INTRODUCTORY BUSINESS STATISTICS

7E

Ronald M. Weiers
Eberly College of Business and Information Technology
Indiana University of Pennsylvania
and
H. John Heinz III College
Carnegie Mellon University

WITH BUSINESS CASES BY

J. Brian Gray
University of Alabama

Lawrence H. Peters
Texas Christian University

Australia • Brazil • Japan • Korea • Mexico • Singapore • Spain • United Kingdom • United States

Introductory Business Statistics, Seventh Edition
Ronald M. Weiers

Vice President of Editorial, Business: Jack W. Calhoun
Publisher: Joe Sabatino
Sr. Acquisitions Editor: Charles McCormick, Jr.
Developmental Editor: Elizabeth Lowry and Suzanna Bainbridge
Editorial Assistant: Nora Heink
Sr. Marketing Communications Manager: Jim Overly
Content Project Manager: Kelly Hillerich
Media Editor: Chris Valentine
Frontlist Buyer, Manufacturing: Miranda Klapper
Compositor: MPS Limited, A Macmillan Company
Sr. Art Director: Stacy Shirley
Internal Designer: Craig Ramsdell
Cover Designer: Patti Hudepohl
Cover Image: © Shutterstock
Photo Acquisition Manager: Don Schlotman
Photo Credits :
B/W Image: Getty Images/PhotoDisc/Chad Baker
Color Image: shutterstock images/Tiberius Dinu

Library of Congress Control Number: 2009943073

International Student Edition ISBN-13: 978-1-111-05885-2
International Student Edition ISBN-10: 1-111-05885-7

Cengage Learning International Offices

Asia
cengageasia.com
tel: (65) 6410 1200

Australia/New Zealand
cengage.com.au
tel: (61) 3 9685 4111

Brazil
cengage.com.br
tel: (011) 3665 9900

India
cengage.co.in
tel: (91) 11 30484837/38

Latin America
cengage.com.mx
tel: +52 (55) 1500 6000

UK/Europe/Middle East/Africa
cengage.co.uk
tel: (44) 207 067 2500

Represented in Canada by Nelson Education, Ltd.
nelson.com
tel: (416) 752 9100/(800) 668 0671

For product information: **www.cengage.com/international**
Visit your local office: **www.cengage.com/global**
Visit our corporate website: **www.cengage.com**

Printed in China by China Translation & Printing Services Limited
2 3 4 5 6 7 13 12 11

To Connor, Madeleine, Hugh, Christina, Aidan,
Mitchell, Owen, Emmett, Mr. Barney Jim,

and

With loving memories of our wonderful son, Bob,
who is swimming with the dolphins off Ocracoke Island

BRIEF CONTENTS

Part 1: Business Statistics: Introduction and Background

1. A Preview of Business Statistics 1
2. Visual Description of Data 15
3. Statistical Description of Data 57
4. Data Collection and Sampling Methods 101

Part 2: Probability

5. Probability: Review of Basic Concepts 133
6. Discrete Probability Distributions 167
7. Continuous Probability Distributions 205

Part 3: Sampling Distributions and Estimation

8. Sampling Distributions 244
9. Estimation from Sample Data 270

Part 4: Hypothesis Testing

10. Hypothesis Tests Involving a Sample Mean or Proportion 311
11. Hypothesis Tests Involving Two Sample Means or Proportions 364
12. Analysis of Variance Tests 411
13. Chi-Square Applications 467
14. Nonparametric Methods 505

Part 5: Regression, Model Building, and Time Series

15. Simple Linear Regression and Correlation 551
16. Multiple Regression and Correlation 600
17. Model Building 644
18. Models for Time Series and Forecasting 687

Part 6: Special Topics

19. Decision Theory 737
20. Total Quality Management 758
21. Ethics in Statistical Analysis and Reporting (Online Chapter)

Appendices

A. Statistical Tables A-1
B. Selected Answers B-1

Index/Glossary I-1

CONTENTS

PART 1: BUSINESS STATISTICS: INTRODUCTION AND BACKGROUND

Chapter 1: A Preview of Business Statistics **1**

1.1 Introduction 2

1.2 Statistics: Yesterday and Today 3

1.3 Descriptive Versus Inferential Statistics 5

1.4 Types of Variables and Scales of Measurement 8

1.5 Statistics in Business Decisions 11

1.6 Business Statistics: Tools Versus Tricks 11

1.7 Summary 12

Chapter 2: Visual Description of Data **15**

2.1 Introduction 16

2.2 The Frequency Distribution and the Histogram 16

2.3 The Stem-and-Leaf Display and the Dotplot 24

2.4 Other Methods for Visual Representation of the Data 28

2.5 The Scatter Diagram 37

2.6 Tabulation, Contingency Tables, and the Excel PivotTable 42

2.7 Summary 48

Integrated Case: Thorndike Sports Equipment (Meet the Thorndikes: See Video Unit One.) 53

Integrated Case: Springdale Shopping Survey 54

Chapter 3: Statistical Description of Data **57**

3.1 Introduction 58

3.2 Statistical Description: Measures of Central Tendency 59

3.3 Statistical Description: Measures of Dispersion 67

3.4 Additional Dispersion Topics 77

3.5 Descriptive Statistics from Grouped Data 83

3.6 Statistical Measures of Association 86

3.7 Summary 90

Integrated Case: Thorndike Sports Equipment 96

Integrated Case: Springdale Shopping Survey 97

Business Case: Baldwin Computer Sales (A) 97

Seeing Statistics Applet 1: Influence of a Single Observation on the Median 99

Seeing Statistics Applet 2: Scatter Diagrams and Correlation 100

Chapter 4: Data Collection and Sampling Methods **101**
4.1 Introduction 102
4.2 Research Basics 102
4.3 Survey Research 105
4.4 Experimentation and Observational Research 109
4.5 Secondary Data 112
4.6 The Basics of Sampling 117
4.7 Sampling Methods 119
4.8 Summary 127

Integrated Case: Thorndike Sports Equipment—Video Unit Two 131
Seeing Statistics Applet 3: Sampling 132

PART 2: PROBABILITY

Chapter 5: Probability: Review of Basic Concepts **133**
5.1 Introduction 134
5.2 Probability: Terms and Approaches 135
5.3 Unions and Intersections of Events 140
5.4 Addition Rules for Probability 143
5.5 Multiplication Rules for Probability 146
5.6 Bayes' Theorem and the Revision of Probabilities 150
5.7 Counting: Permutations and Combinations 156
5.8 Summary 160

Integrated Case: Thorndike Sports Equipment 165
Integrated Case: Springdale Shopping Survey 166
Business Case: Baldwin Computer Sales (B) 166

Chapter 6: Discrete Probability Distributions **167**
6.1 Introduction 168
6.2 The Binomial Distribution 175
6.3 The Hypergeometric Distribution 183
6.4 The Poisson Distribution 187
6.5 Simulating Observations from a Discrete Probability Distribution 194
6.6 Summary 199

Integrated Case: Thorndike Sports Equipment 203

Chapter 7: Continuous Probability Distributions **205**
7.1 Introduction 206
7.2 The Normal Distribution 208
7.3 The Standard Normal Distribution 212
7.4 The Normal Approximation to the Binomial Distribution 223
7.5 The Exponential Distribution 228

7.6 Simulating Observations from a Continuous Probability Distribution 233
7.7 Summary 235

Integrated Case: Thorndike Sports Equipment (Corresponds to Thorndike Video Unit Three) 240
Integrated Case: Thorndike Golf Products Division 240
Seeing Statistics Applet 4: Size and Shape of Normal Distribution 241
Seeing Statistics Applet 5: Normal Distribution Areas 242
Seeing Statistics Applet 6: Normal Approximation to Binomial Distribution 243

PART 3: SAMPLING DISTRIBUTIONS AND ESTIMATION

Chapter 8: Sampling Distributions **244**
8.1 Introduction 245
8.2 A Preview of Sampling Distributions 245
8.3 The Sampling Distribution of the Mean 248
8.4 The Sampling Distribution of the Proportion 254
8.5 Sampling Distributions When the Population Is Finite 257
8.6 Computer Simulation of Sampling Distributions 259
8.7 Summary 262

Integrated Case: Thorndike Sports Equipment 266
Seeing Statistics Applet 7: Distribution of Means: Fair Dice 268
Seeing Statistics Applet 8: Distribution of Means: Loaded Dice 269

Chapter 9: Estimation from Sample Data **270**
9.1 Introduction 271
9.2 Point Estimates 272
9.3 A Preview of Interval Estimates 273
9.4 Confidence Interval Estimates for the Mean: σ Known 276
9.5 Confidence Interval Estimates for the Mean: σ Unknown 281
9.6 Confidence Interval Estimates for the Population Proportion 288
9.7 Sample Size Determination 293
9.8 When the Population Is Finite 298
9.9 Summary 302

Integrated Case: Thorndike Sports Equipment (Thorndike Video Unit Four) 307
Integrated Case: Springdale Shopping Survey 308
Seeing Statistics Applet 9: Confidence Interval Size 309
Seeing Statistics Applet 10: Comparing the Normal and Student *t* Distributions 310
Seeing Statistics Applet 11: Student *t* Distribution Areas 310

PART 4: HYPOTHESIS TESTING

Chapter 10: Hypothesis Tests Involving a Sample Mean or Proportion **311**
10.1 Introduction 312

10.2 Hypothesis Testing: Basic Procedures 317
10.3 Testing a Mean, Population Standard Deviation Known 320
10.4 Confidence Intervals and Hypothesis Testing 329
10.5 Testing a Mean, Population Standard Deviation Unknown 330
10.6 Testing a Proportion 338
10.7 The Power of a Hypothesis Test 346
10.8 Summary 354

Integrated Case: Thorndike Sports Equipment 359
Integrated Case: Springdale Shopping Survey 360
Business Case: Pronto Pizza (A) 361
Seeing Statistics Applet 12: *z*-Interval and Hypothesis Testing 362
Seeing Statistics Applet 13: Statistical Power of a Test 363

Chapter 11: Hypothesis Tests Involving Two Sample Means or Proportions 364

11.1 Introduction 365
11.2 The Pooled-Variances *t*-Test for Comparing the Means of Two Independent Samples 366
11.3 The Unequal-Variances *t*-Test for Comparing the Means of Two Independent Samples 374
11.4 The *z*-Test for Comparing the Means of Two Independent Samples 380
11.5 Comparing Two Means When the Samples Are Dependent 385
11.6 Comparing Two Sample Proportions 391
11.7 Comparing the Variances of Two Independent Samples 397
11.8 Summary 401

Integrated Case: Thorndike Sports Equipment 407
Integrated Case: Springdale Shopping Survey 407
Business Case: Circuit Systems, Inc. (A) 408
Seeing Statistics Applet 14: Distribution of Difference Between Sample Means 410

Chapter 12: Analysis of Variance Tests 411

12.1 Introduction 412
12.2 Analysis of Variance: Basic Concepts 412
12.3 One-Way Analysis of Variance 416
12.4 The Randomized Block Design 429
12.5 Two-Way Analysis of Variance 441
12.6 Summary 457

Integrated Case: Thorndike Sports Equipment (Video Unit Six) 462
Integrated Case: Springdale Shopping Survey 462
Business Case: Fastest Courier in the West 463
Seeing Statistics Applet 15: *F* Distribution and ANOVA 464
Seeing Statistics Applet 16: Interaction Graph in Two-Way ANOVA 465

Chapter 13: Chi-Square Applications 467

13.1 Introduction 468

13.2 Basic Concepts in Chi-Square Testing 468

13.3 Tests for Goodness of Fit and Normality 471

13.4 Testing the Independence of Two Variables 479

13.5 Comparing Proportions from *k* Independent Samples 486

13.6 Estimation and Tests Regarding the Population Variance 489

13.7 Summary 497

Integrated Case: Thorndike Sports Equipment 502

Integrated Case: Springdale Shopping Survey 503

Business Case: Baldwin Computer Sales (C) 503

Seeing Statistics Applet 17: Chi-Square Distribution 504

Chapter 14: Nonparametric Methods 505

14.1 Introduction 506

14.2 Wilcoxon Signed Rank Test for One Sample 508

14.3 Wilcoxon Signed Rank Test for Comparing Paired Samples 513

14.4 Wilcoxon Rank Sum Test for Comparing Two Independent Samples 517

14.5 Kruskal-Wallis Test for Comparing More Than Two Independent Samples 521

14.6 Friedman Test for the Randomized Block Design 525

14.7 Other Nonparametric Methods 530

14.8 Summary 545

Integrated Case: Thorndike Sports Equipment 549

Business Case: Circuit Systems, Inc. (B) 550

PART 5: REGRESSION, MODEL BUILDING, AND TIME SERIES

Chapter 15: Simple Linear Regression and Correlation 551

15.1 Introduction 552

15.2 The Simple Linear Regression Model 553

15.3 Interval Estimation Using the Sample Regression Line 561

15.4 Correlation Analysis 567

15.5 Estimation and Tests Regarding the Sample Regression Line 572

15.6 Additional Topics in Regression and Correlation Analysis 578

15.7 Summary 587

Integrated Case: Thorndike Sports Equipment 595

Integrated Case: Springdale Shopping Survey 596

Business Case: Pronto Pizza (B) 596

Seeing Statistics Applet 18: Regression: Point Estimate for *y* 597

Seeing Statistics Applet 19: Point Insertion Diagram and Correlation 598

Seeing Statistics Applet 20: Regression Error Components 599

Chapter 16: Multiple Regression and Correlation 600
16.1 Introduction 601
16.2 The Multiple Regression Model 602
16.3 Interval Estimation in Multiple Regression 609
16.4 Multiple Correlation Analysis 615
16.5 Significance Tests in Multiple Regression and Correlation 617
16.6 Overview of the Computer Analysis and Interpretation 622
16.7 Additional Topics in Multiple Regression and Correlation 632
16.8 Summary 634

Integrated Case: Thorndike Sports Equipment 639
Integrated Case: Springdale Shopping Survey 640
Business Case: Easton Realty Company (A) 641
Business Case: Circuit Systems, Inc. (C) 643

Chapter 17: Model Building 644
17.1 Introduction 645
17.2 Polynomial Models with One Quantitative Predictor Variable 645
17.3 Polynomial Models with Two Quantitative Predictor Variables 653
17.4 Qualitative Variables 658
17.5 Data Transformations 663
17.6 Multicollinearity 667
17.7 Stepwise Regression 670
17.8 Selecting a Model 676
17.9 Summary 677

Integrated Case: Thorndike Sports Equipment 681
Integrated Case: Fast-Growing Companies 681
Business Case: Westmore MBA Program 682
Business Case: Easton Realty Company (B) 685

Chapter 18: Models for Time Series and Forecasting 687
18.1 Introduction 688
18.2 Time Series 688
18.3 Smoothing Techniques 693
18.4 Seasonal Indexes 702
18.5 Forecasting 708
18.6 Evaluating Alternative Models: *MAD* and *MSE* 713
18.7 Autocorrelation, The Durbin-Watson Test, and Autoregressive Forecasting 715
18.8 Index Numbers 724
18.9 Summary 729

Integrated Case: Thorndike Sports Equipment (Video Unit Five) 735

PART 6: SPECIAL TOPICS

Chapter 19: Decision Theory 737

19.1 Introduction 738

19.2 Structuring the Decision Situation 738

19.3 Non-Bayesian Decision Making 742

19.4 Bayesian Decision Making 745

19.5 The Opportunity Loss Approach 749

19.6 Incremental Analysis and Inventory Decisions 751

19.7 Summary 754

Integrated Case: Thorndike Sports Equipment (Video Unit Seven) 757

Online Appendix to Chapter 19: The Expected Value of Imperfect Information

Chapter 20: Total Quality Management 758

20.1 Introduction 759

20.2 A Historical Perspective and Defect Detection 762

20.3 The Emergence of Total Quality Management 763

20.4 Practicing Total Quality Management 765

20.5 Some Statistical Tools for Total Quality Management 770

20.6 Statistical Process Control: The Concepts 774

20.7 Control Charts for Variables 776

20.8 Control Charts for Attributes 786

20.9 Additional Statistical Process Control and Quality Management Topics 795

20.10 Summary 799

Integrated Case: Thorndike Sports Equipment 803

Integrated Case: Willard Bolt Company 804

Seeing Statistics Applet 21: Mean Control Chart 805

Appendix A: Statistical Tables A-1

Appendix B: Selected Answers B-1

Index/Glossary I-1

Online Chapter 21: Ethics in Statistical Analysis and Reporting

PREFACE

Philosophies and Goals of the Text: A Message to the Student

A book is a very special link between author and reader. In a mystery novel, the author presents the reader with a maze of uncertainty, perplexity, and general quicksand. Intellectual smokescreens are set up all the way to the "whodunit" ending. Unfortunately, many business statistics texts seem to be written the same way—except for the "whodunit" section. This text is specifically designed to be different. Its goals are: (1) to be a clear and friendly guide as you learn about business statistics, (2) to avoid quicksand that could inhibit either your interest or your learning, and (3) to earn and retain your trust in our ability to accomplish goals 1 and 2.

Business statistics is not only relevant to your present academic program, it is also relevant to your future personal and professional life. As a citizen, you will be exposed to, and perhaps may even help generate, statistical descriptions and analyses of data that are of vital importance to your local, state, national, and world communities. As a business professional, you will constantly be dealing with statistical measures of performance and success, as well as with employers who will expect you to be able to utilize the latest statistical techniques and computer software tools—including spreadsheet programs like Excel and statistical software packages like Minitab—in working with these measures.

The chapters that follow are designed to be both informal and informative, as befits an introductory text in business statistics. You will not be expected to have had mathematical training beyond simple algebra, and mathematical symbols and notations will be explained as they become relevant to our discussion. Following an introductory explanation of the purpose and the steps involved in each technique, you will be provided with several down-to-earth examples of its use. Each section has a set of exercises based on the section contents. At the end of each chapter you'll find a summary of what you've read and a listing of equations that have been introduced, as well as chapter exercises, an interesting minicase or two, and in most of the chapters—a realistic business case to help you practice your skills.

Features New to the Seventh Edition

Data Analysis Plus™ 7.0

The Seventh Edition makes extensive use of *Data Analysis Plus™ 7.0*, an updated version of the outstanding add-in that enables Microsoft Excel to carry out practically all of the statistical tests and procedures covered in the text. This excellent software is easy to use, and is available on the premium website that accompanies this text.

The Test Statistics and Estimators Workbooks

The Excel workbooks **Test Statistics** and **Estimators** accompany and are an important complement to ***Data Analysis Plus***™ ***7.0***. These workbooks enable Excel users to quickly perform statistical tests and interval-estimation procedures by simply entering the relevant summary statistics. The workbooks are terrific for solving exercises, checking solutions, and especially for playing "what-if" by trying different inputs to see how they would affect the results. These workbooks, along with **Beta-mean** and three companion workbooks to determine the power of a hypothesis test, accompany ***Data Analysis Plus***™ ***7.0*** and are also available on the premium website at www.cengage.com/login.

Updated Set of 82 Computer Solutions Featuring Complete Printouts and Step-By-Step Instructions for Obtaining Them

Featuring the very latest versions of both Excel and Minitab—Excel 2007 and Minitab 16, respectively—these pieces are located in most of the major sections of the book. Besides providing relevant computer printouts for most of the text examples, they are accompanied by friendly step-by-step instructions written in plain English.

Updated Exercises and Content

The Seventh Edition includes a total of nearly 1600 section and chapter exercises, and more than 300 of them are new or updated. Altogether, there are about 1800 chapter, case, and applet exercises, with about 450 data sets for greater ease and convenience in using the computer. The datasets are in Excel, Minitab, and other popular formats, and are available on the text's premium website. Besides numerous new or updated chapter examples, vignettes, and Statistics in Action items, Chapter 20 (Total Quality Management) has been expanded to include coverage of Process Capability indices and measurement. In response to user preferences and for greater ease of use, the normal distribution table is now cumulative, and it is conveniently located on the rear endsheet of the text.

Continuing Features of *Introductory Business Statistics*

Chapter-Opening Vignettes and Statistics In Action Items

Each chapter begins with a vignette that's both interesting and relevant to the material ahead. Within each chapter, there are also Statistics In Action items that provide further insights into the issues and applications of business statistics in the real world. They include a wide range of topics, including using the consumer price index to time-travel to the (were they *really* lower?) prices in days gone by, and surprisingly-relevant discussion of an odd little car in which the rear passengers faced to the rear. Some of the vignette and Statistics in Action titles:

Get That Cat off the Poll! (p. 116)
Proportions Testing and the Restroom Police (p. 467)
Time-Series-Based Forecasting and the Zündapp (p. 687)

Probabilities, Stolen Lawn Mowers, and the Chance of Rain (p. 138)
The CPI Time Machine (p. 728)
A Sample of Sampling By Giving Away Samples (p. 126)
Gender Stereotypes and Asking for Directions (p. 364)

Extensive Use of Examples and Analogies

The chapters continue to be packed with examples to illustrate the techniques being discussed. In addition to describing a technique and presenting a small-scale example of its application, we will typically present one or more Excel and Minitab printouts showing how the analysis can be handled with popular statistical software. This pedagogical strategy is used so the reader will better appreciate what's going on inside the computer when it's applied to problems of a larger scale.

The Use of Real Data

The value of statistical techniques becomes more apparent through the consistent use of real data in the text. Data sets gathered from such publications as *USA Today*, *Fortune*, *Newsweek*, and *The Wall Street Journal* are used in more than 400 exercises and examples to make statistics both relevant and interesting.

Computer Relevance

The text includes nearly 200 computer printouts generated by Excel and Minitab, and the text's premium website contains data sets for section and chapter exercises, integrated and business cases, and chapter examples. In addition to the new ***Data Analysis Plus™ 7.0*** software and the handy **Test Statistics** and **Estimators** workbooks that accompany it, the Seventh Edition offers the separate collection of 26 Excel worksheet templates generated by the author specifically for exercise solutions and "what-if" analyses based on summary data.

Seeing Statistics Applets

The Seventh Edition continues with the 21 popular interactive java applets, available at the text's premium website. Many of these interesting and insightful applets are customized by their author to specific content and examples in this textbook, and they include a total of 85 applet exercises. The applets are from the award-winning *Seeing Statistics*, authored by Gary McClelland of the University of Colorado, and they bring life and action to many of the most important statistical concepts in the text.

Integrated Cases

At the end of each chapter, you'll find one or both of these case scenarios helpful in understanding and applying concepts presented within the chapter:

(1) Thorndike Sports Equipment Company
The text continues to follow the saga of Grandfather (Luke) and Grandson (Ted) Thorndike as they apply chapter concepts to the diverse opportunities, interesting problems, and assorted dilemmas faced by the Thorndike Sports Equipment Company. At the end of each chapter, the reader has the opportunity to help Luke and Ted apply statistics to their business. The text's premium website offers seven Thorndike video units designed to accompany and reinforce selected written cases.

Viewers will find that they enhance the relevance of the cases as well as provide some entertaining background for the Thorndikes' many statistical adventures.

(2) Springdale Shopping Survey
The Springdale Shopping Survey cases provide the opportunity to apply chapter concepts and the computer to real numbers representing the opinions and behaviors of real people in a real community. The only thing that isn't real is the name of the community. The entire database contains 30 variables for 150 respondents, and is available from the premium website accompanying the text.

Business Cases

The Seventh Edition also provides a set of 12 real-world business cases in 10 different chapters of the text. These interesting and relatively extensive cases feature disguised organizations, but include real data pertaining to real business problems and situations. In each case, the company or organization needs statistical assistance in analyzing their database to help them make more money, make better decisions, or simply make it to the next fiscal year. The organizations range all the way from an MBA program, to a real estate agency, to a pizza delivery service, and these cases and their variants are featured primarily among the chapters in the latter half of the text. The cases have been adapted from the excellent presentations in *Business Cases in Statistical Decision Making*, by Lawrence H. Peters, of Texas Christian University and J. Brian Gray, of the University of Alabama. Just as answers to problems in the real world are not always simple, obvious, and straightforward, neither are some of the solutions associated with the real problems faced by these real (albeit disguised) companies and organizations. However, in keeping with the "Introduction to ..." title of this text, we do provide a few guidelines in the form of specific questions or issues the student may wish to address while using business statistics in helping to formulate observations and recommendations that could be informative or helpful to his or her "client."

Organization of the Text

The text can be used in either a one-term or a two-term course. For one-term applications, Chapters 1 through 11 are suggested. For two-term use, it is recommended that the first term include Chapters 1 through 11, and that the second term include at least Chapters 12 through 18. In either one- or two-term use, the number and variety of chapters allow for instructor flexibility in designing either a course or a sequence of courses that will be of maximum benefit to the student. This flexibility includes the possibility of including one or more of the two remaining chapters, which are in the Special Topics section of the text.

Chapter 1 provides an introductory discussion of business statistics and its relevance to the real world. Chapters 2 and 3 cover visual summarization methods and descriptive statistics used in presenting statistical information. Chapter 4 discusses popular approaches by which statistical data are collected or generated, including relevant sampling methods. In Chapters 5 through 7, we discuss the basic notions of probability and go on to introduce the discrete and continuous probability distributions upon which many statistical analyses depend. In Chapters 8 and 9, we discuss sampling distributions and the vital topic of making estimates based on sample findings.

Chapters 10 through 14 focus on the use of sample data to reach conclusions regarding the phenomena that the data represent. In these chapters, the reader will learn how to use statistics in deciding whether to reject statements that have been made concerning these phenomena. Chapters 15 and 16 introduce methods for obtaining and using estimation equations in describing how one variable tends to change in response to changes in one or more others.

Chapter 17 extends the discussion in the two previous chapters to examine the important issue of model building. Chapter 18 discusses time series, forecasting, and index number concepts used in analyzing data that occur over a period of time. Chapter 19 discusses the role of statistics in decision theory, while Chapter 20 explores total quality management and its utilization of statistics.

At the end of the text, there is a combined index and glossary of key terms, a set of statistical tables, and answers to selected odd exercises. For maximum convenience, immediately preceding the back cover of the text are pages containing the two statistical tables to which the reader will most often be referring: the *t*-distribution and the standard normal, or *z*-distribution.

Ancillary Items

To further enhance the usefulness of the text, a complete package of complementary ancillary items has been assembled, and they are available at the premium website accompanying the text:

Student Premium Website

This website available at www.cengage.com/login, contains *Data Analysis Plus™ 7.0* Excel add-in software and accompanying workbooks, including **Test Statistics** and **Estimators**; *Seeing Statistics* applets; datasets for exercises, cases, and text examples; author-developed Excel worksheet templates for exercise solutions and "what-if" analyses; and the Thorndike Sports Equipment video cases. Also included, in *pdf* format, are Chapter 21, Ethics in Statistical Analysis and Reporting, and the Chapter 19 appendix on the expected value of imperfect information.

Instructor's Resources

The Instructor's Resources are available to qualified adopters and contains *author-generated* complete and detailed solutions to all section, chapter, and applet exercises, integrated cases and business cases; a test bank in Microsoft Word format that includes test questions by section; ExamView testing software, which allows a professor to create exams in minutes; PowerPoint presentations featuring concepts and examples for each chapter; and a set of display *Seeing Statistics* applets based on those in the text and formatted for in-class projection.

Also Available from the Publisher

Available separately from the publisher are other items for enhancing students' learning experience with the textbook. Among them are the following:

Instructor's Solutions Manual (Weiers)

The Instructor's Solutions Manual contains *author-generated* complete and detailed solutions to all section, chapter, and applet exercises, integrated cases and business cases. It is available to qualified adopters and is in Microsoft Word format on the password-protected instructor's website at www.cengage.com/international.

Test Bank (Doug Barrett)

Containing over 2600 test questions, including true-false, multiple-choice, and problems similar to those at the ends of the sections and chapters of the text, the computerized Test Bank makes test creation a cinch. The ExamView program is available from the text's companion website at www.cengage.com/international.

PPTs: (Priscilla Chaffe-Stengel)

The PowerPoint slides contain the chapter learning outcomes, key terms, theoretical overviews, and practical examples to facilitate classroom instruction and student learning. The PowerPoint files are available from the text's companion website at www.cengage.com/international.

Minitab, Student Version for Windows (Minitab, Inc.)

The student version of this popular statistical software package. *Available at a discount when bundled with the text.*

Acknowledgements

Advice and guidance from my colleagues have been invaluable to the generation of the Seventh Edition, and I would like to thank the following individuals for their helpful comments and suggestions:

J. Douglas Barrett
University of North Alabama

David Bush
Villanova University

Priscilla Chaffe-Stengel
California State University-Fresno

Fred Dehner
Rivier College

Jim Ford
University of Delaware

Jeff Grover
Dynamics Research Corporation

Janice Harder
Motlow State Community College

Farid Islam
Utah Valley State College

Yunus Kathawala
Eastern Illinois University

Linda Leighton
Fordham University

Edward Mansfield
University of Alabama

Elizabeth Mayer
St. Bonaventure University

Rich McGowan
Boston College

Patricia Mullins
University of Wisconsin

Alan Olinsky
Bryant University

Deborah J. Rumsey
The Ohio State University

Farhad Saboori
Albright College

Dan Shimshak
University of Massachusetts

Kathy Smith
Carnegie Mellon University

Debra K. Stiver
University of Nevada, Reno

Mark A. Thompson
University of Arkansas at Little Rock

Joseph Van Metre
University of Alabama

I would also like to thank colleagues who were kind enough to serve as reviewers for previous editions of the text: Randy Anderson, *California State University—Fresno*; Leland Ash, *Yakima Valley Community College*; James O. Flynn, *Cleveland State University*; Marcelline Fusilier, *Northwestern State University of Louisiana*; Thomas Johnson, *North Carolina State University*; Mark P. Karscig, *Central Missouri State University*; David Krueger, *Saint Cloud State University*; Richard T. Milam, Jr., *Appalachian State University*; Erl Sorensen, *Northeastern University*; Peter von Allmen, *Moravian College*: R. C. Baker, *University of Texas-Arlington*; Robert Boothe, *Memphis State University*; Raymond D. Brown, *Drexel University*; Shaw K. Chen, *University of Rhode Island*; Gary Cummings, *Walsh College*; Phyllis Curtiss, *Bowling Green State University*; Fred Derrick, *Loyola College*; John Dominguez, *University of Wisconsin—Whitewater*; Robert Elrod, *Georgia State University*; Mohammed A. El-Saidi, *Ferris State University*; Stelios Fotopoulos, *Washington State University*; Oliver Galbraith, *San Diego State University*; Patricia Gaynor, *University of Scranton*; Edward George, *University of Texas—El Paso*; Jerry Goldman, *DePaul University*; Otis Gooden, *Cleveland State University*; Deborah Gougeon, *Appalachian State University*; Jeffry Green, *Ball State University*; Irene Hammerbacher, *Iona College*; Robert Hannum, *University of Denver*; Burt Holland, *Temple University*; Larry Johnson, *Austin Community College*; Shimshon Kinory, *Jersey City State College*; Ron Koot, *Pennsylvania State University*; Douglas Lind, *University of Toledo*; Subhash Lonial, *University of Louisville*; Tom Mathew, *Troy State University—Montgomery*; John McGovern, *Georgian Court College*; Frank McGrath, *Iona College*; Jeff Mock, *Diablo Valley College*; Kris Moore, *Baylor University*; Ryan Murphy, *University of Arizona*; Buddy Myers, *Kent State University*; Leon Neidleman, *San Jose State University*; Julia Norton, *California State University—Hayward*; C. J. Park, *San Diego State University*; Leonard Presby, *William Patterson State College*; Harry Reinken, *Phoenix College*; Vartan Safarian, *Winona State University*; Sue Schou, *Idaho State University*; John Sennetti, *Texas Tech University*; William A. Shrode, *Florida State University*; Lynnette K. Solomon, *Stephen F. Austin State University*; Sandra Strasser, *Valparaiso State University*; Joseph Sukta, *Moraine Valley Community College*; J. B. Spaulding, *University of Northern Texas*; Carol Stamm, *Western Michigan University*; Priscilla Chaffe-Stengel, *California State University—Fresno*; Stan Stephenson, *Southwest Texas State University*; Patti Taylor, *Angelo State University*; Patrick Thompson, *University of Florida—Gainesville*; Russell G. Thompson, *University of Houston*; Susan Colvin-White, *Northwestern State University*; Nancy Williams, *Loyola College*; Dick Withycombe, *University of Montana*; Cliff Young, *University of Colorado at Denver*; and Mustafa Yilmaz, *Northeastern University*.

I would like to thank Vince Taiani for assistance with and permission to use what is known here as the Springdale Shopping Survey computer database. Thanks to Minitab, Inc. for the support and technical assistance they have provided. Thanks to Gary McCelland for his excellent collection of applets for this text, and to Lawrence H. Peters and J. Brian Gray for their outstanding cases and the hands-on experience they have provided to the student. Special thanks to my friend and fellow author Gerry Keller and the producers of *Data Analysis Plus*[TM] *7.0* for their excellent software that has enhanced this edition.

The editorial staff of Cengage Learning is deserving of my gratitude for their encouragement, guidance, and professionalism throughout what has been an arduous, but rewarding task. Among those without whom this project would not have come to fruition are Charles McCormick, Acquisitions Editor; Suzanna

Bainbridge and Elizabeth Lowry, Developmental Editors; Kelly Hillerich, Content Project Manager; Bill Hendee, Vice President of Marketing; Stacy Shirley, Art Director; Eleanora Heink, Editorial Assistant; Suellen Ruttkay, Marketing Coordinator, and Libby Shipp, Marketing Communications Manager. In addition, the world-class editorial skills of Susan Reiland and the detail-orientation of Dr. Jeff Grover, Dr. Debra Stiver, and Dr. Doug Barrett are greatly appreciated.

Last, but certainly not least, I remain extremely thankful to my family for their patience and support through seven editions of this work.

Ronald M. Weiers, Ph.D.
Eberly College of Business and Information Technology
Indiana University of Pennsylvania
and
Adjunct, H. John Heinz III College
Carnegie Mellon University

Using the Computer

In terms of software capability, this edition is the best yet. Besides incorporating Excel 2007, we feature *Data Analysis Plus*™ *7.0* and its primary workbook partners, **Test Statistics** and **Estimators**. The text includes 82 Computer Solutions pieces that show Excel and Minitab printouts relevant to chapter examples, plus friendly step-by-step instructions showing how to carry out each analysis or procedure involved. The Excel materials have been extensively tested with Microsoft Office 2007, but the printouts and instructions will be at least somewhat familiar to users of earlier versions of this spreadsheet software package. The Minitab printouts and instructions pertain to Minitab Release 16, but will be either identical or very similar to those for earlier versions of this dedicated statistical software package. Because operating systems and software versions continually evolve, be sure to get the latest information by visiting the student premium site at: http://www.cengage.com/login.

Minitab

This statistical software is powerful, popular, easy to use, and offers little in the way of surprises—a pleasant departure in an era when we too-often see software crashes and the dreaded "blue screen of death." As a result, there's not much else to be said about this dedicated statistical software. Note that Minitab 16 has excellent graphics that will not be nearly so attractive in some earlier versions.

Excel

This popular spreadsheet software offers a limited number of statistical tests and procedures, but it delivers excellent graphics and it seems to be installed in nearly every computer on the planet. As a result, it gets featured coverage in many of the newer statistics textbooks, including this one. Some special sections with regard to Excel appear below.

Data Analysis/Analysis ToolPak

This is the standard data analysis module in Excel 2007. Within the Data ribbon, Excel's Data Analysis add-in should appear as a menu item at the right. If it does not, load the Data Analysis module as follows: Click the Microsoft Office button at the upper left corner of the screen. Click the **Excel Options** button. Click **Add-Ins.** Be sure that **Excel Add-Ins** appears in the **Manage** box, then click **Go.** In the **Add-Ins Available** box, select **Analysis ToolPak** and click **OK.** The Analysis ToolPak should now be available for use. If you get a prompt indicating that the Analysis ToolPak is not installed on your computer, click **Yes** to install it.

Data Analysis Plus™ 7.0

This outstanding software greatly extends Excel's capabilities to include practically every statistical test and procedure covered in the text, and it is very easy to use. It is available on the premium website and can be automatically installed by means of the startup instructions. Typically, the Excel file STATS will be inserted into the XLstart folder in the Excel portion of your computer's Windows directory. This software is featured in nearly one-third of the Computer Solutions sets of printouts and instructions that appear in the text. After installation, when you click the **Tools** ribbon, the **Data Analysis Plus** item will be among those appearing in the menu below.

Test Statistics and Estimators Workbooks

These Excel workbooks are among those accompanying *Data Analysis Plus*™ *7.0*. They contain worksheets that enable us to carry out procedures or obtain solutions based only on summary information about the problem or situation. This is a real work-saver for solving chapter exercises, checking solutions that have been hand-calculated, or for playing "what-if" by trying different inputs to instantaneously see how they affect the results. These workbooks are typically installed into the same directory where the data files are located.

Other Excel Worksheet Templates

There are 26 Excel worksheet templates generated by the author and carried over from the previous edition. As with the worksheets within the **Test Statistics** and **Estimators** workbooks, they provide solutions based on summary information about a problem or situation. The instructions for using each template are contained within the template itself. When applicable, they are cited within the Computer Solutions items in which the related analyses or procedures appear.

CHAPTER 1

A Preview of Business Statistics

STATISTICS CAN ENTERTAIN, ENLIGHTEN, ALARM

Today's statistics applications range from the inane to the highly germane. Sometimes statistics provides nothing more than entertainment—e.g., a study found that 31% of U.S. adults have re-gifted a present for the holidays.[1] Regarding an actual entertainer, other studies found that the public's "favorable" rating for actor Tom Cruise had dropped from 58% to 35% between 2005 and 2006.[2]

On the other hand, statistical descriptors can be highly relevant to such important matters as corporate ethics and employee privacy. For example, 5% of workers say they use the Internet too much at work, and that decreases their productivity.[3] In the governmental area, U.S. census data can mean millions of dollars to big cities. According to the Los Angeles city council, that city will have lost over $180 million in federal aid because the 2000 census had allegedly missed 76,800 residents, most of whom were urban, minority, and poor.[4]

At a deadly extreme, statistics can also describe the growing toll on persons living near or downwind of Chernobyl, site of the world's worst nuclear accident. Just 10 years following this 1986 disaster, cancer rates in the fallout zone had already nearly doubled, and researchers are now concerned about the possibility of even higher rates with the greater passage of time.[5] In general, statistics can be useful in examining any geographic "cluster" of disease incidence, helping us to decide whether the higher incidence could be due simply to chance variation, or whether some environmental agent or pollutant may have played a role.

[1]Source: Michelle Healy and Veronica Salazar, "Thanks... Again," *USA Today,* December 23, 2008, p. 1D.
[2]Source: Susan Wloszczyna, "In Public's Eyes, Tom's Less of a Top Gun," *USA Today,* May 10, 2006, p. 1D.
[3]Source: Jae Yang and Marcy Mullins, "Internet Usage's Impact on Productivity," *USA Today,* March 21, 2006, p. 1B.
[4]Source: cd13.com. Letter from Los Angeles City Council to U.S. House of Representatives, April 11, 2006.
[5]Allison M. Heinrichs, "Study to Examine Breast Cancer in Europeans," *Pittsburgh Tribune-Review,* April 23, 2006; and world-nuclear.org/info/chernobyl, April 2009.

Anticipating coming attractions

1.1 INTRODUCTION

Timely Topic, Tattered Image

At this point in your college career, toxic dumping, armed robbery, fortune telling, and professional wrestling may all have more positive images than **business statistics.** If so, this isn't unusual, since many students approach the subject believing that it will be either difficult or irrelevant. In a study of 105 beginning students' attitudes toward statistics, 56% either strongly or moderately agreed with the statement, "I am afraid of statistics."[6] (Sorry to have tricked you like that, but you've just been introduced to a statistic, one that you'll undoubtedly agree is neither difficult nor irrelevant.)

Having recognized such possibly negative first impressions, let's go on to discuss statistics in a more positive light. First, regarding ease of learning, the only thing this book assumes is that you have a basic knowledge of algebra. Anything else you need will be introduced and explained as we go along. Next, in terms of relevance, consider the unfortunates of Figure 1.1 and how just the slight change of a single statistic might have considerably influenced each individual's fortune.

What Is Business Statistics?

Briefly defined, business statistics can be described as *the collection, summarization, analysis, and reporting of numerical findings relevant to a business decision or situation.* Naturally, given the great diversity of business itself, it's not surprising that statistics can be applied to many kinds of business settings. We will be examining a wide spectrum of such applications and settings. Regardless of your eventual career destination, whether it be accounting or marketing, finance or politics, information science or human resource management, you'll find the statistical techniques explained here are supported by examples and problems relevant to your own field.

For the Consumer as Well as the Practitioner

As a businessperson, you may find yourself involved with statistics in at least one of the following ways: (1) as a practitioner collecting, analyzing, and presenting

FIGURE 1.1

Some have the notion that statistics can be irrelevant. As the plight of these individuals suggests, nothing could be further from the truth.

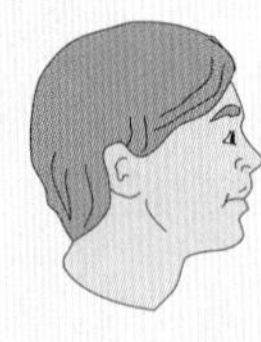

Sidney Sidestreet, former quality assurance supervisor for an electronics manufacturer. The 20 microchips he inspected from the top of the crate all tested out OK, but many of the 14,980 on the bottom weren't quite so good.

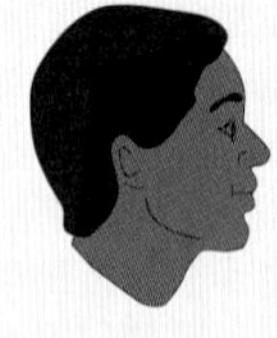

Lefty "H.R." Jones, former professional baseball pitcher. Had an earned-run average of 12.4 last season, which turned out to be his last season.

Walter Wickerbin, former newspaper columnist. Survey by publisher showed that 43% of readers weren't even aware of his column.

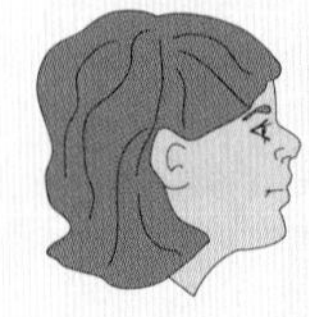

Rhonda Rhodes, former vice president of engineering for a tire manufacturer. The company advertised a 45,000-mile tread life, but tests by a leading consumer magazine found most tires wore out in less than 20,000 miles.

[6]Source: Eleanor W. Jordan and Donna F. Stroup, "The Image of Statistics," *Collegiate News and Views,* Spring 1984, p. 11.

findings based on statistical data or (2) as a consumer of statistical claims and findings offered by others, some of whom may be either incompetent or unethical.

As you might expect, the primary orientation of this text will be toward the "how-to," or practitioner, dimension of business statistics. After finishing this book, you should be both proficient and conversant in most of the popular techniques used in statistical data collection, analysis, and reporting. As a secondary goal, this book will help you protect yourself and your company as a statistical consumer. In particular, it's important that you be able to deal with individuals who arrive at your office bearing statistical advice. Chances are, they'll be one of the following:

1. **Dr. Goodstat** The good doctor has painstakingly employed the correct methodology for the situation and has objectively analyzed and reported on the information he's collected. Trust him, he's OK.
2. **Stanley Stumbler** Stanley means well, but doesn't fully understand what he's doing. He may have innocently employed an improper methodology and arrived at conclusions that are incorrect. In accepting his findings, you may join Stanley in flying blind.
3. **Dr. Unethicus** This character knows what he's doing, but uses his knowledge to sell you findings that he knows aren't true. In short, he places his own selfish interests ahead of both scientific objectivity and your informational needs. He varies his modus operandi and is sometimes difficult to catch. One result is inevitable: When you accept his findings, he wins and you lose.

STATISTICS: YESTERDAY AND TODAY (1.2)

Yesterday

Although statistical data have been collected for thousands of years, very early efforts typically involved simply counting people or possessions to facilitate taxation. This record-keeping and enumeration function remained dominant well into the 20th century, as this 1925 observation on the role of statistics in the commercial and political world of that time indicates:

It is coming to be the rule to use statistics and to think statistically. The larger business units not only have their own statistical departments in which they collect and interpret facts about their own affairs, but they themselves are consumers of statistics collected by others. The trade press and government documents are largely statistical in character, and this is necessarily so, since only by the use of statistics can the affairs of business and of state be intelligently conducted.

Business needs a record of its past history with respect to sales, costs, sources of materials, market facilities, etc. Its condition, thus reflected, is used to measure progress, financial standing, and economic growth. A record of business changes—of its rise and decline and of the sequence of forces influencing it—is necessary for estimating future developments.[7]

Note the brief reference to "estimating future developments" in the preceding quotation. In 1925, this observation was especially pertinent because a transition was in process. Statistics was being transformed from a relatively passive record keeper

[7]Source: Horace Secrist, *An Introduction to Statistical Methods*, rev. ed. New York: Macmillan Company, 1925, p. 1.

and descriptor to an increasingly active and useful business tool, which would influence decisions and enable inferences to be drawn from sample information.

Today

Today, statistics and its applications are an integral part of our lives. In such diverse settings as politics, medicine, education, business, and the legal arena, human activities are both measured and guided by statistics.

Our behavior in the marketplace generates sales statistics that, in turn, help companies make decisions on products to be retained, dropped, or modified. Likewise, auto insurance firms collect data on age, vehicle type, and accidents, and these statistics guide the companies toward charging extremely high premiums for teenagers who own or drive high-powered cars like the Chevrolet Corvette. In turn, the higher premiums influence human behavior by making it more difficult for teens to own or drive such cars. The following are additional examples where statistics are either guiding or measuring human activities.

- Well beyond simply counting how many people live in the United States, the U.S. Census Bureau uses sampling to collect extensive information on income, housing, transportation, occupation, and other characteristics of the populace. The Bureau used to do this by means of a "long form" sent to 1 in 6 Americans every 10 years. Today, the same questions are asked in a 67-question monthly survey that is received by a total of about 3 million households each year. The resulting data are more recent and more useful than the decennial sampling formerly employed, and the data have a vital effect on billions of dollars in business decisions and federal funding.[8]
- According to the International Dairy Foods Association, ice cream and related frozen desserts are consumed by more than 90% of the households in the United States. The most popular flavor is vanilla, which accounts for 30% of sales. Chocolate is a distant second, at 10% of sales.[9]
- On average, U.S. stores lose $35 million each day to shoplifters. The problem becomes even worse when the national economy is weak, and more than 10 million people have been detained for shoplifting in the past five years.[10]

Throughout this text, we will be examining the multifaceted role of statistics as a descriptor of information, a tool for analysis, a means of reaching conclusions, and an aid to decision making. In the next section, after introducing the concept of descriptive versus inferential statistics, we'll present further examples of the relevance of statistics in today's world.

1.1 What was the primary use of statistics in ancient times?

1.2 In what ways can business statistics be useful in today's business environment?

[8]Source: Haya El Nasser, "Rolling Survey for 2010 Census Keeps Data Up to Date," *USA Today*, January 17, 2005, p. 4A.
[9]Source: idfa.org, June 19, 2009.
[10]Source: shopliftingprevention.org, June 19, 2009.

DESCRIPTIVE VERSUS INFERENTIAL STATISTICS (1.3)

As we have seen, statistics can refer to a set of individual numbers or numerical facts, or to general or specific statistical techniques. A further breakdown of the subject is possible, depending on whether the emphasis is on (1) simply describing the characteristics of a set of data or (2) proceeding from data characteristics to making generalizations, estimates, forecasts, or other judgments based on the data. The former is referred to as **descriptive statistics,** while the latter is called **inferential statistics.** As you might expect, both approaches are vital in today's business world.

Descriptive Statistics

In descriptive statistics, we simply summarize and describe the data we've collected. For example, upon looking around your class, you may find that 35% of your fellow students are wearing Casio watches. If so, the figure "35%" is a descriptive statistic. You are not attempting to suggest that 35% of all college students in the United States, or even at your school, wear Casio watches. You're merely describing the data that you've recorded. In the year 1900, the U.S. Postal Service operated 76,688 post offices, compared to just 27,276 in 2007.[11] In 2008, the 1.12 billion common shares of McDonald's Corporation each received a $1.63 dividend on net income of $3.76 per common share.[12] Table 1.1 (page 6) provides additional examples of descriptive statistics. Chapters 2 and 3 will present a number of popular visual and statistical approaches to expressing the data we or others have collected. For now, however, just remember that descriptive statistics are used only to summarize or describe.

Inferential Statistics

In inferential statistics, sometimes referred to as *inductive* statistics, we go beyond mere description of the data and arrive at *inferences* regarding the phenomenon or phenomena for which sample data were obtained. For example, based partially on an examination of the viewing behavior of several thousand television households, the ABC television network may decide to cancel a prime-time television program. In so doing, the network is assuming that millions of other viewers across the nation are also watching competing programs.

Political pollsters are among the heavy users of inferential statistics, typically questioning between 1000 and 2000 voters in an effort to predict the voting behavior of millions of citizens on election day. If you've followed recent presidential elections, you may have noticed that, although they contact only a relatively small number of voters, the pollsters are quite often "on the money" in predicting both the winners and their margins of victory. This accuracy, and the fact that it's not simply luck, is one of the things that make inferential statistics a fascinating and useful topic. (For more examples of the relevance and variety of inferential statistics, refer to Table 1.1.) As you might expect, much of this text will be devoted to the concept and methods of inferential statistics.

[11]Source: Bureau of the Census, U.S. Department of Commerce, *Statistical Abstract of the United States 2009*, p. 690.
[12]Source: McDonald's Corporation, Inc., *2008 Annual Report.*

TABLE 1.1

Some examples of descriptive and inferential statistics.

Descriptive Statistics

- According to the Bureau of the Census, there are 2.2 million U.S. households with a single father and one or more children younger than 18. [p. 1A]
- There have been 82 confirmed or suspected suicides among active-duty service personnel this year, compared to 51 for the same period in 2008. [p. 1A]
- The number of mutual funds peaked at 8305 in 2001, but the combination of bear markets and mergers and acquisitions has driven the number of funds down to 8011. [p. 1B]
- Since March 4, 2009, there have been 190,000 mortgage modifications through President Obama's relief plan; 396,724 homes in payment default; and 607,974 homes in either foreclosure or auction proceedings. [p. 1A]

Inferential Statistics

- In observing a sample of nurses and other healthcare workers who were likely infected with the swine flu, researchers found that only half routinely wore gloves when dealing with patients. [p. 3A]
- In a Zagat survey of diners, Outback Steakhouse had the top-rated steaks in the full-service restaurant category. [p. 7A]
- Survey results revealed that 26% of thirsty golfers order a sports drink when they finish their round and head for the clubhouse. [p. 1C]
- In a survey of U.S. motorists, 33% said their favorite American roadside store was South of the Border, in South Carolina. [p. 1D]

Source: *USA Today*, June 19, 2009. The page references are shown in brackets.

Key Terms for Inferential Statistics

In surveying the political choices of a small number of eligible voters, political pollsters are using a **sample** of voters selected from the **population** of all eligible voters. Based on the results observed in the sample, the researchers then proceed to make inferences on the political choices likely to exist in this larger population of eligible voters. A sample result (e.g., 46% of the sample favor Charles Grady for president) is referred to as a **sample statistic** and is used in an attempt to estimate the corresponding **population parameter** (e.g., the actual, but unknown, national percentage of voters who favor Mr. Grady). These and other important terms from inferential statistics may be defined as follows:

- **Population** Sometimes referred to as the *universe*, this is the entire set of people or objects of interest. It could be all adult citizens in the United States, all commercial pilots employed by domestic airlines, or every roller bearing ever produced by the Timken Company.

A population may refer to things as well as people. Before beginning a study, it is important to clearly define the population involved. For example, in a given study, a retailer may decide to define "customer" as all those who enter her store between 9 A.M. and 5 P.M. next Wednesday.

- **Sample** This is a smaller number (a *subset*) of the people or objects that exist within the larger population. The retailer in the preceding definition may decide to select her sample by choosing every 10th person entering the store between 9 A.M. and 5 P.M. next Wednesday.

A sample is said to be *representative* if its members tend to have the same characteristics (e.g., voting preference, shopping behavior, age, income, educational level) as the population from which they were selected. For example, if 45% of the population consists of female shoppers, we would like our sample to also include 45% females. When a sample is so large as to include all members of the population, it is referred to as a complete *census*.

- **Statistic** This is a measured characteristic of the sample. For example, our retailer may find that 73% of the sample members rate the store as having higher-quality merchandise than the competitor across the street. The sample statistic can be a measure of *typicalness* or central tendency, such as the mean, median, mode, or proportion, or it may be a measure of *spread* or dispersion, such as the range and standard deviation:

The *sample mean* is the arithmetic average of the data. This is the sum of the data divided by the number of values. For example, the mean of $4, $3, and $8 can be calculated as ($4 + $3 + $8)/3, or $5.

The *sample median* is the midpoint of the data. The median of $4, $3, and $8 would be $4, since it has just as many values above it as below it.

The *sample mode* is the value that is most frequently observed. If the data consist of the numbers 12, 15, 10, 15, 18, and 21, the mode would be 15 because it occurs more often than any other value.

The *sample proportion* is simply a percentage expressed as a decimal fraction. For example, if 75.2% is converted into a proportion, it becomes 0.752.

The *sample range* is the difference between the highest and lowest values. For example, the range for $4, $3, and $8 is ($8 − $3), or $5.

The *sample standard deviation,* another measure of dispersion, is obtained by applying a standard formula to the sample values. The formula for the standard deviation is covered in Chapter 3, as are more detailed definitions and examples of the other measures of central tendency and dispersion.

- **Parameter** This is a numerical characteristic of the population. If we were to take a complete census of the population, the parameter could actually be measured. As discussed earlier, however, this is grossly impractical for most business research. The purpose of the sample statistic is to estimate the value of the corresponding population parameter (e.g., the sample mean is used to estimate the population mean). Typical parameters include the population mean, median, proportion, and standard deviation. As with sample statistics, these will be discussed in Chapter 3.

  For our retailer, the actual percentage of the population who rate her store's merchandise as being of higher quality is unknown. (This unknown quantity is the parameter in this case.) However, she may use the sample statistic (73%) as an estimate of what this percentage would have been had she taken the time, expense, and inconvenience to conduct a census of all customers on the day of the study.

EXERCISES

1.3 What is the difference between descriptive statistics and inferential statistics? Which branch is involved when a state senator surveys some of her constituents in order to obtain guidance on how she should vote on a piece of legislation?

1.4 In 2008, Piedmont Natural Gas Corporation sold 50.4 million cubic feet of gas to residential customers, an increase of 3.7% over the previous year. Does this information represent descriptive statistics or inferential statistics? Why? Source: Piedmont Natural Gas Corporation, *Annual Report 2008*, p. 16.

1.5 An article in a fitness magazine described a study that compared the cardiovascular responses of 20 adult subjects for exercises on a treadmill, on a minitrampoline, and jogging in place on a carpeted surface. Researchers found average heart rates were significantly less on the minitrampoline than for the treadmill and stationary jogging. Does this information represent descriptive statistics or inferential statistics? Why?

1.4 TYPES OF VARIABLES AND SCALES OF MEASUREMENT

Qualitative Variables

Some of the variables associated with people or objects are **qualitative** in nature, indicating that the person or object belongs in a category. For example: (1) you are either male or female; (2) you have either consumed Dad's Root Beer within the past week or you have not; (3) your next television set will be either color or black and white; and (4) your hair is likely to be brown, black, red, blonde, or gray. While some qualitative variables have only two categories, others may have three or more. Qualitative variables, also referred to as *attributes*, typically involve counting how many people or objects fall into each category.

In expressing results involving qualitative variables, we describe the percentage or the number of persons or objects falling into each of the possible categories. For example, we may find that 35% of grade-school children interviewed recognize a photograph of Ronald McDonald, while 65% do not. Likewise, some of the children may have eaten a Big Mac hamburger at one time or another, while others have not.

Quantitative Variables

Quantitative variables enable us to determine *how much* of something is possessed, not just whether it is possessed. There are two types of quantitative variables: discrete and continuous.

Discrete quantitative variables can take on only certain values along an interval, with the possible values having gaps between them. Examples of discrete quantitative variables would be the number of employees on the payroll of a manufacturing firm, the number of patrons attending a theatrical performance, or the number of defectives in a production sample. Discrete variables in business statistics usually consist of observations that we can count and often have integer values. Fractional values are also possible, however. For example, in observing the number of gallons of milk that shoppers buy during a trip to a U.S. supermarket, the possible values will be 0.25, 0.50, 0.75, 1.00, 1.25, 1.50, and so on. This is because milk is typically sold in 1-quart containers as well as gallons. A shopper will not be able to purchase a container of milk labeled "0.835 gallons." The distinguishing feature of discrete variables is that gaps exist between the possible values.

Continuous quantitative variables can take on a value at any point along an interval. For example, the volume of liquid in a water tower could be any quantity between zero and its capacity when full. At a given moment, there might be 325,125 gallons, 325,125.41 gallons, or even 325,125.413927 gallons, depending on the accuracy with which the volume can be measured. The possible values that could be taken on would have no gaps between them. Other examples of continuous quantitative variables are the weight of a coal truck, the Dow Jones Industrial Average, the driving distance from your school to your home town, and the temperature outside as you're reading this book. The exact values each of these variables could take on would have no gaps between them.

Scales of Measurement

Assigning a numerical value to a variable is a process called *measurement*. For example, we might look at the thermometer and observe a reading of 72.5 degrees Fahrenheit or examine a box of lightbulbs and find that 3 are broken. The numbers 72.5 and 3 would constitute measurements. When a variable is measured, the result will be in one of the four levels, or *scales*, of measurement—nominal, ordinal, interval, or ratio—summarized in Figure 1.2. The scale to which the measurements belong will be important in determining appropriate methods for data description and analysis.

The Nominal Scale

The **nominal scale** uses numbers only for the purpose of identifying membership in a group or category. Computer statistical analysis is greatly facilitated by the use of numbers instead of names. For example, Entergy Corporation lists four types of domestic electric customers.[13] In its computer records, the company might use "1" to identify residential customers, "2" for commercial customers, "3" for industrial customers, and "4" for government customers. Aside from identification, these numbers have no arithmetic meaning.

The Ordinal Scale

In the **ordinal scale,** numbers represent "greater than" or "less than" measurements, such as preferences or rankings. For example, consider the following

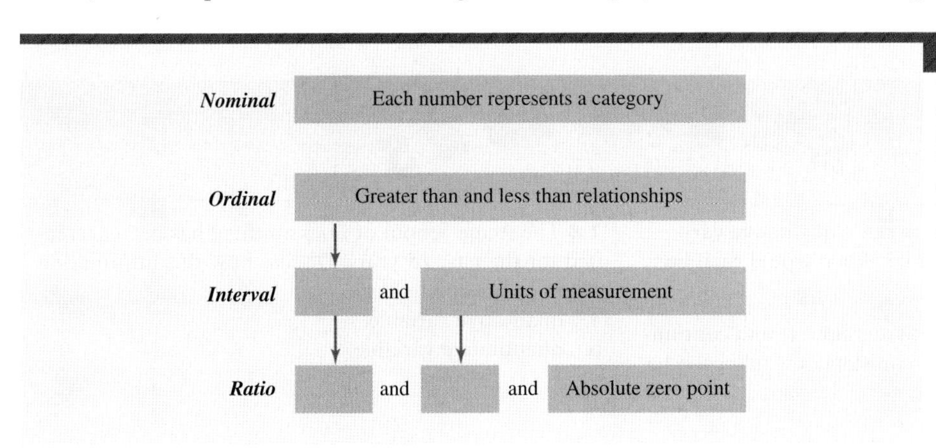

FIGURE 1.2
The methods through which statistical data can be analyzed depend on the scale of measurement of the data. Each of the four scales has its own characteristics.

[13]Source: Entergy Corporation, *2008 Annual Report*.

Women's Tennis Association singles rankings for female tennis players:[14]

1. Dinara Safina
2. Serena Williams
3. Venus Williams
4. Elena Dementieva

In the ordinal scale, numbers are viewed in terms of rank (i.e., greater than, less than), but do not represent distances between objects. For example, we cannot say that the distance between Dinara Safina and Serena Williams is the same as the distance between Serena Williams and Venus Williams. This is because the ordinal scale has no unit of measurement.

The Interval Scale

The **interval scale** not only includes "greater than" and "less than" relationships, but also has a unit of measurement that permits us to describe *how much more or less* one object possesses than another. The Fahrenheit temperature scale represents an interval scale of measurement. We not only know that 90 degrees Fahrenheit is hotter than 70 degrees, and that 70 degrees is hotter than 60 degrees, but can also state that the distance between 90 and 70 is twice the distance between 70 and 60. This is because degree markings serve as the unit of measurement.

In an interval scale, the unit of measurement is arbitrary, and there is no absolute zero level where *none* of a given characteristic is present. Thus, multiples of measured values are not meaningful—e.g., 2 degrees Fahrenheit is not twice as warm as 1 degree. On questionnaire items like the following, business research practitioners typically treat the data as interval scale since the same physical and numerical distances exist between alternatives:

	[ ]	[ ]	[ ]	[ ]	[ ]
Kmart prices are	1	2	3	4	5
	low				high

The Ratio Scale

The **ratio scale** is similar to the interval scale, but has an absolute zero and multiples are meaningful. Election votes, natural gas consumption, return on investment, the speed of a production line, and FedEx Corporation's average daily delivery of 6,900,000 packages during 2008[15] are all examples of the ratio scale of measurement.

1.6 What is the difference between a qualitative variable and a quantitative variable? When would each be appropriate?

1.7 What is the difference between discrete and continuous variables? Under what circumstances would each be applicable?

1.8 The Acme School of Locksmithing has been accredited for the past 15 years. Discuss how this information might be interpreted as a

a. qualitative variable.
b. quantitative variable.

[14]Source: ESPN.com, June 19, 2009.
[15]Source: FedEx Corporation, *2008 Annual Report*, p. 26.

1.9 Jeff Bowlen, a labor relations expert, has collected information on strikes in various industries.

a. Jeff says, "Industry A has been harder hit by strikes than Industry B." In what scale of measurement is this information? Why?
b. Industry C has lost 10.8 days per worker, while Industry D has lost 14.5 days per worker. In what scale of measurement is this information? Why?

1.10 The Snowbird Ski Lodge attracts skiers from several New England states. For each of the following scales of measurement, provide one example of information that might be relevant to the lodge's business.

a. Nominal
b. Ordinal
c. Interval
d. Ratio

STATISTICS IN BUSINESS DECISIONS (1.5)

One aspect of business in which statistics plays an especially vital role is decision making. Every year, U.S. businesses risk billions of dollars in important decisions involving plant expansions, new product development, personnel selection, quality assurance, production techniques, supplier choices, and many others. These decisions almost always involve an element of uncertainty. Competitors, government, technology, and the social and economic environment, along with sometimes capricious consumers and voters, constitute largely uncontrollable factors that can sometimes foil the best-laid plans.

Prior to making decisions, companies often collect information through a series of steps called the *research process*. The steps include: (1) defining the problem in specific terms that can be answered by research, (2) deciding on the type of data required, (3) determining through what means the data will be obtained, (4) planning for the collection of data and, if necessary, selection of a sample, (5) collecting and analyzing the data, (6) drawing conclusions and reporting the findings, and (7) following through with decisions that take the findings into consideration. Business and survey research, discussed more fully in Chapter 4, provides both descriptive and inferential statistics that can improve business decisions in many kinds of situations.

EXERCISES

1.11 Restaurants sometimes provide "customer reaction" cards so that customers can evaluate their dining experience at the establishment. What kinds of decisions might be made on the basis of this information?

1.12 What kinds of statistical data might a burglar alarm company employ in trying to convince urban homeowners to purchase its product?

BUSINESS STATISTICS: TOOLS VERSUS TRICKS (1.6)

The techniques of business statistics are valuable tools for the enhancement of business operations and success. Appropriately, the major emphasis of this text will be to acquaint you with these techniques and to develop your proficiency in using them and interpreting their results.

On the other hand, as suggested earlier, these same techniques can be abused for personal or corporate gain. Improperly used, statistics can become an effective weapon with which to persuade or manipulate others into beliefs or behaviors that we'd like them to adopt. Note too that, even when they are not intentionally misused, the results of statistical research and analyses can depend a lot on when and how they were conducted, as Statistics in Action 1.1 shows.

Unlike many other pursuits, such as defusing torpedoes, climbing mountains, or wrestling alligators, improper actions in business statistics can sometimes work in your favor. (As embezzlers know, this can also be true in accounting.) Naturally, we don't expect that you'll use your knowledge of statistics to manipulate unknowing customers and colleagues, but you should be aware of how *others* may be using statistics in an attempt to manipulate *you*. Remember that one of the key goals of this text is to make you an informed consumer of statistical information generated by others. In general, when you are presented with statistical data or conclusions that have been generated by others, you should ask yourself this key question: *Who carried out this study and analyzed the data, and what benefits do they stand to gain from the conclusions reached?*

1.13 The text claims that a company or organization might actually benefit when one of its employees uses statistics incorrectly. How can this be?

1.14 The headline of an article in your daily newspaper begins "Research Study Reveals. . . ." As a statistics student who wishes to avoid accepting biased results, what single question should be foremost in your mind as you begin reading the article?

1.7 SUMMARY

Business statistics can be defined as the collection, summarization, analysis, and reporting of numerical findings relevant to a business decision or situation. As businesspersons and citizens, we are involved with statistics either as practitioners or as consumers of statistical claims and findings offered by others. Very early statistical efforts primarily involved counting people or possessions for taxation purposes. More recently, statistical methods have been applied in all facets of business as a tool for analysis and reporting, for reaching conclusions based on observed data, and as an aid to decision making.

Statistics can be divided into two branches: descriptive and inferential. Descriptive statistics focuses on summarizing and describing data that have been collected. Inferential statistics goes beyond mere description and, based on sample data, seeks to reach conclusions or make predictions regarding the population from which the sample was drawn. The population is the entire set of all people or objects of interest, with the sample being a subset of this group. A sample is said to be representative if its members tend to have the same characteristics as the larger population. A census involves measuring all people or objects in the population.

The sample statistic is a characteristic of the sample that is measured; it is often a mean, median, mode, proportion, or a measure of variability such as the range or standard deviation. The population parameter is the population characteristic that the sample statistic attempts to estimate.

Variables can be either qualitative or quantitative. Qualitative variables indicate whether a person or object possesses a given attribute, while quantitative

STATISTICS IN ACTION 1.1

High Stakes on the Interstate: Cell Phones and Accidents

Do cell phones contribute to auto accidents? The National Safety Council says they do, and has called for a nation-wide ban on cell phone use while driving. A number of research studies support this view. In one preliminary study, the researchers randomly selected 100 New York motorists who had been in an accident and 100 who had not. Those who had been in an accident were 30% more likely to have a cell phone. In another study, published in *The New England Journal of Medicine,* researchers found that cell phone use while driving quadrupled the chance of having an accident, a risk increase comparable to driving with one's blood alcohol level at the legal limit.

The Cellular Telecommunications Industry Association has a natural stake in this issue. There are more than 270 million cell phone subscribers, tens of thousands are signing up daily, and a high percentage of subscribers use their phones while driving. The association would have a vested interest in dismissing accident studies such as the ones above as limited, flawed, and having research shortcomings.

One thing is certain: More research is on the way. It will be performed by objective researchers as well as by individuals with a vested interest in the results. Future studies, their sponsors, and the interpretation of their results will play an important role in the safety of our highways and the economic vitality of our cellular phone industry.

Sources: "Survey: Car Phone Users Run Higher Risk of Crashes," *Indiana Gazette,* March 19, 1996, p. 10; "Ban Car Phones?", *USA Today,* April 27, 2000, p. 16A; "Get Off the Cell Phone," *Pittsburgh Tribune-Review,* January 29, 2000, p. A6; and "Safety Council Urges Ban on Cell Phone Use While Driving," cnn.com, June 19, 2009.

variables express how much of an attribute is possessed. Discrete quantitative variables can take on only certain values along an interval, with the possible values having gaps between them, while continuous quantitative variables can take on a value at any point along an interval.

When a variable is measured, a numerical value is assigned to it, and the result will be in one of four levels, or scales, of measurement—nominal, ordinal, interval, or ratio. The scale to which the measurements belong will be important in determining appropriate methods for data description and analysis.

By helping to reduce the uncertainty posed by largely uncontrollable factors, such as competitors, government, technology, the social and economic environment, and often unpredictable consumers and voters, statistics plays a vital role in business decision making. Although statistics is a valuable tool in business, its techniques can be abused or misused for personal or corporate gain. This makes it especially important for businesspersons to be informed consumers of statistical claims and findings.

CHAPTER EXERCISES

1.15 In studying the performance of the company's stock investments over the past year, the research manager of a mutual fund company finds that only 43% of the stocks returned more than the rate that had been expected at the beginning of the year.

a. Could this information be viewed as representing the nominal scale of measurement? If so, explain your reasoning. If not, why not?

b. Could this information be viewed as representing the ratio scale of measurement? If so, explain your reasoning. If not, why not?

1.16 For each of the following, indicate the scale of measurement that best describes the information.

a. In 2008, Dell Corporation had approximately 78,000 employees. Source: Fortune, May 4, 2009, p. F-48.

b. *USA Today* reports that the previous day's highest temperature in the United States was 105 degrees in Death Valley, California. Source: *USA Today,* June 19, 2009, p. 12A.

c. An individual respondent answers "yes" when asked if TV contributes to violence in the United States.

d. In a comparison test of family sedans, a magazine rates the Toyota Camry higher than the VW Passat.

1.17 Roger Amster teaches an English course in which 40 students are enrolled. After yesterday's class, Roger questioned the 5 students who always sit in the back of the classroom. Three of the 5 said "yes" when asked if they would like *A Tale of Two Cities* as the next class reading assignment.

a. Identify the population and the sample in this situation.
b. Is this likely to be a representative sample? If not, why not?

1.18 What kinds of statistical data play a role in an auto insurance firm's decision on the annual premium you'll pay for your policy?

1.19 Bill scored 1200 on the Scholastic Aptitude Test and entered college as a physics major. As a freshman, he changed to business because he thought it was more interesting. Because he made the dean's list last semester, his parents gave him $30 to buy a new Casio calculator. For this situation, identify at least one piece of information in the

a. nominal scale of measurement.
b. ordinal scale of measurement.
c. interval scale of measurement.
d. ratio scale of measurement.

1.20 For each of the following, indicate whether the appropriate variable would be qualitative or quantitative. If you identify the variable as quantitative, indicate whether it would be discrete or continuous.

a. Whether you own a Panasonic television set
b. Your status as either a full-time or a part-time student
c. The number of people who attended your school's graduation last year
d. The price of your most recent haircut
e. Sam's travel time from his dorm to the student union
f. The number of students on campus who belong to a social fraternity or sorority

1.21 Most undergraduate business students will not go on to become actual practitioners of statistical research and analysis. Considering this fact, why should such individuals bother to become familiar with business statistics?

1.22 A research firm observes that men are twice as likely as women to watch the Super Bowl on television. Does this information represent descriptive statistics or inferential statistics? Why?

CHAPTER 2

Visual Description of Data

"USA SNAPSHOTS" SET THE STANDARD

When it comes to creative visual displays to summarize data, hardly anything on the planet comes close to *USA Today* and its "USA Snapshots" that appear in the lower-left portion of the front page of each of the four sections of the newspaper. Whether it's "A look at statistics that shape the nation" (section A), "your finances" (section B), "the sports world" (section C), or "our lives" (section D), the visual is apt to be both informative and entertaining.

For example, when the imaginative folks who create "USA Snapshots" get their hands on some numbers, we can expect that practically any related object that happens to be round may end up becoming a pie chart, or that any relevant entity that's rectangular may find itself relegated to duty as a bar chart. An example of this creativity can be seen later in the chapter, in Figure 2.3.

If you're one of the many millions who read *USA Today,* chances are you'll notice a lot of other lively, resourceful approaches to the visual description of information. Complementing their extensive daily fare of news, editorials, and many other items that we all expect a good daily newspaper to present, *USA Today* and the "USA Snapshot" editors set the standard when it comes to reminding us that statistics can be as interesting as they are relevant.

Visualizing the data

LEARNING OBJECTIVES

After reading this chapter, you should be able to:

- Construct a frequency distribution and a histogram.
- Construct relative and cumulative frequency distributions.
- Construct a stem-and-leaf diagram to represent data.
- Visually represent data by using graphs and charts.
- Construct a dotplot and a scatter diagram.
- Construct contingency tables.

2.1 INTRODUCTION

When data have been collected, they are of little use until they have been organized and represented in a form that helps us understand the information contained. In this chapter, we'll discuss how raw data are converted to frequency distributions and visual displays that provide us with a "big picture" of the information collected.

By so organizing the data, we can better identify trends, patterns, and other characteristics that would not be apparent during a simple shuffle through a pile of questionnaires or other data collection forms. Such summarization also helps us compare data that have been collected at different points in time, by different researchers, or from different sources. It can be very difficult to reach conclusions unless we simplify the mass of numbers contained in the original data.

As we discussed in Chapter 1, variables are either quantitative or qualitative. In turn, the appropriate methods for representing the data will depend on whether the variable is quantitative or qualitative. The frequency distribution, histogram, stem-and-leaf display, dotplot, and scatter diagram techniques of this chapter are applicable to quantitative data, while the contingency table is used primarily for counts involving qualitative data.

2.2 THE FREQUENCY DISTRIBUTION AND THE HISTOGRAM

Raw data have not been manipulated or treated in any way beyond their original collection. As such, they will not be arranged or organized in any meaningful manner. When the data are quantitative, two of the ways we can address this problem are the frequency distribution and the histogram. The **frequency distribution** is a table that divides the data values into classes and shows the number of observed values that fall into each class. By converting data to a frequency distribution, we gain a perspective that helps us see the forest instead of the individual trees. A more visual representation, the **histogram** describes a frequency distribution by using a series of adjacent rectangles, each of which has a length that is proportional to the frequency of the observations within the range of values it represents. In either case, we have summarized the raw data in a condensed form that can be readily understood and easily interpreted.

The Frequency Distribution

We'll discuss the frequency distribution in the context of a research study that involves both safety and fuel-efficiency implications. Data are the speeds (miles per hour) of 105 vehicles observed along a section of highway where both accidents and fuel-inefficient speeds have been a problem.

EXAMPLE

Raw Data and Frequency Distribution

Part A of Table 2.1 lists the raw data consisting of measured speeds (mph) of 105 vehicles along a section of highway. There was a wide variety of speeds, and these data values are contained in data file **CX02SPEE**. If we want to learn more from this information by visually summarizing it, one of the ways is to construct a frequency distribution like the one shown in part B of the table.

TABLE 2.1

Raw data and frequency distribution for observed speeds of 105 vehicles.

A. Raw Data

54.2	58.7	71.2	69.7	51.7	75.6	53.7	57.0	82.5	76.8
62.1	67.4	64.6	70.5	48.4	69.0	65.2	64.1	65.8	56.9
82.0	62.0	54.3	73.2	74.5	73.6	76.2	55.1	66.9	62.4
70.7	68.3	62.8	83.5	54.8	68.3	69.4	59.2	60.9	60.4
60.2	75.4	55.4	56.3	77.1	61.2	50.8	67.1	70.6	60.7
56.0	80.7	80.1	56.2	70.1	63.7	70.9	54.6	61.6	63.2
72.2	84.0	76.8	61.6	61.1	80.8	58.3	52.8	74.3	71.4
63.2	57.8	61.9	75.8	80.8	57.3	62.6	75.1	67.6	78.0
52.2	57.6	61.1	66.9	88.5	66.7	61.2	73.9	79.1	70.2
65.2	61.3	69.5	72.8	57.5	71.4	64.7	78.4	67.6	76.3
78.6	66.8	71.1	58.9	61.1					

B. Frequency Distribution (Number of Motorists in Each Category)

Speed (mph)	Number of Motorists
45–under 50	1
50–under 55	9
55–under 60	15
60–under 65	24
65–under 70	17
70–under 75	17
75–under 80	13
80–under 85	8
85–under 90	1

Key Terms

In generating the frequency distribution in part B of Table 2.1, several judgmental decisions were involved, but there is no single "correct" frequency distribution for a given set of data. There are a number of guidelines for constructing a frequency distribution. Before discussing these rules of thumb and their application, we'll first define a few key terms upon which they rely:

Class Each category of the frequency distribution.

Frequency The number of data values falling within each class.

Class limits The boundaries for each class. These determine which data values are assigned to that class.

Class interval The width of each class. This is the difference between the lower limit of the class and the lower limit of the next higher class. When a frequency distribution is to have equally wide classes, the approximate width of each class is

$$\text{Approximate class width} = \frac{\text{Largest value in raw data} - \text{Smallest value in raw data}}{\text{Number of classes desired}}$$

Class mark The midpoint of each class. This is midway between the upper and lower class limits.

Guidelines for the Frequency Distribution

In constructing a frequency distribution for a given set of data, the following guidelines should be observed:

1. The set of classes must be **mutually exclusive** (i.e., a given data value can fall into only one class). There should be no overlap between classes, and limits such as the following would be inappropriate:

   Not allowed, since a value of 60 could fit into either class: 55–60
60–65

   Not allowed, since there's an overlap between the classes: 50–under 55
53–under 58

2. The set of classes must be **exhaustive** (i.e., include all possible data values). No data values should fall outside the range covered by the frequency distribution.
3. If possible, the classes should have equal widths. Unequal class widths make it difficult to interpret both frequency distributions and their graphical presentations.
4. Selecting the number of classes to use is a subjective process. If we have too few classes, important characteristics of the data may be buried within the small number of categories. If there are too many classes, many categories will contain either zero or a small number of values. In general, about 5 to 15 classes will be suitable.
5. Whenever possible, class widths should be round numbers (e.g., 5, 10, 25, 50, 100). For the highway speed data, selecting a width of 2.3 mph for each class would enhance neither the visual attractiveness nor the information value of the frequency distribution.
6. If possible, avoid using **open-end classes.** These are classes with either no lower limit or no upper limit—e.g., 85 mph or more. Such classes may not always be avoidable, however, since some data may include just a few values that are either very high or very low compared to the others.

The frequency distribution in part B of Table 2.1 was the result of applying the preceding guidelines. Illustrating the key terms introduced earlier, we will refer to the "50–under 55" class of the distribution:

- **Class limits** 50–under 55. All values are at least 50, but less than 55.
- **Frequency** 9. The number of motorists with a speed in this category.
- **Class interval** 5. The difference between the lower class limit and that of the next higher class, or 55 minus 50.
- **Class mark** 52.5. The midpoint of the interval; this can be calculated as the lower limit plus half the width of the interval, or $50 + (0.5)(5.0) = 52.5$.

Relative and Cumulative Frequency Distributions

Relative Frequency Distribution. Another useful approach to data expression is the **relative frequency distribution,** which describes the proportion or percentage of data values that fall within each category. The relative frequency distribution for the speed data is shown in Table 2.2; for example, of the 105 motorists, 15 of them (14.3%) were in the 55–under 60 class.

Relative frequencies can be useful in comparing two groups of unequal size, since the actual frequencies would tend to be greater for each class within the larger group than for a class in the smaller one. For example, if a frequency distribution of incomes for 100 physicians is compared with a frequency distribution for 500 business executives, more executives than physicians would be likely to fall into a given class. Relative frequency distributions would convert the groups to the same *size:* 100 percentage points each. Relative frequencies will play an important role in our discussion of probabilities in Chapter 5.

Cumulative Frequency Distribution. Another approach to the frequency distribution is to list the number of observations that are within or below each of the classes. This is known as a **cumulative frequency distribution.** When cumulative frequencies are divided by the total number of observations, the result is a *cumulative relative frequency distribution.* The "Cumulative Relative Frequency (%)" column in Table 2.2 shows the cumulative relative frequencies for the speed data in Table 2.1. Examining this column, we can readily see that 62.85% of the motorists had a speed less than 70 mph.

Cumulative percentages can also operate in the other direction (i.e., "greater than or within"). Based on Table 2.2, we can determine that 90.48% of the 105 motorists had a speed of at least 55 mph.

Speed (mph)	Number of Motorists	Relative Frequency (%)	Cumulative Frequency	Cumulative Relative Frequency (%)
45–under 50	1	0.95	1	0.95
50–under 55	9	8.57	10	9.52
55–under 60	15	14.29	25	23.81
60–under 65	24	22.86	49	46.66
65–under 70	17	16.19	66	62.85
70–under 75	17	16.19	83	79.05
75–under 80	13	12.38	96	91.43
80–under 85	8	7.62	104	99.05
85–under 90	1	0.95	105	100.00

TABLE 2.2

Frequencies, relative frequencies, cumulative frequencies, and cumulative relative frequencies for the speed data of Table 2.1.

The Histogram

The **histogram** describes a frequency distribution by using a series of adjacent rectangles, each of which has a length proportional to either the frequency or the relative frequency of the class it represents. The histogram in part (a) of Figure 2.1 is based on the speed-measurement data summarized in Table 2.1. The lower class limits (e.g., 45 mph, 50 mph, 55 mph, and so on) have been used in constructing the horizontal axis of the histogram.

The tallest rectangle in part (a) of Figure 2.1 is associated with the 60–under 65 class of Table 2.1, identifying this as the class having the greatest number of observations. The relative heights of the rectangles visually demonstrate how the frequencies tend to drop off as we proceed from the 60–under 65 class to the 65–under 70 class and higher.

The Frequency Polygon

Closely related to the histogram, the **frequency polygon** consists of line segments connecting the points formed by the intersections of the class marks with the class frequencies. Relative frequencies or percentages may also be used in constructing the figure. Empty classes are included at each end so the curve will intersect the horizontal axis. For the speed-measurement data in Table 2.1, these are the 40–under 45 and 90–under 95 classes. (*Note:* Had this been a distribution for which the first nonempty class was "0 but under 5," the empty class at the left would have been "−5 but under 0.") The frequency polygon for the speed-measurement data is shown in part (b) of Figure 2.1.

FIGURE 2.1

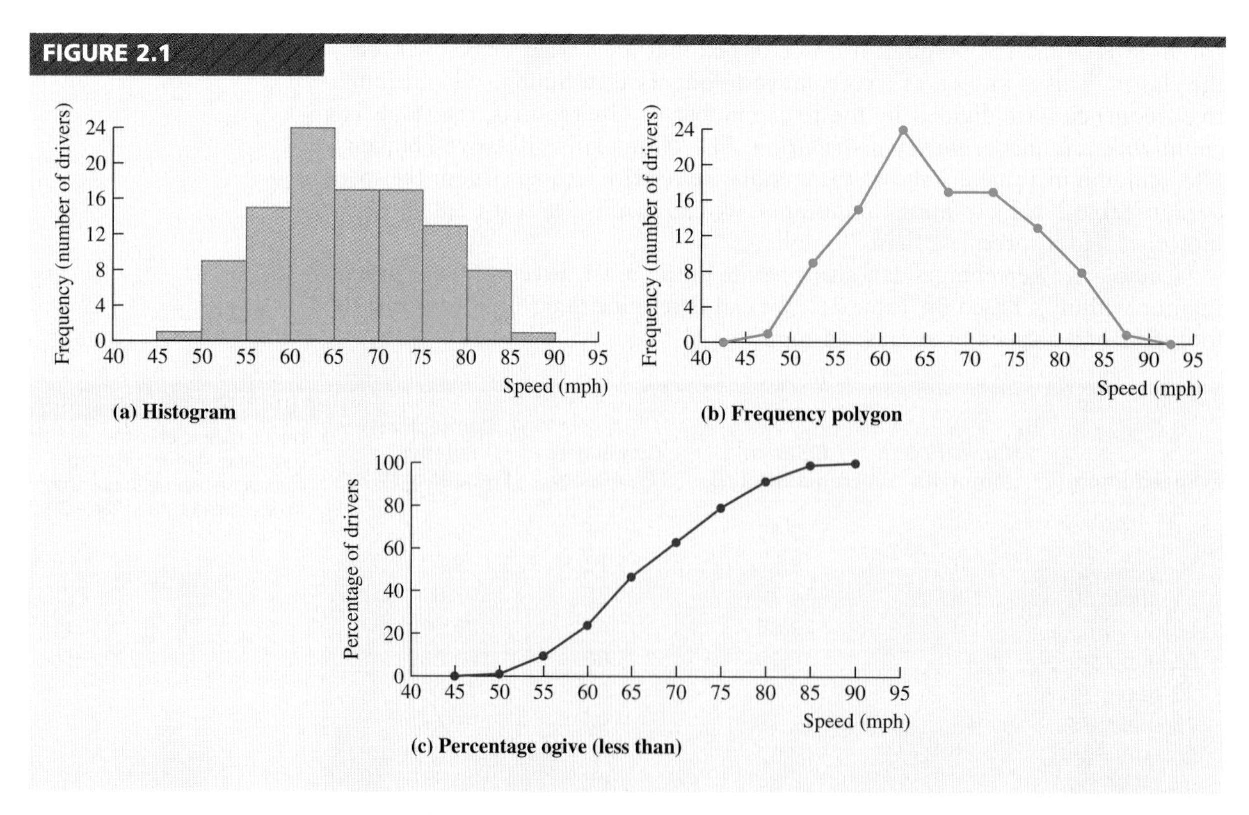

Histogram, frequency polygon, and percentage ogive for the speed-measurement data summarized in Table 2.1

Compared to the histogram, the frequency polygon is more realistic in that the number of observations increases or decreases more gradually across the various classes. The two endpoints make the diagram more complete by allowing the frequencies to taper off to zero at both ends.

Related to the frequency polygon is the **ogive,** a graphical display providing cumulative values for frequencies, relative frequencies, or percentages. These values can be either "greater than" or "less than." The ogive diagram in part (c) of Figure 2.1 shows the percentage of observations that are less than the upper limit of each class.

We can use the computer to generate a histogram as well as the underlying frequency distribution on which the histogram is based. Computer Solutions 2.1 describes the procedures and shows the results when we apply Excel and Minitab in constructing a histogram for the speed data in Table 2.1.

COMPUTER 2.1 SOLUTIONS

The Histogram

EXCEL

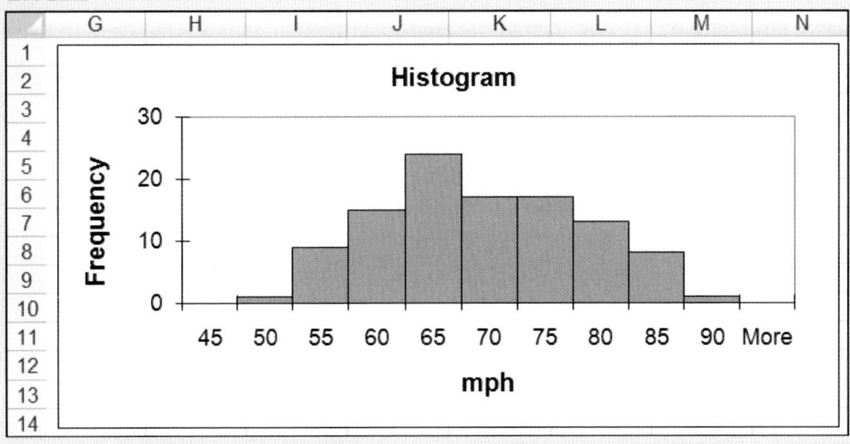

1. Open the Excel data file **CX02SPEE**. The name of the variable (Speed) is in cell A1, and the 105 speeds are in the cells immediately below.
2. Type **Bin** into cell C1. Enter the bin cutoffs (45 to 90, in multiples of 5) into C2:C11. (Alternatively, you can skip this step if you want Excel to generate its default frequency distribution.)
3. From the **Data** ribbon, click **Data Analysis** in the rightmost menu section. Within **Analysis Tools,** click **Histogram.** Click **OK.**
4. Enter the data range **(A1:A106)** into the **Input Range** box. If you entered the bin cutoffs as described in step 2, enter the bin range **(C1:C11)** into the **Bin Range** box. Click to place a check mark in the **Labels** box. (This is because each variable has its name in the first cell of its block.) Select **Output Range** and enter where the output is to begin—this will be cell **E1.**
5. Click to place a check mark into the **Chart Output** box. Click **OK.**
6. Within the chart, click on the word **Bin.** Click again, and type in **mph.** Right-click on any one of the bars in the chart. In the **Format Data Series** menu, click on the **Gap Width** slider and move it leftward to the **No Gap** position. Click **Close.** You can further improve the appearance by clicking on the chart and changing fonts, item locations, such as the key in the lower right, or the background color of the display. In the printout shown here, we have also enlarged the display and moved it slightly to the left.

(continued)

MINITAB

1. Open the Minitab data file **CX02SPEE**. The name of the variable (Speed) is in column C1, and the 105 speeds are in the cells immediately below.

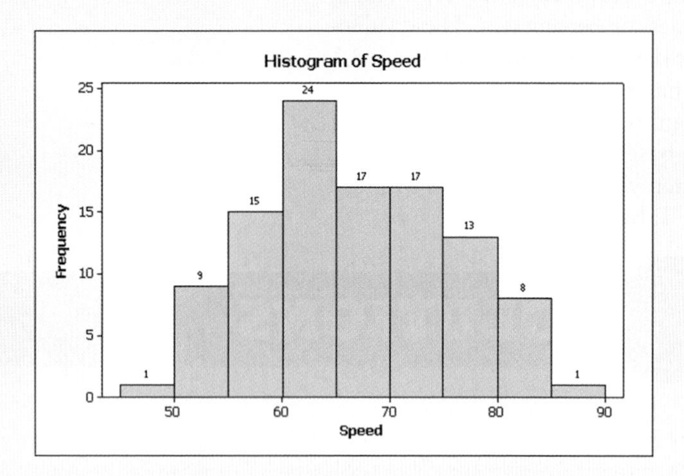

2. Click **Graph.** Click **Histogram.** Select **Simple** and click **OK.** Within the **Histogram** menu, indicate the column to be graphed by entering **C1** into the **Graph variables** box.
3. Click the **Labels** button, then click the **Data Labels** tab. Select **Use y-value labels.** (This will show the counts at the top of the bars.) Click **OK.** Click **OK.**
4. On the graph that appears, double-click on any one of the numbers on the horizontal axis. Click the **Binning** tab. In the **Interval Type** submenu, select **Cutpoint.** In the **Interval Definition** submenu, select **Midpoint/Cutpoint positions** and enter **45:90/5** into the box. (This provides intervals from 45 to 90, with the width of each interval being 5.) Click **OK.**

Excel and Minitab differ slightly in how they describe the classes in a frequency distribution. If you use the defaults in these programs, the frequency distributions may differ slightly whenever a data point happens to have exactly the same value as one of the upper limits because

1. Excel includes the *upper* limit but not the *lower* limit; the Excel bin value of "60" in Computer Solutions 2.1 represents values that are "more than 55, but not more than 60." A particular speed (x) will be in Excel's interval if $55 < x \leq 60$.
2. Minitab includes the *lower* limit, but not the *upper* limit, so a category of 55–60 is "at least 55, but less than 60." Thus, a particular speed (x) will be in Minitab's 55–60 interval if $55 \leq x < 60$.

This was not a problem with the speed data and Computer Solutions 2.1. However, if it had been, and we wanted the Excel frequency distribution to be the same as Minitab's, we could have simply used Excel bin values that were very slightly below those of Minitab's upper limits—for example, a bin value of 44.99 instead of 45.00, 49.99 instead of 50, 54.99 instead of 55, and so on.

EXERCISES

2.1 What is a frequency distribution? What benefits does it offer in the summarization and reporting of data values?

2.2 Generally, how do we go about deciding how many classes to use in a frequency distribution?

2.3 The National Safety Council reports the following age breakdown for licensed drivers in the United States. Source: Bureau of the Census, *Statistical Abstract of the United States 2009,* p. 682.

Age (Years)	Licensed Drivers (Millions)
under 20	10.72
20–under 25	16.84
25–under 35	40.78
35–under 45	45.59
45–under 55	36.70
55–under 65	23.32
65 or over	28.76

Identify the following for the 35–under 45 class: (a) frequency, (b) upper and lower limits, (c) width, and (d) midpoint.

2.4 Using the frequency distribution in Exercise 2.3, identify the following for the 25–under 35 class: (a) frequency, (b) upper and lower limits, (c) width, and (d) midpoint.

2.5 The National Center for Health Statistics reports the following age breakdown of deaths in the United States during 2006. Source: *The New York Times Almanac 2009,* p. 392.

Age (Years)	Number of Deaths (Thousands)
under 15	39.80
15–under 25	34.23
25–under 35	41.93
35–under 45	84.79
45–under 55	183.53
55–under 65	275.30
65–under 75	398.36
75 or over	1389.83

Identify the following for the 45–under 55 class: (a) frequency, (b) upper and lower limits, (c) width, and (d) midpoint.

2.6 Using the frequency distribution in Exercise 2.5, identify the following for the 15–under 25 class: (a) frequency, (b) upper and lower limits, (c) width, and (d) midpoint.

2.7 What is meant by the statement that the set of classes in a frequency distribution must be mutually exclusive and exhaustive?

2.8 For commercial banks in each state, the U.S. Federal Deposit Insurance Corporation has listed their total deposits (billions of dollars) as follows. Source: Bureau of the Census, *Statistical Abstract of the United States 2009,* p. 722.

	Deposits		Deposits		Deposits
AL	74.3	LA	73.0	OH	209.1
AK	6.9	ME	20.0	OK	57.4
AZ	80.0	MD	95.0	OR	47.9
AR	47.4	MA	180.8	PA	259.4
CA	751.0	MI	154.7	RI	25.5
CO	81.3	MN	106.2	SC	64.1
CT	81.0	MS	41.6	SD	63.3
DE	160.2	MO	102.6	TN	107.2
FL	373.9	MT	14.6	TX	450.0
GA	177.9	NE	36.0	UT	181.7
HI	26.8	NV	188.1	VT	9.9
ID	17.7	NH	21.5	VA	182.2
IL	338.9	NJ	222.5	WA	105.7
IN	88.6	NM	21.9	WV	25.9
IA	56.6	NY	722.8	WI	109.7
KS	54.0	NC	206.3	WY	10.6
KY	63.2	ND	13.9		

Construct a frequency distribution and a histogram for these data.

2.9 The accompanying data describe the hourly wage rates (dollars per hour) for 30 employees of an electronics firm:

22.66	24.39	17.31	21.02	21.61	20.97	18.58	16.61
19.74	21.57	20.56	22.16	20.16	18.97	22.64	19.62
22.05	22.03	17.09	24.60	23.82	17.80	16.28	19.34
22.22	19.49	22.27	18.20	19.29	20.43		

Construct a frequency distribution and a histogram for these data.

2.10 The following performance scores have been recorded for 25 job applicants who have taken a pre-employment aptitude test administered by the company to which they applied:

66.6	75.4	66.7	59.2	78.5	80.8	79.9	87.0	94.1
70.2	92.8	86.9	92.8	66.8	65.3	100.8	76.2	87.8
71.0	92.9	97.3	82.5	78.5	72.0	76.2		

Construct a frequency distribution and a histogram for these data.

2.11 During his career in the NHL, hockey great Wayne Gretzky had the following season-total goals for each of his 20 seasons. Source: *The World Almanac and Book of Facts 2000*, p. 954.

51	55	92	71	87	73	52	62	40	54
40	41	31	16	38	11	23	25	23	9

Construct a frequency distribution and a histogram for these data.

2.12 According to the U.S. Department of Agriculture, the distribution of U.S. farms according to value of annual sales is as follows. Source: Bureau of the Census, *Statistical Abstract of the United States 2006*, p. 546.

Annual Sales	Farms (Thousands)
under $10,000	1262
$10,000–under $25,000	256
$25,000–under $50,000	158
$50,000–under $100,000	140
$100,000–under $250,000	159
$250,000–under $500,000	82
$500,000–under $1,000,000	42
$1,000,000 or more	29

Convert this information to a
a. Relative frequency distribution.
b. Cumulative frequency distribution showing "less than or within" frequencies.

2.13 Convert the distribution in Exercise 2.3 to a
a. Relative frequency distribution.
b. Cumulative frequency distribution showing "less than or within" frequencies.

2.14 Convert the distribution in Exercise 2.8 to a
a. Relative frequency distribution.
b. Cumulative relative frequency distribution showing "greater than or within" relative frequencies.

2.15 Using the frequency distribution obtained in Exercise 2.8, convert the information to a "less-than" ogive.

2.16 For the frequency distribution constructed in Exercise 2.11, convert the information to a "less-than" ogive.

(DATA SET) *Note:* Exercises 2.17–2.19 require a computer and statistical software.

2.17 The current values of the stock portfolios for 80 clients of an investment counselor are as listed in data file **XR02017**. Use your computer statistical software to generate a frequency distribution and histogram describing this information. Do any of the portfolio values seem to be especially large or small compared to the others?

2.18 One of the ways to evaluate the ease of navigating a corporate website is to measure how long it takes to reach the link where you can view or download the company's most recent annual report. In randomly selecting and visiting 80 corporate websites, an investor needed an average of 39.9 seconds to reach the point where the company's annual report could be viewed or downloaded. The 80 times are in data file **XR02018**. Generate a frequency distribution and histogram describing this information. Comment on whether any of the corporate sites seemed to require especially long or short times for the annual report to be found.

2.19 A company executive has read with interest the finding that the average U.S. office worker receives about 60 e-mails per day. Assume that an executive, wishing to replicate this study within her own corporation, directs information technology personnel to find out the number of e-mails each of a sample of 100 office workers received yesterday, with the results as provided in the data file **XR02019**. Generate a frequency distribution and histogram describing this information and comment on the extent to which some workers appeared to be receiving an especially high or low number of e-mails. Source: emailstatcenter.com, June 20, 2009.

(2.3) THE STEM-AND-LEAF DISPLAY AND THE DOTPLOT

In this section, we examine two additional methods of showing how the data values are distributed: the stem-and-leaf display and the dotplot. Each of the techniques can be carried out either manually or with the computer and statistical software.

The Stem-and-Leaf Display

The **stem-and-leaf display,** a variant of the frequency distribution, uses a subset of the original digits as class descriptors. The technique is best explained through a few examples. The raw data are the numbers of Congressional bills vetoed during the administrations of seven U.S. presidents, from Johnson to Clinton.[1]

	Johnson	Nixon	Ford	Carter	Reagan	Bush	Clinton
Vetoes	30	43	66	31	78	44	38

In *stem-and-leaf* terms, we could describe these data as follows:

Stem (10's Digit)	Leaf (1's Digit)
3|018	(represents 30, 31, and 38)
4|34	(represents 43 and 44)
5|	(no data values in the 50s)
6|6	(represents 66)
7|8	(represents 78)

The figure to the left of the divider (|) is the **stem,** and the digits to the right are referred to as **leaves.** By using the digits in the data values, we have identified five different categories (30s, 40s, 50s, 60s, and 70s) and can see that there are three data values in the 30s, two in the 40s, one in the 60s, and one in the 70s.

Like the frequency distribution, the stem-and-leaf display allows us to quickly see how the data are arranged. For example, none of these presidents vetoed more than 44 bills, with the exceptions of Presidents Ford and Reagan, who vetoed 66 and 78, respectively. Compared with the frequency distribution, the stem-and-leaf display provides more detail, since it can describe the individual data values as well as show how many are in each group, or stem. Computer Solutions 2.2 shows the procedure and results when Excel and Minitab are used to generate a stem-and-leaf display based on the speed data of Table 2.1.

When the stem-and-leaf display is computer generated, the result may vary slightly from the previous example of the presidential vetoes. For example, Minitab may do the following:

1. Break each stem into two or more lines, each of which will have leaves in a given range. In Computer Solutions 2.2, speeds in the 50s are broken into two lines. The first line includes those with a leaf that is 0 to 4 (i.e., speeds from 50.x to 54.x), the second covers those for which the leaf is from 5 to 9 (i.e., speeds from 55.x to 59.x).
2. Include stem-and-leaf figures for outliers. An *outlier* is a data value very distant from most of the others. When outliers are present, they can be shown separately in a "HI" or a "LO" portion to avoid stretching the display and ending up with a large number of stems that have no leaves.

The stem-and-leaf display shows just two figures for each data value. For example, the data value 1475 could be represented as

1|4 (1000's digit and 100's digit)

or

14|7 (100's digit and 10's digit)

or

147|5 (10's digit and 1's digit)

In a given stem-and-leaf display, only one of these alternatives could be used because (1) a given data value can be represented only once in a display, and

[1]Source: Bureau of the Census, *Statistical Abstract of the United States 2009*, p. 247.

COMPUTER 2.2 SOLUTIONS

The Stem-and-Leaf Display

EXCEL

	A	B	C	D	E	F
1	**Stem & Leaf Display**					
2						
3	**Stems**	**Leaves**				
4	4	->8				
5	5	->0122344445566666777778889				
6	6	->000011111111122222333444555666677777889999				
7	7	->000000111122333445555666678889				
8	8	->000022348				

1. Open the Excel data file **CX02SPEE**. Click on **A1** and drag to **A106** to select the label and data values in cells **A1:A106.** From the **Add-Ins** ribbon, click **Data Analysis Plus** in the leftmost menu section.
2. In the **Data Analysis Plus** menu, click **Stem and Leaf Display.** Click **OK.** Ensure that the **Input Range** box contains the range you specified in step 1. In the **Increment** menu box, select **10** (because the 10's digit is the stem for the display). Click **Labels.** Click **OK.**

MINITAB

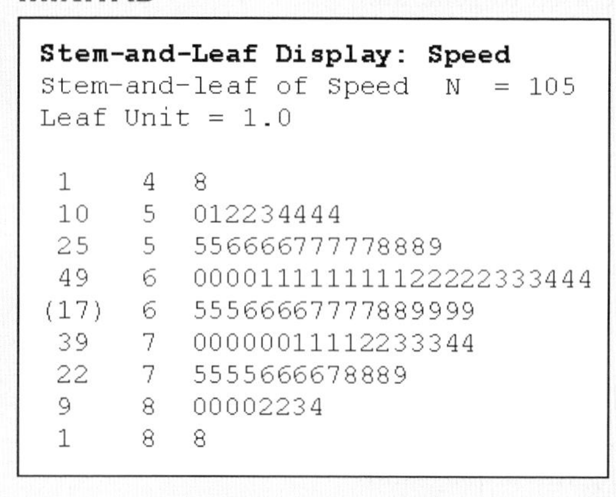

```
Stem-and-Leaf Display: Speed
Stem-and-leaf of Speed  N  = 105
Leaf Unit = 1.0

  1    4  8
 10    5  012234444
 25    5  556666777778889
 49    6  000011111111122222333444
(17)   6  55566667777789999
 39    7  00000011112233344
 22    7  5555666678889
  9    8  00002234
  1    8  8
```

1. Open the Minitab data file **CX02SPEE**. The variable (Speed) is in column C1.
2. Click **Graph.** Click **Stem-And-Leaf.** Enter **C1** into the **Variables** box. Click **OK.**

(2) all of the values must be expressed in terms of the same stem digit and the same leaf digit. If we were to deviate from either of these rules, the resulting display would be meaningless.

When data values are in decimal form, such as 3.4, 2.5, and 4.1, the stem-and-leaf method can still be used. In expressing these numbers, the stem would be the 1's digit, and each leaf would correspond to the first number to the right of the decimal point. For example, the number 3.4 would be converted to a stem of 3 and a leaf of 4. The location of the decimal point would have to be considered during interpretation of the display.

COMPUTER 2.3 SOLUTIONS

The Dotplot

EXCEL

Excel does not currently provide the dotplot display.

MINITAB

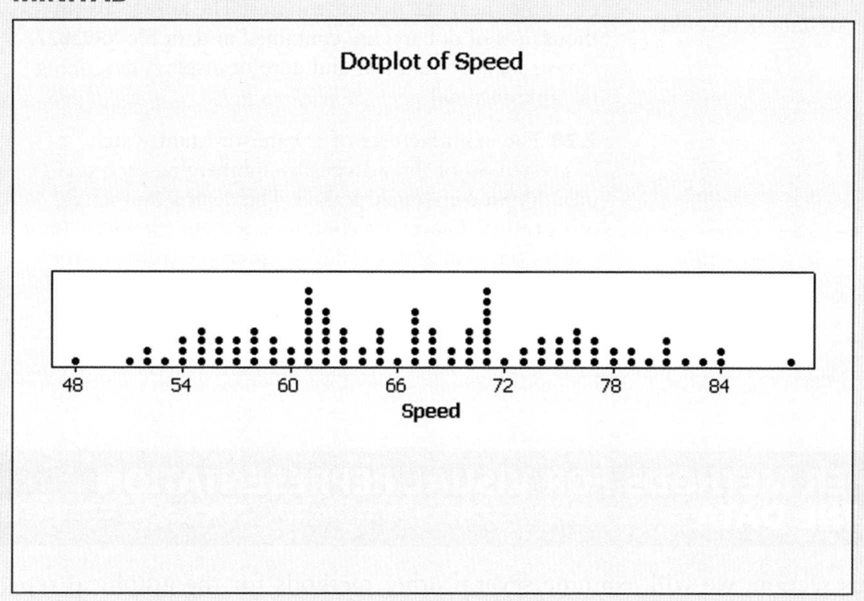

1. Open the Minitab data file **CX02SPEE**. The variable (Speed) is in column C1.
2. Click **Graph.** Click **Dotplot.** Select **One Y Simple.** Click **OK.** Enter **C1** into the **Variables** box. Click **OK.**

The Dotplot

The **dotplot** displays each data value as a dot and allows us to readily see the shape of the distribution as well as the high and low values. Computer Solutions 2.3 shows the procedure and the result when we apply Minitab's dotplot feature to the speed data shown previously in Table 2.1.

EXERCISES

2.20 Construct a stem-and-leaf display for the following data:

15	64	15	34	75	24
81	67	19	25	48	57
69	62	41	46	35	27
72	64	48	51	77	71
21	20	26	42	83	38

2.21 Construct a stem-and-leaf display for the following data:

11	22	28	32	41	46	52	56	59
13	24	28	33	41	46	52	57	60
17	24	29	33	42	46	53	57	61
19	24	30	37	43	48	53	57	61
19	26	30	39	43	49	54	59	63

2.22 In the following stem-and-leaf display for a set of two-digit integers, the stem is the 10's digit, and each leaf is the 1's digit. What is the original set of data?

2|002278
3|011359
4|1344
5|47

2.23 In the following stem-and-leaf display for a set of three-digit integers, the stem is the 100's digit, and each leaf is the 10's digit. Is it possible to determine the exact values in the original data from this display? If so, list the data values. If not, provide a set of data that could have led to this display.

6|133579
7|00257
8|66899
9|0003568

(**DATA SET**) *Note:* Exercises 2.24–2.28 require a computer and statistical software capable of constructing stem-and-leaf and dotplot displays.

2.24 Construct stem-and-leaf and dotplot displays based on the investment portfolio values in the data file (**XR02017**) for Exercise 2.17.

2.25 Construct stem-and-leaf and dotplot displays based on the access times in the data file (**XR02018**) for Exercise 2.18.

2.26 Construct stem-and-leaf and dotplot displays based on the e-mail receipts in the data file (**XR02019**) for Exercise 2.19.

2.27 City tax officials have just finished appraising 60 homes near the downtown area. The appraisals (in thousands of dollars) are contained in data file **XR02027**. Construct stem-and-leaf and dotplot displays describing the appraisal data.

2.28 The manufacturer of a water-resistant watch has tested 80 of the watches by submerging each one until its protective seal leaked. The depths (in meters) just prior to failure are contained in data file **XR02028**. Construct stem-and-leaf and dotplot displays for these data.

(2.4) OTHER METHODS FOR VISUAL REPRESENTATION OF THE DATA

In this section, we will examine several other methods for the graphical representation of data, then discuss some of the ways in which graphs and charts can be used (by either the unwary or the unscrupulous) to deceive the reader or viewer. We will also provide several Computer Solutions to guide you in using Excel and Minitab to generate some of the more common graphical presentations. These are just some of the more popular approaches. There are many other possibilities.

The Bar Chart

Like the histogram, the **bar chart** represents frequencies according to the relative lengths of a set of rectangles, but it differs in two respects from the histogram: (1) the histogram is used in representing quantitative data, while the bar chart represents qualitative data; and (2) adjacent rectangles in the histogram share a common side, while those in the bar chart have a gap between them. Computer Solutions 2.4 shows the Excel and Minitab procedures for generating a bar chart describing the number of U.S. nonrecreational airplane pilot certificates in each of four categories.[2] The relevant data file is **CX02AIR**.

[2]Source: General Aviation Manufacturers Association, *General Aviation Statistical Databook 2008 Edition*, p. 38.

The Line Graph

The **line graph** is capable of simultaneously showing values of two quantitative variables (y, or vertical axis, and x, or horizontal axis); it consists of linear segments connecting points observed or measured for each variable. When x represents time, the result is a time series view of the y variable. Even more information can be presented if two or more y variables are graphed together. Computer Solutions 2.5 shows the Excel and Minitab procedures for generating a line chart showing annual net income for McDonald's Corporation from 2003 through 2008.[3] The pertinent data file is **CX02MCD**.

COMPUTER 2.4 SOLUTIONS

The Bar Chart

EXCEL

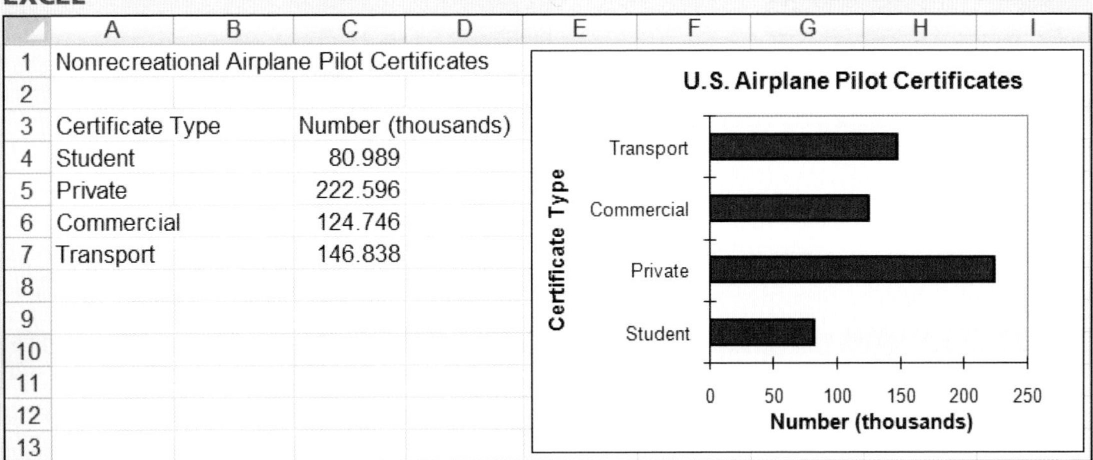

	A	B	C	D
1	Nonrecreational Airplane Pilot Certificates			
2				
3	Certificate Type		Number (thousands)	
4	Student		80.989	
5	Private		222.596	
6	Commercial		124.746	
7	Transport		146.838	
8				
9				
10				
11				
12				
13				

1. Open the Excel data file **CX02AIR**. The labels, **Certificate Type** and **Number (thousands),** have already been entered into **A3** and **C3,** respectively. The name for the first category has already been entered into **A4,** and its corresponding numerical value (thousands of pilots) has been entered into **C4.** We have continued downward, entering category names into column A and corresponding numerical values into column C, until all category names and their values have been entered into columns A and C.
2. Click on **A4** and drag to **C7** to select cells **A4:C7.**
3. From the **Insert** ribbon, click **Bar** from the **Charts** submenu. From the next menu, click the first choice **(Clustered Bar)** from the **2-D Bar** section.
4. To further improve the appearance, click on the chart and drag the borders to expand it vertically and horizontally. If you don't want the little "Legend" box that originally appears, just right-click within it, then click **Delete.** Additional editing can include insertion of the desired labels: Double-click within the chart area. From the **Layout** ribbon, use the **Chart Title** and **Axis Titles** selections within the **Labels** menu.

(continued)

[3]Source: McDonald's Corporation, *2008 Annual Report*, p. 19.

MINITAB

1. Open the Minitab data file CX02AIR. The labels, **Certificate Type** and **Number (thousands),** have already been entered at the top of columns **C1** and **C2,** respectively. The names of the categories have already been entered into **C1,** and the numerical values (thousands of pilots) have been entered into **C2.**
2. Click **Graph.** Click **Bar Chart.** In the **Bars represent** box, select **Values from a table.** Select **One column of values, Simple.** Click **OK.** Within the **Graph variables** menu, enter **C2.** Enter **C1** into the **Categorical variable** box. Click **Scale.** Select the **Axes and Ticks** tab and select **Transpose value and category scales.** Click **OK.** Click **OK.**

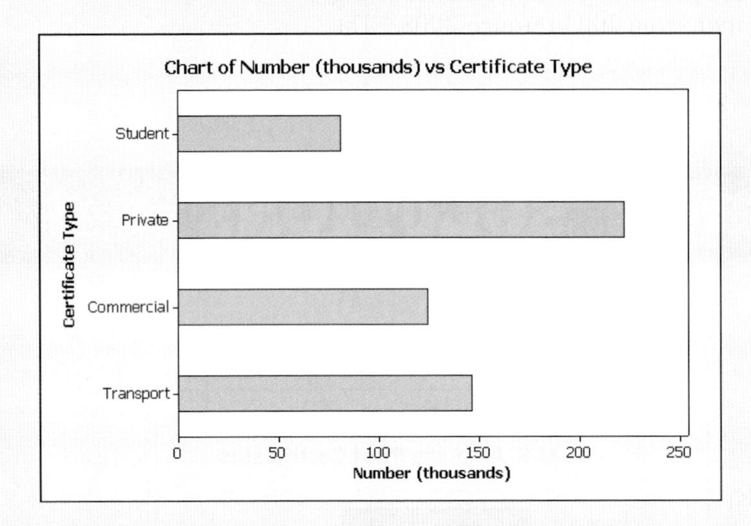

COMPUTER 2.5 SOLUTIONS

The Line Chart

EXCEL

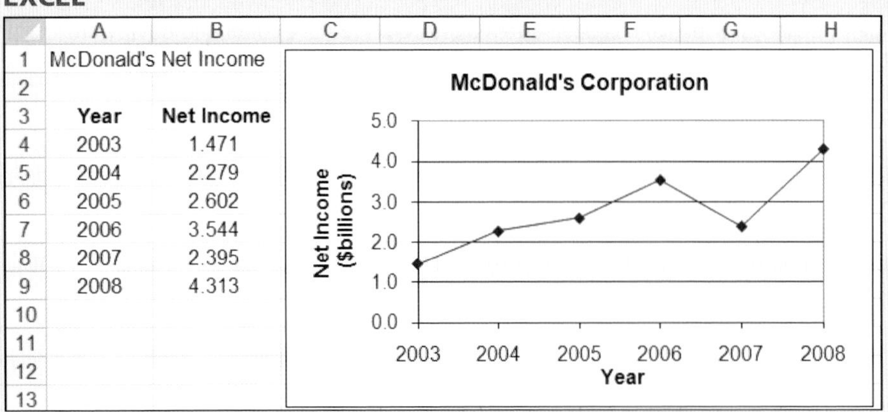

	A	B
1	McDonald's Net Income	
2		
3	Year	Net Income
4	2003	1.471
5	2004	2.279
6	2005	2.602
7	2006	3.544
8	2007	2.395
9	2008	4.313
10		
11		
12		
13		

1. Open the Excel data file **CX02MCD**. The labels and data are already entered, as shown in the printout. Click on **A4** and drag to **B9** to select cells **A4:B9.** (*Note:* Within the two columns, the variable to be represented by the vertical axis should always be in the column at the right.)
2. From the **Insert** ribbon, click **Scatter** from the **Charts** menu. From the **Scatter** submenu, click the fifth option **(Scatter with Straight Lines).**
3. Further appearance improvements can be made by clicking on the chart and dragging the borders to expand it vertically and horizontally. If you don't want the little "Legend" box that originally appears, just right-click within it, then click **Delete.** Additional editing can include insertion of the desired labels: Double-click within the chart area. From the **Layout** ribbon, use the **Chart Title** and **Axis Titles** selections within the **Labels** menu.

MINITAB

1. Open the Minitab data file **CX02MCD**. The labels, **Year** and **Net Income,** have already been entered at the top of columns **C1** and **C2,** respectively. The years and the numerical values have been entered into **C1** and **C2,** respectively.
2. Click **Graph.** Click **Scatterplot.** Select **With Connect Line.** Click **OK.** Enter **C2** into the first line of the **Y variables** box. Enter **C1** into the first line of the **X variables** box. Click **OK.**
3. Double-click on the chart title and the vertical-axis labels and revise as shown.

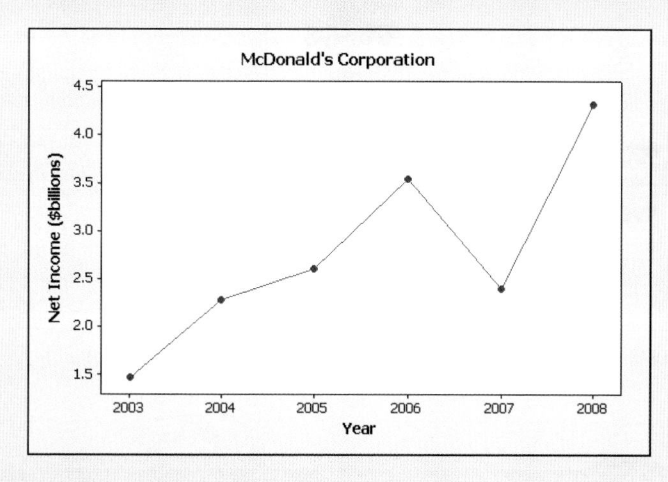

The Pie Chart

The **pie chart** is a circular display divided into sections based on either the number of observations within or the relative values of the segments. If the pie chart is not computer generated, it can be constructed by using the principle that a circle contains 360 degrees. The angle used for each piece of the pie can be calculated as follows:

$$\frac{\text{Number of degrees}}{\text{for the category}} = \frac{\text{Relative value}}{\text{of the category}} \times 360$$

For example, if 25% of the observations fall into a group, they would be represented by a section of the circle that includes (0.25×360), or 90 degrees. Computer Solutions 2.6 shows the Excel and Minitab procedures for generating a pie chart to show the relative importance of four major business segments in contributing to Home Depot Corporation's overall profit.[4] The underlying data file is **CX02HDEP**.

[4]Source: Home Depot Corporation, *2008 Annual Report,* pp. 3, 28.

The Pictogram

Using symbols instead of a bar, the **pictogram** can describe frequencies or other values of interest. Figure 2.2 is an example of this method; it was used by Panamco to describe soft drink sales in Central America over a 3-year period. In the diagram, each truck represents about 12.5 million cases of soft drink products. When setting up a pictogram, the choice of symbols is up to you. This is an important consideration because the right (or wrong) symbols can lend nonverbal or emotional content to the display. For example, a drawing of a sad child with her arm in a cast

FIGURE 2.2

In the pictogram, the symbols represent frequencies or other values of interest. This chart shows how soft drink sales (millions of cases) in Central America increased from 1996 through 1998.

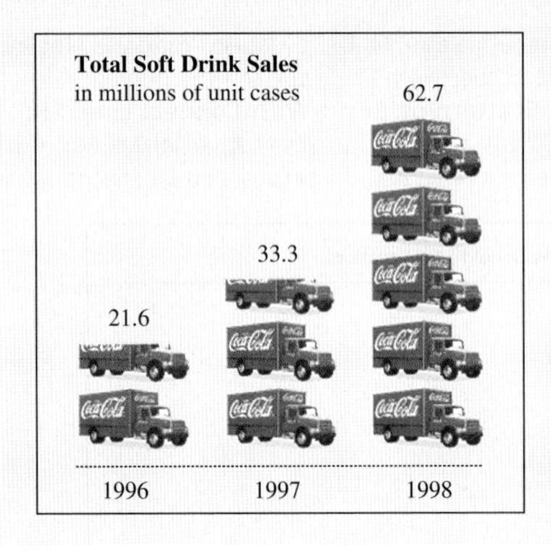

Source: Panamco, *Annual Report 1998*, p. 24.

COMPUTER 2.6 SOLUTIONS

The Pie Chart

EXCEL

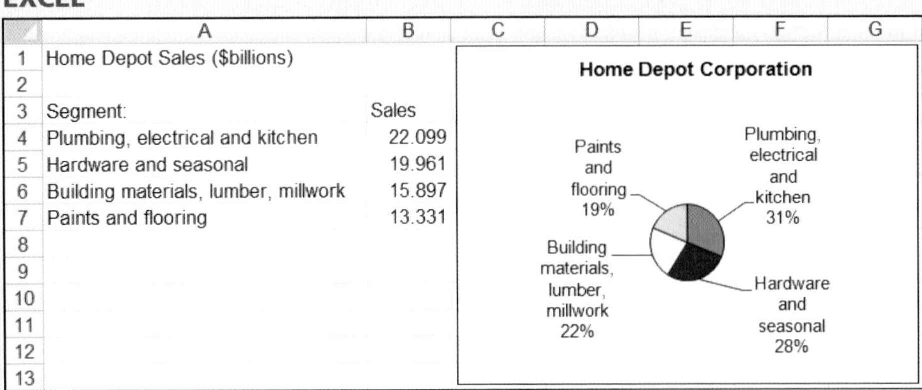

	A	B
1	Home Depot Sales ($billions)	
2		
3	Segment:	Sales
4	Plumbing, electrical and kitchen	22.099
5	Hardware and seasonal	19.961
6	Building materials, lumber, millwork	15.897
7	Paints and flooring	13.331

1. Open the Excel data file **CX02HDEP**. The segment names have already been entered as shown in the display, as have the sales for each segment.
2. Click on **A4** and drag to **B7** to select cells **A4:B7.** From the **Insert** ribbon and its **Charts** menu, click **Pie.** Click on the first option (**Pie**) in the **2-D Pie** menu.
3. Additional editing can include insertion of the desired chart title: Double-click within the chart area. From the **Layout** ribbon, use the **Chart Title** selection within the **Labels** menu.

MINITAB

1. Open the Minitab data file **CX02HDEP**. The labels, **Segment** and **Sales,** have already been entered at the top of columns **C1** and **C2,** respectively. The names of the segments have already been entered into **C1** and the numerical values (sales, billions of dollars) have been entered into **C2.**
2. Click **Graph.** Click **Pie Chart.** Select **Chart values from a table.** Enter **C1** into the **Categorical variable** box. Enter **C2** into the **Summary variables** box. Click **OK.**
3. Double-click on the chart title and revise as shown.

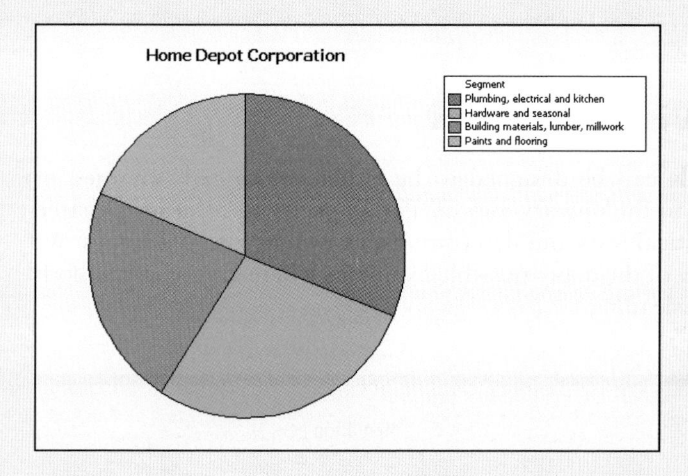

(each symbol representing 10,000 abused children) could help to emphasize the emotional and social costs of child abuse.

The Sketch

Varying in size depending on the frequency or other numerical value displayed, the **sketch** is a drawing or pictorial representation of some symbol relevant to the data. This approach will be demonstrated in part (c) of Figure 2.4.

Other Visuals

The preceding approaches are but a few of the many possibilities for the visual description of information. As noted in the chapter-opening vignette, *USA Today* readers are treated daily to a wide variety of interesting and informative displays, including the "Statistics That Shape Our ... " visuals in the lower-left corner of the front page of each section. One of these displays is shown in Figure 2.3.

FIGURE 2.3

One of the many creative ways *USA Today* and its "USA Snapshots" present informative statistics in an interesting way.

Source: From USA SNAPSHOTS®, "Workers Take Happiness Over Money", *USA Today*, September 20, 2001. Copyright 2001. Reprinted with permission.

The Abuse of Visual Displays

Remember that visuals can be designed to be either emotionally charged or purposely misleading to the unwary viewer. This capacity to mislead is shared by a great many statistical tests and descriptions, as well as visual displays. We will consider just a few of the many possible examples where graphical methods could be viewed as misleading.

FIGURE 2.4

The information presented may be distorted in several ways. Part (a) shows the effect of compressing the data by using a high endpoint for the vertical axis. In part (b), the change is exaggerated by taking a slice from the vertical axis. In part (c), although the sketch representing 2004 is 65.7% higher and 65.7% wider than the 2003 sketch, the increase is distorted because the area of the 2004 sketch is 274.6% as large as that of the 2003 sketch.

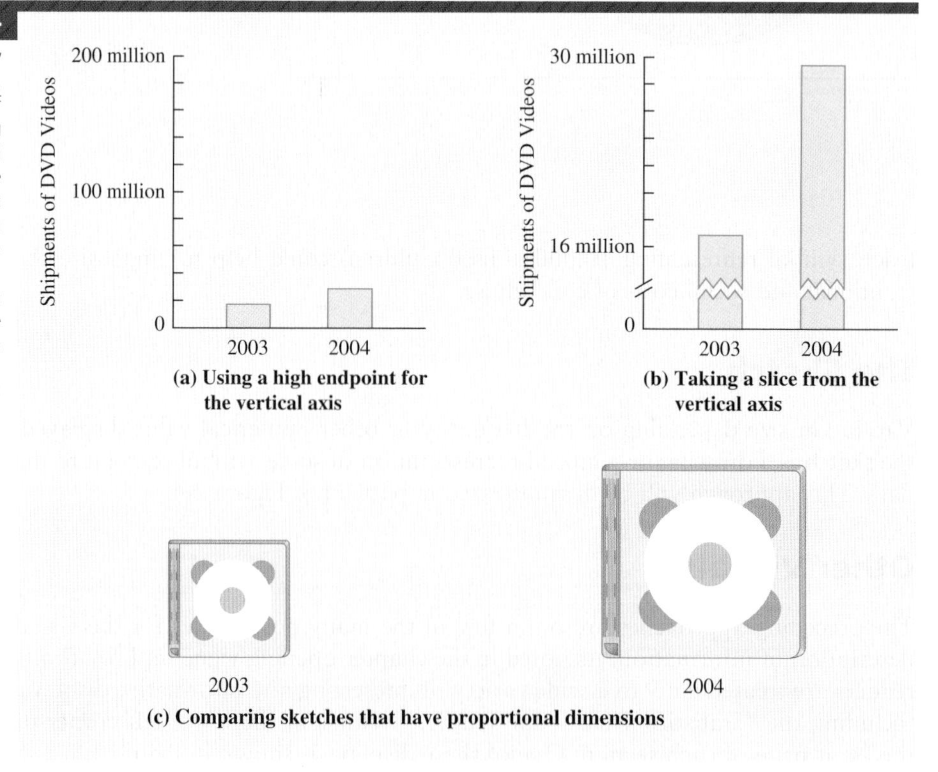

During 2004, 29.0 million DVD videos were shipped, compared to 17.5 million in 2003.[5] Using these data, we can construct the charts in Figure 2.4. In part (a), the vertical axis is given a very high value as its upper limit, and the data are "compressed" so that the increase in shipments isn't very impressive. In part (b) of Figure 2.4, taking a slice from the vertical axis causes the increase in shipments to be more pronounced. Another strategy for achieving the same effect is to begin the vertical axis with a value other than zero.

In part (c) of Figure 2.4, we see how shipments in the two years could be compared using a sketch of a DVD video case. Although the sketch on the right is 65.7% higher than the one on the left, it *looks more than* 65.7% larger. This is because area for each sketch is height times width, and both the height *and* the width are 65.7% greater for the one at the right. Because both height and width are increased, the sketch for 2004 shipments has an area 274.6% (based on 1.657 × 1.657) as large as its 2003 counterpart.

Figure 2.5 is another example of visual distortion by using a nonzero vertical axis. Over the years shown, sales increased by only 1.4%, and this dismal performance is evident from part (a) of the figure. However, in part (b), the starting

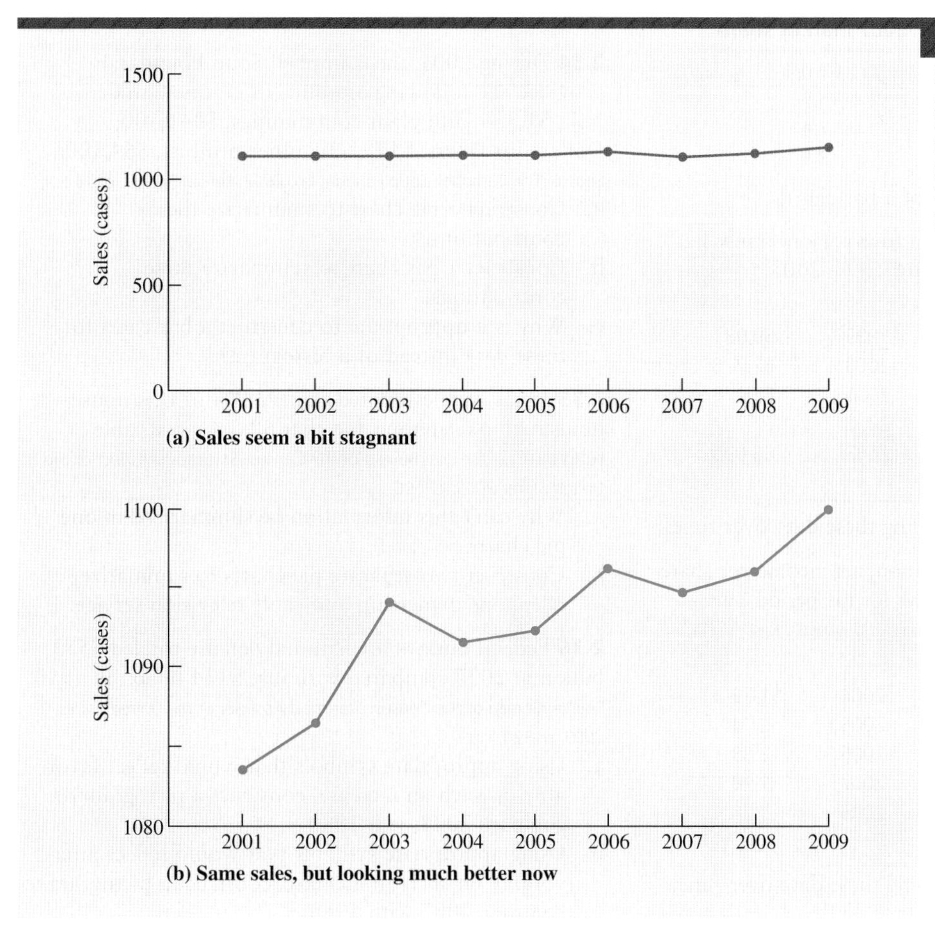

FIGURE 2.5

If the vertical scale is stretched and has a starting point that is greater than zero, the sales for this firm can look a lot better than the 1.4% growth that occurred over the 8-year period.

[5]Data source: Data (but not the visual displays) are from *The World Almanac and Book of Facts 2006*, p. 276.

point for the vertical axis is 1080 cases instead of 0, and the vertical axis is stretched considerably. As a result, it looks as though the company has done very well over the years shown.

EXERCISES

Note: These graphs and charts can be done by hand, but, if possible, use the computer and your statistical software.

2.29 What is the difference between a histogram and a bar chart? For what type of data would each be appropriate?

2.30 The following U.S. market shares for 2007 retail car sales have been reported. Source: *World Almanac and Book of Facts 2009*, p. 122.

Country or Region of Manufacture	2007 market share
North America	69.0%
Japan	15.5
Germany	7.4
Other	8.1

Express the data in the form of a bar chart.

2.31 The median income for a four-person family has been reported as shown here for 1993–2003. Source: *Time Almanac 2006*, p. 627.

1993	$45,161	1999	$59,981
1994	47,012	2000	62,228
1995	49,687	2001	63,278
1996	51,518	2002	62,732
1997	53,350	2003	65,093
1998	56,061		

Construct a line graph describing these data over time.

2.32 For McDonald's Corporation, net income per share of common stock was as follows for the period 1998–2008. Source: McDonald's Corporation, *2005 Annual Report*, p. 1; *2008 Annual Report*, p. 19.

1998	$1.10	2004	$1.79
1999	1.39	2005	2.04
2000	1.46	2006	2.83
2001	1.25	2007	1.98
2002	0.70	2008	3.76
2003	1.15		

Construct a line graph describing these data over time.

2.33 For the period 1998–2008, McDonald's Corporation paid the following dividends (rounded to nearest penny) per share of common stock. Source: McDonald's Corporation, *2005 Annual Report*, p. 1; *2008 Annual Report*, p. 19.

1998	$0.18	2004	$0.55
1999	0.20	2005	0.67
2000	0.22	2006	1.00
2001	0.23	2007	1.50
2002	0.24	2008	1.63
2003	0.40		

Using these data and the data in Exercise 2.32, draw a line graph that describes both net income and dividend per share of common stock over time. (Both lines will be plotted on the same graph.)

2.34 During 2001, the Campbell Soup Foundation provided the following amounts in grants: Camden, N.J., $1,336,700; plant communities, $341,500; Dollars for Doers, $179,600; other projects, $64,100. Source: The Campbell Soup Foundation, *2002 Annual Report*, p. 16.

a. Construct a pie chart to summarize these contributions.
b. Construct a bar chart to summarize these contributions.
c. Why is it appropriate to construct a bar chart for these data instead of a histogram?

2.35 It has been estimated that 92.9% of U.S. households own a telephone and that 68.5% have cable television. Source: Bureau of the Census, *Statistical Abstract of the United States 2009*, p. 696.

a. Why can't this information be summarized in one pie chart?
b. Construct two separate pie charts to summarize telephone ownership and cable television service.

2.36 Federal outlays for national defense totaled $553 billion in 2007 compared with just $134 billion in 1980. Source: Bureau of the Census, *Statistical Abstract of the United States 2009*, p. 324.

a. Using appropriate symbols that would reflect favorably on such an increase, construct a pictogram to compare 2007 with 1980.
b. Using appropriate symbols that would reflect unfavorably on such an increase, construct a pictogram to compare 2007 with 1980.

2.37 For the data in Exercise 2.36, use the sketch technique (and an appropriate symbol of your choice) to compare the 2007 and 1980 expenditures.

THE SCATTER DIAGRAM

2.5

There are times when we would like to find out whether there is a relationship between two quantitative variables—for example, whether sales is related to advertising, whether starting salary is related to undergraduate grade point average, or whether the price of a stock is related to the company's profit per share. To examine whether a relationship exists, we can begin with a graphical device known as the **scatter diagram,** or **scatterplot.**

Think of the scatter diagram as a sort of two-dimensional dotplot. Each point in the diagram represents a pair of known or observed values of two variables, generally referred to as y and x, with y represented along the vertical axis and x represented along the horizontal axis. The two variables are referred to as the **dependent** (y) and **independent** (x) variables, since a typical purpose for this type of analysis is to estimate or predict what y will be for a given value of x.

Once we have drawn a scatter diagram, we can "fit" a line to it in such a way that the line is a reasonable approximation to the points in the diagram. In viewing the "best-fit" line and the nature of the scatter diagram, we can tell more about whether the variables are related and, if so, in what way.

1. A *direct (positive) linear relationship* between the variables, as shown in part (a) of Figure 2.6. The best-fit line is linear and has a positive slope, with both y and x increasing together.
2. An *inverse (negative) linear relationship* between the variables, as shown in part (b) of Figure 2.6. The best-fit line is linear and has a negative slope, with y decreasing as x increases.

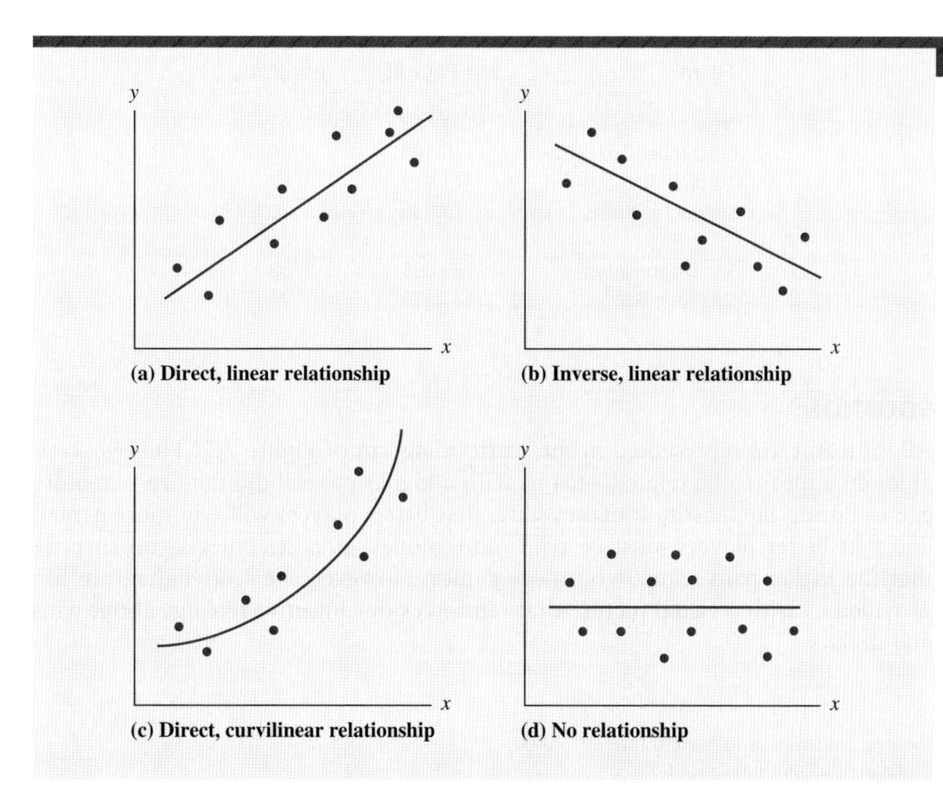

FIGURE 2.6

When two variables are related, as in parts (a) through (c), the relationship can be either direct (positive) or inverse (negative), and either linear or curvilinear. In part (d), there is no relationship at all between the variables.

3. A *curvilinear relationship* between the variables, as shown in part (c) of Figure 2.6. The best-fit line is a curve. As with a linear relationship, a curvilinear relationship can be either direct (positive) or inverse (negative).
4. *No relationship* between the variables, as shown in part (d) of Figure 2.6. The best-fit line is horizontal, with a slope of zero and, when we view the scatter diagram, knowing the value of x is of no help whatsoever in predicting the value of y.

In this chapter, we will consider only *linear* relationships between variables, and there will be two possibilities for fitting a straight line (i.e., a linear equation) to the data. The first possibility is the "eyeball" method, using personal judgment and a ruler applied to the scatter diagram. The second, more accurate, approach is to use the computer and your statistical software to fit a straight line that is mathematically optimum. This technique, known as the "least-squares" method, will be discussed in much more detail in Chapter 15, Simple Linear Regression and Correlation.

EXAMPLE

Best-Fit Equation

In 2006, opening-day Major League Baseball team payrolls ranged from $194.633 million for the New York Yankees to just $14.999 million for the Florida Marlins.[6] During the 162-game season, the Yankees won 97 games; the Marlins, just 78.[7] The payroll and wins data for the Yankees, Marlins, and the other major league teams are contained in the data file CX02BASE. A partial listing of the data for the 30 teams is shown here:

Team	x = Payroll	y = Wins
New York Yankees	$194.633	97
Boston Red Sox	120.100	86
Toronto Blue Jays	71.915	87
Baltimore Orioles	72.586	70
⋮		
San Diego Padres	69.896	88
Colorado Rockies	41.233	76

SOLUTION

All 30 teams are represented in the scatter diagram of Figure 2.7. The two variables do appear to be related—teams with a higher payroll did tend to win more games during the season. It makes sense that better players will win more games, and that better players must be paid more money; thus it comes as no surprise that the higher-paid teams won more games. However, the scatter diagram has provided us with a visual picture that reinforces our intuitive feelings about wins and dollars.

[6]Source: content.usatoday.com, June 21, 2009.
[7]Source: mlb.com, June 21, 2009.

EXAMPLEEXAMPLEEXAMPLEEXAMPLEEXAMPLEEXAMPLEEXA

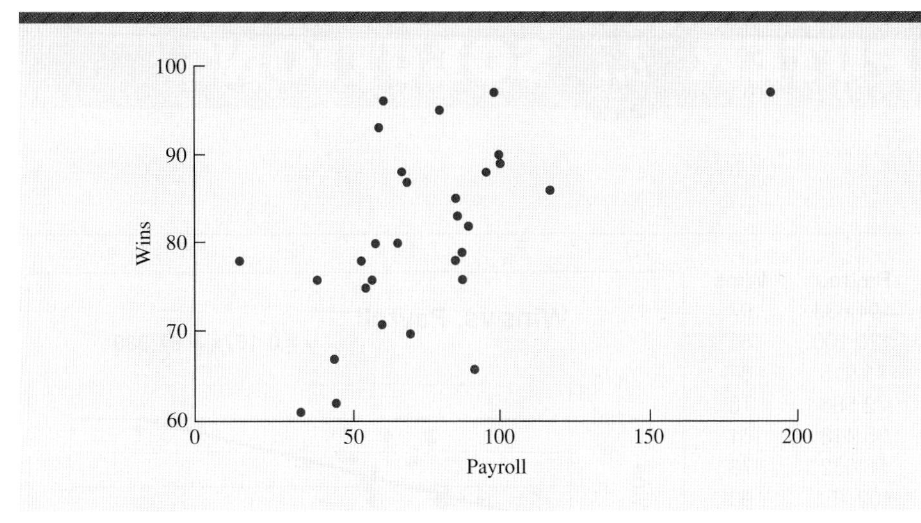

FIGURE 2.7

Scatter diagram of Wins versus Payroll (millions of dollars) for the 30 Major League Baseball teams during the 2006 season. Teams with a higher payroll tended to win more games.

Data sources: content.usatoday.com and mlb.com, June 21, 2009.

We can use Excel or Minitab to easily generate both a scatter diagram for the data and the linear equation that best fits the 30 data points. The procedures and results are shown in Computer Solutions 2.7, and the resulting equation can be written as

$$\text{Wins} = 67.980 + 0.167 \times \text{Payroll}$$

The following are among the important applications for this kind of analysis and best-fit equation:

1. Given a value for x, we can use the equation to estimate a value for y. For example, if a team had a payroll of \$50 million, we estimate that this team would have had 67.980 + 0.167(50), or 76.33 wins during this season. This is obtained simply by substituting x = \$50 million into the equation and calculating an estimated value for y = wins.
2. We can interpret the slope of the equation—in this case, 0.167. Keeping in mind that the slope of the equation is (change in y)/(change in x), we can say that, on average, raising the payroll by an extra \$1.0 million would tend to produce an extra 0.167 wins. Accordingly, an extra \$10 million would tend to produce an extra 1.67 wins.
3. We can interpret the slope in terms of the type of linear relationship between the variables. For the baseball data, the slope is positive (+0.167), reflecting a direct (positive) relationship in which payroll and wins increase and decrease together.
4. We can identify data points where the actual value of y is quite different from the value of y predicted by the equation. In this situation, we might be interested in teams that overperformed (won more games than their payroll would have predicted) or underperformed (won fewer games than their payroll would have predicted). For example, the Oakland Athletics played better than they were paid. Given their payroll of \$62.243 million, the equation would have predicted them to win just 67.980 + 0.167(62.243) = 78.375 games, but they actually won a lot more—93 games.

EXAMPLEEXAMPLEEXAMPLEEXAMPLEEXAMPLE

COMPUTER 2.7 SOLUTIONS

The Scatter Diagram

EXCEL

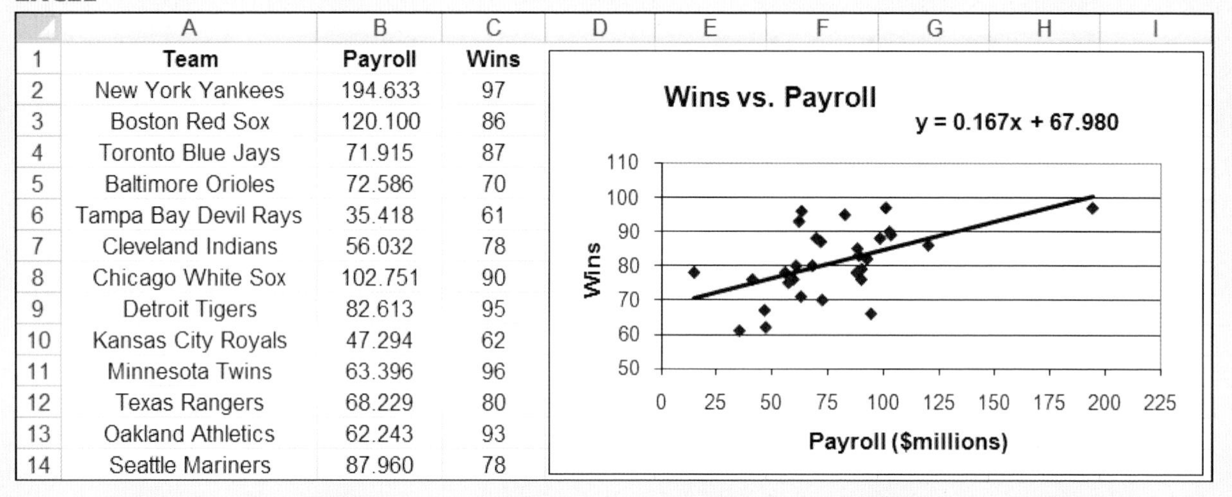

	A	B	C
1	Team	Payroll	Wins
2	New York Yankees	194.633	97
3	Boston Red Sox	120.100	86
4	Toronto Blue Jays	71.915	87
5	Baltimore Orioles	72.586	70
6	Tampa Bay Devil Rays	35.418	61
7	Cleveland Indians	56.032	78
8	Chicago White Sox	102.751	90
9	Detroit Tigers	82.613	95
10	Kansas City Royals	47.294	62
11	Minnesota Twins	63.396	96
12	Texas Rangers	68.229	80
13	Oakland Athletics	62.243	93
14	Seattle Mariners	87.960	78

1. Open the Excel data file **CX02BASE**. The labels and data are already entered, as shown in the printout. Click on **B1** and drag to **C31** to select cells **B1:C31.** (*Note:* As with the line chart, of the two columns, the variable to be represented by the vertical axis should always be in the column to the right.)
2. From the **Insert** ribbon, click **Scatter** from the **Charts** menu. From the **Scatter** submenu, click the first option **(Scatter with only Markers).**
3. This optional step adds the best-fit straight line. Right-click on any one of the points in the scatter diagram. When the menu appears, click **Add Trendline.** In the **Trendline Options** menu, select the **Linear** and **Display Equation on chart** options. **Click Close.**
4. Further appearance improvements can be made by clicking on the chart and dragging the borders to expand it vertically and horizontally. If you don't want the little "Legend" box that originally appears, just right-click within it, then click **Delete.** Additional editing can include insertion of the desired labels: Double-click within the chart area. From the **Layout** ribbon, use the **Chart Title** and **Axis Titles** selections within the **Labels** menu.

MINITAB

The following steps produce the accompanying scatter diagram, complete with the best-fit equation. Along with the equation, the display includes information we will be covering later in the text.

1. Open the Minitab data file **CX02BASE**. The labels, **Payroll** and **Wins,** have already been entered at the top of columns **C2** and **C3,** respectively. The payroll values and the win totals have been entered into **C2** and **C3,** respectively.
2. Click **Stat.** Select **Regression.** Click **Fitted Line Plot.** Enter **C3** into the **Response (Y)** box. Enter **C2** into the **Predictor (X)** box. Click to select the **Linear** model.

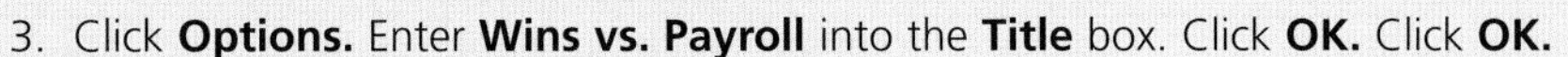

3. Click **Options.** Enter **Wins vs. Payroll** into the **Title** box. Click **OK.** Click **OK.**

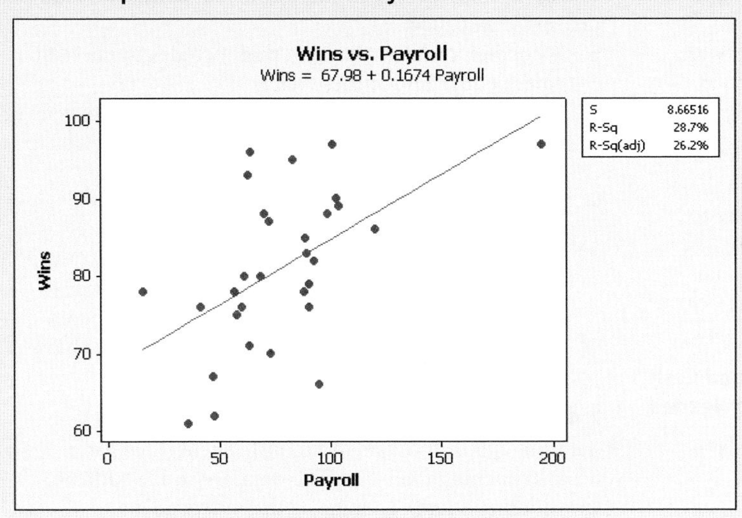

The nature of this chapter has allowed us to only briefly introduce the scatter diagram, the best-fit linear equation, and interpretation of the results. Later in the text, in Chapter 15, Simple Linear Regression and Correlation, we will cover this topic in much greater detail.

EXERCISES

2.38 What is a scatterplot, and for what kind of data is it a useful descriptive device?

2.39 Differentiate between a positive linear relationship and a negative linear relationship between variables.

2.40 When there is no relationship between two variables, what will tend to be the slope of the best-fit straight line through the scatter diagram?

2.41 *PC World* reports the following text and graphics speeds (pages per minute, or ppm) for its top-rated multifunction inkjet printers. Source: *PC World*, July 2009, p. 44.

x = ppm, Plain Text	y = ppm, Graphics	x = ppm, Plain Text	y = ppm, Graphics
30.0	20.0	26.0	17.0
33.0	31.0	7.5	4.5
28.0	23.0	27.0	19.0
32.0	24.0	34.0	33.0
8.4	5.6	33.0	32.0

a. Draw a scatter diagram representing these data.
b. Using the "eyeball" method, fit a straight line to the data.
c. Does there appear to be any relationship between the variables? If so, is the relationship direct or inverse?

2.42 For six local offices of a large tax preparation firm, the following data describe x = service revenues and y = expenses for supplies, freight, and postage during the previous tax preparation season:

x = Service Revenues (Thousands)	y = Supplies, Freight, Postage (Thousands)
$351.4	$18.4
291.3	15.8
325.0	19.3
422.7	22.5
238.1	16.0
514.5	24.6

a. Draw a scatter diagram representing these data.
b. Using the "eyeball" method, fit a straight line to the data.
c. Does there appear to be any relationship between the variables? If so, is the relationship direct or inverse?

2.43 In the 2009 stress tests applied to the nation's biggest banks, the U.S. government determined that 10 of them needed a greater cushion of capital in order to absorb losses in a worst-case economic scenario. The banks, the bailout they received, and the additional cushion they were deemed to need are as shown below.
Source: "Bank Stress-Test Results," *USA Today,* May 8, 2009, p. 3B.

Bank	x = Bailout Received	y = Added Cash Cushion Needed
Bank of America	\$45.0 billion	\$33.9 billion
Wells Fargo	25.0	13.7
GMAC	5.0	11.5
Citigroup	50.0	5.5
Regions	3.5	2.5
SunTrust	4.9	2.2
KeyCorp	2.5	1.8
Morgan Stanley	10.0	1.8
Fifth Third Bancorp	3.4	1.1
PNC	7.6	0.6

a. Draw a scatter diagram representing these data.
b. Using the "eyeball" method, fit a straight line to the data.
c. Does there appear to be any relationship between the variables? If so, is the relationship direct or inverse?

(**DATA SET**) *Note:* Exercises 2.44 and 2.45 require a computer and statistical software.

2.44 For a sample of 100 employees, a personnel director has collected data on x = the number of years an employee has been with the firm and y = the number of shares of company stock the employee owns. The data are in file **XR02044.**
a. Generate a scatter diagram that includes the best-fit linear equation for these data.
b. Does there appear to be any relationship between the variables? If so, is the relationship direct or inverse?
c. Interpret the slope of the equation generated in part (a).
d. Viewing the scatter diagram and equation in part (a), do you think there are any employees who appear to own an unusually high or low amount of stock compared to the ownership that the equation would have predicted?

2.45 A rental management executive has collected data on square footage and monthly rental fee for 80 apartments in her city. The data are in file **XR02045.** Considering square footage as the independent variable and monthly rental fee as the dependent variable:
a. Generate a scatter diagram that includes the best-fit linear equation for these data.
b. Does there appear to be any relationship between the variables? If so, is the relationship direct or inverse?
c. Interpret the slope of the equation generated in part (a).
d. Viewing the scatter diagram and equation in part (a), do you think there are any apartments that seem to have an unusually high or an unusually low rental fee compared to what would be predicted for an apartment of their size? Are there any variables that are not included in the data that might help explain such a differential? If so, what might they be?

2.6 TABULATION, CONTINGENCY TABLES, AND THE EXCEL PivotTable

When some of the variables represent categories, we can apply a popular and useful summarization method called **tabulation,** where we simply count how many people or items are in each category or combination of categories. If each person or item is also described by a quantitative variable (e.g., annual income), we can take the concept one step further and examine how members of one category (e.g., union employees) differ from members of another category (e.g., nonunion employees) in terms of average annual income.

EXAMPLE

Tabulation Methods

For 50 persons observed using a local automated teller machine (ATM), researchers have described the customers according to age category and gender category, and have used a stopwatch to measure how long the customer required to complete his or her transactions at the machine. In this situation, age category and gender are considered to be nominal-scale (category) variables, and time (measured in seconds) is a quantitative variable. The raw data are shown in part A of Table 2.3 and are in data file CX02ATM.

SOLUTIONS

Simple Tabulation

In **simple tabulation,** also known as **marginal** or **one-way tabulation,** we merely count how many people or items are in each category. The simple tabulation in part B of Table 2.3 shows us that the 50 ATM users consisted of 28 males and 22 females. An alternative would have been to express the counts in terms of percentages—e.g., 56% of the subjects were males and 44% were females.

Cross-Tabulation (Contingency Table)

The **cross-tabulation,** also known as the **crosstab** or the **contingency table,** shows how many people or items are in *combinations* of categories. The cross-tabulation in part C of Table 2.3 describes how many of the 50 ATM users were in each age–gender group; for example, 7 of the individuals were males in the youngest age group and 4 were females in the oldest age group. Because there are two category variables involved, part C of Table 2.3 could also be referred to as a **two-way crosstab.** Cross-tabulations help us to identify and examine possible relationships between the variables. For example, the crosstab in part C of Table 2.3 indicates that female ATM users outnumbered men in the youngest age category, but the reverse was true in the middle and older age categories. Given these results, bank management might want to interview customers to learn whether women in the middle and older age categories might have security concerns about the ATM location, lighting, or layout.

In a useful application of the concept of cross-tabulation, we can generate a tabular display that describes how a selected quantitative variable tends to differ from one category to another or from one combination of categories to another. For example, part D of Table 2.3 shows the average time (seconds) required for each of our six categories of ATM users. The average time for all 50 persons was 39.76 seconds, with males in the middle age group having the fastest average time (34.53 seconds) and females in the older age group having the slowest average time (49.95 seconds). From a gender perspective, the average time for males (38.36 seconds) was slightly faster than that for females (41.53 seconds).

EXAMPLEEXAMPLEEXAMPLEEXAMPLEEXAMPLEEXAMPLEEXAMPLE

Cross-tabulations are very basic to many of the data analysis techniques in the remainder of the text, and this section includes two different Computer Solutions in which Excel and Minitab are applied. Computer Solutions 2.8 shows the

TABLE 2.3

Data representing age category, gender, and machine-usage time for 50 persons observed using an automated teller machine (ATM). Age category and gender are nominal-scale (category) variables, and time is a quantitative variable measured in seconds. The coding for the age and gender categories follows: AgeCat = 1 for age < 30, 2 for age 30–60, 3 for age > 60; Gender = 1 for male, 2 for female. The data are in file **CX02ATM**.

A. Raw Data

AgeCat	Gender	Seconds	AgeCat	Gender	Seconds	AgeCat	Gender	Seconds
1	2	50.1	2	2	46.4	2	1	33.2
1	1	53.0	2	1	31.0	2	2	29.1
2	2	43.2	2	1	44.8	2	1	37.7
1	2	34.9	2	1	30.3	2	2	54.9
3	1	37.5	2	1	33.6	1	2	29.4
2	1	37.8	3	2	56.3	3	1	37.1
3	1	49.4	2	2	37.8	3	2	42.9
1	1	50.5	1	2	34.4	1	2	44.3
3	1	48.1	2	1	32.9	3	2	45.0
1	2	27.6	1	1	26.3	2	2	47.6
3	2	55.6	2	1	30.8	1	1	58.1
2	2	50.8	1	1	24.1	1	2	32.0
3	1	43.6	1	2	43.6	2	1	36.9
1	1	35.4	2	2	29.9	1	1	27.5
1	2	37.7	2	1	26.3	3	1	42.3
3	1	44.7	3	1	47.7	2	1	34.1
2	1	39.5	1	2	40.1			

B. Simple Tabulation, by Gender

1 = Male	2 = Female	Total
28	22	50

C. Cross-Tabulation or Contingency Table, by Age Category and Gender

	Gender:		
Age Category:	1 = Male	2 = Female	Total
1 = under 30 yr.	7	10	17
2 = 30–60 yr.	13	8	21
3 = over 60 yr.	8	4	12
Total	28	22	50

D. Mean Usage Times (in Seconds), by Age Category and Gender

	Gender:		
Age Category:	1 = Male	2 = Female	Total
1 = under 30 yr.	39.27	37.41	38.18
2 = 30–60 yr.	34.53	42.46	37.55
3 = over 60 yr.	43.80	49.95	45.85
Total	38.36	41.53	39.76

procedures and the results when Excel and Minitab are used in generating a cross-tabulation, or contingency table, for the data in our ATM example. The underlying data are in **CX02ATM**, and the results correspond to the crosstab in part C of Table 2.3.

COMPUTER 2.8 SOLUTIONS

The Cross-Tabulation

EXCEL

	A	B	C	D	E	F	G
1	AgeCat	Gender	Seconds	Count of Gender	Gender		
2	1	2	50.1	AgeCat	1	2	Grand Total
3	1	1	53.0	1	7	10	17
4	2	2	43.2	2	13	8	21
5	1	2	34.9	3	8	4	12
6	3	1	37.5	Grand Total	28	22	50

1. Open the Excel data file CX02ATM. The labels and data are already entered, as shown in the printout. The variables are age category (1 = <30, 2 = 30–60, and 3 = >60), gender category (1 = male, 2 = female), and time (in seconds). Click on **A1** and drag to **C51** to select cells **A1:C51.**
2. From the **Insert** ribbon and its **Tables** menu, click **PivotTable.** In the **Create PivotTable** menu, click **Select a table or range.** The cells A1:C51 should already be represented in the **Table/Range** box; if not, enter this range into the box. Click to select **Existing Worksheet** and enter **D1** into the box. Click **OK.**
3. Click the **AgeCat** label at the right and drag it into the **Row Labels** rectangle. Click the **Gender** label at the right and drag it into the **Column Labels** rectangle. Click the **Gender** label again and drag it into the **Values** rectangle.
4. Right-click on any one of the data values within the table, then select **Value Field Settings.** In the **Summarize value field by** box, select **Count.** Click **OK.**
5. Click on the **Row Labels** cell and edit by entering **AgeCat.** Click on the **Column labels** cell and edit by entering **Gender.**

MINITAB

```
Tabulated statistics: AgeCat, Gender
 Rows: AgeCat    Columns: Gender

          1   2  All
 1        7  10   17
 2       13   8   21
 3        8   4   12
 All     28  22   50

 Cell Contents:       Count
```

1. Open the Minitab data file CX02ATM. The labels, **AgeCat, Gender,** and **Seconds,** have already been entered at the top of columns **C1, C2,** and **C3,** respectively. The corresponding age categories (1 = <30, 2 = 30–60, and 3 = >60), gender categories (1=male, 2=female), and times (in seconds) have been entered into **C1, C2,** and **C3,** respectively.
2. Click **Stat.** Select **Tables.** Click **Cross Tabulation and Chi-Square.** In the **Categorical variables** section, enter **C1** into the **For rows** box and **C2** into the **For columns** box. In the **Display** portion, click to place a check mark next to **Counts.** Click **OK.**

COMPUTER 2.9 SOLUTIONS

Cross-Tabulation with Cell Summary Information

EXCEL

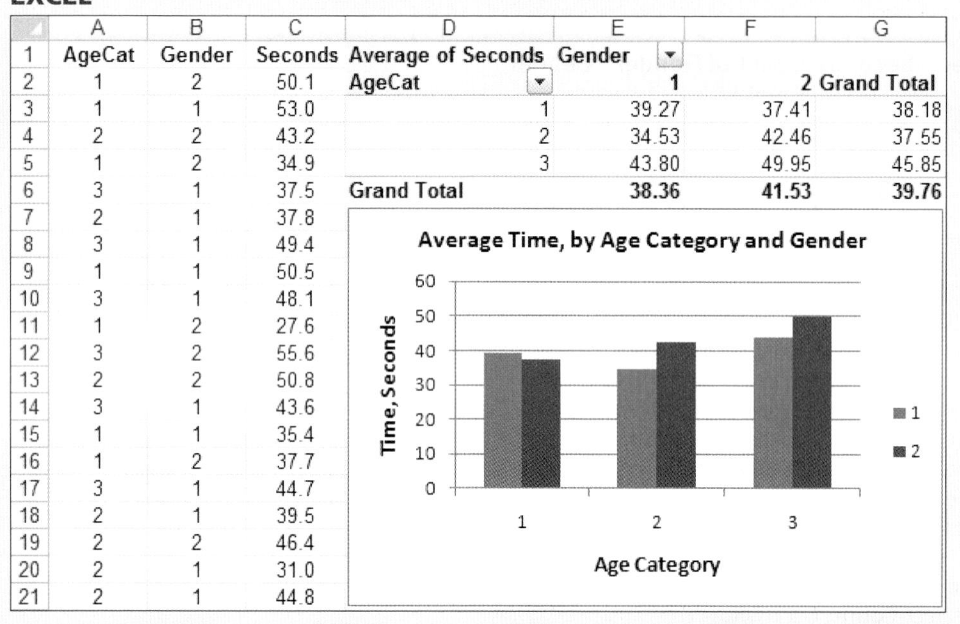

	A	B	C	D	E	F	G
1	AgeCat	Gender	Seconds	Average of Seconds	Gender		
2	1	2	50.1	AgeCat	1	2	Grand Total
3	1	1	53.0	1	39.27	37.41	38.18
4	2	2	43.2	2	34.53	42.46	37.55
5	1	2	34.9	3	43.80	49.95	45.85
6	3	1	37.5	Grand Total	**38.36**	**41.53**	**39.76**
7	2	1	37.8				
8	3	1	49.4				
9	1	1	50.5				
10	3	1	48.1				
11	1	2	27.6				
12	3	2	55.6				
13	2	2	50.8				
14	3	1	43.6				
15	1	1	35.4				
16	1	2	37.7				
17	3	1	44.7				
18	2	1	39.5				
19	2	2	46.4				
20	2	1	31.0				
21	2	1	44.8				

1. Open the Excel data file CX02ATM. Click on **A1** and drag to **C51** to select cells **A1:C51.**
2. From the **Insert** ribbon and its **Tables** menu, click **PivotTable.** In the **Create PivotTable** menu, click **Select a table or range.** The cells A1:C51 should already be represented in the **Table/Range** box; if not, enter this range into the box. Click to select **Existing Worksheet** and enter **D1** into the box. Click **OK.**
3. Click the **AgeCat** label at the right and drag it into the **Row Labels** rectangle. Click the **Gender** label at the right and drag it into the **Column Labels** rectangle. Click the **Seconds** label at the right and drag it into the **Values** rectangle.
4. Right-click on any one of the data values within the table, then select **Value Field Settings.** In the **Summarize value field by** box, select **Average.** Click **OK.** Click on **E3** and drag to **G6** to select cells **E3:G6.** Right-click within the field and select **Number Format.** Select **Number,** and specify **2** in the **Decimal places** box. Click **OK.** Click on the **Row Labels** cell and edit by entering **AgeCat.** Click on the **Column Labels** cell and edit by entering **Gender.**
5. Click on any cell within the table. From the **Insert** ribbon, select **Column** from the **Charts** menu. Select the first option **(Clustered Column)** from the **2-D Column** choices. The chart can now be edited and the desired labels inserted: Double-click within the chart area. From the **Layout** ribbon, use the **Chart Title** and **Axis Titles** selections within the **Labels** menu.

MINITAB

1. Open the Minitab data file CX02ATM. As in Computer Solutions 2.8, the labels, **AgeCat, Gender,** and **Seconds,** have already been entered at the top of columns **C1, C2,** and **C3,** respectively. The corresponding age categories (1 = <30, 2 = 30–60, and 3 = >60), gender categories (1 = male, 2 = female), and times (in seconds) have been entered into **C1, C2,** and **C3,** respectively.
2. Click **Stat.** Select **Tables.** Click **Descriptive Statistics.** In the **Categorical variables** section, enter **C1** into the **For rows** box and **C2** into the **For columns** box.

3. Click on **Display summaries for Associated Variables** and enter **C3** into the **Associated variables** box. Select **Means.** Click **OK.** Click **OK.**

```
Tabulated statistics: AgeCat, Gender
Rows: AgeCat    Columns: Gender

            1      2    All
1       39.27  37.41  38.18
            7     10     17

2       34.53  42.46  37.55
           13      8     21

3       43.80  49.95  45.85
            8      4     12

All     38.36  41.53  39.76
           28     22     50

Cell Contents:   Seconds  :  Mean
                             Count
```

EXERCISES

2.46 When variables are used as the basis for a contingency table, what scale of measurement must they represent?

2.47 Differentiate between simple tabulation and cross-tabulation, and give a real or hypothetical example of each.

Note: Exercises 2.48–2.50 can be done by hand, but computer statistical software is preferable, if available.

2.48 The 30 vehicles operated by a taxi service have been listed according to their engine (1 = diesel, 2 = gasoline), transmission (1 = manual, 2 = automatic), and whether they have air conditioning (1 = no air conditioning, 2 = air conditioning). These characteristics, along with the miles per gallon (mpg) achieved by each vehicle during the past month, are listed here. The data are also provided in file **XR02048.**

a. Construct a simple tabulation in which the counts are according to the type of engine.
b. Construct a cross-tabulation describing the fleet, using type of engine and type of transmission as the categorization variables.
c. Construct a display showing the average mpg according to type of engine and type of transmission. Do the categorization variables seem to be related to mpg? If so, how?

Engine	Trans	AC	mpg	Engine	Trans	AC	mpg
1	1	1	36.4	1	1	2	38.6
1	1	1	37.7	1	1	2	34.8
1	1	1	38.2	1	2	2	30.8
1	1	1	41.8	1	2	2	26.4
1	1	1	34.8	1	2	2	31.6
1	2	1	35.4	1	2	2	28.3
1	2	1	31.3	1	2	2	29.5
1	2	1	32.2	2	1	2	23.0
2	1	1	26.2	2	1	2	21.5
2	1	1	25.5	2	1	2	19.5
2	1	1	27.7	2	2	2	18.3
2	2	1	22.0	2	2	2	19.1
2	2	1	23.3	2	2	2	20.8
1	1	2	34.8	2	2	2	21.8
1	1	2	34.0	2	2	2	18.4

2.49 For the fleet described in Exercise 2.48,

a. Construct a simple tabulation in which the counts are according to the type of engine.

b. Construct a cross-tabulation describing the fleet, using type of engine and whether the vehicle has air conditioning as the categorization variables.

c. Construct a display showing the average mpg according to type of engine and whether the vehicle has air conditioning. Do the categorization variables seem to be related to mpg? If so, how?

2.50 For the fleet described in Exercise 2.48,

a. Construct a simple tabulation in which the counts are according to the type of transmission.

b. Construct a cross-tabulation describing the fleet, using type of transmission and whether the vehicle has air conditioning as the categorization variables.

c. Construct a display showing the average mpg according to type of transmission and whether the vehicle has air conditioning. Do the categorization variables seem to be related to mpg? If so, how?

(DATA SET) *Note:* Exercises 2.51–2.54 require a computer and statistical software.

2.51 A high school guidance counselor has examined the 60 colleges and universities that are on her "highly recommended" list for her school's graduating seniors. The data she has collected on each college or university include highest degree level offered (1 = bachelor's, 2 = master's, 3 = doctorate), primary support for the school (1 = public, 2 = private), type of campus setting (1 = urban, 2 = suburban, 3 = rural), and annual cost for tuition and fees (dollars). The data are in file **XR02051**.

a. Construct a simple tabulation in which the counts are according to the highest degree level offered.

b. Construct a cross-tabulation describing the schools, using the highest degree level offered and whether the school is public or private as the categorization variables.

c. Construct a display showing the average value for tuition and fees according to highest degree level offered and whether the school is public or private. Do the categorization variables seem to be related to the level of tuition and fees? If so, how?

2.52 For the schools described in Exercise 2.51,

a. Construct a simple tabulation in which the counts are according to the type of campus setting.

b. Construct a cross-tabulation describing the schools, using the type of campus setting and whether the school is public or private as the categorization variables.

c. Construct a display showing the average value for tuition and fees according to type of campus setting and whether the school is public or private. Do the categorization variables seem to be related to the level of tuition and fees? If so, how?

2.53 An interdisciplinary research team has examined 35 U.S. cities, reviewing and grading each according to the following performance areas: financial management, capital projects, personnel policies, information technology, and managing for results.[a] The grades awarded were letter grades such as students might receive in school. We have rounded off the "+" and "−" portions of the letter grades so that they are simply A, B, C, D, or F, and the data are coded so that A = 4, B = 3, C = 2, D = 1, and F = 0. We have also included one more variable for each city: its population according to the latest census statistics at the time.[b] The data for all 35 cities are in file **XR02053.** [a]Source: Richard Wolf, "Phoenix Is Managing To Get Everything Right," *USA Today*, January 31, 2000, p. 13A. [b]Source: *The New York Times Almanac 2000*, pp. 248–250.

a. Construct a simple tabulation in which the counts are according to the grade on financial management.

b. Construct a cross-tabulation describing the cities, using grade on financial management and grade on information technology as the categorization variables.

c. Construct a display showing the average population size according to grade on financial management and grade on information technology. Do the categorization variables seem to be related to the level of population? If so, how?

2.54 For the cities described in Exercise 2.53,

a. Construct a simple tabulation in which the counts are according to the grade on personnel policies.

b. Construct a cross-tabulation describing the cities, using the grade on personnel policies and the grade on managing for results as the categorization variables.

c. Construct a display showing the average population according to grade on personnel policies and the grade on managing for results. Do the categorization variables seem to be related to the level of population? If so, how?

(2.7) SUMMARY

- **Frequency distributions and histograms**

To more concisely communicate the information contained, raw data can be visually represented and expressed in terms of statistical summary measures. When data are quantitative, they can be transformed to a frequency distribution or a histogram describing the number of observations occurring in each category.

The set of classes in the frequency distribution must include all possible values and should be selected so that any given value falls into just one category. Selecting the number of classes to use is a subjective process. In general, between 5 and 15 classes are employed. A frequency distribution may be converted to show either relative or cumulative frequencies for the data.

- **Stem-and-leaf displays, dotplots, and other graphical methods**

The stem-and-leaf display, a variant of the frequency distribution, uses a subset of the original digits in the raw data as class descriptors (stems) and class members (leaves). In the dotplot, values for a variable are shown as dots appearing along a single dimension. Frequency polygons, ogives, bar charts, line graphs, pie charts, pictograms, and sketches are among the more popular methods of visually summarizing data. As with many statistical methods, the possibility exists for the purposeful distortion of graphical information.

- **The scatter diagram**

The scatter diagram, or scatterplot, is a diagram in which each point represents a pair of known or observed values of two variables. These are typically referred to as the dependent variable (y) and the independent variable (x). This type of analysis is carried out to fit an equation to the data to estimate or predict the value of y for a given value of x.

- **Tabulation and contingency tables**

When some of the variables represent categories, simple tabulation and cross-tabulation are used to count how many people or items are in each category or combination of categories, respectively. These tabular methods can be extended to include the mean or other measures of a selected quantitative variable for persons or items within a category or combination of categories.

EQUATIONS

Width (Class Interval) of a Frequency Distribution Class

$$\text{Class interval} = \begin{matrix}\text{Lower class limit of}\\ \text{next higher class}\end{matrix} - \begin{matrix}\text{Lower class limit of}\\ \text{the class}\end{matrix}$$

Approximate Width of Classes When Constructing a Frequency Distribution

$$\text{Approximate class width} = \frac{\begin{matrix}\text{Largest value}\\ \text{in raw data}\end{matrix} - \begin{matrix}\text{Smallest value}\\ \text{in raw data}\end{matrix}}{\text{Number of classes desired}}$$

Midpoint of a Class in a Frequency Distribution

$$\text{Class mark} = \text{Lower class limit} + 0.5(\text{class interval})$$

Size of Each Piece of a Pie Chart

$$\begin{matrix}\text{Number of degrees}\\ \text{for the category}\end{matrix} = \begin{matrix}\text{Relative value of the}\\ \text{category}\end{matrix} \times 360$$

CHAPTER EXERCISES

2.55 The National Oceanic and Atmospheric Administration reports the following record high temperatures (degrees Fahrenheit) in 50 U.S. cities. These data are also provided in the file **XR02056.** Source: Bureau of the Census, *Statistical Abstract of the United States 2009,* p. 229.

City	Temperature
Mobile, AL	105
Juneau, AK	90
Phoenix, AZ	122
Little Rock, AR	112
Los Angeles, CA	110
Denver, CO	105
Hartford, CT	102
Wilmington, DE	102
Miami, FL	98
Atlanta, GA	105
Honolulu, HI	95
Boise, ID	111
Chicago, IL	104
Indianapolis, IN	104
Des Moines, IA	108
Wichita, KS	113
Louisville, KY	106
New Orleans, LA	102
Portland, ME	103
Baltimore, MD	105
Boston, MA	102
Detroit, MI	104
Duluth, MN	97
Jackson, MS	107
St. Louis, MO	107
Great Falls, MT	106
Omaha, NE	114
Reno, NV	108
Concord, NH	102
Atlantic City, NJ	106
Albuquerque, NM	107
Buffalo, NY	99
Charlotte, NC	104
Bismarck, ND	111
Cleveland, OH	104
Oklahoma City, OK	110
Portland, OR	107
Pittsburgh, PA	103
Providence, RI	104
Columbia, SC	107
Sioux Falls, SD	110
Nashville, TN	107
Houston, TX	109
(continued)	
Salt Lake City, UT	107
Burlington, VT	101
Norfolk, VA	104
Seattle, WA	100
Charleston, WV	104
Milwaukee, WI	103
Cheyenne, WY	100

a. Construct a stem-and-leaf display for these data.
b. Construct a frequency distribution.
c. Determine the interval width and the class mark for each of the classes in your frequency distribution.
d. Based on the frequency distribution obtained in part (b), draw a histogram and a relative frequency polygon to describe the data.

2.56 The average daily cost to community hospitals for patient stays during 2006 for each of the 50 U.S. states was as follows:

AL	$1274	LA	1288	OH	1773
AK	2455	ME	1627	OK	1379
AZ	1930	MD	2009	OR	2202
AR	1304	MA	1895	PA	1601
CA	2056	MI	1608	RI	1824
CO	1864	MN	1379	SC	1487
CT	1839	MS	1100	SD	774
DE	1634	MO	1687	TN	1349
FL	1595	MT	924	TX	1752
GA	1243	NE	1180	UT	1914
HI	1383	NV	1544	VT	1355
ID	1511	NH	1714	VA	1491
IL	1676	NJ	1910	WA	2298
IN	1699	NM	1738	WV	1152
IA	1093	NY	1602	WI	1566
KS	1155	NC	1362	WY	906
KY	1318	ND	966		

These data are also provided in the file **XR02057.** Source: Bureau of the Census, *Statistical Abstract of the United States 2009,* p. 114.

a. Construct a stem-and-leaf display for these data.
b. Construct a frequency distribution.
c. Determine the interval width and the class mark for each of the classes in your frequency distribution.
d. Based on the frequency distribution obtained in part (b), draw a histogram and a relative frequency polygon.

2.57 The breakdown of U.S. cities having a population of at least 10,000 persons has been reported as follows. Source: Bureau of the Census, *Statistical Abstract of the United States 2009,* p. 34.

Population	Number of Cities
10,000–under 25,000	1510
25,000–under 50,000	684
50,000–under 100,000	430
100,000–under 250,000	191
250,000–under 500,000	37
500,000–under 1,000,000	25
1,000,000 or more	9

a. How many cities have a population of at least 25,000 but less than 500,000?
b. How many cities have a population less than 250,000?
c. How many cities have a population of at least 100,000 but less than 1,000,000? What percentage of cities are in this group?
d. What is the class mark for the 100,000–under 250,000 class?
e. Convert the table to a relative frequency distribution.

2.58 The following stem-and-leaf output has been generated by Minitab. The data consist of two-digit integers.

```
Stem-and-leaf of AGE    N=21
Leaf Unit=1.0

  1    3 3
  4    3 669
  4    4
  8    4 5667
 (6)   5 012233
  7    5 68
  5    6 0
  4    6 779
  1    7 0
```

a. From this display, is it possible to determine the exact values in the original data? If so, list the data values. If not, provide a set of data that could have led to this display.
b. Interpret the numbers in the leftmost column of the output.

2.59 In 2007, unemployment rates in the 50 U.S. states were reported as follows. These data are also provided in the file **XR02058.** Source: Bureau of the Census, *Statistical Abstract of the United States 2009*, p. 373.

	Percent		Percent		Percent
AL	4.0	HI	2.9	MA	4.6
AK	6.2	ID	3.0	MI	7.1
AZ	3.9	IL	5.1	MN	4.6
AR	5.6	IN	4.6	MS	6.1
CA	5.3	IA	3.7	MO	5.0
CO	3.7	KS	4.1	MT	3.6
CT	4.5	KY	5.4	NE	3.1
DE	3.5	LA	4.3	NV	4.6
FL	4.1	ME	4.7	NH	3.6
GA	4.3	MD	3.6	NJ	4.2

(continued)

	Percent		Percent		Percent
NM	3.7	PA	4.3	VT	4.0
NY	4.6	RI	4.9	VA	3.1
NC	4.5	SC	5.6	WA	4.6
ND	3.2	SD	2.9	WV	4.6
OH	5.6	TN	4.6	WI	5.0
OK	4.4	TX	4.3	WY	2.9
OR	5.2	UT	2.6		

a. Construct a stem-and-leaf display for these data.
b. Construct a frequency distribution for these data.
c. Determine the interval width and the class mark for each of the classes in your frequency distribution.
d. Based on the frequency distribution obtained in part (b), draw a histogram and a relative frequency polygon to describe the data.

2.60 The following stem-and-leaf output has been generated by Minitab. The data consist of three-digit integers.

```
Stem-and-leaf of SCORE    N=21
Leaf Unit=10

  2    1 89
  3    2 0
  3    2
  9    2 445555
  9    2
 (2)   2 89
 10    3 0011
  6    3 23
  4    3 455
  1    3
  1    3 8
```

a. From this display, is it possible to determine the exact values in the original data? If so, list the data values. If not, provide a set of data that could have led to this display.
b. Interpret the numbers in the leftmost column of the output.

2.61 During 2007, sales of new, privately owned homes in the total United States and in the western states were broken down into price categories as follows: Source: Bureau of the Census, *Statistical Abstract of the United States 2009*, p. 595.

	Sales (Thousands of Homes)	
Price of Home	Total United States	Western States
under $200,000	268	22
$200,000–under $300,000	227	54
$300,000–under $500,000	186	64
$500,000 or over	94	41

Convert these data to relative frequency distributions, one for the total United States, the other for the western states. Do the results appear to suggest any differences between total U.S. prices and those in the western states?

2.62 A social scientist would like to use a pictogram in comparing the number of violent crimes against persons in 2006 with the comparable figure for 1960. Considering that she can select from a wide variety of symbols to demonstrate these frequencies, suggest several possible symbols that would tend to underscore the violence and/or human tragedy of such crimes.

2.63 For the period 2001–2008, the Bristol-Myers Squibb Company, Inc. reported the following amounts (in billions of dollars) for (1) net sales and (2) advertising and product promotion. The data are also in the file **XR02062**. Source: Bristol-Myers Squibb Company, *Annual Reports, 2005, 2008.*

Year	Net Sales	Advertising/Promotion
2001	$16.612	$1.201
2002	16.208	1.143
2003	18.653	1.416
2004	19.380	1.411
2005	19.207	1.476
2006	16.208	1.304
2007	18.193	1.415
2008	20.597	1.550

For these data, construct a line graph that shows both net sales and expenditures for advertising/product promotion over time. Some would suggest that increases in advertising should be accompanied by increases in sales. Does your line graph support this?

(**DATA SET**) *Note:* Exercises 2.64–2.72 require a computer and statistical software.

2.64 For the law schools discussed in Exercise 2.66, and using the data in file **XR02067**,

a. Construct a scatter diagram where the dependent variable is the percentage of graduates who were employed immediately upon graduation, and the independent variable is the reputation score provided by the lawyers/judges. Do schools given higher scores by the lawyers/judges also seem to have a higher percentage of their graduates employed immediately upon graduation?
b. Fit a linear equation to the scatter diagram. Is the slope positive or is it negative? Is the sign of the slope consistent with your intuitive observation in part (a)?

2.65 For the situation and data described in Exercise 2.67, and using data file **XR02071**,

a. Construct a simple tabulation in which the counts are according to which company supplied the electrical/motor components.
b. Construct a cross-tabulation describing the 100 compressors tested, using electrical/motor supplier and final assembly technician as the categorization variables.
c. Construct a display showing the average pressure (psi) exerted according to electrical/motor supplier and final assembly technician. Which combination of electrical/motor supplier and final assembly technician seems to result in the highest average pressure exerted? The lowest pressure exerted?
d. Based on the crosstab and means in part (c), would it seem that the two technicians might not be equally adept at the final assembly task? Explain your answer.

2.66 Among the information in its top-50 list of the nation's law schools, *U.S. News & World Report* provided two reputation scores (with maximum = 5.0) for each school: one from academicians, the other from lawyers/judges. The magazine also provided the most recent acceptance rate (%), student-faculty ratio (students/faculty), and the percentage of graduates who were employed immediately upon graduation. The data values are listed in file **XR02067**. Source: *U.S. News & World Report, Best Graduate Schools, 2010 edition*, p. 22.

a. Construct a scatter diagram where the variables are the two kinds of reputation scores. Do schools given higher scores by the academicians seem to also receive higher scores from the lawyers/judges?
b. Fit a linear equation to the scatter diagram. Is the slope positive or is it negative? Is the sign of the slope consistent with your intuitive observation in part (a)?

2.67 An air-compressor manufacturer purchases mechanical components from two different suppliers (MechSup = 1 or 2) and electrical/motor components from three different suppliers (ElectSup = 1, 2, or 3), with final assembly being carried out by one of two technicians (Tech = 1 or 2). Following production, finished compressor units are tested to measure how much pressure they can exert (pounds per square inch, or psi). During the past few months, 100 compressors have been tested before shipment to customers, and the resulting data are listed in file **XR02071**.

a. Construct a simple tabulation in which the counts are according to which company supplied the mechanical components.
b. Construct a cross-tabulation describing the 100 compressors tested, using mechanical supplier and electrical/motor supplier as the categorization variables.
c. Construct a display showing the average pressure (psi) according to mechanical supplier and electrical/motor supplier. Which combination of mechanical and electrical/motor suppliers seems to result in the highest pressure exertion? The lowest pressure exertion?
d. Based on the crosstab and means in part (c), would it seem that any of the five suppliers should be examined further with regard to the effect their product might have on the final pressure capabilities of the finished product? Explain your answer.

2.68 If possible with your statistical software, use the dotplot method to compare the 50-state unemployment rates in 1980 with those in 2007. Use the same scale for each dotplot, then comment on whether unemployment appears to have changed in terms of its range (highest minus lowest) or in the general level of unemployment (perhaps a slight shift to the right or to the left) from 1980 to 2007. Data for both years are in the file **XR02066.** Source: Bureau of the Census, *Statistical Abstract of the United States 2009*, p. 373.

2.69 For the nations and household saving rates in Exercise 2.71, and using the data in file **XR02069**,

a. Construct a scatter diagram using the household saving rates in the United States and Germany as the two variables. It doesn't really matter which one you consider to be the independent variable. In the years when U.S. saving rates were higher, did those in Germany seem to be higher as well?
b. Fit a linear equation to the scatter diagram. Is the slope positive or is it negative? Is the sign of the slope consistent with your intuitive observation in part (a)?

2.70 If possible with your statistical software, construct a dotplot for the data of Exercise 2.56 (use data file **XR02057**). Are there any states for which the average daily cost was so high that it would appear to be an outlier? If so, which one(s)?

2.71 The household saving rates (percentage of disposable household income that goes into savings) have been reported for several countries—including the United States, Canada, France, Germany, and Japan—for the years 1980 through 1997. The percentages are listed in data file **XR02069.** Source: Security Industry Association, *1999 Securities Industry Fact Book*, p. 98.

a. Construct a scatter diagram using the household saving rates in the United States and Canada as the two variables. It doesn't really matter which one you consider to be the independent variable. In the years when U.S. saving rates were higher, did those in Canada seem to be higher as well?
b. Fit a linear equation to the scatter diagram. Is the slope positive or is it negative? Is the sign of the slope consistent with your intuitive observation in part (a)?

2.72 If possible with your statistical software, construct a dotplot for the data of Exercise 2.55 (use data file **XR02056**). Are there any cities for which the record high temperature is so high that it would appear to be an outlier? If so, which one(s)?

INTEGRATED CASES

Thorndike Sports Equipment
(Meet the Thorndikes: See Video Unit One.)

Luke Thorndike, founder and current president of Thorndike Sports Equipment, had guided the business through 34 successful years and was now interested in bringing his favorite grandson Ted into the company.

The elder Thorndike, possessed of a sharp but fading wit, begins the meeting with, "Great to see you, Ted. You always were a high-strung kid. Thought you might like to join our tennis racquet division." Ted counters, "Not quite, but you're getting warm, Luke." The Thorndikes have always been a strange bunch.

"Seriously, Ted, I'm getting a little up in years, the microcomputer I bought myself for Christmas is collecting dust, and I think you and your business degree could bring some new blood to the company. I'd like you to be my executive vice president. I've been running this outfit by the seat of my pants for a lot of years now, and the world just seems to be getting too big and too complicated these days. I've got index cards and file folders just about piled up to the ceiling in that little room next door, and there's got to be a better way of making sense of all this information. Maybe I shouldn't have fought the fire department when they wanted to condemn the place."

"Besides all these records piling up, a lot of technical developments are affecting our business—things like composite bicycle wheels, aerodynamic golf balls, and oversize tennis racquets. Just yesterday one of our engineers came in and said she's come up with a new golf ball that will go farther than the conventional design. She seems trustworthy, but we might need to have some numbers to back up our claim if we decide to come out with the product."

After further discussion of the business and the position that Mr. Thorndike has proposed, Ted accepts the offer. As his first official duty, he sets up a test of the new golf ball that's supposed to travel farther than the conventional design. He decides to mix 25 of the new balls with 25 of the old type, have a golf pro hit all 50 of them at a driving range, then measure how far each goes. The results are provided here and are also listed in the data file **THORN02**.

(*continued*)

25 Drives with New Ball (Yards)								
267.5	248.3	265.1	243.6	253.6	232.7	249.2	232.3	252.8
247.2	237.4	223.7	260.4	269.6	256.5	271.4	294.1	256.3
264.3	224.4	239.9	233.5	278.7	226.9	258.8		

25 Drives with Conventional Ball (Yards)								
241.8	255.1	266.8	251.6	233.2	242.7	218.5	229.0	256.3
264.2	237.3	253.2	215.8	226.4	201.0	201.6	244.1	213.5
267.9	240.3	247.9	257.6	234.5	234.7	215.9		

1. Using 10-yard intervals beginning with 200.0–under 210.0, 210.0–under 220.0, on up to 290.0–under 300.0, construct a frequency distribution for the distances traveled by the new ball.
2. Using the same intervals as in part 1, construct a frequency distribution for the distances traveled by the conventional ball.
3. Place the frequency distribution for the new ball next to the one for the conventional ball. Does it appear that the new ball might be more "lively" than the conventional ball?

Springdale Shopping Survey*

The major shopping areas in the community of Springdale include Springdale Mall, West Mall, and the downtown area on Main Street. A telephone survey has been conducted to identify strengths and weaknesses of these areas and to find out how they fit into the shopping activities of local residents. The 150 respondents were also asked to provide information about themselves and their shopping habits. The data are provided in the computer file **SHOPPING**. The variables in the survey were as follows:

A. How Often Respondent Shops at Each Area (Variables 1–3)

	1. Springdale Mall	2. Downtown	3. West Mall
6 or more times/wk.	(1)	(1)	(1)
4–5 times/wk.	(2)	(2)	(2)
2–3 times/wk.	(3)	(3)	(3)
1 time/wk.	(4)	(4)	(4)
2–4 times/mo.	(5)	(5)	(5)
0–1 times/mo.	(6)	(6)	(6)

B. How Much the Respondent Spends during a Trip to Each Area (Variables 4–6)

	4. Springdale Mall	5. Downtown	6. West Mall
$200 or more	(1)	(1)	(1)
$150–under $200	(2)	(2)	(2)
$100–under $150	(3)	(3)	(3)
$ 50–under $100	(4)	(4)	(4)
$ 25–under $50	(5)	(5)	(5)
$ 15–under $25	(6)	(6)	(6)
less than $15	(7)	(7)	(7)

*Source: Materials for this case have been provided courtesy of The Archimedes Group, Indiana, PA. Data are based on actual responses obtained to this subset of the questions included in the survey; town and mall identities have been disguised.

C. General Attitude toward Each Shopping Area (Variables 7–9)

	7. Springdale Mall	8. Downtown	9. West Mall
Like very much	(5)	(5)	(5)
Like	(4)	(4)	(4)
Neutral	(3)	(3)	(3)
Dislike	(2)	(2)	(2)
Dislike very much	(1)	(1)	(1)

D. Which Shopping Area Best Fits Each Description (Variables 10–17)

	Springdale Mall	Downtown	West Mall	No Opinion
10. Easy to return/exchange goods	(1)	(2)	(3)	(4)
11. High quality of goods	(1)	(2)	(3)	(4)
12. Low prices	(1)	(2)	(3)	(4)
13. Good variety of sizes/styles	(1)	(2)	(3)	(4)
14. Sales staff helpful/friendly	(1)	(2)	(3)	(4)
15. Convenient shopping hours	(1)	(2)	(3)	(4)
16. Clean stores and surroundings	(1)	(2)	(3)	(4)
17. A lot of bargain sales	(1)	(2)	(3)	(4)

E. Importance of Each Item in Respondent's Choice of a Shopping Area (Variables 18–25)

	Not Important					Very Important	
18. Easy to return/exchange goods	(1)	(2)	(3)	(4)	(5)	(6)	(7)
19. High quality of goods	(1)	(2)	(3)	(4)	(5)	(6)	(7)
20. Low prices	(1)	(2)	(3)	(4)	(5)	(6)	(7)
21. Good variety of sizes/styles	(1)	(2)	(3)	(4)	(5)	(6)	(7)
22. Sales staff helpful/friendly	(1)	(2)	(3)	(4)	(5)	(6)	(7)
23. Convenient shopping hours	(1)	(2)	(3)	(4)	(5)	(6)	(7)
24. Clean stores and surroundings	(1)	(2)	(3)	(4)	(5)	(6)	(7)
25. A lot of bargain sales	(1)	(2)	(3)	(4)	(5)	(6)	(7)

F. Information about the Respondent (Variables 26–30)

26. Gender: (1) = Male (2) = Female
27. Number of years of school completed:
    (1) = less than 8 years (3) = 12–under 16 years
    (2) = 8–under 12 years (4) = 16 years or more
28. Marital status: (1) = Married (2) = Single or other
29. Number of people in household: ________ persons
30. Age: ________ years

Each respondent in this database is described by 30 variables. As an example of their interpretation, consider row number 1. This corresponds to respondent number 1 and contains the following information.

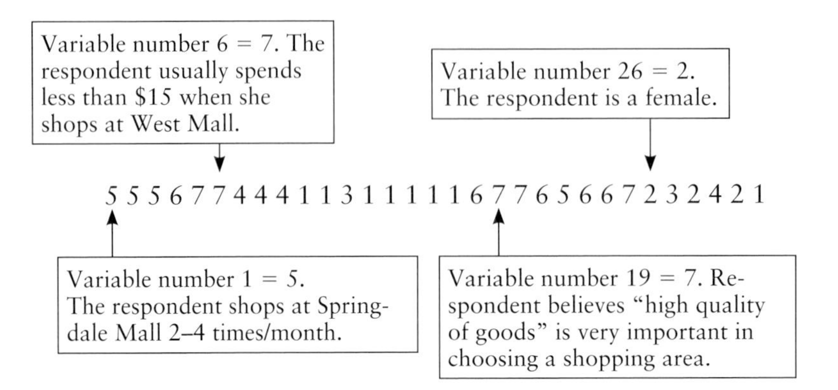

The data from these 150 respondents will be the basis for further analyses later in the text. In applying some of the techniques from this chapter, the following questions could provide insights into the perceptions and behavior of Springdale residents regarding the three shopping areas.

1. Using the data file **SHOPPING**, construct three different frequency distributions, one each for variables 7, 8, and 9. How do the three areas compare in terms of residents' general attitudes toward each?
2. Do people tend to spend differently at the three areas? Construct and compare frequency distributions for variables 4, 5, and 6.
3. To find out more about specific strengths and weaknesses of the areas, set up a frequency distribution for variable 10 (i.e., how many "best-fits" votes did each area get for "easy to return/exchange goods"?). Repeat this for variables 11–17 and interpret the results. Which of the malls seems to be the "bargain mall"? Which area seems to have the most convenient shopping hours?
4. Generate a cross-tabulation in which the categorization variables are variable 26 (gender) and variable 28 (marital status). For each of the four subcategories in the cross-tabulation, have your statistical software include the average for variable 30 (respondent age). Interpret the results.

CHAPTER 3

Statistical Description of Data

EMPLOYEE ABSENTEEISM: SLICK TRICK OR TRULY SICK?

In a survey of 1100 British employers, researchers found quite a discrepancy in the annual sick days used by public- versus private-sector employees. Workers in the public sector used an average of 10.7 sick days per year, compared to just 7.8 sick days per year for those in the private sector. Does the British private sector attract younger and healthier workers, or is it possible that public-sector workers simply take advantage of the greater number of sick days they are generally allowed?

According to a survey of 574 U.S. human resource executives, the answer is the latter—allowing more paid sick days leads to more paid sick days being taken. Human resource directors suspect that only 28% of paid sick leave is actually used because of illness, but they have no way of knowing for sure which absent employees are really sick.

Rosemary Haefner, Senior Career Advisor for CareerBuilder.com, makes an interesting seasonal observation about the contrived absenteeism situation: "With the cold and flu season kicking in, it's a popular time of the year for employees to call in sick. However, the number of those who are actually feeling under the weather may not necessarily match up with unscheduled absences. Twenty percent of workers say they called into work because they just didn't feel like going to the office that day. One in four workers report they feel sick days are equivalent to taking extra vacation days and treat them as such."

The situation can also be viewed from the standpoint of union/management status, and using the median (the sick-days value where just as many workers are higher as lower) instead of the average, or mean. From this perspective, union hourly workers use a median of 6.1 sick days per year, nonunion hourly workers take 4.9, and management uses 4.0.

In the pages ahead, we'll talk more about the mean and the median, as well as several other important numerical measures that describe data. Try not to be absent.

Sources: "Public Sector Sick Days Cost Taxpayers Billions," management-issues.com, June 27, 2009. Del Jones, "Firms Take New Look at Sick Days," *USA Today*, October 8, 1996, p. 8B; and Kate Lorenz, "Why Are People Calling in Sick? What's Your Excuse?" Netscape.com, May 1, 2006.

"I won't eat the mean;
I won't eat the median;
I won't even eat the first decile."

LEARNING OBJECTIVES

After reading this chapter, you should be able to:

- Describe data by using measures of central tendency and dispersion.
- Convert data to standardized values.
- Use the computer to visually represent data with a box-and-whisker plot.
- Determine measures of central tendency and dispersion for grouped data.
- Use the coefficient of correlation to measure association between two quantitative variables.

3.1 INTRODUCTION

In Chapter 2, we saw how raw data are converted into frequency distributions, histograms, and visual displays. We will now examine statistical methods for describing typical values in the data as well as the extent to which the data are spread out. Introduced in Chapter 1, these descriptors are known as **measures of central tendency** and **measures of dispersion:**

- **Measures of central tendency** Numbers describing *typical* data values. Chances are the typical subscriber to *The Wall Street Journal* earns more than the typical subscriber to *Mad* magazine. By using measures of central tendency, we can numerically describe the typical income of members of each group. The primary measures of central tendency discussed in this chapter are the arithmetic mean, weighted mean, median, and mode.
- **Measures of dispersion** Numbers that describe the *scatter* of the data. It's very likely that the variety of heights of centers in the National Basketball Association (most such players are between 6′8″ and 7′4″) is not as wide as the dispersion that exists within the general population, which includes individuals as tall as the NBA centers and persons as short as newborn babies. Measures of dispersion allow us to numerically describe the scatter, or *spread*, of measurements. Among the measures of dispersion discussed in this chapter are the range, quantiles, mean absolute deviation, variance, and standard deviation.

When we discuss measures of central tendency and dispersion, we'll be using two other key terms from Chapter 1: *population* and *sample*. Remember that (1) we often wish to use information about the sample to make inferences about the population from which the sample was drawn, and (2) characteristics of the population are referred to as *parameters*, while characteristics of the sample are referred to as *statistics*.

Chapter 2 also introduced the scatter diagram and best-fit linear equation as a graphical method of examining the possible linear relationship between two quantitative variables. In this chapter, we will extend this to two statistical measures of association: the coefficients of correlation and determination. As with the scatter diagram and best-fit equation, these will be discussed in greater detail in Chapter 15, Simple Linear Regression and Correlation.

STATISTICAL DESCRIPTION: MEASURES OF CENTRAL TENDENCY

3.2

We now discuss methods for representing the data with a single numerical value. This numerical descriptor has the purpose of describing the *typical* observation contained within the data.

The Arithmetic Mean

Defined as the sum of the data values divided by the number of observations, the **arithmetic mean** is one of the most common measures of central tendency. Also referred to as the *arithmetic average* or simply the *mean*, it can be expressed as μ (the population mean, pronounced "myew") or $\bar{x}$ (the sample mean, "x bar").

The population mean (μ) applies when our data represent *all* of the items within the population. The sample mean ($\bar{x}$) is applicable whenever data represent a *sample* taken from the population.

Population mean:

$$\mu = \frac{\sum_{i=1}^{N} x_i}{N} \quad \text{or simply} \quad \mu = \frac{\sum x_i}{N}$$

where μ = population mean
x_i = the ith data value in the population
$\sum$ = the sum of
N = number of data values in the population

The leftmost version of the equation for the population mean (μ) includes complete summation notation. Notation such as the "$i = 1$" and "N" will be included in the remainder of the text only when the nature of the summation is not clear from the context in which it is being used.

Sample mean:

$$\bar{x} = \frac{\sum x_i}{n}$$

where $\bar{x}$ = sample mean
x_i = the ith data value in the sample
$\sum$ = the sum of
n = number of data values in the sample

In determining either a population mean (μ) or a sample mean ($\bar{x}$), the sum of the data values is divided by the number of observations. As an example of the calculation of a population mean, consider the following data for shipments of peanuts from a hypothetical U.S. exporter to five Canadian cities.

City	Peanuts (Thousands of Bags)
Montreal	64.0
Ottawa	15.0
Toronto	285.0
Vancouver	228.0
Winnipeg	45.0

Because these were the only Canadian cities identified as receiving U.S. peanuts from this supplier during this period, they can be considered a population. We can calculate the arithmetic mean (μ) for these data as follows:

$$\mu = \frac{\Sigma x_i}{N} = \frac{64.0 + 15.0 + 285.0 + 228.0 + 45.0}{5} = 127.4 \text{ thousand bags}$$

On the average, each Canadian destination for this U.S. company's peanuts received 127.4 thousand bags of peanuts during the time period involved.

To help understand what the arithmetic mean represents, picture a playground seesaw with markings ranging from 0 at one end to 300 at the other end, with five people of equal weight sitting at positions matching the shipments received by each Canadian city. Assuming a weightless seesaw, the individuals would be perfectly balanced when the pivot bar is located exactly at the 127.4 position, as shown in part (b) of Figure 3.1.

Figure 3.1 also reveals a potential weakness of the mean as a descriptor of typicalness. Notice in part (b) that three of the five cities received relatively moderate shipments, 15.0, 45.0, and 64.0 thousand bags, while the other two cities received 228.0 and 285.0 thousand bags of peanuts. Thus each of the latter received more than the other three cities combined. This results in three of the five cities having shipments that are far below the average, making the arithmetic mean not as typical as we might like. A solution to this difficulty is the median, to be discussed shortly.

FIGURE 3.1

The arithmetic mean, or average, is a mathematical counterpart to the center of gravity on a seesaw. Although the influence of the two values that are more than 200 thousand is not quite as great as that of Arnold, it causes the arithmetic mean to be greater than three of the five data values.

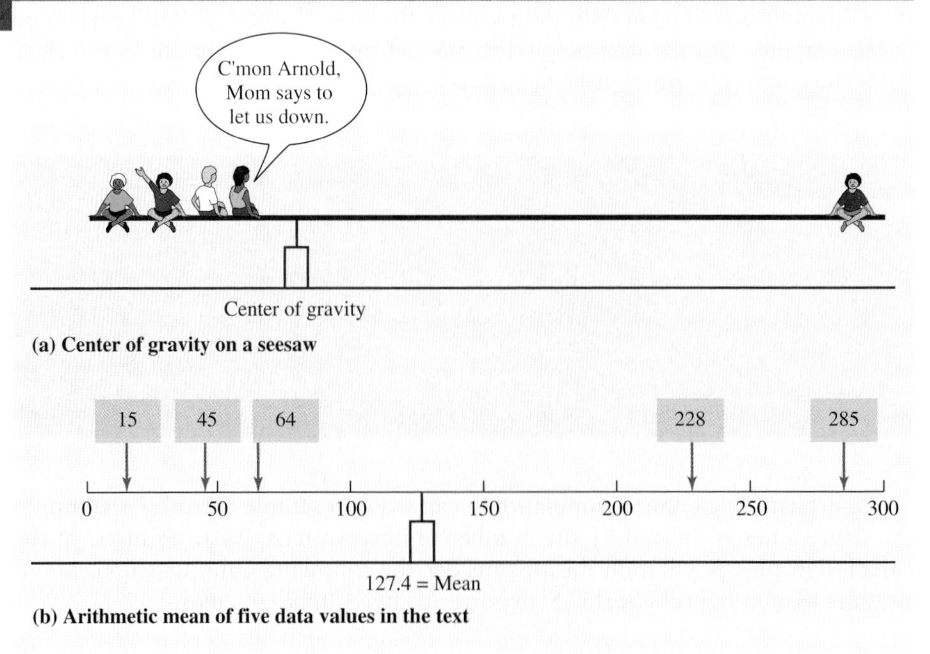

The Weighted Mean

When some values are more important than others, a **weighted mean** (sometimes referred to as a *weighted average*) may be calculated. In this case, each data value is weighted according to its relative importance. The formula for the weighted mean for a population or a sample will be as follows:

Weighted mean, μ_w (for a population) or $\bar{x}_w$ (for a sample):

$$\mu_w \text{ or } \bar{x}_w = \frac{\Sigma w_i x_i}{\Sigma w_i} \quad \text{where } w_i = \text{weight assigned to the } i\text{th data value}, \; x_i = \text{the } i\text{th data value}$$

Continuing with the peanut example, let's assume that shipments to the respective cities will be sold at the following profits per thousand bags: \$15.00, \$13.50, \$15.50, \$12.00, and \$14.00. (*Note:* In this example, we are trying to determine the weighted mean for the *profit;* accordingly, the preceding profit data will constitute the x_i values in our example and will be weighted by the shipment quantities to the five cities.)

The average profit per thousand bags will *not* be (15.00 + 13.50 + 15.50 + 12.00 + 14.00)/5, because the cities did not receive equal quantities of peanuts. A weighted mean must be calculated if we want to find the average profit per thousand bags for all shipments of peanuts from this U.S. exporter to Canada during this period. Using our assumed profit figures, this weighted mean is calculated as

$$\mu_w = \frac{\Sigma w_i x_i}{\Sigma w_i}$$

$$= \frac{64(\$15.00) + 15(\$13.50) + 285(\$15.50) + 228(\$12.00) + 45(\$14.00)}{64 + 15 + 285 + 228 + 45}$$

$$= \$14.04 \text{ per thousand bags}$$

The Median

In a set of data, the **median** is the value that has just as many values above it as below it. For example, the numbers of bags of peanuts (in thousands) shipped by the U.S. exporter to the five Canadian cities were

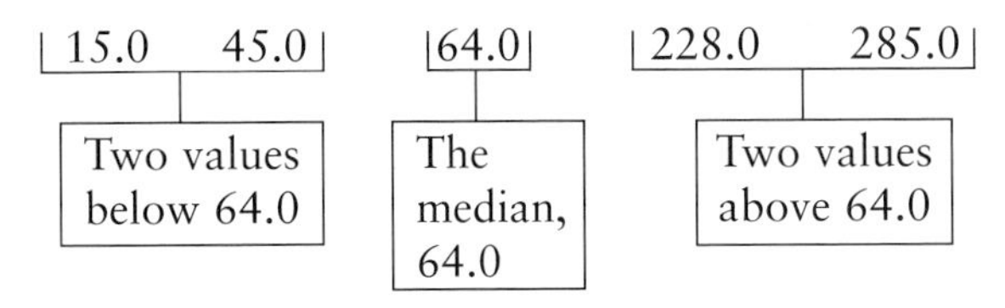

For these data, 64.0 is the value that has just as many values above it as below it. Montreal, which received 64.0 thousand bags of peanuts, is in the median position among the five Canadian cities.

The preceding example had an odd number of data values (5), and the median was the observation with two values above and two below. For some data, however, the number of values will be even. In such cases, the median will be halfway between the *two values* in the middle when the data are arranged in order of size.

For example, Ryder System, Inc. reported the following data for percentage return on average assets over an 8-year period.[1]

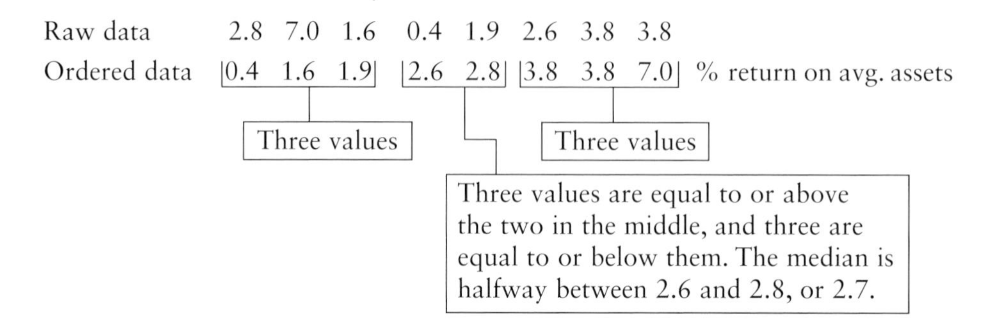

For these data, the median, 2.7, is the value halfway between the two values in the middle of the ordered arrangement, 2.6 and 2.8. Unlike the mean, the median is not influenced by extreme high or low values in the data. For example, if the highest data value had been 600% instead of 7.0%, the median would still be 2.7%.

The Mode

In a set of data, the **mode** is a value that occurs with the greatest frequency. As such, it could be considered the single data value most typical of all the values. Consider again the 8 years of return-on-assets data reported by Ryder System, Inc.:

Ordered data 0.4 1.6 1.9 2.6 2.8 3.8 3.8 7.0 % return on avg. assets

Mode = 3.8

For these data, the mode is 3.8, since it occurs more frequently than any other value. In this case, the mode does not appear to be a very good descriptor of the data, as five of the other six values are all smaller than 3.8.

Depending on the data, there can be more than one mode. For example, if the leftmost data value had been 1.6 instead of 0.4, there would have been two modes, 1.6 and 3.8; if this were the case, the values 1.6 and 3.8 would each have occurred in two different years. When there are two modes, a distribution of values is referred to as **bimodal**.

Comparison of the Mean, Median, and Mode

As we have seen, the mean, median, and mode are alternative approaches to describing the central tendency of a set of values. In deciding which measure to use in a given circumstance, there are a number of considerations:

- The mean gives equal consideration to even very extreme values in the data, while the median tends to focus more closely on those in the middle of the data array. Thus, the mean is able to make more complete use of the data. However, as pointed out earlier, the mean can be strongly influenced by just

[1]Source: Ryder System, Inc., *2005, 2003,* and *1998 Annual Reports.*

one or two very low or high values. A demonstration of how a single observation can affect the mean and median is shown in Seeing Statistics Applet 1, at the end of the chapter.

- There will be just one value for the mean and one value for the median. However, as indicated previously, the data may have more than one mode.
- The mode tends to be less useful than the mean and median as a measure of central tendency. Under certain circumstances, however, the mode can be uniquely valuable. For example, when a television retailer decides how many of each screen size to stock, it would do him little good to know that the mean television set sold has a 38.53-inch screen—after all, there is no such thing as a 38.53-inch television. Knowledge that the mode is 30 inches would be much more useful.

Distribution Shape and Measures of Central Tendency

The relative values of the mean, median, and mode are very much dependent on the shape of the distribution for the data they are describing. As Figure 3.2 shows, distributions may be described in terms of *symmetry* and *skewness*. In a **symmetrical distribution,** such as that shown in part (a), the left and right sides of the distribution are mirror images of each other. The distribution in part (a) has a single mode, is bell shaped, and is known as the normal distribution. It will be discussed in Chapter 6—for now, just note that the values of the mean, median, and mode are equal.

Skewness refers to the tendency of the distribution to "tail off" to the right or left, as shown in parts (b) and (c) of Figure 3.2. In examining the three distribution

FIGURE 3.2

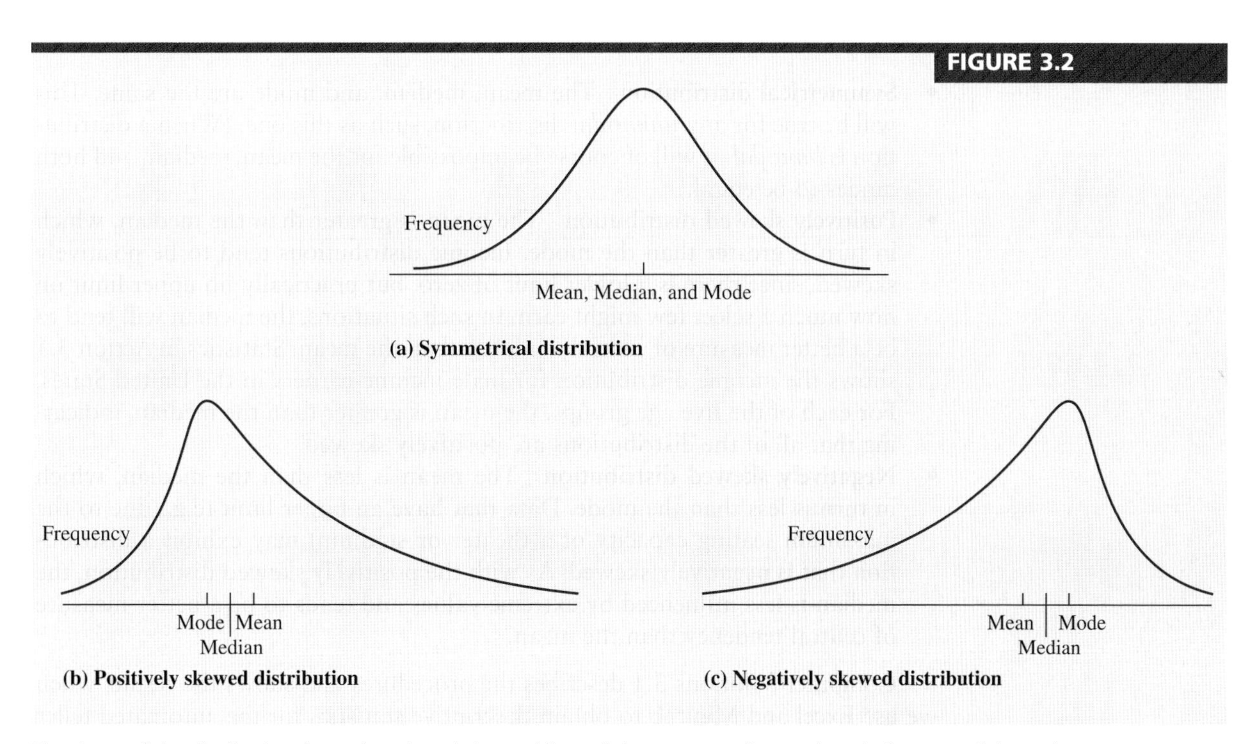

The shape of the distribution determines the relative positions of the mean, median, and mode for a set of data values.

STATISTICS IN ACTION 3.1

U.S. Males Post Skewed Incomes

Data values are skewed positively when the mean exceeds the median, and incomes are especially likely to exhibit this type of distribution. For example, the incomes of males who were 25 or older in 2000 were distributed as follows:

	Age Group				
Income	**25 to <35**	**35 to <45**	**45 to <55**	**55 to <65**	**65 or over**
under $10,000	8.4%	7.7%	8.4%	11.6%	18.7%
$10,000–under $25,000	29.9	21.0	17.7	24.4	45.2
$25,000–under $50,000	40.5	36.7	34.6	30.4	23.5
$50,000–under $75,000	13.8	19.6	20.3	16.7	6.4
$75,000 or over	7.4	15.0	19.0	16.9	6.3
	100.0	100.0	100.0	100.0	100.1*
Median (thousands)	$30.63	$37.09	$41.07	$34.41	$19.17
Mean (thousands)	$42.15	$53.48	$58.46	$61.05	$56.50

*Differs from 100.0 due to rounding.

For each of these distributions, the mean exceeds the median, reflecting that incomes are skewed positively (toward larger incomes) for all of these age groups. In addition, the gap between the median and the mean tends to grow as we proceed from the youngest age group to the oldest. The gap is especially pronounced for 65-or-over males—many of these people are on fixed incomes, but others may have incomes that are quite substantial.

Source: Bureau of the Census, U.S. Department of Commerce, *Statistical Abstract of the United States 2002,* pp. 439–440.

shapes shown in the figure, note the following relationships among the mean, median, and mode:

- **Symmetrical distribution** The mean, median, and mode are the same. This will be true for any *unimodal* distribution, such as this one. (When a distribution is *bimodal,* it will of course be impossible for the mean, median, and both modes to be equal.)
- **Positively skewed distribution** The mean is greater than the median, which in turn is greater than the mode. Income distributions tend to be positively skewed, since there is a lower limit of zero, but practically no upper limit on how much a select few might earn. In such situations, the median will tend to be a better measure of central tendency than the mean. Statistics in Action 3.1 shows the income distribution for male income-earners in the United States. For each of the five age groups, the mean is greater than the median, indicating that all of the distributions are positively skewed.
- **Negatively skewed distribution** The mean is less than the median, which in turn is less than the mode. Data that have an upper limit (e.g., due to the maximum seating capacity of a theater or stadium) may exhibit a distribution that is negatively skewed. As with the positively skewed distribution, the median is less influenced by extreme values and tends to be a better measure of central tendency than the mean.

Computer Solutions 3.1 describes the procedures and shows the results when we use Excel and Minitab to obtain descriptive statistics for the automated teller machine usage times for the 50 customers discussed in Chapter 2. The data are in

COMPUTER 3.1 SOLUTIONS

Descriptive Statistics

EXCEL

	A	B	C	D	E	F
1	**AgeCat**	**Gender**	**Seconds**		*Seconds*	
2	1	2	50.1			
3	1	1	53.0		Mean	39.76
4	2	2	43.2		Standard Error	1.26
5	1	2	34.9		Median	37.80
6	3	1	37.5		Mode	37.80
7	2	1	37.8		Standard Deviation	8.92
8	3	1	49.4		Sample Variance	79.49
9	1	1	50.5		Kurtosis	-0.87
10	3	1	48.1		Skewness	0.23
11	1	2	27.6		Range	34.0
12	3	2	55.6		Minimum	24.1
13	2	2	50.8		Maximum	58.1
14	3	1	43.6		Sum	1987.8
15	1	1	35.4		Count	50

1. Open the Excel data file CX02ATM. The labels and data values have already been entered as shown in the printout.
2. From the **Data** ribbon, click **Data Analysis** in the rightmost menu section. Within **Analysis Tools,** click **Descriptive Statistics.** Click **OK.**
3. In the **Input** section, enter **C1:C51** into the **Input Range** box. In the **Grouped By** section, click to select **Columns.** Because the input range includes the name of the variable in the first row, click to place a check mark in the **Labels in First Row** box.
4. In the **Output Options** section, click to select **Output Range,** then enter **E1** into the **Output Range** box. Click to place a check mark in the **Summary Statistics** box. Click **OK.**
5. Appearance improvements can be made by changing column widths and reducing the number of decimal places shown in the column of descriptive statistics.

MINITAB

```
Descriptive Statistics: Seconds

Variable   N  N*   Mean  SE Mean  StDev  Minimum     Q1  Median     Q3  Maximum
Seconds   50   0  39.76     1.26   8.92    24.10  32.67   37.80  46.70    58.10
```

1. Open the Minitab data file CX02ATM. The times (seconds) for the 50 ATM users are in column **C3.**
2. Click **Stat.** Select **Basic Statistics.** Click **Display Descriptive Statistics.** Enter **C3** into the **Variables** box. To get the printout shown here, click **OK.** [*Note:* As an option, we could easily obtain descriptive statistics for the males compared with those for the females. Just click to place a check mark next to **By variable,** then enter **C2** (the gender variable, where 1 = male, 2 = female) into the **By variable** box.]

file **CX02ATM**. The printouts are typical of what we would expect from most statistical software. Shown below are the printout items that pertain to the measures of central tendency discussed in this section and our interpretation of their values.

- **Sample size = 50** Excel refers to this as "count," while Minitab calls it "*N*". Computer printouts often include only capital letters, so "*N*" can still appear instead of the "*n*" that would be more appropriate for describing the size of a sample. (The N^* term in the Minitab printout shows how many data values were either blank or missing. There are no missing values in this database, so N^* is 0.)
- **Mean = 39.76 seconds and Median = 37.80 seconds** As the printout shows, the average customer required 39.76 seconds to complete his or her transaction, and just as many people had a time longer than 37.80 seconds as had a shorter time. Since the mean is greater than the median, the distribution is positively skewed.
- **Mode = 37.80 seconds** Excel reports the most frequent data value as 37.80 seconds. Minitab can also display the mode, but it has not been selected for this printout.
- **Skewness = 0.23** Excel has quantified the skewness and reported it as +0.23. The sign indicates a positively skewed distribution—that is, one in which the mean exceeds the median.

Not shown in either printout is a measure called the **trimmed mean,** in which statistical software ignores an equal number or percentage of data values at both the high end and the low end of the data before it is calculated. This descriptor can optionally be shown by Minitab, but not by Excel. Minitab's trimmed mean ignores the smallest 5% and the largest 5% of the data values. The purpose of the trimmed mean is to reduce the effect exerted by extreme values in the data.

EXERCISES

3.1 Determine the mean and the median wage rate for the data in Exercise 2.9.

3.2 Using the data in Exercise 2.11, determine the mean and the median number of goals per season that Mr. Gretzky scored during his 20 seasons in the National Hockey League.

3.3 Erika operates a website devoted to providing information and support for persons who are interested in organic gardening. According to the hit counter that records daily visitors to her site, the numbers of visits during the past 20 days have been as follows: 65, 36, 52, 70, 37, 55, 63, 59, 68, 56, 63, 63, 43, 46, 73, 41, 47, 75, 75, and 54. Determine the mean and the median for these data. Is there a mode? If so, what is its value?

3.4 A social scientist for a children's advocacy organization has randomly selected 10 Saturday-morning television cartoon shows and carried out a content analysis in which he counts the number of incidents of verbal or physical violence in each. For the 10 cartoons examined, the counts were as follows: 27, 12, 16, 22, 15, 30, 14, 30, 11, and 21. Determine the mean and the median for these data. Is there a mode? If so, what is its value?

3.5 For 1988 through 2007, the net rate of investment income for U.S. life insurance companies was as follows. What was the mean net rate of investment income for this period? The median? Source: American Council of Life Insurers, *Life Insurers Fact Book 2008,* p. 43.

Year	Percentage	Year	Percentage
1988	9.03	1998	6.95
1989	9.10	1999	6.71
1990	8.89	2000	7.05
1991	8.63	2001	6.31
1992	8.08	2002	5.38
1993	7.52	2003	5.03
1994	7.14	2004	4.80
1995	7.41	2005	4.90
1996	7.25	2006	5.35
1997	7.35	2007	5.71

3.6 The following is a list of closing prices for 15 of the stocks held by an institutional investor. What was the average closing price for this sample of stocks? The median closing price?

$31.69	56.69	65.50	83.50	56.88
72.06	121.44	97.00	42.25	71.88
70.63	35.81	83.19	43.63	40.06

3.7 A reference book lists the following maximum life spans (in years) for animals in captivity: Source: *The World Almanac and Book of Facts 2009*, p. 329.

Beaver	50	Grizzly bear	50
Bison	40	Horse (domestic)	50
Camel	50	Kangaroo	24
Cat (domestic)	37	Moose	27
Chimpanzee	60	Pig	27
Cow	30	Polar bear	45
Dog (domestic)	20	Rabbit	13
Asian elephant	77	Sea lion	34
Giraffe	36	Sheep	20
Goat	18	Tiger	26
Gorilla	54	Zebra	50

What is the mean of these maximum life spans? The median?

3.8 According to a utility company, utility plant expenditures per employee were approximately $50,845, $43,690, $47,098, $56,121, and $49,369 for the years 2005 through 2009. Employees at the end of each year numbered 4738, 4637, 4540, 4397, and 4026, respectively. Using the annual number of employees as weights, what is the weighted mean for annual utility plant investment per employee during this period?

3.9 A student scored 90 on the midterm exam, 78 on the final exam, and 83 on the term project. If these three components are weighted at 35%, 45%, and 20%, respectively, what is the weighted mean for her course performance?

3.10 An observer stands at an overpass and, for each motorcycle that passes below, records the value of $x =$ the number of riders on the motorcycle.

a. What value would you anticipate for the mode of her data? Why?
b. Would you anticipate that the mean would be greater than or less than the mode? Why?
c. Would you anticipate that the mean would be greater than or less than the median? Why?

3.11 In preparation for upcoming contract negotiations, labor and management representatives of the Hoffer Concrete Company have been collecting data on the earnings of a sample of employees (ranging from the president to the lowest paid "go-fer") at similar firms. Assuming that such sample data are available, and that the mean and median have been identified:

a. Which of these measures of central tendency would be likely to best serve the purposes of the management representative? Why?
b. Which of these measures of central tendency would be likely to best serve the purposes of the union representative? Why?

(DATA SET) *Note:* Exercises 3.12–3.15 require a computer and statistical software.

3.12 For the *U.S. News & World Report* top-50 U.S. law schools in Exercise 2.67, find the mean and median scores that the schools received from the academic evaluators. Compare these with the mean and median scores awarded by the lawyer/judge evaluators. Does either set of evaluators seem to provide ratings that are positively skewed or negatively skewed? The data are in file **XR02067**.

3.13 Each of the 100 air compressors described in Exercise 2.71 was tested to measure the amount of pressure it could exert. Find the mean and the median pressure, and comment on possible skewness of the data. The data are in file **XR02071**.

3.14 A personnel administrator has collected data on 100 employees, including gender, age, and number of days absent during the previous year. Compare the mean and median number of days absent for females (coded as gender = 1) versus males (coded as gender = 2) within the company. The data are in file **XR03014**.

3.15 For the data in Exercise 3.14, compare the mean and median ages for female employees (gender code = 1) versus male employees (gender code = 2). The data are in file **XR03014**.

STATISTICAL DESCRIPTION: MEASURES OF DISPERSION (3.3)

Although the mean and other measures are useful in identifying the central tendency of values in a population or a set of sample data, it's also valuable to describe their dispersion, or scatter. As a brief introduction to the concept of dispersion, consider the three distributions in Figure 3.3. Although distributions B and C have the same mean ($\mu = 12.5$ ounces), notice that B has a much wider dispersion

FIGURE 3.3

The dispersion of a distribution refers to the amount of scatter in the values. Distributions A, B, and C describe outputs for a machine that fills cereal boxes. Although B and C have the same mean, B has much greater dispersion and will result in many of the boxes being filled to less than the 12 ounces stated on the package.

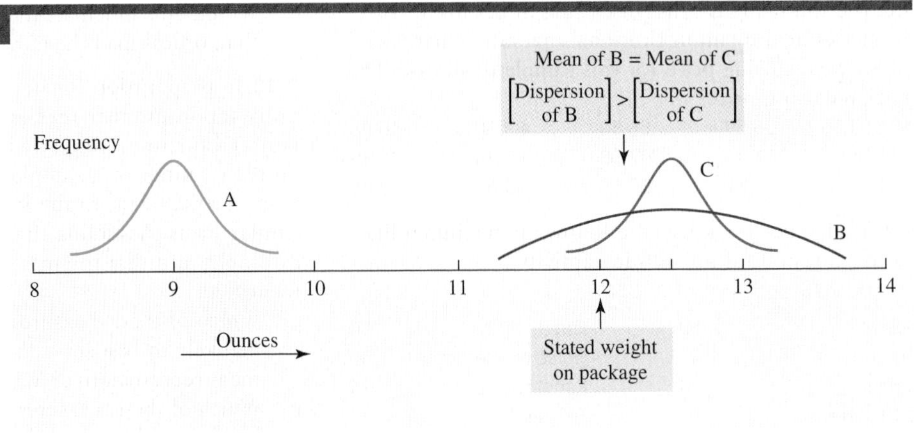

and that both B and C have a much larger mean than A (μ = 9.0 ounces). Considering these three distributions as representing the output of a machine that fills boxes of breakfast cereal, we can make the following observations:

- Machine A is set too low. On the average, it fills boxes to just 9 ounces instead of the 12 ounces advertised on the package. The firm's legal counsel will soon be very busy.
- Machine B delivers an average of 12.5 ounces, but the dispersion is so wide that many of the boxes are being underfilled. The firm's legal counsel will soon be moderately busy.
- Machine C delivers an average of 12.5 ounces, and the dispersion is narrow enough that very few of the boxes are being underfilled. The firm's legal counsel can relax.

As this example shows, measures of dispersion can have great practical value in business. We'll now discuss several of the more common measures, including the range, quantiles, mean absolute deviation, variance, and standard deviation.

Range

The simplest measure of dispersion, the **range** is the difference between the highest and lowest values. Part (a) of Figure 3.4 lists new privately owned housing starts for each of the 50 U.S. states. When the data are arranged in numerical order in part (b) of Figure 3.4, we see a range of 270.2 thousand housing starts (271.4 thousand for California minus the 1.2 thousand for Wyoming).

Although the range is easy to use and understand, it has the weakness of being able to recognize only the extreme values in the data. For example, if we were to eliminate the three highest states, 94% of the states would still be represented, but the range would now be only 71.9 thousand starts (73.1 thousand for Georgia minus the 1.2 thousand for Wyoming). Thus the range would be cut to just one-fourth of its original value despite the unchanged relative positions of the other 47 states.

Ads such as "Earn from $10 to $10,000 a week by sharpening saws at home" take advantage of this aspect of the range. The vast majority of such entrepreneurs may well earn less than $20 per week. However, it takes only one unusually successful person for the range to be so impressive. If there is even one very extreme value, the range can be a large number that is misleading to the unwary.

FIGURE 3.4

(a) Raw data (new housing starts, in thousands)

State	Starts	State	Starts	State	Starts	State	Starts	State	Starts
AL	17.2	HI	7.3	MA	39.2	NM	11.8	SD	2.5
AK	4.0	ID	4.3	MI	37.6	NY	61.9	TN	38.1
AZ	71.8	IL	38.7	MN	28.6	NC	70.7	TX	143.1
AR	9.9	IN	23.0	MS	8.8	ND	2.6	UT	16.5
CA	271.4	IA	5.2	MO	27.2	OH	33.0	VT	4.1
CO	32.8	KS	13.3	MT	2.0	OK	10.7	VA	64.1
CT	24.5	KY	13.8	NE	5.0	OR	11.3	WA	35.5
DE	4.6	LA	18.8	NV	14.0	PA	43.6	WV	1.5
FL	202.6	ME	8.1	NH	17.8	RI	5.4	WI	20.2
GA	73.1	MD	42.1	NJ	55.0	SC	32.8	WY	1.2

(b) Data arranged from smallest to largest

1.2	5.2	13.8	28.6	43.6
1.5	5.4	14.0	32.8	55.0
2.0	7.3	16.5	32.8	61.9
2.5	8.1	17.2	33.0	64.1
2.6	8.8	17.8	35.5	70.7
4.0	9.9	18.8	37.6	71.8
4.1	10.7	20.2	38.1	73.1
4.3	11.3	23.0	38.7	143.1
4.6	11.8	24.5	39.2	202.6
5.0	13.3	27.2	42.1	271.4

(c) Quartiles

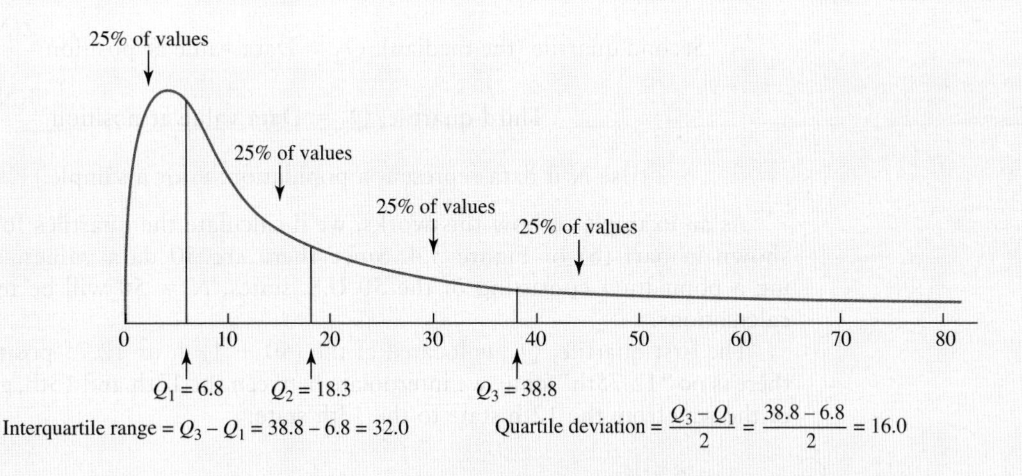

Raw data, ordered data, and quartiles for new privately owned housing starts in the 50 U.S. states
Source: Bureau of the Census, U.S. Department of Commerce, *Construction Reports*, series C20.

The **midrange,** a variant of the range, is the average of the lowest data value and the highest data value. Although it is vulnerable to the effects of extremely low or high outliers in the data, the midrange is sometimes used as a very approximate measure of central tendency. For the data in Figure 3.4, the midrange would be (271.4 + 1.2)/2, or 136.3 thousand housing starts.

Quantiles

We have already seen how the median divides data into two equal-size groups: one with values above the median, the other with values below the median. **Quantiles** also separate the data into equal-size groups in order of numerical value. There are several kinds of quantiles, of which **quartiles** will be the primary topic for our discussion.

PERCENTILES divide the values into 100 parts of equal size, each comprising 1% of the observations. The median describes the 50th percentile.

DECILES divide the values into 10 parts of equal size, each comprising 10% of the observations. The median is the 5th decile.

QUARTILES divide the values into four parts of equal size, each comprising 25% of the observations. The median describes the second quartile, below which 50% of the values fall. (*Note:* In some computer statistical packages, the first and third quartile values may be referred to as *hinges*.)

After values are arranged from smallest to largest, quartiles are calculated similarly to the median. It may be necessary to *interpolate* (calculate a position between) two values to identify the data position corresponding to the quartile.

For N values arranged from lowest to highest:

$$\text{First quartile, } Q_1 = \text{Data value at position } \frac{(N+1)}{4}$$

$$\text{Second quartile (the median), } Q_2 = \text{Data value at position } \frac{2(N+1)}{4}$$

$$\text{Third quartile, } Q_3 = \text{Data value at position } \frac{3(N+1)}{4}$$

(Use N if data represent a population, n for a sample.)

As an example of how this works, we'll calculate the quartiles for the data shown in part (b) of Figure 3.4. Since there are 50 data values, representing a population consisting of the 50 U.S. states, $N = 50$ will be used in the calculations.

The first quartile, Q_1, is located at the $(50 + 1)/4$, or 12.75 position. Since there is no "12.75th" state, we interpolate between the 12th and 13th, going 75% of the way from the 12th state to the 13th state:

Position	x_{12}	$x_{12.75}$	x_{13}
Data value	5.4	6.8	7.3 thousand housing starts

75% of the way from 5.4 to 7.3, $Q_1 = 6.8$
(6.825 rounded to one decimal place)

In calculating Q_2, which is also the median, we identify what would be the $2(50 + 1)/4$, or 25.5th data value. This is halfway between the 25th state (17.8 thousand housing starts) and the 26th state (18.8), or $Q_2 = 18.3$. The third quartile corresponds to the position calculated as $3(50 + 1)/4$, or the 38.25th data value. Interpolating 25% of the way from the 38th state (38.7 thousand housing starts) to the 39th state (39.2), we find $Q_3 = 38.8$ thousand housing starts (38.825 rounded to 38.8). Although the rounding is not strictly necessary, it keeps the quartiles to the same number of decimal places as the original data.

Quartiles offer a special feature that allows us to overcome the inability of the range to consider any but the most extreme values. In particular, there are two useful descriptors that involve quartiles:

The **interquartile range** is the difference between the third quartile and the first quartile, or $Q_3 - Q_1$. In percentile terms, this is the distance between the 75% and 25% values.

The **quartile deviation** is one-half the interquartile range. The quartile deviation is $(Q_3 - Q_1)/2$.

Because they reduce the importance of extreme high and low values, the interquartile range and quartile deviation can be more meaningful than the range as indicators of the dispersion of the data. In addition, they consider more than just the two extreme values, thus making more complete use of the data. Part (c) of Figure 3.4 shows a sketch of the approximate distribution of the 50-state housing-starts data, along with the quartiles, the interquartile range, and the quartile deviation. The distribution is positively skewed (i.e., it tapers off toward a few very high values at the right), and the mean (34.6) is much greater than the median (18.3).

Mean Absolute Deviation (MAD)

With this descriptor, sometimes called the *average deviation* or the *average absolute deviation,* we now consider the extent to which the data values tend to differ from the mean. In particular, the **mean absolute deviation** (*MAD*) is the average of the absolute values of differences from the mean and may be expressed as follows:

Mean absolute deviation (*MAD*) for a population:

$$MAD = \frac{\sum |x_i - \mu|}{N}$$

where μ = population mean
x_i = the ith data value
N = number of data values in the population

(To calculate *MAD* for a sample, substitute n for N and $\bar{x}$ for μ.)

In the preceding formula, we find the sum of the absolute values of the differences between the individual values and the mean, then divide by the number of individual values. The two vertical lines ("| |") tell us that we are concerned with *absolute values,* for which the algebraic sign is ignored. To illustrate how the mean absolute deviation is calculated, we'll examine the following figures, which represent annual research and development (R&D) expenditures by Microsoft Corporation.[2]

[2]Source: Microsoft Corporation, *2005 Annual Report.*

TABLE 3.1

Calculation of mean absolute deviation for annual research and development (R&D) expenditures for Microsoft Corporation. Data are in millions of dollars.

Year	R&D x_i	Deviation from Mean $(x_i - \mu)$	Absolute Value of Deviation from Mean $\lvert x_i - \mu \rvert$
2001	4379	−1868.2	1868.2
2002	6299	51.8	51.8
2003	6595	347.8	347.8
2004	7779	1531.8	1531.8
2005	6184	−63.2	63.2
	31,236	0.0	3862.8
	$= \sum x_i$		$= \sum \lvert x_i - \mu \rvert$

$$\text{Mean: } \mu = \frac{\sum x_i}{N} = \frac{31{,}236}{5} = \$6247.2 \text{ million}$$

$$\text{Mean absolute deviation: } MAD = \frac{\sum \lvert x_i - \mu \rvert}{N} = \frac{3862.8}{5} = \$772.6$$

Year	2001	2002	2003	2004	2005	
R&D	4379	6299	6595	7779	6184	(millions of dollars)

If we consider these data as representing the population of values for these 5 years, the mean will be μ. It is calculated as

$$\mu = \frac{4379 + 6299 + 6595 + 7779 + 6184}{5} = \$6247.2 \text{ million}$$

As shown in Table 3.1, the mean absolute deviation is the sum of the absolute deviations divided by the number of data values, or $3862.8/5 = \$772.6$ million. On the average, annual expenditures on research and development during these 5 years were $772.6 million away from the mean. In the third column of Table 3.1, note that the sum of the deviations is zero. This sum will always be zero, so it isn't of any use in describing dispersion, and we must consider only the *absolute* deviations in our procedure.

Variance and Standard Deviation

Variance

The **variance,** a common measure of dispersion, includes all data values and is calculated by a mathematical formula. For a population, the variance (σ^2, "sigma squared") is the average of squared differences between the N data values and the mean, μ. For a sample variance (s^2), the sum of the squared differences between the n data values and the mean, $\bar{x}$, is divided by $(n - 1)$. These calculations can be summarized as follows:

Variance for a population:

$$\sigma^2 = \frac{\sum (x_i - \mu)^2}{N}$$

where σ^2 = population variance
μ = population mean
x_i = the ith data value
N = number of data values in the population

Variance for a sample:

$$s^2 = \frac{\sum (x_i - \bar{x})^2}{n - 1}$$

where s^2 = sample variance
$\bar{x}$ = sample mean
x_i = the ith data value
n = number of data values in the sample

Using the divisor $(n - 1)$ when calculating the variance for a sample is a standard procedure that makes the resulting sample variance a better estimate of the variance of the population from which the sample was drawn. Actually, for large sample sizes (e.g., $n \geq 30$) the subtraction of 1 from n makes very little difference.

Table 3.2 shows the calculation of the variance for a sample of estimates of highway miles per gallon (mpg) for five sports-utility vehicles. In the table, notice the role played by **residuals.** These are the same as the deviations from the mean discussed in the preceding section. For each observation, the residual is the difference between the individual data value and the mean.

Residual for a data value in a sample $= x_i - \bar{x}$ where $\bar{x}$ = sample mean
x_i = value of the ith observation

For example, the residual for the third observation (21 mpg) is $21.0 - 20.0 = 1.0$ mpg. The residual is an important statistical concept that will be discussed further in later chapters.

TABLE 3.2

Calculation of variance (s^2) and standard deviation (s) for estimated highway fuel economy (mpg) for five 2009 sports-utility vehicles.

Model	Highway mpg (x_i)	x_i^2	Residual $(x_i - \bar{x})$	Residual^2 $(x_i - \bar{x})^2$
Saturn Outlook	23	529	3.0	9.0
Jeep Liberty	21	441	1.0	1.0
Subaru Tribeca	21	441	1.0	1.0
Land Rover LR3	17	289	−3.0	9.0
Porsche Cayenne GTS	18	324	−2.0	4.0
	100 $= \sum x_i$	2024 $= \sum x_i^2$		24.0 $= \sum (x_i - \bar{x})^2$

$$\bar{x} = \frac{\sum x_i}{5} = \frac{100}{5} = 20.0 \text{ miles per gallon}$$

$$s^2 = \frac{\sum (x_i - \bar{x})^2}{n - 1} = \frac{24.0}{5 - 1} = 6.0 \qquad s = \sqrt{6.0} = 2.45$$

Source: U.S. Environmental Protection Agency, *Fuel Economy Guide* 2009.

Standard Deviation

The positive square root of the variance of either a population or a sample is a quantity known as the **standard deviation.** The standard deviation is an especially important measure of dispersion because it is the basis for determining the proportion of data values within certain distances on either side of the mean for certain types of distributions (we will discuss these in a later chapter). The standard deviation may be expressed as

	For a Population	For a Sample
Standard Deviation	$\sigma = \sqrt{\sigma^2}$	$s = \sqrt{s^2}$

For the data in Table 3.2, the sample standard deviation is 2.45 miles per gallon. This is calculated by taking the square root of the sample variance, 6.0.

Variance and Standard Deviation: Additional Notes

- Unless all members of the population or sample have the same value (e.g., 33, 33, 33, 33, and so on), the variance and standard deviation cannot be equal to zero.
- If the same number is added to or subtracted from all the values, the variance and standard deviation are unchanged. For example, the data 24, 53, and 36 will have the same variance and standard deviation as the data 124, 153, and 136. Both the variance and standard deviation depend on the residuals (differences) between the observations and the mean, and adding 100 to each data value will also increase the mean by 100.

Computer Solutions 3.2 describes the procedures and shows the results when we use Excel and Minitab to obtain descriptive statistics for the housing-starts data shown in Figure 3.4. The data are in file **CX03HOUS**. The printout items pertaining to the measures of dispersion in this section and an interpretation of their values follow:

- **Standard Deviation (or StDev) = 49.78** The standard deviation is $s = 49.78$ thousand housing starts during the year. Excel also lists the variance, although this could be readily calculated as the square of the standard deviation.
- **Standard Error (SE Mean) = 7.04** This is the standard error for the mean, a term that will not be discussed until Chapter 8. Disregard it for now.
- **Minimum = 1.20 and Maximum = 271.40** These are the values for the minimum and maximum observations in the data. Excel also provides the range, which is simply 271.40 − 1.20, or 270.20 thousand housing starts.
- **Q1 = 6.82 and Q2 = 38.83** Reported by Minitab, the first and third quartiles are 6.82 and 38.83: 50% of the states had between 6.82 and 38.83 thousand housing starts. Excel does not include the first and third quartiles in its printout of summary statistics, but you can get at least approximate values by using the procedure described in Computer Solutions 3.2. If you are using the Data Analysis Plus add-in with Excel, the first and third quartiles are provided along with the Boxplot; this is demonstrated in Computer Solutions 3.3 in the next section.

Statistical software typically assumes the data are a sample instead of a population and uses (number of observations − 1) in calculating the standard deviation. This is usually not a problem because (1) much of the data in business statistics will

COMPUTER 3.2 SOLUTIONS

Descriptive Statistics

EXCEL

	A	B	C	D
1	**Hstarts**		*Hstarts*	
2	17.2			
3	4.0		Mean	34.65
4	71.8		Standard Error	7.04
5	9.9		Median	18.30
6	271.4		Mode	32.80
7	32.8		Standard Deviation	49.78
8	24.5		Sample Variance	2478.55
9	4.6		Kurtosis	12.20
10	202.6		Skewness	3.27
11	73.1		Range	270.20
12	7.3		Minimum	1.20
13	4.3		Maximum	271.40
14	38.7		Sum	1732.30
15	23.0		Count	50

Open the Excel data file **CX03HOUS**. The steps are the same as for Computer Solutions 3.1, except that we enter **A1:A51** into the **Input Range** box and **C1** into the **Output Range** box. Excel does not include the first and third quartiles in its printout of summary statistics. However, if you have a large number of data values, you can display at least approximate first and third quartiles by dividing the sample size by 4, rounding to the nearest integer if necessary, then entering this number into the **Kth Largest** box as well as the **Kth Smallest** box during step 3.

MINITAB

```
Descriptive Statistics: Hstarts

Variable   N  N*   Mean  SE Mean  StDev  Minimum    Q1  Median     Q3  Maximum
Hstarts   50   0  34.65     7.04  49.78     1.20  6.82   18.30  38.83   271.40
```

Open the Minitab data file **CX03HOUS**. The steps are the same as for Computer Solutions 3.1, except the 50 values for the housing starts are in column **C1.**

actually be sample data, and (2) there will not be an appreciable difference when the number of observations is large. However, if your data do represent a population, and you would like to know the value of σ instead of s, you can multiply s from the printout to get σ, as shown here:

$$\begin{array}{c}\text{Standard deviation}\\ \text{based on } (N) \text{ divisor}\end{array} = \begin{array}{c}\text{Standard deviation}\\ \text{based on } (N-1)\\ \text{divisor}\end{array} \times \sqrt{\frac{N-1}{N}}$$

EXERCISES

3.16 Provide a real or hypothetical example of a situation where the range could be misleading as a measure of dispersion.

3.17 For the sample of computer website hits described in Exercise 3.3, determine the range, the mean absolute deviation, the standard deviation, and the variance.

3.18 For the sample of Saturday-morning cartoon violence counts described in Exercise 3.4, determine the range, the mean absolute deviation, the standard deviation, and the variance.

3.19 During July 2005, the seven most-visited shopping websites (with number of visitors, in millions) were: eBay (61.7), Amazon (42.0), Wal-Mart (22.6), Shopping.com (22.4), Target (20.5), Apple Computer (17.4), and Overstock.com (17.2). Considering these as a population, determine: Source: *The World Almanac and Book of Facts 2006,* p. 390.

a. The mean, median, range, and midrange.
b. The mean absolute deviation.
c. The standard deviation and variance.

3.20 For a sample of 11 employers, the most recent hourly wage increases were 18, 30, 25, 5, 7, 2, 20, 12, 15, 55, and 40 cents per hour. For these sample data, determine:

a. The mean, median, range, and midrange.
b. The mean absolute deviation.
c. The standard deviation and variance.

3.21 According to the U.S. Environmental Protection Agency, a sample of 10 subcompact models shows the following estimated values for highway fuel economy (mpg): 40, 33, 32, 30, 27, 29, 27, 23, 21, and 10. For these sample data, determine:

a. The mean, median, range, and midrange.
b. The mean absolute deviation.
c. The standard deviation and variance.

3.22 For a sample of eight apartment buildings in Chicago, the percentage-occupancy rates during a recent month were as shown below:

50.0%	79.1	61.3	43.9
68.1	70.0	72.5	66.3

For these sample data, determine:

a. The mean, median, range, and midrange.
b. The mean absolute deviation.
c. The standard deviation and variance.

3.23 Determine the first, second, and third quartiles for the data in Exercise 3.20; then calculate the interquartile range and the quartile deviation.

3.24 Determine the first, second, and third quartiles for the data in Exercise 3.21; then calculate the interquartile range and the quartile deviation.

(**DATA SET**) *Note:* Exercises 3.25–3.27 require a computer and statistical software.

3.25 During the 1999 holiday shopping season, Keynote Systems, Inc. visited a number of e-commerce websites to see how long it took for the site to come up on the PC of a consumer using a 56K modem. The average time for Walmart.com was 23.35 seconds. Under the assumption that the company visited Wal-Mart's site on 50 occasions and obtained the sample times (in seconds) in data file **XR03025**, determine and interpret: Source: "How Key Web Sites Handle Holiday Shopping Rush," *USA Today*, November 24, 1999, p. 3B.

a. The mean, median, range, and midrange.
b. The mean absolute deviation.
c. The standard deviation and variance.

3.26 Data file **XR03014** includes the number of absences from work for a sample of 100 employees. For the variable representing number of days absent, determine and interpret the following:

a. The mean, median, range, and midrange.
b. The mean absolute deviation.
c. The standard deviation and variance.

3.27 Data file **XR02028** shows the depths (in meters) that 80 water-resistant watches were able to withstand just before leakage. For the watches in this sample, determine and interpret the following:

a. The mean, median, range, and midrange.
b. The mean absolute deviation.
c. The standard deviation and variance.

ADDITIONAL DISPERSION TOPICS (3.4)

The Box-and-Whisker Plot

Also referred to as a **box plot,** the **box-and-whisker plot** is a graphical device that simultaneously displays several of the measures of central tendency and dispersion discussed previously in the chapter. It highlights the first and third quartiles, the median, and the extreme values in the data, allowing us to easily identify these descriptors. We can also see whether the distribution is symmetrical or whether it is skewed either negatively or positively.

Computer Solutions 3.3 describes the procedures and shows the results when we use Excel and Minitab to generate a box plot for the housing-starts data

COMPUTER 3.3 SOLUTIONS

The Box Plot

EXCEL

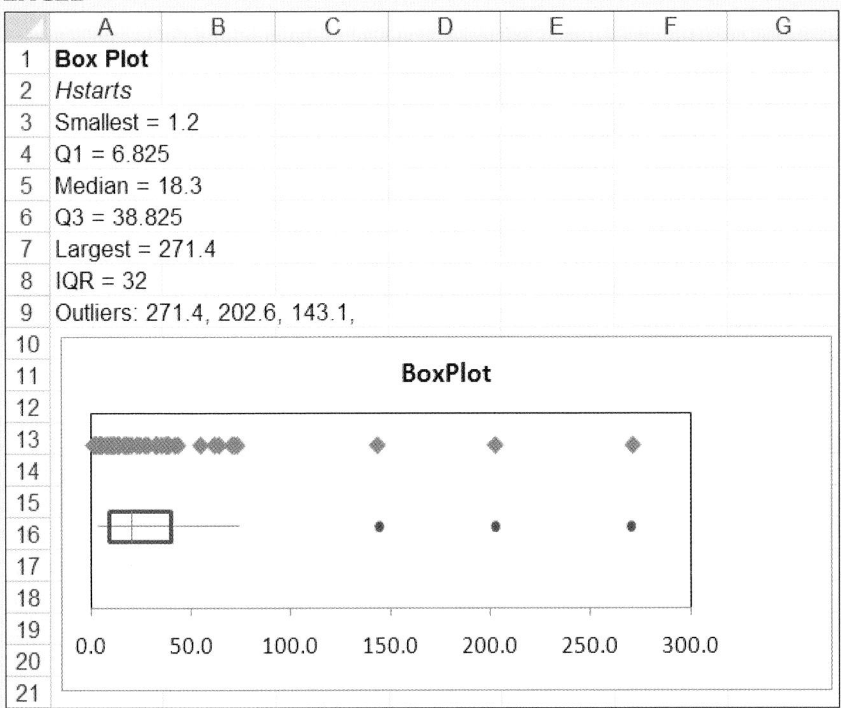

1. Open the Excel data file CX03HOUS. Click on **A1** and drag to **A51** to select the data values in cells **A1:A51.** From the **Add-Ins** ribbon, click **Data Analysis Plus** in the leftmost menu section.
2. In the **Data Analysis Plus** menu, click **Box Plot.** Click **OK.** Ensure that the **Input Range** box contains the range you specified in step 1. Click **Labels.** Click **OK.**
3. Improve the appearance by selecting and rearranging the components of the printout.

(continued)

MINITAB

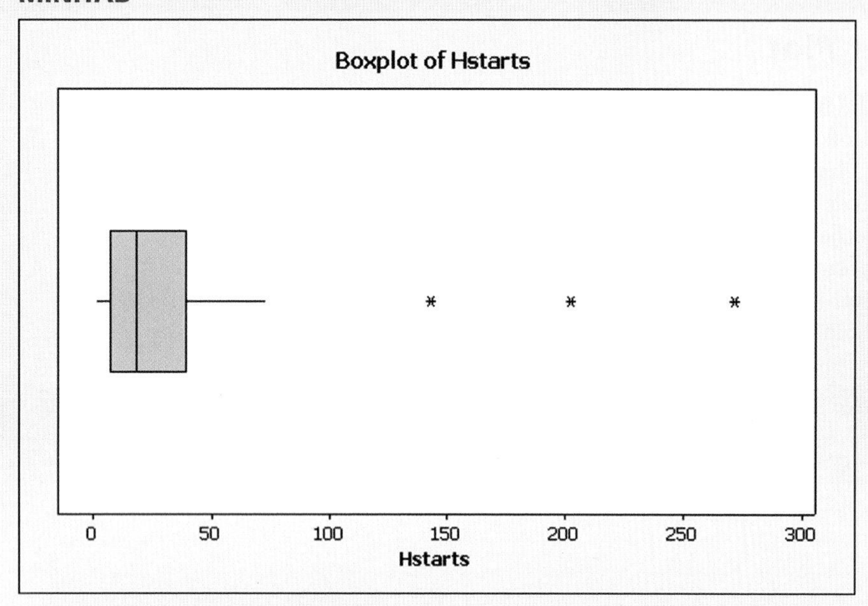

1. Open data file **CX03HOUS**. The 50 values for the housing starts are in column **C1.**
2. Click **Graph.** Click **Boxplot.** Select **One Y Simple** and click **OK.** In the **Graph variables** box, enter **C1.**
3. Click **Scale** and select the **Axes and Ticks** menu. Select **Transpose value and category scales.** (This causes the box plot to be oriented horizontally.) Click **OK.** Click **OK.**

shown in Figure 3.4. The data are in file **CX03HOUS**. The box plot has the following components:

- **Quartiles** The box extends horizontally from the first quartile (25th percentile) to the third quartile (75th percentile). Q_1 and Q_3 are shown at the left and right sides of the box, and the width of the box is the interquartile range.
- **Median** This is shown by the vertical line within the box. It is off-center within the box, being located slightly left of center. Because the median is closer to Q_1 than to Q_3, the distribution is skewed positively.
- **Extreme values** The whisker on the left side of the box extends to the smallest data value, Wyoming's 1.2 thousand housing starts. The whisker on the right side extends to cover all but the three most distant data values, which are considered outliers.

Chebyshev's Theorem

When either a population or a sample has a small standard deviation, individual observations will tend to be closer to the mean. Conversely, a larger standard deviation will result when individual observations are scattered widely about their mean. An early mathematician named Chebyshev set forth a theorem that quantifies this phenomenon. It specifies the minimum percentage of observations that will fall within a given number of standard deviations from the mean, regardless of the shape of the distribution.

CHEBYSHEV'S THEOREM **For either a sample or a population, the percentage of observations that fall within *k* (for $k > 1$) standard deviations of the mean will be at least**

$$\left(1 - \frac{1}{k^2}\right) \times 100$$

To demonstrate this rule, let's examine how it applies to the housing-starts data in Figure 3.4. For these data, values for the population mean and standard deviation are $\mu = 34.6$ and $\sigma = 49.3$, respectively (in thousands of housing starts). Using Chebyshev's theorem, we can find the percentage of values that should fall within $k = 2$ and $k = 3$ standard deviations of the mean. For example, if $k = 2$, this percentage should be at least $[1 - (1/2^2)](100)$, or 75%. Expressed another way, we should find that at least 75% of the states have housing starts that fall in the interval described by $34.6 \pm 2(49.3)$, or from -64.0 to 133.2. The actual percentage of data values falling within this interval is 94%, well above the minimum of 75% predicted by the theorem.

The Empirical Rule

Although Chebyshev's theorem applies to *any* distribution, regardless of shape, there is a rule of thumb that applies *only to distributions that are bell shaped and symmetrical,* like part (a) of Figure 3.2 and part (a) of Figure 3.3. This rule is known as the *empirical rule:*

THE EMPIRICAL RULE **For distributions that are bell shaped and symmetrical:**

- **About 68% of the observations will fall within 1 standard deviation of the mean.**
- **About 95% of the observations will fall within 2 standard deviations of the mean.**
- **Practically all of the observations will fall within 3 standard deviations of the mean.**

Let's consider part (a) of Figure 3.3 (page 68), which happens to be a bell-shaped, symmetrical distribution with a mean of 9.0 and a standard deviation of 0.3. If the filling machine inserts an average of 9.0 ounces of cereal into the boxes, with a standard deviation of 0.3 ounces, then approximately 68% of the boxes will contain $9.0 \pm 1(0.3)$ ounces of cereal, or between 8.7 and 9.3 ounces. Again applying the empirical rule, but using 2 standard deviations instead of 1, approximately 95% of the boxes will contain $9.0 \pm 2(0.3)$ ounces of cereal, or between 8.4 and 9.6 ounces. Practically all of the boxes will contain $9.0 \pm 3(0.3)$ ounces of cereal, or between 8.1 and 9.9 ounces.

Standardized Data

In our discussion of Chebyshev's theorem, we used the standard deviation as a unit of measurement, or measuring stick, for expressing distance from the mean to a data value. **Standardizing** the data is based on this concept and involves

TABLE 3.3

Standardization of estimated highway fuel economy (mpg) data for a sample of five 2009 sports-utility vehicles. Standardizing the data involves expressing each value in terms of its distance from the mean, in standard deviation units.

Original Data			Standardized Data
Model	Highway mpg (x_i)	$x_i - \bar{x}$	$\left[z_i = \frac{(x_i - \bar{x})}{s}\right]$
Saturn Outlook	23.00	3.00	1.22
Jeep Liberty	21.00	1.00	0.41
Subaru Tribeca	21.00	1.00	0.41
Land Rover LR3	17.00	−3.00	−1.22
Porsche Cayenne GTS	18.00	−2.00	−0.82
			0.00 $= \Sigma z_i$

Original Data	Standardized Data
$\bar{x} = 20.00$ mpg	$\bar{z} = 0.000$
$s_x = 2.45$ mpg	$s_z = 1.000$

expressing each data value in terms of its distance (in standard deviations) from the mean. For each observation in a sample:

$$z_i = \frac{x_i - \bar{x}}{s}$$

where z_i = standardized value for the ith observation
$\bar{x}$ = sample mean
x_i = the ith data value
s = sample standard deviation

(When data represent a population, μ replaces $\bar{x}$, and σ replaces s.)

What we're really doing is expressing the data in different units. It's like referring to a 6-footer as someone who is 2 yards tall. Instead of using feet, we're using yards. For example, the mpg estimates for the five sports-utility vehicles in Table 3.2 can be converted as shown in Table 3.3. After conversion, some standardized values are positive, others are negative. This is because we're using the mean as our reference point, and some data values are higher than the mean, others lower.

Standardized values have no units. This is true even though the original data, the standard deviation, and the mean may each represent units such as dollars, hours, or pounds. Because the data values and the standard deviation are in the same units, when one quantity is divided by the other, these units are in both the numerator and denominator, and the units cancel each other out.

A standardized value that is either a large positive number or a large negative number is a relatively unusual occurrence. For example, the estimated 23 mpg for the Saturn Outlook was easily the highest data value for the five vehicles in the sample. A standardized distribution will always have a mean of 0 and a standard deviation of 1. The standard deviation will be either s or σ, depending on whether the data represent a sample or a population. Standardizing the original data values makes it easier to compare two sets of data in which the original measurement units are not the same. Computer Solutions 3.4 describes the procedures and shows the results when we use Excel and Minitab to standardize the data for the five sports-utility vehicles.

COMPUTER 3.4 SOLUTIONS

Standardizing the Data

EXCEL

	A	B	C	D	E
1	MPG	STDMPG		Mean Is:	20.00
2	23	1.22		Std. Dev.Is:	2.45
3	21	0.41			
4	21	0.41			
5	17	-1.22			
6	18	-0.82			

1. Open the Excel data file CX03SUV. The data are entered in column **A,** as shown here. Type the labels **Mean is:** and **Std. Dev. is:** into cells **D1** and **D2,** respectively. Enter the equation **=AVERAGE(A2:A6)** into **E1.** Enter the equation **=STDEV(A2:A6)** into **E2.** (Alternatively, you can find the values for the mean and standard deviation by using a procedure like that of Computer Solutions 3.1 to obtain values for all the descriptive statistics.)
2. Enter the equation **=(A2-20.00)/2.45** into **B2.** Click on cell **B2.** Click on the lower right corner of cell **B2** and the cursor will turn to a "+" sign. Drag the "+" sign down to cell **B6** to fill in the remaining standardized values. Enter the label into cell **B1.**

MINITAB

```
Data Display

Row  MPG    STDMPG
  1   23   1.22474
  2   21   0.40825
  3   21   0.40825
  4   17  -1.22474
  5   18  -0.81650


Descriptive Statistics: MPG, STDMPG

Variable  N  N*    Mean  SE Mean  StDev  Minimum      Q1  Median     Q3
MPG       5   0   20.00     1.10   2.45    17.00   17.50   21.00  22.00
STDMPG    5   0  -0.000    0.447  1.000   -1.225  -1.021   0.408  0.816

Variable  Maximum
MPG         23.00
STDMPG      1.225
```

1. Open the Minitab data file CX03SUV. The five fuel-economy values are in column **C1.**
2. Click **Calc.** Click **Standardize.** Enter **C1** into the **Input column(s)** box. Enter **C2** into the **Store results in** box. Click to select **Subtract mean and divide by std. dev.** Click **OK.**
3. The standardized values are now in **C2,** which we can title by entering **STDMPG** into the label cell above the values.
4. Besides the original and standardized values, the printout shown here includes descriptive statistics for both. (*Note:* The mean and standard deviation of the standardized values are 0 and 1, respectively.) To get the printout: Click **Edit.** Click **Command Line Editor.** Enter **print C1 C2** into the first line. Enter **describe C1 C2** into the second line. Click **Submit Commands.**

The Coefficient of Variation

Expressing the standard deviation as a percentage of the mean, the **coefficient of variation (CV)** indicates the *relative amount of dispersion* in the data. This enables us to easily compare the dispersions of two sets of data that involve different measurement units (e.g., dollars versus tons) or differ substantially in magnitude. For example, the coefficient of variation can be used in comparing the variability in prices of diamonds versus prices of coal or in comparing the variability in sales, inventory, or investment of large firms versus small firms.

In comparing the variability of gold prices and slab zinc prices, for example, standard deviations would tell us that the price of gold varies a great deal more than the price of zinc. However, the dollar value of an ounce of gold is much higher than that of a pound of slab zinc, causing its variability to be exaggerated. Dividing each standard deviation by the mean of its data, we are better able to see how much each commodity varies relative to its own average price. The coefficient of variation is expressed as a percentage and is calculated as follows:

	For a Population	For a Sample
Coefficient of Variation	$CV = \frac{\sigma}{\mu} \times 100$	$CV = \frac{s}{\bar{x}} \times 100$

Suppose that, over a sample of 12 months, the mean price of gold had been $\bar{x}$ = \$810.6 per ounce, with a standard deviation of s = \$49.0 per ounce, while the mean and standard deviation for slab zinc during this same period had been \$1.04 per pound and \$0.08 per pound, respectively. Which commodity would have had the greater relative dispersion, or coefficient of variation?

We calculate the coefficient of variation for each commodity: The CV for gold was $s/\bar{x}$, or \$49.0/\$810.6 = 0.060, and the CV for slab zinc was \$0.08/\$1.04 = 0.077. Expressed as percentages, the coefficients of variation would be 6.0% for gold and 7.7% for slab zinc. Although the standard deviation of monthly gold prices was more than 612 times the standard deviation of zinc prices, from a *relative dispersion* perspective, the variability of gold prices was actually *less than* that for slab zinc.

EXERCISES

3.28 A manufacturing firm has collected information on the number of defects produced each day for the past 50 days. The data have been described by the accompanying Minitab box-and-whisker plot.

a. What is the approximate value of the median? The first and third quartiles?
b. What do the asterisks (*) at the right of the display indicate? What implications might they have for the production supervisor?
c. Does the distribution appear to be symmetric? If not, is it positively skewed or negatively skewed?

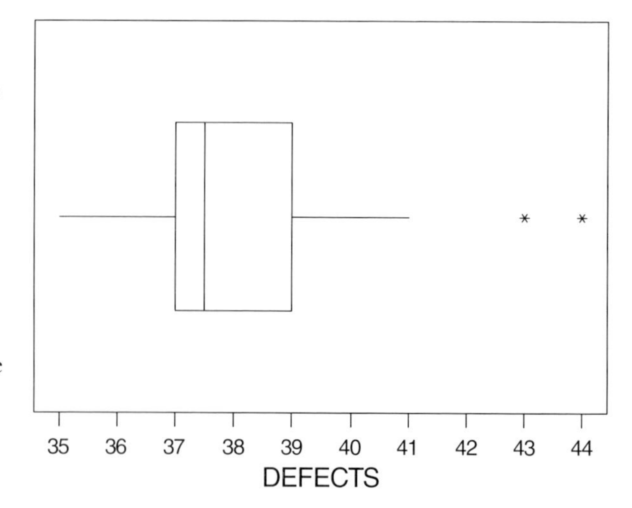

3.29 According to Chebyshev's theorem, what percentage of observations should fall

a. within 2.5 standard deviations of the mean?
b. within 3.0 standard deviations of the mean?
c. within 5.0 standard deviations of the mean?

3.30 For the data in Exercise 3.20, determine the percentage of observations that fall within $k = 1.5$ standard deviation units from the mean. Do the results support Chebyshev's theorem?

3.31 Standardize the data in Exercise 3.21, then identify the percentage of observations that fall within $k = 2.0$ standard deviation units from the mean. Do the results support Chebyshev's theorem?

3.32 The manufacturer of an extended-life lightbulb claims the bulb has an average life of 12,000 hours, with a standard deviation of 500 hours. If the distribution is bell shaped and symmetrical, what is the approximate percentage of these bulbs that will last

a. between 11,000 and 13,000 hours?
b. over 12,500 hours?
c. less than 11,000 hours?
d. between 11,500 and 13,000 hours?

3.33 For college-bound high school seniors from a certain midwestern city, math scores on the Scholastic Aptitude Test (SAT) averaged 480, with a standard deviation of 100. Assume that the distribution of math scores is bell shaped and symmetrical.

a. What is the approximate percentage of scores that were between 380 and 580?
b. What is the approximate percentage of scores that were above 680?
c. Charlie scored 580 on the math portion of the SAT. What is the approximate percentage of students who scored lower than Charlie?
d. Betty scored 680 on the math portion of the SAT. What is the approximate percentage of students who scored higher than Charlie but lower than Betty?

3.34 For data set A, the mean is $1235, with a standard deviation of $140. For data set B, the mean is 15.7 inches, with a standard deviation of 1.87 inches. Which of these two data sets has the greater relative dispersion?

3.35 A sample of 20 customers from Barnsboro National Bank reveals an average savings account balance of $315, with a standard deviation of $87. A sample of 25 customers from Wellington Savings and Loan reveals an average savings account balance of $8350, with a standard deviation of $1800. Which of these samples has the greater relative dispersion?

(**DATA SET**) *Note:* Exercises 3.36–3.38 require a computer and statistical software.

3.36 A resort catering to golfers has reported the average income of golfer households to be $95,000. Assume that this average had been based on the 100 incomes in file **XR03036.** Source: golfwashington.com, June 23, 2009.

a. Construct and interpret the box-and-whisker plot for the data.
b. Standardize the data; then determine the mean and standard deviation of the standardized values.

3.37 In Exercise 3.14, the personnel director has collected data describing 100 employees, with one of the variables being the number of absences during the past year. Using data file **XR03014**, describe the variable, number of absences.

a. Construct and interpret the box-and-whisker plot for the number of absences.
b. Standardize the data; then determine the mean and standard deviation of the standardized values.

3.38 Exercise 3.25 described a study in which e-commerce researchers visited walmart.com on multiple occasions during the holiday shopping season and measured the time (in seconds) it took for the site to come up on the computer screen. Data file **XR03025** contains results that we assumed could have been obtained during this study.

a. Construct and interpret the box-and-whisker plot for the times.
b. Standardize the data; then determine the mean and standard deviation of the standardized values.

DESCRIPTIVE STATISTICS FROM GROUPED DATA (3.5)

The frequency distribution, also referred to as **grouped data,** is a convenient summary of raw data, but it loses some of the information originally contained in the data. As a result, measures of central tendency and dispersion determined from the frequency distribution will be only approximations of the actual values. However, since the original data may not always be available to us, these approximations are sometimes the best we can do. Although other descriptors can be approximated from the frequency distribution, we will concentrate here on the mean and the standard deviation. Techniques for their approximation are

TABLE 3.4

The formulas in the text are used to calculate the approximate mean and standard deviation based only on a frequency distribution for the usage times of the 50 automated teller machine (ATM) customers in data file CX02ATM.

Time (seconds)	(1) Midpoint (m_i)	(2) Frequency (f_i)	(3) $f_i m_i$	(4) m_i^2	(5) $f_i m_i^2$
20–under 25	22.5	1	22.5	506.25	506.25
25–under 30	27.5	7	192.5	756.25	5,293.75
30–under 35	32.5	10	325.0	1056.25	10,562.50
35–under 40	37.5	9	337.5	1406.25	12,656.25
40–under 45	42.5	9	382.5	1806.25	16,256.25
45–under 50	47.5	6	285.0	2256.25	13,537.50
50–under 55	52.5	5	262.5	2756.25	13,781.25
55–under 60	57.5	3	172.5	3306.25	9,918.75
		50	1980.0		82,512.50
		$= \sum f_i = n$	$= \sum f_i m_i$		$= \sum f_i m_i^2$

Approximate mean: $\bar{x} = \frac{\sum f_i m_i}{n} = \frac{1980.0}{50} = 39.6$

Approximate variance: $s^2 = \frac{\sum f_i m_i^2 - n\bar{x}^2}{n-1} = \frac{82{,}512.50 - 50(39.6)^2}{50 - 1}$

$= \frac{82{,}512.50 - 78{,}408}{50 - 1} = \frac{4104.5}{49} = 83.77$

Approximate standard deviation: $s = \sqrt{83.77} = 9.15$

summarized in Table 3.4 and explained subsequently. In this section, we'll assume that all we know about the usage times for the 50 automated teller machine (ATM) customers in data file CX02ATM is the frequency distribution shown in the first three columns of Table 3.4. We will approximate the mean and standard deviation, then compare them with the actual values (which happen to be $\bar{x} = 39.76$ seconds and $s = 8.92$ seconds).

Arithmetic Mean from Grouped Data

When approximating the mean from a frequency distribution, we consider each observation to be located at the midpoint of its class. As an example, for the 30–under 35 class, all 10 persons in this category are assumed to have a time of exactly 32.5 seconds, the midpoint of the class. Using this and other midpoints, we can find the approximate value of the mean.

Approximate value of the sample mean, from grouped data:

$$\bar{x} = \frac{\sum f_i m_i}{n}$$

where $\bar{x}$ = approximate sample mean
f_i = frequency for class i
m_i = midpoint for class i
n = number of data values in the sample

(If data are a population, μ replaces $\bar{x}$ and N replaces n.)

The formula assumes that each midpoint occurs f_i times—e.g., that 22.5 seconds occurs 1 time, 27.5 seconds occurs 7 times, and so on. Applying this formula as shown in Table 3.4, we calculate the mean as approximately 39.6 seconds. Again, this is an approximation—the actual value, computed from the original data, is 39.76 seconds. (*Note:* If each data value *really is* at the midpoint of its class, the calculated mean, variance, and standard deviation will be exact values, not approximations.)

Variance and Standard Deviation from Grouped Data

Variance

As in our previous discussion of the variance, its calculation here will depend on whether the data constitute a population or a sample. The previously determined approximate value for the mean (in this case, estimated as $\bar{x} = 39.6$) is also used. When computing the variance from grouped data, one of the following formulas will apply:

Approximate value of the variance from grouped data:

- **If the data represent a population:**

$$\sigma^2 = \frac{\Sigma f_i m_i^2 - N\mu^2}{N}$$

where μ = approximate population mean
f_i = frequency for class i
m_i = midpoint for class i
N = number of data values in the population

- **If the data represent a sample:**

$$s^2 = \frac{\Sigma f_i m_i^2 - n\bar{x}^2}{n - 1}$$

where $\bar{x}$ = approximate sample mean
f_i = frequency for class i
m_i = midpoint for class i
n = number of data values in the sample

Calculation of the approximate population variance for the ATM customer data is shown in Table 3.4 and results in an estimate of 83.77 for the value of s^2. This differs from the $s^2 = 79.49$ reported by Excel in Computer Solutions 3.1, but remember that our calculation here is an approximation. If the original data are not available, such approximation may be our only choice.

Standard Deviation

The approximate standard deviation from grouped data is the positive square root of the variance. This is true for both a population and a sample. As Table 3.4 shows, the approximate value of s is the square root of 83.77, or 9.15 seconds. Like the variance, this compares favorably with the value that was based on the original data.

EXERCISES

3.39 Using the data in Exercise 3.5, construct a frequency distribution.

a. Determine the approximate values of the mean and standard deviation.
b. Compare the approximate values from part (a) with the actual values.
c. Construct a frequency distribution with twice as many classes as before, then repeat parts (a) and (b). Have the approximations improved?
d. If you were to construct a "frequency distribution" in which each data value was at the midpoint of its own class, what values do you think the "approximations" would have? Explain.

3.40 A sample consisting of 100 employees has been given a manual-dexterity test. Given the accompanying frequency distribution, determine the approximate mean and standard deviation for these data.

3.41 Eighty packages have been randomly selected from a frozen food warehouse, and the age (in weeks) of each package is identified. Given the frequency distribution shown, determine the approximate mean and standard deviation for the ages of the packages in the warehouse inventory.

Data for Exercise 3.40

Score	Number of Persons
5–under 15	7
15–under 25	9
25–under 35	12
35–under 45	14
45–under 55	13
55–under 65	9
65–under 75	8
75–under 85	11
85–under 95	10
95–under 105	7

Data for Exercise 3.41

Age (Weeks)	Number of Packages
0–under 10	25
10–under 20	17
20–under 30	15
30–under 40	9
40–under 50	10
50–under 60	4

3.6 STATISTICAL MEASURES OF ASSOCIATION

In Chapter 2, we introduced the scatter diagram as a method for visually representing the relationship between two quantitative variables, and we also showed how statistical software can be used to fit a linear equation to the points in the scatter diagram. This section introduces a way to numerically measure the strength of the linear relationship between two variables—the coefficient of correlation.

The **coefficient of correlation** (r, with $-1 \leq r \leq +1$) is a number that indicates both the direction and the strength of the linear relationship between the dependent variable (y) and the independent variable (x):

- **Direction of the relationship** If r is positive, y and x are directly related, as in parts (a) and (c) of Figure 3.5. If r is negative, the variables are inversely related, as in parts (b) and (d) of Figure 3.5.
- **Strength of the relationship** The larger the absolute value of r, the stronger the linear relationship between y and x. If $r = -1$ or $r = +1$, the best-fit linear equation will actually include all of the data points. This is the case in parts (a) and (b) of Figure 3.5. However, in parts (c) and (d), the respective equations are less than perfect in fitting the data, resulting in absolute values of r that are less than 1.

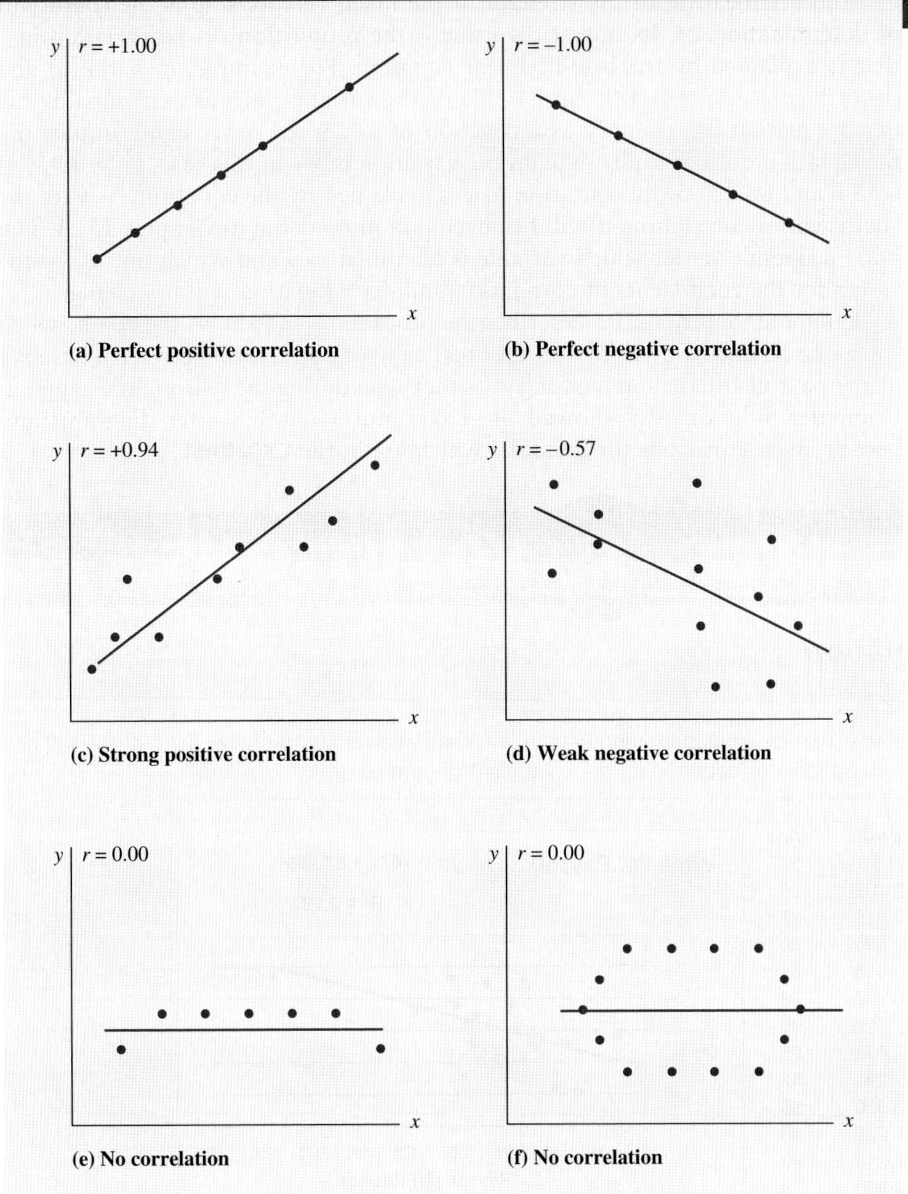

FIGURE 3.5

The stronger the linear relationship between the variables, the greater will be the absolute value of the correlation coefficient, r. When $r = 0$, the slope of the best-fit linear equation will also be 0.

In parts (e) and (f) of Figure 3.5, the slope is zero, r is zero, and there is no linear relationship whatsoever between y and x. In each case, the value of y doesn't depend on the value of x. At the end of the chapter, Seeing Statistics Applet 2 can be used to demonstrate how the value of r changes as the pattern of observations is changed.

In Chapter 15, Simple Linear Regression and Correlation, we will examine the coefficient of correlation and its method of calculation in greater detail. For now, it's important that you simply be aware of this important statistical measure of association. In this chapter, we'll rely on statistical software to do the calculations for us.

Another measure of the strength of the linear relationship is the **coefficient of determination, r^2.** Its numerical value is the proportion of the variation in y that is explained by the best-fit linear equation. For example, in parts (a) and (b) of Figure 3.5, r^2 is 1.00 and 100% of the variation in y is explained by the linear equation describing y as a function of x. On the other hand, in part (c), the equation does not fully explain the variation in y—in this case, $r^2 = 0.94^2 = 0.884$, and 88.4% of the variation in y is explained by the equation. As with the coefficient of correlation, it will be covered in more detail in Chapter 15. *A final note:* It doesn't matter which variable is identified as y and which one as x—the values for the coefficients of correlation and determination will not change.

Computer Solutions 3.5 describes the procedures and shows the results when we use Excel and Minitab to find the coefficient of correlation between a baseball team's payroll and the number of games they won during the season. In Chapter 2, Computer Solutions 2.7 showed the Excel and Minitab scatter diagrams and best-fit linear equations for the data, which are in file CX02BASE.

COMPUTER 3.5 SOLUTIONS

Coefficient of Correlation

EXCEL

The accompanying printout includes the scatter diagram and best-fit linear equation from Computer Solutions 2.7. It also shows the results for each of two options for determining the coefficient of correlation.

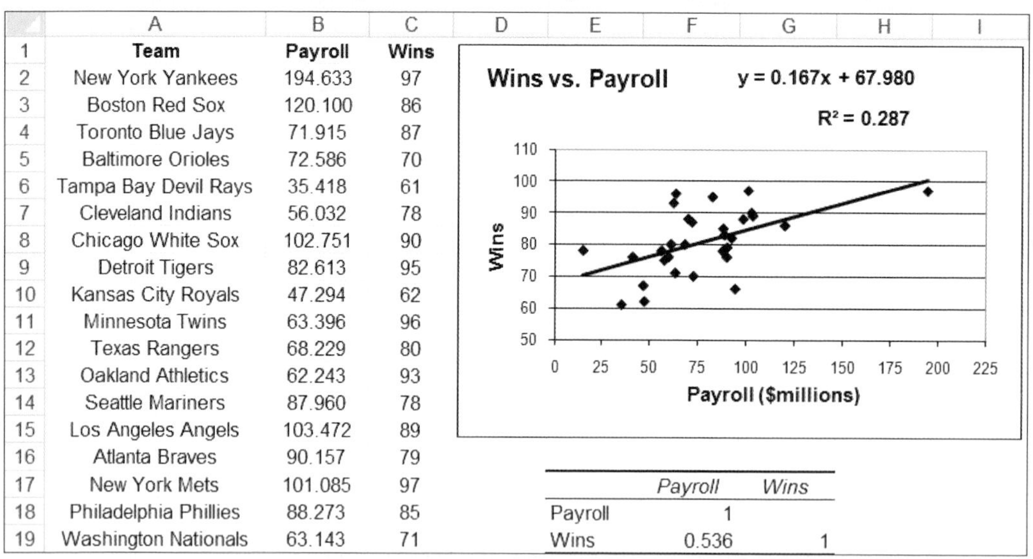

	A	B	C
1	Team	Payroll	Wins
2	New York Yankees	194.633	97
3	Boston Red Sox	120.100	86
4	Toronto Blue Jays	71.915	87
5	Baltimore Orioles	72.586	70
6	Tampa Bay Devil Rays	35.418	61
7	Cleveland Indians	56.032	78
8	Chicago White Sox	102.751	90
9	Detroit Tigers	82.613	95
10	Kansas City Royals	47.294	62
11	Minnesota Twins	63.396	96
12	Texas Rangers	68.229	80
13	Oakland Athletics	62.243	93
14	Seattle Mariners	87.960	78
15	Los Angeles Angels	103.472	89
16	Atlanta Braves	90.157	79
17	New York Mets	101.085	97
18	Philadelphia Phillies	88.273	85
19	Washington Nationals	63.143	71

	Payroll	Wins
Payroll	1	
Wins	0.536	1

Excel Option 1, to simply obtain the coefficient of correlation:

1. Open the Excel data file CX02BASE. The labels and data are already entered, as shown in the printout. Click on **B1** and drag to **C31** to select cells **B1:C31.** From the **Data** ribbon, click **Data Analysis** in the rightmost menu.
2. In the **Analysis Tools** box, click **Correlation.** Click **OK.** Enter **B1:C31** into the **Input range** box. In the **Grouped By** section, click next to **Columns.** Click to place a check mark next to **Labels in First Row.** Enter **E17** into the **Output Range** box. Click **OK.**

(*continued*)

Excel Option 2, to display r^2 on an existing chart like the one from Computer Solutions 2.7:

1. Right-click anywhere on the best-fit line in the chart. Click **Format Trendline.**
2. In the **Trendline Options** menu, select **Display R-squared value on chart.** Click **Close.**

MINITAB

Minitab Option 1, to simply determine the coefficient of correlation:

```
Correlations: Payroll, Wins

Pearson correlation of Payroll and Wins = 0.536
```

1. Open the Minitab data file **CX02BASE**. The labels **Payroll** and **Wins** have already been entered at the top of columns **C2** and **C3,** respectively. The payroll values and the win totals have been entered into **C2** and **C3**, respectively.
2. Click **Stat.** Select **Basic Statistics.** Click **Correlation.** Enter **C2** and **C3** into the **Variables** box. Click **OK.**

Minitab Option 2, to display r^2 along with a scatter diagram and best-fit linear equation:

Generate the Minitab scatter diagram and best-fit linear equation as in the Minitab portion of Computer Solutions 2.7 (not shown here). The chart automatically includes "R-Sq = 28.7%" at the top.

Both software packages report the coefficient of correlation as $r = +0.536$. If we have also generated a scatter diagram and best-fit linear equation, as in Computer Solutions 2.7, the r^2 value (0.287, or 28.7%) is displayed either automatically (Minitab) or as an option (Excel). Interpreting the results, we conclude that there is a positive relationship between a team's payroll and the number of games it won during the season. Further, the size of the payroll explains 28.7% of the variation in the number of wins. From the opposite perspective, (100.0% − 28.7%), or 71.3% of the variation in wins is *not* explained by payroll. This makes sense, as there are probably a great many other variables—for example, level of fan support, quality of coaching, player injuries, strength of schedule—that also influence the number of games a team wins.

EXERCISES

3.42 What is the coefficient of determination and what does it tell us about the relationship between two variables?

3.43 For a set of data, r^2 is 0.64 and the variables x and y are inversely related. What is the numerical value of r?

(DATA SET) *Note:* Exercises 3.44–3.47 require a computer and statistical software.

3.44 In Exercise 3.14, two of the variables examined by the personnel director were age and number of absences during the past year. Using the data in file **XR03014**, generate a scatter diagram (with y = number of absences and x = age), fit a linear equation to the data, then determine and interpret the coefficients of correlation and determination.

3.45 In the *U.S. News & World Report* top-50 U.S. law schools in Exercise 2.67, the listing for each school included two rating scores: one from academic evaluators and another from lawyer/judge evaluators. Using the data in file **XR02067**, generate a scatter diagram with axes of your choice, fit a linear equation to the data, then determine and interpret the coefficients of correlation and determination.

3.46 The U.S. National Center for Health Statistics has compiled data on death rates (deaths per 100,000 people) from heart disease and cancer for the 50 states. The states and their respective death rates from these two diseases are provided in data file **XR03046**. Generate a scatter diagram with axes of your choice, fit a linear equation to the data, then determine and interpret the coefficients of correlation and determination. Source: Bureau of the Census, *Statistical Abstract of the United States 2009*, p. 87.

3.47 A report on the average price of a sample of brand-name prescription drugs with their generic equivalents has found the generics to be considerably cheaper. The drugs and the average prices for their brand-name versus generic versions are listed in data file **XR03047**. Generate a scatter diagram with axes of your choice, fit a linear equation to the data, then determine and interpret the coefficients of correlation and determination. Source: Julie Appleby, "Drug Makers Fight Back As Patents Near Expiration," *USA Today*, November 26, 1999, p. 4B.

(3.7) SUMMARY

- **Measures of central tendency and dispersion**

In addition to summarization by frequency distributions and related visual techniques, raw data may be statistically described through the use of measures of central tendency and dispersion. As with the frequency distribution, this process loses some of the detail in the original data, but offers the benefit of replacing voluminous raw data with just a few numerical descriptors.

Measures of central tendency describe typical values in the data. Four of the most common are the arithmetic mean, weighted mean, median, and mode. Dispersion measures describe the scatter of the data, or how widely the values tend to be spread out. Common measures of dispersion include the range, quantiles, mean absolute deviation, variance, and standard deviation. The midrange is the average of the lowest and highest values and is sometimes used as a very rough measure of central tendency.

Chebyshev's theorem applies to any distribution, regardless of shape, and specifies a minimum percentage of observations that will fall within a given number of standard deviations from the mean. For distributions that are bell shaped and symmetrical, the empirical rule expresses the approximate percentage of observations that will fall within a given number of standard deviations from the mean.

- **Conversion of data to standardized values**

Standardization of the data involves expressing each value in terms of its distance from the mean, in standard deviation units. The resulting standardized distribution will have a mean of 0 and a standard deviation of 1. The coefficient of variation expresses the standard deviation as a percentage of the mean and is useful for comparing the relative dispersion for different distributions.

- **Box-and-whisker plot and estimation from grouped data**

Measures of central tendency and dispersion are part of the output whenever computer statistical packages have been used to analyze the data. In addition, the statistical package is likely to offer further description in the form of information such as the trimmed mean and the box-and-whisker plot. The approximate value of measures of central tendency and dispersion may be calculated from the grouped data of the frequency distribution.

- **The coefficient of correlation**

The coefficient of correlation (r) is an important measure of the strength of the linear relationship between two quantitative variables. The coefficient of correlation is between -1 and $+1$, and the larger the absolute value of r, the stronger the linear relationship between the variables. The coefficient of determination (r^2) is the proportion of the variation in y that is explained by the best-fit linear equation, and r^2 is always ≤ 1.

EQUATIONS

Arithmetic Mean

- Population mean:

$$\mu = \frac{\Sigma x_i}{N}$$

where μ = population mean
x_i = the ith data value in the population
Σ = the sum of
N = number of data values in the population

- Sample mean:

$$\bar{x} = \frac{\Sigma x_i}{n}$$

where $\bar{x}$ = sample mean
x_i = the ith data value in the sample
Σ = the sum of
n = number of data values in the sample

Weighted Mean

Weighted mean, μ_w (for a population) or $\bar{x}_w$ (for a sample):

$$\mu_w \text{ or } \bar{x}_w = \frac{\Sigma w_i x_i}{\Sigma w_i}$$

where w_i = weight assigned to the ith data value
x_i = the ith data value

Range

- Range = Largest data value − Smallest data value
- Midrange = (Largest data value + Smallest data value)/2

Quartiles

In a set of N values arranged from lowest to highest:

- First quartile, Q_1 = Data value at position $\frac{(N + 1)}{4}$
- Second quartile (the median), Q_2 = Data value at position $\frac{2(N + 1)}{4}$
- Third quartile, Q_3 = Data value at position $\frac{3(N + 1)}{4}$

(Use N if data represent a population, n for a sample.)

Interquartile Range and Quartile Deviation

- Interquartile range: third quartile minus first quartile, or $Q_3 - Q_1$
- Quartile deviation: one-half the interquartile range, or $(Q_3 - Q_1)/2$

Mean Absolute Deviation (MAD)

Mean absolute deviation for a population:

$$MAD = \frac{\sum|x_i - \mu|}{N}$$

where μ = population mean
x_i = the ith data value
N = number of data values in the population

(To calculate MAD for a sample, substitute n for N and $\bar{x}$ for μ.)

Variance

- Variance for a population:

$$\sigma^2 = \frac{\sum(x_i - \mu)^2}{N}$$

where σ^2 = population variance
μ = population mean
x_i = the ith data value
N = number of data values in the population

- Variance for a sample:

$$s^2 = \frac{\sum(x_i - \bar{x})^2}{n - 1}$$

where s^2 = sample variance
$\bar{x}$ = sample mean
x_i = the ith data value
n = number of data values in the sample

Standard Deviation

For a Population	For a Sample
$\sigma = \sqrt{\sigma^2}$	$s = \sqrt{s^2}$

Residual

$$\text{Residual for a data value in a sample} = x_i - \bar{x}$$

where $\bar{x}$ = sample mean
x_i = value of the ith observation

Chebyshev's Theorem

For either a sample or a population, the percentage of observations that fall within k (for $k > 1$) standard deviations of the mean will be at least

$$\left(1 - \frac{1}{k^2}\right) \times 100$$

The Empirical Rule

For distributions that are bell shaped and symmetrical:

- About 68% of the observations will fall within 1 standard deviation of the mean.
- About 95% of the observations will fall within 2 standard deviations of the mean.
- Practically all of the observations will fall within 3 standard deviations of the mean.

Standardized Data

$$z_i = \frac{x_i - \bar{x}}{s}$$

where z_i = standardized value for the ith observation
$\bar{x}$ = sample mean
x_i = the ith data value
s = sample standard deviation

(When data represent a population, μ replaces $\bar{x}$ and σ replaces s.)

Coefficient of Variation (CV)

For a Population: $CV = \frac{\sigma}{\mu} \times 100$

For a Sample: $CV = \frac{s}{\bar{x}} \times 100$

Conversion of Computer-Output Standard Deviation Based on (N − 1) Divisor to a Population Standard Deviation Based on (N) Divisor

$$\text{Standard deviation based on } (N) \text{ divisor} = \text{Standard deviation based on } (N-1) \text{ divisor} \times \sqrt{\frac{N-1}{N}}$$

Mean and Variance from Grouped Data

- Approximate value of the mean from grouped data:

$$\mu = \frac{\sum f_i m_i}{N}$$

where μ = approximate population mean
f_i = frequency for class i
m_i = midpoint for class i
N = number of data values in the population

(If data are a sample, $\bar{x}$ replaces μ and n replaces N.)

- Approximate value of the variance from grouped data:
If data represent a population:

$$\sigma^2 = \frac{\sum f_i m_i^2 - N\mu^2}{N}$$

where μ = approximate population mean
f_i = frequency for class i
m_i = midpoint for class i
N = number of data values in the population

If data represent a sample:

$$s^2 = \frac{\sum f_i m_i^2 - n\bar{x}^2}{n-1}$$

where $\bar{x}$ = approximate sample mean
f_i = frequency for class i
m_i = midpoint for class i
n = number of data values in the sample

CHAPTER EXERCISES

Note: For many of the Exercises 3.48–3.64, a computer and statistical software will be desirable and useful. However, any necessary calculations can also be done with the aid of a pocket calculator. For readers using statistical software, keep in mind the file-naming key—for example, the data for Exercise 3.57 will be in data file **XR03057**.

3.48 A dental supplies distributor ships a customer 50 boxes of product A, 30 boxes of B, 60 boxes of C, and 20 boxes of D. The unit shipping costs (dollars per box) for the four products are $5, $2, $4, and $10, respectively. What is the weighted mean for shipping cost per unit?

3.49 The first seven customers of the day at a small donut shop have checks of $1.25, $2.36, $2.50, $2.15, $4.55, $1.10, and $0.95, respectively. Based on the number of customers served each day, the manager of the shop claims that the shop needs an average check of $1.75 per person to stay profitable. Given her contention, has the shop made a profit in serving the first seven customers?

3.50 According to Honda Motor Co., Inc., the exchange rate (yen per U.S. dollar) from 1998 through 2005 was 123, 128, 112, 111, 125, 122, 113, and 108. Determine the mean and median for these data. Is there a mode? If so, what is its numerical value? Source: Honda Motor Co., Ltd., *2005 Annual Report*.

3.51 The accompanying box-and-whisker plot represents the number of gallons of water used by 80 households over a 1-day period. Determine the approximate values for the median, the first and third quartiles, and the range. Does the distribution appear to be skewed? If so, is it positively skewed or negatively skewed?

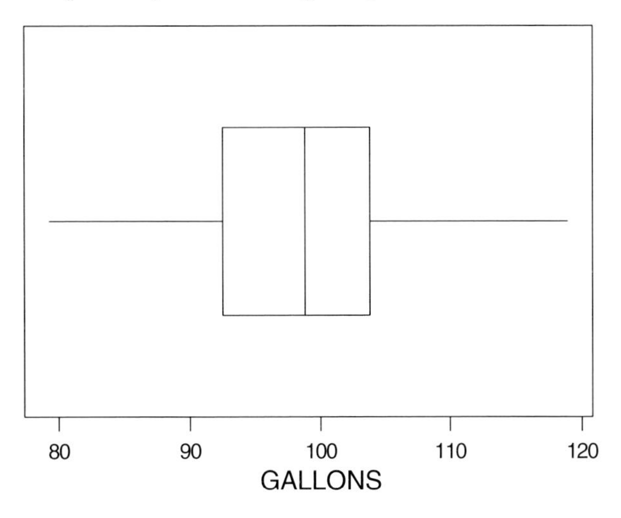

3.52 A quality control supervisor has taken a sample of 16 bolts from the output of a thread-cutting machine and tested their tensile strengths. The results, in tons of force required for breakage, are as follows:

2.20	1.95	2.15	2.08	1.85	1.92
2.23	2.19	1.98	2.07	2.24	2.31
1.96	2.30	2.27	1.89		

a. Determine the mean, median, range, and midrange.
b. Calculate the mean absolute deviation.
c. Calculate the standard deviation and variance.

3.53 For the sample data presented in the dotplot of Exercise 3.55, construct a frequency distribution in which the classes are 10–under 20, 20–under 30, and 30–under 40. If this frequency distribution were all that we knew about the underlying data, what approximate values would be estimated for the mean and standard deviation of these data?

3.54 For a sample of five different years from the period 1960 through 2007, it is found that U.S. work stoppages involving at least 1000 workers occurred 268 times in one of these years, with 424, 235, 145, and 44 work stoppages in the other four. Source: Bureau of the Census, *Statistical Abstract of the United States 2009*, p. 418.

a. Determine the mean, median, range, and midrange.
b. Calculate the mean absolute deviation.
c. Calculate the standard deviation and variance.

3.55 The following dotplot describes the lengths, in pages, of a sample consisting of 20 reports generated by a consulting firm during the past 3 months. Based on the dotplot, what are the values for the median and the first and third quartiles?

3.56 A grocery store owner has found that a sample of customers purchased an average of 3.0 pounds of luncheon meats in the past week, with a sample standard deviation of 0.5 lb.

a. If the store's meat scale is off by 0.1 lb (e.g., a purchase listed as 1 lb actually weighs 1.1 lb), what will be the values of the sample mean and standard deviation after the owner corrects his data for the weighing error?
b. Using the mean and standard deviation you calculated in part (a), and assuming the distribution of purchase amounts was bell shaped and symmetrical, what luncheon meat purchase amount would have been exceeded by only 2.5% of the customers last week?

3.57 Using the frequency distribution in Exercise 3.60, determine the approximate mean and standard deviation for the underlying data.

3.58 The cafeteria manager at a manufacturing plant has kept track of the number of cups of coffee purchased during each of the past 90 working days.

a. Using the following descriptive statistics, construct a box-and-whisker plot for the data. Does the distribution appear to be symmetrical? If not, is it positively skewed or negatively skewed?

```
        N    Mean  Median  Tr Mean  StDev  SE Mean
CUPS   90  125.64  123.50   125.50  30.80     3.25

       Min     Max      Q1       Q3
CUPS  54.00  206.00  102.75   147.25
```

b. If this distribution of daily coffee totals is bell shaped and symmetrical, on approximately what percent of the days were 64 or fewer cups of coffee purchased?

3.59 Use the coefficient of variation to compare the variability of the data in Exercise 3.52 with the variability of the data in Exercise 3.61.

3.60 The frequency distribution for population density (persons per square mile) for the 50 U.S. states is as follows: Source: Bureau of the Census, *Statistical Abstract of the United States 2009*, p. 18.

Population Density	Number of States
0–under 100	27
100–under 200	11

(continued)

Population Density	Number of States
200–under 300	4
300–under 400	1
400–under 500	2
500–under 600	1
600–under 700	0
700–under 800	1
800–under 900	1
900–under 1000	0
1000–under 1100	1
1100–under 1200	1

Does this distribution appear to be symmetrical? If not, is it skewed positively or negatively?

3.61 A law enforcement agency, administering breathalyzer tests to a sample of drivers stopped at a New Year's Eve roadblock, measured the following blood alcohol levels for the 25 drivers who were stopped:

0.00%	0.08%	0.15%	0.18%	0.02%
0.04	0.00	0.03	0.11	0.17
0.05	0.21	0.01	0.10	0.19
0.00	0.09	0.05	0.03	0.00
0.03	0.00	0.16	0.04	0.10

a. Calculate the mean and standard deviation for this sample.
b. Use Chebyshev's theorem to determine the minimum percentage of observations that should fall within $k = 1.50$ standard deviation units of the mean. Do the sample results support the theorem?
c. Calculate the coefficient of variation for these data.

3.62 *Natural History* magazine has published a listing of the maximum speeds (in mph) for a wide variety of animals, including those shown in the table. Source: *The World Almanac and Book of Facts 2009*, p. 330.

Cheetah	70	Grizzly bear	30
Pronghorn antelope	61	Domestic cat	30
Wildebeest	50	Human	27.89
Lion	50	Elephant	25
Coyote	43	Black mamba snake	20
Mongolian wild ass	40	Squirrel	12
Giraffe	32	Spider	1.17
Wart hog	30	Giant tortoise	0.17
White-tailed deer	30	Garden snail	0.03

a. Determine the mean and the median for these data.
b. Is there a mode? If so, what is its numerical value?

3.63 A testing firm has measured the power consumption of 40 dorm-size microwave ovens selected at random from the manufacturer's production line. Given the following box-and-whisker plot, determine the approximate values for the median, the first and third quartiles, and the range. Does the distribution appear to be skewed? If so, is it positively skewed or negatively skewed?

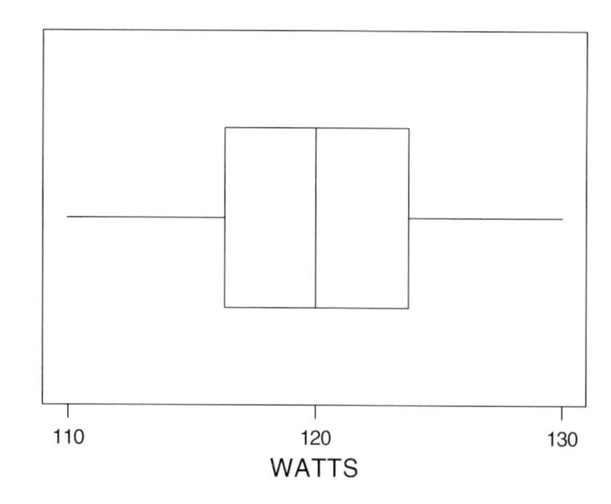

3.64 The 2007 top-grossing feature film was *Spiderman 3*, a product that brought in \$336.5 million at the box office. The gross receipts for this and the rest of the top-20 feature films of that year are shown below. Determine the mean and median for these data. Is there a mode? If so, what is its numerical value? Source: *The New York Times Almanac 2009*, p. 413.

\$336.5	\$322.7	\$319.2	\$309.4	\$292.0
256.4	227.5	219.5	217.3	210.6
206.4	183.1	168.3	148.8	143.5
140.1	134.5	131.9	130.2	127.8

(DATA SET) *Note:* Exercises 3.65–3.69 require a computer and statistical software.

3.65 According to the Energy Information Administration, the average U.S. household spends \$1196 per year for electricity. Assume that this finding could have been based on the electricity expenditures reported by 250 households, with the data in file **XR03065**. Source: eia.doe.gov, June 26, 2009.

a. Confirm the mean and determine the median and the standard deviation of the electricity expenditures.
b. Generate and interpret the box-and-whisker plot for the households from this region.
c. From the results in parts (a) and (b), do there appear to be any "outlier" households? If so, for what reason(s) might energy-conservation officials wish to examine the habits or characteristics of these households?

3.66 *Bride's* magazine reports the cost of the average honeymoon as \$5111. Assume that their findings could have been based on the honeymoon expenditures of a sample consisting of 300 couples whose costs are listed in data file **XR03066**. Source: msnbc.msn.com, June 26, 2009.

a. Confirm the mean and determine the median and the standard deviation of the honeymoon expenditures for this sample.
b. Generate and interpret the box-and-whisker plot for the couples in the sample.
c. From the results in parts (a) and (b), do there appear to be any "outlier" couples who have spent an unusually large amount of money on their honeymoon? If so, what kinds of companies or organizations might have an interest in finding out how to identify such couples before they've made their honeymoon plans?

3.67 During 2005, 1,186,000 college-bound high school seniors took the ACT college admission test. The average score on the mathematics component was 20.7, with a standard deviation of 5.0. We are assuming that the math scores in data file **XR03067** could have been the math scores for a sample of 400 college-bound seniors who took the ACT that year. (*An important note:* The mean and the standard deviation for our sample are close, but not exactly the same as the mean and standard deviation of the population from which the sample was obtained—this is a key concept that is central to Chapter 8, Sampling Distributions.) Based on the assumed sample data in file **XR03067**, Source: nces.ed.gov, June 26, 2009.

a. Determine the mean, the median, and the standard deviation of the math scores in the sample.
b. Generate and interpret the box-and-whisker plot for the math scores in the sample.
c. What math score would a test-taker have to achieve to be higher than 75% of the sample members? To be higher than 25% of the sample members?

3.68 Data for a sample of college players eligible to enter the National Football League include weight (pounds) and time (seconds) required to sprint 40 yards. Using the data listed in file **XR03068**, and with weight as the x (independent) variable and time as the y (dependent) variable, generate a scatter diagram, fit a linear equation to the diagram, then determine and interpret the coefficients of correlation and determination. Source: "Rating the Prospects," *USA Today*, April 14, 2000, p. 19C.

3.69 During a five-year period, the Federal Aviation Administration took administrative actions (letters of correction or warning notices) and levied fines against U.S. airline companies for various safety-related reasons. For each airline involved, data file **XR03069** lists the number of administrative actions taken toward the airline along with the total amount of fines (in millions of dollars) levied against the airline during this time period. Using number of administrative actions as the x (independent) variable and fines as the y (dependent) variable, generate a scatter diagram, fit a linear equation to the diagram, then determine and interpret the coefficients of correlation and determination. Source: "Fines and Administrative Actions Against U.S. Airlines," *USA Today*, March 13, 2000, p. 3B.

INTEGRATED CASES

Thorndike Sports Equipment

When we last visited the Thorndikes, one of the company's engineers had just developed a new golf ball. According to the engineer, the new ball incorporates some changes that may enable it to travel farther than conventional golf balls. The engineer, an avid golfer herself, is very excited about the prospect that she may have developed an innovation that could revolutionize the game of golf. Old Luke says he will be happy only if the new ball revolutionizes his company's net profits.

Some testing has already been done. As reported in the Chapter 2 episode of the Thorndikes, Ted set up a test in which he mixed 25 of the new balls with 25 of the old type, then had a golf pro hit all 50 of them at a driving range. All 50 distances were measured and are listed in data file **THORN02**. For reader convenience, they are also repeated here. Ted has examined the visual descriptions of the data carried out in Chapter 2, but he would like to have some statistical summaries as well.

1. For the 25 drives made with the new balls, calculate statistics that you think would be appropriate in describing the central tendency and the dispersion of the data.
2. Repeat part 1 for the 25 drives made with the conventional golf balls.
3. Ted wants to write a one-paragraph memo to Luke describing the results as they are reflected in parts 1 and 2, but he had to fly to New York this morning. Help Ted by writing a first draft for him. Keep in mind that Luke appreciates plain English, so try to use your own words whenever possible.

25 Drives with New Ball (Yards):

267.5	248.3	265.1	243.6	253.6	232.7	249.2	232.3	252.8
247.2	237.4	223.7	260.4	269.6	256.5	271.4	294.1	256.3
264.3	224.4	239.9	233.5	278.7	226.9	258.8		

25 Drives with Conventional Ball (Yards):

241.8	255.1	266.8	251.6	233.2	242.7	218.5	229.0	256.3
264.2	237.3	253.2	215.8	226.4	201.0	201.6	244.1	213.5
267.9	240.3	247.9	257.6	234.5	234.7	215.9		

Springdale Shopping Survey

This exercise focuses on the Springdale Shopping Survey, the setting and data collection instrument for which were provided at the end of Chapter 2. The data are in the computer file **SHOPPING**.

We will concentrate on variables 18–25, which reflect how important each of eight different attributes is in the respondent's selection of a shopping area. Each of these variables has been measured on a scale of 1 (the attribute is not very important in choosing a shopping area) to 7 (the attribute is very important in choosing a shopping area). The attributes being rated for importance are listed below. Examining the relative importance customers place on these attributes can help a manager "fine-tune" his or her shopping area to make it a more attractive place to shop.

18 Easy to return/exchange goods
19 High quality of goods
20 Low prices
21 Good variety of sizes/styles
22 Sales staff helpful/friendly
23 Convenient shopping hours
24 Clean stores and surroundings
25 A lot of bargain sales

1. Perform the following operations for variables 18–25:
   a. Obtain descriptive statistics, including the mean and median.
   b. Generate a box-and-whisker plot for the variable. Does the distribution appear to be skewed? If so, is the skewness positive or negative?
2. Based on the results for question 1, which attributes seem to be the most important and the least important in respondents' choice of a shopping area?
3. Use their respective coefficients of variation to compare the relative amount of dispersion in variable 29 (number of persons in the household) with that of variable 30 (respondent's age).
4. Determine the coefficient of correlation between variable 29 (number of persons in the household) and variable 30 (age). What percentage of the variation in household size is explained by respondent age?

BUSINESS CASE

Baldwin Computer Sales (A)

Andersen Ross/Photodisc/Getty Images

Baldwin Computer Sales is a small company located in Oldenburg, Washington. The founder of the company, Jonathan Baldwin, began the business by selling computer systems through mail-order at discount prices. Baldwin was one of the first computer mail-order companies to offer a toll-free phone number to their customers for support and trouble-shooting. Although the company has grown over time, many new competitors have entered the computer mail-order marketplace so that Baldwin's share of this market has actually declined.

Five years ago, Bob Gravenstein, a marketing consultant, was contracted by Baldwin to develop long-term marketing plans and strategies for the company. After careful study, he recommended that Baldwin branch out to reach a new and growing segment of the computer sales market: college students. At that time, most universities were providing microcomputer labs and facilities for their students. However, many students were purchasing their own computers and printers so that they could have access to a computer at any time in their apartments or dorms. After graduation, students take the computers, with which they are already familiar, to their new jobs or use them at home after work hours. The percentage of college students owning a computer was increasing rapidly, and Bob Gravenstein recommended that Baldwin Computer Sales take advantage of this marketing opportunity.

The marketing plan developed by Bob Gravenstein worked as follows: Five universities were initially selected for the program with expansion to other universities planned over time. Any student in at least his or her sophomore year at one of these universities was eligible to purchase a discounted JCN-2001 microcomputer system with printer from Baldwin Computer Sales under the program. The JCN-2001 is a private-label, fully compatible system with all of the features of brand-name models. The student makes a small payment each semester at the same time regular tuition payments are due. When the student graduates and finds a job, the payments are increased so that the computer will be paid for within two years after graduation. If the student fails to make payments at any time, Baldwin could repossess the computer system.

The prospect of future sales of computer equipment to these students was also part of the marketing strategy. The JCN-2001 is an entry-level computer system that suffices for most academic and professional uses. Eventually, however, many students who purchased this system would outgrow it and require an upgrade to a more powerful machine with more features. Bob Gravenstein argued that after their good experience with the JCN-2001, these customers would make their future purchases from Baldwin Computer Company.

Today, five years later, Baldwin Computer Sales is still operating the student purchase program and has expanded it to several more universities. There are currently enough data available from the early days of the program for Baldwin to determine whether the program has been successful and whether it should be continued. To discuss the future of the student purchase program, Jonathan Baldwin has called a meeting with Ben Davis, who has been in charge of the program since it began, and Teresa Grimes, the new vice president of marketing.

Baldwin: "I called you both in here today to discuss the student purchase program. As you know, the program has been in place for approximately five years and we need to make decisions about where to go from here. Ben, weren't you telling me last week that we now have enough data to evaluate the program?"

Davis: "Yes, sir. Any student who began the program as a sophomore five years ago should be out of school and employed for at least two years."

Baldwin: "Well, based on your information, would you say the program has been a success or a failure?"

Davis: "That's a tough call, sir. While most of the participants in the program eventually pay their account in full, we have had a high rate of defaults, some while the students were still in school, but mostly after they graduated."

Baldwin: "How much are we losing on those who default?"

Davis: "Each case is different. As I said, some default early and others after they have graduated. There are also the costs of repossession and repair to bring the product back up to resale quality. In many instances we were not able to retrieve the computer systems. Our data suggest that our average loss on each student-customer who defaults is about $1200. On the other hand, our average profit from participants who pay their accounts in full is approximately $750. Overall, we are close to just breaking even."

Grimes: "Ben, have you considered 'qualifying' students for the program, much like a loan officer would qualify someone for a loan?"

Davis: "We had initially thought about doing that, but we didn't believe that there was much information, if any, in the way of a credit history on most college students. Applicants are still requested to provide as much information as possible on their application, including their class, grade point average, work experience, scholarships, and how much of their college expenses were earned through work. However, we weren't sure that this information was particularly useful for screening applicants."

Grimes: "Knowing that we were going to have this discussion today, I had one of my assistants who is well-versed in statistics look over some of those data last week. She has come up with a 'screening test' based only on the information you have been collecting from student applicants. By being more selective about whom we allow to participate in the student purchase program, it may be possible to increase our profit from the program."

Davis: "It would be easy enough to check out her screening test by trying it out on our early data from the program. In those cases, we know whether or not the student actually defaulted."

Baldwin: "Why don't the two of you spend some time on this idea and get back to me next week? At that time I want a recommendation to either discontinue the program, continue it as is, or continue it using this 'screening test' idea. Make sure you have the evidence to back up your recommendation."

Assignment

Ben Davis and Teresa Grimes must analyze the data from the student purchase program and make a recommendation to Mr. Baldwin about the future of the program. The necessary data are contained in the file **BALDWIN**. A description of this data set is given in the Data Description section that follows. Using this data set and other information given in the case, help Ben Davis and Teresa Grimes evaluate the student purchase program and the potential usefulness of the screening test developed by the assistant to Teresa Grimes. The case questions will assist you in your analysis of the data. Use important details from your analysis to support your recommendations.

Data Description

The **BALDWIN** file contains data on all participants in the student purchase program who by now should have either paid in full or defaulted (i.e., those participants who should have graduated and held a job for at least two years). A partial listing of the data is shown here.

Student	School	Default	When	Score
6547	1	1	1	64
4503	2	0	0	58
1219	2	0	0	52
9843	4	1	0	56
6807	1	0	0	47
6386	4	0	0	58
⋮	⋮	⋮	⋮	⋮

SEEING STATISTICS: APPLET 1

Influence of a Single Observation on the Median

The dots in this applet can be viewed as showing the speeds of a sample of nine cars being driven through a residential area. Eight of the values are fixed, but we can change the "speed" of one of them to see how this change affects the median for the sample data. By clicking on the data value for the ninth car (represented by the green dot at the far right), we can drag the value upward or downward and change this car's speed. The median for the data is shown at the top of the applet.

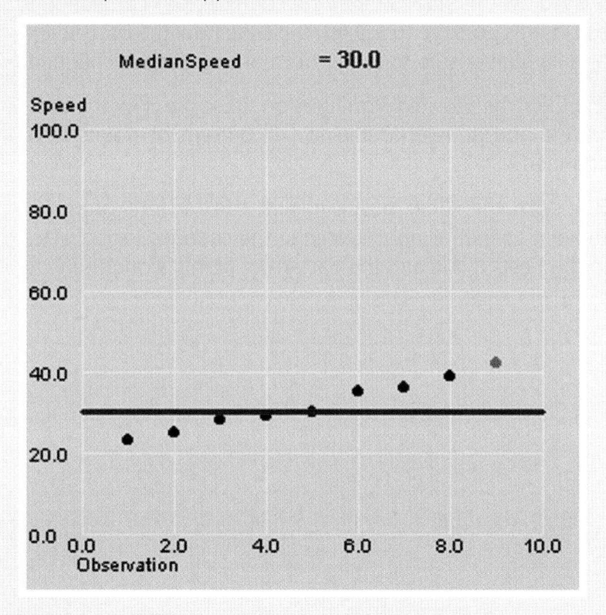

Applet Exercises

1.1 Click on the rightmost green dot and drag it so that the speed of the ninth car is 40 mph. What is the value for the median?

1.2 Click on the rightmost green dot and drag it so that the speed of the ninth car increases from 40 mph to 100 mph. What effect has this had on the value of the median?

1.3 By adjusting the speed of the ninth car, identify the highest and lowest possible values for the median speed for the nine cars in the applet.

1.4 Suppose the ninth car were a jet-powered supersonic vehicle and traveled through the neighborhood at 1000 mph. Although such a speed lies far beyond the scale in the applet, what would be the value of the median? What would be the approximate value for the mean? Which of these two sample descriptors has been more greatly affected by this extreme value in the data?

Try out the Applets: http://www.cengage.com/international

These data are coded as follows:

STUDENT: Student transaction number for identification purposes

SCHOOL: University where the student was enrolled

DEFAULT: 1, in the event of default,
0, if account was paid in full on time

WHEN: 1, if default occurred (or paid in full) before graduation,
0, if default occurred (or paid in full) after graduation

SCORE: Score on screening test based on student applicant information such as his or her class, grade point average, work experience, scholarships, and how much of their college expenses were earned through work

1. Generate appropriate descriptive statistics of the screening test scores for those who did not default on their computer loan. Be sure to include the mean and the third quartile. Do the same for those who did default, then compare the results. Is the mean score on the screening test higher for those who did not default?
2. In the descriptive statistics of the screening test scores for the students who did not default, identify the value of the third quartile and interpret its meaning. If this numerical value had been established as a cutoff for

SEEING STATISTICS: APPLET 2

Scatter Diagrams and Correlation

This applet displays a scatter diagram that we can rearrange by using the slider at the bottom. By clicking on the slider and adjusting its position, we cause the points to gravitate to positions that make the linear relationship between the variables either stronger or weaker. For any given configuration, the best-fit straight line is shown, the coefficient of correlation is displayed at the top of the diagram, and we can reverse the nature of the relationship by clicking on the "Switch Sign" button.

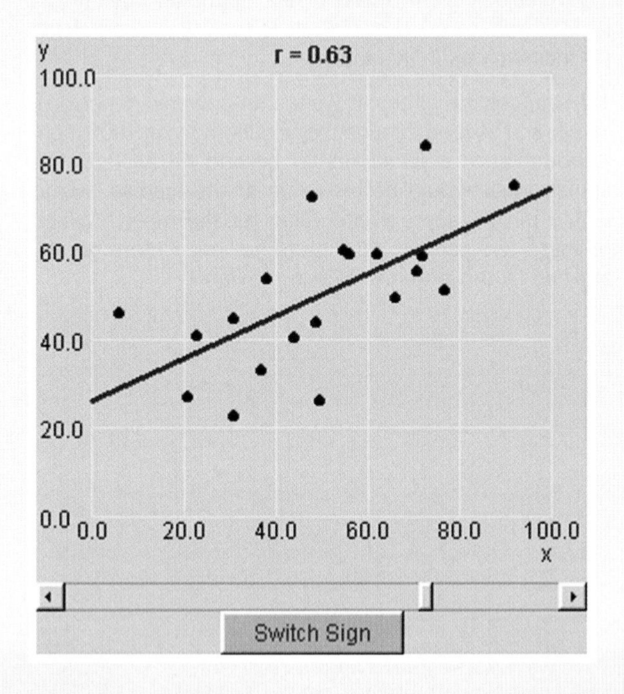

Applet Exercises

2.1 Click on the slider and adjust it so that the best-fit straight line is as close as possible to horizontal. What is the value of the coefficient of correlation?

2.2 Click on the slider and adjust it so that the coefficient of correlation is +1.0. Describe the best-fit line and its relationship to the pattern of points.

2.3 Adjust the slider so that the coefficient of correlation is as close as possible to +0.60. Describe how the best-fit line and its relationship to the pattern of points have changed.

2.4 Click the "Switch Sign" button. Describe how the best-fit line and its relationship to the pattern of points have changed.

2.5 With the slider positioned at the far left, gradually move it to the far right. Describe the changes taking place in the best-fit line and the pattern of points along the way.

Try out the Applets: http://www.cengage.com/international

receiving a computer loan, what percentage of those who repaid would have been denied a loan in the first place? Comment on the advisability of setting a cutoff score that is this high.

3. Construct adjacent dotplots (use box plots if the dotplot feature is not available on your computer statistical package) and visually compare the screening test scores of students who did not default on their computer loan to the scores of those who defaulted. Does the distribution of screening test scores for those who did not default appear to be shifted to the right of the distribution of scores for those who did default?
4. Based on your results and responses to the previous questions, does the screening test appear to be potentially useful as one of the factors in helping Baldwin predict whether a given applicant will end up defaulting on his or her computer loan?

CHAPTER 4

Data Collection and Sampling Methods

U.S. CENSUS: EXTENSIVE, EXPENSIVE, BUT UNDERCOUNTING IS A PROBLEM

Since 1790, at 10-year intervals, the United States has taken a count of the nation's populace. Required by the Constitution, this enumeration is used to allocate seats in the U.S. House of Representatives, to determine election districts, and to distribute more than $1.3 trillion to local and state governments over the next 10 years. In addition, the census shapes important decisions about public health, education, senior services, and transportation. The 2010 census is at least as expensive as it is important. In carrying out this enumeration, the Bureau of the Census will spend more than $14 billion and employ about 1.4 million field operators, more than the entire population of New Hampshire.

Despite such Herculean efforts, undercounting has been a persistent problem. It has been estimated that approximately 5 million people were not included in the 2000 enumeration and that those not counted were especially likely to include blacks, Hispanics, illegal immigrants, and the poor. As a result of undercounting, such communities tend to receive less than proportionate shares of political representation and federal funds.

In adjusting its methodology to better reflect the population and its geographical distribution, the Bureau of the Census is relying partly on sampling approaches similar to those described in this chapter. It has also proposed the controversial technique of allowing citizens to be counted by going online. Changes in enumeration methods can cause great controversy in the political arena, partly because some states (e.g., Texas and California) would benefit from including more of the "uncounteds," while others (e.g., Alabama and Pennsylvania) would tend to suffer.

Thus, although charged with the seemingly straightforward job of determining how many people live in the United States, the Bureau of the Census finds itself caught in the middle of a controversy with significant political overtones. To the extent that federal representation and funding are allocated according to the population distribution, the way in which the official count is determined could greatly influence the fortunes of the individual states affected.

Source: Haya el Nasser, "For 2010 Census, Counting Gets Tougher," usatoday.com, June 27, 2009.

Survey research is a popular method of collecting data.

LEARNING OBJECTIVES

After reading this chapter, you should be able to:

- Describe the types of business research studies and their purposes.
- Explain the difference between primary and secondary data.
- Describe how survey research is conducted.
- Discuss the major sources of error in survey research.
- Describe the experimentation and observational approaches to research.
- Differentiate between internal and external secondary data, and identify sources for each.
- Understand the relevance of the data warehouse and data mining.
- Explain why a sample is generally more practical than a census.
- Differentiate between the probability and nonprobability approaches to sampling, and understand when each approach is appropriate.
- Differentiate between sampling error and nonsampling error.
- Understand that probability samples are necessary to estimate the amount of sampling error.

(4.1) INTRODUCTION

Statistics can be used as (1) a *descriptive* tool for expressing information relevant to business and (2) an *inferential* device for using sample findings to draw conclusions about a population. To use statistics for either of these applications, *data* are required, and such data may be gathered from a wide variety of sources. The techniques presented in subsequent chapters are only as good as the data on which they are based.

In this chapter, we'll examine some of the popular approaches used in collecting statistical data. Our discussion of business and survey research will be on a strictly introductory level, but it should make you more familiar with the sources and procedures used in obtaining the kinds of data seen in this textbook, as well as in the business and general media. We will also discuss the importance of sampling and describe some of the more popular sampling methods used in business research today.

(4.2) RESEARCH BASICS

Types of Studies

Business research studies can be categorized according to the objective involved. Accordingly, we can identify four types of studies: exploratory, descriptive, causal, and predictive.

Exploratory

Often the initial step, **exploratory research** helps us become familiar with the problem situation, identify important variables, and use these variables to form hypotheses that can be tested in subsequent research. The hypothesis will be a statement about a variable or the relationship between variables—for example, "Production will increase if we switch the line locations of operators A and B." The hypothesis may not be true, but it is useful as an assertion that can be examined through the collection of sample data for a test period during which the operators have traded line positions.

Exploratory research can also be of a qualitative nature. One such method is the *focus group interview*, in which a moderator leads a small-group discussion about a topic and the client watches and listens from behind a one-way mirror. The U.S. Environmental Protection Agency has used focus groups to learn what people think and know about their drinking water. Among the results: People expressed distrust in water companies and knew that some contaminants could be in the water, but most people were unable to specify any of the specific contaminants that could be there.[1]

Descriptive

As might be expected, **descriptive research** has the goal of describing something. For example, a survey by the National Association of Home Builders found that the average size of single-family homes ballooned from 1500 square feet in 1970 to nearly 2400 square feet in 2004.[2]

Causal

In **causal research,** the objective is to determine whether one variable has an effect on another. In Suginami City, Japan, there were a record 1710 break-ins during 2002. A few years later, a neighborhood watch group noticed that there seemed to be fewer burglaries on streets where houses had flower boxes in their street-facing windows. Thus began Operation Flower, and the number of burglaries had fallen 80% by 2008. According to city official Kiyotaka Ohyagi, "By planting flowers facing the street, more people will be keeping an eye out while taking care of the flowers or watering them." In Suginami City, Flower Power took on a whole new meaning.[3]

Regarding causal studies, it should be pointed out that statistical techniques alone cannot *prove* causality. Proof must be established on the basis of quantitative findings along with logic. In the case of Suginami City's Operation Flower and the reduction in burglaries, it would seem obvious that causation was not in the *reverse* direction (i.e., lower burglary statistics were not causing flowers to grow). However, we must consider the possibility that one or more other variables might have contributed to the reduced number of break-ins—some of these possibilities could be greater participation in neighborhood watch groups, increased police presence, or the widespread installation of security cameras.

Predictive

Predictive research attempts to forecast some situation or value that will occur in the future. A common variable for such studies is the expected level of future sales. As might be expected, forecasts are not always accurate. For example,

[1]"Consumer Feedback Focus Group Results," U.S. Environmental Protection Agency, epa.gov, July 25, 2006.

[2]David Struckey and Robert W. Ahrens, "The Swelling Dwelling," *USA Today,* April 10, 2006, p. 1A.

[3]"Tokyo Residents Fight Burglars with Flower Power," reuters.com, June 11, 2009.

in 1984, the Semiconductor Industry Association predicted a 22% increase for 1985, a year in which sales actually fell by about 17%.[4] With any forecast, there will tend to be an error between the amount of the forecast and the amount that actually occurs. However, for a good forecasting model, the amount of this error will be consistently smaller than if the model were not being used. Forecasting will be one of the topics in Chapter 18.

The Research Process

A research study begins by defining the problem or question in specific terms that can be answered by research. Relevant variables must also be identified and defined. The steps in the research process include (1) defining the problem, (2) deciding on the type of data required, (3) determining through what means the data will be obtained, (4) planning for the collection of data and, if necessary, selection of a sample, (5) collecting and analyzing the data, (6) drawing conclusions and reporting the findings, and (7) following through with decisions that take the findings into consideration.

Primary versus Secondary Data

Statistical data can be categorized in a number of ways, including primary versus secondary. **Primary data** refer to those generated by a researcher for the specific problem or decision at hand. Survey research (Section 4.3), experimentation, and observational research (both discussed in Section 4.4) are among the most popular methods for collecting primary data.

Secondary data have been gathered by someone else for some other purpose. Secondary data can be **internal** or **external** depending on whether the data were generated from within your firm or organization or by an outside person or group. Some of the many possible sources of secondary data will be described in Section 4.5.

EXERCISES

4.1 What is the difference between primary data and secondary data? Between internal secondary data and external secondary data?

4.2 A published article reports that 60% of U.S. households have cable television. The president of an electronics firm clips the item from the paper and files it under "Cable TV data." Would this material be considered primary data or secondary data?

4.3 A pharmaceutical firm's annual report states that company sales of over-the-counter medicines increased by 48.1%.

a. To the firm itself, would this information be primary data or secondary data?
b. To a competitor, what type of information would this represent?

4.4 Bertram Pearlbinder, owner of an apartment building near the campus of Hightower University, is considering the possibility of installing a large-screen color television in the building's recreation room. For each of the following, indicate whether the associated data are primary or secondary:

a. Bertram interviews the current residents of the building to determine whether they would rather have the color television or a pool table.
b. Bertram, while reading *Newsweek*, notices an article describing the property damage in college dormitories and apartments.
c. One of the tenants tells Bertram that many of the residents are "very excited" about the prospect of getting the big TV.

[4]Source: Otis Port, "A Rosy Forecast for Chips Has Skeptics Hooting," *BusinessWeek*, October 14, 1985, p. 104.

SURVEY RESEARCH

In **survey research,** we communicate with a sample of individuals in order to generalize on the characteristics of the population from which they were drawn. It has been estimated that approximately $20 billion per year is spent on survey research worldwide.[5]

Types of Surveys

The Mail Survey

In a **mail survey,** a mailed questionnaire is typically accompanied by a cover letter and a postage-paid return envelope for the respondent's convenience. A good cover letter will be brief, readable, and will explain who is doing the study, why it's being done, and why it's important for the reader to fill out and return the enclosed questionnaire.

The Personal Interview

In the **personal interview,** an interviewer personally secures the respondent's cooperation and carries out what could be described as a "purposeful conversation" in which the respondent replies to the questions asked of her. The personal interview tends to be relatively expensive compared to the other approaches, but offers a lot of flexibility in allowing the interviewer to explain questions, to probe more deeply into the answers provided, and even to record measurements that are not actually asked of the respondent. For example, along with obtaining responses, the interviewer may record such things as the respondent's gender, approximate age, physical characteristics, and mode of dress.

The Telephone Interview

The **telephone interview** is similar to the personal interview, but uses the telephone instead of personal interaction. Telephone interviews are especially useful for obtaining information on what the respondent is doing at the time of the call (e.g., the television program, if any, being watched). Also, use of the telephone makes it possible to complete a study in a relatively short span of time.

Questionnaire Design

Also referred to as the *data collection instrument*, the **questionnaire** is either filled out personally by the respondent or administered and completed by an interviewer. The questionnaire may contain any of three types of questions: (1) **multiple choice,** in which there are several alternatives from which to choose; (2) **dichotomous,** having only two alternatives available (with "don't know" or "no opinion" sometimes present as a third choice); and (3) **open-ended,** where the respondent is free to formulate his or her own answer and expand on the subject of the question.

In general, multiple choice and dichotomous questions can be difficult to formulate, but data entry and analysis are relatively easily accomplished. The reverse tends to be true for open-ended questions, where a respondent may state or

[5]Source: "Top 25 Research Companies Maintain Position in 2003," esomar.org, July 25, 2006.

write several paragraphs in response to even a very short question. Because they give the respondent an opportunity to more fully express his feelings or describe his behaviors, open-ended questions are especially useful in exploratory research.

Proper wording of each question is important, but often difficult to achieve. A number of problems can arise, including the following: (1) the vocabulary level may be inappropriate for the type of person being surveyed; (2) the respondent may assume a frame of reference other than the one the researcher intended; (3) the question may contain "leading" words or phrases that unduly influence the response; and (4) the respondent may hesitate to answer a question that involves a sensitive topic. Examples of each situation follow:

- **Inappropriate vocabulary level**

  Poor wording "Have you patronized a commercial source of cinematic entertainment within the past month?"

  Problem Vocabulary level will be too high for many respondents.

  Better wording "Have you gone to a movie within the past month?"

- **Confusing frame of reference**

  Poor wording "Are you in better shape than you were a year ago?"

  Problem To what does "shape" refer—physical, financial, emotional?

  Better wording (if the desired frame of reference is physical) "Are you in better physical condition than you were a year ago?"

- **"Leading" words/phrases**

  Poor wording "To help maintain the quality of our schools, hospitals, and public services, do you agree that taxes should be increased next year?"

  Problems "Schools, hospitals, and public services" has an emotional impact. Also, "do you agree" suggests that you *should* agree.

  Better wording "Do you support an increase in taxes next year?"

- **Sensitive topic**

  Poor wording "How much money did you make last year?"

  Problem This question requests detailed, personal information that respondents are hesitant to provide. The question is also very blunt.

  Better wording "Which of the following categories best describes your income last year?"

[ ] under $20,000	[ ] $60,000–$79,999
[ ] $20,000–$39,999	[ ] $80,000–$99,999
[ ] $40,000–$59,999	[ ] $100,000 or more

The preceding represent only a few of the pitfalls that can be encountered when designing a questionnaire. In general, it's a good idea to pretest the questionnaire by personally administering it to a small number of persons who are similar to the eventual sample members.

Sampling Considerations in Survey Research

The sampling techniques presented later in the chapter are readily applicable to survey research. As we will discuss further in Section 4.7, we can make statistical inferences from the sample to the population only if we employ a *probability sample*—one in which every person in the population has either a known or a calculable chance of being included in the sample.

Two of the problems unique to telephone surveys are cell-only households and unlisted numbers. About 16% of all households have wireless or cell-phone service only. Of those who have "landline" telephones, nearly one-third do not have their number listed in the directory. In some locales, the percentage is much higher, with some metropolitan areas having an unlisted percentage up to 60%. Households that are cell-only or unlisted may have different characteristics than those that are listed. For example, studies have found unlisted households are more likely to be urban, young, poor, and mobile compared to households whose number is in the telephone directory.[6]

Random-digit dialing is a technique that has been developed to overcome the problem of unlisted numbers. In this approach, at least the final four digits are randomly selected, either by a computer or from a random number table. The area code and the three-digit exchange number can also be randomly selected if the survey is national in scope. The use of randomly selected numbers results in both listed and unlisted households being included in the sample.

If we are to use information from a few thousand sample members to make inferences about a population numbering in the millions, the sample selection process must be both precise and rigorous. As a result, a number of companies specialize in the generation of samples for survey research clients. Survey Sampling, Inc., a leading supplier of statistical samples for the research industry, is such a company. Each year, Survey Sampling "marks" millions of telephone numbers that have been selected from its Random Digit Sampling Database. Any number that is called is automatically eliminated from consideration for a period of 6 months. One of the purposes of this practice is to help ensure that respondents will be "fresher" and more likely to cooperate.[7]

Mailing lists are readily available for the identification of practically any population a business researcher would ever want to consider. For example, infoUSA, Inc. has a database of 14 million businesses across the United States, and these are coded according to such criteria as geographical location and standard industrial classification (SIC). If you wish to mail a questionnaire to a simple random sample from populations like the 20,338 sporting goods retailers, the 22,114 tanning salons, or the 13,509 siding contractors, you can even obtain the mailing labels directly from the list company.[8]

Either an entire list or a sample from it can be generated on the basis of geographical location, at levels down to individual states and counties. Of particular interest is the ability to select samples of persons from what are known as *compiled* versus *response* lists. The distinction between these two kinds of samples is described here:

COMPILED LIST **A sample consisting of firms or persons who are alike in some way, such as being (1) physicians, (2) homeowners, (3) personal computer owners, or (4) residents of Westmoreland County, Pennsylvania. A compiled list is passive in that you can be placed on one without having done anything.**

(continued)

[6]Source: surveysampling.com, June 28, 2009.
[7]Ibid.
[8]Source: infousa.com, June 28, 2009.

RESPONSE LIST A sample consisting of firms or persons who have engaged in a specified behavior or activity, such as (1) subscribing to *PC Magazine,* (2) contributing to Greenpeace, (3) buying from a mail-order catalog, or (4) applying for an American Express credit card.

Like questionnaire design, survey sampling can itself be the topic of an entire textbook. This is especially true when the design involves either proportionate or disproportionate stratified sampling, two of the more advanced designs discussed later in the chapter.

Errors in Survey Research

Survey research may lead to several different kinds of errors, and it's recommended that you be a skeptical "consumer" of statistical findings generated by this mode of research. These errors may be described as sampling error, response error, and nonresponse error. **Sampling error,** discussed below, is a random error. It can also be described as nondirectional or **nonsystematic,** because measurements exhibiting random error are just as likely to be too high as they are to be too low. On the other hand, **response** and **nonresponse errors** are both of the directional, or **systematic,** type.

Sampling Error

Sampling error occurs because a sample has been taken instead of a complete census of the population. If we have taken a simple random sample, the methods described in Chapter 9 can be used to estimate the likely amount of error between the sample statistic and the population parameter. One of the procedures discussed in Chapter 9 is the determination of the sample size necessary to have a given level of confidence that the sample proportion will not be in error by more than a specified amount.

Response Error

Some respondents may "distort" the truth (to put it kindly) when answering a question. They may exaggerate their income, understate their age, or provide answers that they think are "acceptable." Biased questions can even encourage such response errors, for example, "Shoplifting is not only illegal, but it makes prices higher for everyone. Have you ever shoplifted?"

Nonresponse Error

Not everyone in the sample will cooperate in returning the questionnaire or in answering an interviewer's questions. This would not be a problem, except that those who respond may be different from those who don't. For example, if we're using a mail questionnaire to find out the extent to which people are familiar with the works of William Shakespeare, those who are less literate or less interested in this classic author may also be less likely to complete and return our questionnaires. As a result, we could "measure" a much higher level of interest than actually exists.

EXERCISES

4.5 What are the major approaches to carrying out survey research?

4.6 Provide an example of a survey question that would tend to exceed the vocabulary level of the typical adult.

4.7 Comment on the appropriateness of the question, "Have you ever broken the law and endangered the lives of others by running a red light?"

4.8 What is random-digit dialing and why is it used?

4.9 How does a compiled mailing list differ from a response mailing list?

4.10 Explain what is meant by sampling error, response error, and nonresponse error in survey research.

4.11 A research firm finds that only 33% of those who reported buying Kellogg's Frosted Flakes had actually purchased the product during the period monitored. What type of survey research error does this represent?

4.12 In order to increase the response rate to mail questionnaires, researchers sometimes include a dollar bill or other monetary incentive to reward the respondent for his or her cooperation. Could there be occasions where a relatively large reward—e.g., the inclusion of a $20 bill with a short questionnaire—might end up exerting a biasing influence on the answers the respondent provides? If so, give a real or hypothetical example of such an occasion.

4.13 A company has constructed two prototypes of a new personal digital assistant (PDA) and would like to find out which one of the two looks easier to use. Of the three types of surveys discussed in this section, which one would the company *not* wish to employ?

EXPERIMENTATION AND OBSERVATIONAL RESEARCH (4.4)

Experimentation

In **experiments,** the purpose is to identify cause-and-effect relationships between variables. There are two key variables in an experiment: (1) the *independent* variable, or treatment, and (2) the *dependent* variable, or measurement.

We must also consider what are known as *extraneous* variables—outside variables that are not part of the experiment, but can influence the results. Persons or objects receiving a treatment are said to be in an *experimental* group, while those not exposed are in the *control* group. In symbolic terms, we will refer to the independent variable as "*T*" and each measurement of the dependent variable as "*O*":

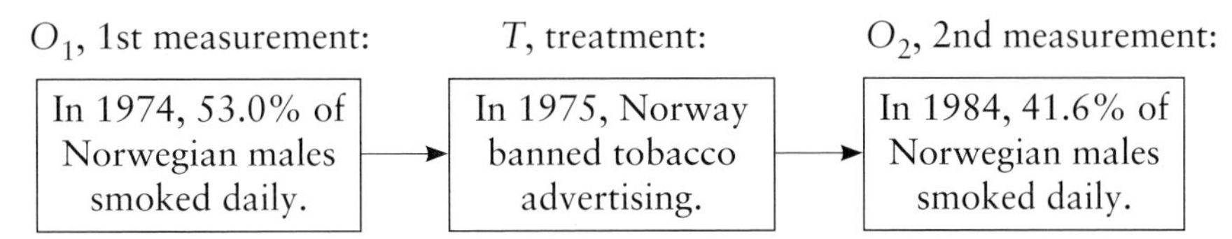

In evaluating experiments, two kinds of validity must be considered: internal and external. We will examine each in terms of the Norwegian ban on tobacco advertising.[9]

Internal Validity

Internal validity refers to whether *T* really made the difference in the measurements obtained. In the preceding experiment, was it really the advertising ban

[9] Source: Jack Burton, "Norway Up in Smoke," *Advertising Age,* April 14, 1986, p. 82.

that made the difference? For example, extraneous variables included a high cigarette tax ($1 to $2 a pack), public service announcements, and extensive educational efforts on the hazards of smoking.

External Validity

Even if *T* *did* make the difference, **external validity** asks whether the results can be generalized to other people or settings. For example, would similar results have been obtained in the United States or Great Britain, or if the tobacco ad ban had been enacted 10 or 20 years earlier instead of in 1975?

Health groups have long studied such phenomena as the effect of cigarette smoking on longevity. However, it is difficult to remove the influence of other variables on this relationship. For example, the heavy smoker is more apt to be a heavy drinker or a compulsive workaholic, while the nonsmoker is more likely to be a nondrinker or light drinker and to be more attentive to other factors (such as exercise and weight control) that tend to increase life span.

It can be useful to design an experiment in which people or objects are randomly placed into either the experimental or control group, then measurements are made on the dependent variable. Consider the following example in which 200 persons are randomly assigned, 100 to each group:

- Experimental group ($n_1 = 100$):

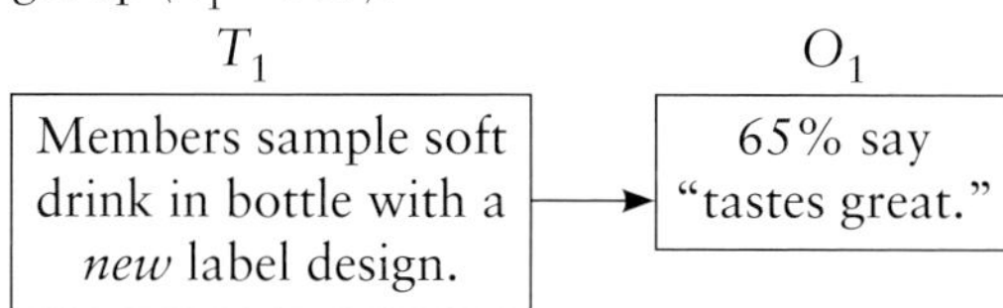

- Control group ($n_2 = 100$):

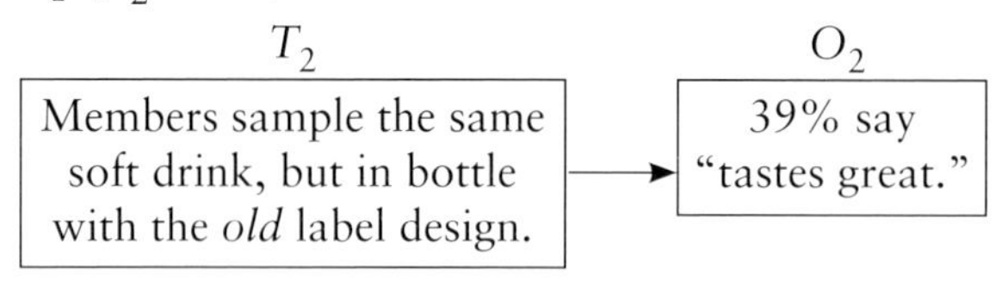

These results suggest that the new label is effective in causing testers to evaluate the soft drink as tasting good. Although this is just one of many possible experimental designs, it shows how an experimental group and a control group can be used in carrying out an experiment. To facilitate internal validity, it's best if persons or objects are randomly assigned to the experimental and control groups. The significance of the difference between the 65% for the experimental group and the 39% for the control group can be obtained using the hypothesis testing methods presented in Chapter 11.

Observational Research

Observation relies on watching or listening, then counting or measuring. For example, observational research allows us to passively watch highway travelers and monitor variables such as speed and adherence to other traffic laws, or the extent to which drivers from one age group, gender group, or vehicle group behave differently from those of other groups. One of the more interesting applications of observational research involves actually contriving a situation that is to be observed—for example, hiring a "mystery shopper" to present retail store personnel with a given problem, product return, or complex request, then have the hired

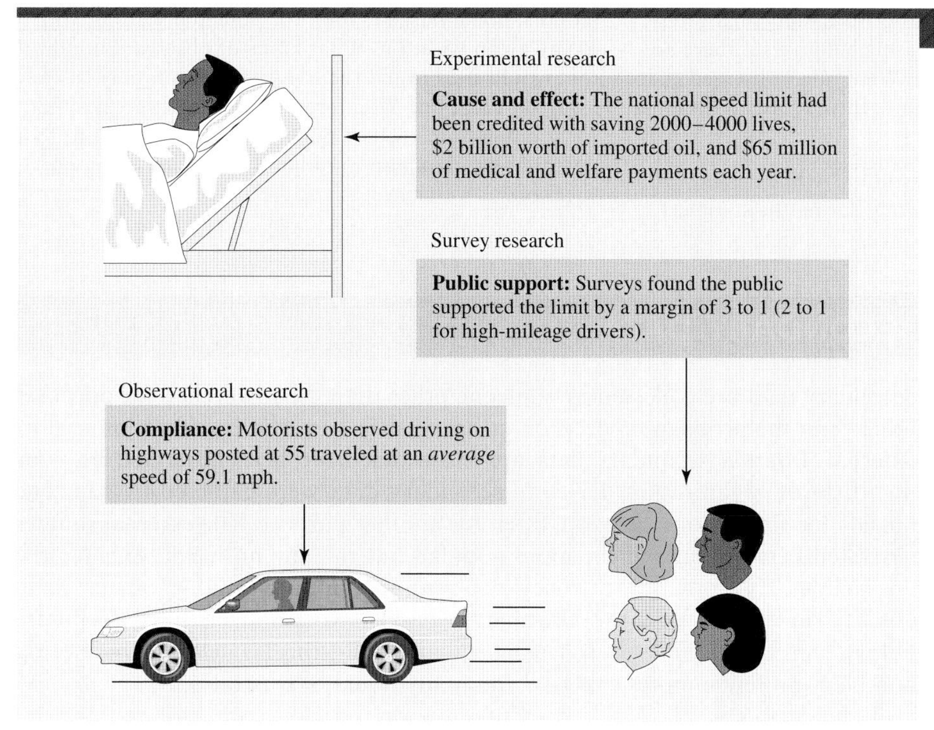

FIGURE 4.1

News media often report research results in the context of an article or story. This figure dramatizes the unusual case of a single article that included the findings of three different types of investigations centered on the same topic: the 55-mph national speed limit that was in effect between 1974 and 1987. Note the contradiction between the support indicated by surveys and the lack of compliance measured in observational studies.

Data source: Geoff Sundstrom, "Speed Limit Retention Is Favored," *Automotive News,* December 3, 1984, p. 6.

shopper submit a report on how he or she was treated, and the extent to which her contrived situation was handled according to corporate policy.

The results of surveys, experiments, and observational studies are often reported in news media. Figure 4.1 shows data from a single article containing elements of all three research methods. The research dealt with the former 55-mph national speed limit and involved a 2-year study by the Transportation Research Board of the National Research Council. The study seems to have included some rather contradictory findings.

EXERCISES

4.14 In studying the effectiveness of monetary rewards on seat belt usage, a firm has been giving out lottery tickets to employees wearing their belts when entering the company parking area. At the end of each week, the holder of the winning ticket gets a free dinner at a local restaurant. Before the experiment, only 36% of employees wore seat belts. After 1 month of the lottery, the usage rate is up to 57%.

a. Identify the dependent and independent variables in this experiment.
b. Is a control group involved in this experiment? If so, who?

4.15 A direct-mail firm tries two different versions of a mailing, one of which includes a 50-cent monetary incentive for the respondent. Responses to the mailing with the incentive were 15% higher than for a similar mailing without it.

a. Identify the dependent and independent variables in this experiment.
b. Is a control group involved in this experiment? If so, who?

4.16 In a health study, the dietary habits of 420,000 men were examined over a 3-year period. For those who ate no fried foods, the death rate was 1208 per 100,000 man-years. For men who consumed fried foods more than 15 times a week, the corresponding rate was only 702. Considering that the relative ages of members of these two groups were not mentioned, critique the internal validity of the experiment. Source: "Science Finds Fried Foods Not Harmful," *Moneysworth* magazine, November 24, 1975, p. 1.

4.17 When traveling, some motorists assume that a restaurant has good food and reasonable prices if there are a lot of trucks parked outside. What other observed characteristics of a restaurant might a traveler use in deciding whether to stop and eat?

4.18 In doing "observational studies" while driving through suburban neighborhoods during the summer months, home burglars look for potential victims who are away on vacation. What kinds of observation indicators would tend to tell the burglars that a family is likely away on vacation? What are some of the strategies that vacationing homeowners might use in attempting to fool potential burglars into thinking they are *not* away on vacation?

4.5 SECONDARY DATA

Secondary data are collected by someone other than the researcher, for purposes other than the problem or decision at hand; such data can be either internal or external. *Internal* secondary data are those that have been generated by your own firm or organization. *External* secondary data were gathered by someone outside the firm or organization, often for another purpose. The external sources described here are among the more popular, but they by no means exhaust the possibilities. As a practical matter, if you're in search of relevant secondary data, don't hesitate to consult the information desk of your university or corporate library. Don't even think about feeling embarrassed—this is one case where being humble for a few minutes can save you many hours of frustration.

Internal Sources

Internal secondary data have traditionally existed in the form of accounting or financial information. Among the many other possibilities are annual reports, as well as more specific reports that may be available from departments such as production, quality control, engineering, marketing, sales, and long-range planning. In short, anything in written form that has ever been generated within the company falls within the realm of internal secondary data. In addition, some information may not even exist in written form, but may be found in the collective memories and observations that executives, managers, and engineers have recorded over the years.

External Sources

Government Agencies

Federal, state, and local governments are extremely prolific generators and purveyors of secondary data, with the federal government, through the U.S. Department of Commerce and its Bureau of the Census, being an especially important source. The following are among the many U.S. Department of Commerce publications that are highly relevant to business research:

- *Statistical Abstract of the United States* Published annually, this volume is an especially extensive collection of social, economic, and political data from a wide variety of sources, both governmental and private.
- *Census of Population* Conducted every 10 years, this is the research for which the Bureau of the Census is probably best known. It provides detailed information on the U.S. population, including age, education, employment, and a broad range of other demographic characteristics.
- *Census of Housing* Also published every 10 years, this report presents voluminous data on such variables as home ownership or rental, type of

structure, number of residents, market value, heating source, major appliances, and much more.
- ***Economic Censuses*** Conducted in years ending in 2 and 7, these studies provide data on retail and wholesale trade, the construction and transportation industries, and the manufacturing and minerals industries. Data are aggregated by national, regional, and local levels.
- ***Survey of Current Business*** A monthly publication containing information on a number of statistical topics, including population, trade, and transportation. It includes over 2500 updated statistical series.

Other key publications of the U.S. Department of Commerce include *Historical Statistics of the United States from Colonial Times to 1970, County Business Patterns, State and Metropolitan Area Data Book,* and the *U.S. Industrial Outlook*. The Department of Labor provides employment and wage information in the *Monthly Labor Review,* while the Federal Reserve System publishes the *Federal Reserve Bulletin,* which includes statistics on banking, industrial production, and international trade. In addition to the staggering quantity of information available from the federal government, published data are also provided by governmental agencies at the state, county, and local levels.

Other Published Sources

This category includes a variety of periodicals, handbooks, and other publications. The following list is a sample of the information available:

- ***Business Periodicals Index*** Published monthly, this is a bibliography of articles in selected business publications, all indexed by subject area.
- ***Reader's Guide to Periodical Literature*** Similar to the *Business Periodicals Index,* but indexes popular, rather than business, publications.
- ***Rand McNally Commercial Atlas and Marketing Guide*** Includes statistical data for each county in the United States and for more than 120,000 cities and towns.

Other published sources include (1) indexes, such as the *New York Times Index* and the *Wall Street Journal Index,* and (2) popular business publications, including *American Demographics, BusinessWeek, Fortune, Sales and Marketing Management, Dun's Business Month, Business Horizons,* and *Industry Week.*

Commercial Suppliers

Information may also be purchased from outside suppliers who specialize in business research. One of the best-known commercial suppliers is the A. C. Nielsen Company, which provides a wide variety of data services for client firms. Although some data are supplied to a number of subscribing firms, studies may also be tailored to the specific needs of a single research customer.

Trade Associations

Trade associations collect and disseminate data that are related to their industry; these groups may be identified in the *Encyclopedia of Associations*. This publication is also an excellent source of private organizations having special interests that may be related to your informational needs. Most such groups generate regular publications that are available upon request or special order.

Data Warehousing and Data Mining

Advancements in computing power, mass data storage, and data analysis techniques have led to a relatively recent and exciting data source—the **data warehouse.** This is a large collection of data from inside and outside the company, put together for the express purpose of using sophisticated analytical techniques to uncover patterns and discover interrelationships within the data. The procedure by which these analytical techniques are applied is known as **data mining**—it is analogous to physically sifting through great quantities of ore in the quest for gold.

The data warehouse typically contains massive information from a variety of sources that pertains to customers, vendors, products, or any other organizational or environmental subjects of interest to the company's decision makers. It is estimated that over 90% of large corporations worldwide either have or are building data warehouses, and that the market for data warehousing hardware and software grew from practically nil in 1994 to more than $15 billion by 2000.[10]

The tools for data mining comprise many of the techniques covered in this text, such as descriptive statistics, cross-tabulations, regression, and correlation, but also include many methods that are far beyond the scope of our text, such as those briefly described here.

Factor analysis A set of techniques for studying interrelationships among variables. A very large set of variables can be reduced to a smaller set of new variables (factors) that are more basic in meaning but contain most of the information in the original set.

Cluster analysis When each of a great many individuals or objects is represented by a set of measurements or variables, cluster analysis categorizes the individuals or objects into groups where the members tend to be similar. In marketing, one application of cluster analysis is to categorize consumers into groups, a procedure associated with what is referred to as *market segmentation.*

Discriminant analysis This technique identifies the variables that best separate members of two or more groups. It can also be used to predict group membership on the basis of variables that have been measured or observed.

Multidimensional scaling Based on a ranking that reflects the extent to which various pairs of objects are similar or dissimilar, multidimensional scaling is a procedure that maps the location of objects in a perceptual space according to dimensions that represent important attributes of the objects. In this approach to multidimensional scaling, one of the biggest challenges is the interpretation of exactly what these dimensions mean.

Data warehouses and data mining can be useful to many kinds of businesses. For example, the descriptive and predictive models developed through data mining can help telecommunications companies answer questions like these: (1) What kinds of customers tend to be profitable or unprofitable to the company? (2) How can we predict whether customers will buy additional products or services, like call waiting? (3) Why do customers tend to call more at certain times? (4) Which products or services are the most profitable?[11] Likewise, companies in the insurance industry can benefit from models that help them identify

[10]Source: Davis, Duane, *Business Research for Decision Making* (Pacific Grove, CA: Duxbury Press, 2000), p. 355.

[11]Source: Groth, Robert, *Data Mining* (Upper Saddle River, NJ: Prentice Hall, 2000), p. 209.

which insurance claims are most likely to be fraudulent, and investment firms can profit from models that help them better differentiate between initial public offerings (IPOs) that should be pursued and those that are better avoided.

SAS, SPSS, and IBM are among those offering data mining software and assistance to a variety of firms and industries. For example, IBM's Intelligent Miner™ software was selected by Bass Export, the United Kingdom's largest beer exporter. As stated by Bass executive Mike Fisher, "Each week, we deliver orders to 23,000 customers. It's vitally important that we understand the buying habits of each of them, the profitability of each individual account, and the profitability of each brand we deliver there."[12] Reuters, a leading provider of financial news and information, has used the SPSS-developed Clementine software to develop models for reducing error rates in its compilation and reporting of the massive amount of data it receives from worldwide sources.[13]

Internet Data Sources

Thousands of database services—some are commercial, but many are free—offer bibliographical and statistical data for personal computer (PC) users. The specific set of available data will depend on your Internet provider, your Web-browsing software, and on the search engines at your disposal for specifying World Wide Web sites for further perusal. The PC and access to the vast electronic "ether" make it very convenient to locate great quantities of information relevant to a particular business situation or decision. The author's seventh-grade teacher had a favorite saying, "The more you know, the less you know you know, because you know there's so much to know." This certainly applies to our familiarity with the Internet and its capabilities, and this section will barely scratch the surface of what's out there.

Corporate News and Annual Reports Online

If you want to read a company's latest annual report, in many cases all you need to do is visit the firm's website and look for the "information for investors" or similar section. There's a good chance that your browser's home page includes a sponsored section where stock quotes (usually with a 20-minute or more delay) are available for companies on the major stock exchanges; if so, current news about a company or its industry is available just by pursuing the news items that accompany the quote.

Secondary Data Online

Many of the sources of published secondary data discussed previously offer information online as well as in hard-copy form. For example, the U.S. Bureau of the Census website (census.gov) can be the starting point for data from sources such as *USA Statistics in Brief* or the *Statistical Abstract of the United States*. Its "State & County QuickFacts" section is particularly interesting and can provide detailed economic and demographic data about the county in which you live. *As with any useful site that you visit, be sure to refer to the "Related Sites" section.* In this case, the Bureau provides links to international statistical agencies, state data centers, more than 100 federal statistical agencies, and to its parent agency, the U.S. Department of Commerce.

[12]Source: Groth, Robert, *Data Mining* (Upper Saddle River, NJ: Prentice Hall, 2000), p. 11.
[13]Ibid.

STATISTICS IN ACTION 4.1

Get That Cat off the Poll!

When a public opinion poll is conducted professionally, respondents are carefully and scientifically selected to ensure that they are representative. However, as with many other human endeavors, public opinion polls are sometimes carried out using methods that are not quite so professional. One unfortunate polling firm was the target of special scrutiny by a federal grand jury in New Haven, Connecticut, for techniques it allegedly employed during some of its studies between 2002 and 2004. According to prosecutors, the polling firm's owners and managers had instructed fieldworkers to fabricate survey results, to complete surveys even if they had to make up the answers themselves, and to meet their response quotas "even if they had to talk to dogs and cats."

Source: Noreen Gillespie, "Polling Firm Accused of Falsifying Results," *The Indiana Gazette,* March 10, 2005, p. 7.

Finding Just About Anything—Those Incredible Search Engines

Tonight Show host Johnny Carson had a segment that he called "Stump the band," where audience members would state a fictitious song title and the band would try to play music that seemed to fit the title. Regardless of whether the band was "stumped," the audience member always received a free dinner or other prize.

If there were ever a contest called "Stump the Internet search engine," it would have few winners. Regardless of how esoteric the information or topic we are seeking, we can use a search engine like Google to obtain a wealth of guidance in leading us toward our goal. A prerequisite to using the search engines effectively is the strategic selection of the key words to be used in the search. It can also be helpful to use the "search within a search" or "power search" features to enter more specific key words to be found within a search that has already been conducted.

Tapping Others' Expertise—Searching Message Boards and Interest Groups

Message boards and interest groups can be viewed as the Internet counterpart to the "brick and mortar" organizations listed in the *Encyclopedia of Associations.* For practically any human interest or pursuit, you'll find some forum with participants who are enthusiastic about the topic and possess high levels of knowledge and experience. Select the "Groups" option at google.com, enter key words that describe your problem or interest, and you'll get a list of postings that contain one or more of the key words. *Note:* There are some privacy issues here—for example, if you've posted to a discussion group, a present or potential employer can search the groups for your name to find out more about *you,* your interests, and the footprints you've made during your electronic travels.

Evaluating Secondary Data

Secondary data should be selected, evaluated, and used with caution, since the information may be irrelevant to your needs, out of date, not expressed in units or levels of aggregation that are compatible with your informational requirements, or may not have been collected or reported objectively. When considering the use of secondary information, be sure to keep in mind both who collected the information and whether the individual or group may have had some stake in the results of their own research. Honesty and reputation are also very important. Statistics in Action 4.1 describes a situation where a company may have been more interested in obtaining respondents than in obtaining facts.

EXERCISES

4.19 What are secondary data? Differentiate between internal and external secondary data.

4.20 Besides accounting and financial information, what are some of the other sources of internal secondary data?

4.21 What kinds of external secondary data are readily available online?

4.22 Briefly, what kinds of information does the U.S. Department of Commerce provide in its *Census of Population?* In its *Census of Housing?*

4.23 What is the *Encyclopedia of Associations,* and how can this publication be useful to the business researcher?

4.24 What is data mining, and how can it be useful to a business or other organization? What are some of the analytical tools it employs?

Note: Exercises 4.25–4.31 require a computer and access to the Internet. Web addresses were accurate at this writing. If a site address is no longer applicable, use a search engine (e.g., google.com) to identify the company or organization's new home page.

4.25 Visit the U.S. Bureau of the Census (census.gov), and find the following descriptive statistics for the county where your college or university is located:

a. Median household income
b. Number of Social Security recipients
c. Per-capita retail sales

4.26 Repeat Exercise 4.25, but obtain the requested descriptive statistics for Dare County, North Carolina.

4.27 Referring to each company's online annual report, find the number of employees and net income in the most recent fiscal year for Coca-Cola, McDonald's, and Microsoft.

4.28 Visit google.com and search message boards for postings that pertain to (a) a professional sports team and (b) a hobby or recreational activity of your choice.

4.29 Using whitepages.com, find out how many telephone listings are in your state for people with the same last name as yours. How many listings are there in your hometown for people named Smith?

4.30 Using one or more search engines and key words of your choice, find (a) a website that reports the average life span of Saint Bernard dogs and (b) a website that advises business professionals on strategies for writing an effective memo.

4.31 Using a search engine and the key words "spy equipment," identify some of the high-tech gadgets that companies or individuals can use in carrying out industrial espionage on other companies or individuals.

THE BASICS OF SAMPLING (4.6)

Selected Terms

The following terms—some of which we have discussed previously in other contexts—are of special relevance to our discussion of samples, the importance of sampling, and the methods by which samples are selected:

POPULATION The set of all possible elements that could theoretically be observed or measured; this is sometimes referred to as the *universe*.

SAMPLE A selected portion from the elements within the population, with these elements actually being measured or observed.

CENSUS The actual measurement or observation of all possible elements from the population; this can be viewed as a "sample" that includes the entire population.

PARAMETER A characteristic of the population, such as the population mean (μ), standard deviation (σ), or population proportion (π).

STATISTIC A characteristic of the sample, such as the sample mean ($\bar{x}$), standard deviation (s), or sample proportion (p). In practice, this is used as an estimate of the (typically unknown) value of the corresponding population parameter.

Why a Sample Instead of a Census?

A Census Can Be Extremely Expensive and Time-Consuming

A complete census of the population is not a problem whenever the population happens to consist of a small number of elements that are easily identified. For example, there are only 29 nuclear electric power generation firms that have 500 or more employees.[14] In this setting, a study involving a complete census would present little difficulty. On the other hand, infoUSA, Inc., lists 77,240 pizza parlors in the United States.[15] Contacting every member of such a large population would require great expenditures of time and money, and a sampling from the list can provide satisfactory results more quickly and at much lower cost.

A Census Can Be Destructive

The Gallo Wine Company, like every other winery, employs wine tasters to ensure the consistency of product quality. Naturally, it would be counterproductive if the tasters consumed *all* of the wine, since none would be left to sell to thirsty customers. Likewise, a firm wishing to ensure that its steel cable meets tensile-strength requirements could not test the breaking strength of its entire output. As in the Gallo situation, the product "sampled" would be lost during the sampling process, so a complete census is out of the question.

In Practice, a Sample Can Be More Accurate Than a Census

Because of the tremendous amounts of time and money required for a complete census of a large population, a sample can actually be more accurate than a census in conducting a real-world study. For example, if we hired 10 people to work 16 hours a day telephoning each of the more than 100 million television households in the United States, it would take years to ask all of them whether they watched the CNN program on political campaigns that aired last night. By the time the last batch of calls was made, nobody would be able to remember whether they had watched a given program that was aired 2 or 3 years earlier.

Sampling can also contribute to accuracy by reducing *nonsampling error*. This occurs because we can now afford to spend more time and attention on the measuring instrument and related procedures. For example, if a survey question is biased ("Most studies find that capital punishment does not deter crime. Do you agree that capital punishment should be abolished?"), asking that same question of everyone in the entire population will only ensure that you get the same biased answers from everyone instead of from just a sample. With some of the money saved by conducting a sample instead of a census, we can fine-tune the measurement procedure through such strategies as pretesting the questionnaire, hiring more-experienced interviewers, or following up on those who did not initially respond. This accuracy advantage of a sample over a census is especially valuable to survey research. As was observed in the vignette at the beginning of this chapter, even the U.S. Bureau of the Census has proposed that its census efforts be supplemented with survey information based on samples.

[14]Source: Bureau of the Census, *Statistical Abstract of the United States 2009*, p. 565.
[15]Source: infousa.com, June 28, 2009.

Although a census is *theoretically* more accurate than a sample, *in practice* we are limited by funding constraints such that a fixed amount of money must be divided between the sampling (or census) process and the measurement process. If we allocate more funds to the sampling or census process, we have less to spend on the assurance of accuracy in the measurement process. Here is a brief review of the concepts of sampling and nonsampling error:

SAMPLING ERROR This is the error that occurs because a sample has been taken instead of a census. For example, the sample mean may differ from the true population mean simply by chance, but is just as likely to be too high as too low. When a probability sample (discussed in the next section) is used, the likely amount of sampling error can be statistically estimated using techniques explained in Chapter 9 and can be reduced by using a larger sample size. For a visual demonstration of sampling error, see Seeing Statistics Applet 3 at the end of the chapter.

NONSAMPLING ERROR This type of error is also referred to as *bias,* because it is a directional error. For example, survey questions may be worded in such a way as to invite responses that are biased in one direction or another. Likewise, a weighing scale may consistently report weights that are too high, or a worn micrometer may tend to report rod diameters larger than those that actually exist. *Nonsampling errors cannot be reduced by simply increasing the size of a sample.* To reduce nonsampling error, it is necessary to take some action that will eliminate the underlying cause of the error. For example, we may need to adjust a machine or replace parts that have become worn through repeated use.

EXERCISES

4.32 What is the difference between a sample and a census? Why can it be advantageous to select a sample instead of carrying out a census?

4.33 Differentiate between sampling error and nonsampling error.

4.34 Provide a real or hypothetical example of a situation in which the sampling process is destructive.

4.35 Differentiate between the terms *parameter* and *statistic*. Which one will be the result of taking a sample?

4.36 To impress the interviewer, many respondents in a survey of 400 persons exaggerated their annual incomes by anywhere from 2% to 50%. Does such exaggeration represent sampling error or nonsampling error? Explain.

4.37 In survey research, what are some of the methods by which we can reduce nonsampling error?

SAMPLING METHODS (4.7)

Sampling methods can be categorized as *probability* or *nonprobability*. The distinction is that with **probability sampling,** each person or element in the population has a known (or calculable) chance of being included in the sample. Such samples also allow us to estimate the maximum amount of sampling error between our sample statistic and the true value of the population parameter being estimated. **Nonprobability sampling** is primarily used in exploratory research studies where there is no intention of making statistical inferences from the sample to the population. Our emphasis in this and the chapters to follow will be on samples of the probability type.

Probability Sampling

In probability sampling, each person or element in the population has some (nonzero) known or calculable chance of being included in the sample. However, every person or element may not have an *equal* chance for inclusion.

The Simple Random Sample

In the **simple random sample,** every person or element in the population has an equal chance of being included in the sample. This type of sample is the equivalent of "drawing names from a hat" but is usually generated using procedures that are more workable than the familiar "hat" analogy.[16]

A practical alternative to placing names in a hat or box is to identify each person or element in the population with a number, then use a random number table to select those who will make up the sample. A portion of the random number table in Appendix A is reproduced here as Table 4.1 and will be used in the example that follows.

EXAMPLE

Simple Random Sampling

A firm has 750 production employees and wishes to select a simple random sample of 40 workers to participate in a quality-assurance training program.

SOLUTION

Using the random digits listed in Table 4.1, the following steps can be taken to determine the 40 individuals to be included in the sample:

1. Identify the employees with numbers ranging from 001 to 750.
2. Arbitrarily select a starting point in the random number table. This can be done by closing your eyes and pointing to a position on the table or by some other relatively random method of your choice. As our starting point for the example, we will begin with digits 11–13 in row 2 of Table 4.1. Three-digit numbers are necessary because the largest employee identification number is three digits.
3. Work your way through the table and identify the first 40 unique (to avoid duplicates, since we're sampling without replacement) numbers that are between 001 and 750.

[16]For practicality, we are using the *operational definition* of the simple random sample. The *theoretical definition* says that, for a given sample size, each possible sample constituency has the same chance of being selected as any other sample constituency. For more details, the interested reader may wish to refer to a classic work on sampling: Leslie Kish, *Survey Sampling,* New York: John Wiley & Sons, 1965, pp. 36–39.

EXAMPLEEXAMPLEEXAMPLEEXAMPLEEXAMPLEE

TABLE 4.1

A portion of the table of random digits from Appendix A. In the example, a simple random sample of 40 is being selected from a population of 750 employees. The digits in row 2, columns 11–13, serve as an arbitrary starting point.

Row	Columns 1–10		Columns 11–20		
1	37220	84861	59998	77311	
2	31618	06840	**451**67	13747	Employee #451 is in sample.
3	53778	71813	**503**06	47415	Employee #503 is in sample.
4	82652	58162	**903**52	10751	903 > 750. 903 is ignored.
5	40372	53109	**769**95	24681	769 > 750. 769 is ignored.
6	24952	30711	**225**35	51397	Employee #225 is in sample.
7	94953	96367	**871**43	71467	871 > 750. 871 is ignored.
8	98972	12203	**907**59	56157	907 > 750. 907 is ignored.
9	61759	32900	**052**99	56687	Employee #052 is in sample.
10	14338	44401	**226**30	06918	Employee #226 is in sample.
⋮					⋮
					The process continues, until all 40 sample members have been selected.

The procedure is summarized in Table 4.1 and begins with the selection of employee #451, who also happens to be at the randomly selected starting point. The next number, 503, also falls within the limit of 750 employees; at this point we have placed individuals #451 and #503 into the sample. Because the next number (903) exceeds 750, it is ignored and we continue down the column.

Computer Solutions 4.1 shows an alternative to using the random number table. Here we show the procedures and results when Excel and Minitab are used to select a simple random sample *from an existing list of data values*. In this case, we are selecting a simple random sample of 10 usage times (seconds) from the 50 ATM customers whose times were examined in Chapters 2 and 3. The data file is **CX02ATM**.

Although the simple random sampling approach just described is a straightforward procedure, it does require that we have a listing of the population members. This is not a problem if we are dealing with telephone directories or members of the *Fortune 500,* but it may not be practical in other situations. For example, it would be both expensive and awkward to take down the names and telephone numbers of all the shoppers at the local mall tomorrow so that we can later narrow them down to a simple random sample of 100 persons.

In some cases, such as when products are coming off an assembly line, we can use a random selection process to determine which ones to sample. For example, if we would like to subject 10% of the units to a detailed inspection, a random number sequence could be followed. As each product comes off the line, we advance to the next random digit in a table of random numbers. If the digit associated with a particular product happens to be a "9" (an arbitrary selection), it and all other units that are associated with the digit 9 receive the detailed inspection. Although it takes a slightly different form, this is a variation of the simple random sampling method, and each unit has the same probability (in this case, 0.10) of being sampled.

COMPUTER 4.1 SOLUTIONS

Simple Random Sampling

These procedures select a simple random sample from a list of existing data values.

EXCEL

	A	B	C	D	E
1	**AgeCat**	**Gender**	**Seconds**		40.1
2	1	2	50.1		50.8
3	1	1	53.0		37.8
4	2	2	43.2		42.9
5	1	2	34.9		40.1
6	3	1	37.5		32.9
7	2	1	37.8		29.4
8	3	1	49.4		49.4
9	1	1	50.5		37.8
10	3	1	48.1		39.5

1. Open the Excel data file CX02ATM. The 50 ATM usage times are in column C, and the variable name (Seconds) is in C1. From the **Data** ribbon, click **Data Analysis** in the rightmost menu. Click **Sampling.** Click **OK.**
2. Enter **C1:C51** into the **Input Range** box. Click to place a check mark into the **Labels** box (because the input range has the name of the variable at the top). Select **Random** and enter **10** into the **Number of Samples** box. Enter **E1** into the **Output Range** box. Click **OK.** The simple random sample of 10 values from the 50 ATM usage times are listed in column E.

MINITAB

```
Data Display

C5
   37.1   46.4   31.0   29.9   47.6   44.7   33.6   42.3   34.9   47.7
```

1. Open the Minitab file CX02ATM. The data are already listed in columns C1 (AgeCat), C2 (Gender), and C3 (Seconds).
2. Click **Calc.** Select **Random Data.** Click **Sample From Columns.** Enter **10** into the **Number of rows to sample:** box. Enter **C3** into the box below the **From columns:** label. Enter **C5** into the **Store samples in** box. (Do not select "Sample with replacement.") Click **OK.** The 10 data values randomly selected from the 50 usage times will be in column C5.

The Systematic Sample

Although similar in concept to the simple random sample, the **systematic sample** is easier to apply in practice. In this approach, we randomly select a starting point between 1 and k, then sample every kth element from the population. For example, in selecting a sample of 100 from a population of 6000, we would select a random number between 1 and 60, then include that person and every 60th person until we have reached the desired sample size. The choice of $k = 60$ is made on the basis that the population is 60 times as large as the desired sample size.

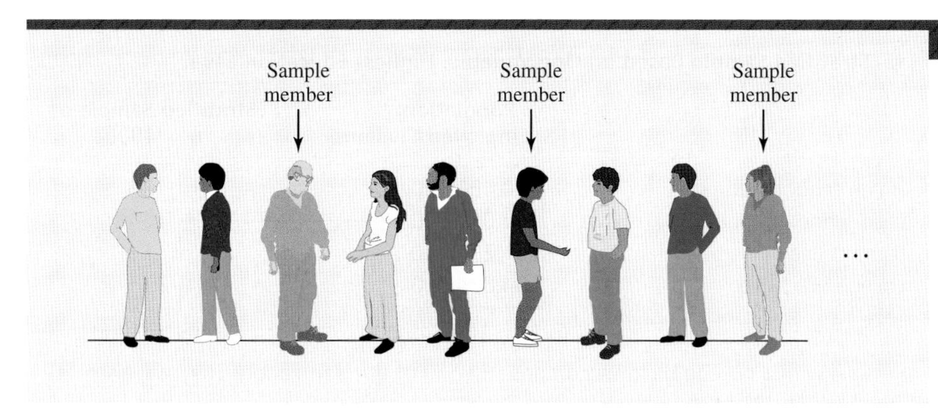

FIGURE 4.2

The systematic sample is a probability sample in which every kth person or element in the population is included in the sample. In this example, $k = 3$.

In the context of a shopping mall, we might simply elect to survey every 10th person passing a given point in the mall during the entire day. Such a process is illustrated in Figure 4.2, except with $k = 3$ as our selection rule. As in the simple random sample, we have a rule to follow, and personal judgment is not allowed to enter the picture. The use of such rules, and their elimination of personal judgment or bias in the selection process, is basic to all probability sampling procedures. If such rules are not followed, we could find ourselves avoiding interviews with mall shoppers who are carrying heavy bags, who appear to be in a hurry, or who in some other way cause us to be either less likely or more likely to interview them.

One potential problem with the systematic sample is **periodicity,** a phenomenon where the order in which the population appears happens to include a cyclical variation in which the length of the cycle is the same as the value of k that we are using in selecting the sample. For example, if daily sales receipts are arranged in order from January 1 to December 31, it would not be wise to select $k = 7$ in choosing a systematic sample. Under these conditions, we could end up with a sample consisting entirely of Mondays, Wednesdays, or some other given day of the week. Likewise, a production machine may have some quirk by which every 14th unit produced happens to be defective. If we were to use $k = 7$ or 14 in selecting our systematic sample, we would end up with a sample that is not representative of production quality. Although such periodicity is not a common problem, the possibility of its existence should be considered when undertaking a systematic sample.

Other Probability Sampling Techniques

A number of more advanced probability sampling methods are also available. Among them are the stratified and cluster sampling approaches.

In the **stratified sample,** the population is divided into layers, or *strata*; then a simple random sample of members from each stratum is selected. Strata members have the same percentage representation in the sample as they do in the population, so this approach is sometimes referred to as the *proportionate* stratified sample. Part A of Table 4.2 is an example of the stratified sampling approach in which the stratification has been done on the basis of education level. The selection of education as the variable on which to stratify could be important if our purpose is to measure the extent to which individuals provide charitable contributions to the performing arts.

A sample can also be stratified on the basis of two or more variables at the same time. Part B of Table 4.2 is similar to part A, except that the stratification is based on two variables. This is referred to as a *two-way* stratified sample since

TABLE 4.2

Examples of stratified sampling using one variable (part A) and two variables (part B) for stratification. In each case, strata members have the same percentage representation in the sample as they do in the population, and sample members from each stratum are selected through the use of probability sampling.

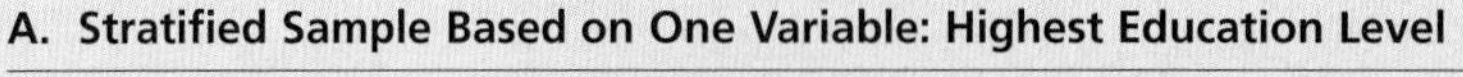

A. Stratified Sample Based on One Variable: Highest Education Level

	Population: U.S. Employed Civilians	Stratified Sample of $n = 1000$
<high school graduate	10,466,000	95
high school graduate	34,011,000	307
some college	31,298,000	283
college graduate	34,874,000	315
	Total = 110,649,000	Total = 1000

Persons with a college diploma comprise 31.5% of the population. They will also comprise 31.5% of the sample.

B. Stratified Sample Based on Two Variables: Education and Gender

	Population: U.S. Employed Civilians		Stratified Sample of $n = 1000$	
	Male	**Female**	**Male**	**Female**
<high school graduate	6,397,000	4,069,000	58	37
high school graduate	18,188,000	15,823,000	164	143
some college	15,613,000	15,685,000	141	142
college graduate	19,033,000	15,841,000	172	143
	Total = 110,649,000		Total = 1000	

Males with some years of college comprise 14.1% of the population. They will also comprise 14.1% of the sample.

Source: Based on data in Bureau of the Census, U.S. Department of Commerce, *Statistical Abstract of the United States 2002*, p. 385.

the sample has been forced to take on the exact percentage breakdown as the population in terms of *two* different measurements or characteristics.

The stratified sample forces the composition of the sample to be the same as that of the population, at least in terms of the stratification variable(s) selected. This can be important if some strata are likely to differ greatly from others with regard to the variable(s) of interest. For example, if you were doing a study to determine the degree of campus support for increased funding for women's sports programs, you might be wise to use a sample that has been stratified according to male/female so that your sample has the same percentage male/female breakdown as the campus population.

The **cluster sample** involves dividing the population into groups (e.g., based on geographic area), then randomly selecting some of the groups and taking either a sample or a census of the members of the groups selected. Such an approach is demonstrated in Figure 4.3. As in the stratified sample, members of the population may not have the *same* probability of inclusion, but these probabilities could be determined if we wished to exert the time and effort.

Nonprobability Sampling

In nonprobability sampling, not every unit in the population has a chance of being included in the sample, and the process involves at least some degree of

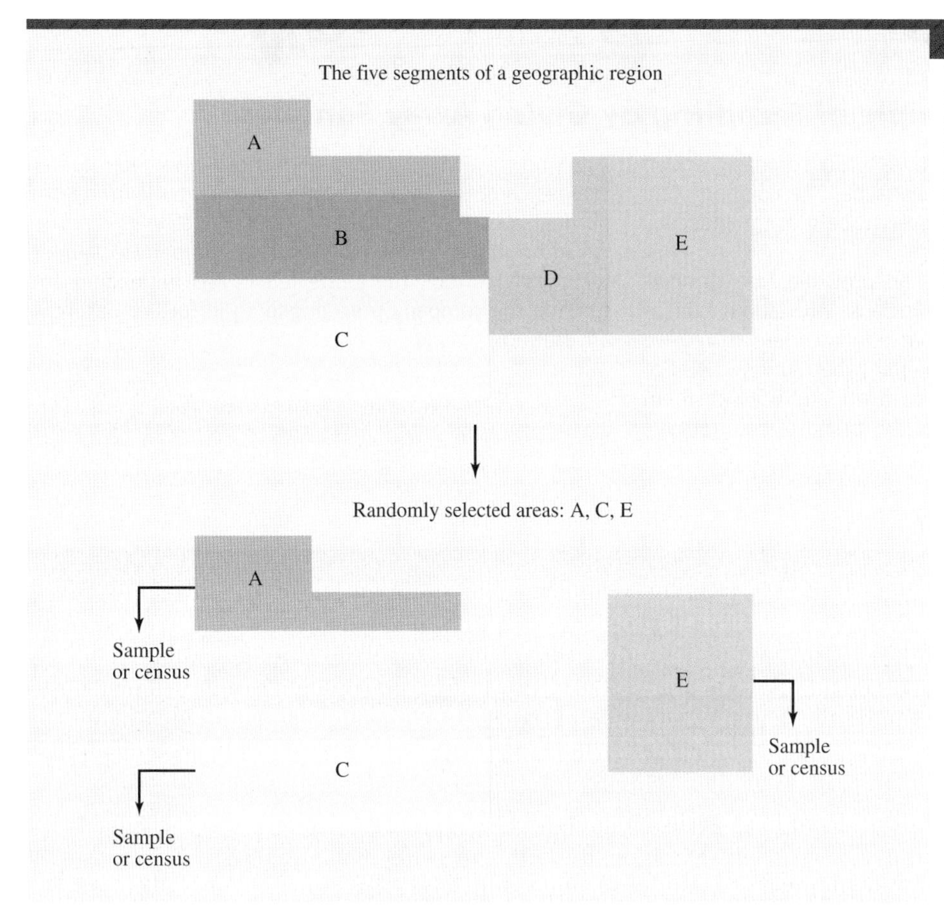

FIGURE 4.3

The cluster sample involves dividing the population into groups or areas, then randomly selecting some of the groups or areas and taking either a sample or a census of each. Shown here is an example of how this approach would apply to a city that has been divided into areas A through E.

personal subjectivity instead of following predetermined, probabilistic rules for selection. Although less important for the purposes of this text, such sampling can be useful in small-scale, exploratory studies where one wishes simply to gain greater familiarity with the population rather than to reach statistical conclusions about its characteristics. We will briefly examine four types of nonprobability samples:

CONVENIENCE SAMPLE Members of such samples are chosen primarily because they are both readily available and willing to participate. For example, a fellow student may ask members of your statistics class to fill out a questionnaire for a study she is doing as an assignment in her sociology course. So-called person-on-the-street interviews by television reporters are often identified as "random," although they are usually conducted at shopping areas near the station or at other locations selected primarily because they happen to be convenient for the interviewer and camera crew.

QUOTA SAMPLE This is similar to the stratified probability sample described previously, except that members of the various strata are *not* chosen through the use of a probability sampling technique. For example, if the sample is to include 30 undergraduates, all 30 could arbitrarily be selected from the marching band. Although it looks good on paper (e.g., the male/female percentage breakdowns are the same for the population and the sample), this type of sample is far inferior to the stratified sample in terms of representativeness.

(*continued*)

STATISTICS IN ACTION 4.2

A Sample of Sampling by Giving Away Samples

A sampling approach used by a drug company ended up giving a federal judge a headache. The company had claimed its brand of pain reliever to be recommended by 70,000 doctors.

It seemed that the firm had provided free samples of its product to 250,000 physicians, with about half of them returning a postcard that entitled them to receive still more free samples. The company then surveyed 404 of those who had returned the postcard, and 65% of them said they recommended use of this brand of pain reliever. Applying the 65% figure to all of those who had returned the postcard, the company came up with its claim that the product was recommended by 70,000 doctors.

Although some might view the company's claim as technically correct, the judge felt it was misleading and forbade the company from engaging in further deceptive advertising.

Source: Based on William Power, "A Judge Prescribes a Dose of Truth To Ease the Pain of Analgesic Ads," *The Wall Street Journal*, May 13, 1987, p. 33.

PURPOSIVE SAMPLE In this sample, members are chosen specifically because they're *not* typical of the population. For example, the maker of a new pocket calculator may submit prototypes to members of an advanced doctoral seminar in mathematics at the Massachusetts Institute of Technology. If these individuals cannot figure out how to operate it, the calculator should obviously be simplified. Conversely, if a sample of individuals known to have IQs below 80 are able to operate it, it should present no problem to the general population.

JUDGMENT SAMPLE This sample is selected on the basis that the researcher *believes* the members to be representative of the population. As a result, the representativeness of such a sample is only as good as the judgment of the person who has selected the sample. The president of a manufacturing firm may really believe that products produced by the 11:00 P.M.–7:00 A.M. shift are typical of those produced by all three work shifts. This belief is all that is necessary for a sample of output from the 11–7 shift to fall into the category of a judgment sample.

As this chapter has pointed out, statistically generalizing to the population on the basis of a nonprobability sample is not valid. Statistics in Action 4.2 describes how a company used a rather bizarre nonprobability sampling approach, then tried to sell the results as accurately describing the population.

EXERCISES

4.38 Attendees at an industrial trade show are given the opportunity to fill out a card that entitles them to receive a free software package from the sponsoring firm. The cards are placed in a bin, and 30 winners will be drawn. Considering the 30 winners as a sample from all of the persons filling out such cards, is this a probability sample or a nonprobability sample? What specific type of probability or nonprobability sample does this procedure represent?

4.39 To obtain a sample of the invoices over the past year, a clerk uses the computer to select all invoice numbers that are a multiple of 13. If the invoice numbers are sequential, beginning with 1, what type of sample does this represent?

4.40 Differentiate between a judgment sample and a convenience sample.

4.41 In what way are the quota sample and the stratified sample similar? In what way are they different? What effect does this difference have on the applications for which such samples might be appropriate?

4.42 What is the difference between a probability sample and a nonprobability sample? Which type is necessary if we wish to statistically generalize from the sample to the population?

4.43 For each of the following situations, would you recommend a sample or a census? Explain your reasoning in each case.

a. A purchasing agent has just received a shipment of shock-resistant watches and wants to find out approximately how far they can be dropped onto a concrete surface without breaking the crystal.
b. General Mills wants to learn the age, gender, and income characteristics of persons who consume Cheerios breakfast cereal.
c. The producers of NBC's *Tonight Show* want to find out what percentage of TV viewers recognize a photo of host Jay Leno.
d. A small company wants to determine whether U.S. companies that manufacture nuclear submarines might be interested in a new technique for purifying air when such craft are submerged.

4.44 Using the random number table in Appendix A, select a simple random sample of 5 from the 50 states of the United States.

4.45 What is meant by *periodicity*, and how can this present a problem when taking a systematic sample? Provide an example of a systematic sampling situation where periodicity could be a factor.

4.46 The manufacturer of a fast-drying, highly effective glue is concerned about the possibility that some users might accidentally glue their fingers together, raising potential legal problems for the firm. Could purposive sampling help determine the extent to which people might use the glue improperly? If so, describe the composition of at least one purposive sample that could be employed for this purpose.

(DATA SET) *Note:* Exercises 4.47–4.49 require a computer and statistical software capable of selecting random or systematic samples.

4.47 Given the 250 household electricity expenditures in Exercise 3.65 (file **XR03065**), generate a simple random sample consisting of 30 data values.

4.48 Given the honeymoon expenditures for the 300 couples in Exercise 3.66 (file **XR03066**), generate a simple random sample consisting of 30 data values.

4.49 A study found that the average American buys 4.6 movie tickets per year, compared to an average of 3.2 in Ireland. Assuming that the data values in file **XR04049** represent the number of movie ticket purchases reported by 100 American respondents. Source: spartatheater.com, June 28, 2009.

a. Select a simple random sample of 20 data values. Compute the sample mean.
b. Select another simple random sample of 20 data values, and compute the mean.
c. To two decimal places, compare the mean of the sample in part (a) with the mean of the sample in part (b). Are the values identical? Would you expect them to be? How do they compare to the mean of the 100 data values from which the samples were drawn? (We will learn more about this topic in Chapter 8, Sampling Distributions.)

SUMMARY (4.8)

Types of business research studies

Business research studies can be categorized according to the objective involved and may be exploratory, descriptive, causal, or predictive. Prior to beginning the study, the problem or question should be defined in specific terms that can be answered by research.

Survey research methods, data, and error sources

In business research, the data collected can be either primary or secondary. Primary data are generated by the researcher for the problem or situation at hand, while secondary data have been collected by someone else for some other purpose. Secondary data can be classified as internal or external, depending on whether the data have been generated from within or outside the researcher's firm or organization. Primary data tend to require more time and expense, but have the advantage of being more applicable to the research problem or situation.

Survey, experimental, and observational research are among the most popular methods for collecting primary data. Surveys may be carried out by mail questionnaires, personal interviews, and telephone interviews. The questionnaire, or data collection instrument, will contain any of three types of questions: multiple choice, dichotomous, and open-ended. In survey research, there are three major sources of error: sampling error, response error, and nonresponse error.

- **Experimental and observational studies**

Experimental studies attempt to identify cause-and-effect relationships between variables. Evaluation of the results of an experimental study involves the assessment of both internal and external validity. Observational studies emphasize what individuals do rather than what they say, with behavior observed and recorded on a data collection instrument resembling the questionnaire used in survey research.

- **Internal secondary data and data mining**

Internal sources of secondary data may take a variety of forms within the researcher's firm or organization, with accounting and financial records being especially useful. The many sources of secondary data include government agencies, publications, commercial suppliers, and trade associations. The data warehouse is a large collection of data put together to apply the sophisticated analytical tools of data mining to identify patterns and interrelationships in the data. More than ever, the Internet is an indispensable source of information and expertise.

- **Sampling can be better than a census**

When a population contains only a small number of elements, a complete census of its members is easily done. In practice, however, most studies involve relatively large populations for which the taking of a sample can provide satisfactory results much more quickly and at considerably lower cost. In some cases, the measurement process can be destructive, necessitating the use of sampling instead of a census. Because of the time and money it saves, a sample can even be more accurate than a census. This occurs because some of the limited funds available for the study can be used to improve the measurement process itself, thereby reducing nonsampling error.

- **Sampling and nonsampling errors**

Error is the difference between the actual value of the population parameter (e.g., μ or π) and the value of the sample statistic (e.g., $\bar{x}$ or p) that is being used to estimate it. Sampling error is a random, nondirectional error that is inevitable whenever a sample is taken instead of a census, but the amount of this error can be estimated whenever a sample is of the probability type. For probability samples, sampling error can be reduced by increasing the sample size. Nonsampling error is a tendency toward error in one direction or the other and can be present even if a complete census is taken.

- **Probability and nonprobability sampling**

In probability sampling, each person or element in the population has some (nonzero) known or calculable chance of being included in the sample. In the simple random sample, each person or element has the same chance for inclusion. Other probability samples discussed in the chapter are the systematic sample, the stratified sample, and the cluster sample.

In nonprobability sampling, not everyone in the population has a chance of being included in the sample, and the process involves at least some degree of personal subjectivity instead of predetermined, probabilistic selection rules.

Such samples can be useful in small-scale, exploratory studies where the researcher does not intend to make statistical generalizations from the sample to the population. Nonprobability techniques discussed in the chapter include convenience, quota, purposive, and judgment sampling.

CHAPTER EXERCISES

4.50 To collect information on how well it is serving its customers, a restaurant places questionnaire cards on the tables, with card deposit boxes located near the exit.

a. Do you think that nonresponse error might influence the results of this effort? In what way(s)?
b. In addition to being a source of information to management, are the cards providing any other benefits? In what way(s)?

4.51 A security officer uses a one-way mirror to watch department store shoppers and identify potential shoplifters. In addition to observing whether a person is attempting to steal merchandise, what other personal or behavioral characteristics can be observed in this setting?

4.52 In an attempt to measure the availability of component parts for a new product line, a manufacturer mails a questionnaire to 15 of the 25 companies that supply these components. Past experience with similar surveys has shown that such companies tend to exaggerate their ability to supply desired components on time and in the quantities desired. Of the companies surveyed, only 8 complete and return the questionnaire. Given this scenario, identify possible sources of sampling error, response error, and nonresponse error.

4.53 Timetech, Inc., has been experimenting with different approaches to improving the performance of its field sales force. In a test involving salespersons in Maine, the company found that giving sales leaders a free snowmobile caused a sales increase of 25% in that state. The experiment in Maine was so successful that company executives are now planning to increase sales nationwide by awarding free snowmobiles to sales leaders in all 50 states. Critique the external validity of this experiment.

4.54 For each of the following report titles, indicate whether the study involved was exploratory, descriptive, causal, or predictive, and explain your reasoning.

a. "The Popularity of Bowling as a Participant Sport in Ohio"
b. "The Effects of TV Violence on Crime in Small Towns"
c. "A Preliminary Investigation of the Market for Pet Foods"
d. "Family Vacationing in the 21st Century"

4.55 The curator of a fossil museum finds there were 1450 visitors in May and 1890 visitors in June. At the beginning of June, he had hired Bert McGruff, a popular local athlete, to serve as a tour guide. Mr. McGruff, pointing out the increase in attendance, has demanded a raise in pay. Considering this situation as an experiment:

a. Identify the dependent and independent variables in this experiment.
b. Are there any extraneous variables that might have influenced this experiment?
c. Given your response to part (b), evaluate the internal validity of the experiment. Does Bert's request for a raise seem to be justified?

4.56 A labor union official has considered three possible negotiating stances that the union might take during upcoming contract negotiations. Members of the rank and file, on the other hand, could differ in terms of the approach they would most like to see. The union official is concerned with properly representing his membership in talks with company representatives. For this situation, formulate a specific question that research might be able to answer.

4.57 A mail survey of junior executives is designed to measure how much time they spend reading *The Wall Street Journal* each day and how they feel about the newspaper's content and layout. Formulate one question of each of the following types that might be included in the questionnaire: (a) multiple-choice; (b) dichotomous; (c) open-ended.

4.58 The Sonic Travel Agency has not kept pace with the current boom in family vacation air travel between the agency's northern city and sunny Florida. For each of the following, indicate whether the associated data are primary or secondary. If the data are secondary, further indicate whether they are internal or external.

a. Sonic's research department conducts a survey of past customers to determine their level of satisfaction with the agency's service.
b. The Florida Bureau of Tourism sends travel agencies across the nation the results of a study describing the characteristics of winter vacationers to the state, including where and how long people tended to stay.
c. To get a better grasp of the problem, the president of Sonic examines the company's annual reports for the past 10 years.

4.59 In general, what are some of the considerations in evaluating the suitability of data that have been generated by someone else? For what types of sources should secondary data be regarded with special skepticism?

4.60 Shirley will be graduating soon and is anticipating job offers from several national-level corporations. Depending on the company she joins, she could end up living in any of four different locations in the United States.

a. What are some of the ways that Shirley could utilize the Internet for information to help her decide which company she would most prefer to join?
b. What are some of the ways that company personnel could use the Internet to find out more about Shirley and her suitability for their company? Would you consider any of these methods to be unethical? If so, which one(s) and why?

4.61 Using either the Internet or the *Encyclopedia of Associations*, identify a group or organization that might be able to supply information on

a. the popularity of bowling.
b. child abuse in the United States.
c. industrial accidents.
d. antique and classic cars.

4.62 When customers purchase goods at a Radio Shack store, they are typically asked for their name and phone number so their purchase information can be stored. This would be one portion of the vast internal and external data available to the company. How might a company like Radio Shack utilize data warehousing and data mining?

Note: Exercises 4.63–4.66 require a computer and access to the Internet. Web addresses were accurate at this writing. If a site address is no longer applicable, use a search engine (e.g., google.com) to identify the company or organization's new home page.

4.63 Visit the U.S. Bureau of the Census (census.gov), and find the following descriptive statistics for Kalamazoo County, Michigan:

a. Of those who are 25 or older, the percentage who are high school graduates
b. The number of persons who are 65 or older
c. The median household income

4.64 Using the U.S. Bureau of the Census (census.gov) as your starting point, use the links it provides and find the most recent population total for Canada.

4.65 One of the capabilities on the Internet is called "reverse lookup." Enter the residential telephone number for someone you know. Evaluate the accuracy of the name and address provided, along with the validity of the map that can be generated as additional information. Was the reverse lookup successful in identifying the person's e-mail address as well?

4.66 Visit google.com and search message boards for postings that pertain to Sony digital cameras. Given the kinds of discussions that take place in such settings, what benefits could Sony obtain by having someone within the company "lurk" (i.e., observe without participating) in one or more groups where digital cameras are being discussed?

4.67 Of a company's 1000 employees, 200 are managers, 700 are factory employees, and 100 are clerical. In selecting a sample of 100 employees, a researcher starts at the beginning of the alphabet and continues until she has collected the names of 20 managers, 70 factory employees, and 10 clerical workers. What type of sampling technique does this represent? Explain your reasoning.

4.68 According to the U.S. Bureau of the Census, 27% of U.S. adults are college graduates. In a stratified sample of 200 U.S. adults, approximately how many persons in each of the following categories should be questioned: (a) college graduate and (b) not a college graduate?
Source: *Statistical Abstract of the United States 2009*, p. 147.

4.69 A college of business has 20 faculty members in Accounting, 30 in Marketing, 20 in Management, 15 in Finance, and 15 in Information Systems. The dean of the college asks her secretary to arbitrarily select 4, 6, 4, 3, and 3 persons from the respective departments to serve on a faculty committee. Is this a probability sample or a nonprobability sample? What specific type of probability or nonprobability sample does this procedure represent?

4.70 Researchers at a university with an enrollment of 12,000 take a census of the entire student population, asking, "Shoplifting is not only illegal, it raises the prices that we all have to pay. Have you ever shoplifted from the university bookstore?" Based on the preceding, discuss the possible presence of (a) sampling error and (b) nonsampling error.

4.71 To test the effectiveness of a new type of plastic handcuffs, a law enforcement agency puts the cuffs on a sample of 30 persons who belong to the local weightlifting club. None is able to break free from the cuffs. What type of sampling technique does this represent? Explain your reasoning.

4.72 A researcher wants to find out to what extent local residents participate in the recreation and entertainment activities available in their community. She selects a simple random sample of numbers from the telephone directory and proceeds to call these numbers between 6 P.M. and 9 P.M. If the line is busy or if there is no answer, she uses the next number on her list as a replacement. Comment on both the randomness and the representativeness of the sample of persons who end up being interviewed.

(DATA SET) *Note:* Exercises 4.73 and 4.74 require a computer and statistical software capable of selecting simple random samples.

4.73 Unknown to a quality assurance technician, the tensile strengths (in pounds per square inch, psi) for all 500 heavy-duty construction bolts in a recent shipment are as listed in file **XR04074**. Because a bolt must be broken to measure its strength, the testing process is destructive. The technician plans to collect a simple random sample of 20 bolts, then measure how much tension each one withstands before it breaks. Generate the simple random sample and compute its mean breaking strength. Assuming that the bolt manufacturer has advertised that, on average, such bolts will withstand 10,000 psi, refer to your sample result in commenting on the manufacturer's claim. (We will discuss the use of sample data in evaluating claims in much greater detail in Chapter 10, Hypothesis Testing.)

4.74 In 2006, the average U.S. production employee worked 38.9 hours per week. An assistant in the human resources department of Acme Eyebolts, Inc., is curious as to how the workers in her company compared to this figure during the first week of this month. The assistant will be reaching her conclusions based on a simple random sample of 30 data values out of the 300 Acme workers' weekly hours listed in file **XR04073**. Help the assistant by generating the simple random sample and computing its mean. Based on the sample, compare the Acme workweek with that of production workers nationally.
Source: *Statistical Abstract of the United States 2009*, p. 409.

INTEGRATED CASE

Thorndike Sports Equipment—Video Unit Two

The first Thorndike video segment takes us back in time, to when Ted got his business degree and first joined his grandfather's company. At the time of this video, Luke isn't much in need of data mining—to reach the things at the bottom of his document heaps, he needs good old-fashioned *shovel* mining.

To get a feel for the industry in which his grandfather operates, Ted wants to find out more about the sporting goods manufacturing business and collect some data on the competition that Thorndike Sports Equipment faces with some of its products, like racquetball and tennis racquets. Please give Ted some help by getting onto the Internet and providing him with the information requested here. Internet search engines and strategically selected key words will come in handy.

1. Approximately how much money do Americans spend on sports equipment annually? If possible, identify which types of sports equipment are the biggest sellers.
2. Who are some of the leading manufacturers of racquetball and tennis racquets? If possible, find out their relative importance in terms of manufacturing volume or retail sales of these products.
3. For two of the companies that manufacture equipment for racquet sports, visit the companies' websites and, if possible, read their most recent annual report. For each company, what have been the trends in overall company sales and profitability over the most recent years reported?
4. Select a company that makes racquetball racquets, then search message boards for comments that group participants may have made about the quality or reputation of racquets carrying that company's brand name.

SEEING STATISTICS: APPLET 3

Sampling

When this applet loads, it contains 100 blue circles. As you click on any 10 of them, their blue covers disappear and each circle changes to either red or green. After you have taken a sample of 10, click on **Show All** and the display at the top will show both the proportion of your sample that were red and the actual proportion of the population that were red. Of course, because of sampling error, it is very likely that your sample proportion will be different from the actual population value. Click the **Reset** button to return the circles to blue so you can try another sample.

Applet Exercises

3.1 Take 20 different samples and for each one, record both the sample proportion and the actual population proportion. How many times was your sample proportion less than the actual population value? Equal to the actual population value? Greater than the actual population value?

3.2 For each of the 20 different samples from Applet Exercise 3.1, calculate the difference between your sample proportion and the actual population proportion. (Remember to retain the positive or negative sign associated with each difference.) What was the average difference for the samples you took?

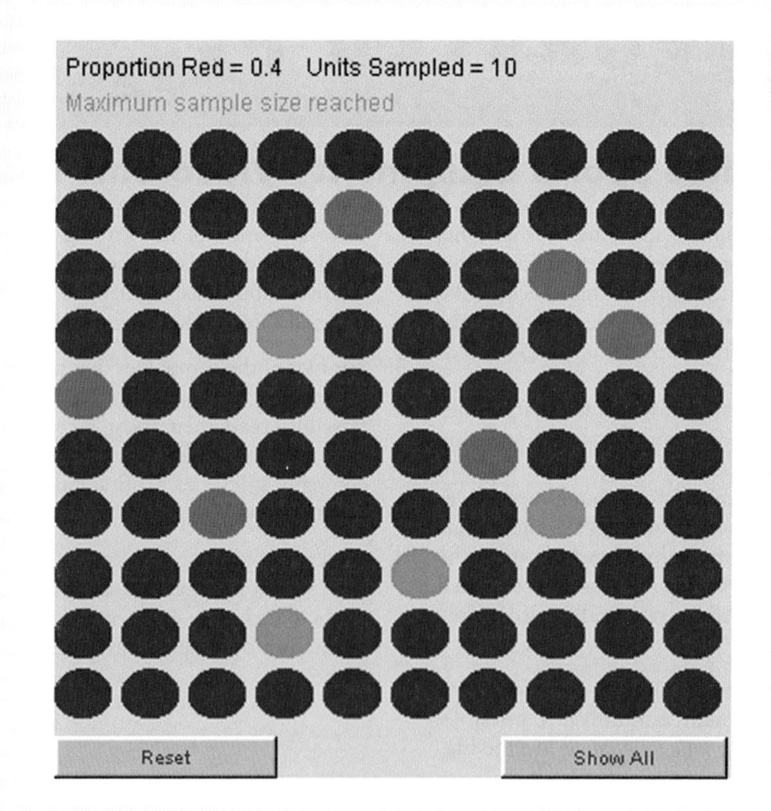

Try out the Applets: http://www.cengage.com/international

CHAPTER 5

Probability: Review of Basic Concepts

BUMPING UP YOUR CHANCES OF GETTING "BUMPED"

Among air travelers, it's common knowledge that airlines overbook flights to make up for no-shows. For the more clever air travelers, it becomes a challenge to get "bumped" from the flight in return for a free ticket, a voucher, or some other monetary goody. During 2007, more than 622,000 passengers agreed to give up their seats on overbooked flights.

There are fewer "no-shows" than in years past, and improved forecasting methods have helped the airline companies more closely predict how many people will need to be "bumped" from a given flight. However, to improve your chances of being bumped and rewarded with free travel or other compensation, experts advise to travel during peak travel periods, book flights that are already nearly full, and favor heavily booked flights on relatively small-bodied aircraft, such as the 727s and 737s. It also helps to arrive early, ask the agent if the flight is fully booked, and tell her you are willing to consider being bumped.

In 2007, about one of every 830 passengers was "bumped" from a flight. To boost your likelihood of becoming a "bumpee," follow these tips and pick up some more from the sources cited below.

Sources: Bestfares.com, July 26, 2006; BudgetTravel.com, July 26, 2006; bts.gov, June 28, 2009.

Facing a low-probability event

Colin Anderson/Blend Images/Getty Images

LEARNING OBJECTIVES

After reading this chapter, you should be able to:

- Understand the basic concept of probability, including the classical, relative frequency, and subjective approaches.
- Interpret a contingency table in terms of the probabilities associated with the two variables.
- Determine the probability that an event will occur.
- Construct and interpret a tree diagram with sequential events.
- Use Bayes' theorem and additional information to revise a probability.
- Determine the number of combinations and the number of possible permutations of *n* objects *r* at a time.

5.1 INTRODUCTION

Uncertainty plays an important role in our daily lives and activities as well as in business. In some cases, uncertainty even helps heighten our interest in these activities. For example, some sports fans refuse to watch the evening news whenever they're going to view a taped replay of a sports event that occurred earlier in the day—the replay is more exciting if they don't know who won the game.

In a business context, investment counselors cannot be sure which of two stocks will deliver the better growth over the coming year, and engineers try to reduce the likelihood that a machine will break down. Likewise, marketers may be uncertain as to the effectiveness of an ad campaign or the eventual success of a new product. The following situations provide other examples of the role of uncertainty in our lives:

- In the last game of the World Series, the home team is down one run with the bases loaded and two out in the bottom of the ninth. The pitcher, whose batting average is only 0.105, is due up. The manager sends in a pinch hitter with a 0.320 batting average to hit for the pitcher.
- Eddie is standing at an intersection, facing the "Don't Walk" light in a city where jaywalking carries a $20 fine. After looking around to be sure there are neither cars nor police, he crosses the street.
- Arthur is 75 years old, has had three heart attacks, and works as a bomb disposal expert with his local police department. When he applies for life insurance, he is turned down as an unacceptable risk.

In each of these cases, chance was involved. The baseball manager selected the batter who was more likely to get a hit. Eddie was trying to reduce the chance that something undesirable would happen to him. Arthur's life insurance company felt that his age, health, and occupation greatly reduced the likelihood of his surviving for many more years.

We'll soon offer several more precise definitions, but for now let's just consider probability as "the chance of something happening." Although humans have been dealing with probabilities ever since cave dwellers discovered their chances for survival would be better if they carried spears, only within the past couple of centuries have probabilities really been a focal point of mathematical attention. Gamblers were among the first to take an interest in probabilities, followed by the insurance industry and its mortality tables that expressed the likelihood that a person of a given age would live an additional year. More recently, probabilities have proven vital in many ways to both business and the social sciences.

The next section considers the three basic approaches to probability. This will be followed by a discussion of unions and intersections of events, concepts that are fundamental to describing situations of uncertainty. Sections 5.4 and 5.5 will examine the rules for calculating probability under different circumstances.

Sometimes, it is useful to *revise* a probability on the basis of additional information that we didn't have before. This will be the subject of Section 5.6. Finally, Section 5.7 will explain how permutations and combinations can help us calculate the number of ways in which things can happen.

PROBABILITY: TERMS AND APPROACHES (5.2)

Basic Terms

The discussion in this and later sections relies heavily on four important, interrelated definitions:

EXPERIMENT An activity or measurement that results in an outcome

SAMPLE SPACE All possible outcomes of an experiment

EVENT One or more of the possible outcomes of an experiment; a subset of the sample space

PROBABILITY A number between 0 and 1 that expresses the chance that an event will occur

The *event* in the preceding definitions could be almost anything of interest to the observer—e.g., rain tomorrow, a 30-point increase in the Dow Jones Industrial Average (DJIA) next week, or even that you'll take time out for a soft drink before finishing this page. As an example application, consider the possibility that the DJIA will increase by at least 30 points during the next week:

- **Experiment** Next week's trading activities on the New York Stock Exchange
- **Sample space** The sample space consists of two possible outcomes: (1) the DJIA goes up by at least 30 points, and (2) it does not.
- **Event** A = the DJIA goes up by at least 30 points next week.
- **Probability** The chance that the DJIA will increase by at least 30 points next week. Represented by $P(A)$, the probability for this and any other event will have the following properties:

$0 \leq P(A) \leq 1$	For any event, the probability will be no less than 0 and no greater than 1.
$P(A) + P(A') = 1$	Either the event will occur (A) or it will not occur (A'). A' is called the **complement** of A.

As indicated above, the probability for any event will be a number between 0 and 1. These extreme values represent events that are either impossible or certain. For example:

$P(B) = 0$ if B = you awake tomorrow morning and find yourself on Jupiter.

$P(C) = 1$ if C = it will rain someplace in the world tomorrow afternoon.

The Classical Approach

The classical approach describes a probability in terms of *the proportion of times that an event can be theoretically expected to occur:*

Classical probability:

For outcomes that are equally likely,

$$\text{Probability} = \frac{\text{Number of possible outcomes in which the event occurs}}{\text{Total number of possible outcomes}}$$

For example, in rolling a 6-sided die, the probability of getting a "three" is

$$\frac{1}{1+1+1+1+1+1}$$

Out of 6 possible (and equally likely) outcomes, there is one in which a "three" occurs.

Likewise, the probability of tossing a coin and having it land "heads" is $\frac{1}{2}$. In this case there are two possible (and equally likely) outcomes, only one of which includes the event "heads." Because we can determine the probability without actually engaging in any experiments, the classical approach is also referred to as the *a priori* ("before the fact") approach. Classical probabilities find their greatest application in games of chance, but they are not so useful when we are engaged in real-world activities where either (1) the possible outcomes are not equally likely or (2) the underlying processes are less well known.

The Relative Frequency Approach

In the relative frequency approach, probability is *the proportion of times an event is observed to occur in a very large number of trials:*

Relative frequency probability:

For a very large number of trials,

$$\text{Probability} = \frac{\text{Number of trials in which the event occurs}}{\text{Total number of trials}}$$

The relative frequency approach to probability depends on what is known as the **law of large numbers:** Over a large number of trials, the relative frequency with which an event occurs will approach the probability of its occurrence for a single trial. For example, if we toss a coin one time, the result will be either heads or tails. With *just one toss,* our relative frequency for heads will be either 0 or 1. If we toss it a great many times, however, the relative frequency will approach the 0.5 that it is theoretically "supposed to be."

On a more serious note, the life insurance industry relies heavily on relative frequencies in determining both candidate acceptability and the premiums that must be charged. Life insurance premiums are based largely on relative frequencies that reflect overall death rates for people of various ages. Table 5.1 is a portion of a mortality table of the type used by the life insurance industry. Of particular interest:

1. The death rate starts out relatively high for newborns, decreases slightly through the younger years, then steadily increases until (at the highest figure

Age	Deaths per 1000 Male	Deaths per 1000 Female	Age	Deaths per 1000 Male	Deaths per 1000 Female	Age	Deaths per 1000 Male	Deaths per 1000 Female
0	4.18	2.89	35	2.11	1.65	70	39.51	22.11
5	0.90	0.76	40	3.02	2.42	75	64.19	38.24
10	0.73	0.68	45	4.55	3.56	80	98.84	65.99
15	1.33	0.85	50	6.71	4.96	85	152.95	116.10
20	1.90	1.05	55	10.47	7.09	90	221.77	190.75
25	1.77	1.16	60	16.08	9.47	95	329.96	317.32
30	1.73	1.35	65	25.42	14.59			

Source: American Council of Life Insurance, *1999 Life Insurance Fact Book,* pp. 182–186.

TABLE 5.1

Life insurance companies use mortality tables to predict how many policy holders in a given age group will die each year. When the death rate is divided by 1000, the result is the probability that a typical policyholder in the age group will pass away during the year.

shown, 95 years) the likelihood of death within the coming year is about one chance out of three.

2. Each death rate can be converted to a probability by dividing by 1000. The resulting probability isn't necessarily the exact probability of death for a specific individual but rather the probability for a "typical" individual in the age group. For example, a typical male policyholder of age 95 would have a 329.96/1000, or a 0.32996, probability of dying within the coming year.
3. Examination of male versus female death rates shows the rate for males is higher *for every age group*. Especially noteworthy is the drastic increase in male death rates between the ages of 10 and 20, a period during which female death rates show a much smaller increase.

The Subjective Approach

The subjective approach to probability is judgmental, representing *the degree to which one happens to believe that an event will or will not happen.* **Subjective probabilities** can also be described as hunches or educated guesses. It has been estimated that wagers in 2000 amounted to $866 billion for all forms of gambling, ranging from casinos to bingo.[1]

Individuals place such bets only if they think they have some chance of winning. In a more negative direction, insurance customers paid $9.4 billion in fire insurance premiums during 2006.[2] They would not have paid such sums if they felt their buildings had a zero probability of catching fire.

As the terms "hunch" and "educated guess" imply, subjective probabilities are neither based on mathematical theory nor developed from numerical analyses of the frequencies of past events. This doesn't make them meaningless, however. For example, the following subjective probabilities might be considered reasonable with regard to your activities today:

Event	Subjective Probability
You will be mugged while reading this page.	Low (Prob. ≤ 0.001)
You'll talk to someone on the phone today.	Medium (Prob. ≥ 0.600)
Your shoes are on the right feet.	High (Prob. ≥ 0.999)

[1]Source: OnPointRadio.org, July 26, 2006.
[2]Source: Bureau of the Census, *Statistical Abstract of the United States 2009,* p. 739.

STATISTICS IN ACTION 5.1

Probabilities, Stolen Lawn Mowers, and the Chance of Rain

If you're a thief in need of money, one solution is to steal a lawn mower and sell it on e-Bay. That's exactly what a German perpetrator did in 2009. However, if you've just had your lawn mower stolen, you're going to be shopping for another one. That's exactly what the victim did. In a highly unlikely scenario, the two crossed paths when the victim perused e-Bay, went to examine the lawn mower he intended to purchase, and noticed it was the very same one he used to own. He called the police, got his mower back, and the thief was promptly arrested. The probability of such a rare turn of events is about as tall as grass in the Sahara.

Although precipitation probabilities have been used since the 1960s, people still don't quite get it when the TV meteorologist says "30% chance of rain today." Some think it will rain all day in just 30% of the local area, while others believe it will rain for 30% of the day. Susan Joslyn, of the Psychology Department at the University of Washington in Seattle, found so much misunderstanding in her sampling of 450 undergraduates in the rainy Pacific Northwest that she fears confusion would be even greater in other parts of the country. In terms of probability, what's the meteorologist really saying? The answer: If there is a 30% chance of rain today, there is a 30% probability that it will rain somewhere within the local area sometime during the day. Keep the umbrella handy.

Sources: "Thief Nabbed Selling Stolen Mower to Owner," reuters.com, June 25, 2009; Doyle Rice, "With a Chance of Misapprehension," *USA Today*, June 24, 2009, p. 7D.

As intelligent human beings, we are the sum total of our experiences and memories, and these enable us to make intuitive estimates of the probabilities of events with which we are somewhat familiar. Such probabilities, though subjective, can be important in drawing conclusions and making decisions. Although you may not be able to precisely quantify the probability that the fast-talking salesperson in the plaid suit is really offering you a sound used car, this shouldn't prevent you from assigning a low probability to such an event. Statistics in Action 5.1 provides two interesting examples of subjective probabilities—one slim, the other often misunderstood.

Probabilities and "Odds"

The term **odds** is sometimes used as a way of expressing the likelihood that something will happen. When someone says "the odds for this outcome are three to one," she is really saying that the chance of the event occurring is three times the chance that it will *not* occur. In this case, the odds might be expressed as "three to one" or "3:1." When you see odds listed in reference to an event, you can convert them to a probability as follows:

Conversion from odds to probability, and vice versa:

1. **If the odds in favor of an event happening are *A* to *B*, or *A*:*B*, the probability being expressed is**

$$\frac{A}{A+B}$$

2. **If the probability of an event is x (with $0 \leq x \leq 1$), the odds in favor of the event are "x to $(1 - x)$," or "x:$(1 - x)$."**

Odds are typically expressed in terms of the lowest applicable integers. For example, "1.5 to 4" would become "3 to 8." Also, the odds *against* the occurrence of an event are the reverse of the odds in favor. For example, if the odds in favor of an event are 2 to 3, the odds against would be 3 to 2.

Like the other probabilities we've discussed, odds can be classical, based on relative frequencies, or even subjective. For example:

- **Classical** In flipping a fair coin, the probability of heads is 0.5. The resulting odds are "0.5 to 0.5," or "1 to 1."
- **Relative frequency** For example, 16.1% of U.S. households have a cell phone with Internet access, a probability of 0.161 for a randomly selected household.[3] If the event is "has cell phone with Internet access," the odds in favor of the household having this device would be 0.161 to 0.839, or approximately 1 to 5.
- **Subjective** In speculation about the upcoming 2000 Major League Baseball season, an oddsmaker assigned odds that each team would end up winning the World Series. The New York Yankees were given odds of 1 to 3 of winning, which translates into a subjective probability of $1/(1 + 3)$, or 0.25. The odds given the Milwaukee Brewers were much less favorable—just 1 to 10,000,000.[4]

1. As discussed earlier, when presenting the odds *against* an event happening, the odds are listed in the opposite order. For example, the published odds *against* the Yankees were "3:1." For purposes of consistency with our previous examples, we've converted the published version to the "odds for" order.
2. In sports events, points are often added to or subtracted from the score of one of the teams (creating a "point spread") in an attempt to convert the odds to 1:1.

EXERCISES

5.1 The president of a computer manufacturing firm states that there is a 70% chance that industry shipments of notebook computers will double in the next 5 years. Is this a classical probability, a relative frequency probability, or a subjective probability?

5.2 If A = "Air France will begin making daily flights to Pocatello, Idaho, next year," identify the sample space and possible events that could occur. Use your judgment in arriving at an approximate probability for the event(s) and complement(s) you've identified.

5.3 If B = "IBM's net profit will increase next year," identify the sample space and possible events that could occur. Use your judgment in arriving at an approximate probability for the event(s) and complement(s) you've identified.

5.4 Electric meters on private homes in a community register the cumulative consumption of kilowatt-hours with a five-digit reading. If a home in the community were randomly selected, it's possible that the right-hand digit (which spins the fastest) would be a "7." Using the concept of classical probability, what is the probability that the digit is a "7"?

5.5 Regarding Exercise 5.4, how might an observer employ the law of large numbers to verify the classical probability that you identified?

5.6 If a die is rolled one time, classical probability would indicate that the probability of a "two" should be $\frac{1}{6}$. If the die is rolled 60 times and comes up "two" only 9 times, does this suggest that the die is "loaded"? Why or why not?

5.7 A newspaper article reported that Massachusetts had become the second state (after Montana) to prohibit sex to be used as a factor in determining insurance premiums. Considering the information in Table 5.1, would such unisex insurance tend to increase or decrease the life insurance premiums previously paid (a) by men? (b) by women? Source: John R. Dorfman, "Proposals for Equal Insurance Fees for Men and Women Spark Battle," *The Wall Street Journal*, August 27, 1987, p. 23.

5.8 It has been reported that about 35% of U.S. adults attended a sports event during the previous year. What are the odds that a randomly selected U.S. adult attended a sports event during the year? Source: Bureau of the Census, *Statistical Abstract of the United States 2009*, p. 747.

5.9 If the odds are 4:7 that an event will occur, what is the corresponding probability?

[3]Source: Adrienne Lewis, "Early Adopter Technologies," *USA Today*, October 17, 2005, p. 2B.
[4]Source: "World Series Odds," *USA Today*, March 14, 2000, p. 11C.

5.3 UNIONS AND INTERSECTIONS OF EVENTS

As discussed previously, an event is one or more of the possible outcomes in an experiment, and the sample space is the entire set of possible outcomes. Sample spaces and events may be portrayed in either a table or a visual display, and these representations can contribute to our discussion of the probability concepts in sections to follow.

The tabular method involves the contingency table, or cross-tabulation, introduced in Chapter 2. This shows either frequencies or relative frequencies for two variables at the same time. The visual mechanism, called a **Venn diagram,** simultaneously displays both the sample space and the possible events within. We will use the contingency table and the Venn diagram in demonstrating several key terms that are important to the consideration of probability:

MUTUALLY EXCLUSIVE EVENTS If one event occurs, the other cannot occur. An event (e.g., *A*) and its complement (*A′*) are always mutually exclusive.

EXHAUSTIVE EVENTS A set of events is *exhaustive* if it includes all the possible outcomes of an experiment. The mutually exclusive events *A* and *A′* are exhaustive because one of them *must* occur. When the events within a set are both mutually exclusive and exhaustive, the sum of their probabilities is 1.0. This is because one of them *must* happen, and they include *all of the possibilities.* The entries within a relative frequency contingency table are mutually exclusive and exhaustive, and their sum will always be 1.0.

INTERSECTION OF EVENTS Two or more events occur at the same time. Such an intersection can be represented by "*A* and *B*," or "*A* and *B* and *C*," depending on the number of possible events involved.

UNION OF EVENTS At least one of a number of possible events occurs. A union is represented by "*A* or *B*," or "*A* or *B* or *C*," depending on the number of events.

The contingency tables in parts (1) and (2) of Table 5.2 show frequencies and relative frequencies describing the sex and age of persons injured by fireworks in 1995. The relative frequencies in part (2) are obtained by dividing the corresponding entries in part (1) by 11,449, the total number of persons injured by fireworks during that year. For example, part (1) shows that 3477 of those injured were males under the age of 15. The relative frequency for persons in this sex/age category is 3477/11,449, or 0.304, as shown in the contingency table in part (2).

Referring to the frequencies in part (1) of Table 5.2, we can identify several that illustrate the terms defined earlier in the section:

- **Mutually exclusive events** The victims were either male (event A, 8913 persons) or female (event A', 2536 persons).
- **Exhaustive events** The four mutually exclusive events (A and B), (A and B'), (A' and B), and (A' and B') are also exhaustive since a victim *must* be in one of these sex/age categories.
- **Intersection of events** There were 3477 victims who were males under the age of 15, the "A and B" category.
- **Union of events** There were 10,162 victims who were either male, or under age 15, or both, the "A or B" category. This category includes everyone except females who were 15 or older and its frequency is calculated as $5436 + 3477 + 1249 = 10{,}162$.

TABLE 5.2

Frequency and relative frequency contingency tables for fireworks victims.

(1) *Frequencies*

			Age		
			B Under 15	B' 15 or Older	
Sex	A	Male	3477	5436	8913
	A'	Female	1249	1287	2536
			4726	6723	11,449

(2) *Relative Frequencies*

			Age		
			B Under 15	B' 15 or Older	
Sex	A	Male	0.304	0.475	0.779
	A'	Female	0.109	0.112	0.221
			0.413	0.587	1.000

Source: U.S. Consumer Product Safety Commission, National Electronic Injury Surveillance System.

Figure 5.1 presents a series of Venn diagrams that also describe the age/sex breakdown of the fireworks victims. The information is the same as that contained in the contingency table of Table 5.2, part (1), but in a more visual form. Examples of calculations for finding the number of persons in the union or the intersection of events are shown in the figure, and the results are identical to those based on the contingency table of frequencies.

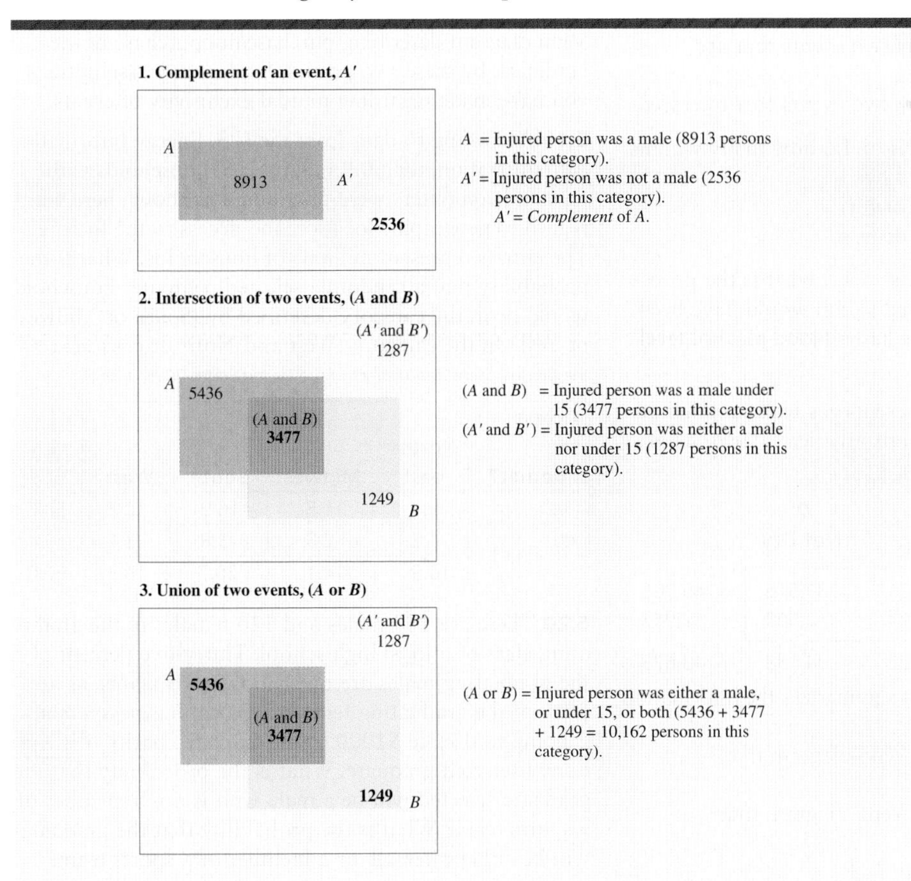

FIGURE 5.1

Venn diagrams describing persons injured by fireworks.

EXERCISES

5.10 A sample space includes the events A, B, and C. Draw a Venn diagram in which the three events are mutually exclusive.

5.11 A sample space includes the events A, B, and C. Draw a Venn diagram in which events A and B are mutually exclusive but events B and C are not mutually exclusive.

5.12 The following contingency table of frequencies is based on a 5-year study of fire fatalities in Maryland. For purposes of clarity, columns and rows are identified by the letters A–C and D–G, respectively. Source: National Fire Protection Association, *The 1984 Fire Almanac*, p. 151.

		Blood Alcohol Level of Victim			
		A	B	C	
	Age	0.00%	0.01–0.09%	≥0.10%	
D	0–19	142	7	6	155
E	20–39	47	8	41	96
F	40–59	29	8	77	114
G	60 or over	47	7	35	89
		265	30	159	454

a. For this table, identify any two events that are mutually exclusive.
b. For this table, identify any two events that intersect.

5.13 Using the table in Exercise 5.12, how many victims were in the category described by:
a. (A and A')? b. (C or F)?
c. (A' and G')? d. (B or G')?

5.14 Using the table in Exercise 5.12, what is the probability that a randomly selected victim would have been at least 60 years old and have had a blood alcohol level of at least 0.10%?

5.15 The following table represents gas well completions during 1986 in North and South America. Source: American Gas Association, *1987 Gas Facts*, p. 50.

		D Dry	*D'* Not Dry	
N	North America	14,131	31,575	45,706
N'	South America	404	2,563	2,967
		14,535	34,138	48,673

a. Draw a Venn diagram that summarizes the information in the table.
b. Identify the region of the Venn diagram that represents (N and D).
c. Identify the region of the Venn diagram that represents (N' and D').
d. Identify the region of the Venn diagram that represents (N or D).
e. Identify the region of the Venn diagram that represents (N' or D').

5.16 Using the table in Exercise 5.15, assume that one well has been selected at random from the 48,673.
a. What is the probability that the well was drilled in North America and was dry?
b. What is the probability that the well was drilled in South America and was not dry?

5.17 The owner of a McDonald's restaurant in France is considering the possibility of opening up a new franchise at the other end of her town. At the same time, the manager of the Pennsylvania Turnpike is deciding whether to recommend raising the speed limit by 10 miles per hour. Draw a single Venn diagram representing their possible decisions. Explain any assumptions you have made regarding mutual exclusivity of events.

5.18 A shopping mall developer and a historical society are the only two bidders for a local historical landmark. The sealed bids are to be opened at next week's city council meeting and the winner announced. Draw a single Venn diagram describing purchase/nonpurchase of the landmark by these two parties. Explain any assumptions you have made regarding mutual exclusivity of events.

5.19 According to data from the U.S. Energy Information Administration, the 60.0 million U.S. households with personal computers were distributed as shown here with regard to geographic location and access to the Internet. The entries represent millions of households. What is the probability that a randomly selected computer household would be in the category described by (South or Midwest or Yes)? In the category (West and No)? Source: Bureau of the Census, *Statistical Abstract of the United States 2006*, p. 638.

Access to Internet?	Region Northeast	Midwest	South	West	
Yes	9.7	11.8	16.9	12.2	50.6
No	1.2	2.3	3.8	2.1	9.4
	10.9	14.1	20.7	14.3	60.0

5.20 There are 100 males and 120 females in the graduating class of a local high school. Thirty-five percent of the graduating males are on school sports teams, as are 30% of the graduating females. A local businessperson is going to donate $1000 to the favorite charity of a randomly selected graduate. What is the probability that the graduate selected will be a male who is not a member of a sports team? What is the probability that the graduate will be either a female or a member of a sports team?

ADDITION RULES FOR PROBABILITY

There are occasions where we wish to determine the probability that one or more of several events will occur in an experiment. The determination of such probabilities involves the use of **addition rules,** and the choice of a given rule will depend on whether the events are mutually exclusive. To illustrate these rules, we will use the two Venn diagrams in Figure 5.2.

When Events Are Mutually Exclusive

When events are *mutually exclusive,* the occurrence of one means that none of the others can occur. In this case, the probability that one of the events will occur is the sum of their individual probabilities. In terms of two events, the rule can be stated as follows:

Rule of addition when events are mutually exclusive:

$$P(A \text{ or } B) = P(A) + P(B)$$

Part (1) of Figure 5.2 is a relative frequency breakdown of colors of sport/compact cars made in North America during the 2004 model year. As the Venn diagram indicates, the possible events are mutually exclusive. For a given vehicle, only one color was applied. Using the letters A and B to identify the colors silver

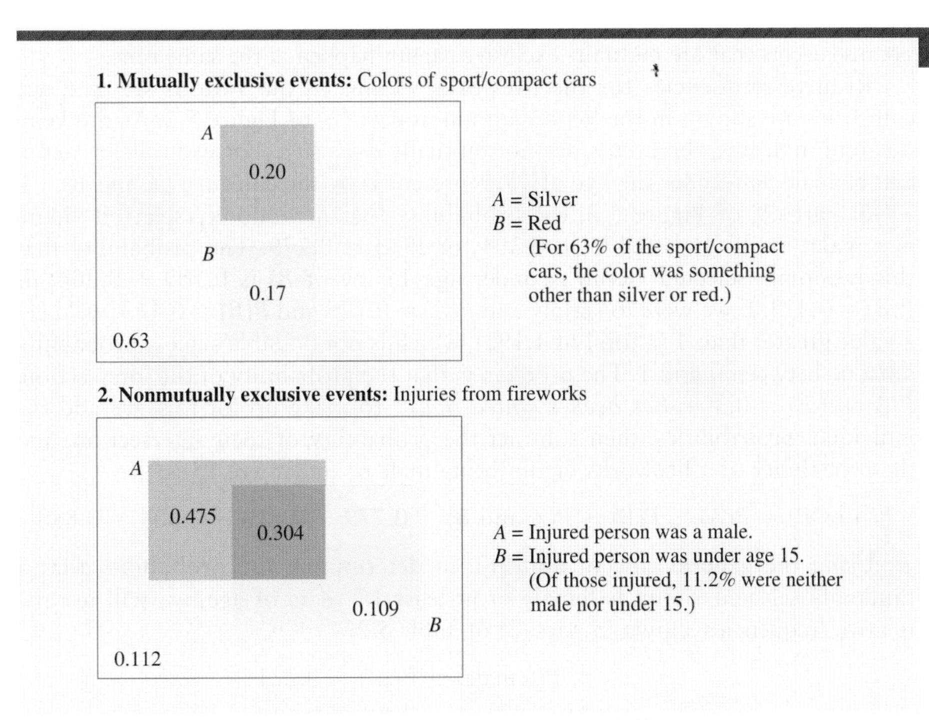

FIGURE 5.2 Venn diagrams showing relative frequencies for (1) colors of sport/compact cars made in North America during the 2004 model year, and (2) persons injured by fireworks.

Sources: Part (1) *Time Almanac 2006*, p. 609. Part (2) U.S. Consumer Product Safety Commission, National Electronic Injury Surveillance System.

and red, respectively, and considering a sport/compact car selected at random from that model year, we can arrive at the following probabilities:

- Probability that the car is either silver or red:

$$P(A \text{ or } B) = P(A) + P(B) = 0.20 + 0.17 = 0.37$$

- Probability that the car is either silver or some color other than silver or red:

  Although we have not specifically identified "neither silver nor red" with a letter designation, it can be expressed as (A' and B'). Using this identification, we can determine the probability that the randomly selected car is either silver or some color other than silver or red:

$$P(A \text{ or } [A' \text{ and } B']) = P(A) + P(A' \text{ and } B') = 0.20 + 0.63 = 0.83$$

When Events Are Not Mutually Exclusive

When events are *not mutually exclusive,* two or more of them can happen at the same time. In this case, the *general rule of addition* can be used in calculating probabilities. In terms of two events, it can be stated as follows:

General rule of addition when events are not mutually exclusive:

$$P(A \text{ or } B) = P(A) + P(B) - P(A \text{ and } B)$$

As befits its name, the general rule of addition can also be applied when the events *are* mutually exclusive. In this case, the final term in the expression will become zero because events that are mutually exclusive cannot happen at the same time.

Relative frequencies for the fireworks victims in the various sex and age categories are shown in the Venn diagram in part (2) of Figure 5.2. As the Venn diagram indicates, the events are not mutually exclusive. For example, a victim can be male *and* under the age of 15, represented by the category (A and B).

In part (2) of Figure 5.2, the probability that a randomly selected victim is a male (event A) is 0.475 + 0.304, or $P(A) = 0.779$. The probability that this randomly selected victim is under age 15 (event B) is 0.109 + 0.304, or $P(B) = 0.413$. If we were to simply add $P(A) = 0.779$ and $P(B) = 0.413$, we'd get a value greater than 1 (a total of 1.192), which is not possible since a probability must be between 0 and 1. The problem is that the probability of the intersection, $P(A \text{ and } B) = 0.304$, *has been counted twice.* To make up for this, we add the individual probabilities, then subtract the probability of their intersection, and the probability of a fireworks victim being male or under age 15 is:

$$P(A \text{ or } B) = P(A) + P(B) - P(A \text{ and } B) = 0.779 + 0.413 - 0.304 = 0.888$$

Using the general rule of addition in determining the probability that a randomly selected victim is female or at least 15 years of age, we will use the relative frequencies shown in part (2) of Table 5.2:

$$P(\text{female}) = P(A') = 0.221$$

$$P(15 \text{ or older}) = P(B') = 0.587$$

$$P(\text{female and } [15 \text{ or older}]) = P(A' \text{ and } B') = 0.112$$

and

$$P(A' \text{ or } B') = P(A') + P(B') - P(A' \text{ and } B')$$
$$= 0.221 + 0.587 - 0.112 = 0.696$$

EXERCISES

5.21 A financial advisor frequently holds investment counseling workshops for persons who have responded to his direct mailings. The typical workshop has 10 attendees. In the past, the advisor has found that in 35% of the workshops, nobody signs up for the advanced class that is offered; in 30% of the workshops, one person signs up; in 25% of the workshops, two people sign up; and in 10% of the workshops, three or more people sign up. The advisor is holding a workshop tomorrow. What is the probability that at least two people will sign up for the advanced class? What is the probability that no more than one person will sign up? Draw a Venn diagram that includes the possible events and their relative frequencies for this situation.

5.22 A survey of employees at a large company found the following relative frequencies for the one-way distances they had to travel to arrive at work:

Number of Miles (One-Way)

	A ≤5	B 6–10	C 11–15	D 16–20	E 21–30	F ≥31
Relative Frequency	0.38	0.25	0.16	0.09	0.07	0.05

a. What is the probability that a randomly selected individual will have to travel 11 or more miles to work?
b. What is the probability that a randomly selected individual will have to travel between 6 and 15 miles to work?
c. Draw a Venn diagram that includes the relative frequency probabilities in the table.
d. Using the letter identifications provided, calculate the following probabilities: $P(A \text{ or } B \text{ or } E)$; $P(A \text{ or } F)$; $P(A' \text{ or } B)$; $P(A \text{ or } B \text{ or } C')$.

5.23 In 2008, McDonald's had 31,967 restaurants systemwide. Of these, 21,328 were operated by franchisees, 6502 by the company, and 4137 by affiliates. What is the probability that a randomly selected McDonald's restaurant is operated by either a franchisee or an affiliate? Source: McDonald's Corporation, *2008 Annual Report*, p. 44.

5.24 For three mutually exclusive events, $P(A) = 0.3$, $P(B) = 0.6$, and $P(A \text{ or } B \text{ or } C) = 1.0$. What is the value of $P(A \text{ or } C)$?

5.25 It has been reported that the 49,600 employees of United Airlines are distributed among the following corporate functions: Source: United Airlines, *Fact Sheet 2008*.

Pilots	13%	(A)
Management	17	(B)
Flight attendants	27	(C)
Mechanics	11	(D)
Other	32	(E)
	100%	

For the sample space consisting of United employees:
a. Draw a Venn diagram representing events A, B, C, D, and E.
b. What is the value of $P(A)$?
c. What is the value of $P(A \text{ or } B)$?
d. What is the value of $P(A \text{ or } D')$?

5.26 In 2003, Entergy Corporation had 2,631,752 electricity customers. Of these, 86.0% were in the residential category (R); 11.8% were commercial (C); 1.6% were industrial (I); and 0.6% were government and municipal (G). Source: Entergy Corporation, *2003 Annual Report*.
a. Draw a Venn diagram representing events R, C, I, and G.
b. What is the value of $P(R)$?
c. What is the value of $P(C \text{ or } G)$?
d. What is the value of $P(R \text{ or } C')$?

5.27 According to the Bureau of Labor Statistics, there are 777,000 men and 588,000 women in the life, physical, and social science occupations category; and 786,000 men and 204,000 women in the farming, fishing, and forestry occupations category. For these 2,355,000 individuals, what is the probability that a randomly selected individual will either be a male or be in the life, physical, and social sciences category? Source: *Encyclopaedia Britannica Almanac 2006*, p. 835.

5.28 Using the information presented in the table in Exercise 5.12, calculate the following probabilities:
a. $P(A \text{ or } D)$ b. $P(B \text{ or } F)$
c. $P(C \text{ or } G)$ d. $P(B \text{ or } C \text{ or } G)$

5.29 Using the information presented in the table in Exercise 5.15, calculate the following probabilities:
a. $P(D \text{ or } N)$ b. $P(D' \text{ or } N')$
c. $P(D \text{ or } N')$ d. $P(D' \text{ or } N)$

5.5 MULTIPLICATION RULES FOR PROBABILITY

While addition rules are used to calculate the probability that *at least one* of several events will occur, in this section we'll consider rules for determining the probability that two or more of the events will *all* occur. These rules involve several important terms:

MARGINAL PROBABILITY The probability that a given event will occur. No other events are taken into consideration. A typical expression is $P(A)$.

JOINT PROBABILITY The probability that two or more events will all occur. A typical expression is $P(A \text{ and } B)$.

CONDITIONAL PROBABILITY The probability that an event will occur, given that another event has already happened. A typical expression is $P(A|B)$, with the verbal description, "the probability of A, given B." A conditional probability may be determined as follows:

Conditional probability of event A, given that event B has occurred:

$$P(A|B) = \frac{P(A \text{ and } B)}{P(B)} \quad (\textit{Note: } \text{This applies only if } P(B) > 0.)$$

The rules that follow are called **multiplication rules;** they determine the probability that two events will *both* happen or that three or more events will *all* happen. There are two multiplication rules, and the one that is applicable will depend on whether the events are *independent* or *dependent:*

INDEPENDENT EVENTS Events are independent when the occurrence of one event has no effect on the probability that another will occur.

DEPENDENT EVENTS Events are dependent when the occurrence of one event changes the probability that another will occur.

When Events Are Independent

When events are independent, their joint probability is the product of their individual probabilities. In the case of two events, the multiplication rule is as follows:

Multiplication rule when events are independent:

$$P(A \text{ and } B) = P(A) \times P(B)$$

To illustrate this rule, we'll use a device called a **tree diagram,** which visually summarizes the occurrence or nonoccurrence of two or more events. The tree diagram is especially useful in visualizing events that occur in sequence. Figure 5.3 illustrates two consecutive tosses of a fair coin. Regardless of the result of the first toss, the probability of heads on the next toss is going to be 0.5; the tree diagram in Figure 5.3 shows all the possibilities, along with their joint probabilities.

For example, the probability of getting heads twice in a row is 0.5×0.5, or 0.25. Likewise, the joint probability of any other given sequence (e.g., $H1 - T2$, $T1 - T2$, or $T1 - H2$) will also be 0.25. Since the diagram includes all four of the possible sequences, the joint probabilities add up to 1.00.

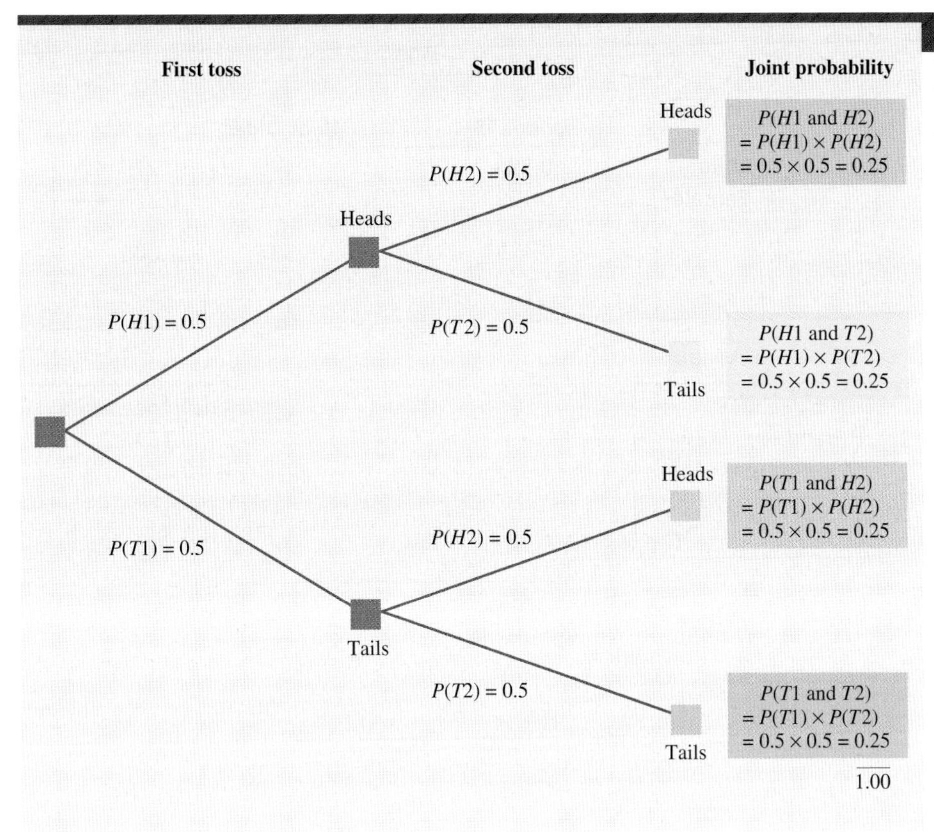

FIGURE 5.3

This tree diagram illustrates two coin tosses, a process involving independent events. The probability of heads on one toss has no effect on the probability of obtaining heads on the next.

When Events Are Not Independent

When events are *not* independent, the occurrence of one will influence the probability that another will take place. Under these conditions, a more general multiplication rule applies:

Multiplication rule when events are not independent:

$$P(A \text{ and } B) = P(A) \times P(B \mid A)$$

Table 5.3 shows marginal, joint, and conditional probabilities for the fireworks victims. It indicates that a fireworks victim is more likely to be male than female. The respective marginal probabilities are $P(A) = 0.779$ and $P(A') = 0.221$. Of the victims, 30.4% were males under the age of 15, and the joint probability for this category is $P(A \text{ and } B) = 0.304$.

The contingency table can be complemented by a tree diagram like the one in Figure 5.4. Using the tree diagram, we can visually describe the two variables and probabilities related to them. The tree diagram in the figure includes the marginal, joint, and conditional probabilities shown in Table 5.3.

The first set of branches in the tree diagram shows the marginal probabilities. These probabilities are based on relative frequencies and have the following values:

$$\text{Probability that the victim is a male} = P(A) = 0.779$$
$$\text{Probability that the victim is a female} = P(A') = 0.221$$

The second set of branches in the tree diagram includes conditional probabilities dependent on the result of each initial branch. For example, given that

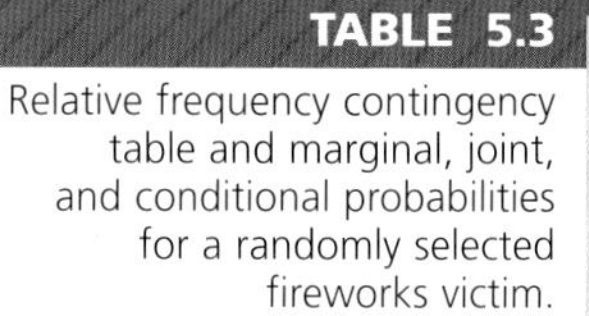

TABLE 5.3

Relative frequency contingency table and marginal, joint, and conditional probabilities for a randomly selected fireworks victim.

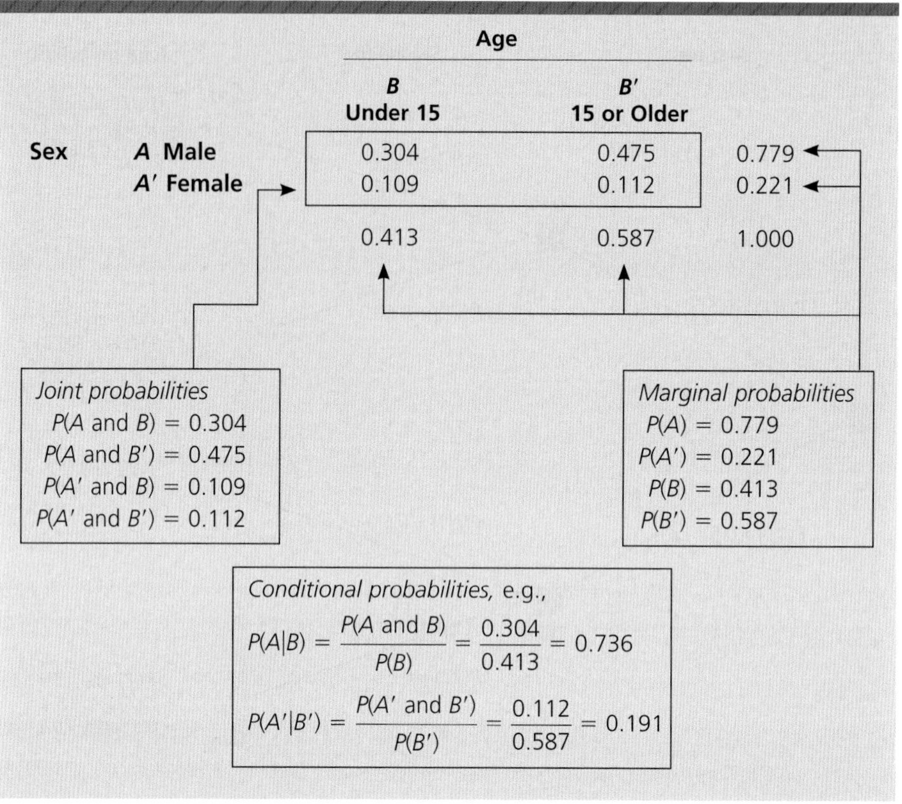

Sex		Age: *B* Under 15	Age: *B′* 15 or Older	
	***A* Male**	0.304	0.475	0.779
	***A′* Female**	0.109	0.112	0.221
		0.413	0.587	1.000

Joint probabilities

$P(A \text{ and } B) = 0.304$

$P(A \text{ and } B') = 0.475$

$P(A' \text{ and } B) = 0.109$

$P(A' \text{ and } B') = 0.112$

Marginal probabilities

$P(A) = 0.779$

$P(A') = 0.221$

$P(B) = 0.413$

$P(B') = 0.587$

Conditional probabilities, e.g.,

$$P(A|B) = \frac{P(A \text{ and } B)}{P(B)} = \frac{0.304}{0.413} = 0.736$$

$$P(A'|B') = \frac{P(A' \text{ and } B')}{P(B')} = \frac{0.112}{0.587} = 0.191$$

FIGURE 5.4

This tree diagram illustrates the sex and age of fireworks accident victims. The events are not independent, so the probabilities in the second set of branches depend on what has happened in the first set.

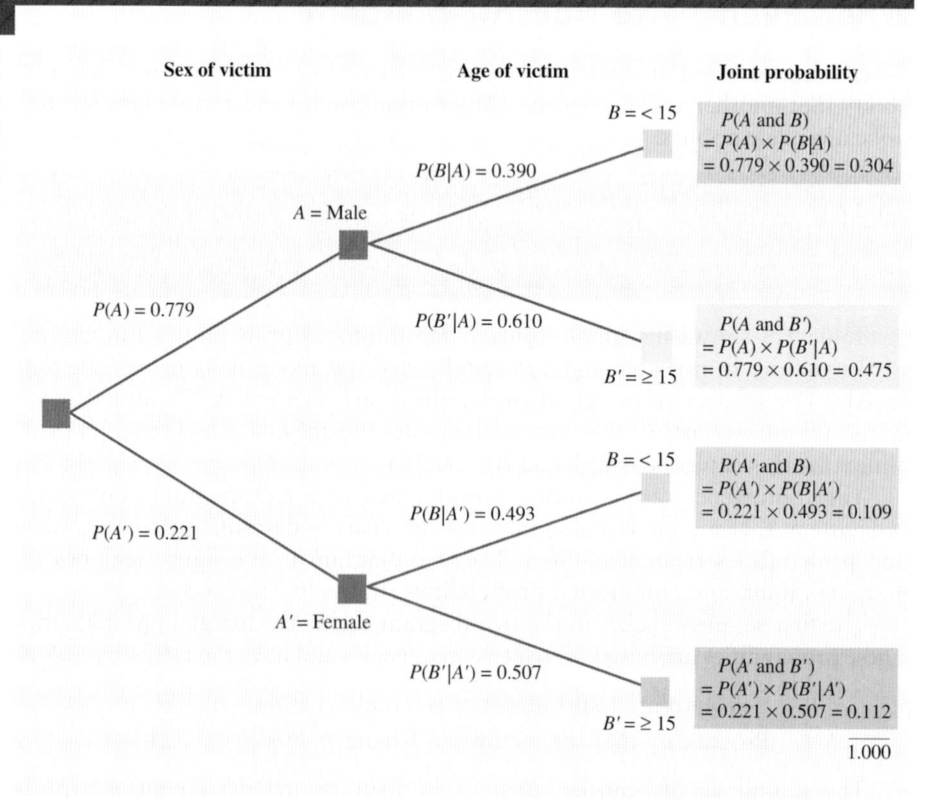

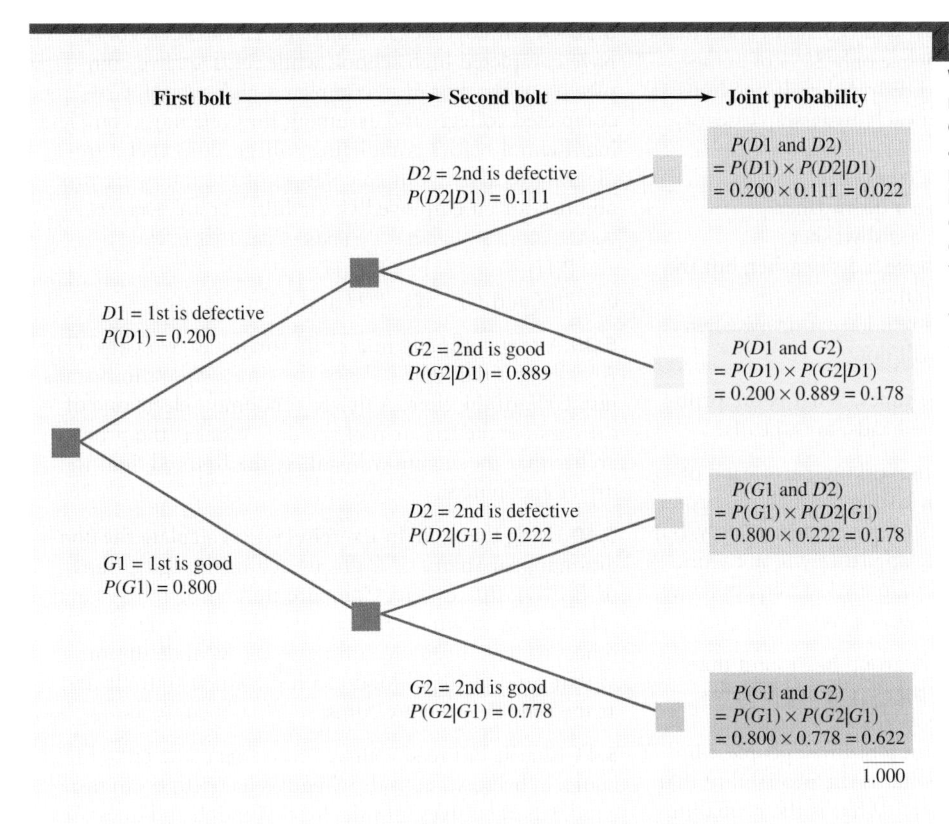

FIGURE 5.5

When sampling without replacement from a bin containing 8 good bolts and 2 defective bolts, the probability that the second bolt is defective will depend on whether or not the first one selected was defective. If the first bolt is defective, the probability that the second one will be defective is $P(D2|D1) = 0.111$.

the victim is a male, the conditional probability that he is under 15 years old is

$$P(B|A) = \frac{P(B \text{ and } A)}{P(A)} = \frac{0.304}{0.779} = 0.390$$

The other conditional probabilities are calculated in a similar manner; then the four different joint probabilities shown in the figure can also be computed. As in Figure 5.3, these are the only possibilities that can occur, so the sum of the joint probabilities is 1.

Events can be dependent when we *sample without replacement*. Figure 5.5 represents the selection of two bolts, without replacement, from a bin of 10 bolts in which 2 are defective. The probability that the first bolt is defective will be $P(D1) = \frac{2}{10}$, or 0.200. If the first bolt *is* defective, then the conditional probability that the second one will be defective is $P(D2|D1) = \frac{1}{9}$, or 0.111. This is because there will be just one defective in the 9 bolts that remain. If the first bolt is *not* defective, the probability that the second one will be defective is $P(D2|G1) = \frac{2}{9}$, or 0.222.

EXERCISES

5.30 What is the difference between a marginal probability and a joint probability?

5.31 It is possible to have a sample space in which $P(A) = 0.7$, $P(B) = 0.6$, and $P(A \text{ and } B) = 0.35$. Given this information, would events A and B be mutually exclusive? Would they be independent?

5.32 If events A and B are independent, will $P(A|B)$ be greater than, less than, or equal to $P(A)$? Explain.

5.33 It has been reported that 57% of U.S. households that rent do not have a dishwasher, while only 28% of homeowner households do not have a dishwasher. If one household is randomly selected from each ownership category, determine the probability that Source: Bureau of the Census, *Statistical Abstract of the United States 2009*, p. 601.

a. neither household will have a dishwasher.
b. both households will have a dishwasher.
c. the renter household will have a dishwasher, but the homeowner household will not.
d. the homeowner household will have a dishwasher, but the renter household will not.

5.34 A fair coin is tossed three times. What is the probability that the sequence will be heads, tails, heads?

5.35 A kitchen appliance has 16 working parts, each of which has a 0.99 probability of lasting through the product's warranty period. The parts operate independently, but if one or more malfunctions, the appliance will not work. What is the probability that a randomly selected appliance will work satisfactorily throughout the warranty period?

5.36 An optometry practitioner group has found that half of those who need vision correction are patients who require bifocal lenses.

a. For a randomly selected group of three people who require vision correction, what is the probability that all three will require bifocals? What is the probability that none of the three will require bifocals?
b. If the three individuals had all been selected from an organization consisting of retirees over age 65, do you believe the joint probabilities calculated in part (a) would still be accurate? If not, why not?

5.37 Through April 30 of the 2009 filing season, 15.8% of all individual U.S. tax returns were prepared by H&R Block. Source: H&R Block, Inc., *Fast Facts 2009*.

a. If two individuals are randomly selected from those filing tax returns during this period, what is the probability that both of their tax returns were prepared by H&R Block?
b. In part (a), what is the probability that neither return was prepared by H&R Block?
c. In part (a), what is the probability that exactly one of the two returns was prepared by H&R Block?

5.38 Of employed U.S. adults age 25 or older, 90.4% have completed high school, while 34.0% have completed college. For H = completed high school, C = completed college, and assuming that one must complete high school before completing college, construct a tree diagram to assist your calculation of the following probabilities for an employed U.S. adult: Source: Bureau of the Census, *Statistical Abstract of the United States 2009*, p. 146.

a. $P(H)$
b. $P(C|H)$
c. $P(H \text{ and } C)$
d. $P(H \text{ and } C')$

5.39 A taxi company in a small town has two cabs. Cab A stalls at a red light 25% of the time, while cab B stalls just 10% of the time. A driver randomly selects one of the cars for the first trip of the day. What is the probability that the engine will stall at the first red light the driver encounters?

5.40 Using the table in Exercise 5.12, calculate the conditional probability of C given each of the age groups, or $P(C|D)$, $P(C|E)$, etc. Compare these probabilities and speculate as to which age groups seem more likely than others to have been (according to the legal definition at that time, 0.10% blood alcohol content) intoxicated at the time they were victims.

5.41 Charlie has read a survey result that says 60% of the adults in his town consider Wendy's hamburgers to taste good. Charlie drives into the local Wendy's and questions a young couple about to enter the restaurant. According to Charlie, there's only a 0.36 (i.e., 0.6×0.6) probability that both of these people will say Wendy's hamburgers taste good. Do you think Charlie is correct? Explain.

5.42 Based on the information in Exercise 5.20, if the student chosen is known to be on a sports team, what is the probability that the student is a female?

5.43 Based on the information in Exercise 5.21, if the advisor has at least one person sign up for the advanced class, what is the probability that at least three people have signed up?

5.44 Based on the information in Exercise 5.22, if a person is known to travel a one-way distance of at least 11 miles to work, determine the probability that he or she drives at least 31 miles to work.

5.6 BAYES' THEOREM AND THE REVISION OF PROBABILITIES

In the 1700s, Thomas Bayes developed a theorem that is an extension of the concept of conditional probability discussed in the previous section. In Bayes' application of conditional probability, the emphasis is on sequential events; in particular, information obtained from a second event is used to revise the probability that a first event has occurred. The theorem will be demonstrated by means of an example, and key terms will be introduced along the way.

EXAMPLE

Bayes' Theorem

Present Information

A dryer manufacturer purchases heating elements from three different suppliers: Argostat, Bermrock, and Thermtek. Thirty percent of the heating elements are supplied by Argostat, 50% by Bermrock, and 20% by Thermtek. The elements are mixed in a supply bin prior to inspection and installation. Based on past experience, 10% of the Argostat elements are defective, compared to only 5% of those supplied by Bermrock, and just 4% of those from Thermtek. An assembly worker randomly selects an element for installation. What is the probability that the element was supplied by Argostat?

SOLUTION

- Events

A_1 = The element was produced by Argostat
A_2 = The element was produced by Bermrock
A_3 = The element was produced by Thermtek

- **Prior probability** This is an initial probability based on the present level of information. On this basis, $P(A_1) = 0.300$ since Argostat supplies 30% of the heating elements.

Additional Information

Upon testing the element before installation, an inspector finds it to be defective.

- Events

B = A tested element is defective
B' = A tested element is not defective

- **Posterior probability** This is a revised probability that has the benefit of additional information. It is a conditional probability and can be expressed as $P(A_1|B)$.

Argostat's quality record is the worst of the three suppliers, so the finding of a defective element would tend to suggest that $P(A_1|B)$ is higher than $P(A_1)$. From an intuitive standpoint, the posterior (revised) probability should be higher than 0.300. But how much higher? As a first step in finding out, consider the tree diagram and joint probabilities in Figure 5.6.

Because the events are dependent, the marginal probabilities in the first branches of the tree diagram are multiplied by the conditional probabilities in the second set of branches to obtain the joint probabilities of the final column. However, there is a problem: In order to construct the tree diagram, we had to use a time sequence that proceeded from the supplier source to the eventual testing of the element.

What we really need is the *reverse* of this tree diagram, that is, proceeding from the present (a known defective) to the past (the probability that it was supplied by Argostat). The joint probabilities of Figure 5.6 can be used in reversing the time sequence. In doing so, it is useful to look at the joint probabilities as relative frequencies out of 1000 cases. Taking this perspective, we observe that in 63 (i.e., 30 + 25 + 8) of the 1000 cases, the element was defective, and in 30 of these the element came from Argostat. In other words, we have just determined that

EXAMPLEEXAMPLEEXAMPLEEXAMPLEEXAMPLEEXAMPLEEXAMPLEEXAMPLEEXA

FIGURE 5.6

Heating-element tree diagram, shown in the order in which the elements are received and found to be either defective or good.

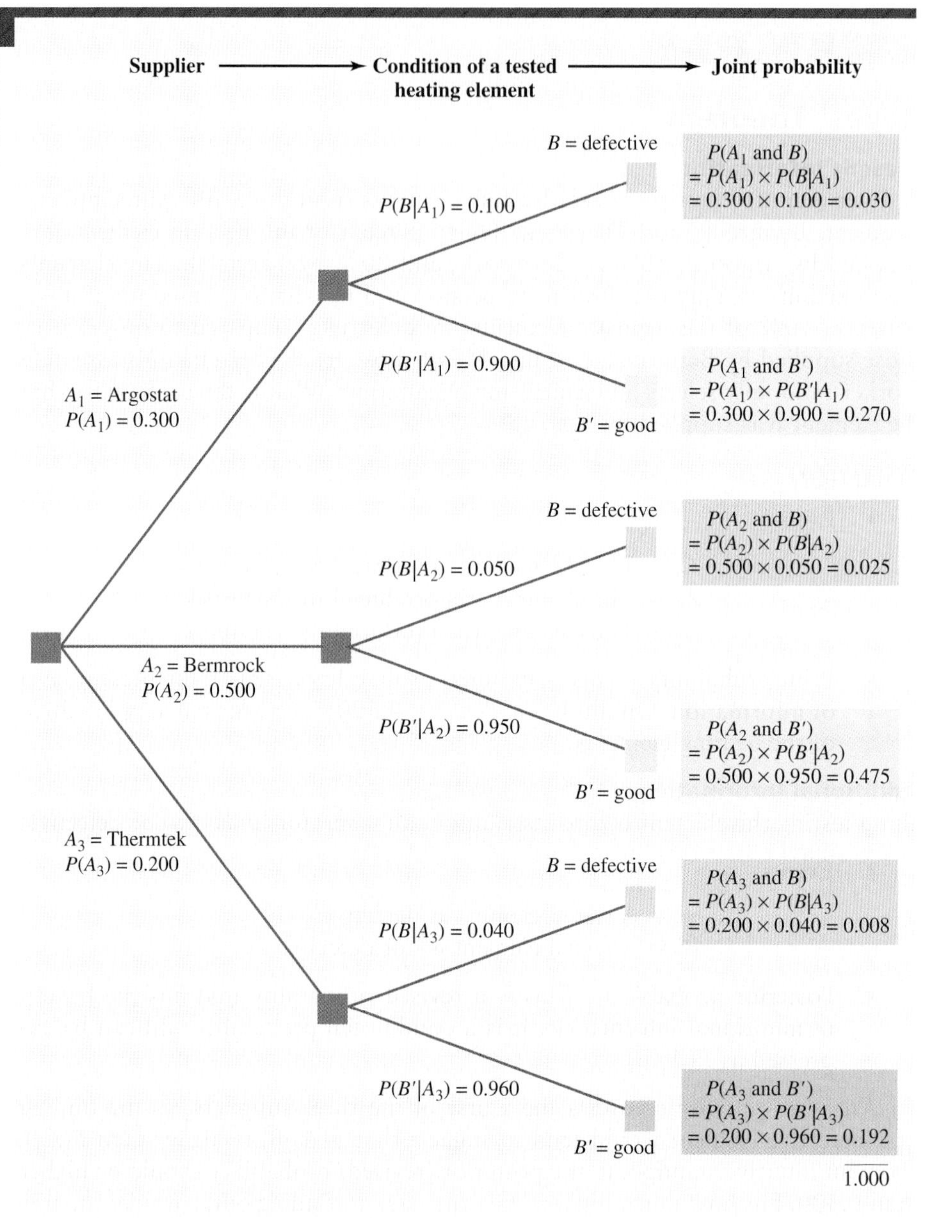

the revised probability, $P(A_1|B)$, is 30/63, or 0.476. This revised probability is among those shown in the *reversed* tree diagram in Figure 5.7.

Probabilities can also be revised by using **Bayes' theorem.** The theorem deals with sequential events, using information obtained about a second event to revise the probability that a first event has occurred.

Bayes' theorem for the revision of probability:

- **Events *A* and *B*** **Probability of *A*, given that event *B* has occurred:**

$$P(A|B) = \frac{P(A \text{ and } B)}{P(B)} = \frac{P(A) \times P(B|A)}{[P(A) \times P(B|A)] + [P(A') \times P(B|A')]}$$

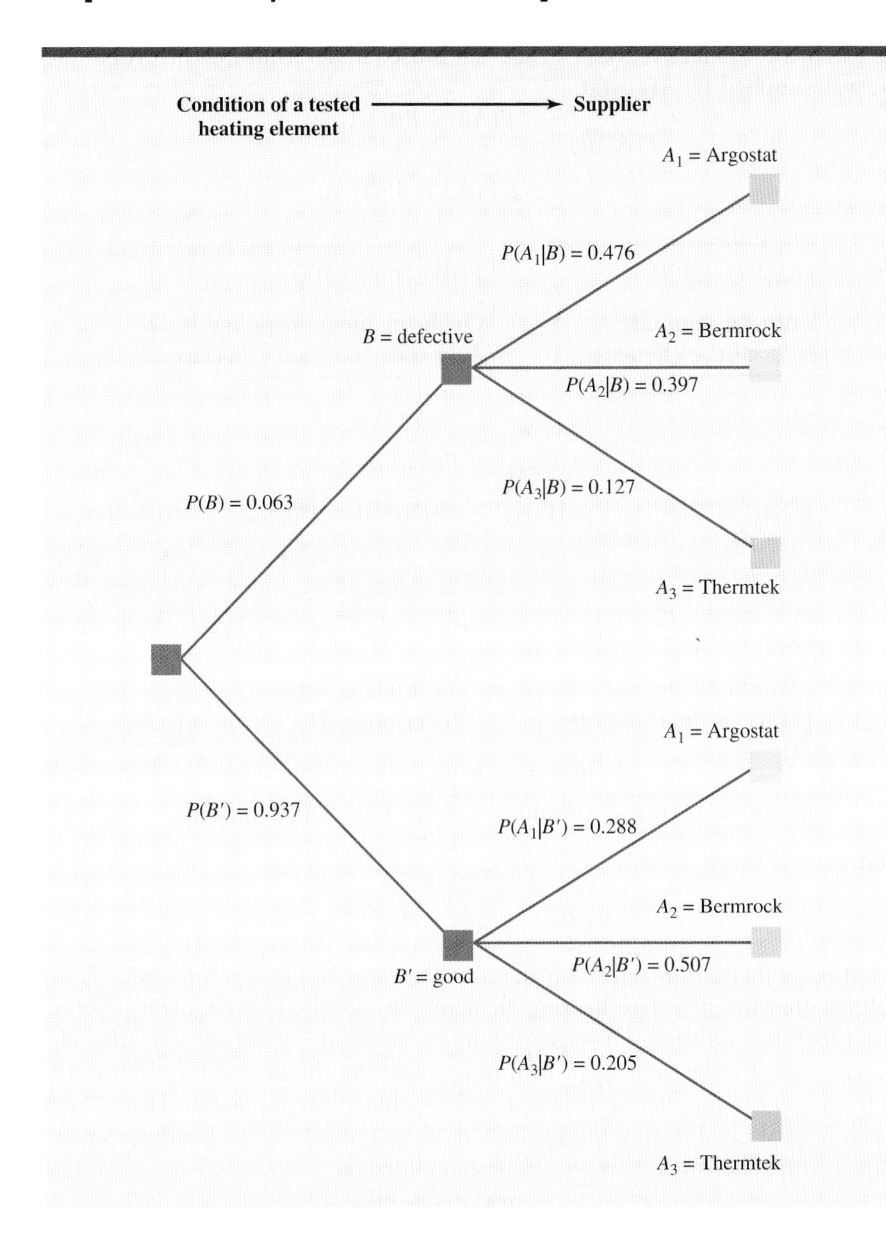

FIGURE 5.7

This reversal of the tree diagram in Figure 5.6 shows that the probability that the defective heating element came from Argostat is 0.476. The prior probability, $P(A_1) = 0.300$, has been converted to a revised probability, $P(A_1|B) = 0.476$.

- **General form** Probability of event A_i, given that event B has occurred:

$$P(A_i|B) = \frac{P(A_i \text{ and } B)}{P(B)} = \frac{P(A_i) \times P(B|A_i)}{\sum_{j=1}^{k} P(A_j) \times P(B|A_j)}$$

where A_i = the ith of k mutually exclusive and exhaustive events

Applying the general form of Bayes' theorem, the heating element had to come from one of $k = 3$ suppliers: Argostat (event A_1), Bermrock (event A_2), or Thermtek (event A_3). Additional information is available in the form of knowledge that the heating element is defective (event B). We must use the general form of the theorem because the number of suppliers is more than two. Given that the heating element is defective, Bayes' theorem is used in determining the probability that it was supplied by Argostat:

$$P(A_i|B) = \frac{P(A_i) \times P(B|A_i)}{\sum_{j=1}^{k} P(A_j) \times P(B|A_j)}$$

$P(A_1) = 0.300$ Argostat supplies 30% of the elements.	$P(B|A_1) = 0.100$ Of the elements supplied by Argostat, 10% are defective.

$$P(A_1|B) = \frac{P(A_1) \times P(B|A_1)}{[P(A_1) \times P(B|A_1)] + [P(A_2) \times P(B|A_2)] + [P(A_3) \times P(B|A_3)]}$$

e.g., $P(A_3) = 0.200$ Thermtek supplies 20% of the elements.	e.g., $P(B|A_3) = 0.040$ Of the elements supplied by Thermtek, 4% are defective.

$$= \frac{0.300 \times 0.100}{[0.300 \times 0.100] + [0.500 \times 0.050] + [0.200 \times 0.040]}$$

and

$$P(A_1|B) = \frac{0.030}{0.063} = 0.476$$

As in our calculations based on the tree diagrams in Figures 5.6 and 5.7, the probability that the defective heating element was supplied by Argostat is found to be 0.476. If the additional information had revealed a *nondefective* heating element, we could use the same approach in determining the probability that it had come from a given supplier. In this case, we would simply replace B with B'. For example, if we wished to find the probability that a nondefective heating element had come from Thermtek, this would be calculated as

$$P(A_3|B') = \frac{P(A_3) \times P(B'|A_3)}{[P(A_1) \times P(B'|A_1)] + [P(A_2) \times P(B'|A_2)] + [P(A_3) \times P(B'|A_3)]}$$

$$= \frac{0.200 \times 0.960}{[0.300 \times 0.900] + [0.500 \times 0.950] + [0.200 \times 0.960]}$$

and

$$P(A_3|B') = \frac{0.192}{0.937} = 0.205$$

For the heating-element example, Table 5.4 shows a contingency table with relative frequencies for suppliers and defectives. Like Table 5.3, Table 5.4 shows marginal, joint, and conditional probabilities. Given that a randomly selected element is defective, we can determine the conditional probability that it was supplied by an individual supplier. This is exactly what we did earlier by using tree diagrams (Figures 5.6 and 5.7) and by using Bayes' theorem for the revision of a probability. In each case, we have taken advantage of the fact that a revised probability can be viewed as a conditional probability.

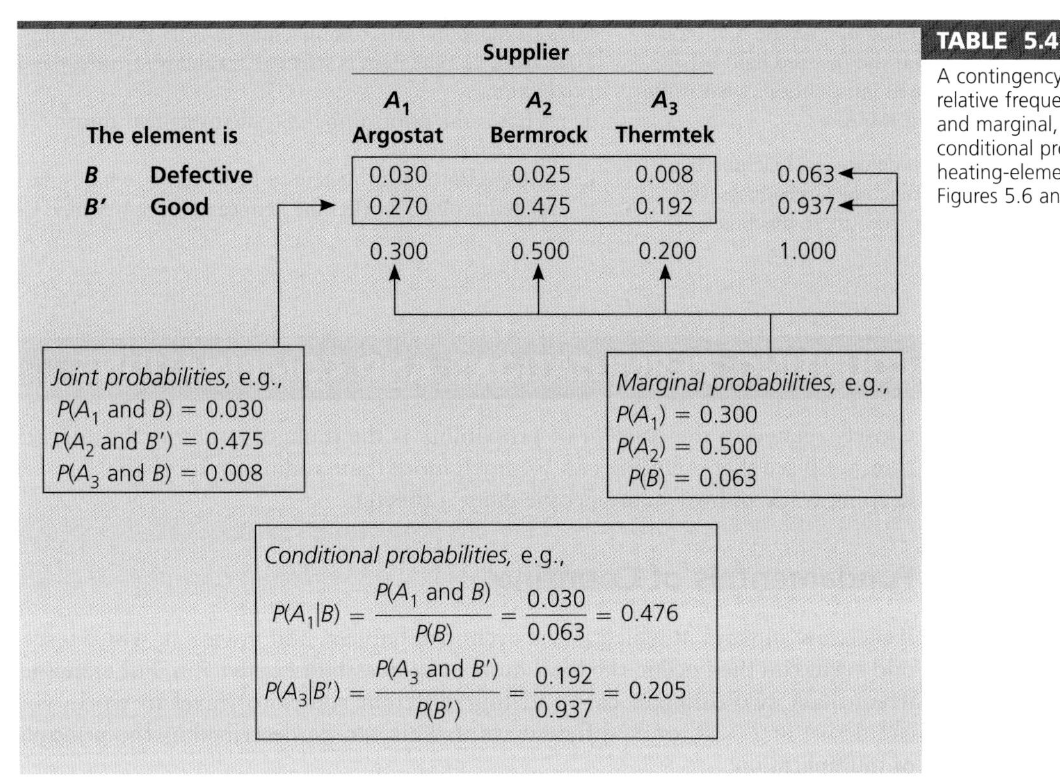

The element is		Supplier A_1 Argostat	A_2 Bermrock	A_3 Thermtek	
B	Defective	0.030	0.025	0.008	0.063
B'	Good	0.270	0.475	0.192	0.937
		0.300	0.500	0.200	1.000

Joint probabilities, e.g.,
$P(A_1 \text{ and } B) = 0.030$
$P(A_2 \text{ and } B') = 0.475$
$P(A_3 \text{ and } B) = 0.008$

Marginal probabilities, e.g.,
$P(A_1) = 0.300$
$P(A_2) = 0.500$
$P(B) = 0.063$

Conditional probabilities, e.g.,

$$P(A_1|B) = \frac{P(A_1 \text{ and } B)}{P(B)} = \frac{0.030}{0.063} = 0.476$$

$$P(A_3|B') = \frac{P(A_3 \text{ and } B')}{P(B')} = \frac{0.192}{0.937} = 0.205$$

TABLE 5.4
A contingency table showing relative frequencies and marginal, joint, and conditional probabilities for the heating-element example of Figures 5.6 and 5.7.

The concept of additional information and revised probabilities is an important part of decision making, an activity in which both statistics and additional information are important. This topic will be covered in greater detail in Chapter 19 and its online appendix.

EXERCISES

5.45 For U.S. live births, $P(\text{boy})$ and $P(\text{girl})$ are approximately 0.51 and 0.49, respectively.[a] According to a newspaper article, a medical process could alter the probabilities that a boy or a girl will be born. Researchers using the process claim that couples who wanted a boy were successful 85% of the time, while couples who wanted a girl were successful 77% of the time.[b] Assuming that the medical process does have an effect on the sex of the child: [a]Source: Bureau of the Census, *Statistical Abstract of the United States 2009*, p. 64. [b]Source: Steve Twedt, "Boy-Girl Selection Available Soon Here," *Pittsburgh Press*, December 31, 1987, pp. 1, 10.

a. Without medical intervention, what is the probability of having a boy?
b. With medical intervention, what is the conditional probability that a couple who wants a boy will have a boy?
c. With medical intervention, what is the conditional probability that a couple who wants a girl will have a girl?

5.46 Using the information in Exercise 5.45, assume that a couple who wanted a girl was randomly placed into either the treatment group (receiving the medical process described) or the control group (no treatment administered) in a test involving the medical procedure.

a. What is the couple's prior probability for being in the treatment group?
b. The couple's newborn child is a girl. Given this additional information, what is the revised probability that the couple was in the treatment group?

5.47 A magician has two coins: One is unbalanced and lands heads 60% of the time; the other is fair and lands heads 50% of the time. A member of the audience randomly selects one of the coins and flips it. The result is heads.

a. What is the prior probability that the fair coin was selected?
b. Given additional information in the form of the single flip that came up heads, what is the revised probability that the coin is the fair one?

5.48 For the information provided in Exercise 5.39, the cabbie finds that his car stalls at the first red light he encounters. Given this additional information, what is the probability that he has selected cab A?

5.49 Machine A produces 3% defectives, machine B produces 5% defectives, and machine C produces 10% defectives. Of the total output from these machines, 60% of the items are from machine A, 30% from B, and 10% from C. One item is selected at random from a day's production.

a. What is the prior probability that the item came from machine C?
b. If inspection finds the item to be defective, what is the revised probability that the item came from machine C?

(5.7) COUNTING: PERMUTATIONS AND COMBINATIONS

Closely related to the concept of probability is the topic of *counting*. In this section, we'll see that counting can be much more than adding up football scores or keeping track of how many people enter a theater.

Fundamentals of Counting

If there are m ways in which a first event can happen, and n ways in which a second event can then occur, the total number of possibilities is $m \times n$. For example, if you have two different cars and four different routes to travel to work, you can arrive in 2×4, or 8, different ways. This can be described as the **principle of multiplication:**

The principle of multiplication:

- **If, following a first event that can happen in n_1 ways, a second event can then happen in n_2 ways, the total number of ways both can happen is $n_1 n_2$.**
- **Alternatively, if each of k independent events can occur in n different ways, the total number of possibilities is n^k.**

As an example of the second approach, assume that a state's auto license plates include 6 different numbers or letters, and that all 26 letters and 10 digits are available. As a result, there can be $(36)^6$, or 2,176,782,336 different license plates. If we include the possibility of blank spaces, this becomes $(37)^6$, or 2,565,726,409, more than enough to satisfy even the most populous state. As a matter of fact, this number of unique plates is more than three times as large as the number of motor vehicles in the entire world.[5]

In an application similar to the selection of letters and numbers for license plates, consonants and vowels can be randomly selected to either create words or fill in blank spaces that are remaining in a word. Statistics in Action 5.2 shows an example in which the computer randomly fills in the remaining three letters of the word "_ _ _ TECH."

In some situations, several events all have different possibilities, but the choice you make for one limits the number of choices for subsequent events. This might occur, for example, if you have five different errands to run and

[5]Source: As of 2005, 862 million motor vehicles were registered throughout the world; *New York Times Almanac 2009*, p. 423.

STATISTICS IN ACTION 5.2

Computer Assistance in the Name Game

When selecting a name for a company or a product, one alternative is to randomly generate combinations of letters, then scan the list for "words" that either look attractive or inspire further ideas. A computer might be programmed to randomly list five-letter words that begin with "k" and end with "x," have a desired order of consonants or vowels, or even include desired sequences like "corp," "tronics," or "electro."

For example, we might generate seven-letter words of the form *cvc*TECH, where c = any of 21 consonants and v = any of 5 vowels. In so forming "_ _ _ TECH," the number of possibilities is 21 × 5 × 21, or 2205. The computer becomes a high-speed variant of the proverbial typewriting monkey who will eventually generate a sentence. As this alphabetized partial listing shows, most of the "words" are not suitable for naming a company, a product, or even a fish. A few of them, however, are fairly reasonable and illustrate the possibilities of this method.

BUGTECH	CAVTECH	DUFTECH	FOZTECH	GAGTECH	HOWTECH
JARTECH	JETTECH	KAPTECH	KUBTECH	LOJTECH	MAZTECH
METTECH	MIPTECH	NESTECH	NOGTECH	PASTECH	QUKTECH
RELTECH	ROSTECH	SAXTECH	SOTTECH	TAGTECH	TOYTECH
VUZTECH	WIJTECH	XOGTECH	YESTECH	ZETTECH	ZIQTECH

must choose the order in which you will do them. In this case, the number of possibilities becomes

$$5 \times 4 \times 3 \times 2 \times 1 = 120$$

There are 5 choices available for the errand to be done first.
Once the 1st errand is chosen, 4 choices are left.
Once the 2nd errand is chosen, 3 choices are left.
Once the 3rd errand is chosen, 2 choices are left.
Once the 4th errand is chosen, only 1 choice remains.

The product (5 × 4 × 3 × 2 × 1) is called a **factorial** and can be described by the symbol "5!." For example, 3! is the same as 3 × 2 × 1. The exclamation point is just a mathematical way of saving space. This approach to counting can be described as the *factorial rule of counting:*

Factorial rule of counting:

$$n! = n \times (n-1) \times (n-2) \times \cdots \times 1$$ **(*Note:* 0! is defined as 1.)**

Permutations

Permutations refer to the number of different ways in which objects can be arranged in order. In a permutation, each item can appear only once, and each order of the items' arrangement constitutes a separate permutation. The number of possible arrangements can be determined as follows:

Number of permutations of n objects taken r at a time:

$$\frac{n!}{(n-r)!}$$

EXAMPLEEXAMPLEEXAMPLEEXAMPLEEXAMP

EXAMPLE

Permutations

Bill has 6 textbooks, but can fit only 4 of them onto a small shelf. If the other 2 books have to sit on the desk, in how many ways can Bill arrange the shelf?

SOLUTION

In this example, there are $n = 6$ objects, taken $r = 4$ at a time, and the number of choices available for the shelf arrangement would be

$$\frac{n!}{(n-r)!} = \frac{6!}{(6-4)!} = \frac{6 \times 5 \times 4 \times 3 \times 2 \times 1}{2 \times 1} = 360$$

The situation can be viewed as follows: (1) the first shelf position can be filled in any of 6 ways, (2) the second shelf position can then be filled in any of 5 ways, (3) the third shelf position can then be filled in any of 4 ways, and (4) the fourth shelf position can then be filled in any of 3 ways. The total number of ways of filling these four shelf positions is thus $6 \times 5 \times 4 \times 3$, or 360. In the equation above, note that the product (2×1) appears in both the numerator and the denominator and is therefore canceled out, reducing the equation to $6 \times 5 \times 4 \times 3 = 360$.

This might seem like a rather large number of possibilities for just 6 books on a 4-book shelf, but each different order of arrangement is regarded as a different permutation. In permutations, the exact order in which the objects are arranged is paramount.

EXAMPLEEXAMPLEEXAMPLEEXAMP

EXAMPLE

Permutations

A tax auditor has 9 different returns to audit, but will have time to examine only 5 of them tomorrow. In how many different orders can tomorrow's task be carried out?

SOLUTION

In this case, there will be $n = 9$ objects, taken $r = 5$ at a time, and the number of different orders for doing tomorrow's work will be

$$\frac{n!}{(n-r)!} = \frac{9!}{(9-5)!} = \frac{9 \times 8 \times 7 \times 6 \times 5 \times 4 \times 3 \times 2 \times 1}{4 \times 3 \times 2 \times 1} = 15{,}120$$

Combinations

Unlike permutations, **combinations** consider only the possible sets of objects, *regardless of the order in which the members of the set are arranged*. The number of possible combinations of n objects taken r at a time will be as follows:

Number of combinations of n objects taken r at a time:

$$\binom{n}{r} = \frac{n!}{r!(n-r)!}$$

In the bookshelf example, there would be 15 different combinations by which Bill could select 4 of the 6 books for storage on the shelf. This is calculated by

$$\binom{n}{r} = \frac{n!}{r!(n-r)!} = \frac{6!}{4!(6-4)!} = \frac{6 \times 5 \times 4 \times 3 \times 2 \times 1}{(4 \times 3 \times 2 \times 1)(2 \times 1)} = 15$$

For greater ease of computation, this calculation can be simplified by taking advantage of the fact that 4! is actually a component of 6!. By making this substitution, the calculation becomes simpler because 4! is canceled out by appearing in both the numerator and denominator:

$$\binom{n}{r} = \frac{n!}{r!(n-r)!} = \frac{6!}{4!(6-4)!} = \frac{6 \times 5 \times 4!}{4!(2 \times 1)} = \frac{6 \times 5}{2 \times 1} = 15$$

In the tax return example, if the tax auditor does not consider the order in which tomorrow's audits are carried out, there will be 126 different compositions for the group of persons whose returns are audited that day. This is calculated as

$$\binom{n}{r} = \frac{n!}{r!(n-r)!} = \frac{9!}{5!(9-5)!} = \frac{9 \times 8 \times 7 \times 6 \times 5!}{5! \times 4!} = \frac{3024}{24} = 126$$

As these examples suggest, for given values of n and r, with $n > r > 1$, the number of permutations will always be greater than the number of combinations. This is because there will be $r!$ permutations for every possible combination. For example, if the auditor selects a given set of 5 returns to audit tomorrow, she can conduct the audits in 5! different orders. Thus, the number of combinations will be just 1/(5!) as large as the number of permutations. This can also be seen by comparing the respective formulas: They are identical except that the formula for combinations has $r!$ in the denominator, while that for permutations does not.

EXERCISES

5.50 A tax accountant has three choices for the method of treating one of a client's deductions. After this choice has been made, there are only two choices for how a second deduction can be treated. What is the total number of possibilities for the treatment of the two deductions?

5.51 A committee consists of eight members, each of whom may or may not show up for the next meeting. Assuming that the members will be making independent decisions on whether or not to attend, how many different possibilities exist for the composition of the meeting?

5.52 Ten prominent citizens have been nominated for a community's "Citizen of the Year" award. First- and second-place trophies are awarded to the two persons receiving the highest and second-highest number of votes. In how many different ways can the trophies be awarded?

5.53 In advertising her oceanfront cottage for summer rental, a property owner can specify (1) whether or not pets are permitted, (2) whether the rent will be \$700, \$900, or \$1200 per week, (3) whether or not children under 10 are allowed, and (4) whether the maximum stay is 1, 2, or 3 weeks. How many different versions of the ad can be generated?

5.54 A state's license plate has 6 positions, each of which has 37 possibilities (letter, integer, or blank). If the buyer of a personalized plate wants his first three positions to be BMW, in how many ways can his request be satisfied?

5.55 When we are considering n objects taken r at a time, with $r > 1$, why will the number of combinations be less than the number of permutations?

5.56 An investment counselor would like to meet with 12 of his clients on Monday, but he has time for only 8 appointments. How many different combinations of the clients could be considered for inclusion into his limited schedule for that day?

5.57 How many different combinations are possible if 6 substitute workers are available to fill 3 openings created by employees planning to take vacation leave next week?

5.58 A roadside museum has 25 exhibits but enough space to display only 10 at a time. If the order of arrangement is considered, how many possibilities exist for the eventual display?

5.59 A sales representative has 35 customers throughout the state and is planning a trip during which 20 will be visited. In how many orders can the visits be made?

5.8 SUMMARY

- **The concept of probability**

Uncertainty, an inherent property of both business and life, can be expressed in terms of probability. For any event, the probability of occurrence will be no less than 0 and no more than 1. All possible outcomes of an experiment are included in the sample space.

- **Classical, subjective, and relative frequency probabilities**

The classical approach describes probability in terms of the proportion of times that an event can be theoretically expected to occur. Because the probability can be determined without actually engaging in any experiments, this is known as the *a priori* approach. In the relative frequency approach, the probability of an event is expressed as the proportion of times it is observed to occur over a large number of trials.

The subjective approach to probability represents the degree to which one believes that an event will happen. Such probabilities can also be described as hunches or educated guesses. Another way to express the probability of an event is to indicate the odds that it will or will not occur.

- **Venn diagrams and contingency tables**

Events are mutually exclusive if, when one occurs, the other(s) cannot. They are exhaustive if they include all of the possible outcomes. An intersection of events represents the occurrence of two or more at the same time, while their union means that at least one of them will happen. The Venn diagram displays both the sample space and the events within. Frequencies or relative frequencies can also be summarized in a contingency table.

- **Determining probabilities**

The addition rules describe the probability that at least one of several events will occur, e.g., $P(A \text{ or } B)$. If the events are mutually exclusive, this probability will be the sum of their individual probabilities. When the events are not mutually exclusive, the general rule of addition applies. This rule takes into account the probabilities that two or more of the events will happen at the same time.

The multiplication rules describe the joint probability that two or more events will occur, e.g., $P(A \text{ and } B)$. When the events are independent, this probability is the product of their individual (marginal) probabilities. When the events are not independent, the general rule of multiplication applies. This rule considers the conditional probability that an event will occur, given that another one has already happened.

- **Bayes' theorem and tree diagrams**

Bayes' theorem, an application of the conditional probability concept, applies to two or more events that occur in sequence. It begins with an initial (prior) probability for an event, then uses knowledge about a second event to calculate a revised (posterior) probability for the original event. The application of Bayes' theorem can be visually described by a tree diagram that shows the possible sequences of events along with their marginal, conditional, and joint probabilities.

- **Permutations and combinations**

Closely related to the concept of probability are the principles of counting. Two such rules are the principle of multiplication and the factorial rule of counting. The concept of permutations is concerned with the number of possible arrangements of n objects taken r at a time. Unlike permutations, combinations consider only the particular set of objects included, regardless of the order in which they happen to be arranged.

EQUATIONS

Classical Approach to Probability

For outcomes that are equally likely,

$$\text{Probability} = \frac{\text{Number of possible outcomes in which the event occurs}}{\text{Total number of possible outcomes}}$$

Relative Frequency Approach to Probability

For a very large number of trials,

$$\text{Probability} = \frac{\text{Number of trials in which the event occurs}}{\text{Total number of trials}}$$

Probabilities and "Odds"

1. If the odds in favor of an event happening are A to B, or $A{:}B$, the probability being expressed is

$$\frac{A}{A + B}$$

2. If the probability of an event is x (with $0 \le x \le 1$), the odds in favor of the event are x to $(1 - x)$, or $x{:}(1 - x)$. (*Note:* In practice, the ratio is converted to a more visually pleasing form in which the numbers are the lowest-possible integers.)

Addition Rule When Events Are Mutually Exclusive

$$P(A \text{ or } B) = P(A) + P(B)$$

Addition Rule When Events Are Not Mutually Exclusive

$$P(A \text{ or } B) = P(A) + P(B) - P(A \text{ and } B)$$

Conditional Probability

$$P(A|B) = \frac{P(A \text{ and } B)}{P(B)}$$ *Note:* This applies only if $P(B) > 0$.

Multiplication Rule When Events Are Independent

$$P(A \text{ and } B) = P(A) \times P(B)$$

Multiplication Rule When Events Are Not Independent

$$P(A \text{ and } B) = P(A) \times P(B|A)$$

Bayes' Theorem for the Revision of a Probability

- **Events A and B** Probability of A, given that event B has occurred:

$$P(A|B) = \frac{P(A \text{ and } B)}{P(B)} = \frac{P(A) \times P(B|A)}{[P(A) \times P(B|A)] + [P(A') \times P(B|A')]}$$

- **General form** Probability of event A_i, given that event B has occurred:

$$P(A_i|B) = \frac{P(A_i \text{ and } B)}{P(B)} = \frac{P(A_i) \times P(B|A_i)}{\sum_{j=1}^{k} P(A_j) \times P(B|A_j)}$$

where A_i = the ith of k mutually exclusive and exhaustive events

The Principle of Multiplication

If, following a first event that can happen in n_1 ways, a second event can then happen in n_2 ways, the total number of ways both can happen is $n_1 n_2$. If each of k independent events can occur in n different ways, the total number of possibilities is n^k.

The Factorial Rule of Counting

$$n! = n \times (n-1) \times (n-2) \times \cdots \times 1 \qquad \textit{Note: } 0! \text{ is defined as } 1.$$

Number of Permutations of n Objects Taken r at a Time

$$\frac{n!}{(n-r)!}$$

Number of Combinations of n Objects Taken r at a Time

$$\binom{n}{r} = \frac{n!}{r!(n-r)!}$$

CHAPTER EXERCISES

5.60 For the situation in Exercise 5.62, what is the probability that Sheila will visit the franchise

a. 5 times without winning a free hamburger?
b. 10 times without winning a free hamburger?
c. 20 times without winning a free hamburger?
d. 40 times without winning a free hamburger?

5.61 It has been estimated that the odds of being struck by lightning in a given year are 1 to 240,000. Using this information: Source: lightningtalks.com, July 27, 2006.

a. Express the odds in terms of "odds against."
b. What is the probability that a randomly selected individual will be struck by lightning this year?
c. What is the probability that the person in part (b) will be struck by lightning at least once during the next 20 years?

5.62 A fast-food chain gives each customer a coupon, the surface of which can be scratched to reveal whether a prize will be received. The odds for winning $1000 per week for life are listed as 1 to 200,000,000, while the odds for winning a free hamburger are 1 to 15. Sheila is going to have lunch at her local franchise of the chain and will receive one coupon during her visit. Determine the probability that she will win

a. $1000 per week for life.
b. a free hamburger.
c. either the lifetime $1000/week or a free hamburger.
d. both the lifetime $1000/week and a free hamburger.

5.63 A firm has two computer systems available for processing telephone orders. At any given time, system A has a 10% chance of being "down," while system B has just a 5% chance of being "down." The computer systems operate independently. For a typical telephone order, determine the probability that

a. neither computer system will be operational.
b. both computer systems will be operational.
c. exactly one of the computer systems will be operational.
d. the order can be processed without delay.

5.64 During fiscal 1995, Ashland Exploration drilled 7 wells in the United States and 3 in Nigeria. One of the wells in Nigeria was productive and 2 of the wells in the United States were productive. The remaining wells were dry. For a well selected at random from those drilled during fiscal 1995, determine the following probabilities: Source: Ashland Oil, Inc., 1995 Fiscal Year Operational Supplement, p. 20.

a. *P*(drilled in the United States)
b. *P*(drilled in the United States or dry)
c. *P*(dry | drilled in Nigeria)

5.65 Dave has been arrested for arson and has been convicted. Edgar is currently being prosecuted for a drug offense. Applying the U.S. Bureau of Justice data in Exercise 5.69, determine the probability that

a. Dave will spend more than a year in jail.
b. Edgar will be convicted.
c. Edgar will spend more than a year in jail.

5.66 Collecting data on traffic accident fatalities, the National Highway Traffic Safety Administration has found that 58.7% of the victims have 0.0% blood alcohol content (BAC), 5.8% of the victims have from 0.01 to 0.07% BAC, and 35.5% of the victims have at least 0.08% BAC. For a randomly selected victim: Source: *Statistical Abstract of the United States 2009*, p. 681.

a. What is the probability that the victim's BAC was at least 0.01%?
b. Given that the victim had been drinking prior to the accident, what is the probability that this victim's BAC was at least 0.08%?

5.67 For the three perpetrators in Exercise 5.69, determine the number of possibilities in which

a. just one person is convicted.
b. exactly two of the three persons are convicted.
c. all three persons are convicted.

5.68 The following relative frequency distribution describes the household incomes for families living in a suburban community:

Income	Relative Frequency
Less than $20,000	0.10
$20,000–under $40,000	0.20
$40,000–under $60,000	0.30
$60,000–under $80,000	0.25
$80,000 or more	0.15

a. For a randomly selected household, what is the probability that its annual income is less than $80,000?
b. If a household is known to have an income of at least $20,000, what is the probability that its annual income is in the $60,000–under $80,000 category?
c. Two households are randomly selected from those whose income is at least $40,000 per year. What is the probability that both of these households are in the $80,000 or more category?

5.69 The U.S. Bureau of Justice released the following probabilities for those arrested for committing various felony crimes in the United States:

	Probability of Being		
Crime	Prosecuted	Convicted	Jailed for >1 Year
Homicide	0.91	0.75	0.73
Assault	0.79	0.64	0.15
Burglary	0.88	0.81	0.28
Arson	0.88	0.72	0.28
Drug offenses	0.78	0.69	0.19
Weapons	0.83	0.70	0.13
Public disorder	0.92	0.85	0.12

Allen has been arrested for burglary, Bill has been arrested for a weapons offense, and Charlie has been arrested on a public-disorder charge. Assuming these individuals are typical perpetrators and the decisions regarding their respective fates are unrelated, determine the probability that Source: U.S. Bureau of Justice, as reported in Sam Meddis, "Felony Arrests: Short Terms," *USA Today*, January 18, 1988, p. 9A.

a. Allen will be jailed for more than a year.
b. either Allen or Bill (or both) will be convicted.
c. none of the three will be jailed for more than a year.
d. Allen and Bill will be convicted, but Charlie will be found innocent.
e. none will be prosecuted.

5.70 An industrial hoist is being used in an emergency job where the weight exceeds the design limits of two of its components. For the amount of weight being lifted, the probability that the upper attachment hook will fail is 0.20. The probability that the lower hook will fail is 0.10. What is the probability that the hoisting job will be successfully completed?

5.71 The "daily number" of a state lottery is a 3-digit integer from 000 to 999.

a. Sam buys a ticket for the number 333. What is the probability that he will win?
b. Is the probability found in part (a) a classical, relative frequency, or subjective probability? Explain your answer.
c. Shirley buys a ticket for the number 418 since it came up yesterday and she thinks it's "hot." Is there a flaw in her logic? Explain.

5.72 Over the years, a realtor estimates that 50% of those who contact her about a listing end up allowing her to provide a guided tour of the property, 20% of those who take a guided tour follow up with a request for more information about the property, and 2% of those who make the follow-up informational request actually end up making an offer on the property. The realtor has just taken a young couple on a guided tour of one of her listed properties. Based on the probability estimates provided above, what is the probability that the young couple will actually make an offer on the property?

5.73 A security service employing 10 officers has been asked to provide 3 persons for crowd control at a local carnival. In how many different ways can the firm staff this event?

5.74 Of the adults in Jefferson County, 10% have had CPR training. For a randomly selected group of adults from this county, how large would the group have to be in order for the probability that at least one group member has had CPR training to be at least 0.90?

5.75 In examining borrower characteristics versus loan delinquency, a bank has collected the following information: (1) 15% of the borrowers who have been employed at their present job for less than 3 years are behind in their payments, (2) 5% of the borrowers who have been employed at their present job for at least 3 years are behind in their payments, and (3) 80% of the borrowers have been employed at their present job for at least 3 years. Given this information:

a. What is the probability that a randomly selected loan account will be for a person in the same job for at least 3 years who is behind in making payments?
b. What is the probability that a randomly selected loan account will be for a person in the same job for less than 3 years or who is behind in making payments?
c. If a loan account is behind, what is the probability that the loan is for a person who has been in the same job for less than 3 years?

5.76 Of the participants in a corporate meeting, 20% have had Red Cross training in cardiopulmonary resuscitation (CPR). During a break in the session, one participant is randomly selected to go across the street and pick up some hoagies from the local deli. A patron at the deli is stricken by a severe heart attack. His survival probability is just 0.40 if he does not receive immediate CPR, but will be 0.90 if CPR can be administered right away. The meeting participant and the stricken person are the only customers in the deli, and none of the employees knows CPR.

a. Without additional information, what is the probability that the stricken patron will survive?
b. Given that the stricken patron survives, what is the probability that the meeting participant knew how to administer CPR?
c. Given that the stricken patron does not survive, what is the probability that the meeting participant knew how to administer CPR?

5.77 When a machine is properly calibrated, 0.5% of its output is defective, but when it is out of adjustment, 6% of the output is defective. The technician in charge of calibration can successfully calibrate the machine in 90% of his attempts. The technician has just made an attempt to calibrate the machine.

a. What is the probability that the machine is in adjustment?
b. The first unit produced after the calibration effort is found to be defective. Given this information, what is the probability that the machine is in adjustment?

5.78 Data from the Federal Bureau of Investigation show that 1 of every 184 motor vehicles was stolen during 2003. Applying this statistic to 5 motor vehicles randomly selected from the nation's vehicle population: Source: Bureau of the Census, *Statistical Abstract of the United States 2009*, pp. 194, 711.

a. What is the probability that none of the 5 motor vehicles will be stolen?
b. What is the probability that all 5 motor vehicles will be stolen?
c. How many possibilities exist in which 2 of the 5 motor vehicles are stolen?

5.79 A computer specialty manufacturer has a two-person technical support staff who work independently of each other. In the past, Tom has been able to solve 75% of the problems he has handled, and Adam has been able to solve 95% of the problems he has handled. Incoming problems are randomly assigned to either Tom or Adam. If a technical problem has just been assigned to the support department, what is the probability that it will be handled by Adam? If it turns out that the problem was solved, what is the probability that it was handled by Adam?

5.80 Avis, Inc., has reported that its fleet consists of 200,000 vehicles. If the vehicles are independent in terms of accident incidence and each has a 0.99999 chance of making it through the year without being in a major accident, what is the probability that the entire fleet will avoid a major accident during the year? Source: Avis.com, July 27, 2006.

5.81 According to Sears, two-thirds of U.S. homeowners have an appliance from Sears. In a randomly selected group of three homeowners, what is the probability that all three would have an appliance from Sears? That at least one of the three would have an appliance from Sears? Source: Sears advertisement insert, *Indiana Gazette*, May 24, 2003.

5.82 A test to compare the taste of 6 soft drinks is being arranged. Each participant will be given 3 of the soft drinks and asked to indicate which one tastes best.

a. How many different possibilities exist for the set of soft drinks that a given participant will taste?
b. If we consider the order in which the participant in part (a) tastes the 3 soft drinks, how many possibilities exist?

5.83 A corporate board of directors consisting of 15 persons is to form a subcommittee of 5 persons to examine an environmental issue currently facing the firm. How many different subcommittees are possible?

5.84 A certain brand of bicycle lock requires that each of four adjacent wheels be rotated to exactly the correct position in order to open the lock and separate the chain to which it is attached. If each wheel is labeled with the digits 0 through 9, how many different possibilities are available to the manufacturer in "programming" the sequence of numbers for opening any given lock?

5.85 The chairperson of the accounting department has three summer courses available: Accounting 201, Accounting 202, and Accounting 305. Twelve faculty members are available for assignment to these courses, and no faculty member can be assigned to more than one course. In how many ways can the chairperson assign faculty to these courses?

INTEGRATED CASES

Thorndike Sports Equipment

Ted Thorndike has been with his grandfather's firm for 5 weeks and is enjoying his job immensely. The elder Thorndike is also pleased—so pleased that he took the morning off to do a little fishing. However, while old Luke is catching some carp, young Ted is catching some flak.

At the other end of this morning's telephone call is an irate customer whose aluminum tennis racquet has cracked in half after just 14 months. The caller is an avid player who gets out on the court about two or three times a week, and he claims the racquet has never been abused in any way. He not only wants Thorndike Sports Equipment to send him another racquet, he also demands that the company reimburse him the $30 he lost after his racquet broke and he was no longer competitive with his opponent.

Ted assures the indignant caller that the company will replace his racquet at no cost, since it is only slightly over the 12-month warranty and the company values his loyalty to its products. Unfortunately, the $30 wager cannot be covered, even though it was a triple-or-nothing bet from the caller's losses in the two previous matches.

On Mr. Thorndike's return to the office, Ted mentions the interesting phone call, and the elder Thorndike is not a bit surprised. He says the firm has been getting a lot of complaints in recent months, mainly from longtime customers who claim their latest Thorndike aluminum racquet didn't hold up as well as the ones they had used in the past.

Speculating, Mr. Thorndike goes on to point out that the company has had two aluminum suppliers for many years, but added a third supplier just a year and a half ago. He suspects that the most recent supplier may be shipping an aluminum alloy that is more brittle and prone to failure. He's not sure, but he thinks the racquets that are failing are more likely than not to be constructed from the aluminum purchased from the newest of the three suppliers.

When the company sends out a replacement racquet, it does so only after it has received the defective one. All of the racquets that have been returned over the past 5 years or so are in a wooden crate in the basement. Mr. Thorndike isn't sure why, but 6 years ago he mandated that each racquet produced have a serial number that identifies when it was made and who supplied the aluminum for its construction. If someone were to ferret through the wooden crate, maybe he could shed some light on this business of customers calling up with broken racquets.

Mr. Thorndike goes on, "Say, Ted, why don't you come in over the weekend and take a look at those broken racquets downstairs, and see if maybe there's something going on here that I should know about? Maybe I should have stayed with the two suppliers I've been using. The prices the new folks were offering seemed too good to pass up, but maybe it was all too good to be true. Net profits from our tennis business aren't going to be very strong if we have to keep on sending people all these free racquets."

As Ted is rummaging through the broken racquets in the basement, what kind of information should he be seeking and how might he structure it in the form of a contingency table? What kinds of probabilities might be useful regarding racquet failure, and what implications could Ted's findings have for helping his grandfather better satisfy the company's racquet customers?

Springdale Shopping Survey

The contingency tables and relative frequency probabilities in this exercise are based on the Springdale Shopping Survey database. Data are in the computer file **SHOPPING**. Information like that gained from the two parts of this exercise could provide helpful insights into the nature of the respondents, their perceptions, and their spending behaviors. In particular, part 2 examines how conditional probabilities related to spending behavior might vary, depending on the gender of the respondent.

1. Based on the relative frequencies for responses to each variable, determine the probability that a randomly selected respondent
   a. [variable 4] spends at least $15 during a trip to Springdale Mall.
   b. [variable 5] spends at least $15 during a trip to Downtown.
   c. [variable 6] spends at least $15 during a trip to West Mall.

   Comparing the preceding probabilities, which areas seem strongest and weakest in terms of the amount of money a shopper spends during a typical shopping visit?
   d. [variable 11] feels that Springdale Mall has the highest-quality goods.
   e. [variable 11] feels that Downtown has the highest-quality goods.
   f. [variable 11] feels that West Mall has the highest-quality goods.

   Comparing the preceding probabilities, which areas are strongest and weakest in terms of the quality of goods offered?

2. Set up a contingency table for the appropriate variables, then determine the following probabilities:
   a. [variables 4 and 26] Given that the random respondent is a female, what is the probability that she spends at least $15 during a trip to Springdale Mall? Is a male more likely or less likely than a female to spend at least $15 during a visit to this area?
   b. [variables 5 and 26] Given that the random respondent is a female, what is the probability that she spends at least $15 during a trip to Downtown? Is a male more likely or less likely than a female to spend at least $15 during a visit to this area?
   c. [variables 6 and 26] Given that the random respondent is a female, what is the probability that she spends at least $15 during a trip to West Mall? Is a male more likely or less likely than a female to spend at least $15 during a visit to this area?

Based on the preceding probabilities, at which shopping areas are males and females most likely and least likely to spend $15 or more during a shopping visit?

BUSINESS CASE

Baldwin Computer Sales (B)

Andersen Ross/Photodisc/Getty Images

In the Baldwin Computer Sales case, visited previously in Chapter 3, one of the areas of concern regarding the student purchase program was whether the probability of defaulting on the computer payments might be related to the university attended by the student. Using the **BALDWIN** data file, first construct an appropriate contingency table of frequencies and determine the overall percentage of students who defaulted on their computer payments. Then address the following:

1. Given that a student attended university number 1, determine the conditional probability that he or she defaulted on the computer payments.
2. Repeat question 1 for each of the other universities. Do any of the conditional probabilities seem especially high or low compared to the others? In general, which school is associated with those most likely to default, and which school is associated with those least likely to default?
3. If a student is randomly chosen from those who have defaulted on their computer payments, determine the revised probability that the student is from the "most likely to default" university identified in question 2.

CHAPTER 6

Discrete Probability Distributions

ATM USERS ARE DISCRETE AS WELL AS DISCREET

Yes, we know automatic teller machine (ATM) customers use discretion as to when and how they use the ATM. For example, many zealously shield the keyboard when entering their personal ID number into the machine.

But ATM users are also discrete: They are members of a discrete probability distribution. For x = the number of persons arriving at the machine over a period of time, x will be a discrete variable—0, 1, 2, 3, and so on.

The next time you're in the vicinity of your local ATM, relax and enjoy a soft drink for 5 minutes while you count how many people arrive at the machine during that time. Do this on 9 more occasions. Now figure out what percent of the time x was equal to 0, what percent of the time x was equal to 1, and so on. Congratulations, you've just created a discrete probability distribution, a type of distribution that is the topic of this chapter.

More specifically, the phenomenon you've been observing tends to be well represented by a theoretical discrete probability distribution called the Poisson distribution, discussed later in the chapter. This distribution is useful in both describing and predicting the number of people arriving at an ATM, a tollbooth, an emergency room, and many other kinds of service facilities over a given period of time.

Stuck in the Friday afternoon discrete probability distribution . . .

LEARNING OBJECTIVES

After reading this chapter, you should be able to:

- Understand both the concept and the applications of a probability distribution for a random variable.
- Determine whether the random variable in a probability distribution is of the discrete type or the continuous type.
- Differentiate among the binomial, hypergeometric, and Poisson discrete probability distributions and their applications.
- Understand what is meant by a Bernoulli process and how this applies to the consecutive trials associated with the binomial distribution.
- Use the appropriate probability distribution in determining the probability that a discrete random variable will have a given value or a value in a given range.

6.1 INTRODUCTION

Chapter 2 introduced the relative frequency distribution as a way of describing the proportion of actual observations that either had one of several possible values or fell into one of a number of different ranges of values. In Chapter 5, we discussed classical probability as a probability that could be theoretically determined without engaging in any experiments [e.g., $P(\text{heads}) = 0.5$ for a coin flip]. In this chapter, we'll see how a probability distribution is really a very special combination of these two concepts.

In general, we can define a **probability distribution** as *the relative frequency distribution that should theoretically occur for observations from a given population*. In business and other contexts, it can be helpful to proceed from (1) a basic understanding of how a natural process seems to operate in generating events to (2) identifying the probability that a given event may occur. By using a probability distribution as a model that represents the possible events and their respective likelihoods of occurrence, we can make more effective decisions and preparations in dealing with the events that the process is generating.

Random Variables: Discrete versus Continuous

A **random variable** is a variable that can take on different values according to the outcome of an experiment. Random variables can be either *discrete* or *continuous*, a distinction discussed briefly in Chapter 1:

DISCRETE RANDOM VARIABLE A random variable that can take on only certain values along an interval, with the possible values having gaps between them. For example, in a given group of five children, the number who got at least one electronic toy for Christmas would be 0, 1, 2, 3, 4, or 5. It could not be a number between any of these values, such as 2.338.

CONTINUOUS RANDOM VARIABLE A random variable that can take on a value at *any* point along an interval. The exact temperature outside as you're reading this book could be 13.568, 78.352, or 83.815 degrees Fahrenheit, or any of an infinity of other values in the range of temperatures where colleges and universities are located. (The temperature examples have been expressed to just three decimal places here, but there would be countless possibilities along the temperature continuum.)

A random variable is described as *random* because we don't know ahead of time exactly what value it will have following the experiment. For example, when we toss a coin, we don't know for sure whether it will land heads or tails. Likewise, when we measure the diameter of a roller bearing, we don't know in advance what the exact measurement will be. In this chapter, the emphasis will be on discrete random variables and their probability distributions. In the next, we'll cover variables of the continuous type.

The Nature of a Discrete Probability Distribution

A **discrete probability distribution** is a listing of all possible outcomes of an experiment, along with their respective probabilities of occurrence. We'll begin our discussion of such distributions with a small-scale example.

EXAMPLE

Discrete Probability Distribution

An experiment is conducted in which a fair coin is flipped twice. The result of the experiment will be the random variable x = the number of times that heads comes up.

SOLUTION

Since the two tosses are independent, the probability of heads remains constant at $P(\text{heads}) = 0.5$ from one toss to the next. Each of four possible sequences is equally likely, and each has a probability of 0.25:

Two tosses of a coin: Possible sequences and their probabilities

Sequence	x = Number of Heads	Probability
HH	2	0.25 = (0.5 × 0.5)
HT	1	0.25
TH	1	0.25
TT	0	0.25
		1.00

Because the possible sequences are mutually exclusive (only one of them can happen for a given series of two tosses), the sum of the joint probabilities is 1.00. Two of the sequences (*HT* and *TH*) involve one head, and we can use the addition rule from Chapter 5 to find the probability that $x = 1$. This can be expressed as $P(HT \text{ or } TH) = P(HT) + P(TH) = 0.25 + 0.25 = 0.50$. We have just generated the following *probability distribution* for the discrete random variable x = number of heads in two coin tosses:

Two tosses of a coin: Discrete probability distribution for the random variable x = number of heads

x	$P(x)$
0	0.25
1	0.50
2	0.25
	1.00

FIGURE 6.1

Discrete probability distribution for the random variable x = number of heads in two consecutive tosses of a coin.

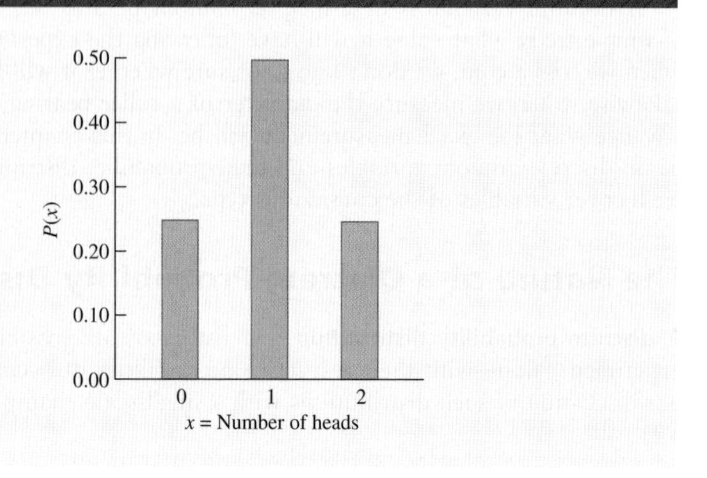

This distribution can also be described graphically, as in Figure 6.1. This discrete probability distribution reflects several important characteristics shared by all such distributions:

Characteristics of a discrete probability distribution:

1. For any value of x, $0 \leq P(x) \leq 1$.
2. The values of x are *exhaustive:* The probability distribution includes all possible values.
3. The values of x are *mutually exclusive:* Only one value can occur for a given experiment.
4. The sum of their probabilities is one, or $\sum P(x_i) = 1.0$.

Although discrete probability distributions are often based on classical probabilities (as in Figure 6.1), they can be the result of relative frequencies observed from past experience or even from subjective judgments made about future events and their likelihood of occurrence. As an example of a discrete probability distribution based on relative frequencies, consider the following situation.

EXAMPLE

Discrete Probability Distribution

A financial counselor conducts investment seminars with each seminar limited to 6 attendees. Because of the small size of the seminar group and the personal attention each person receives, some of the attendees become clients following the seminar. For the past 20 seminars she has conducted, x (for x = the number of attendees who become clients) has had the relative frequency distribution shown in Figure 6.2.

SOLUTION

The probability distribution for x = the number of attendees who become clients can also be described in tabular form. The events and their respective

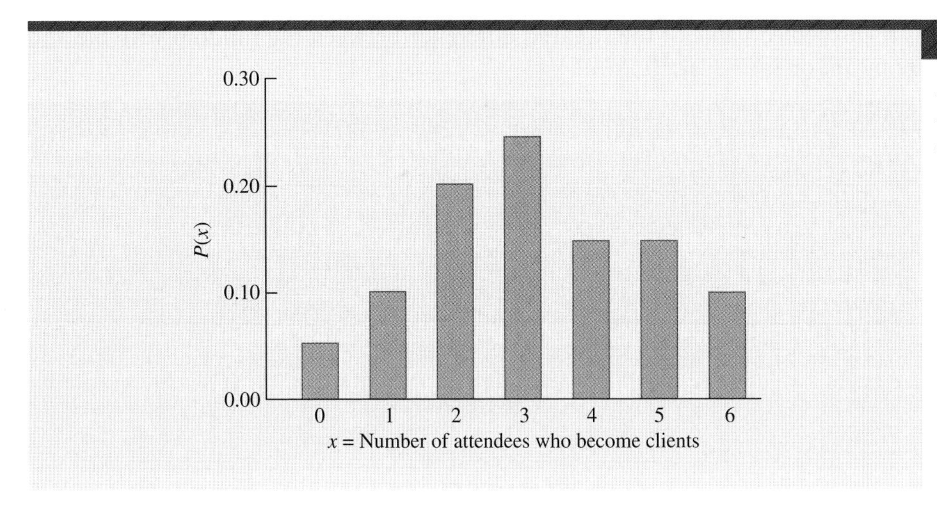

FIGURE 6.2

Relative frequency distribution for the number of attendees in the seminar group who become clients of the seminar leader.

probabilities are as follows:

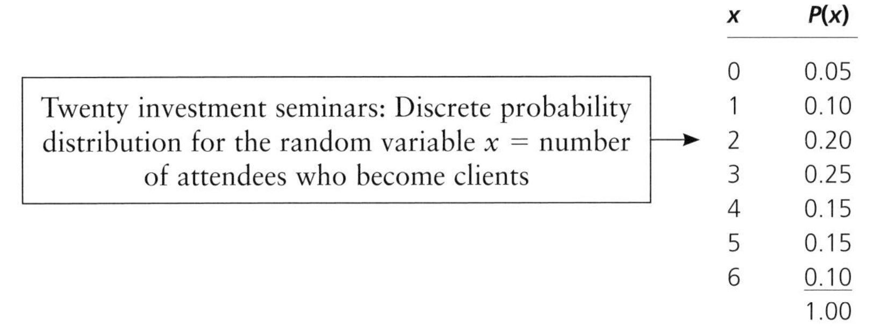

x	$P(x)$
0	0.05
1	0.10
2	0.20
3	0.25
4	0.15
5	0.15
6	0.10
	1.00

As in the classical relative frequency distribution of Figure 6.1, the events in Figure 6.2 are mutually exclusive and exhaustive. Interpreting their relative frequencies as probabilities, as shown in Figure 6.2, the sum of these probabilities will be 1.0. In a given seminar, the probability that nobody will become a client is $P(0) = 0.05$, while the probability that at least 4 will become clients is $P(x \geq 4) = P(x = 4) + P(x = 5) + P(x = 6)$, or $0.15 + 0.15 + 0.10 = 0.40$.

The Mean and Variance of a Discrete Probability Distribution

For the observed data in Chapter 3, we used measures of central tendency and dispersion to describe *typical* observations and *scatter* in the data values, respectively. Similarly, it can be useful to describe a discrete probability distribution in terms of its central tendency and dispersion. For example, the investment counselor of Figure 6.2 may be interested both in how many attendees from a seminar *typically* become clients and in the extent to which the number who become clients *varies*.

The *mean* of a discrete probability distribution for a discrete random variable is called its *expected value*, and this is referred to as $E(x)$, or μ. In the following formula, it can be seen to be a weighted average of all the possible outcomes, with each outcome weighted according to its probability of occurrence. The *variance* is the expected value of the squared difference between the random variable and its mean, or $E[(x - \mu)^2]$:

General formulas for the mean and variance of a discrete probability distribution:

- **Mean:**

$$\mu = E(x) \quad \text{or} \quad \mu = \Sigma x_i P(x_i) \text{ for all possible values of } x$$

- **Variance:**

$$\sigma^2 = E[(x - \mu)^2] \quad \text{or} \quad \sigma^2 = \Sigma(x_i - \mu)^2 P(x_i)$$

for all possible values of x, and the standard deviation is

$$\sigma = \sqrt{\sigma^2}$$

An alternative formula for σ^2 is $\sigma^2 = [\Sigma x_i^2 P(x_i)] - \mu^2$. It provides greater ease of calculation when the data values are large or when there are a great many possible values of x.

To demonstrate how the mean and variance are determined, we will compute these descriptors for the discrete probability distribution in Figure 6.2.

Mean and Variance for x = the Number of Attendees Who Become Clients (Represented by the Discrete Probability Distribution of Figure 6.2)

- Mean:

$$\begin{aligned}\mu = E(x) &= \Sigma x_i P(x_i) \\ &= 0(0.05) + 1(0.10) + 2(0.20) + 3(0.25) + 4(0.15) + 5(0.15) + 6(0.10) \\ &= 3.2 \text{ attendees who become clients}\end{aligned}$$

- Variance:

$$\begin{aligned}\sigma^2 &= E[(x - \mu)^2] = \Sigma(x_i - \mu)^2 P(x_i) \\ &= (0 - 3.2)^2(0.05) + (1 - 3.2)^2(0.10) + (2 - 3.2)^2(0.20) \\ &\quad + (3 - 3.2)^2(0.25) + (4 - 3.2)^2(0.15) + (5 - 3.2)^2(0.15) \\ &\quad + (6 - 3.2)^2(0.10) \\ &= 2.66\end{aligned}$$

and

$$\sigma = \sqrt{2.66} = 1.63 \text{ attendees who become clients}$$

In a given seminar, there will never be exactly $x = 3.2$ attendees who become clients. However, this is the *expected value*, or *mean* of x, and it will tend to be the average result if the probability distribution remains unchanged and a great many such seminars are conducted.

The formulas in this section can be used to determine the mean and variance for *any* discrete probability distribution. For some distributions, however, the mean and variance can be described and computed by using expressions that are both more practical and more easily calculated. As we discuss the distributions to follow, such expressions for the mean and variance will be presented as they become relevant.

Before we move on to examine three particular discrete probability distributions, the binomial, the hypergeometric, and the Poisson, consider the probability distribution represented by a state's three-digit daily lottery number. There are

STATISTICS IN ACTION 6.1

The Discrete Uniform Distribution That Took a Day Off

A feature of the Pennsylvania State Lottery is the "daily number," a three-digit number selected at 7 P.M. each weekday evening. The number is obtained from the first ball that pops from each of three machines containing balls numbered from 0 to 9. The big event is televised, with a senior-citizen witness and a lottery official present to verify the number of each ball.

Since the daily number can range from 000 to 999, the chance of any specific result is just 0.001. Lottery officials rigorously weigh the balls and conduct extensive testing to make sure some numbers don't have a higher probability than others. A number of years ago, an individual was convicted for injecting a latexlike substance that made some of the balls heavier. (The author and many others watching television noticed that some of the bouncing balls seemed a little "tired.") As a result of this one-day nonuniformity of the probability distribution, security was heightened considerably, and officials are now more scrupulous than ever in their efforts to ensure that the probability distribution is indeed uniform.

Despite the uniformity of this discrete probability distribution, some players still insist on playing a "lucky" number based on their birth date, wedding date, or digits from an auto license number. Some even avoid playing yesterday's number because they think it's been "used up" for awhile. Many seem to prefer 333 or 777, although 923, 147, or any other choice would have the same probability of being selected.

1000 possible outcomes (000 to 999) and, because all of the outcomes are equally likely, this is an example of what is called a *discrete uniform distribution.* Attempts to tamper with this uniformity are not well received by state officials. See Statistics in Action 6.1.

EXERCISES

6.1 Why is a random variable said to be "random"?

6.2 Indicate whether each of the following random variables is discrete or continuous.

a. The diameter of aluminum rods coming off a production line
b. The number of years of schooling employees have completed
c. The Dow Jones Industrial Average
d. The volume of milk purchased during a supermarket visit

6.3 Indicate whether each of the following random variables is discrete or continuous.

a. The number of drive-through customers at your local McDonald's today
b. The amount spent by all customers at the local McDonald's today
c. The time, as reported by your digital watch
d. The length of a pedestrian's stride

6.4 The values of x in a discrete probability distribution must be both exhaustive and mutually exclusive. What is meant by these two terms?

6.5 A discrete random variable can have the values $x = 3$, $x = 8$, or $x = 10$, and the respective probabilities are 0.2, 0.7, and 0.1. Determine the mean, variance, and standard deviation of x.

6.6 Determine the mean, variance, and standard deviation of the following discrete probability distribution:

x	0	1	2
$P(x)$	0.60	0.30	0.10

6.7 Determine the mean, variance, and standard deviation of the following discrete probability distribution:

x	0	1	2	3	4
$P(x)$	0.10	0.30	0.30	0.20	0.10

6.8 Ed Tompkins, the assistant dean of a business school, has applied for the position of dean of the school of business at a much larger university. The salary at the new university has been advertised as $200,000, and Ed is very excited about this possibility for making a big career move. He has been told by friends within the administration of the larger university that his resume is impressive and his chances of getting the position he seeks are "about 60%." If Ed stays at his current position, his salary next year will be $120,000. Assuming that his friends have accurately assessed his chances of success, what is Ed's expected salary for next year?

6.9 Scam artists sometimes "stage" accidents by purposely walking in front of slow-moving luxury cars whose drivers are presumed to carry large amounts of liability insurance. The "victim" then offers to accept a small monetary cash settlement so the driver can avoid an insurance claim, a blot on his driving record, and increased insurance premiums. Marlin has just tried to stage such a scheme by walking into a slow-moving Mercedes and faking a minor injury. Assume that the possible outcomes for Marlin are −$500 (i.e., he is arrested and fined for his scam), $0 (the driver recognizes the scam and refuses to pay), and $100 (the driver simply pays up to get rid of Marlin). If the probabilities for these outcomes are 0.1, 0.3, and 0.6, respectively, what is Marlin's expected monetary outcome for trying to scam the Mercedes driver?

6.10 A regional manager visits local fast-food franchises and evaluates the speed of the service. If the manager receives her meal within 45 seconds, the server is given a free movie-admission coupon. If it takes more than 45 seconds, the server receives nothing. Throughout the company's franchises, the probability is 0.60 that a meal will be served within 45 seconds. What is the expected number of coupons a counter employee will receive when serving the regional manager?

6.11 A consultant has presented his client with three alternative research studies that could be conducted. If the client chooses alternative *a*, the consultant will have to hire 1 additional employee. Two new employees will have to be hired if the client chooses *b*, and 5 if the client chooses *c*. The probability that the client will select alternative *a* is 0.1, with $P(b) = 0.5$ and $P(c) = 0.4$. Identify the discrete random variable in this situation and determine its expected value.

6.12 A contractor must pay a $40,000 penalty if construction of an expensive home requires more than 16 weeks. He will receive a bonus of $10,000 if the home is completed within 8 weeks. Based on experience with this type of project, the contractor feels there is a 0.2 chance the home will require more than 16 weeks for completion, and there is a 0.3 chance it will be finished within 8 weeks. If the price of the home is $550,000 before any penalty or bonus adjustment, how much can the buyer expect to pay for her new home when it is completed?

6.13 A music shop is promoting a sale in which the purchaser of a compact disk can roll a die, then deduct a dollar from the retail price for each dot that shows on the rolled die. It is equally likely that the die will come up any integer from 1 through 6. The owner of the music shop pays $5.00 for each compact disk, then prices them at $9.00. During this special promotion, what will be the shop's average profit per compact disk sold?

6.14 Laura McCarthy, the owner of Riverside Bakery, has been approached by insurance underwriters trying to convince her to purchase flood insurance. According to local meteorologists, there is a 0.01 probability that the river will flood next year. Riverside's profits for the coming year depend on whether Laura buys the flood insurance and whether the river floods. The profits (which take into consideration the $10,000 premium for the flood insurance) for the four possible combinations of Laura's choice and river conditions are:

		The River	
		Does Not Flood	Floods
Insurance Decision	No Flood Insurance	$200,000	−$1,000,000
	Get Flood Insurance	$190,000	$200,000

a. If Laura decides not to purchase flood insurance, use the appropriate discrete probability distribution to determine Riverside's expected profit next year.
b. If Laura purchases the flood insurance, what will be Riverside's expected profit next year?
c. Given the results in parts (a) and (b), provide Laura with a recommendation.

6.15 In 2007, nearly 24 million tons of steel mill products went to construction and contracting companies. Transco Steel, a hypothetical manufacturer specializing in the production of steel for this market, is faced with aging production facilities and is considering the possible renovation of its plant. The company has reduced its alternatives to just three: (1) no renovation at all, (2) minor renovation, and (3) major renovation. Whichever strategy it chooses, the company's profits will depend on the size of the construction and contracting industry in future years. Company executives have a subjective probability of 0.3 that the industry will not grow at all, a probability of 0.5 that it will grow moderately, and a probability of 0.2 that it will show a high rate of growth. The company's estimated profits (in millions of dollars) for each combination of renovation strategy and industry growth possibility are:

Source: *Statistical Abstract of the United States 2009*, p. 630.

		Industry Growth		
		None	Moderate	High
Level of Plant Renovation	None	$35	$38	$42
	Minor	$28	$45	$55
	Major	$21	$40	$70

Given the executives' subjective probabilities for the three possible states of industry growth, what will be the company's expected profit if it chooses not to renovate the plant? If it chooses minor renovations to the plant? If it chooses major renovations to the plant? Based on these expected values, provide the company's management with a recommendation as to which level of renovation they should choose.

THE BINOMIAL DISTRIBUTION

6.2

One of the most widely used discrete distributions, the **binomial distribution** deals with consecutive trials, each of which has two possible outcomes. This latter characteristic is the reason for the "bi" part of the descriptor *binomial.* The outcomes can be identified in general terms, such as "success" versus "failure" or "yes" versus "no," or by more specific labels, such as "milk in the refrigerator" versus "no milk in the refrigerator." The binomial distribution relies on what is known as the **Bernoulli process:**

Characteristics of a Bernoulli process:

1. There are two or more consecutive trials.
2. In each trial, there are just two possible outcomes—usually denoted as *success* or *failure.*
3. The trials are statistically independent; that is, the outcome in any trial is not affected by the outcomes of earlier trials, and it does not affect the outcomes of later trials.
4. The probability of a success remains the same from one trial to the next.

Description and Applications

In the binomial distribution, the discrete random variable, x, is the number of successes that occur in n consecutive trials of the Bernoulli process. The distribution, which has a wide variety of applications that will be demonstrated through examples and chapter exercises, can be summarized as follows:

The binomial probability distribution:

The probability of exactly x successes in n trials is

$$P(x) = \frac{n!}{x!(n-x)!}\pi^x(1-\pi)^{n-x}$$

where n = number of trials
x = number of successes
π = the probability of success in any given trial
$(1-\pi)$ = the probability of failure in any given trial

- Mean:

$$\mu = E(x) = n\pi$$

- Variance:

$$\sigma^2 = E[(x-\mu)^2] = n\pi(1-\pi)$$

In the binomial distribution, we can view the n trials as comprising (1) the x trials that are successes and (2) the $(n - x)$ trials that are failures. The second component in the formula, π^x, is the probability that x of x trials will all be successful. The third component, $(1 - \pi)^{n-x}$, is the probability that $(n - x)$ of $(n - x)$

trials will all be failures. The first component is the number of possible combinations for n items taken x at a time; in the context of the binomial distribution, this is the number of possible sequences where there are exactly x successes in n trials.

The binomial distribution is really a *family* of distributions, and the exact *member* of the family is determined by the values of n and π. The following observations may be made regarding the Bernoulli process and the requirement that the probability of success (π) remain unchanged:

1. If **sampling** is done **with replacement** (the person or other item selected from the population is observed, then put back into the population), π will be constant from one trial to the next.
2. If, when **sampling without replacement,** the number of trials (n) is very small compared to the population (N) of such trials from which the sample is taken, as a practical matter π can be considered to be constant from one trial to the next. This would occur, for example, if $n = 5$ interview participants were randomly selected from a population of $N = 1000$ persons. As a rule of thumb, if the population is at least 20 times as large as the number of trials, we will assume the "constant π" assumption has been satisfactorily met.

EXAMPLEEXAMPLEEXAMPLEEXAMPLEEXAMPLEEXAMP

EXAMPLE

Binomial Distribution

Of the 18,000,000 part-time workers in the United States, 20% participate in retirement benefits.[1] A group of 5 people is to be randomly selected from this group, and the discrete random variable is x = the number of persons in the group who participate in retirement benefits.

SOLUTION

This example can be considered a Bernoulli process, with $n = 5$ trials and $\pi = 0.2$. The size of the population ($N = 18{,}000{,}000$) is extremely large compared to the number of trials, so π can be assumed to be constant from one trial to the next. A worker either participates in retirement benefits or does not participate, so there are only two possible outcomes for each of the 5 trials.

What Is the Expected Value of x?

The expected number of people in the group who participate can be calculated as $E(x) = n\pi = 5(0.2)$, or $E(x) = 1.0$. The expectation would be that 1.0 persons in the group would participate in retirement benefits. (In other words, if we were to randomly select a great many such groups, we would expect the average value of x to be 1.0.)

What Is the Probability That the Group Will Include Exactly Two Participants?

Using the binomial formula, with $n = 5$, $\pi = 0.2$, and $(1 - \pi) = (1.0 - 0.2) = 0.8$, the probability of x = exactly 2 participants is

$$P(x) = \frac{n!}{x!(n-x)!}\pi^x (1-\pi)^{n-x}$$

[1]Source: *Statistical Abstract of the United States 2006*, pp. 367, 635.

and

$$P(2) = \frac{5!}{2!(5-2)!}(0.2)^2(0.8)^{5-2} = \boxed{0.205}$$

$$\text{i.e., } \frac{5 \cdot 4 \cdot 3 \cdot 2 \cdot 1}{(2 \cdot 1)(3 \cdot 2 \cdot 1)} = \frac{5 \cdot 4}{2 \cdot 1}$$

For Each of the Possible Values of *x*, What Is the Probability That Exactly This Number of People Out of the Five Will Be Participants?

The values $x = 0$ through $x = 5$ include all of the possibilities for the number of persons in the sample who are participants; their probabilities would be

$$P(0) = \frac{5!}{0!(5-0)!}(0.2)^0(0.8)^5 = \boxed{0.328}$$

(*Note:* Remember that $0! = 1$ and that any nonzero number raised to the zero power is also equal to 1.)

$$P(1) = \frac{5!}{1!(5-1)!}(0.2)^1(0.8)^4 = \boxed{0.410}$$

$$P(2) = \frac{5!}{2!(5-2)!}(0.2)^2(0.8)^3 = \boxed{0.205} \quad \text{(as calculated earlier)}$$

$$P(3) = \frac{5!}{3!(5-3)!}(0.2)^3(0.8)^2 = \boxed{0.051}$$

$$P(4) = \frac{5!}{4!(5-4)!}(0.2)^4(0.8)^1 = \boxed{0.006}$$

$$P(5) = \frac{5!}{5!(5-5)!}(0.2)^5(0.8)^0 = \boxed{0.000} \quad \text{(rounded)}$$

We can verify that $\mu = E(x) = n\pi$, or $5(0.20) = 1.0$, by calculating $E(x) = \Sigma x_i P(x_i)$. Because we have rounded the individual probabilities, the result differs slightly from the actual value, $\mu = 1.000$ persons:

$$\begin{aligned} E(x) &= \Sigma x_i P(x_i) \\ &= 0(0.328) + 1(0.410) + 2(0.205) + 3(0.051) + 4(0.006) + 5(0.000) \\ &= 0.997 \text{ persons (differs from } \mu = 1.000 \text{ due to rounding of the individual probabilities)} \end{aligned}$$

What Is the Probability That the Group Will Include at Least Three Persons Who Are Participants?

Because x can take on only one of the possible values between 0 and 5, the outcomes are mutually exclusive. (Notice that the sum of the six probabilities just calculated will be 1.000.) Thus $P(x \geq 3)$ will be

$$\begin{aligned} P(x \geq 3) &= [P(x = 3) + P(x = 4) + P(x = 5)] \\ &= 0.051 + 0.006 + 0.000 = 0.057 \end{aligned}$$

The binomial distribution for the discrete random variable in this example is shown in Figure 6.3. The distribution is positively skewed. Binomial distributions can have different shapes, including being positively skewed, negatively skewed, or even symmetrical.

FIGURE 6.3

Binomial probability distribution for text example with $n = 5$ and $\pi = 0.20$. Of 18 million part-time workers, only 20% participate in retirement benefits. Because the population is very large compared to the number in the sample, π can be considered as remaining constant from one trial to the next.

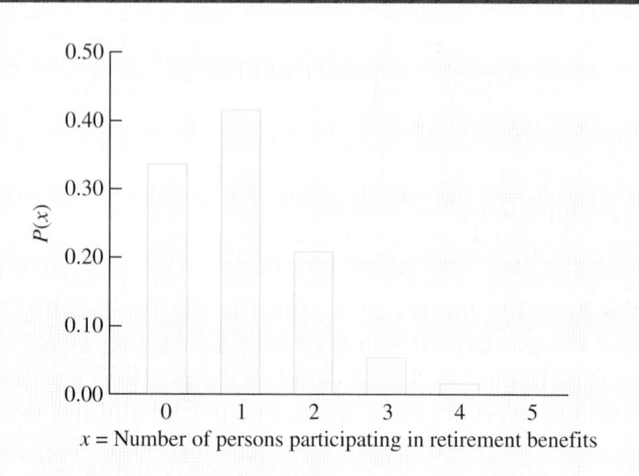

Using the Binomial Tables

As the preceding example indicates, calculating binomial probabilities can be tedious. Fortunately, binomial distribution tables, such as those in Appendix A, make the task a lot easier. We will continue with our example in demonstrating their use.

A given binomial probability distribution depends on the values of π and n, so a separate listing is necessary for various combinations of their values. For the five randomly selected part-time workers, $n = 5$ trials, and we will rely on the individual binomial probabilities table (Table A.1), a portion of which is reproduced here as Table 6.1. The "π" columns in the table range from 0.1 to 0.9 in increments of 0.1. The value of π in our example was $\pi = 0.2$, so this will be the

TABLE 6.1

A portion of the table of *individual* binomial probabilities from Appendix A. For $n = 5$ trials, each entry is $P(x = k)$ for the specified value of the population proportion, π.

Binomial Distribution, Individual Probabilities for x = number of successes in n trials, prob($x = k$)

$n = 5$

k \ π	0.1	0.2	0.3	0.4	0.5	0.6	0.7	0.8	0.9
0	0.5905	0.3277	0.1681	0.0778	0.0313	0.0102	0.0024	0.0003	0.0000
1	0.3281	0.4096	0.3601	0.2592	0.1562	0.0768	0.0284	0.0064	0.0005
2	0.0729	0.2048	0.3087	0.3456	0.3125	0.2304	0.1323	0.0512	0.0081
3	0.0081	0.0512	0.1323	0.2304	0.3125	0.3456	0.3087	0.2048	0.0729
4	0.0005	0.0064	0.0284	0.0768	0.1562	0.2592	0.3601	0.4096	0.3281
5	0.0000	0.0003	0.0024	0.0102	0.0313	0.0778	0.1681	0.3277	0.5905

For $n = 5$ and $\pi = 0.2$, $P(x = 2)$ is 0.2048.

appropriate column for determining the probabilities that we've just calculated the hard way. Briefly summarizing our earlier treatment of this example, $n = 5$, $\pi = 0.2$, and the probabilities calculated for the possible values of x were

x	0	1	2	3	4	5
$P(x)$	0.328	0.410	0.205	0.051	0.006	0.000

Using the individual binomial probabilities in Table 6.1, we refer to the "$\pi = 0.2$" column within the "$n = 5$" section. For example, Table 6.1 shows $P(x = 0)$ to be 0.3277. Likewise, $P(x = 1)$ and $P(x = 2)$ are 0.4096 and 0.2048, respectively.

Appendix A also includes a table for *cumulative* binomial probabilities (Table A.2), a portion of which is reproduced as Table 6.2. Using this table, we find that $P(x \leq 3) = 0.9933$ for $x =$ the number of participants among the five randomly selected workers.

Using the table of cumulative binomial probabilities is the best way to determine "greater than or equal to" probabilities, such as $P(x \geq 2)$. For example, to find $P(x \geq 2)$,

$$P(x \leq 1) + P(x \geq 2) = 1.000$$

and

From the table of cumulative probabilities, $P(x \leq 1) = 0.7373$.

$$P(x \geq 2) = 1.0000 - P(x \leq 1)$$

so

$$P(x \geq 2) = 1.0000 - 0.7373 = 0.2627$$

As an alternative, the individual binomial probabilities in Table 6.1 could also be used in finding $P(x \geq 2)$. Using this table would be more cumbersome,

Binomial Distribution, Cumulative Probabilities for x = number of successes in n trials, prob $(x \leq k)$

$n = 5$

	π	0.1	0.2	0.3	0.4	0.5	0.6	0.7	0.8	0.9
	0	0.5905	0.3277	0.1681	0.0778	0.0313	0.0102	0.0024	0.0003	0.0000
	1	0.9185	0.7373	0.5282	0.3370	0.1875	0.0870	0.0308	0.0067	0.0005
k	2	0.9914	0.9421	0.8369	0.6826	0.5000	0.3174	0.1631	0.0579	0.0086
	3	0.9995	0.9933	0.9692	0.9130	0.8125	0.6630	0.4718	0.2627	0.0815
	4	1.0000	0.9997	0.9976	0.9898	0.9688	0.9222	0.8319	0.6723	0.4095
	5		1.0000	1.0000	1.0000	1.0000	1.0000	1.0000	1.0000	1.0000

For $n = 5$ and $\pi = 0.2$, $P(x \leq 3)$ is 0.9933.

TABLE 6.2

A portion of the table of *cumulative* binomial probabilities from Appendix A. For $n = 5$ trials, each entry is $P(x \leq k)$ for the specified value of the population proportion, π.

however, since we would have to add up each of the probabilities for $x = 2$ through $x = 5$, or

$$\begin{aligned} P(x \geq 2) &= P(x = 2) + P(x = 3) + P(x = 4) + P(x = 5) \\ &= 0.2048 + 0.0512 + 0.0064 + 0.0003 \\ &= 0.2627 \end{aligned}$$

Excel and Minitab can provide either individual or cumulative binomial probabilities. The procedures and the results are shown in Computer Solutions 6.1.

COMPUTER 6.1 SOLUTIONS

Binomial Probabilities

These procedures show how to get binomial probabilities associated with $n = 5$ trials when the probability of a success on any given trial is $\pi = 0.20$.

For each software package:

Procedure I: Individual or cumulative probabilities for $x = 2$ successes.

Procedure II: Complete set of individual or cumulative probabilities for $x = 0$ successes through $x = 5$ successes.

EXCEL

	A	B	C
1	x	P(X = x)	P(X <= x)
2	0	0.3277	0.3277
3	1	0.4096	0.7373
4	2	0.2048	0.9421
5	3	0.0512	0.9933
6	4	0.0064	0.9997
7	5	0.0003	1.0000

I. Excel procedure for just one individual or cumulative binomial probability

1. Click on the cell into which Excel will place the probability. From the **Formulas** ribbon, click on the **Insert Function** symbol at the extreme left. Click **Statistical** in the **Category** box. Click **BINOMDIST** in the **Function** box. Click **OK.**
2. Enter the number of successes **(2)** into the **Number_s** box. Enter the number of trials **(5)** into the **Trials** box. Enter the probability of success on any given trial **(0.20)** into the **Probability_s** box. To obtain the individual probability, $P(x = 2)$, enter **false** into the **Cumulative** box. To obtain the cumulative probability, $P(x \leq 2)$, enter **true** into the **Cumulative** box. Click **OK.**

II. Excel procedure for a complete set of individual or cumulative binomial probabilities

1. Enter the column labels into **A1:C1** as shown in the printout above. Enter the lowest number of successes in the set (in this case, **0**) into **A2.** Enter **1** into **A3.** Click on cell **A2** and highlight cells **A2:A3.** Click at the lower right corner of **A3** (the cursor will turn to a "**+**" sign) and drag downward so cells **A2:A7** will contain the values 0 through 5.
2. To get the set of individual probabilities: Click on cell **B2** (this is where the first individual probability will go). Follow the remainder of the steps in Excel procedure I, except enter **A2** into the **Number_s** (number of successes) box and enter **false** into the **Cumulative** box. The first probability now appears in **B2.** Click to select cell **B2.** Click the lower right corner of **B2** (the cursor turns to a "+" sign) and drag down until cells **B2:B7** contain the set of individual probabilities.
3. To get the set of cumulative probabilities: Click on cell **C2** (this is where the first cumulative probability will go). Follow the remainder of the steps in Excel procedure I, except enter **A2** into the **Number_s** (number of successes) box and enter **true** into the **Cumulative** box. The first probability now appears in **C2.** Click to select cell **C2.**

(continued)

Click the lower right corner of **C2** (the cursor turns to a "+" sign) and drag down until cells **C2:C7** contain the set of cumulative probabilities.

MINITAB

```
Probability Density Function        Cumulative Distribution Function

Binomial with n = 5 and p = 0.2     Binomial with n = 5 and p = 0.2

x  P( X = x )                       x  P( X <= x )
0      0.32768                      0       0.32768
1      0.40960                      1       0.73728
2      0.20480                      2       0.94208
3      0.05120                      3       0.99328
4      0.00640                      4       0.99968
5      0.00032                      5       1.00000
```

I. Minitab procedure for just one individual or cumulative binomial probability

1. Click **Calc.** Select **Probability Distributions.** Click **Binomial.**
2. In the **Binomial Distribution** menu, enter **5** into the **Number of trials** box. Enter **0.20** into the **Probability of success** box. Select **Input constant** and enter the desired *x* value (e.g., **2**) into the box. Select either **Probability** or **Cumulative probability,** depending on whether you want an individual probability or a cumulative probability. Click **OK.**

II. Minitab procedure for a complete set of individual or cumulative binomial probabilities

Enter the desired *x* values (in this case **0** through **5**) into column **C1.** Follow steps 1 and 2 above, except select **Input column** and enter **C1** into the box.

Additional Comments on the Binomial Distribution

Under certain conditions, the binomial distribution can be approximated by other probability distributions. This is especially useful since there is a practical limit to the sheer volume of printed tables that can be included in a statistical appendix. A later section of this chapter discusses how another discrete probability distribution called the *Poisson distribution* can be used to approximate the binomial. Introduced in the next chapter, a continuous distribution called the *normal distribution* is especially useful as a satisfactory substitute in many practical applications.

EXERCISES

6.16 What is necessary for a process to be considered a Bernoulli process?

6.17 When we are sampling without replacement, under what conditions can we assume that the constant π assumption has been satisfactorily met?

6.18 Seven trials are conducted in a Bernoulli process in which the probability of success in a given trial is 0.7. If x = the number of successes, determine the following:

a. $E(x)$ b. σ_x c. $P(x = 3)$
d. $P(3 \leq x \leq 5)$ e. $P(x > 4)$

6.19 Twelve trials are conducted in a Bernoulli process in which the probability of success in a given trial is 0.3. If x = the number of successes, determine the following:

a. $E(x)$ b. σ_x c. $P(x = 3)$
d. $P(2 \leq x \leq 8)$ e. $P(x > 3)$

6.20 A city law-enforcement official has stated that 20% of the items sold by pawn shops within the city have been stolen. Ralph has just purchased 4 items from one of the city's pawn shops. Assuming that the official is correct, and for x = the number of Ralph's purchases that have been stolen, determine the following:

a. $P(x = 0)$ b. $P(2 \leq x)$
c. $P(1 \leq x \leq 3)$ d. $P(x \leq 2)$

6.21 The regional manager in Exercise 6.10 plans to evaluate the speed of service at five local franchises of the restaurant. For the five servers, determine the probability that

a. none will receive a movie coupon.
b. two of the five will receive a coupon.
c. three of the five will receive a coupon.
d. all five will receive a coupon.

6.22 According to the National Marine Manufacturers Association, 50.0% of the population of Vermont were boating participants during the most recent year. For a randomly selected sample of 20 Vermont residents, with the discrete random variable x = the number in the sample who were boating participants that year, determine the following: Source: National Marine Manufacturers Association, *NMMA Boating Industry Report*, p. 3.

a. $E(x)$ b. $P(x \leq 8)$ c. $P(x = 10)$
d. $P(x = 12)$ e. $P(7 \leq x \leq 13)$

6.23 In the town of Hickoryville, an adult citizen has a 1% chance of being selected for jury duty next year. If jury members are selected at random, and the Anderson family includes three adults, determine the probability that

a. none of the Andersons will be chosen.
b. exactly one of the Andersons will be chosen.
c. exactly two of the Andersons will be chosen.
d. all three Andersons will be chosen.

6.24 It has been reported that the overall graduation rate for football players at Division I-A colleges and universities is 65%. Source: "How Football Rates," *USA Today*, December 20, 2005, p. 12C.

a. If five players are randomly selected from the entire population of Division I-A football players, what is the probability that at least 2 of the 5 will graduate?
b. In part (a), what is the probability that all 5 of the players will graduate?

6.25 An association of coffee producers has estimated that 50% of adult workers drink coffee at least occasionally while on the job. Given this estimate:

a. For a randomly selected group of 4 adult workers, what is the probability that no more than 3 of them drink coffee at least occasionally while on the job?
b. In part (a), what is the probability that at least 2 of these individuals drink coffee at least occasionally while on the job?

6.26 The U.S. Department of Labor has reported that 30% of the 2.1 million mathematical and computer scientists in the United States are women. If 3 individuals are randomly selected from this occupational group, and x = the number of females, determine $P(x = 0)$, $P(x = 1)$, $P(x = 2)$, and $P(x = 3)$. Source: Bureau of the Census, *Statistical Abstract of the United States 2002*, p. 381.

6.27 Alicia's schedule includes three Tuesday/Thursday courses in which each professor uses a single coin flip to decide whether to give a short quiz each class day. Alicia is active on the university's student advisory council, works two part-time jobs, and is helping to build a float for the upcoming homecoming parade. As a result of these activities and slightly inadequate planning on her part, she has had very little time to prepare for her Thursday classes being held tomorrow. What is the probability that Alicia will be "lucky" and have no surprise quizzes tomorrow? What is the probability that the worst will occur and she will have surprise quizzes in all three classes? What is the probability that she will escape with minimal damage and have a quiz in only one of the three classes?

6.28 OfficeQuip is a small office supply firm that is currently bidding on furniture and office equipment contracts with four different potential customers who are of comparable size. For each contract, OfficeQuip would gain a profit of $50,000 if that contract were accepted, so the company could make as little as $0 or as much as $200,000. The four potential customers are making independent decisions, and in each case the probability that OfficeQuip will receive the contract is 0.40. When all the decisions have been made, what is the probability that OfficeQuip will receive none of the contracts? Exactly one of the contracts? Exactly two of the contracts? Exactly three of the contracts? All four contracts? Overall, what is OfficeQuip's expected profit in this business-procurement venture?

6.29 Four wheel bearings are to be replaced on a company vehicle, and the mechanic has selected the four replacement parts from a large supply bin. Unknown to the mechanic, 10% of the bearings are defective and will fail within the first 100 miles after installation. What is the probability that the company vehicle will have a breakdown due to defective wheel bearings before it completes tomorrow's 200-mile delivery route?

6.30 It has been estimated that one in five Americans suffers from allergies. The president of Hargrove University plans to randomly select 10 students from the undergraduate population of 1400 to attend a dinner at his home. Assuming that Hargrove students are typical in terms of susceptibility to allergies and that the college president's home happens to contain just about every common allergen to which afflicted persons react, what is the probability that at least 8 of the students will be able to stay for the duration of the dinner event? Source: *World Almanac and Book of Facts 2009*, p. 187.

6.31 Airlines book more seats than are actually available, then "bump" would-be passengers whenever more people show up than there are seats. Through the first quarter of 2009, the rate at which passengers were bumped was 1.40 per 1000 passengers. Assuming that, on average, the probability of any given passenger being bumped is 1.40/1000, or 0.00140: Source: U.S. Bureau of Transportation Statistics, June 2009.

a. Emily is flying nonstop to visit Besco, Inc., for a job interview. What is the probability that she will be bumped?
b. Ten members of the Besco board of directors will be flying nonstop to the firm's annual meeting, and their flights are independent of each other. What is the probability that at least one of them will be bumped?
c. Besco sales personnel made a total of 220 nonstop flights last year, and they experienced a total of four bumpings. Could the company's experience be considered as very unusual?

6.32 Every day, artists at Arnold's House of Fine Figurines produce 5 miniature statues that must be sanded and painted. Past experience has shown that 10% of the statues have a defect that does not show up until after the statue has been sanded and painted. Whenever a statue is found to have this defect, Arnold must pay a specialist $20 to correct the defect and repaint the statue. Describe the probability distribution for the daily amount that Arnold must pay the specialist, and determine the average daily amount the specialist will receive.

THE HYPERGEOMETRIC DISTRIBUTION (6.3)

Description and Applications

The **hypergeometric probability distribution** is similar to the binomial in focusing on the number of successes in a given number of consecutive trials, but there are two major differences: First, *the consecutive trials are not independent;* that is, the outcome in a trial can be affected by the outcomes in earlier trials, and can in turn affect the outcomes in later trials. Second, *the probability of success does not remain constant from one trial to the next.* Unlike the binomial distribution, the hypergeometric distribution does not rely on a Bernoulli process.

The distinction between the binomial and hypergeometric distributions can also be viewed in terms of sampling with replacement versus sampling without replacement. In sampling without replacement, the object selected from the population is not returned to the population. When sampling is done without replacement and the population is not very large, the probability of a success can change quite drastically from one trial to the next.

Situations like these can be analyzed by means of tree diagrams, along with their marginal, conditional, and joint probabilities, but the hypergeometric distribution is a lot easier. The distribution can be described as follows:

The hypergeometric distribution:

The probability of exactly *x* successes in *n* trials that are not independent is

$$P(x) = \frac{\binom{s}{x} \cdot \binom{N-s}{n-x}}{\binom{N}{n}}$$

where N = size of the population
n = size of the sample
s = number of successes in the population
x = number of successes in the sample

• Mean: $$\mu = E(x) = \frac{ns}{N}$$

• Variance: $$\sigma^2 = E[(x-\mu)^2] = \frac{ns(N-s)}{N^2} \cdot \frac{N-n}{N-1}$$

EXAMPLEEXAMPLEEXAMPLEEXAMPLEEXAMPLEEXAMPLEEXAMPLE

EXAMPLE

Hypergeometric Distribution

Twenty businesses located in a small community have filed tax returns for the preceding year. Six of the returns were filled out incorrectly. An Internal Revenue Service (IRS) auditor randomly selects four returns for further examination, and the random variable is x = the number of examined returns that are incorrectly filled out.

What Is the Expected Value of x?

For a sample size of $n = 4$, with $s = 6$ incorrect returns in the population, and a population size of $N = 20$, the expected number of incorrect returns in the sample is

$$E(x) = \frac{ns}{N} = \frac{4(6)}{20} = 1.20 \text{ incorrect returns}$$

What Is the Probability That There Will Be Exactly Two Incorrect Returns Among the Four That Are Chosen for Further Examination?

The quantity to be determined is $P(x = 2)$, with $N = 20$ businesses in the population, $s = 6$ incorrect returns in the population, and $n = 4$ returns randomly selected from the population. Using the hypergeometric formula for $P(x)$:

$$P(x) = \frac{\binom{s}{x} \cdot \binom{N-s}{n-x}}{\binom{N}{n}} \quad \text{and} \quad P(2) = \frac{\binom{6}{2} \cdot \binom{20-6}{4-2}}{\binom{20}{4}} = \frac{\binom{6}{2} \cdot \binom{14}{2}}{\binom{20}{4}}$$

so

$$P(2) = \frac{\dfrac{6!}{2!(6-2)!} \cdot \dfrac{14!}{2!(14-2)!}}{\dfrac{20!}{4!(20-4)!}} = \frac{\dfrac{6 \cdot 5}{2 \cdot 1} \cdot \dfrac{14 \cdot 13}{2 \cdot 1}}{\dfrac{20 \cdot 19 \cdot 18 \cdot 17}{4 \cdot 3 \cdot 2 \cdot 1}} = \frac{1365}{4845} = 0.2817$$

The probability $P(x = 2)$ is 0.2817 that exactly two of the four returns will be incorrect. If we were to repeat this procedure for the other possible values of x, we would find the following probabilities for $x = 0$ through 4 incorrect returns in the sample:

x = Number of incorrect returns:	0	1	2	3	4
$P(x)$:	0.2066	0.4508	0.2817	0.0578	0.0031

As the preceding example illustrates, the calculation of hypergeometric probabilities is rather cumbersome. Unfortunately, with all the possible combinations of the determining characteristics, N, n, and s, tables of hypergeometric probabilities are not a realistic alternative. It is much easier to use the computer, and Computer Solutions 6.2 describes how we can use Excel or Minitab to determine individual or cumulative hypergeometric probabilities.

COMPUTER 6.2 SOLUTIONS

Hypergeometric Probabilities

These procedures show how to get hypergeometric probabilities associated with x = the number of successes in $n = 4$ trials from a population of $N = 20$ when there are $s = 6$ successes within the population. The trials are without replacement. For each software package:

Procedure I: Individual or cumulative probabilities for $x = 2$ successes.

Procedure II: Complete set of individual or cumulative probabilities for $x = 0$ successes through $x = 4$ successes.

EXCEL

	A	B	C
1	x	P(X = x)	P(X <= x)
2	0	0.2066	0.2066
3	1	0.4508	0.6574
4	2	0.2817	0.9391
5	3	0.0578	0.9969
6	4	0.0031	1.0000

I. Excel procedure for obtaining an individual hypergeometric probability

1. Click on the cell into which Excel is to place the individual probability. From the **Formulas** ribbon, click on the **Insert Function** symbol at the extreme left. Click **Statistical** in the **Category** box. Click **HYPGEOMDIST** in the **Function** box. Click **OK.**
2. Enter the number of successes in the sample (**2**) into the **Sample_s** box. Enter the number of trials (**4**) into the **Number_sample** box. Enter the number of successes in the population (**6**) into the **Population_s** box. Enter the size of the population (**20**) into the **Number_pop** box. Click **OK.** The probability, $P(x = 2) = 0.2817$, will appear in the cell selected in step 1.

[Alternatively, you can use Excel worksheet template **tmhyper**.]

II. Excel procedure for a complete set of individual or cumulative hypergeometric probabilities

1. Enter the column labels into **A1:C1** as shown in the printout above. Enter the lowest possible number of successes in the sample (in this case, **0**) into **A2.** Enter **1** into **A3.** Click on cell **A2** and highlight cells **A2:A3.** Click at the lower right corner of **A3** (the cursor will turn to a "+" sign) and drag downward so cells **A2:A6** will contain the values 0 through 4. (Note that the highest possible number of successes in the sample is 4.)
2. To get the set of individual probabilities: Click on cell **B2** (this is where the first individual probability will go). From the **Formulas** ribbon, click on the **Insert Function** symbol at the extreme left. Click **Statistical** in the **Category** box. Click **HYPGEOMDIST** in the **Function** box. Click **OK.** Now follow step 2 in Excel procedure I, except enter **A2** into the **Sample_s** (number of successes in the sample) box. The first individual probability now appears in **B2.**
3. Click to select cell **B2.** Click the lower right corner of **B2** (the cursor turns to a "+" sign) and drag down until cells **B2:B6** contain the set of individual probabilities from $P(x = 0)$ through $P(x = 4)$.
4. To get the set of cumulative probabilities: Copy the value in cell **B2** to cell **C2** (this will be the first cumulative probability). Click on cell **C3** and enter the equation **=C2+B3.** Click the lower right corner of cell **C3** (the cursor turns to a "+" sign) and drag down until cells **C2:C6** contain the set of cumulative probabilities.

(continued)

MINITAB

```
Probability Density Function            Cumulative Distribution Function

Hypergeometric N = 20, M = 6, and n = 4  Hypergeometric N = 20, M = 6, and n = 4

x   P( X = x )                          x   P( X <= x )
0     0.206605                          0   0.20660
1     0.450774                          1   0.65738
2     0.281734                          2   0.93911
3     0.057792                          3   0.99690
4     0.003096                          4   1.00000
```

I. Minitab procedure for just one individual or cumulative hypergeometric probability

1. Click **Calc**. Select **Probability Distributions.** Click **Hypergeometric.**
2. In the **Hypergeometric Distribution** menu, enter **20** into the **Population size** (**N**) box. Enter **6** into the **Event count in population** (**M**) box. Enter **4** into the **Sample size** (**n**) box. Select **Input constant** and enter the desired *x* value (e.g., **2**) into the box. Select either **Probability** or **Cumulative probability,** depending on whether you want an individual probability or a cumulative probability. Click **OK.**

II. Minitab procedure for a complete set of individual or cumulative hypergeometric probabilities

Enter the desired *x* values (in this case, **0 through 4**) into column **C1**. Follow steps 1 and 2 above, except select **Input column** and enter **C1** into the box.

EXERCISES

6.33 Under what circumstances should the hypergeometric distribution be used instead of the binomial distribution?

6.34 Using the hypergeometric distribution, with $N = 4$, $n = 2$, and $s = 3$, determine the following:
a. $P(x = 0)$ b. $P(x = 1)$ c. $P(x = 2)$

6.35 Using the hypergeometric distribution, with $N = 5$, $n = 2$, and $s = 3$, determine the following:
a. $P(x = 0)$ b. $P(x = 1)$ c. $P(x = 2)$

6.36 In a criminal trial, there are 25 persons who have been approved by both parties for possible inclusion in the eventual jury of 12. Of those who have been approved, there are 14 women and 11 men. If the judge forms the final jury of 12 by randomly selecting individuals from the approved listing, what is the probability that at least half of the eventual jurors will be males?

6.37 A computer firm must send a group of three specialists to give a technical presentation to a potential customer. Five such individuals are anxious to attend, but only two are considered to have strong interpersonal skills. Since the specialists have equal levels of seniority, they have agreed to place their names in a hat and allow management to randomly select the three to attend the presentation. What is the probability that there will be at least one person with strong interpersonal skills among the group that is sent?

6.38 Unknown to a rental car office, 3 of the 12 subcompact models now available for rental are subject to a safety recall soon to be announced by the National Highway Traffic Safety Administration. Five subcompacts will be rented today, and the cars will be randomly selected from those available in the pool. What is the probability that exactly one of the recall-affected cars will be rented today? What is the probability that all three of the recall-affected cars will be rented today?

THE POISSON DISTRIBUTION (6.4)

Description and Applications

The **Poisson distribution** is a discrete probability distribution that is applied to events for which the probability of occurrence over a given span of time, space, or distance is extremely small. The discrete random variable, x, is *the number of times* the event occurs over the given span, and x can be 0, 1, 2, 3, and so on, with (theoretically) no upper limit. Besides its ability to approximate the binomial distribution when n is large and π is small, the Poisson distribution tends to describe phenomena like these:

- Customer arrivals at a service point during a given period of time, such as the number of motorists approaching a tollbooth, the number of hungry persons entering a McDonald's restaurant, or the number of calls received by a company switchboard. In this context it is also useful in a management science technique called queuing (waiting-line) theory.
- Defects in manufactured materials, such as the number of flaws in wire or pipe products over a given number of feet, or the number of knots in wooden panels for a given area
- The number of work-related deaths, accidents, or injuries over a given number of production hours
- The number of births, deaths, marriages, divorces, suicides, and homicides over a given period of time

Although it is closely related to the binomial distribution, the Poisson distribution has a number of characteristics that make it unique. Like the binomial distribution, it is a *family* of distributions, but its shape is determined by just *one* descriptor, its mean. In the Poisson distribution, the mean is called λ (the Greek lowercase letter *lambda*) instead of μ. Also, the mean of the distribution is numerically equal to its variance. The distribution can be summarized as follows:

The Poisson distribution:

The probability that an event will occur exactly x times over a given span of time, space, or distance is

$$P(x) = \frac{\lambda^x \cdot e^{-\lambda}}{x!}$$

where λ = the mean, or $E(x)$; the expected number of occurrences over the given span

e = the mathematical constant, 2.71828 (e is the base of the natural logarithm system)

(*Note:* In the Poisson distribution, the mean and the variance are equal.)

1. When using a pocket calculator to make the computation, remember that $2.71828^{-\lambda}$ is the same as 1 divided by 2.71828^{λ}.
2. Appendix A also contains tables for individual and cumulative Poisson probabilities. The application of these tables will be discussed after we use an example in demonstrating how the Poisson distribution works.

EXAMPLE

Poisson Distribution

In an urban county, health care officials anticipate that the number of births this year will be the same as last year, when 438 children were born—an average of 438/365, or 1.2 births per day. Daily births have been distributed according to the Poisson distribution.

SOLUTION

What Is the Mean of the Distribution?

$E(x) = \lambda$ can be expressed in a number of ways, because it reflects here the number of occurrences over a span of time. Accordingly, the "span of time" can be months ($\lambda = 438/12 = 36.5$ births per month), weeks ($\lambda = 438/52 = 8.42$ births per week), or days ($\lambda = 438/365 = 1.2$ births per day). For purposes of our example, we will be using $\lambda = 1.2$ births per day in describing the distribution.

For Any Given Day, What Is the Probability That No Children Will Be Born?

Using the Poisson formula, with $\lambda = 1.2$ births per day, to find $P(x = 0)$:

$$P(x) = \frac{\lambda^x \cdot e^{-\lambda}}{x!}$$

$$\text{and} \quad P(0) = \frac{1.2^0 \cdot e^{-1.2}}{0!} = \frac{1 \cdot 0.30119}{1} = \boxed{0.3012}$$

Calculate Each of the Following Probabilities: $P(x = 1)$, $P(x = 2)$, $P(x = 3)$, $P(x = 4)$, and $P(x = 5)$

Using the same approach as in the preceding calculations, with $\lambda = 1.2$ mean births per day and values of x from 1 to 5 births per day:

$$P(x = 1 \text{ birth}) = \frac{1.2^1 \cdot e^{-1.2}}{1!} = \frac{(1.2000)(0.30119)}{1} = \boxed{0.3614}$$

$$P(x = 2 \text{ births}) = \frac{1.2^2 \cdot e^{-1.2}}{2!} = \frac{(1.4400)(0.30119)}{2 \cdot 1} = \boxed{0.2169}$$

$$P(x = 3 \text{ births}) = \frac{1.2^3 \cdot e^{-1.2}}{3!} = \frac{(1.7280)(0.30119)}{3 \cdot 2 \cdot 1} = \boxed{0.0867}$$

$$P(x = 4 \text{ births}) = \frac{1.2^4 \cdot e^{-1.2}}{4!} = \frac{(2.0736)(0.30119)}{4 \cdot 3 \cdot 2 \cdot 1} = \boxed{0.0260}$$

$$P(x = 5 \text{ births}) = \frac{1.2^5 \cdot e^{-1.2}}{5!} = \frac{(2.4883)(0.30119)}{5 \cdot 4 \cdot 3 \cdot 2 \cdot 1} = \boxed{0.0062}$$

If we were to continue, we'd find that $P(x = 6) = 0.0012$ and $P(x = 7) = 0.0002$, with both rounded to four decimal places. Including these probabilities

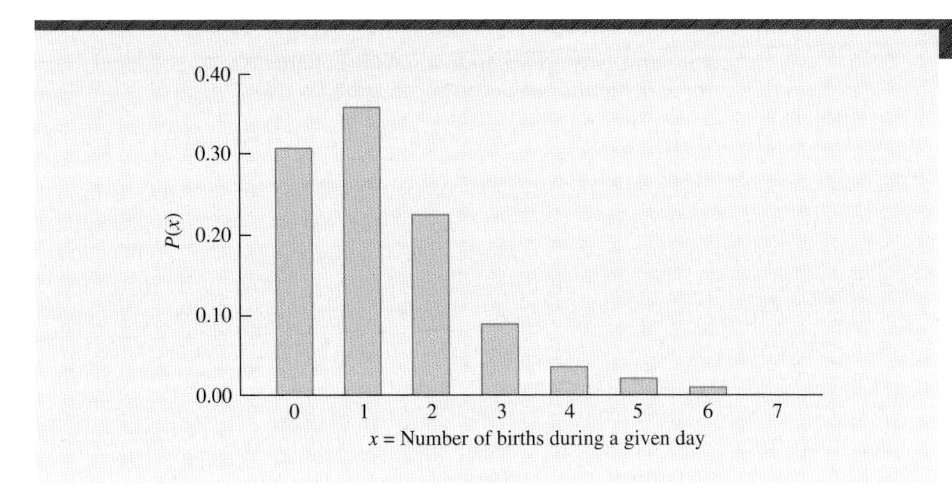

FIGURE 6.4

For the example discussed in the text, the Poisson probability distribution for the number of births per day is skewed to the right. The mean of the distribution is 1.2 births per day, and this descriptor is all that is needed to determine the Poisson probability for each value of x.

with those just calculated, the result would be the following Poisson probability distribution. As Figure 6.4 shows, the distribution is positively skewed.

x	0	1	2	3	4	5	6	7
$P(x)$	0.3012	0.3614	0.2169	0.0867	0.0260	0.0062	0.0012	0.0002

What Is the Probability That No More Than One Birth Will Occur on a Given Day?

Since the events are mutually exclusive, this can be calculated as $P(x = 0) + P(x = 1)$, or $0.3012 + 0.3614$, and $P(x \leq 1) = 0.6626$.

Using the Poisson Tables

The Poisson calculations are much easier than those for the binomial. For example, you need compute $e^{-\lambda}$ only one time for a given value of λ, since the same term is used in finding the probability for each value of x. Using either the individual or the cumulative Poisson distribution table in Appendix A (Tables A.3 and A.4, respectively) makes the job even easier. Table 6.3 shows a portion of Table A.3, while Table 6.4 shows its cumulative counterpart. If the problem on which you are working happens to include one of the λ values listed, you need only refer to the $P(x = k)$ or the $P(x \leq k)$ values in the appropriate table.

In the previous example, λ was 1.2 births per day. Referring to the $\lambda = 1.2$ column of Table 6.3, we see the same probabilities just calculated for $P(x = 1)$ through $P(x = 7)$. For example, in the fifth row of the $\lambda = 1.2$ column, the entry for $P(x = 4)$ is 0.0260. To quickly find the probability that there would be no more than 2 births on a given day, we can refer to the cumulative Poisson probabilities in Table 6.4 and see that $P(x \leq 2)$ is 0.8795.

In some cases, the sum of the individual probabilities in a particular Poisson table will not add up to 1.0000. This is due to (1) rounding and (2) the fact that there is no upper limit to the possible values of x. Some values not listed in the table have $P(x) > 0$, but would appear as 0.0000 because of their extremely small chance of occurrence.

TABLE 6.3

A portion of the table of *individual* Poisson probabilities from Appendix A.

Poisson Distribution, Individual Probabilities for x = number of occurrences, prob ($x = k$)

k \ λ	1.1	1.2	1.3	1.4	1.5	1.6	1.7	1.8	1.9	2.0
0	0.3329	0.3012	0.2725	0.2466	0.2231	0.2019	0.1827	0.1653	0.1496	0.1353
1	0.3662	0.3614	0.3543	0.3452	0.3347	0.3230	0.3106	0.2975	0.2842	0.2707
2	0.2014	0.2169	0.2303	0.2417	0.2510	0.2584	0.2640	0.2678	0.2700	0.2707
3	0.0738	0.0867	0.0998	0.1128	0.1255	0.1378	0.1496	0.1607	0.1710	0.1804
4	0.0203	0.0260	0.0324	0.0395	0.0471	0.0551	0.0636	0.0723	0.0812	0.0902
5	0.0045	0.0062	0.0084	0.0111	0.0141	0.0176	0.0216	0.0260	0.0309	0.0361
6	0.0008	0.0012	0.0018	0.0026	0.0035	0.0047	0.0061	0.0078	0.0098	0.0120
7	0.0001	0.0002	0.0003	0.0005	0.0008	0.0011	0.0015	0.0020	0.0027	0.0034
8	0.0000	0.0000	0.0001	0.0001	0.0001	0.0002	0.0003	0.0005	0.0006	0.0009
9			0.0000	0.0000	0.0000	0.0000	0.0001	0.0001	0.0001	0.0002
10							0.0000	0.0000	0.0000	0.0000

For $\lambda = 1.2$ births per day, the probability that there will be exactly 3 births on a given day is $P(x = 3)$, or 0.0867.

TABLE 6.4

A portion of the table of *cumulative* Poisson probabilities from Appendix A.

Poisson Distribution, Cumulative Probabilities for x = number of occurrences, prob ($x \leq k$)

k \ λ	1.1	1.2	1.3	1.4	1.5	1.6	1.7	1.8	1.9	2.0
0	0.3329	0.3012	0.2725	0.2466	0.2231	0.2019	0.1827	0.1653	0.1496	0.1353
1	0.6990	0.6626	0.6268	0.5918	0.5578	0.5249	0.4932	0.4628	0.4337	0.4060
2	0.9004	0.8795	0.8571	0.8335	0.8088	0.7834	0.7572	0.7306	0.7037	0.6767
3	0.9743	0.9662	0.9569	0.9463	0.9344	0.9212	0.9068	0.8913	0.8747	0.8571
4	0.9946	0.9923	0.9893	0.9857	0.9814	0.9763	0.9704	0.9636	0.9559	0.9473
5	0.9990	0.9985	0.9978	0.9968	0.9955	0.9940	0.9920	0.9896	0.9868	0.9834
6	0.9999	0.9997	0.9996	0.9994	0.9991	0.9987	0.9981	0.9974	0.9966	0.9955
7	1.0000	1.0000	0.9999	0.9999	0.9998	0.9997	0.9996	0.9994	0.9992	0.9989
8			1.0000	1.0000	1.0000	1.0000	0.9999	0.9999	0.9998	0.9998
9							1.0000	1.0000	1.0000	1.0000

For $\lambda = 1.2$ births per day, the probability that there will be no more than 3 births on a given day is $P(x \leq 3)$, or 0.9662.

As with the binomial distribution, Excel and Minitab can provide either individual or cumulative Poisson probabilities. The procedures and the results are shown in Computer Solutions 6.3.

The Poisson Approximation to the Binomial Distribution

When n is relatively large and π (the probability of a success in a given trial) is small, the binomial distribution can be closely approximated by the Poisson distribution. As a rule of thumb, the binomial distribution can be satisfactorily approximated by the Poisson whenever $n \geq 20$ and $\pi \leq 0.05$. Under these conditions, we can just use $\lambda = n\pi$ and find the probability of each value of x using the Poisson distribution.

COMPUTER 6.3 SOLUTIONS

Poisson Probabilities

These procedures show how to get probabilities associated with a Poisson distribution with mean = 1.2.

For each software package:

Procedure I: Individual or cumulative probabilities for $x = 2$ occurrences.

Procedure II: Complete set of individual or cumulative probabilities for $x = 0$ occurrences through $x = 8$ occurrences.

EXCEL

	A	B	C
1	x	P(X = x)	P(X <= x)
2	0	0.3012	0.3012
3	1	0.3614	0.6626
4	2	0.2169	0.8795
5	3	0.0867	0.9662
6	4	0.0260	0.9923
7	5	0.0062	0.9985
8	6	0.0012	0.9997
9	7	0.0002	1.0000
10	8	0.0000	1.0000

I. Excel procedure for just one individual or cumulative Poisson probability

1. Click on the cell into which Excel will place the probability. From the **Formulas** ribbon, click on the **Insert Function** symbol at the extreme left. Click **Statistical** in the **Category** box. Click **POISSON** in the **Function** box. Click **OK.**
2. Enter the number of occurrences (**2**) into the **X** box. Enter the mean of the distribution (**1.2**) into the **Mean** box. To obtain the individual probability, $P(x = 2)$, enter **false** into the **Cumulative** box. To obtain the cumulative probability, $P(x \leq 2)$, enter **true** into the **Cumulative** box. Click **OK.**

(continued)

II. Excel procedure for a complete set of individual or cumulative Poisson probabilities

1. Enter the column labels into **A1:C1** as shown in the printout. Enter the lowest number of occurrences in the set (in this case, **0**) into **A2.** Enter **1** into **A3.** Click on cell **A2** and highlight cells **A2:A3.** Click at the lower right corner of **A3** (the cursor will turn to a "+" sign) and drag downward so cells **A2:A10** will contain the values 0 through 8.
2. To get the set of individual probabilities: Click on cell **B2** (this is where the first individual probability will go). Follow the remainder of the steps in Excel procedure I, except enter **A2** into the **X** (number of occurrences) box and enter **false** into the **Cumulative** box. The first probability now appears in **B2.** Click to select cell **B2.** Click the lower right corner of **B2** (the cursor turns to a "+" sign) and drag down until cells **B2:B10** contain the set of individual probabilities.
3. To get the set of cumulative probabilities: Click on cell **C2** (this is where the first cumulative probability will go). Follow the remainder of the steps in Excel procedure I, except enter **A2** into the **X** (number of occurrences) box and enter **true** into the **Cumulative** box. The first probability now appears in **C2.** Click to select cell **C2.** Click the lower right corner of **C2** (the cursor turns to a "+" sign) and drag down until cells **C2:C10** contain the set of cumulative probabilities.

MINITAB

```
Probability Density Function        Cumulative Distribution Function
Poisson with mean = 1.2             Poisson with mean = 1.2
x   P( X = x )                      x   P( X <= x )
0      0.301194                     0         0.30119
1      0.361433                     1         0.66263
2      0.216860                     2         0.87949
3      0.086744                     3         0.96623
4      0.026023                     4         0.99225
5      0.006246                     5         0.99850
6      0.001249                     6         0.99975
7      0.000214                     7         0.99996
8      0.000032                     8         1.00000
```

I. Minitab procedure for just one individual or cumulative Poisson probability

1. Click **Calc.** Select **Probability Distributions.** Click **Poisson.**
2. In the **Poisson Distribution** menu, enter **1.2** into the **Mean** box. Select **Input constant** and enter the desired *x* value (e.g., **2**) into the box. Select either **Probability** or **Cumulative probability,** depending on whether you want an individual probability or a cumulative probability. Click **OK.**

II. Minitab procedure for a complete set of individual or cumulative binomial probabilities

Enter the desired *x* values (in this case, **0** through **8**) into column **C1.** Follow steps 1 and 2 above, except select **Input column** and enter **C1** into the box.

EXAMPLEEXAMPLE

EXAMPLE

Poisson Approximation to Binomial

Past experience has shown that 1.0% of the microchips produced by a certain firm are defective. A sample of 30 microchips is randomly selected from the firm's production. If x = the number of defective microchips in the sample, determine $P(x = 0)$, $P(x = 1)$, $P(x = 2)$, $P(x = 3)$, and $P(x = 4)$.

SOLUTION

The Binomial Approach

The descriptors for the binomial distribution are $\pi = 0.01$, the chance that any given microchip will be defective, and $n = 30$, the number of microchips in the sample. Although we will skip the actual binomial calculations, we would use $\pi = 0.01$ and $n = 30$ in arriving at the probabilities shown below.

The Poisson Approximation Approach

The only descriptor for the Poisson distribution is $\lambda = n\pi = 30(0.01)$, or $\lambda = 0.3$ microchips expected to be defective in the sample. Referring to the Poisson table for $\lambda = 0.3$, we find the following probabilities for $x = 0$ through $x = 4$. Note the very close similarity between each binomial probability and its Poisson approximation:

	P(*x*), Using the	
x = Number of Defective Microchips in the Sample	Binomial Distribution	Poisson Approximation
0	0.7397	0.7408
1	0.2242	0.2222
2	0.0328	0.0333
3	0.0031	0.0033
4	0.0002	0.0003

EXERCISES

6.39 For a discrete random variable that is Poisson distributed with $\lambda = 2.0$, determine the following:
a. $P(x = 0)$ b. $P(x = 1)$ c. $P(x \leq 3)$
d. $P(x \geq 2)$

6.40 For a discrete random variable that is Poisson distributed with $\lambda = 9.6$, determine the following:
a. $P(x = 7)$ b. $P(x = 9)$ c. $P(x \leq 12)$
d. $P(x \geq 10)$

6.41 In 2006, there were about 490 motor vehicle thefts for every 100,000 registrations. Assuming (1) a Poisson distribution, (2) a community with a comparable theft rate and 1000 registered motor vehicles, and (3) x = the number of vehicles stolen during the year in that community, determine the following: Source: *Statistical Abstract of the United States 2009*, pp. 188, 672.
a. $E(x)$ b. $P(x = 3)$ c. $P(x = 5)$
d. $P(x \leq 8)$ e. $P(3 \leq x \leq 10)$

6.42 Arrivals at a walk-in optometry department in a shopping mall have been found to be Poisson distributed with a mean of 2.5 potential customers arriving per hour. If x = number of arrivals during a given hour, determine the following:
a. $E(x)$ b. $P(x = 1)$ c. $P(x = 3)$
d. $P(x \leq 5)$ e. $P(2 \leq x \leq 6)$

6.43 The U.S. divorce rate has been reported as 3.6 divorces per 1000 population. Assuming that this rate applies to a small community of just 500 people and is Poisson distributed, and that x = the number of divorces in this community during the coming year, determine the following: Source: *New York Times Almanac 2009*, p. 294.
a. $E(x)$ b. $P(x = 1)$ c. $P(x = 4)$
d. $P(x \leq 6)$ e. $P(2 \leq x \leq 5)$

6.44 During the 12 P.M.–1 P.M. noon hour, arrivals at a curbside banking machine have been found to be Poisson distributed with a mean of 1.3 persons per minute. If x = number of arrivals during a given minute, determine the following:
a. $E(x)$ b. $P(x = 0)$ c. $P(x = 1)$
d. $P(x \leq 2)$ e. $P(1 \leq x \leq 3)$

6.45 During the winter heating season in a northern state, Howard's Heating and Cooling Service receives an average of 3.1 emergency service calls per day from heating customers in their area. Howard has increased personnel and equipment resources to better handle such calls, and his company is now able to satisfy a maximum of 8 emergency service calls per day. During a typical day during the heating season, and assuming the daily number of emergency service calls to be Poisson distributed, what is the probability that Howard will receive more emergency service calls tomorrow than his company is able to handle? Considering this probability, would it seem advisable for Howard to increase or to decrease the level of personnel and equipment he devotes to handling such calls?

6.46 Over the past year, a university's computer system has been struck by a virus at an average rate of 0.4 viruses per week. The university's information technology managers estimate that each time a virus occurs, it costs the university $1000 to remove the virus and repair the damages it has caused. Assuming a Poisson distribution, what is the probability that the university will have the good fortune of being virus-free during the upcoming week? During this same week, what is the expected amount of money that the university will have to spend for virus removal and repair?

6.47 According to the Mortgage Bankers Association of America, the foreclosure rate on home mortgages in 2007 was 2%. Assuming that this rate is applicable to a community where 500 homes have mortgages, use the Poisson approximation to the binomial distribution to determine the following for x = the number of foreclosures in this community during the coming year: Source: *Statistical Abstract of the United States 2009*, p. 727.

a. $E(x)$ b. $P(x = 7)$ c. $P(x = 10)$
d. $P(x \leq 12)$ e. $P(8 \leq x \leq 15)$

6.48 The Federal Aviation Administration reports that American Airlines got 6.06 complaints of mishandled baggage per 1000 passengers in 2008. Assuming this rate applies to the next 500 passengers who depart from Portland, Oregon, on American Airlines flights, what is the probability that at least 5 people will have their baggage mishandled? That no more than 3 people will have their baggage mishandled? Source: *The New York Times Almanac 2009*, p. 426.

6.49 Taxicab drivers are more likely to be murdered on the job than members of any other occupation, even police officers and security guards. The annual murder rate for cab drivers is 30.0 homicides per 100,000 workers, compared to just 0.6 per 100,000 for all occupations. Assume that taxicab companies in a large Northeastern city employ a total of 3000 drivers and that the family of any driver murdered on the job receives $50,000 from the city. What is the probability that at least two of the city's taxicab drivers will be murdered on the job during the next year? What is the approximate expected value for the amount of money the city will be providing to families of murdered taxicab drivers during the next year? Source: "Occupational Homicide Rate Is Highest Among Taxi Drivers," *USA Today*, May 2, 2000, p. 3A.

6.50 A manufacturing plant's main production line breaks down an average of 2.4 times per day. Whenever the line goes down, it costs the company $500 in maintenance, repairs, and lost production. What is the probability that the production line will break down at least 3 times tomorrow? What is the approximate expected value for the amount of money that production line breakdowns will cost the company each day?

6.51 The company in Exercise 6.50 has disciplined a worker who was suspected of pilfering tools and supplies from the plant. The very next day, the production line broke down 9 times. Management has confronted the union with accusations of sabotage on the line, but the union president says it's just a coincidence that the production line happened to break down so many times the day after the worker was disciplined. Using probabilities appropriate to your discussion, comment on how much of a coincidence this high number of breakdowns would appear to be.

6.5 SIMULATING OBSERVATIONS FROM A DISCRETE PROBABILITY DISTRIBUTION

Besides describing the exact probability distribution for a discrete random variable, statistical software can also randomly select observations from the distribution itself. For example, we can use Excel or Minitab to randomly select observations from a given Poisson distribution, then display them in the form of a frequency distribution. Computer Solutions 6.4 describes the necessary procedures. Because we are using two different software packages, there will be two different sets of randomly selected data. We will display and discuss the results for Minitab.

Table 6.5 shows the individual $P(x)$ probabilities as computed by Minitab for three different Poisson distributions ($\lambda = 0.1$, $\lambda = 1.2$, and $\lambda = 5.0$). Next to each

COMPUTER 6.4 SOLUTIONS

Simulating Observations from a Discrete Probability Distribution

These procedures simulate 10,000 observations from a Poisson distribution with mean = 1.2. The procedures for simulating observations from a binomial distribution will differ only slightly from those described here.

EXCEL

1. Start with a blank worksheet. From the **Data** ribbon, click **Data Analysis** in the rightmost menu section. Click **Random Number Generation.** Click **OK.**
2. Enter **1** into the **Number of Variables** box. Enter **10000** into the **Number of Random Numbers** box. Within the **Distribution** box, select **Poisson.** Enter the mean (**1.2**) into the **Lambda** box. Select **Output Range** and enter **A1** into the box. Click **OK.** The 10,000 simulated observations will be located in A1:A10000.

MINITAB

1. Click **Calc.** Select **Random Data.** Click **Poisson.** Enter **10000** into the **Number of rows of data to generate** box. Enter **C1** (the destination column) into the **Store in columns** box. Enter **1.2** into the **Mean** box. Click **OK.** The 10,000 simulated observations will be located in **C1.**
2. To obtain the frequencies for the different values of *x*, click **Stat,** select **Tables,** and click **Tally Individual Variables.** Enter **C1** into the **Variables** box and select **Counts.** Click **OK.** The relative frequencies shown in part B of the right side of Table 6.5 are obtained by dividing each of the frequencies by 10,000.

The frequency distributions in Table 6.6 can be generated using the Minitab procedure described previously in Computer Solutions 2.1. As shown in this earlier procedure, we have the option of displaying the frequencies on the chart istelf.

theoretical probability is the relative frequency with which the x value occurred when Minitab randomly selected 10,000 observations from that distribution.

Note that the relative frequencies for the randomly selected x values are very close, but not necessarily the same as the theoretical probabilities listed in the probability distribution. As we discussed in the previous chapter, $P(\text{heads}) = 0.5$ for a coin flip, but flipping a coin 10 times doesn't mean that you're certain to get heads exactly 5 times. However, due to the law of large numbers, observed relative frequencies will more closely approach the theoretical probability distribution as the number of observations grows larger.

Parts A through C of Table 6.6 show Minitab frequency distributions for 10,000 observations randomly selected from the corresponding probability distribution of Table 6.5. In comparing the three frequency distributions of Table 6.6, notice how they become more symmetrical as λ is increased when we proceed from A to C. Although the Poisson distribution is always positively skewed, it becomes more symmetrical as its mean, λ, becomes larger.

The ability to select random observations from a known or assumed probability distribution also has applications in a number of management science areas. Such applications typically involve using one or more such distributions to mathematically represent real-world phenomena of interest. In general, their purpose is to help the management scientist gain insight into how the processes

TABLE 6.5

Besides describing a probability distribution, the computer can also randomly select observations from the distribution itself.

Minitab output. Poisson probability distribution for the discrete random variable, x:

When Minitab randomly selected 10,000 observations from the distribution, the relative frequency with which each x value appeared:

A.

```
Probability Density Function
Poisson with mean = 0.1
x         P( X = x )
0             0.904837
1             0.090484
2             0.004524
3             0.000151
4             0.000004
```

Relative frequency
0.9056
0.0892
0.0050
0.0002
0.0000

B.

```
Probability Density Function
Poisson with mean = 1.2
x         P( X = x )
0             0.301194
1             0.361433
2             0.216860
3             0.086744
4             0.026023
5             0.006246
6             0.001249
7             0.000214
8             0.000032
```

Relative frequency
0.3060
0.3588
0.2085
0.0918
0.0273
0.0063
0.0011
0.0002
0.0000

C.

```
Probability Density Function
Poisson with mean = 5
 x        P( X = x )
 0            0.006738
 1            0.033690
 2            0.084224
 3            0.140374
 4            0.175467
 5            0.175467
 6            0.146223
 7            0.104445
 8            0.065278
 9            0.036266
10            0.018133
11            0.008242
12            0.003434
13            0.001321
14            0.000472
15            0.000157
16            0.000049
```

Relative frequency
0.0067
0.0343
0.0848
0.1424
0.1824
0.1693
0.1433
0.1059
0.0637
0.0345
0.0195
0.0086
0.0029
0.0010
0.0005
0.0001
0.0001

TABLE 6.6

Minitab frequency distributions for 10,000 observations randomly taken from each of the three Poisson probability distributions in Table 6.5.

A. 10,000 random observations from Poisson distribution with mean $\lambda = 0.1$

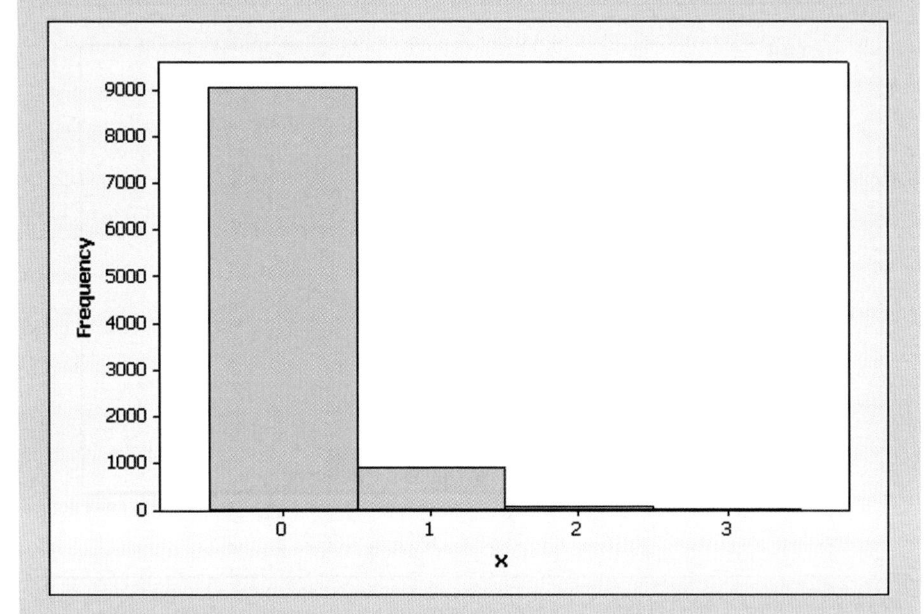

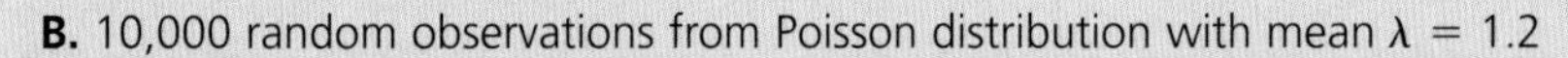

B. 10,000 random observations from Poisson distribution with mean $\lambda = 1.2$

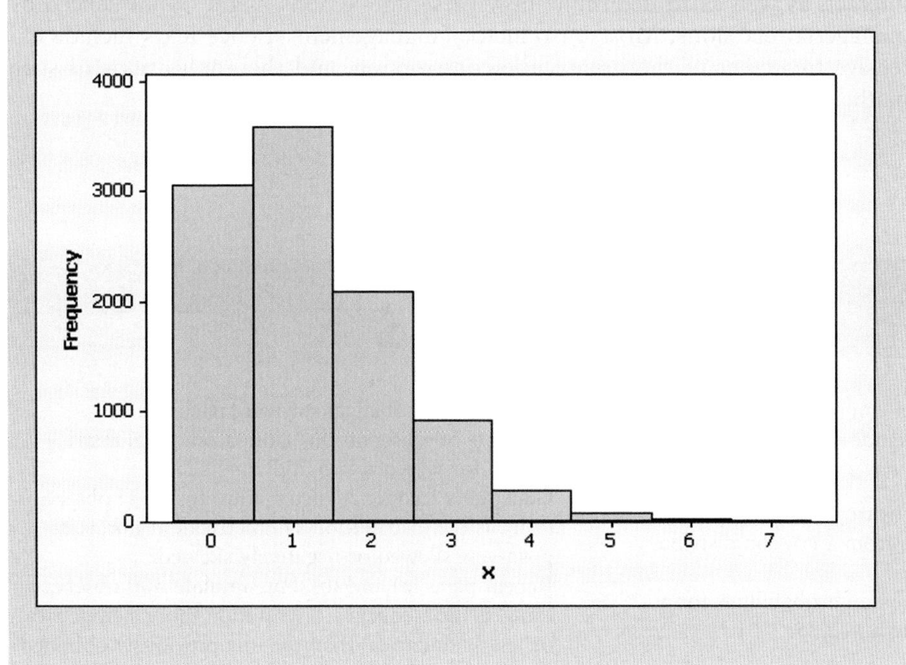

(*continued*)

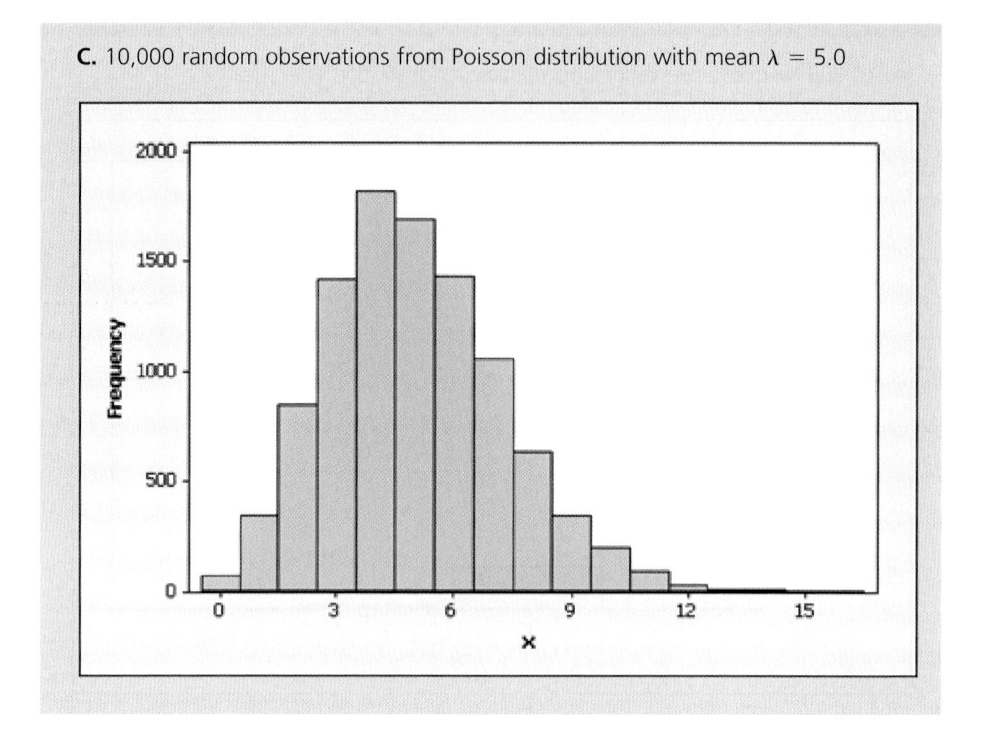

interact, as well as to determine how these interactions might be influenced by managerial decisions. Most introductory management science texts include extensive treatment of the nature, the construction, and the applications of such mathematical models of the real world.

EXERCISES

Note: These exercises require a computer and statistical software capable of selecting random observations from discrete probability distributions.

6.52 Using your statistical software package:

a. Simulate 100 observations from a Poisson distribution having $\lambda = 2.35$.
b. Generate the individual Poisson probabilities for a Poisson distribution having $\lambda = 2.35$.
c. Repeat part (a).
d. Repeat part (b).
e. When part (a) is repeated, the results are now different. However, when part (b) is repeated, the results are the same. What does this say about frequency distributions versus probability distributions?

6.53 Using your statistical software package:

a. Simulate 50 observations from a binomial distribution for which $n = 5$ and $\pi = 0.2$.
b. Generate a frequency distribution for these observations. Is the distribution symmetrical? If not, is it positively skewed or negatively skewed?
c. Repeat parts (a) and (b), but simulate 500 observations. Is the frequency distribution more symmetrical or less symmetrical than the one previously obtained?
d. Comment on how the shape of the distribution would change if you were to continue increasing the number of simulated observations.

SUMMARY (6.6)

Discrete random variable and its probability distribution

A discrete random variable can take on only certain values along a range or continuum, and there are gaps between these possible values. On the other hand, a continuous random variable can have any value along a range or continuum. This chapter has focused on discrete random variables, while the next concentrates on continuous random variables.

The probability distribution for a discrete random variable lists all possible values the variable can assume, along with their respective probabilities of occurrence. There are several types of discrete distributions, each with its own characteristics and applications.

Binomial distribution and the Bernoulli process

The binomial distribution deals with consecutive trials, each of which has only two possible outcomes, generally referred to as *success* or *failure*. This distribution assumes a Bernoulli process, in which the trials are independent of each other, and the probability of success remains constant from one trial to the next.

Hypergeometric distribution

Like the binomial distribution, the hypergeometric distribution deals with consecutive trials, and there are only two possible outcomes for each trial. The trials are not statistically independent, however, and the probability of a success is not constant from one trial to the next.

Poisson distribution

The Poisson distribution applies to events for which the probability of occurrence over a given span of time, space, or distance is extremely small. In this distribution, the random variable is the number of occurrences in the specified span of time, space, or distance. For a Bernoulli process in which the number of trials is large and the probability of success on any given trial is very small, the Poisson distribution can be used as an approximation to the binomial distribution.

Determining probabilities for a discrete random variable

Appendix A contains tables for the binomial and Poisson distributions. However, computer statistical packages enable the user to determine the probabilities associated with a given distribution without the need to refer to such tables.

Simulation of observations from a discrete probability distribution

Some computer packages can randomly select a sample of observations from a probability distribution that has already been specified. In addition to being useful in more advanced management science applications such as simulation models intended to represent real-world phenomena, this capability enables the user to easily view the approximate shape of a given probability distribution. Based on the law of large numbers, the observed relative frequencies will more closely approach the theoretical probability distribution as the number of observations grows larger.

EQUATIONS

Characteristics of a Discrete Probability Distribution

- For any value of x, $0 \leq P(x) \leq 1$.
- The values of x are *exhaustive:* The probability distribution includes all possible values.
- The values of x are *mutually exclusive:* Only one value can occur for a given experiment.
- The sum of their probabilities is 1, or $\Sigma P(x_i) = 1.0$.

General Formulas: Mean and Variance of a Discrete Probability Distribution

- Mean:

$$\mu = E(x) \quad \text{or} \quad \mu = \Sigma x_i P(x_i) \quad \text{for all possible values of } x$$

- Variance:

$$\sigma^2 = E[(x - \mu)^2] \quad \text{or} \quad \sigma^2 = \Sigma(x_i - \mu)^2 P(x_i)$$

for all possible values of x, and the standard deviation is $\sigma = \sqrt{\sigma^2}$

The Binomial Probability Distribution

The probability of exactly x successes in n trials is

$$P(x) = \frac{n!}{x!(n-x)!}\pi^x(1-\pi)^{n-x}$$

where n = number of trials
x = number of successes
π = the probability of success in any given trial
$(1 - \pi)$ = the probability of failure in any given trial

- Mean:

$$\mu = E(x) = n\pi$$

- Variance:

$$\sigma^2 = E[(x - \mu)^2] = n\pi(1 - \pi)$$

The Hypergeometric Distribution

The probability of exactly x successes in n trials that are not independent is

$$P(x) = \frac{\binom{s}{x} \cdot \binom{N-s}{n-x}}{\binom{N}{n}}$$

where N = size of the population
n = size of the sample
s = number of successes in the population
x = number of successes in the sample

- Mean:

$$\mu = E(x) = \frac{ns}{N}$$

- Variance:

$$\sigma^2 = E[(x - \mu)^2] = \frac{ns(N - s)}{N^2} \cdot \frac{N - n}{N - 1}$$

The Poisson Probability Distribution

The probability that an event will occur exactly x times over a given span of time, space, or distance is

$$P(x) = \frac{\lambda^x \cdot e^{-\lambda}}{x!}$$

where λ = the mean, or $E(x)$; the expected number of occurrences over the given span

e = the mathematical constant, 2.71828 (e is the base of the natural logarithm system)

(*Note:* In the Poisson distribution, the mean and the variance are equal.)

CHAPTER EXERCISES

6.54 Twenty percent of the population of Maryland describe themselves as binge drinkers. For a randomly selected group of 5 Maryland residents, use the appropriate statistical table to describe the probability distribution for x = the number in the sample who describe themselves as binge drinkers. Source: *Statistical Abstract of the United States 2009*, p. 130.

6.55 During the 2009 tax filing season, 15.8% of all individual U.S. tax returns were prepared by H&R Block. For a random selection of 3 tax returns, describe the probability distribution for x = the number in the sample whose returns were prepared by H&R Block. Source: H&R Block, Inc., *Fast Facts 2009*.

6.56 A company has installed five intrusion-detection devices in an office building. The devices are very sensitive and, on any given night, each one has a 0.1 probability of mistakenly sounding an alarm when no intruder is present. If two or more of the devices are activated on the same night, the intrusion system automatically sends a signal to the local police. If no intruder is present on a given night, what is the probability that the system will make the mistake of calling the police?

6.57 Bill has to sell three more cars this month in order to meet his quota. Tonight he has after-dinner appointments with five prospective customers, each of whom happens to be interested in a different car. If he has a 30% chance of success with each customer, what is the probability that he will meet his quota by tomorrow morning?

6.58 A mining company finds that daily lost-work injuries average 1.2. If the local union contract has a clause requiring that the mine be shut down as soon as three workers incur lost-work injuries, on what percentage of the days will the mine be operational throughout the day?

6.59 In Exercise 6.57, suppose that three of the customers are interested in the same car, and that they will go elsewhere if it has already been sold. Would it be appropriate to use the binomial distribution under these conditions? Why or why not?

6.60 Unknown to a quality-control inspector, 20% of a very large shipment of electric switches are defective. The inspector has been told to reject the shipment if, in a sample of 15 switches, 2 or more are defective. What is the probability that the shipment will be rejected?

6.61 During 2007, the crash rate for commuter air carriers was 1.0 per 100,000 flying hours. Assuming this rate continues, what is the probability that there will be no more than one crash in the next 50,000 flying hours? Source: *Statistical Abstract of the United States 2009*, p. 662.

6.62 Repeat Exercise 6.29, but under the assumption that there are only 20 bearings in the supply bin, 2 of which are defective.

6.63 The owner of a charter fishing boat has found that 12% of his passengers become seasick during a half-day fishing trip. He has only two beds below deck to accommodate those who become ill. About to embark on a typical half-day trip, he has 6 passengers on board. What is the probability that there will be enough room below deck to accommodate those who become ill?

6.64 A trucking company has found that its trucks average 0.2 breakdowns during round trips from New York to Los Angeles.

a. What is the probability that a single truck will make the complete trip without experiencing a breakdown?
b. If 3 trucks are assigned to a NY/LA round trip, what is the probability that at least 2 of them will make the complete trip without experiencing a breakdown?

6.65 The Bureau of Labor Statistics says that baggage handlers, trash haulers, and coal miners are among the occupations having the highest risk of a debilitating injury or illness. For coal miners working over the course of a year, the injury rate has been reported as 3.6 injuries per 100 coal miners. For a company that employs 200 coal miners, what is the probability that at least 10 of its workers will experience a debilitating injury or illness over the next year? Source: "Workers at Risk," *USA Today*, March 15, 2006, p. 6D.

6.66 A large plumbing contractor buys eight 200-foot rolls from the producer described in Exercise 6.70. Determine the probability that no more than three of the eight rolls will be flawless.

6.67 According to the U.S. Centers for Disease Control and Prevention, Americans 85 years of age and older experience flu-associated respiratory and circulatory illness at an annual rate of approximately 1200 cases per 100,000 persons. If a rural county has 500 seniors in this age group, what is the probability that no more than 5 of them will experience such an illness during the coming year? If, in such a county, 15 seniors in this age group were to come down with such an illness during the coming year, would this be considered rather unusual? Explain your conclusion. Source: Anita Manning, "Tens of Thousands More People Are Hospitalized for Flu Than Expected," *USA Today*, September 15, 2004, p. 5D.

6.68 In 2004, 50% of all U.S. families had a net worth less than \$93,000. What is the probability that exactly three of a randomly selected sample of five families would have a net worth less than \$93,000? Source: *Statistical Abstract of the United States 2009*, p. 459.

6.69 Although they have no commercials, public television stations periodically conduct campaigns during which programming is occasionally interrupted by appeals for subscriber donations. A local public television station has found that, during these breaks, the number of calls averages 3 per minute. It has recently upgraded the switchboard so it can handle as many as 5 calls per minute. The station is now in the middle of a program break. What is the probability that the switchboard will be overloaded during the next minute?

6.70 A producer of copper tubing has found that, on the average, five flaws occur per 2000 feet of tubing. The tubing is sold in continuous rolls of 200 feet each. What is the probability that a randomly selected 200-foot roll will have no flaws?

6.71 A gorilla once made headlines in 400 newspapers and appeared on three national television shows after correctly "selecting" the winner of 13 out of 20 football games. He had made his "choice" for each game by grabbing one of two pieces of paper on which the names of the teams were written. If a great many gorillas had the same opportunity, what percentage of them would be at least this lucky? Source: "Gone Ape," *Pittsburgh Press*, June 8, 1986.

6.72 Seven of the 15 campus police officers available for assignment to the auditorium in which a local politician is to speak have received advanced training in crowd control. If 5 officers are randomly selected for service during the speech, what is the probability that exactly 2 of them will have had advanced training in crowd control? What is the probability that at least 3 of them will have had advanced training?

6.73 It has been estimated that over 90% of the world's large corporations are actively involved in data warehousing. Using the 90% figure, and considering a random sample of 10 large corporations, what is the probability that at least 7 of them are actively involved in data warehousing?

6.74 Fred and Emily are among the 50 finalists in a lottery in which 5 winners will be randomly selected to receive a free vacation in Hawaii. Using the hypergeometric distribution, determine the probability that exactly one of these two individuals will win a free vacation.

6.75 A computer manufacturer that sells notebook computers directly to customers has been having problems with one of its models—it seems that about 10% of the units are inoperable upon arrival to the customer. They are functional when tested at the factory, but there is some problem associated with the shipping/handling process that the company's engineers have not yet been able to identify and correct. A highly valued customer places an emergency order for 8 of these notebooks, and wants them delivered overnight and ready for use in the field by tomorrow. Wanting to be honest with this customer, and wishing to help ensure that the customer receives at least 8 working computers, the manufacturer offers to send 9 units instead of the 8 that the customer needs. The customer accepts this gesture of goodwill and 9 units are sent. What is the probability that the customer will end up with at least 8 working computers?

6.76 A town council includes 10 Democrats and 15 Republicans. In response to a developer's request for a variance that would allow her to build a convenience store in an area of town currently zoned as residential, the mayor is going to randomly select 5 council members for a committee to examine the issue. What is the probability that there will be at least 3 Republicans on the committee?

6.77 A country club has found that the number of complete rounds played by a member during a given outing averages 1.3 rounds. What is the probability that a golfer selected at random will complete at least 2 rounds of golf on his or her next outing?

6.78 A tire manufacturer has advertised that only 2% of its tires have surface blemishes. Further, the manufacturer argues that tires with such blemishes are perfectly safe, and with only the physical appearance suffering very slightly. A consumer magazine randomly purchases 40 of the tires for an upcoming tire performance comparison report. Upon examining the tires, the magazine's test engineers find that 5 of them have surface blemishes. If it were true that just 2% of the tires produced have surface blemishes, what would be the probability of getting 5 or more with such blemishes in a random sample of 40 tires? Considering this probability, comment on the believability of the tire manufacturer's advertising claim.

6.79 J.D. Power and Associates' Initial Quality Study reports that the industry-average problem rate for vehicles is 108 problems per 100 vehicles. The highest-rated brand was Lexus, with a rate of just 84 problems per 100 vehicles. On the other hand, BMW's Mini was reported as having 165 problems per 100 vehicles. Larry is on his way to pick up his brand-new Lexus. At the same time, Mark is on his way to pick up his brand-new Mini. What is the probability that Larry's Lexus will have exactly 2 problems? What is the probability that Mark's Mini will have no more than 1 problem? What is the probability that neither Larry nor Mark will have a problem with his new car? Source: Sharon Silke Carty, "U.S. Autos Power Forward with Gains in Quality Survey," *USA Today*, June 23, 2009, p. 3B.

6.80 It has been estimated that 5% of people who are hospitalized pick up infections while in the hospital, and that 2.2 million people a year get sick and 88,000 die because of these infections. Experts advise patients to be sure that health care personnel have washed their hands, to select surgeons with higher levels of experience, and to minimize the amount of time that they are treated with catheterization. Grove Tree Mountain Hospital anticipates admitting 100 patients this month for surgical and other treatments. Suppose that 14 of the patients in the sample actually end up getting a hospital-borne infection. Treating the 100 patients as a random sample of all hospital patients, and applying the infection rate cited above, what is the probability that a typical hospital would have 14 or more infections in a sample of this size? Considering this probability, comment on whether Grove Tree Mountain might not be a typical hospital in terms of its patients' likelihood of picking up an infection. Source: Anita Manning, "Hospital Infections Jump as Outpatient Care Rises," *USA Today*, March 6, 2000, p. 8D.

INTEGRATED CASE

Thorndike Sports Equipment

Arriving at the office a bit early, Ted finds Luke Thorndike in a fit of rage. Mr. Thorndike is sipping coffee and looking over Ted's notes on the defective racquets that were constructed with aluminum purchased from each of the company's three suppliers.

In rummaging through the basement crate, Ted has found a total of 30 defective racquets. Of these, 5 were made of aluminum supplied by the Snowmet Corporation and 5 were made of aluminum purchased from Barstow Aluminum, Inc. These were the two suppliers from whom Thorndike Sports Equipment had been buying aluminum for many years, and with good results.

The cause of Mr. Thorndike's anger is the total of 20 defective racquets made from aluminum purchased from the company's newest and lowest-priced supplier, Darwood Discount Metals, Inc. Luke recently placed a big order with Darwood, and he is counting the minutes until 10 A.M. so that he can call the West Coast firm, cancel the order, and give Mr. Darwood a big piece of his mind. The arrival of the morning mail only serves to heighten Luke's anger, as three more complaints are received from disgruntled customers demanding immediate replacements for their broken racquets.

Ten o'clock finally rolls around, and Luke places a person-to-person collect call to Mr. Darwood. After berating Mr. Darwood and his products, Luke demands that the order be canceled immediately. Mr. Darwood does not appreciate the elder Thorndike's tirade, but he is quite cognizant of the fact that 10% of his company's profits come from Thorndike Sports Equipment. Though irritated, he patiently tries to reason with Mr. Thorndike.

According to Mr. Darwood, independent metallurgists have conducted a lot of tests in which they found Darwood aluminum to be every bit as good as the product supplied by Snowmet and Barstow. He suggests that the unusually high number of defective racquets found by Luke is merely a fluke. Already operating on a short fuse, Mr. Thorndike responds that he doesn't much care for silly rhymes in the middle of the morning, and he warns Darwood not to load any more aluminum onto Thorndike-bound trucks until he and Ted have a chance to further examine the information Ted collected over the weekend. He promises to call Mr. Darwood with a final decision by 3 P.M., Pacific Time.

Ted spends a very busy morning and skips lunch, but by 2 P.M. he comes up with some data that might prove useful. Most important, he has uncovered a research study in which it was found that 0.8% of all aluminum racquets end up being returned as defective. The number of Snowmet and Barstow racquets in the "defectives" crate is about 1% of those produced. However, of the 1200 racquets made from Darwood aluminum, 20 (about 1.7%) are defective. Ted decides to consider the situation as a binomial probability distribution in which the probability of a defect on a given trial is 0.008, corresponding to the research finding that 0.8% of all the aluminum racquets produced are defective. Using the estimate for the entire industry, the *expected number of defectives* among the 1200 racquets made of Darwood aluminum would be $n\pi = 1200(0.008)$, or 9.6, *but 20 defectives were observed.*

To determine how unusual this result really is, Ted uses Excel and $n = 1200$, $\pi = 0.008$ to find the cumulative probabilities for the number of defects. The results are shown in Table 6.7.

Receiving a telephone call from home and finding out that the water pipes have just burst and his house is flooding, Ted has to flee the office. On his way out, he asks you to use the information he has found, along with Table 6.7, in giving advice to Mr. Thorndike before he makes his call to Darwood.

Table 6.7 Ted Thorndike has used Excel in generating these cumulative probabilities for a binomial distribution in which the number of trials is 1200 and the probability of a defect in any given trial is 0.8%, or 0.008.

	A	B	C	D	E
1	Binomial probabilities for				
2	trials (n), 1200, and				
3	probability of success (pi), 0.008				
4					
5	k	p(x<=k)		k	p(x<=k)
6	0	0.0001		13	0.8927
7	1	0.0007		14	0.9365
8	2	0.0037		15	0.9644
9	3	0.0136		16	0.9811
10	4	0.0373		17	0.9904
11	5	0.0830		18	0.9954
12	6	0.1564		19	0.9979
13	7	0.2574		20	0.9991
14	8	0.3788		21	0.9996
15	9	0.5086		22	0.9998
16	10	0.6331		23	0.9999
17	11	0.7419		24	1.0000
18	12	0.8287			

CHAPTER 7

Continuous Probability Distributions

ATM ARRIVAL TIMES: A CONTINUOUS DISTRIBUTION

In the opening vignette to Chapter 6, we described how the number of automatic teller machine (ATM) customers over a 5-minute period represents a discrete probability distribution, with x = the number of arrivals (0, 1, 2, etc.) during the period.

Without getting into trouble with campus security, stake out your local ATM as you did for the opener to Chapter 6—except now use the chronograph feature of your (or a friend's) watch to record the times between the arrivals of successive customers. For example, x_1 is the time between the arrival of customer A and customer B, x_2 is the time between customers B and C, and so on. What you're doing now is creating a continuous probability distribution, the topic of this chapter.

As a matter of fact, our discussion here is the flip side of the vignette that opened Chapter 6. The inverse of arrivals-per-minute is minutes-between-arrivals. Where the discrete Poisson distribution of Chapter 6 focused on the former, the continuous exponential distribution of this chapter is concerned with the latter. This unique relationship between the Poisson and exponential distributions will be examined later in this chapter. Measuring and predicting time between arrivals can be important to planners and service providers in many real-world situations, especially those dealing with emergency medical services.

Some things look a little like a normal curve.

LEARNING OBJECTIVES

After reading this chapter, you should be able to:

- Understand the nature and the applications of the normal distribution.
- Use the standard normal distribution and *z*-scores to determine probabilities associated with the normal distribution.
- Use the normal distribution to approximate the binomial distribution.
- Understand the nature and the applications of the exponential distribution, including its relationship to the Poisson distribution of Chapter 6.
- Use the computer in determining probabilities associated with the normal and exponential distributions.

7.1 INTRODUCTION

Chapter 6 dealt with probability distributions for *discrete* random variables, which can take on only certain values along an interval, with the possible values having gaps between them. This chapter presents several **continuous probability distributions;** these describe probabilities associated with random variables that are able to assume *any* of an infinite number of values along an interval.

Discrete probability distributions can be expressed as histograms, where the probabilities for the various x values are expressed by the heights of a series of vertical bars. In contrast, continuous probability distributions are smooth curves, where probabilities are expressed as areas under the curves. The curve is a function of x, and $f(x)$ is referred to as a **probability density function.** Since the continuous random variable x can be in an infinitely small interval along a range or continuum, the probability that x will take on any exact value may be regarded as zero. Therefore, we can speak of probabilities only in terms of the probability that x will be within a specified *interval* of values. For a continuous random variable, the probability distribution will have the following characteristics:

The probability distribution for a continuous random variable:

1. **The vertical coordinate is a function of x, described as $f(x)$ and referred to as the probability density function.**
2. **The range of possible x values is along the horizontal axis.**
3. **The probability that x will take on a value between a and b will be the area under the curve between points a and b, as shown in Figure 7.1.**

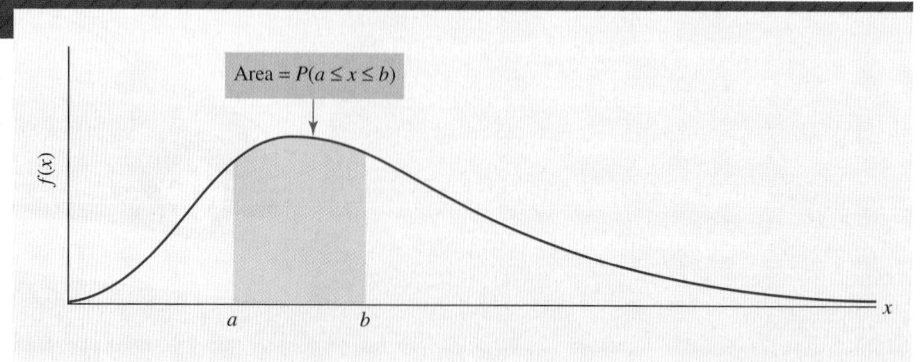

FIGURE 7.1

For a continuous random variable, the probability distribution is described by a curve called the probability density function, $f(x)$. The total area beneath the curve is 1.0, and the probability that x will take on some value between a and b is the area beneath the curve between points a and b.

STATISTICS IN ACTION 7.1

Navigating Normal Curves with Thorndike Sports Equipment

I was flattered but kind of surprised when Dr. Weiers asked Thorndike Sports Equipment to do a Statistics in Action commentary for his textbook. Doesn't he realize we're just imaginary? After all, he made us up in the first place. Anyway, I usually don't spend a lot of time thinking about the normal curve, but preparing this commentary helped open my eyes to how often it comes up.

In many facets of our business, we find a normal curve closely approximates the data we've collected. For example, in the "bounce test" we use in monitoring the liveliness of our golf balls, we find the bounce distances tend to be normally distributed. Although we don't like to talk about it much, we try to make sure our sponsored professionals get balls that are from the most lively 0.5% of the output. My grandson Ted tells me that corresponds to $z = +2.58$ standard deviation units from the mean.

According to our human resources manager, a lot of the aptitude and performance tests we give potential and current employees also tend to result in scores that are distributed according to that classic "bell-shaped" curve. She's been with us for many years, and she finds it very interesting to compare the means and standard deviations today with the ones she got ten or twenty years ago.

Ted says we also rely on the normal distribution in constructing and interpreting the statistical process control charts that have helped Thorndike racquets gain their reputation for consistently high quality. Here at Thorndike Sports Equipment, we greatly appreciate the applications and value of the normal distribution, and we are happy to pass our experience on to you.

Luke Thorndike
President and Founder
Thorndike Sports Equipment

The probability density function, *f*(*x*), for a given continuous distribution is expressed in algebraic terms, and the areas beneath are obtained through the mathematics of calculus. However, tables are provided in the text for readily identifying or calculating such areas.

4. The total area under the curve will be equal to 1.0.

The first continuous probability distribution we will examine is the normal distribution. As suggested by the commentary in Statistics in Action 7.1, this is a very important continuous distribution in both the study of statistics and its application to business. The normal distribution is a bell-shaped, symmetrical curve, the use of which is facilitated by a standard table listing the cumulative areas beneath.

As demonstrated in Chapter 6, binomial probabilities can involve either tedious computations or extensive tables. We will describe a method by which the normal distribution can be used as a convenient approximation for determining the probabilities associated with discrete random variables that are of the binomial distribution.

The next topic will be the exponential distribution. This is related to the (discrete) Poisson distribution of the previous chapter but is a continuous probability distribution describing probabilities associated with the (continuous) time or distance intervals between occurrences of the so-called rare events of a Poisson process.

Finally, we will use Excel and Minitab to randomly select observations from a specified continuous probability distribution. This is similar to the simulated samples obtained from selected discrete probability distributions in Chapter 6.

EXERCISES

7.1 What is the difference between a continuous probability distribution and a discrete probability distribution?

7.2 What is a probability density function and how is it relevant to the determination of probabilities associated with a continuous random variable?

7.3 Why is the total area beneath a probability density function equal to 1.0?

7.4 What is the probability that a continuous random variable will take on any specific value? Explain your answer.

(7.2) THE NORMAL DISTRIBUTION

Description and Applications

The **normal distribution** is the most important continuous distribution in statistics. It occupies this position for three key reasons: (1) many natural and economic phenomena tend to be approximately normally distributed; (2) it can be used in approximating other distributions, including the binomial; and (3) as we will see in Chapter 8, sample means and proportions tend to be normally distributed whenever repeated samples are taken from a given population of any shape. This latter characteristic is related to the *central limit theorem,* a concept also discussed in the next chapter.

The normal distribution is really a family of distributions, each member of which is bell shaped and symmetrical (the left side is a mirror image of the right). As Figure 7.2 shows, (1) the mean, median, and mode are all at the same position on the horizontal axis; (2) the curve is *asymptotic,* approaching the horizontal axis at both ends, but never intersecting with it; and (3) the total area beneath the curve is equal to 1.0. The specific member of the family depends on just two descriptors:

FIGURE 7.2

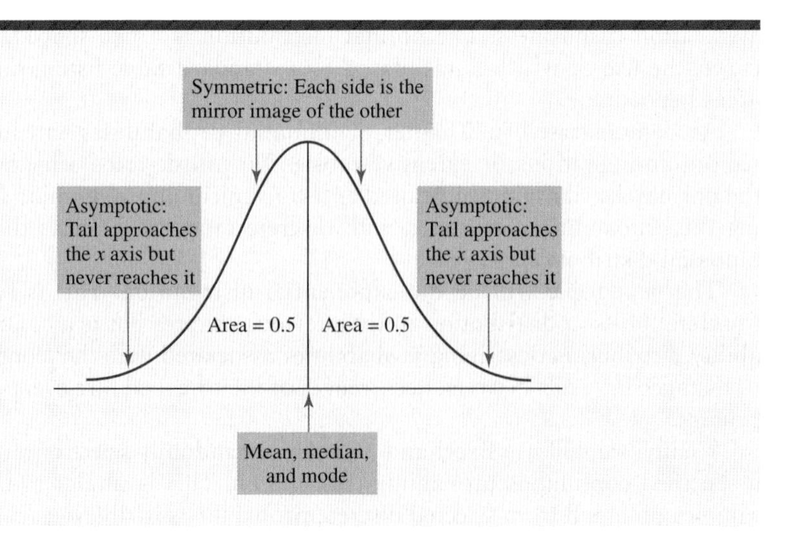

The normal distribution is actually a family of bell-shaped distributions, each of which has these characteristics. The specific member of the family depends on just two descriptors, the mean (μ) and the standard deviation (σ).

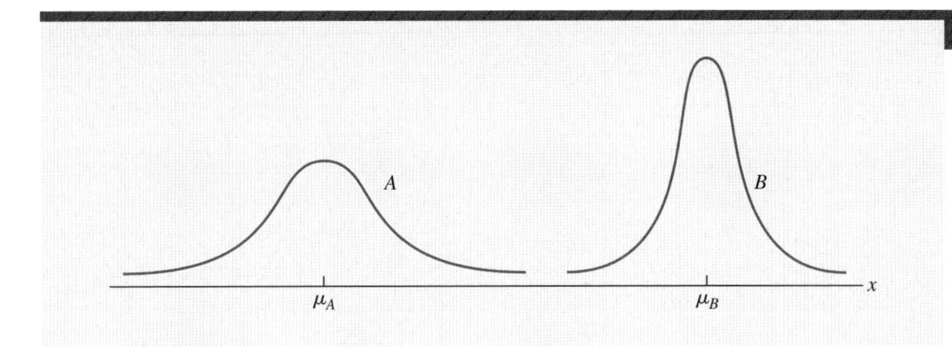

FIGURE 7.3

The shape of a normal curve and its position on the *x* axis will depend on its mean (μ) and standard deviation (σ). Although curve *B* has a greater mean, its standard deviation is smaller.

the mean (μ) and the standard deviation (σ). The normal distribution can be described as follows:

The normal distribution for the continuous random variable, *x* (with $-\infty \le x \le +\infty$):

$$f(x) = \frac{1}{\sigma\sqrt{2\pi}} e^{-(1/2)[(x-\mu)/\sigma]^2}$$

where μ = mean
σ = standard deviation
e = the mathematical constant, 2.71828
π = the mathematical constant, 3.14159

In this equation, π is the geometric constant for the ratio of the circumference of a circle to its diameter, and is not related to the symbol used for a population proportion.

Not all normal curves will look exactly like the one in Figure 7.2. The shape of the curve and its location along the *x* axis will depend on the values of the standard deviation and mean, respectively. You can demonstrate this by using Seeing Statistics Applet 4 at the end of the chapter. Figure 7.3 shows two normal curves representing output diameters for two tubing extrusion machines:

- The average diameter of tubing from machine *B* exceeds that from machine *A*. Looking along the *x* axis, it can be seen that $\mu_B > \mu_A$.
- The variability of the output from machine *A* is greater than that from machine *B*. In symbolic terms, this can be expressed as $\sigma_A > \sigma_B$.

Areas Beneath the Normal Curve

Regardless of the shape of a particular normal curve, the areas beneath it can be described for any interval of our choice. This makes it unnecessary to get involved with calculus and the complexities of the probability density function presented earlier. For *any* normal curve, the areas beneath the curve will be as follows (see also part (a) in Figure 7.4):

- About 68.3% of the area is in the interval $\mu - \sigma$ to $\mu + \sigma$.
- About 95.5% of the area is in the interval $\mu - 2\sigma$ to $\mu + 2\sigma$.
- Nearly all of the area (about 99.7%) is in the interval $\mu - 3\sigma$ to $\mu + 3\sigma$.

FIGURE 7.4

For any normal distribution, the approximate areas shown in part (a) will lie beneath the curve for the intervals shown. In part (b), the areas show how these intervals could apply to the amount of time logged by general-aviation aircraft during the year. Assuming $\mu = 120$ hours and $\sigma = 30$ hours, about 95.5% of the planes logged between 60 and 180 hours of flying time.

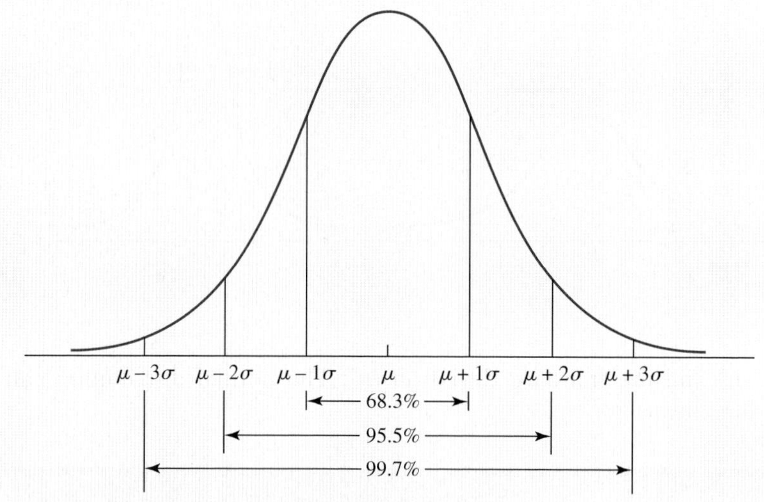

(a) Approximate areas beneath the normal curve

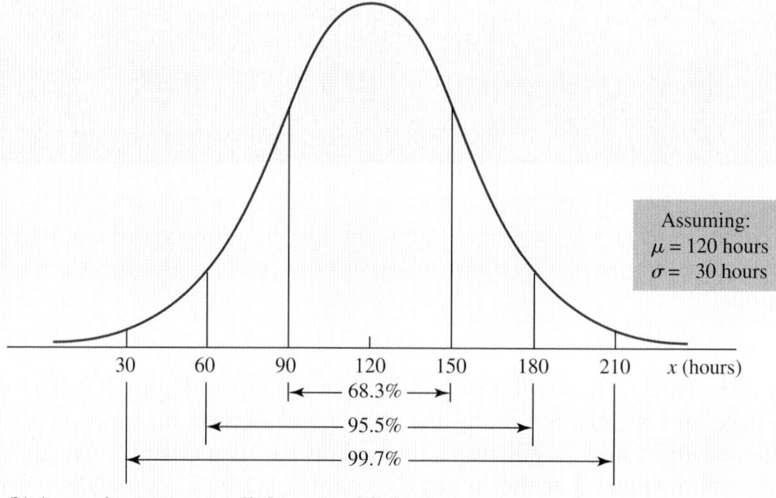

(b) Approximate areas applied to annual flying hours

EXAMPLE

Normal Distribution

The General Aviation Manufacturers Association has reported the average annual number of hours flown by general-aviation aircraft to be approximately 120 hours.[1] For purposes of our example, we will assume a normal distribution and a standard deviation of 30 hours for the flying hours of aircraft within this population.

[1]Source: General Aviation Manufacturers Association, *General Aviation Statistical Databook, 2008 Edition*, pp. 28–29.

SOLUTION

Based on the reported mean, $\mu = 120$ hours, and the assumed standard deviation, $\sigma = 30$ hours, we can use the known properties of the normal distribution to determine that about 95.5% of these aircraft logged between $(120 - 2[30])$ and $(120 + 2[30])$ hours, or between 60 and 180 hours of flying time during that year. Likewise, we could determine that about 68.3% of the aircraft logged between 90 and 150 ($\mu \pm \sigma$) hours and that about 99.7% logged between 30 and 210 ($\mu \pm 3\sigma$) hours. These intervals and approximate areas are shown in part (b) of Figure 7.4.

EXAMPLEEXAMPLEEXAMPLE

EXERCISES

7.5 It has been stated that the normal distribution is really a "family" of distributions. Explain.

7.6 In the normal distribution, the probability that x will exceed $(\mu + 2\sigma)$ is the same as the probability that x will be less than $(\mu - 2\sigma)$. What characteristic of the normal distribution does this reflect?

7.7 Sketch two different normal distributions along a single x axis so that both of the following conditions are satisfied: (a) $\mu_a = \mu_b$ and (b) $\sigma_a > \sigma_b$.

7.8 If x is normally distributed with $\mu = 20.0$ and $\sigma = 4.0$, determine the following:

a. $P(x \geq 20.0)$
b. $P(16.0 \leq x \leq 24.0)$
c. $P(x \leq 12)$
d. $P(x = 22.0)$
e. $P(12.0 \leq x \leq 28.0)$
f. $P(x \geq 16)$

7.9 If x is normally distributed with $\mu = 25.0$ and $\sigma = 5.0$, determine the following:

a. $P(x \geq 25.0)$
b. $P(20.0 \leq x \leq 30.0)$
c. $P(x \leq 30)$
d. $P(x = 26.2)$
e. $P(15.0 \leq x \leq 25.0)$
f. $P(x \geq 15)$

7.10 The Canada Urban Transit Association has reported that the average revenue per passenger trip during a given year was \$1.55. If we assume a normal distribution and a standard deviation of $\sigma = \$0.20$, what proportion of passenger trips produced a revenue of Source: American Public Transit Association, *APTA 2009 Transit Fact Book*, p. 35.

a. less than \$1.55?
b. between \$1.15 and \$1.95?
c. between \$1.35 and \$1.75?
d. between \$0.95 and \$1.55?

7.11 In 2007, the average conventional first mortgage for new single-family homes was \$360,000. Assuming a normal distribution and a standard deviation of $\sigma = \$30,000$, what proportion of the mortgages were Source: Bureau of the Census, *Statistical Abstract of the United States 2009*, p. 726.

a. more than \$360,000?
b. between \$300,000 and \$420,000?
c. between \$330,000 and \$390,000?
d. more than \$270,000?

7.12 In 2009, the average charge for tax preparation by H&R Block, Inc. was \$187. Assuming a normal distribution and a standard deviation of $\sigma = \$20$, what proportion of H&R Block's tax preparation fees were
Source: hrblock.com, July 14, 2009.

a. more than \$187?
b. between \$147 and \$227?
c. between \$167 and \$207?
d. more than \$227?

7.13 It has been reported that the average hotel check-in time, from curbside to delivery of bags into the room, is 12.0 minutes. An Li has just left the cab that brought her to her hotel. Assuming a normal distribution with a standard deviation of 2.0 minutes, what is the probability that the time required for An Li and her bags to get to the room will be:

a. greater than 14.0 minutes?
b. between 10.0 and 14.0 minutes?
c. less than 8.0 minutes?
d. between 10.0 and 16.0 minutes?

7.14 The average American family of four spends \$5000 per year on food prepared at home. Assuming a normal distribution with a standard deviation of \$1000 and a randomly selected American family of four, what is the probability that the family's annual spending for food prepared at home will be: Source: "Business Briefing," *Pittsburgh Tribune-Review*, July 5, 2009, p. E1.

a. more than \$8000?
b. between \$5000 and \$7000?
c. less than \$6000?
d. between \$3000 and \$6000?

7.15 On average, commuters in the Los Angeles, California, area require 30.0 minutes to get to work. Assume a normal distribution with a standard deviation of 5.0 minutes and a randomly selected Los Angeles-area commuter named Jamal. Source: *The World Almanac and Book of Facts 2006*, p. 474.

a. What is the probability that Jamal will require more than 45.0 minutes to get to work on any given day?

b. Jamal has just left home and must attend an important meeting with the CEO in just 25.0 minutes. If the CEO routinely fires employees who are tardy, what is the probability that Jamal will be going to work tomorrow?

7.3 THE STANDARD NORMAL DISTRIBUTION

Description and Applications

Because there is a different normal curve for every possible pair of μ and σ, the number of statistical tables would be limitless if we wished to determine the areas corresponding to possible intervals within all of them. Fortunately, we can solve this dilemma by "standardizing" the normal curve and expressing the original x values in terms of their number of standard deviations away from the mean. The result is referred to as a **standard** (or **standardized**) **normal distribution,** and it allows us to use a single table to describe areas beneath the curve. The key to the process is the ***z*-score:**

The *z*-score for a standard normal distribution:

$$z = \frac{x - \mu}{\sigma}$$

where z = the distance from the mean, measured in standard deviation units
x = the value of x in which we are interested
μ = the mean of the distribution
σ = the standard deviation of the distribution

In the aircraft flying hours example discussed earlier, the mean and (assumed) standard deviation were $\mu = 120$ hours and $\sigma = 30$ hours, respectively. Using the z-score equation, we can convert this distribution into a standard normal distribution in which we have z values instead of x values. For example:

***x* Value in the Original Normal Distribution**	**Corresponding *z* Value in the Standard Normal Distribution**
$x = 120$ hours	$z = \frac{x - \mu}{\sigma} = \frac{120 - 120}{30} = 0.00$
$x = 160$ hours	$z = \frac{x - \mu}{\sigma} = \frac{160 - 120}{30} = 1.33$
$x = 90$ hours	$z = \frac{x - \mu}{\sigma} = \frac{90 - 120}{30} = -1.00$

In Figure 7.5, there are two scales beneath the curve: an x (hours of flying time) scale and a z (standard deviation units) scale. Note that the mean of the z scale is 0.00. Regardless of the mean and standard deviation of any normal distribution, we can use the z-score concept to express the original values in terms of standard deviation multiples from the mean. It is this transformation that allows us to use the standard normal table in determining probabilities associated with any normal distribution. You can use Seeing Statistics Applet 5, at the end of the chapter, to see how the normal distribution areas respond to changes in the beginning and ending z values associated with Figure 7.5.

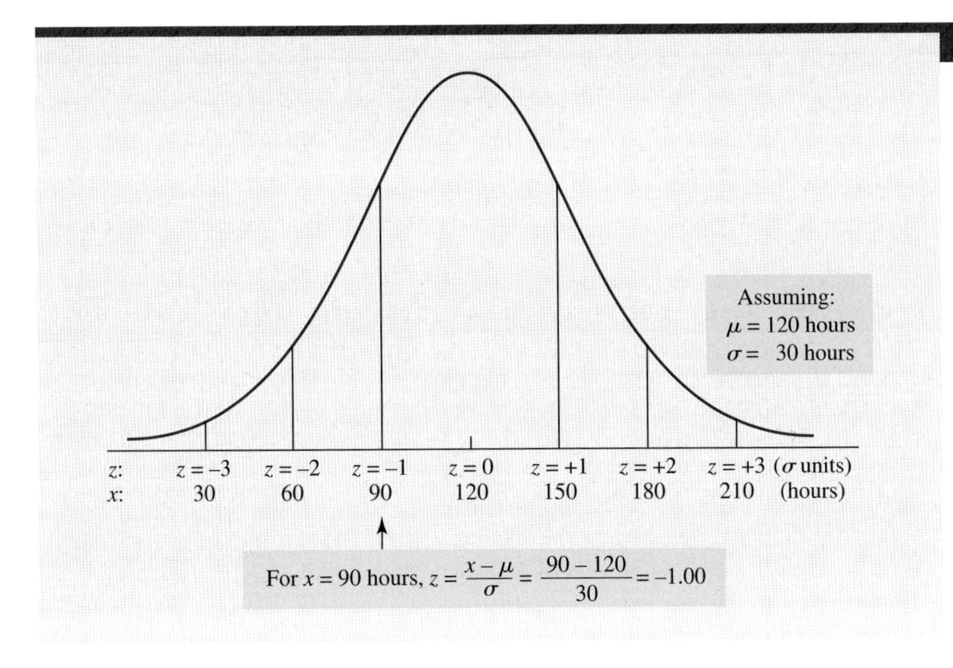

FIGURE 7.5

For the distribution of annual flying hours in part (b) of Figure 7.4, the upper of the two horizontal scales shown here represents the standardized normal distribution, in which x values (hours) have been converted to z-scores (standard deviation units from the mean). The use of z-scores makes it possible to use a single table for all references to the normal distribution, regardless of the original values of μ and σ.

Using the Standard Normal Distribution Table

Regardless of the units involved (e.g., pounds, hours, dollars, and so on) in the original normal distribution, the standard normal distribution converts them into standard deviation distances from the mean. As we proceed through the next example, our original units will be grams.

In demonstrating the use of the standard normal distribution table, we will rely on a situation involving the weight added to balance a generator shaft. For each of the related questions, we will go through these steps:

1. Convert the information provided into one or more z-scores.
2. Use the standard normal table to identify the area(s) corresponding to the z-score(s). A portion of the standard normal distribution table is shown in Table 7.1. The table provides cumulative areas to the z value of interest.
3. Interpret the result in such a way as to answer the original question.

The cumulative standard normal distribution table is on the rear endsheet of the text, and an abbreviated portion of it is shown in Table 7.1. The row labels (e.g., 1.4) refer to the integer and first decimal place of z, and the column labels refer to the second decimal place (e.g., 0.05); thus, we can look in row 1.4 and column 0.05 and see that the cumulative probability (or area) to the left of $z = 1.45$ is 0.9265. Here are a few more examples of how to find cumulative areas for given z values:

- *Cumulative area associated with $z = -1.37$:* Referring to the −1.3 row and the 0.07 column of Table 7.1, we find this probability is 0.0853.
- *Cumulative area associated with $z = -0.08$:* Referring to the −0.0 row and the 0.08 column of Table 7.1, we find this probability is 0.4681.
- *Cumulative area associated with $z = 0.09$:* Referring to the 0.0 row and the 0.09 column of Table 7.1, we see this probability is 0.5359.
- *Cumulative area associated with $z = 1.73$:* Referring to the 1.7 row and the 0.03 column of Table 7.1, we see this probability is 0.9582.

TABLE 7.1

A portion of the cumulative standard normal distribution table from the back endsheet

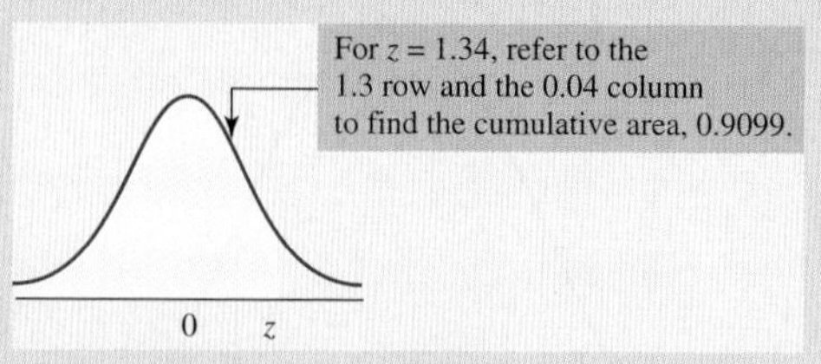

z	0.00	0.01	0.02	0.03	0.04	0.05	0.06	0.07	0.08	0.09
−2.0	0.0228	0.0222	0.0217	0.0212	0.0207	0.0202	0.0197	0.0192	0.0188	0.0183
−1.9	0.0287	0.0281	0.0274	0.0268	0.0262	0.0256	0.0250	0.0244	0.0239	0.0233
−1.8	0.0359	0.0351	0.0344	0.0336	0.0329	0.0322	0.0314	0.0307	0.0301	0.0294
−1.7	0.0446	0.0436	0.0427	0.0418	0.0409	0.0401	0.0392	0.0384	0.0375	0.0367
−1.6	0.0548	0.0537	0.0526	0.0516	0.0505	0.0495	0.0485	0.0475	0.0465	0.0455
−1.5	0.0668	0.0655	0.0643	0.0630	0.0618	0.0606	0.0594	0.0582	0.0571	0.0559
−1.4	0.0808	0.0793	0.0778	0.0764	0.0749	0.0735	0.0721	0.0708	0.0694	0.0681
−1.3	0.0968	0.0951	0.0934	0.0918	0.0901	0.0885	0.0869	0.0853	0.0838	0.0823
−1.2	0.1151	0.1131	0.1112	0.1093	0.1075	0.1056	0.1038	0.1020	0.1003	0.0985
−1.1	0.1357	0.1335	0.1314	0.1292	0.1271	0.1251	0.1230	0.1210	0.1190	0.1170
−1.0	0.1587	0.1562	0.1539	0.1515	0.1492	0.1469	0.1446	0.1423	0.1401	0.1379
−0.9	0.1841	0.1814	0.1788	0.1762	0.1736	0.1711	0.1685	0.1660	0.1635	0.1611
−0.8	0.2119	0.2090	0.2061	0.2033	0.2005	0.1977	0.1949	0.1922	0.1894	0.1867
−0.7	0.2420	0.2389	0.2358	0.2327	0.2296	0.2266	0.2236	0.2206	0.2177	0.2148
−0.6	0.2743	0.2709	0.2676	0.2643	0.2611	0.2578	0.2546	0.2514	0.2483	0.2451
−0.5	0.3085	0.3050	0.3015	0.2981	0.2946	0.2912	0.2877	0.2843	0.2810	0.2776
−0.4	0.3446	0.3409	0.3372	0.3336	0.3300	0.3264	0.3228	0.3192	0.3156	0.3121
−0.3	0.3821	0.3783	0.3745	0.3707	0.3669	0.3632	0.3594	0.3557	0.3520	0.3483
−0.2	0.4207	0.4168	0.4129	0.4090	0.4052	0.4013	0.3974	0.3936	0.3897	0.3859
−0.1	0.4602	0.4562	0.4522	0.4483	0.4443	0.4404	0.4364	0.4325	0.4286	0.4247
−0.0	0.5000	0.4960	0.4920	0.4880	0.4840	0.4801	0.4761	0.4721	0.4681	0.4641
0.0	0.5000	0.5040	0.5080	0.5120	0.5160	0.5199	0.5239	0.5279	0.5319	0.5359
0.1	0.5398	0.5438	0.5478	0.5517	0.5557	0.5596	0.5636	0.5675	0.5714	0.5753
0.2	0.5793	0.5832	0.5871	0.5910	0.5948	0.5987	0.6026	0.6064	0.6103	0.6141
0.3	0.6179	0.6217	0.6255	0.6293	0.6331	0.6368	0.6406	0.6443	0.6480	0.6517
0.4	0.6554	0.6591	0.6628	0.6664	0.6700	0.6736	0.6772	0.6808	0.6844	0.6879
0.5	0.6915	0.6950	0.6985	0.7019	0.7054	0.7088	0.7123	0.7157	0.7190	0.7224
0.6	0.7257	0.7291	0.7324	0.7357	0.7389	0.7422	0.7454	0.7486	0.7517	0.7549
0.7	0.7580	0.7611	0.7642	0.7673	0.7704	0.7734	0.7764	0.7794	0.7823	0.7852
0.8	0.7881	0.7910	0.7939	0.7967	0.7995	0.8023	0.8051	0.8078	0.8106	0.8133
0.9	0.8159	0.8186	0.8212	0.8238	0.8264	0.8289	0.8315	0.8340	0.8365	0.8389
1.0	0.8413	0.8438	0.8461	0.8485	0.8508	0.8531	0.8554	0.8577	0.8599	0.8621
1.1	0.8643	0.8665	0.8686	0.8708	0.8729	0.8749	0.8770	0.8790	0.8810	0.8830
1.2	0.8849	0.8869	0.8888	0.8907	0.8925	0.8944	0.8962	0.8980	0.8997	0.9015
1.3	0.9032	0.9049	0.9066	0.9082	0.9099	0.9115	0.9131	0.9147	0.9162	0.9177
1.4	0.9192	0.9207	0.9222	0.9236	0.9251	0.9265	0.9279	0.9292	0.9306	0.9319
1.5	0.9332	0.9345	0.9357	0.9370	0.9382	0.9394	0.9406	0.9418	0.9429	0.9441
1.6	0.9452	0.9463	0.9474	0.9484	0.9495	0.9505	0.9515	0.9525	0.9535	0.9545
1.7	0.9554	0.9564	0.9573	0.9582	0.9591	0.9599	0.9608	0.9616	0.9625	0.9633
1.8	0.9641	0.9649	0.9656	0.9664	0.9671	0.9678	0.9686	0.9693	0.9699	0.9706
1.9	0.9713	0.9719	0.9726	0.9732	0.9738	0.9744	0.9750	0.9756	0.9761	0.9767
2.0	0.9772	0.9778	0.9783	0.9788	0.9793	0.9798	0.9803	0.9808	0.9812	0.9817

In applications involving the standard normal table, actual areas of interest may lie to the left of the mean or to the right of the mean, and may even require that one cumulative probability be subtracted from another. Because of (1) the variety of possibilities, and (2) the importance of knowing how to use the standard normal table, we will examine a number of applications that relate to the following example.

EXAMPLE

Normal Probabilities

Following their production, industrial generator shafts are tested for static and dynamic balance, and the necessary weight is added to predrilled holes in order to bring each shaft within balance specifications. From past experience, the amount of weight added to a shaft has been normally distributed, with an average of 35 grams and a standard deviation of 9 grams.

SOLUTIONS

What Is the Probability That a Randomly Selected Shaft Will Require Between 35 and 40 Grams of Weight for Proper Balance? [$P(35 \leq x \leq 40)$]

1. Conversion of $x = 40$ grams to z-score:

$$z = \frac{x - \mu}{\sigma} = \frac{40 - 35}{9} = 0.56$$

   Conversion of $x = 35$ grams to z-score:

$$z = \frac{x - \mu}{\sigma} = \frac{35 - 35}{9} = 0.00$$

2. Reference to the standard normal table: Referring to the 0.5 row and the 0.06 column of the table, we find the cumulative area to $z = 0.56$ is 0.7123. Referring to the 0.0 row and the 0.00 column, we can verify that the cumulative area to $z = 0.00$ (the mean) is 0.5000.

3. Solution: The probability that the randomly selected shaft will require between 35 and 40 grams is $0.7123 - 0.5000 = 0.2123$, as illustrated in Figure 7.6.

What Is the Probability That a Randomly Selected Shaft Will Require at Least 50 Grams of Weight for Proper Balance? [$P(x \geq 50)$]

1. Conversion of $x = 50$ grams to z-score:

$$z = \frac{x - \mu}{\sigma} = \frac{50 - 35}{9} = 1.67$$

2. Reference to the standard normal table: Referring to the 1.6 row and the 0.07 column, we find the cumulative area to $z = 1.67$ is 0.9525. Also, we must consider that the total cumulative area (i.e., to z approaching positive infinity) is, by definition, 1.0000.

3. Solution: The probability that the randomly selected shaft will require at least 50 grams is $1.0000 - 0.9525 = 0.0475$, as illustrated in Figure 7.7.

EXAMPLEEXAMPLEEXAMPLEEXAMPLEEXAMPLEEXAMPLEEXAMP

FIGURE 7.6

The probability is 0.2123 that a randomly selected generator shaft will require between 35 and 40 grams of weight in order to achieve proper balance.

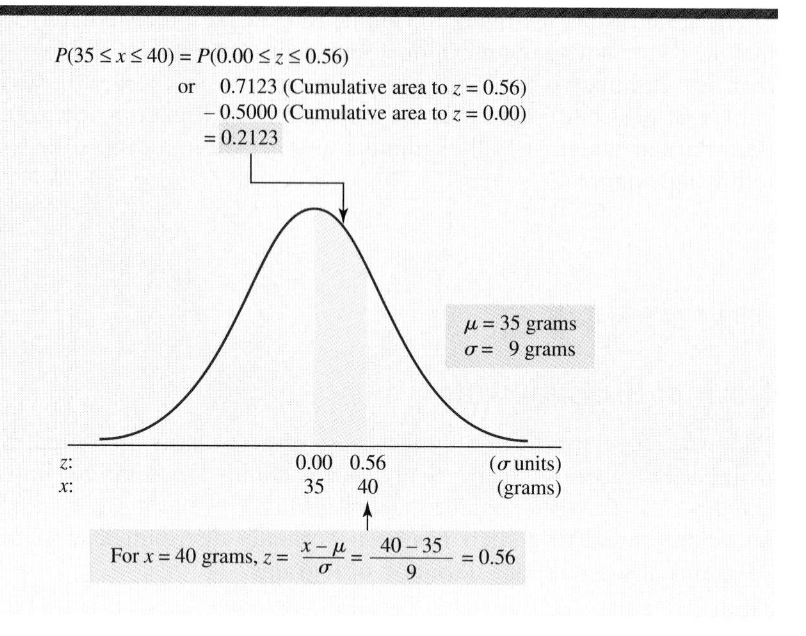

FIGURE 7.7

The probability is 0.0475 that a randomly selected generator shaft will require at least 50 grams for proper balance.

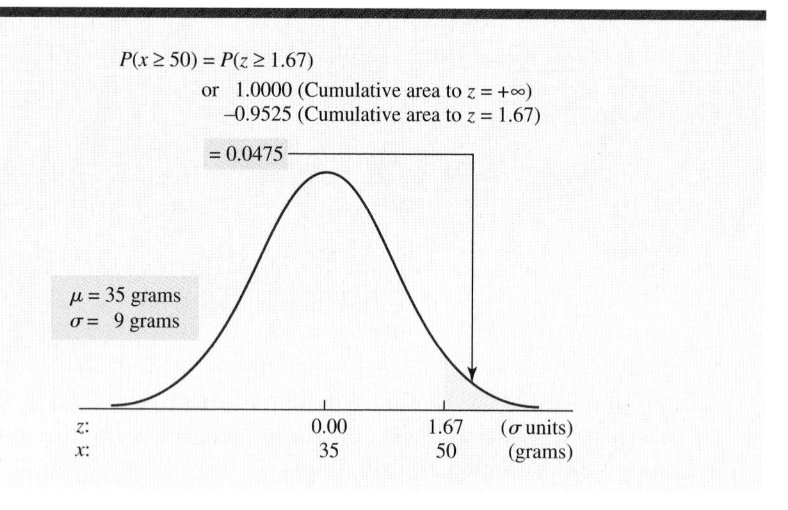

EXAMPLEEXAMPLEEXAMPLEEXAMPLEE

What Is the Probability That a Randomly Selected Shaft Will Require Between 41 and 49 Grams of Weight for Proper Balance? [$P(41 \leq x \leq 49)$]

1. Conversion of $x = 49$ grams to z-score:

$$z = \frac{x - \mu}{\sigma} = \frac{49 - 35}{9} = 1.56$$

   Conversion of $x = 41$ grams to z-score:

$$z = \frac{x - \mu}{\sigma} = \frac{41 - 35}{9} = 0.67$$

2. Reference to the standard normal table: Referring to the 1.5 row and the 0.06 column of the table, we find the cumulative area to $z = 1.56$ is 0.9406. Referring to the 0.6 row and the 0.07 column of the table, the cumulative area to $z = 0.67$ is 0.7486.

3. Solution: The probability that the randomly selected shaft will require between 41 and 49 grams is $0.9406 - 0.7486 = 0.1920$, as illustrated in Figure 7.8.

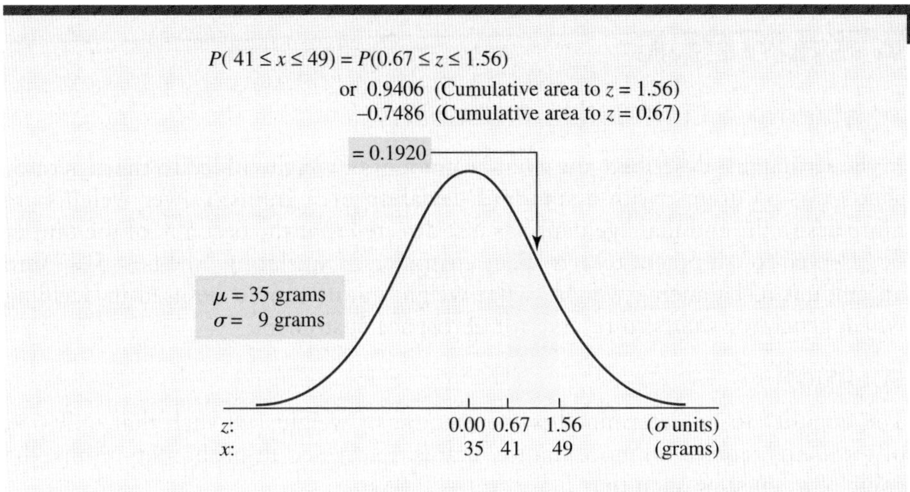

$P(41 \leq x \leq 49) = P(0.67 \leq z \leq 1.56)$
or 0.9406 (Cumulative area to $z = 1.56$)
−0.7486 (Cumulative area to $z = 0.67$)
= 0.1920

$\mu = 35$ grams
$\sigma = 9$ grams

z: 0.00 0.67 1.56 (σ units)
x: 35 41 49 (grams)

FIGURE 7.8

The probability is 0.1920 that a randomly selected generator shaft will require between 41 and 49 grams for proper balance.

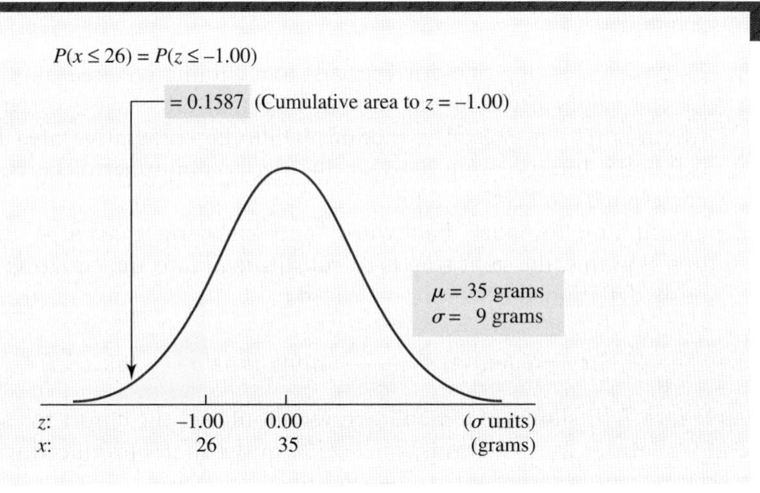

FIGURE 7.9

The probability is 0.1587 that a randomly selected generator shaft will require no more than 26 grams for proper balance.

What Is the Probability That a Randomly Selected Shaft Will Require No More Than 26 Grams of Weight for Proper Balance? [$P(x \leq 26)$]

1. Conversion of $x = 26$ grams to z-score:

$$z = \frac{x - \mu}{\sigma} = \frac{26 - 35}{9} = -1.00$$

2. Reference to the standard normal table: Referring to the −1.0 row and the 0.00 column of the table, we find the cumulative area to $z = -1.00$ is 0.1587.
3. Solution: As Figure 7.9 illustrates, the probability that a randomly selected shaft will require no more than 26 grams is 0.1587.

In the preceding cases, we proceeded from x values to the determination of a corresponding area beneath the standard normal curve. However, it is also possible to specify an area, then identify a range of x values that corresponds to this area.

EXAMPLEEXAMPLEEXAMPLEEXAMPLEEXAMPLEEXAMP

EXAMPLE

x Value for a Cumulative Probability

In the setting just described, the average amount of weight added to the generator shafts was 35 grams, with a standard deviation of 9 grams. Lower weights are more desirable, and management has just directed that the best 5% of the output be reserved for shipment to aerospace customers. Translating "the best 5%" into an amount of balancing weight, what weight cutoff should be used in deciding which generator shafts to reserve for aerospace customers?

SOLUTION

The quantity to be determined is the amount of weight (w) such that only 5% of the shafts require no more than w grams. Expressed in probability terms, the value of w must be such that $P(x \leq w) = 0.05$.

The solution is to find out what value of z corresponds to a cumulative area of 0.05, then convert this z value into w grams. Reversing the steps of the previous examples:

1. Find the value of z that corresponds to a cumulative area of 0.05, or 0.0500: Of the values listed in the standard normal table, we see that the cumulative area to $z = -1.64$ is 0.0505, while the cumulative area to $z = -1.65$ is 0.0495. Interpolating between the two, the number of grams in the cutoff corresponds to $z = -1.645$.
2. Continuing to work backward, and knowing that $\mu = 35$ grams and $\sigma = 9$ grams, the next step is to substitute w into the equation for the previously determined z score, $z = -1.645$:

$$z = \frac{w - \mu}{\sigma} \quad \text{or} \quad -1.645 = \frac{w - 35}{9} \quad \text{and} \quad w = 35 - 9(1.645) = 20.195 \text{ grams}$$

As Figure 7.10 shows, the manufacturer should set the cutoff for aerospace generator shafts at 20.195 grams, since 95% of the shafts produced will require this amount or more in order to achieve proper balance.

FIGURE 7.10

If the best 5% of the output is to be reserved for aerospace customers, shafts requiring no more than 20.195 grams for balancing would be allocated to this industry.

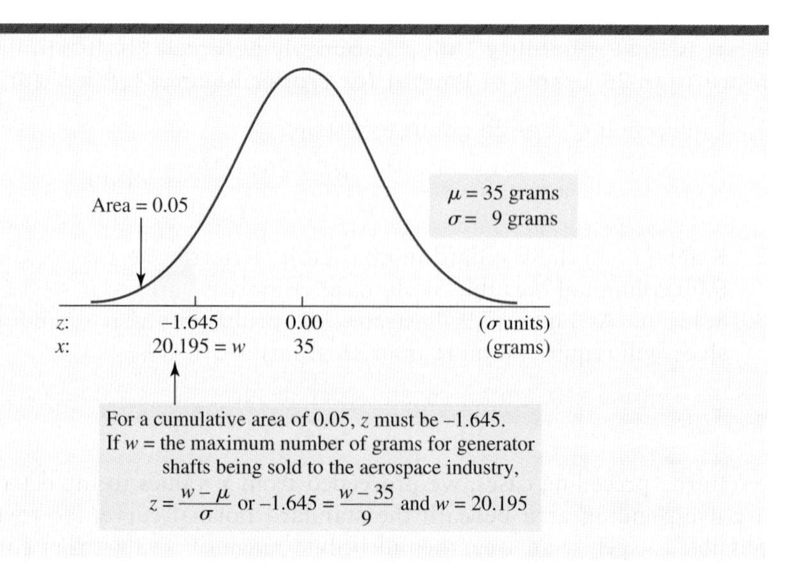

Excel and Minitab can be used to obtain cumulative normal probabilities associated with specified values for x—the procedures and results are shown in Computer Solutions 7.1. These software packages can also be used in doing the reverse—that is, specifying a cumulative probability, then finding the x value associated with it. These procedures and results are shown in Computer Solutions 7.2.

COMPUTER 7.1 SOLUTIONS

Normal Probabilities

These procedures show how to get cumulative probabilities.

For each software package:

Procedure I: Getting the cumulative probability associated with one *x* value.
Procedure II: Getting cumulative probabilities associated with a set of *x* values.

EXCEL

	A	B
1	Normal Distribution	
2	with mean = 35,	
3	std dev. = 9.	
4	x	Area to L.
5	45	0.8667
6	35	0.5000
7	25	0.1333

I. Excel procedure for the cumulative probability associated with one *x* value

1. Click on the cell into which Excel will place the probability. From the **Formulas** ribbon, click on the **Insert Function** symbol at the extreme left. Click **Statistical** in the **Category** box. Click **NORMDIST** in the **Function** box. Click **OK.**
2. Enter the *x* value of interest (e.g., **45**) into the **X** box. Enter the mean of the distribution (e.g., **35**) into the **Mean** box. Enter the standard deviation (e.g., **9**) into the **Standard_dev** box. Enter **True** into the **Cumulative** box. Click **OK.**

II. Excel procedure for cumulative probabilities associated with a set of *x* values

1. Enter the labels into **A4:B4** as shown in the printout above. Enter the *x* values of interest (e.g., **45, 35, 25**) starting with cell **A5.** Click on cell **B5.**
2. From the **Formulas** ribbon, click on the **Insert Function** symbol at the extreme left. Click **Statistical** in the **Category** box. Click **NORMDIST** in the **Function** box. Click **OK.**
3. Enter **A5** into the **X** box. Enter the mean of the distribution (e.g., **35**) into the **Mean** box. Enter the standard deviation (e.g., **9**) into the **Standard_dev** box. Enter **True** into the **Cumulative** box. Click **OK.** The first cumulative probability will appear in cell B5.
4. Click at the lower right corner of **B5** (the cursor will turn to a "+" sign) and drag downward so cells **B5:B7** will contain the set of cumulative probabilities.

(continued)

MINITAB

```
Cumulative Distribution Function
Normal with mean = 35 and standard deviation = 9
 x   P( X <= x )
45      0.866740
35      0.500000
25      0.133260
```

I. Minitab procedure for cumulative probability associated with one *x* value

1. Click **Calc.** Select **Probability Distributions.** Click **Normal.**
2. In the **Normal Distribution** menu, enter the mean of the distribution (e.g., **35**) into the **Mean** box. Enter the standard deviation (e.g., **9**) into the **Standard deviation** box. Select **Input constant** and enter the *x* value of interest (e.g., **45**) into the box. Select **Cumulative.** Click **OK.**

II. Minitab procedure for cumulative probabilities associated with a set of *x* values

Enter the desired *x* values (e.g., **45, 35, 25**) into column **C1.** Follow steps 1 and 2 above, except select **Input column** and enter **C1** into the box.

COMPUTER 7.2 SOLUTIONS

Inverse Normal Probabilities

These procedures show how to get *x* values associated with cumulative normal probabilities.

For each software package:

Procedure I: Getting the *x* value associated with one cumulative probability.
Procedure II: Getting the *x* values associated with a set of cumulative probabilities.

EXCEL

	A	B
1	Normal Distribution	
2	with mean = 35,	
3	std. dev. = 9.	
4	Area to L.	x
5	0.9750	52.64
6	0.5000	35.00
7	0.2500	28.93

I. Excel procedure for the *x* value associated with one cumulative probability

1. Click on the cell into which Excel will place the *x* value. From the **Formulas** ribbon, click on the **Insert Function** symbol at the extreme left. Click **Statistical** in the **Category** box. Click **NORMINV** in the **Function** box. Click **OK.**
2. Enter the cumulative probability of interest (e.g., **0.9750**) into the **Probability** box. Enter the mean of the distribution (e.g., **35**) into the **Mean** box. Enter the standard deviation (e.g., **9**) into the **Standard_dev** box. Click **OK.**

II. Excel procedure for *x* values associated with a set of cumulative probabilities

1. Enter the labels into **A4:B4** as shown in the printout above. Enter the cumulative probabilities of interest (e.g., **0.9750, 0.5000, 0.2500**) starting with cell **A5.** Click on cell **B5.**
2. From the **Formulas** ribbon, click on the **Insert Function** symbol at the extreme left. Click **Statistical** in the **Category** box. Click **NORMINV** in the **Function** box. Click **OK.**
3. Enter **A5** into the **Probability** box. Enter the mean of the distribution (e.g., **35**) into the **Mean** box. Enter the standard deviation (e.g., **9**) into the **Standard_dev** box. Click **OK.** The first cumulative probability will appear in cell B5.
4. Click at the lower right corner of **B5** (the cursor will turn to a "+" sign) and drag downward so cells **B5:B7** will contain the set of cumulative probabilities.

MINITAB

```
Inverse Cumulative Distribution Function
Normal with mean = 35 and standard deviation = 9
P( X <= x )         x
      0.975   52.6397
      0.500   35.0000
      0.250   28.9296
```

I. Minitab procedure for the *x* value associated with one cumulative probability

1. Click **Calc.** Select **Probability Distributions.** Click **Normal.**
2. In the **Normal Distribution** menu, enter the mean of the distribution (e.g., **35**) into the **Mean** box. Enter the standard deviation (e.g., **9**) into the **Standard deviation** box. Select **Input constant** and enter the cumulative probability of interest (e.g., **0.9750**) into the box. Select **Inverse cumulative probability.** Click **OK.**

II. Minitab procedure for the *x* values associated with a set of cumulative probabilities

Enter the cumulative probabilities of interest (e.g., **0.9750, 0.5000, 0.2500**) into column **C1.** Follow steps 1 and 2 above, except select **Input column** and enter **C1** into the box.

EXERCISES

7.16 The normal distribution is really a family of distributions. Is the standard normal distribution also a family of distributions? Explain.

7.17 In the standard normal distribution, approximately what values of z correspond to the
a. first and third quartiles?
b. first and ninth deciles?
c. 23rd and 77th percentiles?

7.18 A continuous random variable, x, is normally distributed with a mean of \$1000 and a standard deviation of \$100. Convert each of the following x values into its corresponding z-score:

a. x = \$1000 b. x = \$750
c. x = \$1100 d. x = \$950
e. x = \$1225

7.19 A continuous random variable, x, is normally distributed with a mean of 200 grams and a standard deviation of 25 grams. Convert each of the following x values into its corresponding z-score:

a. $x = 150$ b. $x = 180$
c. $x = 200$ d. $x = 285$
e. $x = 315$

7.20 Using the standard normal table, find the following probabilities associated with z:
a. $P(0.00 \le z \le 1.25)$
b. $P(-1.25 \le z \le 0.00)$
c. $P(-1.25 \le z \le 1.25)$

7.21 Using the standard normal table, find the following probabilities associated with z:
a. $P(0.00 \le z \le 1.10)$
b. $P(z \ge 1.10)$
c. $P(z \le 1.35)$

7.22 Using the standard normal table, find the following probabilities associated with z:
a. $P(-0.36 \le z \le 0.00)$
b. $P(z \le -0.36)$ c. $P(z \ge -0.43)$

7.23 Using the standard normal table, find the following probabilities associated with z:
a. $P(-1.96 \le z \le 1.27)$
b. $P(0.29 \le z \le 1.00)$
c. $P(-2.87 \le z \le -1.22)$

7.24 Using the standard normal table, determine a z value (to two decimal places) such that the
a. cumulative area to z is 0.7486.
b. cumulative area to z is 0.0735.
c. area between z and positive infinity is 0.3508.
d. area between z and positive infinity is 0.0212.

7.25 Using the standard normal table, determine a z value (to two decimal places) such that the
a. cumulative area to z is 0.70.
b. cumulative area to z is 0.10.
c. area between z and negative infinity is 0.54.
d. area between z and positive infinity is 0.30.

7.26 For the normal distribution described in Exercise 7.10, what is the probability that a randomly selected passenger trip would have produced a revenue
a. between $1.55 and $1.80?
b. over $1.90? c. under $1.25?

7.27 For the normal distribution described in Exercise 7.11, what is the probability that a randomly selected first mortgage would have been for an amount
a. between $305,000 and $325,000?
b. over $325,000? c. under $390,000?

7.28 For the normal distribution described in Exercise 7.12, what is the probability that a randomly selected tax preparation customer would have paid a fee
a. between $150 and $180?
b. under $140? c. over $220?

7.29 For the normal distribution described in Exercise 7.15, what is the probability that Jamal's commuting time to work today will be
a. less than 29 minutes?
b. between 29 and 33 minutes?
c. more than 32 minutes?

7.30 For the normal distribution described in Exercise 7.15, what commuting time will be exceeded on only 10% of Jamal's commuting days?

7.31 For the normal distribution described in Exercise 7.11, what first-mortgage amount would have been exceeded by only 5% of the mortgage customers?

7.32 For the normal distribution described in Exercise 7.12, what tax preparation fee would have been exceeded by 90% of the tax preparation customers?

7.33 It has been reported that households in the West spend an annual average of $6050 for groceries. Assume a normal distribution with a standard deviation of $1500. Source: David Stuckey and Sam Ward, "Home Cooked," *USA Today*, July 25, 2006, p. 1A.
a. What is the probability that a randomly selected Western household spends more than $6350 for groceries?
b. How much money would a Western household have to spend on groceries per year in order to be at the 99th percentile (i.e., only 1% of Western households would spend more on groceries)?

7.34 Andre is a fearless circus performer who gets shot from a special cannon during the grand finale of the show and is supposed to land on a safety net at the other side of the arena. The distance he travels varies, but is normally distributed with a mean of 150 feet and a standard deviation of 10 feet. The landing net is 30 feet long.
a. To maximize Andre's probability of landing on the net, how far away from the cannon should he position the nearest edge of the net?
b. Given the net position in part (a), what is the probability that Andre will be able to return for tomorrow night's show?

7.35 Drying times for newly painted microwave oven cabinets are normally distributed with a mean of 2.5 minutes and a standard deviation of 0.25 minutes. After painting, each cabinet is mated with its electronic modules and mechanical components. The production manager must decide how much time to allow after painting before these other components are installed. If the time is too short, the paint will smudge and the unit will have to be refinished. If the time is too long, production efficiency will suffer. A consultant has concluded that the time delay should be just enough to allow 99.8% of the cabinets to dry completely, with just 0.2% ending up being smudged and sent back for refinishing. Given this information, for what time setting should the production manager set the automatic timer that pauses the production line while each cabinet dries?

7.36 KleerCo supplies an under-hood, emissions-control air pump to the automotive industry. The pump is vacuum powered and works while the engine is operating, cleaning the exhaust by pumping extra oxygen into the exhaust system. If a pump fails before the vehicle in which it is installed has covered 50,000 miles, federal emissions regulations require that it be replaced at no cost to the vehicle owner. The company's current air pump lasts an average of 63,000 miles, with a standard deviation of 10,000 miles. The number of miles a pump operates before becoming ineffective has been found to be normally distributed.

a. For the current pump design, what percentage of the company's pumps will have to be replaced at no charge to the consumer?
b. What percentage of the company's pumps will fail at exactly 50,000 miles?
c. What percentage of the company's pumps will fail between 40,000 and 55,000 miles?
d. For what number of miles does the probability become 80% that a randomly selected pump will no longer be effective?

7.37 The KleerCo company in Exercise 7.36 would like to design a more durable pump so that no more than 2% of original-equipment pumps are returned under free warranty, and the standard deviation continues to be 10,000 miles. What will the new average service life have to be in order for the new pump design to comply with this requirement?

7.38 A battery manufacturer has just begun distribution of its SuperVolt, a 6-volt lantern battery for outdoor enthusiasts. Prototype tests of the new battery found the average lifetime in continuous use to be 8 hours, with a standard deviation of 2 hours. Battery lifetimes in such applications have been approximately normally distributed. Edgar Evans, the consumer reporter for a large metropolitan newspaper, has purchased one of the batteries to take along on a camping trip. On the first day of his trip, Edgar goes fishing in his rowboat, becomes disoriented, and gets lost among the many small islands in the large lake near his camp. Fortunately, Edgar has brought along his trusty flashlight and its new SuperVolt battery. At 9 P.M., Edgar turns on his flashlight and shines it into the air, hoping that someone will see his signal and rescue him. Unfortunately, it is not until 3 A.M. that searchers begin their flight over the lake. If Edgar's light is still shining as they become airborne, they will easily spot it from the air. When Edgar finally gets back to the city, what is the probability that he will have an exciting story to tell about how his flashlight and its SuperVolt battery were two of the heroes in his rescue?

THE NORMAL APPROXIMATION TO THE BINOMIAL DISTRIBUTION (7.4)

The binomial distribution, discussed in Chapter 6, is symmetrical whenever the population proportion, π, is 0.5 and approaches symmetry for values that are close to 0.5. In addition, whenever the number of trials, n, becomes larger, the binomial distribution takes on a bell-shaped appearance that resembles very closely the normal distribution discussed in this chapter.

Although the binomial distribution is discrete and the normal distribution is continuous, the normal distribution is a very good approximation to the binomial whenever both $n\pi$ and $n(1 - \pi)$ are ≥ 5. This can be quite useful when n and π are not among the values listed in the binomial probability tables. As we have seen, such tables are already quite extensive for even relatively small upper limits for n. In addition, calculating binomial probabilities in the absence of such tables is a tedious proposition.

The Binomial Distribution: A Brief Review

The Bernoulli Process

The binomial distribution assumes a *Bernoulli process*, which has the following characteristics:

1. There are two or more consecutive trials.
2. On each trial, there are just two possible outcomes—denoted as "success" or "failure."
3. The trials are statistically independent; that is, the outcome in any trial is not affected by the outcomes of earlier trials, and it does not affect the outcomes of later trials.

4. The probability of a success remains the same from one trial to the next. This is satisfied if either we sample with replacement or the sample is small compared to the size of the population. As a rule of thumb, if we are sampling without replacement and the population is at least 20 times as large as the number of trials, the constant π assumption can be considered to have been met.

The Mean and Standard Deviation

The mean and standard deviation of the binomial distribution are the basis for the normal distribution that will be used in making the approximation. As described in Chapter 6, these are

- Mean:

$$\mu = n\pi$$

- Standard deviation:

$$\sigma = \sqrt{n\pi(1 - \pi)}$$

where n = the number of trials
π = the probability of success on any given trial
x = the number of successes in n trials

Correction for Continuity

Because the binomial distribution has gaps between possible values of x, while the normal distribution is continuous, the normal approximation to the binomial involves a **correction for continuity.** The correction consists of expanding each possible value of the discrete variable, x, by 0.5 in each direction. In continuous distributions like the normal, the probability of x taking on any exact value (e.g., $x = 12.000$) is zero. This is why we must consider each of the discrete values of x as being an *interval.*

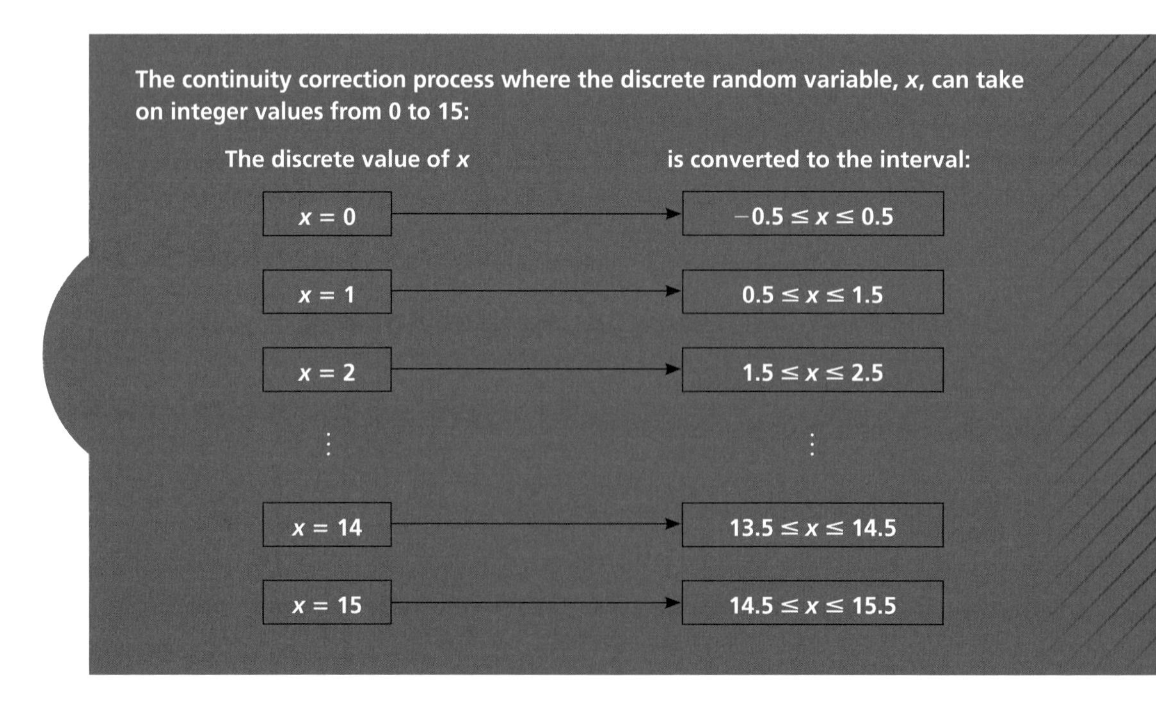

The Approximation Procedure

In using the normal distribution to approximate the binomial, we first determine the mean and standard deviation for the binomial distribution, then use these values as μ and σ in finding the area(s) of interest beneath the standard normal curve. In particular, we can find

1. the probability for an individual value for x, now expressed in terms of the interval from $[x - 0.5]$ to $[x + 0.5]$, or
2. the probability that x will lie within or beyond a given range of values. (*Note:* In this application, the correction for continuity becomes less important whenever x can take on a large number of possible values.)

EXAMPLE

Normal Approximation

This example demonstrates the use of the normal distribution to approximate the binomial. It represents a situation in which both $n\pi$ and $n(1 - \pi)$ are ≥ 5, which is requisite for using the approximation. For purposes of our example, we will assume that 60% of the Americans who exercise each week by walking are women. A randomly selected group comprises 15 persons who exercise each week by walking, with x = the number of women in the group.

SOLUTION

Determine the Mean and Standard Deviation of *x*.
The expected number of females in the group is 9.0.

$$\mu = n\pi \quad \text{or} \quad 15(0.60) = 9$$

$$\sigma = \sqrt{n\pi(1 - \pi)} = \sqrt{15(0.60)(1 - 0.60)} = 1.897$$

What Is the Probability That There Will Be Exactly 11 Females in the Group?
Using the normal distribution to approximate the binomial, with $\mu = 9.0$ and $\sigma = 1.897$, this probability is expressed as $P(10.5 \leq x \leq 11.5)$. The z-score associated with $x = 11.5$ is

$$z = \frac{x - \mu}{\sigma} = \frac{11.5 - 9.0}{1.897} = 1.32$$

The z-score associated with $x = 10.5$ is

$$z = \frac{x - \mu}{\sigma} = \frac{10.5 - 9.0}{1.897} = 0.79$$

Referring to the standard normal table, we find the cumulative area to $z = 1.32$ is 0.9066 and the cumulative area to $z = 0.79$ is 0.7852. As Figure 7.11 shows, $P(10.5 \leq x \leq 11.5)$ is the difference between these areas, or $0.9066 - 0.7852 = 0.1214$.

Note that this approximation is quite close to the actual binomial probability of 0.1268, obtained from the binomial probability tables in Appendix A. Keep in mind, however, that this example was one in which both π and n happened to be available in these tables. This will not always be the case, hence the usefulness of being able to use the normal distribution as an approximation to the binomial.

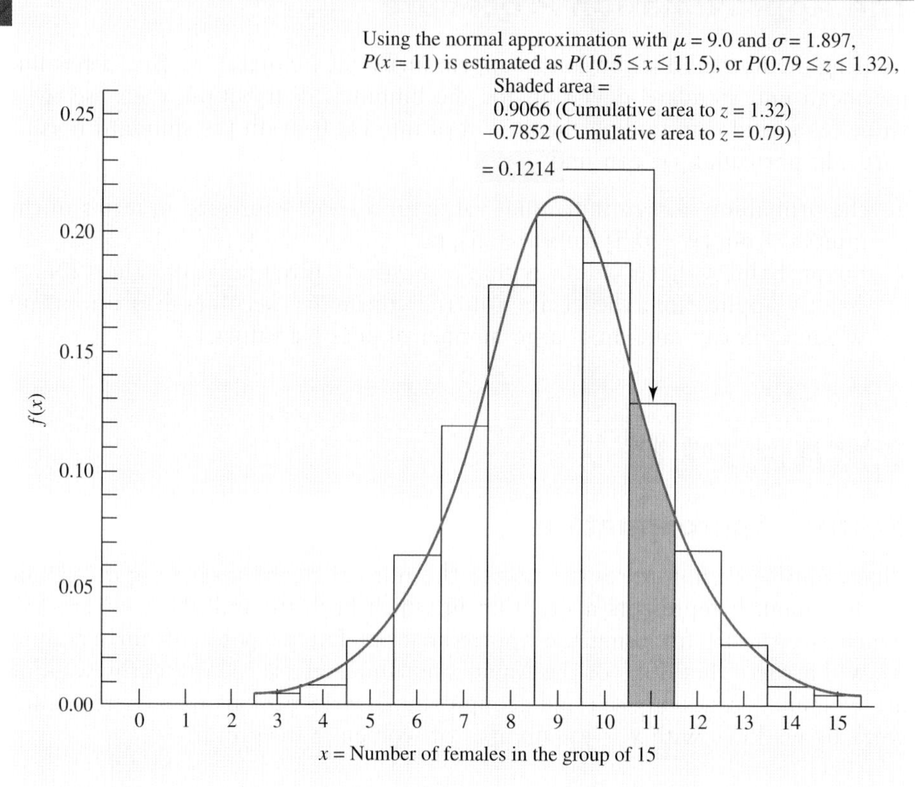

FIGURE 7.11

Normal approximation to a binomial distribution with $n = 15$ and $\pi = 0.60$. The continuity correction represents $x = 11$ as the interval from 10.5 to 11.5, then the standard normal table is used in obtaining the approximate value of 0.1214 for $P(x = 11)$.

What Is the Probability That There Will Be at Least 10 Females in the Group of 15?

Again using the normal distribution to approximate the binomial, the probability of interest is now expressed as $P(x \ge 9.5)$. (The continuity correction has replaced the discrete value of 10 with the interval from 9.5 to 10.5.) The z-score associated with $x = 9.5$ is

$$z = \frac{x - \mu}{\sigma} = \frac{9.5 - 9.0}{1.897} = 0.26$$

Referring to the standard normal table, we see that the cumulative area to $z = 0.26$ is 0.6026. Because the total area beneath the normal curve is 1.0000, the area to the right of $z = 0.26$ is $(1.0000 - 0.6026) = P(x \ge 9.5) = 0.3974$. The normal approximation is again very close to the actual probability (0.4032) obtained from the binomial tables.

The normal approximation to the binomial distribution is quite close whenever both $n\pi$ and $n(1 - \pi)$ are ≥ 5. However, the approximation becomes even better as the value of n increases and whenever π is closer to 0.5. Also, when we are determining probabilities for x being within or beyond a given interval, the correction for continuity becomes less important when x can take on a great many different values. For example, if $n = 1000$ and $\pi = 0.3$, adding and subtracting 0.5 from each x value will make virtually no difference in the results. Refer to Seeing Statistics Applet 6, at the end of the chapter, for a visual demonstration of the normal approximation to the binomial distribution.

EXERCISES

7.39 What is the correction for continuity and why is it employed when the normal distribution is used in approximating the binomial distribution?

7.40 Under what circumstances is it permissible to use the normal distribution in approximating the binomial distribution?

7.41 In a certain binomial distribution, $\pi = 0.25$ and $n = 40$. In using the normal approximation,

a. What are the mean and standard deviation of the corresponding normal distribution?
b. If x = the number of "successes" among the 40 observations, determine the following: $P(x = 8)$, $P(12 \leq x \leq 16)$, $P(10 \leq x \leq 12)$, $P(x \geq 14)$.

7.42 In a certain binomial distribution, $\pi = 0.30$ and $n = 20$. In using the normal approximation,

a. What are the mean and standard deviation of the corresponding normal distribution?
b. If x = the number of "successes" among the 20 observations, determine the following: $P(x = 5)$, $P(4 \leq x \leq 7)$, $P(1 \leq x \leq 5)$, $P(x \geq 7)$.

7.43 Approximately 80% of American families have some form of life insurance coverage. For a randomly selected sample of $n = 15$ families, and the discrete random variable x = the number of families in the sample who are covered by life insurance: Source: American Council of Life Insurers, *2008 Life Insurers Fact Book,* p. 78.

a. Calculate the mean and the standard deviation of the binomial distribution.
b. Using the binomial distribution, calculate the probability that x will be exactly 12 families.
c. Repeat part (b), using the normal approximation to the binomial. (*Hint:* The area of interest under the normal curve will be between $x = 11.5$ and $x = 12.5$.)
d. Using the normal approximation to the binomial, find the probability that at least 10 families in the sample will be covered by some form of life insurance.

7.44 About 40% of new single-family homes completed in the United States during 2007 were heated with electricity. For a randomly selected sample of 20 new single-family homes completed during that year and the discrete random variable, x = the number of homes in this group heated with electricity: Source: *Statistical Abstract of the United States 2009,* p. 593.

a. Calculate the mean and standard deviation of the binomial distribution.
b. Using the binomial distribution tables, determine the probability that x will be exactly 8 homes heated with electricity.
c. Using the binomial distribution tables, determine the probability that x will be at least 6 but no more than 9 homes.
d. Repeat part (b), using the normal approximation to the binomial distribution.
e. Repeat part (c), using the normal approximation to the binomial distribution.

7.45 Of all individual tax returns filed in the United States during the 2009 tax filing season, 15.8% were prepared by H&R Block. For a randomly selected sample of 900 tax returns filed during this period, use the normal approximation to the binomial distribution in determining the probability that between 110 and 140 of these returns were prepared by H&R Block. Source: hrblock.com, July 14, 2009.

7.46 The Electronic Industries Association reports that about 50% of U.S. households have a camcorder. For a randomly selected sample of 800 U.S. households, use the normal approximation to the binomial distribution in determining the probability that at least 410 of these households have a camcorder. Source: *New York Times Almanac 2009,* p. 409.

7.47 Repeat Exercise 7.45, but without using the correction for continuity (i.e., the discrete random variable, x, is not replaced by the interval $x - 0.5$ to $x + 0.5$). Does your answer differ appreciably from that obtained in Exercise 7.45? If not, comment on the need for using the correction for continuity when x can take on a large number of possible values.

7.48 Repeat Exercise 7.46, but without using the correction for continuity (i.e., the discrete random variable, x, is not replaced by the interval $x - 0.5$ to $x + 0.5$). Does your answer differ appreciably from that obtained in Exercise 7.46? If not, comment on the need for using the correction for continuity when x can take on a large number of possible values.

7.5 THE EXPONENTIAL DISTRIBUTION

Description and Applications

Chapter 6 discussed the Poisson distribution, in which the discrete random variable was the number of *rare events* occurring during a given interval of time, space, or distance. For a Poisson process, a distribution called the **exponential distribution** describes the continuous random variable, x = the amount of time, space, or distance between occurrences of these rare events. For example, in the context of arrivals at a service counter, x will be a continuous random variable describing the time between successive arrivals.

As with the Poisson distribution, the exponential distribution has application to the management science topic of queuing, or waiting-line theory. Like the Poisson distribution, it also assumes that the arrivals are independent from one another. The distribution can be summarized as follows:

The exponential distribution:

- For x = the length of the interval between occurrences:

  $f(x) = \lambda e^{-\lambda x}$ for $x > 0$ and $\lambda > 0$ where λ = the mean and variance of a Poisson distribution

  $1/\lambda$ = the mean and standard deviation of the corresponding exponential distribution

  e = the mathematical constant, 2.71828, the base of the natural logarithm system

- Areas beneath the curve:

  $P(x \geq k) = e^{-\lambda k}$ where k = the time, space, or distance until the next occurrence

Like the Poisson distribution, the exponential distribution is really a family of distributions, and the particular member of the family is determined by just one descriptor: its mean. Actually, the mean of the exponential distribution is the *inverse* of that of the Poisson. For example, if the average number of arrivals at a tollbooth is $\lambda = 5$ persons per minute, the mean of the corresponding exponential distribution will be $1/\lambda = 1/5$, or 0.20 minutes between persons. Note that the Poisson random variable is discrete (number of persons), whereas the random variable for the exponential distribution is continuous, in this case, *time*.

It is important that we be consistent in our use of physical-measurement units when we invert λ to arrive at the mean of the corresponding exponential distribution. For example, if arrivals tend to occur at the rate of 15 vehicles per hour, we can express this as 15 arrivals divided by 60 minutes, or 0.25 arrivals per minute. If we then consider $\lambda = 0.25$ arrivals per minute as the Poisson mean, the mean of the corresponding exponential distribution must also be based on minutes, and $1/\lambda$ will be $1/0.25$, or 4 minutes between arrivals.

EXAMPLE

Exponential Distribution

Calls to the "911" emergency number of a large community have been found to be Poisson distributed with an average of 10 calls per hour.

SOLUTIONS

Determine the Mean and Standard Deviation of the Corresponding Exponential Distribution

The mean of the Poisson distribution is $\lambda = 10$ calls per hour, and the mean of the corresponding exponential distribution is the inverse of this, $1/\lambda = 1/10$, or 0.10 hours between calls. The standard deviation is numerically equal to the mean.

For greater clarity in answering the questions that follow, we will express the means of the Poisson and exponential distributions using minutes instead of hours. Accordingly, the mean of the Poisson distribution is 10 calls per 60 minutes, or $\lambda = 1/6$ calls per minute. The mean of the corresponding exponential distribution is $1/\lambda = 6$ minutes between calls.

What Is the Probability That the Next Call Will Occur at Least 5 Minutes from Now?

Referring to the continuous probability distribution of Figure 7.12, this probability corresponds to the area to the right of $k = 5$ minutes on the x axis. Applying the formula for determining areas beneath the curve:

$$P(x \geq k) = e^{-\lambda k} \quad \text{where } k = 5 \text{ minutes}$$

or

$$P(x \geq 5) = e^{-(1/6)(5)} = 2.71828^{-0.833} \quad \text{or} \quad 0.4347$$

There is a 0.4347 chance that at least 5 minutes will elapse before the next call is received. Alternatively, the probability that the next call will occur *within* the next 5 minutes is (1.000 − 0.4347), or 0.5653. This is because (1) the total area beneath the curve is 1.0, and (2) the exponential distribution is continuous

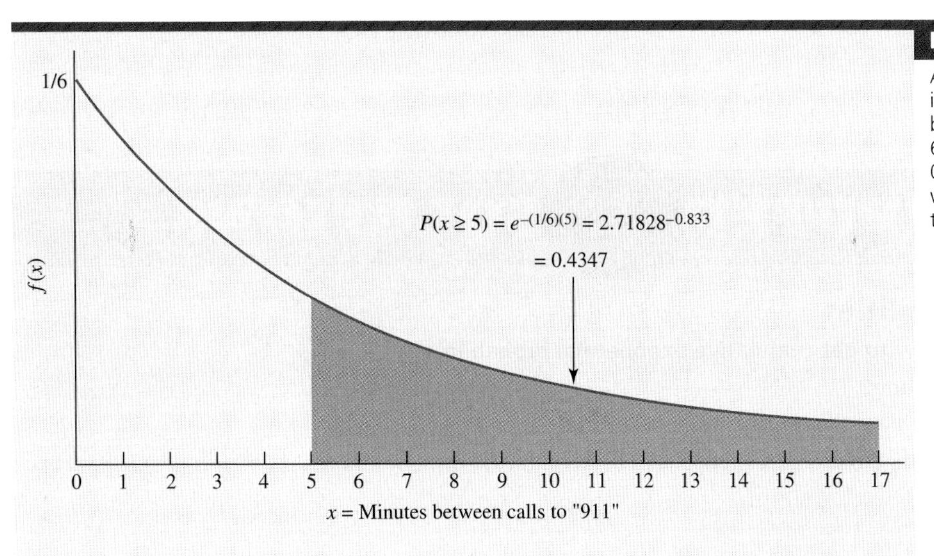

FIGURE 7.12

An exponential distribution in which the mean time between occurrences is 6 minutes. The probability is 0.4347 that at least 5 minutes will elapse before the next call to the "911" switchboard.

and the probability is zero that a continuous random variable will take on any exact value [i.e., $P(x = 5)$ is 0.0].

What Is the Probability That the Next Call Will Occur Between 3 and 8 Minutes from Now?

The area beneath the curve in the interval $3 \le x \le 8$ can be determined by subtracting the area representing $x \ge 8$ from the area representing $x \ge 3$. This is shown in Figure 7.13 and is calculated as

$$P(3 \le x \le 8) = P(x \ge 3) - P(x \ge 8) = e^{-(1/6)(3)} - e^{-(1/6)(8)}$$
$$= 2.71828^{-0.500} - 2.71828^{-1.333} \quad \text{or} \quad 0.6065 - 0.2637 = 0.3428$$

Excel and Minitab can be used to obtain cumulative exponential probabilities associated with specified values for x—the procedures and results are shown in Computer Solutions 7.3. These software packages can also be used in doing the reverse—that is, specifying a cumulative probability, then finding the x value associated with it. These procedures and results are shown in Computer Solutions 7.4.

FIGURE 7.13

For the exponential distribution in which the mean time between calls is 6 minutes, the probability is 0.3428 that between 3 and 8 minutes will elapse before the next call to the "911" switchboard.

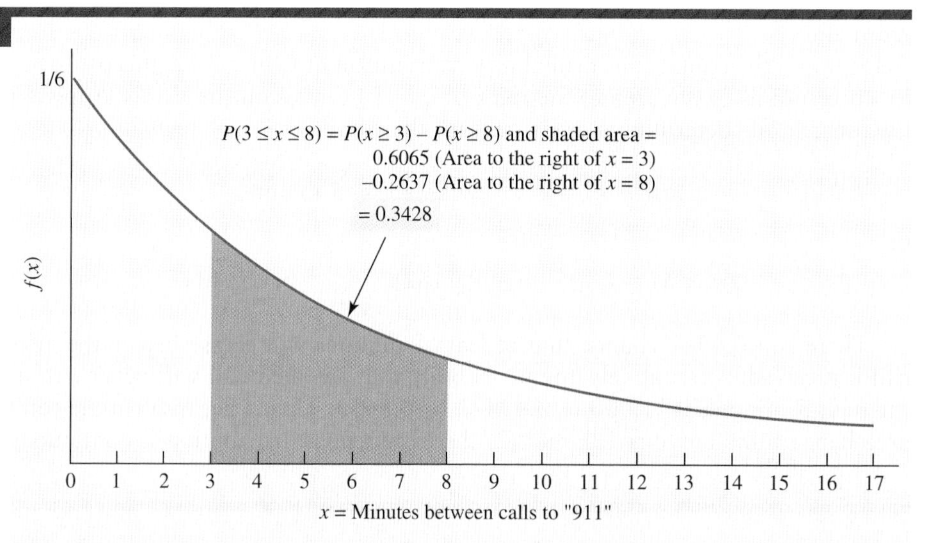

COMPUTER 7.3 SOLUTIONS

Exponential Probabilities

These procedures show how to get cumulative exponential probabilities.

For each software package:

Procedure I: Getting the cumulative probability associated with one x value.

Procedure II: Getting cumulative probabilities associated with a set of x values.

EXCEL

	A	B	C
1	Exponential Distribution		
2	with mean = 6.0.		
3	x	Area to L.	
4	7.0	0.6886	
5	6.0	0.6321	
6	5.0	0.5654	

I. Excel procedure for the cumulative probability associated with one *x* value

1. Click on the cell into which Excel will place the probability. From the **Formulas** ribbon, click on the **Insert Function** symbol at the extreme left. Click **Statistical** in the **Category** box. Click **EXPONDIST** in the **Function** box. Click **OK.**
2. Enter the *x* value of interest (e.g., **5**) into the **X** box. Enter the *inverse* of the exponential mean (e.g., if the mean is 6 minutes between calls, enter 1/6, or **0.16667**) into the **Lambda** box. Enter **True** into the **Cumulative** box. Click **OK.**

II. Excel procedure for cumulative probabilities associated with a set of *x* values

1. Enter the labels into **A3:B3** as shown in the printout above. Enter the *x* values of interest (e.g., **7, 6, 5**) starting with cell **A4.** Click on cell **B4.**
2. From the **Formulas** ribbon, click on the **Insert Function** symbol at the extreme left. Click **Statistical** in the **Category** box. Click **EXPONDIST** in the **Function** box. Click **OK.**
3. Enter **A4** into the **X** box. Enter the *inverse* of the exponential mean (e.g., 1/6, or **0.16667**) into the **Lambda** box. Enter **True** into the **Cumulative** box. Click **OK.** The first cumulative probability will appear in cell B4.
4. Click at the lower right corner of **B4** (the cursor will turn to a "+" sign) and drag downward so cells **B4:B6** will contain the set of cumulative probabilities.

MINITAB

```
Cumulative Distribution Function
Exponential with mean = 6
x  P( X <= x )
7     0.688597
6     0.632121
5     0.565402
```

I. Minitab procedure for cumulative probability associated with one *x* value

1. Click **Calc.** Select **Probability Distributions.** Click **Exponential.**
2. In the **Exponential Distribution** menu, enter the mean of the distribution (e.g., **6**) into the **Mean** or **Scale** box. Select **Input constant** and enter the *x* value of interest (e.g., **5**) into the box. Select **Cumulative probability.** Click **OK.**

II. Minitab procedure for cumulative probabilities associated with a set of *x* values

Enter the desired *x* values (e.g., **7, 6, 5**) into column **C1.** Follow steps 1 and 2 above, except select **Input column** and enter **C1** into the box.

COMPUTER 7.4 SOLUTIONS

Inverse Exponential Probabilities

For Minitab only. Excel does not currently offer inverse exponential probabilities.

MINITAB

```
Inverse Cumulative Distribution Function
Exponential with mean = 6
P( X <= x )          x
      0.975   22.1333
      0.500    4.1589
      0.250    1.7261
```

I. Minitab procedure for the *x* value associated with one cumulative probability

1. Click **Calc.** Select **Probability Distributions.** Click **Exponential.**
2. In the **Exponential Distribution** menu, enter the mean of the distribution (e.g., **6**) into the **Mean** or **Scale** box. Select **Input constant** and enter the cumulative probability of interest (e.g., **0.8000**) into the box. Select **Inverse cumulative probability.** Click **OK.**

II. Minitab procedure for the *x* values associated with a set of cumulative probabilities

Enter the cumulative probabilities of interest (e.g., **0.9750, 0.5000, 0.2500**) into column **C1.** Follow steps 1 and 2 above, except select **Input column** and enter **C1** into the box.

EXERCISES

7.49 What is the relationship between the Poisson distribution and the exponential distribution?

7.50 Every day, drivers arrive at a tollbooth. If the Poisson distribution were applied to this process, what would be an appropriate random variable? What would be the exponential-distribution counterpart to this random variable?

7.51 The main switchboard at the Home Shopping Network receives calls from customers. If the Poisson distribution were applied to this process, what would be an appropriate random variable? What would be the exponential-distribution counterpart to this random variable?

7.52 A random variable is Poisson distributed with $\lambda = 1.5$ occurrences per hour. For the corresponding exponential distribution, and x = hours until the next occurrence, identify the mean of x and determine the following:

a. $P(x \geq 0.5)$ b. $P(x \geq 1.0)$
c. $P(x \geq 1.5)$ d. $P(x \geq 2.0)$

7.53 A random variable is Poisson distributed with $\lambda = 0.02$ occurrences per minute. For the corresponding exponential distribution, and x = minutes until the next occurrence, identify the mean of x and determine the following:

a. $P(x \geq 30.0)$ b. $P(x \geq 40.0)$
c. $P(x \geq 50.0)$ d. $P(x \geq 60.0)$

7.54 A random variable is Poisson distributed with $\lambda = 0.50$ arrivals per minute. For the corresponding exponential distribution, and x = minutes until the next arrival, identify the mean of x and determine the following:
a. $P(x \leq 0.5)$ b. $P(x \leq 1.5)$
c. $P(x \geq 2.5)$ d. $P(x \geq 3.0)$

7.55 The owner of a self-service carwash has found that customers take an average of 8 minutes to wash and dry their cars. Assuming that the self-service times tend to be exponentially distributed, what is the probability that a customer will require more than 10 minutes to complete the job?

7.56 A taxi dispatcher has found that successive calls for taxi service are exponentially distributed, with a mean time between calls of 5.30 minutes. The dispatcher must disconnect the telephone system for 3 minutes in order to have the push-button mechanism repaired. What is the probability that a call will be received while the system is out of service?

7.57 During 2008, U.S. general aviation pilots had 1.20 fatal crashes per 100,000 flying hours. Harriet Arnold is president of Arnold's Flying Service, a company that operates a total of 50 sightseeing planes based in 20 regions of the United States. Altogether, the planes in this fleet are in the air about 40,000 hours per year. Ever since a fatal crash last year involving one of the company's planes that took off under questionable weather conditions, the company has been getting close scrutiny from both government safety officials and the national media. Harriet believes her planes are just as safe as anyone else's, but is concerned that even one more fatal accident within the coming year might be too much for the now-struggling company to overcome. Assuming an exponential distribution for x = thousands of flying hours between fatal crashes, what is the probability that Arnold's Flying Service will not experience a fatal crash until at least a year from today? Until at least 2 years from today? What number of flying hours would be associated with a 90% probability of experiencing no crashes within that number of flying hours from today? Source: General Aviation Manufacturers Association, *General Aviation Statistical Databook, 2008 Edition*, p. 58.

7.58 The Bureau of Labor Statistics reports the fatality rate for workers in the rail transportation industry is 0.006 fatalities per 200,000 worker-hours. During 2007, rail transportation workers put in a total of 545 million worker-hours. Assuming that (1) an exponential distribution applies for the number of worker-hours between fatalities in this industry, and (2) the annual 545 million worker-hours of activity continues in the industry, what is the expected number of worker-hours until the next fatality occurs in this industry? If a given rail transportation company requires 1,000,000 worker-hours each year, what is the probability that the next fatal injury in this company will not occur until at least 30 years from now? Source: Bureau of Labor Statistics, *Census of Fatal Occupational Injuries, 2007*.

SIMULATING OBSERVATIONS FROM A CONTINUOUS PROBABILITY DISTRIBUTION (7.6)

Although practically all statistical packages can provide probabilities like those in the Computer Solutions earlier in this chapter, some can also randomly select observations from the continuous distribution itself. For example, Computer Solutions 7.5 lists the procedures and shows the results when Excel and Minitab are used in randomly selecting 10,000 observations from a normal distribution. Both sets of simulated observations are from a normal distribution examined earlier in the chapter—in this situation, the continuous random variable is x = the amount of weight added so that a generator shaft would have proper static and dynamic balance. In the probability distribution, the average amount of weight added is $\mu = 35$ grams and the standard deviation is $\sigma = 9$ grams.

In each of the printouts in Computer Solutions 7.5, the mean and standard deviation of the simulated observations are very close to the actual mean and standard deviation of the theoretical distribution from which the 10,000 observations were drawn. In addition to differing slightly from the theoretical values, the Excel and Minitab sample means and standard deviations also differ slightly

COMPUTER 7.5 SOLUTIONS

Simulating Observations from a Continuous Probability Distribution

These procedures simulate 10,000 observations from a normal distribution with mean = 35 and standard deviation = 9. The procedures for simulating observations from an exponential distribution will differ only slightly from those described here.

EXCEL

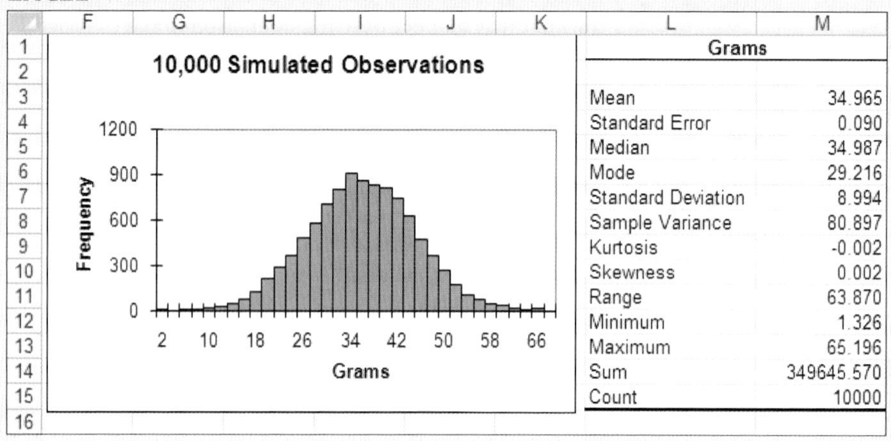

Grams	
Mean	34.965
Standard Error	0.090
Median	34.987
Mode	29.216
Standard Deviation	8.994
Sample Variance	80.897
Kurtosis	-0.002
Skewness	0.002
Range	63.870
Minimum	1.326
Maximum	65.196
Sum	349645.570
Count	10000

1. Start with a blank worksheet. From the **Data** ribbon, click **Data Analysis** in the rightmost menu section. Click **Random Number Generation.** Click **OK.**
2. Enter **1** into the **Number of Variables** box. Enter **10000** into the **Number of Random Numbers** box. Select **Normal** within the **Distribution** box. Enter **35** into the **Mean** box. Enter **9** into the **Standard Deviation** box. Select **Output Range** and enter **A1** into the box. Click **OK.** The 10,000 simulated observations will be located in A1:A10000.
3. The Excel descriptive statistics and histogram above can be generated using the procedures in Computer Solutions 2.1 and 3.1.

MINITAB

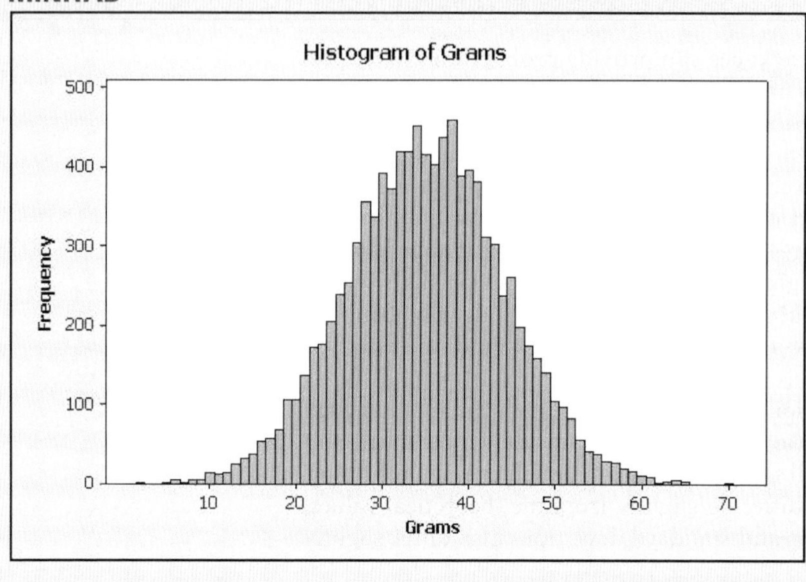

```
Descriptive Statistics: Grams
Variable        N  N*     Mean  SE Mean  StDev  Minimum      Q1  Median      Q3
Grams       10000   0   34.995    0.091  9.078    2.055  28.835  35.020  41.079

Variable  Maximum
Grams      70.164
```

1. Click **Calc.** Select **Random Data.** Click **Normal.** Enter **10000** into the **Number of rows of data to generate** box. Enter **C1** (the destination column) into the **Store in columns** box. Enter **35** into the **Mean** box. Enter **9** into the **Standard deviation** box. Click **OK.** The 10,000 simulated observations will be located in column C1.
2. The Minitab descriptive statistics and histogram shown above can be generated using the procedures in Computer Solutions 2.1 and 3.1.

from each other. This is because each package generated its own sample of 10,000 observations, and each sample had a different set of observations from the other.

According to the law of large numbers, histograms like those in Computer Solutions 7.5 will more closely approach the theoretical distribution from which they were drawn whenever the number of observations becomes larger. The topic of how samples can differ from each other, and from the population from which they were drawn, will be an important topic in the next two chapters.

EXERCISES

7.59 A computer statistical package has simulated 1000 random observations from a normal distribution with $\mu = 50$ and $\sigma = 10$. Sketch the approximate box-and-whisker display for the resulting data.

7.60 A computer statistical package has simulated 2000 random observations from a normal distribution with $\mu = 80$ and $\sigma = 20$. Sketch the approximate box-and-whisker display for the resulting data.

7.61 If a computer statistical package were to simulate 2000 random observations from a normal distribution with $\mu = 150$ and $\sigma = 25$, what percentage of these observations would you expect to have a value of 140 or less? Do you think the actual number in the "≤140" range would equal the expected number in this range? If so, why? If not, why not?

7.62 If a computer statistical package were to simulate 1000 random observations from a normal distribution with $\mu = 200$ and $\sigma = 50$, what percentage of these observations would you expect to have a value of 300 or more? Do you think the actual number in the "≥300" range would equal the expected number in this range? If so, why? If not, why not?

SUMMARY (7.7)

● Nature and applications of the normal distribution

Continuous probability distributions describe probabilities associated with random variables that can take on any value along a given range or continuum and for which there are no gaps between these possible values. The curve $f(x)$ is a function of the continuous random variable, and $f(x)$ is called the probability density function. For any specified interval of values, the probability that x will assume a value within the interval is the area beneath the curve between the two points describing the interval. The total area beneath the curve is 1.0, and the probability that x will take on any specific value is 0.0.

The most important continuous distribution is the normal distribution. This is because many natural and economic phenomena tend to be approximately normally distributed; it can be used to approximate other distributions, such as the binomial; and the distribution of sample means and proportions tends to be normally distributed whenever repeated samples are taken from a given population of any shape.

The normal distribution is a symmetrical, bell-shaped curve, and the mean, median, and mode are located at the same position. The curve is asymptotic in that each end approaches the horizontal (x) axis, but never reaches it. The normal distribution is actually a family of such distributions, and the specific member of the family is determined by the mean (μ) and standard deviation (σ). Regardless of the shape of a particular normal curve, about 68.3% of the area is within the interval $\mu \pm \sigma$, 95.5% within $\mu \pm 2\sigma$, and 99.7% within $\mu \pm 3\sigma$.

● Standard normal distribution and *z*-scores

When the original x values are expressed in terms of the number of standard deviations from the mean, the result is the standard (or standardized) normal distribution in which the z value (or z-score) replaces x. The standard normal distribution has a mean of 0 and a standard deviation of 1, and cumulative areas beneath its curve are readily available from the standard normal table.

● Normal distribution to approximate the binomial distribution

Although the binomial distribution is discrete and the normal distribution is continuous, the normal distribution is a good approximation to the binomial whenever both $n\pi$ and $n(1 - \pi)$ are ≥ 5. The normal approximation is useful whenever suitable binomial tables are not available, and makes it unnecessary to engage in the tedious computations that the binomial distribution would otherwise require.

Because the binomial distribution has gaps between possible values of x, the normal approximation to the binomial involves a correction for continuity, which consists of "expanding" each possible value of x by 0.5 in each direction. This correction becomes less important whenever the number of trials is large and we are finding the probability that x lies within or beyond a given range of values.

● Exponential distribution

For a Poisson process, the exponential distribution describes the continuous random variable, x = the amount of time, space, or distance between occurrences of the rare events of interest. The exponential distribution is a family of distributions, all of which are positively skewed, and the specific member is determined by the value of the mean.

● Computer-generated probabilities and simulations

As with discrete probability distributions, computer statistical packages can determine the probabilities associated with possible values of x. In the case of continuous distributions, this includes finding the exact area beneath a specified portion of the curve, or probability density function, and eliminates the need for calculations or table references. Some packages, such as Excel and Minitab, can also simulate the random selection of observations from the distribution itself.

EQUATIONS

The Normal Distribution

$$f(x) = \frac{1}{\sigma\sqrt{2\pi}} e^{-(1/2)[(x-\mu)/\sigma]^2}$$

where μ = mean
σ = standard deviation
e = the mathematical constant, 2.71828
π = the mathematical constant, 3.14159

The *z*-Score for a Standard Normal Distribution

$$z = \frac{x - \mu}{\sigma}$$

where z = the distance from the mean, measured in standard deviation units
x = the value of x in which we are interested
μ = the mean of the distribution
σ = the standard deviation of the distribution

The Normal Approximation to the Binomial Distribution

- Mean:

$$\mu = n\pi$$

- Standard deviation:

$$\sigma = \sqrt{n\pi(1 - \pi)}$$

where n = the number of trials
π = the probability of success on any given trial
x = the number of successes in n trials

(*Note*: This approximation is used when $n\pi$ and $n(1 - \pi)$ are both ≥ 5.)

The Exponential Distribution

- For x = the length of the interval between occurrences:

$$f(x) = \lambda e^{-\lambda x} \quad \text{for } x > 0 \text{ and } \lambda > 0$$

where λ = the mean and variance of a Poisson distribution
$1/\lambda$ = the mean and standard deviation of the corresponding exponential distribution
e = the mathematical constant, 2.71828, the base of the natural logarithm system

- Areas beneath the curve:

$$P(x \geq k) = e^{-\lambda k}$$

where k = the time, space, or distance until the next occurrence

CHAPTER EXERCISES

7.63 A food expert claims that there is no difference between the taste of two leading soft drinks. In a taste test involving 200 persons, however, 55% of the subjects say they like soft drink A better than B. If the food expert is correct, we would expect that 50% of the persons would have preferred soft drink A. Under the assumption that the food expert is correct, what would be the probability of a "taste-off" in which 110 or more of the subjects preferred soft drink A?

7.64 According to a recent survey, 35% of the adult citizens of Mayville have a charge account at Marcy's Department Store. Mr. Marcy's nephew is going to trial for embezzlement, and nine jurors are to be randomly selected from the adult citizenry. What is the probability that at least four of the jurors will be charge-account holders?

7.65 Of the 1.0 million persons of voting age in Maine, 72% voted in the 2008 presidential election. For a randomly selected group of 30 Maine residents who were of voting age at that time, what is the probability that at least 20 of these persons voted in the presidential election? Source: *nonprofitvote.org*, July 15, 2009.

7.66 In Exercise 7.63, the maker of soft drink A claims its product to be "superior" to soft drink B. Given the results obtained in Exercise 7.63, do they appear to have a solid basis for their claim?

7.67 A public relations agency tells its client that 80% of the residents in a 50-mile radius believe the company is "an industry leader." Skeptical, the company commissions a survey in which just 320 of the 500 persons interviewed felt the company was an industry leader. *If the public relations agency's claim is correct,* what is the probability that interviews with 500 people would yield 320 or fewer who felt the company was an industry leader?

7.68 During an annual heating season, the average gas bill for customers in a New England community heating their homes with gas was $457. Assuming a normal distribution and a standard deviation of $80:

a. What proportion of homes heating with gas had a gas bill over $382?
b. What proportion of homes heating with gas had a gas bill between $497 and $537?
c. What amount was exceeded by only 2.5% of the homes heating with gas?
d. What amount was exceeded by 95% of the homes heating with gas?

7.69 Given the results of Exercise 7.67, evaluate the claim made by the public relations agency.

7.70 It has been reported that the average monthly cell phone bill is $50. Assuming a normal distribution and a standard deviation of $10, what is the probability that a randomly selected cell phone subscriber's bill last month was less than $35? More than $70? Source: *Statistical Abstract of the United States 2009*, p. 706.

7.71 A researcher is studying a particular process and has identified the following times, in minutes, between consecutive occurrences: 1.8, 4.3, 2.3, 2.1, 5.6, 1.5, 4.5, 1.8, 2.4, 2.6, 3.2, 3.5, 1.6, 2.1, and 1.9. Given this information, what would be the researcher's best estimate for the mean of the exponential distribution? What would be her best estimate for the mean of the corresponding Poisson distribution?

7.72 During 2004, the average number of flying hours for aircraft operated by regional airlines in the United States was 2389. Assume a normal distribution and a standard deviation of 300 hours. Source: *Statistical Abstract of the United States 2006*, p. 701.

a. What proportion of aircraft flew more than 2200 hours during the year?
b. What proportion of aircraft flew between 2000 and 2400 hours during the year?
c. What number of annual flying hours was exceeded by only 15% of the aircraft?
d. What number of annual flying hours was exceeded by 75% of the aircraft?

7.73 A safety researcher has found that fourth to sixth graders engaging in unorganized activities on school grounds experienced injuries at the rate of 2.4 injuries per 100 thousand student-days. On the average, how many student-days elapse between injuries to fourth to sixth graders in unorganized activities at school, and what is the probability that the next such injury will occur before 45 thousand student-days have passed?

7.74 During fiscal 2008, the average daily volume for FedEx Corporation was 7,000,000 packages per day. Assuming a normal distribution and a standard deviation of 800,000 packages per day, on what proportion of the days was the volume between 6,000,000 and 6,500,000 packages? Source: *FedEx Corporation, 2008 Annual Report*, p. 28.

7.75 The number of defects in rolls of aluminum sheet average 3.5 defects per 100 feet. The company president is accompanying overseas visitors on a tour of the plant, and they have just arrived at the aluminum-sheet machine. If they remain at the machine long enough to observe 32 feet of aluminum being produced, what is the probability that they will have left the area before the next defect occurs?

7.76 For itemized tax returns in the $60,000–$75,000 income group for the most recent year reported, the average charitable contribution was $1935. Assume a normal distribution and a standard deviation of $400. Source: *Statistical Abstract of the United States 2002*, p. 360.

a. For a randomly selected household from this income group, what is the probability that the household made charitable contributions of at least $1600?
b. For a randomly selected household from this income group, what is the probability that the household made charitable contributions of between $2200 and $2400?
c. What level of contributions was exceeded by only 20% of the itemized returns in this income group?
d. What level of contributions was exceeded by 40% of the itemized returns in this income group?

7.77 The U-Drive car rental corporation has found that the cars in the company's rental fleet experience punctures at the rate of 1.25 punctures per 10,000 miles, and the number of punctures per distance traveled is Poisson distributed. Ed and Harriet are taking advantage of a U-Drive promotion in which there is no mileage charge. During their vacation, they plan to drive from Boston to Los Angeles and back, a total distance of 6164 miles.

a. What is the probability that Ed and Harriet will not have to change any tires during their vacation?
b. While traveling west, what is the probability that Ed and Harriet will experience a punctured tire before they make it to Denver, 2016 miles away?
c. For what mileage is the probability 0.80 that Ed and Harriet will experience a puncture before they reach this distance?

7.78 The trainer for a professional football team has found that, on the average, his team experiences 2.3 sprained ankles per 1000 game minutes. If a game lasts 60 minutes, and the number of game minutes between ankle sprains is exponentially distributed, determine the probability that the next ankle sprain will occur

a. during the team's next game.
b. sometime during the team's next ten games.
c. at the moment the 2-minute warning is sounded in the next game.

7.79 Bonnie Rogers, the customer relations manager for a large computer manufacturer, has recently studied the duration of telephone calls received by the company's technical support department. The call lengths are exponentially distributed, with a mean duration of 8 minutes. Experience has taught Bonnie that lengthy calls usually can be more efficiently handled by a technical support manager, rather than by the less experienced personnel who receive the initial technical inquiry. To avoid having the higher-paid managers virtually take over the task of handling technical support, Bonnie would like to have only the longest 10% of the calls redirected to a technical support manager. How long must a technical support call last before it qualifies for redirection to a technical support manager?

7.80 The precooked weight of hamburgers at a gourmet hamburger restaurant has been normally distributed with a mean of 5.5 ounces and a standard deviation of 0.15 ounces. A dining-column journalist from the local newspaper enters the restaurant and orders a hamburger. What is the probability that he will receive a hamburger that had a precooked weight less than 5.3 ounces? If, in a separate visit to the restaurant, the journalist is accompanied by three colleagues who also order a hamburger, what is the probability that at least 2 of these 4 customers will receive a hamburger that had a precooked weight greater than 5.7 ounces?

7.81 In Exercise 7.85, the company would like to put in a greater weight "cushion" to help protect itself from consumer advocates. If the company wants to have just 2% of the packages contain less than 20 ounces, to what average weight must the filling machine be set?

7.82 State Police Academy graduates from a large graduating class have scored an average of 81.0 on a proficiency test, and the scores are normally distributed with $\sigma = 8.5$. Graduates are randomly paired and assigned to barracks throughout the state. Madeleine has scored 83.0 on the test. What is the probability that she will be paired with someone whose score was within 5.0 points of her own?

7.83 Discount Micros, a computer mail-order company, has received 2000 desktop computers from a major manufacturer. Unknown to the mail-order company, 250 of the computers have a hard disk that was improperly installed at the factory. A consulting firm purchases 40 computers from Discount Micros. If the order is filled by random selection from the 2000 computers just received, what is the probability that the shipment to the consulting firm will include no more than 2 defective computers? At least 1, but no more than 3 defective computers?

7.84 The "20-ounce" package of mozzarella cheese distributed by a food company actually has a weight that is normally distributed, with $\mu = 21$ ounces and $\sigma = 0.2$ ounces.

a. What is the probability that a randomly selected package of this cheese will actually contain at least 20.5 ounces?
b. What is the probability that a randomly selected package of this cheese will actually contain between 20.5 and 21.3 ounces?
c. A shopper randomly selects 8 of these packages. What is the probability that 3 or more of them will contain at least 21.2 ounces of cheese?

7.85 Boxes are filled with sugar by a machine that is set to deliver an average of 20.3 ounces. The weights are normally distributed with a standard deviation of 0.3 ounces. If a consumer advocate buys a package of the product and then proceeds to weigh the contents, what is the probability that the package she has purchased will have a content weight between 20 and 21 ounces? If the consumer advocate plans to purchase 100 boxes of the product, then file suit against the company if more than 5 of the boxes have content weights below the 20-ounce stated weight on the package, what is the probability that she will end up filing suit against the company?

7.86 The mileage death rate for motorcycle riders has been estimated to be about 39 deaths per 100 million miles of motorcycle travel. A national motorcycling association has 1200 members who travel a total of 2 million miles each year on their motorcycles. What is the probability that the next highway fatality experienced by the association will not occur until more than a year from now? Until more than 2 years from now?
Source: USAToday.com, July 30, 2006.

7.87 A storage warehouse in a relatively remote part of a county is protected by a homemade burglary detection system. Once actuated by an intruder, the system sounds a horn and flashes a light for 15 minutes, then saves battery power by shutting down and resetting itself until the next intrusion is detected. Police in three patrol cars have this part of the county as one of their responsibilities, and they routinely drive by the warehouse facility during the hours from dark to dawn. On average, a patrol car passes the warehouse every 20 minutes, and the time between patrol car arrivals is exponentially distributed. It's the middle of the night and a burglar has just broken into the storage warehouse and set off the alarm. What is the probability that the alarm will shut off before the next police patrol car drives by?

INTEGRATED CASES

Thorndike Sports Equipment (Corresponds to Thorndike Video Unit Three)

Since Thorndike "Graph-Pro" racquetball racquets were introduced several years ago, no attempt has ever been made to offer a lightweight version. Ted has done some research and found that some of the other manufacturers offer a lighter version of their standard racquet. In addition, some of the racquetball players he has talked to have said a lighter racquet would probably improve their game.

Talking with production engineers, Ted learns that the racquets coming off the line don't all weigh the same anyway. According to the chief engineer, the racquets weigh an average of 240 grams, but the weights "vary all over the lot," and he takes advantage of the situation to press for a high-tech $700,000 machine that would produce racquets of equal quality while holding this variation within narrower limits.

Old Luke isn't too keen on buying a $700,000 machine, especially since Ted might have a better solution. Ted has dusted off the literature that accompanied the racquet-production machine and found that the machine generates output that is approximately normally distributed, with a standard deviation of 10 grams. This is a little sloppy these days, but maybe Thorndike Sports Equipment can turn this weakness into a strength.

Conversing further with Luke, talking to three sporting-goods retailers, and interviewing players at five different racquetball facilities, Ted comes back to the office with a terrific idea. Asking Luke to sit down, and bringing him some coffee, Ted begins his pitch: "Say, grandfather, why don't we take the racquets at the lighter end of the range and label them 'Graph-Pro Light,' with the ones more in the middle being 'Graph-Pro Regular,' and the ones at the heavier end can be 'Graph-Pro Stout'?"

Luke responds that just because this works for the beer companies, this doesn't mean it will work for Thorndike Sports Equipment. However, he's willing to give it a try. After accompanying Ted on another round of conversations with retailers and players, the elder Thorndike tells Ted that for now, he'd like to see 15% of the production consist of "Graph-Pro Light," with 80% "Graph-Pro Regular," and 5% "Graph-Pro Stout."

Ted's next task is to tell the production people how to go about selecting which racquets are to be put into these three different categories. He's also trying to follow his grandfather's directive to come up with a better name for the "Stout" model, but he figures he can do that after lunch.

Thorndike Golf Products Division

During an 8-hour shift of golf-ball production, one golf ball is randomly selected from each 2 minutes' worth of output. The ball is then tested for "liveliness" by rolling it down a grooved, stainless-steel surface. At the bottom, it strikes a smooth iron block and bounces backward, up a gradually sloped, grooved surface inscribed with distance markings. The higher the ball gets on the rebound surface, the more lively it is.

The production machine is set so that the average rebound distance will be $\mu = 30.00$ inches, and the standard deviation is $\sigma = 2.00$ inches. The scores are normally distributed.

SEEING STATISTICS: APPLET 4

Size and Shape of Normal Distribution

This applet has two normal curves, a blue one and a red one. The blue one stays fixed with a mean of 0 and a standard deviation of 1. Using the sliders at the bottom, we can change the mean and standard deviation of the red curve and observe how these changes affect its position and its shape. Moving the top slider to the left or right will decrease or increase the mean of the red normal curve, and moving the bottom slider to the left or right will decrease or increase its standard deviation.

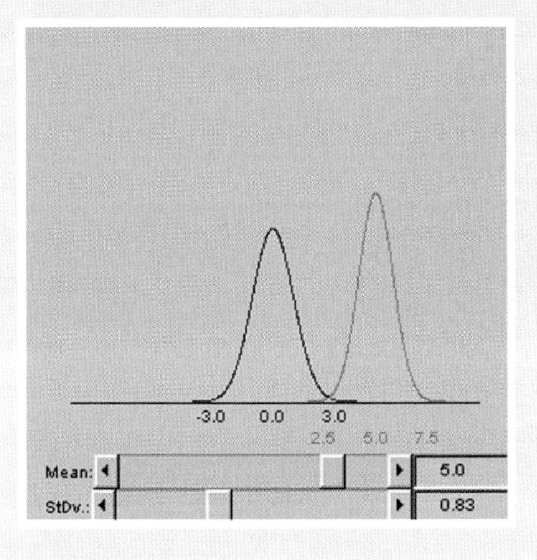

Applet Exercises

4.1 Position the top slider at the far left, then gradually move it to the far right. What effect does this have on the mean of the distribution?

4.2 Using the bottom slider, change the standard deviation so that it is greater than 1. How does the shape of the red curve compare to that of the blue curve?

4.3 Using the two sliders, position the red curve so that its mean is as small as possible and its shape is as narrow as possible. Compare the numerical values of the mean and standard deviation to those of the blue curve (mean = 0, standard deviation = 1).

Try out the Applets: http://www.cengage.com/international

1. A driving range has just purchased 100 golf balls. Use the computer and the binomial distribution in determining the individual and cumulative probabilities for x = the number of balls that scored at least 31.00 inches on the "bounce" test.
2. Repeat part (1), but use the normal approximation to the binomial distribution. Do the respective probabilities appear to be very similar?
3. Given the distribution of bounce test scores described above, what score value should be exceeded only 5% of the time? For what score value should only 5% of the balls do more poorly?
4. What is the probability that, for three consecutive balls, all three will happen to score below the lower of the two values determined in part (3)?

Two hundred forty golf balls have been subjected to the bounce test during the most recent 8-hour shift. From the 1st through the 240th, their scores are provided in computer data file **CDB07** as representing the continuous random variable, BOUNCE. The variable BALL is a sequence from 1 to 240.

5. Using the computer, generate a line graph in which BOUNCE is on the vertical axis and BALL is on the horizontal axis.
6. On the graph obtained in part (5), draw two horizontal lines—one at each of the BOUNCE scores determined in part (3).
7. Examining the graph and the horizontal lines drawn in part (6), does it appear that about 90% of the balls have BOUNCE scores between the two horizontal lines you've drawn, as the normal distribution would suggest when μ = 30.00 and σ = 2.00?
8. Comparing the BOUNCE scores when BALL = 1 through 200 to those when BALL = 201 through 240, does it appear that the process may have changed in some way toward the end of the work shift? In what way? Does it appear that the machine might be in need of repair or adjustment? If so, in what way should the adjustment alter the process as it appeared to exist at the end of the work shift?

Note: Items (5) through (8) contain elements of the topic known as *statistical process control*. It is an important technique that will be examined more completely in Chapter 20.

SEEING STATISTICS: APPLET 5

Normal Distribution Areas

This applet allows us to conveniently find the area corresponding to a selected interval beneath a normal curve that has been specified. We can specify the mean by typing it into the box at the lower left (be sure to press the Return or Enter key while the cursor is still in the text box). Likewise, we can specify the standard deviation of the normal curve by using the text box at the lower right.

With the distribution described by its mean and standard deviation, we can click on the shaded portion of the curve and change the left and right endpoints for the interval. The resulting z values and the probability corresponding to the shaded area are shown at the top of the applet.

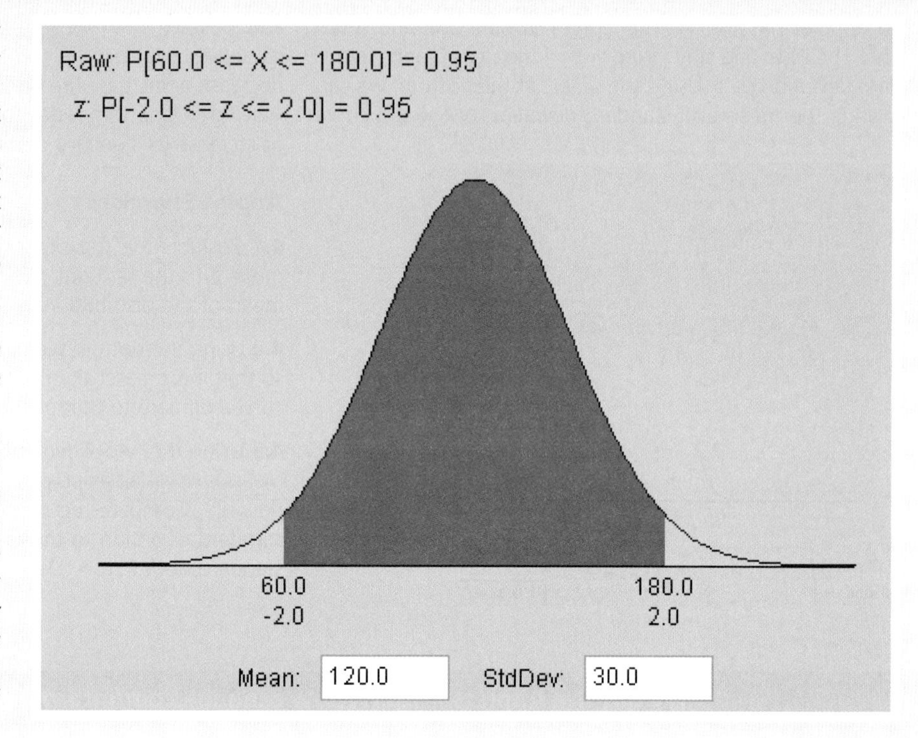

Applet Exercises

5.1 Using the text boxes, ensure that the mean is 120 and the standard deviation is 30, as in the distribution of flying hours in Figure 7.5. (Don't forget to hit the **Enter** or **Return** key while the cursor is still in the text box.) Next, ensure that the left and right boundaries of the shaded area are 60 and 180, respectively; if necessary, click at the left and right edges of the shaded area and drag them. Observing the display at the top, what is the probability that a randomly selected plane will have flown between 60 and 180 hours during the year? What values of z correspond to 60 hours and 180 hours, respectively?

5.2 With the mean and standard deviation set at 120 and 30, respectively, drag the edges of the shaded area so that its left boundary is at 120 and its right boundary is at 180. What is the probability that a randomly selected plane will have flown between 120 and 180 hours during the year? What values of z correspond to 120 hours and 180 hours, respectively?

5.3 By dragging the left and right edges of the shaded area, determine the probability that a randomly selected plane will have flown between 130 and 160 hours during the year. What values of z correspond to 130 hours and 160 hours, respectively?

5.4 Drag the left and right edges of the shaded area so that the left boundary corresponds to $z = 0$ and the right boundary corresponds to $z = +1.0$. What probability is now associated with the shaded area?

SEEING STATISTICS: APPLET 6

Normal Approximation to Binomial Distribution

This applet demonstrates the normal approximation to the binomial distribution. We specify the desired binomial distribution by entering the sample size (n) and the population proportion (π, shown in the applet as **pi**). We enter the k value of interest into the text box at the lower left. Once again, when using the text boxes, remember to press **Enter** or **Return** while the cursor is still within the box. Once we have specified n, π, and k, we see two results: (1) the binomial probability $P(x \leq k)$, shown in the **Prob** text box, and (2) the corresponding approximation using the normal distribution. Please note that probabilities shown in the applet may sometimes differ very slightly from those we might calculate using the pocket calculator and printed tables. This is because the applet is using more exact values than we are able to show within printed tables like the binomial and normal tables in the text.

Applet Exercises

6.1 With $n = 15$ and $\pi = 0.6$, as in the example in Section 7.4, what is the probability that there are no more than $k = 9$ females in the sample of 15 walkers? How does this compare with the corresponding probability using the normal approximation to the binomial distribution?

6.2 With $n = 5$ and $\pi = 0.6$, what is the actual binomial probability that there are no more than $k = 3$ females in the sample of 5? How does this compare with the corresponding probability using the normal approximation? What effect has the smaller sample size had on the closeness of the approximation?

6.3 With $n = 100$ and $\pi = 0.6$, what is the actual binomial probability that there are no more than $k = 60$ females in the sample of 100? How does this compare with the corresponding probability using the normal approximation? What effect has the larger sample size had on the closeness of the approximation?

6.4 Repeat Applet Exercise 6.1 for k values of 6 through 12, in each case identifying the actual binomial probability that there will be no more than k females in the sample. For each value of k, also note the normal approximation result and its closeness to the actual binomial probability.

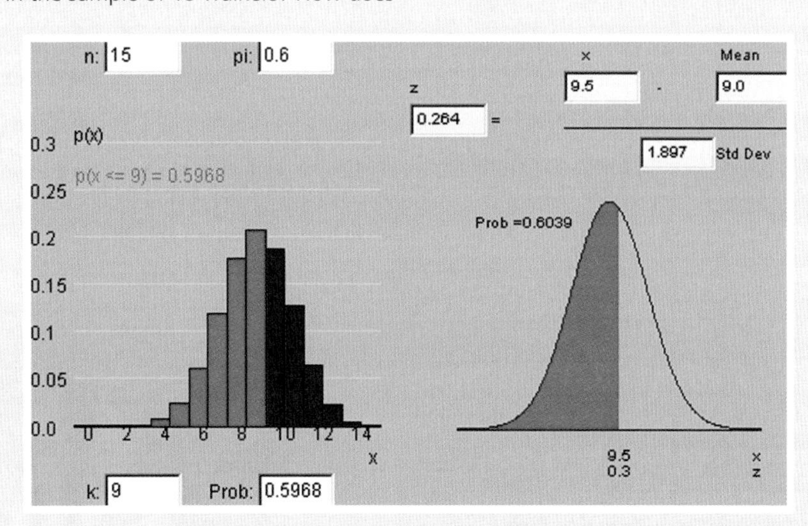

CHAPTER 8

Sampling Distributions

PROCESS CONTROL AND THE SAMPLING DISTRIBUTION OF THE MEAN

When an industrial process is operating properly, it is said to be "in control." However, random variations from one sample to the next can lead to sample means that differ both from each other and from the assumed population mean, μ. Applying the concepts introduced in this chapter, we can state that if the process is in control, the means of samples of $n \geq 30$ should tend to be normally distributed, with $\mu_{\bar{x}} = \mu$ and $\sigma_{\bar{x}} = \sigma/\sqrt{n}$.

As long as the process is in control, there is a 95.5% chance that a sample mean will be within $z = \pm 2$ standard errors of the population mean, a 99.73% chance that it will be within $z = \pm 3$ standard errors of the population mean, and so on. Using this approach, "control limits" can be established to monitor the production process.

For example, if a filling machine is known to have a process standard deviation of $\sigma = 0.1$ ounces and is set to fill boxes with 12.0 ounces of product, the 95.5% ("two-sigma") control limits for the means of samples of $n = 30$ would be

$$12.0 \pm 2.0\left(\frac{0.1}{\sqrt{30}}\right) = 12.0 \pm 0.037 \text{ ounces}$$

or from 11.963 to 12.037 ounces

If sample means consistently tend to fall outside these limits, this suggests that the population mean may have "drifted" from the desired 12.0 ounces and that a filling-quantity adjustment is in order. Such use of what is called an $\bar{x}$ control chart will be examined further in Chapter 20, Total Quality Management.

Samples tend to vary in their proximity to the population target.

LEARNING OBJECTIVES

After reading this chapter, you should be able to:

- Understand that the sample mean or the sample proportion can be considered a random variable.
- Understand and determine the sampling distribution of means for samples from a given population.
- Understand and determine the sampling distribution of proportions for samples from a given population.
- Explain the central limit theorem and its relevance to the shape of the sampling distribution of a mean or proportion.
- Determine the effect on the sampling distribution when the samples are relatively large compared to the population from which they are drawn.

8.1 INTRODUCTION

In the two preceding chapters, we discussed populations in terms of (1) whether they represented discrete versus continuous random variables and (2) their specific probability distributions and the shape of these distributions. In both chapters, we emphasized entire populations. More often, however, we must use sample data to gain insight into the characteristics of a population. Using information from a sample to reach conclusions about the population from which it was drawn is known as **inferential statistics.** The purpose of this chapter is to examine how sample means and proportions from a population will tend to be distributed, and to lay the groundwork for the inferential statistics applications in the chapters that follow.

To underscore the pivotal importance of this chapter, let's specifically consider it in terms of the chapters that immediately precede and follow:

- In Chapter 7, we had access to the population mean, and we made probability statements about individual x values taken from the population.
- In this chapter, we again have access to the population mean, but we will be making probability statements about the means of samples taken from the population. (With this chapter, the sample mean itself is now considered as a random variable!)
- In Chapter 9, we will lack access to the population mean, but we will begin using sample data as the basis from which to make probability statements about the true (but unknown) value of the population mean.

In essence, this chapter is a turning point where we begin the transition from (a) knowing the population and making probability statements *about elements taken from it,* to (b) not knowing the population, but making probability statements *about the population* based on sample data taken from it.

8.2 A PREVIEW OF SAMPLING DISTRIBUTIONS

As suggested in Section 8.1, we're now going to escalate to a new level and *consider the sample mean itself as a random variable.* If we were to take a great many samples of size n from the same population, we would end up with a great many different values for the sample mean. The resulting collection of sample means could then be viewed as a new random variable with its own mean and standard deviation. The probability distribution of these sample means is called the distribution of sample means, or the **sampling distribution of the mean.**

For any specified population, the sampling distribution of the mean is defined as *the probability distribution of the sample means for all possible samples of that particular size.* Table 8.1 provides an introduction to the concept of the sampling distribution of the mean. Referring to parts A–C of Table 8.1:

- **Part A** *The population.* Each of the four members of the population has a given number of bottles of Diet Pepsi in his or her refrigerator. Bill has 1, Carl has 1, Denise has 3, and Ed has 5. The probability distribution of the discrete random variable, x = number of bottles of Diet Pepsi in the refrigerator, is shown in this portion of Table 8.1, and $\mu = (1 + 1 + 3 + 5)/4$, or $\mu = 2.5$ bottles of Diet Pepsi.
- **Part B** *All possible simple random samples of $n = 2$ from this population,* along with the probability that each possible sample might be selected. With the simple random sampling technique, each sample has the same probability

TABLE 8.1

Although the population is very small, parts (A) through (C) show the basic principles underlying the sampling distribution of the mean. The discrete random variable is x = the number of bottles of Diet Pepsi in the person's refrigerator.

A. *The Population, Its Probability Distribution, and Its Mean*

Person	x
Bill	1
Carl	1
Denise	3
Ed	5

Mean: $\mu = (1 + 1 + 3 + 5)/4 = 2.5$

B. *All possible simple random samples of $n = 2$*

Sample	Mean of this Sample	Probability of Selecting this Sample
Bill, Carl	$\bar{x} = (1 + 1)/2 = 1.0$	1/6
Bill, Denise	$\bar{x} = (1 + 3)/2 = 2.0$	1/6
Bill, Ed	$\bar{x} = (1 + 5)/2 = 3.0$	1/6
Carl, Denise	$\bar{x} = (1 + 3)/2 = 2.0$	1/6
Carl, Ed	$\bar{x} = (1 + 5)/2 = 3.0$	1/6
Denise, Ed	$\bar{x} = (3 + 5)/2 = 4.0$	1/6

C. *The Probability Distribution of the Sample Means (the Sampling Distribution of the Mean)*

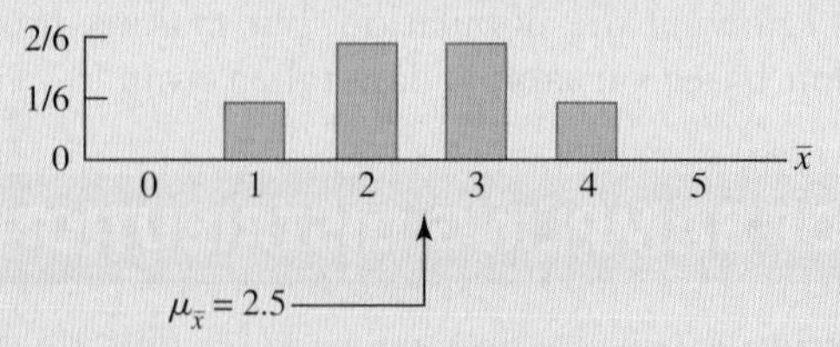

$\mu_{\bar{x}} = 2.5$

The mean of the sample means, $\mu_{\bar{x}}$, is calculated as

$$\mu_{\bar{x}} = \frac{1}{6}(1.0) + \frac{2}{6}(2.0) + \frac{2}{6}(3.0) + \frac{1}{6}(4.0) = 2.5 \text{ bottles}$$

of selection. The number of combinations of 4 elements taken 2 at a time is 6, the number of different simple random samples possible for $n = 2$. In this portion of the table, we also have the mean of each possible sample.

- **Part C** *The probability distribution of the sample means* (i.e., the sampling distribution of the mean), along with $\mu_{\bar{x}}$ = the mean of the sampling distribution. Note that $\mu_{\bar{x}}$ is the same as μ, or 2.5 bottles.

The sampling distribution of the mean has four important characteristics:

1. The sampling distribution of the mean will have the same mean as the original population from which the samples were drawn [i.e., $E(\bar{x}) = \mu_{\bar{x}} = \mu$].
2. The standard deviation of the sampling distribution of the mean is referred to as the **standard error of the mean,** or $\sigma_{\bar{x}}$. For the sampling distribution in Table 8.1, $\sigma_{\bar{x}}$ can be calculated in the same way as the standard deviations in Chapter 6, as the positive square root of its variance:

$$\sigma_{\bar{x}}^2 = E[(\bar{x} - \mu)^2] = (1.0 - 2.5)^2\left(\frac{1}{6}\right) + (2.0 - 2.5)^2\left(\frac{2}{6}\right) + (3.0 - 2.5)^2\left(\frac{2}{6}\right) + (4.0 - 2.5)^2\left(\frac{1}{6}\right) = 0.917$$

and

$$\sigma_{\bar{x}} = \sqrt{0.917} = 0.958 \text{ bottles}$$

3. If the original population is normally distributed, the sampling distribution of the mean will also be normal.
4. If the original population is *not* normally distributed, the sampling distribution of the mean will be approximately normal for large sample sizes and will more closely approach the normal distribution as the sample size increases. Known as the *central limit theorem,* this result will be an especially important topic in the next section.

When a discrete random variable is the result of a Bernoulli process, with x = the number of successes in n trials, the result can be expressed as a sample proportion, p:

$$p = \frac{\text{Number of successes}}{\text{Number of trials}} = \frac{x}{n}$$

Like the sample mean, the sample proportion can also be considered a random variable that will take on different values as the procedure leading to the sample proportion is repeated a great many times. The resulting probability distribution is called the sampling distribution of the proportion, and it will also have an expected value (π, the probability of success on any given trial) and a standard deviation (σ_p). These will be discussed in Section 8.4.

EXERCISES

8.1 What is the difference between a probability distribution and a sampling distribution?

8.2 What is the difference between a standard deviation and a standard error?

8.3 In a simple random sample of 1000 households, 150 households happen to own a barbecue grill. Based on the characteristics of the population, the expected number of grill owners in the sample was 180. What are the values of π, p, and n?

8.4 For a population of five individuals, television ownership is as follows:

	x = Number of Television Sets Owned
Allen	2
Betty	1
Chuck	3
Dave	4
Eddie	2

a. Determine the probability distribution for the discrete random variable, x = number of television sets owned. Calculate the population mean and standard deviation.
b. For the sample size $n = 2$, determine the mean for each possible simple random sample from the five individuals.
c. For each simple random sample identified in part (b), what is the probability that this particular sample will be selected?
d. Combining the results of parts (b) and (c), describe the sampling distribution of the mean.
e. Repeat parts (b)–(d) using $n = 3$ instead of $n = 2$. What effect does the larger sample size have on the mean of the sampling distribution? On the standard error of the sampling distribution?

8.5 Given the following probability distribution for an infinite population with the discrete random variable, x:

x	1	2
$P(x)$	0.5	0.5

a. Determine the mean and the standard deviation of x.
b. For the sample size $n = 2$, determine the mean for each possible simple random sample from this population.
c. For each simple random sample identified in part (b), what is the probability that this particular sample will be selected?
d. Combining the results of parts (b) and (c), describe the sampling distribution of the mean.
e. Repeat parts (b)–(d) using a sample size of $n = 3$ instead of $n = 2$. What effect does the larger sample size have on the mean of the sampling distribution? On the standard error of the sampling distribution?

8.6 The operator of a museum exhibit has found that 30% of the visitors donate the dollar that has been requested to help defray costs, with the others leaving without paying. Three visitors are on their way into the exhibit. Assuming that they will be making independent decisions on whether to donate the dollar that has been requested:
a. Determine the mean and standard deviation for x = the amount paid by individual visitors.
b. Determine the mean and the standard deviation for the sampling distribution of $\bar{x}$ for samples consisting of 3 visitors each.

(8.3) THE SAMPLING DISTRIBUTION OF THE MEAN

When the Population Is Normally Distributed

When a great many simple random samples of size n are drawn from a population that is normally distributed, the sample means will also be normally distributed. *This will be true regardless of the sample size.* In addition, the standard error of the distribution of these means will be smaller for larger values of n. In summary:

Sampling distribution of the mean, simple random samples from a normally distributed population:

Regardless of the sample size, the sampling distribution of the mean will be normally distributed, with

- **Mean:**

$$E(\bar{x}) = \mu_{\bar{x}} = \mu$$

- **Standard error:**

$$\sigma_{\bar{x}} = \frac{\sigma}{\sqrt{n}}$$

where μ **= population mean**
σ **= population standard deviation**
n **= sample size**

The standard deviation (referred to as the standard error) of the sample means will be smaller than that of the original population. This is because each sample mean is the average of several values from the population, and relatively large or small values of x are being combined with values that are less extreme.

EXAMPLE

Sampling Distribution

In Chapter 7, we reported that the average annual number of hours flown by general-aviation aircraft was 120 hours.[1] For purposes of our discussion here, we will continue to assume that the flying hours for these aircraft are normally distributed with a standard deviation of 30 hours.

SOLUTION

Assuming a normal distribution with $\mu = 120.0$ hours and $\sigma = 30.0$ hours, part (a) of Figure 8.1 shows the shape of the population distribution, while part (b) shows the sampling distribution of the mean for simple random samples of $n = 9$ aircraft each. Notice how the sampling distribution of the mean is much narrower than the distribution of flying hours for the original population. While the population x values have a standard deviation of 30.0 hours, the sample means have a standard error of just $30.0/\sqrt{9}$, or 10.0 hours.

In part (c) of Figure 8.1, we see how the sampling distribution of the mean becomes even narrower when the sample size is increased. For simple random samples of size $n = 36$, the sample means have a standard error of only $30.0/\sqrt{36}$, or 5.0 hours.

EXAMPLEEXAMPLEEXAMPLEEXAMPL

Just as we can use the z-score to convert a normal distribution to a standard normal distribution, we can also convert a normally distributed sampling distribution of means to a standard normal distribution. The z-score formula for doing so is very similar to the one introduced in the previous chapter:

z-score for the sampling distribution of the mean, normally distributed population:

$$z = \frac{\bar{x} - \mu}{\sigma_{\bar{x}}}$$

where z = distance from the mean, measured in standard error units

$\bar{x}$ = value of the sample mean in which we are interested

μ = population mean

$\sigma_{\bar{x}}$ = standard error of the sampling distribution of the mean, or $\sigma/\sqrt{n}$

[1]Source: General Aviation Manufacturers Association, *General Aviation Statistical Databook, 2008 Edition*, pp. 28–29.

FIGURE 8.1

For a normal population, the sampling distribution of $\bar{x}$ will also be normal regardless of the sample size. As n increases, the standard error of the sampling distribution will become smaller.

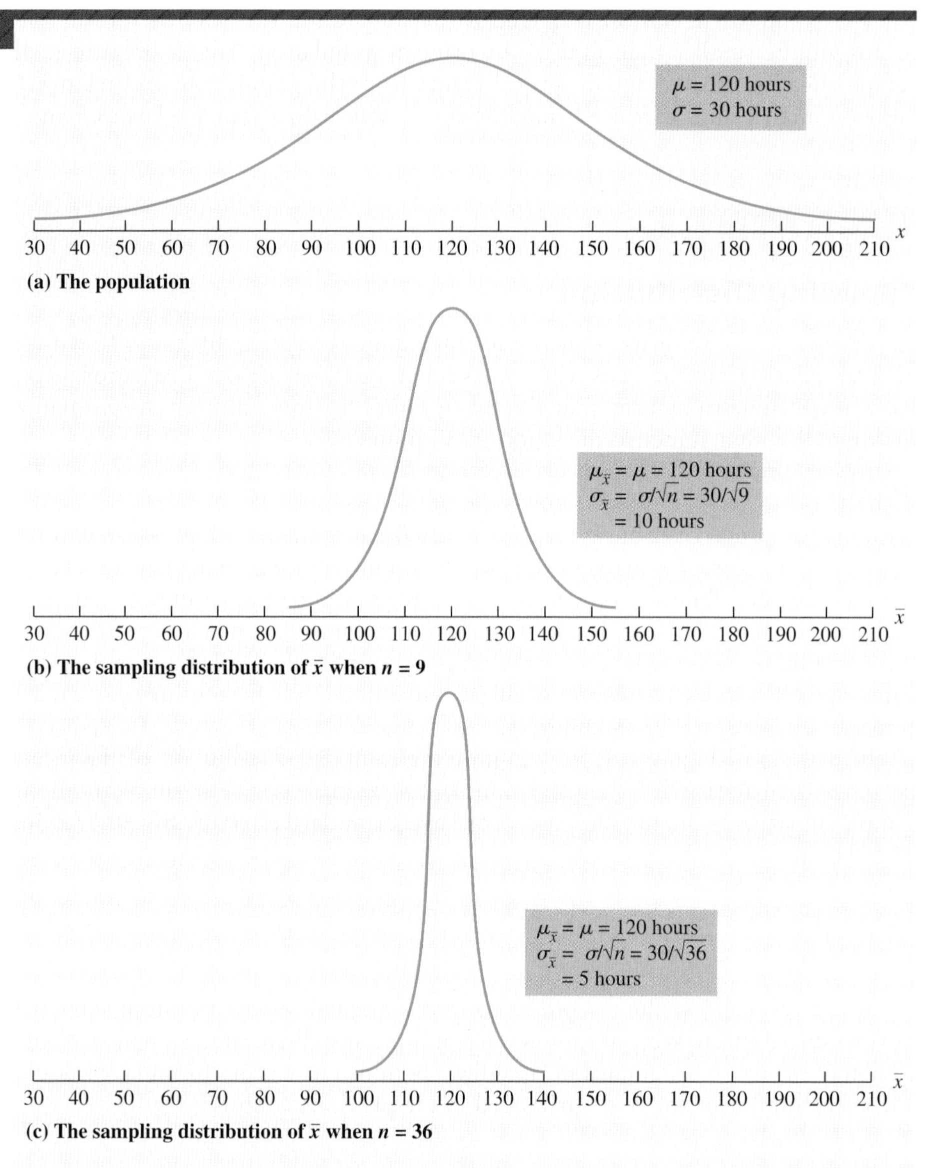

EXAMPLEEXAMPLEEXAMP

EXAMPLE

z-Score

For this example, we have a population of general-aviation aircraft, with the assumption of $\mu = 120.0$ flying hours and $\sigma = 30.0$ flying hours.

SOLUTION

Two important notes:

1. This is the same population of aircraft we discussed in Chapter 7. In Chapter 7, we were making probability statements about *individual planes* selected from

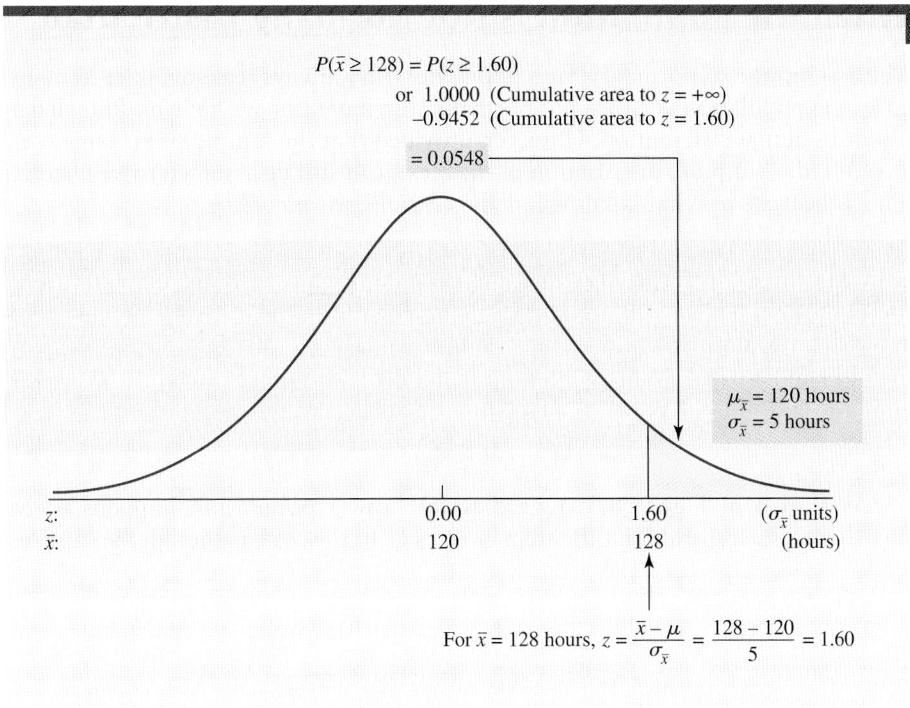

$P(\bar{x} \geq 128) = P(z \geq 1.60)$

or 1.0000 (Cumulative area to $z = +\infty$)

−0.9452 (Cumulative area to $z = 1.60$)

= 0.0548

$\mu_{\bar{x}} = 120$ hours

$\sigma_{\bar{x}} = 5$ hours

z: 0.00 1.60 ($\sigma_{\bar{x}}$ units)

$\bar{x}$: 120 128 (hours)

For $\bar{x} = 128$ hours, $z = \frac{\bar{x} - \mu}{\sigma_{\bar{x}}} = \frac{128 - 120}{5} = 1.60$

FIGURE 8.2

For a simple random sample of $n = 36$ from the aircraft flying-hours population, the probability is just 0.0548 that the sample mean will be 128 hours or more.

the population. Here we are making probability statements about *the means of samples of planes* taken from the original population.

2. In Chapter 7, we dealt with a population consisting of individual planes. Here we are considering a population consisting of sample means—this is what is meant by the sampling distribution of the mean.

For a Simple Random Sample of 36 Aircraft, What Is the Probability That the Average Flying Time for the Aircraft in the Sample Was at Least 128.0 Hours?

As calculated previously, the standard error of the sampling distribution of the mean for samples of this size is $\sigma_{\bar{x}} = 30.0/\sqrt{36}$, or 5.0 hours. We can now proceed to calculate the z-score corresponding to a sample mean of 128.0 hours:

$$z = \frac{\bar{x} - \mu}{\sigma_{\bar{x}}} = \frac{128.0 - 120.0}{5.0} = \frac{8.0}{5.0} = 1.60$$

Referring to the standard normal table, the cumulative area to $z = 1.60$ is 0.9452. Since the total area beneath the curve is 1.0000, the probability that $\bar{x}$ will be at least 128.0 hours is $1.0000 - 0.9452$, or 0.0548, as shown in Figure 8.2.

As with our use of the z-score in finding areas beneath the standard normal curve in the preceding chapter, many other possibilities are available for identifying probabilities associated with the sample mean. For example, you may wish to verify on your own that $P(123.0 \leq \bar{x} \leq 129.0) = 0.2384$ for a simple random sample of $n = 36$ aircraft.

When the Population Is Not Normally Distributed

The assumption of normality for a population isn't always realistic, since in many cases the population is either not normally distributed or we have no knowledge about its actual distribution. However, provided that the sample size is large (i.e., $n \geq 30$), the sampling distribution of the mean can still be assumed to be normal. This is because of what is known as the *central limit theorem:*

Central limit theorem. For large, simple random samples from a population that is not normally distributed, the sampling distribution of the mean will be approximately normal, with the mean $\mu_{\bar{x}} = \mu$ and the standard error $\sigma_{\bar{x}} = \sigma/\sqrt{n}$. As the sample size ($n$) is increased, the sampling distribution of the mean will more closely approach the normal distribution.

As shown in Figure 8.3, regardless of how a population happens to be distributed, the central limit theorem enables us to proceed as though the samples

FIGURE 8.3

When populations are not normally distributed, the central limit theorem states that the sampling distribution of the mean will be approximately normal for large simple random samples.

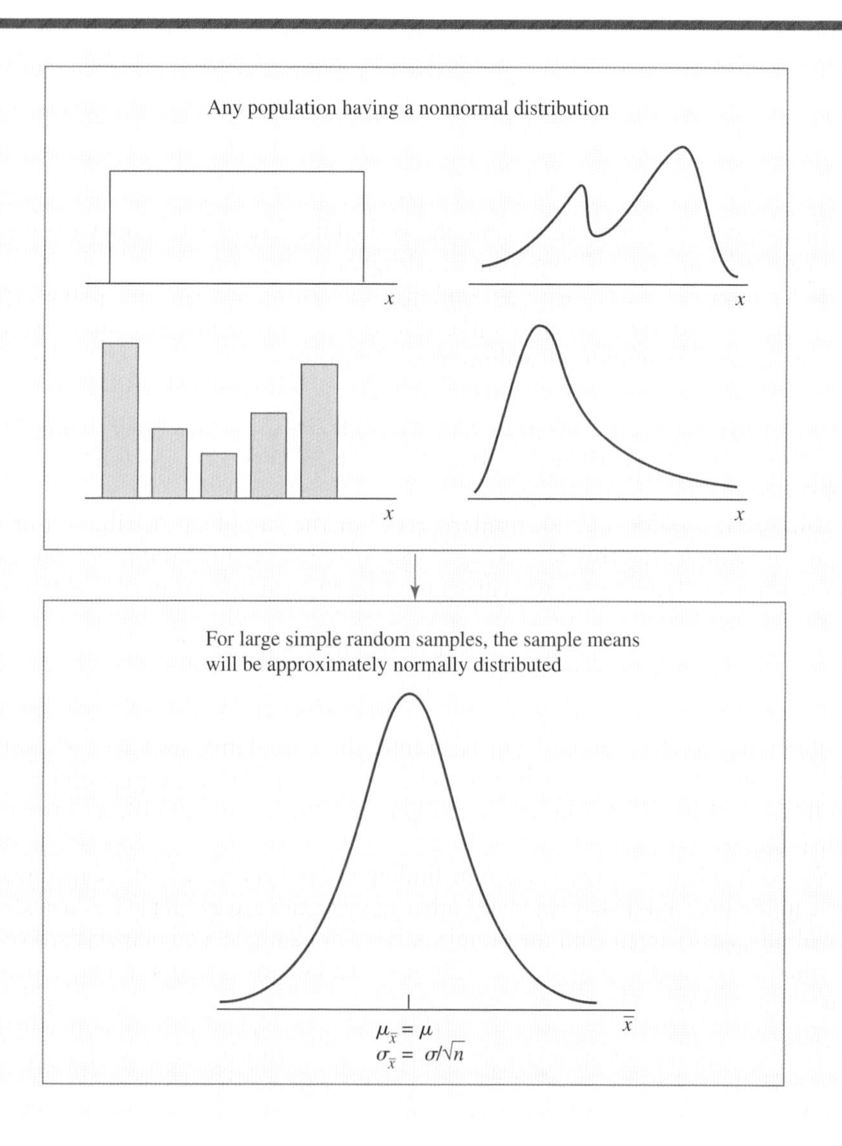

were being drawn from a population that is normally distributed.[2] The only requirement here is that the sample size be sufficiently large. In Chapter 9, we will consider the constraint that σ is known. As a practical matter, when the sample size is at least 30, the "sufficiently large" requirement is assumed to have been met. As the definition of the central limit theorem indicates, the approximation to the normal distribution becomes even better for larger samples. Refer to Seeing Statistics Applets 7 and 8, at the end of the chapter, for visual demonstrations of the sampling distribution of the mean for fair and unfair dice, respectively.

The central limit theorem is basic to the concept of statistical inference because it permits us to draw conclusions about the population based strictly on sample data, and without having any knowledge about the distribution of the underlying population. Such inferences are important to the topics of the next three chapters, and they also play a key role in the application of statistics to total quality management. This chapter's opening vignette is a process-control application to be examined in more detail in Chapter 20.

EXERCISES

8.7 A random variable is normally distributed with mean $\mu = \$1500$ and standard deviation $\sigma = \$100$. Determine the standard error of the sampling distribution of the mean for simple random samples with the following sample sizes:

a. $n = 16$ b. $n = 100$
c. $n = 400$ d. $n = 1000$

8.8 For a random variable that is normally distributed, with $\mu = 200$ and $\sigma = 20$, determine the probability that a simple random sample of 4 items will have a mean that is

a. greater than 210.
b. between 190 and 230.
c. less than 225.

8.9 For a random variable that is normally distributed, with $\mu = 80$ and $\sigma = 10$, determine the probability that a simple random sample of 25 items will have a mean that is

a. greater than 78.
b. between 79 and 85.
c. less than 85.

8.10 Employees in a large manufacturing plant worked an average of 62.0 hours of overtime last year, with a standard deviation of 15.0 hours. For a simple random sample of $n = 36$ employees and $x =$ the number of overtime hours worked last year, determine the z-score corresponding to each of the following sample means:

a. $\bar{x} = 55.0$ hours b. $\bar{x} = 60.0$ hours
c. $\bar{x} = 65.0$ hours d. $\bar{x} = 47.0$ hours

8.11 For the manufacturing plant discussed in Exercise 8.10, the union president and the human resources director jointly select a simple random sample of 36 employees to engage in a discussion with regard to the company's work rules and overtime policies. What is the probability that the average number of overtime hours last year for members of this sample would have been less than 65.0 hours? Between 55.0 and 65.0 hours?

8.12 Shirley Johnson, a county political leader, is looking ahead to next year's election campaign and the possibility that she might be forced to support a property tax increase to help combat projected shortfalls in the county budget. For the dual purpose of gaining voter support and finding out more about what her constituents think about taxes and other important issues, she is planning to hold a town meeting with a simple random sample of 40 homeowners from her county. For the more than 100,000 homes in Shirley's county, the mean value is \$190,000 and the standard deviation is \$50,000. What is the probability that the mean value of the homes owned by those invited to her meeting will be greater than \$200,000?

[2]The central limit theorem applies to any distribution that has a nonzero variance. If there is zero variance in the population, sample means will all have the same value, hence they cannot be normally distributed under this (rare) condition.

8.13 An industrial crane is operated by four electric motors working together. For the crane to perform properly, the four motors have to generate a total of at least 380 horsepower. Engineers from the plant that produces the electric motors have stated that motors off their assembly line average 100 horsepower, with a standard deviation of 10 horsepower. The four electric motors on the crane have just been replaced with four brand new motors from the motor manufacturer. A safety engineer is concerned that the four motors might not be powerful enough to properly operate the crane. Stating any additional assumptions you are using, does the engineer's concern appear to be justified?

8.14 From past experience, an airline has found the luggage weight for individual air travelers on its trans-Atlantic route to have a mean of 80 pounds and a standard deviation of 20 pounds. The plane is consistently fully booked and holds 100 passengers. The pilot insists on loading an extra 500 pounds of fuel whenever the total luggage weight exceeds 8300 pounds. On what percentage of the flights will she end up having the extra fuel loaded?

8.15 A hardware store chain has just received a truckload of 5000 electric drills. Before accepting the shipment, the purchasing manager insists that 9 of the drills be randomly selected for testing. He intends to measure the maximum power consumption of each drill and reject the shipment if the mean consumption for the sample is greater than the 300 watts listed on the product label. Unknown to the purchasing manager, the drills on the truck require an average of 295 watts, with a standard deviation of 12 watts. Stating any additional assumptions you are using, find the probability that the truckload of drills will be rejected.

8.16 The tensile strength of spot welds produced by a robot welder is normally distributed, with a mean of 10,000 pounds per square inch and a standard deviation of 800 pounds per square inch. For a simple random sample of $n = 4$ welds, what is the probability that the sample mean will be at least 10,200 pounds per square inch? Less than 9900 pounds per square inch?

8.17 At a department store catalog orders counter, the average time that a customer has to wait before being served has been found to be approximately exponentially distributed, with a mean (and standard deviation) of 3.5 minutes. For a simple random sample of 36 recent customers, invoke the central limit theorem and determine the probability that their average waiting time was at least 4.0 minutes.

8.18 It has been reported that the average U.S. teenager sends 80 text messages per day. For purposes of this exercise, we will assume the daily number of text messages sent is normally distributed with a standard deviation of 15.0 messages. For a randomly selected group of 10 teenagers, and considering these persons to be a simple random sample of all U.S. teens, what is the probability that the group will send at least 900 text messages (i.e., a sample mean of at least 90 messages per teen) next Wednesday? Source: healthnews.com, May 27, 2009.

8.19 The average length of a hospital stay in general or community hospitals in the United States is 5.5 days. Assuming a population standard deviation of 2.5 days and a simple random sample of 50 patients, what is the probability that the average length of stay for this group of patients will be no more than 6.5 days? If the sample size had been 8 patients instead of 50, what further assumption(s) would have been necessary in order to solve this problem? Source: Bureau of the Census, *Statistical Abstract of the United States 2009*, p. 115.

8.20 According to the Newspaper Association of America, the average visitor to online newspaper sites spends 45 minutes per month reading online news content. Assuming a population standard deviation of 10 minutes and a simple random sample of 30 online newspaper readers, what is the probability that members of this group will average at least 40 minutes reading online newspapers during the coming month? Source: niemanlab.org, July 17, 2009.

8.4 THE SAMPLING DISTRIBUTION OF THE PROPORTION

As mentioned earlier, the proportion of successes in a sample consisting of n trials will be

$$\text{Sample proportion} = p = \frac{\text{Number of successes}}{\text{Number of trials}} = \frac{x}{n}$$

Whenever both $n\pi$ and $n(1 - \pi)$ are ≥ 5, the normal distribution can be used to approximate the binomial, an application examined in Chapter 7. When these conditions are satisfied and the procedure leading to the sample outcome is repeated a great many times, the sampling distribution of the proportion (p) will be approximately normally distributed, with this approximation becoming better for larger values of n and for values of π that are closer to 0.5.

Like the sampling distribution of the mean, the sampling distribution of the proportion has an expected value and a standard error. Also, as the size of the sample becomes larger, the standard error becomes smaller. For simple random samples, the sampling distribution of the proportion can be described as follows:

- **Sampling distribution of the proportion, p:**

$$\text{Mean} = E(p) = \pi$$

$$\text{Standard error} = \sigma_p = \sqrt{\frac{\pi(1-\pi)}{n}} \quad \text{where } \pi = \text{population proportion}, \; n = \text{sample size}$$

- **z-score for a given value of p:**

$$z = \frac{p-\pi}{\sigma_p} \quad \text{where } p = \text{the sample proportion value of interest}$$

EXAMPLE

Sampling Distribution, Proportion

According to the Bureau of the Census, 85.2% of adult residents of Oklahoma have completed high school.[3]

SOLUTION

What Is the Probability That No More Than 80% of the Persons in a Simple Random Sample of 200 Adult Residents of Oklahoma Have Finished High School (i.e., the Probability That $p \leq 0.800$)?

The expected value of the sampling distribution of the proportion is $\pi = 0.852$. This, along with the sample size of $n = 200$, is used in determining the standard error of the sampling distribution and the z-score corresponding to a sample proportion of $p - 0.800$:

$$\text{Standard error} = \sigma_p = \sqrt{\frac{\pi(1-\pi)}{n}} = \sqrt{\frac{0.852(1-0.852)}{200}} = 0.0251$$

$$z = \frac{p-\pi}{\sigma_p} = \frac{0.800-0.852}{0.0251} = -2.07$$

Referring to the standard normal distribution table, the cumulative area to $z = -2.07$ is 0.0192. As shown in Figure 8.4, the probability that no more than 80% of these 200 people are high school graduates is 0.0192.

[3]Source: Bureau of the Census, *Statistical Abstract of the United States 2009*, p. 149.

FIGURE 8.4

The population proportion (π) of adult residents of Oklahoma who have finished high school is 0.852. For a simple random sample of 200 adults from that state, the probability is 0.0192 that the proportion of sample members who are high school graduates will be no more than 0.800.

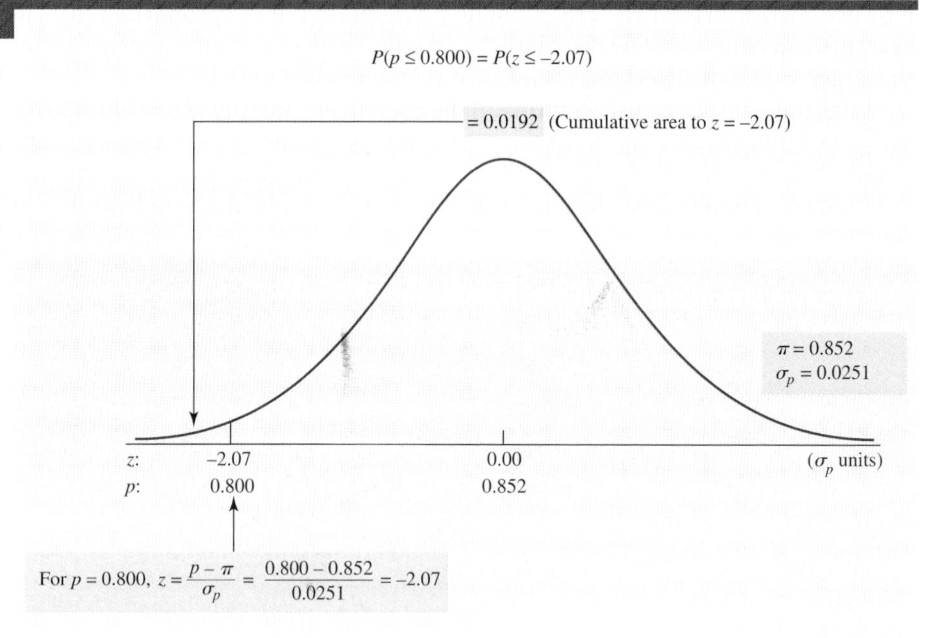

Although the binomial distribution is discrete and the normal continuous, we have not used the correction for continuity in reaching this solution. As Chapter 7 explained, such correction makes little difference in the results when sample sizes are large and the number of successes can take on a large number of possible values. Because of the sample sizes typically involved when dealing with the sampling distribution of the proportion, this correction is generally not employed in these applications and will not be necessary in the exercises that follow.

EXERCISES

8.21 According to the Federal Reserve System, 46% of U.S. households have credit-card debt. For simple random samples of size n, with $n > 30$, describe the sampling distribution of p = the proportion of households in the sample who have credit-card debt. Source: *Statistical Abstract of the United States 2009*, p. 719.

8.22 Shirley Johnson, the county political leader discussed in Exercise 8.12, is a Republican. According to voter registration records, 60% of the homeowners in her county are also Republicans. For the simple random sample of 40 homeowners invited to attend her town meeting, describe the sampling distribution of p = the proportion of Republicans among those invited to the meeting.

8.23 A simple random sample with $n = 300$ is drawn from a binomial process in which $\pi = 0.4$. Determine the following probabilities for p = the proportion of successes:

a. $P(p \geq 0.35)$
b. $P(0.38 \leq p \leq 0.42)$
c. $P(p \leq 0.45)$

8.24 A simple random sample with $n = 200$ is drawn from a binomial process in which $\pi = 0.6$. Determine the following probabilities for p = the proportion of successes:

a. $P(0.50 \leq p)$
b. $P(0.65 \leq p \leq 0.70)$
c. $P(p \geq 0.70)$

8.25 It has been reported that 40% of U.S. workers employed as purchasing managers are females. In a simple random sample of U.S. purchasing managers, 70 out of the 200 are females. Given this information: Source: Bureau of the Census, *Statistical Abstract of the United States 2009*, p. 384.

a. What is the population proportion, π?
b. What is the sample proportion, p?
c. What is the standard error of the sample proportion?
d. In the sampling distribution of the proportion, what is the probability that a sample of this size would result in a sample proportion at least as large as that found in part (b)?

8.26 According to the Investment Company Institute, 40% of U.S. households have an Individual Retirement Account (IRA). Assuming the population proportion to be $\pi = 0.40$ and that a simple random sample of 400 households has been selected: Source: Investment Company Institute, *2008 Investment Company Fact Book*, p. 87.

a. What is the expected value of p = the proportion in the sample having an IRA?
b. What is the standard error of the sampling distribution of the proportion?
c. What is the probability that at least 35% of those in the sample will have an IRA?
d. What is the probability that between 38% and 45% of those in the sample will have an IRA?

8.27 A mail-order discount outlet for running shoes has found that 20% of the orders sent to new customers end up being returned for refund or exchange because of size or fit problems experienced by the customer. The 100-member Connor County Jogging Society has just found out about this mail-order source and each member has ordered a pair of running shoes. The group-orders manager of the mail-order firm has promised the club a 15% discount on its next order if 10 or more of the pairs ordered must be returned because of size or fit problems. What is the probability that the club will get 15% off on its next order?

8.28 The Templovee Company supplies temporary workers to industry and has found that 58% of the positions it is asked to fill require a knowledge of Microsoft Excel. The firm currently has 25 individuals available who know Microsoft Excel. If Templovee Company receives 50 inquiries for temporary workers next week, how likely is it that the company will not have enough Microsoft Excel-skilled people on hand?

8.29 The absentee rate for bus drivers in a large school district has been 5% in recent years. During the past few months, the school district and the bus drivers' union have been engaged in negotiations for a new contract. Negotiations are at an impasse over several key issues, and the drivers already have worked 6 months past the termination of their previous labor contract with the district. Yesterday's negotiating session was particularly contentious, with members of both sides engaging in heated debate and name-calling. Of the 400 bus drivers who serve the school district, 60 of them called in sick this morning, an absentee rate of 15%. The school district has accused the union of collusion in carrying out a work stoppage and the union has responded that the school district is just trying to make a big deal out of more people than usual happening to be sick on a given day. Given the historical absentee rate, what is the probability that a daily absentee rate would happen to be 15% or more? Comment on the competing claims of the school district and the bus drivers' union.

8.30 According to the Federal Trade Commission (FTC), about 3.4% of items come up at the wrong price when scanned at the retail checkout counter. Freda Thompson, director of investigations for her state's consumer affairs office, sets up a study in which she has undercover shoppers (sometimes known as "mystery shoppers") randomly select and purchase 600 items from Wal-Mart department stores across her state. Suppose that researchers find 15 of the 600 items have been incorrectly priced by the checkout scanner, an error rate of just 2.5%. If Wal-Mart's overall rate of scanning errors is the same as the rate cited by the FTC, what is the probability that no more than 2.5% of the items purchased by Freda's investigators would be incorrectly priced when they are scanned? Given the assumed sample result, comment on whether Wal-Mart might be typical of other companies in terms of their error rate for scanning retail items. Source: nist.gov, July 17, 2009.

SAMPLING DISTRIBUTIONS WHEN THE POPULATION IS FINITE (8.5)

When sampling is without replacement and from a finite population, the techniques we've discussed must be modified slightly. The purpose is to arrive at a corrected (reduced) value of the standard error for the sampling distribution. To better appreciate why this is necessary, consider the possibility of selecting a sample of 990 persons from a population of just 1000. Under these conditions, there would be practically no sampling error because the sample would include almost everyone in the population.

Whether we are dealing with the sampling distribution of the mean ($\bar{x}$) or the sampling distribution of the proportion (p), the same correction factor is applied. This factor depends on the sample size (n) versus the size of the population (N) and, as a rule of thumb, should be applied whenever the sample is at least 5% as large as the population. When $n < 0.05N$, the correction will have very little effect.

- **Standard error for the sample mean when sampling without replacement from a finite population:**

$$\sigma_{\bar{x}} = \frac{\sigma}{\sqrt{n}} \cdot \sqrt{\frac{N-n}{N-1}}$$

where n = sample size
N = population size
σ = population standard deviation

- **Standard error for the sample proportion when sampling without replacement from a finite population:**

$$\sigma_p = \sqrt{\frac{\pi(1-\pi)}{n}} \cdot \sqrt{\frac{N-n}{N-1}}$$

where n = sample size
N = population size
π = population proportion

Note that the first term in each of the preceding formulas is the same as we would have used earlier in calculating the standard error. The second term in each case is the **finite population correction factor,** and its purpose is to reduce the standard error according to how large the sample is compared to the population.

EXAMPLE

Finite Populations

Of the 629 passenger vehicles imported by a South American country in a recent year, 117 were Volvos. A simple random sample of 300 passenger vehicles imported during that year is taken.

SOLUTION

What Is the Probability That at Least 15% of the Vehicles in This Sample Will Be Volvos?

For the 629 vehicles imported, the population proportion of Volvos is $\pi = 117/629$, or 0.186. The sample size ($n = 300$) is more than 5% of the size of the population ($N = 629$), so the finite population correction factor is included in the calculation of the standard error of the proportion:

This is the same as would have been calculated in Section 8.4.

The finite population correction term, this will be smaller than 1.0 so the standard error will be less than if an infinite population were involved.

$$\sigma_p = \sqrt{\frac{\pi(1-\pi)}{n}} \cdot \sqrt{\frac{N-n}{N-1}}$$

$$= \sqrt{\frac{0.186(1-0.186)}{300}} \cdot \sqrt{\frac{629-300}{629-1}} = 0.0225 \cdot 0.7238 = 0.0163$$

EXAMPLEEXAMPLEEXAMPLEEXAMPLEEXAMP

Note that the correction factor has done its job. The resulting standard error is 72.38% as large as it would have been for an infinite population. Having obtained $\sigma_p = 0.0163$, we can now determine the z-score that corresponds to the sample proportion of interest, $p = 0.15$:

$$z = \frac{p - \pi}{\sigma_p} = \frac{0.15 - 0.186}{0.0163} = -2.21$$

Using the standard normal distribution table, we see the cumulative area to $z = -2.21$ is 0.0136. The total area beneath the standard normal curve is 1.0000, so the probability that at least 15% of the cars in the sample will be Volvos is $1.0000 - 0.0136$, or 0.9864.

For problems where we happen to be dealing with a sample mean, the same finite population correction approach is used. As with the sample proportion, the correction term will be less than 1.0 in order to reduce the value of the standard error. Again, whenever $n < 0.05N$, the correction will have little effect. For example, at the extreme for this rule of thumb, if $n = 100$ and $N = 2000$, the correction term will have a value of 0.975. The resulting standard error will be very close to that which would have been determined under the assumption of an infinite population.

EXERCISES

8.31 What is meant by a *finite population* and how can this affect the sampling distribution of a mean or a proportion?

8.32 Under what circumstances is it desirable to use the finite population correction factor in describing the sampling distribution of a mean or a proportion?

8.33 Compared to the situation in which the population is either extremely large or infinite in size, what effect does the use of the finite population term tend to have on the resulting value of the standard error?

8.34 A population of 500 values is distributed such that $\mu = \$1000$ and $\sigma = \$400$. For a simple random sample of $n = 200$ values selected without replacement, describe the sampling distribution of the mean.

8.35 A civic organization includes 200 members, who have an average income of $58,000, with a standard deviation of $10,000. A simple random sample of $n = 30$ members is selected to participate in the annual fund-raising drive. What is the probability that the average income of the fund-raising group will be at least $60,000?

8.36 Of the 217 airports available for public use in Belgium, 132 are paved. For a simple random sample of $n = 50$, what is the probability that at least 22 of the airports in the sample will be paved? Source: General Aviation Manufacturers Association, *General Aviation Statistical Databook, 2008 Edition*, p. 48.

8.37 A firm's receiving department has just taken in a shipment of 300 generators, 20% of which are defective. The quality control inspector has been instructed to reject the shipment if, in a simple random sample of 40, 15% or more are defective. What is the probability that the shipment will be rejected?

COMPUTER SIMULATION OF SAMPLING DISTRIBUTIONS (8.6)

Using the computer, we can simulate the collection of simple random samples from a population or a probability distribution. For example, Computer Solutions 8.1 describes the procedures for using Excel or Minitab to generate a simple random sample of observations from a Poisson probability distribution with a mean of 4.

COMPUTER 8.1 SOLUTIONS

Sampling Distributions and Computer Simulation

These procedures simulate 200 samples from a Poisson distribution with a mean of 4.0. In the first simulation, the size of each sample is n = 5. In the second simulation, the size of each sample is n = 30. The Excel results are displayed in Figure 8.5 and the Minitab results would be similar.

EXCEL

1. Start with a blank worksheet. From the **Data** ribbon, click **Data Analysis** in the rightmost menu section. Click **Random Number Generation.** Click **OK.**
2. Enter **5** into the **Number of Variables** box. Enter **200** into the **Number of Random Numbers** box. Within the **Distribution** box, select **Poisson.** Enter **4.0** into the **Lambda** box. Select **Output Range** and enter **A2** into the dialogue box. Click **OK.** The 200 simulated samples will be located in A2:E201, with each row representing one of the 200 samples.
3. Click on cell **F2** and enter the formula **=AVERAGE(A2:E2).** Click at the lower right corner of cell **F2** and drag downward to fill the remaining cells in column F. These values are the 200 sample means. Click on cell **F1** and enter the title for the F column as **Means.**
4. Determine the mean of the sample means by clicking on cell **G2** and entering the formula **=AVERAGE(F2:F201).** Determine the standard deviation of the sample means by clicking on cell **H2** and entering the formula **=STDEV(F2:F201).**
5. Generate the histogram of means when n = 5: Click on cell **J1** and enter the title for the J column as **Bin.** Enter **1.0** into **J2** and **1.2** into **J3.** Select **J2:J3,** then click at the lower right corner of cell **J3** and drag downward to fill, so that **7.0** will be in cell **J32.**
6. From the **Data** ribbon, click **Data Analysis** in the rightmost menu section. Within **Analysis Tools,** click **Histogram.** Click **OK.** Enter **F1:F201** into **Input Range.** Enter **J1:J32** into **Bin Range.** Select **Labels.** Select **Output Range** and enter **K1** into the dialogue box. Select **Chart Output.** Click **OK.** In fine-tuning the histogram, eliminate the gaps between the bars by right-clicking on one of the bars, selecting the **Format Data Series** menu, then clicking on the **Gap Width** slider and moving it leftward to the **No Gap** position, and clicking **Close.** The histogram is shown in part (b) of Figure 8.5.
7. For samples with n = 30, the procedure for generating 200 samples, their means, and the histogram of the sample means will be similar to the steps shown above. For example, in step 2, we would enter **30** into the **Number of Variables** box and enter **200** into the **Number of Random Numbers** box. For each sample, the sample values would be in columns A through AD and the sample means would be in column AE.

MINITAB

1. Click **Calc**. Select **Random Data.** Click **Poisson.** Enter **200** into the **Number of rows of data to generate** box. Enter **C1–C5** (the destination columns) into the **Store in columns** box. Enter **4.0** into the **Mean** box. Click **OK.** The 200 simulated samples will be located in columns C1–C5, with each row constituting a separate sample.
2. Determine the sample means: Click **Calc.** Click **Row Statistics.** Select **Mean.** Enter **C1–C5** into the **Input Variables** box. Enter **C6** into the **Store result in** box. Click **OK.** Enter **Means** as the label of column **C6.**
3. Generate the histogram for the 200 sample means: Click **Graph.** Click **Histogram.** Select **Simple** and Click **OK.** Enter **C6** into the **Graph variables** box. Click **OK.**
4. For samples with n = 30, the procedure for generating 200 samples, their means, and the histogram of the sample means will be similar to the steps shown above. For example, the destination in step 1 will be columns C1–C30, and the means will be located in column C31.

FIGURE 8.5

Although the population is a Poisson distribution with a mean of 4.0, the sampling distribution of the mean still approaches the normal distribution as the Excel-simulated samples become larger. The simulations represented in parts (b) and (c) were generated using the Excel procedures in Computer Solutions 8.1. The Minitab results would be similar.

(a) The population from which the Excel-simulated samples were drawn

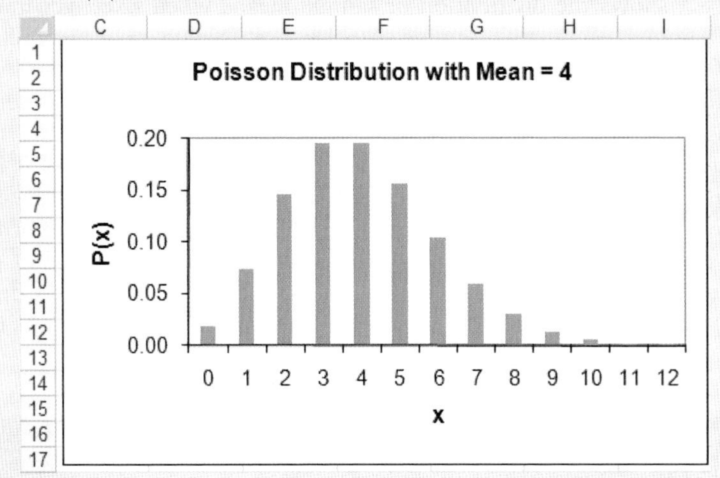

(b) Distribution of means for 200 simulated samples, each with $n = 5$

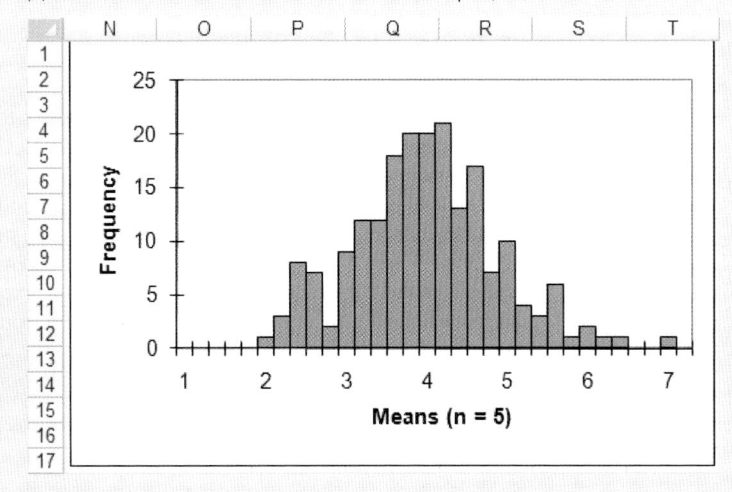

(c) Distribution of means for 200 simulated samples, each with $n = 30$

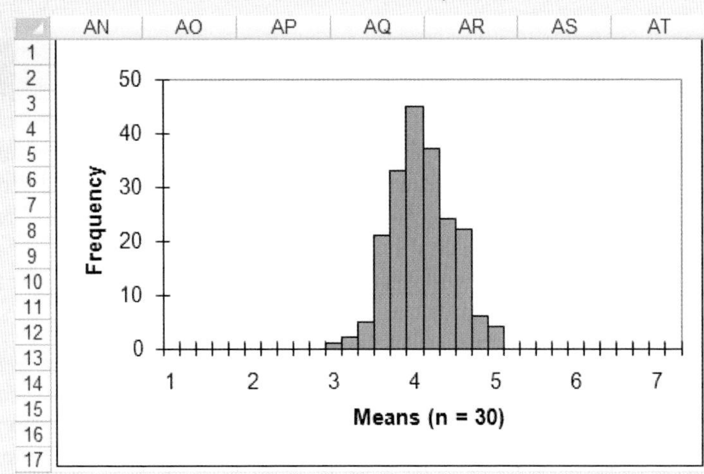

In part (a) of Figure 8.5, it is evident that the Poisson distribution from which the simulated samples are to be taken is neither continuous nor symmetrical. Part (b) of Figure 8.5 shows the distribution of sample means when Excel generated 200 simulated samples, each with a sample size of $n = 5$. In part (c) of Figure 8.5, each of the 200 simulated samples is larger, with $n = 30$. Note that the distribution of sample means not only becomes narrower, but it also more closely resembles a normal distribution.

The probability distribution in part (a) of Figure 8.5 has a mean of 4.0. Because it is a Poisson distribution, its variance is also 4.0, and its standard deviation is the positive square root of the variance, or 2.0. According to the central limit theorem, samples of $n = 30$ from this population should have a mean of 4.0 and a standard error of $\sigma/\sqrt{n} = 2.0/\sqrt{30} = 0.365$. For the 200 samples described in part (c) of Figure 8.5, the mean was 4.009 and the standard error was 0.3759, both very close to the values predicted by the theorem.

EXERCISES

Note: These exercises require Excel, Minitab, or other statistical software capable of simulating simple random samples from a population or probability distribution.

8.38 Using the procedures in Computer Solutions 8.1 as a general guide, simulate 150 simple random samples of size $n = 2$ from an exponential distribution with a mean of 6. Determine the mean and the standard deviation of the 150 sample means, then compare these with their theoretically expected values. Given the nature of the population distribution, would we expect the distribution of sample means to be at least approximately normally distributed? Generate a histogram of the sample means and comment on the shape of the distribution.

8.39 Repeat Exercise 8.38, but with $n = 30$ observations in each sample.

8.40 Using the procedures in Computer Solutions 8.1 as a general guide, simulate 150 simple random samples of size $n = 4$ from a normal distribution with a mean of 80 and a standard deviation of 20. Determine the mean and the standard deviation of the 150 sample means, then compare these with their theoretically expected values. Given the nature of the population distribution, would we expect the distribution of sample means to be at least approximately normally distributed? Generate a histogram of the sample means and comment on the shape of the distribution.

8.41 Repeat Exercise 8.40, but with $n = 30$ observations in each sample.

8.7 SUMMARY

- **The sample mean or sample proportion considered as a random variable**

The two preceding chapters dealt with populations representing discrete and continuous random variables, and with the probability distributions for these variables. The purpose of this chapter is to examine how sample means and proportions from a population will tend to be distributed, and to lay the foundation for the inferential statistics of the chapters that follow. Inferential statistics involves the use of sample information to draw conclusions about the population from which the sample was drawn.

- **Sampling distribution of the sample mean or the sample proportion**

If we were to take a great many samples of a given size from the same population, the result would be a great many different values for the sample mean or sample proportion. Such a process involves what is referred to as a sampling distribution of the sample mean or the sample proportion, and the sample mean or proportion is the random variable of interest. The standard deviation of the sampling distribution of the mean or proportion is referred to as the standard error of the sampling distribution. As the sample size becomes larger, the standard error of the sampling distribution will decrease.

● The central limit theorem

If a population is normally distributed, the sampling distribution of the mean will also be normal. Regardless of how the population values are distributed, the central limit theorem holds that the sampling distribution of the mean will be approximately normal for large samples (e.g., $n \geq 30$) and will more closely approach the normal distribution as the sample size increases. Whenever both $n\pi$ and $n(1 - \pi)$ are ≥ 5, the normal distribution can be used to approximate the binomial, and the sampling distribution of the proportion will be approximately normally distributed, with the approximation improving for larger values of n and for values of π that are closer to 0.5.

Just as we can use the z-score to convert a normal distribution to the standard normal distribution, the same approach can be used to convert a normal sampling distribution to the standard normal distribution. In the process, the z-score can be used to determine probabilities associated with the sample mean or the sample proportion.

● When the sample is large compared to the population

When sampling is without replacement and from a finite population, calculation of the standard error of the sampling distribution may require multiplying by a correction term called the finite population correction factor. The purpose of the correction is to reduce the value of the standard error according to how large the sample is compared to the population from which it was drawn. As a rule of thumb, the finite population correction is made whenever the sample is at least 1/20th the size of the population.

● Using the computer to demonstrate the central limit theorem

Some computer statistical packages can generate simple random samples from a specified distribution. The chapter uses Excel and Minitab to demonstrate the central limit theorem by showing how the distribution of means of simulated samples from a nonnormal population tends to approach the normal distribution as the sample size is increased.

EQUATIONS

Sampling Distribution of the Mean: Simple Random Samples from a Normally Distributed Population

- Regardless of the sample size, the sampling distribution of the mean will be normally distributed, with
- Mean:

$$E(\bar{x}) = \mu_{\bar{x}} = \mu$$

- Standard error:

$$\sigma_{\bar{x}} = \frac{\sigma}{\sqrt{n}}$$ where μ = population mean
σ = population standard deviation
n = sample size

- z-score for the sampling distribution of the mean:

$$z = \frac{\bar{x} - \mu}{\sigma_{\bar{x}}}$$ where z = distance from the mean, measured in standard error units
$\bar{x}$ = value of the sample mean in which we are interested
$\sigma_{\bar{x}}$ = standard error of the sampling distribution of the mean, or $\sigma/\sqrt{n}$

- According to the central limit theorem, the sampling distribution of the mean will be approximately normal even when the samples are drawn from a nonnormal population, provided that the sample size is sufficiently large. It is common practice to apply the above formulas when $n \geq 30$.

Sampling Distribution of the Proportion, p

- Mean = $E(p) = \pi$
- Standard error:

$$\sigma_p = \sqrt{\frac{\pi(1 - \pi)}{n}}$$ where π = population proportion
n = sample size

- z-score for a given value of p:

$$z = \frac{p - \pi}{\sigma_p}$$ where p = sample proportion value of interest

Standard Error of the Sampling Distribution: Sampling without Replacement from a Finite Population

- Standard error for the sample mean:

$$\sigma_{\bar{x}} = \frac{\sigma}{\sqrt{n}} \cdot \sqrt{\frac{N - n}{N - 1}}$$ where n = sample size
N = population size
σ = population standard deviation

- Standard error for the sample proportion:

$$\sigma_p = \sqrt{\frac{\pi(1 - \pi)}{n}} \cdot \sqrt{\frac{N - n}{N - 1}}$$ where n = sample size
N = population size
π = population proportion

CHAPTER EXERCISES

8.42 For the situation described in Exercise 8.43, suppose that only 49% of the registered voters in the sample favor the candidate over her strongest opponent. Given this result:

a. If the campaign manager's claim is correct, what is the probability that the sample proportion would be no more than 0.49 for a sample of this size?

b. Based on your answer to part (a), speculate on whether the campaign manager's claim might be mistaken.

8.43 The campaign manager for a political candidate claims that 55% of registered voters favor the candidate over her strongest opponent. Assuming that this claim is true, what is the probability that in a simple random sample of 300 voters, at least 60% would favor the candidate over her strongest opponent?

8.44 For a given sample size, the standard error of the sample mean is 10 grams. With other factors unchanged, to what extent must the sample size be increased if the standard error of the sample mean is to be just 5 grams?

8.45 When a production machine is properly calibrated, it requires an average of 25 seconds per unit produced, with a standard deviation of 3 seconds. For a simple random sample of $n = 36$ units, the sample mean is found to be $\bar{x} = 26.2$ seconds per unit.

a. What z-score corresponds to the sample mean of $\bar{x} = 26.2$ seconds?
b. When the machine is properly calibrated, what is the probability that the mean for a simple random sample of this size will be at least 26.2 seconds?
c. Based on the probability determined in part (b), does it seem likely that the machine is properly calibrated? Explain your reasoning.

8.46 As a project for their high school mathematics class, four class members purchase packages of a leading breakfast cereal for subsequent weighing and comparison with the 12-ounce label weight. The students are surprised when they weigh the contents of their four packages and find the average weight to be only 11.5 ounces. When they write to the company, its consumer relations representative explains that "the contents of the '12-ounce' packages average 12.2 ounces and have a standard deviation of 0.4 ounces. The amount of cereal will vary from one box to the next, and your sample is very small—this explains why your average weight could be just 11.5 ounces." If the spokesperson's information is correct, how unusual would it be for these students to get a sample mean as small as the one they obtained? Be sure to state any assumptions you have made in obtaining your answer.

8.47 Based on past experience, a telemarketing firm has found that calls to prospective customers take an average of 2.0 minutes, with a standard deviation of 1.5 minutes. The distribution is positively skewed, since persons who actually become customers require more of the caller's time than those who are not home, who simply hang up, or who say they're not interested. Albert has been given a quota of 220 calls for tomorrow, and he works an 8-hour day. Assuming that his list represents a simple random sample of those persons who could be called, what is the probability that Albert will meet or exceed his quota?

8.48 The manufacturer of a travel alarm clock claims that, on the average, its clocks deviate from perfect time by 30 seconds per month, with a standard deviation of 10 seconds. Engineers from a consumer magazine purchase 40 of the clocks and find that the average clock in the sample deviated from perfect accuracy by 34 seconds in one month.

a. If the manufacturer's claim is correct (i.e., $\mu = 30$ seconds, $\sigma = 10$ seconds), what is the probability that the average deviation from perfect accuracy would be 34 seconds or more?
b. Based on your answer to part (a), speculate on the possibility that the manufacturer's claim might not be correct.

8.49 In 2009, the average fee paid by H&R Block tax preparation customers was \$187. Assume that the standard deviation of fees was \$60 but that we have no idea regarding the shape of the population distribution.
Source: hrblock.com, July 14, 2009.

a. What additional assumption about the population would be needed in order to use the standard normal table in determining the probability that the mean fee for a simple random sample of 5 customers was less than \$170?
b. What is the probability that the mean fee for a simple random sample of 36 customers was less than \$170? What role does the central limit theorem play in making it possible to determine this probability?

8.50 Based on past experience, 20% of the contacts made by a firm's sales representatives result in a sale being made. Charlie has contacted 100 potential customers but has made only 10 sales. Assume that Charlie's contacts represent a simple random sample of those who could have been called upon. Given this information:

a. What is the sample proportion, p = proportion of contacts that resulted in a sale being made?
b. For simple random samples of this size, what is the probability that $p \le 0.10$?
c. Based on your answer to (b), would you tend to accept Charlie's explanation that he "just had a bad day"?

8.51 Given the following probability distribution for an infinite population with the discrete random variable, x:

x	0	1	2	3
$P(x)$	0.2	0.1	0.3	0.4

a. Determine the mean and the standard deviation of x.
b. For the sample size $n = 2$, determine the mean for each possible simple random sample from this population.
c. For each simple random sample identified in part (b), what is the probability that this particular sample will be selected?
d. Combining the results of parts (b) and (c), describe the sampling distribution of the mean.

8.52 Employees at a corporate headquarters own an average of 240 shares of the company's common stock, with a standard deviation of 40 shares. For a simple random sample consisting of 5 employees, what assumption would be required if we are to use the standard normal distribution in determining $P(\bar{x} \geq 220)$?

8.53 In the 2000 census, the so-called "long form" received by one of every six households contained 52 questions, ranging from your occupation and income all the way to whether you had a bathtub. According to the U.S. Census Bureau, the mean completion time for the long form is 38 minutes. Assuming a standard deviation of 5 minutes and a simple random sample of 50 persons who filled out the long form, what is the probability that their average time for completion of the form was more than 45 minutes? Source: Haya El Nasser, "Census Forms Can Be Filed by Computer," *USA Today*, February 10, 2000, p. 4A.

8.54 The overall pass rate for law school graduates taking the Maryland bar exam has been reported as 76%. Assume that a certain Maryland law school has had 400 of its most recent graduates take the Maryland bar exam, but only 60% passed. When asked about these results, the dean of this university's law school claims his graduates are just as good as others who have taken the Maryland bar exam, and the low pass rate for the recent graduates was just "one of those statistical fluctuations that happen all the time." If this university's overall pass rate for the Maryland bar exam were really 76%, what would be the probability of a simple random sample of 400 having a pass rate of 60% or less? Based on this probability, comment on the dean's explanation. Source: *U.S. News & World Report, Best Graduate Schools, 2010 Edition,* p. 22.

8.55 Seventy percent of the registered voters in a large community are Democrats. The mayor will be selecting a simple random sample of 20 registered voters to serve on an advisory panel that helps formulate policy for the community's parks and recreation facilities. What is the probability that the constituency of the committee will be no more than 50% Democrats?

8.56 A lighting vendor has described its incandescent bulbs as having normally distributed lifetimes with a mean of 2000 hours and a standard deviation of 100 hours. The vendor is faced with an especially demanding industrial customer who insists on conducting a longevity test before committing to a very large order. The customer's purchasing department plans to test a simple random sample of 16 bulbs and to place the large order only if the sample mean is at least 2050 hours. Given the longevity of the vendor's bulbs and the test the potential customer will be carrying out, what is the probability that the vendor will get the contract?

8.57 Of 500 vehicles sold by an auto dealership last month, 200 have a defect that will require a factory recall for dealer correction. For a simple random sample of $n = 50$ of last month's customers, what is the probability that no more than 15 will be affected by the recall?

8.58 In performing research on extrasensory perception, researchers sometimes place two individuals in different locations, then have person A (the "sender") view each item in a series of stimuli while person B (the "receiver") tries to correctly identify the image person A is viewing. In carrying out such a study, a researcher enlists the help of Amy and Donald as volunteers to "send" and "receive," respectively. The researcher randomly shows Amy one of two pictures—either a sunny sky or a stormy sky—and this procedure is carried out for a total of 200 exposures at 20-second intervals. If Donald is simply guessing, he will tend to be correct 50% of the time. In this particular study, Donald was successful in 108 out of 200 attempts to identify which picture Amy was viewing. If someone has absolutely no extrasensory capabilities and is simply guessing, what is the probability of guessing correctly on at least 108 out of 200 attempts? Given the probability you've calculated, would you suggest that Donald drop his plan to list himself in the yellow pages as a psychic phenomenon?

INTEGRATED CASE

Thorndike Sports Equipment

As he does on the first business day of each month, Ted has just opened up the suggestion box that Luke placed in the plant to attract ideas from the employees. Along with the usual monthly suggestions that old Luke attempt activities that are either inappropriate for his age or physiologically impossible, Ted finds a message from a production worker who is very concerned about the company's racquetball racquet machine:

Dear Thorndike management:

I am getting very worried about the racquetball racquet machine. Some of the older workers say it was here 15 years ago when they started, and I think it is just about worn out. I know new machines are expensive, so I thought I would collect some numbers that might help you out in making a decision to replace it.

I know that the weight of the racquets coming off the machine is a big deal, since all of the racquets are weighed and then you decide which ones will be "Light," "Regular," and whatever you've decided to call the heavier model. Anyway, I've been doing some research. On my own time, honest. From time to time during the day, I take the 30 racquets that have just come from the machine and weigh them myself.

You've told us that some racquets weigh more than others, so I can understand that one racquet might weigh 245 grams, the very next one might weigh 230 grams, and then the next one off the line might be all the way up to 250 grams. Like I said before, I think the machine is worn out. The first 30 racquets I weighed yesterday weighed an average of 236.5 grams, the next 30 weighed an average of 243.1 grams. I think the machine is going crazy, and you should either get it fixed or replace it.

My friends would get pretty upset if they knew I wrote this suggestion, since buying a brand-new machine could mean that nobody gets a Christmas bonus this year. But I'm a young guy and I'd like for you to be successful so that I can have some job security and take care of my family. If you end up having a problem with the weights of your racquets, I might end up being out of a job.

I'd like to find out if maybe I'm missing something here, and if the machine really is OK or not. So my friends don't find out about this suggestion, would you please put together a reply and duct-tape it to the underside of the pay phone next to the candy machine? I'll plan on picking it up after the first work shift on the 15th of next month.

Discarding the suggestions that Luke would probably find offensive, Ted mentions the young worker's letter regarding the racquetball racquet machine. Luke asks if the machine is still working the same as when the decision was made to offer racquets in three different weight categories. Ted replies that it seems to be. The average weight is still 240 grams, the standard deviation is still 10 grams, although he isn't too sure whether the weights are still normally distributed. Luke asks Ted to generate a reply to the worker's suggestion and tape it to the underside of the pay phone as requested.

SEEING STATISTICS: APPLET 7

Distribution of Means: Fair Dice

This applet simulates the rolling of dice when the dice are *fair*—i.e., for each die, the outcomes (integers 1 through 6) are equally likely. You can select the size of each sample (n = 1, 3, or 12 dice at a time), and the applet will update the relative frequencies with which the sample means have occurred. The three versions of this applet are shown and described here.

SAMPLE SIZE = 1

Each time you click the **Roll One Set** button, you are rolling *a single die*. **N** shows the number of samples you have taken so far, and the chart shows the relative frequency with which each of the possible sample means has occurred. Naturally, since each sample is only $n = 1$, the result in each sample is also the mean of each sample. You can make things happen much faster by clicking either the **Roll 10 Sets** button or the **Roll 100 Sets** button. In the latter cases, you are shown the means for the 10 most recent samples.

Sample Size = 1

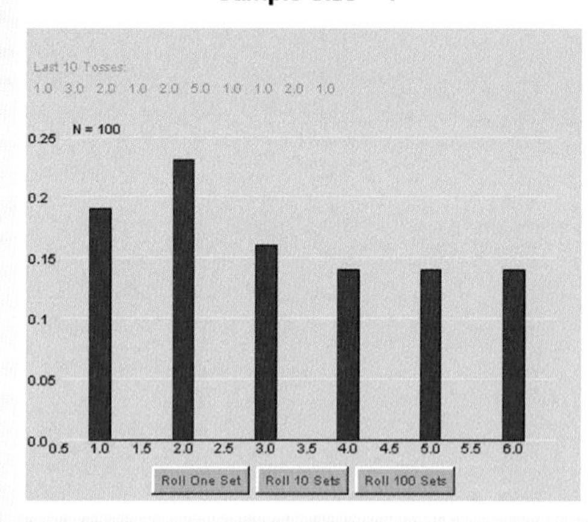

Sample Size = 3

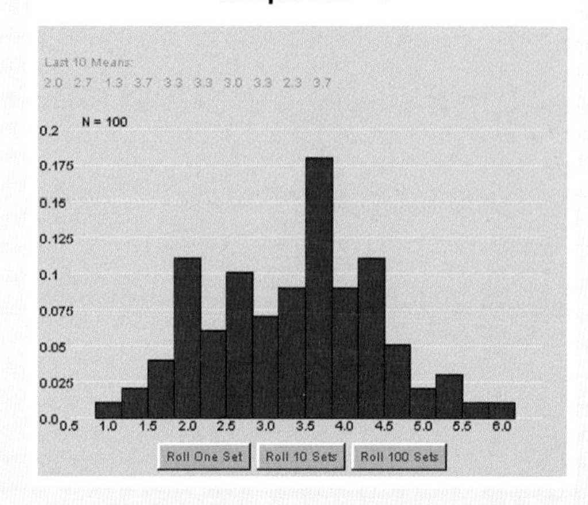

SAMPLE SIZE = 3

Each time you click the **Roll One Set** button, you are now rolling *three dice*. **N** shows the number of samples (or sets) you have taken so far, and the chart is updated to show the relative frequency with which each of the possible sample means has occurred.

SAMPLE SIZE = 12

Each time you click the **Roll One Set** button, you are rolling *twelve dice*. The procedure is the same as above, except each sample mean is now the mean for a set of 12 dice.

Sample Size = 12

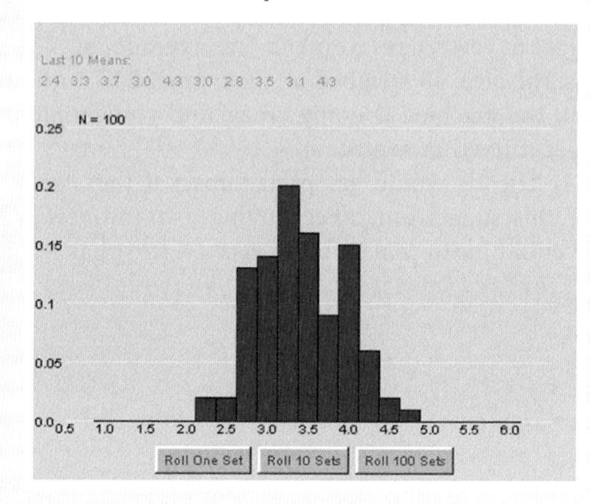

Applet Exercises

7.1 Generate 3000 rolls of a single die: Select the **Sample Size = 1** applet version, and click the **Roll 100 Sets** button 30 times. Describe the shape of the distribution of the sample results in the relative frequencies chart. Is the shape similar to what you might expect?

7.2 Generate 3000 samples, each one representing a set of three dice: Select the **Sample Size = 3** applet version and click the **Roll 100 Sets** button 30 times. Comment on how the distribution of sample means has changed from the distribution obtained in Applet Exercise 7.1.

7.3 Generate 3000 samples, each one representing a set of twelve dice: Select the **Sample Size = 12** applet version and click the **Roll 100 Sets** button 30 times. How has the increase in sample size affected the shape of the distribution of sample means compared to those obtained in Applet Exercises 7.1 and 7.2?

7.4 If there were an additional applet version that allowed each sample to consist of 100 dice, and we generated 2000 of these samples, what do you think the distribution of sample means would tend to look like? In general, as the sample sizes become larger, what shape will the distribution of sample means tend to approach?

Try out the Applets: http://www.cengage.com/international

Distribution of Means: Loaded Dice

This applet simulates the rolling of *loaded* dice—i.e., for each die, the outcomes (integers 1 through 6) are most definitely NOT equally likely, and you should never gamble with people who use these things. As with the "honest" dice in Applet 7, you can select the size of each sample (n = 1, 3, or 12 dice at a time), and the applet will update the relative frequencies with which the sample means have occurred. The same features and procedures apply as in Applet 7, but you will find the results to be quite different. Examples of the three versions of the "loaded" applet are shown here:

Sample Size = 1

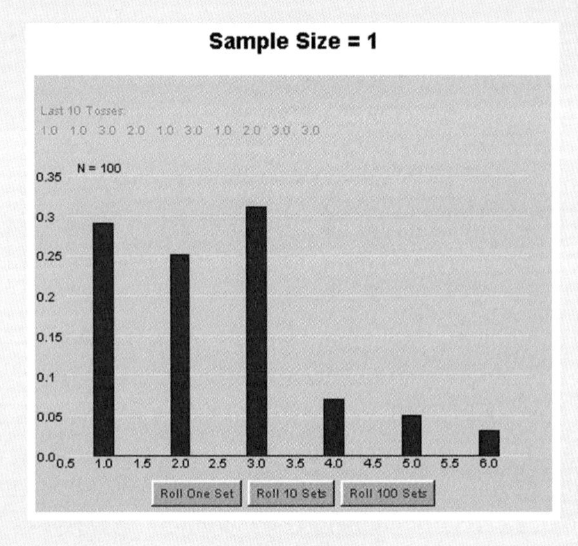

Sample Size = 3

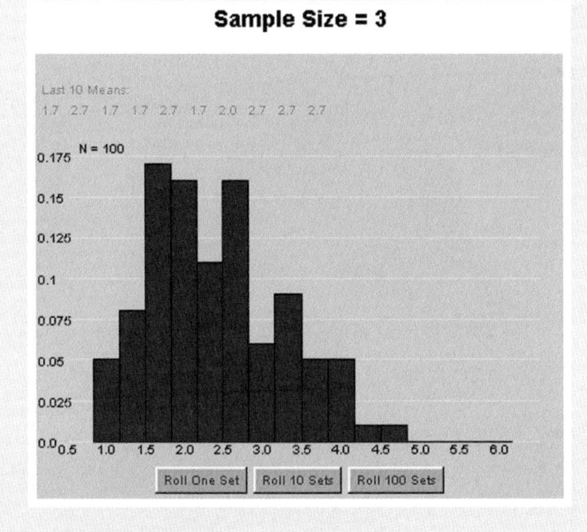

Sample Size = 12

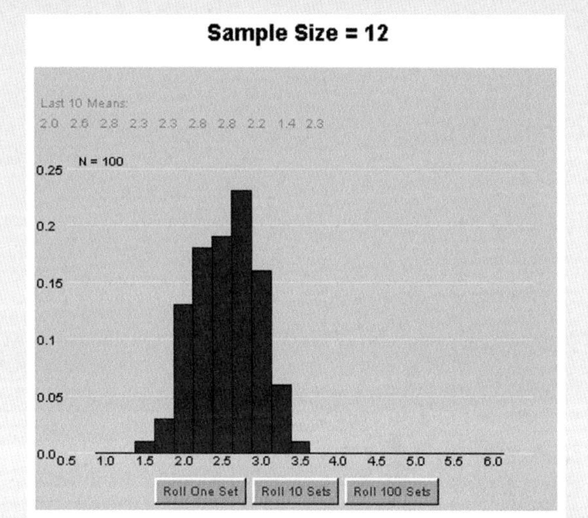

Applet Exercises

8.1 Generate 3000 rolls of a single die: Select the **"Sample Size = 1"** applet version, and click the **"Roll 100 Sets"** button 30 times. Describe the shape of the distribution of the sample results in the relative frequencies chart. Does the shape look a little peculiar? Can you tell which faces of the dice might be "loaded"?

8.2 Generate 3000 samples, each one representing a set of three dice: Select the **"Sample Size = 3"** applet version, and click the **"Roll 100 Sets"** button 30 times. Comment on how the distribution of sample means has changed from the distribution obtained in Applet Exercise 8.1.

8.3 Generate 3000 samples, each one representing a set of twelve dice: Select the **"Sample Size = 12"** applet version, and click the **"Roll 100 Sets"** button 30 times. How has the increase in sample size affected the shape of the distribution of sample means compared to those obtained in Applet Exercises 8.1 and 8.2?

8.4 If there were an additional applet version that allowed each sample to consist of 100 dice, and we generated 2000 of these samples, what do you think the distribution of sample means would tend to look like? In general, as the sample sizes become larger, what shape will the distribution of sample means tend to approach?

CHAPTER 9

Estimation from Sample Data

WORK SAMPLING

In the workplace, production experts sometimes conduct studies to find out how much time workers spend doing various job activities. This information can be used in establishing or updating standards for performance, as well as for comparing and evaluating worker performance. Among the approaches for determining how much time a worker spends at various activities is a technique known as *work sampling*. Compared to alternative methods (such as following the person and timing him or her with a stopwatch), work sampling is unobtrusive in that the behavior observed is not influenced by the observation process itself.

In work sampling, a worker is observed at randomly selected points along an interval of time; then the proportion of these observations that involve each selected activity is determined. For example, if we want to determine how much of the time a secretary spends keyboarding, we can observe the person at random times during a typical day or week, then calculate the proportion of time that he or she happens to be keyboarding. If the person were observed to be keyboarding in 100 of 280 random observations, the sample proportion would be 100/280, or 0.357.

Using this information, along with estimation techniques described later in this chapter, we could then arrive at an interval estimate reflecting the likely range of values within which the true population proportion lies. When you finish Section 9.6, you might like to pay a short visit back here and verify that the 90% confidence interval for the population proportion is from 0.310 to 0.404, and that we are 90% confident that the person spends somewhere between 31.0% and 40.4% of his or her time clicking away at the keyboard.

Making an inference based on sample data

- Explain the difference between a point estimate and an interval estimate for a population parameter.
- Use the standard normal distribution in constructing a confidence interval for a population mean or proportion.
- Use the *t* distribution in constructing a confidence interval for a population mean.
- Decide whether the standard normal distribution or the *t* distribution should be used in constructing a given confidence interval.
- Determine how large a simple random sample must be in order to estimate a population mean or proportion at specified levels of accuracy and confidence.
- Use Excel and Minitab to construct confidence intervals.

LEARNING OBJECTIVES

After reading this chapter, you should be able to:

INTRODUCTION (9.1)

In Chapter 8, we began with a population having a known mean (μ) or proportion (π); then we examined the sampling distribution of the corresponding sample statistic ($\bar{x}$ or p) for samples of a given size. In this chapter, we'll be going in the opposite direction—based on sample data, we will be making estimates involving the (unknown) value of the population mean or proportion. As mentioned previously, the use of sample information to draw conclusions about the population is known as *inferential statistics*.

To repeat a very important point, this chapter completes a key transition discussed at the beginning of Chapter 7:

- In Chapter 7, we had access to the population mean and we made probability statements *about individual x values taken from the population.*
- In Chapter 8, we again had access to the population mean, but we invoked the central limit theorem and began making probability statements *about the means of samples taken from the population.* (Beginning with Chapter 8, the sample mean itself is considered as a random variable.)
- In this chapter, we again lack access to the population mean, but we will begin using sample data as the basis from which to make probability statements *about the true (but unknown) value of the population mean.* As in Chapter 8, we will be relying heavily on the central limit theorem.

In the following sections, we will use sample data to make both point and interval estimates regarding the population mean or proportion. While the **point estimate** is a single number that estimates the exact value of the population parameter of interest (e.g., μ or π), an **interval estimate** includes a range of possible values that are likely to include the actual population parameter. When the interval estimate is associated with a degree of confidence that it actually includes the population parameter, it is referred to as a **confidence interval.**

Point and interval estimates can also be made regarding the *difference* between two population means ($\mu_1 - \mu_2$) or proportions ($\pi_1 - \pi_2$). These involve data from *two* samples, and they will be discussed in the context of the hypothesis-testing procedures of Chapter 11.

Whenever sample data are used for estimating a population mean or proportion, sampling error will tend to be present because a sample has been taken instead of a census. As a result, the observed sample statistic ($\bar{x}$ or p) will differ from the actual value of the population parameter (μ or π). Assuming a simple random sampling of elements from the population, formulas will be presented for determining how large a sample size is necessary to ensure that such sampling error is not likely to exceed a given amount.

EXERCISES

9.1 Differentiate between a point estimate and an interval estimate for a population parameter.

9.2 What is meant by inferential statistics, and what role does it play in estimation?

9.3 What is necessary for an interval estimate to be a confidence interval?

9.2 POINT ESTIMATES

An important consideration in choosing a sample statistic as a point estimate of the value of a population parameter is that the sample statistic be an **unbiased estimator.** An estimator is *unbiased* if the expected value of the sample statistic is the same as the actual value of the population parameter it is intended to estimate. Three important point estimators introduced in the chapter are those for a population mean (μ), a population variance (σ^2), and a population proportion (π).

As Chapter 8 showed, the expected value of the sample mean is the population mean, and the expected value of the sample proportion is the population proportion. As a result, $\bar{x}$ and p are unbiased estimators of μ and π, respectively. Table 9.1 presents a review of the applicable formulas. Note that the divisor in the formula for the sample variance (s^2) is $(n - 1)$. Using $(n - 1)$ as the divisor in calculating the variance of the sample results in s^2 being an unbiased estimate of the (unknown) population variance, σ^2. The positive square root of s^2, the sample standard deviation (s), will *not* be an unbiased estimate of the population standard deviation (σ). In practice, however, s is the most frequently used estimator of its population counterpart, σ.

TABLE 9.1

An estimator is unbiased if its expected value is the same as the actual value of the corresponding population parameter. Listed here are unbiased point estimators for a population mean, a population variance, and a population proportion.

Population Parameter	Unbiased Estimator	Formula
Mean, μ	$\bar{x}$	$\bar{x} = \frac{\Sigma x_i}{n}$
Variance, σ^2	s^2	$s^2 = \frac{\Sigma(x_i - \bar{x})^2}{n - 1}$
Proportion, π	p	$p = \frac{x \text{ successes}}{n \text{ trials}}$

EXERCISES

9.4 What is meant when a sample statistic is said to be an *unbiased estimator*?

9.5 When calculating the sample variance, what procedure is necessary to ensure that s^2 will be an unbiased estimator of σ^2? Will s be an unbiased estimator of σ?

9.6 During the month of July, an auto manufacturer gives its production employees a vacation period so it can tool up for the new model run. In surveying a simple random sample of 200 production workers, the personnel director finds that 38% of them plan to vacation out of state for at least one week during this period. Is this a point estimate or an interval estimate? Explain.

9.7 A simple random sample of 8 employees is selected from a large firm. For the 8 employees, the number of days each was absent during the past month was found to be 0, 2, 4, 2, 1, 7, 3, and 2, respectively.

a. What is the point estimate for μ, the mean number of days absent for the firm's employees?
b. What is the point estimate for σ^2, the variance of the number of days absent?

9.8 The average annual U.S. per capita consumption of iceberg lettuce has been estimated as 20.2 pounds. The annual per capita consumption 2 years earlier had been estimated as 21.3 pounds. Could either or both of these consumption figures be considered a point estimate? Could the difference between the two consumption figures be considered an interval estimate? Explain your reasoning in both cases. Source: Bureau of the Census, *Statistical Abstract of the United States 2009,* p. 136.

A PREVIEW OF INTERVAL ESTIMATES (9.3)

When we know the values of the population mean and standard deviation, we can (if either the population is normally distributed or n is large) use the standard normal distribution in determining the proportion of sample means that will fall within a given number of standard error ($\sigma_{\bar{x}}$) units of the known population mean. This is exactly what we did in Chapter 8.

It is typical of inferential statistics that we must use the mean ($\bar{x}$) and standard deviation (s) of a single sample as our best estimates of the (unknown) values of μ and σ. However, this does not prevent us from employing $\bar{x}$ and s in constructing an estimated sampling distribution for all means having this sample size. This is the basis for the construction of an interval estimate for the population mean.

When we apply the techniques of this chapter and establish the sample mean as the midpoint of an interval estimate for the population mean, the resulting interval may or may not include the actual value of μ. For example, in Figure 9.1, six of the seven simple random samples from the same population led to an interval estimate that included the true value of the population mean.

In Figure 9.1, the mean of sample number 1 ($\bar{x}_1$) is slightly greater than the population mean (μ), and the interval estimate based on this sample actually includes μ. For sample 3, taken from the same population, the estimation interval does not include μ. In Figure 9.1, we can make these observations because the value of μ is known. In practice, however, we will not have the benefit of knowing the actual value of the population mean. Therefore, we will not be able to say *with complete certainty* that an interval based on our sample result will actually include the (unknown) value of μ.

The interval estimate for the mean simply describes a range of values that is likely to include the actual population mean. This is also the case for our use of

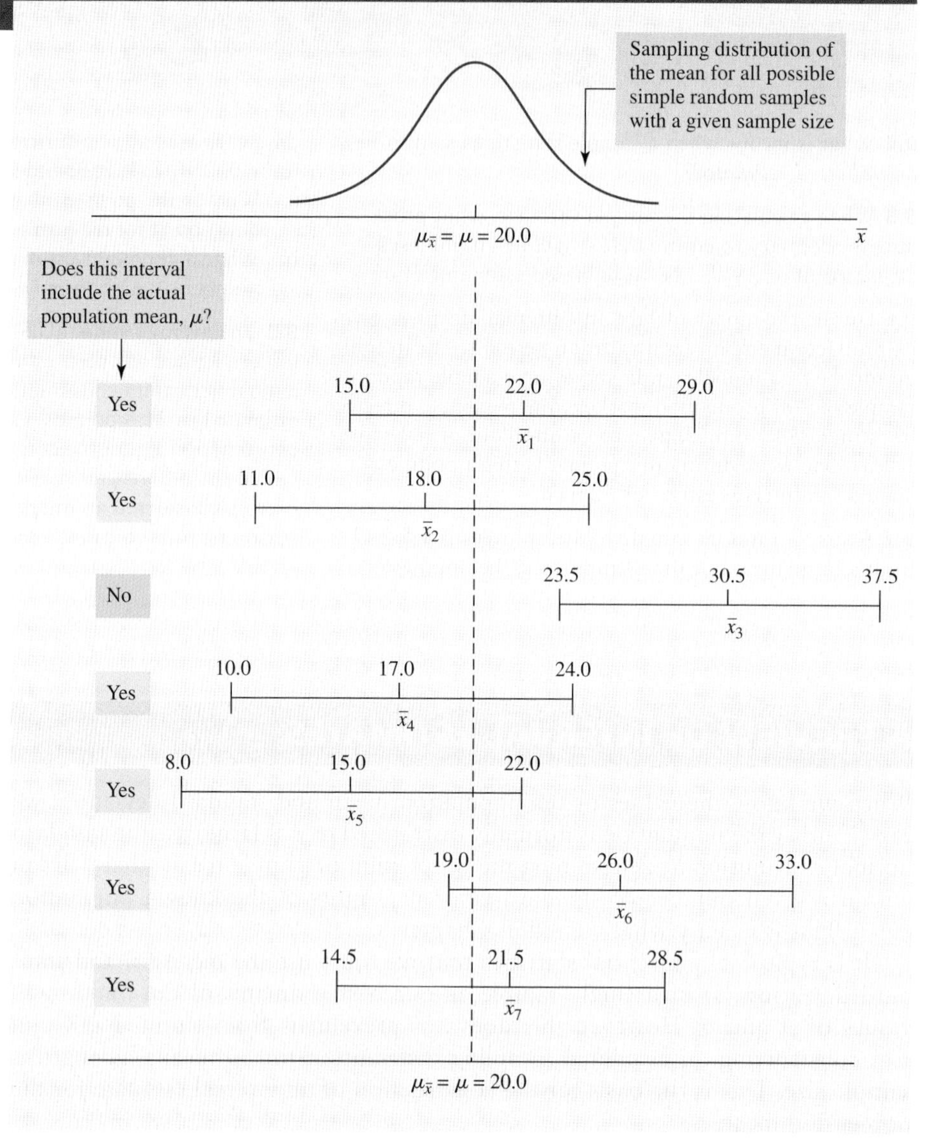

FIGURE 9.1

Examples of seven different interval estimates for a population mean, with each interval based on a separate simple random sample from the population. Six of the seven interval estimates include the actual value of μ.

the sample proportion (p) to estimate the population proportion (π), as well as for our construction of an interval estimate within which the actual value of π is likely to fall.

The following terms are of great importance in interval estimation:

INTERVAL ESTIMATE **A range of values within which the actual value of the population parameter may fall.**

INTERVAL LIMITS **The lower and upper values of the interval estimate.**

CONFIDENCE INTERVAL **An interval estimate for which there is a specified degree of certainty that the actual value of the population parameter will fall within the interval.**

CONFIDENCE COEFFICIENT For a confidence interval, the proportion of such intervals that would include the population parameter if the process leading to the interval were repeated a great many times.

CONFIDENCE LEVEL Like the confidence coefficient, this expresses the degree of certainty that an interval will include the actual value of the population parameter, but it is stated as a percentage. For example, a 0.95 confidence coefficient is equivalent to a 95% confidence level.

ACCURACY The difference between the observed sample statistic and the actual value of the population parameter being estimated. This may also be referred to as *estimation error* or *sampling error.*

To illustrate these and several other terms discussed so far, we have provided their values in the following example, which is typical of published statistical findings. The methods by which the values were determined will become apparent in the sections to follow.

EXAMPLE

Interval Estimates

"In our simple random sample of 2000 households, we found the average income to be $\bar{x}$ = \$65,000, with a standard deviation, s = \$12,000. Based on these data, we have 95% confidence that the population mean is somewhere between \$64,474 and \$65,526."

SOLUTION

- **Point estimate of μ** \$65,000
- **Point estimate of σ** \$12,000
- **Interval estimate of μ** \$64,474 to \$65,526
- **Lower and upper interval limits for μ** \$64,474 and \$65,526
- **Confidence coefficient** 0.95
- **Confidence level** 95%
- **Accuracy** For 95% of such intervals, the sample mean would not differ from the actual population mean by more than \$526.

EXAMPLEEXAMPLEEXAMPLEEXA

When constructing a confidence interval for the mean, a key consideration is whether we know the actual value of the population standard deviation (σ). As Figure 9.2 shows, this will determine whether the normal distribution or the t distribution (see Section 9.5) will be used in determining the appropriate interval. Figure 9.2 also summarizes the procedure for constructing the confidence interval for the population proportion, a technique that will be discussed in Section 9.6.

FIGURE 9.2

This figure provides an overview of the methods for determining confidence interval estimates for a population mean or a population proportion and indicates the chapter section in which each is discussed. Key assumptions are reviewed in the figure notes.

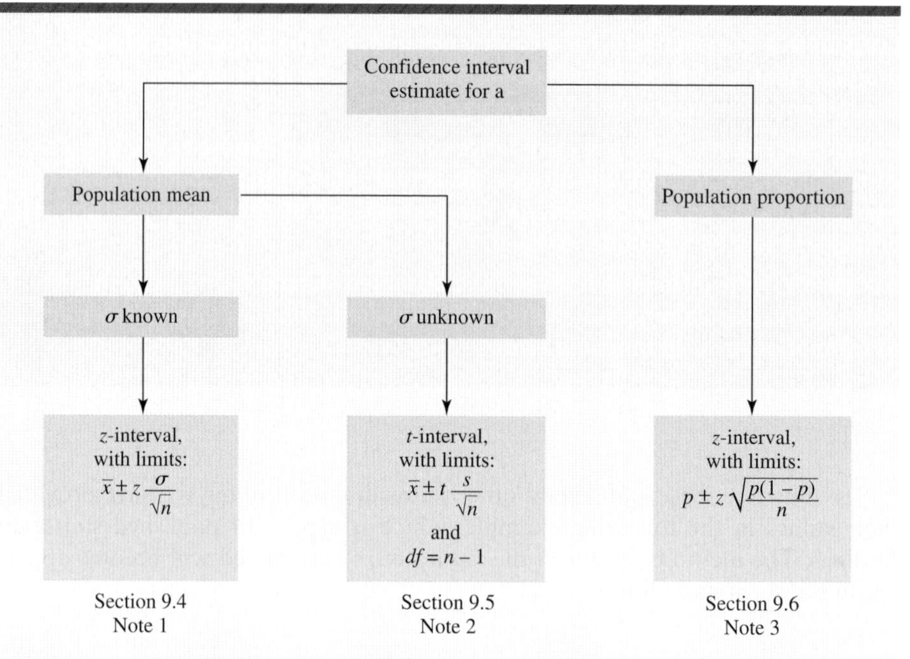

[1]If the population is not normally distributed, n should be at least 30 for the central limit theorem to apply.

[2]When σ is unknown, but the underlying population can be assumed to be approximately normally distributed, use of the t distribution is a *necessity* when $n < 30$. Use of the t distribution is also appropriate, however, when σ is unknown and the sample sizes are larger. Most computer statistical packages routinely use the t-interval *for all sample sizes* when s is used to estimate σ.

[3]Assumes that both np and $n(1 - p)$ are ≥ 5. The normal distribution as an approximation to the binomial improves as n becomes larger and for values of p that are closer to 0.5.

EXERCISES

9.9 Exactly what is meant by the *accuracy* of a point estimate?

9.10 A population is approximately normally distributed and the sample size is to be $n = 40$. What additional factor must be considered in determining whether to use the standard normal distribution in constructing the confidence interval for the population mean?

9.11 "In surveying a simple random sample of 1000 employed adults, we found that 450 individuals felt they were underpaid by at least $3000. Based on these results, we have 95% confidence that the proportion of the population of employed adults who share this sentiment is between 0.419 and 0.481." For this summary statement, identify the

a. point estimate of the population proportion.
b. confidence interval estimate for the population proportion.
c. confidence level and the confidence coefficient.
d. accuracy of the sample result.

9.4 CONFIDENCE INTERVAL ESTIMATES FOR THE MEAN: σ KNOWN

If we don't know the value of the population mean, chances are that we also do not know the value of the population standard deviation. However, in some cases, usually industrial processes, σ may be known while μ is not. If the population cannot be assumed to be normally distributed, the sample size must be at least 30 for the central limit theorem to apply. When these conditions are met, the confidence interval for the population mean will be as follows:

Confidence interval limits for the population mean, σ known:

$$\bar{x} \pm z\frac{\sigma}{\sqrt{n}}$$

where $\bar{x}$ = sample mean
σ = population standard deviation
n = sample size
z = z value corresponding to the level of confidence desired (e.g., $z = 1.96$ for the 95% confidence level)
$\sigma/\sqrt{n}$ = standard error of the sampling distribution of the mean

This application assumes that either (1) the underlying population is normally distributed or (2) the sample size is $n \geq 30$. Also, an alternative way of describing the z value is to refer to it as $z_{\alpha/2}$, with $\alpha/2$ being the area to the right. For example, $z_{0.025}$ would be 1.96. The most commonly used confidence intervals and the corresponding z values typically used to obtain them are shown below:

Confidence Level	z
90%	1.645
95%	1.96
98%	2.33
99%	2.58

EXAMPLE

z-Interval, Mean

From past experience, the population standard deviation of rod diameters produced by a machine has been found to be $\sigma = 0.053$ inches. For a simple random sample of $n = 30$ rods, the average diameter is found to be $\bar{x} = 1.400$ inches. The underlying data are in file **CX09RODS**.

SOLUTION

What Is the 95% Confidence Interval for the Population Mean, μ?

Although we don't know the value of μ, $\bar{x} = 1.400$ inches is our best estimate of the population mean diameter. As Figure 9.3 shows, the sampling distribution of the mean will have a standard error of $\sigma/\sqrt{n}$, or $0.053/\sqrt{30}$. For the standard normal distribution, 95% of the area will fall between $z = -1.96$ and $z = +1.96$. We are able to use the standard normal distribution table because $n \geq 30$ and the central limit theorem can be invoked. As a result, the 95% confidence interval for the (unknown) population mean can be calculated as

$$\bar{x} \pm z\frac{\sigma}{\sqrt{n}} = 1.400 \pm 1.96\frac{0.053}{\sqrt{30}} \quad \text{or between 1.381 and 1.419 inches}$$

Figure 9.3 shows the midpoint ($\bar{x} = 1.400$ inches) for the 95% confidence interval for the mean, along with the lower and upper limits for the confidence

EXAMPLEEXAMPLEEXAMPLEEXAMPLEEXAN

FIGURE 9.3

Construction of the 95% confidence interval for the population mean, based on a sample of 30 rods for which the average diameter is 1.400 inches. From past experience, the population standard deviation is known to be $\sigma = 0.053$ inches. Because σ is known, the normal distribution can be used in determining the interval limits. We have 95% confidence that μ is between 1.381 and 1.419 inches.

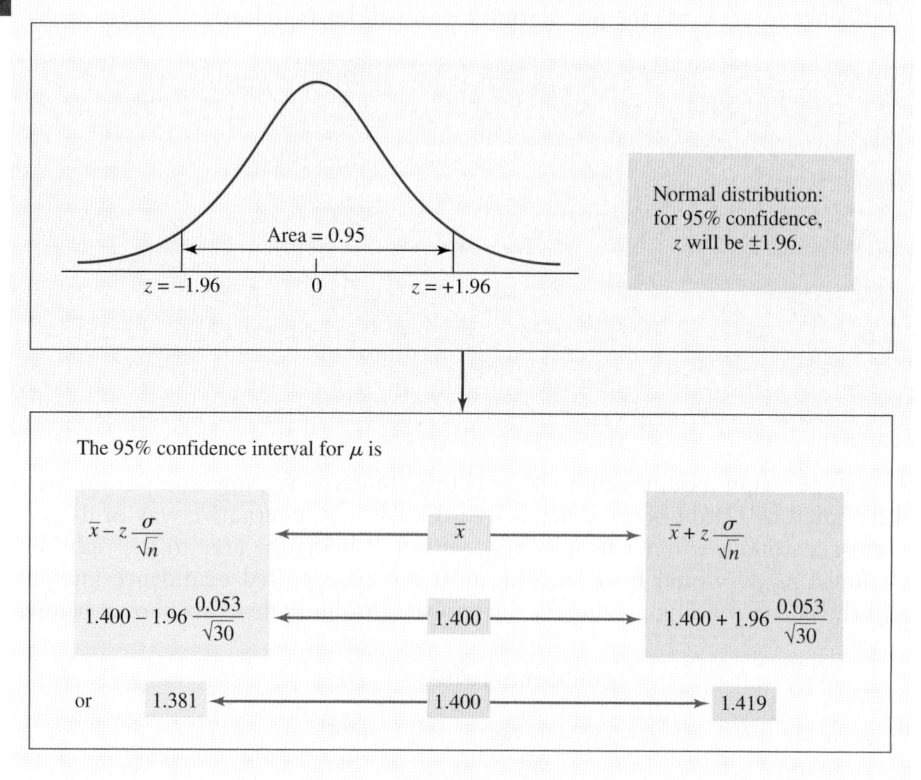

interval. Based on our calculations, the 95% confidence interval for the population mean is from 1.381 to 1.419 inches.

More precisely, 95% of such intervals constructed in this way would include the population mean. Since we have taken only one sample and constructed just one interval, it is technically correct to say we have 95% confidence that this particular interval contains the population mean. Although the logic may be tempting, this is *not* the same as saying the probability is 0.95 that *this particular interval* will include the population mean.

With other factors unchanged, a higher confidence level will require a wider confidence interval. Likewise, a lower confidence level will lead to a narrower confidence interval. In other words, the more certain we wish to be that the interval estimate contains the population parameter, the wider the interval will have to be. Refer to Seeing Statistics Applet 9, at the end of the chapter, to visually demonstrate how the width of the confidence interval changes when higher and lower levels of confidence are specified.

Computer Solutions 9.1 shows how we can use Excel or Minitab to generate a confidence interval for the mean when the population standard deviation is known or assumed. In this case, we are replicating the 95% confidence interval shown in Figure 9.3, and the 30 data values are in file **CX09RODS**. If we use an Excel procedure based on summary statistics, it can be interesting to examine “what-if” scenarios to instantly see how changes in the specified confidence level would change the width of the confidence interval.

COMPUTER 9.1 SOLUTIONS

Confidence Interval for Population Mean, σ Known

These procedures show how to construct a confidence interval for the population mean when the population standard deviation is known.

EXCEL

	A	B	C
1	z-Estimate: Mean		
2			
3			*diameter*
4	Mean		1.400
5	Standard Deviation		0.052
6	Observations		30
7	SIGMA		0.053
8	LCL		1.381
9	UCL		1.419

Excel confidence interval for μ based on raw data and σ known

1. For example, using the 30 rod diameters (in Excel file **CX09RODS**) on which Figure 9.3 is based: The label and 30 data values are in **A1:A31.** From the **Add-Ins** ribbon, click **Data Analysis Plus** in the leftmost menu section. Click **Z-Estimate: Mean.** Click **OK.**
2. Enter **A1:A31** into the **Input Range** box. Enter the known population standard deviation **(0.053)** into the **Standard Deviation (SIGMA)** box. Click **Labels,** since the variable name is in the first cell within the field. The desired confidence level as a decimal fraction is 0.95, so the corresponding alpha value is 1 − 0.95 = 0.05. Enter **0.05** into the **Alpha** box. Click **OK.** The confidence interval results will be as shown above.

Excel confidence interval for μ based on summary statistics and σ known

1. For example, with $\bar{x} = 1.400$, $\sigma = 0.053$, and $n = 30$, as in Figure 9.3: Open the **ESTIMATORS** workbook.
2. Using the arrows at the bottom left, select the ***z*-Estimate_Mean** worksheet. Enter the sample mean **(1.4),** the known sigma **(0.053),** the sample size **(30),** and the desired confidence level as a decimal fraction **(0.95).**

(*Note:* As an alternative, you can use Excel worksheet template **TMZINT**. The steps are described within the template.)

MINITAB

Confidence interval for μ based on raw data and σ known

```
One-Sample Z: diameter

The assumed standard deviation = 0.053

Variable   N     Mean     StDev  SE Mean          95% CI
diameter  30  1.40000  0.05196  0.00968  (1.38103, 1.41897)
```

1. For example, using the data (in Minitab file **CX09RODS**) on which Figure 9.3 is based, with the 30 data values in column **C1:** Click **Stat.** Select **Basic Statistics.** Click **1-Sample Z.**
2. Select **Samples in columns** and enter **C1** into the box. Enter the known population standard deviation **(0.053)** into the **Standard deviation** box. The **Perform hypothesis test** box should be left blank—we will not be doing hypothesis testing until the next chapter.

(continued)

3. Click **Options.** Enter the desired confidence level as a percentage **(95.0)** into the **Confidence Level** box. Within the **Alternative** box, select **not equal.** Click **OK.** Click **OK.** The printout also includes the sample mean (1.400), the known sigma (0.053), the sample size (30), and the standard error of the mean (calculated as $0.053/\sqrt{30} = 0.00968$). Although the sample standard deviation is shown, it is not used in the construction of this confidence interval.

Confidence interval for μ based on summary statistics and σ known

Follow the procedure in the previous steps for raw data, but in step 2 select **Summarized data** and enter **30** into the **Sample size** box and **1.4** into the **Mean** box.

EXERCISES

9.12 What role does the central limit theorem play in the construction of a confidence interval for the population mean?

9.13 In using the standard normal distribution to construct a confidence interval for the population mean, what two assumptions are necessary if the sample size is less than 30?

9.14 The following data values are a simple random sample from a population that is normally distributed, with $\sigma^2 = 25.0$: 47, 43, 33, 42, 34, and 41. Construct and interpret the 95% and 99% confidence intervals for the population mean.

9.15 A simple random sample of 30 has been collected from a population for which it is known that $\sigma = 10.0$. The sample mean has been calculated as 240.0. Construct and interpret the 90% and 95% confidence intervals for the population mean.

9.16 A simple random sample of 25 has been collected from a normally distributed population for which it is known that $\sigma = 17.0$. The sample mean has been calculated as 342.0, and the sample standard deviation is $s = 14.9$. Construct and interpret the 95% and 99% confidence intervals for the population mean.

9.17 The administrator of a physical therapy facility has found that postoperative performance scores on a knee flexibility test have tended to follow a normal distribution with a standard deviation of 4. For a simple random sample of ten patients who have recently had knee surgery, the scores are as follows: 101, 92, 94, 88, 52, 93, 76, 84, 72, and 98. Construct and interpret the 90% and 95% confidence intervals for the population mean.

9.18 In testing the heat resistance of electrical components, safety engineers for an appliance manufacturer routinely subject wiring connectors to a temperature of 450 degrees Fahrenheit, then record the amount of time it takes for the connector to melt and cause a short circuit. Past experience has shown the standard deviation of failure times to be 6.4 seconds. In a simple random sample of 40 connectors from a very large production run, the mean time until failure was found to be 35.5 seconds. Construct and interpret the 99% confidence interval for μ = the mean time until failure for all of the connectors from the production run.

9.19 An assembly process includes a torque wrench device that automatically tightens compressor housing bolts; the device has a known process standard deviation of $\sigma = 3$ lb-ft in the torque applied. A simple random sample of 35 nuts is selected, and the average torque to which they have been tightened is 150 lb-ft. What is the 95% confidence interval for the average torque being applied during the assembly process?

9.20 A machine that stuffs a cheese-filled snack product can be adjusted for the amount of cheese injected into each unit. A simple random sample of 30 units is selected, and the average amount of cheese injected is found to be $\bar{x} = 3.5$ grams. If the process standard deviation is known to be $\sigma = 0.25$ grams, construct the 95% confidence interval for μ = the average amount of cheese being injected by the machine.

9.21 In Exercise 9.20, if the sample size had been $n = 5$ instead of $n = 30$, what assumption would have to be made about the population distribution of filling weights in order to use z values in constructing the confidence interval?

(**DATA SET**) *Note:* Exercises 9.22 and 9.23 require a computer and statistical software.

9.22 For one of the tasks in a manufacturing process, the mean time for task completion has historically been 35.0 minutes, with a standard deviation of 2.5 minutes. Workers have recently complained that the machinery used in the task is wearing out and slowing down. In response to the complaints, plant engineers have measured the time required for a sample consisting of 100 task operations. The 100 sample times, in minutes, are in data file **XR09022**. Using the mean for this sample, and assuming that the population standard deviation has remained unchanged at 2.5 minutes, construct the 95% confidence interval for the population mean. Is 35.0 minutes within the confidence interval? Interpret your "yes" or "no" answer in terms of whether the mean time for the task may have changed.

9.23 Sheila Johnson, a state procurement manager, is responsible for monitoring the integrity of a wide range of products purchased by state agencies. She is currently examining a sample of paint containers recently received from a long-time supplier. According to the supplier, the process by which the cans are filled involves a small amount of variation from one can to the next, and the standard deviation is 0.25 fluid ounces. The 40 cans in Sheila's sample were examined to determine how much paint they contained, and the results (in fluid ounces) are listed in data file **XR09023**. Using the mean for this sample, and assuming that the population standard deviation is 0.25 fluid ounces, construct the 90% confidence interval for the population mean volume for the cans of paint provided by the supplier. If the labels on the paint cans say the mean content for such containers is 100.0 fluid ounces, would your confidence interval tend to support this possibility?

CONFIDENCE INTERVAL ESTIMATES FOR THE MEAN: σ UNKNOWN

9.5

It is rare that we know the standard deviation of a population but have no knowledge about its mean. For this reason, the techniques of the previous section are much less likely to be used in practice than those discussed here. Whenever the population standard deviation is unknown, it must be estimated by the sample standard deviation, s. For such applications, there is a continuous distribution called the Student's t distribution.

The Student's *t* Distribution

Description

Also referred to as simply the *t distribution,* this distribution is really a family of continuous, unimodal, bell-shaped distributions. It was developed in the early 1900s by W. S. Gosset, who used the pen name "Student" because his company did not permit employees to publish their research results. The t distribution is the probability distribution for the random variable $t = (\bar{x} - \mu)/(s/\sqrt{n})$. It has a mean of zero, but its shape is determined by what is called the number of **degrees of freedom (*df*)**. For confidence interval applications, the specific member of the family is determined by $df = n - 1$.

The term *degrees of freedom* refers to the number of values that remain *free* to vary once some information about them is already known. For example, if four items have a mean of 10.0, and three of these items are known to have values of 8, 12, and 7, there is no choice but for the fourth item to have a value of 13. In effect, one degree of freedom has been lost.

The t distribution tends to be flatter and more spread out than the normal distribution, especially for very small sample sizes. Figure 9.4 compares the approximate shape of a standard normal distribution with that of a t distribution for which $df = 6$. The t distribution converges to the normal distribution as the sample size (and df) increases, and as the number of degrees of freedom approaches *infinity*, the two distributions are actually identical. As with our use of z previously in this chapter, t represents distance in terms of standard error units. Seeing Statistics Applet 10, at the end of the chapter, can be used in seeing how the shape of the t distribution responds to different df values.

FIGURE 9.4

A comparison of the approximate shape of the standard normal distribution with that of a *t* distribution having 6 degrees of freedom. The shape of the *t* distribution is flatter and more spread out, but approaches that of the standard normal distribution as the number of degrees of freedom increases.

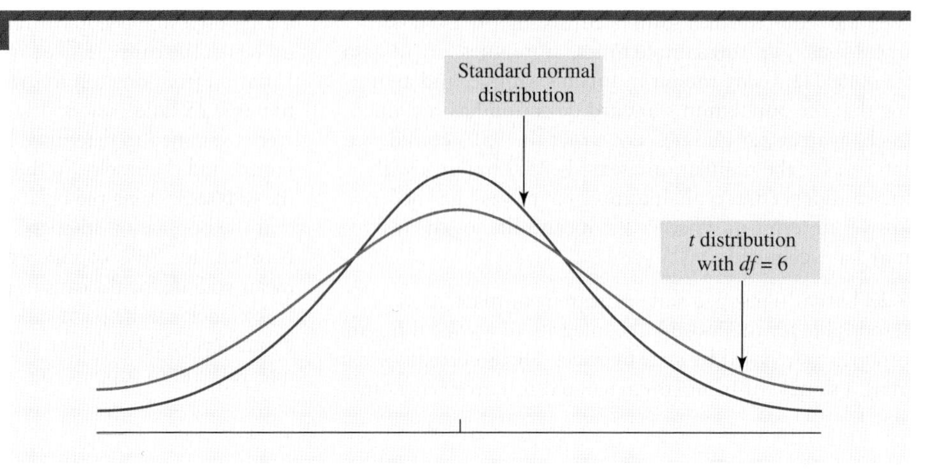

EXAMPLEEXAMPLEEXAMPLEEXAMPLEEXAMPLEEXAMPLEEXAM

EXAMPLE

t Distribution Table

Using the *t* Distribution Table. A table for *t* values that correspond to selected areas beneath the *t* distribution appears on the pages immediately preceding the back cover. A portion of the table is reproduced as Table 9.2. In general, it is used in the same way as the standard normal table, but there are two exceptions: (1) the areas provided are for the right tail only, and (2) it is necessary to refer to the appropriate degrees of freedom (df) row in finding the appropriate *t* value.

SOLUTION

For a Sample Size of $n = 15$, What *t* Values Would Correspond to an Area Centered at $t = 0$ and Having an Area beneath the Curve of 95%?

The area of interest beneath the curve can be expressed as 0.95, so the total area in both tails combined will be $(1.00 - 0.95)$, or 0.05. Since the curve is symmetrical, the area in just one tail will be $0.05/2$, or 0.025. The number of degrees of freedom will be the sample size minus 1, or $df = n - 1$, or $15 - 1 = 14$. Referring to the 0.025 column and the $df = 14$ row of the table, we find that the value of *t* corresponding to a right-tail area of 0.025 is $t = +2.145$. Because the curve is symmetrical, the value of *t* for a left-tail area of 0.025 will be $t = -2.145$.

Note that these values of *t* ($t = \pm 2.145$) are farther apart than the *z* values ($z = \pm 1.96$) that would have led to a 95% area beneath the standard normal curve. Remember that the shape of the *t* distribution tends to be flatter and more spread out than that of the normal distribution, especially for small samples.

For a Sample Size of $n = 99$, What *t* Values Would Correspond to an Area Centered at $t = 0$ and Having an Area beneath the Curve of 90%?

In this case, the proportion of the area beneath the curve is 0.90, so each tail will have an area of $(1.00 - 0.90)/2$, or 0.05. Therefore, we will refer to the 0.05 column in the *t* table. Subtracting 1 from the sample size of 99, $df = 99 - 1$, or 98. Using the 0.05 column and the $df = 98$ row, the corresponding *t* value

TABLE 9.2

A portion of the Student's t distribution table. The t distribution is really a family of symmetric, continuous distributions with a mean of $t = 0$. The specific member of the distribution depends on the number of degrees of freedom, or df. As df increases, the t distribution approaches the normal distribution, and the t values in the infinity row are identical to the z values for the standard normal distribution.

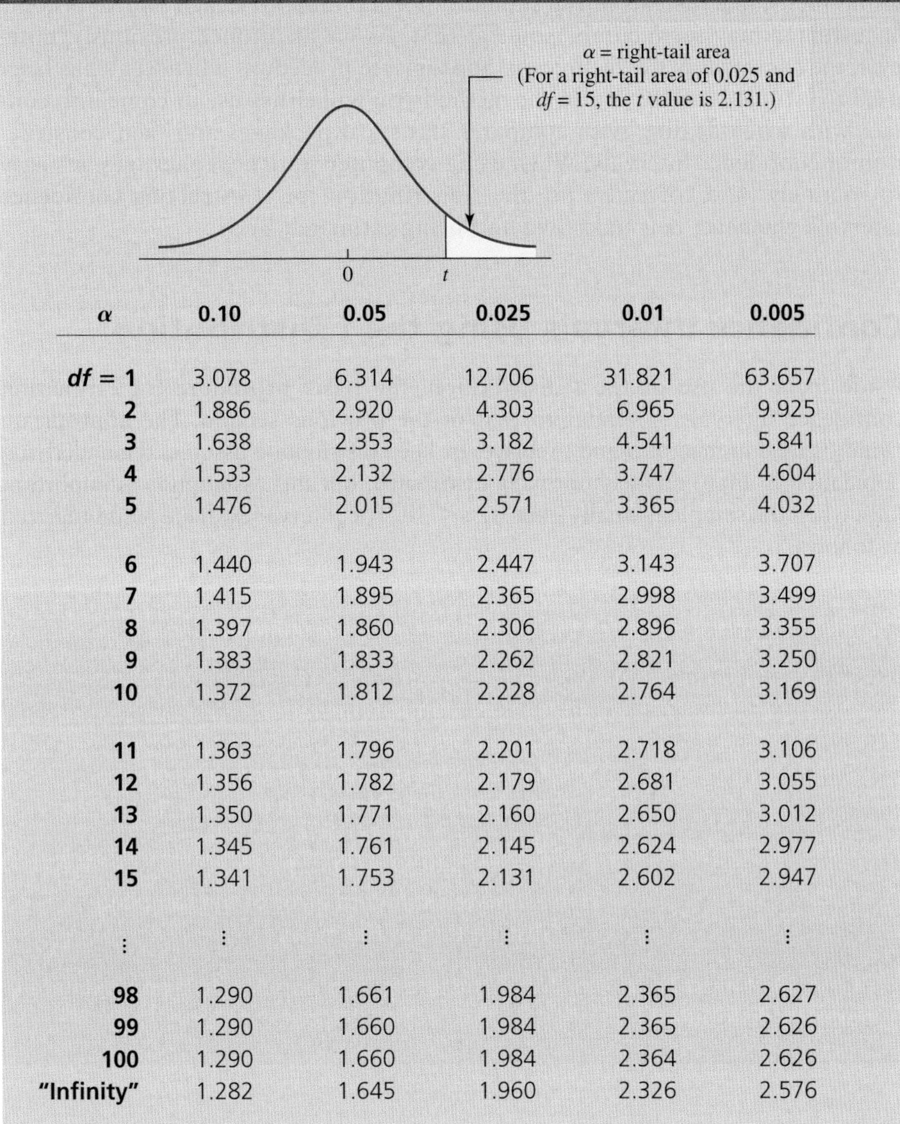

α	0.10	0.05	0.025	0.01	0.005
df = 1	3.078	6.314	12.706	31.821	63.657
2	1.886	2.920	4.303	6.965	9.925
3	1.638	2.353	3.182	4.541	5.841
4	1.533	2.132	2.776	3.747	4.604
5	1.476	2.015	2.571	3.365	4.032
6	1.440	1.943	2.447	3.143	3.707
7	1.415	1.895	2.365	2.998	3.499
8	1.397	1.860	2.306	2.896	3.355
9	1.383	1.833	2.262	2.821	3.250
10	1.372	1.812	2.228	2.764	3.169
11	1.363	1.796	2.201	2.718	3.106
12	1.356	1.782	2.179	2.681	3.055
13	1.350	1.771	2.160	2.650	3.012
14	1.345	1.761	2.145	2.624	2.977
15	1.341	1.753	2.131	2.602	2.947
⋮	⋮	⋮	⋮	⋮	⋮
98	1.290	1.661	1.984	2.365	2.627
99	1.290	1.660	1.984	2.365	2.626
100	1.290	1.660	1.984	2.364	2.626
"Infinity"	1.282	1.645	1.960	2.326	2.576

is $t = +1.661$. Since $t = +1.661$ corresponds to a right-tail area of 0.05, $t = -1.661$ will correspond to a left-tail area of 0.05. This is due to the symmetry of the distribution, and the distance from $t = -1.661$ to $t = +1.661$ will include 90% of the area beneath the curve. You can use Seeing Statistics Applet 11, at the end of the chapter, to further examine areas beneath the t distribution curve.

Should you encounter a situation in which the number of degrees of freedom exceeds the $df = 100$ limit of the t distribution table, just use the corresponding z value for the desired level of confidence. These z values are listed in the $df = infinity$ row in the t distribution table.

Using the t table instead of the standard normal table to which you've become accustomed may seem cumbersome at first. As we mentioned previously, however, the t-interval is the technically appropriate procedure whenever s has been used to estimate σ. This is also the method you will either use or come into contact with when dealing with computer statistical packages and their construction of confidence intervals. When using computer statistical packages, it's easy to routinely (and correctly) use the t distribution for constructing confidence intervals whenever σ is unknown and being estimated by s.

Confidence Intervals Using the *t* Distribution

Aside from the use of the t distribution, the basic procedure for estimating confidence intervals is similar to that of the previous section. The appropriate t value is used instead of z, and s replaces σ. The t distribution assumes the underlying population is approximately normally distributed, but this assumption is important only when the sample is small—that is, $n < 30$. The interval estimate is summarized as follows:

Confidence interval limits for the population mean, σ unknown:

$$\bar{x} \pm t\frac{s}{\sqrt{n}}$$

where $\bar{x}$ = sample mean
s = sample standard deviation
n = sample size
t = t value corresponding to the level of confidence desired, with $df = n - 1$ (e.g., $t = 2.201$ for 95% confidence, $n = 12$, and $df = 12 - 1 = 11$)
$s/\sqrt{n}$ = estimated standard error of the sampling distribution of the mean

(If $n < 30$, it must be assumed that the underlying population is approximately normally distributed.)

EXAMPLEEXAMPLEEXAMPLEEX

EXAMPLE

t-Interval, Mean

A simple random sample of $n = 90$ manufacturing employees has been selected from those working throughout a state. The average number of overtime hours worked last week was $\bar{x} = 8.46$ hours, with a sample standard deviation of $s = 3.61$ hours. The underlying data are in file CX09OVER.

SOLUTION

What Is the 98% Confidence Interval for the Population Mean, μ?

The first step in determining the appropriate value of t is to identify the column of the t distribution table to which we must refer. Since the confidence level is 98%, the right-tail area of interest is $(1.00 - 0.98)/2$, or 0.01. For this sample size, the number of degrees of freedom will be $90 - 1$, or $df = 89$. Referring

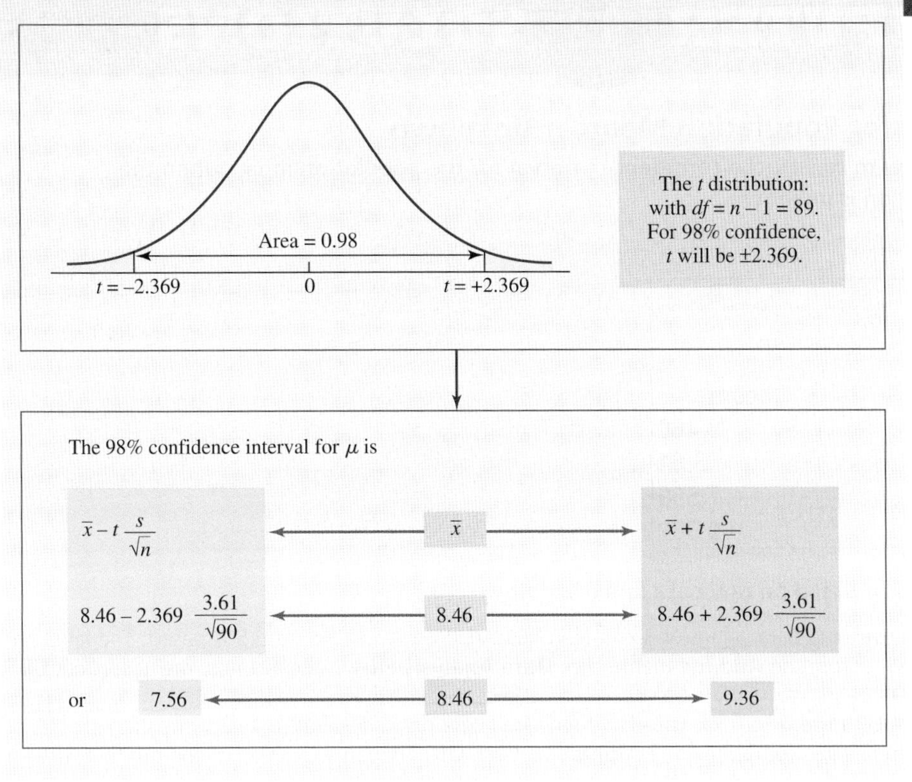

FIGURE 9.5

Although the sample size in this example is relatively large (n = 90), the t distribution was used in constructing the 98% confidence interval for the population mean. This is because the population standard deviation is unknown and is being estimated by the sample standard deviation (s = 3.61 hours).

to the 0.01 column and the df = 89 row, we determine t = +2.369. Due to the symmetry of the t distribution, 98% of the area beneath the curve will be between $t = -2.369$ and $t = +2.369$. For the results summarized in Figure 9.5, the underlying calculations for the 98% confidence interval are

$$\bar{x} \pm t\frac{s}{\sqrt{n}} = 8.46 \pm 2.369\frac{3.61}{\sqrt{90}} \quad \text{or between 7.56 and 9.36 hours}$$

MPLEEXAMPLE

If n is so large (e.g., $n > 101$) that df exceeds the finite limits of the t table, just use the infinity row of the table. In the preceding example, if n were >101, we would refer to the 0.01 column and the df = infinity row and obtain a t value of 2.326. This is the same as using z instead of t. *For such large samples, the z and t distributions are similar enough that the z will be a very close approximation.*

Computer Solutions 9.2 shows how we can use Excel or Minitab to generate a confidence interval for the mean when the population standard deviation is unknown. In this case, we are replicating the 98% confidence interval shown in Figure 9.5, and the 90 data values are in file **CX09OVER**. Once again, if we use an Excel procedure based on summary statistics, it can be interesting to examine "what-if" scenarios to instantly see how changes in the specified confidence level would change the width of the confidence interval.

COMPUTER 9.2 SOLUTIONS

Confidence Interval for Population Mean, σ Unknown

These procedures show how to construct a confidence interval for the population mean when the population standard deviation is unknown.

EXCEL

	A	B	C	D
1	t-Estimate: Mean			
2				
3				*hours*
4	Mean			8.460
5	Standard Deviation			3.61
6	LCL			7.559
7	UCL			9.362

Excel confidence interval for μ based on raw data and σ unknown

1. For example, using the 90 overtime data values (in Excel file CX09OVER) on which Figure 9.5 is based: The label and 90 data values are in **A1:A91.** From the **Add-Ins** menu, click **Data Analysis Plus** in the leftmost menu section. Click ***t*-Estimate: Mean.** Click **OK.**
2. Enter **A1:A91** into the **Input Range** box. Click **Labels.** The desired confidence level as a decimal fraction is 0.98, so enter the corresponding alpha value **(0.02)** into the **Alpha** box. Click **OK.** The lower portion of the printout lists the lower and upper limits for the 98% confidence interval.

Excel confidence interval for μ based on summary statistics and σ unknown

1. Open the ESTIMATORS workbook.
2. Using the arrows at the bottom left, select the ***t*-Estimate_Mean** worksheet. Enter the sample mean **(8.46),** the sample standard deviation **(3.61),** the sample size **(90),** and the desired confidence level as a decimal fraction **(0.98).**

(*Note:* As an alternative, you can use the Excel worksheet template TMTINT. The steps are described within the template.)

MINITAB

Confidence interval for μ based on raw data and σ unknown

```
One-Sample T: hours

Variable   N   Mean  StDev  SE Mean       98% CI
hours     90  8.460  3.610    0.381  (7.559, 9.362)
```

1. For example, using the data (in Minitab file CX09OVER) on which Figure 9.5 is based, with the 90 data values in column **C1:** Click **Stat.** Select **Basic Statistics.** Click **1-Sample t.**
2. Select **Samples in columns** and enter **C1** into the box. The **Perform hypothesis test** box should be left blank—we will not be doing hypothesis testing until the next chapter.
3. Click **Options.** Enter the desired confidence level as a percentage **(98.0)** into the **Confidence Level** box. Within the **Alternative** box, select **not equal.** Click **OK.** Click **OK.**

Confidence interval for μ based on summary statistics and σ unknown

Follow the procedure in the previous steps for raw data, but in step 2 select **Summarized data,** enter **90** into the **Sample size** box, **8.46** into the **Mean** box, and **3.61** into the **Standard deviation** box.

EXERCISES

9.24 When the t distribution is used in constructing a confidence interval based on a sample size of less than 30, what assumption must be made about the shape of the underlying population?

9.25 Why are the t values listed in the df = infinity row of the t distribution table identical to the z values that correspond to the same right-tail areas of the standard normal distribution? What does this indicate about the relationship between the t and standard normal distributions?

9.26 In using the t distribution table, what value of t would correspond to an upper-tail area of 0.025 for 19 degrees of freedom?

9.27 In using the t distribution table, what value of t would correspond to an upper-tail area of 0.10 for 28 degrees of freedom?

9.28 For $df = 25$, determine the value of A that corresponds to each of the following probabilities:

a. $P(t \geq A) = 0.025$
b. $P(t \leq A) = 0.10$
c. $P(-A \leq t \leq A) = 0.99$

9.29 For $df = 85$, determine the value of A that corresponds to each of the following probabilities:

a. $P(t \geq A) = 0.10$
b. $P(t \leq A) = 0.025$
c. $P(-A \leq t \leq A) = 0.98$

9.30 Given the following observations in a simple random sample from a population that is approximately normally distributed, construct and interpret the 90% and 95% confidence intervals for the mean:

67 79 71 98 74 70 59 102 92 96

9.31 Given the following observations in a simple random sample from a population that is approximately normally distributed, construct and interpret the 95% and 99% confidence intervals for the mean:

66	34	59	56	51	45	38	58	52	52
50	34	42	61	53	48	57	47	50	54

9.32 A consumer magazine has contacted a simple random sample of 33 owners of a certain model of automobile and asked each owner how many defects had to be corrected within the first 2 months of ownership. The average number of defects was $\bar{x} = 3.7$, with a standard deviation of 1.8 defects.

a. Use the t distribution to construct a 95% confidence interval for μ = the average number of defects for this model.
b. Use the z distribution to construct a 95% confidence interval for μ = the average number of defects for this model.
c. Given that the population standard deviation is not known, which of these two confidence intervals should be used as the interval estimate for μ?

9.33 The service manager of Appliance Universe has recorded the times for a simple random sample of 50 refrigerator service calls taken from last year's service records. The sample mean and standard deviation were 25 minutes and 10 minutes, respectively.

a. Construct and interpret the 95% confidence interval for the mean.
b. It's quite possible that the population of such times is strongly skewed in the positive direction—that is, some jobs, such as compressor replacement, might take 3 or 4 hours. If this were true, would the interval constructed in part (a) still be appropriate? Explain your answer.

9.34 An automobile rental agency has the following mileages (in thousands) for a simple random sample of 20 cars rented last year. Given this information, and assuming the data are from a population that is approximately normally distributed, construct and interpret the 90% confidence interval for the population mean:

55	35	65	64	69	37	88	80	39	61
54	50	74	92	59	50	38	59	29	60 miles

9.35 One of the most popular products sold by a manufacturer of electrical fuses is the standard 30-ampere fuse used by electrical contractors in the construction industry. The company has tested a simple random sample of 16 fuses from its production output and found the amperages at which they "blew" to be as shown here. Given this information, and assuming the data are from a normally distributed population, construct and interpret the 95% confidence interval for the population mean amperage these fuses will withstand.

30.6	30.2	27.7	28.5	29.0	27.5	28.9	28.1
30.3	30.8	28.5	30.3	30.0	28.5	29.0	28.2 amperes

9.36 The author of an entry-level book on using Microsoft Word has carried out a test in which 35 novices were provided with the book, a computer, and the assignment of converting a complex handwritten document with text and tables into a Microsoft Word file. The novices required an average time of 105 minutes, with a standard deviation of 20 minutes. Construct and interpret the 90% confidence interval for the population mean time it would take all such novices to complete this task.

9.37 An office equipment manufacturer has developed a new photocopy machine and would like to estimate the average number of $8^1/_2$-by-11 copies that can be made using a single bottle of toner. For a simple random sample of 20 bottles of toner, the average was 1535 pages, with a standard deviation of 30 pages. Making and stating whatever assumptions you believe are necessary, construct and interpret the 95% confidence interval for the population mean.

9.38 The average capacity usage for iPhone users has been estimated as 400 megabytes per month. Assuming this finding to be based on a simple random sample of 80 iPhone users, with a sample standard deviation of $s = 90$ megabytes per month, construct and interpret the 95% confidence interval for the population mean usage per month. Given this confidence interval, would it seem very unusual if another sample of this size were to have a mean of 350.0 megabytes per month? Source: Leslie Cauley, "IPhone Gulps AT&T Network Capacity," *USA Today,* June 17, 2009, p. 1B.

9.39 According to Nielsen/NetRatings, the average visitor to amazon.com spends 19.7 minutes at the site. Assuming this finding to be based on a simple random sample of 20 visitors to the site, with a sample standard deviation of $s = 4.0$ minutes, and from a population that is approximately normally distributed, construct and interpret the 98% confidence interval for the population mean. Given this confidence interval, would it seem very unusual if another sample of this size were to have a mean visiting time of 18.5 minutes? Source: *The New York Times Almanac 2009,* p. 812.

(DATA SET) *Note:* Exercises 9.40 and 9.41 require a computer and statistical software.

9.40 Sallie Mae researchers report that the average college junior has a credit card debt of $2912. Under the assumption that this finding was based on the sample data in file **XR09040**, construct and interpret the 99% confidence interval for the population mean credit card debt for college juniors. Based on this confidence interval, would it seem very unusual if another sample of this size were to result in a mean of $3150? Source: Sandra Block, "Credit Card Reform Swipes Easy Plastic from College Students," *USA Today,* May 26, 2009, p. 3B.

9.41 According to a Kelton Research survey, the average U.S. woman owns 15 pairs of shoes. Under the assumption that this finding was based on the sample data in file **XR09041**, construct and interpret the 95% confidence interval for the population mean number of pairs of shoes owned by U.S. women. Based on this confidence interval, would it seem very unusual if another sample of this size were to result in a sample mean of 13.2 pairs of shoes? Source: Anne R. Carey and Sam Ward, "Who Owns More Shoes?", *USA Today,* July 8, 2009, p. 1A.

9.6 CONFIDENCE INTERVAL ESTIMATES FOR THE POPULATION PROPORTION

Determining a confidence interval estimate for the population proportion requires that we use the sample proportion (p) for two purposes: (1) as a point estimate of the (unknown) population proportion, π, and (2) in combination with the sample size (n) in estimating the standard error of the sampling distribution of the sample proportion for samples of this size.

The technique of this section uses the normal distribution as an approximation to the binomial distribution. This approximation is considered satisfactory whenever np and $n(1 - p)$ are both ≥ 5, and becomes better for large values of n and whenever p is closer to 0.5. The midpoint of the confidence interval is the sample proportion, and the lower and upper confidence limits are determined as follows:

Confidence interval limits for the population proportion:

$$p \pm z\sqrt{\frac{p(1-p)}{n}}$$

where p = sample proportion = $\dfrac{\text{number of successes}}{\text{number of trials}}$

n = sample size

z = z value corresponding to desired level of confidence (e.g., $z = 1.96$ for 95% confidence)

$\sqrt{\dfrac{p(1-p)}{n}}$ = estimated standard error of the sampling distribution of the proportion

EXAMPLE

z-Interval, Proportion

In an April 2007 NBC News/*Wall Street Journal* poll, 1008 adults were randomly sampled from across the United States. In response to the question, "All in all, do you think that things in the nation are generally headed in the right direction, or do you feel that things are off on the wrong track?" 22% responded "headed in the right direction."[1]

SOLUTION

What Is the 95% Confidence Interval for the Population Proportion Who Would Have Answered "Headed in the Right Direction" to the Question Posed?

The sample proportion $p = 0.22$ is our point estimate of π and the midpoint of the interval. Since the confidence level is to be 95%, z will be ± 1.96. The resulting confidence interval, shown in Figure 9.6, is

$$p \pm z\sqrt{\frac{p(1-p)}{n}} = 0.22 \pm 1.96\sqrt{\frac{0.22(1-0.22)}{1008}} = 0.194 \text{ to } 0.246$$

From these results, we have 95% confidence that the population proportion is somewhere between 0.194 and 0.246. Expressed in terms of percentage points, the interval would be from 19.4% to 24.6% for the percentage of the population who would have said "headed in the right direction," and the interval width would be (24.6 − 19.4), or 5.2 percentage points.

The example in this section involves a political poll of the citizenry. Such polls are very important in a democracy, and Statistics in Action 9.1 helps explain why.

[1] Source: online.wsj.com, July 21, 2009.

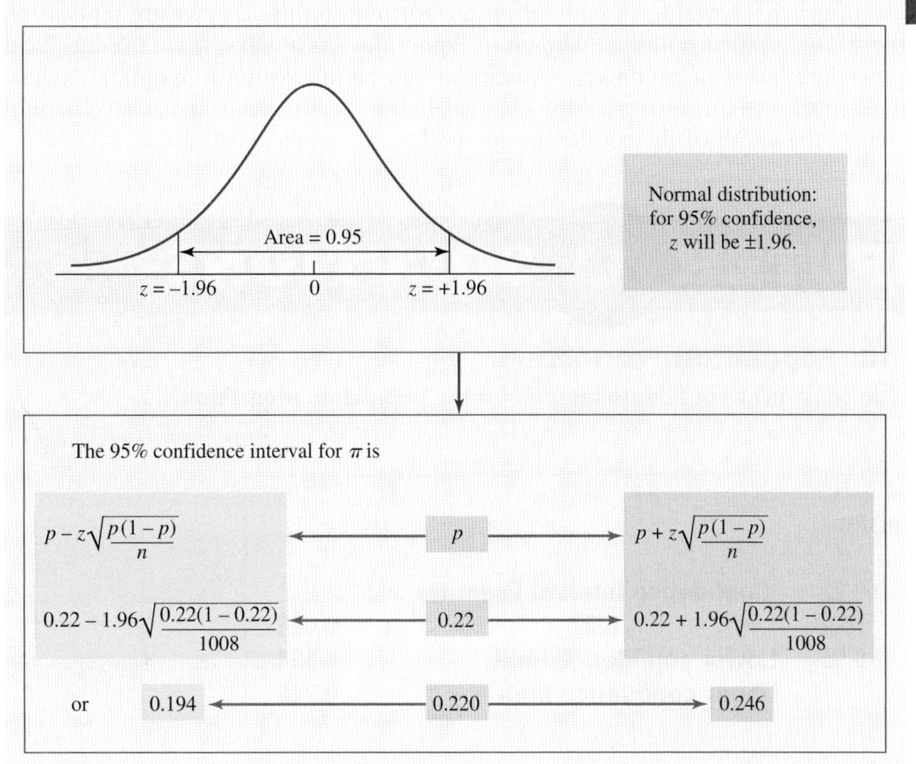

FIGURE 9.6

The 95% confidence interval for a population proportion, based on a political poll having a sample proportion of $p = 0.22$ and a sample size of $n = 1008$. We have 95% confidence that π is between 0.194 and 0.246.

STATISTICS IN ACTION 9.1

Political Polls: Important in a Democracy

President Barack Obama, elected during probably the worst economic climate since the Great Depression, had a 55% approval rating after 6 months in office. At this point in his presidential tenure, he was in 10th place out of the 12 presidents who have served since the end of World War II.

Although the presidential approval rating seems to be just another number cited by media reporters, it is actually an important statistic. A president must have a strong approval rating to retain the support of his own party and attract support from members of the opposition party. As Republican pollster Whit Ayres puts it, "Approval ratings are absolutely critical for a president to achieve his agenda."

At the 6-month point, President Obama had been experiencing a downward slide in approval ratings that was hindering his ability to champion a variety of agenda issues. At this juncture of his presidency, more people disapproved than approved of his stance on the economy, taxes, healthcare, and the federal budget deficit. Such negativity makes it difficult for any president to get his agenda through Congress. According to H. W. Brands, of the University of Texas–Austin, "It's important if a president is trying to accomplish some big stuff legislatively... Members of Congress are somewhat reluctant to tangle with a president who seems to have the backing of the American people."

In a democracy, presidents and other political leaders rely heavily on pollsters to advise them as to which path the citizens wish to take. In this respect, the typical political leader "leads" by looking in the rearview mirror to see where his followers are heading, then rapidly adjusting his own direction so as to continue marching at the front of the ideological parade.

When you see politician approval ratings and political poll results, remember that your elected leaders are looking at them as well—and developing their platforms and policies accordingly.

When a political pollster contacts you, even though you are one of a relatively small sample of potential respondents, your opinions during the survey can be every bit as important as your constitutional voice every 4 years behind the election-booth curtain.

Source: Susan Page, "Polls Can Affect President's Hold on Party," *USA Today*, July 21, 2009, p. 6A.

Computer Solutions 9.3 shows how we can use Excel or Minitab to generate a confidence interval for a population proportion. In this case, we are replicating the 95% confidence interval shown in Figure 9.6. As always, if we use an Excel procedure based on summary statistics, it can be interesting to examine "what-if" scenarios to instantly see how changes in the specified confidence level would change the width of the confidence interval.

COMPUTER SOLUTIONS 9.3

Confidence Interval for Population Proportion

These procedures show how to construct a confidence interval for the population proportion.

EXCEL

	A	B	C	D	E
1	z-Estimate of a Proportion				
2					
3	Sample proportion	0.22	Confidence Interval Estimate		
4	Sample size	1008	0.22	±	0.026
5	Confidence level	0.95	Lower confidence limit		0.194
6			Upper confidence limit		0.246

Excel confidence interval for π based on summary statistics

1. For example, with $n = 1008$ and $p = 0.22$, as in Figure 9.6: Open the **ESTIMATORS** workbook.
2. Using the arrows at the bottom left, select the ***z*-Estimate_Proportion** worksheet. Enter the sample proportion **(0.22),** the sample size **(1008),** and the desired confidence level as a decimal fraction **(0.95).** The confidence interval appears as shown here.

(*Note:* As an alternative, you can use Excel worksheet template **TMPINT**. The steps are described within the template.)

Excel confidence interval for π based on raw data

1. For example, if we had 20 data values that were coded (1 = female, 2 = male), with the label and data values in A1:A21: From the **Add-Ins** ribbon, click **Data Analysis Plus.** Click ***Z*-Estimate: Proportion.** Click **OK.**
2. Enter **A1:A21** into the **Input Range** box. Enter **1** into the **Code for Success** box. Click **Labels.** The desired confidence level as a decimal fraction is 0.95, so enter the corresponding alpha value **(0.05)** into the **Alpha** box. Click **OK.**

MINITAB

Minitab confidence interval for π based on summary statistics

```
Test and CI for One Proportion

Sample    X      N  Sample p         95% CI
1       222   1008  0.220238  (0.194655, 0.245821)

Using the normal approximation.
```

1. This interval is based on the summary statistics for Figure 9.6: Click **Stat.** Select **Basic Statistics.** Click **1 Proportion.** Select **Summarized Data.** Enter the sample size (in this case, **1008**) into the **Number of Trials** box. Multiply the sample proportion (0.22) times the sample size (1008) to get the number of "successes" or "events" as (0.22)(1008) = 221.76. Round to the nearest integer and enter the result **(222)** into the **Number of Events** box. The **Perform hypothesis test** box should be left blank.
2. Click **Options.** Enter the desired confidence level as a percentage **(95.0)** into the **Confidence Level** box. Within the **Alternative** box, select **not equal.** Click to select **Use test and interval based on normal distribution.** Click **OK.** Click **OK.**

Minitab confidence interval for π based on raw data

1. For example, if column C1 contains 20 data values that are coded (1 = female, 2 = male): Click **Stat.** Select **Basic Statistics.** Click **1 Proportion.** Select **Samples in columns** and enter **C1** into the dialogue box.
2. Follow step 2 in the summary-information procedure. *Note:* Minitab will select the larger of the two codes (i.e., 2 = male) as the "success" or "event" and provide the sample proportion and the confidence interval for the population proportion of males. To obtain the results for females, just recode the data so females will have the higher code number: Click **Data.** Select **Code.** Click **Numeric to Numeric.** Enter **C1** into both the **Code data from columns** box and the **Store coded data in columns** box. Enter **1** into the **Original values** box. Enter **3** into the **New** box. Click **OK.** The new codes will be (3 = female, 2 = male).

EXERCISES

9.42 Under what conditions is it appropriate to use the normal approximation to the binomial distribution in constructing the confidence interval for the population proportion?

9.43 A pharmaceutical company found that 46% of 1000 U.S. adults surveyed knew neither their blood pressure nor their cholesterol level. Assuming the persons surveyed to be a simple random sample of U.S. adults, construct a 95% confidence interval for π = the population proportion of U.S. adults who would have given the same answer if a census had been taken instead of a survey.

9.44 An airline has surveyed a simple random sample of air travelers to find out whether they would be interested in paying a higher fare in order to have access to e-mail during their flight. Of the 400 travelers surveyed, 80 said e-mail access would be worth a slight extra cost. Construct a 95% confidence interval for the population proportion of air travelers who are in favor of the airline's e-mail idea.

9.45 In response to media inquiries and concerns expressed by groups opposed to violence, the president of a university with over 25,000 students has agreed to survey a simple random sample of her students to find out whether the student body thinks the school's "Plundering Pirate" mascot should be changed to one that is less aggressive in name and appearance. Of the 200 students selected for participation in the survey, only 20% believe the school should select a new and more kindly mascot. Construct a 90% confidence interval for the population proportion of students who believe the mascot should be changed. Based on the sample findings and associated confidence interval, comment on the credibility of a local journalist's comment that "over 50%" of the students would like a new mascot.

9.46 In examining a simple random sample of 100 sales invoices from several thousand such invoices for the previous year, a researcher finds that 65 of the invoices involved customers who bought less than $2000 worth of merchandise from the company during that year. Construct a 90% confidence interval for the proportion of all sales invoices that were for customers buying less than $2000 worth of merchandise during the year.

9.47 It has been estimated that 48% of U.S. households headed by persons in the 35–44 age group own mutual funds. Assuming this finding to be based on a simple random sample of 1000 households headed by persons in this age group, construct a 95% confidence interval for π = the population proportion of such households that own mutual funds. Source: Investment Company Institute, *Investment Company Fact Book 2008,* p. 72.

9.48 A study by Careerbuilder.com found that 20% of companies check out job candidates' profiles on social networking sites like Facebook and MySpace before deciding whether to employ them. Assuming that the survey included a simple random sample of 1200 companies, construct a 90% confidence interval for π = the population proportion of companies that check social-networking sites before offering employment. Source: Carrie-Ann Skinner, "Employers Admit Checking Facebook Before Hiring," pcworld.com, September 14, 2008.

9.49 According to Nielsen Media Research viewership data, the top television broadcast of all time was the last episode of *M*A*S*H*, which aired on February 28, 1983, and was viewed by an estimated 60.2% of all TV households. Assuming this estimate was based on a simple random sample of 1800 TV households, what is the 95% confidence interval for π = the proportion of all TV households who viewed the last episode of *M*A*S*H*? Source: *The World Almanac and Book of Facts 2009,* p. 293.

9.50 In a major industry where well over 100,000 manufacturing employees are represented by a single union, a simple random sampling of $n = 100$ union members finds that 57% of those in the sample intend to vote for the new labor contract negotiated by union and management representatives.

a. What is the 99% confidence interval for π = the population proportion of union-represented employees who intend to vote for the labor contract?
b. Based on your response to part (a), does contract approval by the union appear to be a "sure thing"? Why or why not?

9.51 Repeat Exercise 9.50, but assume that the sample size was $n = 900$ instead of $n = 100$.

9.52 A 2009 Pew Internet & American Life Project survey of 2253 U.S. adults found that 69% were using the Internet to help them find jobs, bargains, benefits, financing, and to otherwise cope with the economic recession. Assuming this to be a simple random sample of U.S. adults, construct and interpret the 90% confidence interval for the population proportion of U.S. adults who were using the Internet to cope with the recession. Source: Lee Rainie and Aaron Smith, "The Internet and the Recession," pewinternet.org, July 15, 2009.

9.53 According to a National Organization for Youth Safety Survey of 605 drivers aged 16–20, 68% have sent a text message while driving. Assuming this to be a simple random sample of drivers from this age group, construct and interpret the 90% confidence interval for the population proportion of drivers aged 16–20 who have texted while driving. Source: Anne R. Carey and Karl Gelles, "Young Drivers Busy Behind the Wheel," *USA Today*, June 16, 2009, p. 1A.

9.54 A Zogby poll found that 75% of Americans can name all of The Three Stooges—Moe, Curly, and Larry. Assuming this finding to be based on a simple random sample of 1200 Americans, construct and interpret the 95% confidence interval for the population proportion of Americans who can name all three Stooges. Source: Brit Hume, "Zogby Poll: Most Americans Can Name Three Stooges, But Not Three Branches of Government," foxnews.com, August 15, 2006.

9.55 Estelle McCarthy, a candidate for state office in New Jersey, has been advised that she must get at least 65% of the union vote in her state. A recent political poll of likely voters included 800 union-member respondents, and 60% of them said they intended to vote for Ms. McCarthy. Based on the survey, construct and interpret the 95% confidence interval for the population proportion of likely-voter union members who intend to vote for Ms. McCarthy. Is the 65% level of support within the confidence interval? Given your answer to the preceding question, comment on the possibility that Ms. McCarthy might not succeed in obtaining the level of union support she needs.

(DATA SET) *Note:* Exercises 9.56 and 9.57 require a computer and statistical software.

9.56 In the documentation that accompanies its products that are returned for in-warranty service, a manufacturer of electric can openers asks the customer to indicate the reason for the return. The codes and return problem categories are: (1) "doesn't work," (2) "excessive noise," and (3) "other." Data file **XR09056** contains the problem codes for a simple random sample of 300 product returns. Based on this sample, construct and interpret the 95% confidence interval for the population proportion of returns that were because the product "doesn't work."

9.57 An investment counselor has purchased a large mailing list consisting of 50,000 potential investors. Before creating a brochure to send to members of the list, the counselor mails a questionnaire to a small simple random sampling of them. In one of the questions, the respondent is asked, "Do I think of myself as someone who enjoys taking risks?" The response codes are (1) "Yes" and (2) "No." The results for the 200 investors who answered this question are represented by the codes listed in data file **XR09057**. Based on this sample, construct and interpret the 99% confidence interval for the population proportion of investors on the counselor's mailing list who think of themselves as someone who enjoys taking risks.

SAMPLE SIZE DETERMINATION (9.7)

In our interval estimates to this point, we have taken our results, including a stated sample size, then constructed a confidence interval. In this section, we'll proceed in the opposite direction—in other words, we will decide in advance the desired confidence interval width, then work backward to find out how large a sample size is necessary to achieve this goal. Central to our discussion in this section is the fact that the maximum likely sampling error (accuracy) is one-half the width of the confidence interval.

Estimating the Population Mean

To show how the necessary sample-size equation is obtained, we will consider a case in which σ is known. Of especial importance is that the distance from the midpoint to the upper confidence limit can be expressed as either (1) the maximum likely sampling error (e) or (2) z times the standard error of the sampling distribution. Since the two quantities are the same, we can set up the following equation and solve for n:

$$e = z\frac{\sigma}{\sqrt{n}} \longrightarrow \boxed{\text{Solving for } n} \longrightarrow n = \frac{z^2 \cdot \sigma^2}{e^2}$$

Required sample size for estimating a population mean:

$$n = \frac{z^2 \cdot \sigma^2}{e^2}$$

where n = required sample size
z = z value for which $\pm z$ corresponds to the desired level of confidence
σ = known (or, if necessary, estimated) value of the population standard deviation
e = maximum likely error that is acceptable

One way of estimating an *unknown* σ is to use a relatively small-scale pilot study from which the sample standard deviation is used as a point estimate of the population standard deviation. A second approach is to estimate σ by using the results of a similar study done at some time in the past. A third method is to estimate σ as 1/6 the approximate range of data values.

EXAMPLEEXAMPLEEXAMPLEEXAMPLEEXAMPLE

EXAMPLE

Sample Size, Estimating a Mean

A state politician would like to determine the average amount earned during summer employment by state teenagers during the past summer's vacation period. She wants to have 95% confidence that the sample mean is within \$50 of the actual population mean. Based on past studies, she has estimated the population standard deviation to be σ = \$400.

SOLUTION

What Sample Size Is Necessary to Have 95% Confidence That $\bar{x}$ Will Be Within \$50 of the Actual Population Mean?

For this situation, 95% confidence leads to a z value of 1.96, e = the \$50 maximum likely error that is acceptable, and the estimated value of σ is \$400. The necessary sample size will be

$$n = \frac{z^2 \cdot \sigma^2}{e^2} = \frac{1.96^2 \cdot 400^2}{50^2} = 245.9 \text{ persons, rounded up to } 246$$

Since we can't include a fraction of a person in the sample, we round up to n = 246 to ensure 95% confidence in being within \$50 of the population mean. Whenever the calculated value of n is not an integer, it is a standard (though slightly conservative) practice to round up to the next integer value.

Note that if we cut the desired maximum likely error in half, the necessary sample size will quadruple. This is because the e term in the denominator is squared. The desire for extremely accurate results can lead to sample sizes that grow very rapidly in size (and expense) as the specified value for e is reduced.

Estimating the Population Proportion

As when estimating the population mean, the maximum likely error (e) in estimating a population proportion will be one-half of the eventual confidence interval width. Likewise, the distance from the midpoint of the confidence interval to the upper confidence limit can be described in two ways. Setting them equal and solving for n gives the following result:

$$e = z\sqrt{\frac{p(1-p)}{n}} \longrightarrow \boxed{\text{Solving for } n} \longrightarrow n = \frac{z^2 p(1-p)}{e^2}$$

Required sample size for estimating a population proportion:

$$n = \frac{z^2 p(1-p)}{e^2}$$

where n = required sample size
z = z value for which $\pm z$ corresponds to the desired level of confidence
p = the estimated value of the population proportion (As a conservative strategy, use $p = 0.5$ if you have no idea as to the actual value of π.)
e = maximum likely error that is acceptable

In applying the preceding formula, we should first consider whether the true population proportion is likely to be either much less than or much greater than 0.5. If we have absolutely no idea as to what π might be, using $p = 0.5$ is the conservative strategy to follow. This is because the required sample size, n, is proportional to the value of $p(1 - p)$, and this value is largest whenever $p = 0.5$.

If we are totally uncertain regarding the actual population proportion, we may wish to conduct a pilot study to get a rough idea of its value. If we can estimate the population proportion as being either much less or much more than 0.5, we can obtain the desired accuracy with a smaller sample than would have otherwise been necessary.

If the population proportion is believed to be within a range, such as "between 0.20 and 0.40," we should use as our estimate the value that is closest to 0.5. For example, if we believe the population proportion is somewhere between 0.20 and 0.40, it should be estimated as 0.40 when calculating the required sample size.

EXAMPLE

Sample Size, Estimating a Proportion

A tourist agency researcher would like to determine the proportion of U.S. adults who have ever vacationed in Mexico and wishes to be 95% confident that the sampling error will be no more than 0.03 (3 percentage points).

SOLUTION

Assuming the Researcher Has No Idea Regarding the Actual Value of the Population Proportion, What Sample Size Is Necessary to Have 95% Confidence That the Sample Proportion Will Be within 0.03 (3 Percentage Points) of the Actual Population Proportion?

For the 95% level of confidence, the z value will be 1.96. The maximum acceptable error is $e = 0.03$. Not wishing to make an estimate, the researcher will use $p = 0.5$ in calculating the necessary sample size:

$$n = \frac{z^2 p(1-p)}{e^2} = \frac{1.96^2(0.5)(1-0.5)}{0.03^2} = 1067.1 \text{ persons, rounded up to } 1068$$

If the Researcher Believes the Population Proportion Is No More Than 0.3, and Uses $p = 0.3$ as the Estimate, What Sample Size Will Be Necessary?

Other factors are unchanged, so z remains 1.96 and e is still specified as 0.03. However, the $p(1-p)$ term in the numerator will be reduced due to the assumption that the population proportion is no more than 0.3. The required sample size will now be

$$n = \frac{z^2 p(1-p)}{e^2} = \frac{1.96^2(0.3)(1-0.3)}{0.03^2} = 896.4 \text{ persons, rounded up to } 897$$

As in determining the necessary size for estimating a population mean, lower values of e lead to greatly increased sample sizes. For example, if the researcher estimated the population proportion as being no more than 0.3, but specified a maximum likely error of 0.01 instead of 0.03, he would have to include *nine times as many* people in the sample (8068 instead of 897).

Computer Solutions 9.4 shows how we can use Excel to determine the necessary sample size for estimating a population mean or proportion. With these procedures, it is very easy to examine "what-if" scenarios and instantly see how changes in confidence level or specified maximum likely error will affect the required sample size.

EXERCISES

9.58 "If we want to cut the maximum likely error in half, we'll have to double the sample size." Is this statement correct? Why or why not?

9.59 In determining the necessary sample size in making an interval estimate for a population mean, it is necessary to first make an estimate of the population standard deviation. On what bases might such an estimate be made?

9.60 From past experience, a package-filling machine has been found to have a process standard deviation of 0.65 ounces of product weight. A simple random sample is to be selected from the machine's output for the purpose of determining the average weight of product being packed by the machine. For 95% confidence that the sample mean will not differ from the actual population mean by more than 0.1 ounces, what sample size is required?

9.61 Based on a pilot study, the population standard deviation of scores for U.S. high school graduates taking a new version of an aptitude test has been estimated as 3.7 points. If a larger study is to be undertaken, how large a simple random sample will be necessary to have 99% confidence that the sample mean will not differ from the actual population mean by more than 1.0 point?

9.62 A consumer agency has retained an independent testing firm to examine a television manufacturer's claim that its 50-inch console model consumes just 110 watts of electricity. Based on a preliminary study, the population standard deviation has been estimated as 11.2 watts for these sets. In undertaking a larger study, and using a simple random sample, how many sets must be tested for the firm to be 95% confident that its sample mean does not differ from the actual population mean by more than 3.0 watts?

9.63 A national political candidate has commissioned a study to determine the percentage of registered voters who intend to vote for him in the upcoming election. To have 95% confidence that the sample percentage will be within 3 percentage points of the actual population percentage, how large a simple random sample is required?

9.64 Suppose that Nabisco would like to determine, with 95% confidence and a maximum likely error of 0.03, the proportion of first graders in Pennsylvania who had Nabisco's Spoon-Size Shredded Wheat for breakfast at least once last week. In determining the necessary size of a simple random sample for this purpose:

a. Use 0.5 as your estimate of the population proportion.

b. Do you think the population proportion could really be as high as 0.5? If not, repeat part (a) using an estimated proportion that you think would be more likely to be true. What effect does your use of this estimate have on the sample size?

9.65 The Chevrolet dealers of a large county are conducting a study to determine the proportion of car owners in the county who are considering the purchase of a new car within the next year. If the population proportion is believed to be no more than 0.15, how many owners must be included in a simple random sample if the dealers want to be 90% confident that the maximum likely error will be no more than 0.02?

9.66 In Exercise 9.65, suppose that (unknown to the dealers) the actual population proportion is really 0.35. If they use their estimated value ($\pi \leq 0.15$) in determining the sample size and then conduct the study, will their maximum likely error be greater than, equal to, or less than 0.02? Why?

9.67 In reporting the results of their survey of a simple random sample of U.S. registered voters, pollsters claim 95% confidence that their sampling error is no more than 4 percentage points. Given this information only, what sample size was used?

COMPUTER 9.4 SOLUTIONS

Sample Size Determination

These procedures determine the necessary sample size for estimating a population mean or a population proportion.

EXCEL

	A	B	C	D
1	**Sample size required for estimating a**			
2	**population mean:**			
3				
4	Estimate for sigma:			400.00
5	Maximum likely error, e:			50.00
6				
7	Confidence level desired:			0.95
8	alpha = (1 - conf. level desired):			0.05
9	The corresponding z value is:			1.960
10				
11	The required sample size is n =			245.9

Sample size for estimating a population mean, using Excel worksheet template TMNFORMU

To determine the necessary sample size for estimating a population mean within $50 and with 95% confidence, assuming a population standard deviation of $400: Open Excel worksheet **TMNFORMU**. Enter the estimated sigma **(400)**, the

(continued)

maximum likely error **(50)**, and the specified confidence level as a decimal fraction **(0.95)**. The required sample size (in cell D11) should then be rounded *up* to the nearest integer (246). This procedure is also described within the worksheet template.

Caution Do not save any changes when exiting Excel.

Sample size for estimating a population proportion, using Excel worksheet template TMNFORPI

To determine the necessary sample size for estimating a population proportion within 0.03 (3 percentage points) and with 95% confidence: Open Excel worksheet **TMNFORPI**. Enter the estimate for pi **(0.50)**, the maximum likely error **(0.03)**, and the specified confidence level as a decimal fraction **(0.95)**. The required sample size (in cell D11) should then be rounded *up* to the nearest integer (1068). (*Note:* If you have knowledge about the population and can estimate pi as either less than or greater than 0.50, use your estimate and the necessary sample size will be smaller. Otherwise, be conservative and use 0.50 as your estimate.) This procedure is also described within the worksheet template.

Caution Do not save any changes when exiting Excel.

9.8 WHEN THE POPULATION IS FINITE

Whenever sampling is without replacement and from a finite population, it may be necessary to modify slightly the techniques for confidence interval estimation and sample size determination in the preceding sections. As in Chapter 8, the general idea is to reduce the value of the standard error of the estimate for the sampling distribution of the mean or proportion. As a rule of thumb, the methods in this section should be applied whenever the sample size (n) is at least 5% as large as the population. When $n < 0.05N$, there will be very little difference in the results.

Confidence Interval Estimation

Whether we are dealing with interval estimation for a population mean (μ) or a population proportion (π), the confidence intervals will be similar to those in Figure 9.2. The only difference is that the "$\pm$" term will be multiplied by the "finite population correction factor" shown in Table 9.3. As in Chapter 8, this correction depends on the sample size (n) and the population size (N).

As an example of how this works, we'll consider a situation such as that in Section 9.5, where a confidence interval is to be constructed for the population mean, and the sample standard deviation (s) is used as an estimate of the population standard deviation (σ). In this case, however, the sample will be relatively large compared to the size of the population.

EXAMPLEEXAMPLEE

EXAMPLE

Interval Estimates, Finite Population

According to the Bureau of the Census, the population of Hinsdale County, Colorado, is 838 persons.[2] For purposes of our example, assume that a researcher has interviewed a simple random sample of 400 persons and found that their average number of years of formal education is $\bar{x} = 11.5$ years, with a standard deviation of $s = 4.3$ years.

[2] Source: *The World Almanac and Book of Facts 2009*, p. 611.

TABLE 9.3

Summary of confidence interval formulas when sampling without replacement from a finite population. As a rule of thumb, they should be applied whenever the sample is at least 5% as large as the population. The formulas and terms are similar to those in Figure 9.2 but include a "finite population correction factor," the value of which depends on the relative sizes of the sample (n) and population (N).

Confidence Interval Estimate for the Population Mean, σ Known	
Infinite Population	**Finite Population**
$\bar{x} \pm z\frac{\sigma}{\sqrt{n}}$	$\bar{x} \pm z\left(\frac{\sigma}{\sqrt{n}} \cdot \sqrt{\frac{N-n}{N-1}}\right)$
Confidence Interval Estimate for the Population Mean, σ Unknown	
Infinite Population	**Finite Population**
$\bar{x} \pm t\frac{s}{\sqrt{n}}$	$\bar{x} \pm t\left(\frac{s}{\sqrt{n}} \cdot \sqrt{\frac{N-n}{N-1}}\right)$
Confidence Interval Estimate for the Population Proportion	
Infinite Population	**Finite Population**
$p \pm z\sqrt{\frac{p(1-p)}{n}}$	$p \pm z\left(\sqrt{\frac{p(1-p)}{n}} \cdot \sqrt{\frac{N-n}{N-1}}\right)$

SOLUTION

Considering That the Population Is Finite and $n \geq 0.05N$, What Is the 95% Confidence Interval for the Population Mean?

Since the number of degrees of freedom ($df = n - 1 = 400 - 1$, or 399) exceeds the limits of our t distribution table, the t distribution and normal distribution can be considered to be practically identical, and we can use the infinity row of the t table. The appropriate column in this table will be 0.025 for a 95% confidence interval, and the entry in the infinity row of this column is a t value of 1.96.

Since s is being used to estimate σ, and the sample is more than 5% as large as the population, we will use the "σ unknown" formula of the finite population expressions in Table 9.3. The 95% confidence interval for the population mean can be determined as

> The finite population correction term: This will be smaller than 1.0 so the standard error will be less than if an infinite population were involved.

$$\bar{x} \pm t\left(\frac{s}{\sqrt{n}} \cdot \sqrt{\frac{N-n}{N-1}}\right)$$

$$= 11.5 \pm 1.96\left(\frac{4.3}{\sqrt{400}} \cdot \sqrt{\frac{838-400}{838-1}}\right)$$

$$= 11.5 \pm 1.96(0.215 \cdot 0.723) = 11.5 \pm 0.305$$

or from 11.195 to 11.805

As expected, the finite correction term (0.723) is less than 1.0 and leads to a 95% confidence interval that is narrower than if an infinite population had been assumed. (*Note:* If the population had been considered infinite, the resulting interval would have been wider, with lower and upper limits of 11.079 and 11.921 years, respectively.)

EXAMPLEEXAMPLEEXAMPLEEXAMPLEEXAMPLE

Sample Size Determination

As in confidence interval estimation, the rule of thumb is to change our sample size determination procedure slightly whenever we are sampling without replacement from a finite population and the sample is likely to be at least 5% as large as the population. Although different in appearance, the following formulas are applied in the same way that we used their counterparts in Section 9.7.

If you were to substitute an N value of *infinity* into each of the following equations, you would find that the right-hand term in the denominator of each would be eliminated, and the result would be an expression exactly the same as its counterpart in Section 9.7.

Required sample size for estimating the mean of a finite population:

$$n = \frac{\sigma^2}{\frac{e^2}{z^2} + \frac{\sigma^2}{N}}$$

where n = required sample size
N = population size
z = z value for which $\pm z$ corresponds to the desired level of confidence
σ = known (or, if necessary, estimated) value of the population standard deviation
e = maximum likely error that is acceptable

Required sample size, estimating the proportion for a finite population:

$$n = \frac{p(1-p)}{\frac{e^2}{z^2} + \frac{p(1-p)}{N}}$$

where n = required sample size
N = population size
z = z value for which $\pm z$ corresponds to the desired level of confidence
p = the estimated value of the population proportion (As a conservative strategy, use $p = 0.5$ if you have no idea as to the actual value of π.)
e = maximum likely error that is acceptable

EXAMPLE

Sample Size, Finite Population

The Federal Aviation Administration (FAA) lists 12,290 pilots holding commercial helicopter certificates.[3] Suppose the FAA wishes to question a simple random sample of these individuals to find out what proportion are interested in switching jobs within the next 3 years. Assume the FAA wishes to have 95% confidence that the sample proportion is no more than 0.04 (i.e., 4 percentage points) away from the true population proportion.

[3]Source: General Aviation Manufacturers Association, *General Aviation Statistical Databook, 2008 Edition,* p. 38.

EXAMPLEEXAMPLEEXAMP

SOLUTION

Considering That the Population Is Finite, What Sample Size Is Necessary to Have 95% Confidence That the Sample Proportion Will Not Differ from the Population Proportion by More Than 0.04?

Since the actual population proportion who are interested in switching jobs has not been estimated, we will be conservative and use $p = 0.5$ in deciding on the necessary sample size. For the 95% confidence level, z will be 1.96. Applying the finite population formula, with $N = 12{,}290$, the number of pilots who should be included in the sample is

$$n = \frac{p(1-p)}{\frac{e^2}{z^2} + \frac{p(1-p)}{N}} = \frac{0.5(1-0.5)}{\frac{0.04^2}{1.96^2} + \frac{0.5(1-0.5)}{12{,}290}} = 572.3$$

Had the population been considered infinite, the required sample size would have been calculated as in Section 9.7. This would have resulted in $n = 600.25$, rounded up to 601. By recognizing that the population is finite, we are able to achieve the desired confidence level and maximum error with a sample size that includes only 573 pilots instead of 601.

EXERCISES

9.68 As a rule of thumb, under what conditions should the finite population correction be employed in determining confidence intervals and calculating required sample sizes?

9.69 Compared to situations where the population is either infinite or very large compared to the sample size, what effect will the finite population correction tend to have on

a. the width of a confidence interval?
b. the required size of a sample?

9.70 The personnel manager of a firm with 200 employees has selected a simple random sample of 40 employees and examined their health-benefit claims over the past year. The average amount claimed during the year was $260, with a standard deviation of $80. Construct and interpret the 95% confidence interval for the population mean. Was it necessary to make any assumptions about the shape of the population distribution? Explain.

9.71 Of 1200 undergraduates enrolled at a university, a simple random sample of 600 have been surveyed to measure student support for a $5 activities fee increase to help fund women's intercollegiate athletics at the NCAA division 1A level. Of those who were polled, 55% supported the fee increase. Construct and interpret the 95% and 99% confidence intervals for the population proportion. Based on your results, comment on the possibility that the fee increase might lose when it is voted on at next week's university-wide student referendum.

9.72 A local environmental agency has selected a simple random sample of 16 homes to be tested for tap-water lead. Concentrations of lead were found to have a mean of 12 parts per billion and a standard deviation of 4 parts per billion. Considering that the homes were selected from a community in which there are 100 homes, construct and interpret the 95% confidence interval for the population mean. Based on your results, comment on the possibility that the average lead concentration in this community's homes might exceed the Environmental Protection Agency's recommended limit of 15 parts per billion of lead.

9.73 A simple random sample is to be drawn from a population of 800. In order to have 95% confidence that the sampling error in estimating π is no more than 0.03, what sample size will be necessary?

9.74 A simple random sample is to be drawn from a population of 2000. The population standard deviation has been estimated as being 40 grams. In order to have 99% confidence that the sampling error in estimating μ is no more than 5 grams, what sample size will be necessary?

9.75 There are 100 members in the U.S. Senate. A political scientist wants to estimate, with 95% confidence and within 3 percentage points, the percentage who own stock in foreign companies. How many senators should be interviewed? Explain any assumptions you used in obtaining your recommended sample size.

9.76 A transportation company operates 200 trucks and would like to use a hidden speed monitor device to record the maximum speed at which a truck is operated during the period that the device is installed. The trucks are driven primarily on interstate highways, and the company wants to estimate the average maximum speed for its fleet with 90% confidence and within 2 miles per hour. Using (and explaining) your own estimate for the population standard deviation, determine the number of trucks on which the company should install the hidden speed-recording device.

9.77 A research firm supports a consumer panel of 2000 households that keep written diaries of their weekly grocery expenditures. The firm would like to estimate, with 95% confidence and within 4 percentage points, the percentage of its panel households who would be interested in providing more extensive information in return for an extra $50 per week remuneration. How many of the households should be surveyed? Explain any assumptions you used in obtaining your recommended sample size.

9.78 A university official wants to estimate, with 99% confidence and within $2, the average amount that members of fraternities and sororities spend at local restaurants during the first week of the semester. If the total fraternity/sorority membership is 300 people, how many members should be included in the sample? Use (and explain) your own estimate for the population standard deviation.

9.79 A quality-management supervisor believes that no more than 5% of the items in a recent shipment of 2000 are defective. If she wishes to determine, within 1 percentage point and with 99% confidence, the percentage of defective items in the shipment, how large a simple random sample would be necessary?

9.9 SUMMARY

- **Inferential statistics: Point and interval estimates for a population parameter**

Chapter 8 examined the sampling distribution of a sample mean or a sample proportion from a known population. In this chapter, the emphasis has been on the estimation of an unknown population mean (μ) or proportion (π) on the basis of sample statistics. Point estimates involve using the sample mean ($\bar{x}$) or proportion (p) as the single best estimate of the value of the population mean or proportion. Interval estimates involve a range of values that may contain the actual value of the population parameter. When interval estimates are associated with a degree of certainty that they really do include the true population parameter, they are referred to as confidence intervals.

- **Constructing a confidence interval for a population mean or a population proportion**

The procedure appropriate to constructing an interval estimate for the population mean depends largely on whether the population standard deviation is known. Figure 9.2 summarizes these procedures and their underlying assumptions. Although the t-interval is often associated with interval estimates based on small samples, it is appropriate for larger samples as well. Using computer statistical packages, we can easily and routinely apply the t distribution for interval estimates of the mean whenever σ is unknown, even for very large sample sizes.

A trade-off exists between the degree of confidence that an interval contains the population parameter and the width of the interval itself. The more certain we wish to be that the interval estimate contains the parameter, the wider the interval will have to be.

- **Sample size determination**

Accuracy, or sampling error, is equal to one-half of the confidence interval width. The process of sample size determination anticipates the width of the eventual confidence interval, then determines the required sample size that will limit the maximum likely sampling error to an acceptable amount.

- **When the sample is large compared to the population**

As in Chapter 8, when sampling is without replacement from a finite population, it is appropriate to use a finite population correction factor whenever the sample is at least 5% of the size of the population. Such corrections are presented for both interval estimation and sample size determination techniques within the chapter.

- **Computer-generated confidence intervals**

Most computer statistical packages are able to construct confidence interval estimates of the types discussed in the chapter. Examples of Excel and Minitab outputs are provided for a number of chapter examples in which such confidence intervals were developed.

EQUATIONS

Confidence Interval Limits for the Population Mean, σ Known

$$\bar{x} \pm z\frac{\sigma}{\sqrt{n}}$$

where
- $\bar{x}$ = sample mean
- σ = population standard deviation
- n = sample size
- z = z value for desired confidence level
- $\sigma/\sqrt{n}$ = standard error of the sampling distribution of the mean

(Assumes that either (1) the underlying population is normally distributed or (2) the sample size is $n \geq 30$.)

Confidence Interval Limits for the Population Mean, σ Unknown

$$\bar{x} \pm t\frac{s}{\sqrt{n}}$$

where
- $\bar{x}$ = sample mean
- s = sample standard deviation
- n = sample size
- t = t value corresponding to the level of confidence desired, with $df = n - 1$
- $s/\sqrt{n}$ = estimated standard error of the sampling distribution of the mean

(If $n < 30$, this requires the assumption that the underlying population is approximately normally distributed.)

Confidence Interval Limits for the Population Proportion

$$p \pm z\sqrt{\frac{p(1-p)}{n}}$$

where
- p = sample proportion = $\frac{\text{number of successes}}{\text{number of trials}}$
- n = sample size
- z = z value corresponding to desired level of confidence (e.g., $z = 1.96$ for 95% confidence)
- $\sqrt{\frac{p(1-p)}{n}}$ = estimated standard error of the sampling distribution of the proportion

Required Sample Size for Estimating a Population Mean

$$n = \frac{z^2 \cdot \sigma^2}{e^2}$$

where n = required sample size
z = z value for which $\pm z$ corresponds to the desired level of confidence
σ = known (or, if necessary, estimated) value of the population standard deviation
e = maximum likely error that is acceptable

Required Sample Size for Estimating a Population Proportion

$$n = \frac{z^2 p(1-p)}{e^2}$$

where n = required sample size
z = z value for desired level of confidence
p = estimated value of the population proportion (if not estimated, use $p = 0.5$)
e = maximum likely error that is acceptable

Confidence Interval Estimates When the Population Is Finite

- For the population mean, σ known:

$$\bar{x} \pm z\left(\frac{\sigma}{\sqrt{n}} \cdot \sqrt{\frac{N-n}{N-1}}\right)$$

where n = sample size
N = population size

- For the population mean, σ unknown:

$$\bar{x} \pm t\left(\frac{s}{\sqrt{n}} \cdot \sqrt{\frac{N-n}{N-1}}\right)$$

- For the population proportion:

$$p \pm z\left(\sqrt{\frac{p(1-p)}{n}} \cdot \sqrt{\frac{N-n}{N-1}}\right)$$

Required Sample Size for Estimating the Mean of a Finite Population

$$n = \frac{\sigma^2}{\dfrac{e^2}{z^2} + \dfrac{\sigma^2}{N}}$$

where n = required sample size
N = population size
z = z value for desired level of confidence
σ = known (or estimated) value of the population standard deviation
e = maximum likely error that is acceptable

Required Sample Size for Estimating the Proportion for a Finite Population

$$n = \frac{p(1-p)}{\dfrac{e^2}{z^2} + \dfrac{p(1-p)}{N}}$$

where n = required sample size
N = population size
z = z value for desired level of confidence
p = the estimated population proportion (if not estimated, use $p = 0.5$)
e = maximum likely error that is acceptable

CHAPTER EXERCISES

9.80 In a survey of 500 U.S. adults, 45% of them said that lounging at the beach was their "dream vacation." Assuming this to be a simple random sample of U.S. adults, construct and interpret the 95% and 99% confidence intervals for the proportion of U.S. adults who consider lounging at the beach to be their dream vacation.

9.81 In Exercise 9.83, a small-scale preliminary survey has indicated that no more than 20% of the television households will tune in to the first episode of the miniseries. Given this information, how large must the sample be?

9.82 For the following simple random sample of household incomes (thousands of dollars) from a large county, construct and interpret the 90% and 95% confidence intervals for the population mean. The data are also in file **XR09087**.

58.3	50.0	58.1	33.5	51.1	38.1	42.3	60.4	55.8	46.2
40.4	52.5	51.3	47.5	48.5	59.3	40.9	37.1	39.1	43.6
55.3	42.3	48.2	42.8	61.1	34.7	35.5	52.9	44.7	51.5

9.83 There are approximately 113 million television households in the United States. A ratings service would like to know, within 5 percentage points and with 95% confidence, the percentage of these households who tune in to the first episode of a network miniseries. How many television households must be included in the sample? Source: *The World Almanac and Book of Facts 2009*, p. 291.

9.84 For a new process with which the production personnel have little experience, neither the standard deviation nor the mean of the process is known. Twenty different simple random samples, each with $n = 50$, are to be drawn from the process, and a 90% confidence interval for the mean is to be constructed for each sample. What is the probability that at least 2 of the confidence intervals will not contain the population mean?

9.85 A torque wrench used in the final assembly of cylinder heads has a process standard deviation of 5.0 lb-ft. The engineers have specified that a process average of 135 lb-ft is desirable. For a simple random sample of 30 nuts that the machine has recently tightened, the sample mean is 137.0 lb-ft. Construct and interpret the 95% confidence interval for the current process mean. Discuss the possibility that the machine may be in need of adjustment to correct the process mean.

9.86 There were 904 new Subway Restaurants franchises opened during 2002. Suppose that Subway wished to survey a simple random sample of the new franchisees to find out what percentage of them were totally pleased with their relationship with the company. If Subway wanted to have 90% confidence in being within 3 percentage points of the population percentage who are pleased, how many of the new franchisees would have to be included in the sample? Source: Subway.com. June 13, 2003.

9.87 The accompanying data represent one-way commuting times (minutes) for a simple random sample of 15 persons who work at a large assembly plant. The data are also in file **XR09082**. Assuming an approximately normal distribution of commuting times for those who work at the plant, construct and interpret the 90% and 95% confidence intervals for the mean.

21.7	26.8	33.1	27.9	23.5
39.0	28.0	24.7	28.4	28.9
30.0	33.6	33.3	34.1	35.1

9.88 In Exercise 9.86, suppose Subway has carried out the study, using the sample size determined in that exercise, and 27.5% of the franchisees say they are pleased with their relationship with Subway. Construct and interpret the 95% confidence interval for the population percentage.

9.89 Working independently, each of two researchers has devised a sampling plan to be carried out for the purpose of constructing a 90% confidence interval for the mean of a certain population. What is the probability that neither of their confidence intervals will include the population mean?

9.90 A research firm wants to be 90% confident that a population percentage has been estimated to within 3 percentage points. The research manager calculates the necessary sample size with 0.5 as his estimate of the population proportion. A new business school graduate who has just joined the firm questions the research manager further, and they agree that the population proportion is no more than 0.3. If interviews cost $10 each, how much money has the new graduate just saved the company?

9.91 In a destructive test of product quality, a briefcase manufacturer places each of a simple random sample of the day's production in a viselike device and measures how many pounds it takes to crush the case. From past experience, the standard deviation has been found to be 21.5 pounds. For 35 cases randomly selected from today's production, the average breaking strength was 341.0 pounds. Construct and interpret the 99% confidence interval for the mean breaking strength of the briefcases produced today.

9.92 In a survey of 1320 executives who oversee corporate data systems, 24% said they had experienced losses caused by computer viruses during the past year. Assuming the executives were a simple random sample of all such executives, construct and interpret the 90% confidence interval for the population proportion who were monetarily harmed by computer viruses that year.

9.93 The Colgate-Palmolive Company has 36,600 employees. If the company wishes to estimate, within 2 percentage points and with 99% confidence, the percentage of employees who are interested in participating in a new stock option benefits program, how large a simple random sample will be necessary? Source: Colgate-Palmolive Company, *2008 Annual Report,* p. 2.

9.94 In a work-sampling study, an industrial engineer has observed the activities of a clerical worker on 121 randomly selected times during a workweek. On 32 of these occasions, the employee was talking on the telephone. For an 8-hour day, what are the upper and lower 95% confidence limits for the number of minutes this employee talks on the phone?

9.95 In Exercise 9.97, suppose the tool manufacturer has carried out the study, using the sample size determined in that exercise, and 39.0% of the machinery rebuilding and repairing companies are interested in the new tool design. Construct and interpret the 95% confidence interval for the population percentage.

9.96 A researcher would like to determine, within 3 percentage points and with 90% confidence, the percentage of Americans who have a certain characteristic. If she feels certain that the percentage is somewhere between 20% and 40%, how many persons should be included in the sample?

9.97 There are 1254 machinery rebuilding and repairing companies in the United States. A tool manufacturer wishes to survey a simple random sample of these firms to find out what proportion of them are interested in a new tool design. If the tool manufacturer would like to be 95% confident that the sample proportion is within 0.01 of the actual population proportion, how many machinery rebuilding and repairing companies should be included in the sample? Source: infousa.com, July 21, 2009.

9.98 The makers of Count Chocula breakfast cereal would like to determine, within 2 percentage points and with 99% confidence, the percentage of U.S. senior citizens who have Count Chocula for breakfast at least once a week. What sample size would you recommend?

9.99 A consultant conducts a pilot study to estimate a population standard deviation, then determines how large a simple random sample will be necessary to have a given level of confidence that the difference between $\bar{x}$ and μ will be within the maximum error specified by her client. The necessary sample size has been calculated as $n = 100$. If the client suddenly decides that the maximum error must be only one-fourth that originally specified, what sample size will now be necessary?

9.100 For a process having a known standard deviation, a simple random sample of 35 items is selected. If the width of the 95% confidence interval is identified as y, express the width of the 99% confidence interval as a multiple of y.

9.101 An airline would like to determine, within 3 percentage points and with 95% confidence, the percentage of next month's customers who judge the courtesy of its employees as being "very good to excellent." What sample size would you recommend?

9.102 A research firm has found that 39% of U.S. adults in the over-$75,000 income category work at least 51 hours per week. Assuming this was a simple random sample of 500 adults in this income group, construct and interpret the 95% and 99% confidence intervals for the proportion who work at least 51 hours per week. For each of the confidence intervals, identify and explain the maximum likely error in the study.

9.103 To gain information about competitors' products, companies sometimes employ "reverse engineering," which consists of buying the competitor's product, then taking it apart and examining the parts in great detail. Engaging in this practice, a bicycle manufacturer intends to buy two or more of a leading competitor's mountain bikes and measure the tensile strength of the crossbar portion of the frame. Past experience has shown these strengths to be approximately normally distributed with a standard deviation of 20 pounds per square inch (psi). If the bike purchaser wants to have 90% confidence that the sampling error will be no more than 5 psi, how many of the competitor's mountain bikes should be purchased for destructive testing?

9.104 The activities director of a large university has surveyed a simple random sample of 100 students for the purpose of determining approximately how many students to expect at next month's awards ceremony to be held in the gymnasium. Forty of the students said they plan to attend. What are the upper and lower 95% confidence limits for the number of the university's 10,000 students who plan to attend the awards ceremony?

9.105 A researcher has estimated that U.S. college students spend an average of 17.2 hours per week on the Internet. Assuming a simple random sample of 500 college students and a sample standard deviation of 1.4 hours per week, construct and interpret the 99% confidence interval for the population mean.

9.106 A truck loaded with 8000 electronic circuit boards has just pulled into a firm's receiving dock. The supplier claims that no more than 3% of the boards fall outside the most rigid level of industry performance specifications. In a simple random sample of 300 boards from this shipment, 12 fall outside these specifications. Construct the 95% confidence interval for the percentage of all boards in this shipment that fall outside the specifications, then comment on whether the supplier's claim would appear to be correct.

9.107 A researcher, believing π to be no more than 0.40, calculates the necessary sample size for the confidence level and maximum likely error he has specified. Upon completing the study, he finds the sample proportion to be 0.32. Is the maximum likely error greater than, equal to, or less than that originally specified? Explain.

9.108 A survey of business travelers found that 40% of those surveyed utilize hotel exercise facilities during their stay. Under the assumption that a simple random sample of 1000 business travelers were surveyed, construct and interpret the 90% and 95% confidence intervals for the proportion of business travelers who use their hotel's exercise facilities.

(DATA SET) *Note:* Exercises 9.109–9.111 require a computer and statistical software.

9.109 It has been estimated that the average dinner check at Morton's, the world's largest chain of upscale steakhouses, is $97 per person. Such a finding could have been based on data like the 800 sample checks in file **XR09109**. Using the data in this file, construct and interpret the 95% confidence interval for the population mean. Source: Bruce Horovitz, "Struggling Restaurant Chains Try Cheap Eats to Lure Diners," usatoday.com, July 21, 2009; and mortons.com, July 21, 2009.

9.110 According to the Internal Revenue Service, the average deduction for charitable contributions in the most recent tax year reported was $4388. Curious to see how his legislative district compares, a state senator surveys a simple random sample of 200 taxpayers from the district, with the data as shown in file **XR09110**. Using the data in this file, construct and interpret the 90% confidence interval for the population mean charitable-gifts contribution for taxpayers from the district. Is $4388 within the confidence interval? Given the answer to the preceding question, comment on whether the district's taxpayers might not be typical of those in the nation as a whole in terms of their tax-deductible charitable contributions. Source: Bureau of the Census, *Statistical Abstract of the United States 2009*, p. 364.

9.111 To avoid losing part of their federal highway fund allocation, state safety administrators must ensure that interstate speed limits are adequately enforced within their state. In an upcoming test, federal researchers will be randomly selecting and clocking a very large sample of vehicles on a given section of the state's portion of an interstate highway that has historically had a relatively high accident rate. In anticipation of the upcoming study, state administrators randomly select and clock 100 vehicles along this route, obtaining the speeds shown in data file **XR09111**. Construct and interpret the 95% confidence interval for the population mean vehicle speed along this stretch of highway. Based on this interval, comment on whether the mean speed for the population of vehicles using this part of the highway might be 70 mph, the cutoff above which federal highway funds become endangered.

INTEGRATED CASES

Thorndike Sports Equipment (Thorndike Video Unit Four)

Seeing the fishing pole in his grandfather's office, Ted Thorndike's first thought is that old Luke is going to go fishing again and leave him to manage the store. He is quite surprised to learn the fishing pole is actually an inspiration for a new series of ads that Luke has in mind.

The elder Thorndike explains, "Ted, this fishing pole is made of graphite, the same stuff that goes into our Graf-Pro racquetball racquets. It's so flexible and strong that it can be bent so the two ends actually touch each other. They even show this in the ads." Although Luke realizes that you can't do exactly the same thing with a racquetball racquet, he'd

(continued)

like to put some of his racquets into a horizontal mounting device, then see how much weight they'll take before they break.

If the amount of weight is impressive enough, Luke plans to include this kind of test in the television advertisements he's planning for the firm's racquetball racquets. However, he wants to be careful not to brag about the racquet being able to hold *too* much weight, since the firm could get into trouble with the government and other truth-in-advertising advocates.

He asks Ted to set up a test in which racquets are mounted horizontally, then the weight on the end is gradually increased until they break. Based on the test results, a weight value would be selected such that the average racquet would almost certainly be able to withstand this amount. Although accuracy is important, Ted has been instructed not to break more than 15 or 20 racquets in coming up with an average for all the racquets.

For 20 racquets subjected to this severe test, the weight (in pounds) at which each one failed was as follows. The data are also in file **THORN09**.

221	228	223	218	218
208	220	217	224	225
224	222	229	215	221
217	230	236	222	234

Ted believes it's reasonable to assume the population of breaking strengths is approximately normally distributed. Because of Luke's concern about being able to support the advertising claim, he wants to be very conservative in estimating the population mean for these breaking strengths. Ted needs some help in deciding how conservative he would like to be, and in coming up with a number that can be promoted in the ads.

Springdale Shopping Survey

The case in Chapter 2 listed 30 questions asked of 150 respondents in the community of Springdale. The coding key for the responses was also provided in that earlier exercise. The data are in file **SHOPPING**. In this exercise, some of the estimation techniques presented in the chapter will be applied to the survey results. You may assume that these respondents represent a simple random sample of all potential respondents within the community and that the population is large enough that application of the finite population correction would not make an appreciable difference in the results.

Managers associated with shopping areas like these find it useful to have point estimates regarding variables describing the characteristics and behaviors of their customers. In addition, it is helpful for them to have some idea as to the likely accuracy of these estimates. Therein lies the benefit of the techniques presented in this chapter and applied here.

1. Item C in the description of the data collection instrument lists variables 7, 8, and 9, which represent the respondent's general attitude toward each of the three shopping areas. Each of these variables has numerically equal distances between the possible responses, and for purposes of analysis they may be considered to be of the interval scale of measurement.
   a. Determine the point estimate, then construct the 95% confidence interval for μ_7 = the average attitude toward Springdale Mall. What is the maximum likely error in the point estimate of the population mean?
   b. Repeat part (a) for μ_8 and μ_9, the average attitudes toward Downtown and West Mall, respectively.
2. Given the breakdown of responses for variable 26 (sex of respondent), determine the point estimate, then construct the 95% confidence interval for π_{26} = the population proportion of males. What is the maximum likely error in the point estimate of the population proportion?
3. Given the breakdown of responses for variable 28 (marital status of respondent), determine the point estimate, then construct the 95% confidence interval for π_{28} = the population proportion in the "single or other" category. What is the maximum likely error in the point estimate of the population proportion?

SEEING STATISTICS: APPLET 9

Confidence Interval Size

This applet allows us to construct and view z-intervals for the population mean by using the slider to specify the confidence level. As in Figure 9.3, the sample mean is 1.400 inches, the sample size is 30, and the population standard deviation is known to be 0.053 inches.

Note that the confidence interval limits shown in the graph may sometimes differ slightly from those we would calculate using the pocket calculator and our standard normal distribution table. This is because the applet is using more exact values for z than we are able to show within printed tables like the one in the text.

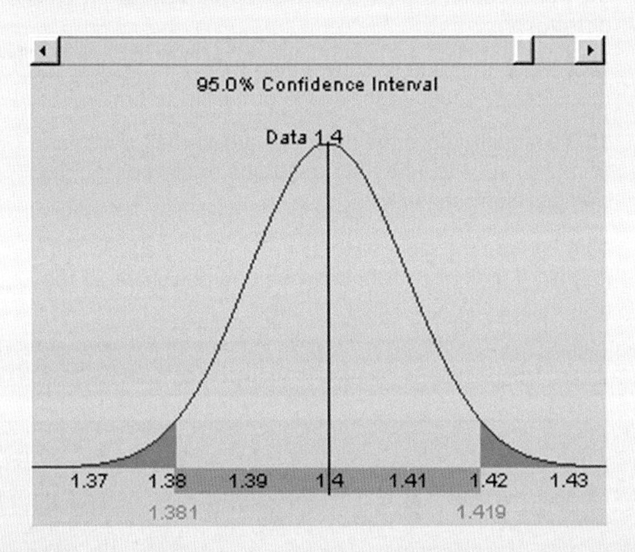

Applet Exercises

9.1 With the slider positioned so as to specify a 95% confidence interval for μ, what are the upper and lower confidence limits?

9.2 Move the slider so that the confidence interval is now 99%. Describe how the increase in the confidence level has changed the width of the confidence interval.

9.3 Move the slider so that the confidence interval is now 80%. Describe how the decrease in the confidence level has changed the width of the confidence interval.

9.4 Position the slider at its extreme left position, then gradually move it to the far right. Describe how this movement changes the confidence level and the width of the confidence interval.

Try out the Applets: http://www.cengage.com/international

SEEING STATISTICS: APPLET 10

Comparing the Normal and Student *t* Distributions

In this applet, we use a slider to change the number of degrees of freedom and shape for the Student *t* distribution and then observe how the resulting shape compares to that of the standard normal distribution. The standard normal distribution is fixed and shown in red, and the Student *t* distribution is displayed in blue.

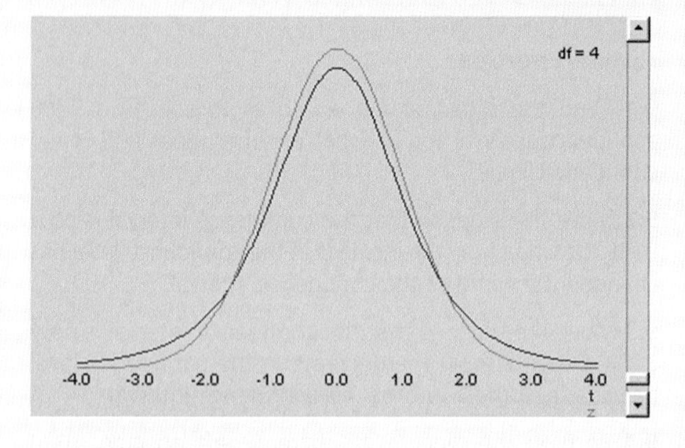

Applet Exercises

10.1 Move the slider so that $df = 5$. Describe the shape of the *t* distribution compared to that of the standard normal distribution.

10.2 Move the slider downward so that $df = 2$. How has this decrease changed the shape of the *t* distribution?

10.3 Gradually move the slider upward so that *df* increases from 2 to 10. Describe how the shape of the *t* distribution changes along the way.

10.4 Position the slider so that $df = 2$, then gradually move it upward until $df = 100$. Describe how the shape of the *t* distribution changes along the way.

Try out the Applets: http://www.cengage.com/international

SEEING STATISTICS: APPLET 11

Student *t* Distribution Areas

In this applet, we use a slider to change the number of degrees of freedom for the *t* distribution, and text boxes allow us to change the *t* value or the two-tail probability for a given *df*. When changing a text-box entry, be sure the cursor is still within the box before pressing the enter or return key.

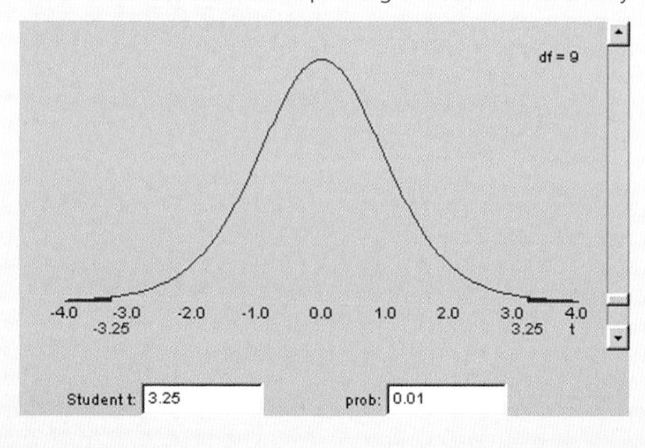

Applet Exercises

11.1 With the slider set so that $df = 9$ and the left text box containing $t = 3.25$, what is the area beneath the curve between $t = -3.25$ and $t = +3.25$?

11.2 Gradually move the slider upward until $df = 89$. What effect does this have on the *t* value shown in the text box?

11.3 Position the slider so that $df = 2$, then gradually move it upward until $df = 100$. Describe how the value in the *t* text box and the shape of the *t* distribution change along the way.

11.4 With the slider set so that $df = 9$, enter 0.10 into the two-tail probability text box at the right. What value of *t* now appears in the left text box? To what right-tail area does this correspond? Verify the value of *t* for $df = 9$ and this right-tail area by using the *t* table that precedes the rear endsheet of the book.

Try out the Applets: http://www.cengage.com/international

CHAPTER 10

Hypothesis Tests Involving a Sample Mean or Proportion

FAT-FREE OR REGULAR PRINGLES: CAN TASTERS TELL THE DIFFERENCE?

When the makers of Pringles potato chips came out with new Fat-Free Pringles, they wanted the fat-free chips to taste just as good as their already successful regular Pringles. Did they succeed? In an independent effort to answer this question, *USA Today* hired registered dietitian Diane Wilke to give 44 people a chance to see whether they could tell the difference between the two kinds of Pringles. Each tester was given two bowls of chips—one containing Fat-Free Pringles, the other containing regular Pringles—and nobody was told which was which.

On average, if the two kinds of chips really taste the same, we'd expect such testers to have a 50% chance of correctly identifying the bowl containing the fat-free chips. However, 25 of the 44 testers (56.8%) successfully identified the bowl with the fat-free chips.

Does this result mean that Pringles failed in its attempt to make the products taste the same, or could the difference between the observed 56.8% and the theoretical 50% have happened just by chance? Actually, if the chips really taste the same and we were to repeat this type of test many times, pure chance would lead to about 1/5 of the tests yielding a sample percentage at least as high as the 56.8% observed here. Thus, this particular test would not allow us to rule out the possibility that the chips taste the same. After reading Sections 10.3 and 10.6 of this chapter, you'll be able to verify how we reached this conclusion. For now, just trust us and read on. Thanks.

Source: Beth Ashley, "Taste Testers Notice Little Difference Between Products," *USA Today,* September 30, 1996, p. 6D. Interested readers may also refer to Fiona Haynes, "Do Low-Fat Foods Really Taste Different?", lowfatcooking.about.com, July 23, 2009.

Is the taster correct significantly more than half the time?

LEARNING OBJECTIVES

After reading this chapter, you should be able to:

- Describe the meaning of a null and an alternative hypothesis.
- Transform a verbal statement into appropriate null and alternative hypotheses, including the determination of whether a two-tail test or a one-tail test is appropriate.
- Describe what is meant by Type I and Type II errors, and explain how their probabilities can be reduced in hypothesis testing.
- Carry out a hypothesis test for a population mean or a population proportion, interpret the results of the test, and determine the appropriate business decision that should be made.
- Determine and explain the *p*-value for a hypothesis test.
- Explain how confidence intervals are related to hypothesis testing.
- Determine and explain the power curve for a hypothesis test and a given decision rule.
- Determine and explain the operating characteristic curve for a hypothesis test and a given decision rule.

10.1 INTRODUCTION

In statistics, as in life, nothing is as certain as the presence of uncertainty. However, just because we're not 100% sure of something, that's no reason why we can't reach some conclusions that are highly likely to be true. For example, if a coin were to land heads 20 times in a row, we might be wrong in concluding that it's unfair, but we'd still be wise to avoid engaging in gambling contests with its owner. In this chapter, we'll examine the very important process of reaching conclusions based on sample information—in particular, of evaluating hypotheses based on claims like the following:

- Titus Walsh, the director of a municipal transit authority, claims that 35% of the system's ridership consists of senior citizens. In a recent study, independent researchers find that only 23% of the riders observed are senior citizens. Should the claim of Walsh be considered false?
- Jackson T. Backus has just received a railroad car of canned beets from his grocery supplier, who claims that no more than 20% of the cans are dented. Jackson, a born skeptic, examines a random sample from the shipment and finds that 25% of the cans sampled are dented. Has Mr. Backus bought a batch of botched beets?

Each of the preceding cases raises a question of "believability" that can be examined by the techniques of this chapter. These methods represent *inferential statistics,* because information from a sample is used in reaching a conclusion about the population from which the sample was drawn.

Null and Alternative Hypotheses

The first step in examining claims like the preceding is to form a **null hypothesis,** expressed as H_0 ("H sub naught"). The null hypothesis is a statement about the value of a population parameter and is put up for testing in the face of numerical evidence. The null hypothesis is either rejected or fails to be rejected.

The null hypothesis tends to be a "business as usual, nothing out of the ordinary is happening" statement that practically invites you to challenge its truthfulness. In the philosophy of hypothesis testing, the null hypothesis is assumed to be true unless we have statistically overwhelming evidence to the contrary. In other words, it gets the benefit of the doubt.

The **alternative hypothesis,** H_1 ("H sub one"), is an assertion that holds *if* the null hypothesis is false. For a given test, the null and alternative hypotheses include all possible values of the population parameter, so either one or the other must be false.

There are three possible choices for the set of null and alternative hypotheses to be used for a given test. Described in terms of an (unknown) population mean (μ), they might be listed as shown below. Notice that each null hypothesis has an *equality* term in its statement (i.e., "=," "≥," or "≤").

Null Hypothesis	Alternative Hypothesis	
H_0: μ = \$10	H_1: $\mu \neq$ \$10	(μ is \$10, or it isn't.)
H_0: $\mu \geq$ \$10	H_1: $\mu <$ \$10	(μ is at least \$10, or it is less.)
H_0: $\mu \leq$ \$10	H_1: $\mu >$ \$10	(μ is no more than \$10, or it is more.)

Directional and Nondirectional Testing

A *directional* claim or assertion holds that a population parameter is *greater than* (>), *at least* (≥), *no more than* (≤), or *less than* (<) some quantity. For example, Jackson's supplier claims that no more than 20% of the beet cans are dented.

A *nondirectional* claim or assertion states that a parameter is *equal* to some quantity. For example, Titus Walsh claims that 35% of his transit riders are senior citizens.

Directional assertions lead to what are called **one-tail tests,** where a null hypothesis can be rejected by an extreme result in one direction only. A nondirectional assertion involves a **two-tail test,** in which a null hypothesis can be rejected by an extreme result occurring in either direction.

Hypothesis Testing and the Nature of the Test

When formulating the null and alternative hypotheses, the nature, or purpose, of the test must also be taken into account. To demonstrate how (1) directionality versus nondirectionality and (2) the purpose of the test can guide us toward the appropriate testing approach, we will consider the two examples at the beginning of the chapter. For each situation, we'll examine (1) the claim or assertion leading to the test, (2) the null hypothesis to be evaluated, (3) the alternative hypothesis, (4) whether the test will be two-tail or one-tail, and (5) a visual representation of the test itself.

Titus Walsh

1. Titus' assertion: "35% of the riders are senior citizens."
2. Null hypothesis: H_0: $\pi = 0.35$, where π = the population proportion. The null hypothesis is identical to his statement since he's claimed an exact value for the population parameter.
3. Alternative hypothesis: H_1: $\pi \neq 0.35$. If the population proportion is not 0.35, then it must be some other value.

FIGURE 10.1

Hypothesis tests can be two-tail (a) or one-tail (b), depending on the purpose of the test. A one-tail test can be either left-tail (not shown) or right-tail (b).

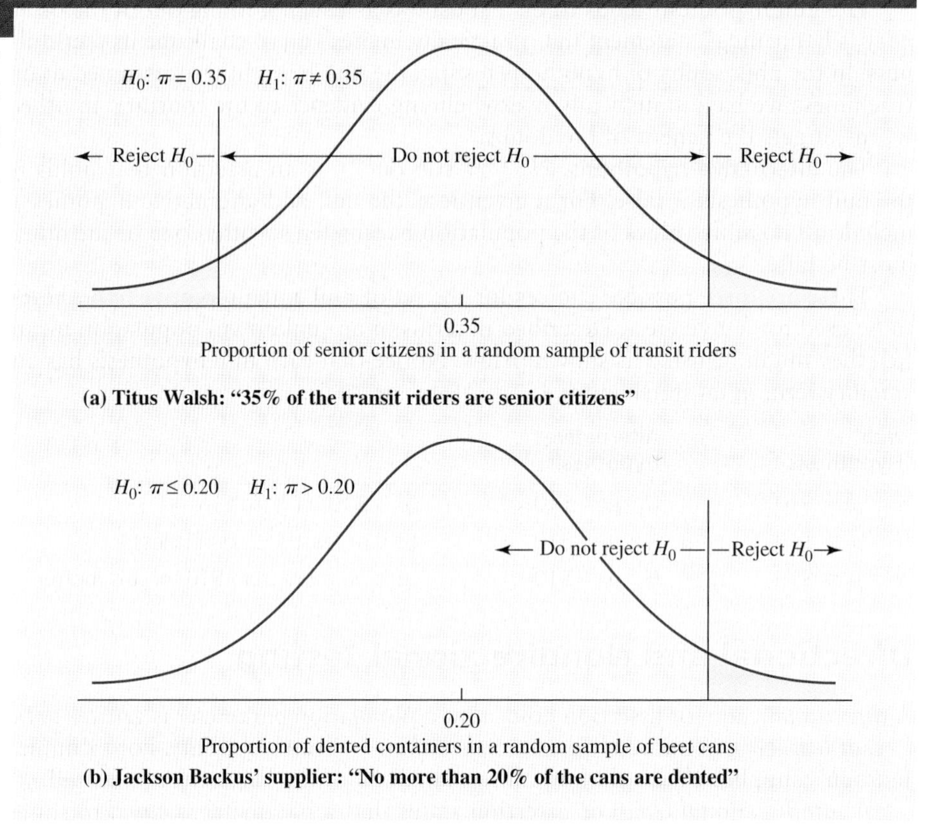

4. A two-tail test is used because the null hypothesis is nondirectional.
5. As part (a) of Figure 10.1 shows, $\pi = 0.35$ is at the center of the hypothesized distribution, and a sample with either a very high proportion or a very low proportion of senior citizens would lead to rejection of the null hypothesis. Accordingly, there are *reject* areas at both ends of the distribution.

Jackson T. Backus

1. Supplier's assertion: "No more than 20% of the cans are dented."
2. Null hypothesis: H_0: $\pi \leq 0.20$, where π = the population proportion. In this situation, the null hypothesis happens to be the same as the claim that led to the test. This is not always the case when the test involves a directional claim or assertion.
3. Alternative hypothesis: H_1: $\pi > 0.20$. Jackson's purpose in conducting the test is to determine whether the population proportion of dented cans could really be greater than 0.20.
4. A one-tail test is used because the null hypothesis is directional.
5. As part (b) of Figure 10.1 shows, a sample with a very high proportion of dented cans would lead to the rejection of the null hypothesis. A one-tail test in which the rejection area is at the right is known as a **right-tail test.** Note that in part (b) of Figure 10.1, the center of the hypothesized distribution is identified as $\pi = 0.20$. This is the highest value for which the null hypothesis could be true. From Jackson's standpoint, this may be viewed as somewhat conservative, but remember that the null hypothesis tends to get the benefit of the doubt.

TABLE 10.1

Categories of verbal statements and typical null and alternative hypotheses for each.

A. Verbal Statement Is an Equality, "="

Example: "Average tire life [is] 35,000 miles."

H_0: $\mu = 35{,}000$ miles

H_1: $\mu \neq 35{,}000$ miles

B. Verbal Statement is "≥" or "≤" (not > or <)

Example: "Average tire life [is at least] 35,000 miles."

H_0: $\mu \geq 35{,}000$ miles

H_1: $\mu < 35{,}000$ miles

Example: "Average tire life [is no more than] 35,000 miles."

H_0: $\mu \leq 35{,}000$ miles

H_1: $\mu > 35{,}000$ miles

In directional tests, the directionality of the null and alternative hypotheses will be in opposite directions and will depend on the purpose of the test. For example, in the case of Jackson Backus, Jackson was interested in rejecting H_0: $\pi \leq 0.20$ only if evidence suggested π to be higher than 0.20. As we proceed with the examples in the chapter, we'll get more practice in formulating null and alternative hypotheses for both nondirectional and directional tests. Table 10.1 offers general guidelines for proceeding from a verbal statement to typical null and alternative hypotheses.

Errors in Hypothesis Testing

Whenever we reject a null hypothesis, there is a chance that we have made a mistake—i.e., that we have rejected a true statement. Rejecting a true null hypothesis is referred to as a **Type I error,** and our probability of making such an error is represented by the Greek letter **alpha (α).** This probability, which is referred to as the **significance level** of the test, is of primary concern in hypothesis testing.

On the other hand, we can also make the mistake of failing to reject a false null hypothesis—this is a **Type II error.** Our probability of making it is represented by the Greek letter **beta (β).** Naturally, if we either fail to reject a true null hypothesis or reject a false null hypothesis, we've acted correctly. The probability of rejecting a false null hypothesis is called the **power of the test,** and it will be discussed in Section 10.7. The four possibilities are shown in Table 10.2. In hypothesis testing, there is a necessary trade-off between Type I and Type II errors: For a given sample size, reducing the probability of a Type I error increases the probability of a Type II error, and vice versa. The only sure way to avoid accepting false claims is to never accept *any* claims. Likewise, the only sure way to avoid rejecting true claims is to never reject any claims. Of course, each of these extreme approaches is impractical, and we must usually compromise by accepting a reasonable risk of committing either type of error.

TABLE 10.2

A summary of the possibilities for mistakes and correct decisions in hypothesis testing. The probability of incorrectly rejecting a true null hypothesis is α, the significance level. The probability that the test will correctly reject a false null hypothesis is $(1 - \beta)$, the power of the test.

Hypothesis Test Says	The Null Hypothesis (H_0) Is Really: True	False
"Do Not Reject H_0"	Correct decision.	Incorrect decision (Type II error). Probability of making this error is β.
"Reject H_0"	Incorrect decision (Type I error). Probability of making this error is α, the *significance level*.	Correct decision. Probability $(1 - \beta)$ is the *power of the test.*

EXERCISES

10.1 What is the difference between a null hypothesis and an alternative hypothesis? Is the null hypothesis always the same as the verbal claim or assertion that led to the test? Why or why not?

10.2 For each of the following pairs of null and alternative hypotheses, determine whether the pair would be appropriate for a hypothesis test. If a pair is deemed inappropriate, explain why.

a. H_0: $\mu \geq 10$, H_1: $\mu < 10$
b. H_0: $\mu = 30$, H_1: $\mu \neq 30$
c. H_0: $\mu > 90$, H_1: $\mu \leq 90$
d. H_0: $\mu \leq 75$, H_1: $\mu \leq 85$
e. H_0: $\bar{x} \geq 15$, H_1: $\bar{x} < 15$
f. H_0: $\bar{x} = 58$, H_1: $\bar{x} \neq 58$

10.3 For each of the following pairs of null and alternative hypotheses, determine whether the pair would be appropriate for a hypothesis test. If a pair is deemed inappropriate, explain why.

a. H_0: $\pi \geq 0.30$, H_1: $\pi < 0.35$
b. H_0: $\pi = 0.72$, H_1: $\pi \neq 0.72$
c. H_0: $\pi \leq 0.25$, H_1: $\pi > 0.25$
d. H_0: $\pi \geq 0.48$, H_1: $\pi > 0.48$
e. H_0: $\pi \leq 0.70$, H_1: $\pi > 0.70$
f. H_0: $p \geq 0.65$, H_1: $p < 0.65$

10.4 The president of a company that manufactures central home air conditioning units has told an investigative reporter that at least 85% of its homeowner customers claim to be "completely satisfied" with the overall purchase experience. If the reporter were to subject the president's statement to statistical scrutiny by questioning a sample of the company's residential customers, would the test be one-tail or two-tail? What would be the appropriate null and alternative hypotheses?

10.5 On CNN and other news networks, guests often express their opinions in rather strong, persuasive, and sometimes frightening terms. For example, a scientist who strongly believes that global warming is taking place will warn us of the dire consequences (such as rising sea levels, coastal flooding, and global climate change) she foresees if we do not take her arguments seriously. If the scientist is correct, and the world does not take her seriously, would this be a Type I error or a Type II error? Briefly explain your reasoning.

10.6 Many law enforcement agencies use voice-stress analysis to help determine whether persons under interrogation are lying. If the sound frequency of a person's voice changes when asked a question, the presumption is that the person is being untruthful. For this situation, state the null and alternative hypotheses in verbal terms, then identify what would constitute a Type I error and a Type II error in this situation.

10.7 Following a major earthquake, the city engineer must determine whether the stadium is structurally sound for an upcoming athletic event. If the null hypothesis is "the stadium is structurally sound," and the alternative hypothesis is "the stadium is not structurally sound," which type of error (Type I or Type II) would the engineer *least* like to commit?

10.8 A state representative is reported as saying that about 10% of reported auto thefts involve owners whose cars have not really been stolen, but who are trying to defraud their insurance company. What null and alternative hypotheses would be appropriate in evaluating the statement made by this legislator?

10.9 In response to the assertion made in Exercise 10.8, suppose an insurance company executive were to claim the percentage of fraudulent auto theft reports to be "no more than 10%." What null and alternative hypotheses would be appropriate in evaluating the executive's statement?

10.10 For each of the following statements, formulate appropriate null and alternative hypotheses. Indicate whether the appropriate test will be one-tail or two-tail, then sketch a diagram that shows the approximate location of the "rejection" region(s) for the test.

a. "The average college student spends no more than $300 per semester at the university's bookstore."
b. "The average adult drinks 1.5 cups of coffee per day."
c. "The average SAT score for entering freshmen is at least 1200."
d. "The average employee put in 3.5 hours of overtime last week."

10.11 In administering a "field sobriety" test to suspected drunks, officers may ask a person to walk in a straight line or close his eyes and touch his nose. Define the Type I and Type II errors in terms of this setting. Speculate on physiological variables (besides the drinking of alcoholic beverages) that might contribute to the chance of each type of error.

10.12 In the judicial system, the defense attorney argues for the null hypothesis that the defendant is innocent. In general, what would be the result if judges instructed juries to

a. *never* make a Type I error?
b. *never* make a Type II error?
c. compromise between Type I and Type II errors?

10.13 Regarding the testing of pharmaceutical companies' claims that their drugs are safe, a U.S. Food and Drug Administration official has said that it's "better to turn down 1000 good drugs than to approve one that's unsafe." If the null hypothesis is H_0: "The drug is not harmful," what type of error does the official appear to favor?

HYPOTHESIS TESTING: BASIC PROCEDURES (10.2)

There are several basic steps in hypothesis testing. They are briefly presented here and will be further explained through examples that follow.

1. *Formulate the null and alternative hypotheses.* As described in the preceding section, the null hypothesis asserts that a population parameter is *equal to*, *no more than*, or *no less than* some exact value, and it is evaluated in the face of numerical evidence. An appropriate alternative hypothesis covers other possible values for the parameter.
2. *Select the significance level.* If we end up rejecting the null hypothesis, there's a chance that we're wrong in doing so—i.e., that we've made a Type I error. The significance level is the maximum probability that we'll make such a mistake. In Figure 10.1, the significance level is represented by the shaded area(s) beneath each curve. For two-tail tests, the level of significance is the sum of *both* tail areas. In conducting a hypothesis test, we can choose any significance level we desire. In practice, however, levels of 0.10, 0.05, and 0.01 tend to be most common—in other words, if we reject a null hypothesis, the maximum chance of our being wrong would be 10%, 5%, or 1%, respectively. This significance level will be used to later identify the critical value(s).
3. *Select the test statistic and calculate its value.* For the tests of this chapter, the **test statistic** will be either z or t, corresponding to the normal and t distributions, respectively. Figure 10.2 shows how the test statistic is selected. An important consideration in tests involving a sample mean is whether the population standard deviation (σ) is known. As Figure 10.2 indicates, the **z-test** (normal distribution and test statistic, z) will be used for hypothesis tests involving a sample proportion.

FIGURE 10.2

An overview of the process of selecting a test statistic for single-sample hypothesis testing. Key assumptions are reviewed in the figure notes.

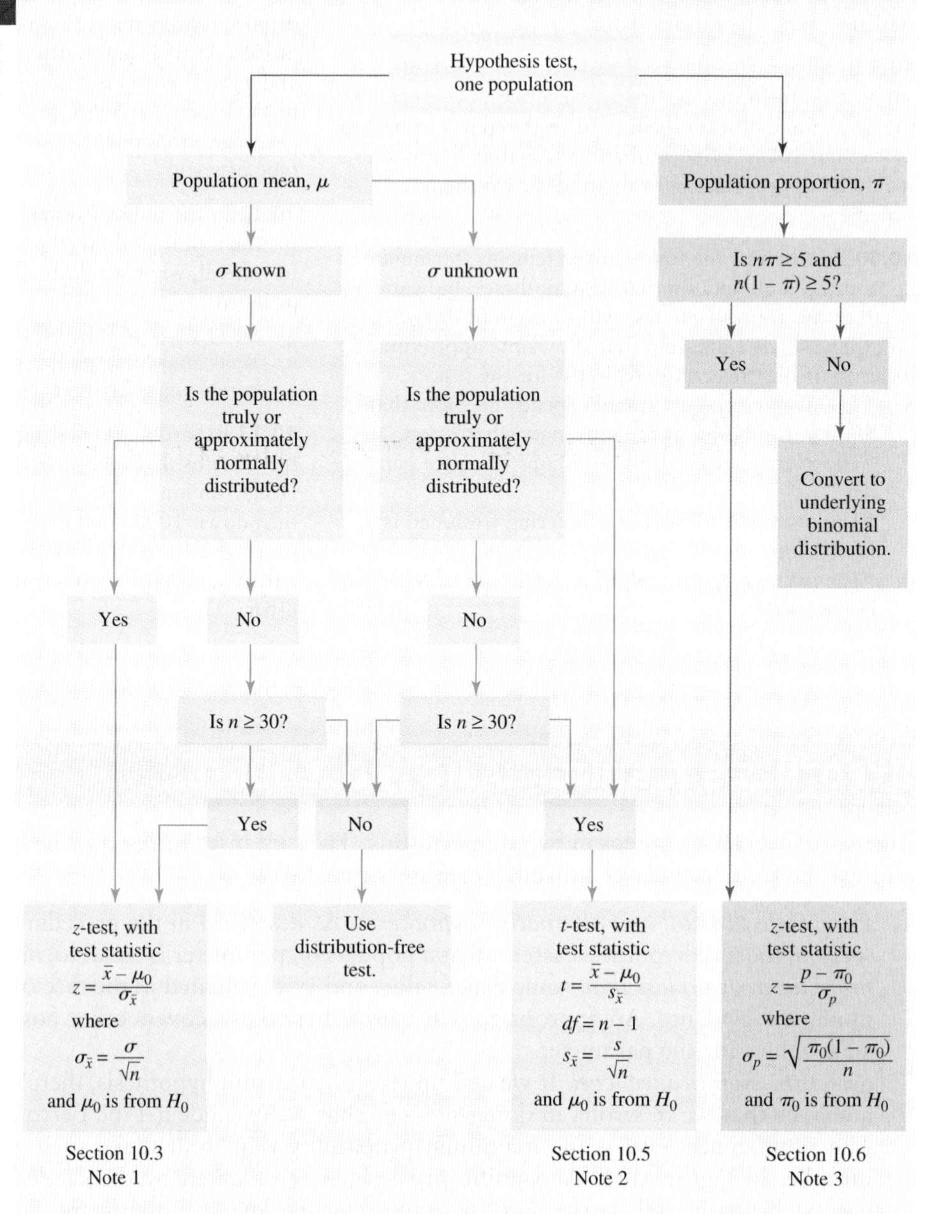

[1] The z distribution: If the population is not normally distributed, n should be ≥ 30 for the central limit theorem to apply. The population σ is usually not known.

[2] The t distribution: For an unknown σ, and when the population is approximately normally distributed, the t-test is appropriate *regardless of the sample size*. As n increases, the normality assumption becomes less important. If $n < 30$ and the population is not approximately normal, nonparametric testing (e.g., the sign test for central tendency, in Chapter 14) may be applied. The t-test is "robust" in terms of not being adversely affected by slight departures from the population normality assumption.

[3] When $n\pi \geq 5$ and $n(1 - \pi) \geq 5$, the normal distribution is considered to be a good approximation to the binomial distribution. If this condition is not met, the exact probabilities must be derived from the binomial distribution. Most practical business settings involving proportions satisfy this condition, and the normal approximation is used in this chapter.

4. *Identify critical value(s) for the test statistic and state the decision rule.* The **critical value(s)** will bound rejection and nonrejection regions for the null hypothesis, H_0. Such regions are shown in Figure 10.1. They are determined from the significance level selected in step 2. In a one-tail test, there will be one critical value since H_0 can be rejected by an extreme result in just one direction. Two-tail tests will require two critical values since H_0 can be rejected by an extreme result in either direction. If the null hypothesis were really true, there would still be some probability (the significance level, α) that the test statistic would be so extreme as to fall into a rejection region. The rejection and nonrejection regions can be stated as a **decision rule** specifying the conclusion to be reached for a given outcome of the test (e.g., "Reject H_0 if $z > 1.645$, otherwise do not reject").
5. *Compare calculated and critical values and reach a conclusion about the null hypothesis.* Depending on the **calculated value** of the test statistic, it will fall into either a rejection region or the nonrejection region. If the calculated value is in a rejection region, the null hypothesis will be rejected. Otherwise, the null hypothesis cannot be rejected. Failure to reject a null hypothesis does not constitute proof that it is true, but rather that we are unable to reject it at the level of significance being used for the test.
6. *Make the related business decision.* After rejecting or failing to reject the null hypothesis, the results are applied to the business decision situation that precipitated the test in the first place. For example, Jackson T. Backus may decide to return the entire shipment of beets to his distributor.

EXERCISES

10.14 A researcher wants to carry out a hypothesis test involving the mean for a sample of size $n = 18$. She does not know the true value of the population standard deviation, but is reasonably sure that the underlying population is approximately normally distributed. Should she use a z-test or a t-test in carrying out the analysis? Why?

10.15 A research firm claims that 62% of women in the 40–49 age group save in a 401(k) or individual retirement account. If we wished to test whether this percentage could be the same for women in this age group living in New York City and selected a random sample of 300 such individuals from New York, what would be the null and alternative hypotheses? Would the test be a z-test or a t-test? Why?

10.16 In hypothesis testing, what is meant by the *decision rule*? What role does it play in the hypothesis-testing procedure?

10.17 A manufacturer informs a customer's design engineers that the mean tensile strength of its rivets is at least 3000 pounds. A test is set up to measure the tensile strength of a sample of rivets, with the null and alternative hypotheses, H_0: $\mu \geq 3000$ and H_1: $\mu < 3000$. For each of the following individuals, indicate whether the person would tend to prefer a numerically very high (e.g., $\alpha = 0.20$) or a numerically very low (e.g., $\alpha = 0.0001$) level of significance to be specified for the test.

a. The marketing director for a major competitor of the rivet manufacturer.
b. The rivet manufacturer's advertising agency, which has already made the "at least 3000 pounds" claim in national ads.

10.18 It has been claimed that no more than 5% of the units coming off an assembly line are defective. Formulate a null hypothesis and an alternative hypothesis for this situation. Will the test be one-tail or two-tail? Why? If the test is one-tail, will it be left-tail or right-tail? Why?

10.3 TESTING A MEAN, POPULATION STANDARD DEVIATION KNOWN

Situations can occur where the population mean is unknown but past experience has provided us with a trustworthy value for the population standard deviation. Although this possibility is more likely in an industrial production setting, it can sometimes apply to employees, consumers, or other nonmechanical entities.

In addition to the assumption that σ is known, the procedure of this section assumes either (1) that the sample size is large ($n \geq 30$), or (2) that, if $n < 30$, the underlying population is normally distributed. These assumptions are summarized in Figure 10.2. If the sample size is large, the central limit theorem assures us that the sample means will be approximately normally distributed, regardless of the shape of the underlying distribution. The larger the sample size, the better this approximation becomes. Because it is based on the normal distribution, the test is known as the *z*-test, and the test statistic is as follows:

Test statistic, *z*-test for a sample mean:

$$z = \frac{\bar{x} - \mu_0}{\sigma_{\bar{x}}}$$

where $\sigma_{\bar{x}}$ = standard error for the sample mean, $= \sigma/\sqrt{n}$
$\bar{x}$ = sample mean
μ_0 = hypothesized population mean
n = sample size

The symbol μ_0 is the value of μ that is assumed for purposes of the hypothesis test.

Two-Tail Testing of a Mean, σ Known

EXAMPLE

Two-Tail Test

When a robot welder is in adjustment, its mean time to perform its task is 1.3250 minutes. Past experience has found the standard deviation of the cycle time to be 0.0396 minutes. An incorrect mean operating time can disrupt the efficiency of other activities along the production line. For a recent random sample of 80 jobs, the mean cycle time for the welder was 1.3229 minutes. The underlying data are in file **CX10WELD**. Does the machine appear to be in need of adjustment?

SOLUTION

Formulate the Null and Alternative Hypotheses

H_0: $\mu = 1.3250$ minutes The machine is in adjustment.
H_1: $\mu \neq 1.3250$ minutes The machine is out of adjustment.

In this test, we are concerned that the machine might be running at a mean speed that is either too fast or too slow. Accordingly, the null hypothesis could be

EXAMPLEEXAMPLEEXAMPLEEXAMP

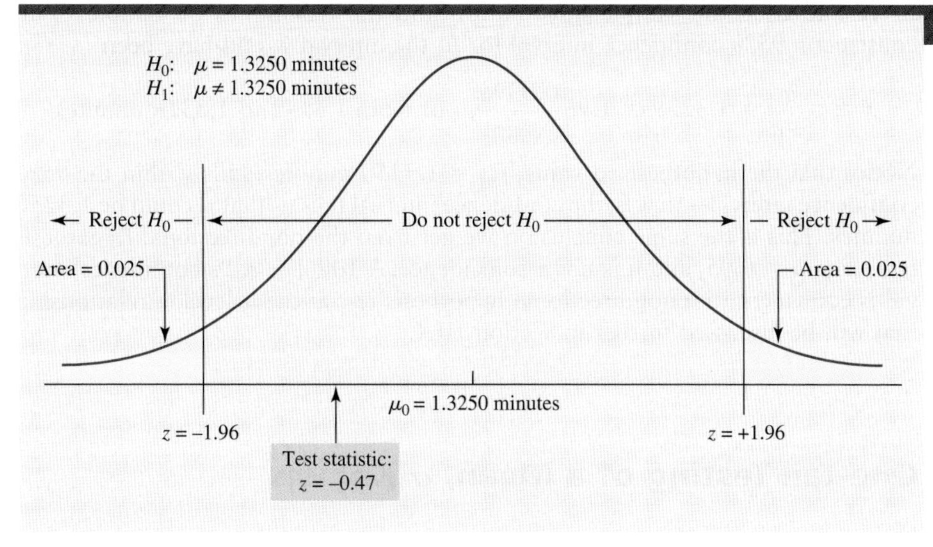

FIGURE 10.3

When the robot welder is in adjustment, the mean cycle time is 1.3250 minutes. This two-tail test at the 0.05 level of significance indicates that the machine is not out of adjustment.

rejected by an extreme sample result in either direction. The hypothesized value for the population mean is $\mu_0 = 1.3250$ minutes, shown at the center of the distribution in Figure 10.3.

Select the Significance Level

The significance level used will be $\alpha = 0.05$. If the machine is running properly, there is only a 0.05 probability of our making the mistake of concluding that it requires adjustment.

Select the Test Statistic and Calculate Its Value

The population standard deviation (σ) is known and the sample size is large, so the normal distribution is appropriate and the test statistic will be z, calculated as

$$z = \frac{\bar{x} - \mu_0}{\sigma_{\bar{x}}} = \frac{1.3229 - 1.3250}{0.0396/\sqrt{80}} = \frac{-0.0021}{0.00443} = -0.47$$

Identify Critical Values for the Test Statistic and State the Decision Rule

For a two-tail test using the normal distribution and $\alpha = 0.05$, $z = -1.96$ and $z = +1.96$ will be the respective boundaries for lower and upper tails of 0.025 each. These are the critical values for the test, and they identify the rejection and nonrejection regions shown in Figure 10.3. The decision rule can be stated as "Reject H_0 if calculated $z < -1.96$ or $> +1.96$, otherwise do not reject."

Compare Calculated and Critical Values and Reach a Conclusion for the Null Hypothesis

The calculated value, $z = -0.47$, falls within the nonrejection region of Figure 10.3. At the 0.05 level of significance, the null hypothesis cannot be rejected.

Make the Related Business Decision

Based on these results, the robot welder is not in need of adjustment. The difference between the hypothesized population mean, $\mu_0 = 1.3250$ minutes, and the observed sample mean, $\bar{x} = 1.3229$, is judged to have been merely the result of chance variation.

EXAMPLEEXAMPLEEXAMPLEEXAMPLEEXAMPLEEXAMPLE

If we had used the sample information and the techniques of Chapter 9 to construct a 95% confidence interval for μ, the interval would have been

$$\bar{x} \pm z\frac{\sigma}{\sqrt{n}} = 1.3229 \pm 1.96\frac{0.0396}{\sqrt{80}}, \quad \text{or from 1.3142 to 1.3316 minutes}$$

Notice that the hypothesized value, $\mu_0 = 1.3250$ minutes, falls within the 95% confidence interval—that is, the confidence interval tells us that μ could be 1.3250 minutes. This is the same conclusion we get from the nondirectional hypothesis test using $\alpha = 0.05$, and it is not a coincidence. A $100(1 - \alpha)$% confidence interval is equivalent to a nondirectional hypothesis test at the α level, a relationship that will be discussed further in Section 10.4.

One-Tail Testing of a Mean, σ Known

EXAMPLE

One-Tail Test

The lightbulbs in an industrial warehouse have been found to have a mean lifetime of 1030.0 hours, with a standard deviation of 90.0 hours. The warehouse manager has been approached by a representative of Extendabulb, a company that makes a device intended to increase bulb life. The manager is concerned that the average lifetime of Extendabulb-equipped bulbs might not be any greater than the 1030 hours historically experienced. In a subsequent test, the manager tests 40 bulbs equipped with the device and finds their mean life to be 1061.6 hours. The underlying data are in file **CX10BULB**. Does Extendabulb really work?

SOLUTION

Formulate the Null and Alternative Hypotheses

The warehouse manager's concern that Extendabulb-equipped bulbs might not be any better than those used in the past leads to a directional test. Accordingly, the null and alternative hypotheses are:

H_0: $\mu \leq 1030.0$ hours Extendabulb is no better than the present system.

H_1: $\mu > 1030.0$ hours Extendabulb really does increase bulb life.

At the center of the hypothesized distribution will be the highest possible value for which H_0 could be true, $\mu_0 = 1030.0$ hours.

Select the Significance Level

The level chosen for the test will be $\alpha = 0.05$. If Extendabulb really has no favorable effect, the maximum probability of our mistakenly concluding that it does will be 0.05.

Select the Test Statistic and Calculate Its Value

As in the previous test, the population standard deviation (σ) is known and the sample size is large, so the normal distribution is appropriate and the test statistic will be z. It is calculated as

$$z = \frac{\bar{x} - \mu_0}{\sigma_{\bar{x}}} = \frac{1061.6 - 1030.0}{90.0/\sqrt{40}} = 2.22$$

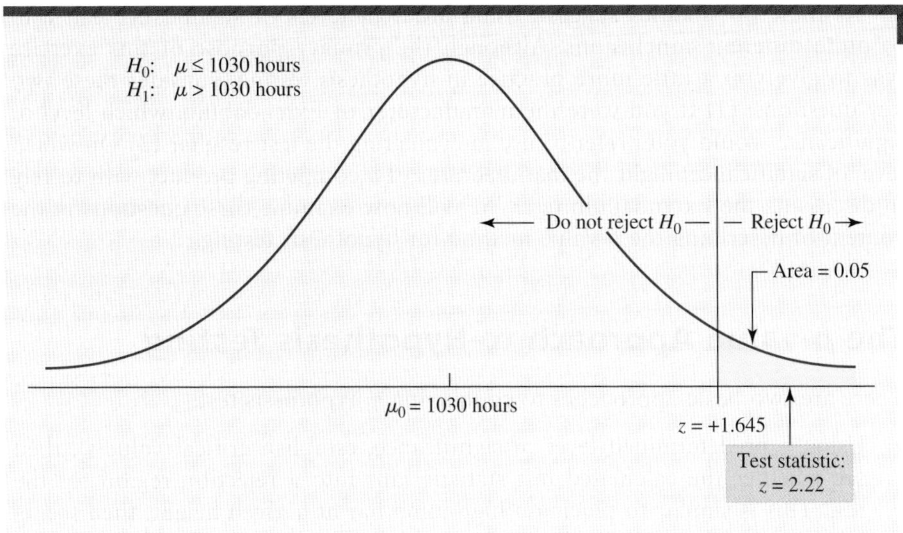

H_0: $\mu \leq 1030$ hours
H_1: $\mu > 1030$ hours

Do not reject H_0 | Reject H_0

Area = 0.05

$\mu_0 = 1030$ hours

$z = +1.645$

Test statistic: $z = 2.22$

FIGURE 10.4

The warehouse manager is concerned that Extendabulb might not increase the lifetime of lightbulbs. This right-tail test at the 0.05 level suggests otherwise.

Select the Critical Value for the Test Statistic and State the Decision Rule

For a right-tail z-test in which $\alpha = 0.05$, $z = +1.645$ will be the boundary separating the nonrejection and rejection regions. This critical value for the test is included in Figure 10.4. The decision rule can be stated as "Reject H_0 if calculated $z > +1.645$, otherwise do not reject."

Compare Calculated and Critical Values and Reach a Conclusion for the Null Hypothesis

The calculated value, $z = +2.22$, falls within the rejection region of the diagram in Figure 10.4. At the 0.05 level of significance, the null hypothesis is rejected.

Make the Related Business Decision

The results suggest that Extendabulb does increase the mean lifetime of the bulbs. The difference between the mean of the hypothesized distribution, $\mu_0 = 1030.0$ hours, and the observed sample mean, $\bar{x} = 1061.6$, is judged too great to have occurred by chance. The firm may wish to incorporate Extendabulb into its warehouse lighting system.

Other Levels of Significance

This test was conducted at the 0.05 level, but would the conclusion have been different if other levels of significance had been used instead? Consider the following possibilities:

- For the **0.05 level of significance** at which the test was conducted. The critical z is $+1.645$, and the calculated value, $z = 2.22$, exceeds it. The null hypothesis is rejected, and we conclude that Extendabulb does increase bulb life.
- For the **0.025 level of significance.** The critical z is $+1.96$, and the calculated value, $z = 2.22$, exceeds it. The null hypothesis is rejected, and we again conclude that Extendabulb increases bulb life.
- For the **0.005 level of significance.** The critical z is $+2.58$, and the calculated value, $z = 2.22$, does not exceed it. The null hypothesis is not rejected, and we conclude that Extendabulb does *not* increase bulb life.

As these possibilities suggest, using different levels of significance can lead to quite different conclusions. Although the primary purpose of this exercise was to give you a little more practice in hypothesis testing, consider these two key questions: (1) If you were the manufacturer of Extendabulb, which level of significance would you prefer to use in evaluating the test results? (2) On which level of significance might the manufacturer of a competing product wish to rely in discussing the Extendabulb test? We will now examine these questions in the context of describing the *p*-value method for hypothesis testing.

The *p*-value Approach to Hypothesis Testing

There are two basic approaches to conducting a hypothesis test:

- Using a predetermined level of significance, establish critical value(s), then see whether the calculated test statistic falls into a rejection region for the test. This is similar to placing a high-jump bar at a given height, then seeing whether you can clear it.
- Determine the exact level of significance associated with the calculated value of the test statistic. In this case, we're identifying the most extreme critical value that the test statistic *would be capable of exceeding*. This is equivalent to your jumping as high as you can with no bar in place, then having the judges tell you how high you *would have cleared* if there had been a crossbar.

In the two tests carried out previously, we used the first of these approaches, making the hypothesis test a "yes–no" decision. In the Extendabulb example, however, we did allude to what we're about to do here by trying several different significance levels in our one-tail test examining the ability of Extendabulb to increase the lifetime of lightbulbs.

We saw that Extendabulb showed a significant improvement at the 0.05 and 0.025 levels, but was not shown to be effective at the 0.005 level. In our high-jumping analogy, we might say that Extendabulb "cleared the bar" at the 0.05 level, cleared it again when it was raised to the more demanding 0.025 level, but couldn't quite make the grade when the bar was raised to the very demanding 0.005 level of significance. In summary:

- **0.05 level** Extendabulb significantly increases bulb life (e.g., "clears the high-jump bar").
- **0.025 level** Extendabulb significantly increases bulb life ("clears the bar").
- ***p*-value level** Extendabulb just barely shows significant improvement in bulb life ("clears the bar, but lightly touches it on the way over").
- **0.005 level** Extendabulb shows no significant improvement in bulb life ("insufficient height, fails to clear").

As suggested by the preceding, and illustrated in part (a) of Figure 10.5, there is some level of significance (the ***p*-value**) where the calculated value of the test statistic is exactly the same as the critical value. For a given set of data, the *p*-value is sometimes referred to as the *observed* level of significance. It is the lowest possible level of significance at which the null hypothesis can be rejected. (*Note:* The lowercase *p* in "*p*-value" is not related to the symbol for the sample proportion.) For the Extendabulb test, the calculated value of the test statistic was $z = 2.22$. For a critical $z = +2.22$, the right-tail area can be found using the normal distribution table at the back of the book.

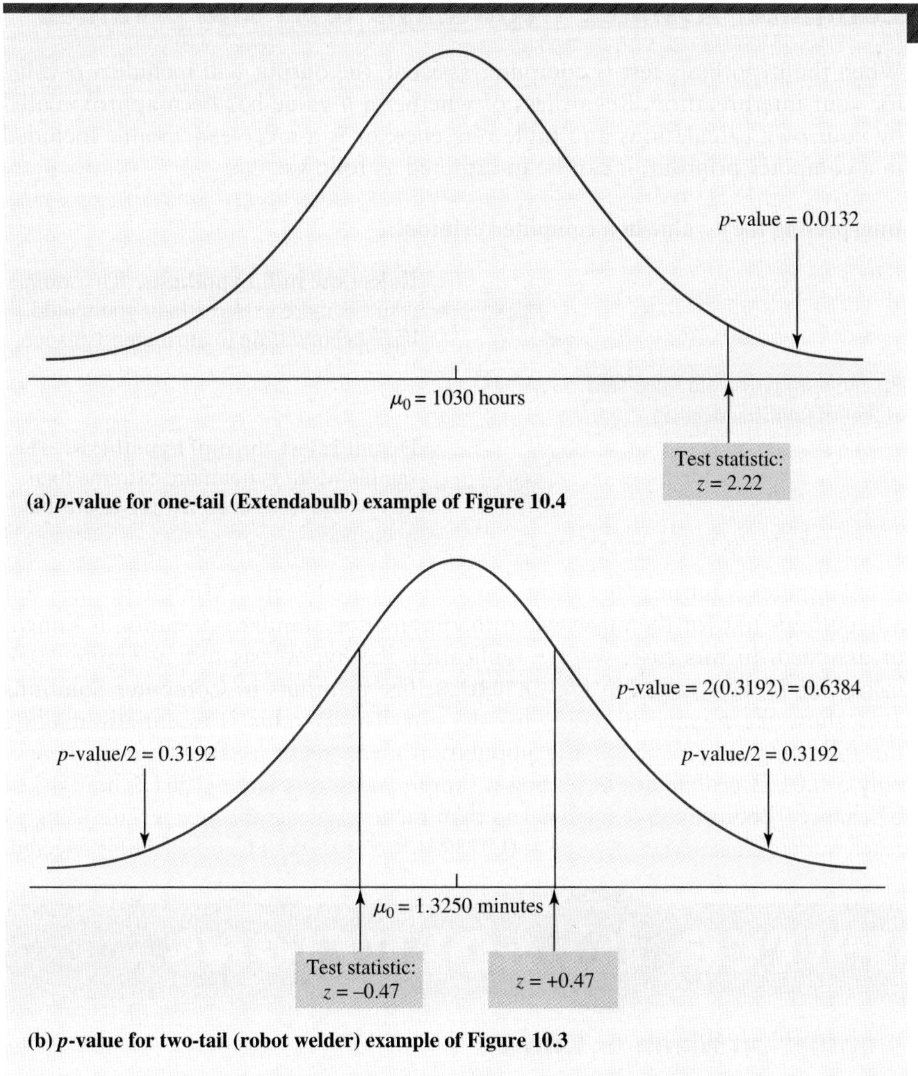

FIGURE 10.5

The *p*-value of a test is the level of significance where the observed value of the test statistic is exactly the same as a critical value for that level. These diagrams show the *p*-values, as calculated in the text, for two of the tests performed in this section. When the hypothesis test is two-tail, as in part (b), the *p*-value is the sum of two tail areas.

Referring to the normal distribution table, we see that 2.22 standard error units to the right of the mean includes a cumulative area of 0.9868, leaving (1.0000 − 0.9868), or 0.0132, in the right-tail area. This identifies the most demanding level of significance that Extendabulb could have achieved. If we had originally specified a significance level of 0.0132 for our test, the critical value for z would have been exactly the same as the value calculated. Thus, the p-value for the Extendabulb test is found to be 0.0132.

The Extendabulb example was a one-tail test—accordingly, the p-value was the area in just one tail. For two-tail tests, such as the robot welder example of Figure 10.3, the p-value will be the sum of both tail areas, as shown in part (b) of Figure 10.5. The calculated test statistic was $z = -0.47$, resulting in a cumulative area of 0.3192 in the left tail of the distribution. Since the robot welder test was two-tail, the 0.3192 must be multiplied by 2 to get the p-value of 0.6384.

Computer-Assisted Hypothesis Tests and *p*-values

When the hypothesis test is computer-assisted, the output will include a p-value for your interpretation. Regardless of whether a p-value has been approximated by your own calculations and table reference, or is a more exact value included in a computer printout, it can be interpreted as follows:

Interpreting the p-value in a computer printout:

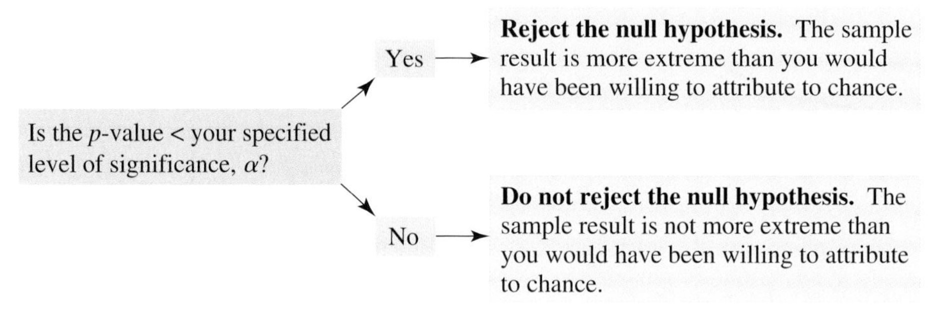

Computer Solutions 10.1 shows how we can use Excel or Minitab to carry out a hypothesis test for the mean when the population standard deviation is known or assumed. In this case, we are replicating the hypothesis test in Figure 10.4, using the 40 data values in file CX10BULB. The printouts in Computer Solutions 10.1 show the p-value (0.0132) for the test. This p-value is essentially making the following statement: "If the population mean really is 1030 hours, there is only a 0.0132 probability of getting a sample mean this large (1061.6 hours) just by chance." Because the p-value is less than the level of significance we are using to reach our conclusion (i.e., p-value $= 0.0132$ is $< \alpha = 0.05$), H_0: $\mu \leq 1030$ is rejected.

COMPUTER 10.1 SOLUTIONS

Hypothesis Test for Population Mean, σ Known

These procedures show how to carry out a hypothesis test for the population mean when the population standard deviation is known.

EXCEL

	A	B	C	D
1	Z-Test: Mean			
2				hours
3	Mean			1061.61
4	Standard Deviation			93.60
5	Observations			40
6	Hypothesized Mean			1030
7	SIGMA			90
8	z Stat			2.221
9	P(Z<=z) one-tail			0.0132
10	z Critical one-tail			1.645
11	P(Z<=z) two-tail			0.0264
12	z Critical two-tail			1.96

Excel hypothesis test for μ based on raw data and σ known

1. For example, for the 40 bulb lifetimes (file **CX10BULB**) on which Figure 10.4 is based, with the label and 40 data values in **A1:A41:** From the **Add-Ins** ribbon, click **Data Analysis Plus.** Click ***Z*-Test: Mean.** Click **OK.**
2. Enter **A1:A41** into the **Input Range** box. Enter the hypothesized mean **(1030)** into the **Hypothesized Mean** box. Enter the known population standard deviation **(90.0)** into the **Standard Deviation (SIGMA)** box. Click **Labels,** since the variable name is in the first cell within the field. Enter the level of significance for the test **(0.05)** into the **Alpha** box. Click **OK.** The printout includes the *p*-value for this one-tail test, 0.0132.

Excel hypothesis test for μ based on summary statistics and σ known

1. For example, with $\bar{x} = 1061.6$, $\sigma = 90.0$, and $n = 40$, as in Figure 10.4: Open the **TEST STATISTICS** workbook.
2. Using the arrows at the bottom left, select the **z-Test_Mean** worksheet. Enter the sample mean **(1061.6),** the known sigma **(90.0),** the sample size **(40),** the hypothesized population mean **(1030),** and the level of significance for the test **(0.05).**

(*Note:* As an alternative, you can use Excel worksheet template **TMZTEST**. The steps are described within the template.)

MINITAB

Minitab hypothesis test for μ based on raw data and σ known

```
One-Sample Z: hours

Test of mu = 1030 vs > 1030
The assumed standard deviation = 90

                                           95% Lower
Variable    N     Mean   StDev   SE Mean       Bound     Z       P
hours      40   1061.6    93.6      14.2      1038.2  2.22   0.013
```

1. For example, using the data (file **CX10BULB**) on which Figure 10.4 is based, with the 40 data values in column **C1:** Click **Stat.** Select **Basic Statistics.** Click **1-Sample Z.**
2. Select **Samples in Columns** and enter **C1** into the box. Enter the known population standard deviation **(90.0)** into the **Standard deviation** box. Select **Perform hypothesis test** and enter the hypothesized population mean **(1030)** into the **Hypothesized mean:** box.
3. Click **Options**. Enter the desired confidence level as a percentage **(95.0)** into the **Confidence Level** box. Within the **Alternative** box, select **greater than**. Click **OK.** Click **OK.** By default, this test also provides the lower boundary of the 95% confidence interval (unless another confidence level has been specified).

Minitab hypothesis test for μ based on summary statistics and σ known

Follow the procedure in steps 1 through 3, above, but in step 2 select **Summarized data** and enter **40** and **1061.6** into the **Sample size** and **Mean** boxes, respectively.

EXERCISES

10.19 What is the central limit theorem, and how is it applicable to hypothesis testing?

10.20 If the population standard deviation is known, but the sample size is less than 30, what assumption is necessary to use the *z*-statistic in carrying out a hypothesis test for the population mean?

10.21 What is a *p*-value, and how is it relevant to hypothesis testing?

10.22 The p-value for a hypothesis test has been reported as 0.03. If the test result is interpreted using the $\alpha = 0.05$ level of significance as a criterion, will H_0 be rejected? Explain.

10.23 The p-value for a hypothesis test has been reported as 0.04. If the test result is interpreted using the $\alpha = 0.01$ level of significance as a criterion, will H_0 be rejected? Explain.

10.24 A hypothesis test is carried out using the $\alpha = 0.01$ level of significance, and H_0 cannot be rejected. What is the most accurate statement we can make about the p-value for this test?

10.25 For each of the following tests and z values, determine the p-value for the test:

a. Right-tail test and $z = 1.54$
b. Left-tail test and $z = -1.03$
c. Two-tail test and $z = -1.83$

10.26 For each of the following tests and z values, determine the p-value for the test:

a. Left-tail test and $z = -1.62$
b. Right-tail test and $z = 1.43$
c. Two-tail test and $z = 1.27$

10.27 For a sample of 35 items from a population for which the standard deviation is $\sigma = 20.5$, the sample mean is 458.0. At the 0.05 level of significance, test H_0: $\mu = 450$ versus H_1: $\mu \neq 450$. Determine and interpret the p-value for the test.

10.28 For a sample of 12 items from a normally distributed population for which the standard deviation is $\sigma = 17.0$, the sample mean is 230.8. At the 0.05 level of significance, test H_0: $\mu \leq 220$ versus H_1: $\mu > 220$. Determine and interpret the p-value for the test.

10.29 A quality-assurance inspector periodically examines the output of a machine to determine whether it is properly adjusted. When set properly, the machine produces nails having a mean length of 2.000 inches, with a standard deviation of 0.070 inches. For a sample of 35 nails, the mean length is 2.025 inches. Using the 0.01 level of significance, examine the null hypothesis that the machine is adjusted properly. Determine and interpret the p-value for the test.

10.30 In the past, patrons of a cinema complex have spent an average of \$5.00 for popcorn and other snacks, with a standard deviation of \$1.80. The amounts of these expenditures have been normally distributed. Following an intensive publicity campaign by a local medical society, the mean expenditure for a sample of 18 patrons is found to be \$4.20. In a one-tail test at the 0.05 level of significance, does this recent experience suggest a decline in spending? Determine and interpret the p-value for the test.

10.31 Following maintenance and calibration, an extrusion machine produces aluminum tubing with a mean outside diameter of 2.500 inches, with a standard deviation of 0.027 inches. As the machine functions over an extended number of work shifts, the standard deviation remains unchanged, but the combination of accumulated deposits and mechanical wear causes the mean diameter to "drift" away from the desired 2.500 inches. For a recent random sample of 34 tubes, the mean diameter was 2.509 inches. At the 0.01 level of significance, does the machine appear to be in need of maintenance and calibration? Determine and interpret the p-value for the test.

10.32 A manufacturer of electronic kits has found that the mean time required for novices to assemble its new circuit tester is 3 hours, with a standard deviation of 0.20 hours. A consultant has developed a new instructional booklet intended to reduce the time an inexperienced kit builder will need to assemble the device. In a test of the effectiveness of the new booklet, 15 novices require a mean of 2.90 hours to complete the job. Assuming the population of times is normally distributed, and using the 0.05 level of significance, should we conclude that the new booklet is effective? Determine and interpret the p-value for the test.

(DATA SET) *Note:* Exercises 10.33 and 10.34 require a computer and statistical software.

10.33 According to bankrate.com, the average cost to remodel a home office is \$10,526. Assuming a population standard deviation of \$2000 and the sample of home office conversion prices charged for 40 recent jobs performed by builders in a region of the United States, examine whether the mean price for home office conversions for builders in this region might be different from the average for the nation as a whole. The underlying data are in file **XR10033**. Identify and interpret the p-value for the test. Using the 0.05 level of significance, what conclusion will be reached?
Source: bankrate.com, July 23, 2009.

10.34 A machine that fills shipping containers with driveway filler mix is set to deliver a mean fill weight of 70.0 pounds. The standard deviation of fill weights delivered by the machine is known to be 1.0 pounds. For a recent sample of 35 containers, the fill weights are listed in data file **XR10034**. Using the mean for this sample, and assuming that the population standard deviation has remained unchanged at 1.0 pounds, examine whether the mean fill weight delivered by the machine might now be something other than 70.0 pounds. Identify and interpret the p-value for the test. Using the 0.05 level of significance, what conclusion will be reached?

CONFIDENCE INTERVALS AND HYPOTHESIS TESTING (10.4)

In Chapter 9, we constructed confidence intervals for a population mean or proportion. In this chapter, we sometimes carry out nondirectional tests for the null hypothesis that the population mean or proportion could have a given value. Although the purposes may differ, the concepts are related.

In the previous section, we briefly mentioned this relationship in the context of the nondirectional test summarized in Figure 10.3. Consider this nondirectional test, carried out at the $\alpha = 0.05$ level:

1. Null and alternative hypotheses: H_0: $\mu = 1.3250$ minutes and H_1: $\mu \neq 1.3250$ minutes.
2. The standard error of the mean: $\sigma_{\bar{x}} = \sigma/\sqrt{n} = 0.0396/\sqrt{80}$, or 0.00443 minutes.
3. The critical z values for a two-tail test at the $\alpha = 0.05$ level are $z = -1.96$ and $z = +1.96$.
4. Expressing these z values in terms of the sample mean, critical values for $\bar{x}$ would be calculated as $1.325 \pm 1.96(0.00443)$, or 1.3163 minutes and 1.3337 minutes.
5. The observed sample mean was $\bar{x} = 1.3229$ minutes. This fell within the acceptable limits and we were not able to reject H_0.

Based on the $\alpha = 0.05$ level, the nondirectional hypothesis test led us to conclude that H_0: $\mu = 1.3250$ minutes was believable. The observed sample mean (1.3229 minutes) was close enough to the 1.3250 hypothesized value that the difference could have happened by chance.

Now let's approach the same situation by using a 95% confidence interval. As noted previously, the standard error of the sample mean is 0.00443 minutes. Based on the sample results, the 95% confidence interval for μ is $1.3229 \pm 1.96(0.00443)$, or from 1.3142 minutes to 1.3316 minutes. In other words, we have 95% confidence that the population mean is somewhere between 1.3142 minutes and 1.3316 minutes. If someone were to suggest that the population mean were actually 1.3250 minutes, we would find this believable, since 1.3250 falls within the likely values for μ that our confidence interval represents.

The nondirectional hypothesis test was done at the $\alpha = 0.05$ level, the confidence interval was for the 95% confidence level, and the conclusion was the same in each case. As a general rule, we can state that *the conclusion from a nondirectional hypothesis test for a population mean at the α level of significance will be the same as the conclusion based on a confidence interval at the $100(1 - \alpha)$% confidence level.*

When a hypothesis test is nondirectional, this equivalence will be true. This exact statement cannot be made about confidence intervals and *directional* tests—although they can also be shown to be related, such a demonstration would take us beyond the purposes of this chapter. Suffice it to say that confidence intervals and hypothesis tests are both concerned with using sample information to make a statement about the (unknown) value of a population mean or proportion. Thus, it is not surprising that their results are related.

By using Seeing Statistics Applet 12, at the end of the chapter, you can see how the confidence interval (and the hypothesis test conclusion) would change in response to various possible values for the sample mean.

EXERCISES

10.35 Based on sample data, a confidence interval has been constructed such that we have 90% confidence that the population mean is between 120 and 180. Given this information, provide the conclusion that would be reached for each of the following hypothesis tests at the $\alpha = 0.10$ level:

a. H_0: $\mu = 170$ versus H_1: $\mu \neq 170$
b. H_0: $\mu = 110$ versus H_1: $\mu \neq 110$
c. H_0: $\mu = 130$ versus H_1: $\mu \neq 130$
d. H_0: $\mu = 200$ versus H_1: $\mu \neq 200$

10.36 Given the information in Exercise 10.27, construct a 95% confidence interval for the population mean, then reach a conclusion regarding whether μ could actually be equal to the value that has been hypothesized. How does this conclusion compare to that reached in Exercise 10.27? Why?

10.37 Given the information in Exercise 10.29, construct a 99% confidence interval for the population mean, then reach a conclusion regarding whether μ could actually be equal to the value that has been hypothesized. How does this conclusion compare to that reached in Exercise 10.29? Why?

10.38 Use an appropriate confidence interval in reaching a conclusion regarding the problem situation and null hypothesis for Exercise 10.31.

10.5 TESTING A MEAN, POPULATION STANDARD DEVIATION UNKNOWN

The true standard deviation of a population will usually be unknown. As Figure 10.2 shows, the **t-test** is appropriate for hypothesis tests in which the sample standard deviation (s) is used in estimating the value of the population standard deviation, σ. The t-test is based on the t distribution (with number of degrees of freedom, $df = n - 1$) and the assumption that the population is approximately normally distributed. As the sample size becomes larger, the assumption of population normality becomes less important.

As we observed in Chapter 9, the t distribution is a *family* of distributions (one for each number of degrees of freedom, df). When df is small, the t distribution is flatter and more spread out than the normal distribution, but for larger degrees of freedom, successive members of the family more closely approach the normal distribution. As the number of degrees of freedom approaches *infinity,* the two distributions become identical.

Like the z-test, the t-test depends on the sampling distribution for the sample mean. The appropriate test statistic is similar in appearance, but includes s instead of σ, because s is being used to estimate the (unknown) value of σ. The test statistic can be calculated as follows:

Test statistic, *t*-test for a sample mean:

$$t = \frac{\bar{x} - \mu_0}{s_{\bar{x}}}$$

where $s_{\bar{x}}$ = estimated standard error for the sample mean, $= s/\sqrt{n}$
$\bar{x}$ = sample mean
μ_0 = hypothesized population mean
n = sample size

Two-Tail Testing of a Mean, σ Unknown

EXAMPLE

Two-Tail Test

The credit manager of a large department store claims that the mean balance for the store's charge account customers is $410. An independent auditor selects a random sample of 18 accounts and finds a mean balance of $\bar{x}$ = $511.33 and a standard deviation of s = $183.75. The sample data are in file **CX10CRED**. If the manager's claim is not supported by these data, the auditor intends to examine all charge account balances. If the population of account balances is assumed to be approximately normally distributed, what action should the auditor take?

SOLUTION

Formulate the Null and Alternative Hypotheses

H_0: $\mu = \$410$ The mean balance is actually $410.

H_1: $\mu \neq \$410$ The mean balance is some other value.

In evaluating the manager's claim, a two-tail test is appropriate since it is a nondirectional statement that could be rejected by an extreme result in either direction. The center of the hypothesized distribution of sample means for samples of $n = 18$ will be $\mu_0 = \$410$.

Select the Significance Level

For this test, we will use the 0.05 level of significance. The sum of the two tail areas will be 0.05.

Select the Test Statistic and Calculate Its Value

The test statistic is $t = (\bar{x} - \mu_0)/s_{\bar{x}}$, and the t distribution will be used to describe the sampling distribution of the mean for samples of $n = 18$. The center of the distribution is $\mu_0 = \$410$, which corresponds to $t = 0.000$. Since the population standard deviation is unknown, s is used to estimate σ. The sampling distribution has an estimated standard error of

$$s_{\bar{x}} = \frac{s}{\sqrt{n}} = \frac{\$183.75}{\sqrt{18}} = \$43.31$$

and the calculated value of t will be

$$t = \frac{\bar{x} - \mu_0}{s_{\bar{x}}} = \frac{\$511.33 - \$410.00}{\$43.31} = 2.340$$

Identify Critical Values for the Test Statistic and State the Decision Rule

For this test, $\alpha = 0.05$, and the number of degrees of freedom will be $df = (n - 1)$, or $(18 - 1) = 17$. The t distribution table at the back of the book provides one-tail areas, so we must identify the boundaries where each tail area is one-half of α, or 0.025. Referring to the 0.025 column and 17th row of the table, we find the critical values for the test statistic to be $t = -2.110$ and $t = +2.110$. (Although the "-2.110" is not shown in the table, we can identify this as the left-tail boundary because the distribution is symmetrical.) The rejection and nonrejection areas

EXAMPLEEXAMPLEEXAMPLEEXAMPLEEXAMPLEEXAMPLEEXAMPLEEXAMPLEEXAMP

FIGURE 10.6

The credit manager has claimed that the mean balance of his charge customers is $410, but the results of this two-tail test suggest otherwise.

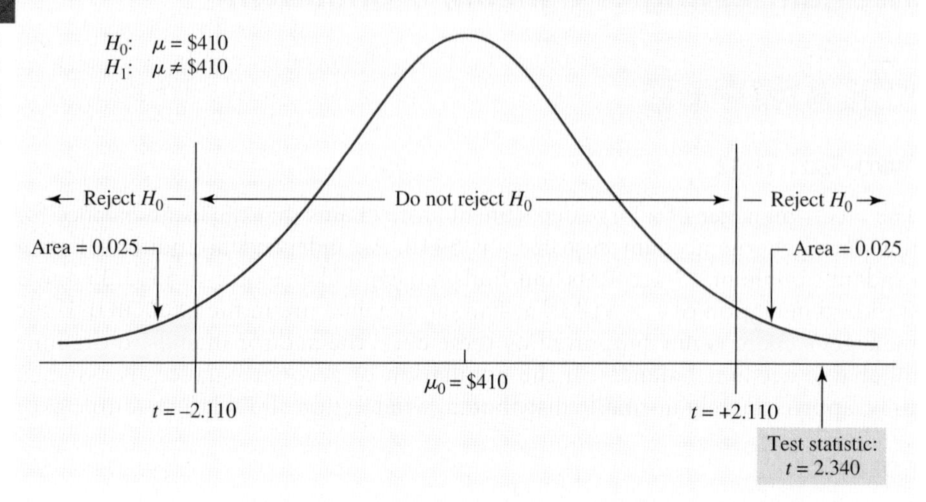

are shown in Figure 10.6, and the decision rule can be stated as "Reject H_0 if the calculated t is either < -2.110 or $> +2.110$, otherwise do not reject."

Compare the Calculated and Critical Values and Reach a Conclusion for the Null Hypothesis

The calculated test statistic, $t = 2.340$, exceeds the upper boundary and falls into this rejection region. H_0 is rejected.

Make the Related Business Decision

The results suggest that the mean charge account balance is some value other than $410. The auditor should proceed to examine all charge account balances.

One-Tail Testing of a Mean, σ Unknown

EXAMPLE

One-Tail Test

The Chekzar Rubber Company, in financial difficulties because of a poor reputation for product quality, has come out with an ad campaign claiming that the mean lifetime for Chekzar tires is at least 60,000 miles in highway driving. Skeptical, the editors of a consumer magazine purchase 36 of the tires and test them in highway use. The mean tire life in the sample is $\bar{x} = 58{,}341.69$ miles, with a sample standard deviation of $s = 3632.53$ miles. The sample data are in file **CX10CHEK**.

SOLUTION

Formulate the Null and Alternative Hypotheses

Because of the directional nature of the ad claim and the editors' skepticism regarding its truthfulness, the null and alternative hypotheses are

H_0: $\mu \geq 60{,}000$ miles The mean tire life is at least 60,000 miles.
H_1: $\mu < 60{,}000$ miles The mean tire life is under 60,000 miles.

Select the Significance Level

For this test, the significance level will be specified as 0.01.

Select the Test Statistic and Calculate Its Value

The test statistic is $t = (\bar{x} - \mu_0)/s_{\bar{x}}$, and the t distribution will be used to describe the sampling distribution of the mean for samples of $n = 36$. The center of the distribution is the lowest possible value for which H_0 could be true, or $\mu_0 = 60{,}000$ miles. Since the population standard deviation is unknown, s is used to estimate σ. The sampling distribution has an estimated standard error of

$$s_{\bar{x}} = \frac{s}{\sqrt{n}} = \frac{3632.53 \text{ miles}}{\sqrt{36}} = 605.42 \text{ miles}$$

and the calculated value of t will be

$$t = \frac{\bar{x} - \mu_0}{s_{\bar{x}}} = \frac{58{,}341.69 - 60{,}000.00}{605.42} = -2.739$$

Identify the Critical Value for the Test Statistic and State the Decision Rule

For this test, α has been specified as 0.01. The number of degrees of freedom is $df = (n - 1)$, or $(36 - 1) = 35$. The t distribution table is now used in finding the value of t that corresponds to a one-tail area of 0.01 and $df = 35$ degrees of freedom. Referring to the 0.01 column and 35th row of the table, we find this critical value to be $t = -2.438$. (Although the value listed is positive, remember that the distribution is symmetrical, and we are looking for the left-tail boundary.) The rejection and nonrejection regions are shown in Figure 10.7, and the

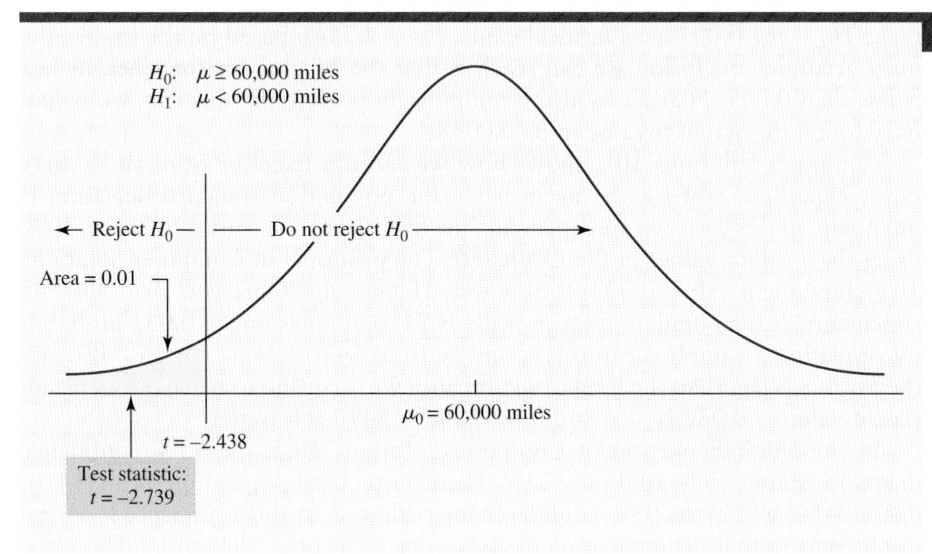

FIGURE 10.7
The Chekzar Rubber Company has claimed that, in highway use, the mean lifetime of its tires is at least 60,000 miles. At the 0.01 level in this left-tail test, the claim is not supported.

decision rule can be stated as "Reject H_0 if the calculated t is less than -2.438, otherwise do not reject."

Compare the Calculated and Critical Values and Reach a Conclusion for the Null Hypothesis

The calculated test statistic, $t = -2.739$, is less than the critical value, $t = -2.438$, and falls into the rejection region of the test. The null hypothesis, H_0: $\mu \geq 60{,}000$ miles, must be rejected.

Make the Related Business Decision

The test results support the editors' doubts regarding Chekzar's ad claim. The magazine may wish to exert either readership or legal pressure on Chekzar to modify its claim.

Compared to the t-test, the z-test is a little easier to apply if the analysis is carried out by pocket calculator and references to a statistical table. (There are lesser "gaps" between areas listed in the normal distribution table compared to values provided in the t table.) Also, courtesy of the central limit theorem, results can be fairly satisfactory when n is large and s is a close estimate of σ.

Nevertheless, the t-test remains the appropriate procedure whenever σ is unknown and is being estimated by s. In addition, this is the method you will either use or come into contact with when dealing with computer statistical packages handling the kinds of analyses in this section. For example, with Excel, Minitab, SPSS, SAS, and others, we can routinely (and correctly) apply the t-test whenever s has been used to estimate σ.

An important note when using statistical tables to determine p-values: For t-tests, the p-value can't be determined as exactly as with the z-test, because the t table areas include greater "gaps" (e.g., the 0.005, 0.01, 0.025 columns, and so on). However, we can *narrow down* the t-test p-value to a range, such as "between 0.01 and 0.025."

For example, in the Chekzar Rubber Company t-test of Figure 10.7, the calculated t statistic was $t = -2.739$. We were able to reject the null hypothesis at the 0.01 level (critical value, $t = -2.438$), and would also have been able to reject H_0 at the 0.005 level (critical value, $t = -2.724$). Based on the t table, the most accurate conclusion we can reach is that the p-value for the Chekzar test is less than 0.005. Had we used the computer in performing this test, we would have found the actual p-value to be 0.0048.

Computer Solutions 10.2 shows how we can use Excel or Minitab to carry out a hypothesis test for the mean when the population standard deviation is unknown. In this case, we are replicating the hypothesis test shown in Figure 10.6, using the 18 data values in file **CX10CRED**. The printouts in Computer Solutions 10.2 show the p-value (0.032) for the test. This p-value represents the following statement: "If the population mean really is \$410, there is only a 0.032 probability of getting a sample mean this far away from \$410 just by chance." Because the p-value is less than the level of significance we are using to reach a conclusion (i.e., p-value $= 0.032$ is $< \alpha = 0.05$), H_0: $\mu = \$410$ is rejected.

In the Minitab portion of Computer Solutions 10.2, the 95% confidence interval is shown as \$420.0 to \$602.7. The hypothesized population mean (\$410) does not fall within the 95% confidence interval; thus, at this confidence level, the results suggest that the population mean is some value other than \$410. This same conclusion was reached in our two-tail test at the 0.05 level of significance.

COMPUTER 10.2 SOLUTIONS

Hypothesis Test for Population Mean, σ Unknown

These procedures show how to carry out a hypothesis test for the population mean when the population standard deviation is unknown.

EXCEL

	A	B	C	D
1	t-Test: Mean			
2				*balance*
3	Mean			511.33
4	Standard Deviation			183.750
5	Hypothesized Mean			410
6	df			17
7	t Stat			2.3396
8	P(T<=t) one-tail			0.0159
9	t Critical one-tail			1.7396
10	P(T<=t) two-tail			0.0318
11	t Critical two-tail			2.1098

Excel hypothesis test for μ based on raw data and σ unknown

1. For example, for the credit balances (file **CX10CRED**) on which Figure 10.6 is based, with the label and 18 data values in **A1:A19:** From the **Add-Ins** ribbon, click **Data Analysis Plus.** Click ***t*-Test: Mean.** Click **OK.**
2. Enter **A1:A19** into the **Input Range** box. Enter the hypothesized mean **(410)** into the **Hypothesized Mean** box. Click **Labels.** Enter the level of significance for the test **(0.05)** into the **Alpha** box. Click **OK.** The printout shows the *p*-value for this two-tail test, 0.0318.

Excel hypothesis test for μ based on summary statistics and σ unknown

1. For example, with $\bar{x} = 511.33$, $s = 183.75$, and $n = 18$, as in Figure 10.6: Open the **TEST STATISTICS** workbook.
2. Using the arrows at the bottom left, select the ***t*-Test_Mean** worksheet. Enter the sample mean **(511.33),** the sample standard deviation **(183.75),** the sample size **(18),** the hypothesized population mean **(410),** and the level of significance for the test **(0.05).**

(*Note:* As an alternative, you can use Excel worksheet template **TMTTEST**. The steps are described within the template.)

MINITAB

Minitab hypothesis test for μ based on raw data and σ unknown

```
One-Sample T: balance

Test of mu = 410 vs not = 410


Variable   N   Mean   StDev  SE Mean        95% CI          T      P
balance   18  511.3   183.8     43.3  (420.0, 602.7)    2.34  0.032
```

1. For example, using the data (file **CX10CRED**) on which Figure 10.6 is based, with the 18 data values in column **C1:** Click **Stat.** Select **Basic Statistics.** Click **1-Sample t.**
2. Select **Samples in Columns** and enter **C1** into the box. Select **Perform hypothesis test** and enter the hypothesized population mean **(410)** into the **Hypothesized mean:** box.

(*continued*)

3. Click **Options**. Enter the desired confidence level as a percentage **(95.0)** into the **Confidence Level** box. Within the **Alternative** box, select **not equal.** Click **OK.** Click **OK.**

Minitab hypothesis test for μ based on summary statistics and σ unknown

Follow the procedure in steps 1 through 3, above, but in step 1 select **Summarized data** and enter **18, 511.33,** and **183.75** into the **Sample size, Mean,** and **Standard deviation** boxes, respectively.

EXERCISES

10.39 Under what circumstances should the t-statistic be used in carrying out a hypothesis test for the population mean?

10.40 For a simple random sample of 40 items, $\bar{x} = 25.9$ and $s = 4.2$. At the 0.01 level of significance, test H_0: $\mu = 24.0$ versus H_1: $\mu \neq 24.0$.

10.41 For a simple random sample of 15 items from a population that is approximately normally distributed, $\bar{x} = 82.0$ and $s = 20.5$. At the 0.05 level of significance, test H_0: $\mu \geq 90.0$ versus H_1: $\mu < 90.0$.

10.42 The average age of passenger cars in use in the United States is 9.0 years. For a simple random sample of 34 vehicles observed in the employee parking area of a large manufacturing plant, the average age is 10.4 years, with a standard deviation of 3.1 years. At the 0.01 level of significance, can we conclude that the average age of cars driven to work by the plant's employees is greater than the national average? Source: polk.com, August 9, 2006.

10.43 The average length of a flight by regional airlines in the United States has been reported as 464 miles. If a simple random sample of 30 flights by regional airlines were to have $\bar{x} = 479.6$ miles and $s = 42.8$ miles, would this tend to cast doubt on the reported average of 464 miles? Use a two-tail test and the 0.05 level of significance in arriving at your answer. Source: Bureau of the Census, *Statistical Abstract of the United States 2009*, p. 664.

10.44 The International Coffee Association has reported the mean daily coffee consumption for U.S. residents as 1.65 cups. Assume that a sample of 38 people from a North Carolina city consumed a mean of 1.84 cups of coffee per day, with a standard deviation of 0.85 cups. In a two-tail test at the 0.05 level, could the residents of this city be said to be significantly different from their counterparts across the nation? Source: coffeeresearch.org, August 8, 2006.

10.45 Taxco, a firm specializing in the preparation of income tax returns, claims the mean refund for customers who received refunds last year was \$150. For a random sample of 12 customers who received refunds last year, the mean amount was found to be \$125, with a standard deviation of \$43. Assuming that the population is approximately normally distributed, and using the 0.10 level in a two-tail test, do these results suggest that Taxco's assertion may be accurate?

10.46 The new director of a local YMCA has been told by his predecessors that the average member has belonged for 8.7 years. Examining a random sample of 15 membership files, he finds the mean length of membership to be 7.2 years, with a standard deviation of 2.5 years. Assuming the population is approximately normally distributed, and using the 0.05 level, does this result suggest that the actual mean length of membership may be some value other than 8.7 years?

10.47 A scrap metal dealer claims that the mean of his cash sales is "no more than \$80," but an Internal Revenue Service agent believes the dealer is untruthful. Observing a sample of 20 cash customers, the agent finds the mean purchase to be \$91, with a standard deviation of \$21. Assuming the population is approximately normally distributed, and using the 0.05 level of significance, is the agent's suspicion confirmed?

10.48 During 2008, college work-study students earned a mean of \$1478. Assume that a sample consisting of 45 of the work-study students at a large university was found to have earned a mean of \$1503 during that year, with a standard deviation of \$210. Would a one-tail test at the 0.05 level suggest the average earnings of this university's work-study students were significantly higher than the national mean? Source: Bureau of the Census, *Statistical Abstract of the United States 2009*, p. 178.

10.49 According to the Federal Reserve Board, the mean net worth of U.S. households headed by persons 75 years or older is \$640,000. Suppose a simple random sample of 50 households in this age group is obtained from a certain region of the United States and is found to have a mean net worth of $615,000, with a standard deviation of $120,000. From these sample results, and using the 0.05 level of significance in a two-tail test, comment on whether the mean net worth for all the region's households in this age category might not be the same as the mean value reported for their counterparts across the nation. Source: Federal Reserve Board, *Changes in U.S. Family Finances from 2004 to 2007*, p. A11.

10.50 Using the sample results in Exercise 10.49, construct and interpret the 95% confidence interval for the population mean. Is the hypothesized population mean ($640,000) within the interval? Given the presence or absence of the $640,000 value within the interval, is this consistent with the findings of the hypothesis test conducted in Exercise 10.49?

10.51 It has been reported that the average life for halogen lightbulbs is 4000 hours. Learning of this figure, a plant manager would like to find out whether the vibration and temperature conditions that the facility's bulbs encounter might be having an adverse effect on the service life of bulbs in her plant. In a test involving 15 halogen bulbs installed in various locations around the plant, she finds the average life for bulbs in the sample is 3882 hours, with a standard deviation of 200 hours. Assuming the population of halogen bulb lifetimes to be approximately normally distributed, and using the 0.025 level of significance, do the test results tend to support the manager's suspicion that adverse conditions might be detrimental to the operating lifespan of halogen lightbulbs used in her plant? Source: Cindy Hall and Gary Visgaitis, "Bulbs Lasting Longer," *USA Today*, March 9, 2000, p. 1D.

10.52 In response to an inquiry from its national office, the manager of a local bank has stated that her bank's average service time for a drive-through customer is 93 seconds. A student intern working at the bank happens to be taking a statistics course and is curious as to whether the true average might be some value other than 93 seconds. The intern observes a simple random sample of 50 drive-through customers whose average service time is 89.5 seconds, with a standard deviation of 11.3 seconds. From these sample results, and using the 0.05 level of significance, what conclusion would the student reach with regard to the bank manager's claim?

10.53 Using the sample results in Exercise 10.52, construct and interpret the 95% confidence interval for the population mean. Is the hypothesized population mean (93 seconds) within the interval? Given the presence or absence of the 93 seconds value within the interval, is this consistent with the findings of the hypothesis test conducted in Exercise 10.52?

10.54 The U.S. Census Bureau says the 52-question "long form" received by 1 in 6 households during the 2000 census takes a mean of 38 minutes to complete. Suppose a simple random sample of 35 persons is given the form, and their mean time to complete it is 36.8 minutes, with a standard deviation of 4.0 minutes. From these sample results, and using the 0.10 level of significance, would it seem that the actual population mean time for completion might be some value other than 38 minutes? Source: Haya El Nasser, "Census Forms Can Be Filed by Computer," *USA Today*, February 10, 2000, p. 4A.

10.55 Using the sample results in Exercise 10.54, construct and interpret the 90% confidence interval for the population mean. Is the hypothesized population mean (38 minutes) within the interval? Given the presence or absence of the 38 minutes value within the interval, is this consistent with the findings of the hypothesis test conducted in Exercise 10.54?

(DATA SET) *Note:* Exercises 10.56–10.58 require a computer and statistical software.

10.56 The International Council of Shopping Centers reports that the average teenager spends $57 during a shopping trip to the mall. The promotions director of a local mall has used a variety of strategies to attract area teens to his mall, including live bands and "teen-appreciation days" that feature special bargains for this age group. He believes teen shoppers at his mall respond to his promotional efforts by shopping there more often and spending more when they do. Mall management decides to evaluate the promotions director's success by surveying a simple random sample of 45 local teens and finding out how much they spent on their most recent shopping visit to the mall. The results are listed in data file **XR10056**. Use a suitable hypothesis test in examining whether the mean mall shopping expenditure for teens in this area might be higher than for U.S. teens as a whole. Identify and interpret the p-value for the test. Using the 0.025 level of significance, what conclusion do you reach? Source: icsc.org, July 23, 2009.

10.57 According to the Insurance Information Institute, the mean annual expenditure for automobile insurance for U.S. motorists is $817. Suppose that a government official in North Carolina has surveyed a simple random sample of 80 residents of her state, and that their auto insurance expenditures for the most recent year are in data file **XR10057**. Based on these data, examine whether the mean annual auto insurance expenditure for motorists in North Carolina might be different from the $817 for the country as a whole. Identify and interpret the p-value for the test. Using the 0.05 level of significance, what conclusion do you reach? Source: iii.org, July 23, 2009.

10.58 Using the sample data in Exercise 10.57, construct and interpret the 95% confidence interval for the population mean. Is the hypothesized population mean ($817) within the interval? Given the presence or absence of the $817 value within the interval, is this consistent with the findings of the hypothesis test conducted in Exercise 10.57?

10.6 TESTING A PROPORTION

Occasions may arise when we wish to compare a sample proportion, p, with a value that has been hypothesized for the population proportion, π. As we noted in Figure 10.2, the theoretically correct distribution for dealing with proportions is the binomial distribution. However, the normal distribution is a good approximation when $n\pi \geq 5$ and $n(1 - \pi) \geq 5$. The larger the sample size, the better this approximation becomes, and for most practical settings, this condition is satisfied. When using the normal distribution for hypothesis tests of a sample proportion, the test statistic is as follows:

Test statistic, *z*-test for a sample proportion:

$$z = \frac{p - \pi_0}{\sigma_p}$$

where p = sample proportion
π_0 = hypothesized population proportion
n = sample size
σ_p = standard error of the distribution of the sample proportion

$$\sigma_p = \sqrt{\frac{\pi_0(1 - \pi_0)}{n}}$$

Two-Tail Testing of a Proportion

Whenever the null hypothesis involves a proportion and is nondirectional, this technique is appropriate. To demonstrate how it works, consider the following situation.

EXAMPLE

Two-Tail Test

The career services director of Hobart University has said that 70% of the school's seniors enter the job market in a position directly related to their undergraduate field of study. In a sample consisting of 200 of the graduates from last year's class, 66% have entered jobs related to their field of study. The underlying data are in file **CX10GRAD**, with values coded as 1 = no job in field, 2 = job in field.

SOLUTION

Formulate the Null and Alternative Hypotheses

The director's statement is nondirectional and leads to null and alternative hypotheses of

H_0: $\pi = 0.70$ The proportion of graduates entering jobs in their field is 0.70.

H_1: $\pi \neq 0.70$ The proportion is some value other than 0.70.

Select the Significance Level

For this test, the 0.05 level will be used. The sum of the two tail areas will be 0.05.

EXAMPLEEXAMPLEEXAMPLEEXAMPLEEXAMPLEEX

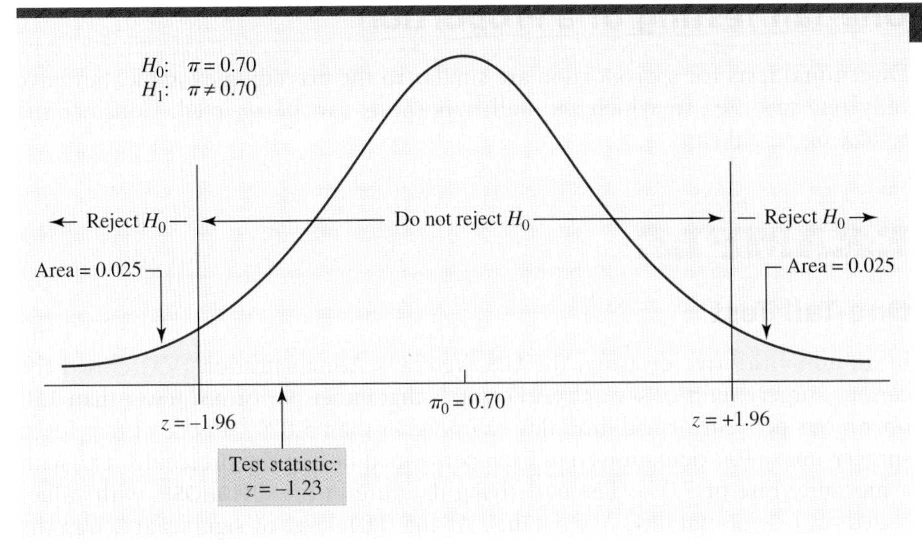

FIGURE 10.8

In this two-tail test involving a sample proportion, the sample result leads to nonrejection of the career services director's claim that 70% of a university's seniors enter jobs related to their field of study.

Select the Test Statistic and Calculate Its Value

The test statistic will be z, the number of standard error units from the hypothesized population proportion, $\pi_0 = 0.70$, to the sample proportion, $p = 0.66$. The standard error of the sample proportion is

$$\sigma_p = \sqrt{\frac{\pi_0(1 - \pi_0)}{n}} = \sqrt{\frac{0.70(1 - 0.70)}{200}} = 0.0324$$

and the calculated value of z will be

$$z = \frac{p - \pi_0}{\sigma_p} = \frac{0.66 - 0.70}{0.0324} = -1.23$$

Identify Critical Values for the Test Statistic and State the Decision Rule

Since the test is two-tail and the selected level of significance is 0.05, the critical values will be $z = -1.96$ and $z = +1.96$. The decision rule can be stated as "Reject H_0 if calculated z is either < -1.96 or $> +1.96$, otherwise do not reject."

Compare the Calculated and Critical Values and Reach a Conclusion for the Null Hypothesis

The calculated value of the test statistic, $z = -1.23$, falls between the two critical values, placing it in the nonrejection region of the distribution shown in Figure 10.8. The null hypothesis is not rejected.

Make the Related Decision

Failure to reject the null hypothesis leads us to conclude that the proportion of graduates who enter the job market in careers related to their field of study could indeed be equal to the claimed value of 0.70. If the career services director has been making this claim to her students or their parents, this analysis would suggest that her assertion not be challenged.

One-Tail Testing of a Proportion

Directional tests for a proportion are similar to the preceding example, but have only one tail area in which the null hypothesis can be rejected. Consider the following actual case.

EXAMPLE

One-Tail Test

In an administrative decision, the U.S. Veterans Administration (VA) closed the cardiac surgery units of several VA hospitals that either performed fewer than 150 operations per year or had mortality rates higher than 5.0%.[1] In one of the closed surgery units, 100 operations had been performed during the preceding year, with a mortality rate of 7.0%. The underlying data are in file **CX10HOSP**, with values coded as 1 = nonfatality, 2 = fatality. At the 0.01 level of significance, was the mortality rate of this hospital significantly greater than the 5.0% cutoff point? Consider the hospital's performance as representing a sample from the population of possible operations it might have performed if the patients had been available.

SOLUTION

Formulate the Null and Alternative Hypotheses

The null hypothesis makes the assumption that the "population" mortality rate for the hospital cardiac surgery unit is really no greater than 0.05, and that the observed proportion, $p = 0.07$, was simply due to chance variation.

H_0: $\pi \leq 0.05$ The true mortality rate for the unit is no more than 0.05.

H_1: $\pi > 0.05$ The true mortality rate is greater than 0.05.

The center of the hypothesized distribution, $\pi_0 = 0.05$, is the highest possible value for which the null hypothesis could be true.

Select the Significance Level

The significance level has been specified as $\alpha = 0.01$. If the null hypothesis were really true, there would be no more than a 0.01 probability of incorrectly rejecting it.

Select the Test Statistic and Calculate Its Value

The test statistic will be z, calculated as $z = (p - \pi_0)/\sigma_p$. The standard error of the sample proportion and the calculated value of the test statistic are

$$\sigma_p = \sqrt{\frac{\pi_0(1 - \pi_0)}{n}} = \sqrt{\frac{0.05(1 - 0.05)}{100}} = 0.02179$$

and

$$z = \frac{p - \pi_0}{\sigma_p} = \frac{0.07 - 0.05}{0.02179} = 0.92$$

Identify the Critical Value for the Test Statistic and State the Decision Rule

For the 0.01 level, the critical value of z is $z = +2.33$. The decision rule can be stated as "Reject H_0 if calculated $z > +2.33$, otherwise do not reject."

[1]Source: "VA Halts Some Heart Operations," *Indiana Gazette*, January 13, 1987.

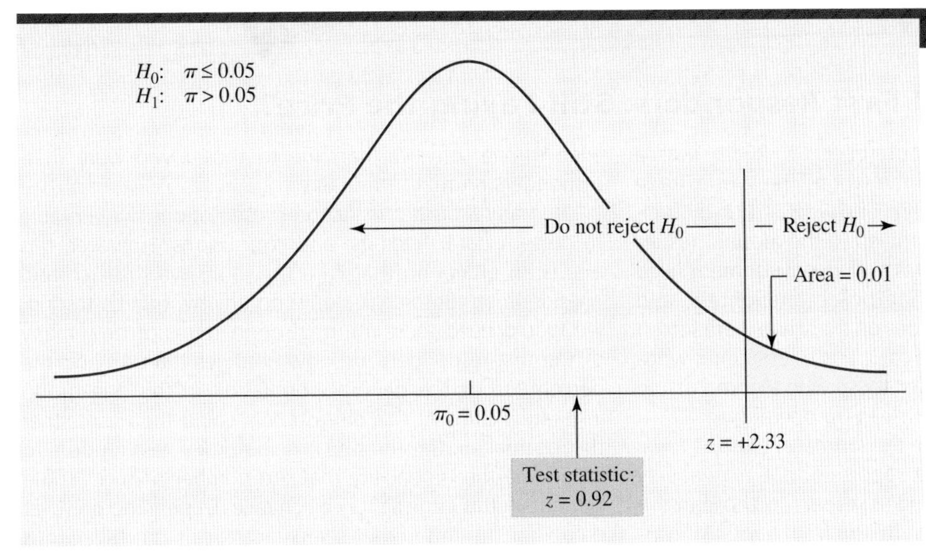

FIGURE 10.9

The U.S. Veterans Administration closed a number of cardiac surgery units because they either performed fewer than 150 operations or had a mortality rate over 5.0% during the previous year. For one of the hospitals, there was a mortality rate of 7.0% in 100 operations, but this relatively high mortality rate could have been due to chance variation.

Compare the Calculated and Critical Values and Reach a Conclusion for the Null Hypothesis

Since the calculated value, $z = 0.92$, is less than the critical value, it falls into the nonrejection region of Figure 10.9 and the null hypothesis, H_0: $\pi \le 0.05$, cannot be rejected.

Make the Related Business Decision

The cardiac surgery mortality rate for this hospital could have been as high as 0.07 merely by chance, and closing it could not be justified strictly on the basis of a "significantly greater than 0.05" guideline. [*Notes:* (1) The VA may have been striving for some lower population proportion not mentioned in the article, and (2) because the cardiac unit did not meet the minimum requirement of 150 operations per year, the VA would have closed it anyway.]

Statistics in Action 10.1 addresses a serious issue but can provide no conclusions. Use the Internet, find the latest data, and you can do your own testing.

Computer Solutions 10.3 shows how we can use Excel or Minitab to carry out a hypothesis test for a proportion. In this case, we are replicating the hypothesis test shown in Figure 10.8, using summary information. The printouts in Computer Solutions 10.3 show the p-value (0.217) for the test. This p-value represents the following statement: "If the population proportion really is 0.70, there is a 0.217 probability of getting a sample proportion this far away from 0.70 just by chance." The p-value is not less than the level of significance we are using to reach a conclusion (i.e., p-value $= 0.217$ is not less than $\alpha = 0.05$), and H_0: $\pi = 0.70$ cannot be rejected.

In the Minitab portion of Computer Solutions 10.3, the 95% confidence interval is shown as (0.594349 to 0.725651). The hypothesized population proportion (0.70) falls within the 95% confidence interval and, at this confidence level, it appears that the population proportion could be 0.70. This relationship between confidence intervals and two-tail hypothesis tests was discussed in Section 10.4.

STATISTICS IN ACTION 10.1

9/11 First Responders: Still Paying the Price?

As most everyone knows, September 11, 2001, was a sad day for both America and the world. The World Trade Center Twin Towers in New York City were struck by fuel-filled, hijacked aircraft and crumbled to the ground, resulting in both catastrophic fatalities and an environment filled with toxic smoke and dust. The police, firefighters, and other emergency-response heroes on the scene were forced to breathe a mixture of air and toxic dusts containing who-knows-what. The same goes for those who worked at the site during search, rescue, and clean-up operations.

In August 2009, an article in the *Journal of Occupational and Environmental Medicine* reported that a relatively unusual number of first-responders, especially those in the under-45 age group, are now suffering from a form of cancer that has historically been considered a disease of the elderly. Of the 28,252 responders studied, eight cases of multiple myeloma were diagnosed between September 11, 2001, and September 10, 2007, and half of these were in persons now under the age of 45.

According to the study authors, the expected number of cases of this disease among the under-45 responders would be 1.2, but the observed number was 4.0. Was this relatively high number of diagnoses among the under-45 due to chance variation, or did that toxic cloud inflict what medical professionals might call a major bodily insult in the form of a ticking time bomb that will eventually cause even greater physical damage to those who were first-responders on that fateful day?

In many *Statistics in Action* presentations, we are able to present data, findings, and conclusions. However, this one is an exception that calls for your own research and consideration. Before we could do any hypothesis testing, we would need more information in terms of how many of the first-responders are now younger than 45 and the percentage of the general population in this age group who are victims of this disease. We could then use either the normal or (preferably) the binomial distribution to determine the probability that the number of cases could have been this large simply by chance. Depending on what data your future Internet searching reveals on this serious matter, you may gain insights by using either the binomial distribution or the normal distribution approximation.

As a business statistics student, and as a concerned and caring citizen, you may find it relevant to use the Internet and other sources to keep abreast of these developments and be aware of the after-effects being experienced by the 9/11 first-responders. Perhaps more than any example we might provide within these pages, this one exemplifies that statistics are more than mere numbers.

Sources: Amanda Gardner, "9/11 Responders May Be at Raised Myeloma Risk," medicine.net, August 14, 2009.

COMPUTER 10.3 SOLUTIONS

Hypothesis Test for a Population Proportion

These procedures show how to carry out a hypothesis test for the population proportion.

EXCEL

	A	B	C	D
1	z-Test of a Proportion			
2				
3	Sample proportion	0.66	z Stat	-1.23
4	Sample size	200	P(Z<=z) one-tail	0.1085
5	Hypothesized proportion	0.70	z Critical one-tail	1.645
6	Alpha	0.05	P(Z<=z) two-tail	0.2170
7			z Critical two-tail	1.960

Excel hypothesis test for π based on summary statistics

1. For example, with $n = 200$ and $p = 0.66$, as in Figure 10.8: Open the TEST STATISTICS workbook.
2. Using the arrows at the bottom left, select the **z-Test_Proportion** worksheet. Enter the sample proportion **(0.66)**, the sample size **(200)**, the hypothesized population proportion **(0.70)**, and the level of significance for the test **(0.05)**. The *p*-value for this two-tail test is shown as 0.2170.

(*Note:* As an alternative, you can use Excel worksheet template TMPTEST. The steps are described within the template.)

Excel hypothesis test for π based on raw data

1. For example, using data file CX10GRAD, with the label and 200 data values in **A1:A201** and data coded as 1 = no job in field, 2 = job in field: From the **Add-Ins** ribbon, click **Data Analysis Plus.** Click ***Z*-Test: Proportion.** Click **OK.**
2. Enter **A1:A201** into the **Input Range** box. Enter **2** into the **Code for Success** box. Enter **0.70** into the **Hypothesized Proportion** box. Click **Labels.** Enter the level of significance for the test **(0.05)** into the **Alpha** box. Click **OK.**

MINITAB

Minitab hypothesis test for π based on summary statistics

```
Test and CI for One Proportion

Test of p = 0.7 vs p not = 0.7


Sample     X    N  Sample p            95% CI            Z-Value  P-Value
1        132  200  0.660000  (0.594349, 0.725651)          -1.23    0.217

Using the normal approximation.
```

1. This interval is based on the summary statistics for Figure 10.8: Click **Stat.** Select **Basic Statistics.** Click **1 Proportion.** Select **Summarized Data.** Enter the sample size **(200)** into the **Number of Trials** box. Multiply the sample proportion (0.66) times the sample size (200) to get the number of "events" or "successes" (0.66)(200) = 132. (Had this not been an integer, it would have been necessary to round to the nearest integer.) Enter the number of "successes" **(132)** into the **Number of events** box. Select **Perform hypothesis test** and enter the hypothesized population proportion **(0.70)** into the **Hypothesized proportion** box.
2. Click **Options.** Enter the desired confidence level as a percentage **(95.0)** into the **Confidence Level** box. Within the **Alternative** box, select **not equal.** Click to select **Use test and interval based on normal distribution.** Click **OK.** Click **OK.**

Minitab hypothesis test for π based on raw data

1. For example, using file CX10GRAD, with column C1 containing the 200 assumed data values (coded as 1 = no job in field, 2 = job in field): Click **Stat.** Select **Basic Statistics.** Click **1 Proportion.** Select **Samples in columns** and enter **C1** into the dialog box. Select **Perform hypothesis test** and enter the hypothesized proportion **(0.70)** into the **Hypothesized Proportion** box.
2. Follow step 2 in the summary-information procedure, above. *Note:* Minitab will select the larger of the two codes (i.e., 2 = job in field) as the "success" and provide the sample proportion and the confidence interval for the population proportion of graduates having jobs in their fields. To obtain the results for those not having jobs in their fields, just recode the data so graduates without jobs in their fields will have the higher code number: Click **Data.** Select **Code.** Click **Numeric to Numeric.** Enter **C1** into both the **Code data from columns** box and the **Store coded data in columns** box. Enter **1** into the **Original values** box. Enter **3** into the **New** box. Click **OK.** The new codes will be 3 = no job in field, 2 = job in field.

Had we used the computer to perform the test summarized in Figure 10.9, we would have found the p-value for this one-tail test to be 0.179. The p-value for the test says, "If the population proportion really is 0.05, there is a 0.179 probability of getting a sample proportion this large (0.07) just by chance." The p-value (0.179) is not less than the level of significance we are using as a criterion ($\alpha = 0.01$), so we are unable to reject H_0: $\pi \leq 0.05$.

EXERCISES

10.59 When carrying out a hypothesis test for a population proportion, under what conditions is it appropriate to use the normal distribution as an approximation to the (theoretically correct) binomial distribution?

10.60 For a simple random sample, $n = 200$ and $p = 0.34$. At the 0.01 level, test H_0: $\pi = 0.40$ versus H_1: $\pi \neq 0.40$.

10.61 For a simple random sample, $n = 1000$ and $p = 0.47$. At the 0.05 level, test H_0: $\pi \geq 0.50$ versus H_1: $\pi < 0.50$.

10.62 For a simple random sample, $n = 700$ and $p = 0.63$. At the 0.025 level, test H_0: $\pi \leq 0.60$ versus H_1: $\pi > 0.60$.

10.63 A simple random sample of 300 items is selected from a large shipment, and testing reveals that 4% of the sampled items are defective. The supplier claims that no more than 2% of the items in the shipment are defective. Carry out an appropriate hypothesis test and comment on the credibility of the supplier's claim.

10.64 The director of admissions at a large university says that 15% of high school juniors to whom she sends university literature eventually apply for admission. In a sample of 300 persons to whom materials were sent, 30 students applied for admission. In a two-tail test at the 0.05 level of significance, should we reject the director's claim?

10.65 According to the human resources director of a plant, no more than 5% of employees hired in the past year have violated their preemployment agreement not to use any of five illegal drugs. The agreement specified that random urine checks could be carried out to ascertain compliance. In a random sample of 400 employees, screening detected at least one of these drugs in the systems of 8% of those tested. At the 0.025 level, is the human resources director's claim credible? Determine and interpret the p-value for the test.

10.66 It has been claimed that 65% of homeowners would prefer to heat with electricity instead of gas. A study finds that 60% of 200 homeowners prefer electric heating to gas. In a two-tail test at the 0.05 level of significance, can we conclude that the percentage who prefer electric heating may differ from 65%? Determine and interpret the p-value for the test.

10.67 In the past, 44% of those taking a public accounting qualifying exam have passed the exam on their first try. Lately, the availability of exam preparation books and tutoring sessions may have improved the likelihood of an individual's passing on his or her first try. In a sample of 250 recent applicants, 130 passed on their first attempt. At the 0.05 level of significance, can we conclude that the proportion passing on the first try has increased? Determine and interpret the p-value for the test.

10.68 Heritage Union has said that 66% of U.S. adults have purchased life insurance. Suppose that for a random sample of 50 adults from a given U.S. city, a researcher finds that only 56% of them have purchased life insurance. At the 0.05 level in a one-tail test, is this sample finding significantly lower than the 66% reported by Heritage Union? Determine and interpret the p-value for the test. Source: reuters.com, September 23, 2008.

10.69 According to the National Association of Home Builders, 55% of new single-family homes built during 2005 had a fireplace. Suppose a nationwide homebuilder has claimed that its homes are "a cross section of America," but a simple random sample of 600 of its single-family homes built during that year included only 50.0% that had a fireplace. Using the 0.05 level of significance in a two-tail test, examine whether the percentage of sample homes having a fireplace could have differed from 55% simply by chance. Determine and interpret the p-value for the test. Source: National Association of Homebuilders, *2007 Housing Facts, Figures, and Trends*, p. 13.

10.70 Based on the sample results in Exercise 10.69, construct and interpret the 95% confidence interval for the population proportion. Is the hypothesized proportion (0.55) within the interval? Given the presence or absence of the 0.55 value within the interval, is this consistent with the findings of the hypothesis test conducted in Exercise 10.69?

10.71 According to the U.S. Bureau of Labor Statistics, 7.0% of female hourly workers who are 16 to 24 years old are being paid minimum wage or less. (Note that some workers in some industries are exempt from the minimum wage requirement of the Fair Labor Standards Act and, thus, could be legally earning less than the "minimum" wage.) A prominent politician is interested in how young working women within her county compare to this national percentage, and selects a simple random sample of 500 female hourly workers who are 16 to 24 years old. Of the women in the sample, 42 are being paid minimum wage or less. From these sample results, and using the 0.10 level of significance, could the politician conclude that the percentage of young female hourly workers who are low-paid in her county might be the same as the percentage of young women who are low-paid in the nation as a whole? Determine and interpret the p-value for the test. Source: *Statistical Abstract of the United States 2009,* p. 413.

10.72 Using the sample results in Exercise 10.71, construct and interpret the 90% confidence interval for the population proportion. Is the hypothesized population proportion (0.07) within the interval? Given the presence or absence of the 0.07 value within the interval, is this consistent with the findings of the hypothesis test conducted in Exercise 10.71?

10.73 Brad Davenport, a consumer reporter for a national cable TV channel, is working on a story evaluating generic food products and comparing them to their brand-name counterparts. According to Brad, consumers claim to like the brand-name products better than the generics, but they can't even tell which is which. To test his theory, Brad gives each of 200 consumers two potato chips—one generic, the other a brand name—and asks them which one is the brand-name chip. Fifty-five percent of the subjects correctly identify the brand-name chip. At the 0.025 level, is this significantly greater than the 50% that could be expected simply by chance? Determine and interpret the p-value for the test.

10.74 It has been reported that 80% of taxpayers who are audited by the Internal Revenue Service end up paying more money in taxes. Assume that auditors are randomly assigned to cases, and that one of the ways the IRS oversees its auditors is to monitor the percentage of cases that result in the taxpayer paying more taxes. If a sample of 400 cases handled by an individual auditor has 77.0% of those she audited paying more taxes, is there reason to believe her overall "pay more" percentage might be some value other than 80%? Use the 0.10 level of significance in reaching a conclusion. Determine and interpret the p-value for the test. Source: Sandra Block, "Audit Red Flags You Don't Want to Wave," *USA Today,* April 11, 2000, p. 3B.

10.75 Based on the sample results in Exercise 10.74, construct and interpret the 90% confidence interval for the population proportion. Is the hypothesized proportion (0.80) within the interval? Given the presence or absence of the 0.80 value within the interval, is this consistent with the findings of the hypothesis test conducted in Exercise 10.74?

(DATA SET) *Note:* Exercises 10.76–10.78 require a computer and statistical software.

10.76 According to the National Collegiate Athletic Association (NCAA), 41% of male basketball players graduate within 6 years of enrolling in their college or university, compared to 56% for the student body as a whole. Assume that data file **XR10076** shows the current status for a sample of 200 male basketball players who enrolled in New England colleges and universities 6 years ago. The data codes are 1 = left school, 2 = still in school, 3 = graduated. Using these data and the 0.10 level of significance, does the graduation rate for male basketball players from schools in this region differ significantly from the 41% for male basketball players across the nation? Identify and interpret the p-value for the test. Source: "NCAA Basketball 'Reforms' Come Up Short," *USA Today,* April 1, 2000, p. 17A.

10.77 Using the sample results in Exercise 10.76, construct and interpret the 90% confidence interval for the population proportion. Is the hypothesized proportion (0.41) within the interval? Given the presence or absence of the 0.41 value within the interval, is this consistent with the findings of the hypothesis test conducted in Exercise 10.76?

10.78 Website administrators sometimes use analysis tools or service providers to "track" the movements of visitors to the various portions of their site. Overall, the administrator of a political action website has found that 35% of the visitors who visit the "Environmental Issues" page go on to visit the "Here's What You Can Do" page. In an effort to increase this rate, the administrator places a photograph of an oil-covered sea otter on the Issues page. Of the next 300 visitors to the Issues page, 40% also visit the Can Do page. The data are in file **XR10078**, coded as 1 = did not go on to visit Can Do page and 2 = went on to visit Can Do page. At the 0.05 level, is the 40% rate significantly greater than the 35% that had been occurring in the past, or might this higher rate be simply due to chance variation? Identify and interpret the p-value for the test.

10.7 THE POWER OF A HYPOTHESIS TEST

Hypothesis Testing Errors and the Power of a Test

As discussed previously in the chapter, incorrect conclusions can result from hypothesis testing. As a quick review, the mistakes are of two kinds:

- **Type I error, rejecting a true hypothesis:**

$$\alpha = \text{probability of rejecting } H_0 \text{ when } H_0 \text{ is true}$$

or

$$\alpha = P(\text{reject } H_0 \mid H_0 \text{ true})$$
$$\alpha = \textit{the level of significance of a test}$$

- **Type II error, failing to reject a false hypothesis:**

$$\beta = \text{probability of failing to reject } H_0 \text{ when } H_0 \text{ is false}$$

or

$$\beta = P(\text{fail to reject } H_0 \mid H_0 \text{ false})$$
$$1 - \beta = \text{probability of rejecting } H_0 \text{ when } H_0 \text{ is false}$$
$$1 - \beta = \textit{the power of a test}$$

In this section, our focus will be on $(1 - \beta)$, the power of a test. As mentioned previously, there is a trade-off between α and β: For a given sample size, reducing α tends to increase β, and vice versa; with larger sample sizes, however, *both* α and β can be decreased for a given test.

In wishing people luck, we sometimes tell them, "Don't take any wooden nickels." As an analogy, the power of a hypothesis test is the probability that the test will correctly reject the "wooden nickel" represented by a false null hypothesis. In other words, $(1 - \beta)$, the *power of a test*, is the probability that the test will respond correctly by rejecting a false null hypothesis.

The Power of a Test: An Example

As an example, consider the Extendabulb test, presented in Section 10.3 and illustrated in Figure 10.4. The test can be summarized as follows:

- Null and alternative hypotheses:

H_0: $\mu \le 1030$ hours	Extendabulb is no better than the previous system.
H_1: $\mu > 1030$ hours	Extendabulb does increase bulb life.

- Significance level selected: 0.05
- Calculated value of test statistic:

$$z = \frac{\bar{x} - \mu_0}{\sigma/\sqrt{n}} = \frac{1061.6 - 1030.0}{90/\sqrt{40}} = 2.22$$

- Critical value for test statistic: $z = +1.645$
- Decision rule: Reject H_0 if calculated $z > +1.645$, otherwise do not reject.

For purposes of determining the power of the test, we will first convert the critical value, $z = +1.645$, into the equivalent mean bulb life for a sample of this size. This will be 1.645 standard error units to the right of the mean of the hypothesized distribution (1030 hours). The standard error for the distribution of sample means is $\sigma_{\bar{x}} = \sigma/\sqrt{n} = 90/\sqrt{40}$, or 14.23 hours. The critical z value can now be converted into a critical sample mean:

$$\text{Sample mean, } \bar{x} \text{ corresponding to critical } z = +1.645 = 1030.00 + 1.645(14.23) = 1053.41 \text{ hours}$$

and the decision rule, "Reject H_0 if calculated test statistic is greater than $z = +1.645$" can be restated as "*Reject H_0 if sample mean is greater than 1053.41 hours.*"

The power of a test to correctly reject a false hypothesis depends on the true value of the population mean, a quantity that we do not know. At this point, we will assume that the true mean has a value that would *cause* the null hypothesis to be false, then the decision rule of the test will be applied to see whether this "wooden nickel" is rejected, as it should be.

As an arbitrary choice, the true mean life of Extendabulb-equipped bulbs will be assumed to be $\mu = 1040$ hours. The next step is to see how the decision rule, "Reject H_0 if the sample mean is greater than 1053.41 hours," is likely to react. In particular, interest is focused on the probability that the decision rule will correctly reject the false null hypothesis that the mean is no more than 1030 hours.

As part (a) of Figure 10.10 shows, the distribution of sample means is centered on $\mu = 1040$ hours, the true value assumed for bulb life. The standard error of the distribution of sample means remains the same, so the *spread* of the sampling distribution is unchanged compared to that in Figure 10.4. In part (a) of Figure 10.10, however, the entire distribution has been "shifted" 10 hours to the right.

If the true mean is 1040 hours, the shaded portion of the curve in part (a) of Figure 10.10 represents the power of the hypothesis test—that is, the probability that it will correctly reject the false null hypothesis. Using the standard error of the sample mean, $\sigma_{\bar{x}} = 14.23$ hours, we can calculate the number of standard error units from 1040 to 1053.41 hours as

$$z = \frac{\bar{x} - \mu}{\sigma_{\bar{x}}} = \frac{1053.41 - 1040.00}{14.23} = \frac{13.41}{14.23} = 0.94 \text{ standard error units to the right of the population mean}$$

From the normal distribution table, we find the cumulative area to $z = +0.94$ is 0.8264. Since the total area beneath the curve is 1.0000, we can calculate the shaded area as $1.0000 - 0.8264$, or 0.1736. Thus, if the true mean life of Extendabulb-equipped bulbs is 1040 hours, there is a 0.1736 probability that a sample of 40 bulbs will have a mean in the "reject H_0" region of our test and that we will correctly reject the false null hypothesis that μ is no more than 1030 hours. For a true mean of 1040 hours, the power of the test is 0.1736.

FIGURE 10.10

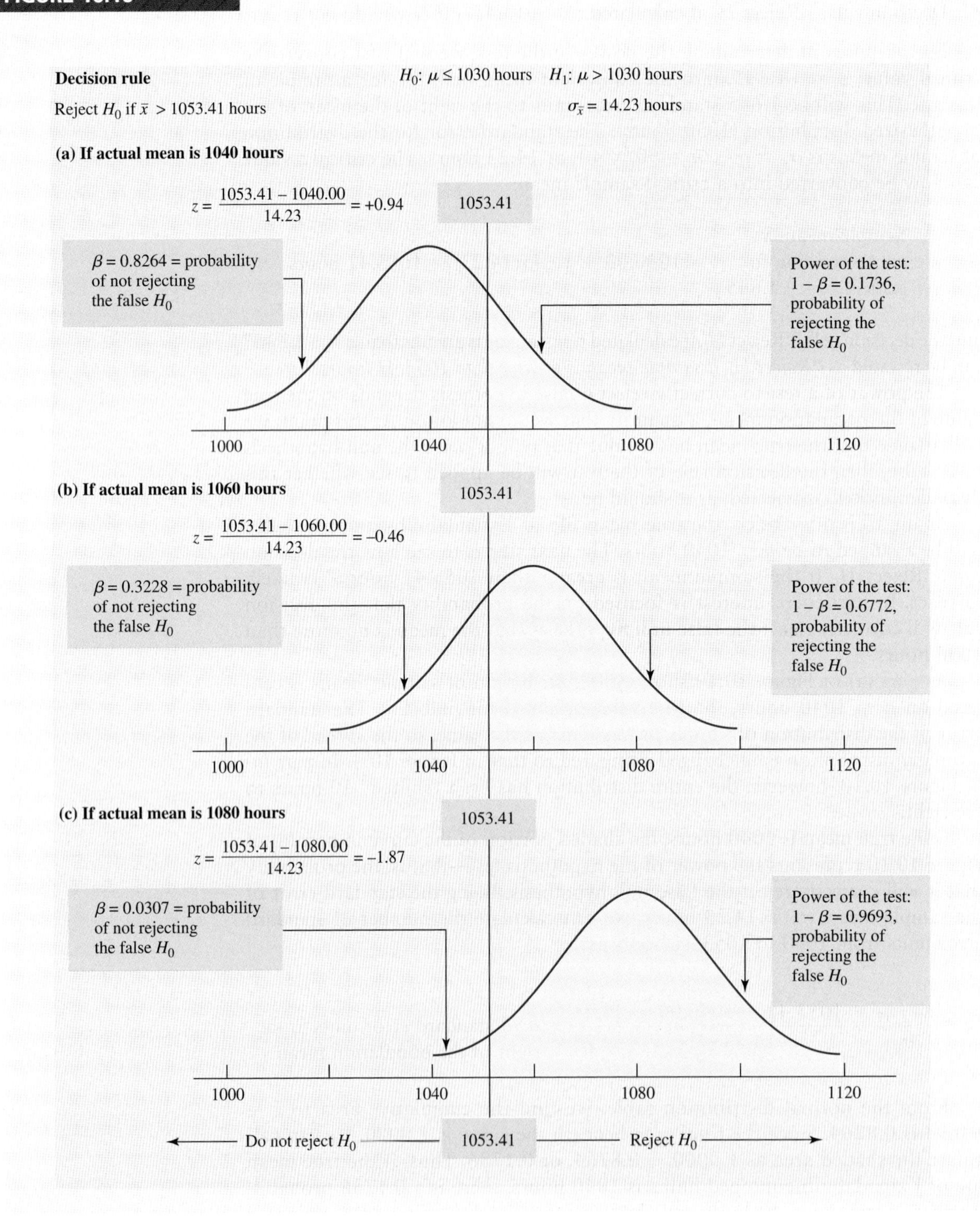

The power of the test $(1 - \beta)$ is the probability that the decision rule will correctly reject a false null hypothesis. For example, if the population mean were really 1040 hours (part a), there would be a 0.1736 probability that the decision rule would correctly reject the null hypothesis that $\mu \leq 1030$.

The Power Curve for a Hypothesis Test

One-Tail Test

In the preceding example, we arbitrarily selected one value ($\mu = 1040$ hours) that would make the null hypothesis false, then found the probability that the decision rule of the test would correctly reject the false null hypothesis. In other words, we calculated the power of the test $(1 - \beta)$ for just one possible value of the actual population mean. If we were to select many such values (e.g., $\mu = 1060$, $\mu = 1080$, $\mu = 1100$, and so on) for which H_0 is false, we could calculate a corresponding value of $(1 - \beta)$ for each of them.

For example, part (b) of Figure 10.10 illustrates the power of the test whenever the Extendabulb-equipped bulbs are assumed to have a true mean life of 1060 hours. In part (b), the power of the test is 0.6772. This is obtained by the same approach used when the true mean life was assumed to be 1040 hours, but we are now using $\mu = 1060$ hours instead of 1040.

The diagram in part (c) of Figure 10.10 repeats this process for an assumed value of 1080 hours for the true population mean. Notice how the shape of the distribution is the same in diagrams (a), (b), and (c), but that the distribution itself shifts from one diagram to the next, reflecting the new true value being assumed for μ.

Calculating the power of the test $(1 - \beta)$ for several more possible values for the population mean, we arrive at the **power curve** shown by the lower line in Figure 10.11. (The upper line in the figure will be discussed shortly.) As Figure 10.11 shows, the power of the test becomes stronger as the true population mean exceeds 1030 by a greater margin. For example, our test is almost

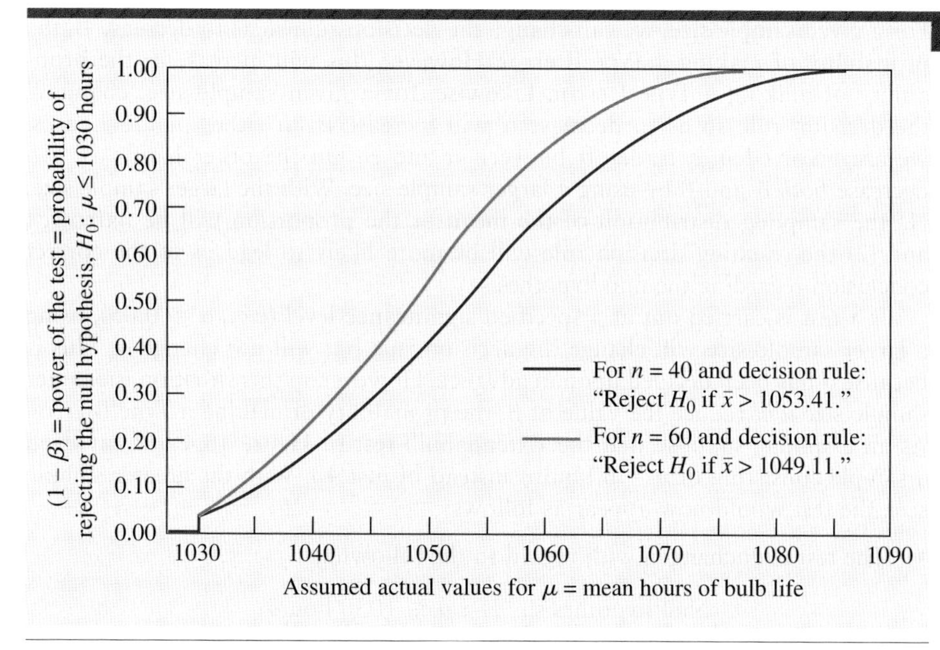

FIGURE 10.11

The power curve for the Extendabulb test of Figure 10.4 shows the probability of correctly rejecting H_0: $\mu \le 1030$ hours for a range of actual population means for which the null hypothesis would be false. If the actual population mean were 1030 hours or less, the power of the test would be zero because the null hypothesis is no longer false. The lower line represents the power of the test for the original sample size, $n = 40$. The upper line shows the increased power if the hypothesis test had been for a sample size of 60.

Note: As the specified actual value for the mean becomes smaller and approaches 1030 hours, the power of the test approaches 0.05. This occurs because (1) the mean of the hypothesized distribution for the test was set at the highest possible value for which the null hypothesis would still be true (i.e., 1030 hours), and (2) the level of significance selected in performing the test was $\alpha = 0.05$.

certain (probability = 0.9949) to reject the null hypothesis whenever the true population mean is 1090 hours. In Figure 10.11, the power of the test drops to zero whenever the true population mean is 1030 hours. This would also be true for all values lower than 1030 as well, because such actual values for the mean would result in the null hypothesis actually being true—in such cases, it would not be possible to reject a false null hypothesis, since the null hypothesis would not be false.

A complement to the power curve is known as the **operating characteristic (*OC*) curve.** Its horizontal axis would be the same as that in Figure 10.11, but the vertical axis would be identified as β instead of $(1 - \beta)$. In other words, the operating characteristic curve plots the probability that the hypothesis test will *not* reject the null hypothesis for each of the selected values for the population mean.

Two-Tail Test

In two-tail tests, the power curve will have a zero value when the assumed population mean equals the hypothesized value, then will increase toward 1.0 in both directions from that assumed value for the mean. In appearance, it will somewhat resemble an upside-down normal curve. The basic principle for power curve construction will be the same as for the one-tail test: Assume different population mean values for which the null hypothesis would be false, then determine the probability that an observed sample mean would fall into a rejection region originally specified by the decision rule of the test.

The Effect of Increased Sample Size on Type I and Type II Errors

For a given sample size, we can change the decision rule so as to decrease β, the probability of making a Type II error. However, this will increase α, the probability of making a Type I error. Likewise, for a given sample size, changing the decision rule so as to decrease α will increase β. In either of these cases, we are involved in a trade-off between α and β. On the other hand, we can decrease *both* α and β by using a larger sample size. With the larger sample size, (1) the sampling distribution of the mean or the proportion will be narrower, and (2) the resulting decision rule will be more likely to lead us to the correct conclusion regarding the null hypothesis.

If a test is carried out at a specified significance level (e.g., $\alpha = 0.05$), using a larger sample size will change the decision rule but will not change α. This is because α has been decided upon in advance. However, in this situation the larger sample size *will* reduce the value of β, the probability of making a Type II error. As an example, suppose that the Extendabulb test of Figure 10.4 had involved a sample consisting of $n = 60$ bulbs instead of just 40. With the greater sample size, the test would now appear as follows:

- The test is unchanged with regard to the following:

Null hypothesis: H_0: $\mu \leq 1030$ hours
Alternative hypothesis: H_1: $\mu > 1030$ hours
Population standard deviation: $\sigma = 90$ hours
Level of significance specified: $\alpha = 0.05$

- The following are changed as the result of $n = 60$ instead of $n = 40$:

  The standard error of the sample mean, $\sigma_{\bar{x}}$, is now

$$\frac{\sigma}{\sqrt{n}} = \frac{90}{\sqrt{60}} = 11.62 \text{ hours}$$

  The critical z of $+1.645$ now corresponds to a sample mean of

$$1030.00 + 1.645(11.62) = 1049.11 \text{ hours}$$

  The decision rule becomes, "Reject H_0 if $\bar{x} > 1049.11$ hours."

With the larger sample size and this new decision rule, if we were to repeat the process that led to Figure 10.10, we would find the following values for the power of the test. In the accompanying table, they are compared with those reported in Figure 10.10, *with each test using its own decision rule* for the 0.05 level of significance.

		Power of the Test	
		With $n = 60$ prob($\bar{x} > 1049.11$)	With $n = 40$ prob($\bar{x} > 1053.41$)
True Value of	1040	0.2177	0.1736
Population	1060	0.8264	0.6772
Mean	1080	0.9961	0.9693

For example, for $n = 60$ and the decision rule shown,

$$z = \frac{\bar{x} - \mu}{\sigma_{\bar{x}}} = \frac{1049.11 - 1080.00}{11.62} = -2.66$$

with the normal curve area to the right of $z = -2.66$ equal to 0.9961.

As these figures indicate, the same test with $n = 60$ (and the corresponding decision rule) would have a greater probability of correctly rejecting the false null hypothesis for each of the possible population means in Figure 10.10. If the Extendabulb example had remained the same, but only the sample size were changed, the $n = 60$ sample size would result in much higher $(1 - \beta)$ values than the ones previously calculated for $n = 40$. This curve is shown by the upper line in Figure 10.11. Combining two or more power curves into a display similar to Figure 10.11 can reveal the effect of various sample sizes on the susceptibility of a test to Type II error. Seeing Statistics Applet 13, at the end of the chapter, allows you to do "what-if" analyses involving a nondirectional test and its power curve values.

Computer Solutions 10.4 shows how we can use Excel or Minitab to determine the power of the test and generate a power curve like the ones in Figure 10.11. With Excel, we specify the assumed population means one at a time, while Minitab offers the capability of finding power values for an entire set of assumed population means at the same time. In either case, the procedure is much more convenient than using a pocket calculator and statistical tables, as was done in generating the values shown in Figure 10.10.

COMPUTER 10.4 SOLUTIONS

The Power Curve for a Hypothesis Test

These procedures show how to construct the power curve for a hypothesis test. The example used is the Extendabulb test described in Figures 10.4, 10.10, and 10.11.

EXCEL

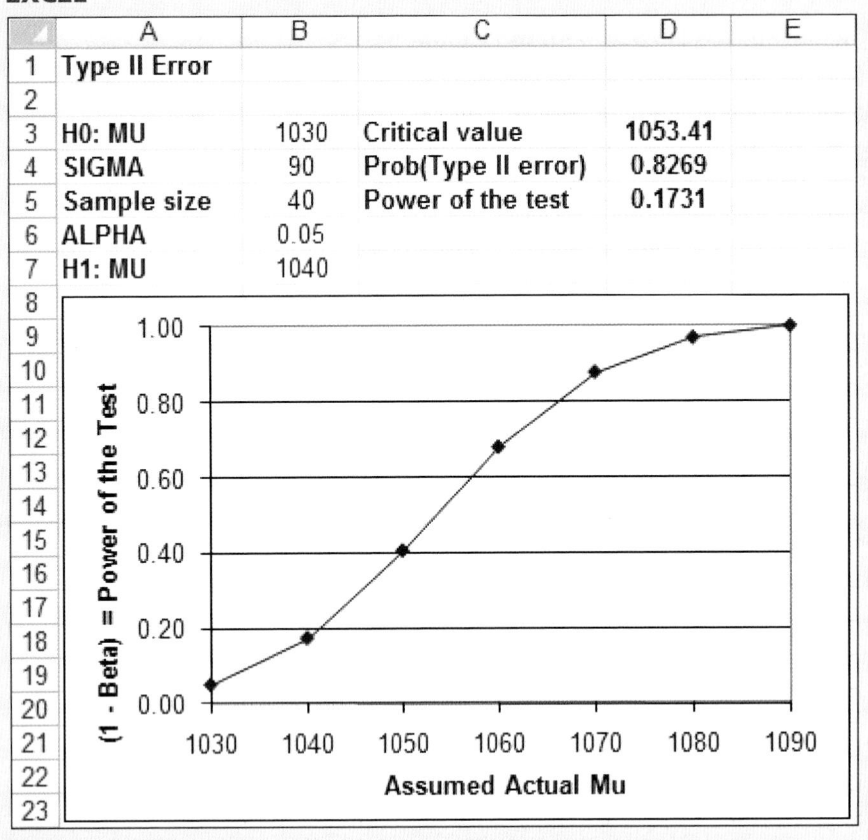

Excel procedure for obtaining the power curve for a hypothesis test

1. Open the **BETA-MEAN** workbook. Using the arrows at the bottom left, select the **Right-tail Test** worksheet.
2. Enter **1030** as the hypothesized population mean, **H0: MU.** Enter **90** as **SIGMA, 40** as the **Sample size,** and **0.05** as **ALPHA.** Enter **1040** as the value for **H1:MU,** the assumed population mean for which we want the power of the test. Excel automatically calculates the critical value for the sample mean as 1053.41 hours, and the power of the test appears in row 5 as 0.1731. This is comparable to part (a) of Figure 10.10, but is slightly more accurate because its calculation does not rely upon printed tables and the inherent gaps within them. We now have one point on the power curve for $n = 40$: Power = 0.1731 on the vertical axis for an Assumed Actual μ of 1040 hours on the horizontal axis. This result is shown above, in the upper portion of the Excel printout.
3. Repeat step 2 to obtain the power of the test for each of several other assumed actual values for μ—for example, 1030, 1050, 1060, 1070, 1080, and 1090. The resulting points can then be plotted as shown in the lower portion of the Excel printout.

MINITAB

Minitab procedure for obtaining the power curve for a hypothesis test

```
Power and Sample Size

1-Sample Z Test

Testing mean = null (versus > null)
Calculating power for mean = null + difference
Alpha = 0.05  Assumed standard deviation = 90

               Sample
Difference       Size      Power
        10         40   0.173064
        20         40   0.405399
        30         40   0.678437
        40         40   0.878205
        50         40   0.969174
        60         40   0.994937
```

1. Click **Stat.** Select **Power and Sample Size.** Click **1-Sample Z.** Enter **40** into the **Sample Sizes** box. Enter **10 20 30 40 50 60** (separated by spaces) into the **Differences** box. These are the differences between the hypothesized population mean (1030) and the selected values for the assumed actual mean (i.e., 1040, 1050, 1060, 1070, 1080, and 1090). Enter **90** into the **Standard deviation** box.
2. Click **Options.** Within the **Alternative Hypothesis** box, select **Greater than.** Enter **0.05** into the **Significance level** box. Click **OK.** Click **OK.** The printout shows the power of the test for sample sizes of 40 and for assumed actual population means of 1040 (difference from 1030 is 10), 1050, 1060, 1070, 1080, and 1090. It corresponds to the values plotted in Figure 10.11. Minitab will also generate a power curve graph (not shown here).

EXERCISES

10.79 What is a power curve and how is it applicable to hypothesis testing?

10.80 What is an operating characteristic curve and how is it related to the power curve for a test?

10.81 A hypothesis test has been set up and is to be conducted at the $\alpha = 0.05$ level of significance. If the sample size is doubled, what will be the effect on α? On β?

10.82 For the test described in Exercise 10.31, if the true population mean is really 2.520 inches, what is the probability that the inspector will correctly reject the false null hypothesis that $\mu = 2.500$ inches?

10.83 For the test described in Exercise 10.32, assume that the true population mean for the new booklet is $\mu = 2.80$ hours. Under this assumption, what is the probability that the false null hypothesis, H_0: $\mu \geq 3.00$ hours, will be rejected?

10.84 Using assumed true population means of 2.485, 2.490, 2.495, 2.500, 2.505, 2.510, and 2.515 inches, plot the power curve for the test in Exercise 10.31.

10.85 Using assumed true population means of 2.80, 2.85, 2.90, 2.95, and 3.00 hours, plot the power curve for the test in Exercise 10.32.

10.86 Using assumed true population percentages of 2%, 3%, 4%, 5%, 6%, and 7%, plot the power curve for the test in Exercise 10.63.

10.87 When a thread-cutting machine is operating properly, only 2% of the units produced are defective. Since the machine was bumped by a forklift truck, however, the quality-control manager is concerned that it may require expensive downtime and a readjustment. The manager would like to set up a right-tail test at the 0.01 level of significance, then identify the proportion of defectives in a sample of 400 units produced. His null hypothesis is H_0: $\pi \leq 0.02$, with H_1: $\pi > 0.02$.

a. What critical value of z will be associated with this test?
b. For the critical z determined in part (a), identify the sample proportion that this z value represents, then use this value of p in stating a decision rule for the test.
c. What is the probability that the quality-control manager will fail to reject a false H_0 if the actual population proportion of defectives is $\pi = 0.02$? If $\pi = 0.03$? If $\pi = 0.04$? If $\pi = 0.05$? If $\pi = 0.06$?
d. Making use of the probabilities calculated in part (c), plot the power and operating characteristic curves for the test.

10.88 Plot the operating characteristic curve that corresponds to the power curve constructed for Exercise 10.84.

10.89 Plot the operating characteristic curve that corresponds to the power curve constructed for Exercise 10.86.

10.8 SUMMARY

- **Hypothesis testing: null and alternative hypotheses**

Hypothesis testing is a method of using sample information to evaluate a claim that has been converted into a set of statements called the null and alternative hypotheses. The null hypothesis can be either rejected or not rejected; if it is not rejected, the alternative hypothesis, which asserts the null hypothesis to be wrong, must be rejected. The null hypothesis is given the benefit of the doubt and is assumed to be true unless we are faced with statistically overwhelming evidence to the contrary.

- **Directional and nondirectional hypothesis tests**

A verbal statement from which a hypothesis test originates may be either directional or nondirectional. Accordingly, the appropriate hypothesis test will also be either directional (one-tail test) or nondirectional (two-tail test). The object of the hypothesis test may be a single population mean or proportion, as discussed in this chapter, or the difference between two population means or proportions (covered in the next chapter).

- **Type I and Type II errors**

When we reject or fail to reject the null hypothesis, there is the possibility that our conclusion may be mistaken. Rejecting a true null hypothesis is known as a Type I error, while failing to reject a false null hypothesis results in a Type II error. Alpha (α), the probability of incorrectly rejecting a true null hypothesis, is also known as the level of significance of the test. Beta (β) is the probability of incorrectly failing to reject a false null hypothesis. With sample size and type of test (one-tail versus two-tail) being equal, reducing α tends to increase β, and vice versa, though random samples of larger size can simultaneously reduce both α and β.

● The hypothesis testing procedure

Hypothesis testing involves (1) formulating the null and alternative hypotheses, (2) selecting the significance level (α) to be used in the test, (3) selecting the test statistic and calculating its value, (4) identifying the critical value(s) for the test statistic and stating the decision rule, (5) comparing calculated and critical value(s) for the test statistic and reaching a conclusion about the null hypothesis, and (6) making the related business decision for the situation that precipitated the hypothesis test in the first place. In a two-tail test, there will be two critical values for the test statistic; in a one-tail test, a single critical value.

● Tests involving a sample mean or proportion

When testing a sample mean, an important consideration in the selection of the test statistic (z or t) is whether the population standard deviation (σ) is known. If other assumptions are met, the t-test is appropriate whenever σ is unknown. Figure 10.2 summarizes the test statistic selection process and related assumptions.

If the calculated test statistic falls into a rejection region defined by the critical value(s), the difference between the sample mean or proportion and its hypothesized population value is judged to be too great to have occurred by chance. In such a case, the null hypothesis is rejected.

● The *p*-value for a hypothesis test

The p-value for a hypothesis test is the level of significance where the calculated value of the test statistic is exactly the same as a critical value, and is another way of expressing the level of significance (α) of the test. When the hypothesis test is computer assisted, the computer output will include a p-value for use in interpreting the results. The p-value is the lowest level of significance at which the null hypothesis can be rejected.

● Confidence intervals and nondirectional tests

Although confidence intervals and hypothesis tests have different purposes, the concepts are actually related. In the case of a nondirectional test for a population mean, the conclusion at the α level of significance will be the same as that reached on the basis of a confidence interval constructed for the $100(1 - \alpha)\%$ confidence level.

● Power and operating characteristic curves for a test

Another measure of the performance of a hypothesis test is the power of the test. This is the probability $(1 - \beta)$ that the test will avoid a Type II error by correctly rejecting a false null hypothesis. When the power of the test (vertical axis) is plotted against a range of population parameter values for which the null hypothesis would be false (horizontal axis), the result is known as the power curve for the test. A complement to the power curve is the operating characteristic (OC) curve, which has the same horizontal axis, but uses β instead of $(1 - \beta)$ for the vertical axis. The OC curve expresses the probability that the test will make a Type II error by incorrectly failing to reject the null hypothesis for a range of population parameter values for which the null hypothesis would be false.

EQUATIONS

Testing a Mean, Population Standard Deviation Known

z-test, with test statistic:

$$z = \frac{\bar{x} - \mu_0}{\sigma_{\bar{x}}}$$

where $\sigma_{\bar{x}}$ = standard error for the sample mean, $= \sigma/\sqrt{n}$
$\bar{x}$ = sample mean
μ_0 = hypothesized population mean
n = sample size

If the population is not normally distributed, n should be ≥ 30 for the central limit theorem to apply. The population σ is usually not known.

Testing a Mean, Population Standard Deviation Unknown

t-test, with test statistic:

$$t = \frac{\bar{x} - \mu_0}{s_{\bar{x}}}$$

where $s_{\bar{x}}$ = estimated standard error for the sample mean, $= s/\sqrt{n}$
$\bar{x}$ = sample mean
μ_0 = hypothesized population mean
n = sample size and $df = n - 1$

For an unknown σ, and when the population is approximately normally distributed, the t-test is appropriate *regardless of the sample size.* As n increases, the normality assumption becomes less important. If $n < 30$ and the population is not approximately normal, nonparametric testing (e.g., the sign test for central tendency, in Chapter 14) may be applied. The t-test is "robust" in terms of not being adversely affected by slight departures from the population normality assumption.

Testing a Proportion

z-test, with test statistic:

$$z = \frac{p - \pi_0}{\sigma_p}$$

where p = sample proportion
π_0 = hypothesized population proportion
n = sample size
σ_p = standard error of the distribution of the sample proportion

$$\sigma_p = \sqrt{\frac{\pi_0(1 - \pi_0)}{n}}$$

When $n\pi \geq 5$ and $n(1 - \pi) \geq 5$, the normal distribution is considered to be a good approximation to the binomial distribution. If this condition is not met, the exact probabilities must be derived from the binomial table. Most practical business settings involving proportions satisfy this condition, and the normal approximation is used in the chapter.

Confidence Intervals and Hypothesis Testing

Confidence intervals and hypothesis testing are related. For example, in the case of a nondirectional test for a population mean, the conclusion at the α level of significance will be the same as that reached on the basis of a confidence interval constructed for the $100(1 - \alpha)\%$ confidence level.

CHAPTER EXERCISES

10.90 The average U.S. family includes 3.13 persons. To determine whether families in her city tend to be representative in size compared to those across the United States, a city council member selects a simple random sample of 40 families. She finds the average number of persons in a family to be 3.40, with a standard deviation of 1.10. Source: Bureau of the Census, *Statistical Abstract of the United States 2009*, p. 52.

a. Using a nondirectional hypothesis test and a level of significance of your choice, what conclusion would be reached?
b. For the significance level used in part (a), construct the appropriate confidence interval and verify the conclusion reached in part (a).

10.91 Before the hiring of an efficiency expert, the mean productivity of a firm's employees was 45.4 units per hour, with a standard deviation of 4.5 units per hour. After incorporating the changes recommended by the expert, it was found that a sample of 30 workers produced a mean of 47.5 units per hour. Using the 0.01 level of significance, can we conclude that the mean productivity has increased?

10.92 The average U.S. household has $235,600 in life insurance. A local insurance agent would like to see how households in his city compare to the national average, and selects a simple random sample of 30 households from the city. For households in the sample, the average amount of life insurance is $\bar{x} = \$245{,}800$, with $s = \$25{,}500$. Source: reuters.com, September 23, 2008.

a. Using a nondirectional hypothesis test and a level of significance of your choice, what conclusion would be reached?
b. For the significance level used in part (a), construct the appropriate confidence interval and verify the conclusion reached in part (a).

10.93 Before accepting a large shipment of bolts, the director of an elevator construction project checks the tensile strength of a simple random sample consisting of 20 bolts. She is concerned that the bolts may be counterfeits, which bear the proper markings for this grade of bolt, but are made from inferior materials. For this application, the genuine bolts are known to have a tensile strength that is normally distributed with a mean of 1400 pounds and a standard deviation of 30 pounds. The mean tensile strength for the bolts tested is 1385 pounds. Formulate and carry out a hypothesis test to examine the possibility that the bolts in the shipment might not be genuine.

10.94 Technical data provided with a room dehumidifier explain that units tested from the production line consume a mean of 800 watts of power, with a standard deviation of 12 watts. The superintendent of a large office complex, upon evaluating 30 of the units recently purchased and finding their mean power consumption to be 805 watts, claims that her shipment does not meet company specifications. At the 0.05 level of significance, is the superintendent's complaint justified?

10.95 Strands of human hair absorb elements from the bloodstream and provide a historical record of both health and the use or nonuse of chemical substances. Hair grows at the rate of about one-half inch per month, and a person with long hair might be accused or absolved on the basis of a segment of hair that sprouted many months or years ago. By separately analyzing sections of the hair strand, scientists can even approximate the periods of time during which drug use was heavy, moderate, light, or altogether absent. If a transit employee provides a strand of hair for drug testing, state the null and alternative hypotheses in verbal terms, then identify what would constitute a Type I error and a Type II error in this situation. Source: omegalabs.com, July 24, 2009.

10.96 A regional office of the Internal Revenue Service randomly distributes returns to be audited to the pool of auditors. Over the thousands of returns audited last year, the average amount of extra taxes collected was $356 per audited return. One of the auditors, Jeffrey Jones, is suspected of being too lenient with persons whose returns are being audited. For a simple random sample of 30 of the returns audited by Mr. Jones last year, an average of $322 in extra taxes was collected, with a standard deviation of $90. Based on this information and a hypothesis test of your choice, do the suspicions regarding Mr. Jones appear to be justified?

10.97 For each of the following situations, determine whether a one-tail test or a two-tail test would be appropriate. Describe the test, including the null and alternative hypotheses, then explain your reasoning in selecting it.

a. A machine that has not been serviced for several months is producing output in which 5% of the items are defective. The machine has just been serviced and quality should now be improved.
b. In a speech during her campaign for reelection, a Republican candidate claims that 55% of registered Democrats in her county intend to vote for her.
c. Of those who have bought a new car in the past, a dealer has found that 70% experience three or more mechanical problems in the first four months of

ownership. Unhappy with this percentage, the dealer has heavily revised the procedure by which predelivery mechanical checks are carried out.

10.98 In a taste comparison test, it was found that 58 of 100 persons preferred the chunky version of a peanut butter over the creamy type. An interested observer would like to determine whether this proportion (0.58) is significantly greater than the (0.50) proportion that would tend to result from chance. Using the 0.05 level of significance, what conclusion will be reached?

10.99 Historically, Shop-Mart has gotten an average of 2000 hours of use from its G&E fluorescent lightbulbs. Because its fixtures are attached to the ceiling, the bulbs are rather cumbersome to replace, and Shop-Mart is looking into the possibility of switching to Phipps bulbs, which cost the same. A sample of Phipps bulbs lasted an average of 2080 hours, and the p-value in a right-tail test (H_0: $\mu \leq 2000$ versus H_1: $\mu > 2000$) is 0.012. Shop-Mart has shared these results with both G&E and Phipps.

a. Give Shop-Mart a recommendation. In doing so, use "0.012" in a sentence that would be understood by a Shop-Mart executive who has had no statistical training but is comfortable with probabilities as used by weather forecasters.
b. In interpreting these results, what level of significance might G&E like to use in reaching a conclusion? Explain.
c. In interpreting these results, what level of significance might Phipps like to use in reaching a conclusion? Explain.
d. If the test had been two-tail (i.e., H_0: $\mu = 2000$ versus H_1: $\mu \neq 2000$) instead of right-tail, would the p-value still be 0.012? If not, what would the p-value be? Explain.

10.100 The administrator of a local hospital has told the governing board that 30% of its emergency room patients are not really in need of emergency treatment (i.e., the problems could just as easily have been handled by an appointment with their family physician). In checking a random sample of 400 emergency room patients, a board member finds that 35% of those treated were not true emergency cases. Using an appropriate hypothesis test and the 0.05 level, evaluate the administrator's statement.

10.101 A consumer agency suspects that a pet food company may be underfilling packages for one of its brands. The package label states "1600 grams net weight," and the president of the company claims the average weight is at least 1600 grams. For a simple random sample of 35 boxes collected by the consumer agency, the mean and standard deviation were 1591.7 grams and 18.5 grams, respectively.

a. At the 0.05 level of significance, what conclusion should be reached by the consumer agency? Would the president of the company prefer to use a different level of significance in reaching a conclusion? Explain.
b. Use the decision rule associated with part (a) and a range of selected assumed values for μ in constructing the power curve for the test.

10.102 A 1931 issue of *Time* magazine contained an advertisement for *Liberty* magazine. According to a study cited by *Liberty*, it was found that 15% of *Liberty* families had a "mechanical refrigerator," compared to just 8% for all U.S. families. Assuming that the study included a sample of 120 *Liberty* families, and using the 0.01 level of significance, could the percentage of *Liberty* families owning this modern (for 1931) convenience be higher than for the nation overall? Source: *Time*, August 10, 1931, p. 49.

10.103 In an interview with a local newspaper, a respected trial lawyer claims that he wins at least 75% of his court cases. Bert, a skeptical statistics student, sets up a one-tail test at the 0.05 level of significance to evaluate the attorney's claim. The student plans to examine a random sample of 40 cases tried by the attorney and determine the proportion of these cases that were won. The null and alternative hypotheses are H_0: $\pi \geq 0.75$ and H_1: $\pi < 0.75$. Using the techniques of this chapter, Bert sets up a hypothesis test in which the decision rule is "Reject H_0 if $z < -1.645$, otherwise do not reject." What is the probability that Bert will make a Type II error (fail to reject a false null hypothesis) if the attorney's true population proportion of wins is actually

a. $\pi = 0.75$? b. $\pi = 0.70$? c. $\pi = 0.65$?
d. $\pi = 0.60$? e. $\pi = 0.55$?
f. Making use of the probabilities calculated in parts (a) through (e), describe and plot the power curve for Bert's hypothesis test.

10.104 During 2006, 3.0% of all U.S. households were burglary victims. For a simple random sample of 300 households from a certain region, suppose that 18 households were victimized by burglary during that year. Apply an appropriate hypothesis test and the 0.05 level of significance in determining whether the region should be considered as having a burglary problem greater than that for the nation as a whole. Source: Bureau of the Census, *Statistical Abstract of the United States 2009*, p. 195.

10.105 A national chain of health clubs says the mean amount of weight lost by members during the past month was at least 5 pounds. Skeptical of this claim, a consumer advocate believes the chain's assertion is an exaggeration. She interviews a random sample of 40 members, finding their mean weight loss to be 4.6 pounds, with a standard deviation of 1.5 pounds. At the 0.01 level of significance, evaluate the health club's contention.

10.106 A state transportation official claims that the mean waiting time at exit booths from a toll road near the capitol is no more than 0.40 minutes. For a sample of 35 motorists exiting the toll road, it was found that the mean waiting time was 0.46 minutes, with a standard deviation of 0.16 minutes. At the 0.05 level of significance, can we reject the official's claim?

10.107 An exterminator claims that no more than 10% of the homes he treats have termite problems within 1 year after treatment. In a sample of 100 homes, local officials find that 14 had termites less than 1 year after being treated. At the 0.05 level of significance, evaluate the credibility of the exterminator's statement.

(**DATA SET**) *Note:* Exercises 10.108–10.113 require a computer and statistical software.

10.108 Before installing a high-tech laser exhibit near the exit area, a museum of American technology found the average contribution by patrons was $2.75 per person. For a sample of 30 patrons following installation of the new exhibit, the contributions are as listed in data file **XR10108**. Based on the sample data, and using the 0.025 level of significance, can we conclude that the new exhibit has increased the average contribution of exhibit patrons? Identify and interpret the p-value for the test.

10.109 According to the U.S. Department of Justice, 25% of violent crimes involve the use of a weapon. We will assume that data file **XR10109** contains a sample of the crime information for a given city, with data for the 400 crimes coded as 1 = crime involved a weapon and 2 = crime did not involve a weapon. Using the 0.05 level in a two-tail test, was the percentage of violent crimes involving a weapon in this city significantly different from the 25% in the nation as a whole? Identify and interpret the p-value for the test. Source: Bureau of the Census, *Statistical Abstract of the United States 2009*, p. 194.

10.110 Based on the sample results in Exercise 10.109, construct and interpret the 95% confidence interval for the population proportion. Is the hypothesized proportion (0.25) within the interval? Given the presence or absence of the 0.25 value within the interval, is this consistent with the findings of the hypothesis test conducted in Exercise 10.109?

10.111 In making aluminum castings into alternator housings, an average of 3.5 ounces per casting must be trimmed off and recycled as a raw material. A new manufacturing procedure has been proposed to reduce the amount of aluminum that must be recycled in this way. For a sample of 35 castings made with the new process, data file **XR10111** lists the weights of aluminum trimmed and recycled. Based on the sample data, and using the 0.01 level, has the new procedure significantly reduced the average amount of aluminum trimmed and recycled? Identify and interpret the p-value for the test.

10.112 Use the decision rule associated with Exercise 10.111 and a range of selected assumed values for μ to construct the power curve for the test.

10.113 In the past, the mean lifetime of diesel engine injection pumps has been 12,000 operating hours. A new injection pump is available that is promoted as lasting longer than the old version. In a test of 50 of the new pumps, the lifetimes are as listed in data file **XR10113**. Based on the sample data, and using the 0.025 level of significance, examine the possibility that the new pumps might have a mean life that is no more than that of the old design. Identify and interpret the p-value for the test.

INTEGRATED CASES

Thorndike Sports Equipment

The first item on Ted Thorndike's agenda for today is a meeting with Martha Scott, a technical sales representative with Cromwell Industries. Cromwell produces a racquet-stringing machine that is promoted as being the fastest in the industry. In their recent telephone conversation, Martha offered Ted the chance to try the machine for a full week, then return it to Cromwell if he decides it doesn't offer a worthwhile improvement.

Currently, the Thorndike racquetball and tennis racquets are strung using a type of machine that has been around for about 10 years, and the Thorndikes have been very pleased with the results. When efficiency experts visited the plant last year, they found it took an average of 3.25 minutes to string a racquetball racquet and an average of 4.13 minutes for a tennis racquet. Both distributions were found to be approximately normal.

Ms. Scott comes by, explains the operation of the machine, and assists in its installation. Because its controls are similar in function and layout to the older model, no operator training is necessary. During the 1-week trial, the new machine will be working side-by-side with older models in the racquet-stringing department.

After Ms. Scott leaves, Luke and Ted Thorndike discuss the machine further. Luke indicates that the oldest of the current machines is about due for replacement and, if the Cromwell model is really faster, he will buy the Cromwell to replace it. It's possible that Cromwell models would also be purchased as other stringing machines approach the end of their operating lifetimes and are retired. Luke cautions Ted that the Cromwell must be purchased only if it is indeed faster than the current models—although Ms. Scott seems to have an honest face, he's never trusted anybody whose briefcase contains brochures. The Cromwell must be quicker in stringing both racquetball racquets and tennis racquets. Otherwise, the firm will continue purchasing and using the current model.

Evaluating the Cromwell model, Ted measures the exact time required for each racquet in separate samples consisting of 40 racquetball racquets and 40 tennis racquets. The times are given here and are also in data file **THORN10**.

Since Luke seems quite adamant against buying something that isn't any faster than the current models, Ted must be very confident that any increased speeds measured are not simply due to chance. Using a computer statistical package to analyze the sample times, do the respective tests appear to warrant purchase of the Cromwell machine, or should it be returned to Cromwell at the end of the week?

40 Racquetball Racquets:

2.97	3.43	3.10	2.79	3.41	3.22	3.36	3.46
3.11	3.26	3.36	3.48	3.09	3.06	2.67	3.13
2.97	3.09	3.18	2.71	3.41	2.83	3.20	3.32
2.70	3.17	3.16	3.16	3.24	2.91	3.34	2.97
3.14	3.36	3.89	3.21	3.13	2.92	2.94	3.18

minutes

40 Tennis Racquets:

3.77	3.97	4.08	4.36	3.85	4.65	3.89	3.99
3.51	3.43	4.31	3.84	3.86	3.50	4.23	3.88
3.91	4.47	4.27	3.70	3.80	4.54	4.46	3.97
4.49	3.74	4.17	3.68	4.09	4.27	3.84	3.51
3.71	3.77	3.90	4.14	4.03	4.61	3.63	4.62

minutes

Springdale Shopping Survey

The Springdale computer database exercise of Chapter 2 listed 30 shopping-related questions asked of 150 respondents in the community of Springdale. The data are in file **SHOPPING**. The coding key for these responses was also provided in the Chapter 2 exercise. In this section, hypothesis testing will be used to examine the significance of some of the survey results.

From a managerial perspective, the information gained from this database exercise would be useful in determining whether the mean attitude toward a given area differed significantly from the "neutral" level, in observing and comparing the mean attitude scores exhibited toward the three shopping areas, and in examining the strengths and weaknesses of a manager's shopping area relative to the other two areas. If such a study were repeated from time to time, management could observe the extent to which the overall attitude toward their shopping area was becoming stronger or weaker in the eyes of the consumer, thus helping them select business strategies that could retain and take advantage of their area's strengths and either correct or minimize the effect of its weaknesses.

1. Item C in the description of the data collection instrument lists variables 7, 8, and 9, which represent the respondent's general attitude toward each of the three shopping areas. Each of these variables has numerically equal distances between the possible responses, and for purposes of analysis they may be considered to be of the interval scale of measurement.
   a. Calculate the mean attitude score toward the Springdale Mall shopping area and compute the sample standard deviation. Does this area seem to be well regarded by the respondents?
   b. In a two-tail test at the 0.10 level, compare the sample mean calculated in part (a) to the 3.0 "neutral" value. Use the null hypothesis H_0: $\mu = 3.0$ and the alternative hypothesis H_1: $\mu \neq 3.0$.
   c. Generate the 90% confidence interval for the population mean and use this interval in verifying the conclusion obtained in part (b).
   d. What is the p-value associated with the hypothesis test?
   e. Repeat parts (a) through (d) for the Downtown shopping area.
   f. Repeat parts (a) through (d) for the West Mall shopping area.
2. If Springdale Mall offered exactly the same benefits as the other two shopping areas, we would expect exactly one-third of those who expressed an opinion to select it as the area best fitting the description for variable number 10 ("Easy to return/exchange goods"). In testing whether Springdale Mall differs significantly from this expected proportion, the null and alternative hypotheses will be H_0: $\pi = 0.333$ and H_1: $\pi \neq 0.333$. Carry out the following analyses for Springdale Mall:
   a. Analyzing the data for variable 10, use the preceding null and alternative hypotheses in testing the null hypothesis at the 0.05 level.
   b. Determine the p-value for the test conducted in part (a).
   c. Repeat parts (a) and (b) for variables 11–17.
   d. Based on the preceding analyses, identify the principal strengths and weaknesses exhibited by Springdale Mall.

BUSINESS CASE

Pronto Pizza (A)

© Adam Crowley/Photodisc/Getty Images

Pronto Pizza is a family-owned pizza restaurant in Vinemont, a small town of 20,000 people in upstate New York. Antonio Scapelli started the business 30 years ago as Antonio's Restaurant with just a few thousand dollars. Antonio, his wife, and their children, most of whom are now grown, operate the business. Several years ago, one of Antonio's sons, Tony Jr., graduated from NYU with an undergraduate degree in business administration. After graduation, he came back to manage the family business. Pronto Pizza was one of the earliest pizza restaurants to offer pizza delivery to homes. Fortunately, Tony had the foresight to make this business decision a few years ago. At the same time, he changed the restaurant's name from Antonio's to Pronto Pizza to emphasize the pizza delivery service. The restaurant has thrived since then and has become one of the leading businesses in the area. While many of their customers still "dine in" at the restaurant, nearly 90% of Pronto's current business is derived from the pizza delivery service.

Recently, one of the national chains of fast-food pizza delivery services found its way to Vinemont, New York. In order to attract business, this new competitor guarantees delivery of its pizzas within 30 minutes after the order is placed. If the delivery is not made within 30 minutes, the customer receives the order without charge. Before long, there were signs that this new pizza restaurant was taking business away from Pronto Pizza. Tony realized that Pronto Pizza would have to offer a similar guarantee in order to remain competitive.

After a careful cost analysis, Tony determined that to offer a guarantee of 29 minutes or less, Pronto's average delivery time would have to be 25 minutes or less. Tony thought that this would limit the percentage of "free pizzas" under the guarantee to about 5% of all deliveries, which he had figured to be the break-even point for such a promotion. To be sure of Pronto's ability to deliver on a promise of 29 minutes or less, Tony knew that he needed to collect data on Pronto's pizza deliveries.

Pronto Pizza's delivery service operates from 4:00 P.M. to midnight every day of the week. After an order for a pizza is phoned in, one of the two cooks is given the order for preparation. When the crust is prepared and the ingredients have been added, the pizza is placed on the belt of the conveyor oven. The speed of the conveyor is set so that pizzas come out perfectly, time after time. Once the pizza is ready and one of Pronto's drivers is available to make the delivery, the pizza is taken in a heat-insulated bag to the customer. Pronto uses approximately five to six drivers each night for deliveries. Most of the drivers hired by Pronto Pizza are juniors and seniors at the local high school.

Given the large number of deliveries made each evening, Tony knew that he could not possibly monitor every single delivery. He had thought of the possibility of having someone else collect the data, but given the importance of accurate data, he decided to make all of the measurements himself. This, of course, meant taking a random sample of, rather than all, deliveries over some time period. Tony decided to monitor deliveries over the course of a full month. During each hour of delivery service operation, he randomly selected a phoned-in order. He then carefully measured the time required to prepare the order and the amount of time that the order had to wait for a delivery person to become available. Tony would then go with the delivery person to accurately measure the delivery time. After returning, Tony randomly selected an order placed during the next hour and repeated the process. At the end of the month, Tony had collected data on 240 deliveries.

Once the data were available, Tony knew there were several issues that should be addressed. He was committed to going with the 29-minute delivery guarantee unless the data strongly indicated that the true average delivery time was greater than 25 minutes. How would he make this decision? Tony also realized that there were three components that could affect pizza delivery times: the preparation time, the waiting time for an available driver, and the travel time to deliver the pizza to the customer. Tony hoped that he had collected sufficient data to allow him to determine how he might improve the delivery operation by reducing the overall delivery time.

Assignment

Tony has asked you for some assistance in interpreting the data he has collected. In particular, he needs to know whether the true average delivery time for Pronto Pizza might be more than 25 minutes. Use the data in the file **PRONTO** in answering his question. A description of this data set is given in the accompanying Data Description section. Based on your examination of the data, provide Tony with some suggestions that would help him in making his decision about the 29-minute delivery guarantee and in improving his pizza delivery service. The case questions will assist you in your analysis of the data. Use important details from your analysis to support your recommendations.

SEEING STATISTICS: APPLET 12

z-Interval and Hypothesis Testing

When this applet loads, it shows the hypothesized population mean for the nondirectional z-test discussed in this section as "Null Hyp 1.325." It also shows the sample mean ("Sample Mean 1.3229") along with the resulting 95% confidence interval for the population mean. In this example, we took only one sample. However, by moving the slider at the bottom of the graph, we can see what the confidence interval *would have been* if our sample mean had been lower or higher than the 1.3229 minutes we actually obtained.

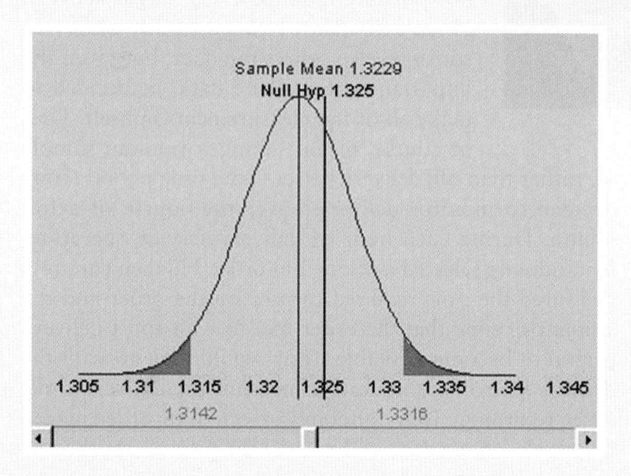

Note that the confidence interval limits shown in the graph may sometimes differ slightly from those we would calculate using the pocket calculator and our standard normal distribution table. This is because the applet is using more exact values for z than we are able to show within printed tables like the one in the text.

Applet Exercises

12.1 When the applet initially loads, identify the sample mean and describe the 95% confidence interval for μ. Based on this confidence interval, would it seem believable that the true population mean might be 1.325 minutes?

12.2 Using the slider, increase the sample mean to approximately 1.330 minutes. What is the 95% confidence interval for μ? Based on this confidence interval, would it seem believable that the true population mean might be 1.325 minutes?

12.3 Using the slider, decrease the sample mean to approximately 1.310 minutes. Based on this confidence interval, would it seem believable that the true population mean might be 1.325 minutes?

Try out the Applets: http://www.cengage.com/international

Data Description

The **PRONTO** file contains the data collected by Tony Scapelli over the past month on pizza deliveries. Data recorded for each delivery order are as shown in the table below.

The variables are defined as follows:

DAY: Day of the week (1 = Monday, 7 = Sunday)

HOUR: Hour of the day (4–11 P.M.)

PREP_TIME: Time required (in minutes) to prepare the order

WAIT_TIME: Time (in minutes) from completing preparation of the order until a delivery person was available to deliver the order

TRAVEL_TIME: Time (in minutes) it took the car to reach the delivery location

DISTANCE: Distance (in miles) from Pronto Pizza to the delivery location

(*Note:* You will first have to help Tony by creating a new variable called TOT_TIME, which represents the total amount of time from the call being received to the delivery being made. TOT_TIME is the total of PREP_TIME, WAIT_TIME, and TRAVEL_TIME. It is the time to which the guarantee would be applied and is the time of interest in the questions below.)

1. Is there sufficient evidence in the data to conclude that the average time to deliver a pizza, from the time of the phone call to the time of delivery, is greater

DAY	HOUR	PREP_TIME	WAIT_TIME	TRAVEL_TIME	DISTANCE
5	4	14.86	3.08	6.02	2.5
5	5	14.84	13.81	5.47	3.3
5	6	15.41	9.91	8.99	4.9
5	7	16.34	2.08	7.98	3.8
5	8	15.19	2.69	9.01	4.9
⋮	⋮	⋮	⋮	⋮	⋮

SEEING STATISTICS: APPLET 13

Statistical Power of a Test

This applet allows "what-if" analyses of the power of a nondirectional test with null and alternative hypotheses H_0: $\mu = 10$ versus H_1: $\mu \neq 10$. There are three sliders, allowing us to choose the α level for the test (left slider), the sample size (right slider), and the actual value of μ (identified as "Alt Hyp" and controlled by the bottom slider). As we move any of the sliders, we can immediately see the power of the test in the display at the top of the graph. Recall that the power of the test is $(1 - \beta)$, and that it is the probability of rejecting a false null hypothesis.

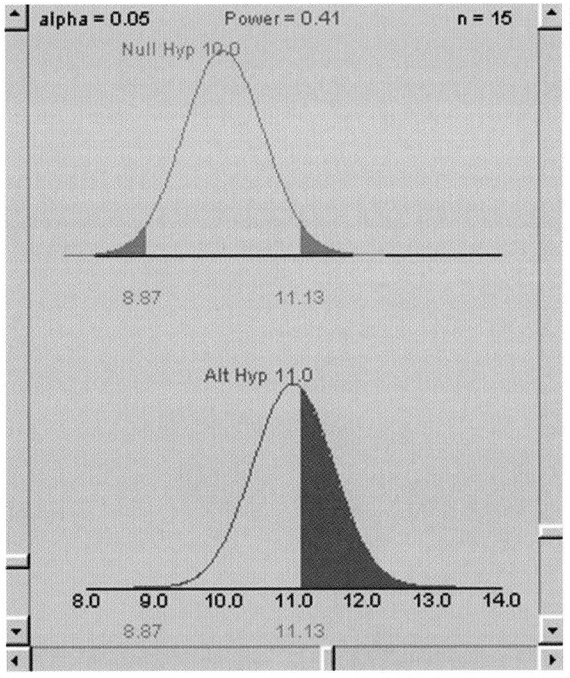

The distribution in the upper portion of the chart shows the hypothesized population mean (H_0: $\mu = 10$) as well as the lower and upper critical values for the sample mean in a nondirectional test for the combination of α and n being used. The sum of the blue areas in the lower distribution show, for a given α, actual μ, and n, the probability that $\bar{x}$ will fall outside the critical values and cause us to reject the null hypothesis.

alpha = 0.05 Power = 0.41 n = 15
Null Hyp 10.0
8.87 11.13
Alt Hyp 11.0
8.0 9.0 10.0 11.0 12.0 13.0 14.0
8.87 11.13

Applet Exercises

13.1 Using the left and right sliders, set up the test so that $\alpha = 0.10$ and $n = 20$. Now move the bottom slider so that the actual μ is as close as possible to 10 without being equal to 10 (e.g., 10.01). What is the value for the power of the test? Is this the value you would expect?

13.2 Set the left, right, and bottom sliders so that $\alpha = 0.05$, $n = 15$, and $\mu =$ the same value you selected in Applet Exercise 13.1. Now move the left slider upward and downward to change the α level for the test. In what way(s) do these changes in α affect the power of the test?

13.3 Set the left and bottom sliders so that $\alpha = 0.05$ and $\mu = 11.2$. Now move the right slider upward and downward to change the sample size for the test. Observe the power of the test for each of the following values for n: 2, 10, 20, 40, 60, 80, and 100.

13.4 Set the right and bottom sliders so that $n = 20$ and $\mu = 11.2$. Now move the left slider upward and downward to change the α level for the test. Observe the power of the test for each of the following values for α: 0.01, 0.02, 0.05, 0.10, 0.20, 0.30, 0.40, and 0.50.

Try out the Applets: http://www.cengage.com/international

than 25 minutes? Using a level of significance of your choice, perform an appropriate statistical analysis and attach any supporting computer output. What is the p-value for the test?

2. Based on the data Tony has already collected, approximately what percentage of the time will Pronto Pizza fail to meet its guaranteed time of 29 minutes or less? Explain how you arrived at your answer and discuss any assumptions that you are making. Does it appear that they will meet their requirement of failing to meet the guarantee 5% of the time or less?
3. Compare the average delivery times for different days of the week. Does the day of the week seem to have an effect on the average time a customer will have to wait for his or her pizza? Attach any supporting output for your answer.
4. Compare the average delivery times for different hours of the day. Does the time of day seem to have an effect on the average time a customer will have to wait for his or her pizza? Attach any supporting output for your answer.
5. Based on your analysis of the data, what action (or actions) would you recommend to the owners of Pronto Pizza to improve their operation? Attach any supporting output that led to your conclusion.

CHAPTER 11

Hypothesis Tests Involving Two Sample Means or Proportions

GENDER STEREOTYPES AND ASKING FOR DIRECTIONS

Many of us have heard the stereotypical observation that men absolutely refuse to stop and ask for directions, preferring instead to wander about until they "find their own way." Is this just a stereotype, or is there really something to it? When Lincoln Mercury was carrying out research prior to the design and introduction of its in-vehicle navigational systems, the company surveyed men and women with regard to their driving characteristics and navigational habits.

According to the survey, 61% of the female respondents said they would stop and ask for directions once they figured out they were lost. On the other hand, only 42% of the male respondents said they would stop and ask for directions under the same circumstances. For purposes of our discussion, we'll assume that Lincoln Mercury surveyed 200 persons of each gender.

Is it possible that the lower rate of direction-asking for men—a sample proportion of just 0.42 (84 out of 200) versus 0.61 (122 out of 200)—may have occurred simply by chance variation? Actually, if the population proportions were really the same, there would be only a 0.0001 probability of the direction-asking proportion for men being this much lower just by chance. Section 11.6 of this chapter will provide some guidance regarding a statistical test with which you can compare these gender results and verify the conclusion for yourself. There will be no need to stop at Section 11.5 to ask for directions.

Source: Jamie LaReau, "Lincoln Uses Gender Stereotypes to Sell Navigation System," *Automotive News*, May 30, 2005, p. 32.

Comparing two samples the old-fashioned way

Reza Estakhrian/Stone/Getty Images

- Select and use the appropriate hypothesis test in comparing the means of two independent samples.
- Test the difference between sample means when the samples are not independent.
- Test the difference between proportions for two independent samples.
- Determine whether two independent samples could have come from populations having the same standard deviation.

LEARNING OBJECTIVES

After reading this chapter, you should be able to:

INTRODUCTION (11.1)

One of the most useful applications of business statistics involves comparing two samples to examine whether a difference between them is (1) significant or (2) likely to have been due to chance. This lends itself quite well to the analysis of data from experiments such as the following:

EXAMPLE

Comparing Samples

A local YMCA, in the early stages of telephone solicitation to raise funds for expanding its gymnasium, is testing two appeals. Of 100 residents approached with appeal A, 21% pledged a donation. Of 150 presented with appeal B, 28% pledged a donation.

SOLUTION

Is B really a superior appeal, or could its advantage have been merely the result of chance variation from one sample to the next? Using the techniques in this chapter, we can reach a conclusion on this and similar questions. The approach will be very similar to that in Chapter 10, where we dealt with one sample statistic (either $\bar{x}$ or p), its standard error, and the level of significance at which it differed from its hypothesized value.

EXAMPLEEXAMPLEEXAMPLE

Sections 11.2–11.4 and 11.6 deal with the comparison of two means or two proportions from independent samples. **Independent samples** are those for which the selection process for one is not related to that for the other. For example, in an experiment, independent samples occur when persons are randomly assigned to the experimental and control groups. In these sections, a hypothesis-testing procedure very similar to that in Chapter 10 will be followed. As before, either one or two critical value(s) will be identified, and then a decision rule will be applied to see if the calculated value of the test statistic falls into a rejection region specified by the rule.

When comparing two independent samples, the null and alternative hypotheses can be expressed in terms of either the population parameters (μ_1 and μ_2, or π_1 and π_2) or the sampling distribution of the difference between the sample statistics ($\bar{x}_1$ and $\bar{x}_2$, or p_1 and p_2). These approaches to describing the null and alternative hypotheses are equivalent and are demonstrated in Table 11.1.

TABLE 11.1

When comparing means from independent samples, null and alternative hypotheses can be expressed in terms of the population parameters (on the left) or described by the mean of the sampling distribution of the difference between the sample statistics (on the right). This also applies to testing two sample proportions. For example, H_0: $\pi_1 = \pi_2$ is the same as H_0: $\mu_{(p_1-p_2)} = 0$.

Null and Alternative Hypotheses Expressed in Terms of the Population Means		Hypotheses Expressed in Terms of the Sampling Distribution of the Difference Between the Sample Means
Two-tail test:		
H_0: $\mu_1 = \mu_2$	or	H_0: $\mu_{(\bar{x}_1-\bar{x}_2)} = 0$
H_1: $\mu_1 \neq \mu_2$		H_1: $\mu_{(\bar{x}_1-\bar{x}_2)} \neq 0$
Left-tail test:		
H_0: $\mu_1 \geq \mu_2$	or	H_0: $\mu_{(\bar{x}_1-\bar{x}_2)} \geq 0$
H_1: $\mu_1 < \mu_2$		H_1: $\mu_{(\bar{x}_1-\bar{x}_2)} < 0$
Right-tail test:		
H_0: $\mu_1 \leq \mu_2$	or	H_0: $\mu_{(\bar{x}_1-\bar{x}_2)} \leq 0$
H_1: $\mu_1 > \mu_2$		H_1: $\mu_{(\bar{x}_1-\bar{x}_2)} > 0$

Section 11.5 is concerned with the comparison of means for two dependent samples. Samples are **dependent** when the selection process for one is related to the selection process for the other. A typical example of dependent samples occurs when we have before-and-after measures of the same individuals or objects. In this case we are interested in only *one variable:* the difference between measurements for each person or object.

In this chapter, we present three different methods for comparing the means of two independent samples: the pooled-variances t-test (Section 11.2), the unequal-variances t-test (Section 11.3), and the z-test (Section 11.4). Figure 11.1 summarizes the procedure for selecting which test to use in comparing the sample means.

As shown in Figure 11.1, an important factor in choosing between the pooled-variances t-test and the unequal-variances t-test is whether we can assume the population standard deviations (and, hence, the variances) might be equal. Section 11.7 provides a hypothesis-testing procedure by which we can actually test this possibility. However, for the time being, we will use a less rigorous standard—that is, based on the sample standard deviations, whether it *appears that* the population standard deviations might be equal.

11.2 THE POOLED-VARIANCES *t*-TEST FOR COMPARING THE MEANS OF TWO INDEPENDENT SAMPLES

Situations can arise where we'd like to examine whether the difference between the means of two independent samples is large enough to warrant rejecting the possibility that their population means are the same. In this type of setting, the alternative conclusion is that the difference between the sample means is small enough to have occurred by chance, and that the population means really could be equal.

FIGURE 11.1

Selecting the test statistic for hypothesis tests comparing the means of two independent samples.

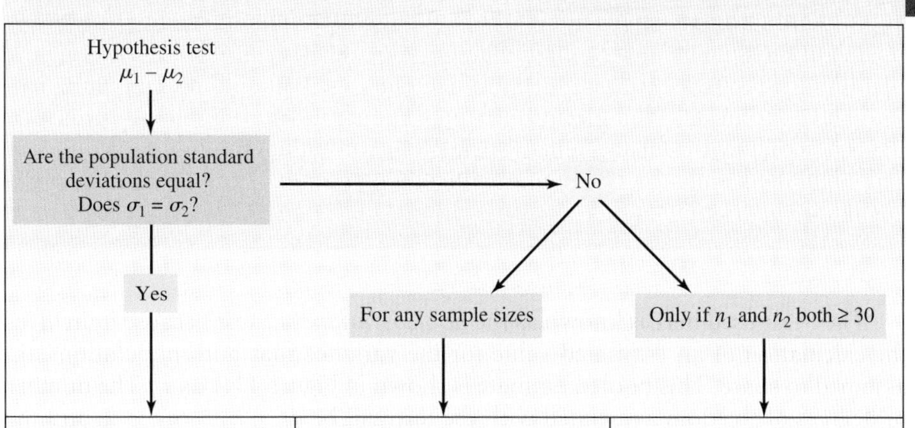

Compute the pooled estimate of the common variance as

$$s_p^2 = \frac{(n_1 - 1)s_1^2 + (n_2 - 1)s_2^2}{n_1 + n_2 - 2}$$

and perform a pooled-variances t-test where

$$t = \frac{(\bar{x}_1 - \bar{x}_2) - (\mu_1 - \mu_2)_0}{\sqrt{s_p^2\left(\frac{1}{n_1} + \frac{1}{n_2}\right)}}$$

$df = n_1 + n_2 - 2$

and $(\mu_1 - \mu_2)_0$ is from H_0

Test assumes samples are from normal populations with equal standard deviations.

Section 11.2
See Note 1

Perform an unequal-variances t-test where

$$t = \frac{(\bar{x}_1 - \bar{x}_2) - (\mu_1 - \mu_2)_0}{\sqrt{\frac{s_1^2}{n_1} + \frac{s_2^2}{n_2}}}$$

and

$$df = \frac{[(s_1^2/n_1) + (s_2^2/n_2)]^2}{\frac{(s_1^2/n_1)^2}{n_1 - 1} + \frac{(s_2^2/n_2)^2}{n_2 - 1}}$$

and $(\mu_1 - \mu_2)_0$ is from H_0

Test assumes samples are from normal populations. When either test can be applied to the same data, the unequal-variances t-test is preferable to the z-test—especially when doing the test with computer assistance.

Section 11.3
See Note 2

A z-test approximation can be performed where

$$z = \frac{(\bar{x}_1 - \bar{x}_2) - (\mu_1 - \mu_2)_0}{\sqrt{\frac{s_1^2}{n_1} + \frac{s_2^2}{n_2}}}$$

with s_1^2 and s_2^2 as estimates of σ_1^2 and σ_2^2,

and $(\mu_1 - \mu_2)_0$ is from H_0

The central limit theorem prevails and there are no limitations on the population distributions. This test may also be more convenient for solutions based on pocket calculators.

Section 11.4
See Note 3

1. Section 11.7 describes a procedure for testing the null hypothesis that $\sigma_1 = \sigma_2$. However, when using a computer and statistical software, it may be more convenient to simply bypass this assumption and apply the unequal-variances t-test described in Section 11.3.
2. This test involves a corrected df value that is smaller than if $\sigma_1 = \sigma_2$ had been assumed. When using a computer and statistical software, this test can be used routinely instead of the other two tests shown here. The nature of the df expression makes hand calculations somewhat cumbersome. The normality assumption becomes less important for larger sample sizes.
3. When sample sizes are large (each $n \geq 30$), the z-test is a useful alternative to the unequal-variances t-test, and may be more convenient when hand calculations are involved.
4. For each test, Computer Solutions within the chapter describe the Excel and Minitab procedures for carrying out the test. Procedures based on data files and those based on summary statistics are included.

Our use of the *t*-test assumes that (1) the (unknown) population standard deviations are equal, and (2) the populations are at least approximately normally distributed. Because of the central limit theorem, the assumption of population normality becomes less important for larger sample sizes. Although it is often associated only with small-sample tests, the *t* distribution is appropriate when the population standard deviations are unknown, *regardless of how large or small the samples happen to be.*

The *t*-test used here is known as the *pooled-variances t*-test because it involves the calculation of an estimated value for the variance that both populations are assumed to share. This pooled estimate is shown in Figure 11.1 as s_p^2. The number of degrees of freedom associated with the test will be $df = n_1 + n_2 - 2$, and the test statistic is calculated as follows:

Test statistic for comparing the means of two independent samples, σ_1 and σ_2 assumed to be equal:

$$t = \frac{(\bar{x}_1 - \bar{x}_2) - (\mu_1 - \mu_2)_0}{\sqrt{s_p^2\left(\frac{1}{n_1} + \frac{1}{n_2}\right)}}$$

where $\bar{x}_1$ and $\bar{x}_2$ = means of samples 1 and 2
$(\mu_1 - \mu_2)_0$ = the hypothesized difference between the population means
n_1 and n_2 = sizes of samples 1 and 2
s_1 and s_2 = standard deviations of samples 1 and 2
s_p = pooled estimate of the common standard deviation

with

$$s_p^2 = \frac{(n_1 - 1)s_1^2 + (n_2 - 1)s_2^2}{n_1 + n_2 - 2} \quad \text{and} \quad df = n_1 + n_2 - 2$$

Confidence interval for $\mu_1 - \mu_2$:

$$(\bar{x}_1 - \bar{x}_2) \pm t_{\alpha/2}\sqrt{s_p^2\left(\frac{1}{n_1} + \frac{1}{n_2}\right)}$$

with

$$\alpha = (1 - \text{confidence coefficient})$$

The numerator of the *t*-statistic includes $(\mu_1 - \mu_2)_0$, the hypothesized value of the difference between the population means. The hypothesized difference is generally zero in tests like those in this chapter. The term in the denominator of the *t*-statistic is the estimated standard error of the difference between the sample means. It is comparable to the standard error of the sampling distribution for the sample mean discussed in Chapter 10. Also shown is the confidence interval for the difference between the population means.

EXAMPLE

Pooled-Variances *t*-Test

Entrepreneurs developing an accounting review program for persons preparing to take the Certified Public Accountant (CPA) examination are considering two possible formats for conducting the review sessions. A random sample of 10 students are trained using format 1, and then their number of errors is recorded for a prototype examination. Another random sample of 12 individuals are trained according to format 2, and their errors are similarly recorded for the same examination. For the 10 students trained with format 1, the individual performances are 11, 8, 8, 3, 7, 5, 9, 5, 1, and 3 errors. For the 12 students trained with format 2, the individual performances are 10, 11, 9, 7, 2, 11, 12, 3, 6, 7, 8, and 12 errors. These data are in file CX11CPA.

SOLUTION

Since the study was not conducted with directionality in mind, the appropriate test will be two-tail. The null hypothesis is H_0: $\mu_1 = \mu_2$, and the alternative hypothesis is H_1: $\mu_1 \neq \mu_2$. The null and alternative hypotheses may also be expressed as follows:

- **Null hypothesis**

  H_0: $\mu_{(\bar{x}_1-\bar{x}_2)} = 0$ The two review formats are equally effective.

- **Alternative hypothesis**

  H_1: $\mu_{(\bar{x}_1-\bar{x}_2)} \neq 0$ The two review formats are not equally effective.

In comparing the performances of the two groups, the 0.10 level of significance will be used. Based on these data, the 10 members of group 1 made an average of 6.000 errors, with a sample standard deviation of 3.127. The 12 students trained with format 2 made an average of 8.167 errors, with a standard deviation of 3.326. The sample standard deviations do not appear to be very different, and we will assume that the population standard deviations could be equal. (As noted previously, this rather informal inference can be replaced by the separate hypothesis test in Section 11.7.) In applying the pooled-variances *t*-test, the pooled estimate of the common variance, s_p^2, and the test statistic, *t*, can be calculated as

$$s_p^2 = \frac{(10 - 1)(3.127)^2 + (12 - 1)(3.326)^2}{10 + 12 - 2} = 10.484$$

and

$$t = \frac{(6.000 - 8.167) - 0}{\sqrt{10.484\left(\frac{1}{10} + \frac{1}{12}\right)}} = -1.563$$

For the 0.10 level of significance, the critical values of the test statistic will be $t = -1.725$ and $t = +1.725$. These are based on the number of degrees of freedom, $df = (n_1 + n_2 - 2)$, or $(10 + 12 - 2) = 20$, and the specification that

FIGURE 11.2

In this two-tail pooled-variances *t*-test, we are not able to reject the null hypothesis that the two accounting review formats could be equally effective.

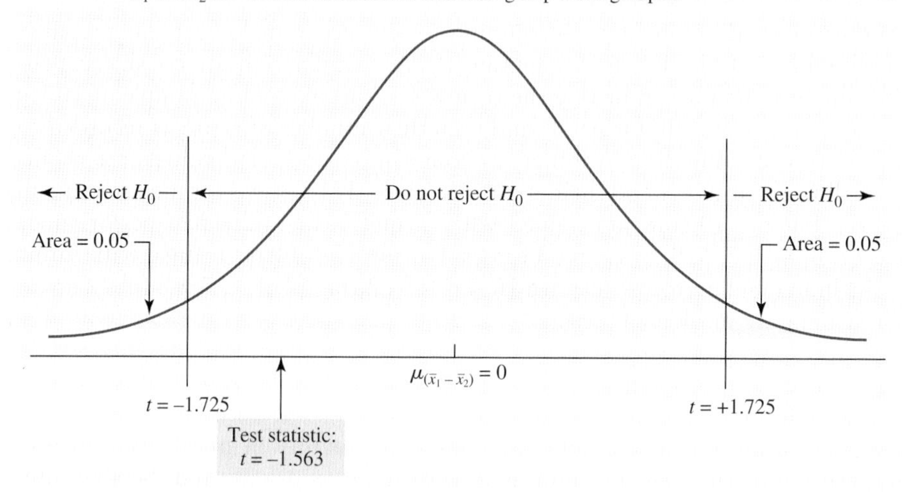

the two tail areas must add up to 0.10. The decision rule is to reject the null hypothesis (i.e., conclude that the population means are not equal for the two review formats) if the calculated test statistic is either less than $t = -1.725$ or greater than $t = +1.725$.

As Figure 11.2 shows, the calculated test statistic, $t = -1.563$, falls into the nonrejection region of the test. At the 0.10 level, we must conclude that the review formats are equally effective in training individuals for the CPA examination. For this level of significance, the observed difference between the groups' mean errors is judged to have been due to chance.

Based on the sample data, we will also determine the 90% confidence interval for $(\mu_1 - \mu_2)$. For $df = 20$ and $\alpha = 0.10$, this is

$$(\bar{x}_1 - \bar{x}_2) \pm t_{\alpha/2}\sqrt{s_p^2\left(\frac{1}{n_1} + \frac{1}{n_2}\right)} = (6.000 - 8.167) \pm 1.725\sqrt{10.484\left(\frac{1}{10} + \frac{1}{12}\right)}$$

$$= -2.167 \pm 2.392, \text{ or from } -4.559 \text{ to } +0.225$$

The hypothesized difference (zero) is contained within the 90% confidence interval, so we are 90% confident the population means could be the same. As discussed in Chapter 10, a nondirectional test at the α level of significance and a $100(1 - \alpha)\%$ confidence interval will lead to the same conclusion. Computer Solutions 11.1 shows Excel and Minitab procedures for the pooled-variances *t*-test.

COMPUTER 11.1 SOLUTIONS

Pooled-Variances *t*-Test for $(\mu_1 - \mu_2)$, Population Variances Unknown but Assumed Equal

These procedures show how to use the pooled-variances *t*-test to compare the means of two independent samples. The population variances are unknown, but assumed to be equal.

EXCEL

	D	E	F
1	t-Test: Two-Sample Assuming Equal Variances		
2			
3		*format1*	*format2*
4	Mean	6.000	8.167
5	Variance	9.778	11.061
6	Observations	10	12
7	Pooled Variance	10.48333	
8	Hypothesized Mean Difference	0	
9	df	20	
10	t Stat	-1.563	
11	P(T<=t) one-tail	0.067	
12	t Critical one-tail	1.325	
13	P(T<=t) two-tail	0.134	
14	t Critical two-tail	1.725	

Excel pooled-variances *t*-test for $(\mu_1 - \mu_2)$, based on raw data

1. For the sample data (Excel file CX11CPA) on which Figure 11.2 is based, with the label and 10 data values for format 1 in column A, and the label and 12 data values for format 2 in column B: From the **Data** ribbon, click **Data Analysis.** Click ***t*-Test: Two-Sample Assuming Equal Variances.** Click **OK.**
2. Enter **A1:A11** into the **Variable 1 Range** box and **B1:B13** into the **Variable 2 Range** box. Enter **0** into the **Hypothesized Mean Difference** box. Click to select **Labels.** Specify the significance level for the test by entering **0.10** into the **Alpha** box. Select **Output Range** and enter **D1** into the box. Click **OK.** The results will be as shown above. This is a two-tail test, so we refer to the 0.134 *p*-value. (Excel doesn't ask whether our test is one-tail or two-tail, but it provides critical values and *p*-values for both.)
3. To obtain a confidence interval for $(\mu_1 - \mu_2)$, it will be necessary to refer to the procedure described below. It is based on summary statistics for the samples.

Excel Unstacking Note: When this analysis is based on raw data, Excel requires that the two samples be in two distinctly separate fields (e.g., two separate columns). If the data are in one column and the subscripts identifying group membership are in another column, it will be necessary to "unstack" the data. For example, if the data were in column A and the subscripts (1 and 2, to denote prep formats 1 and 2) were in column B, and each column had a label at the top: Click and drag to select the numbers and labels in columns A and B. From the **Data** ribbon, click **Sort.** In the **Sort by** box, select the category variable in column B (i.e., **Format**). Click **OK.** The data for sample 1 will now be listed directly above the data for sample 2—from here, we need only select each grouping of data (e.g., scores) and either move or copy it to its own column. The result will be one column for the scores of group 1 and an adjacent column for the scores of group 2.

Excel pooled-variances *t*-test and confidence interval for $(\mu_1 - \mu_2)$, based on summary statistics

1. Using the summary statistics associated with CX11CPA: Open the TEST STATISTICS workbook.
2. Using the arrows at the bottom left, select the ***t*-Test_2 Means (Eq-Var)** worksheet.

(continued)

3. Enter the sample means, variances, and sizes into the appropriate cells. Enter the hypothesized difference (0, in this case) and the desired alpha level for the test. The calculated *t*-statistic and a two-tail *p*-value will be shown at the right.
4. To obtain a confidence interval for $(\mu_1 - \mu_2)$, follow steps 1–3, but open the **ESTIMATORS** workbook and select the ***t*-Estimate_2 Means (Eq-Var)** worksheet. Enter the desired confidence level as a decimal fraction (e.g., 0.90).

Note: As an alternative, you can use Excel worksheet template **TMT2POOL**. It simultaneously conducts the test and reports a confidence interval. The steps are described within the template.

MINITAB

Minitab pooled-variances *t*-test and confidence interval for $(\mu_1 - \mu_2)$, based on raw data

```
Two-Sample T-Test and CI: format1, format2

Two-sample T for format1 vs format2

          N   Mean  StDev  SE Mean
format1  10   6.00   3.13     0.99
format2  12   8.17   3.33     0.96

Difference = mu (format1) - mu (format2)
Estimate for difference:  -2.17
90% CI for difference:  (-4.56, 0.22)
T-Test of difference = 0 (vs not =): T-Value = -1.56  P-Value = 0.134  DF = 20
Both use Pooled StDev = 3.2378
```

1. For example, using the data (Minitab file **CX11CPA**) on which Figure 11.2 is based, with the data values for format 1 in column **C1** and the data values for format 2 in column **C2:** Click **Stat.** Select **Basic Statistics.** Click **2-Sample *t*.**
2. Select **Samples in different columns.** Enter **C1** into the **First** box and **C2** into the **Second** box. Click to select **Assume equal variances.** (*Note:* If all the data had been in a single column [i.e., "stacked"], it would have been necessary to select the **Samples in one column** option, then to specify the column containing the data and the column containing the subscripts identifying group membership.)
3. Click **Options.** Enter the desired confidence level as a percentage (e.g., **90.0**) into the **Confidence Level** box. Enter the hypothesized difference (**0**) into the **Test difference** box. Within the **Alternative** box, select **not equal.** Click **OK.** Click **OK.**

Minitab pooled-variances *t*-test and confidence interval for $(\mu_1 - \mu_2)$, based on summary data

Follow steps 1 through 3 above, but select **Summarized data** in step 2 and insert the appropriate summary statistics into the **Sample size, Mean,** and **Standard deviation** boxes for each sample.

EXERCISES

11.1 "When comparing two sample means, the *t*-test should be used only when the sample sizes are less than 30." Comment.

11.2 An educator is considering two different videotapes for use in a half-day session designed to introduce students to the basics of economics. Students have been randomly assigned to two groups, and they all take the same written examination after viewing the videotape. The scores are summarized here. Assuming normal populations with equal standard deviations, does it appear that the two videotapes could be equally effective? What is the most accurate statement that could be made about the *p*-value for the test?

Videotape 1: $\bar{x}_1 = 77.1$ $\quad s_1 = 7.8$ $\quad n_1 = 25$

Videotape 2: $\bar{x}_2 = 80.0$ $\quad s_2 = 8.1$ $\quad n_2 = 25$

11.3 Using independent random samples, a researcher is comparing the number of hours of television viewed last week for high school seniors versus sophomores. The results are shown here. Assuming normal populations with equal standard deviations, does it appear that the average number of television hours per week could be equal for these two populations? What is the most accurate statement that could be made about the p-value for the test?

Seniors:	$\bar{x}_1 = 3.9$ hours	$s_1 = 1.2$ hours	$n_1 = 32$
Sophomores:	$\bar{x}_2 = 3.5$ hours	$s_2 = 1.4$ hours	$n_2 = 30$

11.4 An ambulance service located at the edge of town is responsible for serving a large office building in the downtown area. Testing different routes for getting from the ambulance station to the office building, a driver finds that five trips using route A take an average of 5.9 minutes, with a standard deviation of 1.4 minutes; six trips using route B take an average of 4.2 minutes, with a standard deviation of 1.8 minutes. Assuming normal populations with equal standard deviations, and using the 0.10 level, is there a significant difference between the two routes? Construct and interpret the 90% confidence interval for the difference between the population means.

11.5 A maintenance supervisor is comparing the standard version of an instructional booklet with one that has been claimed to be superior. An experiment is conducted in which 26 technicians are divided into two groups, provided with one of the booklets, then given a test a week later. For the 13 using the standard version, the average exam score was 72.0, with a standard deviation of 9.3. For the 13 given the new version, the average score was 80.2, with a standard deviation of 10.1. Assuming normal populations with equal standard deviations, and using the 0.05 level of significance, does the new booklet appear to be better than the standard version?

11.6 A sample of 40 investment customers serviced by an account manager are found to have had an average of \$23,000 in transactions during the past year, with a standard deviation of \$8500. A sample of 30 customers serviced by another account manager averaged \$28,000 in transactions, with a standard deviation of \$11,000. Assuming the population standard deviations are equal, use the 0.05 level of significance in testing whether the population means could be equal for customers serviced by the two account managers. Using the appropriate statistical table, what is the most accurate statement we can make about the p-value for this test? Construct and interpret the 95% confidence interval for the difference between the population means.

11.7 Comparing dexterity-test scores of workers on the day shift versus those on the night shift, the production manager of a large electronics plant finds that a sample of 37 workers from the day shift have an average score of 73.1, with a standard deviation of 12.3. For 42 workers from the night shift, the average score was 77.3, with a standard deviation of 8.4. Assuming the population standard deviations are equal, use the 0.05 level of significance in comparing the average scores for the two shifts. Using the appropriate statistical table, what is the most accurate statement we can make about the p-value for this test? Construct and interpret the 95% confidence interval for the difference between the population means.

11.8 Sheila Smith, the manager of a large resort's main hotel, has been receiving complaints from some guests that they are not being provided with prompt service upon approaching the front desk. In particular, she is concerned that desk staff might be providing female guests with less prompt service than their male counterparts. In observing a sample of 34 male guests, she finds it takes an average of 15.2 seconds, with a standard deviation of 5.9 seconds, for them to be greeted after their arrival at the front desk. For a sample of 39 female guests, the mean and standard deviation are 17.4 seconds and 6.4 seconds, respectively. Assuming the population standard deviations to be equal, use the 0.05 level of significance in examining whether the population mean time for serving female guests might actually be no greater than that for serving male guests. Using the appropriate statistical table, what is the most accurate statement we can make about the p-value for the test?

11.9 Media observers have been examining the number of minutes devoted to business and financial news during the half-hour evening news broadcasts of two local television channels. For each channel, they have randomly selected 10 weekday broadcasts and observed the number of minutes spent on business and financial news during that broadcast. The times measured in these independent samples are shown here. Assuming normal populations with equal standard deviations, use the 0.10 level of significance in testing whether the population means might actually be the same. Using the appropriate statistical table, what is the most accurate statement we can make about the p-value for the test? Construct and interpret the 90% confidence interval for the difference between the population means.

Channel 2: 3.8 2.7 4.9 3.4 3.7 4.5 4.2 2.8 3.5 4.6 minutes
Channel 4: 3.6 4.0 4.5 5.2 4.8 4.3 5.7 3.5 3.7 5.8 minutes

11.10 In a test of the effectiveness of a new battery design, 16 battery-powered music boxes are randomly provided with either the old design or the new version. Hours of playing time before battery failure were as follows:

8 boxes, new battery type:	3.3, 6.4, 3.9, 5.4, 5.1, 4.6, 4.9, 7.2 hrs
8 boxes, old battery type:	4.2, 2.9, 4.5, 4.9, 5.0, 5.1, 3.2, 4.0 hrs

Assuming normal populations with equal standard deviations, use the 0.05 level to determine whether the new battery could be better than the old design. Using the appropriate statistical table, what is the most accurate statement we can make about the p-value for this test?

11.11 A nutritionist has noticed a FoodFarm ad stating the company's peanut butter contains less fat than that produced by a major competitor. She purchases 11 8-ounce jars of each brand and measures the fat content of each. The 11 FoodFarm jars had an average of 31.3 grams of fat, with a standard deviation of 2.1 grams. The 11 jars from the other company had an average of 33.2 grams of fat, with a standard deviation of 1.8 grams. Assuming normal populations with equal standard deviations, use the 0.05 level of significance in examining whether FoodFarm's ad claim could be valid. What is the most accurate statement that could be made about the p-value for this test?

(DATA SET) *Note:* Exercises 11.12–11.17 require a computer and statistical software.

11.12 A study published in the *Archives of Internal Medicine* examined the prevalence of so-called "difficult" patients who ask for unneeded prescriptions, unnecessarily complain, or otherwise cause extraordinary frustrations for their medical provider. Researchers found the mean age of doctors having the greatest problems with difficult patients was 41, while the mean age of doctors having the least problems with difficult patients was 46. Assume that data file **XR11012** contains the ages for doctors in each of these patient-difficulty samples. In a suitable one-tail test at the 0.01 level, was the mean age for doctors having the greatest problems with difficult patients significantly less than that for those having the least problems with difficult patients? Identify and interpret the p-value for the test. Source: Rita Rubin, "'Difficult Patients Can Test Doctors' Patience," *USA Today*, February 24, 2009, p. 5D.

11.13 It has been claimed that flattening ("rolling") the barrel of a graphite baseball bat can stretch the fibers and result in a pitched baseball being returned faster than with a regulation bat. This was a controversial topic during the 2009 baseball College World Series. Assume that data file **XR11013** describes the results of a controlled test measuring the return speeds of balls pitched to a flattened bat and a regulation bat, respectively. Based on these sample results, and using a suitable one-tail test at the 0.05 level, was the mean return speed for the flattened bat significantly greater than for the regulation bat? Identify and interpret the p-value for the test. Source: Andy Gardiner, "NCAA on Guard for 'Rolled' Bats," *USA Today*, June 17, 2009, p. 8C.

11.14 Comparing the number of Facebook "friends" for men and women, observers have speculated about whether the mean number of friends could be the same for each group. Assume that data file **XR11014** lists the number of Facebook friends for independent samples of male and female Facebook users. Based on the results of a two-tail test at the 0.05 level, comment on whether the difference between the sample means could have simply occurred by chance. Identify and interpret the p-value for the test. Source: "Primates on Facebook," economist.com, February 26, 2009.

11.15 Using the sample results in Exercise 11.14, construct and interpret the 95% confidence interval for the difference between the population means. Is the hypothesized difference (0.00) within the interval? Given the presence or absence of the 0.00 value within the interval, is this consistent with the findings of the hypothesis test conducted in Exercise 11.14?

11.16 An engineer has measured the hardness scores for a sample of conveyor-belt support bearings that have been hardened by two different methods. The first method is used by her company, and the second method is known to be used by a number of other companies in the industry. With the resulting data in file **XR11016**, use the 0.05 level of significance in comparing the mean hardness scores of the two samples, and comment on the possibility that the difference between the sample means could have occurred by chance. Identify and interpret the p-value for the test.

11.17 Using the sample results in Exercise 11.16, construct and interpret the 95% confidence interval for the difference between the population means. Is the hypothesized difference (0.00) within the interval? Given the presence or absence of the 0.00 value within the interval, is this consistent with the findings of the hypothesis test conducted in Exercise 11.16?

(11.3) THE UNEQUAL-VARIANCES t-TEST FOR COMPARING THE MEANS OF TWO INDEPENDENT SAMPLES

When the population standard deviations are unknown and are not assumed to be equal, pooling the sample standard deviations into a single estimate of their common population value is no longer applicable. As a result, s_1 and s_2 must be used to estimate their respective population standard deviations, σ_1 and σ_2. The test assumes the populations to be at least approximately normally distributed, an assumption that becomes less important for larger sample sizes.

In the unequal-variances t-test, the t-statistic expression is straightforward, but the df formula is a little more complex—it is a correction formula that provides a *df* value that is smaller than its counterpart in the preceding section. For accuracy, it's best to maintain a lot of decimal places if you are computing *df* with a pocket calculator. If we are using the computer and statistical software, the unequal-variances *t*-test presents no computational difficulties, and is the preferred method for comparing the means of two independent samples, regardless of the sample sizes. The test statistic, *df*, and confidence interval expressions for this test are shown here:

Unequal-variances *t*-test for comparing the means of two independent samples, σ_1 and σ_2 unknown and not assumed to be equal:

$$t = \frac{(\bar{x}_1 - \bar{x}_2) - (\mu_1 - \mu_2)_0}{\sqrt{\frac{s_1^2}{n_1} + \frac{s_2^2}{n_2}}}$$

where $\bar{x}_1$ and $\bar{x}_2$ = means of samples 1 and 2
$(\mu_1 - \mu_2)_0$ = hypothesized difference between the population means
n_1 and n_2 = sizes of samples 1 and 2
s_1 and s_2 = standard deviations of samples 1 and 2

with

$$df = \frac{[(s_1^2/n_1) + (s_2^2/n_2)]^2}{\frac{(s_1^2/n_1)^2}{n_1 - 1} + \frac{(s_2^2/n_2)^2}{n_2 - 1}}$$

Confidence interval for $\mu_1 - \mu_2$:

$$(\bar{x}_1 - \bar{x}_2) \pm t_{\alpha/2}\sqrt{\frac{s_1^2}{n_1} + \frac{s_2^2}{n_2}}$$

with

$$\alpha = (1 - \text{confidence coefficient})$$

EXAMPLE

Unequal-Variances *t*-Test

The makers of Graphlex, a graphite additive for engine oil, have conducted an experimental study to determine the effectiveness of their product in improving the fuel efficiency of automobiles. In cooperation with the Metropolitan Cab Company, they've randomly divided the company's cabs into two groups of equal size. Graphlex was added to the engine oil of the 45 cabs in the experimental group, while the 45 cabs in the control group continued to operate with the usual lubricant. Drivers were not informed of the experiment. After 1 month, fuel efficiency records were examined. For the 45 cabs using Graphlex, the average cab achieved 18.94 miles per gallon (mpg), with a standard deviation of 3.90 mpg. For the 45 cabs not using Graphlex, the average mpg was 17.51, with a standard deviation of 2.87 mpg. The underlying data are in file CX11MPG. Graphlex is preparing a national advertising campaign to promote its ability to improve fuel efficiency.

EXAMPLEEXAMPLEEXAMPLEEXAM

FIGURE 11.3

The makers of Graphlex claim their graphite oil additive improves the fuel efficiency of automobiles. In this right-tail unequal-variances test at the 0.05 level, the results indicate they may be correct.

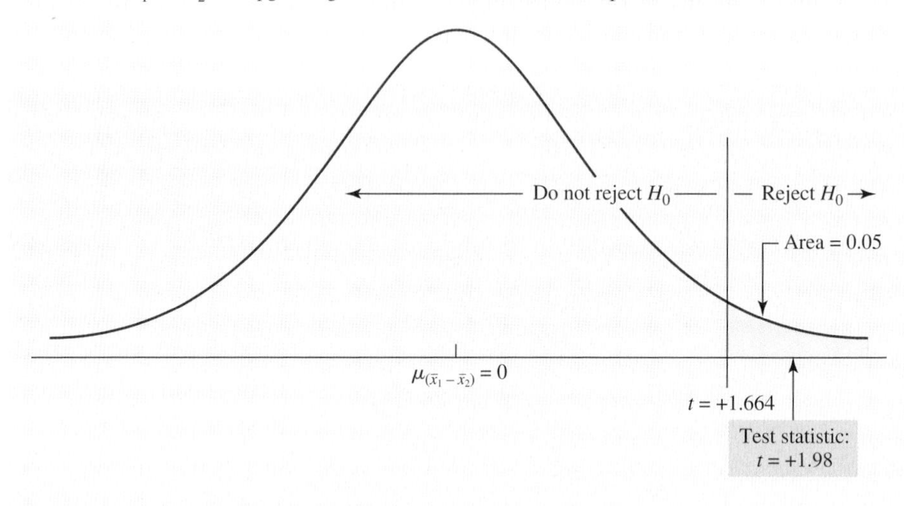

SOLUTION

The results will be evaluated at the 0.05 level of significance. Since the purpose of the study was to determine whether cabs using Graphlex (group 1) get better fuel economy than cabs without it (group 2), the null and alternative hypotheses are H_0: $\mu_1 \leq \mu_2$, and H_1: $\mu_1 > \mu_2$. Expressed in terms of the sampling distribution of the difference between sample means, the null and alternative hypotheses can be stated as follows:

- **Null hypothesis**

  H_0: $\mu_{(\bar{x}_1-\bar{x}_2)} \leq 0$ mpg with Graphlex is no higher than with conventional oil.

- **Alternative hypothesis**

  H_1: $\mu_{(\bar{x}_1-\bar{x}_2)} > 0$ mpg with Graphlex is higher.

For these data, the values for the test statistic (t) and the number of degrees of freedom (df) are calculated as

$$t = \frac{(\bar{x}_1 - \bar{x}_2) - (\mu_1 - \mu_2)_0}{\sqrt{\dfrac{s_1^2}{n_1} + \dfrac{s_2^2}{n_2}}} = \frac{(18.94 - 17.51) - 0}{\sqrt{\dfrac{3.90^2}{45} + \dfrac{2.87^2}{45}}} = 1.98$$

and

$$df = \frac{\left[(s_1^2/n_1) + (s_2^2/n_2)\right]^2}{\dfrac{(s_1^2/n_1)^2}{n_1 - 1} + \dfrac{(s_2^2/n_2)^2}{n_2 - 1}} = \frac{[(3.90^2/45) + (2.87^2/45)]^2}{\dfrac{(3.90^2/45)^2}{45 - 1} + \dfrac{(2.87^2/45)^2}{45 - 1}} = 80.85, \text{ rounded to } 81$$

For the 0.05 level of significance, $df = 81$, and a right-tail test, the critical value of the test statistic is $t = +1.664$. The decision rule is, "Reject H_0 if the calculated test statistic is greater than $t = +1.664$, otherwise do not reject." As Figure 11.3 shows, the calculated test statistic exceeds the critical value, the null hypothesis is rejected, and we conclude that Graphlex really works.

As we saw in Chapter 10, different levels of significance can lead to different conclusions. For example, at the 0.01 level (critical $t = +2.373$), the null hypothesis would not have been rejected. While the Graphlex Company would likely stress the additive's effectiveness by relying on the 0.05 level of significance, the manufacturer of a competing brand would tend to prefer a more demanding test (e.g., $\alpha = 0.01$) in order to boast that Graphlex has no effect.

You may have noticed that the degrees of freedom value for this test, $df = 81$, was near the upper part of the df range of the t distribution table. Should you encounter a test in which df happens to be over 100, just use the "infinity" row of the table to determine the critical value(s) of the test statistic. This row represents the normal distribution, toward which the t distribution converges as df becomes larger.

Based on the sample data, we will also determine the 90% confidence interval for $(\mu_1 - \mu_2)$. For $df = 81$ and $\alpha = 0.10$, this will be

$$(\bar{x}_1 - \bar{x}_2) \pm t_{\alpha/2}\sqrt{\frac{s_1^2}{n_1} + \frac{s_2^2}{n_2}} = (18.94 - 17.51) \pm 1.664\sqrt{\frac{3.90^2}{45} + \frac{2.87^2}{45}}$$

$$= 1.43 \pm 1.20, \quad \text{or from } 0.23 \text{ to } 2.63$$

Computer Solutions 11.2 shows Excel and Minitab procedures for the unequal-variances t-test.

COMPUTER 11.2 SOLUTIONS

Unequal-Variances t-Test for $(\mu_1 - \mu_2)$, Population Variances Unknown and Not Equal

These procedures show how to use the unequal-variances *t*-test to compare the means of two independent samples. The population variances are unknown and not assumed to be equal.

EXCEL

	D	E	F
1	t-Test: Two-Sample Assuming Unequal Variances		
2			
3		*Expgrp*	*Ctrlgrp*
4	Mean	18.94	17.51
5	Variance	15.22	8.24
6	Observations	45	45
7	Hypothesized Mean Difference	0	
8	df	81	
9	t Stat	1.98	
10	P(T<=t) one-tail	0.026	
11	t Critical one-tail	1.664	
12	P(T<=t) two-tail	0.051	
13	t Critical two-tail	1.990	

Excel unequal-variances *t*-test for $(\mu_1 - \mu_2)$, based on raw data

1. For the sample data (Excel file **CX11MPG**) on which Figure 11.3 is based, with the label and 45 mpg values for the Graphlex cabs in column A, and the label and 45 data values for the standard cabs in column B: From the **Data** ribbon, click **Data Analysis.** Click ***t*-Test: Two-Sample Assuming Unequal Variances.** Click **OK.**

(continued)

2. Enter **A1:A46** into the **Variable 1 Range** box and **B1:B46** into the **Variable 2 Range** box. Enter **0** into the **Hypothesized Mean Difference** box. Click to select **Labels.** Specify the significance level by entering **0.05** into the **Alpha** box. Select **Output Range** and enter **D1** into the box. Click **OK.** This is a one-tail test, so we refer to the 0.026 *p*-value.
3. To obtain a confidence interval for $(\mu_1 - \mu_2)$, it will be necessary to refer to the procedure described next. It is based on summary statistics for the samples.

Note: For an analysis based on raw data, Excel requires the two samples to be in two distinctly separate fields (e.g., two separate columns). If the data are in one column and the subscripts identifying group membership are in another column, see the Excel Unstacking Note in Computer Solutions 11.1.

Excel unequal-variances *t*-test and confidence interval for $(\mu_1 - \mu_2)$, based on summary statistics

1. Using the summary statistics associated with CX11MPG (the test is summarized in Figure 11.3): Open the TEST STATISTICS workbook.
2. Using the arrows at the bottom left, select the ***t*-Test_2 Means (Uneq-Var)** worksheet.
3. Enter the sample means, variances, and sizes into the appropriate cells. Enter the hypothesized difference (0, in this case) and the desired alpha level for the test. The calculated *t*-statistic and a one-tail *p*-value will be shown at the right.
4. To get a confidence interval for $(\mu_1 - \mu_2)$, follow steps 1–3, above, but open the ESTIMATORS workbook and select the ***t*-Estimate_2 Means (Uneq-Var)** worksheet. Enter the desired confidence level as a decimal fraction (e.g., 0.90).

Note: As an alternative, you can use Excel worksheet template TMT2UNEQ. It simultaneously conducts the test and reports a confidence interval. The steps are described within the template.

MINITAB

Minitab unequal-variances *t*-test and confidence interval for $(\mu_1 - \mu_2)$, based on raw data

```
Two-Sample T-Test and CI: Expgrp, Ctrlgrp

Two-sample T for Expgrp vs Ctrlgrp

          N   Mean  StDev  SE Mean
Expgrp   45  18.94   3.90     0.58
Ctrlgrp  45  17.51   2.87     0.43

Difference = mu (Expgrp) - mu (Ctrlgrp)
Estimate for difference:  1.429
90% lower bound for difference:  0.496
T-Test of difference = 0 (vs >): T-Value = 1.98  P-Value = 0.026  DF = 80
```

1. For example, using the data (Minitab file CX11MPG) on which Figure 11.3 is based, with the data values for the Graphlex cabs in column **C1** and the data values for the standard cabs in column **C2:** Click **Stat.** Select **Basic Statistics.** Click **2-Sample *t*.**
2. Select **Samples in different columns.** Enter **C1** into the **First** box and **C2** into the **Second** box. Do NOT select the "Assume equal variances" option. (*Note:* If all the data had been in a single column [i.e., "stacked"], it would have been necessary to select the **Samples in one column** option, then to specify the column containing the data and the column containing the subscripts identifying group membership.)
3. Click **Options.** Enter the desired confidence level as a percentage (e.g., **90.0**) into the **Confidence Level** box. Enter the hypothesized difference (**0**) into the **Test difference** box. Within the **Alternative** box, select **greater than.** Click **OK.** Click **OK.**

Minitab unequal-variances *t*-test and confidence interval for $(\mu_1 - \mu_2)$, based on summary data

Follow steps 1 through 3 above, but select **Summarized data** in step 2 and insert the appropriate summary statistics into the **Sample size, Mean,** and **Standard deviation** boxes for each sample.

EXERCISES

11.18 In two independent samples from populations that are normally distributed, $\bar{x}_1 = 35.0$, $s_1 = 5.8$, $n_1 = 12$ and $\bar{x}_2 = 42.5$, $s_2 = 9.3$, $n_2 = 14$. Using the 0.05 level of significance, test H_0: $\mu_1 = \mu_2$ versus H_1: $\mu_1 \neq \mu_2$.

11.19 In two independent samples, $\bar{x}_1 = 165.0$, $s_1 = 21.5$, $n_1 = 40$ and $\bar{x}_2 = 172.9$, $s_2 = 31.3$, $n_2 = 32$. Using the 0.10 level of significance, test H_0: $\mu_1 \geq \mu_2$ versus H_1: $\mu_1 < \mu_2$.

11.20 In two independent samples, $\bar{x}_1 = 125.0$, $s_1 = 21.5$, $n_1 = 40$, and $\bar{x}_2 = 116.4$, $s_2 = 10.8$, $n_2 = 35$. Using the 0.025 level of significance, test H_0: $\mu_1 \leq \mu_2$ versus H_1: $\mu_1 > \mu_2$.

11.21 Fred and Martina, senior agents at an airline security checkpoint, carry out advanced screening procedures for hundreds of randomly selected passengers per day. For a random sample of 30 passengers recently processed by Fred, the mean processing time was 124.5 seconds, with a standard deviation of 20.4 seconds. For a random sample of 36 passengers recently processed by Martina, the corresponding mean and standard deviation were 133.0 seconds and 38.7 seconds, respectively. Using the 0.05 level of significance, can we conclude that the population mean processing times for Fred and Martina could be the same? Using the appropriate statistical table, what is the most accurate statement we can make about the p-value for the test? Construct and interpret the 95% confidence interval for the difference between the population means.

11.22 A tire company is considering switching to a new type of adhesive designed to improve tire reliability in high-temperature and overload conditions. In laboratory "torture" tests with temperatures and loads 90% higher than the maximum normally encountered in the field, 15 tires constructed with the new adhesive run an average of 65 miles before failure, with a standard deviation of 14 miles. For 18 tires constructed with the conventional adhesive, the mean mileage before failure was 53 miles, with a standard deviation of 22 miles. Assuming normal populations and using the 0.05 level of significance, can we conclude that the new adhesive is superior to the old under such test conditions? What is the most accurate statement that could be made about the p-value for this test?

11.23 The credit manager for Braxton's Department Store, in examining the accounts for various types of customers served by the establishment, has noticed that the mean outstanding balance for a sample of 20 customers from the local ZIP code is \$375, with a standard deviation of \$75. For a sample of 25 customers from a nearby ZIP code, the mean and standard deviation were \$425 and \$143, respectively. Assuming normal populations and using the 0.05 level of significance, examine whether credit customers from the two ZIP codes might have the same mean outstanding balance. What is the most accurate statement that could be made about the p-value for this test? Construct and interpret the 95% confidence interval for the difference between the population means. Is the hypothesized difference (0.00) within the interval? Given the presence or absence of the 0.00 value within the interval, is this consistent with the findings of the hypothesis test?

11.24 Safety engineers at a manufacturing plant are evaluating two brands of 50-ampere electrical fuses for possible purchase and use in the plant. One of the performance characteristics they are considering is how long the fuse will carry 50 amperes before it blows. In a sample of 35 fuses from the Shockley Fuse Company, the mean time was found to be 240 milliseconds, with a standard deviation of 50 milliseconds. In comparable tests of a sample of 30 fuses from the Fusemaster Corporation, the mean time was 221 milliseconds, with a standard deviation of 28 milliseconds. Using the 0.10 level of significance, examine whether the population mean times for fuses from the two companies might be the same. What is the most accurate statement that could be made about the p-value for this test? Construct and interpret the 90% confidence interval for the difference between the population means. Is the hypothesized difference (0.00) within the interval? Given the presence or absence of the 0.00 value within the interval, is this consistent with the findings of the hypothesis test?

11.25 According to a national Gallup poll, men visit the doctor's office an average of 3.8 times per year, while women visit an average of 5.8 times per year. In a similar poll conducted in a Midwest county, a sample of 50 men visited the doctor an average of 2.2 times in the past year, with a standard deviation of 0.6 visits. For a sample of 40 women from the same county, the mean and standard deviation were 3.9 and 0.9, respectively. In a two-tail test at the 0.05 level, test whether the observed difference between $\bar{x}_1$ and $\bar{x}_2$ is significantly different from the $(3.8 - 5.8) = -2.0$ visits per year that was found for the nation as a whole. Construct and interpret the 95% confidence interval for $(\mu_1 - \mu_2)$ for the county. Is the hypothesized difference (-2.0) within the interval? Given the presence or absence of the -2.0 value within the interval, is this consistent with the findings of the hypothesis test? Source: Cindy Hall and Marcy E. Mullins, "Doctors See Women More Often," *USA Today,* May 2, 2000, p. 7D.

11.26 One of the measures of the effectiveness of a stimulus is how much the viewer's pulse rate increases on exposure to it. In testing a lively new music theme for its television commercials, an advertising agency shows ads with the new music to a sample of 25 viewers. Their mean pulse rate increase is 20.5 beats per minute, with a standard deviation of 7.4. For a comparable sample of 25 viewers seeing the same ads with the previous music theme, the mean pulse rate increase is 16.4 beats per minute, with a standard deviation of 4.9. Assuming normal populations and using the 0.025 level of significance, can we conclude that the new music theme is better than the old in terms of increasing the pulse rate of viewers? What is the most accurate statement that could be made about the p-value for this test?

(**DATA SET**) *Note:* Exercises 11.27–11.30 require a computer and statistical software.

11.27 According to a Yankelovich poll, women spend an average of 19.1 hours shopping during the month of December, compared to 12.7 hours for men. Assuming that file **XR11027** contains the survey data underlying these results, use the 0.01 level in examining whether the sample mean for women is significantly higher than that for men. Identify and interpret the p-value for the test.
Source: Anne R. Carey and Gary Visgaitis, "'Tis the Season, Health," *USA Today 1999 Snapshot Calendar,* December 8.

11.28 When companies are designing a new product, one of the steps typically taken is to see how potential buyers react to a picture or prototype of the proposed product. The product-development team for a notebook computer company has shown picture A to a large sample of potential buyers and picture B to another, asking each person to indicate what they "would expect to pay" for such a product. The data resulting from the two pictures are provided in file **XR11028**. Using the 0.05 level of significance, determine whether the prototypes might not really differ in terms of the price that potential buyers would expect to pay. Identify and interpret the p-value for the test. Construct and interpret the 95% confidence interval for the difference between the population means. Is the hypothesized difference (0.00) within the interval? Given the presence or absence of the 0.00 value within the interval, is this consistent with the findings of the hypothesis test?

11.29 It has been reported that the average visitor from Japan spent \$3120 during a trip to the United States, while the average for a visitor from the United Kingdom was \$2654. Assuming that file **XR11029** contains the survey data underlying these results, use the 0.05 level in examining whether the sample mean for Japanese visitors is significantly higher than that for visitors from the United Kingdom. Identify and interpret the p-value for the test.
Source: *The World Almanac and Book of Facts 2009,* p. 127.

11.30 During May 2009, visitors to usatoday.com spent an average of 12.2 minutes per visit, compared to 11.0 minutes for visitors to washingtonpost.com. Assuming that file **XR11030** contains the sample data underlying these results, use the 0.01 level of significance in examining whether the population mean visiting times for the two sites might really be the same. Identify and interpret the p-value for the test. Construct and interpret the 99% confidence interval for the difference between the population means. Is the hypothesized difference (0.00) within the interval? Given the presence or absence of the 0.00 value within the interval, is this consistent with the findings of the hypothesis test?
Source: Jennifer Saba, "Average Time Spent on Top 30 Newspaper Web Sites Declines," editorandpublisher.com, June 22, 2009.

11.4 THE *z*-TEST FOR COMPARING THE MEANS OF TWO INDEPENDENT SAMPLES

The z-test approximation is included in Figure 11.1 and presented here as an alternative to the unequal-variances t-test whenever both n_1 and n_2 are ≥ 30. Besides requiring no assumptions about the shape of the population distributions, it offers the advantages of slightly greater simplicity and avoidance of the cumbersome df correction formula used in the unequal-variances t-test; thus, it can be useful to those who are not relying on a computer and statistical software. This test has been popular for many years as a method for comparing the means of two large, independent samples when σ_1 and σ_2 are unknown, and of two independent samples of any size when σ_1 and σ_2 are known and the two populations are normally distributed. Like the unequal-variances t-test, the z-test approximation does not assume the population standard deviations are equal, and s_1 and s_2 are used to estimate their respective population standard deviations, σ_1 and σ_2.

z-test approximation for comparing the means of two independent samples, σ_1 and σ_2 unknown, and each $n \geq 30$:

$$z = \frac{(\bar{x}_1 - \bar{x}_2) - (\mu_1 - \mu_2)_0}{\sqrt{\frac{s_1^2}{n_1} + \frac{s_2^2}{n_2}}}$$

where $\bar{x}_1$ and $\bar{x}_2$ = means of samples 1 and 2
$(\mu_1 - \mu_2)_0$ = hypothesized difference between the population means
n_1 and n_2 = sizes of samples 1 and 2
s_1 and s_2 = standard deviations of samples 1 and 2

Confidence interval for $\mu_1 - \mu_2$:

$$(\bar{x}_1 - \bar{x}_2) \pm z_{\alpha/2}\sqrt{\frac{s_1^2}{n_1} + \frac{s_2^2}{n_2}}$$

with

$$\alpha = (1 - \text{confidence coefficient})$$

EXAMPLE

z-Test

A university's placement center has collected data comparing the starting salaries of graduating students with surnames beginning with the letters A through M with those whose surnames begin with N through Z. For a sample of 30 students in the A–M category, the average starting salary was \$37,233.33, with a standard deviation of \$3475.54. For a sample of 36 students with surnames beginning with N–Z, the average starting salary was \$35,855.81, with a standard deviation of \$2580.02. The underlying data are in file CX11GRAD.

SOLUTION

For this study, the null hypothesis is that there is no difference between the population means, or H_0: $\mu_1 = \mu_2$. Because the intent of the test is nondirectional, the null hypothesis can be rejected by an extreme difference in either direction, and the alternative hypothesis is H_1: $\mu_1 \neq \mu_2$. For testing the null hypothesis, we'll use the 0.02 level of significance. As described in Table 11.1, the null and alternative hypotheses can also be stated as follows:

- **Null hypothesis**

  H_0: $\mu_{(\bar{x}_1 - \bar{x}_2)} = 0$ The starting salaries are the same for both populations.

- **Alternative hypothesis**

  H_1: $\mu_{(\bar{x}_1 - \bar{x}_2)} \neq 0$ The starting salaries are not the same.

For these data, the calculated value of the test statistic, z, can be computed as

$$z = \frac{(\bar{x}_1 - \bar{x}_2) - 0}{\sqrt{\frac{s_1^2}{n_1} + \frac{s_2^2}{n_2}}} = \frac{(37{,}233.33 - 35{,}855.81) - 0}{\sqrt{\frac{3475.54^2}{30} + \frac{2580.02^2}{36}}} = 1.80$$

For the 0.02 level of significance, the critical values will be $z = -2.33$ and $z = +2.33$. The decision rule will be to reject the null hypothesis of equal population means if the calculated z is either less than -2.33 or greater than $+2.33$, as shown in Figure 11.4. Because the calculated test statistic, $z = 1.80$, falls into the nonrejection region of the diagram, the null hypothesis cannot be rejected at

EXAMPLEEXAMPLEEXAMPLEEXAMPLEEXAMPLEEXAMPLEEXAN

FIGURE 11.4

This is an application of the z-test in comparing two sample means. From the results, we are not able to reject the possibility that graduates with surnames beginning with A–M receive the same starting salaries as graduates whose names begin with N–Z.

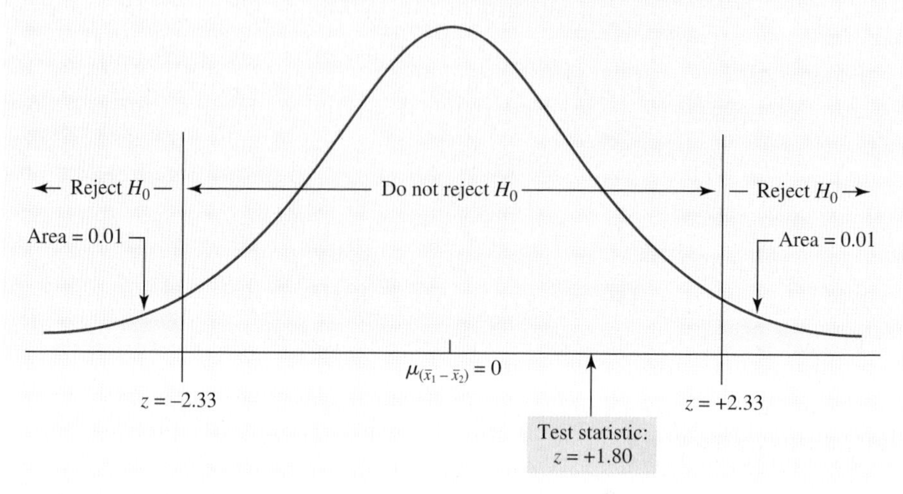

LEEXAMPLE

the 0.02 level of significance. From this analysis, we cannot conclude that people with surnames in the first part of the alphabet receive different starting salaries than persons whose names are in the latter portion.

The approximate p-value for this test can be determined by finding the area to the right of the calculated test statistic, $z = 1.80$, then (because it is a nondirectional test) multiplying this quantity by 2. Referring to the normal distribution table, we find the cumulative area to $z = 1.80$ is 0.9641. Subtracting 0.9641 from 1.0000 gives the right-tail area as 0.0359. The approximate p-value for this two-tail test will be 2(0.0359), or 0.0718.

Based on the sample data, we will also determine the 98% confidence interval for $(\mu_1 - \mu_2)$. This corresponds to $\alpha = 0.02$ and, to the best accuracy possible using the normal table with $z = 2.33$, the interval will be

$$(\bar{x}_1 - \bar{x}_2) \pm z_{\alpha/2}\sqrt{\frac{s_1^2}{n_1} + \frac{s_2^2}{n_2}}$$

$$= (37{,}233.33 - 35{,}855.81) \pm 2.33\sqrt{\frac{3475.54^2}{30} + \frac{2580.02^2}{36}}$$

$$= 1377.52 \pm 1785.99 \quad \text{or from } -408.47 \text{ to } +3163.51$$

The hypothesized difference (zero) is contained within the 98% confidence interval, so we are 98% confident the population means could be the same. As we discussed in Chapter 10, a nondirectional test at the α level of significance and a $100(1 - \alpha)\%$ confidence interval will lead to the same conclusion. Computer Solutions 11.3 describes Excel procedures for the z-test and confidence interval for the difference between population means. The computer-generated confidence interval differs very slightly from ours because our z-value (2.33) contained only the standard two decimal places from the normal distribution table.

Seeing Statistics Applet 14, at the end of the chapter, allows you to visually examine the sampling distribution of the difference between sample means and how it responds to the changes you select.

COMPUTER 11.3 SOLUTIONS

The z-Test for $(\mu_1 - \mu_2)$

These procedures show how to use the *z*-test to compare the means of two independent samples. Each sample size should be ≥30. Excel offers the *z*-test; Minitab does not.

EXCEL

	D	E	F
1	z-Test: Two Sample for Means		
2			
3		*GroupAM*	*GroupNZ*
4	Mean	37233.33	35855.81
5	Known Variance	12079378.29	6656503.20
6	Observations	30	36
7	Hypothesized Mean Difference	0	
8	z	1.797	
9	P(Z<=z) one-tail	0.036	
10	z Critical one-tail	2.054	
11	P(Z<=z) two-tail	0.072	
12	z Critical two-tail	2.326	

Excel *z*-test for $(\mu_1 - \mu_2)$, based on raw data

1. Open Excel data file **CX11GRAD**. It corresponds to Figure 11.4 and has the label and 30 salary values for the A–M group in column A, and the label and 36 salary values for the N–Z group in column B.
2. Excel does this test assuming the population variances are known. To estimate these using the sample variances, first compute the variances of the two samples: Select any available cell and enter **=VAR(A2:A31)** and hit **return.** Select another available cell, enter **=VAR(B2:B37),** and hit **return.** From the **Data** ribbon, click **Data Analysis.** Click **z-Test: Two Sample for Means.** Click **OK.**
3. Enter **A1:A31** into the **Variable 1 Range** box and **B1:B37** into the **Variable 2 Range** box. Enter **0** into the **Hypothesized Mean Difference** box. Click to select **Labels.** Specify the significance level by entering **0.02** into the **Alpha** box.
4. Enter the variances for columns A and B, obtained in step 2, into the **Variable 1 Variance (known)** and **Variable 2 Variance (known)** boxes, respectively. Select **Output Range** and enter **D1** into the box. Click **OK.** This is a two-tail test, so we refer to the 0.072 *p*-value.
5. To get a confidence interval for $(\mu_1 - \mu_2)$, it will be necessary to refer to the Excel worksheet template and procedure described below. It is based on summary statistics for the samples.

Note: For an analysis based on raw data, Excel requires the two samples to be in two distinctly separate fields (e.g., two separate columns). If the data are in one column and the subscripts identifying group membership are in another column, see the Excel Unstacking Note in Computer Solutions 11.1.

Excel *z*-test and confidence interval for $(\mu_1 - \mu_2)$, based on summary data

	A	B	C	D	E
1	**2-Sample Z-Test Comparing Means (based on summary statistics)**				
2					
3	Hypothesized Difference (mu1 - mu2)	0.000			
4					
5	*Summary of Sample Data:*		*Calculated Values:*		
6	Mean for Sample 1, xbar1	37233.33	Std. Error =	766.517	
7	Std. Dev. For Sample 1, s1	3475.54	z =	1.797	
8	Size of Sample 1, n1	30			
9	Mean for Sample 2, xbar2	35855.81	*p-value if the test is:*		
10	Std. Dev. For Sample 2, s2	2580.02	Left-Tail	**Two-Tail**	Right-Tail
11	Size of Sample 2, n2	36	0.964	**0.072**	0.036
12					
13	Confidence level desired	0.98			
14	Lower confidence limit	-405.7			
15	Upper confidence limit	3160.7			

(continued)

1. Open Excel worksheet TMZ2TEST. Enter the hypothesized difference between the population means **(0)** into cell B3. Enter the summary statistics for the example on which Figure 11.4 is based into cells B6:B11. Enter the desired confidence level as a decimal fraction (e.g., **0.98**) into cell B13.
2. Refer to the *p*-value that corresponds to the type of test being conducted. The test in Figure 11.4 is a two-tail test, so the *p*-value is in cell D11, or 0.072.

EXERCISES

11.31 Under what conditions is it appropriate to use the *z*-test as an approximation to the unequal-variances *t*-test when comparing two sample means?

11.32 For the following independent random samples, use the *z*-test and the 0.01 level of significance in testing H_0: $\mu_1 = \mu_2$ versus H_1: $\mu_1 \neq \mu_2$.

$\bar{x}_1 = 33.5 \quad s_1 = 6.4 \quad n_1 = 31$
$\bar{x}_2 = 27.6 \quad s_2 = 2.7 \quad n_2 = 30$

11.33 For the following independent random samples, use the *z*-test and the 0.05 level of significance in testing H_0: $\mu_1 \leq \mu_2$ versus H_1: $\mu_1 > \mu_2$.

$\bar{x}_1 = 85.2 \quad s_1 = 9.6 \quad n_1 = 40$
$\bar{x}_2 = 81.7 \quad s_2 = 4.1 \quad n_2 = 32$

11.34 Repeat Exercise 11.20 using the *z*-test approximation to the unequal-variances *t*-test.

11.35 Repeat Exercise 11.21 using the *z*-test approximation to the unequal-variances *t*-test.

11.36 According to the U.S. Bureau of Labor Statistics, personal expenditures for entertainment fees and admissions averaged $349 per person in the Northeast and $420 in the West. Assuming that these data involved (1) sample sizes of 800 and 600 and (2) standard deviations of $215 and $325, respectively, use a *z*-test and the 0.01 level of significance in testing the difference between these means. Determine and interpret the *p*-value for the test. Construct and interpret the 99% confidence interval for the test. Source: U.S. Bureau of Labor Statistics, *Consumer Expenditure Survey,* Interview Survey, annual.

11.37 Surveys of dog owners in two adjacent towns found the average veterinary-related expense total for dog owners in town A during the preceding year was $240, with a standard deviation of $55, while the comparable results in nearby town B were $225 and $32, respectively. Assuming a sample size of 30 dog owners surveyed in each town, use a *z*-test and the 0.05 level of significance in determining whether the population means might actually be the same. Determine and interpret the *p*-value for the test; then construct and interpret the 95% confidence interval for the difference between the population means.

11.38 A study of 102 patients who had digestive-tract surgery found that those who chewed a stick of gum for 15 minutes at mealtimes ended up being discharged after an average of 4.4 days in the hospital, compared to an average of 5.2 days for their counterparts who were not provided with gum. Assuming (1) a sample size of 51 for each group and (2) sample standard deviations of 1.3 and 1.9 days, respectively, use a *z*-test and the 0.01 level of significance in examining whether the mean hospital stay for the gum-chewing patients could have been this much lower simply by chance. Determine and interpret the *p*-value for the test. Source: Jennifer Bails, "Chewing Gum Might Help Digestive-Tract Patients Get Home Sooner," *Pittsburgh Tribune-Review,* October 15, 2005, p. A4.

11.39 A study found that women of normal weight missed an average of 3.4 days of work due to illness during the preceding year, compared to an average of 5.2 days for women who were considered overweight. Assuming (1) a sample size of 30 for each group and (2) sample standard deviations of 2.5 and 3.6 days, respectively, use a *z*-test and the 0.05 level of significance in examining whether the mean number of absence days for the overweight group could have been this much higher simply by chance. Determine and interpret the *p*-value for the test. Source: Nanci Hellmich, "Heavy Workers, Hefty Price," *USA Today,* September 9, 2005, p. 7D.

11.40 According to a psychologist at the Medical College of Virginia, soccer players who "head" the ball 10 or more times a game are at risk for brain damage that could lower their intellectual abilities. In tests involving young male soccer players, those who typically headed the ball 10 or more times a game recorded a mean IQ score of 103, compared to a mean of 112 for those who usually headed the ball once or less a game. Assuming 30 players in each sample, with sample standard deviations of 17.4 and 14.5 IQ points for the "headers" and "nonheaders," respectively, use an appropriate one-tail z-test and the 0.01 level of significance in reaching a conclusion as to whether frequent "heading" of the ball by soccer players might lower intellectual performance. Determine and interpret the p-value for the test. Source: Marilyn Elias, "Heading Soccer Ball Can Lower IQ," *USA Today*, August 14, 1995, p. 1D.

(DATA SET) *Note:* Exercises 11.41–11.43 require a computer and statistical software.

11.41 In their 2003 and 2008 studies on how long wireless customers have to spend on hold before speaking with a customer service representative, J.D. Power and Associates found the mean times to be 3.3 minutes and 4.4 minutes, respectively. Assume that data file **XR11041** contains the data, in minutes, for times on hold during the 2003 and 2008 studies. Use a one-tail test and the 0.05 level of significance in concluding whether the population mean time for 2008 could be greater than that for 2003. Determine and interpret the p-value for the test. Source: J.D. Power and Associates press release, "Customer Hold Times for Wireless Phone Customers Reach an All-Time High," jdpower.com, August 14, 2008.

11.42 In planning the processes to be incorporated into a new manufacturing facility, engineers have proposed two possible assembly procedures for one of the phases in the production sequence. Sixty of the eventual production line workers have taken part in preliminary tests, with data representing productivity (units produced in 1 hour) for 30 workers using procedure A and 30 using procedure B. Given the data in file **XR11042**, use the 0.10 level of significance in determining whether the two procedures might be equally efficient. Identify and interpret the p-value for the test, then construct and interpret the 90% confidence interval for the difference between the population means.

11.43 A study has been conducted to examine the effectiveness of a new experimental program for preparing high school students for the Scholastic Aptitude Test. Eighty students have been randomly divided into two groups of 40. The eventual SAT scores for those exposed to the new program and the conventional program are listed in data file **XR11043**. Use a one-tail test and the 0.025 level of significance in concluding whether the experimental program might be better than the conventional method in preparing students for the SAT. Determine and interpret the p-value for the test.

COMPARING TWO MEANS WHEN THE SAMPLES ARE DEPENDENT (11.5)

In previous comparisons of two sample means, the samples were independent from each other. That is, the selection process for one sample was not related to the selection process for the other. However, there may be times when we wish to test hypotheses involving samples that are *not* independent. For example, we may wish to examine the before-and-after productivity of individual employees after a change in their workstation layout, or compare the before-and-after reading speeds of individual participants in a speed-reading course. In such cases, we do not really have two different samples of persons, but rather *before* and *after* measurements for the *same* individuals. As a result, there will be just one variable: the difference recorded for each individual.

Tests in which the samples are not independent are also referred to as **paired observations,** or *matched pairs*, and they are essentially the same as those discussed in Chapter 10 for the mean of a single sample. The variable under consideration in this case is $d = (x_1 - x_2)$, where x_1 and x_2 are the before and after measurements, respectively. As in the tests for one sample mean, the null and alternative hypotheses will be one of the following, with the test statistic calculated as shown here:

Null Hypothesis	Alternative Hypothesis	Type of Test
H_0: $\mu_d = 0$	H_1: $\mu_d \neq 0$	Two-tail
H_0: $\mu_d \geq 0$	H_1: $\mu_d < 0$	Left-tail
H_0: $\mu_d \leq 0$	H_1: $\mu_d > 0$	Right-tail

Test statistic for comparing the means for paired observations:

$$t = \frac{\bar{d}}{s_d/\sqrt{n}}$$

where d = for each individual or test unit, $(x_1 - x_2)$, the difference between the two measurements

$\bar{d}$ = the average difference, $= \sum d_i / n$

n = number of pairs of observations

s_d = the standard deviation of d, or $\sqrt{\frac{\sum d_i^2 - n\bar{d}^2}{n-1}}$

$df = n - 1$

Confidence interval for μ_d:

$$\bar{d} \pm t_{\alpha/2} \frac{s_d}{\sqrt{n}}$$

EXAMPLE

Dependent Samples

Exploring ways to increase office productivity, a company vice president has ordered 12 ergonomic keyboards for distribution to a sample of secretarial employees. If the keyboards substantially increase productivity, she plans to replace all of the firm's current keyboards with the new models. Prior to delivery of the keyboards, each of the 12 sample members types a standard document on his or her old keyboard, and the number of words per minute is measured. After receiving the new keyboards and spending a few weeks becoming familiar with their operation, each employee then types the same document using the ergonomic model. Table 11.2 shows the number of words per minute each of the 12 persons typed in each test. The data are also in file CX11TYPE.

TABLE 11.2

For dependent samples, only one variable is tested: the difference between measurements. For each of 12 individuals, the typing speed for generating a standard document is shown before and after learning to use an ergonomic keyboard.

Person	x_1, Words/Minute with Old Keyboard	x_2, Words/Minute with New Keyboard	Difference $d = (x_1 - x_2)$	d^2
1	25.5	43.6	−18.1	327.61
2	59.2	69.9	−10.7	114.49
3	38.4	39.8	−1.4	1.96
4	66.8	73.4	−6.6	43.56
5	44.9	50.2	−5.3	28.09
6	47.4	53.9	−6.5	42.25
7	41.6	40.3	1.3	1.69
8	48.9	58.0	−9.1	82.81
9	60.7	66.9	−6.2	38.44
10	41.0	66.5	−25.5	650.25
11	36.1	27.4	8.7	75.69
12	34.4	33.7	0.7	0.49
			−78.7 $= \sum d$	1407.33 $= \sum d^2$

SOLUTION

Because the vice president doesn't want to replace the current stock of keyboards unless the ergonomic model is clearly superior, the burden of proof is on the new model, and a one-tail test is appropriate. The 0.025 level will be used to examine whether the new keyboard has significantly increased typing speeds. For each person in the sample, the difference in typing speed between the first and second measurements is $d = (x_1 - x_2)$ words per minute. The null and alternative hypotheses will be as follows:

- **Null hypothesis**

  $H_0\text{:}\ \mu_d \geq 0$ Typing with the ergonomic keyboard is no faster than with the current keyboard.

- **Alternative hypothesis**

  $H_1\text{:}\ \mu_d < 0$ The ergonomic keyboard is faster.

The sample mean and standard deviation for d are calculated as in Chapter 3, and can be expressed as

$$\bar{d} = \frac{-18.1 - 10.7 - 1.4 - 6.6 - 5.3 - 6.5 + 1.3 - 9.1 - 6.2 - 25.5 + 8.7 + 0.7}{12}$$
$$= -6.558$$

$$s_d = \sqrt{\frac{\Sigma d_i^2 - n\bar{d}^2}{n - 1}} = \sqrt{\frac{1407.33 - 12(-6.558)^2}{12 - 1}} = 9.001$$

and the test statistic is calculated as

$$t = \frac{\bar{d}}{s_d/\sqrt{n}} = \frac{-6.558}{9.001/\sqrt{12}} = -2.524$$

The number of degrees of freedom for the test is $df = (n - 1)$, or $(12 - 1) = 11$. For the 0.025 level of significance in a left-tail test, the critical value for the test statistic will be $t = -2.201$. This is obtained by referring to the $\alpha = 0.025$ column and $df = 11$ row of the table. The decision rule is, "Reject the null hypothesis if the calculated test statistic is less than $t = -2.201$, otherwise do not reject."

As Figure 11.5 shows, the calculated test statistic is less than the critical value and falls into the rejection region for the test. As a result, the null hypothesis is rejected, and we conclude that the ergonomic keyboard does increase typing speeds. Following through with the intent of her test, the vice president should order them for all secretarial personnel.

Based on the sample data, we will also determine the 95% confidence interval for μ_d. This corresponds to $\alpha = 0.05$. With $df = 12 - 1 = 11$ and $t = 2.201$, the interval will be

$$\bar{d} \pm t_{\alpha/2}\frac{s_d}{\sqrt{n}} = -6.558 \pm 2.201\frac{9.001}{\sqrt{12}}$$
$$= -6.558 \pm 5.719, \quad \text{or from } -12.277 \text{ to } -0.839$$

FIGURE 11.5

A summary of the hypothesis test for the paired observations in Table 11.2. At the 0.025 level in a one-tail test, we conclude that the ergonomic keyboard increases typing speeds.

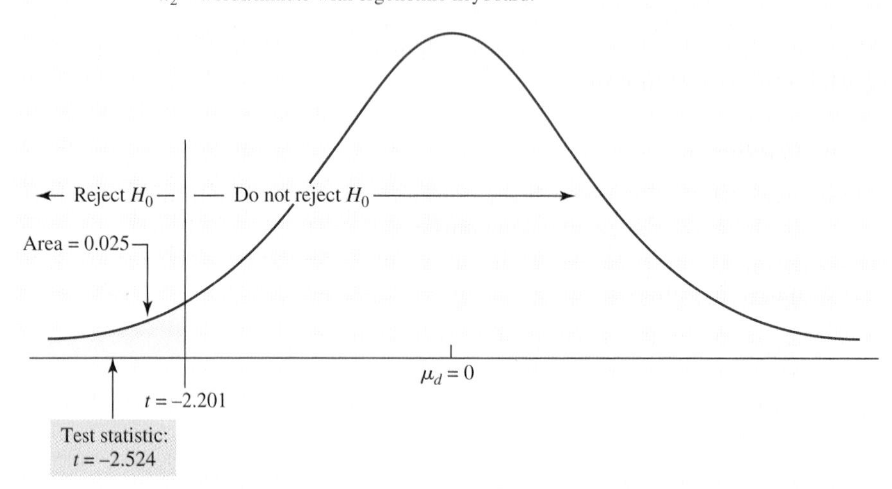

Computer Solutions 11.4 describes Excel and Minitab procedures for the t-test and confidence interval when comparing the means of dependent samples. In a one-tail test, Minitab will provide either an upper or a lower confidence limit for μ_d, depending on the directionality of the test.

COMPUTER 11.4 SOLUTIONS

Comparing the Means of Dependent Samples

These procedures show how to use a *t*-test to compare sample means when the samples are not independent.

EXCEL

	D	E	F
1	t-Test: Paired Two Sample for Means		
2			
3		*Old*	*New*
4	Mean	45.41	51.97
5	Variance	144.12	231.80
6	Observations	12	12
7	Pearson Correlation	0.807	
8	Hypothesized Mean Difference	0	
9	df	11	
10	t Stat	-2.524	
11	P(T<=t) one-tail	0.014	
12	t Critical one-tail	2.201	
13	P(T<=t) two-tail	0.028	
14	t Critical two-tail	2.593	

Excel *t*-test for comparing the means of dependent samples, based on raw data

1. For the sample data (Excel file CX11TYPE) on which Figure 11.5 is based, with the label and 12 "old keyboard" data values in column A, and the label and 12 "new keyboard" data values in column B: From the **Data** ribbon, click **Data Analysis.** Click ***t*-Test: Paired Two-Sample For Means.** Click **OK.**
2. Enter **A1:A13** into the **Variable 1 Range** box and **B1:B13** into the **Variable 2 Range** box. Enter **0** into the **Hypothesized Mean Difference** box. Click to select **Labels**. Specify the significance level by entering **0.025** into the **Alpha** box. Select **Output Range** and enter **D1** into the box. Click **OK.** This is a one-tail test, so we refer to the 0.014 *p*-value.
3. To obtain a confidence interval for $(\mu_1 - \mu_2)$, it will be necessary to refer to the procedure described below. It is based on summary statistics for the samples.

Excel *t*-test for comparing the means of dependent samples, based on summary statistics

1. Using the summary statistics associated with CX11TYPE (the test is summarized in Figure 11.5): Open the TEST STATISTICS workbook.
2. Using the arrows at the bottom left, select the ***t*-Test_Mean** worksheet.
3. For $d = x_1 - x_2$, enter the mean of d **(−6.558),** the standard deviation of d **(9.001),** and the number of pairs **(12)** into the appropriate cells. Enter the hypothesized difference (**0**) and the desired alpha level for the test **(0.025).** The calculated *t*-statistic and a one-tail *p*-value will be shown at the right.
4. To get a confidence interval for μ_d, follow steps 1–3, but open the ESTIMATORS workbook and select the ***t*-Estimate_Mean** worksheet. Enter the desired confidence level as a decimal fraction (e.g., **0.95**).

Note: As an alternative, you can use Excel worksheet template TMTTEST. The steps are described within the template.

MINITAB

Minitab *t*-test for comparing the means of dependent samples, based on raw data

```
Paired T-Test and CI: Old, New

Paired T for Old - New

              N    Mean   StDev  SE Mean
Old          12   45.41   12.00     3.47
New          12   51.97   15.22     4.40
Difference   12   -6.56    9.00     2.60

95% upper bound for mean difference: -1.89
T-Test of mean difference = 0 (vs < 0): T-Value = -2.52  P-Value = 0.014
```

1. For example, using the data (Minitab file CX11TYPE) on which Figure 11.5 is based, with the "old keyboard" data values in column **C1** and the "new keyboard" data values in column **C2:** Click **Stat.** Select **Basic Statistics.** Click **Paired *t*.**
2. Select **Samples in columns** and enter **C1** into the **First sample** box and **C2** into the **Second sample** box.
3. Click **Options.** Enter the desired confidence level as a percentage (e.g., **95.0**) into the **Confidence Level** box. Enter the hypothesized difference (**0**) into the **Test mean** box. Within the **Alternative** box, select **less than.** Click **OK.** Click **OK.**

Minitab *t*-test for comparing the means of dependent samples, based on summary data

Follow steps 1 through 3 above, but select **Summarized data (differences)** in step 2 and insert the appropriate summary statistics (number of pairs, the mean of d, and the standard deviation of d) into the **Sample size, Mean,** and **Standard deviation** boxes.

EXERCISES

11.44 A pharmaceutical firm has checked the cholesterol levels for each of 30 male patients, then provided them with fish-oil capsules to take on a daily basis. The cholesterol levels are rechecked after a 1-month period. Does this study involve independent samples or dependent samples?

11.45 Each of 20 consumers is provided with a package containing two different brands of instant coffee. A week later, they are asked to rate the taste of each coffee on a scale of 1 (poor taste) to 10 (excellent taste). Is this an example of independent samples or dependent samples?

11.46 A university president randomly selects 10 tenured faculty from the College of Arts and Sciences and 10 tenured faculty from the College of Business. Each faculty member is then asked to rate his or her job satisfaction on a scale of 1 (very dissatisfied) to 10 (very satisfied). Would this be an example of independent samples or dependent samples?

11.47 A trucking firm is considering the installation of a new, low-restriction engine air filter for its long-haul trucks, but doesn't want to make the switch unless the new filter can be shown to improve the fuel economy of these vehicles. A test is set up in which each of 10 trucks makes the same run twice—once with the old filtration system and once with the new version. Given the sample results shown below, use the 0.05 level of significance in determining whether the new filtration system could be superior.

Truck Number	Current Filter	New Filter
1	7.6 mpg	7.3 mpg
2	5.1	7.2
3	10.4	6.8
4	6.9	10.6
5	5.6	8.8
6	7.9	8.7
7	5.4	5.7
8	5.7	8.7
9	5.5	8.9
10	5.3	7.1

11.48 In an attempt to measure the emotional effect of a proposed billboard ad, an advertising agency checks the pulse rate of 10 persons before and after they are shown a photograph of the billboard. The agency believes that an effective billboard will increase the pulse rate of those who view it. In its test, the agency found the mean change in pulse rate was +5.7 beats per minute, with a standard deviation of 1.6. Using the 0.01 level of significance, examine whether the billboard stimulus could meet the agency's criterion for effectiveness.

11.49 The students in an aerobics class have been weighed both before and after the 5-week class, with the following results:

Person Number	Weight Before	Weight After
1	198 lb	194 lb
2	154	151
3	124	126
4	110	104
5	127	123
6	162	155
7	141	129
8	180	165

Using the 0.05 level of significance, evaluate the effectiveness of the program. Using the appropriate statistical table, what is the most accurate statement we can make about the p-value for this test?

(DATA SET) *Note:* Exercises 11.50 and 11.51 require a computer and statistical software.

11.50 For a special pre–New Year's Eve show, a radio station personality has invited a small panel of prominent local citizens to help demonstrate to listeners the adverse effect of alcohol on reaction time. The reaction times (in seconds) before and after consuming four drinks are in data file **XR11050**. At the 0.005 level, has the program host made his point? Identify and interpret the p-value for the test.

11.51 A plant manager has collected productivity data for a sample of workers, intending to see whether there is a difference in the number of units they produce on Monday versus Thursday. The results (units produced) are in data file **XR11051**. Using the 0.01 level of significance, evaluate the null hypothesis that there is no difference in worker productivity between the two days. Identify and interpret the p-value for the test, then construct and interpret the 99% confidence interval for the mean difference in productivity between the days.

COMPARING TWO SAMPLE PROPORTIONS (11.6)

The comparison of sample proportions from two independent samples is a frequent subject for statistical analysis. The following are but a few of the possibilities:

- Comparing the percentage of defective parts between shipments provided by two different suppliers.
- Determining whether the proportion of headache sufferers getting relief from a new medication is significantly greater than for those using aspirin.
- Comparing the enlistment percentage of high school seniors who have viewed version A of a recruiting film versus those seeing version B.

In this section, tests assume that both sample sizes are large (each $n \geq 30$). In addition, n_1p_1, $n_1(1 - p_1)$, n_2p_2, and $n_2(1 - p_2)$ should all be ≥ 5. (These requirements are necessary in order that the normal distribution used here will be a close approximation to the binomial distribution.) As in the comparison of means from independent samples, tests involving proportions can be either nondirectional or directional. Possible null and alternative hypotheses are similar to those summarized in Table 11.1.

Unlike the previous sections, our choice of test statistic will depend on the hypothesized difference between the population proportions, $(\pi_1 - \pi_2)_0$. In the vast majority of practical applications, the hypothesized difference will be zero and the appropriate test statistic will be the first of the two alternatives shown below. Accordingly, that will be our emphasis in this section. The confidence interval for $(\pi_1 - \pi_2)$ is not affected.

Test statistic for comparing proportions of two independent samples:

1. **When the hypothesized difference is zero (the usual case):**

$$z = \frac{(p_1 - p_2)}{\sqrt{\bar{p}(1 - \bar{p})\left(\frac{1}{n_1} + \frac{1}{n_2}\right)}}$$

with $\bar{p} = \dfrac{n_1p_1 + n_2p_2}{n_1 + n_2}$

where p_1 and p_2 = the sample proportions
n_1 and n_2 = the sample sizes
$\bar{p}$ = pooled estimate of the population proportion

2. **When the hypothesized difference is $(\pi_1 - \pi_2)_0 \neq 0$:**

$$z = \frac{(p_1 - p_2) - (\pi_1 - \pi_2)_0}{\sqrt{\frac{p_1(1 - p_1)}{n_1} + \frac{p_2(1 - p_2)}{n_2}}}$$

In either case, confidence interval for $(\pi_1 - \pi_2)$:

$$(p_1 - p_2) \pm z_{\alpha/2}\sqrt{\frac{p_1(1 - p_1)}{n_1} + \frac{p_2(1 - p_2)}{n_2}}$$

EXAMPLE

Sample Proportions

In a 10-year study sponsored by the National Heart, Lung and Blood Institute, 3806 middle-age men with high cholesterol levels but no known heart problems were divided into two groups. Members of the first group received a new drug designed to lower cholesterol levels, while the second group received daily

EXAMPLEEXAMP

dosages of a placebo. Besides lowering cholesterol levels, the drug appeared to be effective in reducing the incidence of heart attacks. During the 10 years, 155 of those in the first group suffered a heart attack, compared to 187 in the placebo group.[1] Assume the underlying data are in file **CX11HRT**, coded as 1 = did not have a heart attack, and 2 = had a heart attack.

SOLUTION

If we assume the 3806 participants were randomly divided into two groups, there would have been 1903 men in each group. Under this assumption, the sample proportions for heart attacks within the two groups are $p_1 = 155/1903$, or $p_1 = 0.0815$, and $p_2 = 187/1903$, or $p_2 = 0.0983$. Since the intent of the study was to evaluate the effectiveness of the new drug, the hypothesis test will be directional. In terms of the population proportions, the null and alternative hypotheses are $H_0\colon \pi_1 \geq \pi_2$ and $H_1\colon \pi_1 < \pi_2$. The hypotheses can also be expressed as

- **Null hypothesis**

  $H_0\colon\ \mu_{(p_1-p_2)} \geq 0$ — Users of the new drug are at least as likely to experience a coronary.

- **Alternative hypothesis**

  $H_1\colon\ \mu_{(p_1-p_2)} < 0$ — Users of the new drug are less likely to experience a coronary.

In testing the null hypothesis, we will use the 0.05 level of significance. The pooled estimate of the (assumed equal) population proportions is calculated as

$$\bar{p} = \frac{n_1p_1 + n_2p_2}{n_1 + n_2} = \frac{(1903)(0.0815) + (1903)(0.0983)}{1903 + 1903} = 0.0899$$

The calculated value of the test statistic, z, is

$$z = \frac{p_1 - p_2}{\sqrt{\bar{p}(1-\bar{p})\left(\dfrac{1}{n_1} + \dfrac{1}{n_2}\right)}} = \frac{0.0815 - 0.0983}{\sqrt{0.0899(1 - 0.0899)\left(\dfrac{1}{1903} + \dfrac{1}{1903}\right)}} = -1.81$$

For the 0.05 level in this left-tail test, the critical value of z will be $z = -1.645$. The decision rule is, "Reject H_0 if the calculated test statistic is < -1.645, otherwise do not reject."

As Figure 11.6 shows, the calculated test statistic, $z = -1.81$, is less than the critical value and falls into the rejection region. At the 0.05 level of significance, the null hypothesis is rejected, and we conclude that the new medication is effective.

Using the normal distribution table, we find the cumulative area to $z = -1.81$ to be 0.0351. This is the approximate p-value for the test.

Based on the sample data, we will also determine the 90% confidence interval for $(\pi_1 - \pi_2)$. With $z = 1.645$, this will be

$$(p_1 - p_2) \pm z_{\alpha/2}\sqrt{\frac{p_1(1-p_1)}{n_1} + \frac{p_2(1-p_2)}{n_2}}$$

$$= (0.0815 - 0.0983) \pm 1.645\sqrt{\frac{0.0815(1 - 0.0815)}{1903} + \frac{0.0983(1 - 0.0983)}{1903}}$$

$$= -0.0168 \pm 0.0152, \quad \text{or from } -0.0320 \text{ to } -0.0016$$

[1]Source: "News from the World of Medicine," *Reader's Digest,* May 1984, p. 222.

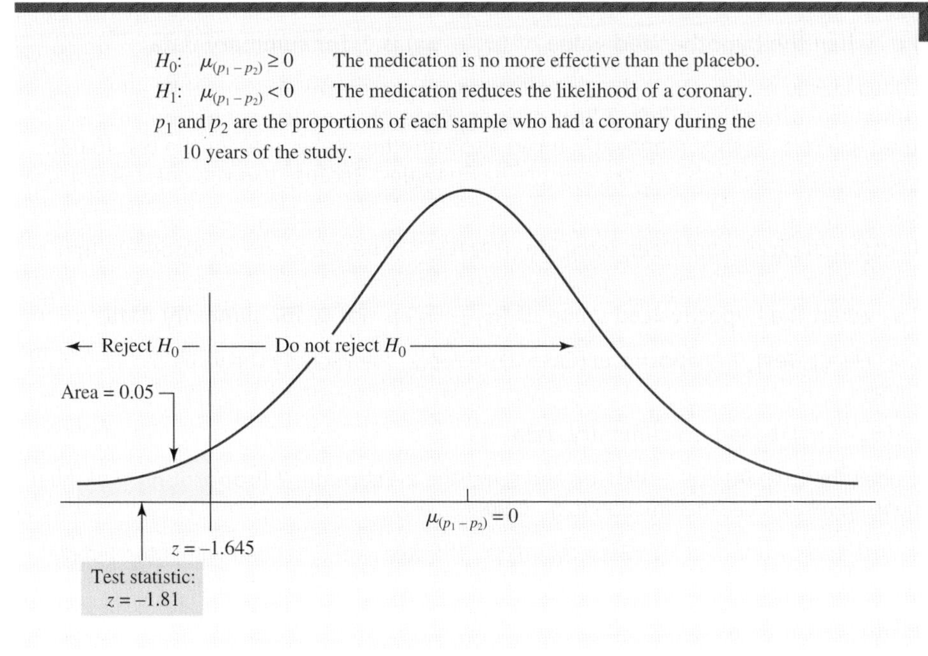

FIGURE 11.6

At the 0.05 level of significance, the medication that was the subject of this study appears to have been effective in reducing the incidence of heart attacks in middle-age men.

Computer Solutions 11.5 describes Excel and Minitab procedures for the z-test and confidence interval when comparing the proportions from two independent samples. In a one-tail test, Minitab will provide either an upper or a lower confidence limit for $(\pi_1 - \pi_2)$, depending on the directionality of the test.

COMPUTER 11.5 SOLUTIONS

The *z*-Test for Comparing Two Sample Proportions

These procedures show how to use the *z*-test to compare proportions from two independent samples.

EXCEL—USING SUMMARY DATA

	A	B	C	D	E
1	z-Test of the Difference Between Two Proportions (Case 1)				
2					
3		Sample 1	Sample 2	z Stat	-1.812
4	Sample proportion	0.0815	0.0983	P(Z<=z) one-tail	0.035
5	Sample size	1903	1903	z Critical one-tail	1.645
6	Alpha	0.05		P(Z<=z) two-tail	0.070
7				z Critical two-tail	1.960

(continued)

Excel comparison of p_1 and p_2 when the hypothesized value of $(\pi_1 - \pi_2)$ is 0, for summary data

1. For the summary data described in Figure 11.6: Open the **TEST STATISTICS** workbook.
2. Using the arrows at the bottom left, select the ***z*-Test_2 Proportions (Case 1).** For each sample, enter the sample proportion and the sample size, as shown above. You can also enter the alpha level for the test **(0.05).** Along with the *z* statistic, the printout lists both one-tail and two-tail *p*-values. Because this is a one-tail test, the *p*-value is 0.035.

Note: As an alternative, you can use Excel worksheet template **TM2PTEST**. The steps are described within the template.

Excel comparison of p_1 and p_2 when the hypothesized value of $(\pi_1 - \pi_2)$ is not 0, for summary data

Follow the procedure above, but select ***z*-Test_2 Proportions (Case 2)** in step 2, then specify the value of $(\pi_1 - \pi_2)$ associated with the null hypothesis.

Excel confidence interval for $(\pi_1 - \pi_2)$, based on summary data

Follow the first procedure, but open the **ESTIMATORS** workbook and select the ***z*-Estimate_2 Proportions** worksheet, then specify the desired confidence level as a decimal fraction (e.g., **0.90**).

Note: As an alternative, you can use Excel worksheet template **TM2PTEST**. The steps are described within the template.

EXCEL—USING RAW DATA

Excel comparison of p_1 and p_2 for raw data, regardless of the hypothesized value of $(\pi_1 - \pi_2)$

1. For the sample data (Excel file **CX11HRT**) on which Figure 11.6 is based, with the labels and data values for the drug and placebo groups in columns A and B, and coded as 1 = did not have a heart attack and 2 = had a heart attack: Click on any cell within the data field. From the **Add-Ins** ribbon, click **Data Analysis Plus.** Click ***Z*-Test: Two Proportions.** Click **OK.**
2. Enter **A1:A1904** into the **Variable 1 Range** box. Enter **B1:B1904** into the **Variable 2 Range** box. Enter **2** into the **Code for Success** box. Enter the hypothesized difference (in this case, **0**) into the **Hypothesized Difference** box. Click **Labels.** Enter **0.05** into the **Alpha** box. Click **OK.**

Excel confidence interval for $(\pi_1 - \pi_2)$, based on raw data

Follow the preceding procedure, but select ***Z*-Estimate: Two Proportions** from the Data Analysis Plus menu. Specify the desired confidence interval by entering the appropriate value for alpha. For example, to get a 95% confidence interval, input **0.05** into the **Alpha** box in step 2.

MINITAB—USING SUMMARY DATA

Minitab comparison of p_1 and p_2 when the hypothesized value of $(\pi_1 - \pi_2)$ is 0, for summary data

```
Test and CI for Two Proportions

Sample     X     N  Sample p
1        155  1903  0.081450
2        187  1903  0.098266

Difference = p (1) - p (2)
Estimate for difference:   -0.0168156
90% upper bound for difference:   -0.00493936
Test for difference = 0 (vs < 0):   Z = -1.81   P-Value = 0.035
```

1. Using the summary statistics associated with Figure 11.6: Click **Stat.** Select **Basic Statistics.** Click **2 Proportions.** Select **Summarized Data.** For sample 1, enter the sample size (**1903**) into the **Trials** box. Multiply the sample

proportion (0.0815) times the sample size (1903) to get the number of "successes" (0.0815)(1903) = 155.1, rounded to **155,** and enter this into the **Events** box for sample 1. Repeat this for sample 2, entering **1903** into the **Trials** box and (0.0983)(1903) = 187.1, rounded to **187** into the **Events** box.

2. Click **Options.** Enter the desired confidence level as a percentage **(90.0)** into the **Confidence Level** box. Enter the hypothesized difference between the population proportions **(0)** into the **Test difference** box. Within the **Alternative** box, select **less than.** Click to select **Use pooled estimate of p for test.** Click **OK.** Click **OK.**

Minitab comparison of p_1 and p_2 when the hypothesized value of $(\pi_1 - \pi_2)$ is not 0, for summary data

Follow the previous procedure, but in step 2 do NOT select "Use pooled estimate of p for test." Specify the hypothesized nonzero difference between the population proportions.

MINITAB—USING RAW DATA

Minitab comparison of p_1 and p_2 when the hypothesized value of $(\pi_1 - \pi_2)$ is 0, for raw data

1. For the sample data (Minitab file CX11HRT) on which Figure 11.6 is based, with the data for the drug group in column C1, the data for the placebo group in column C2, and data coded as 1 = did not have a heart attack and 2 = had a heart attack: Click **Stat.** Select **Basic Statistics.** Click **2 Proportions.** Select **Samples in different columns.** Enter **C1** into the **First** box and **C2** into the **Second** box. Note that Minitab will select the larger of the two codes (i.e., 2 = had heart attack) as the "success" or "event."
2. Click **Options**. Enter the desired confidence level as a percentage **(90.0)** into the **Confidence Level** box. Enter the hypothesized difference between the population proportions **(0)** into the **Test difference** box. Within the **Alternative** box, select **less than.** Click to select **Use pooled estimate of *p* for test.** Click **OK.** Click **OK.**

Minitab comparison of p_1 and p_2 when the hypothesized value of $(\pi_1 - \pi_2)$ is not 0, for raw data

Follow the previous procedure, but in step 2 do NOT select "Use pooled estimate of p for test." Specify the hypothesized nonzero difference between the population proportions.

EXERCISES

11.52 Summary data for two independent samples are $p_1 = 0.36$, $n_1 = 150$, and $p_2 = 0.29$, $n_2 = 100$. Use the 0.025 level of significance in testing H_0: $\pi_1 \leq \pi_2$ versus H_1: $\pi_1 > \pi_2$.

11.53 Summary data for two independent samples are $p_1 = 0.31$, $n_1 = 400$, and $p_2 = 0.38$, $n_2 = 500$. Use the 0.05 level of significance in testing H_0: $\pi_1 = \pi_2$ versus H_1: $\pi_1 \neq \pi_2$.

11.54 A bank manager has been presented with a new brochure that was designed to be more effective in attracting current customers to a personal financial counseling session that would include an analysis of additional banking services that could be advantageous to both the bank and the customer. The manager's assistant, who created the new brochure, randomly selects 400 current customers, then randomly chooses 200 to receive the standard brochure that has been used in the past, with the other 200 receiving the promising new brochure that he has developed. Of those receiving the standard brochure, 35% call for more information about the counseling session, while 42% of those receiving the new brochure call for more information. Using the 0.10 level of significance, is it possible that the superior performance of the new brochure was just due to chance and that the new brochure might really be no better than the old one?

11.55 In examining the ability of users to complete a variety of information-seeking tasks on their mobile devices, Nielsen Norman Group assigned sample members to get the answers to a variety of questions like, "How many calories are there typically in a slice of thin-crust pizza?" The success rate for achieving such tasks was 75% for those using touch-screen phones like the iPhone to access the mobile Internet, compared to 80% for persons accessing websites on a conventional personal computer. Assuming these percentages to be based on independent samples of 500 persons each, and using the 0.01 level of significance, can we conclude that the population success rate with a conventional PC is greater than that when using the mobile Internet and touch-screen phone? Source: "Study: Mobile Web a Throwback to '90s," *USA Today*, July 20, 2009, p. 3B.

11.56 In a study by the United Dairy Industry Association, 42% of 655 persons age 35–44 said they seldom or never ate cottage cheese. For 455 individuals age 45–54, the corresponding percentage was 34%. At the 0.01 level, can we reject the possibility that the population percentages could be equal for these two groups? Source: United Dairy Industry Association, *Cottage Cheese: Attitudes & Usage*, p. 21.

11.57 During the course of a year, 38.0% of the 213 merchant ships lost were general cargo carriers, while 43.8% of the 219 ships lost during the following year were in this category. Using a two-tail test at the 0.05 level, examine whether this difference could have been the result of chance variation from one year to the next. Source: Bureau of the Census, *Statistical Abstract of the United States 1996*, p. 657.

11.58 Of 200 subjects approached by interviewer A, 45 refused to be interviewed. Of 120 approached by interviewer B, 42 refused an interview. At the 0.05 level of significance, can we reject the possibility that the interviewers are equally capable of obtaining interviews?

11.59 Community National Bank is using an observational study to examine the utilization of its new 24-hour banking machine. Of 300 males who used the machine last week, 42% made two or more transactions before leaving. Of 250 female users during the same period, 50% made at least two transactions while at the machine. At the 0.10 level, do males and females differ significantly in terms of making multiple transactions? Determine and interpret the p-value for this test, then construct and interpret the 90% confidence interval for $\pi_1 - \pi_2$.

11.60 A telephone sales solicitor, trying to decide between two alternative sales pitches, randomly alternated between them during a day of calls. Using approach A, 20% of 100 calls led to requests for the mailing of additional product information. For approach B in another 100 calls, only 14% led to requests for the product information mailing. At the 0.05 level, can we conclude that the difference in results was due to chance? Determine and interpret the p-value for this test, then construct and interpret the 95% confidence interval for $\pi_1 - \pi_2$.

11.61 In a preliminary study, the U.S. Veterans Affairs Department found that 30.9% of the 81 soldiers who were near an accidental nerve gas release just after the 1991 Persian Gulf War had muscle and bone ailments, compared to a rate of 23.5% for the 52,000 Gulf veterans who were not near that area. Use an appropriate one-tail test and the 0.10 level of significance in examining the difference between these two rates. Determine and interpret the p-value for the test. Source: John Diamond, "Some Veterans Suffer Bone, Muscle Ailments," *The Indiana Gazette*, January 22, 1997, p. 5.

11.62 According to the ICR Research Group, 63% of Americans in the 18–34 age group say they are comfortable filing income tax returns electronically, compared to just 49% of those who are 55–64. Using the 0.025 level of significance, and assuming there were 200 persons surveyed from each age group, examine whether Americans in the 18–34 age group might be more comfortable with electronic filing than their counterparts in the 55–64 group. Determine and interpret the p-value for the test. Source: Anne R. Carey and Genevieve Lynn, "E-Filing: No Problem," *USA Today*, March 22, 2000, p. 1B.

(DATA SET) *Note:* Exercises 11.63–11.65 require a computer and statistical software.

11.63 An American Express survey of small-business owners found that 71% of female owners feel stressed by their work/life balance, compared to 62% of their male counterparts. Assume that file **XR11063** contains the underlying sample data for female and male small-business owners, respectively, with the data coded so that 1 = feels no stress in work/life balance and 2 = feels stress in work/life balance. Using the 0.01 level of significance, evaluate the null hypothesis that female and male small-business owners might really be equally likely to feel stress in their work/life balance. Identify and interpret the p-value for the test, then construct and interpret the 99% confidence interval for the difference between the population proportions. Source: Jeff Cornwall, "Small Business Owners Still Struggle with Balance," smartbiz.com, May 30, 2007.

11.64 Attempting to improve the quality of services provided to customers, the owner of a chain of high-fashion department stores randomly selected a number of clerks for special training in customer relations. Of this group, only 10% were the subject of complaints to the store manager during the 3 months following the training. On the other hand, 15% of a sample of untrained clerks were mentioned in customer complaints to the manager during this same period. The data are in file **XR11064**, with data for each group coded as 1 = not mentioned in a complaint and 2 = mentioned in a complaint. Using the 0.05 level of significance, does the training appear to be effective in reducing the incidence of customer dissatisfaction with sales personnel? Identify and interpret the p-value for the test.

11.65 A study by the National Marine Manufacturers Association found that 12.2% of those who participated in sailing were females age 25–34. Of those who participated in horseback riding during the same period, 14.7% were females in this age group. Assume that file **XR11065** contains the underlying data for each activity group, coded as 1 = not a female in this age group and 2 = a female in this age group. Using the 0.10 level of significance, test whether the population proportion could be equal for females age 25–34 participating in each of these activities. Identify and interpret the p-value for the test, then construct and interpret the 90% confidence interval for the difference between the population proportions. Source: National Marine Manufacturers Association, *The Boating Market: A Sports Participation Study*, p. 40.

COMPARING THE VARIANCES OF TWO INDEPENDENT SAMPLES (11.7)

There are occasions when it is useful to compare the variances of two independent samples. For example, we might be interested in whether one manufacturing process differs from another in terms of the amount of variation among the units produced. We can examine two different portfolio strategies to determine whether there is significantly more variation in the performances of the investments in one of the portfolios than in the other. We can also compare the variances of two independent samples to determine the permissibility of using the pooled-variances t-test of Section 11.2, which assumes that the standard deviations (and, thus, the variances as well) of the respective populations are equal.

The test in this section involves the F distribution. Like the t distribution, it is a family of distributions and is continuous. Unlike the t distribution, however, its exact shape is determined by *two* different degrees of freedom instead of just a single value. From a theoretical standpoint, the F distribution is the sampling distribution of s_1^2/s_2^2 that would result if two samples were repeatedly drawn from the same, normally distributed population.

In terms of the hypothesis-testing procedure introduced in Chapter 10, the test can be described as follows:

1. *Formulate null and alternative hypotheses.* The null and alternative hypotheses are $H_0: \sigma_1^2 = \sigma_2^2$ and $H_1: \sigma_1^2 \neq \sigma_2^2$.
2. *Select the significance level, α.* There are three F distribution tables in Appendix A (Table A.6, Parts A–C). They represent upper-tail areas of 0.05, 0.025, and 0.01, respectively. Since these are one-tail areas, they represent $\alpha = 0.10$, $\alpha = 0.05$, and $\alpha = 0.02$ for our two-tail test.
3. *Calculate the test statistic.* The calculated test statistic is

$$F = \frac{s_1^2}{s_2^2} \quad \text{or} \quad \frac{s_2^2}{s_1^2}, \quad \text{whichever is larger}$$

4. *Identify the critical value of the test statistic and state the decision rule.* Although the test is nondirectional (i.e., $H_1: \sigma_1^2 \neq \sigma_2^2$), there will be just one critical value of F. This is because we have selected the larger of the two ratios in step 3.

   The critical value of F will be

   $F(\alpha/2, \nu_1, \nu_2)$ where α = specified level of significance: 0.10, 0.05, or 0.02

   $\nu_1 = (n - 1)$, where n is the size of the sample that had the larger variance

   $\nu_2 = (n - 1)$, where n is the size of the sample that had the smaller variance

   The critical value is found by consulting the F table that corresponds to $\alpha/2$ (0.05, 0.025, or 0.01), with ν_1 = the number of degrees of freedom associated with the numerator of the F ratio, and ν_2 = the number of degrees of freedom for the denominator. If ν_1 or ν_2 happens to be one of the larger values not included in the table, interpolate between the listed entries. The decision rule is, "Reject H_0 if calculated $F >$ critical F, otherwise do not reject."

5. *Compare the calculated and critical values and reach a conclusion.* If the calculated F exceeds the critical F, we are not able to assume that the population variances are equal.
6. *Make the related decision.* This will depend on the purpose for the test. For example, if the variance for one investment portfolio strategy differs significantly from the variance for another, we may wish to pursue the one for which the variance is lower. The ability to assume equal variances would also allow us to compare two sample means with the pooled-variances t-test of Section 11.2. However, keep in mind that the unequal-variances t-test of Section 11.3 can routinely be applied without having to go through the inconvenience of the test for variance equality.

EXAMPLEEXAMPLEEXAMPLEEXAMPLEEXAMPLEEXAMPLE

EXAMPLE

Comparing Variances

A sample of 9 technicians exposed to the standard version of a training film required an average of 31.4 minutes to service a compressor system, with a standard deviation of 14.5 minutes. For 7 technicians viewing an alternative version of the film, the average time required was 22.3 minutes, with a standard deviation of 10.2 minutes. If the sampled populations are approximately normally distributed, can σ_1^2 and σ_2^2 be assumed to be equal? The underlying data are in file **CX11TECH**.

SOLUTION

Formulate Null and Alternative Hypotheses

The null and alternative hypotheses are H_0: $\sigma_1^2 = \sigma_2^2$ and H_1: $\sigma_1^2 \neq \sigma_2^2$.

Select the Significance Level, α

For this test, the level of significance will be $\alpha = 0.02$.

Calculate the Test Statistic

The calculated F statistic will be

$$\frac{s_1^2}{s_2^2} = \frac{14.5^2}{10.2^2} \quad \text{or} \quad \frac{s_2^2}{s_1^2} = \frac{10.2^2}{14.5^2}, \qquad \text{whichever is larger}$$

Since the first ratio is larger, the calculated F is

$$\frac{14.5^2}{10.2^2} = 2.02$$

Identify the Critical Value of the Test Statistic and State the Decision Rule

The sample associated with the numerator of the F statistic is the one having the larger variance. It had a sample size of 9, so ν_1 is $9 - 1$, or $\nu_1 = 8$. The sample associated with the denominator of the F statistic is the one having the smaller variance. It had a sample size of 7, so $\nu_2 = 7 - 1$, or $\nu_2 = 6$.

The specified level of significance was $\alpha = 0.02$, and the critical value of F will be $F(\alpha/2, \nu_1, \nu_2) = F(0.01, 8, 6)$. Referring to the F table with an upper-tail area of 0.01, with $\nu_1 = 8$ and $\nu_2 = 6$, we find the critical value to be 8.10. The decision rule is, "Reject H_0 if calculated $F > 8.10$, otherwise do not reject."

Compare the Calculated and Critical Values and Reach a Conclusion

As the test summary in Figure 11.7 shows, the calculated F (2.02) does not exceed the critical value (8.10), so the null hypothesis of equal population variances is not rejected.

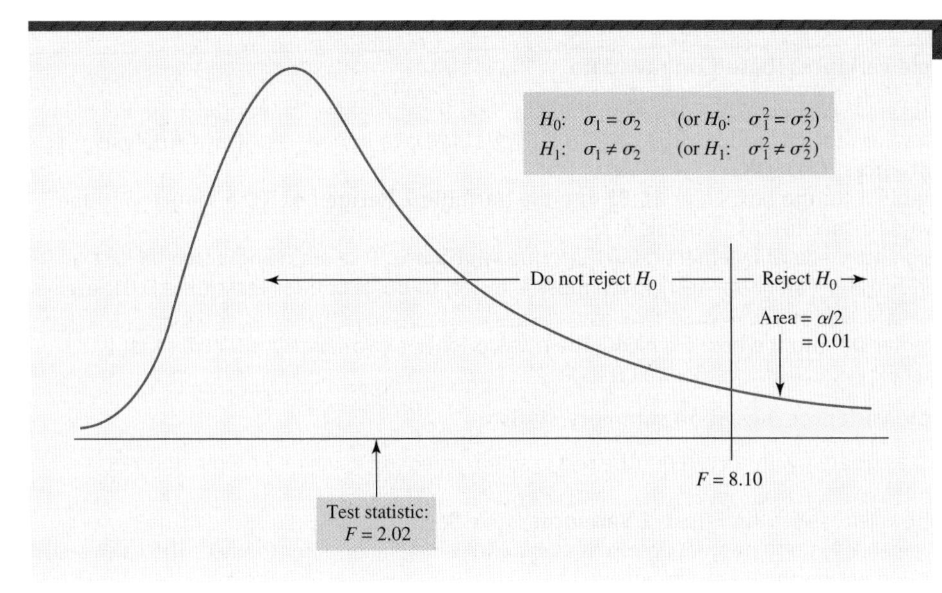

FIGURE 11.7

The F distribution can be used to test the null hypothesis that the population variances are equal. In this test at the $\alpha = 0.02$ level, the calculated F was less than the critical F, so the population variances can be assumed to be equal.

Make the Related Decision

At the 0.02 level of significance, we conclude that the variances in service times associated with the two different technician-training films could be equal. If we were to compare the means of the two samples, the population standard deviations could be assumed to be equal and it would be permissible to apply the pooled-variances t-test of Section 11.2. Computer Solutions 11.6 describes Excel and Minitab procedures for testing whether two population variances could be equal.

EXAMPLE

COMPUTER 11.6 SOLUTIONS

Testing for the Equality of Population Variances

These procedures show how to use the F-test to compare the variances of two independent samples.

EXCEL

	D	E	F
1	F-Test Two-Sample for Variances		
2			
3		Method1	Method2
4	Mean	31.40	22.30
5	Variance	210.20	104.03
6	Observations	9	7
7	df	8	6
8	F	2.02	
9	P(F<=f) one-tail	0.2035	
10	F Critical one-tail	8.10	

(continued)

Excel *F*-test comparing sample variances, based on raw data

1. For Excel data file **CX11TECH** on which Figure 11.7 is based, the label and 9 data values for method 1 are in column A, and the label and 7 data values for method 2 are in column B. From the **Data** ribbon, click **Data Analysis.** Click ***F*-Test Two-Sample for Variances.** Click **OK.**
2. Enter **A1:A10** into the **Variable 1 Range** box. Enter **B1:B8** into the **Variable 2 Range** box. (The sample with the greater variance should always be specified as "Variable 1.")
3. Select **Labels.** Specify the significance level by entering **0.01** into the **Alpha** box. (We are doing a two-tail test at the 0.02 level, but Excel performs only a one-tail test, so we must tell Excel to do its one-tail test at the 0.01 level to get comparable results.)
4. Select **Output Range** and enter **D1** into the box. Click **OK.** The *p*-value for our two-tail test is 0.407—that is, double the 0.2035 value that Excel shows for its one-tail test.

Excel *F*-test comparing sample variances, based on summary statistics

1. Using the summary statistics associated with **CX11TECH** (the test is summarized in Figure 11.7): Open the **TEST STATISTICS** workbook.
2. Using the arrows at the bottom left, select the **F-Test_2 Variances** worksheet.
3. Enter the sample sizes and variances into the appropriate cells along with an alpha level for the test. The calculated *F*-statistic and a two-tail *p*-value will be shown at the right.

MINITAB

Minitab *F*-test comparing sample variances, based on raw data

Test and CI for Two Variances: Method1, Method2

Method	DF1	DF2	Test Statistic	P-Value
F Test (normal)	8	6	2.02	0.407

1. Using Minitab data file **CX11TECH** on which Figure 11.7 is based, with the data for method 1 in column **C1** and the data for method 2 in column **C2:** Click **Stat.** Select **Basic Statistics.** Click **2 Variances.**
2. Select **Samples in different columns.** Enter **C1** into the **First** box and **C2** into the **Second** box. Click **OK.** A portion of the printout, including the two-tail *p*-value, is shown here.

Minitab *F*-test for comparing sample variances, based on summary data

Follow steps 1 and 2 above, but select **Sample variances** in step 2 and insert the appropriate summary statistics for each sample into the **Sample size** and **Variance** boxes.

EXERCISES

11.66 For two samples from what are assumed to be normally distributed populations, the sample sizes and standard deviations are $n_1 = 10$, $s_1 = 23.5$, $n_2 = 9$, and $s_2 = 10.4$. At the 0.10 level of significance, test the null hypothesis that the population variances are equal. Would your conclusion be different if the test had been conducted at the 0.05 level? At the 0.02 level?

11.67 For two samples from what are assumed to be normally distributed populations, the sample sizes and standard deviations are $n_1 = 28$, $s_1 = 103.1$, $n_2 = 41$, and $s_2 = 133.5$. At the 0.10 level of significance, test the null hypothesis that the population variances are equal. Would your conclusion be different if the test had been conducted at the 0.05 level? At the 0.02 level?

11.68 Given the information in Exercise 11.10, and using the 0.02 level of significance in comparing the sample standard deviations, were we justified in assuming that the population standard deviations are equal? Would your conclusion change if the standard deviation of playing times for music boxes equipped with the new battery design had been 20% larger than the value calculated in Exercise 11.10?

11.69 Given the information in Exercise 11.11, and using the 0.05 level of significance in comparing the sample standard deviations, were we justified in assuming that the population standard deviations were equal? Would your conclusion change if the standard deviation for the FoodFarm jars had been 3.0 grams?

11.70 According to the National Association of Homebuilders, the average life expectancies of a dishwasher and a microwave oven are about the same: 9 years. Assume that their finding was based on a sample of $n_1 = 60$ dishwashers and $n_2 = 40$ microwave ovens, and that the corresponding sample standard deviations were $s_1 = 3.0$ years and $s_2 = 3.7$ years. Using the 0.02 level of significance, examine whether the population standard deviations for the lifetimes of these two types of appliances could be the same. Source: National Association of Homebuilders, *Study of Life Expectancy of Home Components*, February 2007, p. 5.

11.71 In conducting her annual review of suppliers, a purchasing agent has collected data on a sample of orders from two of her company's leading vendors. On average, the 24 shipments from company 1 have arrived 3.4 days after the order was placed, with a standard deviation of 0.4 days. The 30 shipments from company 2 arrived an average of 3.6 days after the order was placed, with a standard deviation of 0.7 days. The average time for shipments to be received is about the same, regardless of supplier, but the purchasing agent is concerned about company 2's higher variability in shipping time. Using the 0.025 level of significance in a one-tail test, should the purchasing agent conclude that company 2's higher standard deviation in shipping times is due to something other than chance?

(DATA SET) *Note:* Exercises 11.72–11.74 require a computer and statistical software.

11.72 A researcher has observed independent samples of females and males, recording how long each person took to complete his or her shopping at a local mall. The respective times, in minutes, are listed in file **XR11072**. Using the 0.025 level of significance in a one-tail test, would females appear to exhibit more variability than males in the length of time shopping in this mall?

11.73 For independent samples of customers of Internet providers A and B, the ages are as listed in file **XR11073**. Using the 0.05 level of significance in a nondirectional test, examine whether the variation in ages could be the same for customers of the two providers.

11.74 An investment analyst has data for the past 8 years on each of two mutual funds, with the annual rates of return for each listed in file **XR11074**. On average, each fund has had an excellent annual rate of return over the time period for his data, but the analyst is concerned that a mutual fund with greater variation in rate of return tends to involve greater risk for his investment clients. Using the 0.05 level in a nondirectional test, examine whether the two mutual funds have differed significantly in their performance variability.

SUMMARY (11.8)

• Comparing sample means or proportions

One of the most useful applications of business statistics involves comparing two samples to examine whether a difference between them is (1) significant or (2) more likely due to chance variation from one sample to the next. As with testing hypotheses for just a single sample, comparing the means or proportions from two samples requires comparing the calculated value of a test statistic with its critical value(s), then deciding whether the null hypothesis should be rejected.

• Independent versus dependent samples

Independent samples are those for which the selection process for one is not related to the selection process for the other. An example of independent samples occurs when subjects are randomly assigned to the experimental and control groups of an experiment. Samples are dependent when the selection process for one is related to the selection process for the other. A typical example of dependent samples occurs with before-and-after measurements for the same individuals or test units. In this case, there is really only one variable: the difference between the two measurements recorded for each individual or test unit.

● **Tests for comparing means of independent samples**

When comparing the means of independent samples, the t-test is applicable whenever the population standard deviations are unknown. The *pooled-variances* t-test is used whenever the population standard deviations are assumed to be equal, regardless of the sample size. The population standard deviations are typically unknown.

The chapter describes the *unequal-variances* t-test to be used if the population standard deviations are unknown and cannot be assumed to be equal. The unequal-variances t-test involves a special correction formula for the number of degrees of freedom (df). The z-test can be used as a close approximation to the unequal-variances t-test when the population standard deviations are not assumed to be equal, but samples are large (each $n \geq 30$).

● **Comparing proportions from independent samples**

Comparing proportions from two independent samples involves a z-test. When samples are large and other conditions described in the chapter are met, the normal distribution is a close approximation to the binomial.

● **Comparing variances for independent samples**

A special hypothesis test can be applied to test the null hypothesis that the population variances are equal for two independent samples. Based on the F distribution, it has many possible applications, including determining whether the pooled-variances t-test is appropriate for a given set of data.

EQUATIONS

Pooled-variances t-test for comparing the means of two independent samples, σ_1 and σ_2 unknown and assumed to be equal

- $$t = \frac{(\bar{x}_1 - \bar{x}_2) - (\mu_1 - \mu_2)_0}{\sqrt{s_p^2\left(\frac{1}{n_1} + \frac{1}{n_2}\right)}}$$

where $\bar{x}_1$ and $\bar{x}_2$ = means of samples 1 and 2
$(\mu_1 - \mu_2)_0$ = hypothesized difference between the population means
n_1 and n_2 = sizes of samples 1 and 2
s_1 and s_2 = standard deviations of samples 1 and 2

with $$s_p^2 = \frac{(n_1 - 1)s_1^2 + (n_2 - 1)s_2^2}{n_1 + n_2 - 2}$$ and $df = n_1 + n_2 - 2$

- Confidence interval for $\mu_1 - \mu_2$:

$$(\bar{x}_1 - \bar{x}_2) \pm t_{\alpha/2}\sqrt{s_p^2\left(\frac{1}{n_1} + \frac{1}{n_2}\right)}$$

Unequal-variances t-test for comparing the means of two independent samples, σ_1 and σ_2 unknown and not assumed to be equal

- $$t = \frac{(\bar{x}_1 - \bar{x}_2) - (\mu_1 - \mu_2)_0}{\sqrt{\frac{s_1^2}{n_1} + \frac{s_2^2}{n_2}}}$$

where $\bar{x}_1$ and $\bar{x}_2$ = means of samples 1 and 2
$(\mu_1 - \mu_2)_0$ = hypothesized difference between the population means
n_1 and n_2 = sizes of samples 1 and 2
s_1 and s_2 = standard deviations of samples 1 and 2

with $$df = \frac{\left[(s_1^2/n_1) + (s_2^2/n_2)\right]^2}{\frac{(s_1^2/n_1)^2}{n_1 - 1} + \frac{(s_2^2/n_2)^2}{n_2 - 1}}$$

- Confidence interval for $\mu_1 - \mu_2$:

$$(\bar{x}_1 - \bar{x}_2) \pm t_{\alpha/2}\sqrt{\frac{s_1^2}{n_1} + \frac{s_2^2}{n_2}}$$

z-test approximation for comparing the means of two independent samples, σ_1 and σ_2 unknown, and each $n \geq 30$

- $$z = \frac{(\bar{x}_1 - \bar{x}_2) - (\mu_1 - \mu_2)_0}{\sqrt{\frac{s_1^2}{n_1} + \frac{s_2^2}{n_2}}}$$

where $\bar{x}_1$ and $\bar{x}_2$ = means of samples 1 and 2
$(\mu_1 - \mu_2)_0$ = hypothesized difference between the population means
n_1 and n_2 = sizes of samples 1 and 2
s_1 and s_2 = standard deviations of samples 1 and 2

- Confidence interval for $\mu_1 - \mu_2$:

$$(\bar{x}_1 - \bar{x}_2) \pm z_{\alpha/2}\sqrt{\frac{s_1^2}{n_1} + \frac{s_2^2}{n_2}}$$

Comparing proportions from two independent samples

- z-test, with test statistic

1. When the hypothesized difference is zero (the usual case):

$$z = \frac{(p_1 - p_2)}{\sqrt{\bar{p}(1 - \bar{p})\left(\frac{1}{n_1} + \frac{1}{n_2}\right)}}$$

where p_1 = observed proportion, sample 1
p_2 = observed proportion, sample 2
n_1 = sample size, sample 1
n_2 = sample size, sample 2
$\bar{p}$ = pooled estimate of the population proportion

with $\bar{p} = \frac{n_1 p_1 + n_2 p_2}{n_1 + n_2}$

2. When the hypothesized difference is $(\pi_1 - \pi_2)_0 \neq 0$:

$$z = \frac{(p_1 - p_2) - (\pi_1 - \pi_2)_0}{\sqrt{\frac{p_1(1 - p_1)}{n_1} + \frac{p_2(1 - p_2)}{n_2}}}$$

- Confidence interval for $(\pi_1 - \pi_2)$:

$$(p_1 - p_2) \pm z_{\alpha/2}\sqrt{\frac{p_1(1 - p_1)}{n_1} + \frac{p_2(1 - p_2)}{n_2}}$$

Comparing the means when samples are dependent

- t-test, with test statistic:

$$t = \frac{\bar{d}}{s_d/\sqrt{n}}$$

where d = for each individual or test unit, $(x_1 - x_2)$, the difference between the two measurements
$\bar{d}$ = the average difference, $= \sum d_i/n$
n = number of pairs of observations
s_d = the standard deviation of d, or $\sqrt{\frac{\sum d_i^2 - n\bar{d}^2}{n - 1}}$
$df = n - 1$

- Confidence interval for μ_d:

$$\bar{d} \pm t_{\alpha/2}\frac{s_d}{\sqrt{n}}$$

Comparing variances from two independent samples

- F-test, with test statistic:

$$F = \frac{s_1^2}{s_2^2} \quad \text{or} \quad \frac{s_2^2}{s_1^2}, \qquad \text{whichever is larger}$$

- Critical value of F:

$F(\alpha/2, \nu_1, \nu_2)$ where α = specified level of significance for a nondirectional test: 0.10, 0.05, or 0.02
$\nu_1 = (n - 1)$, where n is the size of the sample that had the larger variance
$\nu_2 = (n - 1)$, where n is the size of the sample that had the smaller variance

CHAPTER EXERCISES

Reminder: Be sure to refer to Figure 11.1 in selecting the testing procedure for an exercise. When either test can be applied to the same data, the unequal-variances t-test is preferable to the z-test—especially when doing the test with computer assistance.

11.75 Suspecting that television repair shops tend to charge women more than they do men, Emily disconnected the speaker wire on her portable television and took it to a sample of 12 shops. She was given repair estimates that averaged \$85, with a standard deviation of \$28. Her friend John, taking the same set to another sample of 9 shops, was provided with an average estimate of \$65, with a standard deviation of \$21. Assuming normal populations with equal standard deviations, use the 0.05 level in evaluating Emily's suspicion. Using the appropriate statistical table, what is the approximate p-value for this test?

11.76 The manufacturer of a small utility trailer is interested in comparing the amount of fuel used in towing the trailer with that required to overcome the weight and wind resistance of a rooftop carrier. Over a standard test route at highway speeds, 8 trips with the trailer resulted in an average fuel consumption of 0.28 gallons, with a standard deviation of 0.03. Under similar conditions, 10 trips with the rooftop carrier led to an average consumption of 0.35 gallons, with a standard deviation of 0.05 gallons. Assuming normal populations with equal standard deviations, use the 0.01 level in examining whether pulling the trailer uses significantly less fuel than driving with the rooftop carrier.

11.77 A compressor manufacturer is testing two different designs for an air tank. Testing involves pumping air into a tank until it bursts, then noting the air pressure just prior to tank failure. Four tanks of design A are found to fail at an average of 1400 pounds per square inch (psi), with a standard deviation of 250 psi. Six tanks of design B fail at an average of 1620 psi, with a standard deviation of 230 psi. Assuming normal populations with equal standard deviations, use the 0.10 level of significance in comparing the two designs. Construct and interpret the 90% confidence interval for the difference between the population means.

11.78 An industrial engineer has designed a new workstation layout that she claims will increase production efficiency. For a sample of 8 workers using the new workstation layout, the average number of units produced per hour is 36.4, with a standard deviation of 4.9. For a sample of 6 workers using the standard layout, the average number of units produced per hour is 30.2, with a standard deviation of 6.2. Assuming normal populations with equal standard deviations, use the 0.025 level of significance in testing the engineer's claim.

11.79 Bob Techster is very serious about winning this year's All-American Soap Box Derby competition. Since there is no wind-tunnel time available at the aerospace firm where his father works, Bob is using a local hill to test the effectiveness of an aerodynamic design change. In 10 runs with the standard car, the average time was 22.5 seconds, with a standard deviation of 1.3 seconds. With the design change, 12 runs led to an average time of 21.1 seconds, with a standard deviation of 0.9 seconds. Assuming normal populations with equal standard deviations, use the 0.01 level in helping Bob determine whether the modifications are really effective in improving his racer's speed. Using the appropriate statistical table, what is the approximate p-value for this test?

11.80 The developer of a new welding rod claims that spot welds using his product will have greater strength than conventional welds. For 45 welds using the new rod, the average tensile strength is 23,500 pounds per square inch, with a standard deviation of 600 pounds. For 40 conventional welds on the same materials, the average tensile strength is 23,140 pounds per square inch, with a standard deviation of 750 pounds. Use the 0.01 level in testing the claim of superiority for the new rod. Using the appropriate statistical table, what is the approximate p-value for this test?

11.81 Independent random samples of vehicles traveling past a given point on an interstate highway have been observed on Monday versus Wednesday. For 16 cars observed on Monday, the average speed was 59.4 mph, with a standard deviation of 3.7 mph. For 20 cars observed on Wednesday, the average speed was 56.3 mph, with a standard deviation of 4.4 mph. At the 0.05 level, and assuming normal populations, can we conclude that the average speed for all vehicles was higher on Monday than on Wednesday? What is the most accurate statement that can be made about the p-value for this test?

11.82 A sample of 25 production employees has been tested twice on a standard test of manual dexterity. The average change in the time required to finish the test was a decrease of 1.5 minutes, with a standard deviation of 0.3 minutes. At the 0.05 level, can we conclude that the average production employee will complete the test more quickly the second time he or she takes it?

11.83 For a sample of 48 finance majors, the average time spent reading each issue of the campus newspaper is 19.7 minutes, with a standard deviation of 7.3 minutes. The corresponding figures for a sample of 40 management information systems majors are 16.3 and 4.1 minutes. Using the 0.01 level of significance, test the null hypothesis that the population means are equal. Using the appropriate statistical table, what is the approximate p-value for this test? Construct and interpret the 99% confidence interval for the difference between the population means.

11.84 Two machines are supposed to be producing steel bars of approximately the same length. A sample of 35 bars from one machine has an average length of 37.013 inches, with a standard deviation of 0.095 inches. For 38 bars produced by the other machine, the corresponding figures are 36.974 inches and 0.032 inches. Using the 0.05 level of significance, test the null hypothesis that the population means are equal. Using the appropriate statistical table, what is the approximate p-value for this test? Construct and interpret the 95% confidence interval for the difference between the population means.

11.85 Observing a sample of 35 morning customers at a convenience store, a researcher finds their average stay in the store is 3.0 minutes, with a standard deviation of 0.9 minutes. For a sample of 43 afternoon customers, the mean and standard deviation were 3.5 and 1.8 minutes, respectively. Using the 0.05 level of significance, determine whether the population means could be equal. Construct and interpret the 95% confidence interval for the difference between the population means.

11.86 In an experiment comparing a new long-distance golf ball with the conventional design, each of 20 golfers hits one drive with each ball. The average golfer in the sample hit the new ball 9.3 yards farther, with a standard deviation of 10.5 yards. At the 0.005 level, evaluate the effectiveness of the new ball in increasing distance. Using the appropriate statistical table, what is the approximate p-value for this test?

11.87 The idling speed of 14 gasoline-powered generators is measured with and without an oil additive that is designed to lower friction. With the additive installed, the mean change in speed was +23 revolutions per minute (rpm), with a standard deviation of 3.5 rpm. At the 0.05 level of significance, is the additive effective in increasing engine rpm?

11.88 A motel manager, concerned with customer theft of towels, decided that the theft rate might be reduced by changing from white, imprinted towels to a drab green version. Of the 120 guests provided with the white towels, 35% took at least one towel with them when they checked out. Of the 160 guests given the drab green towels, only 25% checked out with one or more towels in their possession. At the 0.01 level of significance, can we conclude that the manager's idea is effective in reducing the rate of towel theft?

11.89 An Internal Revenue Service manager is comparing the results of recent taxpayer interviews conducted by two auditors. Of 200 taxpayers audited by Ms. Smith, 30% had to pay additional taxes. Of 100 audited by Mr. Burke, only 19% paid additional taxes. At the 0.01 level of significance, can the manager conclude that Ms. Smith is a more effective auditor than Mr. Burke?

11.90 A pharmaceutical manufacturer has come up with a new drug intended to provide greater headache relief than the old formula. Of 250 patients treated with the previous medication, 130 reported "fast relief from headache pain." Of 200 individuals treated with the new formula, 128 said they got "fast relief." At the 0.05 level, can we conclude that the new formula is better than the old? Using the appropriate statistical table, what is the approximate p-value for this test?

11.91 In tests of a speed-reading course, it is found that 10 subjects increased their reading speed by an average of 200 words per minute, with a standard deviation of 75 words. At the 0.10 level, does this sample information suggest that the course is really effective in improving reading speed? Using the appropriate statistical table, what is the approximate p-value for this test?

11.92 A maintenance engineer has been approached by a supplier who promises less downtime for machines using its new molybdenum-based super lubricant. Of 120 machines maintained with the new lubricant, only 20% were out of service for more than an hour during the month of the test, while 30% of the 100 machines using the previous lubricant were down for more than an hour. Using the 0.05 level in an appropriate test, evaluate the supplier's contention. Using the appropriate statistical table, what is the approximate *p*-value for this test?

11.93 According to the National Center for Education Statistics, 88% of elementary, middle-school, and secondary-school students from families earning $75,000 or more per year use a computer at school, compared to 81% from families earning less than $20,000. Using a one-tail test at the 0.01 level, and assuming that the percentages are from independent samples of 500 families each, is the percentage of more-affluent students using computers at school significantly greater than for those from lower-income families? Determine and interpret the *p*-value for the test. Source: Bureau of the Census, *Statistical Abstract of the United States 2009*, p. 163.

11.94 A study by Experian found that 20% of consumers in New Jersey have 10 or more credit cards, compared to 11% in Tennessee. In a two-tail test at the 0.05 level of significance, and assuming the percentages are from independent samples of 100 consumers each, can we conclude that New Jersey and Tennessee consumers might not differ in terms of the percentage who have 10 or more credit cards? Determine and interpret the *p*-value for the test, then construct and interpret the 95% confidence interval for the difference between the population proportions. Source: "1 in 7 Americans Carry 10 or More Credit Cards," moneycentral.msn.com, July 25, 2009.

11.95 Given the information in Exercise 11.75, and using the 0.05 level of significance in comparing the sample standard deviations, were we justified in assuming that the population standard deviations were equal? Would your conclusion change if the standard deviation of the estimates received by Emily had been $35?

11.96 Given the information in Exercise 11.76, and using the 0.01 level of significance in comparing the sample standard deviations, were we justified in assuming that the population standard deviations were equal? Would your conclusion change if the standard deviation of the data values with the rooftop carrier had been 0.07 gallons?

(**DATA SET**) *Note:* Exercises 11.97–11.101 require a computer and statistical software.

11.97 Three years after receiving their degrees, graduates of a university's MBA program have reported their annual salary rates, with a portion of the data listed in file **XR11097**. Graduates in one of the groups represented in the data are employed by consulting firms, while another group consists of graduates who are with national-level corporations. Considering these groups as data from independent samples, use the 0.05 level in determining whether the sample means differ significantly from each other. Identify and interpret the *p*-value for the test, then generate and interpret the 95% confidence interval for the difference between the population means.

11.98 A supermarket manager is examining whether a new plastic bagging configuration in the produce area might make it more efficient for customers who are bagging their own fruits and vegetables. She observes a sample of customers using the new system and a sample of customers using the standard system, and she notes how many seconds it took each of them to unroll and open the plastic bag in preparation for bagging the veggies. The data are in file **XR11098**. Considering these as data from independent samples, use the 0.025 level of significance in examining whether the population mean for the new bagging system might be less than that for the conventional system. Identify and interpret the *p*-value for the test.

11.99 A company that makes an athletic shoe designed for basketball has stated in its advertisements that the shoe increases the jumping ability of players who wear it. The general manager of a professional team has conducted a test in which each player's vertical leap is measured with the current shoe versus the new model, with the data as listed in file **XR11099**. At the 0.05 level of significance, evaluate the claim that has been made by the shoe company. Identify and interpret the *p*-value for the test.

11.100 The data in file **XR11100** are the weights (in grams) for random samples of grain packages filled by two different filling machines. The machines have a fine adjustment for the mean amount of fill, but the standard deviations are inherent in the design of the grain delivery mechanism. Based on these data, and using the 0.01 level of significance, is there reason to conclude that the population standard deviations might not be equal for the quantities being delivered from the two machines? Identify and interpret the *p*-value for the test.

11.101 An anthropologist studying personal advertisements in a Utica, New York, newspaper has observed whether the advertiser included a mention of an interest in the outdoors in his or her ads. According to the anthropologist, citing the outdoors "may be taken to imply not only good life habits, but also sound character." Overall, 58% of men and 62% of women mentioned the outdoors in their ad. Assuming that the data are in file **XR11101**, coded as 1 = did not mention the outdoors and 2 = mentioned the outdoors, use the 0.10 level of significance in examining whether the population percentages for men and women who mention the outdoors might be the same. Identify and interpret the *p*-value for the test, then construct and interpret the 90% confidence interval for the difference between the population proportions. Source: Karen S. Peterson, "Personal Ads Get Back To Nature," *USA Today*, November 23, 1999, p. 1D.

INTEGRATED CASES

Thorndike Sports Equipment

The Thorndikes have submitted a bid to be the sole supplier of swimming goggles for the U.S. Olympic team. OptiView, Inc. has been supplying the goggles for many years, and the Olympic committee has said it will switch to Thorndike only if the Thorndike goggles are found to be significantly better in a standard leakage test.

For purposes of fairness, the committee has purchased 16 examples from each manufacturer in the retail marketplace. This is to avoid the possibility that either manufacturer might supply goggles that have been specially modified for the test. Testing involves installing the goggles on a surface that simulates the face of a swimmer, then submitting them to increasing water pressure (expressed in meters of water depth) until the goggles leak. The greater the number of meters before leakage, the better the quality of the goggles.

Both companies have received copies of the test results and have an opportunity to offer their respective comments before the final decision is made. Ted Thorndike has just received his company's copy of the results, rounded to the nearest meter of water depth. The data are also listed in file **THORN11**.

Thorndike Goggles (meters)

82	117	91	95	110	81	101	108
106	114	106	95	101	92	94	108

OptiView Goggles (meters)

73	95	83	106	70	103	86	100
92	108	94	77	109	90	107	73

1. Based on analysis of these data, formulate a commentary that Ted Thorndike might wish to make to the committee.
2. Based on analysis of these data, formulate a commentary that OptiView might wish to make to the committee.
3. What would be your recommendation to the committee?

Springdale Shopping Survey

Item C of the Springdale Shopping Survey, introduced at the end of Chapter 2, describes variables 7–9 of the survey. These variables represent the general attitude respondents have toward each of the three shopping areas, and range from 5 (like very much) to 1 (dislike very much). Samples involving consumer groups often differ in their results, and managers find it useful to determine whether the differences could be due to sampling variation or whether there is "something going on" regarding the attitudes, perceptions, and behaviors of one consumer group versus another.

Variable 7 = Attitude toward Springdale Mall
Variable 8 = Attitude toward Downtown
Variable 9 = Attitude toward West Mall

Part I: Attitude Comparisons Based on Marital Status of Respondents

1. For variable 7 (attitude toward Springdale Mall), carry out an appropriate hypothesis test to determine whether married persons (code = 1 on variable 28, marital status) have a different mean attitude than unmarried persons (code = 2 on variable 28). Interpret the resulting computer printout, including the *p*-value for the test.
2. Repeat step 1 for variable 8 (attitude toward Downtown).
3. Repeat step 1 for variable 9 (attitude toward West Mall).
4. Comment on the extent to which attitudes toward each of the shopping areas differ between married and unmarried respondents.

Part II: Attitude Comparisons Based on Gender of Respondent

Repeat Part I, using variable number 26 (gender of respondent) instead of variable 28 (marital status of respondent) as the basis on which the groups are identified.

BUSINESS CASE

Circuit Systems, Inc. (A)

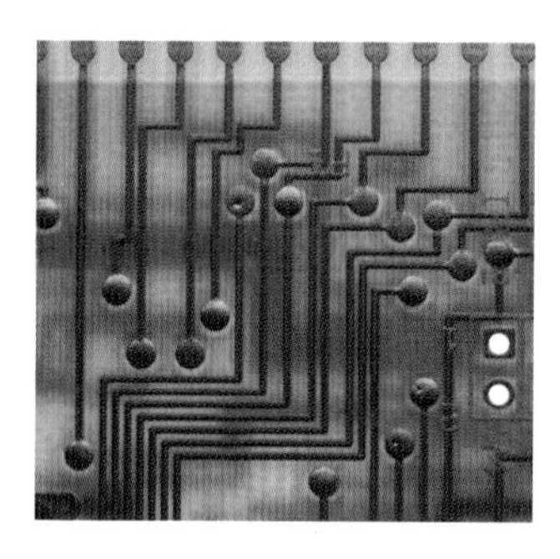

Circuit Systems, Inc., located in northern California, is a company that produces integrated circuit boards for the microcomputer industry. In addition to salaried management and office staff personnel, Circuit Systems currently employs approximately 250 hourly production workers involved in the actual assembly of the circuit boards. These hourly employees earn an average of $11.00 per hour.

Thomas Nelson, the Director of Human Resources at Circuit Systems, has been concerned with hourly employee absenteeism within the company. Presently, each hourly employee earns 18 days of paid sick leave per year. Thomas has found that many of these employees use most or all of their sick leave well before the year is over. After an informal survey of employee records, Thomas is convinced that while most hourly employees make legitimate use of their sick leave, there are many who view paid sick leave as "extra" vacation time and "call in sick" when they want to take off from work. This has been a source of conflict between the hourly production workers and management. The problem is due in part to a restrictive vacation policy at Circuit Systems in which hourly employees receive only one week of paid vacation per year in addition to a few paid holidays. With only one week of paid vacation and a few paid holidays, the hourly production employees work a 50-week year, not counting paid sick leave.

In an effort to save money and increase productivity, Thomas has developed a two-point plan that was recently approved by the president of Circuit Systems. To combat the abuse of paid sick leave, hourly workers will now be allowed to convert unused paid sick leave to cash on a "three-for-one" basis, i.e., each unused day of sick leave can be converted into an additional one-third of a day's pay. An hourly employee could earn up to an additional six days of pay each year if he or she does not take any paid sick leave during the year. Even though a worker could gain more time off by dishonestly "phoning in sick," Thomas hopes that the majority of hourly employees will view this approved conversion of sick leave into extra pay as a more acceptable alternative. In the second part of his plan, Thomas is instituting a voluntary exercise program for hourly employees to improve their overall health. At an annual company expense of $200 for each hourly employee who participates, Circuit Systems will subsidize membership in a local health club. In return, the participating employee is required to exercise at least three times per week outside of regular working hours to maintain his or her free membership. Circuit Systems believes that, in the long term, an investment in employees' physical well-being may increase their productivity as well as reduce the company's future health insurance premiums. In discussions with hourly employees, Thomas has found that many of them approve of the exercise program and are willing to participate.

Many of the supervisors that Thomas has spoken with believe that the paid sick leave conversion and the exercise program may help in curbing the absenteeism problem, but others do not give it much hope for succeeding and think the cost would outweigh any benefits. The president of Circuit Systems agreed to give the proposal a one-year trial period. At the end of the trial period, Thomas must evaluate the new anti-absenteeism plan, present the results, and make a recommendation to either continue or discontinue the plan.

Assignment

Over the next year, during which time the sick leave conversion and exercise program are in place, Thomas Nelson has maintained data on employee absences, use of the sick leave conversion privilege, participation in the exercise program, and other pertinent information. He has also gone back to collect data from the year prior to starting the new program in order to better evaluate the new program. His complete data are in the file CIRCUIT. A description of this data set is given in the Data Description section.

Using this data set and other information given in the case, help Thomas Nelson evaluate the new program to determine whether it is effective in reducing the average cost of absenteeism by hourly employees, thereby increasing worker productivity. In particular, you need to compare this year's data to last year's data to determine whether there has been a reduction in the average cost of absenteeism per hourly production worker by going to the new program. The case questions will assist you in your analysis of the data. Use important details from your analysis to support your recommendation.

Data Description

The CIRCUIT file contains data for the past two years on the 233 hourly production employees in the company who were with the company for that entire period of time. A partial listing of the data is shown here.

Employee	Hourly Pay	Sick Leave Last Year	Sick Leave This Year	Exercise Program
6631	$10.97	3.50	2.00	0
7179	11.35	24.00	12.50	0
2304	10.75	18.00	12.75	0
9819	10.96	21.25	14.00	0
4479	10.59	16.50	11.75	0
1484	11.41	16.50	9.75	1
⋮	⋮	⋮	⋮	⋮

These data are coded as follows:

Employee: Employee ID number.

Hourly Pay: Hourly pay of the employee in both years. Unfortunately, due to economic conditions, there were no pay raises last year.

Sick Leave Last Year: Actual number of days of sick leave taken by the employee last year before the new program started.

Sick Leave This Year: Actual number of days of sick leave taken by the employee this year under the new program.

Exercise Program: 1, if participating in the exercise program.
0, if not participating.

1. Using the method presented in the chapter for comparing the means of paired samples, compare the two years in terms of days missed before and after the new program was implemented. On this basis alone, does it appear that the program has been effective in reducing the number of days missed?
2. Keeping in mind that the goal of the program is to reduce the *cost* of absenteeism, you will need to create two new variables for each employee: (1) the cost of paid absences last year, and (2) the cost associated with absences this year. *A few hints on creating these variables for each person: For* (1), *assume an 8-hour workday and consider both the person's daily pay and his or her number of absence days. For* (2), *assume an 8-hour workday and keep in mind that the total cost associated with absenteeism for each person must include the cost of paid absences, the extra pay for unused sick leave (if any), and health club membership (if applicable). You might call these new variables Cost_Before and Cost_After.* Use these new variables in repeating the procedure you followed in Question 1, then discuss the results and make a recommendation to Mr. Nelson regarding the effectiveness and possible continuation of the new program.
3. Using the new variables you created in Question 2, use this year's costs and an appropriate statistical test in evaluating the effectiveness of the exercise program. Discuss the results and make a recommendation to Mr. Nelson regarding the effectiveness and possible continuation of the company-paid health club memberships.

SEEING STATISTICS: APPLET 14

Distribution of Difference Between Sample Means

This applet has three parts:

1. The upper portion shows two normally distributed populations. For purposes of simplicity in the applet, their standard deviations are the same. By moving the slider at the top of the applet, we can shift one of the population curves back and forth, thus changing the difference between the population means. By moving the upper of the two sliders at the right, we can change the standard deviations of the populations.
2. The center portion of the applet shows the sampling distribution of the means for simple random samples taken from each of the populations at the top. To reduce complexity within the applet, the sample sizes are the same. By moving the lower of the two sliders at the right, we can change the sizes of the samples.
3. The bottom portion of the applet shows the sampling distribution of $(\bar{x}_1 - \bar{x}_2)$, the difference between the sample means. The standard error of this sampling distribution is shown at the lower right as "sigma," and the applet calculates it using the formula provided in text Section 11.4, but with σ_1 and σ_2 replacing s_1 and s_2, respectively. Despite the small sample sizes, this is appropriate because the populations are normal and the population standard deviations are known. In real-world applications, we almost never know both population standard deviations, but don't let this detract from your use of this very interesting applet.

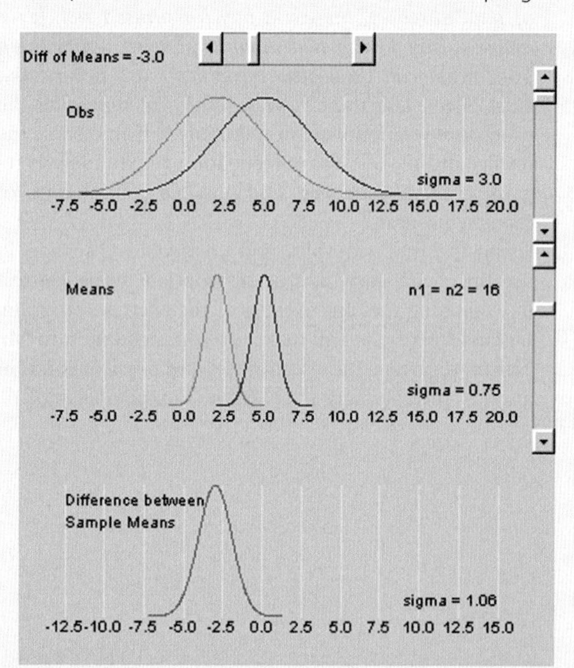

Applet Exercises

14.1 Set the top slider so that the difference between the population means is −3.0. Set the slider at the upper right so that the standard deviation of each population is 2.5. Set the slider at the lower right so that $n_1 = n_2 = 20$.

a. Is there very much overlap between the two population curves at the top of the applet?
b. Is there very much overlap between the two sampling distribution curves in the center part of the applet?
c. Viewing the bottom portion of the applet, and assuming that a sample is going to be taken from each of the two populations, does it seem very likely that $(\bar{x}_1 - \bar{x}_2)$ will be greater than zero?

14.2 Repeat Applet Exercise 14.1, but with the top slider set so the difference between the sample means is +0.5.

14.3 Repeat Applet Exercise 14.1, but with the top slider set so the difference between the sample means is +3.0.

14.4 Use the top slider to gradually change the difference between the population means from −0.5 to 4.5. Describe how this affects the graphs in the three portions of the applet.

14.5 Use the upper right slider to gradually increase the population standard deviations from 1.1 to 3.0. Describe how this affects the graphs in the three portions of the applet.

14.6 Use the lower right slider to gradually increase the sample sizes from 2 to 20. Describe how this affects the graphs in the three portions of the applet.

CHAPTER 12

Analysis of Variance Tests

SYNERGY, ANOVA, AND THE THORNDIKES

Synergy refers to a situation where the effect of the whole differs from the sum of the effects of its parts. Plaids may look nice, stripes may look nice; but, luckily, most of us have someone in our lives who prevents us from wearing our nice plaid pants with our nice striped shirt. On the other hand, some things go together, each enhancing the effect of the other—ocean surf and a sunny sky come to mind here.

In statistics, we sometimes want to examine whether two factors might be interacting in either a very positive or a very negative way, thus having a synergistic effect on a measurement of interest. The technique involved is called *two-way analysis of variance,* and it's the topic of Section 12.5 of this chapter.

In our Thorndike Sports Equipment minicase for this chapter, Luke and Ted have sponsored a super-duper racquetball player and are trying to determine which combination of string and racquet best suits his power game. By the time you get to the Thorndikes and their latest dilemma, you'll have read the chapter and should be able to help them out.

Some things interact well together.

LEARNING OBJECTIVES

After reading this chapter, you should be able to:

- Describe the general approach by which analysis of variance is applied and the type of applications for which it is used.
- Understand the relationship between analysis of variance and the design of experiments.
- Differentiate between the one-way, randomized block, and two-way analysis of variance techniques and their respective purposes.
- Arrange data into a format that facilitates their analysis by the appropriate analysis of variance procedure.
- Use the one-way, randomized block, and two-way analysis of variance methods in testing appropriate hypotheses relative to experimental data.
- Appreciate that computer assistance is especially important in analysis of variance tests and be able to interpret computer outputs for these tests.

12.1 INTRODUCTION

Among the tests in Chapter 11 was the comparison of the means for two independent samples. This chapter introduces **analysis of variance (ANOVA),** a set of techniques that allow us to compare *two or more* sample means at the same time.

The availability of ANOVA as a technique for data analysis is an important consideration in the design of experiments. We will discuss this further in the next section, along with a presentation of the basic concepts underlying ANOVA.

In Sections 12.3–12.5, we will examine three of the most widely used techniques involving ANOVA. Although similar in some respects, they differ in terms of their specific purposes and procedures. In each section, we will generate and interpret computer-assisted results that correspond to the small-scale, hand-calculated examples presented for each of these ANOVA procedures. Computer assistance is especially useful in ANOVA because the calculations can be quite extensive even for small amounts of data.

12.2 ANALYSIS OF VARIANCE: BASIC CONCEPTS

Analysis of Variance and Experimentation

In the late 1700s, the British Navy began issuing daily rations of fruit to sailors. This was in response to the finding that vitamin C helped to prevent scurvy, then a common disease. More recently, you may have used a toothpaste containing stannous fluoride when you brushed your teeth this morning. This is a common toothpaste ingredient, as dental studies have found stannous fluoride to be effective in reducing tooth decay. The rationing of fruit to the sailors and the approval of stannous fluoride by the American Dental Association were decisions based on the results of *experiments*. Since ANOVA is closely related to experimentation, we'll introduce a few terms that are important to the remainder of the chapter:

EXPERIMENT A study or investigation designed for the purpose of examining the effect that one variable has on the value of another variable.

DEPENDENT VARIABLE The variable for which a value is measured or observed. In ANOVA, the dependent variable will be a quantitative variable—for example, soft drink consumption, examination score, or the time required to type a document.

INDEPENDENT VARIABLE A variable that is observed or controlled for the purpose of determining its effect on the value of the dependent variable. In ANOVA, the independent variable can be qualitative (e.g., marital status) or quantitative (e.g., age group). The following terms are of particular relevance to the independent variable:

1. An independent variable is referred to as a **factor**, and one or more factors may be involved in a given study.
2. The experiment may involve different **factor levels** (categories).
3. Each specific level of a factor (or, in multiple-factor experiments, the intersection of a level of one factor with a level of another factor) is referred to as a **treatment**.

When there is only one factor in an experiment, *factor levels* and *treatments* are synonymous. Therefore, when dealing with an experiment that involves just one factor, we will use the terms *factor levels* and *treatments* interchangeably.

Our emphasis in the chapter will be on **designed experiments,** in which we actually assign treatments to persons or test units on a random basis. However, the techniques may also be applied when we wish to compare treatments that cannot be randomly assigned. For example, we might like to compare incomes for high school, college, and graduate school graduates, but we obviously cannot randomly assign these treatments to individuals. In such cases, our analysis would have to be based on data that already exist.

As has been suggested, the starting point for ANOVA is often an experiment in which we try to determine whether various levels of a given factor might be having different effects on something we're observing or measuring. For example, a lock manufacturer might test four different lock designs to determine whether they differ significantly in the amount of force each will withstand just prior to breakage:

Factor: Lock Design

	Level 1	Level 2	Level 3	Level 4	
Measurements: Breaking Strength, Pounds, 3 Locks of Each Design	1050	957	1008	1235	
	1023	1114	849	1093	
	981	1056	972	1110	
Mean	1018.0	1042.3	943.0	1146.0	pounds

In this experiment, the dependent variable is "breaking strength," the independent variable (or factor) is "lock design," and the factor levels (or treatments) are the four different designs being tested. Again, since there is just one factor, each factor level can also be referred to as a treatment. This is an example of a **balanced experiment,** one in which an equal number of persons or test units receives each treatment. There are three test units (locks) in each treatment group.

In its most basic application, ANOVA can determine whether the sample means ($\bar{x}_1$, $\bar{x}_2$, $\bar{x}_3$, and $\bar{x}_4$) differ significantly from one another or whether the differences observed could have been due to chance. For example, in the experiment just presented, can we conclude that the populations of these four lock designs are equally strong?

Variation Between and Within the Groups

In the preceding experiment, the variation in the breaking strengths can be viewed in terms of (1) variation *between* the groups, reflecting the effect of the factor levels (the four types of lock design), and (2) variation *within* the groups, which represents random error from the sampling process. Comparing these two kinds of variation is the basis of ANOVA.

To show how these two kinds of variation are applicable to ANOVA, we'll rely on two computer-assisted examples. Each will consist of three independent samples represented in dotplot diagrams, and each can be viewed as a separate experiment that involved three treatments. Based on each set of plots, you'll be asked to judge whether the samples could have come from populations that had the same value for the population mean.

Part A. As a first step in this "you be the judge" scenario, consider the three independent samples represented in part A of Table 12.1. From these plots (without peeking at the table caption), would it appear that all three samples could have come from populations having the same value for the population mean (i.e., $\mu_1 = \mu_2 = \mu_3$)?

TABLE 12.1

The basis for ANOVA lies in the amount of variation *between* the samples compared to the amount of variation *within* them. For the three independent samples in part A, we would tend to reject H_0: $\mu_1 = \mu_2 = \mu_3$. For the three independent samples in part B, it appears that the population means could be equal.

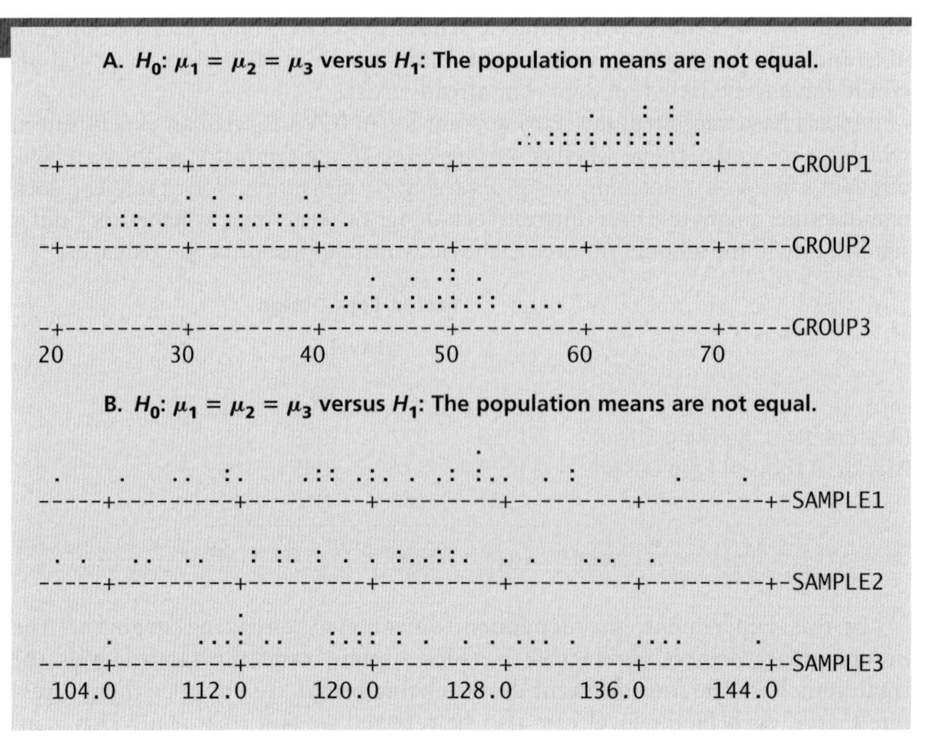

Part B. Next, consider the three independent samples represented in part B of Table 12.1. Does it seem possible that all three of these samples might have come from populations having the same value for the population mean (i.e., $\mu_1 = \mu_2 = \mu_3$)?

In part A, it does not appear that the population means could be the same. Intuitively, this is because the variation *between* the samples is rather large compared to the amount of variation *within* the samples themselves. As an analogy, the data points in part A somewhat resemble the likely outcome if a single-barrel shotgun were to be fired in three different directions.

Looking at part B, the same intuition would tell us that these samples don't appear to be very different from each other. Stated another way, the variation *between* the samples is small relative to the variation *within* the samples. Although the sample means are not shown in the dotplots, they would be very close together compared to the amount of variation within the samples themselves. Using the shotgun analogy, part B could well be the result of having fired a single-barrel shotgun three times in the *same* direction.

If the data in part A of Table 12.1 were the result of applying three different treatments in a one-factor experiment, we would tend to conclude that the treatments are not equally effective. Taking a similar view toward a second experiment described by the data in part B, we would tend to conclude that the treatments in this experiment might be equally effective.

Although ANOVA quantifies variation between versus within samples, the basic process is an extension of the intuitive conclusions drawn from parts A and B of Table 12.1. The most basic approach to ANOVA is the situation in which there is just one factor that operates on two or more different levels, as with the four lock designs discussed previously. Since there is just one factor (lock design), this is referred to as *one-way analysis of variance*. It is the subject of the next section.

EXERCISES

12.1 What is meant by a designed experiment?

12.2 "A factor level is the same as a treatment." Is this statement always true? If not, what conditions are necessary for it to be true?

12.3 What is necessary for an experiment to be "balanced"?

12.4 Explain the basic role that between-sample variation and within-sample variation play in carrying out an analysis of variance.

12.5 Differentiate between the independent and dependent variables in an experiment.

12.6 A university president collects data showing the number of absences over the past academic year for a random sample of 6 professors in the College of Engineering. She does the same for a random sample of 9 professors in the College of Business and for a random sample of 8 professors in the College of Fine Arts. Does this represent a designed experiment? Explain.

12.7 Twenty accounting students are randomly assigned to two different sections of an intermediate accounting class. Each section ends up consisting of 10 students. In one of the sections, computer-assisted instruction and review software is utilized; in the other section, it is not. All students are given the same final examination at the end of the semester. Does this represent a designed experiment? Explain.

12.8 From each of four suppliers, a quality-control technician collects a random sample of 10 rivets, then measures the number of pounds each will withstand before it fails. Does this represent a designed experiment? Explain.

12.9 For the experiment described in Exercise 12.6, identify the dependent and independent variables. Indicate whether each variable is quantitative or qualitative.

12.10 For the experiment described in Exercise 12.7, identify the dependent and independent variables. Indicate whether each variable is quantitative or qualitative.

12.11 For the experiment described in Exercise 12.8, identify the dependent and independent variables. Indicate whether each variable is quantitative or qualitative.

12.12 The accompanying dotplots represent independent samples. Would you tend to reject the null hypothesis that $\mu_A = \mu_B$? Explain.

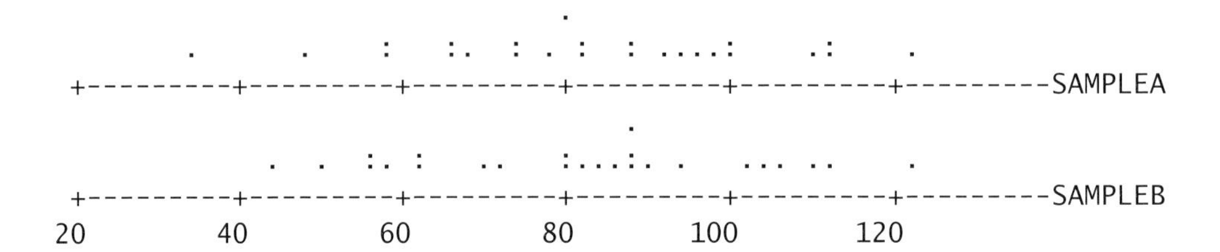

12.3 ONE-WAY ANALYSIS OF VARIANCE

Purpose

The **one-way analysis of variance** examines two or more independent samples to determine whether their population means could be equal. Since the treatments are randomly assigned to all of the persons or other test units in the experiment, this is also referred to as the *one-factor, completely randomized design*. When there are just two samples, it becomes the equivalent of the two-sample, pooled-variances *t*-test of Chapter 11. We will examine this equivalence later in this section.

The null and alternative hypotheses are

H_0: $\mu_1 = \mu_2 = \cdots = \mu_t$ for treatments 1 through t

H_1: The population means are not equal.

Model and Assumptions

The preceding null and alternative hypotheses represent one of the possibilities in viewing the one-way ANOVA. It can also be described in terms of the following model in which each individual observation is considered to be the sum of three separate components:

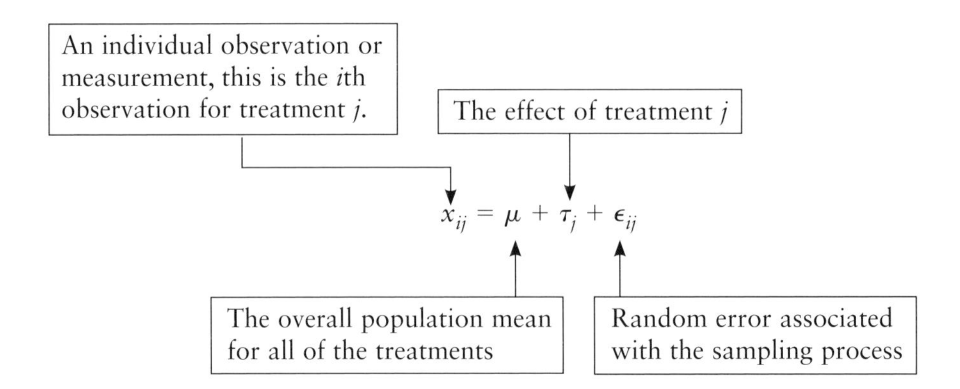

According to this model, each measurement (x_{ij}) associated with treatment j is the sum of the actual population mean for all of the treatments (μ), the effect of treatment j (τ_j), and a sampling error (ϵ_{ij}). The null and alternative hypotheses shown previously can now be presented in an equivalent form that is relevant to the model described here:

H_0: $\tau_j = 0$ for treatments $j = 1$ through t
(Each treatment has no effect.)

H_1: $\tau_j \neq 0$ for at least one of the $j = 1$ through t treatments
(One or more treatments has an effect.)

These null and alternative hypotheses are equivalent to those expressed in terms of the population means. If the effect of each treatment (τ_j, $j = 1$ through t) is zero, then the population means for the treatments ($\mu_1, \mu_2, \ldots, \mu_t$) will be equal. In terms of the ($x_{ij} = \mu + \tau_j + \epsilon_{ij}$) model, if these equivalent null hypotheses cannot be rejected, we must conclude that each population mean could be equal to the overall mean, μ.

Regarding the terms *effective, equally effective*, and *effect:* It's possible that all of the treatments might be very *effective* in influencing the dependent variable (e.g., all of the lock designs tested could be virtually unbreakable). To this extent, μ in the ($x_{ij} = \mu + \tau_j + \epsilon_{ij}$) model would be an extremely high number of pounds before breakage. However, the purpose of the one-way ANOVA is to determine whether the treatments could be *equally effective*. If we are able to conclude that the treatments could be equally effective, this is equivalent to the conclusion that each treatment *effect* (τ_j) could be zero.

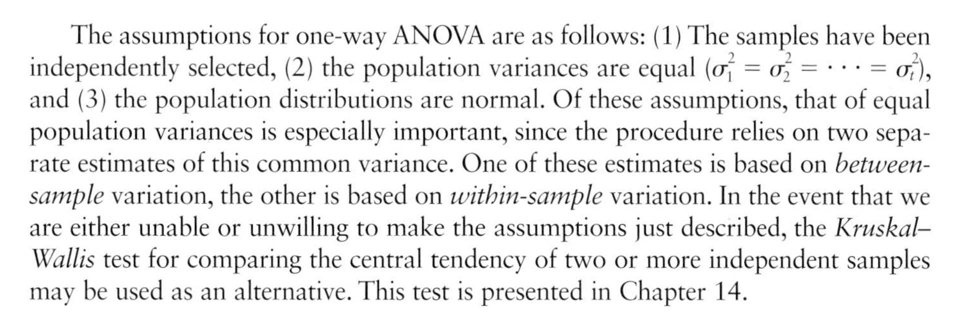

The assumptions for one-way ANOVA are as follows: (1) The samples have been independently selected, (2) the population variances are equal ($\sigma_1^2 = \sigma_2^2 = \cdots = \sigma_t^2$), and (3) the population distributions are normal. Of these assumptions, that of equal population variances is especially important, since the procedure relies on two separate estimates of this common variance. One of these estimates is based on *between-sample* variation, the other is based on *within-sample* variation. In the event that we are either unable or unwilling to make the assumptions just described, the *Kruskal–Wallis* test for comparing the central tendency of two or more independent samples may be used as an alternative. This test is presented in Chapter 14.

Procedure

The procedure for carrying out the one-way ANOVA is summarized in Table 12.2, and its steps are generally similar to those for hypothesis tests performed in Chapters 10 and 11. Whether done by hand calculations or with the assistance of a computer statistical package (a much preferred approach), the procedure can be described according to the process shown in parts A through D of Table 12.2.

Part A: The Null and Alternative Hypotheses

The null and alternative hypotheses are expressed in terms of the equality of the population means for all of the treatment groups.

Part B: The Format of the Data to Be Analyzed

The data can be listed in tabular form, as shown, with a separate column for each of the t treatments. The number of observations in the columns ($n_1, n_2, n_3, \ldots, n_t$) need not be equal. For greater convenience in the formulas to follow, N is shown

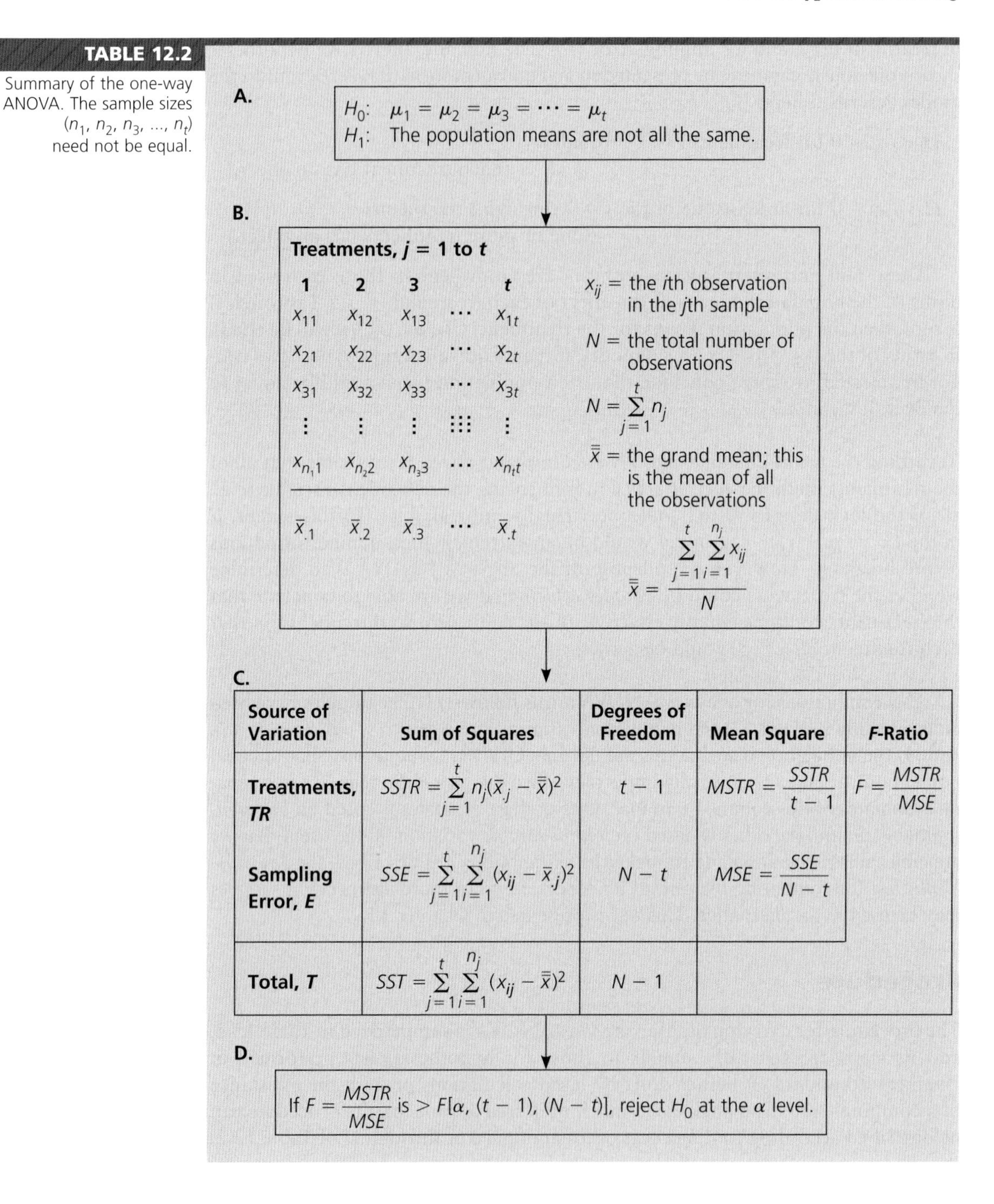

A.

H_0: $\mu_1 = \mu_2 = \mu_3 = \cdots = \mu_t$
H_1: The population means are not all the same.

B.

Treatments, j = 1 to t

1	2	3		t
x_{11}	x_{12}	x_{13}	$\cdots$	x_{1t}
x_{21}	x_{22}	x_{23}	$\cdots$	x_{2t}
x_{31}	x_{32}	x_{33}	$\cdots$	x_{3t}
$\vdots$	$\vdots$	$\vdots$	$\vdots$	$\vdots$
$x_{n_1 1}$	$x_{n_2 2}$	$x_{n_3 3}$	$\cdots$	$x_{n_t t}$
$\bar{x}_{.1}$	$\bar{x}_{.2}$	$\bar{x}_{.3}$	$\cdots$	$\bar{x}_{.t}$

x_{ij} = the ith observation in the jth sample

N = the total number of observations

$$N = \sum_{j=1}^{t} n_j$$

$\bar{\bar{x}}$ = the grand mean; this is the mean of all the observations

$$\bar{\bar{x}} = \frac{\sum_{j=1}^{t}\sum_{i=1}^{n_j} x_{ij}}{N}$$

C.

Source of Variation	Sum of Squares	Degrees of Freedom	Mean Square	*F*-Ratio
Treatments, *TR*	$SSTR = \sum_{j=1}^{t} n_j(\bar{x}_{.j} - \bar{\bar{x}})^2$	$t - 1$	$MSTR = \frac{SSTR}{t-1}$	$F = \frac{MSTR}{MSE}$
Sampling Error, *E*	$SSE = \sum_{j=1}^{t}\sum_{i=1}^{n_j} (x_{ij} - \bar{x}_{.j})^2$	$N - t$	$MSE = \frac{SSE}{N-t}$	
Total, *T*	$SST = \sum_{j=1}^{t}\sum_{i=1}^{n_j} (x_{ij} - \bar{\bar{x}})^2$	$N - 1$		

D.

If $F = \frac{MSTR}{MSE}$ is $> F[\alpha, (t-1), (N-t)]$, reject H_0 at the α level.

TABLE 12.2

Summary of the one-way ANOVA. The sample sizes ($n_1, n_2, n_3, \ldots, n_t$) need not be equal.

in Table 12.2 as the total number of observations in all of the samples combined. Likewise, the **grand mean** ($\bar{\bar{x}}$) is the mean of all of the observations that have been recorded.

In the format shown for listing the x_{ij} data values, note that the mean for each treatment (j = 1 through t treatments) has a dot within the subscript. For example, the mean for treatment 1 is expressed as $\bar{x}_{.1}$. This indicates that this mean is for all values of x_{i1}; in other words, it represents the entire column of x_{i1} values. The dot can be considered a "wildcard" subscript that includes all values to which it applies; its use will be clarified in the numerical example that follows this discussion.

Part C: The Calculations for One-Way ANOVA

Part C of Table 12.2 describes the specific computations necessary to carry out one-way ANOVA. Although these computations are not as imposing as they might seem, it remains advisable to use a computer statistical package if one is available. Each of these quantities is associated with a specific source of variation within the sample data, and they correspond to the $x_{ij} = \mu + \tau_j + \epsilon_{ij}$ model discussed previously.

The Sum of Squares Terms: Quantifying the Two Sources of Variation

- **Treatments, *TR*** *SSTR* is the **sum of squares** value reflecting variation between individual treatment means and the overall mean for all treatments ($\bar{\bar{x}}$). Weighted according to the sample sizes for the respective treatment groups, *SSTR* expresses the amount of variation that is attributable to the treatments.
- **Sampling error, *E*** *SSE* is the sum of the squared differences between observed values and the means for their respective treatment groups; *SSE* expresses the amount of variation due to sampling error.
- **Total variation, *T*** *SST* is the total amount of variation, or $SST = SSTR + SSE$.

Making the Amounts from the Two Sources of Variation Comparable. *MSTR* is the **mean square** for the between-group variation. It is obtained by dividing *SSTR* by an appropriate number of degrees of freedom $(t - 1)$ so *MSTR* will be comparable to *MSE* in the calculation of the test statistic.

MSE is the *mean square* for within-group variation. It is obtained by dividing *SSE* by an appropriate number of degrees of freedom $(N - t)$, again so that *MSTR* and *MSE* will be comparable in the calculation of the test statistic.

Part D: The Test Statistic, the Critical Value, and the Decision Rule

The Test Statistic. The *F*-ratio, *MSTR/MSE*, is the test statistic upon which we rely in reaching a conclusion. *MSTR* has estimated the common variance (σ^2) based on variation *between* the treatment means. *MSE* has estimated the common variance based on variation *within* the treatment groups themselves, and the test statistic, *F*, is the ratio of these separate estimates of common variance:

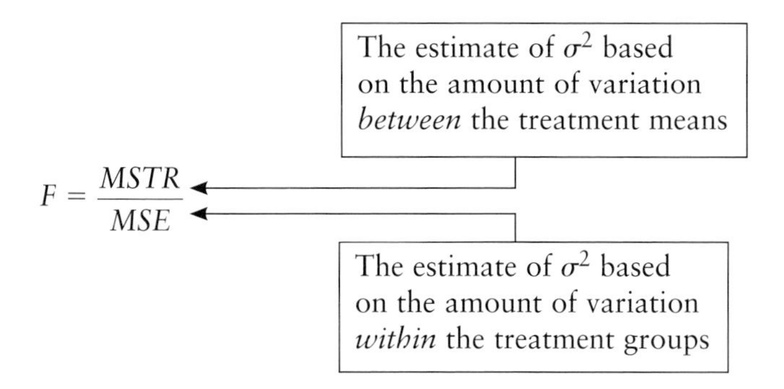

The Critical Value and the Decision Rule. The test is right-tail and, for a given level of significance (α), we will reject H_0: $\mu_1 = \mu_2 = \cdots = \mu_t$ if the calculated value of *F* is greater than $F[\alpha, (t - 1), (N - t)]$. In referring to the *F* distribution table, we use the $v_1 = (t - 1)$ column and the $v_2 = (N - t)$ row in identifying the critical value. By using Seeing Statistics Applet 15, at the end of the chapter, you can demonstrate how the F distribution and its probabilities respond to changes in v_1 and v_2.

The F distribution is used in this test because, as discussed in Chapter 11, it is the sampling distribution for the ratio of the two sample variances whenever two random samples are repeatedly drawn from the same, normally distributed population. To the extent that the calculated F is large, we will tend toward the conclusion that the population means might not be the same. Naturally, the specific conclusion we reach will depend on (1) the level of significance (α) we have selected and (2) our comparison of the calculated F with the critical F for this level of significance.

EXAMPLEEXAMPLEEXAMP

EXAMPLE

One-way ANOVA Procedure

An accounting firm has developed three methods to guide its seasonal employees in preparing individual income tax returns. In comparing the effectiveness of these methods, a test is set up in which each of 10 seasonal employees is randomly assigned to use one of the three methods in preparing a hypothetical income tax return. The preparation times (in minutes) are shown in Table 12.3 and are in data file CX12ACCT. At the $\alpha = 0.025$ level, can we conclude that the three methods could be equally effective?

SOLUTION

As Table 12.3 shows, the three treatment groups are not equal in size, with $n_1 = 4$, $n_2 = 3$, and $n_3 = 3$, but this presents no problem in the analysis. The

TABLE 12.3

Underlying data and the results of preliminary calculations for a one-way analysis of variance to determine whether three tax-preparation methods could be equally effective in guiding tax preparers through the completion of a hypothetical return contrived for the purpose of the experiment. The three treatments were randomly assigned to 10 tax preparers, and data are the number of minutes each person required for completion of the return.

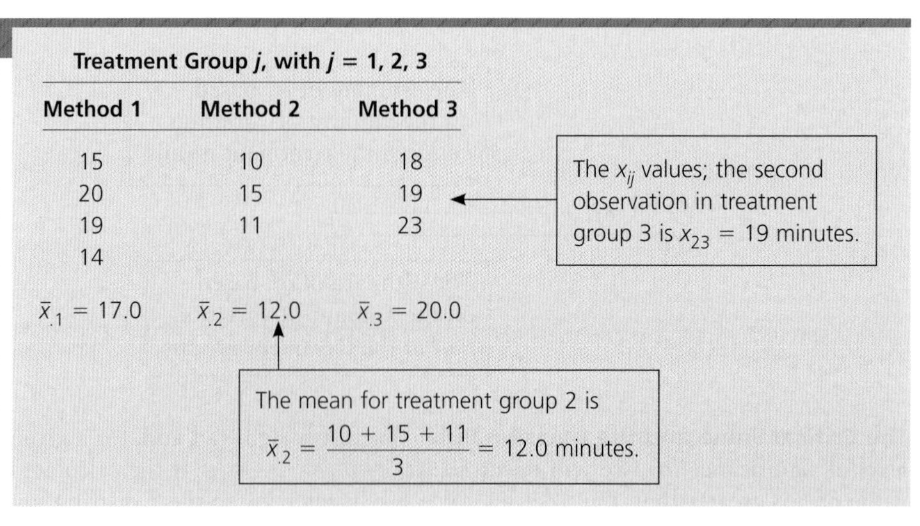

Treatment Group j, with j = 1, 2, 3		
Method 1	Method 2	Method 3
15	10	18
20	15	19
19	11	23
14		
$\bar{x}_{.1} = 17.0$	$\bar{x}_{.2} = 12.0$	$\bar{x}_{.3} = 20.0$

The x_{ij} values; the second observation in treatment group 3 is $x_{23} = 19$ minutes.

The mean for treatment group 2 is

$$\bar{x}_{.2} = \frac{10 + 15 + 11}{3} = 12.0 \text{ minutes.}$$

total number of observations in all treatment groups is as described in part B of Table 12.2, or

$$N = \sum_{j=1}^{t} n_j = 4 + 3 + 3 \quad \text{and} \quad N = 10$$

The grand mean ($\bar{\bar{x}}$) is the mean of the $N = 10$ observations. It is obtained by dividing the sum of all the observations by N, or

$$\bar{\bar{x}} = \frac{\sum_{j=1}^{t}\sum_{i=1}^{n_j} x_{ij}}{N}$$

and

$$\bar{\bar{x}} = \frac{15 + 20 + 19 + 14 + 10 + 15 + 11 + 18 + 19 + 23}{10} = 16.4 \text{ minutes}$$

Having carried out the preceding calculations, we can now proceed with those described in part C of Table 12.2.

Treatment Sum of Squares, *SSTR*

This quantity expresses the amount of variation attributable to the treatments, and each treatment mean is compared to the overall, or grand mean, during the calculation of *SSTR*:

$$\begin{aligned} SSTR &= \sum_{j=1}^{t} n_j (\bar{x}_{.j} - \bar{\bar{x}})^2 \\ &= 4(17.0 - 16.4)^2 + 3(12.0 - 16.4)^2 + 3(20.0 - 16.4)^2 \end{aligned}$$

and

$$\boxed{SSTR = 98.4}$$

Error Sum of Squares, *SSE*

This quantity represents the amount of variation *within* the treatment groups, or sampling error. Notice that in this calculation, each observation is being compared with the mean of the treatment group of which it is a member, and that each treatment group "contributes" to the value of *SSE*.

$$\begin{aligned} SSE &= \sum_{j=1}^{t}\sum_{i=1}^{n_j} (x_{ij} - \bar{x}_{.j})^2 \\ &= \begin{bmatrix} (15.0 - 17.0)^2 \\ +(20.0 - 17.0)^2 \\ +(19.0 - 17.0)^2 \\ +(14.0 - 17.0)^2 \end{bmatrix} + \begin{bmatrix} (10.0 - 12.0)^2 \\ +(15.0 - 12.0)^2 \\ +(11.0 - 12.0)^2 \end{bmatrix} + \begin{bmatrix} (18.0 - 20.0)^2 \\ +(19.0 - 20.0)^2 \\ +(23.0 - 20.0)^2 \end{bmatrix} \end{aligned}$$

and

$$\boxed{SSE = 54.0}$$

Total Sum of Squares, *SST*

To apply one-way ANOVA to the testing of our null hypothesis, it isn't really necessary to use the formula in Table 12.2 to compute *SST*. In practice, we could just calculate $SST = SSTR + SSE$. The calculation of *SST* is shown here only to demonstrate that the total variation is the sum of variation due to the treatments

($SSTR$) plus that due to sampling error (SSE). In the expressions that follow, note that the terms are similar to those of the SSE computation, except that each observation is now being compared to the grand mean, $\bar{\bar{x}} = 16.4$ minutes:

$$SST = \sum_{j=1}^{t} \sum_{i=1}^{n_j} (x_{ij} - \bar{\bar{x}})^2$$

$$= \begin{bmatrix} (15.0 - 16.4)^2 \\ +(20.0 - 16.4)^2 \\ +(19.0 - 16.4)^2 \\ +(14.0 - 16.4)^2 \end{bmatrix} + \begin{bmatrix} (10.0 - 16.4)^2 \\ +(15.0 - 16.4)^2 \\ +(11.0 - 16.4)^2 \end{bmatrix} + \begin{bmatrix} (18.0 - 16.4)^2 \\ +(19.0 - 16.4)^2 \\ +(23.0 - 16.4)^2 \end{bmatrix}$$

and

$$\boxed{SST = 152.4 \quad \text{or} \quad SST = SSTR + SSE}$$

Treatment Mean Square (*MSTR*) and Error Mean Square (*MSE*)

$SSTR$ and SSE must now be divided by their respective degrees of freedom so that (1) they will be comparable and (2) each will be a separate estimate of the common variance that the treatment group populations are assumed to share. Recall that we have $t = 3$ treatments and a total of $N = 10$ observations.

The estimate of σ^2 that is based on the *between-treatment* variation is

$$MSTR = \frac{SSTR}{t - 1} = \frac{98.4}{3 - 1} \quad \text{and} \quad \boxed{MSTR = 49.20}$$

The estimate of σ^2 that is based on the *within-treatment*, or sampling-error variation is

$$MSE = \frac{SSE}{N - t} = \frac{54.0}{10 - 3} \quad \text{and} \quad \boxed{MSE = 7.71}$$

The Test Statistic, *F*

The test statistic is the ratio of the two estimates for σ^2, or

$$F = \frac{MSTR}{MSE} = \frac{49.20}{7.71} \quad \text{and} \quad \boxed{F = 6.38}$$

At this point, we have generated all of the information described in part C of Table 12.2, and the results can be summarized as shown in the standard one-way ANOVA table in Table 12.4. Depending on your computer statistical package, the items in Table 12.4 might be the only information provided, and you might have to draw conclusions on the basis of statistical tables as we are now about to do.

TABLE 12.4

Given the data in Table 12.3 and the computations in the text, the result is this standard format describing the results of the one-way analysis of variance test. The format is similar to that shown in part C of Table 12.2. Depending on your computer statistical package, this may be all the information that you will be provided.

Variation Source	Sum of Squares	Degrees of Freedom	Mean Square	*F*
Treatments[1]	98.40	2	49.20	6.38
Error[2]	54.00	7	7.71	
Total	152.40	9		

[1] This source may also be described as "between-group" variation.
[2] This source may also be described as "within-group" variation.

The Critical Value of F and the Decision

The calculated F might seem large enough to suggest that the population means might not be the same, but the key question is, *Is it large enough that we are able to reject the null hypothesis at the $\alpha = 0.025$ level of significance?* To find out, we must first look up the critical value of F from the $\alpha = 0.025$ F distribution table in Appendix A:

$$\text{Critical Value of } F = F(\alpha, v_1, v_2)$$

The df associated with the numerator of F is

$$v_1 = (t - 1) \quad \text{or} \quad (3 - 1) = 2$$

The df associated with the denominator of F is

$$v_2 = (N - t) \quad \text{or} \quad (10 - 3) = 7$$

Thus, for $\alpha = 0.025$, $t = 3$ treatments, and a total of $N = 10$ observations, the critical F is $F(0.025, 2, 7) = 6.54$.

As Figure 12.1 shows, the calculated value ($F = 6.38$) does not exceed the critical value (6.54), and we are not able to reject the null hypothesis, H_0: $\mu_1 = \mu_2 = \mu_3$. At the 0.025 level of significance, the training methods could be equally effective.

Using the 0.05, 0.025, and 0.01 F distribution tables in Appendix A, we can narrow down the exact p-value for the test:

α	Calculated F		Critical F Value, $F(\alpha, 2, 7)$	Decision
0.01	6.38	is not >	9.55	Cannot reject H_0
0.025	6.38	is not >	6.54	Cannot reject H_0
0.05	6.38	exceeds	4.74	Reject H_0

Although the null hypothesis would be rejected at the 0.05 level, we are not able to reject it at the 0.025 level. Based on our statistical tables, the most accurate statement we can make about the p-value for the test is that it is somewhere between 0.025 and 0.05.

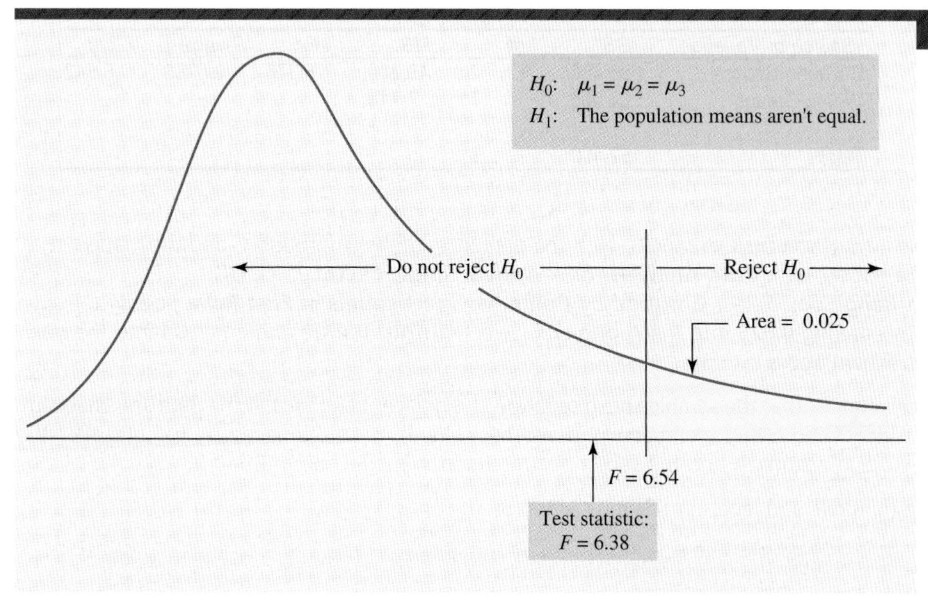

FIGURE 12.1

Results of the one-way analysis of variance for the data in Table 12.3 and summary figures in Table 12.4. The test was conducted at the 0.025 level, and the calculated F does not exceed the critical value for this level of significance, so we are unable to reject the null hypothesis that the population means are equal.

Computer Outputs and Interpretation

The printouts in Computer Solutions 12.1 show the Excel and Minitab results for the ANOVA problem we've just carried out by pocket calculator. Although their formats differ slightly, the content of each output is very similar to the summary results shown in Table 12.4.

In the Excel portion of Computer Solutions 12.1, the same quantities calculated in our example appear in the Excel output, along with the p-value of 0.0265. Among the values listed is the total sum of squares (152.400), with separate entries for its between-group (98.400) and within-group (54.000) components. These are equivalent to the *SST*, *SSTR*, and *SSE* terms that we have been using in describing the respective sources of variation.

The Minitab portion of Computer Solutions 12.1 shows the same quantities as Excel, including the p-value for the test. The respective p-values differ only due to rounding. Note that "Factor" is used in representing the variation

COMPUTER 12.1 SOLUTIONS

One-Way Analysis of Variance

EXCEL

	A	B	C	D	E	F	G	H	I	J
1	Method1	Method2	Method3	Anova: Single Factor						
2	15	10	18							
3	20	15	19	SUMMARY						
4	19	11	23	*Groups*	*Count*	*Sum*	*Average*	*Variance*		
5	14			Method1	4	68	17	8.667		
6				Method2	3	36	12	7		
7				Method3	3	60	20	7		
8										
9										
10				ANOVA						
11				*Source of Variation*	*SS*	*df*	*MS*	*F*	*P-value*	*F crit*
12				Between Groups	98.400	2	49.200	6.378	0.0265	6.542
13				Within Groups	54.000	7	7.714			
14										
15				Total	152.400	9				

Excel one-way ANOVA

1. For the sample data in Table 12.3 (Excel file **CX12ACCT**), with the data in columns A through C and their labels in the first row: From the **Data** ribbon, click **Data Analysis.** Click **Anova: Single Factor.** Click **OK.**
2. Enter **A1:C5** into the **Input Range** box. Select **Grouped By Columns.** Select **Labels in First Row.** Specify the significance level for the test by entering **0.025** into the **Alpha** box. Select **Output Range** and enter **D1** into the box. Click **OK.** The results are as shown above.

Excel Unstacking Note: Excel requires that the samples be in separate and adjacent columns. If the data are "stacked" in one column, with subscripts identifying group membership in another column, it will be necessary to "unstack" the data. For example, if the data were in column A and the subscripts (methods) were in column B:

1. Click and drag to select the numbers and labels in columns A and B.

2. From the **Data** ribbon, click **Sort.** In the **Sort by** box, select the category variable in column B (i.e., **Method**). Click **OK.** The data for sample 1 will now be listed directly above the data for sample 2, which in turn will be listed above the data for sample 3—from here, we need only select each grouping of data and either move or copy it to its own column.

MINITAB

Minitab one-way ANOVA

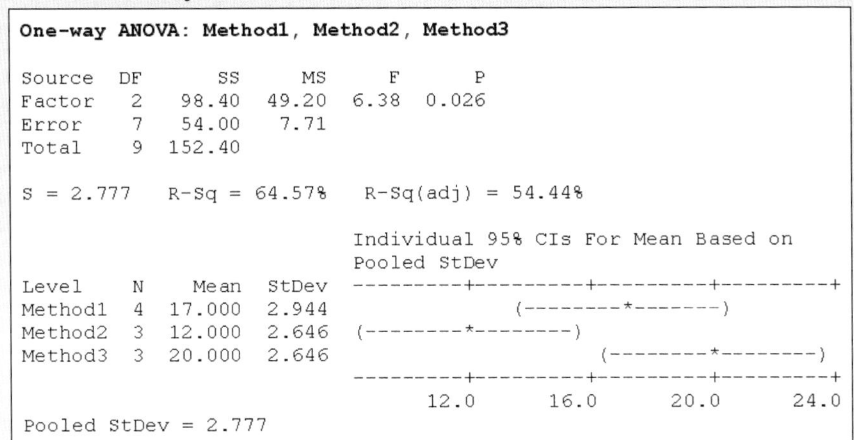

```
One-way ANOVA: Method1, Method2, Method3

Source  DF      SS     MS     F      P
Factor   2   98.40  49.20  6.38  0.026
Error    7   54.00   7.71
Total    9  152.40

S = 2.777   R-Sq = 64.57%   R-Sq(adj) = 54.44%

                                Individual 95% CIs For Mean Based on
                                Pooled StDev
Level    N    Mean  StDev  ---------+---------+---------+---------+
Method1  4  17.000  2.944              (--------*-------)
Method2  3  12.000  2.646  (--------*--------)
Method3  3  20.000  2.646                    (--------*--------)
                           ---------+---------+---------+---------+
                                 12.0      16.0      20.0      24.0
Pooled StDev = 2.777
```

1. For the sample data in Table 12.3 (Minitab file **CX12ACCT**), with the data in columns C1 through C3: Click **Stat.** Select **ANOVA.** Click **One-Way (Unstacked).**
2. Enter **C1-C3** into the **Responses (in separate columns)** box. Click **OK.**

Note: If all the data had been in a single column [i.e., "stacked"], it would have been necessary to select the **One-way** option in step 1, then to specify the column containing the data and the column containing the subscripts in step 2.

due to the treatments. As pointed out previously, when there is just one factor, treatments and factor levels are synonymous. Beyond the standard listing of one-way ANOVA summary information, Minitab provides further information in the form of:

1. The pooled estimate of the population standard deviation, "Pooled StDev = 2.777." Keeping in mind that the model assumes the treatment-group populations have the same variance, this part of the output indicates that 2.777^2 is the estimated value for the variance for each of the populations from which the treatment groups were drawn.

   Note that $2.777^2 = 7.71$, the value listed for the error mean square in the "Error" row and "MS" column of the Minitab output. (*MSE* is equivalent to s_p^2, the pooled estimate of σ^2.) As discussed previously, one-way ANOVA uses the error mean square as one of the two estimates of the common variance that the treatment-group populations are assumed to share.
2. The 95% confidence interval for the population mean corresponding to each treatment. The Minitab output shows each of these confidence intervals in a visual format, but they are based on the pooled estimate of σ, which is

"Pooled StDev = 2.777." Each of these confidence intervals is calculated using the t distribution and the same df that is associated with the error mean square, MSE:

One-way ANOVA, confidence interval for the mean, treatment j:
The confidence interval for μ_j is

$$\bar{x}_j \pm t\left(\frac{s_p}{\sqrt{n_j}}\right)$$

where $s_p = \sqrt{MSE}$, the pooled estimate of σ
n_j = number of persons or test units in treatment group j
$\bar{x}_j$ = sample mean for treatment group j
t = from the t distribution, the t value for the desired confidence level, using the same df as associated with MSE

For example, the mean for the 4 persons in treatment group 1 ("Method 1") was 17.0 minutes, and the number of degrees of freedom associated with MSE was $df = 7$. Using the pooled estimate for the population standard deviation (2.777 minutes), the 95% confidence interval for μ_1 is

$$\bar{x}_1 \pm t\left(\frac{s_p}{\sqrt{n_1}}\right) \quad \text{where } df = 7 \text{ degrees of freedom associated with } MSE\text{, and } t = 2.365 \text{ for a 95\% interval}$$

$$= 17.000 \pm t\left(\frac{2.777}{\sqrt{4}}\right)$$

$$= 17.000 \pm 2.365(1.389), \quad \text{or from 13.72 to 20.28 minutes}$$

We have 95% confidence that the population mean for treatment 1 is somewhere from 13.72 to 20.28 minutes; the confidence intervals for μ_2 and μ_3 shown in the Minitab portion of Computer Solutions 12.1 were determined through the same procedure. The simultaneous display of confidence intervals such as those in the Minitab printout helps us to see at a glance the extent to which one or more treatments might differ from the rest.

One-Way ANOVA and the Pooled-Variances *t*-Test

As mentioned previously, when there are just two treatments, one-way ANOVA and the pooled-variances t-test are equivalent. Each is testing whether two population means could be the same, and each relies on a pooled estimate for the variance that the populations are assumed to share. In addition, both tests assume the population is (at least approximately) normally distributed and the samples are independent.

With two treatment groups, one-way ANOVA is testing H_0: $\mu_1 = \mu_2$ versus H_1: $\mu_1 \neq \mu_2$. The test is nondirectional, since H_0 can be rejected by an extreme difference in either direction. In a nondirectional, pooled-variances t-test, the null and alternative hypotheses are exactly the same as in one-way ANOVA with two treatment groups. In each case, the null hypothesis can be rejected by an extreme result in either direction. The tests are equivalent and their results will provide the same p-value.

EXERCISES

Note: The computer is highly recommended when using ANOVA techniques. For exercises involving raw data, data files are shown in parentheses.

12.13 What is the purpose of the one-way ANOVA?

12.14 What assumptions are required in using the one-way ANOVA?

12.15 Why is the F distribution applicable to ANOVA?

12.16 For the one-way ANOVA model, $x_{ij} = \mu + \tau_j + \epsilon_{ij}$, explain the meaning of each term in the equation, then use the appropriate terms from the equation in presenting and explaining the null and alternative hypotheses to be examined.

12.17 In terms of the assumptions required for using ANOVA, what is the special relevance of the value calculated for the error mean square, MSE?

12.18 On snow-covered roads, winter tires enable a car to stop in a shorter distance than if summer tires were installed. In terms of the additive model for one-way ANOVA, and for an experiment in which the mean stopping distances on a snow-covered road are measured for each of five brands of winter tires, differentiate between the use of the terms "effectiveness" and "effect."

12.19 In using ANOVA to compare the means of several samples, what value would be expected for the calculated F-statistic if the population means really are the same? Explain.

12.20 In a one-way ANOVA, there are three independent samples, with $n_1 = 8$, $n_2 = 10$, and $n_3 = 7$. The calculated F-statistic is $F = 3.95$. At the 0.05 level of significance, what conclusion would be reached? Based on the F distribution tables, what is the most accurate statement that can be made about the p-value for this test?

12.21 In a one-way ANOVA, there are two independent samples, with $n_1 = 20$ and $n_2 = 15$. The calculated F-statistic is $F = 3.60$. At the 0.05 level of significance, what conclusion would be reached? Based on the F distribution tables, what is the most accurate statement that can be made about the p-value for this test?

12.22 A one-way ANOVA has been conducted for an experiment in which there are three treatments and each treatment group consists of 10 persons. The results include the sum of squares terms shown here. Based on this information, construct the ANOVA table of summary findings and use the 0.025 level of significance in testing whether all of the treatment effects could be zero.

$$SSTR = 252.1 \qquad SSE = 761.1$$

12.23 For the experiment described in Exercise 12.6, what null and alternative hypotheses would be appropriate for the one-way ANOVA to examine the data? Given the following data, what conclusion would be reached at the 0.05 level of significance? From the F distribution tables, what is the approximate p-value for this test? (Use data file **XR12023**.)

Absences, Sample Members from the College of

Engineering	Business	Fine Arts
8	5	9
10	7	10
6	6	10
8	7	9
4	7	7
8	6	5
	8	13
	8	7
	1	

12.24 For the experiment described in Exercise 12.7, what null and alternative hypotheses would be appropriate for the one-way ANOVA to examine the data? If the data are as shown here, what conclusion would be reached at the 0.025 level of significance? From the F distribution tables, what is the approximate p-value for this test? (Use data file **XR12024**.)

Final Exam Scores of Students for Whom Computers and Software Were

Used	73	82	72	75	88	73	74	81	86	83
Not Used	78	67	71	72	69	81	64	79	73	73

12.25 For the experiment described in Exercise 12.8, what null and alternative hypotheses would be appropriate for the one-way ANOVA to examine the data? If the data are as shown here, what conclusion would be reached at the 0.01 level of significance? (Use data file **XR12025**.)

Breaking Strengths (Pounds) of 10 Rivets from

Supplier A	517	484	463	452	502
	447	481	500	485	566
Supplier B	479	499	488	430	482
	457	424	488	526	455
Supplier C	435	443	480	465	435
	430	465	514	463	510
Supplier D	526	537	443	505	468
	533	481	477	490	470

12.26 For the following data from independent samples, could the null hypothesis that the population means are equal be rejected at the 0.05 level? (Use data file **XR12026**.)

Sample 1	Sample 2	Sample 3	Sample 4
15.2	10.2	14.9	11.5
12.2	12.1	13.0	13.1
15.0	8.5	10.8	11.5
14.6	8.1	10.7	6.3
9.7	10.9	13.3	12.0
	13.6	10.6	8.7
	8.5		11.9

12.27 A study has been carried out to compare the United Way contributions made by clerical workers from three local corporations. A sample of clerical workers has been randomly selected from each firm, and the dollar amounts of their contributions are as follows. (Use data file **XR12027**.)

Firm 1	Firm 2	Firm 3
199	108	162
236	104	86
167	153	160
263	218	135
254	210	207
	96	201

a. What are the null and alternative hypotheses for this test?
b. Use ANOVA and the 0.05 level of significance in testing the null hypothesis identified in part (a).

12.28 Students in a large section of a biology class have been randomly assigned to one of two graduate students for the laboratory portion of the class. A random sample of final examination scores has been selected from students supervised by each graduate student, with the following results. (Use data file **XR12028**.)

Grad Student A	78	78	71	89	80	93	73	76		
Grad Student B	74	81	65	73	80	63	71	64	50	80

a. What are the null and alternative hypotheses for this test?
b. Use ANOVA and the 0.01 level of significance in testing the null hypothesis identified in part (a).

12.29 Each of fifteen 60-watt lightbulbs has been randomly placed into an outlet for which the voltage is 3 volts below the line voltage, 2 volts below the line voltage, or equal to the line voltage. The following data are the lifetimes of the bulbs, expressed in days of continuous use. (Use data file **XR12029**.)

Three Volts Below Line	58	63	46	57	51
Two Volts Below Line	46	59	51	46	42
Equal to Line Voltage	52	48	38	48	42

a. What are the null and alternative hypotheses for this test?
b. Use ANOVA and the 0.01 level of significance in testing the null hypothesis identified in part (a).
c. For each sample, construct the 95% confidence interval for the population mean.

12.30 Safety researchers, interested in determining whether the occupancy of a vehicle might be related to the speed at which the vehicle is driven, have observed the following speed (mph) measurements for two random samples of vehicles. (Use data file **XR12030**.)

Driver Alone	64	50	71	55	67	61	80	56	59	74		
At Least One Passenger	44	52	54	48	69	67	54	57	58	51	62	67

a. What are the null and alternative hypotheses for this test?
b. Use ANOVA and the 0.025 level of significance in testing the null hypothesis identified in part (a).
c. For each sample, construct the 95% confidence interval for the population mean.

12.31 For the following summary table for a one-way ANOVA, fill in the missing items (indicated by asterisks), identify the null and alternative hypotheses, then use the 0.05 level of significance in reaching a conclusion regarding the null hypothesis.

Variation Source	Sum of Squares	Degrees of Freedom	Mean Square	*F*
Treatments	6752.0	2	3376.0	***
Error	30178.0	***	***	
Total	36930.0	29		

12.32 For the following summary table for a one-way ANOVA, fill in the missing items (indicated by asterisks), identify the null and alternative hypotheses, then use the 0.025 level of significance in reaching a conclusion regarding the null hypothesis.

Variation Source	Sum of Squares	Degrees of Freedom	Mean Square	*F*
Treatments	665.0	4	***	***
Error	***	60	***	
Total	3736.3	***		

12.33 A researcher has used a standard test in measuring the job-satisfaction scores for employees randomly selected from three departments of a large firm. Interpret the results summarized in the following Minitab output.

```
Analysis of Variance
Source    DF        SS        MS        F        p
Factor     2      69.9      35.0     2.45    0.092
Error     92    1311.2      14.3
Total     94    1381.1
                                 Individual 95% CIs for Mean
                                 Based on Pooled StDev
Level      N      Mean     StDev   --------+---------+---------+--------
DEPT1     30    17.640     3.458                (---------*--------)
DEPT2     25    15.400     3.841   (---------*---------)
DEPT3     40    16.875     3.956              (------*-------)
                                   --------+---------+---------+--------
Pooled StDev =   3.775                   15.0      16.5      18.0
```

12.34 A large investment firm claims that no discrepancy exists between the average incomes of its male and female investment counselors. Random samples consisting of 15 male counselors and 17 female counselors have been selected, and the results examined through the use of ANOVA. In terms of this situation, interpret the components of the following Minitab output.

```
Analysis of Variance
Source    DF        SS        MS        F        p
Factor     1     142.8     142.8     2.70    0.111
Error     30    1586.0      52.9
Total     31    1728.8
                                 Individual 95% CIs for Mean
                                 Based on Pooled StDev
Level      N      Mean     StDev   -------+---------+---------+---------
MALES     15    47.654     6.963             (----------*----------)
FEMALES   17    43.420     7.530   (---------*---------)
                                   -------+---------+---------+---------
Pooled StDev =   7.271                  42.0      45.5      49.0
```

12.35 Given the summary information in Exercise 12.33, verify the calculation of the 95% confidence interval for each of the treatment means.

12.36 Given the summary information in Exercise 12.34, verify the calculation of the 95% confidence interval for each of the treatment means.

12.37 In general, how do the assumptions, procedures, and results of one-way ANOVA compare with those for the pooled-variances t-test of Chapter 11?

12.38 Use the pooled-variances t-test of Chapter 11 in comparing the sample means for the data of Exercise 12.28. Is the conclusion consistent with the one reached in that exercise? Explain.

12.39 Use the pooled-variances t-test of Chapter 11 in comparing the sample means for the data of Exercise 12.30. Is the conclusion consistent with the one reached in that exercise? Explain.

THE RANDOMIZED BLOCK DESIGN (12.4)

Purpose

In the one-way, or completely randomized, ANOVA of the previous section, treatments are randomly assigned to all of the persons or other test units in the experiment. As a result, the composition of the treatment groups may be such that certain kinds of people or test units are overrepresented in some treatment groups and underrepresented in others, simply by chance. If the characteristics of the participants or test units have a strong influence on the measurements we obtain, we may be largely measuring the differing group compositions rather than the effects of the treatments.

For example, let's assume we have randomly selected 12 citizens from a small community and these persons are to participate in an experiment intended to compare the night-vision effectiveness of four different headlamp designs. If we have treatment groups of equal size and randomly assign the treatments, as shown in part A of Table 12.5, it's likely that the representation of older drivers would not be exactly the same in all four groups. This would reduce our ability to compare the headlamp designs, since night vision tends to decrease with age. In this situation, the value of the variable that we really want to measure (i.e., the distance at which a headlamp enables a suburban traffic sign to be read) is being strongly influenced by another variable (age category) that has not been considered in the experiment.

In the **randomized block** design presented in this section, persons or test units are first arranged into similar groups, or **blocks,** before the treatments are assigned. This allows us to reduce the amount of *error* variation. For example, in

TABLE 12.5

In the one-way, or completely randomized, design of part A, our ability to compare the night-vision effectiveness of headlamp treatments may be reduced as the result of chance differences in the age-category compositions of the groups. The randomized block design in part B ensures that groups will be comparable in terms of the age categories of their members.

Measured: The distance (yards) at which a traffic sign can be read at night.
Treatments: Four headlamp designs, 1, 2, 3, and 4.
Age categories: *Y*, driver is under 30 years; *M*, 30–60 years; *O*, >60 years.

A. *Completely Randomized Approach*

1. Twelve drivers are randomly selected from the community (e.g., 5 *Y*, 3 *M*, and 4 *O*).
2. One of the four treatments is randomly assigned to each person. With random assignment, the treatment groups could end up like this:

Members of Treatment Group

1	2	3	4
Y	*M*	*Y*	*O*
Y	*O*	*M*	*Y*
O	*O*	*Y*	*M*

For night-vision measurement, treatment 2 would be at a distinct disadvantage and treatment 3 would have a distinct advantage.

B. *Randomized Block Approach*

1. Four members from each age group are randomly selected from the community (4 *Y*, 4 *M*, and 4 *O*).
2. For each age category, treatments are randomly assigned to the members. Each treatment group will include one person from each age category:

	Members of Treatment Group			
	1	**2**	**3**	**4**
4 *Y*, treatments randomly assigned →	*Y*	*Y*	*Y*	*Y*
4 *M*, treatments randomly assigned →	*M*	*M*	*M*	*M*
4 *O*, treatments randomly assigned →	*O*	*O*	*O*	*O*

the night-vision experiment just described, use of the randomized block design would ensure that the treatment groups are comparable in terms of the age categories of their members. Exerting this control over the age-category variable (now referred to as a *blocking* variable) allows us to better compare the effectiveness of the headlamp designs, or treatments. The resulting experiment would have a format like that shown in part B of Table 12.5.

Although we are controlling, or *blocking*, one variable, our primary concern lies in testing whether the population means could be the same for all of the treatment groups. Accordingly, the null and alternative hypotheses are

$$H_0\text{:}\quad \mu_1 = \mu_2 = \cdots = \mu_t \text{ for treatments 1 through } t$$
$$H_1\text{:}\quad \text{The population means are not equal.}$$

Model and Assumptions

As with the one-way ANOVA of the previous section, the null and alternative hypotheses can also be expressed in terms of an equation in which each individual observation is considered to be the sum of several components. For the randomized block design, these components include both treatment and block effects:

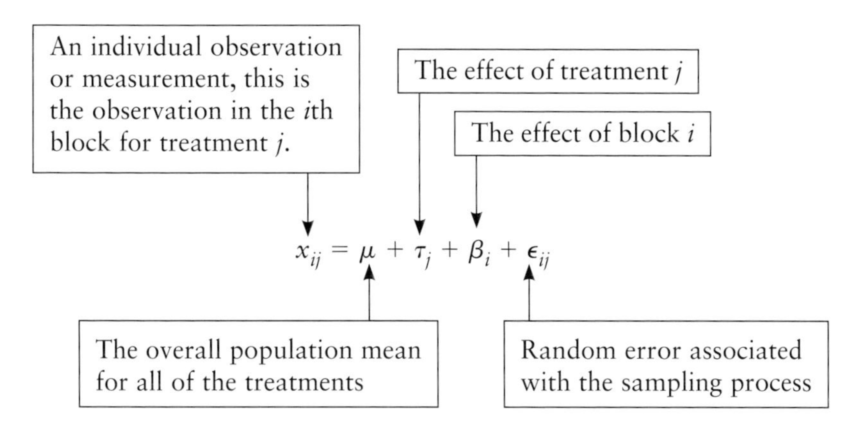

In this model, a measurement (x_{ij}) associated with block i and treatment j is the sum of the actual population mean for all of the treatments (μ), the effect of treatment j (τ_j), the effect of block i (β_i), and a sampling error (ϵ_{ij}). We are only *controlling for* the effect of the blocking variable, not attempting to examine its influence. Thus, the equivalent null and alternative hypotheses are expressed only in terms of the treatments:

H_0: $\tau_j = 0$ for treatments $j = 1$ through t (Each treatment has no effect.)

H_1: $\tau_j \neq 0$ for at least one of the $j = 1$ through t treatments. (One or more treatments has an effect.)

In the randomized block design, there is just one observation or measurement for each block-treatment combination. For example, as part B of Table 12.5 shows, headlamp treatment 1 would be administered to one person in the under-30 category, one person age 30–60, and one person over 60. Thus, each combination of block level and treatment represents a sample of one person or test unit.

The randomized block design assumes (1) for each block-treatment combination (i.e., each combination of the values of i and j), the sample of size 1 has been randomly selected from a population in which the x_{ij} values are normally distributed; (2) the variances are equal for the x_{ij} values in these populations; and (3) there is no interaction between the blocks and the treatments. In the randomized block design, **interaction** is present *when the effect of a treatment depends on the block to which it has been administered.* For example, in the headlamp experiment in part B of Table 12.5, the performance of a given headlamp design relative to the others should be approximately the same, regardless of which age group is considered. The presence or absence of such interactions will be the subject of a later discussion. As with one-way ANOVA, an alternative method is available if one or more of these assumptions cannot be made. In the case of the randomized block design, this is the *Friedman test* of Chapter 14.

Procedure

The procedure for carrying out the randomized block design, which is summarized in Table 12.6, is generally similar to the one-way ANOVA described in Table 12.2. The following descriptions correspond to parts A through D of Table 12.6.

Part A: The Null and Alternative Hypotheses

The null and alternative hypotheses are expressed in terms of the equality of the population means for all of the treatment groups.

TABLE 12.6

Summary of the randomized block design.

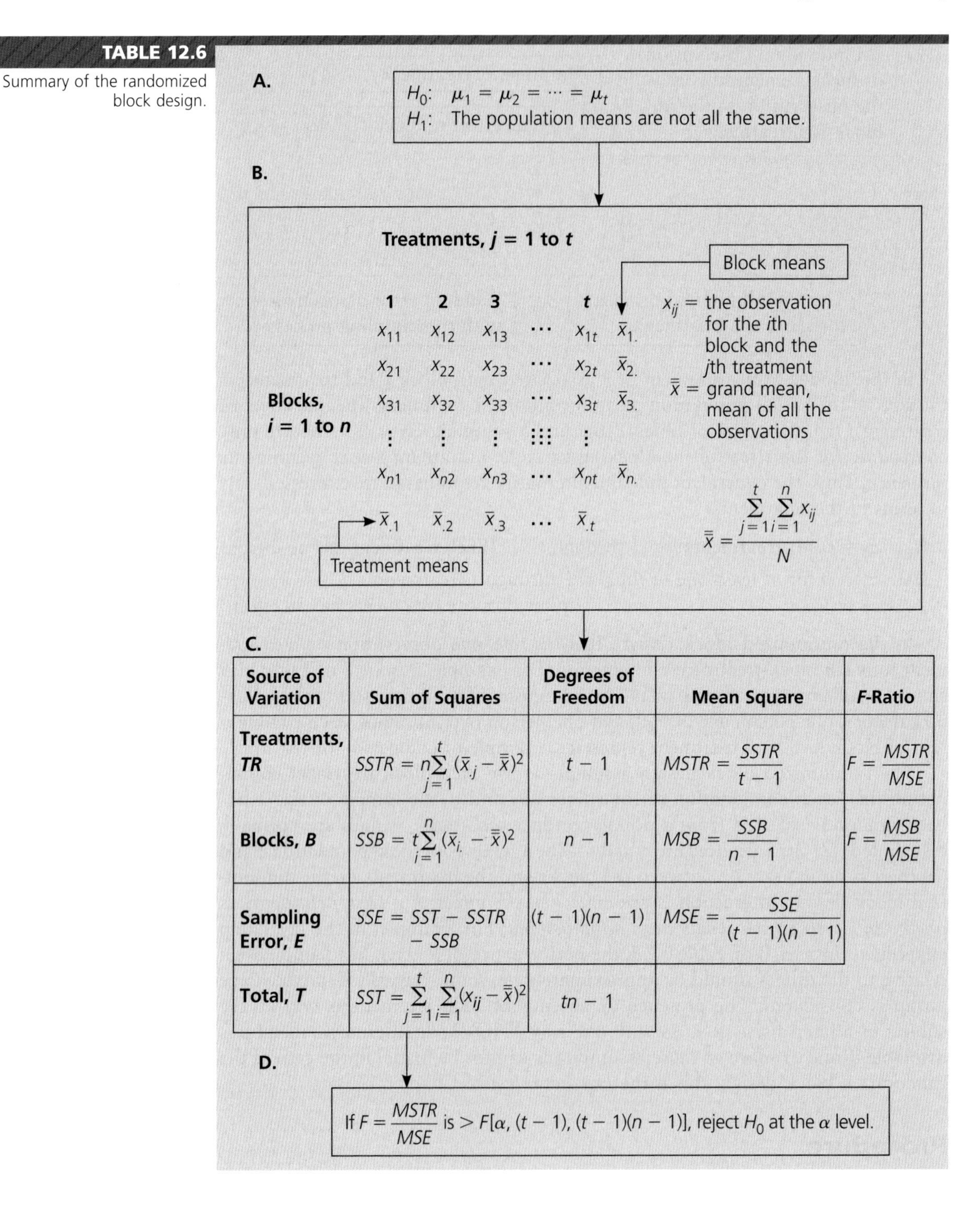

Source of Variation	Sum of Squares	Degrees of Freedom	Mean Square	*F*-Ratio
Treatments, *TR*	$SSTR = n\sum_{j=1}^{t}(\bar{x}_{.j} - \bar{\bar{x}})^2$	$t - 1$	$MSTR = \frac{SSTR}{t-1}$	$F = \frac{MSTR}{MSE}$
Blocks, *B*	$SSB = t\sum_{i=1}^{n}(\bar{x}_{i.} - \bar{\bar{x}})^2$	$n - 1$	$MSB = \frac{SSB}{n-1}$	$F = \frac{MSB}{MSE}$
Sampling Error, *E*	$SSE = SST - SSTR - SSB$	$(t-1)(n-1)$	$MSE = \frac{SSE}{(t-1)(n-1)}$	
Total, *T*	$SST = \sum_{j=1}^{t}\sum_{i=1}^{n}(x_{ij} - \bar{\bar{x}})^2$	$tn - 1$		

Part B: The Format of the Data to Be Analyzed

The data can be listed in tabular form, as shown, with a separate column for each of the t treatments and a separate row for each of the n blocks. Each cell (combination of a block i and a treatment j) contains just one observation. As with one-way, or completely randomized, ANOVA, the grand mean ($\bar{\bar{x}}$) is the mean of all of the observations that have been recorded.

In addition to listing a format for the x_{ij} values, part B of Table 12.6 shows the block means and the treatment means. For example, $\bar{x}_{1.}$ is the mean for block 1, with the dot (.) in the subscript indicating the mean is across all of the treatments. Likewise, $\bar{x}_{.3}$ is the mean for treatment 3, with the dot portion of the subscript indicating the mean is across all of the blocks.

Part C: The Calculations for the Randomized Block Design

Part C of Table 12.6 describes the specific computations for the randomized block design; each quantity is associated with a specific source of variation within the sample data. They correspond to the $x_{ij} = \mu + \tau_j + \beta_i + \epsilon_{ij}$ model discussed previously, and their calculation and interpretation are similar to their counterparts in the one-way, or completely randomized, design.

The Sum of Squares Terms: Quantifying the Sources of Variation

- **Treatments, *TR*** *SSTR* is the sum of squares that reflects the amount of variation between the treatment means and the grand mean, $\bar{\bar{x}}$. *SSB* is the sum of squares that reflects the amount of variation between the block means and the grand mean, $\bar{\bar{x}}$.
- **Sampling error, *E*** *SSE* is the sum of squares that reflects the total amount of variation that is due to sampling error. It is most easily calculated by first determining *SST* then subtracting *SSTR* and *SSB*.
- **Total variation, *T*** *SST* is the sum of squares that reflects the total amount of variation in the data, with each data value being compared to the grand mean, then the differences are squared and summed.

Making the Amounts from the Sources of Variation Comparable. *MSTR*, *MSB*, and *MSE* are the mean squares for treatments, blocks, and error, respectively. As part C of Table 12.6 shows, each is obtained by dividing the corresponding sum of squares by an appropriate value for *df*.

Part D: The Test Statistic, the Critical Value, and the Decision Rule

The Test Statistic. The test statistic is *MSTR*/*MSE*. As with the one-way or completely randomized design, *MSTR* has estimated the common variance (σ^2) based on variation *between* the treatment means, while *MSE* has estimated the common variance based on variance *within* the treatment groups. The test statistic is the ratio of these separate estimates of the common variance.

The Critical Value and the Decision Rule. The test is right-tail and, for a given level of significance (α), we will reject H_0: $\mu_1 = \mu_2 = \cdots = \mu_t$ if the calculated value of F is greater than $F[\alpha, (t-1), (t-1)(n-1)]$. In referring to the F distribution table, we use the $v_1 = (t-1)$ column and the $v_2 = (t-1)(n-1)$ row in identifying the critical value.

EXAMPLE

Randomized Block ANOVA Procedure

To illustrate the application of the randomized block design, we will use an example that corresponds to our introductory discussion for this procedure. Computer outputs and their interpretation will then be presented.

EXAMPLEEXAMP

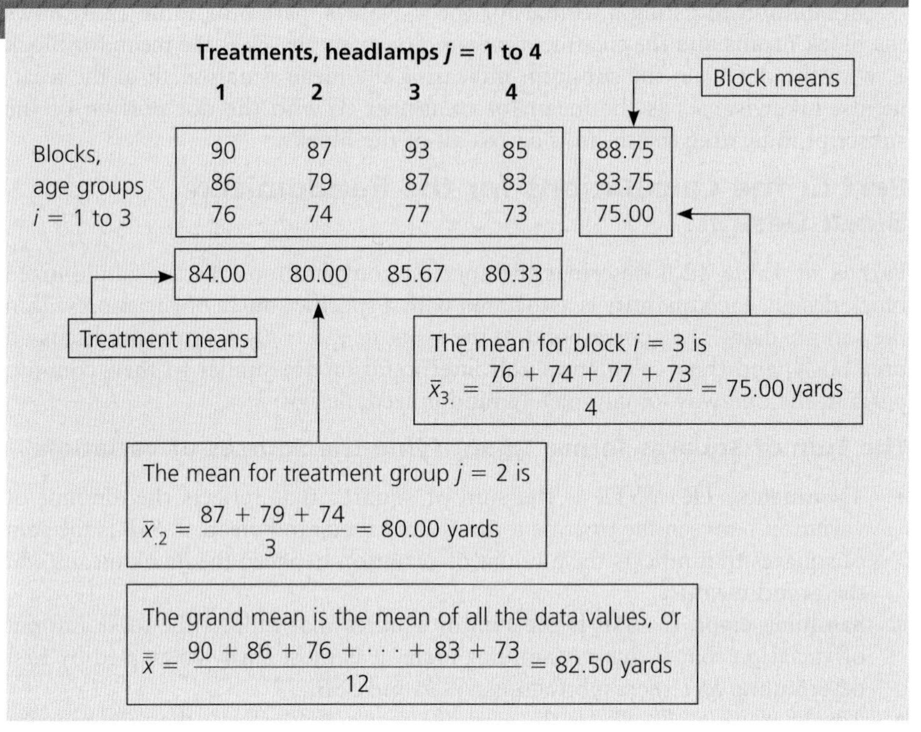

Blocks, age groups i = 1 to 3	Treatments, headlamps j = 1 to 4: 1	2	3	4	Block means
	90	87	93	85	88.75
	86	79	87	83	83.75
	76	74	77	73	75.00
Treatment means	84.00	80.00	85.67	80.33	

The mean for block i = 3 is

$$\bar{x}_{3.} = \frac{76 + 74 + 77 + 73}{4} = 75.00 \text{ yards}$$

The mean for treatment group j = 2 is

$$\bar{x}_{.2} = \frac{87 + 79 + 74}{3} = 80.00 \text{ yards}$$

The grand mean is the mean of all the data values, or

$$\bar{\bar{x}} = \frac{90 + 86 + 76 + \cdots + 83 + 73}{12} = 82.50 \text{ yards}$$

TABLE 12.7

The data and the results of preliminary calculations for the randomized block design to determine whether four headlamp designs could be equally effective. The blocks are according to age group, with i = 1 for the four under-30 drivers, i = 2 for the four 30–60 drivers, and i = 3 for the four over-60 drivers. Data are the distance (in yards) at which a suburban traffic sign can be read.

The new-product development team for an automotive headlamp firm has four different headlamp designs under consideration. A test is set up to compare their effectiveness in night-driving conditions, and the measurement of interest is the distance at which a suburban traffic sign can be read by the driver. Recognizing that younger drivers tend to have better night vision than older drivers, the team has planned the experiment so that *age group* will be a blocking variable. Four persons from each age group (or block) are selected, then the treatments are randomly assigned to members from each block. When each person is subjected to one of the headlamp designs, the distance at which the traffic sign can be read is measured, with the results in Table 12.7. At the 0.05 level, could the headlamp designs be equally effective? The data are also in file **CX12LITE**.

SOLUTION

For this test, the null hypothesis is H_0: $\mu_1 = \mu_2 = \mu_3 = \mu_4$ for the four headlamp designs, or treatments. According to the null hypothesis, the treatment population means are the same—i.e., the four headlamp designs are equally effective. The alternative hypothesis holds that the population means are not equal and that the headlamp designs are not equally effective. Preliminary calculations, such as the block means, treatment means, and the grand mean are shown in Table 12.7. We will now proceed with the remaining calculations, described in part C of Table 12.6.

Treatment Sum of Squares, *SSTR*

The treatment sum of squares is calculated in a manner analogous to that of the one-way ANOVA method. In the randomized block design, the treatment groups are of equal size, and in this experiment each group has n = 3 persons.

$$SSTR = n\sum_{j=1}^{t}(\bar{x}_{.j} - \bar{\bar{x}})^2$$

$$= 3\begin{bmatrix}(84.00 - 82.50)^2 + (80.00 - 82.50)^2 \\ +(85.67 - 82.50)^2 + (80.33 - 82.50)^2\end{bmatrix} \text{ and } \boxed{SSTR = 69.77}$$

Block Sum of Squares, *SSB*

The sum of squares comparing block means with the overall mean is calculated as

$$SSB = t\sum_{i=1}^{n}(\bar{x}_{i.} - \bar{\bar{x}})^2$$

$$= 4[(88.75 - 82.50)^2 + (83.75 - 82.50)^2 + (75.00 - 82.50)^2]$$

and

$$\boxed{SSB = 387.50}$$

Total Sum of Squares, *SST*

SST considers the individual data values. In this sum of squares, each x_{ij} value is compared to the grand mean, $\bar{\bar{x}}$:

$$SST = \sum_{j=1}^{t}\sum_{i=1}^{n}(x_{ij} - \bar{\bar{x}})^2$$

$$= \begin{bmatrix}(90 - 82.50)^2 \\ +(86 - 82.50)^2 \\ +(76 - 82.50)^2\end{bmatrix} + \begin{bmatrix}(87 - 82.50)^2 \\ +(79 - 82.50)^2 \\ +(74 - 82.50)^2\end{bmatrix}$$

$$+ \begin{bmatrix}(93 - 82.50)^2 \\ +(87 - 82.50)^2 \\ +(77 - 82.50)^2\end{bmatrix} + \begin{bmatrix}(85 - 82.50)^2 \\ +(83 - 82.50)^2 \\ +(73 - 82.50)^2\end{bmatrix} \text{ and } \boxed{SST = 473.00}$$

Error Sum of Squares, *SSE*

This quantity is most easily computed by first calculating the preceding quantities, then using the relationship $SST = SSTR + SSB + SSE$, or

$$SSE = SST - SSTR - SSB = 473.00 - 69.77 - 387.50 \text{ and } \boxed{SSE = 15.73}$$

Treatment Mean Square (*MSTR*) and Error Mean Square (*MSE*)

SSTR and *SSE* are now divided by their respective degrees of freedom so that (1) they will be comparable and (2) each will be a separate estimate of the common variance that the treatment group populations are assumed to share. Recall that we have $t = 4$ treatments and $n = 3$ blocks.

The estimate of σ^2 that is based on the *between-treatment* variation is

$$MSTR = \frac{SSTR}{t - 1} = \frac{69.77}{4 - 1} \text{ and } \boxed{MSTR = 23.26}$$

The estimate of σ^2 that is based on the *within-treatment*, or sampling-error variation is

$$MSE = \frac{SSE}{(t - 1)(n - 1)} = \frac{15.73}{(4 - 1)(3 - 1)} \text{ and } \boxed{MSE = 2.62}$$

TABLE 12.8

From the data in Table 12.7 and the computations in the text, we can construct this standard table describing the results of the randomized block ANOVA test. The format is similar to that shown in part C of Table 12.6. Because our calculations were rounded to two decimal places, the results shown here will differ slightly from those in the corresponding computer printouts.

Variation Source	Sum of Squares	Degrees of Freedom	Mean Square	F
Treatments	69.77	3	23.26	8.88
Blocks	387.50	2	193.75	73.95
Error	15.73	6	2.62	
Total	473.00	11		

EXAMPLEEXAMPLEEXAMPLEEXAMPLEEXAMPLEEXAMPLE

The Test Statistic, *F*

The test statistic is the ratio of the two estimates for σ^2, or

$$F = \frac{MSTR}{MSE} = \frac{23.26}{2.62} \quad \text{and} \quad \boxed{F = 8.88}$$

We have now generated the information described in part C of Table 12.6, and the results are summarized in the randomized block ANOVA table shown in Table 12.8. Note that Table 12.8 includes the listing of the mean square [$MSB = SSB/(n - 1) = 193.75$] and the F-ratio ($F = MSB/MSE = 73.95$) for the blocking variable, neither of which was calculated previously. These were omitted from our previous calculations because our intent in the randomized block design is only to *control* for the effect of the blocking variable, not to investigate its effect. In Section 12.5, we will examine the simultaneous effects of *two* independent variables in a method called *two-way analysis of variance*.

The Critical Value of *F* and the Decision

The test is being carried out at the $\alpha = 0.05$ level, and the critical value of F can be found from Appendix A as

$$\text{Critical value of } F = F(\alpha, v_1, v_2)$$

The df associated with the numerator of F is

$$v_1 = (t - 1) \quad \text{or} \quad (4 - 1) = 3$$

The df associated with the denominator of F is

$$\begin{aligned} v_2 &= (t - 1)(n - 1) \\ &= (4 - 1)(3 - 1) = 6 \end{aligned}$$

For $\alpha = 0.05$, $v_1 = 3$ and $v_2 = 6$, the critical F is $F(0.05, 3, 6) = 4.76$. The calculated value ($F = 8.88$) exceeds the critical value, and, at the 0.05 level, we are able to reject the null hypothesis that the population means are equal. At this level of significance, our conclusion is that the headlamp treatments are not equally effective.

Using the F distribution tables in Appendix A, we can narrow down the p-value for the test. For the 0.05, 0.025, and 0.01 levels, the respective critical values are 4.76, 6.60, and 9.78, and the calculated F ($F = 8.88$) falls between the critical values for the 0.025 and 0.01 levels. Based on our statistical tables, the most accurate statement we can make about the p-value for the test is that it is somewhere between 0.025 and 0.01.

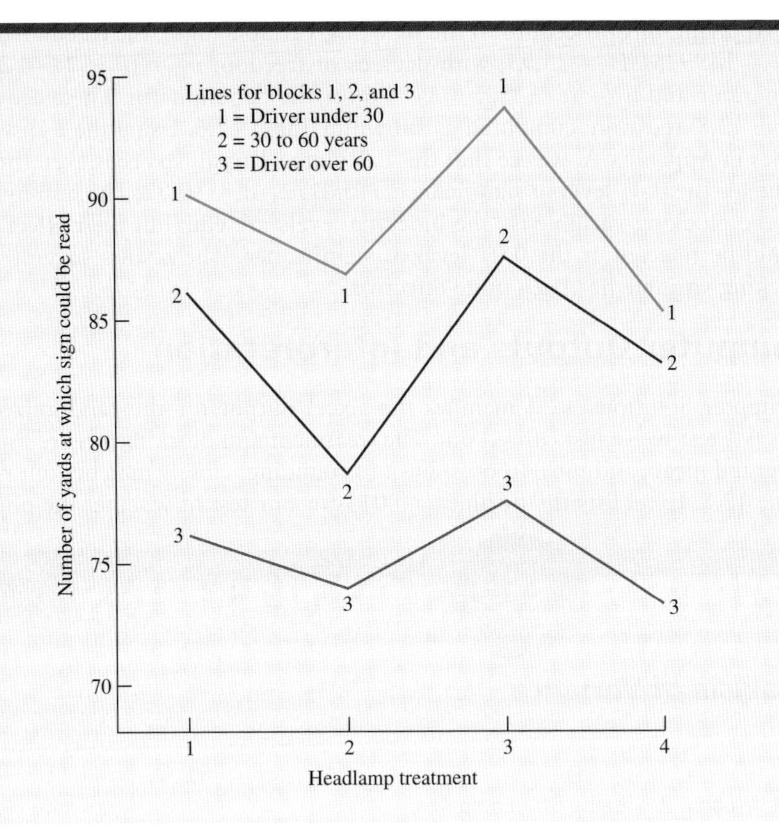

FIGURE 12.2

This multiple plot visually summarizes the data in Table 12.7. The display helps show why blocking according to age group would be appropriate for this type of study.

Figure 12.2 is a multiple plot of the measurements according to the treatments and blocks. Each line connects the distances measured for persons in one of the age-group blocks. Regardless of the headlamp design, the distance at which the traffic sign could be read was longest for the youngest person exposed to that treatment, and shortest for the oldest person exposed to the treatment. This further demonstrates the appropriateness of using the randomized block design with age group as a blocking variable.

Testing the Effectiveness of the Blocking Variable

As mentioned earlier, our primary intent is to *control* for the effect of the blocking variable, not to measure its effect. If we *do* wish to examine the effectiveness of the blocking variable, we can use this procedure:

Hypotheses for testing the effectiveness of the blocking variable:

H_0: The levels of the blocking variable are equal in their effect.
H_1: At least one level has an effect different from the others.

Calculated value of $F = MSB/MSE$, computed as shown in Table 12.6
Critical value of $F = F(\alpha, v_1, v_2)$
where $v_1 = (n - 1)$, *df* associated with the numerator, and
$v_2 = (t - 1)(n - 1)$, *df* associated with the denominator
and n = number of blocks, t = number of treatments

Decision rule: If calculated $F > F(\alpha, v_1, v_2)$, reject H_0 at the α level.

Applying this procedure to the blocking variable (age), the calculated F-ratio is $F = MSB/MSE = 73.95$, with degrees of freedom $v_1 = (n - 1) = 2$ for the numerator and $v_2 = (t - 1)(n - 1) = 6$ for the denominator. These are listed in Table 12.8 and were computed as shown in Table 12.6. Using the F distribution tables in Appendix A, we see that this calculated F easily exceeds even the critical value for the 0.01 level of significance, $F(0.01, 2, 6) = 10.92$. The null hypothesis for the blocking variable is "the blocking levels are equal in their effect," and it would be rejected at the 0.01 level of significance. The results suggest that the blocking variable has been quite effective.

Computer Outputs and Interpretation

Computer Solutions 12.2 includes the Excel output for the randomized block ANOVA test we've just carried out. The treatment and block means are listed, as is the grand mean, and the output (subject to rounding) closely matches our results in Table 12.8. Excel lists the p-value as 0.013 for our test of whether the population

COMPUTER 12.2 SOLUTIONS

Randomized Block Analysis of Variance

EXCEL

	A	B	C	D	E	F	G
1		Lamp 1	Lamp 2	Lamp 3	Lamp 4		
2	Age block 1	90	87	93	85		
3	Age block 2	86	79	87	83		
4	Age block 3	76	74	77	73		
5							
6	Anova: Two-Factor Without Replication						
7							
8	SUMMARY	Count	Sum	Average	Variance		
9	Age block 1	4	355	88.750	12.250		
10	Age block 2	4	335	83.750	12.917		
11	Age block 3	4	300	75.000	3.333		
12							
13	Lamp 1	3	252	84.000	52.000		
14	Lamp 2	3	240	80.000	43.000		
15	Lamp 3	3	257	85.667	65.333		
16	Lamp 4	3	241	80.333	41.333		
17							
18							
19	ANOVA						
20	Source of Variation	SS	df	MS	F	P-value	F crit
21	Rows	387.500	2	193.750	73.421	6.05E-05	5.143
22	Columns	69.667	3	23.222	8.800	0.013	4.757
23	Error	15.833	6	2.639			
24							
25	Total	473.000	11				

Excel randomized block ANOVA

1. For the sample data in Table 12.7 (Excel file **CX12LITE**), with the data and labels in columns A through E, as shown above: From the **Data** ribbon, click **Data Analysis.** Click **Anova: Two-Factor Without Replication.** Click **OK.**

2. Enter **A1:E4** into the **Input Range** box. Select **Labels.** Specify the significance level for the test by entering **0.05** into the **Alpha** box. Select **Output Range** and enter **A6** into the box. Click **OK.** The results are as shown on page 438.

MINITAB

Minitab randomized block ANOVA

```
Two-way ANOVA: Distance versus Block, Treatmnt

Source    DF       SS       MS      F      P
Block      2  387.500  193.750  73.42  0.000
Treatmnt   3   69.667   23.222   8.80  0.013
Error      6   15.833    2.639
Total     11  473.000

S = 1.624   R-Sq = 96.65%   R-Sq(adj) = 93.86%
```

1. For the sample data in Table 12.7 (Minitab file **CX12LITE**), with the distances in C1, the block identifications in C2, and the treatment identifications in C3: Click **Stat.** Select **ANOVA.** Click **Two-way.**
2. Enter **C1** into the **Response** box. Enter **C2** into the **Row factor** box. Enter **C3** into the **Column factor** box. Click **OK.**

means could be equal for the four treatments. Note that Excel has also listed an F-ratio (73.421) and a p-value (6.05×10^{-5}, or 0.0000605) for a test of the block effects. In terms of our model, this would be a test of (H_0: $\beta_i = 0$, $i = 1$ through n) versus (H_1: $\beta_i \neq 0$ for at least one of the n blocks, or age categories).

Although the Excel printout indicates that the blocking variable is exerting an extremely strong influence, remember that our purpose was only to *control* for the blocking variable in examining the effect of the four headlamps, or treatments. That the blocking variable is so highly influential comes as no surprise, since its impact on night vision is the reason why the randomized block design was used in the first place.

In examining the Excel and Minitab printouts, our purpose in this randomized block ANOVA was not to examine the impact of the blocking variable (age category) as an independent variable. Also, our model does not include any consideration of possible interaction between blocks and treatments, as will the two-way ANOVA in the next section.

Minitab can include confidence intervals for the treatment means, as was the case for the one-way ANOVA printout in the Minitab portion of Computer Solutions 12.1. With the randomized block design, however, the confidence interval for a treatment mean would not be very meaningful. Unlike the one-way, completely randomized design, we have tampered with the randomness of the treatment groups by placing drivers into blocks consisting of persons whose ages are similar.

Randomized Block ANOVA and the Dependent-Samples *t*-Test

In the previous section, we pointed out that one-way ANOVA with two treatment groups is equivalent to the two-tail, pooled-variances t-test in Chapter 11. When there are just two treatments, the randomized block ANOVA is also equivalent to one of the tests in Chapter 11, in this case the two-tail, dependent-samples t-test. In applying the randomized block ANOVA to data representing dependent samples, each pair of x_1 and x_2 values would represent a separate block. Because the tests are equivalent when they compare two treatments, they will provide the same p-value when applied to a given set of matched pairs.

EXERCISES

Note: The computer is highly recommended when using ANOVA techniques. For exercises involving raw data, data files are shown in parentheses.

12.40 What is the purpose of the randomized block design?

12.41 What assumptions are required in using the randomized block design?

12.42 How does the randomized block design differ from the one-way, completely randomized design?

12.43 For the randomized block ANOVA model, $x_{ij} = \mu + \tau_j + \beta_i + \epsilon_{ij}$, explain the meaning of each term in the equation, then use the appropriate terms from the equation in presenting and explaining the possible sets of null and alternative hypotheses that can be tested.

12.44 In the randomized block design, what benefit is gained from blocking?

12.45 In one-way ANOVA, we were able to determine the confidence interval for the population value of each treatment mean. Why is this not appropriate for the treatments being examined in the randomized block design?

12.46 For a randomized block experiment in which there are three treatments and two blocks, the calculated value of $F = MSTR/MSE$ is 16.1. Using the 0.05 level of significance, what conclusion would be reached? Based on the F distribution tables, what is the most accurate statement that could be made about the p-value for the test?

12.47 For a randomized block experiment in which there are five treatments and four blocks, the calculated value of $F = MSTR/MSE$ is 3.75. Using the 0.05 level of significance, what conclusion would be reached? Based on the F distribution tables, what is the most accurate statement that could be made about the p-value for the test?

12.48 For a randomized block experiment having three treatments and three blocks, the results include the following sum of squares terms. Based on this information, construct the appropriate table of ANOVA summary findings and use the 0.05 level of significance in testing whether all of the treatment effects could be zero.

$$SSTR = 394.0 \qquad SSB = 325.0 \qquad SSE = 180.0$$

12.49 A randomized block design has three different income groups as blocks, and members of each block have been randomly assigned to the treatment groups shown here. For these data, use the 0.01 level in determining whether the treatment effects could both be zero. Using the 0.01 level, evaluate the effectiveness of the blocking variable. (Use data file **XR12049**.)

		Treatment	
		1	2
Block	A	46	31
	B	37	26
	C	44	35

12.50 A randomized block design has five different age groups as blocks, and members of each block have been randomly assigned to the treatment groups shown here. For these data, use the 0.05 level in determining whether the treatment effects could all be zero. Using the 0.01 level, evaluate the effectiveness of the blocking variable. (Use data file **XR12050**.)

		Treatment			
		1	2	3	4
Block	A	51.2	50.3	47.2	42.0
	B	41.0	37.6	37.0	35.7
	C	57.5	56.9	54.7	49.2
	D	51.2	49.3	46.9	50.9
	E	36.9	34.6	37.2	33.2

12.51 The continuous operating lifetime of a battery depends very much on the electrical demands of the device in which the battery is installed. In a test of three battery brands, four examples of each brand are randomly selected. Three examples of four different toys are also randomly selected, randomly equipped with one of the battery brands, then each is operated until it stops. The times, in hours, are shown here. At the 0.025 level of significance, test whether the battery treatment effects could all be zero. Using the 0.01 level, evaluate the effectiveness of the blocking variable. (Use data file **XR12051**.)

		Hours of Continuous Performance When Installed in	
		"Jumper Rabbit"	"Drummer Duck"
Battery Brand	A	6.1	7.7
	B	5.6	5.7
	C	3.4	4.3
		"Talking Teeth"	"Toddling Turtle"
Battery Brand	A	7.4	5.5
	B	6.9	4.9
	C	6.4	5.0

12.52 Three racquetball players, one from each skill level, have been randomly selected from the membership list of a health club. Using the same ball, each person hits five serves, one with each of five racquets, and using the racquets in a random order. Each serve is clocked with a radar gun, and the results are shown here. With player skill level as a blocking variable, use the 0.025 level of significance in determining whether the treatment effects of the five racquets could all be zero. Using the 0.01 level, evaluate the effectiveness of the blocking variable. (Use data file **XR12052**.)

		Player Skill Level		
		Beginner	Intermediate	Advanced
Racquet Model	*A*	73 mph	64 mph	83 mph
	B	63	72	89
	C	51	54	72
	D	56	81	86
	E	69	90	97

12.53 Two different machine control-button configurations are being tested. Recognizing that some operators are faster than others, management uses seven operators as blocks, then has each operator carry out a standard operation using each control-button layout. Each operator proceeds from one treatment to the next in a random order, and the times for completion of the operation are shown here. At the 0.05 level, determine whether the treatment effects of the two layouts could be zero. (Use data file **XR12053**.)

		Control-Button Configuration	
		1	2
Operator	*A*	5	6 seconds
	B	9	6
	C	11	8
	D	13	10
	E	10	7
	F	9	10
	G	12	9

12.54 Given the following summary table for a randomized block design, fill in the missing items (indicated by asterisks), identify the null and alternative hypotheses in terms of the treatment effects, then use the 0.025 level of significance in reaching a conclusion regarding the null hypothesis.

Variation Source	Sum of Squares	Degrees of Freedom	Mean Square	*F*
Treatments	30.89	2	***	***
Blocks	80.22	2	***	***
Error	7.11	***	***	
Total	118.22	8		

12.55 Given the following summary table for a randomized block design, fill in the missing items (indicated by asterisks), identify the null and alternative hypotheses in terms of the treatment effects, then use the 0.05 level of significance in reaching a conclusion regarding the null hypothesis.

Variation Source	Sum of Squares	Degrees of Freedom	Mean Square	*F*
Treatments	35.33	***	***	***
Blocks	134.40	2	***	***
Error	16.27	8	***	
Total	186.00	14		

12.56 The following Minitab output summarizes the results of a randomized block ANOVA in which the blocks consisted of 4 different income categories, the treatments were 3 different kinds of appeals for contributions to a national charity, and the variable measured (dollars) was the amount contributed to the charity. State the null and alternative hypotheses in terms of the treatment effects, then use the 0.025 level of significance in reaching a conclusion regarding the null hypothesis.

```
Analysis of Variance for Dollars
Source      DF      SS      MS       F       P
Block        3   75613   25204    9.08   0.012
Treatment    2   36422   18211    6.56   0.031
Error        6   16663    2777
Total       11  128698
```

12.57 Compare the randomized block ANOVA with two treatments to the dependent-samples *t*-test of Chapter 11.

12.58 Apply the dependent-samples *t*-test to the data in Exercise 12.49. Is the conclusion consistent with the one reached in that exercise? Explain.

12.59 Apply the dependent-samples *t*-test to the data in Exercise 12.53. Is the conclusion consistent with the one reached in that exercise? Explain.

TWO-WAY ANALYSIS OF VARIANCE (12.5)

Purpose

In the one-way ANOVA of Section 12.3, our emphasis was on determining the effect of a single factor (tax-preparation method) on the dependent variable, or measurement. In the randomized block ANOVA of Section 12.4, we included a blocking variable (age category), but this was only for the purpose of exerting improved control over the examination of the single factor of interest (headlamp design). This section considers **two-way analysis of variance,** a method that simultaneously examines

(1) the effect of *two* factors on the dependent variable, along with (2) the effects of *interactions* between the different levels of these two factors. Two-way ANOVA is the most basic form of the **factorial experiment,** one in which there are two or more factors and the treatments represent all possible combinations of their levels.

Unlike the randomized block ANOVA, two-way ANOVA examines *interactions* between different levels of the factors, or independent variables. In a two-factor experiment, interaction exists *when the effect of a level for one factor depends on which level of the other factor is present.* Medical professionals are always concerned about the interactive effects of drugs, whether prescribed or unprescribed. For example, the slight drowsiness that would be caused by either a small amount of alcohol *or* a small amount of cough medication can become major drowsiness when *both* alcohol and cough medication have been consumed. So-called binary chemical-warfare weapons also exemplify the phenomenon of interaction. Although each of the separate components is harmless by itself, they become a deadly combination when the weapon is activated and they are mixed together.

In the randomized block example of the preceding section, there was no need to consider interactions between the factor levels and the blocks. As Figure 12.2 showed, the rank order of viewing distance according to age group remained the same for all four headlamp designs. Regardless of the headlamp design considered, the youngest driver had the longest vision, the oldest driver the shortest. However, one or more interaction effects would have been present if there had been any *crossovers*—for example, if the oldest driver exposed to headlamp design 4 had been able to see farther than the youngest driver who was exposed to that headlamp. The concept of interaction will be further examined in the context of the two-factor ANOVA example of this section.

Model and Assumptions

In the two-way ANOVA, random assignments are made such that two or more persons or other test units are subjected to each possible combination of the factor levels. The number of persons or test units within each of these combinations (or cells) is referred to as r = the number of replications, with $r \geq 2$. In this section, we will consider only the *balanced design*, where there is an equal number of replications (r) within each combination of factor levels. In other words, within each combination of levels, there will be $k = 1$ through r observations. As with the other ANOVA designs we've examined, each individual observation is considered to be the sum of several components:

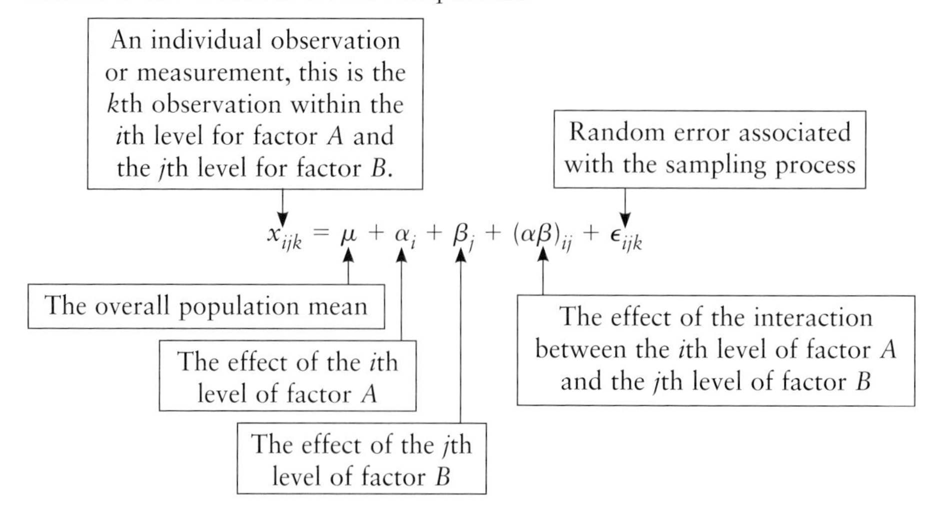

In this model, the kth observation for level i of factor A and level j of factor B is x_{ijk}, and it is viewed as the sum of the overall population mean (μ), the effect of the ith level of factor A (α_i), the effect of the jth level of factor B (β_j), the interaction effect between level i of factor A and level j of factor B [$(\alpha\beta)_{ij}$], plus random error due to sampling (ϵ_{ijk}). There are three sets of null and alternative hypotheses to be tested.

1. Testing for *main effects*, factor A:

   H_0: $\alpha_i = 0$ for each level of factor A, with $i = 1$ through a.
   (No level of factor A has an effect.)

   H_1: $\alpha_i \neq 0$ for at least one value of i, with $i = 1$ through a.
   (At least one level of factor A has an effect.)

2. Testing for *main effects*, factor B:

   H_0: $\beta_j = 0$ for each level of factor B, with $j = 1$ through b.
   (No level of factor B has an effect.)

   H_1: $\beta_j \neq 0$ for at least one value of j, with $j = 1$ through b.
   (At least one level of factor B has an effect.)

3. Testing for *interaction effects* between levels of factors A and B:

   H_0: $(\alpha\beta)_{ij} = 0$ for each combination of i and j.
   (There are no interaction effects.)

   H_1: $(\alpha\beta)_{ij} \neq 0$ for at least one combination of i and j.
   (At least one combination of i and j has an effect.)

The assumptions for the two-way ANOVA design are similar to those for the randomized block design, with *blocks* being replaced by *factor levels*. There are $a \times b$ factor-level combinations, or cells, and it is assumed that the r observations in each cell have been drawn from normally distributed populations with equal variances. The assumption of *no interactions* no longer applies. If one or more of these assumptions cannot be made, the *Friedman test* of Chapter 14 can be applied.

Procedure

Table 12.9 summarizes the procedure for carrying out the two-way ANOVA design. Although the process is generally similar to the randomized block method of Table 12.6, there are now three sets of null and alternative hypotheses to be tested, interaction effects will be examined, and each observation is identified with three subscripts instead of two. The following descriptions correspond to parts A–D of Table 12.9.

Part A: The Null and Alternative Hypotheses

The null and alternative hypotheses are expressed in terms of the main effects (factors A and B) and interaction effects (combinations of levels of these factors).

Part B: The Format of the Data to Be Analyzed

The data can be listed in tabular form, as shown, with each cell identified as a combination of the ith level of factor A with the jth level of factor B. Each cell contains r observations, or replications. For each level of each factor, a mean is calculated. For example, $\bar{x}_{2..}$ is the mean for all observations that received the second level of factor A. Likewise, $\bar{x}_{.1.}$ is the mean for all observations that received the first level of factor B. As in previous analyses, the grand mean ($\bar{\bar{x}}$) is the mean of all the observations that have been recorded.

TABLE 12.9

Summary of the two-way ANOVA design.

A. *Null and Alternative Hypotheses to Be Tested*

1. Main effects, factor *A*:
   H_0: $\alpha_i = 0$ for each value of i, with $i = 1$ through a
   H_1: $\alpha_i \neq 0$ for at least one value of i
2. Main effects, factor *B*:
   H_0: $\beta_j = 0$ for each value of j, with $j = 1$ through b
   H_1: $\beta_j \neq 0$ for at least one value of j
3. Interaction effects:
   H_0: $(\alpha\beta)_{ij} = 0$ for each combination of i and j
   H_1: $(\alpha\beta)_{ij} \neq 0$ for at least one combination of i and j

B. *Data Format*

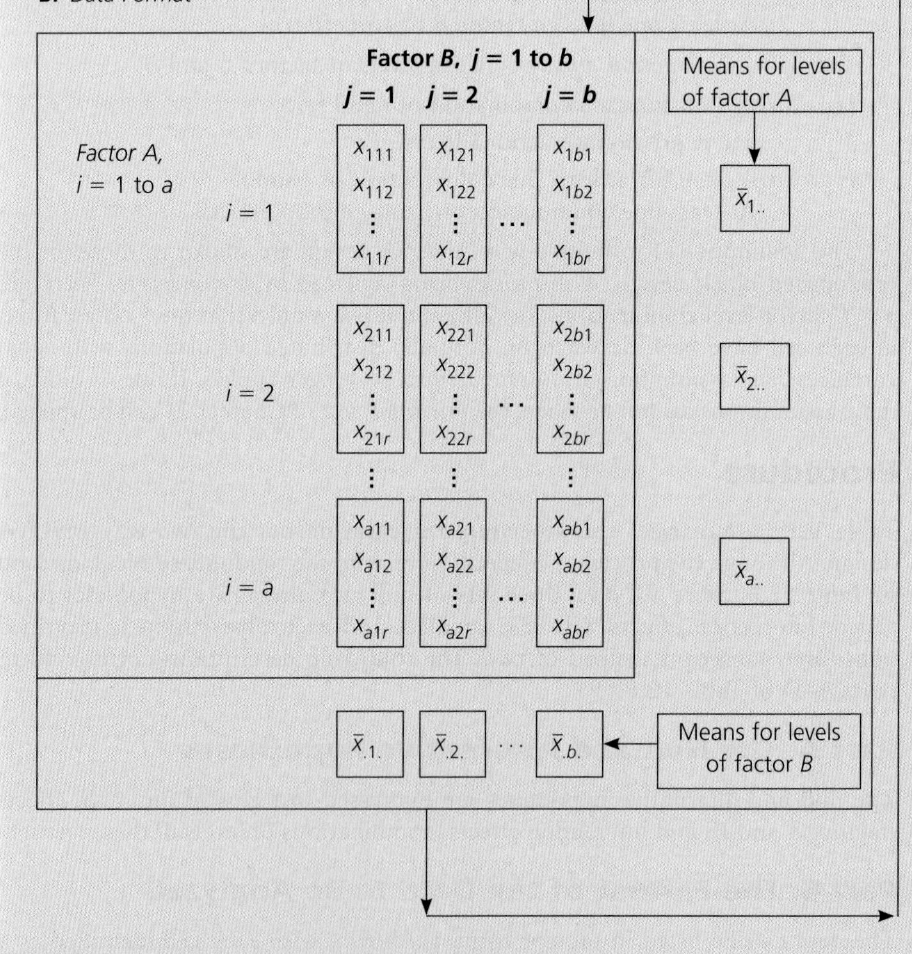

C. *Calculations*

Source of Variation	Sum of Squares	Degrees of Freedom	Mean Square	*F*-Ratio
Factor A	$SSA = rb\sum_{i=1}^{a}(\bar{x}_{i..} - \bar{\bar{x}})^2$	$a - 1$	$MSA = \frac{SSA}{a-1}$	$F = \frac{MSA}{MSE}$
Factor B	$SSB = ra\sum_{j=1}^{b}(\bar{x}_{.j.} - \bar{\bar{x}})^2$	$b - 1$	$MSB = \frac{SSB}{b-1}$	$F = \frac{MSB}{MSE}$
Interaction between A and B	$SSAB = SST - SSA - SSB - SSE$	$(a-1)(b-1)$	$MSAB = \frac{SSAB}{(a-1)(b-1)}$	$F = \frac{MSAB}{MSE}$
Sampling Error, E	$SSE = \sum_{i=1}^{a}\sum_{j=1}^{b}\sum_{k=1}^{r}(x_{ijk} - \bar{x}_{ij.})^2$	$ab(r-1)$	$MSE = \frac{SSE}{ab(r-1)}$	
Total, T	$SST = \sum_{i=1}^{a}\sum_{j=1}^{b}\sum_{k=1}^{r}(x_{ijk} - \bar{\bar{x}})^2$	$abr - 1$		

D. *Critical values and decisions*

1. Main effects, factor *A:*

   Reject H_0: all $\alpha_i = 0$, if $F = \frac{MSA}{MSE}$ is $> F[\alpha, (a-1), ab(r-1)]$.

2. Main effects, factor *B:*

   Reject H_0: all $\beta_j = 0$, if $F = \frac{MSB}{MSE}$ is $> F[\alpha, (b-1), ab(r-1)]$.

3. Interaction effects:

   Reject H_0: all $(\alpha\beta)_{ij} = 0$, if $F = \frac{MSAB}{MSE}$ is $> F[\alpha, (a-1)(b-1), ab(r-1)]$.

Part C: The Calculations for the Two-Way ANOVA Design

Part C of Table 12.9 describes the specific computations, with each quantity being associated with a specific source of variation within the sample data. They correspond to the $x_{ijk} = \mu + \alpha_i + \beta_j + (\alpha\beta)_{ij} + \epsilon_{ijk}$ model discussed previously.

The Sum of Squares Terms: Quantifying the Sources of Variation

- **Variation due to the factors** *SSA* is the sum of squares reflecting variation caused by the levels of factor *A*. *SSB* is the sum of squares reflecting variation caused by the levels of factor *B*.
- **Variation due to interactions between factor levels** *SSAB* is the sum of squares reflecting variation caused by interactions between the levels of factors *A* and *B*. This is most easily calculated by first computing the other sum of squares terms, then $SSAB = SST - SSA - SSB - SSE$.
- **Sampling error, *E*** *SSE* is the sum of squares reflecting variation due to sampling error. In this calculation, each data value is compared to the mean of its own cell.
- **Total variation, *T*** *SST* is the sum of squares reflecting the overall variation in the data, with each observation being compared to the grand mean, then the differences squared and summed.

Making the Amounts from the Sources of Variation Comparable. *MSA*, *MSB*, *MSAB*, and *MSE* are the mean squares for factors *A* and *B*, interactions between *A* and *B*, and error, respectively. As part C of the table shows, each is obtained by dividing the corresponding sum of squares by the number of degrees of freedom (*df*) associated with this sum of squares.

Part D: Test Statistics, Critical Values, and Decision Rules

For each null hypothesis to be tested, a separate test statistic is calculated. The numerator and denominator are separate estimates of the variance that the cell populations are assumed to share. For each null hypothesis, the critical value of *F* will depend on the level of significance that has been selected, and on the number of degrees of freedom associated with the numerator and denominator of the *F* statistic. In testing each H_0, the values of v_1 and v_2 are shown in the table.

If a calculated *F* exceeds $F[\alpha, v_1, v_2]$, the corresponding null hypothesis will be rejected.

EXAMPLE

Two-way ANOVA Procedure

An aircraft firm is considering three different alloys for use in the wing construction of a new airplane. Each alloy can be produced in four different thicknesses (1 = thinnest, 4 = thickest). Two test samples are constructed for each combination of alloy type and thickness, then each of the 24 test samples is subjected to a laboratory device that severely flexes it until failure occurs. For each test sample, the number of flexes before failure is recorded, with the results shown in Table 12.10. At the 0.05 level of significance, examine (1) whether the alloy thickness has an effect on durability, (2) whether the alloy type has an effect on durability, and (3) whether durability is influenced by interactions between alloy thickness and alloy type. The data are in file CX12WING.

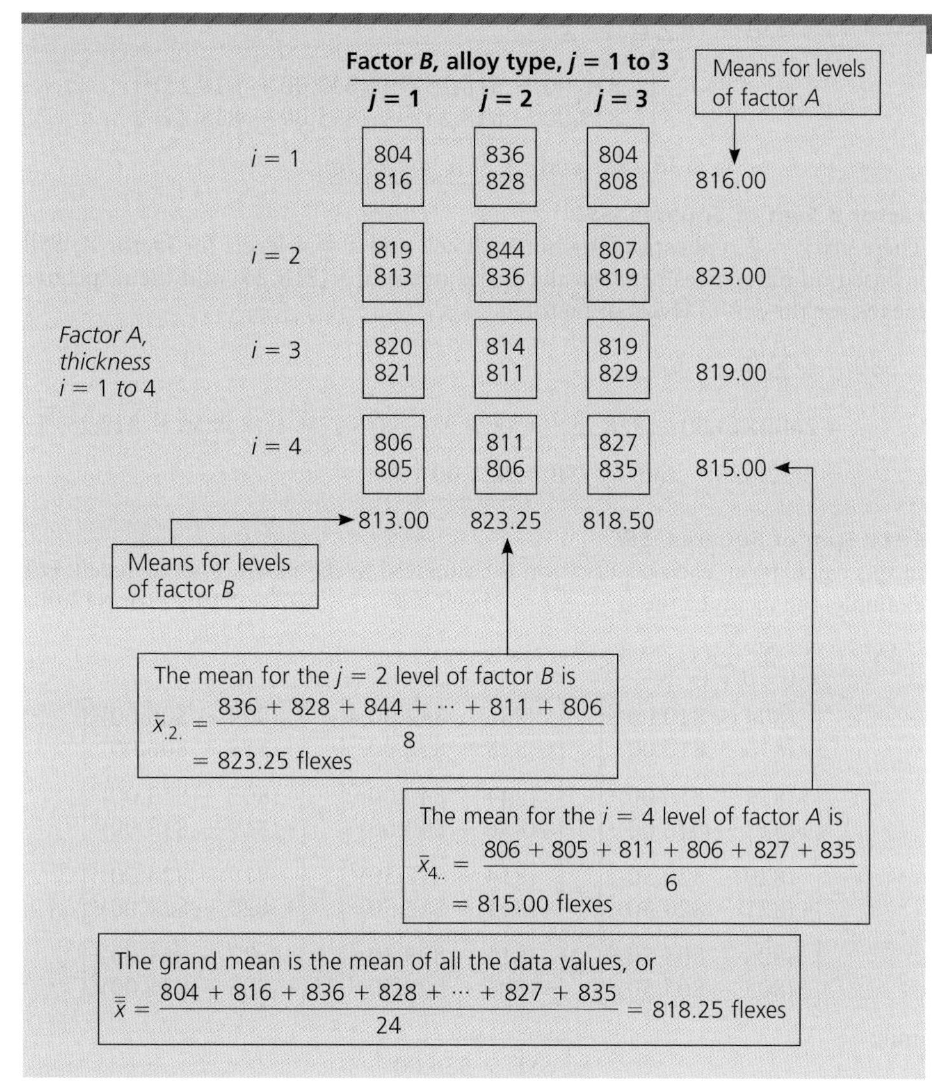

Factor A, thickness, i = 1 to 4	Factor B, alloy type, j = 1	j = 2	j = 3	Means for levels of factor A
i = 1	804, 816	836, 828	804, 808	816.00
i = 2	819, 813	844, 836	807, 819	823.00
i = 3	820, 821	814, 811	819, 829	819.00
i = 4	806, 805	811, 806	827, 835	815.00
Means for levels of factor B	813.00	823.25	818.50	

The mean for the $j = 2$ level of factor B is

$$\bar{x}_{.2.} = \frac{836 + 828 + 844 + \cdots + 811 + 806}{8} = 823.25 \text{ flexes}$$

The mean for the $i = 4$ level of factor A is

$$\bar{x}_{4..} = \frac{806 + 805 + 811 + 806 + 827 + 835}{6} = 815.00 \text{ flexes}$$

The grand mean is the mean of all the data values, or

$$\bar{\bar{x}} = \frac{804 + 816 + 836 + 828 + \cdots + 827 + 835}{24} = 818.25 \text{ flexes}$$

TABLE 12.10

Data and example calculations for a two-way ANOVA design to examine main and interactive effects of factor A (alloy thickness) and factor B (alloy type). Each cell is a combination of factor levels i and j, and contains $r = 2$ observations or replications.

SOLUTION

The null and alternative hypotheses for this analysis are as shown in part A of Table 12.9. In terms of these data, the three null hypotheses hold that (1) no level of factor A (thickness) has an effect, (2) no level of factor B (alloy type) has an effect, and (3) there is no interaction between levels of factor A and levels of factor B.

In Table 12.10, there are 4 levels of factor A and 3 levels of factor B, leading to $4 \times 3 = 12$ combinations, or cells. Within each cell, there are $r = 2$ observations, or replications. For example, the cell for ($i = 2, j = 3$) contains the following observations: $x_{231} = 807$ flexes and $x_{232} = 819$ flexes. Also shown in Table 12.10 are examples of the calculation of the means for the levels of both factors. The dot in the subscripts has the same meaning as in previous analyses—e.g., $\bar{x}_{.2.}$ indicates that this mean is for all observations that received the $j = 2$ level of factor B. The calculations that follow correspond to part C of Table 12.9.

Factor *A* Sum of Squares, *SSA*

There are $r = 2$ replications within each cell and $b = 3$ levels for factor B. SSA is based on differences between the grand mean ($\bar{\bar{x}} = 818.25$) and the respective means for the $a = 4$ levels of factor A:

$$SSA = rb\sum_{i=1}^{a}(\bar{x}_{i..} - \bar{\bar{x}})^2$$
$$= 2(3)\begin{bmatrix}(816.00 - 818.25)^2 + (823.00 - 818.25)^2 \\ +(819.00 - 818.25)^2 + (815.00 - 818.25)^2\end{bmatrix}$$
$$= 6(38.75) \quad \text{and} \quad \boxed{SSA = 232.50}$$

Factor *B* Sum of Squares, *SSB*

There are $r = 2$ replications within each cell and $a = 4$ levels for factor A. SSB is based on differences between the grand mean ($\bar{\bar{x}} = 818.25$) and the respective means for the $b = 3$ levels of factor B.

$$SSB = ra\sum_{j=1}^{b}(\bar{x}_{.j.} - \bar{\bar{x}})^2$$
$$= 2(4)[(813.00 - 818.25)^2 + (823.25 - 818.25)^2 + (818.50 - 818.25)^2]$$
$$= 8(52.625) \quad \text{and} \quad \boxed{SSB = 421.00}$$

Error Sum of Squares, *SSE*

In this calculation, each observation is compared to the mean of its own cell. For example, the mean of the ($i = 2$, $j = 3$) cell is $\bar{x}_{23.} = (807 + 819)/2$, or 813.00.

$$SSE = \sum_{i=1}^{a}\sum_{j=1}^{b}\sum_{k=1}^{r}(x_{ijk} - \bar{x}_{ij.})^2$$
$$= \begin{bmatrix}(804 - 810.00)^2 \\ +(816 - 810.00)^2\end{bmatrix} + \begin{bmatrix}(836 - 832.00)^2 \\ +(828 - 832.00)^2\end{bmatrix} + \begin{bmatrix}(804 - 806.00)^2 \\ +(808 - 806.00)^2\end{bmatrix}$$
$$+ \begin{bmatrix}(819 - 816.00)^2 \\ +(813 - 816.00)^2\end{bmatrix} + \begin{bmatrix}(844 - 840.00)^2 \\ +(836 - 840.00)^2\end{bmatrix} + \begin{bmatrix}(807 - 813.00)^2 \\ +(819 - 813.00)^2\end{bmatrix}$$
$$+ \begin{bmatrix}(820 - 820.50)^2 \\ +(821 - 820.50)^2\end{bmatrix} + \begin{bmatrix}(814 - 812.50)^2 \\ +(811 - 812.50)^2\end{bmatrix} + \begin{bmatrix}(819 - 824.00)^2 \\ +(829 - 824.00)^2\end{bmatrix}$$
$$+ \begin{bmatrix}(806 - 805.50)^2 \\ +(805 - 805.50)^2\end{bmatrix} + \begin{bmatrix}(811 - 808.50)^2 \\ +(806 - 808.50)^2\end{bmatrix} + \begin{bmatrix}(827 - 831.00)^2 \\ +(835 - 831.00)^2\end{bmatrix}$$

and

$$\boxed{SSE = 334.00}$$

Total Sum of Squares, *SST*

This calculation compares each observation with the grand mean ($\bar{\bar{x}} = 818.25$), with the differences squared and summed:

$$SST = \sum_{i=1}^{a}\sum_{j=1}^{b}\sum_{k=1}^{r}(x_{ijk} - \bar{\bar{x}})^2$$
$$= \begin{bmatrix}(804 - 818.25)^2 \\ +(816 - 818.25)^2\end{bmatrix} + \begin{bmatrix}(836 - 818.25)^2 \\ +(828 - 818.25)^2\end{bmatrix} + \begin{bmatrix}(804 - 818.25)^2 \\ +(808 - 818.25)^2\end{bmatrix}$$
$$+ \begin{bmatrix}(819 - 818.25)^2 \\ +(813 - 818.25)^2\end{bmatrix} + \begin{bmatrix}(844 - 818.25)^2 \\ +(836 - 818.25)^2\end{bmatrix} + \begin{bmatrix}(807 - 818.25)^2 \\ +(819 - 818.25)^2\end{bmatrix}$$
$$+ \begin{bmatrix}(820 - 818.25)^2 \\ +(821 - 818.25)^2\end{bmatrix} + \begin{bmatrix}(814 - 818.25)^2 \\ +(811 - 818.25)^2\end{bmatrix} + \begin{bmatrix}(819 - 818.25)^2 \\ +(829 - 818.25)^2\end{bmatrix}$$
$$+ \begin{bmatrix}(806 - 818.25)^2 \\ +(805 - 818.25)^2\end{bmatrix} + \begin{bmatrix}(811 - 818.25)^2 \\ +(806 - 818.25)^2\end{bmatrix} + \begin{bmatrix}(827 - 818.25)^2 \\ +(835 - 818.25)^2\end{bmatrix}$$

and

$$\boxed{SST = 3142.50}$$

Interaction Sum of Squares, *SSAB*

Having calculated the other sum of squares terms, we can obtain *SSAB* by subtracting the other terms from *SST*, or

$$SSAB = SST - SSA - SSB - SSE$$
$$= 3142.50 - 232.50 - 421.00 - 334.00 \quad \text{and} \quad \boxed{SSAB = 2155.00}$$

The Mean Square Terms

As part C of Table 12.9 shows, each sum of squares term is divided by the number of degrees of freedom with which it is associated. There are $a = 4$ levels for factor A, $b = 3$ levels for factor B, and $r = 2$ replications per cell, and the mean square terms are as follows:

- Factor A:

$$MSA = \frac{SSA}{a-1} = \frac{232.50}{4-1} \quad \text{and} \quad \boxed{MSA = 77.50}$$

- Factor B:

$$MSB = \frac{SSB}{b-1} = \frac{421.00}{3-1} \quad \text{and} \quad \boxed{MSB = 210.50}$$

- Interaction, AB:

$$MSAB = \frac{SSAB}{(a-1)(b-1)} = \frac{2155.00}{(4-1)(3-1)} \quad \text{and} \quad \boxed{MSAB = 359.17}$$

- Error, E:

$$MSE = \frac{SSE}{ab(r-1)} = \frac{334.00}{4(3)(2-1)} \quad \text{and} \quad \boxed{MSE = 27.83}$$

The Test Statistics, the Critical Values, and the Decisions

There are three sets of null and alternative hypotheses to be evaluated. In each case, the calculated F is compared to the critical F, listed in the F distribution table and described in part D of Table 12.9; then a decision is reached. The denominator of the F-ratio for each test is MSE, which has $ab(r-1)$, or $4(3)(2-1) =$ 12 degrees of freedom.

1. **Testing for main effects, factor A** The calculated F is

$$\frac{MSA}{MSE} = \frac{77.50}{27.83} = 2.78$$

The critical F for the 0.05 level is

$$F[0.05, (a-1), ab(r-1)]$$

or

$$F(0.05, 3, 12) = 3.49$$

Because the calculated value of F (2.78) does not exceed the critical value (3.49), H_0: all $\alpha_i = 0$ cannot be rejected. Our conclusion is that none of the thicknesses has any effect on the number of flexes a test unit will withstand before it fails.

2. **Testing for main effects, factor *B*** The calculated F is

$$\frac{MSB}{MSE} = \frac{210.50}{27.83} = 7.56$$

The critical F for the 0.05 level is

$$F[0.05, (b - 1), ab(r - 1)]$$

or

$$F(0.05, 2, 12) = 3.89$$

In this case, the calculated F (7.56) is greater than the critical value (3.89), and H_0: all $\beta_j = 0$ is rejected. At least one of the alloys has an effect on the number of flexes a test unit will withstand before it fails. The calculated F also exceeds the critical values for both the 0.025 level (5.10) and 0.01 level (6.93), so the null hypothesis would be rejected at these levels as well. From the F table listings, the most accurate statement we can make about the p-value for this test is that the p-value is less than 0.01.

3. **Testing for interaction effects between levels of factors *A* and *B*** The calculated F is

$$\frac{MSAB}{MSE} = \frac{359.17}{27.83} = 12.91$$

The critical F for the 0.05 level is

$$F[0.05, (a - 1)(b - 1), ab(r - 1)]$$

or

$$F(0.05, 6, 12) = 3.00$$

In the test for interaction effects, the calculated F (12.91) exceeds the critical value (3.00) and H_0: all $(\alpha\beta)_{ij} = 0$ is rejected. The factors are not operating independently, and there is some relationship between the levels of thickness (factor A) and alloy type (factor B) in determining how many flexes a test unit will withstand before failure. The calculated F also exceeds the critical values for both the 0.025 level (3.73) and the 0.01 level (4.82), so the null hypothesis would be rejected at these levels as well. As with the test for the main effects of factor B, the most accurate statement we can make based on our tables is that the p-value for this test is less than 0.01.

The summary findings for the preceding analysis are shown in Table 12.11, a standard format for displaying the results of a two-way ANOVA. As suggested from the listings in the table, it is evident that a great deal of the variability has come from interactions between the levels of factor A (thickness) and factor B (alloy type).

TABLE 12.11
Summary results for the two-way ANOVA performed on the data shown in Table 12.10.

Variation Source	Sum of Squares	Degrees of Freedom	Mean Square	*F*
Factor *A*	232.50	3	77.50	2.78
Factor *B*	421.00	2	210.50	7.56
Interaction, *AB*	2155.00	6	359.17	12.91
Error	334.00	12	27.83	
Total	3142.50	23		

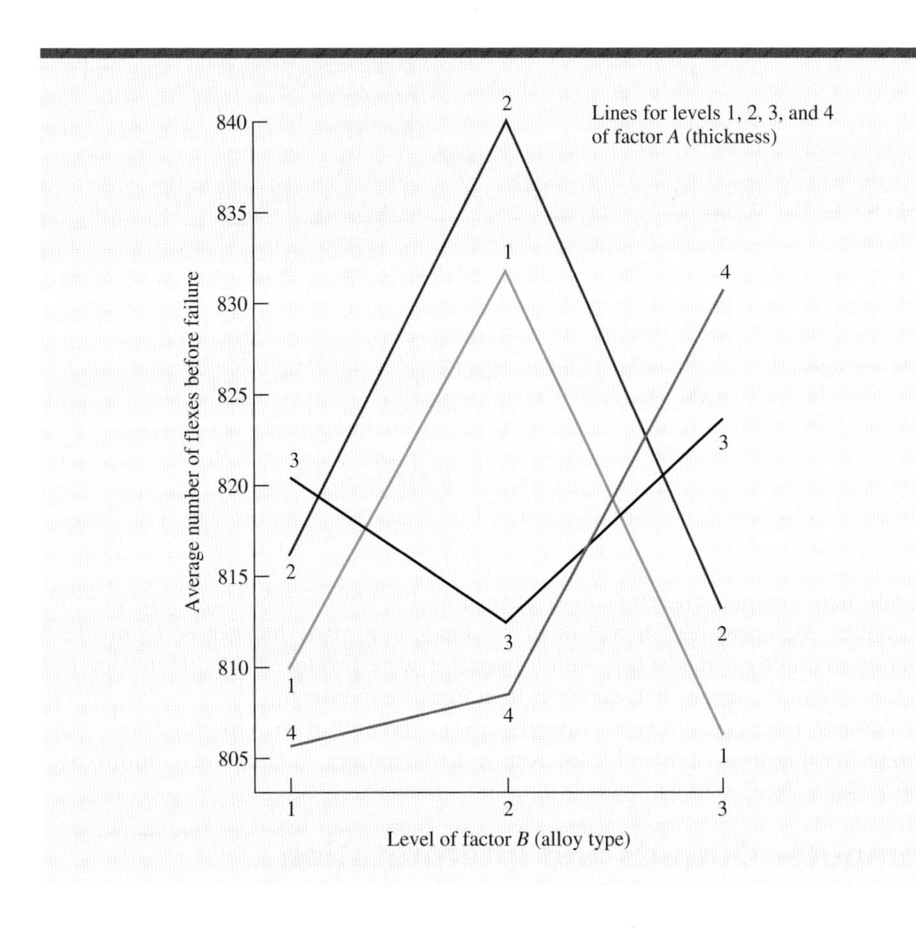

FIGURE 12.3
Interaction between levels of factor *A* (thickness) and factor *B* (alloy type) is present. Thicknesses 1 and 2 seem best for alloy 2, but thickness 4 seems best for alloy 3.

To obtain further insight into the nature of the interactions between the factor levels, let's first consider the plots shown in Figure 12.3. In this graph, the vertical axis represents the average number of flexes for the $r = 2$ test units within each cell. The horizontal axis represents the levels for factor *B*, alloy type. Thicknesses 1 and 2 seem relatively durable for alloy 2, but these thicknesses lead to early failure for alloys 1 and 3. According to this figure, the longest-lasting combination is thickness 2 and alloy 2.

Figure 12.4 is similar to Figure 12.3, except that we now have the levels of factor *A* (thickness) as the horizontal axis and three lines that represent levels 1, 2, and 3 of factor *B* (alloy type). Keeping in mind that the test unit becomes thicker from thickness = 1 to thickness = 4, some interesting observations can be made. Referring to the lines in the plot, alloy 2 lasts longest when manufactured at thickness 2 but fails more quickly when it is at thickness levels 3 and 4. Similarly, alloy 1 turns in

FIGURE 12.4

Another perspective regarding interaction between levels of factor A (thickness) and factor B (alloy type). The thicker the test unit, the longer alloy 3 seems to last, but the reverse seems true for alloy 2.

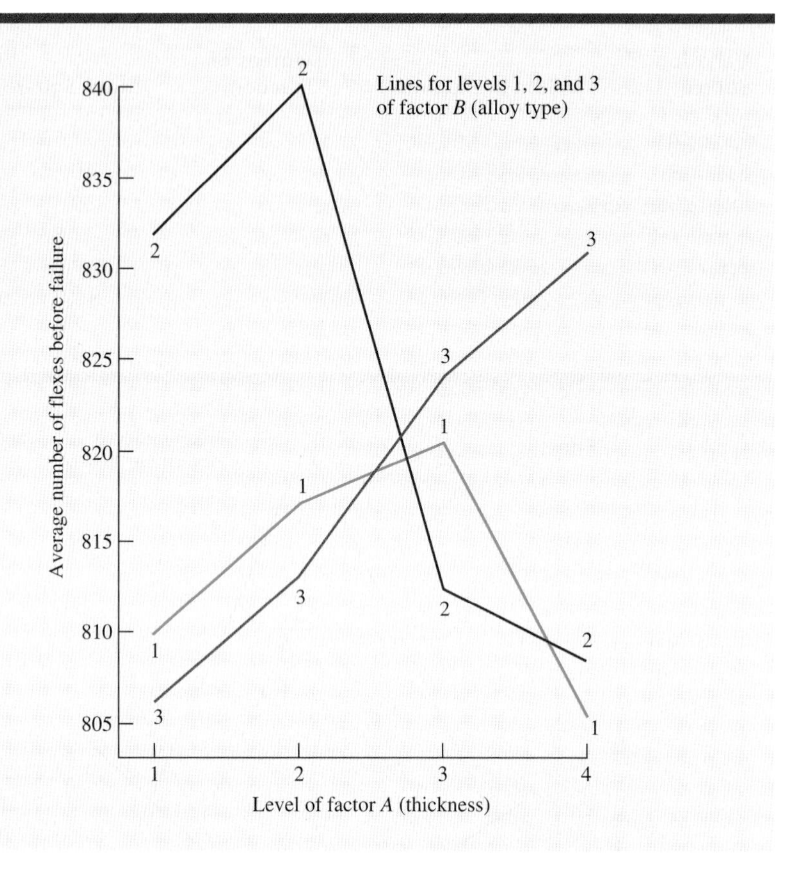

the best performance when it is made at thickness 3. On the other hand, the durability of alloy 3 progressively increases as it is made in greater thicknesses. Figure 12.4 suggests that alloy 3 might last even longer if it were produced at a greater level of thickness than the levels that were included in this experiment.

You can use Seeing Statistics Applet 16, at the end of the chapter, to see how graphs like Figures 12.3 and 12.4 respond to changes in the row, column, and interaction effects.

Computer Outputs and Interpretation

The Excel portion of Computer Solutions 12.3 shows the Excel data configuration and partial output for the two-way ANOVA we've just carried out. Excel provides the count, sum, mean, and variance for each factor level and cell, but only the means are shown in our printout.

Note that Excel has identified factor *A* as "Sample" (rows) and factor *B* as "Columns." Although this is consistent with the format of the data in Table 12.10, it's not carved in stone that tabular displays of the data must have the levels of factor *A* represented by rows and the levels of factor *B* represented by columns, so be careful. As long as you use and interpret the subscripts consistently, it doesn't really matter whether the levels of a given factor are associated with the rows or with the columns in a given display of the data.

The Excel printout in Computer Solutions 12.3 also shows exact p-values for the respective hypothesis tests that were carried out earlier. In testing for the main effects of factor A, we were unable to reject H_0: all $\alpha_i = 0$ at the 0.05 level of significance. This is consistent with the p-value of 0.0864 reported by Excel for this test. For the test of the main effects of factor B, H_0: all $\beta_j = 0$ can be rejected at the 0.0075 level. For the test of the interaction effects between levels of factors A and B, H_0: all $(\alpha\beta)_{ij} = 0$ can be rejected at an even more significant level, 0.0001.

The Minitab portion of Computer Solutions 12.3 shows results comparable to the Excel printout. As an option, Minitab can also display the 95% confidence intervals for the population means associated with the factor levels. These intervals are calculated using a procedure similar to that described in Section 12.3.

COMPUTER 12.3 SOLUTIONS

Two-Way Analysis of Variance

EXCEL

	A	B	C	D	E	F	G
1		Alloy 1	Alloy 2	Alloy 3			
2	Thickness1	804	836	804			
3		816	828	808			
4	Thickness 2	819	844	807			
5		813	836	819			
6	Thickness 3	820	814	819			
7		821	811	829			
8	Thickness 4	806	811	827			
9		805	806	835			
10							
11	Anova: Two-Factor With Replication						
12							
13	ANOVA						
14	*Source of Variation*	*SS*	*df*	*MS*	*F*	*P-value*	*F crit*
15	Sample	232.50	3	77.500	2.784	0.0864	3.490
16	Columns	421.00	2	210.500	7.563	0.0075	3.885
17	Interaction	2155.00	6	359.167	12.904	0.0001	2.996
18	Within	334.00	12	27.833			
19							
20	Total	3142.50	23				

Excel two-way ANOVA

1. For the sample data in Table 12.10 (Excel file CX12WING), with the data and labels in columns A through E, as shown above: From the **Data** ribbon, click **Data Analysis.** Click **Anova: Two-Factor With Replication.** Click **OK.**
2. Enter **A1:D9** into the **Input Range** box. Enter **2** into the **Rows per sample** box. Specify the significance level for the test by entering **0.05** into the **Alpha** box. Select **Output Range** and enter **A11** into the box. Click **OK.** The data and a portion of the printout are shown above.

(continued)

MINITAB

Minitab two-way ANOVA

```
Two-way ANOVA: Flexes versus Thicknes, Alloytyp

Source       DF      SS       MS      F      P
Thicknes      3   232.5   77.500   2.78  0.086
Alloytyp      2   421.0  210.500   7.56  0.007
Interaction   6  2155.0  359.167  12.90  0.000
Error        12   334.0   27.833
Total        23  3142.5

S = 5.276   R-Sq = 89.37%   R-Sq(adj) = 79.63%
```

1. For the sample data in Table 12.10 (Minitab file **CX12WING**), with the performances (number of flexes) in C1, the thickness identifications in C2, and the alloy identifications in C3: Click **Stat.** Select **ANOVA.** Click **Two-way.**
2. Enter **C1** into the **Response** box. Enter **C2** into the **Row factor** box. Enter **C3** into the **Column factor** box. To get a display of the 95% confidence intervals for the population means associated with the factor levels (not shown here), click to select the **Display means** boxes. Click **OK.**

Each confidence interval is constructed by employing the t distribution, with s_p (computed as the square root of MSE) used as an estimate of σ. These confidence intervals are calculated as follows:

Two-way ANOVA, confidence interval for the mean, level k of factor M: The confidence interval for μ_{kM} is

$$\bar{x}_{kM} \pm t\left(\frac{s_p}{\sqrt{n_{kM}}}\right)$$

where $s_p = \sqrt{MSE}$, the pooled estimate of σ
n_{kM} = the total number of persons or test units receiving level k of factor M
$\bar{x}_{kM}$ = the mean for all persons or test units receiving level k of factor M
t = from the t distribution, the t value for the desired confidence level, using the same df as associated with MSE

For example, the mean for level 1 of factor A (thickness) was 816.00 flexes, and a total of 6 test units was subjected to this level of factor A. The pooled estimate for the population standard deviation is $s_p = \sqrt{MSE} = \sqrt{27.8}$, or $s_p = 5.27$. Using the t distribution with the $df = 12$ associated with MSE, the 95% confidence interval for μ_{1A}, the population mean associated with level 1 of factor A (thickness) is

$$\bar{x}_{1A} \pm t\left(\frac{s_p}{\sqrt{n_{1A}}}\right)$$

$$= 816.00 \pm t\left(\frac{5.27}{\sqrt{6}}\right)$$ where df = 12 degrees of freedom associated with MSE, and $t = 2.179$ for a 95% interval

$$= 816.00 \pm 2.179(2.15),$$ or from 811.32 to 820.68 flexes

We have 95% confidence that the population mean for level 1 of factor A is somewhere from 811.32 to 820.68 flexes. The other confidence intervals for the population means associated with the factor levels can be similarly calculated.

EXERCISES

Note: The computer is highly recommended when using ANOVA techniques. For exercises involving raw data, data files are shown in parentheses.

12.60 What is the purpose of two-way analysis of variance?

12.61 What assumptions are required in using the two-way ANOVA?

12.62 Why are there more sets of null and alternative hypotheses that can be tested in two-way ANOVA compared to the one-way and randomized block designs?

12.63 How is two-way ANOVA similar to the randomized block design? How does it differ?

12.64 What are *main effects* and *interactive effects* in the two-way ANOVA?

12.65 In the two-way ANOVA, what is meant by the term *replications*?

12.66 In a two-way ANOVA experiment, factor A is operating on 4 levels, factor B is operating on 3 levels, and there are 3 replications per cell. How many treatments are there in this experiment? Explain.

12.67 For the two-way ANOVA model, $x_{ijk} = \mu + \alpha_i + \beta_j + (\alpha\beta)_{ij} + \epsilon_{ijk}$, explain the meaning of each of the six terms in the equation, then use the appropriate terms from the equation in presenting and explaining the three sets of null and alternative hypotheses that can be tested.

12.68 In a two-way ANOVA experiment, factor A is operating on 3 levels, factor B is operating on 2 levels, and there are 2 replications per cell. If $MSA/MSE = 5.35$, $MSB/MSE = 5.72$, and $MSAB/MSE = 6.75$, and using the 0.05 level of significance, what conclusions would be reached regarding the respective null hypotheses for this experiment?

12.69 In a two-way ANOVA experiment, factor A is operating on 4 levels, factor B is operating on 3 levels, and there are 3 replications per cell. If $MSA/MSE = 3.54$, $MSB/MSE = 5.55$, and $MSAB/MSE = 12.40$, and using the 0.025 level of significance, what conclusions would be reached regarding the respective null hypotheses for this experiment?

12.70 For a two-way ANOVA in which factor A operates on 3 levels and factor B operates on 4 levels, there are 2 replications within each cell. Given the following sum of squares terms, construct the appropriate table of ANOVA summary findings and use the 0.05 level in examining the null and alternative hypotheses associated with the experiment.

$SSA = 89.58$ $SSB = 30.17$
$SSAB = 973.08$ $SSE = 29.00$

12.71 Given the following data for a two-way ANOVA, identify the sets of null and alternative hypotheses, then use the 0.05 level in testing each null hypothesis. (Use data file **XR12071**.)

		Factor *B*		
		1	**2**	**3**
Factor A	1	13 10	19 15	19 17
	2	15 15	22 19	16 17

12.72 Given the following data for a two-way ANOVA, identify the sets of null and alternative hypotheses, then use the 0.05 level in testing each null hypothesis. (Use data file **XR12072**.)

		Factor *B*		
		1	**2**	**3**
Factor A	1	152 151	158 154	160 160
	2	158 154	164 158	152 155
	3	160 161	147 150	147 146

12.73 Given the following data for a two-way ANOVA, identify the sets of null and alternative hypotheses, then use the 0.05 level in testing each null hypothesis. (Use data file **XR12073**.)

		Factor *B*			
		1	**2**	**3**	**4**
Factor A	1	58 64 61	63 66 61	64 68 59	66 68 72
	2	59 64 59	56 61 56	58 60 63	60 59 62
	3	64 62 61	57 58 51	55 56 52	61 58 64

12.74 Given the following summary table for a two-way ANOVA, fill in the missing items (indicated by asterisks), identify the sets of null and alternative hypotheses to be examined, then use the 0.05 level in reaching a conclusion regarding each null hypothesis.

Variation Source	Sum of Squares	Degrees of Freedom	Mean Square	*F*
Factor *A*	40.50	***	***	***
Factor *B*	154.08	1	***	***
Interaction, *AB*	70.17	2	***	***
Error	31.50	6	***	
Total	296.25	11		

12.75 A study has been undertaken to examine the effect of background music and assembly method on the productivity of workers on a production line for electronic circuit boards. Each of 8 workers has been randomly assigned to one of 4 cells, as shown here. So that each worker will hear only the intended type of music, headphones are used for the background music in the experiment. Data are the number of circuit boards assembled in 1 hour. Using the 0.05 level of significance in testing for the main and interactive effects, what conclusions should be drawn from this experiment? Determine the 95% confidence interval for the population mean for each factor level. (Use data file **XR12075**.)

		Music: Classical	Music: Rock
Assembly Method	1	29 32	49 45
	2	35 32	41 40

12.76 For Exercise 12.75, and using the levels of the assembly method factor as the horizontal axis and "units produced" as the vertical axis, plot and connect the cell means for the cells associated with the "classical music" level of the background music factor. On the same graph, plot and connect the cell means for the cells associated with the "rock music" level of the background music factor. Do the plots indicate the presence of interaction between the factor levels? Explain.

12.77 Each of 12 undergraduate students has been randomly assigned to one of the 6 cells shown here. The purpose of the study is to test whether factor *A* (if a shopping bag is being carried) and factor *B* (mode of dress) have main and interactive effects regarding the number of seconds it takes to get waited on at a jewelry counter of an upscale department store. Students have entered the store at times that have been randomly scheduled for the day of the experiment. Using the 0.025 level of significance in testing for the main and interactive effects, what conclusions should be drawn from this experiment? Determine the 95% confidence interval for the population mean for each factor level. (Use data file **XR12077**.)

		Mode of Dress: Sloppy	Casual	Dressy
Shopping Bag Carried?	Yes	41 39	24 29	14 19
	No	52 49	16 16	27 21

12.78 For Exercise 12.77, and using the levels of the "shopping bag carried" factor as the horizontal axis and "seconds" as the vertical axis, plot and connect the cell means for the cells associated with the "sloppy" level of the dress factor. On the same graph, plot and connect the cell means for the cells associated with the "casual" level of the dress factor. Do a third plot, connecting the cell means for cells associated with the "dressy" level of the dress factor. Do the plots indicate the presence of interaction between the factor levels?

12.79 An experiment has been conducted to examine main and interactive effects of factor *A* (keyboard configuration, 3 levels) and factor *B* (word processing package, 2 levels). Each cell consists of two replications, representing the number of minutes each of two secretaries randomly assigned to that cell required to type a standard document. Interpret the accompanying computer output for this study, including your conclusion (use the 0.05 level) for each null hypothesis associated with the experiment.

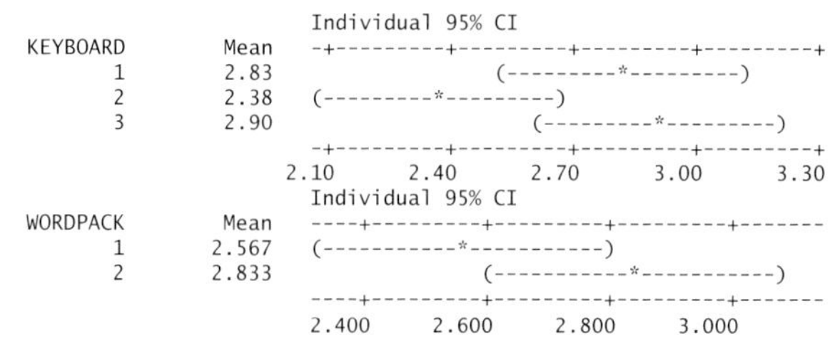

```
Two-Way Analysis of Variance

Analysis of Variance for MINUTES
Source        DF       SS        MS          F        P
KEYBOARD       2   0.6450    0.3225       5.53    0.043
WORDPACK       1   0.2133    0.2133       3.66    0.104
Interaction    2   0.9117    0.4558       7.82    0.021
Error          6   0.3500    0.0583
Total         11   2.1200

                       Individual 95% CI
KEYBOARD      Mean   -+---------+---------+---------+---------+
     1        2.83                 (---------*---------)
     2        2.38    (---------*---------)
     3        2.90                      (---------*---------)
                     -+---------+---------+---------+---------+
                    2.10      2.40      2.70      3.00      3.30
                       Individual 95% CI
WORDPACK      Mean   ----+---------+---------+---------+-------
     1       2.567    (-----------*-----------)
     2       2.833                 (-----------*-----------)
                     ----+---------+---------+---------+-------
                       2.400     2.600     2.800     3.000
```

12.80 Through separate calculations, verify the confidence interval for each of the factor-level means shown in the output for Exercise 12.79.

SUMMARY

12.6

- **ANOVA and experiments**

The chapter goes beyond the two-sample tests of Chapter 11 to introduce analysis of variance (ANOVA), a procedure capable of comparing the means of two or more independent samples at the same time. The starting point for ANOVA is often an experiment in which the goal is to determine whether various levels of an independent variable, or factor, might be exerting different effects on the dependent variable, or measurement. When there is only one factor in an experiment, each factor level can be referred to as a *treatment*.

- **The ANOVA procedure**

Basic to ANOVA is the comparison of variation between samples versus the amount of variation within the samples. The test statistic is an F-ratio in which the numerator reflects variation between the samples and the denominator reflects the variation within them. If the calculated F-statistic exceeds the critical F for a given test, the null hypothesis of equal population means is rejected. The numerator and denominator of the F-statistic are separate estimates of the variance that the populations are assumed to share. Further assumptions are that the samples have been independently selected and that the population distributions are normal.

- **One-way ANOVA**

In the one-way ANOVA, there is just one factor, and the null hypothesis is that the population means are equal for the respective treatments, or factor levels. Treatments are randomly assigned to the persons or test units in the experiment, so this method is also referred to as the completely randomized ANOVA. The null hypothesis can also be expressed in terms of a model where each measurement is assumed to be the sum of three components: an overall mean, the effect of a treatment, and a random error due to sampling. When there are only two treatments, the one-way ANOVA is equivalent to the nondirectional pooled-variances t-test of Chapter 11.

- **Randomized block ANOVA**

In the randomized block design, persons or test units are first arranged into similar groups, or *blocks,* then the treatments are randomly assigned. Blocking helps reduce error variation by ensuring that the treatment groups are comparable in terms of the blocking variable. This design assumes that no interactions exist between the blocks and the treatments. Although the blocking variable can be viewed as representing a second independent variable, it is introduced solely for the purpose of controlling error variation, and the major purpose continues to be the examination of the effects of the factor levels, or treatments. When there are just two treatments, the randomized block design is equivalent to the nondirectional, dependent-samples t-test of Chapter 11.

- **Two-way ANOVA**

In two-way ANOVA, there are two factors, each of which operates on two or more levels. In this design, it is no longer appropriate to refer to the factor levels as treatments, since each combination of their levels constitutes a separate treatment. The two-way ANOVA examines the main effects of the levels of both factors as well as interactive effects associated with the combinations of their levels. Accordingly, significance tests include those for the main effects of each factor and for interactive effects between the combinations of factor levels.

- **Computer-assisted ANOVA**

A computer statistical package is especially useful in ANOVA procedures, since calculations become quite extensive for even small amounts of data. Along with small-scale examples for demonstration purposes, the chapter includes the corresponding Excel and Minitab printouts and interpretation for each analysis.

EQUATIONS

One-Way Analysis of Variance

This ANOVA procedure assumes that (1) the samples have been independently selected, (2) the population variances are equal ($\sigma_1^2 = \sigma_2^2 = \cdots = \sigma_t^2$) for the t treatments, and (3) the population distributions are normal. The null and alternative hypotheses, data format, calculation formulas, and decision rule appear in Table 12.2.

One-Way Analysis of Variance, Confidence Interval for a Treatment Mean

For treatment j, the confidence interval for μ_j is

$$\bar{x}_j \pm t\left(\frac{s_p}{\sqrt{n_j}}\right)$$

where $s_p = \sqrt{MSE}$, the pooled estimate of σ

n_j = number of persons or test units in treatment group j

$\bar{x}_j$ = sample mean for treatment group j

t = from the t distribution, the t value for the desired confidence level, using the same df associated with MSE

Randomized Block Design

The randomized block design introduces a blocking variable to reduce error variation, but the intent of the design is not to measure the effect of this variable. Accordingly, the significance test usually considers only the treatment variable. Assumptions for the randomized block design are that (1) the one observation in each treatment-block combination (cell) has been randomly selected from a normal population, (2) the variances are equal for the values in the respective populations, and (3) there is no interaction between the blocks and the treatments. The null and alternative hypotheses, data format, calculation formulas, and decision rule appear in Table 12.6. The process of blocking persons or test units into homogeneous blocks interferes with the randomness of the treatment groups, and we are not able to calculate confidence intervals for the treatment means. For those who do wish to examine the effectiveness of the blocking variable, the procedure is discussed in the chapter and briefly summarized below:

Hypotheses for testing the effectiveness of the blocking variable:

H_0: The levels of the blocking variable are equal in their effect.
H_1: At least one level has an effect different from the others.

Calculated value of $F = MSB/MSE$, computed as shown in Table 12.6
Critical value of $F = F(\alpha, v_1, v_2)$

where $v_1 = (n - 1)$, df associated with the numerator
$v_2 = (t - 1)(n - 1)$, df associated with the denominator
and n = number of blocks
t = number of treatments

Decision rule: If calculated $F > F(\alpha, v_1, v_2)$, reject H_0 at the α level.

Two-Way Analysis of Variance

In this design, there are two factors, and the significance tests involve both main effects and interactive effects. There are r replications within each of the factor-level combinations, or cells. It is assumed that (1) the r observations within each cell have been drawn from a normal population, and (2) the variances of the respective populations are equal. The null and alternative hypotheses, data format, calculation formulas, and decision rules appear in Table 12.9.

Two-Way Analysis of Variance, Confidence Interval for the Mean of a Factor Level

For level k of factor M, the confidence interval for μ_{kM} is

$$\bar{x}_{kM} \pm t\left(\frac{s_p}{\sqrt{n_{kM}}}\right)$$

where $s_p = \sqrt{MSE}$, the pooled estimate of σ

n_{kM} = total number of persons or test units receiving level k of factor M

$\bar{x}_{kM}$ = mean for all persons or test units receiving level k of factor M

t = from the t distribution, the t value for the desired confidence level, using the same df associated with MSE

CHAPTER EXERCISES

Note: The computer is highly recommended when using ANOVA techniques. For exercises involving raw data, data files are shown in parentheses.

12.81 Compare the respective purposes of one-way, randomized block, and two-way analysis of variance. In general, under what circumstances would each method be used?

12.82 Since the t-tests of Chapter 11 and the ANOVA techniques of this chapter are both involved with the comparison of sample means, why can't the techniques always be used interchangeably?

12.83 The data in an experiment have been examined using ANOVA, and a confidence interval has been constructed for the mean associated with each level for a factor. Which ANOVA technique (one-way, randomized block, or two-way) has *not* been used in this analysis? Explain.

12.84 The personnel director for a large firm selects a random sample consisting of 100 clerical employees, then finds out whether they have been with the firm for more than 5 years and how many shares of the company's stock they own. The purpose is to see if longevity within the firm has an influence over the number of shares of stock that are held; ANOVA is to be used in analyzing the data. Identify the independent and dependent variables and explain whether this is a designed experiment.

12.85 Three different faceplate designs have been selected for the radio intended for a new luxury automobile, and safety engineers would like to examine the extent to which their operation will be relatively intuitive to the motorist. Nine drivers are selected and randomly placed into one of three groups (faceplate 1, faceplate 2, and faceplate 3). Each driver is then asked to insert a music CD and play track number 18, and engineers measure how long it takes for this task to be accomplished. ANOVA is to be used in analyzing the data. Identify the independent and dependent variables, and explain whether this is a designed experiment.

12.86 Given the situation described in Exercise 12.85, which ANOVA procedure from this chapter would be appropriate for the analysis of the resulting data?

12.87 Given the situation described in Exercise 12.85, suppose the groups are contrived so that each group contains one driver who is under 21 years of age, one driver who is between 22 and 60, and one driver who is over 60. If the three groups were constructed in this manner, would the ANOVA procedure associated with Exercise 12.86 still be applicable? Why or why not?

12.88 A firm that specializes in preparing recent law school graduates for the state bar exam has formulated two alternatives to their current preparation course.

To examine the relative effectiveness of the three possible courses, they have set up a study in which they randomly assigned a group of recent graduates so that five of them take each preparation program. For the five who have taken alternative program A, their eventual scores on the bar exam are 82, 75, 57, 74, and 74. For the five who have taken alternative program B, their eventual scores on the bar exam are 73, 63, 57, 59, and 76. For the five who have taken the current program, their eventual scores on the bar exam are 74, 71, 66, 78, and 61. (The data are also in file **XR12088**.) At the 0.025 level of significance, can the firm conclude that the three preparation programs might be equally effective in getting recent graduates ready for the state bar exam?

12.89 An industrial sales manager, testing the effectiveness of three different sales presentations, randomly selected a presentation to be used when making the next sales call on 14 customers. The numbers of units purchased as a result of each sales call are shown here. Use the 0.05 level in determining whether the three sales presentations could be equally effective. (Use data file **XR12089**.)

Presentation		
1	2	3
7	13	14
9	10	15
11	11	12
7	13	10
	14	13

12.90 Four different brands of brake shoes have been installed on 12 city transit buses, with each brand installed on 3 buses selected at random from the 12. The number of thousands of miles before the lining required replacement was as follows for each bus. (Use data file **XR12090**.)

Brand of Brake Lining			
1	2	3	4
23	19	23	17
20	11	25	13
17	16	30	19

a. From the variation between the samples, what is the estimated value of the common variance, σ^2, that is assumed to be shared by the four populations?
b. From the variation within the samples, what is the estimated value of the common variance, σ^2, that is assumed to be shared by the four populations?
c. At the 0.01 level of significance, could the four types of brake linings be equally durable?

12.91 In an attempt to compare the assessments provided by the four assessors it employs, a municipal official sends each assessor to view the same five homes. Their visits to the homes are in a random order, and the assessments they provide are as shown here. Use the 0.025 level in comparing the assessor effects. (Use data file **XR12091**.)

		Assessor Number			
		1	2	3	4
	A	40	48	55	53 thousand dollars
	B	49	46	52	50
Home	*C*	35	47	51	48
Assessed	*D*	60	54	70	72
	E	100	81	109	88

12.92 Given the following data from three independent samples, use the 0.025 level in determining whether the population means could be the same. (Use data file **XR12092**.)

Sample		
1	2	3
6	7	14
9	20	18
18	11	23
12	15	17
13	23	27
10	16	

12.93 A testing agency is evaluating three different brands of bathroom scales and has selected random samples of each brand. For brand A, a test object was found to weigh 204, 202, 197, 204, and 205 pounds on the five scales sampled. For four scales of brand B, the test object weighed 201, 199, 196, and 203 pounds, while for six scales of brand C, the object weighed 195, 197, 192, 196, 198, and 196 pounds. Using the 0.025 level of significance, determine whether the three brands could have the same population mean for this test object. (Use data file **XR12093**.)

12.94 Four different alloy compositions have been used in manufacturing metal rods, with the hardness measurements shown here. Use the 0.05 level of significance in testing whether the population means might be equal, then construct the 95% confidence interval for each population mean. (Use data file **XR12094**.)

Alloy Composition			
1	2	3	4
42	52	45	46
51	54	48	54
46	45	39	63
	50		56
			53

12.95 An investor has consulted four different financial advisors with regard to the expected annual rate of return for each of three portfolio possibilities she is considering. The financial advisors have been chosen because they are known to range from very conservative (advisor A) to very optimistic (advisor D). The advisors' respective estimates for the three portfolios are shown here. Use the 0.05 level in comparing the portfolios. (Use data file **XR12095**.)

		Estimated Annual Rate of Return for Portfolio		
		1	2	3
Advisor	*A*	8%	8	5
	B	12	10	8
	C	8	11	10
	D	15	12	11

12.96 Researchers have obtained and tested samples of four different brands of nylon rope that are advertised as having a breaking strength of 100 pounds. Given the breaking strengths (in pounds) shown here, use the 0.025 level in comparing the brands. (Use data file **XR12096**.)

Brand A	Brand B	Brand C	Brand D
103.1	111.6	109.0	118.0
108.9	117.8	111.8	115.8
106.7	109.8	113.0	114.2
114.3	110.1	109.7	117.3
113.3	118.3	108.6	113.8
110.5	116.7	114.7	110.6

12.97 A state law enforcement agency has come up with three different methods for publicizing burglary-prevention measures during vacation periods. Recognizing that there are more burglaries in larger communities than in smaller communities, three communities have been randomly selected from each size category, then assigned to a treatment group. The accompanying results are the number of home burglaries during the test month for communities in which the respective publicity methods were employed. Use the 0.05 level in comparing the publicity techniques. (Use data file **XR12097**.)

		Publicity Method		
		1	2	3
Community Size	*Small*	14	13	8 burglaries
	Medium	17	15	13
	Large	27	23	17

12.98 Interested in comparing the effectiveness of four different driving strategies, a government agency has equipped a compact car with a fuel-consumption meter that measures every 0.01 gallon of gasoline consumed. Each of five randomly selected drivers applies each strategy in negotiating the same test course. The order in which each driver applies the strategies is randomly determined, and the fuel-consumption data are shown here. Use the 0.05 level in comparing the driving strategies. (Use data file **XR12098**.)

		Driving Strategy			
		1	2	3	4
Driver	*A*	21	34	25	38
	B	23	29	20	32
	C	28	33	26	37
	D	25	28	18	24
	E	19	26	23	16

hundredths of a gallon

12.99 A magazine publisher is studying the influence of type style and darkness on the readability of her publication. Each of 12 persons has been randomly assigned to one of the cells in the experiment, and the data are the number of seconds each person requires to read a brief test item. For these data, use the 0.05 level of significance in drawing conclusions about the main and interactive effects in the experiment. Determine the 95% confidence interval for the population mean for each factor level. (Use data file **XR12099**.)

		Type Darkness		
		Light	Medium	Dark
Type Style	1	29 32	23 28	26 30
	2	29 31	26 23	23 24

12.100 Three different heat-treatment methods are being tested against four different zinc-coating techniques, with data representing the number of pounds required for a test probe to penetrate the layer of zinc coating. Metal samples have been randomly assigned to the factor levels, with two replications in each cell. For the data shown here, use the 0.01 level of significance in drawing conclusions from this experiment. Determine the 95% confidence interval for the population mean for each factor level. (Use data file **XR12100**.)

		Zinc-Coating Technique			
		1	2	3	4
Heat Treatment Method	1	663 695	736 740	736 675	747 736
	2	726 759	778 759	690 650	697 710
	3	774 745	671 690	756 774	774 813

12.101 Three different point-of-purchase displays are being considered for lottery ticket sales at a chain of convenience stores. Two different locations within the store are also being considered. Eighteen stores for which lottery ticket sales have been comparable are randomly assigned to the six cells shown here, with three stores receiving each treatment. Data are the number of lottery tickets sold during the day of the experiment. For these data, use the 0.025 level of significance in drawing conclusions about the main and interactive effects in the experiment. Determine the 95% confidence interval for the population mean for each factor level. (Use data file **XR12101**.)

		Type of Display		
Position		1	2	3
Next to Cash Register	1	43, 39, 40	39, 38, 43	57, 60, 49
Mounted on Entrance Door	2	53, 46, 51	58, 55, 50	47, 42, 46

INTEGRATED CASES

Thorndike Sports Equipment (Video Unit Six)

Ted and Luke Thorndike have been approached by several local racquetball players, each of whom would like to be sponsored by Thorndike Sports Equipment during regional tournaments to be held during the next calendar year. In the past, Luke has resisted the sponsorship possibility. His philosophy has been, "Our racquets are the best in the business. If they're smart, they'll use a Thorndike racquet; and if we're smart, we won't give it to them for free."

It wasn't easy, but Ted finally convinced Luke of the benefits to be gained by exposing racquet club members and tournament spectators to high-caliber players who rely on Thorndike racquets and related sports equipment. According to Ted, this exposure is easily worth a few racquets, a couple of pairs of court shoes, and a half-dozen "Thorndike Power" T-shirts.

The Thorndikes compromise between Luke's original idea of sponsoring nobody and Ted's proposal to sponsor at least three players. Their eventual selection is Kermit Clawson, an outstanding young player who finished third in the regional championship tournament last year. Kermit and the Thorndikes have reached agreement on all of the details of the sponsorship arrangement but must decide on which combination of racquet and string would be best.

Kermit is a power player who relies heavily on the speed of his serves and return shots. The Thorndike line currently includes four "power" racquets, each of which can be strung with one of three types of string most often used by strong hitters like Kermit. Ted sets up a test in which Kermit will hit two "power serves" with each combination of racquet and string. The 24 serves will be made in a random order, and each will be clocked with a radar gun. Besides helping Kermit select the combination he likes best, the test will also help the Thorndikes study what Ted has referred to as the "main" and "interactive" effects of the racquets and strings. The data from the test are shown here and are provided in file **THORN12**.

From these data, which combination of racquet and string would you recommend that Kermit use? Also, employ the appropriate ANOVA procedure in helping Ted respond to Luke's statement that "the racquets are all the same, the strings are all the same, and it doesn't matter which string goes into which racquet."

		Racquet			
		1	2	3	4
	1	105, 108	113, 109	114, 112	109, 108 mph
String	*2*	110, 108	109, 107	118, 114	113, 113
	3	113, 112	114, 109	114, 114	110, 107

Springdale Shopping Survey

Item C of the Springdale Shopping Survey, introduced at the end of Chapter 2, describes variables 7–9 for the survey. These variables represent the general attitude respondents have toward each of the three shopping areas, and they range from 5 (like very much) to 1 (dislike very much). The data are in file **SHOPPING**.

Variable 7 = Attitude toward Springdale Mall
Variable 8 = Attitude toward Downtown
Variable 9 = Attitude toward West Mall

The information gained from this database exercise would be useful in determining whether the population mean attitudes toward the three shopping areas could be the same for men and women, for persons from various education levels, or for persons who differ on the basis of other criteria. Such knowledge could help managers to better understand and serve their customers.

Part I: Attitude Comparisons Based on the Gender of the Respondent

1. For variable 7 (attitude toward Springdale Mall), compare the mean scores for respondents according to variable 26 (gender of respondent). At the 0.05 level of significance, could the population means of these groups be the same? What is the p-value for the test?
2. Repeat step 1 for variable 8 (attitude toward Downtown).
3. Repeat step 1 for variable 9 (attitude toward West Mall).

Part II: Attitude Comparisons Based on Education Level

1. For variable 7 (attitude toward Springdale Mall), compare the mean scores for respondents according to variable 27 (education category). At the 0.05 level of significance, could the population means of these groups be the same? What is the p-value for the test?
2. Repeat step 1 for variable 8 (attitude toward Downtown).
3. Repeat step 1 for variable 9 (attitude toward West Mall).

Part III: Some Further Comparisons

Given the number of classificatory variables in this database, a great many other comparisons can be made among sample means. For example, variables 10–17 indicate which shopping area the respondent feels is best in terms of possessing a specific, desirable attribute. As just one example, compare the means for variable 7 (attitude toward Springdale Mall) for groups classified according to variable 11 (shopping area having the highest-quality goods). Depending on your time, curiosity, and computer access, you may wish to try some of the many other classificatory variables in selecting groups and comparing attitude scores.

BUSINESS CASE

Fastest Courier in the West

Ryan McVay/Photodisc/Getty Images

The law firm of Adams, Babcock, and Connors is located in the Dallas–Fort Worth metroplex. Randall Adams is the senior and founding partner in the firm. John Babcock has been a partner in the firm for the past eight years, and Blake Connors became a partner just last year. The firm employs two paralegal assistants and three secretaries. In addition, Bill Davis, the newly hired office manager, is in charge of day-to-day operations and manages the financial affairs of the law firm.

A major aspect of the law firm's business is the preparation of contracts and other legal documents for its clients. A courier service is employed by the firm to deliver legal documents to its many clients as they are scattered throughout the metroplex. The downtown centers of Dallas and Fort Worth are separated by a distance of approximately 30 miles. With the large sizes of these cities and their associated heavy traffic, a trip by car from the southwest side of Fort Worth to the northeast side of Dallas can easily take longer than an hour. Due to the importance of the legal documents involved, their timely delivery is a high priority. At a recent partners' meeting, the topic of courier delivery came up.

Adams: "Recently, we have received a couple of complaints from some of our best clients about delayed contract deliveries. I spent the better part of an hour yesterday afternoon trying to calm down old man Dixon. He claims that if those contracts had arrived any later, his deal with the Taguchi Group would have fallen through."

Connors: "Well, it wasn't our fault. Anne had the contracts all typed and proofread before nine in the morning."

Adams: "No, no. Everything was handled fine on our end. The delay was the courier's fault. Something about a delay in the delivery. . . ."

Babcock: "Metro Delivery has always done a good job for us in the past. I am sure that these are just a few unusual incidents."

(continued)

SEEING STATISTICS: APPLET 15

F Distribution and ANOVA

When this applet loads, it shows an *F* distribution diagram similar to the one in Figure 12.1. The components include the degrees of freedom for the numerator ($v_1 = 2$ is shown as "df1 = 2"), the degrees of freedom for the denominator ($v_2 = 7$ is shown as "df2 = 7"), and the critical value of *F* (6.54) associated with the specified level of significance for the test ("prob = 0.025").

By moving the left and right sliders, we can change the number of degrees of freedom associated with the numerator and denominator of the *F*-ratio, and the shape of the curve changes accordingly. Likewise, we can use the text boxes to enter a desired value for either the *F*-ratio or the probability.

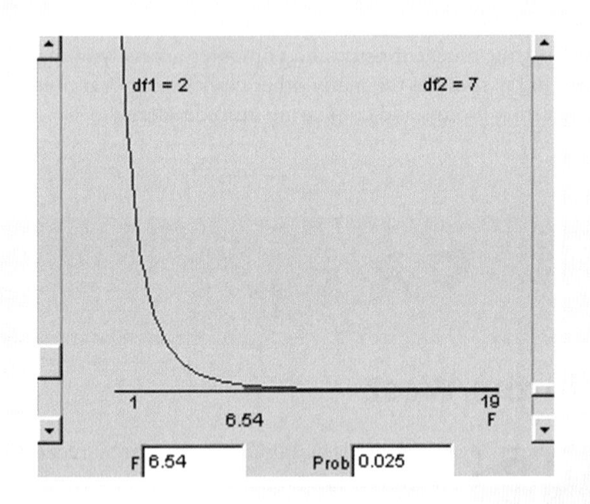

Note that the *F*-ratio and probability values in the text boxes will sometimes differ slightly from those we would look up in our *F* distribution tables in the back of the book. Besides being more accurate because they do not rely on printed tables, these text boxes can also provide a wider variety of values for *F*-ratios and their related upper-tail probabilities than would be available from our printed tables.

Applet Exercises

15.1 Using the left slider, increase the number of degrees of freedom for the numerator of the *F*-ratio to df1 = 5, then to df1 = 10. What happens to the shape of the *F* distribution curve?

15.2 Use the left and right sliders to set the degrees of freedom back to df1 = 2 and df2 = 7. Next, use the right slider to set the number of degrees of freedom for the denominator to df2 = 10, then to df2 = 15. What happens to the shape of the *F* distribution curve?

15.3 Use the left and right sliders to set the degrees of freedom back to df1 = 2 and df2 = 7. Next, use the left text box to increase the *F* value to 9.55. (Be sure to press the **Return** or **Enter** key after changing the text box entry.) What effect does this have on the probability?

15.4 Use the left and right sliders to set the degrees of freedom back to df1 = 2 and df2 = 7, and use the left text box to return the *F* value to 6.54. (Be sure to press the **Return** or **Enter** key after changing the text box entry.) Next, use the right text box to change the probability to 0.01. What effect does this have on the *F* value?

Try out the Applets: http://www.cengage.com/international

Connors: "On the other hand, it could be an indication that their service is slipping."

Adams: "In any event, we cannot afford to offend our clients. No one is perfect, but it only takes one or two bad incidents to lose important clients. At least two new courier services have opened in the metroplex during the last two years. I hear good things about them from some of my friends. The question is, Should we keep using Metro or consider using one of these other services?"

Connors: "How would you suggest that we make this decision?"

Babcock: "Why not give each one a trial period and choose the best performer?"

Adams: "Great idea! But how would you decide who's best?"

Babcock: "Well, obviously the choice boils down to picking the fastest courier service. Given our recent problem, we also want to avoid the infrequent, but costly, delayed deliveries. Delivery cost is an important secondary criterion."

Connors: "Why not let our new office manager run this little 'contest' for a few weeks? Bill normally keeps fairly detailed information about contract deliveries anyway. As the need for deliveries arises, he can rotate through all three couriers."

Adams: "Let's be sure not to let any of the couriers know about the contest; otherwise, we may not see their

SEEING STATISTICS: APPLET 16

Interaction Graph in Two-Way ANOVA

This applet shows interaction graphs similar to those in Figures 12.3 and 12.4, but it gives us the opportunity to see how the interaction graphs change in response to changes we make in the row, column, and interaction effects. The top slider changes the difference between the row means ("Row"), the middle slider changes the difference between the column means ("Col"), and the bottom slider changes the interaction effects ("RxC").

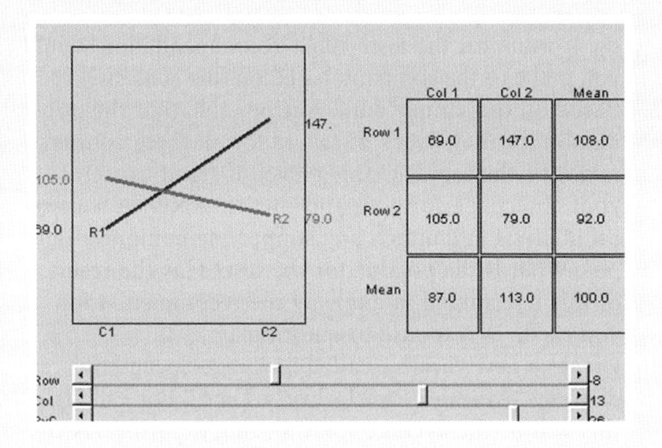

Applet Exercises

16.1 If they are not already in this position, center all three sliders so that the value at the far right of each slider scale is a zero. Next, slide the top slider to the right and set it at +10 to increase the difference between the row means. Now slide it further to the right, to +20. Describe the nature of the graph and what happens when the difference between the row means is increased.

16.2 Center all three sliders so that the value at the far right of each slider scale is a zero. Next, slide the middle slider to the right and set it at +10 to increase the difference between the column means. Now slide it further to the right, to +20. Describe the nature of the graph and what happens when the difference between the column means is increased.

16.3 Center all three sliders so that the value at the far right of each slider scale is a zero. Next, set each of the top two sliders to −20 and describe the appearance of the graph. Now set each of the top two sliders to +20. In what way(s) has the graph changed, and in what way(s) has it remained the same?

16.4 With each of the top two sliders set at +20, move the bottom slider across its range of movement, from far left to far right. Describe how the graph responds to this movement.

16.5 Set the sliders so that the following conditions will exist simultaneously: (a) Row 1 will have a greater effect (higher mean) than Row 2; (b) Column 2 will have a greater effect than Column 1; and (c) the strongest positive interaction effect will be from the combination of Row 1 and Column 2. (*Note:* There are a great many possibilities, but just select three slider positions of your choice that will cause all three of these conditions to be present at the same time.)

Try out the Applets: http://www.cengage.com/international

typical performance. We'll take up this topic again after Bill has collected and analyzed some data and is ready to make a presentation."

During the past month, Bill Davis has kept detailed records of the deliveries made by each of three courier services: DFW Express, Carborne Carrier, and Metro Delivery (the courier presently used by the law firm). Due to the importance of the documents delivered, a courier is required to phone the law office as soon as a delivery has been made at its destination. For each delivery, Bill's data set contains the courier used, the pickup time, the delivery time, the mileage of the delivery, and the cost of the delivery. Each of the courier services charges a flat fee plus a mileage charge for each delivery. These charges vary from courier to courier.

Assignment

As office manager, Bill Davis is responsible for making the decision as to which courier service will be given the exclusive contract. Using the data set stored in the file **COURIER**, assist Bill in choosing among the three courier services and defending his choice to the partners of the firm by performing a statistical analysis of the data. Write a short report to the partners giving your final recommendation. The case questions will assist you in your analysis of the data. Use important details from your analysis of the data to support your recommendation.

(*continued*)

Data Description

For each delivery, Bill Davis' data set contains the courier used, the pickup time, the delivery time, the mileage of the delivery, and the cost of the delivery. The first few entries in the database are shown here. The entire database is in the file **COURIER**, and it contains information on 182 courier deliveries.

Courier	Pickup Time	Delivery Time	Mileage	Cost
1	13	14	7	16.55
3	20	51	20	26.50
2	22	33	12	19.00
3	11	47	19	25.60
3	17	18	8	15.70
⋮	⋮	⋮	⋮	⋮

The variables are defined as follows:

Courier:	1 = DFW Express 2 = Carborne Carrier 3 = Metro Delivery
Pickup Time:	Time in minutes from when the order is phoned in until a courier agent arrives
Delivery Time:	Time in minutes that it takes for the documents to be delivered to the destination from the firm
Mileage:	Distance in miles from the law firm to the destination
Cost:	Charge for the delivery. Each of the courier services charges a flat fee plus a mileage charge. These charges vary from courier to courier.

1. Based on the sample data, is it possible that the population mean delivery *times* for all three couriers might be the same? Using a level of significance of your choice, perform an appropriate ANOVA statistical analysis and attach any supporting computer output. What is the *p*-value for the test? What recommendation would you make to the company based on this analysis?
2. Based on the sample data, is it possible that the population mean delivery *distances* for all three couriers might be the same? Using a level of significance of your choice, perform an appropriate ANOVA statistical analysis and attach any supporting computer output. What is the *p*-value for the test? Has the result of this test altered in any way the recommendation you made in response to question 1?
3. Create a new variable called "Average Speed" by dividing "Mileage" by "Delivery Time." For each trip, this will be the average speed during the trip. Based on the sample data, is it possible that the population mean *average speeds* for all three couriers might be the same? Using a level of significance of your choice, perform an appropriate ANOVA statistical analysis and attach any supporting computer output. What is the *p*-value for the test? On the basis of this analysis, what recommendation would you make to the firm?
4. Overall, considering the analyses performed above, do you feel that the evidence is sufficiently conclusive that the firm should switch from their present courier? Present a brief written recommendation to the firm.

CHAPTER 13

Chi-Square Applications

PROPORTIONS TESTING AND THE RESTROOM POLICE

On the basis of observations in selected restrooms across the nation, researchers have discovered that many Americans don't wash their hands after using the facilities. Among their findings, obtained by peering from stalls or pretending to be combing their hair, are that 75% of 1576 baseball fans at an Atlanta Braves home game, 81% of 789 visitors to Chicago's Museum of Science and Industry, and 72% of 515 persons observed at New York's Penn Station washed their hands.

Could these three percentages—75%, 81%, and 72%—differ from each other just by chance (sampling error), or can we conclude that the populations involved really don't exercise the same level of restroom fastidiousness? Actually, there's only a 0.0003 probability (the *p*-value) that such great differences among the sample percentages could have occurred simply by chance. In Section 13.5 of this chapter, you'll learn about the test we used to get this result, and you'll be able to verify it for yourself.

Note: The study was sponsored by the American Society for Microbiology, a group that feels very strongly about washing our hands to reduce the spread of those little micro-critters that give us flu, colds, and other ailments. *Another note:* It's not at all unusual for organizations to package a rather straightforward message in the context of reporting the results of an interesting survey or other research effort to which newspaper editors are more likely to be attracted.

Source: American Society for Microbiology and Harris Interactive, Inc., *A Survey of Handwashing Behavior*, September 2007.

Observed data often differ from expected data. The runner-up in the 2009 Nathan's Famous Hot Dog Eating Contest was the slender gentleman above, who ate more than 64 hot dogs in just 10 minutes.

LEARNING OBJECTIVES

After reading this chapter, you should be able to:

- Explain the nature of the chi-square distribution.
- List and understand the general procedures involved in chi-square testing.
- Apply the chi-square distribution in testing whether a sample could have come from a population having a specified probability distribution.
- Apply the chi-square distribution in testing whether two nominal-scale (category) variables could be independent.
- Apply the chi-square distribution in comparing the proportions of two or more independent samples.
- Determine the confidence interval for, and carry out hypothesis tests for a population variance.

13.1 INTRODUCTION

The chapter introduces a new probability distribution called the **chi-square distribution.** Using this distribution, along with sample data and frequency counts, we will be able to examine (1) whether a sample could have come from a given type of population distribution, (2) whether two nominal variables could be independent of each other, and (3) whether two or more independent samples could have the same population proportion. The chi-square distribution will also be used in constructing confidence intervals and carrying out hypothesis tests regarding the value of a population variance. As in previous chapters, relevant computer outputs for examples will be presented and interpreted.

13.2 BASIC CONCEPTS IN CHI-SQUARE TESTING

The Chi-Square Distribution

Like the t and F distributions covered previously, the chi-square (χ^2, "chi" rhymes with "sky") distribution is a sampling distribution. Specifically, when samples of size n are drawn from a normal population, the chi-square distribution is the sampling distribution of $\chi^2 = (n - 1)s^2/\sigma^2$.

The chi-square distribution is a *family* of probability distributions. As with the t distribution, the specific member depends on a value for the number of degrees of freedom. (Recall that the F distribution is dependent on *two* separate df values, one associated with the numerator of F, the other with the denominator.) The chi-square distribution is skewed positively, but as df increases, it approaches the shape of the normal distribution. In Figure 13.1, which shows the chi-square curve for selected df values, note the extreme positive skewness that is present when df takes on smaller values. Since $(n - 1)$, s^2, and σ^2 will always be ≥ 0, χ^2 will always be ≥ 0 as well. By using Seeing Statistics Applet 17, at the end of the chapter, you can observe how the chi-square distribution shape and probabilities change in response to changes in df.

As with the t distribution, it isn't practical to include a table of areas beneath the curve for every possible member of this family of probability distributions. The χ^2 table in Appendix A lists the values of χ^2 corresponding to selected right-tail areas for various values of df. A portion of this table is shown as Table 13.1. In

FIGURE 13.1

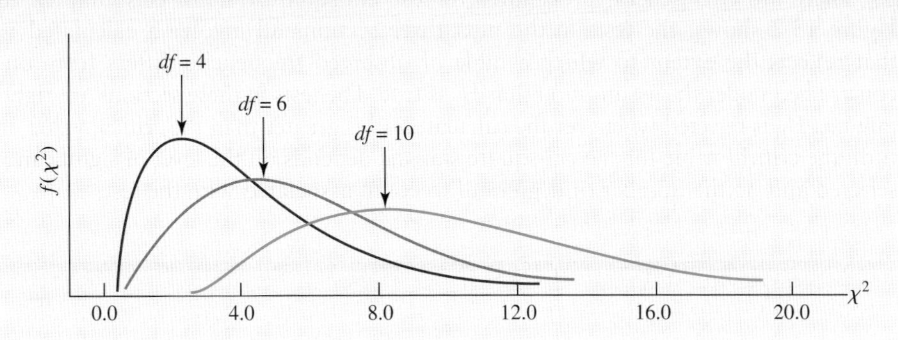

The chi-square distribution is really a family of distributions, and the specific member depends on the number of degrees of freedom. For small values of *df*, there is extreme positive skewness, but the curve approaches the normal distribution as *df* becomes larger.

TABLE 13.1

A portion of the chi-square table from Appendix A. Like the *t* distribution, the chi-square distribution is really a family of continuous distributions, but the shape of the chi-square distribution is positively skewed.

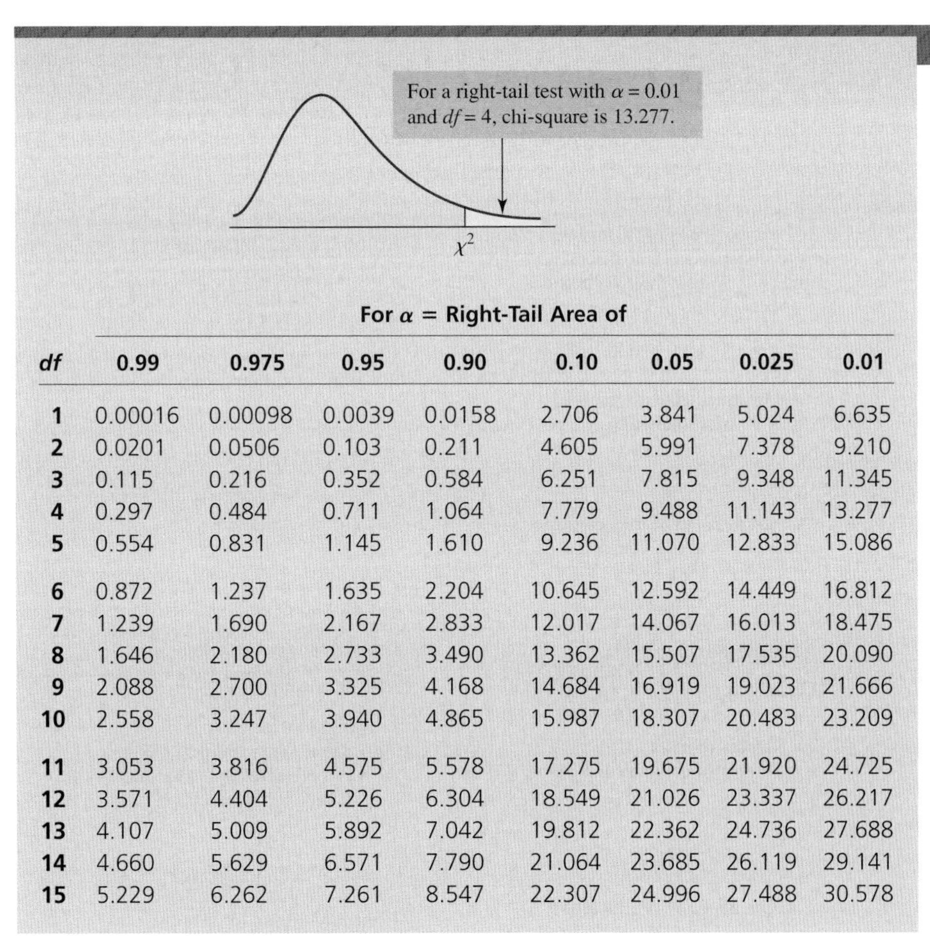

	For α = Right-Tail Area of							
df	**0.99**	**0.975**	**0.95**	**0.90**	**0.10**	**0.05**	**0.025**	**0.01**
1	0.00016	0.00098	0.0039	0.0158	2.706	3.841	5.024	6.635
2	0.0201	0.0506	0.103	0.211	4.605	5.991	7.378	9.210
3	0.115	0.216	0.352	0.584	6.251	7.815	9.348	11.345
4	0.297	0.484	0.711	1.064	7.779	9.488	11.143	13.277
5	0.554	0.831	1.145	1.610	9.236	11.070	12.833	15.086
6	0.872	1.237	1.635	2.204	10.645	12.592	14.449	16.812
7	1.239	1.690	2.167	2.833	12.017	14.067	16.013	18.475
8	1.646	2.180	2.733	3.490	13.362	15.507	17.535	20.090
9	2.088	2.700	3.325	4.168	14.684	16.919	19.023	21.666
10	2.558	3.247	3.940	4.865	15.987	18.307	20.483	23.209
11	3.053	3.816	4.575	5.578	17.275	19.675	21.920	24.725
12	3.571	4.404	5.226	6.304	18.549	21.026	23.337	26.217
13	4.107	5.009	5.892	7.042	19.812	22.362	24.736	27.688
14	4.660	5.629	6.571	7.790	21.064	23.685	26.119	29.141
15	5.229	6.262	7.261	8.547	22.307	24.996	27.488	30.578

using the χ^2 table, we need only refer to the upper-tail area of interest, then use the appropriate *df* row in identifying the χ^2 value. For example, if the right-tail area of interest is 0.05 and $df = 10$, the corresponding χ^2 value will be at the intersection of the $\alpha = 0.05$ column and the $df = 10$ row, $\chi^2 = 18.307$. Although most of our tests will be right-tail, those in Section 13.6 can be two-tail as well. This is why our chi-square table also includes columns for right-tail areas of 0.99, 0.975, 0.95, and 0.90.

There are two general procedures for the chi-square tests of this chapter. As Figure 13.2 shows, the tests in the next three sections all involve a calculated χ^2 that reflects the extent to which a table of observed frequencies differs from one constructed under the assumption that H_0 is true. The tests in these sections are right-tail; i.e., H_0 is rejected whenever the calculated χ^2 is greater than the critical value. In turn, the critical value for the chi-square statistic depends on the level of significance selected and on the number of degrees of freedom associated with the test.

FIGURE 13.2

General procedures for chi-square tests discussed in the chapter.

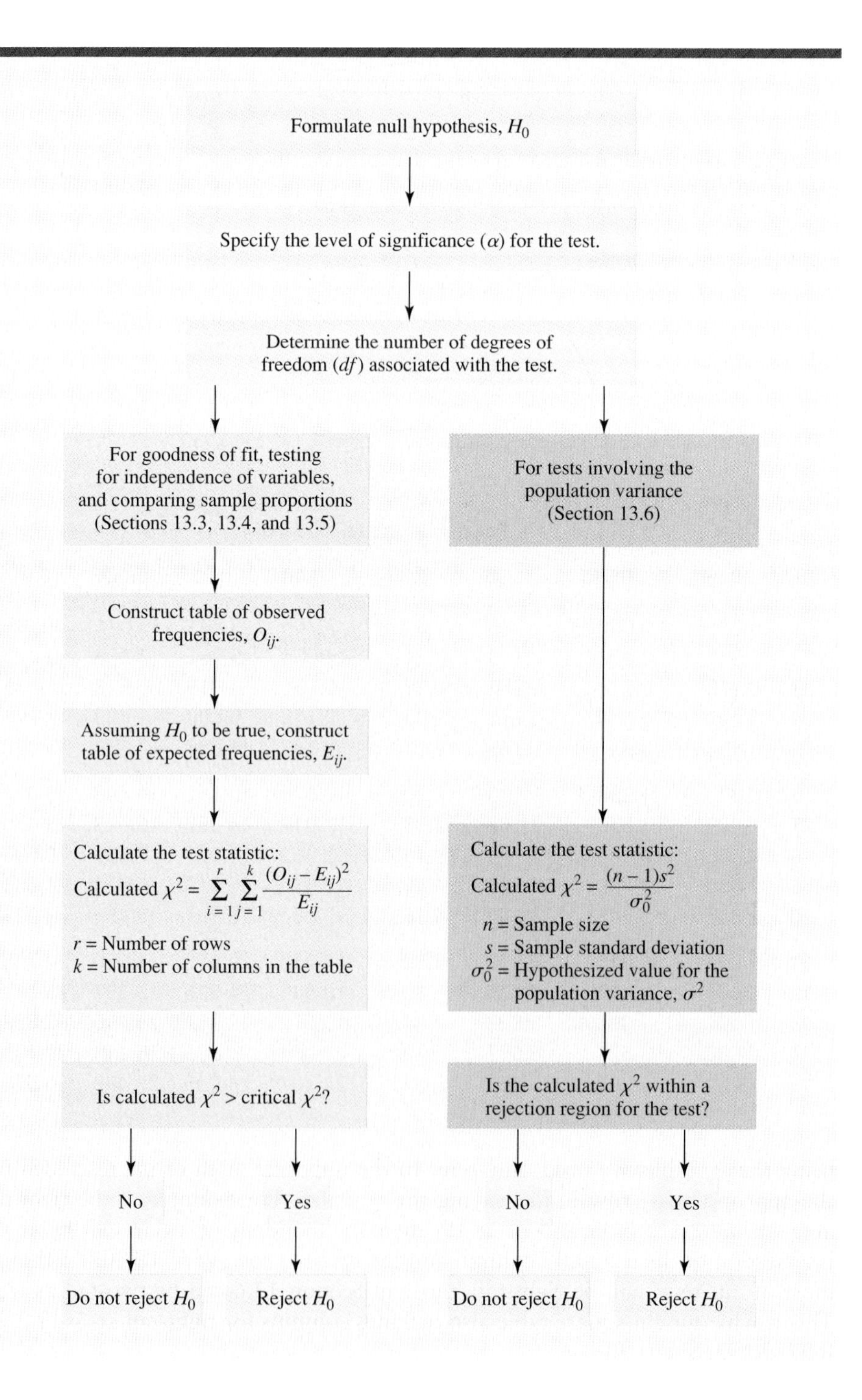

Sections 13.3, 13.4, and 13.5 each present a specific variant of the general formula shown in Figure 13.2. The frequency tables in these sections will differ in the number of rows they contain, with the tests in Sections 13.3, 13.4, and 13.5 having 1, ≥2, and 2 rows, respectively.

In using the chi-square distribution to construct confidence intervals and carry out hypothesis tests involving a population variance, the topics of Section 13.6, there are no observed versus expected tables to be compared. In this application, the chi-square statistic is used in a manner that is more analogous to our use of the *t*-statistic in Chapters 9 and 10, where we constructed confidence intervals and carried out tests involving a population mean.

EXERCISES

13.1 For what kinds of tests can chi-square analysis be used?

13.2 Is the chi-square distribution a continuous distribution or a discrete distribution? Explain.

13.3 In what way are the chi-square and normal distributions related?

13.4 Sketch the approximate shape of the chi-square curve when $df = 2$ and when $df = 100$.

13.5 Why can the chi-square statistic never be negative?

13.6 For $df = 5$ and the constant A, identify the value of A such that

a. $P(\chi^2 > A) = 0.90$
b. $P(\chi^2 > A) = 0.10$
c. $P(\chi^2 > A) = 0.95$
d. $P(\chi^2 > A) = 0.05$
e. $P(\chi^2 < A) = 0.975$
f. $P(\chi^2 < A) = 0.025$

13.7 For $df = 8$ and the constant A, identify the value of A such that

a. $P(\chi^2 > A) = 0.90$
b. $P(\chi^2 > A) = 0.10$
c. $P(\chi^2 > A) = 0.95$
d. $P(\chi^2 > A) = 0.05$
e. $P(\chi^2 < A) = 0.975$
f. $P(\chi^2 < A) = 0.025$

13.8 For $df = 10$ and the constants A and B, identify the values of A and B such that the tail areas are equal and

a. $P(A < \chi^2 < B) = 0.80$
b. $P(A < \chi^2 < B) = 0.90$
c. $P(A < \chi^2 < B) = 0.95$
d. $P(A < \chi^2 < B) = 0.98$

13.9 For $df = 15$ and the constants A and B, identify the values of A and B such that the tail areas are equal and

a. $P(A < \chi^2 < B) = 0.80$
b. $P(A < \chi^2 < B) = 0.90$
c. $P(A < \chi^2 < B) = 0.95$
d. $P(A < \chi^2 < B) = 0.98$

TESTS FOR GOODNESS OF FIT AND NORMALITY (13.3)

Purpose

In **goodness-of-fit tests,** chi-square analysis is applied for the purpose of examining whether sample data could have been drawn from a population having a specified probability distribution. For example, we may wish to determine from sample data (1) whether a random number table differs significantly from its hypothesized discrete uniform distribution, (2) whether a set of data could have come from a population having a normal distribution, or (3) whether the distribution of contributions to a local public television station differs significantly from those for such stations across the nation. With regard to item 2, we have already discussed a number of techniques which assume that sample data have come from a population that is either normally distributed or approximately so. This section provides an Excel-based, chi-square goodness-of-fit test designed specifically to examine whether sample data could have come from a normally distributed population.

Procedure

The procedure first requires that we arrange the sample data into categories that are mutually exclusive and exhaustive. We then count the number of data values in each of these categories and use these counts in constructing the table of observed frequencies. In the context of the general procedure shown in Figure 13.2, this table of observed frequencies will have one row and k columns. Since there is just one row, we will simply refer to each **observed frequency** as O_j, with $j = 1$ through k.

The null and alternative hypotheses are the key to the next step. Again, these are

H_0: The sample is from the specified population.

H_1: The sample is not from the specified population.

Under the assumption that the null hypothesis is true, we construct a table of **expected frequencies** (E_j, with $j = 1$ through k) that are based on the probability distribution from which the sample is assumed to have been drawn. The test statistic, χ^2, is a measure of the extent to which these tables differ. From an intuitive standpoint, a large amount of discrepancy between the frequencies that are observed and those that are expected tends to cause us to reject the null hypothesis. The test statistic is calculated as follows:

Chi-square test for goodness-of-fit:

$$\text{Calculated } \chi^2 = \sum_{j=1}^{k} \frac{(O_j - E_j)^2}{E_j}$$

where k = number of categories, or cells in the table
O_j = observed frequency in cell j
E_j = expected frequency in cell j

and $df = k - 1 - m$, with m = the number of population parameters (e.g., μ or σ) that must be estimated from sample data in order to carry out the goodness-of-fit test.

The test is upper-tail, and the critical value of χ^2 will be that associated with the level of significance (α) and degrees of freedom (df) for the test. Regarding the determination of df, the $(k - 1)$ portion of the df expression is required because we have already lost one degree of freedom by arranging the frequencies into categories. Since we know the total number of observations in all cells (the sample size), we need only know the contents of $(k - 1)$ cells to determine the count in the kth cell. We must also further reduce df by 1 for each population parameter that has been estimated in order to carry out the test. Each such estimation represents additional help in our construction of the table of expected frequencies, and the purpose of the reductions beyond $(k - 1)$ is to correct for this assistance. This aspect of the df determination will be further discussed in the context of the examples that follow.

As a rule of thumb, it is important that the total sample size and selection of categories be such that *each expected frequency* (E_j) *is at least* 5. This is because the chi-square distribution is continuous, while the counts on which the test statistic is based are discrete, and the approximation will be unsatisfactory whenever one or more of the expected frequencies is very small. There are two ways to avoid this problem: (1) use an overall sample size that is large enough that each expected frequency will be at least 5, or (2) combine adjacent cells so that the result will be a cell in which the expected frequency is at least 5.

EXAMPLE

Poisson Distribution

Automobiles leaving the paint department of an assembly plant are subjected to a detailed examination of all exterior painted surfaces. For the most recent 380 automobiles produced, the number of blemishes per car is summarized here—for example, 242 of the automobiles had 0 blemishes, 94 had 1 blemish, and so on. The underlying data are in file **CX13BLEM**.

Blemishes per car:	0	1	2	3	4
Number of cars:	242	94	38	4	2

Using the 0.05 level of significance, could these data have been drawn from a population that is Poisson distributed?

SOLUTION

Formulate the Null and Alternative Hypotheses

The null hypothesis is H_0: the sample was drawn from a population that is Poisson distributed. The alternative hypothesis is that it was not drawn from such a population.

Specify the Level of Significance (α) for the Test

The $\alpha = 0.05$ level has been specified.

Construct the Table of Observed Frequencies, O_j

The observed frequencies are as listed earlier and are repeated below.

Blemishes per car:	0	1	2	3	4	
Observed frequency, O_j:	242	94	38	4	2	380

Assuming the Null Hypothesis to Be True, Construct the Table of Expected Frequencies, E_j

To construct the table of expected frequencies, we must first use the mean of the sample data to estimate the mean of the Poisson population that is being hypothesized. The sample mean is simply the total number of blemishes divided by the total number of cars, or:

$$\bar{x} = \frac{0(242) + 1(94) + 2(38) + 3(4) + 4(2) \text{ blemishes}}{380 \text{ cars}}$$

$$\text{and} \quad \bar{x} = 190/380 = 0.5 \text{ blemishes per car}$$

Using $\bar{x} = 0.5$ blemishes per car as our estimate of the true population mean (λ), we now refer to the "$\lambda = 0.5$" portions of the individual and cumulative tables of

Poisson probabilities in Appendix A. For a Poisson distribution with $\lambda = 0.5$, and x = the number of blemishes per car:

x = number of blemishes	$P(x) \cdot n$	Expected number of cars with x blemishes
0	0.6065 · 380 =	230.470
1	0.3033 · 380 =	115.254
2	0.0758 · 380 =	28.804
3	0.0126 · 380 =	4.788
4	0.0016 · 380 =	0.608
5 or more	0.0002 · 380 =	0.076
	1.0000	380.000

The final three categories ($x = 3$, $x = 4$, and $x \geq 5$ blemishes) have expected frequencies that are less than 5.0. To satisfy the requirement that all expected frequencies be at least 5.0, we will combine these into a new category called "3 or more blemishes," and it will have an expected frequency of 4.788 + 0.608 + 0.076, or 5.472 cars. After this merger, the expected frequencies will be as shown below.

Blemishes per car:	0	1	2	≥3	
Expected frequency, E_j:	230.470	115.254	28.804	5.472	380.000

The observed frequencies must be based on the same categories used in the expected frequencies, so we need to express the observed frequencies in terms of the same categories (0, 1, 2, and ≥3) used above. The observed frequencies can now be as shown below.

Blemishes per car:	0	1	2	≥3	
Observed frequency, O_j:	242	94	38	6	380

Determine the Calculated Value of the χ^2 Test Statistic

Over the $k = 4$ cells, the calculated chi-square is 7.483, computed as follows:

$$\text{Calculated } \chi^2 = \sum_{j=1}^{k} \frac{(O_j - E_j)^2}{E_j}$$

$$= \frac{(242 - 230.470)^2}{230.470} + \frac{(94 - 115.254)^2}{115.254} + \frac{(38 - 28.804)^2}{28.804} + \frac{(6 - 5.472)^2}{5.472}$$

$$= 7.483$$

Identify the Critical Value of the Chi-Square Statistic

The number of degrees of freedom is $df = k - 1 - m$, with $k = 4$ categories and $m = 1$ parameter (the mean) being estimated from sample data, and $df = 4 - 1 - 1$, or 2. For the 0.05 level of significance and $df = 2$, the critical chi-square value is 5.991.

NOTE

Regarding tests involving the Poisson distribution, in which the mean and the variance are equal, $df = k - 1 - m$ will have either $m = 0$ or $m = 1$ parameters being estimated. If λ is given in the null hypothesis (e.g., H_0: "The sample was drawn from a Poisson distribution having $\lambda = 1.8$"), m will be 0. If it is necessary to estimate λ based on your sample data (e.g., H_0: "The sample was drawn from a population that is Poisson distributed"), m will be 1. Just because λ and σ^2 are equal in this distribution doesn't mean that they must both be counted in the df expression if λ has to be estimated based on the sample data.

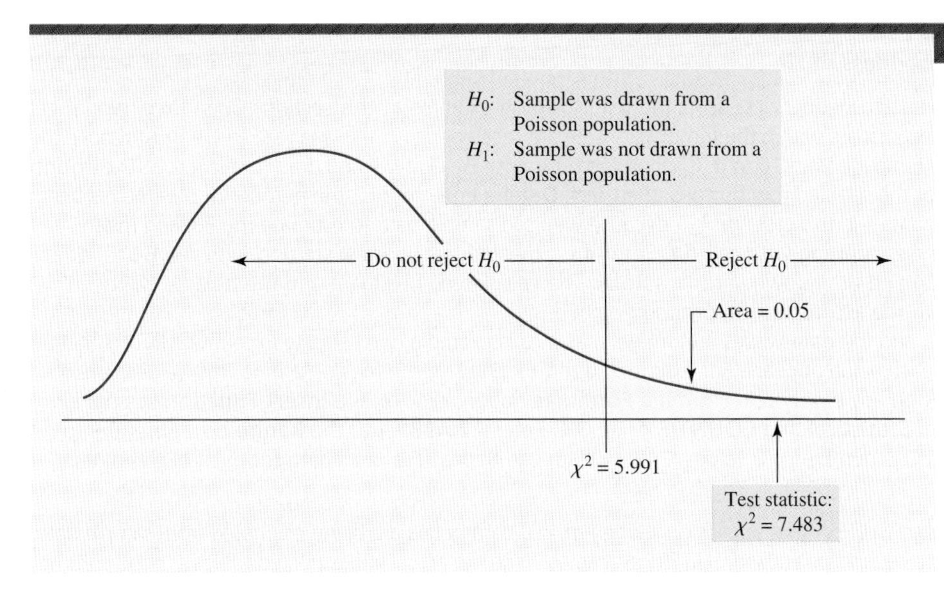

FIGURE 13.3

Given the frequency distribution for the number of paint blemishes on 380 automobiles, the null hypothesis is that the sample actually was drawn from a Poisson population. Using the chi-square goodness-of-fit test, we are able to reject the null hypothesis at the 0.05 level of significance.

Compare the Calculated and Critical Values of Chi-Square

The calculated chi-square (7.483) exceeds the critical value (5.991), and the null hypothesis can be rejected at the 0.05 level. At the 0.05 level of significance, we conclude that the sample was not drawn from a population that is Poisson distributed.

The calculated chi-square also exceeds 7.378, the critical value for the 0.025 level of significance. However, it does not exceed 9.210, the critical value for the 0.01 level. Based on the chi-square table, the most accurate statement we can make about the exact p-value for the test is that it is somewhere between 0.025 and 0.01. The test is summarized in Figure 13.3.

EXAMPLEEXAMPLE

Computer Solutions 13.1 shows how we can use Excel to carry out goodness-of-fit tests like the one summarized in Figure 13.3. Keep in mind that Excel's CHITEST function should be used only if $m = 0$ population parameters have been estimated on the basis of sample data, so it would not be applicable to this example.

COMPUTER 13.1 SOLUTIONS

Chi-Square Test for Goodness of Fit

EXCEL

	A	B	C	D	E
1	Chi-Square Goodness-of-Fit Test			no. of cells, k =	4
2	Cell Frequencies:			no. of parameters estimated, m =	1
3	Observed (Oj):	Expected (Ej):	(Oj-Ej)^2/Ej:	df = k - 1 - m =	2
4	242	230.470	0.577	calculated chi-square =	7.483
5	94	115.254	3.919	p-value =	0.024
6	38	28.804	2.936		
7	6	5.472	0.051		

(continued)

Excel Chi-Square Test for Goodness of Fit

1. Given the observed and expected frequencies, as in the paint-blemish example in this section: Open Excel worksheet template TMCHIFIT. The directions are within the template. There are 4 categories in this analysis, so we must reduce the number of categories in the template from 10 to 4. Click on the button for Excel row **7** and drag to select rows **7 through 12.** Right-click on any one of the selected row buttons, then click **Delete.** Enter into E2 the number of parameters being estimated based on sample data—in this case, $m = \mathbf{1}$ because we used sample data to estimate the Poisson mean.
2. Enter the observed frequencies, starting with cell **A4.** Enter the expected frequencies, starting with cell **B4.** Click at the lower right corner of **C4** and drag downward to autofill. The calculated chi-square and the *p*-value will be in cells E4 and E5, as shown in the accompanying printout.

   If we have only raw data and the data values are integers, as in this example, we can use Excel's COUNTIF function to count how many times each value appears. For example, using CX13BLEM, the file containing 380 values consisting of 0, 1, 2, 3, or 4: Click on any empty cell. From the **Formulas** ribbon, click on the **Insert Function** symbol at the extreme left, then select **Statistical** and **COUNTIF.** Click **OK.** Enter **A2:A381** into the **Range** box. Enter **0** into the **Criteria** box. Click **OK.** The number of 0 values (242) will appear. Repeat the preceding for each of the other codes (1, 2, 3, and 4).

Note: Excel offers a function called CHITEST that applies *only* when $m = 0$ parameters are being estimated from sample data. *If* this had been the case here, we could have applied CHITEST as follows: With the actual and expected values as shown above, click on an empty cell. From the **Formulas** ribbon, click on the **Insert Function** symbol. Select **Statistical** and **CHITEST.** Click **OK.** Enter **A4:A7** into the **Actual_range** box. Enter **B4:B7** into the **Expected_range** box. The *p*-value is provided.

(Minitab does not provide this goodness-of-fit test.)

EXAMPLEEXAMPLEEXAMPLEEXAMPLEEXAMPLEEXAMP

EXAMPLE

Test for Normality

A researcher in the department of education has collected data consisting of a simple random sample of 300 SAT scores of high school seniors in her state who took the college entrance examination last year. For this sample, $\bar{x} = 945.04$, $s = 142.61$, and the underlying data are in file CX13SAT. Based on these sample data, use the 0.01 level of significance in determining whether the sample could have been drawn from a population in which the scores are normally distributed.

SOLUTION

Although this test could be performed with tables and pocket calculator, we will use the procedure described in Computer Solutions 13.2, and we will discuss it in terms of the printout shown there. The null hypothesis is that the sample came from a normal distribution, and the test will be conducted using the 0.01 level of significance. Whether the test is carried out by hand calculations or using the computer, the key components will be those shown in Computer Solutions 13.2:

1. The expected frequencies can be defined in terms of area under the normal curve. For example, as shown in cells B8 and C8, 2.28% of the observations (or 300[0.0228] = 6.825 observations) would be expected to be at least two standard deviations below the mean.
2. If we were to arrange the data in order of size, we would find that 8 observations were actually at least two standard deviations below the mean.
3. Using the same procedure as in steps 1 and 2, we could obtain the probabilities, expected frequencies, and actual frequencies for the other five groups (intervals). The expected frequencies are shown in cells C8:C13 and the observed frequencies are shown in cells D8:D13.

4. We can use the same formula as in Figure 13.2 and the previous example to obtain the calculated value of the chi-square statistic, or calculated chi-square = 15.24. This could be calculated as

$$\sum_{j=1}^{k} \frac{(O_j - E_j)^2}{E_j} = \frac{(8 - 6.825)^2}{6.825} + \frac{(28 - 40.772)^2}{40.772} + \cdots + \frac{(11 - 6.825)^2}{6.825} = 15.24$$

5. We have used the sample mean and standard deviation to estimate their population counterparts, so the value of df is $k - 1 - m$, or $6 - 1 - 2 = 3$. The critical value of chi-square for the 0.01 level of significance and $df = 3$ is 11.345.
6. The calculated chi-square (15.24) exceeds the critical value (11.345) and, at the 0.01 level of significance, we reject the null hypothesis that these data were drawn from a normal distribution. At this level of significance, we conclude that the disparity between the expected and observed frequencies is simply too great to have happened by chance. The exact p-value for the test is shown in cell B17 as 0.0016.

EXAMPLEEXAMPLEEXAMPLEEXAMPLE

COMPUTER 13.2 SOLUTIONS

Chi-Square Goodness-of-Fit Test for Normality

EXCEL

	A	B	C	D
1	**Chi-Squared Test of Normality**			
2		*SAT*		
3	Mean	945.04		
4	Standard deviation	142.61		
5	Observations	300		
6				
7	Intervals	Probability	Expected	Observed
8	(z <= -2)	0.0228	6.825	8
9	(-2 < z <= -1)	0.1359	40.772	28
10	(-1 < z <= 0)	0.3413	102.404	127
11	(0 < z <= 1)	0.3413	102.404	94
12	(1 < z <= 2)	0.1359	40.772	32
13	(z > 2)	0.0228	6.825	11
14				
15	chi-squared Stat	15.24		
16	df	3		
17	p-value	0.0016		
18	chi-squared Critical	11.345		

Excel Chi-Square Goodness-of-Fit Test for Normality (requires $n \geq 32$)

1. For example, testing the data in Excel file **CX13SAT**, which contains 300 Scholastic Aptitude Test (SAT) scores with label and data values in A1:A301: From the **Add-Ins** ribbon, click **Data Analysis Plus.** Click **Chi-Squared Test of Normality.** Click **OK.**
2. Enter **A1:A301** into the **Input Range** box. Click **Labels.** Enter **0.01** into the **Alpha** box. Click **OK.** The observed and expected frequencies, the chi-square statistic, and the p-value will be as shown above. The p-value is quite small, and we can conclude that the sample data did not come from a normally distributed population.

(Minitab does not provide the chi-square goodness-of-fit test for normality. Other Minitab tests for normality are presented in Chapters 14 and 15.)

EXERCISES

13.10 In carrying out a chi-square goodness-of-fit test, what are the "k" and "m" terms in the "$df = k - 1 - m$" expression and why is each term present?

13.11 If a table of expected frequencies differs very little from the frequencies that were observed, would the calculated chi-square be large or small? Why?

13.12 For the null hypothesis, H_0: "The data were drawn from a Poisson distribution with $\lambda = 7.0$," how many degrees of freedom would be associated with the test if there are 8 cells in the table of frequencies to be analyzed?

13.13 For the null hypothesis, H_0: "The data were drawn from a uniform continuous distribution," how many degrees of freedom would be associated with the test if there are 5 cells in the table of frequencies to be analyzed?

13.14 Sample data have been collected, and the null hypothesis to be tested is, H_0: "The sample was drawn from a normal population." If the analysis is based on a categorization that includes 5 cells:

a. How many degrees of freedom will be associated with the test?
b. For a test at the 0.05 level, what is the critical value of chi-square?
c. If the calculated value of the chi-square statistic is 8.13, what conclusion would be reached regarding the null hypothesis?

13.15 Sample data have been collected, and the null hypothesis to be tested is H_0: "The sample was drawn from a normal population in which $\mu = 130$." If the analysis is based on a categorization that includes 8 cells:

a. How many degrees of freedom will be associated with the test?
b. For a test at the 0.05 level, what is the critical value of chi-square?
c. If the calculated value of the chi-square statistic is 11.25, what conclusion would be reached regarding the null hypothesis?

13.16 According to the Bureau of the Census, 18.1% of the U.S. population lives in the Northeast, 21.9% in the Midwest, 36.7% in the South, and 23.3% in the West. In a random sample of 200 recent calls to a national 800-number hotline, 39 of the calls were from the Northeast, 55 from the Midwest, 60 from the South, and 46 from the West. At the 0.05 level, can we conclude that the geographical distribution of hotline callers could be the same as the U.S. population distribution? Source: Bureau of the Census, factfinder.census.gov, July 30, 2009.

13.17 From the one-day work absences during the past year, the personnel director for a large firm has identified the day of the week for a random sample of 150 of the absences. Given the following observed frequencies, and for $\alpha = 0.01$, can the director conclude that one-day absences during the various days of the week are not equally likely?

	Monday	Tuesday	Wednesday	Thursday	Friday	
Absences	42	18	24	27	39	150

13.18 It has been reported that 8.7% of U.S. households do not own a vehicle, with 33.1% owning 1 vehicle, 38.1% owning 2 vehicles, and 20.1% owning 3 or more vehicles. The data for a random sample of 100 households in a resort community are summarized in the frequency distribution below. At the 0.05 level of significance, can we reject the possibility that the vehicle-ownership distribution in this community differs from that of the nation as a whole? Source: planetforward.org, July 30, 2009.

Number of Vehicles Owned	Number of Households
0	20
1	35
2	23
3 or more	22
	100

13.19 For the data provided in Exercise 13.18, use the 0.05 level in testing whether the sample could have been drawn from a Poisson population with $\lambda = 1.5$.

13.20 Employees in a production department assemble the products, then submit them in batches of 25 for quality inspection. The number of defectives in each batch of a random sample of 50 recently produced batches is shown below. At the 0.01 level, test whether the population of x = the number of defectives in a batch could be Poisson distributed with $\lambda = 2.0$.

x = Number of Defectives	Number of Batches
0	5
1	15
2	14
3	10
4	4
5	2
	50

13.21 Given the information in Exercise 13.20, use the 0.05 level in testing whether the number of defectives in each batch of $n = 25$ could be binomially distributed with $\pi = 0.1$.

13.22 Approximately 13.2% of U.S. drivers are younger than age 25, with 37.7% in the 25–44 age group, and 49.1% in the 45-and-over category. For a random sample of 200 fatal accidents in her state, a safety expert finds that 42 drivers were under 25 years old, 80 were 25–44 years old, and 78 were at least 45 years old. At the 0.05 level, test whether the age distribution of drivers involved in fatal accidents within the state could be the same as the age distribution of all U.S. drivers.
Source: *The World Almanac and Book of Facts 2006*, p. 117.

(**DATA SET**) *Note:* Exercises 13.23–13.26 require a computer and statistical software.

13.23 For a random sample of 200 U.S. motorists, the mileages driven last year are in data file **XR13023**. Use the 0.01 level of significance in determining whether the mileages driven by the population of U.S. motorists could be normally distributed.

13.24 For the previous 300 plywood panels from a lumber company's production line, the data in file **XR13024** list the number of surface defects per panel. Based on this information, use the 0.05 level of significance in examining whether the data could have come from a Poisson distribution.

13.25 For Exercise 13.24, use the 0.05 level of significance in examining whether the data could have come from a normal distribution.

13.26 The outstanding balances for 500 credit-card customers are listed in file **XR13026**. Using the 0.10 level of significance, examine whether the data could have come from a normal distribution.

TESTING THE INDEPENDENCE OF TWO VARIABLES (13.4)

Purpose

When data involve two nominal-scale variables, with each variable having two or more categories, chi-square analysis can be used to test whether or not there might be some relationship between the variables. For example, a campaign manager might be interested in whether party affiliation could be related to educational level, or an insurance underwriter may wish to determine whether different occupational groups tend to suffer different kinds of on-the-job injuries. The purpose of chi-square analysis in such applications is not to identify the exact nature of a relationship between nominal variables; the goal of this technique is simply to test whether or not the variables could be independent of each other.

Procedure

The starting point for the chi-square **test of variable independence** is the **contingency table.** Introduced in Chapter 2, this is a table where the rows represent categories of one variable, the columns represent categories of another variable, and entries are the frequencies of occurrence for various row and column combinations. A contingency table has r rows and k columns, where $r \geq 2$ and $k \geq 2$. The null and alternative hypotheses are

H_0: The variables are independent of each other.
H_1: The variables are not independent of each other.

As in the preceding section, there will be a table of observed frequencies and a table of expected frequencies, and the amount of disparity between these tables will be compared during the calculation of the χ^2 test statistic. The observed frequencies will reflect a cross-classification for members of a single sample, and the table of expected frequencies will be constructed under the assumption that the null hypothesis is true. The test statistic, calculated as shown below, will be compared to a critical χ^2 for a given level of significance using an appropriate value for df.

Chi-square test for the independence of two variables:

$$\text{Calculated } \chi^2 = \sum_{i=1}^{r}\sum_{j=1}^{k}\frac{(O_{ij} - E_{ij})^2}{E_{ij}}$$

where r = number of rows in the contingency table
k = number of columns in the contingency table
O_{ij} = observed frequency in row i, column j
E_{ij} = expected frequency in row i, column j
$df = (r - 1)(k - 1)$

The rationale for $df = (r - 1)(k - 1)$ is that this quantity represents the number of cell frequencies that are free to vary. Given that we know the sample size (the total number of observations in the table), we need only know the contents of $(r - 1)$ rows and $(k - 1)$ columns in order to completely fill in the cells within the $r \times k$ table. As in the preceding section, we should combine rows or columns whenever necessary, so that each E_{ij} value will be at least five. Again, this is because we are using a continuous distribution in analyzing *count* data, which are discrete.

The calculation of the E_{ij} values in the table of expected frequencies relies on (1) the assumption that H_0 is true and the variables really are independent, and (2) the joint and marginal probability concepts presented in Chapter 5.

Expected frequencies when testing for the independence of variables:

The expected frequency in row *i*, column *j* is

$$E_{ij} = p_j(n_i)$$

where p_j = for all the observations, the proportion that are in column j
n_i = the number of observations in row i

In the above expression, p_j is the marginal probability that a randomly selected sample member will be in column j. If the variables are really independent, this probability should be applicable regardless of which row is considered, and the expected number of observations in the ij cell will be the product of p_j times n_i.

EXAMPLEEXAMPLEE

EXAMPLE

Test for Independence

A traffic-safety researcher has observed 500 vehicles at a stop sign in a suburban neighborhood and recorded (1) the type of vehicle (car, sport utility vehicle [SUV], pickup truck) and (2) driver behavior at the stop sign (complete stop, near stop, "ran" the stop sign). Her results are summarized in part A of Table 13.2. At the 0.05 level of significance, could there be some relationship between driver

TABLE 13.2

Observed and expected frequencies in testing the independence of two nominal-scale variables.

A. *Observed Frequencies, O_{ij}*

		Behavior at Stop Sign			
		Stopped	Coasted	Ran It	
Type of Vehicle	Car	183	107	60	350
	SUV	54	27	19	100
	Pickup	14	20	16	50
		251	154	95	500

B. *Expected Frequencies, E_{ij}*

		Behavior at Stop Sign			
		Stopped	Coasted	Ran It	
Type of Vehicle	Car	175.7	107.8	66.5	350.0
	SUV	50.2	30.8	19.0	100.0
	Pickup	25.1	15.4	9.5	50.0
		251.0	154.0	95.0	500.0

$E_{31} = \frac{251}{500} \times 50$, or 25.1

behavior and the type of vehicle being driven? The underlying data are in file **CX13STOP**.

SOLUTION

As part A of Table 13.2 shows, the table of observed frequencies reflects intersections of the various categories of the two variables. For example, 183 persons driving cars were seen to come to a complete stop at the sign. The contingency table has $r = 3$ rows and $k = 3$ columns. Let's proceed through the steps in the hypothesis-testing procedure.

Formulate the Null Hypothesis

The null hypothesis is H_0: driver behavior and type of vehicle are independent. The alternative hypothesis is that a relationship exists between the variables.

Specify the Level of Significance (α) for the Test

The $\alpha = 0.05$ level has been specified.

Construct the Table of Observed Frequencies, O_{ij}

Part A of Table 13.2 shows the observed frequencies recorded by the researcher.

Assuming the Null Hypothesis to Be True, Construct the Table of Expected Frequencies, E_{ij}

Each E_{ij} value is calculated by the formula presented earlier. In the entire sample of 500 observations, 251 of the 500 were in column 1 (those who stopped at the sign). Under the assumption that the variables are independent, we would expect that the proportion of the persons in each row (vehicle type) who stopped at the sign would also be 251/500. For rows 1 through 3, the expected frequencies in column 1 can be calculated as

- Row $i = 1$, column $j = 1$:

$$E_{ij} = p_j(n_i) = \frac{251}{500}(350) \quad \text{and} \quad E_{11} = 175.7$$

- Row $i = 2$, column $j = 1$:

$$E_{ij} = p_j(n_i) = \frac{251}{500}(100) \quad \text{and} \quad E_{21} = 50.2$$

- Row $i = 3$, column $j = 1$:

$$E_{ij} = p_j(n_i) = \frac{251}{500}(50) \quad \text{and} \quad E_{31} = 25.1$$

The expected frequencies within columns 2 and 3 are calculated using the same approach. For example, the expected frequency for the "car—ran it" cell (1st row, 3rd column) is $(95/500) \times 350$, or 66.5. If the variables were really independent, and $95/500 = 19.0\%$ of all drivers "ran" the stop sign, it would be expected that 19.0% of those driving each type of vehicle would have run through the stop sign.

Determine the Calculated Value of the χ^2 Test Statistic

The calculated chi-square is 12.431, computed as follows:

$$\text{Calculated } \chi^2 = \sum_{i=1}^{r}\sum_{j=1}^{k} \frac{(O_{ij} - E_{ij})^2}{E_{ij}}$$

$$= \frac{(183.0 - 175.7)^2}{175.7} + \frac{(107.0 - 107.8)^2}{107.8} + \frac{(60.0 - 66.5)^2}{66.5}$$

$$+ \frac{(54.0 - 50.2)^2}{50.2} + \frac{(27.0 - 30.8)^2}{30.8} + \frac{(19.0 - 19.0)^2}{19.0}$$

$$+ \frac{(14.0 - 25.1)^2}{25.1} + \frac{(20.0 - 15.4)^2}{15.4} + \frac{(16.0 - 9.5)^2}{9.5}$$

$$= 12.431$$

Identify the Critical Value of the Chi-Square Statistic

The table includes $r = 3$ rows and $k = 3$ columns, so the number of degrees of freedom for the test is $df = (r - 1)(k - 1)$, or $(3 - 1)(3 - 1) = 4$. For the 0.05 level of significance and $df = 4$, the critical value of chi-square is 9.488.

Compare the Calculated and Critical Values of Chi-Square

The calculated chi-square (12.431) exceeds the critical value (9.488), so the null hypothesis can be rejected at the 0.05 level. Had the test been done at the 0.025 level (critical chi-square = 11.143), it would have been rejected at that level as well. It would not have been rejected at the 0.01 level (critical chi-square = 13.277). Based on the chi-square table, the most accurate statement we can make about the exact p-value for the test is that it is somewhere between 0.025 and 0.01. From this analysis it appears that driving behavior at the stop sign is not independent of the type of vehicle being driven.

EXAMPLEEXAMPLE

Computer Solutions 13.3 shows how we can use Excel and Minitab to carry out a test for the independence of variables like the one in this section. The analysis can be based on either summary information or raw data.

COMPUTER 13.3 SOLUTIONS

Chi-Square Test for Independence of Variables

EXCEL

	A	B	C	D	E
1	**Contingency Table**				
2		*Column 1*	*Column 2*	*Column 3*	TOTAL
3	*Row 1*	183	107	60	350
4	*Row 2*	54	27	19	100
5	*Row 3*	14	20	16	50
6	TOTAL	251	154	95	500
7					
8	chi-squared Stat			12.431	
9	df			4	
10	p-value			0.014	
11	chi-squared Critical			9.488	

Excel Chi-Square Test for Independence of Variables, from Summary Information

1. Enter the observed frequencies into adjacent columns. For example, the 9 observed frequencies in Table 13.2 would be in cells A1:C3. From the **Add-Ins** ribbon, click **Data Analysis Plus.** Click **Contingency Table.** Click **OK.**
2. Enter **A1:C3** into the **Input Range** box. Do not select **Labels.** Enter **0.05** into the **Alpha** box. Click **OK.** The observed frequencies, along with the chi-square statistic and the *p*-value, will be as shown above.

Note: As an alternative, you can use Excel worksheet template **TMCHIVAR**. The procedure is described within the template.

Excel Chi-Square Test for Independence of Variables, from Raw Data

1. The underlying data for Table 13.2 are in Excel file **CX13STOP**, with vehicle codes in column A (1 = car, 2 = SUV, and 3 = pickup) and behavior codes in column B (1 = stopped, 2 = coasted, and 3 = ran it). The labels are in cells A1 and B1.
2. From the **Add-Ins** ribbon, click **Data Analysis Plus.** Click **Contingency Table (Raw Data).** Click **OK.**
3. Enter **A1:B501** into the **Input Range** box. Click **Labels.** Enter **0.05** into the **Alpha** box. Click **OK.** The observed frequencies, along with the chi-square statistic and the *p*-value, are provided.

(continued)

MINITAB

```
Chi-Square Test: C1, C2, C3

Expected counts are printed below observed counts
Chi-Square contributions are printed below expected counts

           C1      C2      C3  Total
    1     183     107      60    350
       175.70  107.80   66.50
        0.303   0.006   0.635

    2      54      27      19    100
        50.20   30.80   19.00
        0.288   0.469   0.000

    3      14      20      16     50
        25.10   15.40    9.50
        4.909   1.374   4.447

Total     251     154      95    500

Chi-Sq = 12.431, DF = 4, P-Value = 0.014
```

Minitab Chi-Square Test for Independence of Variables, from Summary Information

1. Enter the 9 observed frequencies into adjacent columns, in the same layout as in Table 13.2—e.g., the C1 column will consist of 183, 54, and 14. Click **Stat.** Select **Tables.** Click **Chi-Square Test (Two-Way Table in Worksheet).**
2. Enter **C1-C3** into the **Columns containing the table** box. Click **OK.** The observed frequencies, expected frequencies, chi-square statistic, and *p*-value will be as shown above.

Minitab Chi-Square Test for Independence of Variables, from Raw Data

1. The underlying data for Table 13.2 are in Minitab file CX13STOP, with vehicle codes in column C1 (1 = car, 2 = SUV, and 3 = pickup) and behavior codes in column C2 (1 = stopped, 2 = coasted, and 3 = ran it).
2. Click **Stat.** Select **Tables.** Click **Cross Tabulation and Chi-Square.**
3. In the **Categorical variables** menu, enter **C1** into the **For rows** box and **C2** into the **For columns** box. In the **Display** menu, select **Counts.**
4. Click **Chi-Square.** In the **Display** menu, select **Chi-Square analysis** and **Expected cell counts.** Click **OK.** Click **OK.**

EXERCISES

13.27 In conducting a chi-square test, why is it advisable that each expected frequency be at least 5.0? If the expected frequency in a cell happens to be less than 5.0, what should be done in order to carry out the analysis?

13.28 In carrying out a chi-square test for the independence of variables, what is the procedure for determining the number of degrees of freedom to be used in the test?

13.29 For a contingency table with r rows and k columns, determine the df for the test if

a. $r = 3, k = 4$ b. $r = 2, k = 3$
c. $r = 4, k = 5$ d. $r = 5, k = 3$
e. $r = 3, k = 7$ f. $r = 3, k = 3$

13.30 In testing the independence of two variables described in a contingency table, determine the critical value of chi-square if the test is to be conducted at the

a. $\alpha = 0.05$ level and $df = 3$
b. $\alpha = 0.01$ level and $df = 5$
c. $\alpha = 0.10$ level and $df = 2$
d. $\alpha = 0.025$ level and $df = 4$

13.31 In testing the independence of two variables described in a contingency table, determine the critical value of chi-square if the test is to be conducted at the

a. $\alpha = 0.025$ level and $df = 5$
b. $\alpha = 0.05$ level and $df = 8$
c. $\alpha = 0.01$ level and $df = 6$
d. $\alpha = 0.10$ level and $df = 4$

13.32 In a test of the independence of two variables, one of the variables has two possible categories and the other has three possible categories. What will be the critical value of chi-square if the test is to be carried out at the 0.025 level? At the 0.05 level?

13.33 A researcher has observed 100 shoppers from three different age groups entering a large discount store and noted the nature of the greeting received by the shopper. Given the results shown here, and using the 0.025 level of significance, can we conclude that the age category of the shopper is independent of the nature of the greeting he or she receives upon entering the store? Based on the chi-square table, what is the most accurate statement that can be made about the p-value for the test?

		Shopper Age Category (years)			
		21 or less	22–50	51 or more	
Greeting	Cool	16	12	5	33
	Friendly	8	20	6	34
	Hearty	6	14	13	33
		30	46	24	100

13.34 A research organization has collected the following data on household size and telephone ownership for 200 U.S. households. At the 0.05 level, are the two variables independent? Based on the chi-square table, what is the most accurate statement that can be made about the p-value for the test?

		Telephones Owned			
		≤1	2	≥3	
Persons in the Household	≤2	49	18	13	80
	3–4	40	27	21	88
	≥5	11	13	8	32
		100	58	42	200

13.35 Researchers in a California community have asked a sample of 175 automobile owners to select their favorite from three popular automotive magazines. Of the 111 import owners in the sample, 54 selected *Car and Driver,* 25 selected *Motor Trend,* and 32 selected *Road & Track*. Of the 64 domestic-make owners in the sample, 19 selected *Car and Driver,* 22 selected *Motor Trend,* and 23 selected *Road & Track*. At the 0.05 level, is import/domestic ownership independent of magazine preference? Based on the chi-square table, what is the most accurate statement that can be made about the p-value for the test?

13.36 A pharmaceutical firm, studying the selection of "name brand" versus "generic equivalent" on prescription forms, has been given a sample of 150 recent prescriptions submitted to a local pharmacy. Of the 44 under-40 patients in the sample, 16 submitted a prescription form with the "generic equivalent" box checked. Of the 52 patients in the 40–60 age group, 28 submitted a prescription form specifying "generic equivalent," and for the 54 patients in the 61-or-over age group, 32 submitted a prescription form specifying "generic equivalent." At the 0.025 level, is age group independent of name-brand/generic specification? Based on the chi-square table, what is the most accurate statement that can be made about the p-value for the test?

13.37 Customers of the Sky Mountain Grocery chain are routinely asked at the checkout whether they prefer paper or plastic bags for their purchases. In a recent study, researchers observed the type of bag specified and surveyed the customer for other information, including his or her level of education. For the 175 persons in the sample, bag selection and education levels were as shown below. At the 0.01 level, is bag selection independent of education level? Based on the chi-square table, what is the most accurate statement that can be made about the p-value for the test?

		Education Level				
		High School	Some College	College Grad	Graduate Study	
Bag Selection	Paper	14	13	34	2	63
	Plastic	17	19	19	3	58
	No Preference	8	28	13	5	54
		39	60	66	10	175

13.38 Upon leaving an assembly area, production items are examined and some of them are found to be in need of either further work or total scrapping. Tags on a sample of 150 items that failed final inspection show both the recommended action and the identity of the inspector who examined the item. The summary information for the sample is shown below. At the 0.10 level, is the recommended action independent of the inspector? Based on the chi-square table, what is the most accurate statement that can be made about the p-value for the test?

		Inspector			
		A	B	C	
Recommended Action	Major Rework	20	14	13	47
	Minor Rework	18	16	23	57
	Scrap	16	21	9	46
		54	51	45	150

13.39 Collecting data for eight games, a basketball coach has compiled the following contingency table for the quarter of the game versus the result of "one-and-one" free-throw attempts in which a second shot can be attempted if the first one is made. At the 0.025 level, can he conclude that the quarter of the game is independent of the result of the free-throw opportunity?

		Quarter				
		1	2	3	4	
Points Scored	0	6	12	10	17	45
	1	21	17	19	25	82
	2	26	24	11	15	76
		53	53	40	57	203

(DATA SET) *Note:* Exercises 13.40 and 13.41 require a computer and statistical software.

13.40 The manager of a television station employs three different weather reporters (coded as 1–3) and has surveyed viewers to find out whether education level (coded as 1–4) might be related to the weather reporter the viewer prefers. For the results listed in data file **XR13040**, use the 0.05 level of significance in concluding whether weather reporter preference and education level might be related.

13.41 For a random sample of returns audited by the IRS, the data in file **XR13041** describe the results according to the income category of the audited party (coded as 1 = low, 2 = medium, 3 = moderate, and 4 = high) and the type of IRS personnel who examined the return (coded as 1 = revenue agent, 2 = tax auditor, and 3 = service center). At the 0.05 level, are we able to conclude that the category of return is independent of the type of IRS examiner to which a return was assigned?

13.5 COMPARING PROPORTIONS FROM *k* INDEPENDENT SAMPLES

Purpose

In Chapter 11, we presented a test for comparing two sample proportions to examine whether or not the population proportions could be the same. By using chi-square analysis, we can compare two or more sample proportions at the same time. For example, a politician might like to determine whether persons in different age groups are equally likely to have voted in the most recent election, or a production manager could be interested in whether the percentage of defective products differs significantly among workers in a given department.

Procedure

For all practical purposes, the **test for equality of proportions,** the procedure in this section, is really just a special case for the independence of two variables. The only differences are that here (1) each column consists of observations for an independent sample, while in the previous section the entire table represented just one sample; and (2) each table (observed and expected) will always have $r = 2$ rows. The O_{ij} and E_{ij} tables, the calculated χ^2, and the determination of df are all handled the same way as in Section 13.4. The null and alternative hypotheses are

H_0: $\pi_1 = \pi_2 = \pi_3 = \cdots = \pi_k$, for the $j = 1$ through k populations.

H_1: At least one of the π_j values differs from the others.

EXAMPLE

Comparing Proportions

According to the Bureau of the Census, 17% of Americans moved to a different house during the 1980–1981 period. This compares to 16% during 1990–1991 and 14% during 2000–2001.[1] For our example, we will assume that 1000 persons were surveyed during the two earlier studies and that 2000 persons were surveyed in the most recent one. Using the 0.05 level of significance, is it possible

[1]Bureau of the Census, *Statistical Abstract of the United States 2009*, p. 35.

EXAMPLEEXAMPLEEXAMP

that the percentage of Americans who moved in each of these three periods was really the same, but the three surveys showed differences only because of sampling error?

SOLUTION

As mentioned previously, the appropriate procedure is just a special case of the contingency table method in the preceding section, and we proceed in the same manner. Although the test can be performed by tables and pocket calculator, we will use the procedure described in Computer Solutions 13.4, and we will discuss it in terms of the printout shown there. The null hypothesis is H_0: $\pi_1 = \pi_2 = \pi_3$, that the population proportions are equal. Whether the test is carried out by hand calculations or using the computer, the key components are shown below and in the Excel printout in Computer Solutions 13.4:

1. The column totals are 1000, 1000, and 2000, respectively, as shown in cells B5:D5.
2. For each column, we obtain the number of "movers" by multiplying the column total by the percentage of people who moved that year. These frequencies are shown in row 3 of the printout.
3. For each column, we obtain the number of "didn't move" observations by subtracting the number of "movers" from the column total.
4. Overall, 15.25% of the respondents had moved. This is obtained by (170 + 160 + 280)/4000 = 0.1525, and 0.1525 is our estimate of the common proportion for all three groups that is expressed in the null hypothesis. Based on what we've done so far, we can construct the following tables for the observed and expected frequencies:

	1980–1981	1990–1991	2000–2001	
Observed Frequencies				
Moved	170	160	280	610
Didn't move	830	840	1720	3390
Total	1000	1000	2000	4000
Expected Frequencies				
Moved	152.5	152.5	305.0	610
Didn't move	847.5	847.5	1695.0	3390
Total	1000	1000	2000	4000

5. Using the same formula as for the contingency table analyses in the previous section, we obtain the calculated value of the chi-square statistic:

$$\sum_{i=1}^{r}\sum_{j=1}^{k}\frac{(O_{ij}-E_{ij})^2}{E_{ij}} = \frac{(170-152.5)^2}{152.5} + \frac{(160-152.5)^2}{152.5} + \cdots + \frac{(1720-1695)^2}{1695}$$
$$= 5.223$$

6. The calculated chi-square (5.223) does not exceed the critical value (5.992 for the 0.05 level and $df = 2$) and, at the 0.05 level of significance, we are not able to reject the null hypothesis that the proportions of Americans who moved during these time periods were the same. As shown in Computer Solutions 13.4, the exact p-value for the test is 0.073.

COMPUTER 13.4 SOLUTIONS

Chi-Square Test Comparing Proportions from Independent Samples

EXCEL

	A	B	C	D	E
1	**Contingency Table**				
2		*1980-81*	*1990-91*	*2000-01*	TOTAL
3	*moved*	170	160	280	610
4	*didn't*	830	840	1720	3390
5	TOTAL	1000	1000	2000	4000
6					
7	chi-squared Stat			5.223	
8	df			2	
9	p-value			0.073	
10	chi-squared Critical			5.992	

Whether using Excel or Minitab, this procedure is the same as that for testing the independence of two variables, described in Computer Solutions 13.3, except that at least one of the two variables will have just two categories. Using Excel and Data Analysis Plus, the results are as shown above.

Note: As an alternative, you can use Excel worksheet template **TMCHIPRO**. The procedure is described within the template.

EXERCISES

13.42 For the following data obtained from three independent samples, use the 0.05 level in testing H_0: $\pi_1 = \pi_2 = \pi_3$ versus H_1: "At least one population proportion differs from the others."

$$n_1 = 100, \quad p_1 = 0.20$$
$$n_2 = 120, \quad p_2 = 0.25$$
$$n_3 = 200, \quad p_3 = 0.18$$

13.43 For the following data obtained from four independent samples, use the 0.025 level in testing H_0: $\pi_1 = \pi_2 = \pi_3 = \pi_4$ versus H_1: At least one population proportion differs from the others.

$$n_1 = 150, \quad p_1 = 0.30$$
$$n_2 = 80, \quad p_2 = 0.20$$
$$n_3 = 140, \quad p_3 = 0.25$$
$$n_4 = 50, \quad p_4 = 0.26$$

13.44 For three independent samples, each with $n = 100$, the respective sample proportions are 0.30, 0.35, and 0.25. Use the 0.05 level in testing whether the three population proportions could be the same.

13.45 An investment firm survey included the finding that 52% of 150 clients describing themselves as "very aggressive" investors said they were optimistic about the near-term future of the stock market, compared to 46% of 100 describing themselves as "moderate" and 38% of 100 describing themselves as "conservative." Use the 0.01 level in testing whether the three population proportions could be the same.

13.46 According to a study by the Pew Research Center, 75% of adults 18–30 said they went online daily, compared with 40% of those 65–74, and just 16% for those 75 or older. Using the 0.01 level of significance, and assuming that a random sample of 200 persons was surveyed from each age group, could the population proportions be equal for those who say they go online daily from the three groups? Source: "Poll Finds Generation Gap Biggest Since Vietnam War," *USA Today*, June 29, 2009, p. 7A.

13.47 It has been reported that 18.3% of all U.S. households were heated by electricity in 1980, compared to 27.4% in 1995 and 31.5% in 2005. At the 0.05 level, and assuming a sample size of 1000 U.S. households for each year, test whether the population percentages could be equal for these years. Source: *Statistical Abstract of the United States 1996*, p. 722; *Statistical Abstract of the United States 2009*, p. 610.

13.48 In analyzing the consumption of cottage cheese by members of various occupational groups, the United Dairy Industry Association found that 326 of 837 professionals seldom or never ate cottage cheese, versus 220 of 489 white-collar workers and 522 of 1243 blue-collar workers. Assuming independent samples, use the 0.05 level in testing the null hypothesis that the population proportions could be the same for the three occupational groups. Source: United Dairy Industry Association, *Cottage Cheese Attitudes & Usage*, p. 25.

(**DATA SET**) *Note:* Exercises 13.49 and 13.50 require a computer and statistical software.

13.49 The movie complex at a shopping mall shows three movies simultaneously on the same evening. On a recent Friday evening, each movie drew a capacity crowd. A sample of the evening's movie patrons have been coded according to whether they purchased snacks (1 = purchased snacks, 2 = did not purchase snacks) and which movie they attended (1 = the "G" movie, 2 = the "PG" movie, and 3 = the "R" movie). The data are in file **XR13049**. At the 0.025 level, could the percentage buying snacks be the same for all three movies?

13.50 An experiment has been conducted to compare the ease of use of several pocket calculators, with subjects randomly provided with one of four calculator designs. The subjects have been coded according to which one of the four calculators they tried (codes = 1–4) and whether they thought the calculator was "easy to use" (1 = easy to use, 2 = not easy to use). The data are in file **XR13050**. At the 0.01 level of significance, test whether the calculators could be equally easy to use.

ESTIMATION AND TESTS REGARDING THE POPULATION VARIANCE (13.6)

Besides enabling us to carry out the tests presented in earlier sections of the chapter, the chi-square distribution is the basis for both estimation and hypothesis testing regarding the population variance. Like the population mean, the population variance will typically be unknown; it must be estimated by a sample statistic, and there will be uncertainty with regard to its actual value. In this section, we will quantify this uncertainty by using sample information and the chi-square distribution in (1) constructing a confidence interval for σ^2 and (2) conducting hypothesis tests regarding its value.

The Confidence Interval for a Population Variance

The sample variance (s^2) is a point estimate of the population variance, and as such it is our single best estimate as to the value of σ^2. Just as it can be desirable to have an interval estimate for μ, it can also be useful to have an interval estimate for σ^2. For example, a production machine may be readily adjustable with regard to the mean of its output (e.g., a diameter or weight), but due to mechanical wear on bearings and other components, the amount of variability in its output may both increase and accelerate as the "wearing-out" process persists. At some point, a decision may have to be made regarding either the replacement or the overhaul of the machine, and such a decision could rely heavily on the variability that is present in a sample of output.

The chi-square distribution was described earlier as the sampling distribution for $\chi^2 = (n - 1)s^2/\sigma^2$ when random samples of size n are repeatedly drawn from a normal population. Of particular importance is the fact that we can transform the preceding equation and have $\sigma^2 = (n - 1)s^2/\chi^2$. This will be the statistic of interest in our determination of the confidence interval for σ^2. An important assumption for both the construction of confidence intervals and the carrying out of hypothesis tests for σ^2 is that *the population must be normally distributed.*

While s^2 is the point estimate of σ^2, and a quantity from which we proceed to determine the lower and upper confidence limits, the confidence interval is not $s^2 \pm$ a fixed quantity. This is because, unlike the t and z distributions, the chi-square distribution is asymmetric. As Figure 13.1 showed, it is positively skewed. Thus, the description of the confidence interval for σ^2 requires separate calculations for its upper and lower limits:

Confidence interval limits for the variance of a normal population:

There is $100(1 - \alpha)\%$ confidence that σ^2 is within the interval

$$\frac{(n-1)s^2}{\chi^2_U} \le \sigma^2 \le \frac{(n-1)s^2}{\chi^2_L}$$

where s^2 = sample variance
n = sample size
χ^2_U = chi-square value where upper-tail area = $\alpha/2$
χ^2_L = chi-square value where lower-tail area = $\alpha/2$
$df = n - 1$

For the $100(1 - \alpha)\%$ confidence interval and $df = n - 1$, χ^2_L is the chi-square value for which the area in the lower, or left, tail is $\alpha/2$, while χ^2_U is the chi-square value for which the area in the upper, or right, tail is $\alpha/2$. Notice that χ^2_L is within the expression for the *upper* confidence limit, and χ^2_U is within that for the *lower* confidence limit. Because they are in the *denominators* of their respective expressions, their roles are the reverse of what intuition might suggest.

For example, if $df = 5$ and we are constructing a 90% confidence interval, the lower confidence limit for σ^2 would have $\chi^2_U = 11.070$ in its denominator, and the upper confidence limit for σ^2 would have $\chi^2_L = 1.145$ in its denominator. Shown in Figure 13.4, along with their respective tail areas, these values would be used in determining the confidence limits for σ^2.

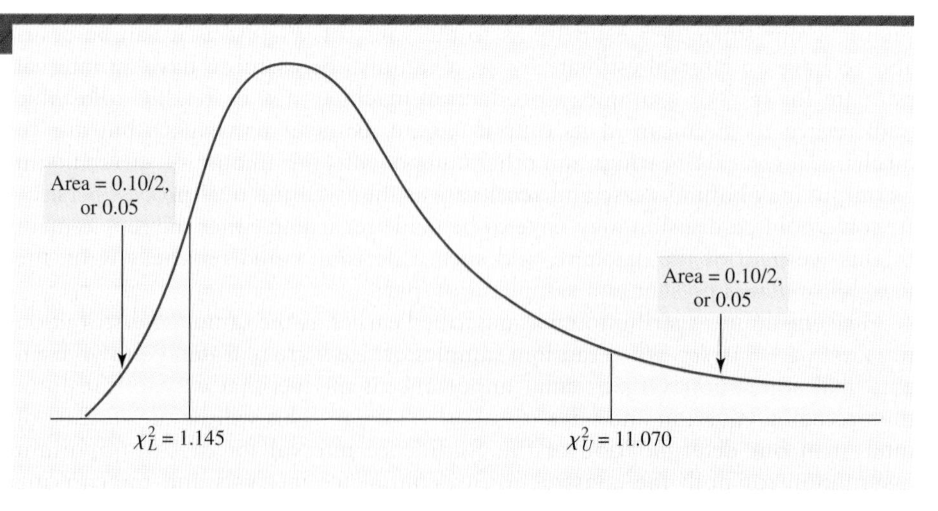

FIGURE 13.4

Chi-square values used in determining the 90% confidence limits for σ^2 when $df = 5$. Because they appear in the denominators of their respective confidence limit expressions, χ^2_U is associated with the *lower* confidence limit for σ^2, and χ^2_L is associated with the *upper* confidence limit for σ^2.

EXAMPLE

Confidence Interval for a Population Variance

A total quality management (TQM) supervisor is interested in examining variability in the amount of product being inserted by a filling machine. From a sample of $n = 15$ packages, the average weight of the contents was 412.72 grams, with a standard deviation of 5.30 grams. The underlying data are in file CX13FILL. If the filling weights are from a normally distributed population, what is the 98% confidence interval for the population variance?

SOLUTION

From this sample, the point estimate of σ^2 is s^2, or $5.30^2 = 28.09$. However, since the chi-square distribution is not symmetrical, $s^2 = 28.09$ will not be at the midpoint of the confidence interval. For the 98% confidence interval, $\alpha = 0.02$, and the area in each tail beneath the chi-square curve will be $\alpha/2$, or 0.01. The next step is to determine the value of χ^2_L that corresponds to a lower-tail area of 0.01 and the value of χ^2_U that corresponds to an upper-tail area of 0.01. These can be found by referring to chi-square values listed in Appendix A, or

- For a lower-tail area of 0.01 and $df = 14$:

$$\chi^2_L = 4.660$$

- For an upper-tail area of 0.01 and $df = 14$:

$$\chi^2_U = 29.141$$

The 98% confidence interval for σ^2 can now be found through separate calculations for its two limits:

- The lower 98% confidence limit for σ^2 is

$$\frac{(n-1)s^2}{\chi^2_U} = \frac{(15-1)(5.30)^2}{29.141} = 13.495$$

- The upper 98% confidence limit for σ^2 is

$$\frac{(n-1)s^2}{\chi^2_L} = \frac{(15-1)(5.30)^2}{4.660} = 84.391$$

The TQM supervisor will have 98% confidence that the population variance is somewhere between 13.495 and 84.391. Since σ is the square root of σ^2, we can also say that he has 98% confidence that the population standard deviation is somewhere between $\sqrt{13.495}$ and $\sqrt{84.391}$, or from 3.674 to 9.186 grams.

Hypothesis Tests for the Population Variance

As with the construction of a confidence interval for σ^2, hypothesis testing also requires that the population be normally distributed. As with previous hypothesis tests, we can carry out either nondirectional or directional tests for σ^2. Regardless of the directionality of the test, the test statistic will be the same.

Test statistic for tests regarding a population variance:

$$\chi^2 = \frac{(n-1)s^2}{\sigma_0^2}$$

where s^2 = sample variance
n = sample size
σ_0^2 = a hypothesized value for σ^2
$df = n - 1$

Although the test statistic is the same for any test, the null and alternative hypotheses, the critical value(s), and the decision rule will depend on the type of test being conducted. The possibilities are summarized in Table 13.3, where the level of significance for each test has been identified as α.

TABLE 13.3

Hypothesis tests for the population variance. Each test is at the α level, the test statistic is $\chi^2 = (n-1)s^2/\sigma_0^2$, and $df = n - 1$.

A. *Two-tail test*

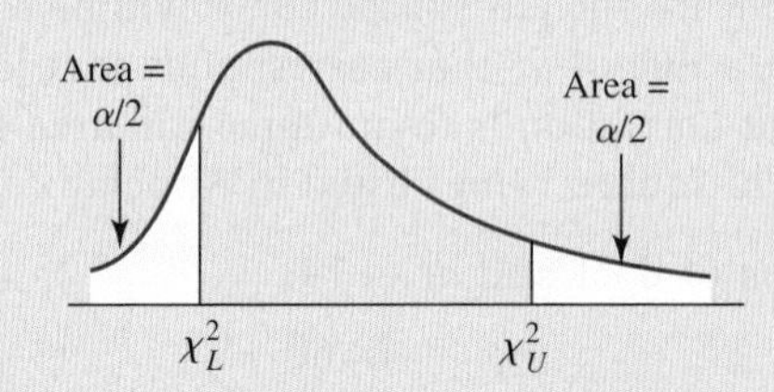

Null and alternative hypotheses: H_0: $\sigma^2 = \sigma_0^2$ and H_1: $\sigma^2 \neq \sigma_0^2$
Critical values: χ_L^2, corresponding to lower-tail area of $\alpha/2$
χ_U^2, corresponding to upper-tail area of $\alpha/2$
Decision rule: Reject H_0 if calculated χ^2 is either $< \chi_L^2$ or $> \chi_U^2$; otherwise, do not reject.

B. *Upper-tail test*

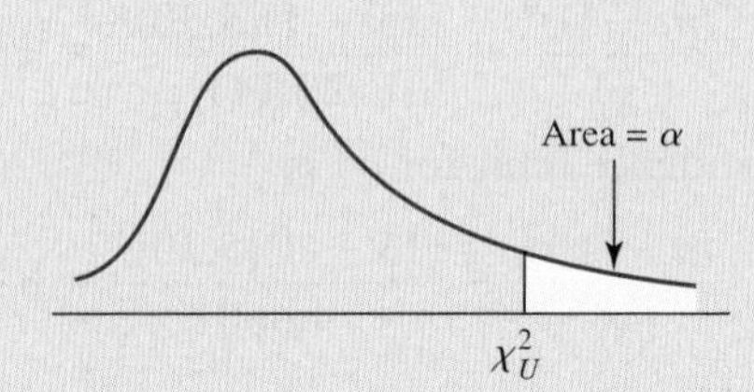

Null and alternative hypotheses: H_0: $\sigma^2 \leq \sigma_0^2$ and H_1: $\sigma^2 > \sigma_0^2$
Critical value: χ_U^2, corresponding to upper-tail area of α
Decision rule: Reject H_0 if calculated $\chi^2 > \chi_U^2$; otherwise, do not reject.

C. *Lower-tail test*

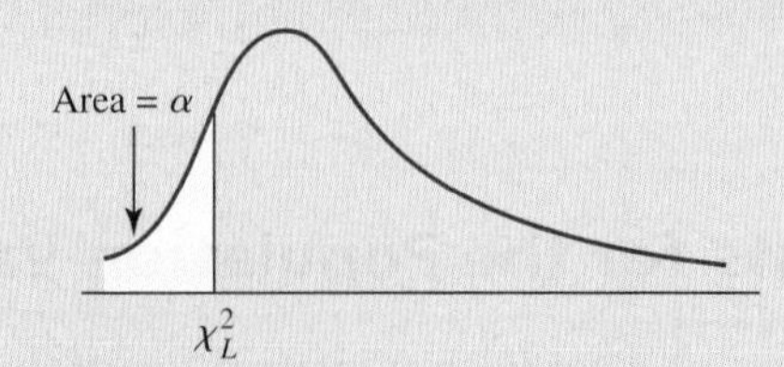

Null and alternative hypotheses: H_0: $\sigma^2 \geq \sigma_0^2$ and H_1: $\sigma^2 < \sigma_0^2$
Critical value: χ_L^2, corresponding to lower-tail area of α
Decision rule: Reject H_0 if calculated $\chi^2 < \chi_L^2$; otherwise, do not reject.

EXAMPLE

Hypothesis Test for a Population Variance

A computer manufacturer requires that springs purchased for installation beneath its keyboard keys have no more than $\sigma = 5$ grams of variability in the amount of force required for key depression. From a shipment of several thousand springs, a random sample of 20 springs is tested, and the standard deviation is 7.78 grams. The underlying data are in file **CX13KEYS**. Assuming a normal distribution for the force required to compress the springs, use the 0.01 level of significance in examining whether the shipment meets the manufacturer's specifications.

SOLUTION

The sample variance is $7.78^2 = 60.53$, and the value of σ^2 associated with the test is $\sigma_0^2 = 5^2 = 25.00$. The test is one-tail, since the computer manufacturer wishes to know whether the population variance exceeds the amount that is acceptable. The null and alternative hypotheses are

$$H_0\text{: } \sigma^2 \leq 25.00 \quad \text{and} \quad H_1\text{: } \sigma^2 > 25.00$$

The test statistic is χ^2. For $\sigma_0^2 = 25.00$, $s^2 = 60.53$, and $n = 20$, the calculated value of the test statistic is

$$\text{Calculated } \chi^2 = \frac{(n-1)s^2}{\sigma_0^2} = \frac{(20-1)(60.53)}{25.00} = 46.00$$

As part B of Table 13.3 shows, this combination of null and alternative hypotheses calls for an upper-tail test. Referring to the $df = 19$ row of the chi-square table, we find that $\chi_U^2 = 36.191$ is the chi-square value that corresponds to an upper-tail area of 0.01. The calculated value (46.00) falls into the "rejection" region shown in Figure 13.5. At the 0.01 level of significance, we reject H_0 and conclude that the shipment of springs does not meet the computer manufacturer's specifications.

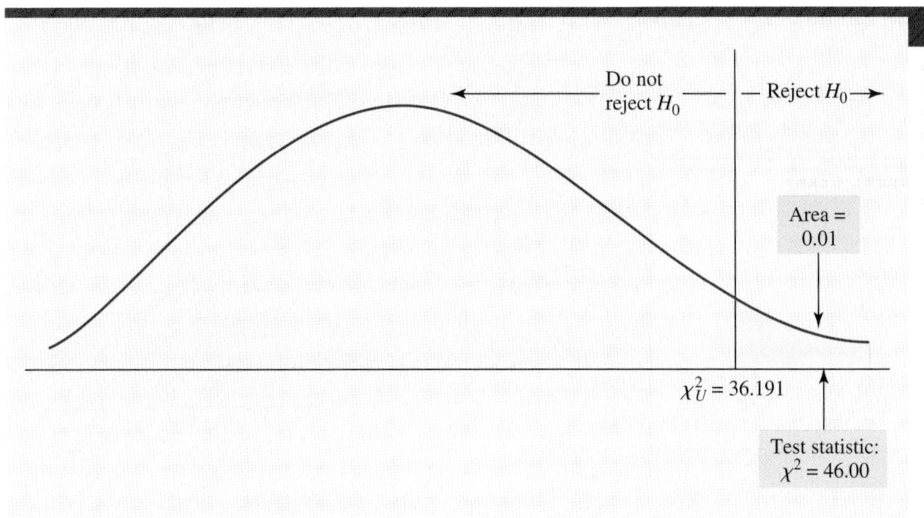

FIGURE 13.5

An upper-tail test for a population variance. The calculated value of the χ^2 test statistic allows us to reject the null hypothesis that the population variance is no more than 25.00.

Computer Solutions 13.5 shows the procedure and results when we use Excel and Minitab to generate a confidence interval for a population variance. In Computer Solutions 13.6, we use Excel and Minitab to carry out a hypothesis test for a population variance.

COMPUTER 13.5 SOLUTIONS

Confidence Interval for a Population Variance

EXCEL

	A	B	C	D
1	Chi-Squared Estimate of a Variance			
2				
3	Sample variance	28.09	Confidence Interval Estimate	
4	Sample size	15	Lower confidence limit	13.495
5	Confidence level	0.98	Upper confidence limit	84.383

Excel confidence interval for a population variance, based on summary information

1. The Excel data file for the filling machine example earlier in this section is CX13FILL. If we were not provided with summary information in this example, we would first use Excel to find the sample variance. In this file, the data values are in cells A2:A16. Just click on any empty cell and enter the formula **=VAR(A2:A16).** The variance is 5.30^2, or 28.09.
2. Open the ESTIMATORS workbook.
3. Using the arrows at the bottom left, select **Chi-Squared-Estimate_Variance.** Enter the sample variance **(28.09),** the sample size **(15),** and the desired confidence level as a decimal fraction **(0.98)** into the appropriate cells. The upper and lower confidence limits for the population variance are provided as shown above. (These limits are slightly more accurate than our text example, in which we used table-based chi-square values in our calculations.)

Excel confidence interval for a population variance, based on raw data

1. Open Excel file CX13FILL. From the **Add-Ins** ribbon, click **Data Analysis Plus.** Select **Chi-Squared Estimate: Variance.** Click **OK.**
2. Enter **A1:A16** into the **Input Range** box. Click **Labels.** For a 98% confidence interval, enter **0.02** into the **Alpha** box. Click **OK.**

MINITAB

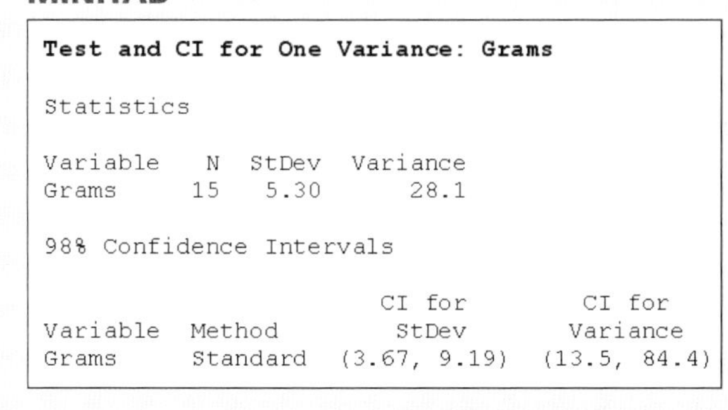

```
Test and CI for One Variance: Grams

Statistics

Variable   N  StDev  Variance
Grams     15   5.30      28.1

98% Confidence Intervals

                          CI for        CI for
Variable  Method          StDev       Variance
Grams     Standard  (3.67, 9.19)  (13.5, 84.4)
```

Minitab confidence interval for a population variance, based on raw data

1. Open Minitab file CX13FILL. Click **Stat,** select **Basic Statistics,** and click **1 Variance.**
2. Select **Samples in columns** and enter **C1** into the box. Click **Options** and enter **98** into the **Confidence Level** box. Click **OK.** Click **OK.** The 98% confidence interval will be as shown at bottom of previous page.

Minitab confidence interval for a population variance, based on summary data

This will be identical to the procedure for raw data, but in step 2, select **Enter variance** and **Summarized data,** then enter **15** into the **Sample size** box and **28.09** into the **Sample variance** box.

COMPUTER 13.6 SOLUTIONS

Hypothesis Test for a Population Variance

EXCEL

	A	B	C	D
1	Chi-squared Test of a Variance			
2				
3	Sample variance	60.53	Chi-squared Stat	46.003
4	Sample size	20	P(CHI<=chi) one-tail	0.0005
5	Hypothesized variance	25	chi-squared Critical one-tail	36.191
6	Alpha	0.01	P(CHI<=chi) two-tail	0.001
7			chi-squared Critical two-tail	6.844
8				38.582

Excel hypothesis test for a population variance, based on summary information

1. The Excel data file for the keyboard spring force example in this section is CX13KEYS. If we were not provided with summary information in this example, we would first use Excel to find the sample variance. In this file, the data values are in cells A2:A21. Just click on any empty cell and enter the formula **=VAR(A2:A21).** The variance is 7.78^2, or 60.53.
2. Open the TEST STATISTICS workbook.
3. Using the arrows at the bottom left, select **Chi-squared Test_Variance.** Enter the sample variance **(60.53),** the sample size **(20),** the hypothesized value for the population variance **(25.0)** and the level of significance **(0.01)** for the test into the appropriate cells.

Excel hypothesis test for a population variance, based on raw data

1. Open Excel file CX13KEYS. From the **Add-Ins** ribbon, click **Data Analysis Plus.** Select **Chi-Squared Test: Variance.** Click **OK.**
2. Enter **A1:A21** into the **Input Range** box. Click **Labels.** Enter **25.0** into the **Hypothesized Variance** box and **0.01** into the **Alpha** box. Click **OK.**

(continued)

MINITAB

```
Test and CI for One Variance

Null hypothesis         Sigma-squared = 25
Alternative hypothesis  Sigma-squared > 25

Statistics

 N  StDev  Variance
20   7.78      60.5

95% One-Sided Confidence Intervals

          Lower
          Bound
            for   Lower Bound
Method    StDev  for Variance
Standard   6.18          38.2

Tests

Method    Chi-Square  DF  P-Value
Standard       46.00  19    0.000
```

Minitab hypothesis test for a population variance, based on summary information

1. Click **Stat,** select **Basic Statistics,** and click **1 Variance.**
2. Select **Sample Variance**, then enter **20** into the **Sample size** box and **60.53** into the **Sample variance** box. Select **Perform hypothesis test** and enter **25** into the **Hypothesized variance** box.Click **Options** and select **Greater than** from the **Alternative** menu. Click **OK.** Click **OK.** The results of the hypothesis test will appear as shown above.

Minitab hypothesis test for a population variance, based on raw data

Open Minitab file CX13KEYS. The procedure will now be the same as that for summary data but, in step 2, select **Samples in columns** and enter **C1** into the box.

EXERCISES

13.51 In applying the chi-square statistic to estimation and tests for a population variance, why must the population be normally distributed?

13.52 In previous chapters, confidence intervals have been expressed in terms of a sample statistic plus or minus a given expression—e.g., $\bar{x} \pm t(s/\sqrt{n})$. However, the confidence interval for σ^2 cannot be expressed as s^2 plus or minus a comparable term. Why is this not possible?

13.53 A random sample of 30 observations has been drawn from a normal population, and the sample variance is found to be $s^2 = 23.8$. Determine the 95% confidence interval for σ^2.

13.54 A random sample of 10 observations has been drawn from a normal population, and the sample variance is found to be $s^2 = 19.5$. Determine the 98% confidence interval for σ^2.

13.55 A random sample of 20 observations has been drawn from a normal population, and the sample variance is found to be $s^2 = 4.53$. Determine the 95% confidence interval for σ^2.

13.56 A pharmaceutical company has specified that the variation in the quantity of active ingredient in its leading prescription medication must be such that the population standard deviation must be no more than 1.20 micrograms. The ingredient content of the pills is normally distributed. In a random sample of 20 pills, the standard deviation is found to be $s = 1.32$ micrograms. Based on the sample data, use the $\alpha = 0.05$ level in testing H_0: $\sigma \leq 1.20$ micrograms versus H_1: $\sigma > 1.20$ micrograms.

13.57 During the early stages of production for a single-cylinder engine designed for use in portable generators, the manufacturer randomly selects 12 engines from the end of the assembly line and subjects them to a battery of tests. In the horsepower tests, the engines are found to have an output standard deviation of $s = 0.18$ horsepower. Assuming a normal distribution of horsepower outputs, determine the 90% confidence interval for σ.

13.58 An instrument maker has randomly selected and submitted 20 of its electronic blood-pressure measurement units to a medical laboratory for testing. At the lab, each unit measures the "blood pressure" of a testing device that simulates a human arm; within the device the pressure is a known constant. For the instruments submitted for testing, the standard deviation of the readings is 1.4 millimeters. Assuming a normal distribution of measurement values, determine the 95% confidence interval for the population standard deviation of instruments produced by this manufacturer.

13.59 A random sample of $n = 30$ is drawn from a population that is normally distributed, and the sample variance is $s^2 = 41.5$. Use $\alpha = 0.05$ in testing H_0: $\sigma^2 = 29.0$ versus H_1: $\sigma^2 \neq 29.0$.

13.60 A random sample of $n = 12$ is drawn from a population that is normally distributed, and the sample variance is $s^2 = 19.3$. Use $\alpha = 0.025$ in testing H_0: $\sigma^2 \leq 9.4$ versus H_1: $\sigma^2 > 9.4$.

13.61 According to industry specifications, the standard deviation of gasket thicknesses for a given application must be no more than 3 thousandths of an inch. In a random sample of 25 gaskets produced by one of the firms, the sample standard deviation is found to be 3.7 thousandths of an inch. Assuming the population of thicknesses is normally distributed, and using $\alpha = 0.025$, examine whether the firm could be in violation of the industry specification. Based on the chi-square table, what is the most accurate statement that can be made about the p-value for the test?

(**DATA SET**) *Note:* Exercises 13.62 and 13.63 require a computer and statistical software.

13.62 In past studies, a fast-food franchise has found the standard deviation of weights for quarter-pound hamburgers to be approximately 0.28 ounces after cooking. A consultant has suggested a new procedure for hamburger preparation and cooking that she believes will reduce the variability in the cooked weight of the hamburgers. In a test of the new method, 61 hamburgers are prepared, cooked, and weighed, with the resulting cooked weights listed in data file **XR13062**. At the 0.05 level, and assuming a normal distribution for the weights, examine the possibility that the new procedure is actually no better than the old one.

13.63 Given the information in Exercise 13.62 and data file **XR13062**, determine the 98% confidence interval for the population variance of the cooked weights for hamburgers prepared using the consultant's recommended cooking procedure.

SUMMARY (13.7)

● **The chi-square distribution**

The chapter introduces a probability distribution called the chi-square distribution. When samples of size n are drawn from a normal population, the chi-square distribution is the sampling distribution of $\chi^2 = (n - 1)s^2/\sigma^2$. The chi-square distribution is also a family of distributions, with the specific member of the family depending on the number of degrees of freedom, and with the determination of df depending on the particular test or procedure being carried out. The chi-square distribution has extreme positive skewness for smaller values of df, but approaches the normal distribution as df becomes larger.

● **Applications and procedures for chi-square tests**

The chi-square distribution, along with sample data and frequency counts, is used to examine (1) whether a sample could have come from a given type of population distribution (e.g., the normal distribution), (2) whether two nominal variables could be independent of each other, and (3) whether two or more independent samples could have the same population proportion. In these applications, chi-square analysis involves the comparison of a table of observed frequencies (O_{ij}) with a table of expected frequencies (E_{ij}) that has been constructed under the assumption that the null hypothesis is true.

The calculated value of the chi-square statistic represents the extent to which these tables differ. If the calculated chi-square exceeds the critical value, H_0 is rejected. In these tests, it is important that the total sample size and the categories be such that each E_{ij} value is at least 5. This is because the chi-square distribution is continuous and is being used to analyze discrete (*counts*) data. If one or more expected frequencies do not meet this minimum requirement, we should either increase the total sample size or combine adjacent rows or columns in the table.

- **Confidence intervals and testing for a population variance**

The chi-square distribution is also used in constructing confidence intervals and carrying out hypothesis tests regarding the value of a population variance. In these applications, an important assumption is that the population is normally distributed.

EQUATIONS

Chi-Square Test for Goodness of Fit

$$\text{Calculated } \chi^2 = \sum_{j=1}^{k} \frac{(O_j - E_j)^2}{E_j}$$

where k = number of categories, or cells in the table
O_j = observed frequency in cell j
E_j = expected frequency in cell j

In this test, $df = k - 1 - m$, with m = the number of population parameters (e.g., μ or σ) that must be estimated from sample data in order to carry out the goodness-of-fit test. The test is upper-tail, and the critical value of χ^2 will be that associated with the level of significance (α) and degrees of freedom (df) for the test. Each E_j must be at least 5. If not, either the total sample size should be increased, or cells with $E_j < 5$ can be combined with adjacent cells.

Chi-Square Test for the Independence of Two Variables

$$\text{Calculated } \chi^2 = \sum_{i=1}^{r} \sum_{j=1}^{k} \frac{(O_{ij} - E_{ij})^2}{E_{ij}}$$

where r = number of rows in the contingency table
k = number of columns in the contingency table
O_{ij} = observed frequency in row i, column j
E_{ij} = expected frequency in row i, column j

The expected frequency in row i, column j is

$$E_{ij} = p_j(n_i)$$

where p_j = for all the observations, the proportion that are in column j
n_i = number of observations in row i

In this test, $df = (r - 1)(k - 1)$, the number of cell frequencies that are free to vary. Given that we know the sample size (the total number of observations in the table), we need only know the contents of $(r - 1)$ rows and $(k - 1)$ columns in order to completely fill in the cells within the $r \times k$ table. Each E_{ij} must be at least 5. If not, either the total sample size should be increased, or adjacent rows or columns can be combined until each E_{ij} is ≥ 5.

Chi-Square Test for the Equality of Population Proportions

This procedure is a special case of the test for independence above, but the columns represent independent samples from the respective populations, and there will be just two rows in the contingency table.

Confidence Interval Limits for the Variance of a Normal Population

There is $100(1 - \alpha)\%$ confidence that σ^2 is within the interval

$$\frac{(n-1)s^2}{\chi^2_U} \leq \sigma^2 \leq \frac{(n-1)s^2}{\chi^2_L}$$

where s^2 = sample variance
n = sample size
χ^2_U = chi-square value where upper-tail area = $\alpha/2$
χ^2_L = chi-square value where lower-tail area = $\alpha/2$

In this test, $df = n - 1$. The population must be normally distributed.

Test Statistic for Hypothesis Tests Regarding a Population Variance

$$\chi^2 = \frac{(n-1)s^2}{\sigma_0^2}$$

where s^2 = sample variance
n = sample size
σ_0^2 = a hypothesized value for σ^2

Regardless of the test, $df = n - 1$. The test statistic is the same for any test, but the null and alternative hypotheses, the critical value(s), and the decision rule will depend on the type of test being conducted. The possibilities are summarized in Table 13.3, and the population must be normally distributed.

CHAPTER EXERCISES

13.64 The personnel director of a large firm has summarized a random sample of last year's absentee reports in the accompanying contingency table. At the 0.01 level of significance, test the independence of age versus length of absence.

		Number of Days Absent				
		1	2–4	5–7	>7	
Age Group	<25 years	30	8	3	4	45
	26–40 years	15	12	5	7	39
	>40 years	18	22	11	10	61
		63	42	19	21	145

13.65 Frank Gerbel, a local insurance representative, has hired a part-time assistant to make "cold" calls on potential customers. The assistant has been given a quota of 20 such calls to make each day, and a call is deemed a success if the potential customer agrees to set up an appointment to discuss his or her insurance needs with Mr. Gerbel. Over the past 50 days, the distribution of the number of successful calls made by the assistant is as shown below. Using the 0.05 level of significance, test the goodness of fit between these data and a Poisson distribution with $\lambda = 1.4$.

Number of Successful Calls	Number of Days
0	17
1	15
2	9
3	4
4	1
5	3
6	1
	50

13.66 For the most recent year reported, the Federal Bureau of Investigation has listed the following data for robberies in the United States. Source: *Statistical Abstract of the United States 2006,* p. 201.

Street or highway	179,000 robberies
Bank or commercial establishment	70,000
Gas station or convenience store	37,000
Residence	57,000

The police chief in a medium-sized community has data indicating that the respective numbers of these robberies in his community during the same year were 23, 9, 7, and 11. At the 0.05 level, can the distribution of robberies in this community be considered as having differed significantly from that experienced by the United States as a whole?

13.67 The following data are the number of persons who were waiting in line at a checkout counter at 100 randomly selected times during the week. At the 0.05 level, test whether the population of these values could be Poisson distributed.

x = Number of Persons	Number of Observations
0	24
1	24
2	18
3	20
4	10
5	4
	100

13.68 A newspaper reporter, collecting data for a feature article on her state lottery system, has found the 200 digits most recently selected to be distributed as shown below. Based on this information, and using the 0.10 level of significance, can she conclude that the digits have not been drawn from the appropriate discrete uniform distribution?

Digit	0	1	2	3	4	5	6	7	8	9	
Frequency	17	18	24	25	19	16	14	23	24	20	200

13.69 According to the Bureau of the Census, the U.S. population includes 37.7 million persons who were born in Europe, Asia, the Caribbean, Central America, South America, or in another area outside the United States. A percentage breakdown of the regions of birth is shown here. If a random sample of 300 foreign-born persons were to be selected from a major U.S. city, with the observed frequencies shown in the rightmost column of the table, could we conclude that the distribution of birth regions for the foreign-born persons in this city could be the same as the national-level distribution in the adjacent column? Use the 0.05 level of significance in reaching a conclusion. Source: Bureau of the Census, *Statistical Abstract of the United States 2009,* p. 44.

Birth Region	National-Level Percentage	Number of Persons in Sample
Europe	11.8%	43
Asia	26.3	87
Caribbean	9.1	30
Central America	38.9	102
South America	6.8	25
Other	7.1	13
	100.0	300

13.70 A national public television network has found that 35% of its contributions are for less than \$20, with 45% for \$20–\$50, and 20% for more than \$50. In a random sample of 200 of the contributions to a local station, 42% were for less than \$20, 43% were for \$20–\$50, and 15% were over \$50. At the 0.05 level, does the local station's distribution of contributions differ significantly from that experienced nationally?

13.71 Twenty-five percent of the employees of a large firm have been with the firm less than 10 years, 40% have been with the firm for 10 to 20 years, and 35% have been with the firm for more than 20 years. Management claims to have randomly selected 20 employees to participate in a drawing for a new car. Of the 20 employees chosen, 3 have fewer than 10 years with the firm, 5 have been there between 10 and 20 years, and 12 have been with the firm for over 20 years. At the 0.025 level of significance, evaluate management's claim of random selection. Using the chi-square table in Appendix A, what is the most accurate statement that can be made about the exact p-value for the test?

13.72 The following contingency table describes types of collisions versus driving environments for a random sample of two-vehicle accidents that occurred in a given region last year. At the 0.01 level of significance, can we conclude that the type of collision is independent of whether the accident took place in a rural versus an urban setting?

		Type of Collision			
		Angle	**Rear-end**	**Other**	
Driving Environment	**Urban**	40	30	72	142
	Rural	6	12	15	33
		46	42	87	175

13.73 Of the students at a large university, 30% are freshmen, 25% are sophomores, 25% are juniors, and 20% are seniors. The representation of the four classes in the

100-member marching band is found to be 24%, 28%, 19%, and 29%. At the 0.05 level of significance, does the composition of the marching band differ significantly from the class distribution of the university? Using the chi-square table in Appendix A, what is the most accurate statement that can be made about the exact p-value for the test?

13.74 Three different instructional techniques are being considered for training mechanics to perform a difficult emissions-reduction adjustment on fuel-injection engines. Each of 60 mechanics is randomly assigned to receive one of the training techniques. For the 20 trained by technique A, 70% are able to perform the adjustment successfully on the first try. For mechanics trained by techniques B and C, the respective percentages are 50% and 65%. At the 0.05 level, can we conclude that the three techniques are equally effective?

13.75 Taking air samples at random times during a 2-month period, an environmental group finds the standard deviation of ozone concentration measurements in the vicinity of a chemical plant to be 0.03 parts per million (ppm). The group believes the population of ozone measurements is normally distributed, and they have decided to use the 0.03 ppm figure as a baseline for the variability in the values. They will continue taking air samples, with 12 samples taken at random times during each week. If they find that the standard deviation of observations for any given week is abnormally higher or lower than 0.03 ppm, they will conclude that something unusual is happening at the plant. Given that the group is estimating σ as 0.03, and assuming that they are relying on a 98% confidence interval, how high or low would a weekly standard deviation have to be for them to conclude that unusual events are taking place at the plant?

13.76 A local bank has recently installed an automatic teller machine (ATM). At a regional meeting, an ATM specialist tells the bank manager that usage times will be normally distributed and the variability in the amount of time customers require to use the machine will start out relatively high, then eventually settle down to a standard deviation of 0.8 minutes. The manager, skeptical that the variability could be this great, has a study done in which a random sample of 30 persons is timed while using the machine. Analysis of the data shows a standard deviation of 0.61 minutes. At the 0.05 level, test whether the standard deviation of usage times for this bank's customers might actually be less than the amount predicted at the regional meeting.

13.77 According to the Bureau of the Census, 36% of U.S. adults in the $20,001–$35,000 income group participated in adult education during the most recent reporting year. Data for this and other groups are shown here. Source: *Statistical Abstract of the United States 2009*, p. 184.

Income Group	Percentage Who Participated in Adult Education
$20,001–$35,000	36%
$35,001–$50,000	42%
$50,001–$75,000	48%

Assuming independent samples of 1000 for each of these income groups, use the 0.10 level of significance in testing the null hypothesis that $\pi_1 = \pi_2 = \pi_3$.

13.78 Over the past year, an audio store has sold stereo systems produced by three different manufacturers. Of the 400 units from company A, 12% were returned for service during the warranty period. This compares to 15% of 300 units from company B, and 16% of 500 from company C. Assuming that the store's sales are a representative sample of the stereo systems produced by the three manufacturers, use the 0.025 level of significance in testing whether the population percentages of warranty service returns could be equal.

13.79 Union members in a large firm work in three departments: assembly, maintenance, and shipping/receiving. Twenty of a random sample of 30 assembly workers say they intend to ratify the new contract negotiated by union leaders and management. This compares to 30 of 40 maintenance workers and 25 of 50 shipping/receiving workers who say they will ratify the contract. At the 0.025 level of significance, test the null hypothesis that the population proportions are the same for all three departments.

13.80 An aspirin manufacturer has designed the child-proof cap on its bottles so that it must be pressed down before it can be turned and removed. A consumer magazine has randomly purchased 15 bottles of the company's aspirin for analysis in a feature article on pain relievers and their packaging. Among the analyses is a test of how much force is required to depress the cap on each bottle. The standard deviation for the bottles made by this manufacturer is found to be 7.2 grams. From the magazine's findings, and assuming a normal distribution, determine the 95% confidence interval for the population standard deviation of the force required to depress the firm's aspirin bottle caps.

13.81 A campus newspaper is examining the extent to which various segments of the university community favor the Student Government Association's recommendation that the school make the transition from Division II to Division I NCAA football. A random sample of 100 students is chosen from each of the major colleges. Of persons surveyed from the College of Business, 47% favor the move. Corresponding percentages for the College of Fine Arts, the College of Science and Technology, and the College of Humanities and Social Sciences are 62%, 45%, and 39%. At the 0.05 level, test whether the population percentages could be the same for students from these four colleges within the university.

13.82 A casting machine is supposed to produce units for which the standard deviation of weights is 0.4 ounces. According to the maker of the machine, it should be shut down and thoroughly examined for wear if the standard deviation becomes greater than the 0.4 ounces that has been specified. In a random sample of 30 units recently produced by the machine, the standard deviation is found to be 0.82 ounces. Given this information, and assuming a normally distributed population of weights, would you feel very secure in advising management that the machine should be shut down? Use the approximate p-value of the test for assistance in formulating your answer.

(**DATA SET**) *Note:* Exercises 13.83–13.88 require a computer and statistical software.

13.83 At the beginning of last year, a power company asked each of its industrial customers to estimate how much electricity it would require during the year. At the end of the year, the power company recorded the extent to which a customer's estimate was over or under the amount of electricity actually used. In data file **XR13083**, customers are described according to their size (1 = small, 2 = medium, 3 = large) and the extent to which their usage differed compared to their estimate (1 = used more than 110% of estimate, 2 = within 10% of estimate, and 3 = used less than 90% of estimate). At the 0.025 level of significance, test whether customer size might be independent of estimation accuracy.

13.84 The data in file **XR13084** represent the Scholastic Aptitude Test (SAT) scores for a random sample of 200 high school seniors who have taken the exam. Using the 0.025 level of significance, test whether the sample scores could have come from a population that is normally distributed.

13.85 For a random sample of households in a county, file **XR13085** lists the number of TV sets owned by each of the households in the study. Using the 0.025 level of significance, test whether x = the number of television sets per household could be Poisson distributed in the county's households.

13.86 Researchers have coded a random sample of households according to household income (1 = under \$30,000, 2 = \$30,000–\$49,999, 3 = \$50,000–\$69,999, and 4 = \$70,000 or more) and number of vehicles owned (1 = 0, 2 = 1 or 2, and 3 = 3 or more). The data are in file **XR13086**. Using the 0.05 level of significance, test whether household income is independent of the number of vehicles owned.

13.87 Safety researchers in a government agency believe that too much variability in the speeds of vehicles on urban sections of interstate highways can contribute to accidents by causing a greater level of interaction between vehicles traveling in the same direction. They believe that a standard deviation in excess of 5 mph is undesirable. Observing a random sample of vehicles on an urban portion of the interstate highway in their locale, they find the speeds to be as listed in file **XR13087**. At the 0.025 level, and assuming a normal distribution of vehicle speeds, could they be mistaken in their conclusion that too much variability exists in the speeds of vehicles passing this location?

13.88 Given the information in Exercise 13.87 and data file **XR13087**, determine the 95% confidence interval for the population variance of the speeds of vehicles passing this location on the urban interstate.

INTEGRATED CASES

Thorndike Sports Equipment

Ted Thorndike is trying to close a sale with the Alvindale Chipmunks, a minor-league professional baseball team. Although his grandfather isn't very thrilled about any kind of deal with somebody with a name like "Chipmunks," Ted feels this could mark the beginning of some really high-level business with professional baseball teams, all the way up to the majors.

The Chipmunks, who are relatively high tech for a minor-league team, have been using a Thorndike competitor's automatic pitching machine for batting practice and are looking for a replacement. The problem is that the machine is very erratic. Although it has not "beaned" any players yet, they are concerned about the amount of vertical variability it exhibits in its pitches. Tom Johnson, principal owner of the Chipmunks, did a study in which he set up a target at home plate, then measured the heights at which the balls hit the target.

Mr. Johnson found the heights were normally distributed, as he has discovered is the case for pitching machines throughout the industry. However, he feels that a standard deviation of 3.5 inches in the heights of the pitches is just too "wild," even for the minor-league level. He has offered Ted Thorndike the opportunity to supply a replacement machine. The only catch to the sale is that Ted must prove that the Thorndike "Rapid Robot" machine is significantly better than the machine the Chipmunks are now using. Mr. Johnson has specified that the Thorndike pitcher will be purchased only if it passes the following test: Based on 30 pitches by Rapid Robot, the standard deviation of the vertical heights must be significantly less than 3.5 inches, and at the 0.01 level of significance.

All of this talk about vertical standard deviation is over Luke's head, but he tells Ted to go ahead and see if he can swing a deal with the Chipmunks. Ted is confident that the Thorndike Rapid Robot can meet Mr. Johnson's specifications, so he and Mr. Robot make the trip to Alvindale.

Ted and Mr. Johnson set up the Thorndike Rapid Robot machine at the team's practice facility, along with a target at home plate that will reveal exactly where each pitch lands. Measured from ground level at home plate, the 30 pitches are the following distances from the ground. The data are also in file **THORN13**.

26.1	26.2	27.2	26.9	25.8
26.8	25.3	29.9	27.1	26.3
28.1	31.4	27.3	23.5	27.0
29.0	31.3	24.2	28.7	30.6
26.9	28.9	25.4	26.9	22.7
32.1	30.1	26.6	25.5	27.5 inches

1. Given Mr. Johnson's decision rule, is this a two-tail or a one-tail hypothesis test? What are the null and alternative hypotheses?
2. When Mr. Johnson has completed his analysis of the data represented by the pitches in the test session, will he sign Thorndike's Rapid Robot as a mechanical member of the Chipmunks?

Springdale Shopping Survey

These analyses involve the following variables from the data file **SHOPPING**, the Springdale Shopping Survey introduced in Chapter 2:

1, 2, and 3 = Frequency of shopping at Springdale Mall, Downtown, and West Mall, respectively
4, 5, and 6 = Typical amount spent during a trip to Springdale Mall, Downtown, and West Mall, respectively
17 = The area identified as best for bargain sales
26 = Gender of the respondent
27 = Education level of the respondent
28 = Marital status of the respondent

Information like that gained from this database exercise could provide helpful insights into the nature of the respondents themselves as well as the possibility that some respondent characteristics might not be independent of other characteristics. Only a few of the many possible comparisons have been specified in this exercise.

Part I: Respondent Characteristics and the Area Judged Best for Bargains

1. Construct and use the 0.05 level in testing the contingency table for variables 26 and 17. Can we conclude that the gender of the respondent is independent of the shopping area judged best for bargains?
2. Repeat step 1 for variables 27 and 17. Can we conclude that the education level of the respondent is independent of the shopping area judged best for bargains?

Part II: Respondent Characteristics

1. Construct and use the 0.05 level in testing the contingency table for variables 26 and 27. Can we conclude that the gender of the respondent is independent of the education level of the respondent?
2. Construct and use the 0.05 level in testing the contingency table for variables 26 and 28. Can we conclude that the gender of the respondent is independent of the marital status of the respondent?

BUSINESS CASE

Baldwin Computer Sales (C)

In the Baldwin Computer Sales case, visited previously in Chapters 3 and 5, one of the areas of concern regarding the student purchase program was whether defaulting on the computer payments is independent of the university attended by the student. Using the **BALDWIN** data file, apply the appropriate chi-square test from this chapter to determine whether these variables could be independent. Use a level of significance of your choice in reaching a conclusion, and identify and interpret the p-value for the test. What, if any, recommendations would you make to Baldwin management on the basis of this test?

Andersen Ross/Photodisc/Getty Images

SEEING STATISTICS: APPLET 17

Chi-Square Distribution

When this applet loads, it shows a chi-square distribution similar to those in Figure 13.1. By using the slider at the right, we can change the degrees of freedom for the distribution and observe how this changes the distribution shape. We can also use the text boxes to enter either a chi-square value or a probability value, thus providing results similar to those in Table 13.1. Note that the chi-square and probability values in this applet will sometimes differ slightly from those in Table 13.1 and the corresponding chi-square table in the back of the text. Besides being more accurate because they do not rely on printed tables, these text boxes can also provide a wider variety of chi-square and probability values than are available from our printed tables.

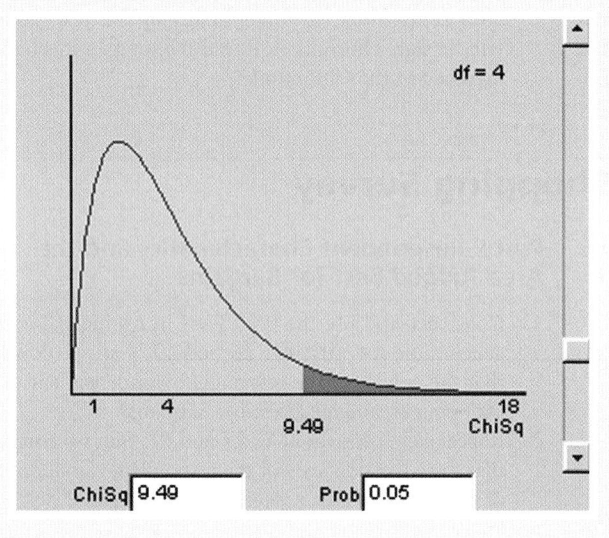

Applet Exercises

17.1 Using the slider, increase the number of degrees of freedom. What happens to the shape of the distribution?

17.2 Using the "ChiSq" text box, enter a larger chi-square value. (Be sure to press the Return or Enter key after changing the text box entry.) What effect does this have on the probability?

17.3 Using the "Prob" text box, enter a numerically higher probability. (Be sure to press the Return or Enter key after changing the text box entry.) What effect does this have on the chi-square value?

CHAPTER 14

Nonparametric Methods

PRESIDENTIAL SURNAMES: A NONRANDOM SEQUENCE?

In a series of numbers, names, or other data, is it possible that the series might exhibit some nonrandom tendencies? This chapter provides a procedure that allows us to test just that: the runs test for randomness. A "run" occurs when data of one type (e.g., above the median versus below the median) happens one or more times in a row; and if we have either too many or too few runs, this suggests the data might be other than random.

This technique is a handy way to examine whether a series of data might have come under the influence of one or more outside forces. It doesn't tell us exactly what the forces might be; it just sends a message that some force, sinister or otherwise, appears to be at work on the data.

Have there been nonrandom tendencies in the sequence of surnames for U.S. presidents? In the series below, also listed in file CX14PRES, the last names of the first 44 presidents are converted to a "1" (name begins with A–M) or a "2" (name begins with N–Z):

George Washington

2 1 1 1 1 1 1 2 1 2 2 2 1 2 1 1 1 1 1 1 1 1

1 1 1 2 2 2 1 1 1 2 2 1 1 1 2 1 1 2 1 1 1 2

Barack Obama

Using Minitab and applying the runs test for randomness, we find 17 streaks or runs in the observed series, compared to 20.1 runs that would be expected from a random series. Although our series does appear to have some long "clusters," the *p*-value is 0.275 and we are not able to reject the null hypothesis that the series is random.

You can verify these results for yourself when you get to the runs test later in the chapter. While you're at it, check out Exercise 14.60. It gives you a chance to statistically test the series of Super Bowl winners (American Conference versus National Conference), a series that many football fans are convinced is decidedly nonrandom. Is it really?

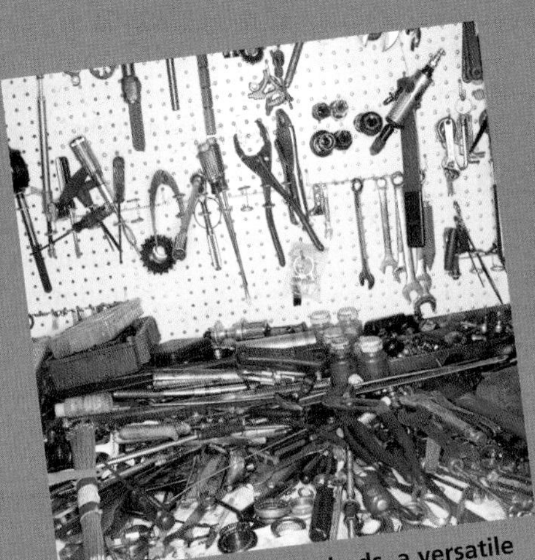

Like nonparametric methods, a versatile set of tools...

LEARNING OBJECTIVES

After reading this chapter, you should be able to:

- Differentiate between nonparametric and parametric hypothesis tests.
- Explain the advantages and disadvantages of nonparametric versus parametric testing.
- Determine when a nonparametric hypothesis test should be used instead of its parametric counterpart.
- Apply each of the nonparametric methods in the chapter to tests of hypotheses for which they are appropriate.

(14.1) INTRODUCTION

Chapters 10 and 11 discussed t-tests for examining hypotheses about either one or two samples. Chapter 12 extended this idea to include one-way analysis of variance and more than two samples, and also introduced the randomized block ANOVA design. This chapter gives alternative approaches to these tests, describes the conditions under which such alternatives are necessary, and examines each test through the use of a computer statistical package.

Nonparametric Testing

A **nonparametric test** is one that makes no assumptions about the specific shape of the population from which a sample is drawn. This is unlike most of our previous tests, which assumed that a population was either normally distributed or approximately so. When two or more populations were being compared, another typical assumption was that their population variances were equal. Also, most parametric tests require data to be of the interval or ratio scale of measurement, while many nonparametric techniques have no such requirement. In Chapter 13, the chi-square tests for goodness of fit, independence, and comparison of sample proportions were nonparametric methods having a variety of applications, but there are many more procedures in the realm of nonparametric testing.

A nonparametric test should be used instead of its parametric counterpart whenever

1. data are of the nominal or ordinal scale of measurement, or
2. data are of the interval or ratio scale of measurement but one or more other assumptions, such as the normality of the underlying population distribution, are not met.

Both nonparametric and parametric testing rely on the basic principles of hypothesis testing. Figure 14.1 shows five of the nonparametric tests in this chapter, along with their parametric counterparts from earlier in the text.

Advantages and Disadvantages of Nonparametric Testing

Compared to parametric tests, nonparametric testing has the following advantages and disadvantages:

Advantages

1. Fewer assumptions about the population. Most important, the population need not be normally distributed or approximately so. Nonparametric tests do not assume the population has *any* specific distribution.

FIGURE 14.1

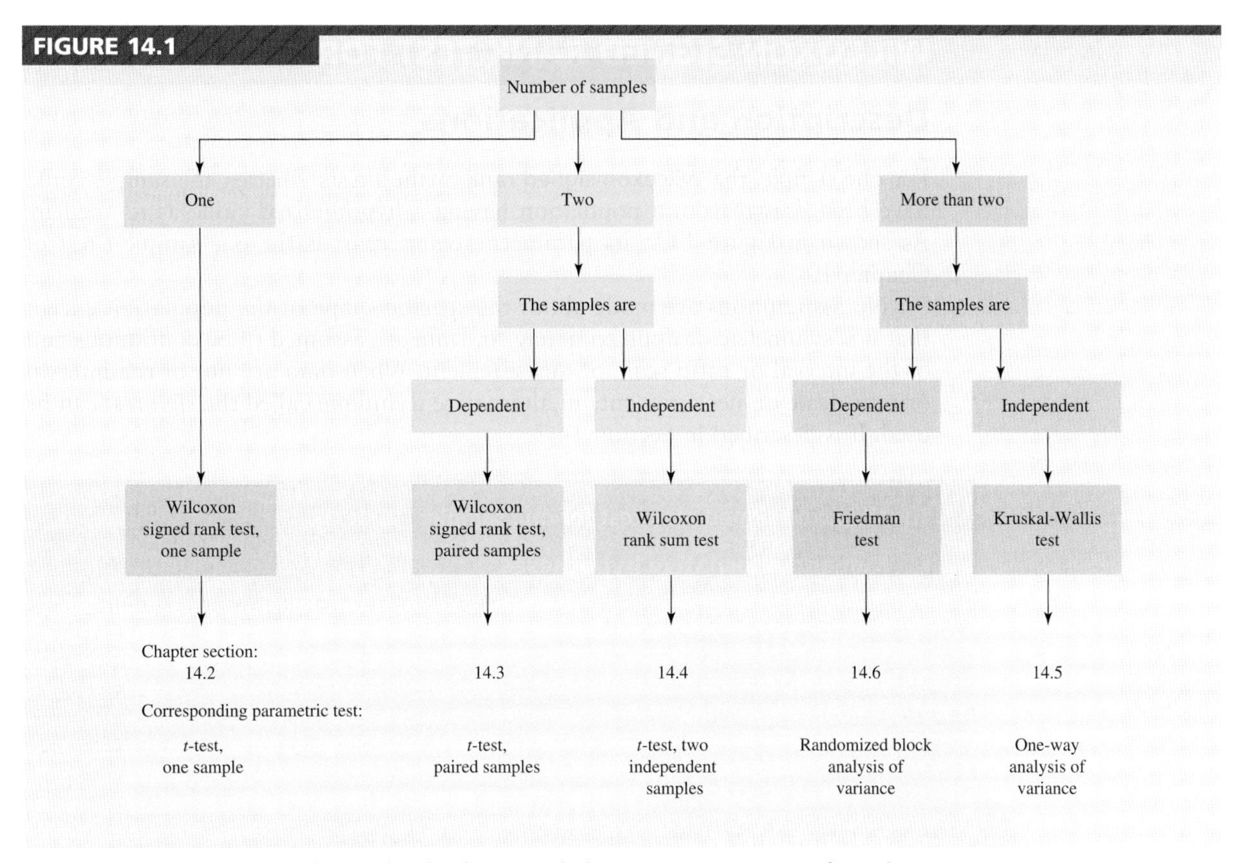

Five of the nonparametric tests discussed in the chapter, with the parametric counterpart for each

2. The techniques can be applied when sample sizes are very small.
3. Samples with data of the nominal and ordinal scales of measurement can be tested.

Disadvantages

1. Compared to a parametric test, the information in the data is used less efficiently, and the power of the test will be lower. For this reason, a parametric test is preferable whenever its assumptions have been met.
2. Nonparametric testing places greater reliance on statistical tables, if a computer statistical package or spreadsheet is not being used.

EXERCISES

14.1 What key feature distinguishes nonparametric from parametric tests?

14.2 What are some advantages and disadvantages of nonparametric testing?

14.3 In general, when should a nonparametric technique be used instead of its parametric counterpart?

14.4 Referring to Chapter 1 as necessary, differentiate among the nominal, ordinal, interval, and ratio scales of measurement.

14.2 WILCOXON SIGNED RANK TEST FOR ONE SAMPLE

Description and Applications

For one sample, the **Wilcoxon signed rank** method tests whether the sample could have been drawn from a population having a hypothesized value as its median. As shown in Figure 14.1, its parametric counterpart is the one-sample *t*-test of Chapter 10.

No assumptions are made about the specific shape of the population except that it is symmetric or approximately so. Data are assumed to be continuous and of the interval or ratio scale of measurement. When data are not of the interval or ratio scale of measurement, an alternative technique called the *sign test* can be used. It is described in Section 14.7.

The Wilcoxon signed rank test for one sample:

- **Null and alternative hypotheses**

	Two-tail Test	Left-tail Test	Right-tail Test
Null hypothesis, H_0:	$m = m_0$	$m \geq m_0$	$m \leq m_0$
Alternative hypothesis, H_1:	$m \neq m_0$	$m < m_0$	$m > m_0$

where m = the population median
m_0 = a value that has been specified

- **Test statistic, W**

1. **For each of the observed values, calculate $d_i = x_i - m_0$.**
2. **Ignoring observations where $d_i = 0$, rank the $|d_i|$ values so the smallest $|d_i|$ will have a rank of 1. If there are ties, assign each of the tied rows the average of the ranks they are occupying.**
3. **For observations where $x_i > m_0$, list the rank in the $R+$ column.**
4. **The test statistic is the sum of the $R+$ column, $W = \sum R+$.**

- **Critical value of W** **The Wilcoxon signed rank table in Appendix A lists lower and upper critical values for various levels of significance, with n = the number of observations for which $d_i \neq 0$. The rejection region will be in either one or both tails, depending on the null hypothesis being tested. For n values beyond the range of this table, a z-test approximation can be used.**

EXAMPLE

Wilcoxon Signed Rank Test, One Sample

An environmental activist believes her community's drinking water contains at least the 40.0 parts per million (ppm) limit recommended by health officials for a certain metal. In response to her claim, the health department samples and analyzes drinking water from a sample of 11 households in the community. The results are residue levels of 39.0, 20.2, 40.0, 32.2, 30.5, 26.5, 42.1, 45.6, 42.1, 29.9, and 40.9 ppm. The data are also in file CX14WAT. At the 0.05 level of significance, can we conclude that the community's drinking water might equal or exceed the 40.0 ppm recommended limit?

TABLE 14.1

Data and calculations for Wilcoxon signed rank test. The median of the hypothesized population distribution is $m_0 = 40.0$ ppm.

Household	Observed Concentration x_i	$d_i = x_i - m_0$	$\lvert d_i \rvert$	Rank	$R+$	$R-$
A	39.0 ppm	−1.0	1.0	2		2
B	20.2	−19.8	19.8	10		10
C	40.0	0.0	0.0	—		
D	32.2	−7.8	7.8	6		6
E	30.5	−9.5	9.5	7		7
F	26.5	−13.5	13.5	9		9
G	42.1	2.1	2.1	3.5	3.5	
H	45.6	5.6	5.6	5	5	
I	42.1	2.1	2.1	3.5	3.5	
J	29.9	−10.1	10.1	8		8
K	40.9	0.9	0.9	1	1	
					13.0	42.0

$W = \Sigma R+ = 13.0$

SOLUTION

We will use a left-tail test in evaluating the activist's statement. The null and alternative hypotheses are

$$H_0\colon \quad m \geq 40.0 \text{ ppm}$$
$$H_1\colon \quad m < 40.0 \text{ ppm}$$

Table 14.1 shows the data and the calculations needed to conduct the Wilcoxon signed rank test. For each household, $d_i = x_i - m_0$ is the difference between the observed concentration and the specified population median ($m_0 = 40.0$ ppm) in the null hypothesis. The absolute values of these differences are placed in the $|d_i|$ column, then the nonzero $|d_i|$ values are ranked.

The smallest nonzero $|d_i|$ value is the 0.9 for household K. This household gets a rank of 1 in the "Rank" column. Household A has the next smallest $|d_i|$, so it gets a rank of 2. Households G and I are tied, each having $|d_i| = 2.1$. Since these two $|d_i|$ values are taking up ranks 3 and 4, each is assigned the average of the two ranks, or $(3 + 4)/2 = 3.5$. For household C, $|d_i| = 0.0$, and this row is omitted from further analysis.

For those households having $x_i > 40.0$, the rank is placed in the $R+$ column. Although it isn't used in the analysis, the corresponding $R-$ column is also listed, since its inclusion can help us avoid errors in the assignment of the ranks. The sum of both columns is shown, and the test statistic W is the sum of the $R+$ column, or $W = 13.0$.

For n = the number of observations for which $d_i \neq 0$, the Wilcoxon signed rank statistic will be between $W = 0$ (if all of the d_i values are negative) and $W = n(n + 1)/2$ (if all of the d_i values are positive). The largest possible value, $n(n + 1)/2$, is the formula for the sum of consecutive integers from 1 to n. In this test, $n = 10$ nonzero differences, and W must have a value between 0 and 55.0. (Note that $\Sigma R+ = 13.0$ plus $\Sigma R- = 42.0$ is 55.0, the maximum value that W could have attained.) $W = 13.0$ is relatively low within these possible values.

TABLE 14.2

A portion of the table of critical values for the Wilcoxon signed rank test in Appendix A. Lower and upper critical values are provided.

Two-Tail Test: One-Tail Test:	$\alpha = 0.20$ $\alpha = 0.10$	$\alpha = 0.10$ $\alpha = 0.05$	$\alpha = 0.05$ $\alpha = 0.025$	$\alpha = 0.02$ $\alpha = 0.01$	$\alpha = 0.01$ $\alpha = 0.005$
$n = 4$	1, 9	0, 10	0, 10	0, 10	0, 10
5	3, 12	1, 14	0, 15	0, 15	0, 15
6	4, 17	3, 18	1, 20	0, 21	0, 21
7	6, 22	4, 24	3, 25	1, 27	0, 28
8	9, 27	6, 30	4, 32	2, 34	1, 35
9	11, 34	9, 36	6, 39	4, 41	2, 43
10	15, 40	11, 44	9, 46	6, 49	4, 51
11	18, 48	14, 52	11, 55	8, 58	6, 60
12	22, 56	18, 60	14, 64	10, 68	8, 70
⋮	⋮	⋮	⋮	⋮	⋮

Sources: Adapted from Roger C. Pfaffenberger and James H. Patterson, *Statistical Methods for Business and Economics,* (Homewood, Ill.: Richard D. Irwin, Inc., 1987), p. 1176; and R. L. McCornack, "Extended Tables of the Wilcoxon Matched Pairs Signed Rank Statistics," *Journal of the American Statistical Association,* 60 (1965), pp. 864–871.

EXAMPLE

The critical value of W can be found from the table of critical values for the Wilcoxon signed rank test in Appendix A. A portion of this table is shown in Table 14.2. Upper and lower critical values are listed, but we need only a lower critical value for our left-tail test. For $n = 10$ nonzero differences and $\alpha = 0.05$, the critical value is 11. The test statistic, $W = 13.0$, exceeds 11, and the null hypothesis cannot be rejected at the 0.05 level. At the 0.05 level, we are unable to reject the possibility that the city's water supply might have at least 40.0 ppm of the metal.

The Normal Approximation

When the number of observations for which $d_i \neq 0$ is $n \geq 20$, a z-test will be a close approximation to the Wilcoxon signed rank test. This is possible because the W distribution approaches a normal curve as n becomes larger. The format for this approximation is shown below.

z-test approximation to the Wilcoxon signed rank test:

Test statistic:

$$z = \frac{W - \frac{n(n+1)}{4}}{\sqrt{\frac{n(n+1)(2n+1)}{24}}}$$

where W = sum of the $R+$ ranks
n = number of observations for which $d_i \neq 0$

Although our example has only $n = 10$ nonzero differences, we'll use it to demonstrate the procedure for using the normal approximation. Because the sample size is smaller than the more desirable $n \geq 20$, this approximation will be "rougher" than if n were 20 or larger.

Substituting $W = 13.0$ and $n = 10$ into this expression, $z = -14.5/9.81$, or $z = -1.48$. For a left-tail test at the 0.05 level, the critical value of z will be -1.645 (from the normal distribution table). Since $z = -1.48$ falls to the right of this critical value, the null hypothesis cannot be rejected. We are unable to reject the possibility that the city's water supply might have at least 40.0 ppm of the metal.

Computer Solutions 14.1 shows how we can use Excel and Minitab in applying the Wilcoxon signed rank test to the data in Table 14.1. The Excel method uses the z-test approximation, and is best used when there are at least 20 nonzero differences. The "estimated median" shown by Minitab is computed through a special algorithm, and is not necessarily the same as the median for the sample.

EXERCISES

14.5 What is the parametric counterpart to the Wilcoxon signed rank test for one sample? Compare the assumptions involved in using the respective tests.

14.6 Given the following randomly selected observations, use the Wilcoxon signed rank test and the 0.05 level of significance in examining whether the population median could be equal to 4.9. Using the appropriate statistical table, what is the most accurate statement that can be made about the p-value for the test?

2.9 2.7 5.2 5.3 4.5 3.2 3.1 4.8

14.7 Given the following randomly selected observations, use the Wilcoxon signed rank test and the 0.01 level of significance in examining whether the population median could be greater than 37.0. Using the appropriate statistical table, what is the most accurate statement that can be made about the p-value for the test?

34.6 40.0 33.8 47.7 41.4 40.2 47.0
39.5 36.1 48.1 39.1 45.0 45.7 46.6

14.8 According to the director of a county tourist bureau, there is a median of 10 hours of sunshine per day during the summer months. For a random sample of 20 days during the past three summers, the number of hours of sunshine has been recorded as shown below. Use the 0.05 level in evaluating the director's claim.

8 9 8 10 9 7 7 9 7 7
9 8 11 9 10 7 8 11 8 12 hours

14.9 As a guideline, a local Internal Revenue Service office recommends that auditors try to spend 15 minutes or less with taxpayers who are called in for audit interviews. In a recent random sample of 22 interviews, the interview times were as shown below. Using the 0.025 level, do these data suggest that the median interview time for auditors from this office exceeds the recommended guideline? Using the appropriate statistical table, what is the most accurate statement we can make about the p-value for the test?

12 15 16 16 17 10 22 13 20 20 20
19 24 12 11 18 23 23 13 24 24 12

(**DATA SET**) *Note:* Exercises 14.10 and 14.11 require a computer and statistical software.

14.10 The placement director for a university claims that last year's graduates had a median starting salary of $35,000. For a random sample of graduates, the starting salaries, in thousands of dollars, are listed in data file **XR14010**. At the 0.05 level, could the director's claim be true? Identify and interpret the p-value for the test.

14.11 The median age for a male at the time of his first marriage has been reported as 27.5 years. In a given region, a random sample of males' first marriages showed the ages listed in data file **XR14011**. Using the 0.05 level of significance, test whether the median age at which males in this region are first married could be the same as that for the nation as a whole. Identify and interpret the p-value for the test. Source: *The New York Times Almanac 2009*, p. 298.

COMPUTER 14.1 SOLUTIONS

Wilcoxon Signed Rank Test for One Sample

This hypothesis test is for the median of a population.

EXCEL

	A	B	C	D
1	Wilcoxon Signed Rank Sum Test			
2				
3	Difference		*ppm - median0*	
4	T+		13	
5	T-		42	
6	Observations (for test)		10	
7	z Stat		-1.478	
8	P(Z<=z) one-tail		0.0697	
9	z Critical one-tail		1.645	
10	P(Z<=z) two-tail		0.1394	
11	z Critical two-tail		1.960	

Excel Wilcoxon Signed Rank Test for One Sample, Using the *z*-Test Approximation

1. The data in Table 14.1 are also in Excel file CX14WAT, with the label (ppm) in A1 and the data in A2:A12. Enter the median associated with H_0 **(40)** into **B2.** Click at the bottom right corner of B2, then drag and fill through B12 so that 40 will be in all of these cells. Enter the label **(median0)** into **B1.**
2. From the **Add-Ins** ribbon, click **Data Analysis Plus.** Click **Wilcoxon Signed Rank Sum Test.** Click **OK.**
3. Enter **A1:A12** into the **Variable 1 Range** box. Enter **B1:B12** into the **Variable 2 Range** box. Click **Labels.** Enter **0.05** into the **Alpha** box. Click **OK.** The printout includes the number of nonzero differences, the sum of the ranks for the positive differences and the negative differences, the *z*-statistic, and the *p*-value for this one-tail test (0.0697).

The *z*-test approximation here is applicable when there are at least 20 nonzero differences. For the data in file CX14WAT, this requirement is not met, but we have applied the approximation here to demonstrate the steps involved.

Note: As an alternative, you can use Excel worksheet template TMWILSR1. The procedure is described within the template.

MINITAB

Minitab Wilcoxon Signed Rank Test for One Sample

```
Wilcoxon Signed Rank Test: ppm

Test of median = 40.00 versus median < 40.00

          N for   Wilcoxon              Estimated
      N   Test   Statistic       P         Median
ppm  11     10        13.0   0.077          35.85
```

1. With the Table 14.1 data values in column C1 of Minitab file CX14WAT: Click **Stat.** Select **Nonparametrics.** Click **1-Sample Wilcoxon.**
2. Enter **C1** into the **Variables** box. Select **Test median** and enter **40** into the box. Select **less than** within the **Alternative** box. Click **OK.** The printout includes the *p*-value (0.077) as well as a restatement of the null and alternative hypotheses. Unlike the *z*-test method, this Minitab test can be used even with small data sets.

WILCOXON SIGNED RANK TEST FOR COMPARING PAIRED SAMPLES

14.3

Description and Applications

The Wilcoxon signed rank test can also be used for paired samples. In this context, it is the nonparametric counterpart to the paired-samples *t*-test of Chapter 11.

As in Section 14.2, the technique assumes the data are continuous and of the interval or ratio scale of measurement. In this application, the measurement of interest is the difference between paired observations, or $d_i = x_i - y_i$. Depending on the type of test, the null and alternative hypotheses will be as follows:

	Two-tail Test	Left-tail Test	Right-tail Test
Null hypothesis, H_0:	$m_d = 0$	$m_d \geq 0$	$m_d \leq 0$
Alternative hypothesis, H_1:	$m_d \neq 0$	$m_d < 0$	$m_d > 0$

where m_d = the population median of $d_i = x_i - y_i$

The population of *d* values is assumed to be symmetric or nearly so, but need not be normally distributed or have any other specific shape. Applying the Wilcoxon signed rank test to paired samples is nearly identical to its use for one sample, discussed in the preceding section. Because the two approaches are so similar, this one will be described through the use of an example problem.

EXAMPLE

Wilcoxon Signed Rank Test, Paired Samples

Two computer software packages are being considered for use in the inventory control department of a small manufacturing firm. The firm has selected 12 different computing tasks that are typical of the kinds of jobs such a package would have to perform, then recorded the number of seconds each package required to complete each task. The results are shown in Table 14.3, which includes the difference (d_i) in the number of seconds the two packages required for each task. The data are also in file **CX14TIME**. At the 0.10 level, can we conclude that the median difference (m_d) for the population of such tasks might be zero?

SOLUTION

The background for this test does not imply that one package is any faster or slower than the other, so a two-tail test is used. The null and alternative hypotheses are

H_0: $m_d = 0$ The population median of $d_i = x_i - y_i$ is zero.

H_1: $m_d \neq 0$ The population median of $d_i = x_i - y_i$ is not zero.

As Table 14.3 shows, the difference (d_i) is calculated for each pair of x_i and y_i values, the absolute values for d_i are obtained, then the $|d_i|$ values are ranked. Finally, the ranks associated with positive differences are listed in the $R+$ column and added to get the observed value of the test statistic, *W*.

EXAMPLEEXAMPLEEXAMPLEEXAMPLEEXAMPLE

TABLE 14.3

Data and calculations for the application of the Wilcoxon signed rank test to the comparison of paired samples. Prior to the test, there was no reason to believe one software package took more time than the other, so the test is two-tail.

Computing Task	Time Required for Software Packages x and y						
	x_i	y_i	$d_i = x_i - y_i$	$\lvert d_i \rvert$	Rank	$R+$	$R-$
A	24.0	23.1	0.9	0.9	1	1	
B	16.7	20.4	−3.7	3.7	4		4
C	21.6	17.7	3.9	3.9	5	5	
D	23.7	20.7	3.0	3.0	2.5	2.5	
E	37.5	42.1	−4.6	4.6	6		6
F	31.4	36.1	−4.7	4.7	7		7
G	14.9	21.8	−6.9	6.9	10		10
H	37.3	40.3	−3.0	3.0	2.5		2.5
I	17.9	26.0	−8.1	8.1	11		11
J	15.5	15.5	0.0	0.0	—		
K	29.0	35.4	−6.4	6.4	9		9
L	19.9	25.5	−5.6	5.6	8		8
						8.5	57.5

$W = \Sigma R+ = 8.5$

The smallest nonzero $|d_i|$ value is $|d_i| = 0.9$ seconds for computing task A, and it gets a rank of 1. Tasks D and H are tied, each having $|d_i| = 3.0$ seconds. These two tasks are occupying ranks 2 and 3, so each is assigned the average of the two ranks, or 2.5. Because $|d_i| = 0.0$ for task J, it is dropped from the analysis.

As Table 14.3 shows, the test statistic is $W = \Sigma R+ = 8.5$. For $n = 11$ nonzero differences and $\alpha = 0.10$, the Wilcoxon signed rank table (see the portion shown in Table 14.2) gives lower and upper critical values for W of 14 and 52, respectively. The observed value, $W = 8.5$, falls outside these limits, and the null hypothesis is rejected. Based on these data, the software packages are not equally rapid in handling computing tasks like those in the sample. At the 0.10 level, we are able to conclude that the population median for $d_i = x_i - y_i$ is not equal to zero and that package x is faster than package y in handling computing tasks like the ones sampled.

The z-test approximation described in the preceding section can also be used for paired samples. Again, it is desirable that the number of nonzero differences be at least 20, but we can use this method to get a very rough approximation if n is smaller. Using the normal approximation in this case, the observed z is $z = -2.18$. This is based on $W = 8.5$, $n = 11$, and the z expression in the preceding section. Using the normal distribution table, we find that 0.0146 of the area falls in the left tail. Since the test is two-tail, the approximate p-value is 2(0.0146), or 0.0292.

Computer Solutions 14.2 shows how we can use Excel and Minitab in applying the Wilcoxon signed rank test to the paired data in Table 14.3. The Excel method uses the z-test approximation, and is best used when there are at least

COMPUTER 14.2 SOLUTIONS

Wilcoxon Signed Rank Test for Comparing Paired Samples

This hypothesis test applies to paired samples, such as before-and-after comparisons for the same items or subjects.

EXCEL

	A	B	C
1	Wilcoxon Signed Rank Sum Test		
2			
3	Difference		*X - Y*
4	T+		8.5
5	T-		57.5
6	Observations (for test)		11
7	z Stat		-2.178
8	P(Z<=z) one-tail		0.0147
9	z Critical one-tail		1.282
10	P(Z<=z) two-tail		0.0294
11	z Critical two-tail		1.645

Excel Wilcoxon Signed Rank Test for Paired Samples, Using the *z*-test Approximation

1. The data in Table 14.3 are in adjacent columns in Excel file **CX14TIME**, with the labels in A1 and B1, and the data in A2:B13. From the **Add-Ins** ribbon, click **Data Analysis Plus.** Click **Wilcoxon Signed Rank Sum Test.** Click **OK.**
2. Enter **A1:A13** into the **Variable 1 Range** box. Enter **B1:B13** into the **Variable 2 Range** box. Click **Labels.** Enter **0.10** into the **Alpha** box. Click **OK.** The printout includes the number of nonzero differences, the sum of the ranks for the positive differences and the negative differences, the *z*-statistic, and the *p*-value for this two-tail test (0.0294).

The *z*-test approximation here is applicable when there are at least 20 nonzero differences. For the data in file **CX14TIME**, this requirement is not met, but we have applied the approximation here to demonstrate the steps involved.

Note: As an alternative, you can use Excel worksheet template **TMWILSR1**. The procedure is described within the template.

MINITAB

Minitab Wilcoxon Signed Rank Test for Paired Samples

```
Wilcoxon Signed Rank Test: DIFF

Test of median = 0.000000 versus median not = 0.000000

            N for  Wilcoxon             Estimated
       N    Test  Statistic       P        Median
DIFF  12      11        8.5   0.033        -3.100
```

1. With the Table 14.3 data values in columns C1 and C2 of Minitab file **CX14TIME**, we must first calculate the differences and insert them into C3: Click **Calc.** Click **Calculator.** Enter **C1-C2** into the **Expression** box. Enter **C3** into the **Store result in variable** box. Click **OK.** Type the label **(DIFF)** at the top of column C3.
2. With the differences in C3, click **Stat.** Select **Nonparametrics.** Click **1-Sample Wilcoxon.**
3. Enter **C3** into the **Variables** box. Select **Test median** and enter **0** into the box. Select **not equal** within the **Alternative** box. Click **OK.** Unlike the *z*-test method, this Minitab test is not an approximation, and it can be used even with small data sets.

20 nonzero differences. The "estimated median" shown by Minitab is computed through a special algorithm, and is not necessarily the same as the median for the sample. The Excel and Minitab p-values differ slightly from each other, as well as from our p-value that was based on the normal distribution table and its inherent gaps between listed z values.

EXERCISES

14.12 What is the parametric counterpart to the Wilcoxon signed rank test for paired samples? Compare the assumptions involved in using the respective tests.

14.13 For nine paired samples, the data are as listed in the table. For d = the difference between paired observations $(x_1 - x_2)$, use the 0.10 level of significance in testing H_0: $m_d \leq 0$ versus H_1: $m_d > 0$.

Sample:	1	2	3	4	5	6	7	8	9
x_1	7.8	4.0	4.2	9.8	8.7	4.3	8.7	8.1	8.9
x_2	9.7	2.1	3.7	8.2	6.6	5.7	5.6	7.7	8.6

14.14 For ten paired samples, the data are as listed here. For d = the difference between paired observations $(x_1 - x_2)$, use the 0.10 level of significance in testing H_0: $m_d = 0$ versus H_1: $m_d \neq 0$.

Sample:	1	2	3	4	5	6	7	8	9	10
x_1	12.5	11.2	14.4	8.9	10.5	12.3	11.8	12.4	13.9	8.8
x_2	12.7	11.2	15.6	8.7	11.8	12.2	11.3	13.4	14.4	8.3

14.15 Tom Frederick is the computer support manager for a large company whose employees have been complaining about "spam," which many of us know as unwanted e-mail to solicit our money or our attention. Tom asked a sample of 9 employees to keep track of the number of spam messages they received during the previous week. He then installed a spam "filter" into the e-mail system to block some of the spam by identifying key words that often appear in such messages. During the week following installation of the filter, the number of spam messages received by each of the 9 employees was again counted, with the results shown here. At the 0.05 level of significance, can we conclude that Tom's filtering system is effective in reducing the weekly number of spam messages an employee receives?

Employee:	1	2	3	4	5	6	7	8	9
Spam Messages, Before Filter	28	27	24	18	23	28	25	20	28
Spam Messages, After Filter	19	25	25	19	20	21	18	21	22

14.16 A researcher studying the purchase habits of eight married couples has found they spent the amounts shown below for clothing. At the 0.05 level, can we conclude that the population of husbands and wives do not spend equally in this purchase category?

	1	2	3	4	5	6	7	8
Husband	$300	240	230	250	230	300	470	230
Wife	$380	375	180	390	400	480	470	320

14.17 A new procedure is being tested for the assembly of electronic circuit boards. For each of five employees, the time required to assemble a board is measured for both the current procedure and the method that has been proposed. Given the times listed in the table, use the 0.025 level of significance in testing whether the new method could be an improvement over the one currently used. Using the appropriate statistical table, what is the most accurate statement that can be made about the p-value for the test?

		Current	Proposed
Person	1	66.4	70.0
	2	39.8	35.4
	3	50.5	43.7
	4	65.1	59.0
	5	55.6	49.9 seconds

(DATA SET) *Note:* Exercises 14.18 and 14.19 require a computer and statistical software.

14.18 Early in the 1995–1996 NHL hockey season, players and coaches observed with great interest an apparent drop in the number of goals scored so far during that season, compared to scoring over the same number of games during the previous season. For each team, the number of goals in the 1995–1996 season versus the same stage of the 1994–1995 season are listed in data file **XR14018**. At the 0.05 level, could the drop in goals be simply due to chance, or does it appear that something may have happened between seasons to reduce teams' scoring power? Identify and interpret the p-value for the test. Source: Kevin Allen, "Some Players, Coaches Applaud Trend Toward Lower-Scoring Games," *USA Today,* November 12, 1996, p. 5C.

14.19 A company has randomly selected a number of employees to test the effectiveness of a speed-reading course. If the course is found to significantly increase reading speed, it will be offered to all of the firm's clerical employees. For each of the employees selected for the test, reading speeds (words per minute) before and after the course are listed in data file **XR14019**. Using the 0.01 level of significance in evaluating the effect of the course, should the firm offer the program to all of its clerical employees? Identify and interpret the p-value for the test.

WILCOXON RANK SUM TEST FOR COMPARING TWO INDEPENDENT SAMPLES

14.4

Description and Applications

The **Wilcoxon rank sum test** compares two independent samples and is a nonparametric counterpart to the two-sample pooled t-test of Chapter 11. The test assumes that the data are at least of the ordinal scale of measurement, that the samples are independent and randomly selected, and that the populations have approximately the same shape.

When doubts exist as to whether data meet the requirements of the interval scale of measurement, or if it cannot be assumed that the populations are normal, with equal variances, the Wilcoxon rank sum test should be used instead of the pooled-variances t-test for independent samples. The sample sizes need not be equal, and the test is as shown:

Wilcoxon rank sum test for comparing two independent samples:

- **Null and alternative hypotheses**

	Two-tail Test	Left-tail Test	Right-tail Test
H_0:	$m_1 = m_2$	$m_1 \geq m_2$	$m_1 \leq m_2$
H_1:	$m_1 \neq m_2$	$m_1 < m_2$	$m_1 > m_2$

where m_1 and m_2 are the population medians.

- **Test statistic, W**
  1. Designate the smaller of the two samples as sample 1. If the sample sizes are equal, either sample may be designated as sample 1.
  2. Rank the combined data values as if they were from a single group. The smallest data value gets a rank of 1, the next smallest, 2, and so on. In the event of a tie, each of the tied values gets the average rank that the values are occupying.
  3. List the ranks for data values from sample 1 in the R_1 column and the ranks for data values from sample 2 in the R_2 column.
  4. The observed value of the test statistic is $W = \Sigma R_1$.

- **Critical value of W** The Wilcoxon rank sum table in Appendix A lists lower and upper critical values for the test, with n_1 and n_2 as the number of observations in the respective samples. The rejection region will be in either one or both tails, depending on the null hypothesis being tested. For n_1 and n_2 values beyond the range of this table, a z-test approximation is used in carrying out the test.

EXAMPLE

Wilcoxon Rank Sum Test

In evaluating the flexibility of rubber tie-down lines, an inspector selects a random sample of the straps and counts the number of 360-degree twists each will withstand before breaking. For 7 lines from one production lot, the number of turns before breaking is 112, 105, 83, 102, 144, 85, and 50. For 10 lines from a second production lot, the number of turns before breaking is 91, 183, 141, 219, 105, 138, 146, 84, 134, and 106. The data are also in file **CX14TIE**. At the 0.05 level, can it be concluded that the two production lots have the same median flexibility?

SOLUTION

Prior to the testing, there was no reason to believe that one lot was either more flexible or less flexible than the other, so the test is two-tail. The null and alternative hypotheses are H_0: $m_1 = m_2$ and H_1: $m_1 \neq m_2$, where m_1 and m_2 are the medians of the respective populations.

Combining the data values and ranking them, the rankings are shown in Table 14.4. The smallest value, 50, gets a rank of 1 and the next smallest, 83, gets the rank of 2. This continues until the largest value, 219, has been assigned the rank of 17. Two values are tied at 105, and each is given the rank of 7.5.

The sum of ranks of values from sample 1 is $W = \Sigma R_1 = 44.5$, the observed value of the test statistic. The smaller of the two samples has already been designated as sample 1. If this had not been done, the identification of the samples would have to be reversed in order to carry out the analysis.

A portion of the Wilcoxon rank sum table in Appendix A is provided in Table 14.5. For $n_1 = 7$, $n_2 = 10$, and a two-tail test at the 0.05 level, the lower and upper critical values of W are 43 and 83, respectively. Since the calculated value, $W = 44.5$, falls between these limits, the null hypothesis cannot be rejected at the 0.05 level. Our conclusion is that the median flexibility of the two lots could be the same.

TABLE 14.4

Data and calculations for Wilcoxon rank sum comparison of two independent samples. The sample with the smaller number of observations is designated as sample 1.

Sample 1 and Ranks		Sample 2 and Ranks	
112	10	91	5
105	7.5	183	16
83	2	141	13
102	6	219	17
144	14	105	7.5
85	4	138	12
50	1	146	15
	44.5	84	3
		134	11
		106	9
			108.5

$W = \Sigma R_1 = 44.5$

TABLE 14.5

A portion of the Wilcoxon rank sum table from Appendix A. Lower and upper limits are provided, and, if sample sizes are unequal, sample 1 will be the one with the smaller number of observations.

$\alpha = 0.025$ (One-Tail) or $\alpha = 0.05$ (Two-Tail)

	n_1	3	4	5	6	7	8	9	10
	3	5, 16	6, 18	6, 21	7, 23	7, 26	8, 28	8, 31	9, 33
	4	6, 18	11, 25	12, 28	12, 32	13, 35	14, 38	15, 41	16, 44
	5	6, 21	12, 28	18, 37	19, 41	20, 45	21, 49	22, 53	24, 56
	6	7, 23	12, 32	19, 41	26, 52	28, 56	29, 61	31, 65	32, 70
n_2	7	7, 26	13, 35	20, 45	28, 56	37, 68	39, 73	41, 78	43, 83
	8	8, 28	14, 38	21, 49	29, 61	39, 73	49, 87	51, 93	54, 98
	9	8, 31	15, 41	22, 53	31, 65	41, 78	51, 93	63, 108	66, 114
	10	9, 33	16, 44	24, 56	32, 70	43, 83	54, 98	66, 114	79, 131

Note: n_1 is the smaller of the two samples—i.e., $n_1 \leq n_2$.
Sources: F. Wilcoxon and R. A. Wilcox, *Some Approximate Statistical Procedures* (New York: American Cyanamid Company, 1964), pp. 20–23.

The Normal Approximation

When $n_1 \geq 10$ and $n_2 \geq 10$, a normal approximation to the Wilcoxon rank sum test can be used. The format is as follows:

z-test approximation to the Wilcoxon rank sum test:

Test statistic:

$$z = \frac{W - \frac{n_1(n+1)}{2}}{\sqrt{\frac{n_1 n_2 (n+1)}{12}}}$$

where W = sum of the R_1 ranks
$n = n_1 + n_2$

Although our example falls short of the $n_1 \geq 10$, $n_2 \geq 10$ rule of thumb, it will be used to demonstrate the normal approximation. The results will be more "approximate" than if the rule of thumb had been met.

Substituting $W = 44.5$, $n_1 = 7$, $n_2 = 10$, and $n = 17$ into the above expression, we find that $z = -1.81$. For a two-tail test at the 0.05 level, the critical z values are -1.96 and $+1.96$. The calculated z is between these limits, and the null hypothesis cannot be rejected.

The approximate p-value for this two-tail test is twice the area under the normal curve to the left of $z = -1.81$. Using the normal distribution table, we find the approximate p-value to be 2(0.0351), or 0.0702.

Computer Solutions 14.3 shows how we can use Excel and Minitab in applying the Wilcoxon rank sum test to compare the independent samples in Table 14.4. The Excel method uses the z-test approximation, and is best used when each sample size is at least 10. Note that the Minitab procedure calls for the Mann-Whitney test, which is equivalent to the Wilcoxon test described in the text.

COMPUTER 14.3 SOLUTIONS

Wilcoxon Rank Sum Test for Two Independent Samples

This hypothesis test compares the medians of two independent samples.

EXCEL

	A	B	C	D
1	Wilcoxon Rank Sum Test			
2				
3			Rank Sum	Observations
4	*LOT 1*		44.5	7
5	*LOT 2*		108.5	10
6	z Stat		-1.81	
7	P(Z<=z) one-tail		0.0355	
8	z Critical one-tail		1.645	
9	P(Z<=z) two-tail		0.071	
10	z Critical two-tail		1.960	

Excel Wilcoxon Rank Sum Test for Two Independent Samples, Using the *z*-test Approximation

1. The data in Table 14.4 are in adjacent columns in Excel file **CX14TIE**, with labels in A1 and B1, and data in A2:A8 and B2:B11. From the **Add-Ins** ribbon, click **Data Analysis Plus.** Click **Wilcoxon Rank Sum Test.** Click **OK.**
2. Enter **A1:A8** into the **Variable 1 Range** box. Enter **B1:B11** into the **Variable 2 Range** box. Click **Labels.** Enter **0.05** into the **Alpha** box. Click **OK.** The printout includes the *p*-value for this two-tail test (0.071).

The *z*-test approximation here is applicable when each sample size is at least 10. For the data in file **CX14TIE**, this requirement is not met, but we have applied the approximation here to demonstrate the steps involved.

Note: As an alternative, you can use Excel worksheet template **TMWILRS2**. The procedure is described within the template.

MINITAB

Minitab Wilcoxon Rank Sum Test for Two Independent Samples

```
Mann-Whitney Test and CI: LOT 1, LOT 2

        N  Median
LOT 1   7  102.00
LOT 2  10  136.00

Point estimate for ETA1-ETA2 is -34.00
95.5 Percent CI for ETA1-ETA2 is (-78.00,2.99)
W = 44.5
Test of ETA1 = ETA2 vs ETA1 not = ETA2 is significant at 0.0790
The test is significant at 0.0788 (adjusted for ties)
```

1. With the Table 14.4 data values in columns C1 and C2 of Minitab file **CX14TIE**: Click **Stat.** Select **Nonparametrics.** Click **Mann-Whitney,** the equivalent test that Minitab performs.
2. Enter **C1** into the **First Sample** box. Enter **C2** into the **Second Sample** box. Select **not equal** within the **Alternative** box. Click **OK.** Unlike the *z*-test method, this Minitab test can be used even with small data sets.

EXERCISES

14.20 Differentiate between the Wilcoxon signed rank test and the Wilcoxon rank sum test, and explain the circumstances under which each should be applied.

14.21 What is the parametric counterpart to the Wilcoxon rank sum test? Compare the assumptions involved in using the respective tests.

14.22 For the following random samples from two populations, use the 0.05 level of significance in testing H_0: $m_1 = m_2$ versus H_1: $m_1 \neq m_2$.

Sample 1: 40 34 53 28 41
Sample 2: 29 31 52 29 20 31 26

14.23 For the following random samples from two populations, use the 0.025 level of significance in testing H_0: $m_1 \geq m_2$ versus H_1: $m_1 < m_2$.

Sample 1: 46.8 48.0 72.6 55.1 43.6
53.0 41.8 45.5
Sample 2: 56.4 59.3 59.1 58.4 57.9
46.4 49.6 50.2 55.8 53.8

14.24 Twelve family pets have been randomly placed into two groups. Group 1 receives the standard version of a flea powder, while group 2 receives a modified formula. Four days following treatment, researchers count the number of fleas collected from each pet during a 2-minute period. The results are shown below. At the 0.025 level, is the new formula more effective than the old?

Number of fleas after old formula:
17 16 17 4 8 15
Number of fleas after new formula:
6 1 1 2 7 5

(**DATA SET**) *Note:* Exercises 14.25 and 14.26 require a computer and statistical software.

14.25 A shop owner has collected random samples of sales receipts for credit purchases made with MasterCard and Visa. The amounts are listed in data file **XR14025**. At the 0.05 level, can we conclude that the shop's median sale for MasterCard is the same as that for Visa? Identify and interpret the p-value for the test.

14.26 A word-processing firm has narrowed down its choice of a printer to two major brands. Using a 20-page document that is typical of the firm's work, a number of printers of each brand are timed while they generate a letter-quality copy of the document. The times are listed in data file **XR14026**. At the 0.05 level, can we conclude that the two brands are equally fast? Identify and interpret the p-value for the test.

KRUSKAL-WALLIS TEST FOR COMPARING MORE THAN TWO INDEPENDENT SAMPLES (14.5)

Description and Applications

An extension of the Wilcoxon rank sum technique of the previous section, the **Kruskal-Wallis test** compares more than two independent samples. It is the nonparametric counterpart to the one-way analysis of variance. However, unlike one-way ANOVA, it does not assume that samples have been drawn from normally distributed populations with equal variances. The scale of measurement of the data must be at least ordinal, and the samples are assumed to be randomly selected from their respective populations.

In this test, the null hypothesis is that the medians of the k populations are the same, or H_0: $m_1 = m_2 = \cdots = m_k$. The test is one-tail and is carried out as follows:

Kruskal-Wallis test for comparing more than two independent samples:

- **Null and alternative hypotheses**

  H_0: $m_1 = m_2 = \cdots = m_k$ for the j = 1 through k populations. (The population medians are equal.)

  H_1: At least one m_j differs from the others. (The population medians are not equal.)

(continued)

- **Test statistic, *H***

  1. Rank the combined data values as if they were from a single group. The smallest data value gets a rank of 1, the next smallest, 2, and so on. In the event of a tie, each of the tied values gets their average rank.
  2. Add the ranks for data values from each of the *k* groups, obtaining ΣR_1, ΣR_2, through ΣR_k.
  3. The calculated value of the test statistic is

$$H = \frac{12}{n(n+1)}\left[\frac{(\Sigma R_1)^2}{n_1} + \frac{(\Sigma R_2)^2}{n_2} + \cdots + \frac{(\Sigma R_k)^2}{n_k}\right] - 3(n+1)$$

where $n_1, n_2, \ldots, n_k$ = the respective sample sizes for the *k* samples

$$n = n_1 + n_2 + \cdots + n_k$$

- **Critical value of *H*** The distribution of *H* is closely approximated by the chi-square distribution whenever each sample size is at least 5 and, for α = the level of significance for the test, the critical *H* is the chi-square value for which $df = k - 1$ and the upper-tail area is α. If the calculated *H* exceeds the critical value, the null hypothesis is rejected. Otherwise, it cannot be rejected. If one or more of the sample sizes is < 5, either a computer statistical package or special tables must be used for the test.

EXAMPLEEXAMPLEEXAMPLEEXAMPLEEXAMPLEEXAMPLE

EXAMPLE

Kruskal-Wallis Test

Each of three aerospace companies has randomly selected a group of technical staff workers to participate in a training conference sponsored by a supplier firm. The three companies have sent 6, 5, and 7 employees, respectively. At the beginning of the session, a preliminary test is given, and the scores are shown in Table 14.6. The data are also in file CX14TECH. At the 0.05 level, can we conclude that the median scores for the three populations of technical staff workers could be the same?

SOLUTION

As in the Wilcoxon rank sum test, the combined data values are ranked, with the smallest value receiving a rank of 1. The process is summarized in Table 14.6, which also shows the sum of the ranks for each sample. There is one tie, and each person who scored 67 on the test receives a rank of 7.5.

The sums of the ranks are $\Sigma R_1 = 32$, $\Sigma R_2 = 45$, and $\Sigma R_3 = 94$. There is a total of $n = 18$ observations in the $k = 3$ samples, with $n_1 = 6$, $n_2 = 5$, and $n_3 = 7$. Substituting these into the expression for the test statistic, *H*, we have

$$H = \frac{12}{n(n+1)}\left[\frac{(\Sigma R_1)^2}{n_1} + \frac{(\Sigma R_2)^2}{n_2} + \cdots + \frac{(\Sigma R_k)^2}{n_k}\right] - 3(n+1)$$

$$= \frac{12}{18(18+1)}\left[\frac{32^2}{6} + \frac{45^2}{5} + \frac{94^2}{7}\right] - 3(18+1) = 7.490$$

Technical Staff from Firm 1		Technical Staff from Firm 2		Technical Staff from Firm 3	
Test Score	Rank	Test Score	Rank	Test Score	Rank
67	7.5	64	5	75	15
57	1	73	13	61	3
62	4	72	12	76	16
59	2	68	9	71	11
70	10	65	6	78	17
67	7.5		$\Sigma R_2 = 45$	74	14
	$\Sigma R_1 = 32$			79	18
					$\Sigma R_3 = 94$

TABLE 14.6

Exam scores and ranks for members of the three groups of technical staff workers. The Kruskal-Wallis test is an extension of the Wilcoxon rank sum test, and the ranks are assigned the same way.

The critical value of H is the chi-square statistic corresponding to an upper tail area of $\alpha = 0.05$ and $df = k - 1 = 2$. Referring to the chi-square table in Appendix A, this is found to be 5.991. The calculated value, $H = 7.490$, exceeds 5.991, and the null hypothesis is rejected at the 0.05 level of significance. At this level, we conclude that the three populations do not have the same median.

The calculated H also exceeds the critical value for the 0.025 level of significance (chi-square = 7.378), but does not exceed that for the 0.01 level (chi-square = 9.210). Based on the table in Appendix A, the most accurate statement we can make about the p-value for the test is that it is between 0.025 and 0.01.

AMPLEEXAMPLEEXAMPLE

Computer Solutions 14.4 shows how we can use Excel and Minitab in applying the Kruskal-Wallis test to compare the three independent samples in Table 14.6. Minitab lists the same H value (7.49) we've just obtained, along with an "adjusted" H value (7.50) that corrects for the tie that occurred during the ranking process. Although it would not change our conclusion in this example, it is preferable to use the adjusted H value when it is available.

EXERCISES

14.27 Differentiate between the Kruskal-Wallis test and the one-way analysis of variance in terms of their assumptions and the circumstances under which each should be applied.

14.28 For the following independent and random samples, use the 0.05 level of significance in testing whether the population medians could be equal.

Sample 1	Sample 2	Sample 3
31.3	29.4	36.0
30.7	20.8	37.7
35.4	22.2	31.0
36.1	24.9	28.4
30.3	21.4	31.7
25.5		24.1
		32.6

14.29 For the following independent and random samples, use the 0.10 level of significance in testing whether the population medians could be equal.

Sample 1	Sample 2	Sample 3	Sample 4
313	361	197	382
240	316	203	302
262	149	244	260
218	325	108	370
236	171	102	190
253			382

14.30 A sample of five law firms has been selected from each of three major cities for the purpose of obtaining a quote for legal services on a relatively routine business contract. The quotes provided by the fifteen firms are as shown in the accompanying table. At the 0.05 level

COMPUTER 14.4 SOLUTIONS

Kruskal-Wallis Test for Comparing More Than Two Independent Samples

This hypothesis test compares the medians of more than two independent samples.

EXCEL

	A	B	C
1	Kruskal-Wallis Test		
2			
3	Group	Rank Sum	Observations
4	*Firm1*	32	6
5	*Firm2*	45	5
6	*Firm3*	94	7
7	H Stat		7.4896
8	df		2
9	p-value		0.0236
10	chi-squared Critical		5.9915

Excel Kruskal-Wallis Test for Comparing More Than Two Independent Samples

1. The data in Table 14.6 are in adjacent columns in Excel file **CX14TECH**, with the labels in A1:C1, and the data in A2:C8. From the **Add-Ins** ribbon, click **Data Analysis Plus.** Click **Kruskal Wallis Test**. Click **OK.**
2. Enter **A1:C8** into the **Input Range** box. Click **Labels.** Enter **0.05** into the **Alpha** box. Click **OK.**

MINITAB

Minitab Kruskal-Wallis Test for Comparing More Than Two Independent Samples

```
Kruskal-Wallis Test: Score versus Firm

Kruskal-Wallis Test on Score

Firm       N  Median  Ave Rank      Z
1          6   64.50       5.3  -2.34
2          5   68.00       9.0  -0.25
3          7   75.00      13.4   2.49
Overall   18               9.5

H = 7.49  DF = 2  P = 0.024
H = 7.50  DF = 2  P = 0.024  (adjusted for ties)
```

1. The data in Table 14.6 must be stacked, as in Minitab file **CX14TECH**. With the test scores in column C1 and the firm identification (1, 2, or 3) in column C2: Click **Stat.** Select **Nonparametrics.** Click **Kruskal-Wallis.**
2. Enter **C1** into the **Response** box. Enter **C2** into the **Factor** box. Click **OK.** The printout will include two *p*-values, including one that is adjusted for ties.

of significance, does it seem that the population median bids from the cities' firms could be the same?

City A	City B	City C
452	439	477
469	461	460
430	487	496
403	431	484
456	401	423

14.31 In testing three different rubber compounds, a tire manufacturer finds the tread life of tires made from each to be as shown below. At the 0.05 level, could the three compounds deliver the same median tread life?

Design 1:	34	38	33	30	30 thousand miles
Design 2:	46	43	39	46	36
Design 3:	48	39	33	35	41

14.32 In an agricultural test, each of four organic compounds is applied to a sample of plants. At the end of 4 weeks, the heights of the plants are as shown here. At the 0.025 level, are the compounds equally effective in promoting plant growth?

Formula 1:	19	18	20	20	18 inches	
Formula 2:	9	13	20	16	13	18
Formula 3:	14	8	8	17	8	
Formula 4:	10	13	12	19	18	11

(**DATA SET**) *Note:* Exercises 14.33 and 14.34 require a computer and statistical software.

14.33 For three random samples of employees, each sample consisting of employees from a given age group, the data in file **XR14033** show the number of absences over the past 6 months. At the 0.10 level, can we conclude that the populations of such employee groups have the same median number of days missed? Identify and interpret the p-value for the test.

14.34 The Environmental Protection Agency measures city fuel economy by placing the vehicle on a laboratory dynamometer and determining the amount of fuel consumed during a standard speed-distance schedule that simulates an urban trip. File **XR14034** lists the miles per gallon data for three different automobile models. Can we conclude at the 0.10 level that the three models have the same median fuel economy? Identify and interpret the p-value for the test.

FRIEDMAN TEST FOR THE RANDOMIZED BLOCK DESIGN (14.6)

Description and Applications

The **Friedman test** is an extension of the Wilcoxon signed rank test for paired samples and is the nonparametric counterpart to the randomized block ANOVA design of Chapter 12. While randomized block ANOVA requires that observations be from normal populations with equal variances, the Friedman test makes no such demands. The Friedman test is applicable regardless of the shapes of the populations from which the individual data values are obtained. Unlike the randomized block ANOVA, the Friedman test can also be extended to the examination of ordinal data instead of being limited to data from the interval or ratio scales of measurement.

Like the randomized block ANOVA, the Friedman test has the goal of comparing treatments, though it is concerned with their medians rather than their means, and persons or test units are arranged into homogeneous blocks before the treatments are randomly assigned to members of each. The purpose of the blocks is to reduce error variation that might otherwise be present as the result of chance differences from one treatment group to the next with regard to the characteristics represented by the blocking variable. The randomized block ANOVA and the Friedman test differ in their assumptions and procedures, but their purposes and applications are similar.

As in the randomized block ANOVA of Chapter 12, the data can be summarized in a $b \times t$ table, where each column represents one of the t treatment groups and each row represents one of the b blocks. In each cell (combination of block and treatment), there will be one observation. Within each block (row), the observations are ranked from lowest (rank = 1) to highest (rank = t). If there are ties, each of the tied cells is assigned the average of the ranks they are occupying.

The null and alternative hypotheses, the test statistic, and the decision rule are as follows:

Friedman test for the randomized block design:

- **Null and alternative hypotheses**

  H_0: $m_1 = m_2 = \cdots = m_t$ for the $j = 1$ through t treatments. (The population medians are equal.)

  H_1: At least one m_j differs from the others. (The population medians are not equal.)

- **Test statistic**

  $$F_r = \frac{12}{bt(t+1)} \sum_{j=1}^{t} R_j^2 - 3b(t+1)$$

  where b = number of blocks
  t = number of treatments
  R_j = sum of the ranks for treatment j

- **Critical value and decision rule** For a test at the α level of significance, the critical value is the chi-square statistic with $df = (t - 1)$ and an upper-tail area of α. If F_r exceeds the critical chi-square, reject H_0; otherwise, do not reject.

For the Friedman test, either the number of blocks or the number of treatments should be ≥ 5 for the chi-square distribution to be a good approximation. If both b and t are <5, it is advisable either to use a computer statistical package or to rely on a set of special tables that are available in references such as Siegel (1956).[1] It is also desirable that the number of ties be small relative to the total number of observations, or cells. This presents no problem if the test is computer assisted, as most computer statistical packages provide results that are corrected for ties.

EXAMPLE

Friedman Test

The maker of a stain remover is testing the effectiveness of four different formulations for a new product. An experiment has been conducted in which each formula was randomly applied to one of four fabric pieces stained with the same household or food product. Six common types of stains were used as the blocks in the experiment, and each measurement represents the research director's subjective judgment on a scale of 1 to 10 using the following criteria:

1	2	3	4	5	6	7	8	9	10
☐	☐	☐	☐	☐	☐	☐	☐	☐	☐
"Poor"		"Fair"		"Good"		"Excellent"			"Outstanding"

[1]Source: See S. Siegel, *Nonparametric Statistics for the Behavioral Sciences* (New York: McGraw-Hill, 1956).

EXAMPLEEXAMPLEEXAMPLEEXAMPLEEXAMP

TABLE 14.7

Part A shows the underlying data for a Friedman test in which the blocks are types of stains and the treatments are four different formulations for a stain remover. The ratings reflect how well the stain was removed, with numerically higher numbers representing greater effectiveness. In part B, the scores within each block have been converted into ranks.

A. *Ratings Data*

		Stain-Remover Formula			
		1	**2**	**3**	**4**
Type of Stain	**Creosote**	2	7	3	6
	Crayon	9	10	7	5
	Motor Oil	4	6	1	4
	Grape Juice	9	7	4	5
	Ink	6	8	4	3
	Coffee	9	4	2	6

B. *After Converting to Ranks, with 1 = Lowest Rating and 4 = Highest Rating*

		Stain-Remover Formula			
		1	**2**	**3**	**4**
Type of Stain	**Creosote**	1	4	2	3
	Crayon	3	4	2	1
	Motor Oil	2.5	4	1	2.5
	Grape Juice	4	3	1	2
	Ink	3	4	2	1
	Coffee	4	2	1	3
	Sum of the Ranks:	17.5	21.0	9.0	12.5

The fabric samples having each type of stain are presented to the research director in a random order, and the ratings that resulted are shown in part A of Table 14.7. The data are also in file **CX14FAB**. At the 0.05 level, examine whether the stain-remover formulas could be equally effective in removing stains from this type of fabric.

SOLUTION

Although there are equal numerical distances from one rating category to the next, the presence of verbal descriptors along the way detracts from our ability to assume the resulting data are in the interval scale of measurement. For example, although they are shown equidistant along the scale, doubts may exist as to whether the distance between "Fair" and "Good" is the same as the distance between "Good" and "Excellent." It's therefore best that the ratings in part A of Table 14.7 be treated as ordinal (ranks) data.

The first step in the analysis is to convert the ratings into ranks within blocks. The results of this conversion are shown in part B of Table 14.7. For example, part A reveals that formula 4 has the lowest rating for removing ink stains, and it is assigned a rank value of 1 within that block, or row. Within the same block, formula 3 has the next lowest rating for ink stains, and it is assigned a rank value of 2. Formula 1 has the next to the highest rating on ink stains, and it is assigned a rank value of 3. Finally, formula 2 has the highest rating for ink stains and is assigned a rank value of 4. The ranks within the other blocks are assigned using the same procedure.

In the motor oil block, formulas 1 and 4 had identical ratings. The rank values they are occupying include ranks 2 and 3, so each is assigned the average of these ranks or $(2 + 3)/2 = 2.5$. As part B of Table 14.7 shows, this is the only tie that exists in the data.

Next, we calculate the F_r test statistic. The calculation relies on the sums of the ranks shown in part B of Table 14.7. These are $R_1 = 17.5$, $R_2 = 21.0$, $R_3 = 9.0$, and $R_4 = 12.5$. For $b = 6$ blocks and $t = 4$ treatments, the calculated value of the test statistic is

$$F_r = \frac{12}{bt(t + 1)} \sum_{j=1}^{t} R_j^2 - 3b(t + 1)$$

$$= \frac{12}{6(4)(4 + 1)}(17.5^2 + 21.0^2 + 9.0^2 + 12.5^2) - 3(6)(4 + 1)$$

$$= \frac{12}{120}(984.5) - 90 = 8.45$$

For a test at the 0.05 level of significance, the critical value of F_r is the chi-square statistic with $df = (t - 1) = 3$ and an upper-tail area of 0.05. Referring to the chi-square table, we find the critical value to be 7.815. For the 0.05 level of significance, the calculated value of F_r (8.45) exceeds the critical value (7.815), and we are able to reject the null hypothesis that the treatment medians could be equal. In other words, at least one of them is better or worse than the others.

Had the test been carried out at the 0.025 level of significance, the calculated value (8.45) would not have exceeded the critical value for this level (9.348), and the null hypothesis could not have been rejected. Based on the chi-square table, the most accurate statement we can make about the p-value for this test is that it is somewhere between 0.05 and 0.025.

Computer Solutions 14.5 shows how we can use Excel and Minitab in applying the Friedman test to the randomized block data in part A of Table 14.7. Minitab provides two p-values: one that ignores ties (0.038) and one that is adjusted for the one tie that was present in the ratings (0.035). Two values are similarly given for the calculated test statistic, identified in the printout as "S."

EXERCISES

14.35 Compare the Friedman test with the randomized block ANOVA in terms of (a) their respective null and alternative hypotheses, and (b) the assumptions required in order to use each test.

14.36 The randomized block design has been used in comparing the effectiveness of three treatments, with the data as shown in the table. Using the 0.05 level of significance, can we conclude that the treatments are equally effective?

		Treatment		
		1	2	3
Block	1	80	75	72
	2	60	70	60
	3	53	45	50
	4	72	65	49
	5	84	82	75

COMPUTER 14.5 SOLUTIONS

Friedman Test for the Randomized Block Design

This hypothesis test compares the medians of more than two dependent samples.

EXCEL

	A	B	C
1	Friedman Test		
2			
3	Group		Rank Sum
4	*Formula1*		17.5
5	*Formula2*		21
6	*Formula3*		9
7	*Formula4*		12.5
8	Fr Stat		8.45
9	df		3
10	p-value		0.0376
11	chi-squared Critical		7.815

Excel Friedman Test for Comparing More Than Two Dependent Samples

1. The data in Table 14.7 are in adjacent columns in Excel file **CX14FAB**, with the labels in A1:D1, and the data in A2:D7. The columns must represent the treatments and the rows must represent the blocks, as in part A of Table 14.7. From the **Add-Ins** ribbon, click **Data Analysis Plus.** Click **Friedman Test.** Click **OK.**
2. Enter **A1:D7** into the **Input Range** box. Click **Labels.** Enter **0.05** into the **Alpha** box. Click **OK.** The printout includes the calculated test statistic (8.45) and the *p*-value for the test (0.0376).

MINITAB

Minitab Friedman Test for Comparing More Than Two Dependent Samples

```
Friedman Test: Rating versus Formula blocked by StainTyp

S = 8.45  DF = 3  P = 0.038
S = 8.59  DF = 3  P = 0.035 (adjusted for ties)

                  Est   Sum of
Formula   N   Median    Ranks
1         6    6.438     17.5
2         6    7.188     21.0
3         6    3.438      9.0
4         6    4.688     12.5

Grand median = 5.438
```

1. The data in Table 14.7 must be stacked, as in Minitab file **CX14FAB**. With the formula (coded as 1–4) in C1, the stain type (1–6) in C2, and the performance rating in C3: Click **Stat.** Select **Nonparametrics.** Click **Friedman.**
2. Enter **C3** into the **Response** box. Enter **C1** into the **Treatment** box. Enter **C2** into the **Blocks** box. Click **OK.** The printout includes a *p*-value that is adjusted for ties.

14.37 A researcher has obtained the following salary figures for the mayor, police chief, and fire chief for a sample of U.S. cities. Assuming these to be a random sample of major cities in the United States, and using the 0.10 level of significance, can we conclude that the median national salaries for these positions of responsibility could be the same? Using the appropriate statistical table, what is the most accurate statement we can make about the p-value for the test?

City	Mayor	Police Chief	Fire Chief
A	$105,000	$89,356	$93,516
B	120,000	118,750	128,476
C	117,400	106,408	115,888
D	126,991	95,000	95,000
E	120,670	87,108	104,961
F	100,000	139,456	125,626

14.38 An extermination firm is testing several brands of pesticide spray, all of which claim to be effective against ants. Under controlled conditions, each spray is used on 10 ants of the species listed here. The measurement in each cell is the number of seconds until all 10 ants are dead. Given these data, and using the 0.05 level of significance, can the firm conclude the brands are equally effective? Using the appropriate statistical table, what is the most accurate statement that can be made about the p-value for the test?

	Pesticide Spray			
	1	2	3	4
Fire Ants	10	14	13	17
Bulldog Ants	12	16	8	19
Honey Ants	17	14	15	20
Carpenter Ants	15	12	14	18
Weaver Ants	14	11	10	15
Janitor Ants	13	15	17	17 seconds

(**DATA SET**) *Note:* Exercises 14.39 and 14.40 require a computer and statistical software.

14.39 Three movie critics have each rated a number of current movies on a scale of 1 = poor to 10 = excellent. Given the ratings in file **XR14039**, use the 0.025 level of significance in comparing the critics. Does it appear that they may be relying on different value systems in assigning their ratings? Identify and interpret the p-value for the test.

14.40 To evaluate whether three of its slopes should be classified as equally difficult, a ski resort sets up an experiment in which three skiers from each skill category (A = beginner to F = expert) are randomly selected to make one run on the slope to which they have been assigned. Given the times (minutes) in file **XR14040**, and using the 0.05 level of significance, can it be concluded that the slopes could be equally difficult? Identify and interpret the p-value for the test.

14.7 OTHER NONPARAMETRIC METHODS

The nonparametric methods covered earlier are just the tip of the iceberg, as a great many other tests have been devised for specific applications. This section presents three such techniques.

Sign Test for Comparing Paired Samples

The **sign test** is used for the same purposes as the Wilcoxon signed rank test of Sections 14.2 and 14.3, but it assumes that data are ordinal instead of interval or ratio. In the sign test, the difference between a data value and the hypothesized median (one-sample test) or the difference between two data values (paired-samples test) is replaced with a plus (+) or a minus (−) sign indicating the direction of the difference. We'll discuss this test in the context of comparing paired samples.

The sign test relies on the binomial distribution and the fact that, if m_d, the population median for $d_i = x_i - y_i$, is actually zero, $P(+)$ will be 0.5 for any pair of observations that are not tied. The method we are about to present involves the binomial tables, and it is exact. For larger samples, however, it can be approximated through the use of the normal distribution. The normal approximation will be discussed later.

The sign test for comparing paired samples:

- **The test statistic**
  1. For each pair of values, calculate $d_i = x_i - y_i$ and record the sign of d_i.
  2. n = the number of pairs for which $d_i \neq 0$.
  3. The test statistic is T = the number of pairs for which $d_i > 0$.
- **Determining the p-value for the test** For n = the number of nonzero differences, use the binomial tables (with the number of trials = n, and the probability of a "success," $\pi = 0.5$) to identify probabilities associated with s = the number of successes for the type of test being conducted:
  - Two-tail test with H_0: $m_d = 0$ versus H_1: $m_d \neq 0$.

    $$p\text{-value} = P(s \leq T) + P(s \geq [n - T])$$
  - Right-tail test with H_0: $m_d \leq 0$ versus H_1: $m_d > 0$.

    $$p\text{-value} = P(s \geq T)$$
  - Left-tail test with H_0: $m_d \geq 0$ versus H_1: $m_d < 0$.

    $$p\text{-value} = P(s \leq T)$$
- **Decision rule** For a test at the α level, reject H_0 if p-value $< \alpha$; otherwise, do not reject.
- **Carrying out the sign test for one sample** In this application, each value in the y_i column will be m_0, the value that has been hypothesized for the median. Otherwise, the procedure is similar to the procedure for comparing paired observations above.

EXAMPLE

Sign Test

To demonstrate how the sign test works, we'll apply it to the paired samples described earlier in Table 14.3, and the test will be carried out at the 0.10 level of significance. As in our earlier analysis, the data are in file CX14TIME, the test is two-tail, and the null and alternative hypotheses are H_0: $m_d = 0$ and H_1: $m_d \neq 0$.

SOLUTION

As Table 14.8 shows, the sign test converts the $d_i = x_i - y_i$ values to plus or minus, depending on whether x_i is greater than or less than y_i for each pair of observations. For computing task J, $d_i = 0$, and this pair of observations is omitted from further analysis. For the $n = 11$ nonzero differences that remain, $T = 3$ of these differences are positive.

In determining the p-value for the test, we can use the cumulative binomial probabilities for $n = 11$ trials and $\pi = 0.5$ for "success" (a positive difference) on each trial. For convenience, these probabilities have been reproduced in Table 14.9, which also summarizes the determination of the p-value for the test. If the null hypothesis were true, the probability of achieving $s \leq 3$ successes would be

$$P(s \leq 3) = 0.1133 \quad \text{as shown in Table 14.9}$$

EXAMPLEEXAMPLEEXAMPLEEXAMPLEEXA

TABLE 14.8

Using the sign test to compare the paired samples of Table 14.3.

Computing Task	Time Required for Software Packages x and y: x_i	y_i	The Sign of $d_i = x_i - y_i$
A	24.0	23.1	+
B	16.7	20.4	–
C	21.6	17.7	+
D	23.7	20.7	+
E	37.5	42.1	–
F	31.4	36.1	–
G	14.9	21.8	–
H	37.3	40.3	–
I	17.9	26.0	–
J	15.5	15.5	0
K	29.0	35.4	–
L	19.9	25.5	–

TABLE 14.9

The underlying binomial probabilities for a sign test of the data in Tables 14.3 and 14.8.

Binomial probabilities for 11 nonzero differences

```
Cumulative Distribution Function
Binomial with n = 11 and p = 0.5

      x        P( X <= x)
   0.00          0.0005
   1.00          0.0059
   2.00          0.0327
   3.00          0.1133
   4.00          0.2744
   5.00          0.5000
   6.00          0.7256
   7.00          0.8867
   8.00          0.9673
   9.00          0.9941
  10.00          0.9995
  11.00          1.0000
```

For $n = 11$, $\pi = 0.50$, the exact probability of three or fewer "+" differences is 0.1133.

For $n = 11$, $\pi = 0.50$, the exact probability of eight or more "+" differences is $1 - 0.8867 = 0.1133$.

Since $T = 3$ and the test is two-tail, the exact p-value is $0.1133 + 0.1133$, or 0.2266.

However, the test is two-tail and $P(s \leq 3) = 0.1133$ represents only the left-tail component of the p-value. The right-tail component will be its mirror image (recall that the binomial distribution is symmetric when $\pi = 0.5$), and this is the probability that $s \geq (n - T)$, or $P(s \geq 8)$. This component can be calculated as

$$P(s \geq 8) = 1 - P(s \leq 7)$$
$$= 1.0000 - 0.8867 = 0.1133 \quad \text{also shown in Table 14.9}$$

and the p-value for the two-tail test is

$$P(s \leq 3) + P(s \geq 8) = 0.1133 + 0.1133 = 0.2266$$

This analysis had the goal of using the $\alpha = 0.10$ level in testing whether m_d could be zero. Since the p-value (0.2266) is not less than α (0.10), the null hypothesis cannot be rejected at the 0.10 level. Based on the sign test, we are not able to conclude that the software packages differ with respect to the time they require for various computing tasks.

EXAMPLEEXAMPLE

If the number of nonzero differences is $n \geq 10$, the normal approximation to the binomial distribution can be used in carrying out the sign test. As discussed in Chapter 7, the normal distribution is a close approximation to the binomial whenever both $n\pi$ and $n(1 - \pi)$ are ≥ 5. Since the null hypothesis in the sign test always assumes that $\pi = 0.5$, the approximation will be satisfactory as long as $n \geq 10$. The procedure for the normal approximation is described below.

The sign test, using the normal approximation to the binomial distribution:

For n = the number of nonzero differences, T = number of (+) differences, the test statistic is

$$z = \frac{T - 0.5n}{0.5\sqrt{n}}$$

The critical value(s) for z will be similar to other z-tests. In the sign-test application, the normal approximation to the binomial distribution may be used whenever $n \geq 10$.

The rightmost term in the numerator of the z-statistic is the expected number of successes in n trials, and the denominator is the standard deviation of the number of successes. In Chapter 7, the standard deviation was presented as the square root of $[n\pi(1 - \pi)]$, and the denominator in the z-statistic is its equivalent when $\pi = 0.5$.

In the example just presented, we carried out the sign test on $n = 11$ nonzero differences by referring to the binomial tables. Since n was ≥ 10, we could have used the normal approximation instead. Applying the normal approximation method to this two-tail test, the first step is to compare the observed and expected values of T. The calculated value of the test statistic will be

$$z = \frac{T - 0.5n}{0.5\sqrt{n}} = \frac{3.0 - 0.5(11)}{0.5\sqrt{11}} = -1.51$$

For $\alpha = 0.10$ and a two-tail test, the critical values of z are $z = -1.645$ and $z = +1.645$. The calculated value ($z = -1.51$) falls within the "do not reject" region bounded by these values, and H_0: $m_d = 0$ cannot be rejected. This is the same conclusion we reached when we used the exact binomial probabilities in carrying out the test. When the number of nonzero differences is very large, the normal approximation will very closely approach the exact binomial distribution results in the sign test.

The sign test can also be used for tests involving a single sample. In this case, we just compare each observed value to the hypothesized value for the median. The observations are represented as x_i, and the hypothesized value for the median is represented as y_i. All of the values in the y_i column will be identical (and equal to m_0, the value that has been hypothesized for the median). The procedure is

COMPUTER 14.6 SOLUTIONS

Sign Test for Comparing Paired Samples

This hypothesis test applies to paired samples, such as before-and-after comparisons for the same items or subjects.

EXCEL

	A	B	C	D	E
1	**Sign Test for Testing One Sample or for Comparing Paired Samples**				
2					
3	n, the Number of NonZero Differences:	11	*Calculated Values:*		
4	T, the Number of Differences with di > 0:	3	*p-value if the test is:*		
5			Left-Tail	**Two-Tail**	Right-Tail
6			0.1133	**0.2266**	0.9673

Excel Sign Test for Paired Samples

1. The data in Table 14.3 (repeated in Table 14.8) are in adjacent columns in Excel file **CX14TIME**, with labels in A1 and B1, and data in A2:A13 and B2:B13. Enter the label **(Difference)** into **C1.** Select **C2** and enter the equation, **=A2-B2.** Click on the lower right corner of **C2,** then autofill downward so the differences will be in **C2:C13.** Count the number of positive, negative, and zero differences.
2. Open Excel worksheet template **TMSIGN**. In this worksheet, enter the number of nonzero differences into **B3.** Enter the number of positive differences into **B4.** The *p*-value for the test is based on the binomial distribution and is automatically calculated by Excel. For this two-tail test, the *p*-value is reported as 0.2266.

Note: As an alternative, the *z*-test approximation is applicable when there are at least 10 pairs of values. This test is available via the **Add-Ins, Data Analysis Plus, Sign Test** sequence of menu selections. (*Note:* Select one of the cells in the data field before proceeding.)

MINITAB

Minitab Sign Test for Paired Samples

```
Sign Test for Median: DIFF

Sign test of median =  0.00000 versus not = 0.00000

      N  Below  Equal  Above        P  Median
DIFF  12     8      1      3  0.2266  -4.150
```

1. With the Table 14.3 (and Table 14.8) data values in columns C1 and C2 of file **CX14TIME**, we must first calculate the differences and insert them into C3, as described in step 1 of Computer Solutions 14.2.
2. Click **Stat.** Select **Nonparametrics.** Click **1-Sample Sign.**
3. Enter **C3** into the **Variables** box. Select **Test median** and enter **0** into the box. Select **not equal** within the **Alternative** box. Click **OK.**

then carried out using the methods just described. If n = the number of nonzero differences is <10, the binomial tables should be used, while if $n \geq 10$, it is satisfactory to use the normal approximation.

Computer Solutions 14.6 shows how we can use Excel and Minitab in applying the sign test to compare the paired samples in Table 14.8. The p-value

(0.2266) is greater than for the Wilcoxon signed rank test for the same data (0.03). This is because the sign test has made less efficient use of the data, assuming it to be ordinal instead of interval.

The Runs Test for Randomness

The **runs test** evaluates the *randomness* of a series of observations by analyzing the number of *runs* it contains. A *run* is the consecutive appearance of one or more observations that are similar. For example, suppose a portion of a random number table included the series 2, 7, 6, 8, 9, 6, 7, 6, 6, 4. The first digit (2) is below the supposed median of 4.5 and constitutes one run. The middle eight digits are all above 4.5 and are another run. The final digit (4) is below 4.5 and is a third run. For a listing of digits from a supposed random number table, the presence of either very few runs or a great many runs would cause us to doubt whether the "random" digits are truly random.

If data are nominal, runs can also be counted. In a series of coin flips, we might observe H H T T T H T. In recording the order in which males and females arrive at work, part of the sequence could be F M M M F F M. Each of these sequences contains four runs. The runs test is carried out as follows:

Runs test:

- **Null and alternative hypotheses**

  H_0: The sequence is random, and H_1: The sequence is not random.

- **Procedure**
  - **For nominal data with two categories**
    1. Determine n_1 and n_2, the number of observations of each type.
    2. Count the number of runs, T.
  - **For ordinal, interval, or ratio data**
    1. Determine the median, m, of the data values.
    2. Identify each data value with a plus (+) if $x_i \geq m$ and with a minus (−) if $x_i < m$.
    3. Determine n_1 and n_2, the number of (+) and (−) observations.
    4. Count the number of runs, T.

- **Test statistic**

$$z = \frac{T - \left(\dfrac{2n_1n_2}{n} + 1\right)}{\sqrt{\dfrac{2n_1n_2(2n_1n_2 - n)}{n^2(n-1)}}}$$

where T = the number of runs
n_1 = the number of observations of the first type
n_2 = the number of observations of the second type
n = the total number of observations, $n_1 + n_2$

In the numerator of the expression above, the quantity in parentheses is the number of runs that would be expected if the observations were randomly arranged. If the number of runs is significantly less than expected, this indicates that similar kinds of observations tend to occur in *clusters,* as would happen if a sports team tended to win and lose in streaks. This may also occur if there is an upward or downward trend from earlier values to later ones. If the number of runs is significantly more than expected, this reflects that the data values tend to reverse direction more often than chance processes would allow.

When data are of the interval or ratio scale of measurement, the mean can be used instead of the median. Since the median can apply to any data that are not nominal, however, we will use it in our presentation and examples. If a sequence has a great many values and a computer is not being used, it can be time-saving to use the more easily calculated mean instead of the median.

EXAMPLE

Runs Test for Randomness

A political activist claims to have "randomly" stopped persons at a street corner and asked them to sign his petition and give their age. During his first hour on the street, 30 people signed the document and gave their age, and the order is as shown in Table 14.10. The data are also in file **CX14AGE**. At the 0.05 level of significance, evaluate the randomness of the ages for this sequence of 30 respondents.

SOLUTION

In conducting this test, the null hypothesis is H_0: The ages are in a random order. The age values, shown in Table 14.10, have a median of 44 years. Each age is converted to a plus (+) if it is 44 or higher and to a minus (−) if it is less than 44. This process provides the sequence of (+) and (−) symbols in the lower part of the table. For clarity, each of the $T = 10$ runs is enclosed in a box. There are $n_1 = 15$ (+) symbols, $n_2 = 15$ (−) symbols, and the total sample size is $n = 15 + 15 = 30$. The z-statistic is

The actual number of runs → T; The expected number of runs → $\left(\frac{2n_1n_2}{n} + 1\right)$

$$z = \frac{T - \left(\frac{2n_1n_2}{n} + 1\right)}{\sqrt{\frac{2n_1n_2(2n_1n_2 - n)}{n^2(n-1)}}} = \frac{10 - \left(\frac{2(15)(15)}{30} + 1\right)}{\sqrt{\frac{2(15)(15) \cdot [2(15)(15) - 30]}{30^2(30-1)}}}$$

$$= \frac{10 - 16}{2.691} = -2.23$$

For a two-tail test at the 0.05 level of significance, the critical z values are $z = -1.96$ and $z = +1.96$. The observed value, $z = -2.23$, is outside these limits, and the null hypothesis is rejected. At the 0.05 level, the ages do not occur in a random order. That the number of runs (10) is much less than the expected number (16) is not necessarily the petitioner's fault, since persons of similar ages may tend to be found together or to frequent the street at different times of the day. For example, older persons may tend to be more prevalent during the early morning and noontime hours.

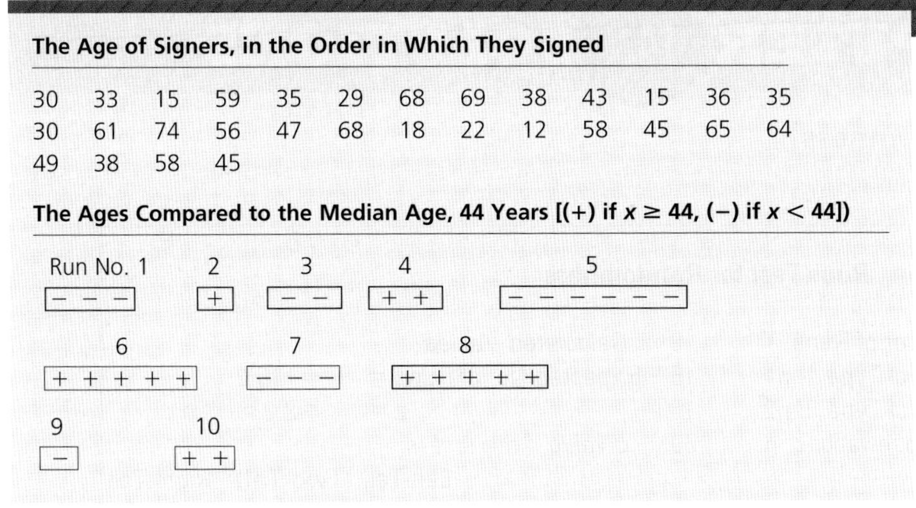

The Age of Signers, in the Order in Which They Signed

30	33	15	59	35	29	68	69	38	43	15	36	35
30	61	74	56	47	68	18	22	12	58	45	65	64
49	38	58	45									

The Ages Compared to the Median Age, 44 Years [(+) if $x \geq 44$, (−) if $x < 44$])

Run No.	Signs
1	− − −
2	+
3	− −
4	+ +
5	− − − − − −
6	+ + + + +
7	− − −
8	+ + + + +
9	−
10	+ +

TABLE 14.10

In this sequence of persons signing a petition, the median age is 44 years. The sequence contains $T = 10$ runs versus an expected value of 16 runs, and the null hypothesis that the order is random is rejected at the 0.05 level of significance.

Using the standard normal table, we see the cumulative area to $z = -2.23$ is 0.0129. Because the test is two-tail, the p-value is twice this area, or $2(0.0129) = 0.0258$.

Computer Solutions 14.7 shows how we can use Excel and Minitab in applying the runs test for randomness to the sequence in Table 14.10. With Excel, we manually count the runs, then apply Excel worksheet template **TMRUNS**. Minitab reports both the actual and the expected number of runs. The actual number of runs differs greatly from the expected number, and the p-value for the test is just 0.026. Using the 0.05 level of significance that has been specified, the sequence of ages is judged to be other than random.

Kolmogorov-Smirnov Test for Normality

The **Kolmogorov-Smirnov test for normality** is an alternative to the chi-square test for normality discussed in Chapter 13. Compared to the chi-square method, this test does not require any minimum values for expected frequencies and can be used with small as well as large sample sizes. The data must be a random sample from the population, and the variable being observed or measured has to be at least interval in its scale of measurement.

The population parameters are not specified, and the sample mean and standard deviation are used to estimate their population counterparts. The null and alternative hypotheses are

H_0: The sample was drawn from a normal distribution.

H_1: The sample was not drawn from a normal distribution.

This test involves two cumulative relative frequencies: one that is derived from the actual data observed, the other constructed under the assumption that the null hypothesis is true, and relying on the mean and standard deviation of the sample data. A comparison is made between these cumulative

COMPUTER 14.7 SOLUTIONS

Runs Test for Randomness

EXCEL

	A	B	C	D	E
1	**Z-Test Approximation , Runs Test for Randomness**				
2					
3	Number of Runs, T:	10	*Calculated Values:*		
4	Number of Observs. of Type 1, n1:	15	Expected No. of Runs =	16.0	
5	Number of Observs. of Type 2, n2:	15	z =	-2.230	
6	Total Number of Observations, n:	30	*p-value if the test is:*		
7			Left-Tail	**Two-Tail**	Right-Tail
8			0.0129	**0.0258**	0.9871

Excel Runs Test for Randomness

1. The ages in Table 14.10 are in Excel file **CX14AGE**, with the label in A1 and the data values in A2:A31. Click on cell B2 and enter the equation **=A2−43.999.** Click on the bottom right corner of cell B2 and autofill downward to cell B31. An examination of the sequence of values in B2:B31 will reveal the same series of signs shown in the lower portion of Table 14.10. Count the number of runs—this will be $T = 10$. Count the number of "+" values in the sequence—this will be $n_1 = 15$. Count the number of "−" values—this will be $n_2 = 15$.
2. Open Excel worksheet template **TMRUNS**. Enter T **(10)** into **B3.** Enter n_1 **(15)** into **B4.** Enter n_2 **(15)** into **B5.** The expected number of runs, the calculated z-statistic, and the two-tail p-value for the test will be displayed.

MINITAB

Minitab Runs Test for Randomness

```
Runs Test: Age

Runs test for Age

Runs above and below K = 43.999

The observed number of runs = 10
The expected number of runs = 16
15 observations above K, 15 below
P-value = 0.026
```

1. With the Table 14.10 sequence of ages in column C1 of **CX14AGE**: Click **Stat.** Select **Nonparametrics.** Click **Runs Test.**
2. Enter **C1** into the **Variables** box. Select **Above and below:** and enter **43.999** into the box. Click **OK.** The p-value (0.026) is comparable to that obtained with the Excel worksheet template above.

relative frequencies, and the test statistic is the maximum amount of divergence between them. The procedure and test statistic are as follows:

Kolmogorov-Smirnov test for normality:

1. **Data values are arranged in order, from smaller to larger values of x.**
2. **Calculated test statistic:**

$$D = \max|F_i - E_i|$$

where F_i = observed cumulative relative frequency for the ith data value
E_i = expected cumulative relative frequency for the ith data value

3. **If $D >$ the critical value listed in the critical values for the Kolmogorov-Smirnov test table in Appendix A for sample size = n and level of significance = α, the null hypothesis will be rejected. Otherwise, it cannot be rejected.**

EXAMPLE

Kolmogorov-Smirnov Test for Normality

A random sample consists of just 5 data values: 438, 424, 213, 181, and 137, which are also in file CX14XDAT. Using the Kolmogorov-Smirnov test for normality and the 0.10 level of significance, could these data have been drawn from a population that is normally distributed?

SOLUTION

This sample size is much too small for us to apply the chi-square test for normality, and it might even seem too small for us to be able to apply any test at all. But the Kolmogorov-Smirnov test for normality is applicable even for a sample this small. For this test, the null hypothesis is that the sample was drawn from a normal population and the alternative hypothesis is that it was not.

The test is easily carried out. All we have to do is arrange the data in increasing order, then compare the observed and the expected cumulative relative frequencies.

Table 14.11 summarizes the calculations involved in the test. The leftmost column shows the data values arranged in increasing order of size, and the F_i column lists the observed cumulative relative frequencies for each of the data values. For example, 3 of the 5 observations (60%, for a cumulative relative frequency of 0.6000) were no greater than 213.

The entries in the E_i column are based on a normal distribution with μ = 278.6 and σ = 141.8. Each data value is converted to a z-score, then we use the standard normal table to find the area to the left. Table 14.11 shows an example calculation for the expected cumulative relative frequency for 424, or $P(x \leq 424)$ for a normal distribution in which $\mu = 278.6$ and $\sigma = 141.8$. The z-score is $z = 1.03$, and the standard normal table is used in identifying the area to the left, which is 0.8485.

The value of the test statistic, $D = \max|F_i - E_i|$, is 0.2772. Referring to Appendix A and the critical values for the Kolmogorov-Smirnov test table, we

EXAMPLEEXAMPLEEXAMPLEEXAMPLEEXAMPLEEXAMPLE

TABLE 14.11

The Kolmogorov-Smirnov test for normality can be applied even for very small samples. In this example, there are just 5 data values, but we can still test whether the sample could have been drawn from a normal population. The sample mean and standard deviation are used in constructing the expected cumulative relative frequencies.

Data Value	Relative Frequency	Cumulative Relative Frequencies: F_i = Observed	Cumulative Relative Frequencies: E_i = Expected	$\lvert F_i - E_i \rvert$
137	0.2000	0.2000	0.1587	0.0413
181	0.2000	0.4000	0.2451	0.1549
213	0.2000	0.6000	0.3228	0.2772
424	0.2000	0.8000	0.8485	0.0485
438	0.2000	1.0000	0.8686	0.1314

$D = \max|F_i - E_i| = 0.2772$

In a normal distribution with $\mu = 278.6$ and $\sigma = 141.8$, $P(x \leq 424)$ is found by first calculating

$$z = \frac{424 - 278.6}{141.8} = 1.03,$$

then referring to the standard normal table. The area to the left of $z = 1.03$ is 0.8485.

AMPLEEXAMPLEEXAMPLE

find the critical value to be 0.315 for a test at the 0.10 level with $n = 5$. Since the observed value of D (0.2772) does not exceed the critical value (0.315), we are not able to reject the null hypothesis. The data could have been drawn from a normal distribution.

Computer Solutions 14.8 shows how we can use Minitab in applying the Kolmogorov-Smirnov test for normality to the data values in Table 14.11. The printout includes a calculated D statistic that is comparable to the one in Table 14.11, but with greater accuracy, because our calculations were based on the normal distribution table, which has inherent gaps between the printed z values.

Spearman Coefficient of Rank Correlation

For paired observations in which the two variables are at least ordinal, we can use the **Spearman coefficient of rank correlation** (r_s) to measure the strength and direction of the relationship between them. For example, the Spearman coefficient of rank correlation could be applied if two different investment advisers were to rank 10 stocks from best to worst. If they provided identical rankings, r_s would be $+1$. If their rankings were exactly opposite, r_s would be -1. For variables x and y, a positive correlation indicates that higher ranks for x tend to be associated with higher ranks for y, and a negative correlation indicates that lower ranks for x tend to be associated with higher ranks for y. When the original data are not in the form of ranks, they are converted to ranks for purposes of this analysis. The Spearman coefficient of rank correlation has a parametric counterpart that was discussed in Chapter 2 and will be discussed further in the next chapter. The formula below is exact when there are no ties, and is a very close approximation when the number of ties is small relative to the number of pairs.

Spearman coefficient of rank correlation:

Test statistic:

$$r_s = 1 - \frac{6(\sum d_i^2)}{n(n^2 - 1)}$$

where $\sum d_i^2$ = the sum of the squared differences between the ranks

n = the number of pairs of observations

COMPUTER 14.8 SOLUTIONS

The Kolmogorov-Smirnov Test for Normality

This test examines whether a set of data could have come from a normal population.

EXCEL

Excel does not provide the Kolmogorov-Smirnov test for normality. As an alternative, the chi-square test for normality (Chapter 13) can be used when $n \geq 32$.

MINITAB

Minitab Kolmogorov-Smirnov Test for Normality

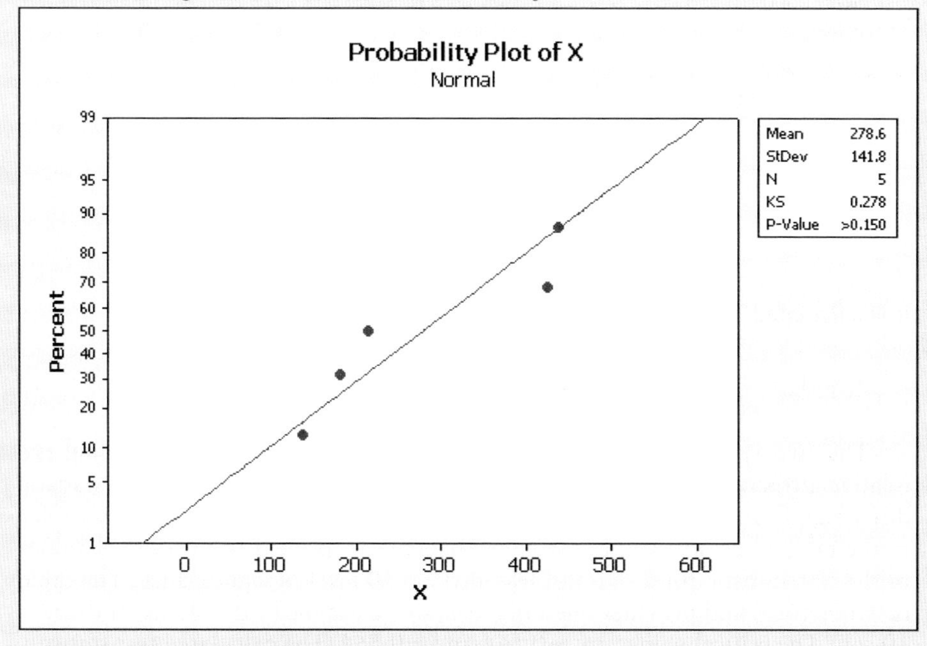

1. The data in the leftmost column of Table 14.11 are in column C1 of file CX14XDAT. Click **Stat.** Select **Basic Statistics.** Click **Normality Test.**
2. Enter **C1** into the **Variable** box. Select **Kolmogorov-Smirnov.** Click **OK.**
3. In the normal probability plot that is generated, data that are actually from a normal distribution are more likely to fall along the straight line that is fitted. The output also includes the D statistic (shown as "KS") and an approximate p-value ($p > 0.15$) for the test. Because ($p > 0.15$) is not less than the 0.10 level for our test, we conclude that the data could have come from a normal distribution.

Significance test for ρ_s = the population value of r_s:

For $n \leq 30$, the rejection region(s) for the tests of H_0: $\rho_s = 0$, H_0: $\rho_s \geq 0$, and H_0: $\rho_s \leq 0$ are in the table of critical values for r_s in Appendix A.

For larger values of n, the normal distribution approximation can be used, and the test statistic will be $z = r_s\sqrt{n-1}$.

EXAMPLEEXAMPLEEXAMPLEEXAMPLEEXAMPLEEXAMPLE

EXAMPLE

Spearman Rank Correlation

Two sportswriters have ranked 10 Olympic sports according to how interesting they are to watch in person, with the results shown here. Calculate and interpret the Spearman coefficient of rank correlation, and use the 0.10 level of significance in testing whether its true value could be something other than 0.

Olympic Sport	Arnold's Ranking, x_i	Barbara's Ranking, y_i	$d_i = x_i - y_i$	d_i^2
Track & field	1	2	−1	1
Basketball	2	4	−2	4
Volleyball	3	5	−2	4
Swimming	4	1	3	9
Boxing	5	8	−3	9
Ice skating	6	3	3	9
Weight lifting	7	6	1	1
Wrestling	8	7	1	1
Judo	9	10	−1	1
Equestrian	10	9	1	1

$\Sigma d_i^2 = 40$

For these data, the Spearman coefficient of rank correlation is $r_s = 0.758$, and it is calculated as follows:

$$r_s = 1 - \frac{6(\Sigma d_i^2)}{n(n^2 - 1)} = 1 - \frac{6(40)}{10(10^2 - 1)} = 1 - \frac{240}{990} = 0.758$$

The two sportswriters seem to be somewhat in agreement regarding the relative attractiveness of the various sports, but does the calculated test statistic differ significantly from 0 at the 0.10 level of significance? To find out, we refer to the table of critical values for r_s in Appendix A to identify the critical values of the test statistic for a two-tail test at the 0.10 level of significance. The table is symmetrical and provides one-tail areas, so we use the 0.05 column of the n = 10 row. The critical values are −0.564 and +0.564. The calculated test statistic, r_s = 0.758, falls into the upper rejection region and we are able to reject the possibility that the population value for the Spearman coefficient of rank correlation could be 0. You may wish to further use the table in verifying that the most accurate statement we can make about the p-value for this two-tail test is that it is somewhere between 0.01 and 0.02.

Computer Solutions 14.9 shows the Excel and Minitab procedures for obtaining the Spearman coefficient of rank correlation. Although each printout has the same correlation value, note that Excel has used the normal approximation in its p-value and critical-value calculations. We've used the Excel printout here as an example, but this approach is best used when the number of sample pairs is at least 30.

COMPUTER 14.9 SOLUTIONS

Spearman Coefficient of Rank Correlation

EXCEL

	A	B	C	D
1	**Spearman Rank Correlation**			
2				
3	*Arnold and Barbara*			
4	Spearman Rank Correlation			0.758
5	z Stat			2.273
6	P(Z<=z) one tail			0.012
7	z Critical one tail			1.282
8	P(Z<=z) two tail			0.023
9	z Critical two tail			1.645

Using Excel to obtain Spearman coefficient of rank correlation

1. Open Excel file CX14SPEAR. From the **Add-Ins** ribbon, click **Data Analysis Plus.** Select **Correlation (Spearman)** and click **OK.**
2. Enter **B1:B11** into the **Variable 1 Range** box and **C1:C11** into the **Variable 2 Range** Box. Select **Labels** and enter **0.10** into the **Alpha** box. Click **OK.**

Note: The Spearman coefficient of rank correlation has been calculated as described in the text, but the Excel p-values and critical values are based on the normal approximation. This approximation works best when there are at least 30 pairs of values, but it is shown here for demonstration purposes.

MINITAB

```
Correlations: Arnold, Barbara

Pearson correlation of Arnold and Barbara = 0.758
P-Value = 0.011
```

Using Minitab to obtain Spearman coefficient of rank correlation

1. Open Minitab workbook CX14SPEAR. Click **Stat,** select **Basic Statistics,** and click **Correlation.**
2. Enter **C2** and **C3** into the **Variables** box, select **Display p-values,** and click **OK.**

EXERCISES

14.41 What role does the binomial distribution play in the sign test, and when is it permissible to use the normal approximation in carrying out the test?

14.42 Whether testing a single sample or comparing paired samples, how should one go about deciding whether to use the sign test versus the Wilcoxon signed rank test?

14.43 For $n = 8$ nonzero differences between paired observations, there are 5 (+) differences for $d = x_1 - x_2$. Use the 0.05 level of significance in testing H_0: $m_d = 0$ versus H_1: $m_d \neq 0$.

14.44 For $n = 6$ nonzero differences in a one-sample test, there are 5 (+) differences for $d = x - 20$. Use the 0.10 level of significance in testing H_0: $m_d \leq 20$ versus H_1: $m_d > 20$.

14.45 Using the data of Exercise 14.13 and the 0.10 level, apply the sign test to compare the two samples. How do the results compare to those of that exercise? If they differ, what does this imply about the respective tests?

14.46 Using the data of Exercise 14.8 and the 0.05 level, apply the sign test to compare the data with the claimed median. How do the results compare to those of that exercise? If they differ, what does this imply about the respective tests?

14.47 In the runs test for randomness, what is a "run"?

14.48 How many runs are there in each of the following series?

a. 1 1 0 0 0 1 0 0 1 0 1 1 0 0 0 0 1 1 0 1 1 0 1 0 1 1
1 0 0 1

b. 0 1 0 1 1 0 1 1 1 0 0 0 1 0 1 1 0 0 1 0 1 0 0 1 0
1 1 0 0 1

c. 1 1 1 0 0 0 1 1 1 0 0 0 1 1 0 0 1 1 0 0 1 0 1 0 1
0 0 1 0 1

d. 1 1 1 1 1 1 1 1 1 1 1 1 1 1 0 0 0 0 0 0 0 0 0 0 0
0 0 0 0 0

14.49 From 1789 through 1870, the last names of associate justices appointed to the U.S. Supreme Court were as coded below. Last names beginning with A–M are shown as a "1," those beginning with N–Z as a "2." Compare the observed number of runs with the expected number, then interpret what this difference might mean in terms of the sequence of the appointees. At the 0.10 level of significance, test the randomness of this series.

Source: *Time Almanac 2003*, p. 91.

2 2 1 1 1 1 2 1 2 1 1 1 2 1 2 2 2 1
1 2 1 1 1 1 2 2 1 1 1 1 2 1 1 1 2 1

14.50 Franchise headquarters claims it randomly selected the local franchises that received surprise visits during the past year. The sequence of visits is as shown below for male (M) versus female (F) managers. Using the 0.10 level of significance, evaluate the randomness of the sequence.

F M M M F M F M F M F F M
M F M F M M M F M M F

14.51 A random sample of recent LSAT (Law School Admission Test) scores submitted by applicants to a university's law program are as listed below. Using the 0.10 level of significance, apply the Kolmogorov-Smirnov test for normality in examining whether the scores could have come from a normally distributed population.

142 161 142 149 147 145 164 170

14.52 Product testers for a radio manufacturer routinely select a sample of units from the production line and test circuit integrity by subjecting each unit to extreme vibrations that are approximately triple the most violent shaking they are ever likely to encounter in the field. For the most recent five radios tested, the number of minutes until failure was found to be 15, 79, 68, 10, and 17 minutes. Using the 0.05 level of significance, apply the Kolmogorov-Smirnov test for normality in examining whether the population of failure times might be normally distributed.

14.53 A local supermarket has contracted for floor cleaning and polishing to be done each evening by a small work crew using mechanized equipment. After the crew has performed the service for several weeks and settled into a routine, the supermarket manager has timed the most recent jobs to get a better idea as to the nature of the distribution of times for performing the task. Over the past week, the cleaning and polishing times have been 61, 44, 57, 49, 59, 43, and 42 minutes. Using the 0.10 level of significance, apply the Kolmogorov-Smirnov test for normality in examining whether the population of cleaning times might be normally distributed.

14.54 What is the Spearman coefficient of rank correlation, what range of values can it have, and how is its numerical value interpreted?

14.55 Two travel editors have each ranked 10 popular vacation spots according to the value they represent for the money. The sum of the squared differences between their rankings is 28. Determine and interpret the Spearman coefficient of rank correlation. At the 0.05 level, does r_s differ significantly from 0?

14.56 For the data of Exercise 14.16, convert the husbands' and wives' expenditures to ranks, then determine and interpret the Spearman coefficient of rank correlation. At the 0.05 level, does r_s differ significantly from 0?

14.57 A Republican and a Democrat candidate for political office have each rated the 12 most recent pieces of legislation in their state from 1 (most beneficial to the state) to 12 (least beneficial). Given the data shown here, determine and interpret the Spearman coefficient of rank correlation. At the 0.02 level, does r_s differ significantly from 0?

Legislation	A	B	C	D	E	F	G	H	I	J	K	L
Republican's ranking	1	2	3	4	5	6	7	8	9	10	11	12
Democrat's ranking	3	4	6	9	8	5	10	7	12	11	1	2

(DATA SET) *Note:* Exercises 14.58–14.63 require a computer and statistical software.

14.58 Testing the effect of a diet and exercise regimen on subjects' cholesterol levels, a sports medicine clinic enlists 40 volunteers whose cholesterol levels are measured before and after the 1-month program. The before and after data are listed in data file **XR14058**. Using the sign test and the 0.025 level of significance, examine the effectiveness of the diet and exercise program. Identify and interpret the p-value for the test.

14.59 According to R.L. Polk & Company, the median age of automobiles and trucks on the road in the United States is about 9 years. The ages for a random sample of vehicles in the executive parking lot of a metal fabrication company are listed in file **XR14059**. Using the sign test and the 0.05 level of significance, examine whether the median age for cars driven to work by the company's executives might be less than the median reported by R.L. Polk. Identify and interpret the p-value for the test. Source: theautochannel.com, July 31, 2009.

14.60 For the first 43 Super Bowl games, the winner is listed in file **XR14060** according to "1" (American Conference) or "2" (National Conference). At the 0.05 level of significance, can this sequence be considered as other than random? Identify and interpret the p-value for the test. Source: NFL.com, July 31, 2009.

14.61 For the first 43 Super Bowl games, the winning margins are listed in file **XR14061**. At the 0.10 level of significance, can this sequence be considered as other than random? Identify and interpret the p-value for the test. Source: NFL.com, July 31, 2009.

14.62 For the set of winning margins in Exercise 14.61, use either the Kolmogorov-Smirnov test or the chi-square test for normality and the 0.05 level of significance in examining whether the data could have come from a normal distribution.

14.63 Last year's rates of return for a random sample of mutual funds are listed in file **XR14063**. Using either the Kolmogorov-Smirnov test or the chi-square test for normality and the 0.10 level of significance, examine whether the data could have come from a normal distribution.

SUMMARY (14.8)

Nonparametric versus parametric testing

A nonparametric test is one that makes no assumptions about the specific shape of the population from which a sample is drawn. In Chapter 13, the chi-square tests for goodness of fit, independence, and comparison of sample proportions were nonparametric, but chi-square tests of hypotheses regarding the population variance were parametric. (Recall that chi-square tests for σ^2 required that the population be normally distributed.) This chapter covers a variety of other nonparametric techniques, most of which have a parametric counterpart covered in other chapters of the text.

When to use a nonparametric test

A nonparametric test is used instead of its parametric counterpart (1) when the data are of the nominal or ordinal scale of measurement or (2) when the data are of the interval or ratio scale of measurement, but one or more other assumptions, such as the normality of the underlying population distribution, are not met. Unlike parametric methods, nonparametric techniques can be applied even when the sample sizes are very small.

Nonparametric tests: types and applications

The Wilcoxon signed rank test is applied to a single sample in testing whether the population median could be a hypothesized value. In this application, it is the nonparametric counterpart to the single-sample t-test. With paired samples, the Wilcoxon signed rank test is the nonparametric counterpart to the paired-samples t-test, and tests whether the population median for the difference between paired observations could be zero. The Wilcoxon rank sum test compares the medians of two independent samples and is the nonparametric counterpart to the two-sample, pooled-variances t-test.

The Kruskal-Wallis test compares the medians of more than two independent samples and is the nonparametric counterpart to the one-way analysis of variance. Like the randomized block ANOVA, the Friedman test compares the effectiveness of treatments, but it is concerned with their medians rather than their means. Also, it can examine data that are ordinal (ranks) and makes no assumptions regarding the populations from which data values have been drawn.

The sign test is used for the same purposes as the Wilcoxon signed rank test, but data need only be of the ordinal scale of measurement. In this test, each observation (one-sample test) or difference between paired observations (paired samples) is expressed only as a plus (+) or a minus (−) to indicate the direction of the difference. Because it assumes that data are ordinal instead of interval, it makes less efficient use of the data than does the Wilcoxon signed rank test.

The runs test evaluates the randomness of a series of observations by analyzing the number of "runs" it contains. A run is the consecutive appearance of one or more observations that are similar in some way. If the number of runs differs significantly from the number expected, we conclude that some pattern exists in the sequence of data values.

The Kolmogorov-Smirnov test for normality can be used for samples of any size, including those that are too small for the chi-square method presented in Chapter 13. The test is based on the comparison of two cumulative relative frequencies: one that is derived from the actual data, the other constructed under the assumption that the sample was actually drawn from a normal distribution.

For paired observations where the two variables are at least ordinal, the Spearman coefficient of rank correlation can be applied to the ranks in determining the strength and direction of the relationship between the variables.

EQUATIONS

This is where you would ordinarily find a neat, concise listing of the equations introduced in the chapter. However, for this rather unique chapter, such a listing would be neither neat nor concise. The chapter's many equations tend to be inseparable from the detailed procedures used in obtaining values for the terms within them. For this reason, we are foregoing our usual summary in favor of recommending that you refer to the chapter sections for the techniques and their respective equation/procedure summaries.

CHAPTER EXERCISES

14.64 Two different paste wax formulations are being compared to see if they hold up equally well in repeated washing at an automated car wash. On each of 10 test vehicles, formulas A and B are applied to opposite sides of the roof, and a coin flip is used in determining the side that receives formula A. The number of washings before water no longer "beads" on the surface is found to be as listed here. At the 0.05 level, can we conclude that the wax formulations are equally effective? Using the appropriate statistical table, what is the most accurate statement that can be made about the p-value for the test?

		Vehicle									
		1	2	3	4	5	6	7	8	9	10
Washings Survived by Wax Formulation	*A*	12	14	15	14	14	11	14	10	8	15
	B	11	19	17	18	11	16	17	13	13	12

14.65 Comparing the lifetimes of two brands of penlight batteries, a researcher has randomly selected 10 batteries of brand A and 10 of brand B, then subjected each battery to the same usage test. Given the lifetimes listed below and using the 0.05 level of significance, can we conclude that brand A is superior to brand B? Using the appropriate statistical table, what is the most accurate statement that can be made about the p-value for the test?

Brand A	93.4	119.1	94.3	90.0	66.0
	121.4	92.5	111.6	87.6	98.0 minutes
Brand B	59.2	65.5	81.5	92.5	71.7
	69.4	83.4	92.5	93.3	81.7 minutes

14.66 It has been reported that the median price of a home purchased in Portland, Oregon, during 2007 was \$295,200. For a random sample of homes purchased in a certain area of the city during that year, the prices were as listed below. Using the 0.025 level of significance, test whether the median purchase price of all homes sold in that area of the city during 2007 could have been less than \$295,200. Using the appropriate statistical table, what is the most accurate statement that can be made about the p-value for the test? Source: Bureau of the Census, *Statistical Abstract of the United States 2009*, p. 596.

\$300.3	271.4	270.3	295.1	252.9
287.0	287.3	298.1	257.9	313.9 thousand

14.67 The resting pulse rates of 6 randomly selected adults have been measured before and after a 4-week aerobic fitness program, with the results listed below. At the 0.05 level, can we conclude that the program has had a significant effect? Using the appropriate statistical table, what is the most accurate statement that can be made about the p-value for the test?

		Resting Pulse Rate	
		Before	**After**
Person	1	86	85
	2	81	82
	3	77	73
	4	85	79
	5	77	67
	6	77	73 beats/minute

14.68 In a fast-food restaurant, it's desirable to have seats that are comfortable but not *too* comfortable. Patrons sitting at a seat that is too comfortable may tend to linger and tax the capacity of the restaurant. In an experiment to compare seat designs by observing how long people with the same food order stay in the restaurant, each of 18 persons is given a hamburger, fries, and shake, then timed to see how much time they require to eat and leave. The times are shown here. At the 0.05 level, do the seat designs differ significantly in terms of the time that patrons stay? Using the appropriate statistical table, what is the most accurate statement we can make about the p-value for the test?

Time, six patrons and seat design 1:
6 10 9 5 5 8 minutes

Time, six patrons and seat design 2:
13 10 8 13 9 12

Time, six patrons and seat design 3:
10 15 10 8 15 15

14.69 A consumer magazine is evaluating five brands of trash compactors for their effectiveness in reducing the volume of typical household products that are discarded. In the experiment, each block consists of three bags containing identical mixtures of the kinds of trash described, and each measurement is the volume (cubic feet) to which a given compactor reduced the load it was assigned. At the 0.025 level, can the magazine conclude the compactors are equally effective in reducing the volume of household trash items? Using the appropriate statistical table, what is the most accurate statement that can be made about the p-value for the test?

	Compactor					
	1	**2**	**3**	**4**	**5**	
Cans and Bottles	1.2	1.3	1.4	1.6	1.5	
Cardboard Boxes	1.5	1.5	1.8	1.8	1.7	
Newspapers and Magazines	1.7	1.9	2.1	2.4	2.3	ft^3

14.70 The manager of a photography studio has carried out an experiment to compare the attractiveness of four different poster-size photographs for use in the display window. During 23 business days of the previous month, she randomly selected one of the four photographs for display; the data below show the number of customers for each day that a given photograph was displayed in the window. At the 0.01 level of significance, examine whether the poster-size photos could be equally effective in attracting customers into the store. Using the appropriate statistical table, what is the most accurate statement that can be made about the p-value for the test?

Photo 1	Photo 2	Photo 3	Photo 4
14	39	19	28
39	19	27	21
45	25	22	32
48	37	15	29
26	18	31	42
30	45		48

14.71 Three different O-ring shapes are being considered for use in a deep-sea submersible vehicle that will be remote-operated from the surface vessel. Because the sealing effectiveness depends partially on the temperature of the surrounding sea water, a randomized block design has been set up in which each design will be pressure-tested at five temperatures. The temperatures have been selected to represent the maximum of approximately 86° Fahrenheit at surface locations in the tropical climates to the near 32° temperatures found at the greatest depths. For each seal-temperature cell, the measurement in the table is the pressure (thousands of pounds per square inch) recorded just prior to seal failure. At the 0.025 level of significance, could the respective O-ring shapes be equally effective? Using the appropriate statistical table, what is the most accurate statement that can be made about the p-value for the test?

		O-Ring Shape			
		1	2	3	
	86	18.5	18.3	17.4	
	75	14.5	15.8	10.6	
Temperature,	*65*	13.5	14.9	11.9	
Degrees	*55*	11.9	11.5	10.5	
Fahrenheit	*45*	10.7	9.7	7.6	
	33	8.7	5.8	6.2	thousands of lb. per square inch

14.72 Prior to taking an aerobics class, each of 12 health-club members is asked to rate the instructor on a scale in which 3 = "excellent," 2 = "good," and 1 = "poor." Following the class, each student is again asked to rate the instructor. Given the following before-and-after ratings, use the 0.05 level in determining whether there has been a significant improvement in the ratings received by the instructor.

	Student											
	1	**2**	**3**	**4**	**5**	**6**	**7**	**8**	**9**	**10**	**11**	**12**
Precourse Rating	1	3	2	2	2	2	1	3	1	2	2	1
Postcourse Rating	2	2	3	3	3	2	3	3	3	2	3	2

14.73 Selecting a position that ensures that you will not be conspicuous, stand or park near a local convenience store and observe 35 consecutive individuals entering the store. Jot down whether each is male versus female, then use the runs test and the 0.05 level of significance in determining whether this sequence of customers might be other than random.

14.74 An interviewer has been told to randomly select respondents and to include x = whether or not they are a college graduate among the data recorded. The categories of the first 30 persons interviewed, with G = college graduate and N = nongraduate, are shown in the sequence below. At the 0.10 level, evaluate the randomness of the sequence.

G N G N G N N G N G G N N N G
N G G N N G N N G G N G N N G

14.75 During her interview with an urban firm, Betty was told by the personnel director that she would typically require "no more than 20 minutes" to drive to work from the apartment complex to which she has recently moved. During the first 2 weeks of work, her travel times have been as listed below. Use the 0.05 level of significance in evaluating the personnel director's claim. Using the appropriate statistical table, what is the most accurate statement that can be made about the p-value for the test?

24.7 19.5 22.6 21.4 18.0
22.2 17.1 27.1 24.9 23.1 min

14.76 For a random sample of drivers using an automated toll booth that requires exact change, the following data show the elapsed time (in seconds) from the moment the vehicle stopped to the moment it began to move away following payment of the toll: 6, 5, 7, 4, and 17. Using the 0.10 level of significance, apply the Kolmogorov-Smirnov test for normality in examining whether the times could have come from a normally distributed population.

14.77 Two investment counselors have evaluated the 15 major companies in an industry according to their potential as a "safe investment." Both counselors agree that company C is the best firm in which to invest. These and the remaining ranks are shown in the table. Use the Spearman coefficient of rank correlation in examining the direction and the strength of the agreement between the two counselors. At the 0.05 level, does r_s differ significantly from 0?

Firm	A	B	C	D	E	F	G	H	I	J	K	L	M	N	O
Counselor 1	2	14	1	8	7	13	3	12	9	4	10	15	5	11	6
Counselor 2	3	12	1	5	6	13	7	10	11	2	9	14	8	15	4

(DATA SET) *Note:* Exercises 14.78–14.86 require a computer and statistical software.

14.78 To help ensure that packages contain at least the stated weight of 12 ounces, a cereal manufacturer wants the filling machine set to 12.5 ounces. For a random sample of boxes, the weights (in ounces) are listed in file **XR14078**. At the 0.10 level of significance, does the filling machine seem to have "drifted" from the desired setting? Identify and interpret the p-value for the test.

14.79 In the past, telephone solicitors for a home improvement firm have been paid an hourly wage. A small-scale experiment has been set up in which five solicitors are given further incentive in the form of 1% of the eventual sale. Data file **XR14079** shows the number of sales originating from each solicitor for the day before and the day after the test policy was put into place. At the 0.05 level, has the incentive policy increased performance? Identify and interpret the *p*-value for the test.

14.80 Two applicants for a component assembly position with an electronics firm have been asked to complete each of six tasks that their jobs would typically involve. For each assigned task, file **XR14080** shows the times (minutes) required by each of the applicants. At the 0.10 level, could the applicants be equally qualified? Identify and interpret the *p*-value for the test.

14.81 A new taxi driver will be making regular trips from the airport to a downtown hotel. She has heard conflicting opinions as to which of two routes takes less time. When she randomly selects which route to take for each of 10 trips, the five times (minutes) for each route are as listed in file **XR14081**. Using the 0.05 level, examine whether the two routes could be equally time efficient. Identify and interpret the *p*-value for the test.

14.82 Comparing the attitudes of customers at three restaurants, a researcher selects a random sample of customers from each and asks each person to rate the restaurant on a scale of 50 to 100, with 50 = "poor" and 100 = "excellent." Given the results in file **XR14082**, and using the 0.05 level of significance, examine whether the population median customer ratings for the three restaurants could be the same. Identify and interpret the *p*-value for the test.

14.83 Reacting to a skeptical reader who doubts that so-called wine connoisseurs can really tell the difference between cheap and expensive wines, the dining editor of a city newspaper sets up a test in which each of five connoisseurs is asked to taste three unidentified wines, then rate each one on a scale of 1 ("poor") to 10 ("excellent"). Unknown to the raters, wine 1 is $1.50 per gallon Elderberry, wine 2 is $8 per liter Rhine, and wine 3 is $35 per liter Cabernet Sauvignon. Their ratings are in file **XR14083**. Considering the wine categories as "treatments" and the tasters as "blocks," and using the 0.05 level of significance, advise the dining editor as to how to respond to this reader when he writes his next column dealing with wine. Identify and interpret the *p*-value for the test.

14.84 Five computer operators have been asked to rate four different workstation arrangements in terms of comfort with 1 = very uncomfortable to 10 = very comfortable. Their ratings are in file **XR14084**. At the 0.05 level, could the workstation arrangements be equally comfortable? Identify and interpret the *p*-value for the test.

14.85 The weights (in ounces) of the first 34 units in a production run of cast-iron flywheels are in file **XR14085**. Using the 0.10 level of significance, test the randomness of this sequence of measurements. Identify and interpret the *p*-value for the test.

14.86 Apply either the chi-square or the Kolmogorov-Smirnov test for normality to the data described in Exercise 14.85. At the 0.05 level of significance, could the data have come from a normal population?

INTEGRATED CASE

Thorndike Sports Equipment

Ted and Luke Thorndike have just returned from lunch with Candice Ergonne, president of a chain of 550 health clubs located across the United States. The clubs have a quality image and are known for featuring equipment and facilities that are second to none. The Thorndikes are rightfully pleased that Ms. Ergonne has invited their proposal for supplying exercise bicycles to the 120 new clubs that are to be opened during the next year.

After further negotiations, the deal is practically closed, but some fine-tuning remains with regard to the seat design for the "ThornBike 2000" model that has been selected. Ms. Ergonne recognizes that some club members require wider seats than others but doesn't want to call attention to this fact by having differing seat widths from one bike to the next. She would much prefer a "one size fits all" approach to the seating decision.

In an experiment to evaluate four seat designs, Ms. Ergonne supplies seven persons who vary widely in size, while the Thorndikes provide the seats to be tested. Each of the individuals spends 5 minutes riding on each of the four seat designs, with the order determined randomly. The subject then rates the comfort of the seat on a scale of 1 to 10, with 1 = "poor" and 10 = "excellent." The results are as shown below. The data are also in file **THORN14**.

From these results, and using the 0.10 level of significance, can Ted Thorndike reject the null hypothesis that the four seating designs could be equally comfortable?

		Seat Design			
		1	2	3	4
	1	3	9	5	7
	2	7	8	6	8
	3	8	10	9	7
Rider	*4*	4	8	3	6
	5	6	8	5	10
	6	9	5	7	6
	7	8	10	7	9

BUSINESS CASE

Circuit Systems, Inc. (B)

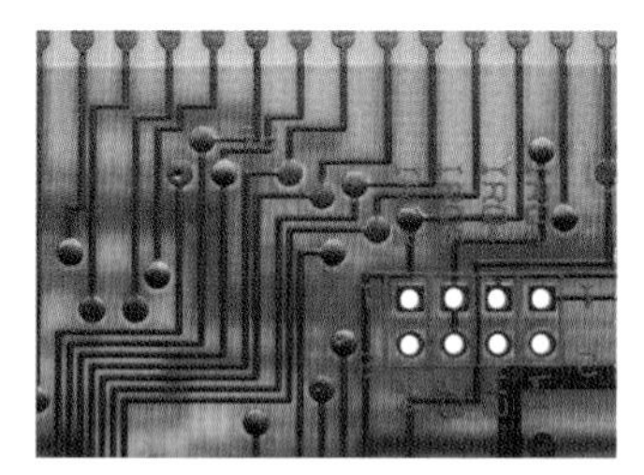
Photodisc/Getty Images

In Chapter 11, we visited Circuit Systems, Inc., a company that was concerned about the effectiveness of its new program for reducing the cost of absenteeism among hourly workers. Besides allowing employees to be paid for unused sick leave, the program also pays $200 toward membership in a local health club for those employees who wish to join. Thomas Nelson, who introduced the new program, has collected data describing the 233 employees who have been with the company for both the year preceding the new program and the year following its implementation. The information Mr. Nelson has collected is in the **CIRCUIT** data file, and the variables and description of his program are as described in Circuit Systems, Inc. (A), in Chapter 11.

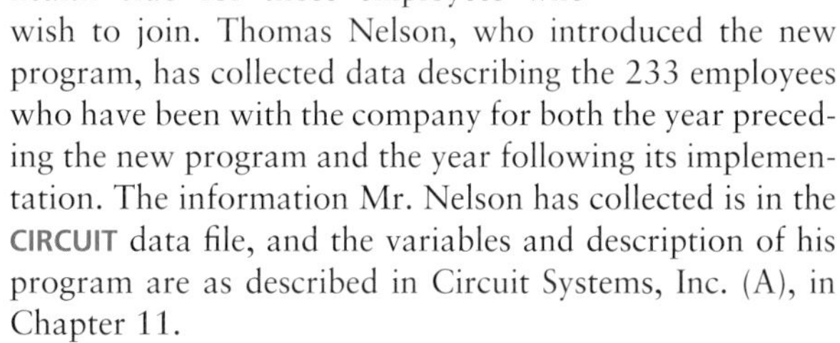

1. Repeat Question 1 from the Circuit Systems (A) case, but use the Wilcoxon signed rank test for paired samples instead of the dependent-samples t-test in comparing the two years in terms of days missed before and after the new program was implemented.
2. Repeat Question 2 from the Circuit Systems (A) case, but use the Wilcoxon signed rank test for paired samples. Be sure to keep in mind that the goal of the program is to reduce the *cost* of absenteeism, and that you will need to either create two new variables or utilize the ones you created in the earlier case. See the discussion and hints in the Circuit Systems (A) case questions regarding these variables.
3. Repeat Question 3 from the Circuit Systems (A) case, but use an appropriate nonparametric statistical test from this chapter in evaluating the effectiveness of the exercise program.

CHAPTER 15

Simple Linear Regression and Correlation

WINNING FOR DOLLARS

In sports, some teams are referred to as "the best that money can buy." If team payrolls and wins during the 2006 Major League Baseball season are any indication, there may be some truth to that adage. Although the highest-paid team didn't win the World Series, there was a very strong relationship between the size of a team's payroll and the number of wins it achieved during the regular season.

This relationship can be described by a linear equation fitted to the data using a technique called *simple linear regression*. This popular method for data analysis involves both the *scatterplot* discussed in Chapter 2 and the *coefficient of correlation* that was briefly introduced in Chapter 3. Using linear regression and correlation, we can make interesting observations like these: (1) variation in team payrolls explains approximately 29% of the variation in team wins; (2) on average, each additional win cost an extra $5.99 million in team payroll; and (3) the Minnesota Twins, one of the lowest-paid teams in baseball, pleased their fans and owners by winning about 17 more games than their payroll level would have predicted. Later in the chapter, in Statistics in Action 15.2, we'll delve a little further into how regression and correlation analysis applies to the payrolls and win totals of the 2006 Major League Baseball season.

Some slopes are a little steeper than others.

LEARNING OBJECTIVES

After reading this chapter, you should be able to:

- Explain the individual terms in the simple linear regression model, and describe the assumptions that the model requires.
- Determine the least-squares regression equation, and make point and interval estimates for the dependent variable.
- Determine and interpret the value of the coefficient of correlation.
- Describe the meaning of the coefficient of determination.
- Construct confidence intervals and carry out hypothesis tests involving the slope of the regression line.
- Test the significance of the correlation coefficient.
- Use residual analysis in examining the appropriateness of the linear regression model and the extent to which underlying assumptions are met.

(15.1) INTRODUCTION

In Section 3.6 of Chapter 3, we briefly introduced the coefficient of correlation as a way of examining whether two interval- or ratio-scale variables could be related, and in Section 13.4 of Chapter 13, we used chi-square analysis to examine whether two nominal-scale variables could be related. Although these approaches can suggest that a relationship exists, they do not reveal *in exactly what way* the variables are related. The techniques of this chapter are applicable to two interval- or ratio-scale variables and describe both the nature and the strength of the relationship between the variables.

Regression analysis provides a "best-fit" mathematical equation for the values of the two variables. The equation may be linear (a straight line) or curvilinear, but we will be concentrating on the linear type. **Correlation analysis** measures the strength of the relationship between the variables.

The topic of this chapter is known as *simple* linear regression and correlation because there are just two variables, y and x. These are called the **dependent (y)** and **independent (x) variables,** since a typical purpose for this type of analysis is to estimate or predict what y will be for a given value of x. In the next chapter we'll cover *multiple* linear regression and correlation, in which two or more independent variables are used in estimating a dependent variable. The following are examples of dependent and independent variables that might be the subject of simple regression and correlation:

Dependent Variable (y)	Independent Variable (x)
Sales	Advertising
Starting salary	Grade point average
Product recalls	Quality assurance expenditures
Cost of a magazine ad	Circulation of the magazine

One way to examine whether two variables might be linearly related is to construct a scatter diagram, as we did in Section 2.5 of Chapter 2, then use the "eyeball method" to fit a straight line roughly approximating the points in the diagram. In the next section, we will introduce the simple linear regression model and employ a more quantitative approach that relies on what is known as the least-squares criterion.

THE SIMPLE LINEAR REGRESSION MODEL 15.2

Model and Assumptions

The *simple linear regression model* is a linear equation having a y-intercept and a slope, with estimates of these population parameters based on sample data and determined by standard formulas. The model is described in terms of the population parameters as follows:

The simple linear regression model:

$y_i = \beta_0 + \beta_1 x_i + \epsilon_i$ where y_i = a value of the dependent variable, y
x_i = a value of the independent variable, x
β_0 = the y-intercept of the regression line
β_1 = the slope of the regression line
ϵ_i = random error, or residual

1. For a given value of x, the expected value of y is given by the linear equation, $\mu_{y.x} = \beta_0 + \beta_1 x_i$. The term $\mu_{y.x}$ can be stated as "the mean of y, given a specific value of x."
2. The difference between the actual value of y and the expected value of y is the error, or residual, $\epsilon_i = y_i - (\beta_0 + \beta_1 x_i)$.

According to this model, the y-intercept for the population of (x_i, y_i) pairs is β_0 and the slope is β_1. The ϵ_i term is the random error, or **residual,** for the ith observation or measurement—this residual is the difference between the actual value (y_i) and the expected value, $\mu_{y.x} = \beta_0 + \beta_1 x_i$, from the regression line. The y values may be scattered above and below the regression line, but the *expected* value of y for a given value of x will be given by the linear equation, $\mu_{y.x} = \beta_0 + \beta_1 x_i$.

Three assumptions underlie the simple linear regression model:

1. For any given value of x, the y values are normally distributed with a mean that is on the regression line, $\mu_{y.x} = \beta_0 + \beta_1 x_i$.
2. Regardless of the value of x, the standard deviation of the distribution of y values about the regression line is the same. The assumption of equal standard deviations about the regression line is called **homoscedasticity.**
3. The y values are statistically independent of each other. For example, if a given y value happens to exceed $\mu_{y.x} = \beta_0 + \beta_1 x_i$, this does not affect the probability that the next y value observed will also exceed $\mu_{y.x} = \beta_0 + \beta_1 x_i$.

Figure 15.1 shows the variation of y values above and below a population regression line. There is a "family" of such distributions (one for each possible x value). Each distribution has $\mu_{y.x}$ as its mean, and the standard deviations are the same $(\sigma_{y.x} = \sigma)$.

The three assumptions can also be expressed in terms of the error, or residual, component (ϵ_i) in the simple linear regression model: (1) For any given value of x, the population of ϵ_i values will be normally distributed with a mean of zero and a standard deviation of σ; (2) this standard deviation will be the same regardless of the value of x; and (3) the ϵ_i values are statistically independent of each other.

FIGURE 15.1

For any given value of x, the y values are assumed to be normally distributed about the population regression line and to have the same standard deviation, σ. The regression line based on sample data is an estimate of this "true" line. Likewise, $s_{y.x}$ is our sample estimate of σ.

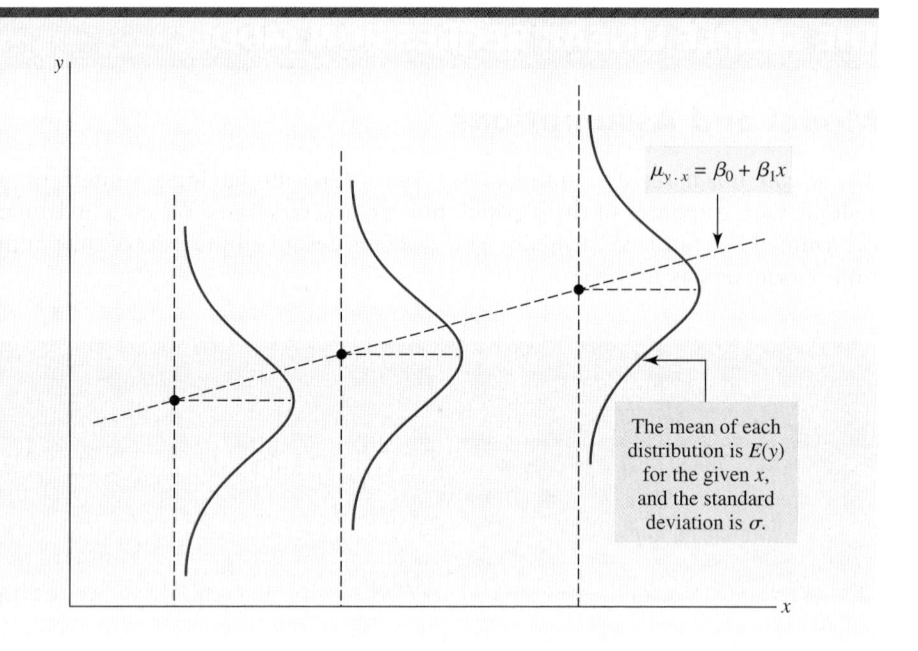

Based on the sample data, the y-intercept and slope of the population regression line can be estimated. The result is the sample regression line:

Sample regression line:

$\hat{y} = b_0 + b_1x$ where $\hat{y}$ = the estimated value of the dependent variable (y) for a given value of x

b_0 = the y-intercept; this is the value of y where the line intersects the y-axis whenever $x = 0$.

b_1 = the slope of the regression line

x = a value for the independent variable

The cap (ˆ) over the y indicates that it is an estimate of the (unknown) "true" value of y. The equation is completely described by the y-intercept (b_0) and slope (b_1), which are sample estimates of their population counterparts, β_0 and β_1, respectively. An infinite number of possible equations can be fitted to a given scatter diagram, and each equation will have a unique combination of values for b_0 and b_1. However, only one equation will be the "best fit" as defined by the least-squares criterion we are going to use.

The Least-Squares Criterion

The **least-squares criterion** requires that the sum of the squared deviations between y values in the scatter diagram and y values predicted by the equation be minimized. In symbolic terms:

Least-squares criterion for determining the best-fit equation:

The equation must be such that $\sum(y_i - \hat{y}_i)^2$ is minimized

where y_i = the observed value of y for the given value of x
$\hat{y}_i$ = the predicted value of y for that x value, as determined from the regression equation

To show how the least-squares criterion works, consider parts (a) and (b) of Figure 15.2. In part (a) the sum of the squared deviations between observed and predicted y values is 100.0, while in part (b) the sum is only 67.0. According to the least-squares criterion, the line in part (b) is a better fit to the data than the line in part (a).

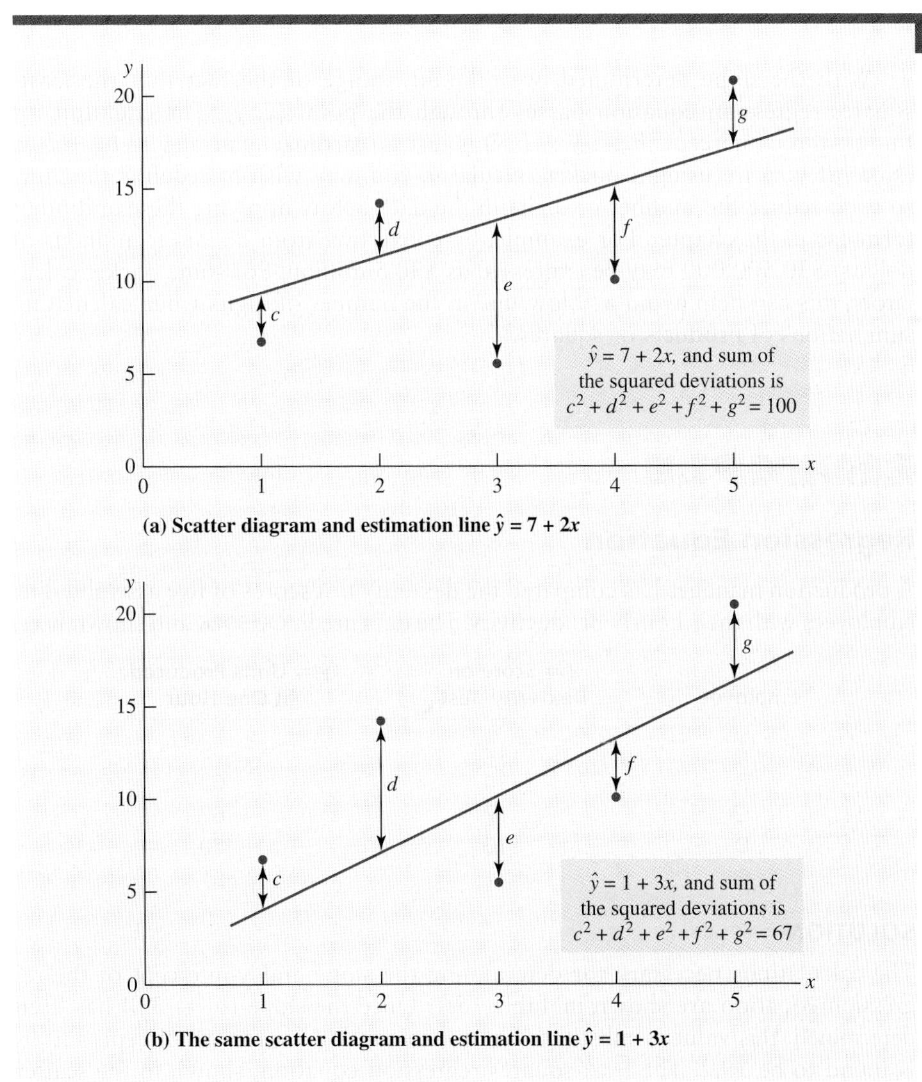

FIGURE 15.2

Using the least-squares criterion, the line fitted in part (b) is a better fit to the data than the line in part (a).

Determining the Least-Squares Regression Line

Equations have been developed for proceeding from a set of data to the least-squares regression line. They are based on the methods of calculus and provide values for b_0 and b_1 such that the least-squares criterion is met. The least-squares regression line may also be referred to as the *least-squares regression equation* or as simply the **regression line:**

Least-squares regression line, $\hat{y} = b_0 + b_1 x$:

- **Slope**

$$b_1 = \frac{\left(\sum x_i y_i\right) - n\bar{x}\bar{y}}{\left(\sum x_i^2\right) - n\bar{x}^2}$$

where n = number of data points

- **y-intercept**

$$b_0 = \bar{y} - b_1\bar{x}$$

With the slope determined, we take advantage of the fact that the least-squares regression equation passes through the point $(\bar{x}, \bar{y})$. The equation for finding the y-intercept ($b_0 = \bar{y} - b_1\bar{x}$) is just a rearrangement of $\bar{y} = b_0 + b_1\bar{x}$. (*Note:* If you are using a pocket calculator, you may wish to redefine the units so as to reduce the number of digits in the data before applying these and other formulas in the chapter. For example, by converting from dollars to millions of dollars, \$30,500,000 may be expressed as \$30.5 million. For some pocket calculators, this can help avoid a "blow-up" in the number of digits when calculating summations of products or squares.)

EXAMPLEEXAMPLEEXAMPLEEXAMPLE

EXAMPLE

Regression Equation

A production manager has compared the dexterity test scores of five assembly-line employees with their hourly productivity. The data are in CX15DEX and shown here.

Employee	x = Score on Dexterity Test	y = Units Produced in One Hour
A	12	55
B	14	63
C	17	67
D	16	70
E	11	51

SOLUTION

The calculations necessary for determining the slope and y-intercept of the regression equation are shown in Table 15.1. Once the slope ($b_1 = 3.0$) has been determined, this value is substituted into the equation for the y-intercept, and b_0 is found to be 19.2. The least-squares regression equation, shown in the scatter diagram of Figure 15.3, is $\hat{y} = 19.2 + 3.0x$.

TABLE 15.1

Data and calculations for determining the least-squares regression line for the example involving dexterity test score (x) and units produced per hour (y).

Data and Preliminary Calculations

Employee	x_i = Score on Dexterity Test	y_i = Units Produced in One Hour	$x_i y_i$	x_i^2	y_i^2
A	12	55	660	144	3025
B	14	63	882	196	3969
C	17	67	1139	289	4489
D	16	70	1120	256	4900
E	11	51	561	121	2601
	70	306	4362	1006	18,984
	$\sum x_i$	$\sum y_i$	$\sum x_i y_i$	$\sum x_i^2$	$\sum y_i^2$

$\bar{x} = 70/5 = 14.0$ $\qquad \bar{y} = 306/5 = 61.2$

Calculations for Slope and *y*-Intercept of Least-Squares Regression Line

$$\text{slope, } b_1 = \frac{(\sum x_i y_i) - n\bar{x}\bar{y}}{(\sum x_i^2) - n\bar{x}^2} = \frac{4362 - 5(14.0)(61.2)}{1006 - 5(14.0)^2} = \frac{78.0}{26.0} = 3.0$$

$$y\text{-intercept, } b_0 = \bar{y} - b_1\bar{x} = 61.2 - 3.0(14.0) = 61.2 - 42.0 = 19.2$$

The least-squares regression line is $\hat{y} = 19.2 + 3.0x$

where $\hat{y}$ = estimated units produced per hour
x = score on manual dexterity test

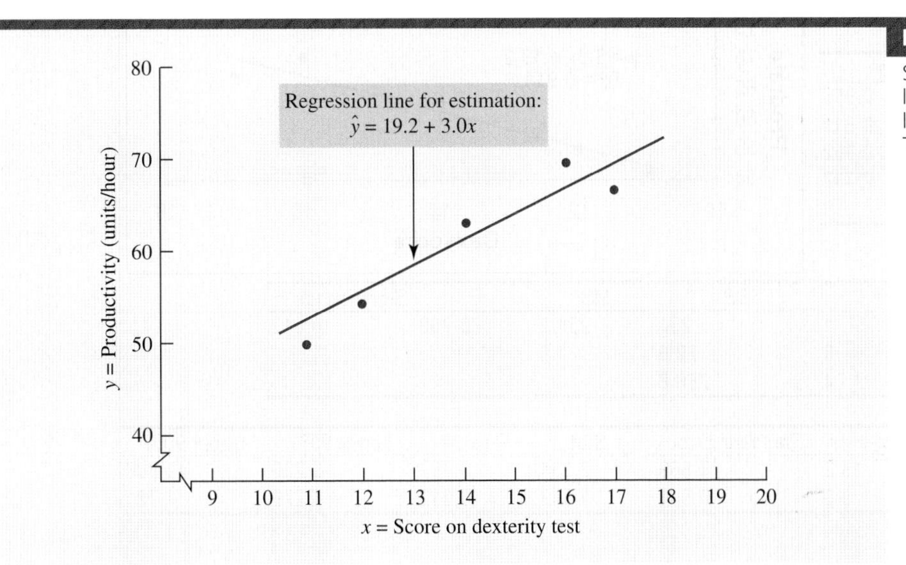

FIGURE 15.3

Scatter diagram and least-squares regression line for the data of Table 15.1.

The slope of the regression line is positive, suggesting a direct relationship between dexterity test score and productivity. The value of the slope ($b_1 = 3.0$) indicates that each one-point increase in the dexterity test score will increase the estimated productivity by 3.0 units per hour.

Point Estimates Using the Regression Line

Making point estimates based on the regression line is simply a matter of substituting a known or assumed value of x into the equation, then calculating the estimated value of y. For example, if a job applicant were to score $x = 15$ on the manual dexterity test, we would predict this person would be capable of producing 64.2 units per hour on the assembly line. This is calculated as shown below. You can also use Seeing Statistics Applet 18, at the end of the chapter, to demonstrate how different inputs for the dexterity test score will produce different point estimates for productivity.

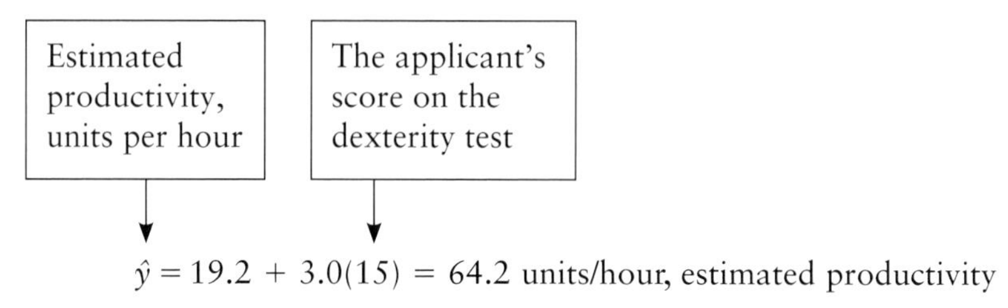

$$\hat{y} = 19.2 + 3.0(15) = 64.2 \text{ units/hour, estimated productivity}$$

Computer Solutions 15.1 describes the procedures and shows the results when Excel and Minitab are used in performing a simple linear regression analysis on

COMPUTER 15.1 SOLUTIONS

Simple Linear Regression

EXCEL

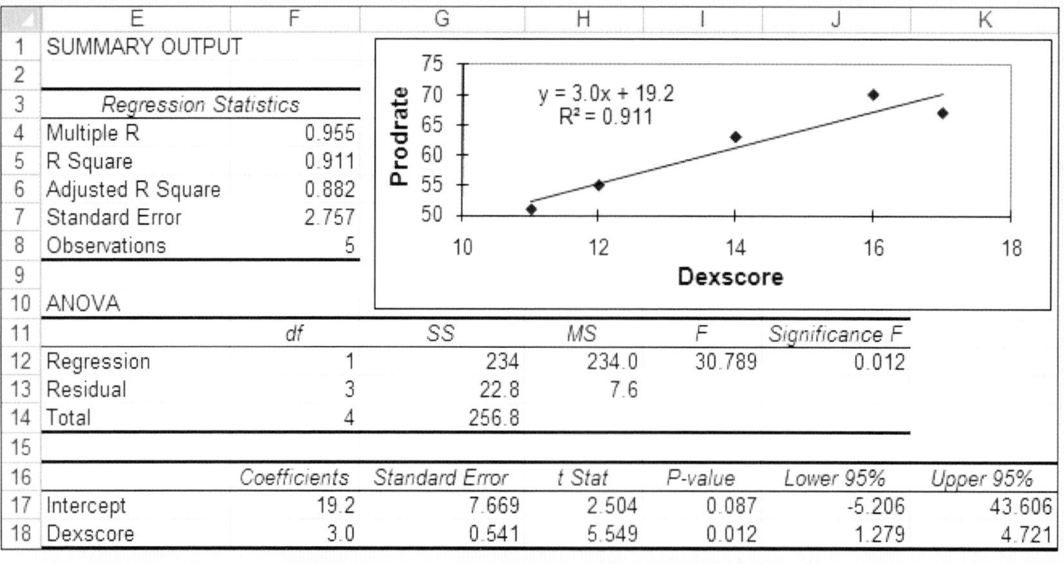

	E	F	G	H	I	J	K
1	SUMMARY OUTPUT						
2							
3	*Regression Statistics*						
4	Multiple R	0.955					
5	R Square	0.911					
6	Adjusted R Square	0.882					
7	Standard Error	2.757					
8	Observations	5					
9							
10	ANOVA						
11		*df*	*SS*	*MS*	*F*	*Significance F*	
12	Regression	1	234	234.0	30.789	0.012	
13	Residual	3	22.8	7.6			
14	Total	4	256.8				
15							
16		*Coefficients*	*Standard Error*	*t Stat*	*P-value*	*Lower 95%*	*Upper 95%*
17	Intercept	19.2	7.669	2.504	0.087	-5.206	43.606
18	Dexscore	3.0	0.541	5.549	0.012	1.279	4.721

Excel Simple Linear Regression

1. The data for the primary example in this chapter are in Excel file **CX15DEX**. The dexterity test scores (Dexscore) are in A2:A6, the production rates (Prodrate) are in B2:B6, and the labels are in A1 and B1. From the **Data** ribbon, click **Data Analysis.** Select **Regression.** Click **OK.**
2. Enter **B1:B6** into the **Input Y Range** box. Enter **A1:A6** into the **Input X Range** box. Select **Labels.** Enter **E1** into the **Output Range** box. Check the **Line Fit Plots** box. Click **OK.**
3. Right-click within the legend descriptions to the right of the chart and click **Delete.** Right-click on any one of the linearly-arranged squares that describe predicted values for the data points and click **Delete.**
4. Right-click on any one of the data points in the scatter diagram and click **Add Trendline.** Select the **Linear** option. Click to select the **Display Equation on chart** and **Display R-squared value on chart** options. Click **Close.** Make formatting adjustments to the chart.

MINITAB

Minitab Simple Linear Regression

```
Regression Analysis: Prodrate versus Dexscore

The regression equation is Prodrate = 19.2 + 3.00 Dexscore

Predictor     Coef   SE Coef      T      P
Constant    19.200     7.669   2.50  0.087
Dexscore    3.0000    0.5407   5.55  0.012

S = 2.75681    R-Sq = 91.1%    R-Sq(adj) = 88.2%

Analysis of Variance

Source          DF      SS      MS      F      P
Regression       1  234.00  234.00  30.79  0.012
Residual Error   3   22.80    7.60
Total            4  256.80
```

1. Using Minitab file **CX15DEX**, with dexterity test scores in C1 and production rates in C2: Click **Stat.** Select **Regression.** Click **Regression.** Enter **C2** (Prodrate) into the **Response** box. Enter **C1** (Dexscore) into the **Predictors** box. Click **OK.**
2. Alternatively, we can generate a scatter diagram and line plot (not shown): Click **Stat.** Select **Regression.** Click **Fitted Line Plot.** Enter **C2** (Prodrate) into the **Response** box. Enter **C1** (Dexscore) into the **Predictor** box. Select **Linear.** Click **OK.**
3. Another alternative is to use the equation to estimate *y* for a given value of *x*: In step 1, click **Options.** Enter the desired *x* value (e.g., **15**) into the **Prediction intervals for new observations** box. Click **OK.** The estimated value of *y* (not shown here) will be 64.2.

the productivity data in our example. We will be referring back to these printouts in later sections of the chapter. In the Excel printout, the intercept (19.2) and slope (3.0) are shown in cells F17:F18. Minitab shows the intercept and slope in the context of the least-squares equation.

EXERCISES

15.1 Explain each term in the linear regression model, $y_i = \beta_0 + \beta_1 x_i + \epsilon_i$.

15.2 What assumptions are required in using the linear regression model?

15.3 In the linear regression equation, $\hat{y} = b_0 + b_1 x$, why is the term at the left given as $\hat{y}$ instead of simply y?

15.4 What is the least-squares criterion, and what does it have to do with obtaining a regression line for a given set of data?

15.5 A scatter diagram includes the data points ($x = 2$, $y = 10$), ($x = 3$, $y = 12$), ($x = 4$, $y = 20$), and ($x = 5$, $y = 16$). Two regression lines are proposed: (1) $\hat{y} = 10 + x$, and (2) $\hat{y} = 8 + 2x$. Using the least-squares criterion, which of these regression lines is the better fit to the data? Why?

15.6 Determine the least-squares regression line for the data in Exercise 15.5, then calculate the sum of the squared deviations between y and $\hat{y}$.

15.7 A scatter diagram includes the data points ($x = 3$, $y = 8$), ($x = 5$, $y = 18$), ($x = 7$, $y = 30$), and ($x = 9$, $y = 32$). Two regression lines are proposed: (1) $\hat{y} = 5 + 3x$, and (2) $\hat{y} = -2 + 4x$. Using the least-squares criterion, which of these regression lines is the better fit to the data? Why?

15.8 Determine the least-squares regression line for the data in Exercise 15.7, then calculate the sum of the squared deviations between y and $\hat{y}$.

15.9 For a sample of 8 employees, a personnel director has collected the following data on ownership of company stock versus years with the firm.

x = Years	y = Shares
6	300
12	408
14	560
6	252
9	288
13	650
15	630
9	522

a. Determine the least-squares regression line and interpret its slope.
b. For an employee who has been with the firm 10 years, what is the predicted number of shares of stock owned?

15.10 The following data represent x = boat sales and y = boat trailer sales from 1995 through 2000. Source: Bureau of the Census, *Statistical Abstract of the United States 1999*, p. 269 and *Statistical Abstract 2002*, p. 755.

Year	Boat Sales (Thousands)	Boat Trailer Sales (Thousands)
1995	649	207
1996	619	194
1997	596	181
1998	576	174
1999	585	168
2000	574	159

a. Determine the least-squares regression line and interpret its slope.
b. Estimate, for a year during which 500,000 boats are sold, the number of boat trailers that would be sold.
c. What reasons might explain why the number of boat trailers sold per year is less than the number of boats sold per year?

15.11 A media analyst studying popular movies has generated the accompanying table, which shows gross ticket sales during the initial 2 weeks following release compared to total gross sales during the movie's run. Given these data, determine the least-squares equation for predicting total gross ticket sales and interpret its slope. What would be the estimated total gross sales for a film that had $100 million in ticket sales during the first 2 weeks of its run?

After 2 Weeks ($ millions)	Total Gross Ticket Sales During Run ($ millions)
104.5	207.9
125.0	250.3
64.7	191.8
88.2	205.0
75.4	175.1

(**DATA SET**) *Note:* Exercises 15.12–15.14 require a computer and statistical software.

15.12 For the top law firms in the world in terms of profit per equity partner, data file **XR15012** lists the number of equity partners and the gross revenue ($ millions) for the most recent fiscal year. Given these data, determine the least-squares equation for predicting gross revenue on the basis of the number of equity partners, then interpret its slope. What would be the estimated gross revenue for a firm with 200 equity partners? Source: law.com, August 8, 2009.

15.13 The Florida ecosystem is heavily reliant on wet and dry seasons, and is particularly dependent on the winter and spring rainfall that helps reduce damage due to wildfires. Data file **XR15013** lists the number of acres burned and the number of inches of rainfall during January–May for 1988–2000. Given these data, determine the least-squares equation for predicting acreage burned during the January–May period as a function of rainfall during that period, then interpret its slope. What would be the estimated number of acres burned during January–May if there were 18 inches of rainfall during the same period? Source: "Rainfall, Fires Linked In Florida," *USA Today*, June 9, 2000, p. 4A.

15.14 For the leading health maintenance organization (HMO) firms, data file **XR15014** lists the number of persons enrolled in the HMO along with the number of plans the firm administers. Given these data, determine the least-squares equation for predicting the total enrollment as a function of the number of plans administered. What would be the estimated total enrollment for an HMO firm that administered 20 plans? Source: *The Wall Street Journal Almanac 1999*, p. 565.

INTERVAL ESTIMATION USING THE SAMPLE REGRESSION LINE (15.3)

In the preceding example, we found that an applicant scoring $x = 15$ on the dexterity test would be predicted to produce 64.2 units per hour. This is just a point estimate, however, and there will be some uncertainty as to how high or low the individual's productivity might actually be. Using the techniques of this section, we can draw some conclusions regarding the upper and lower limits within which this individual's actual productivity is likely to fall.

The Standard Error of Estimate

To develop interval estimates for the dependent variable, we must first determine the **standard error of estimate,** $s_{y.x}$. This is a standard deviation describing the dispersion of data points above and below the regression line. The formula for the standard error of estimate is shown below and is very similar to that for determining a sample standard deviation, s.

Standard error of estimate

$$s_{y.x} = \sqrt{\frac{\Sigma(y_i - \hat{y}_i)^2}{n-2}}$$

where y_i = each observed value of y in the data
$\hat{y}_i$ = the value of y that would have been estimated from the regression line
n = the number of data points

The number of degrees of freedom associated with $s_{y.x}$ is $n - 2$, since both the y-intercept b_0 and the slope b_1 are estimated from sample data. For the dexterity test example of Table 15.1 and Figure 15.3, the standard error of estimate can be calculated as shown below. In each row, the predicted value ($\hat{y}$) is estimated by the regression line, $\hat{y} = 19.2 + 3.0x$, and the deviation between the observed and predicted values is the *residual*. Each residual is the vertical distance from the regression line to an observed data point.

Employee	x_i = Test Score	Number of Units/Hour Actual (y_i)	Number of Units/Hour Predicted $(\hat{y}_i)$	Error, or Residual, $(y_i - \hat{y}_i)$	Error, or Residual, Squared, $(y_i - \hat{y}_i)^2$
A	12	55	55.2	−0.2	0.04
B	14	63	61.2	1.8	3.24
C	17	67	70.2	−3.2	10.24
D	16	70	67.2	2.8	7.84
E	11	51	52.2	−1.2	1.44
					$\Sigma(y_i - \hat{y}_i)^2 = 22.80$

Having calculated the sum of the squared deviations (i.e., squared residuals) about the regression line as $\Sigma(y_i - \hat{y}_i)^2 = 22.80$, we may now calculate the standard error of estimate as

$$s_{y.x} = \sqrt{\frac{\Sigma(y_i - \hat{y}_i)^2}{n - 2}} = \sqrt{\frac{22.80}{5 - 2}} = \sqrt{7.6} = 2.757 \text{ units/hour}$$

This is the same value provided in the Excel and Minitab portions of Computer Solutions 15.1, shown previously.

As with any standard deviation, larger values of the standard error of estimate reflect a greater amount of *scatter* in the data. If every data point in the scatter diagram were to fall exactly on the regression line, the standard error of estimate would be zero. In this event, there would be no variability at all above and below the regression line.

Confidence Interval for the Mean of *y* Given a Specific *x* Value

Given a specific value of x, we can make two kinds of interval estimates regarding y: (1) a confidence interval for the (unknown) true mean of y, and (2) a prediction interval for an individual y observation. These applications are examined in this section and the next.

For a given value of x, the best point estimate we can make about the mean of y is obtained from the regression line, or $\mu_{y.x} = \hat{y} = b_0 + b_1x$. However, the regression line itself is based on sample data, and b_0 is an estimate of the true y-intercept (β_0) and b_1 is an estimate of the true slope (β_1). As such, both b_0 and b_1 are subject to error. Estimates improve when they are based on an x value that is closer to $\bar{x}$. [Remember that the regression line for a set of data always includes the point $(\bar{x}, \bar{y})$.] The following formula also considers the number of data points, since regression lines based on fewer data points (smaller n) tend to result in greater errors of estimation:

Confidence interval for the mean of y, given a specific value of x:

$$\hat{y} \pm t(s_{y.x})\sqrt{\frac{1}{n} + \frac{(x \text{ value} - \bar{x})^2}{(\Sigma x_i^2) - (\Sigma x_i)^2/n}}$$

where $\hat{y}$ = the estimated value of y for the given value of x
n = the number of data points
$s_{y.x}$ = the standard error of estimate
t = t value for confidence level desired and $df = n - 2$
x value = the given value of x

EXAMPLE

Confidence Interval

For persons scoring $x = 15$ on the dexterity test, what is the 95% confidence interval for their mean productivity?

SOLUTION

The x value given is $x = 15$, and the standard error of estimate has been calculated as $s_{y.x} = 2.757$ units/hour. The correction factor (the square root symbol and the terms within) will end up being slightly larger than if we were making an interval estimate based on an x value closer to the mean of x.

For $x = 15$, $\hat{y} = 19.2 + 3.0(15)$, or 64.2 units per hour. This will be the midpoint of the interval. The mean of x is $\bar{x} = 14.0$, there are $n = 5$ data points, and the Σx_i^2 and Σx_i terms have been calculated in Table 15.1 as 1006 and 70, respectively. For the 95% level of confidence and $df = n - 2 = 3$, $t = 3.182$ and the 95% confidence interval can now be calculated as

$$\hat{y} \pm t(s_{y.x})\sqrt{\frac{1}{n} + \frac{(x \text{ value} - \bar{x})^2}{(\Sigma x_i^2) - (\Sigma x_i)^2/n}}$$

$$= 64.2 \pm 3.182(2.757)\sqrt{\frac{1}{5} + \frac{(15.0 - 14.0)^2}{1006 - (70)^2/5}} = 64.2 \pm 8.773\sqrt{0.2385}$$

$$= 64.2 \pm 8.773(0.488) = 64.2 \pm 4.281, \quad \text{or between } 59.919 \text{ and } 68.481$$

Based on these calculations, we have 95% confidence that the mean productivity for persons scoring $x = 15$ on the dexterity test will be between 59.919 and 68.481 units per hour.

EXAMPLEEXAMPLEEXAMPLEEXAMPLEEXAMPLE

Prediction Interval for an Individual y Observation

For a given value of x, the estimation interval for an individual y observation is called the **prediction interval.** For the dexterity-test regression line, we'll demonstrate this concept by constructing a prediction interval for the productivity of an individual job applicant who has achieved a given score on the dexterity test. The formula is similar to that just presented, except that a 1 is included beneath the square root symbol:

Prediction interval for an individual y, given a specific value of x:

$$\hat{y} \pm t(s_{y.x})\sqrt{1 + \frac{1}{n} + \frac{(x \text{ value} - \bar{x})^2}{\left(\sum x_i^2\right) - \left(\sum x_i\right)^2/n}}$$

where $\hat{y}$ = estimated value of y for the given value of x
n = number of data points
$s_{y.x}$ = the standard error of estimate
t = t value for the confidence level desired and $df = n - 2$
x value = the given value of x

EXAMPLE

Prediction Interval

A prospective employee has scored $x = 15$ on the dexterity test. What is the 95% prediction interval for his productivity?

SOLUTION

For $x = 15$, y is estimated as $\hat{y} = 19.2 + 3.0(15)$, or $\hat{y} = 64.2$. As in the preceding example, this will be the midpoint of the interval. Applying the prediction interval formula, the interval is described by

$$\hat{y} \pm t(s_{y.x})\sqrt{1 + \frac{1}{n} + \frac{(x \text{ value} - \bar{x})^2}{\left(\sum x_i^2\right) - \left(\sum x_i\right)^2/n}}$$

$$= 64.2 \pm 3.182(2.757)\sqrt{1 + \frac{1}{5} + \frac{(15.0 - 14.0)^2}{1006 - (70)^2/5}}$$

$$= 64.2 \pm 8.773\sqrt{1.2385}$$

$$= 64.2 \pm 8.773(1.113)$$

$$= 64.2 \pm 9.764, \quad \text{or between 54.436 and 73.964}$$

For this applicant, we have 95% confidence that his productivity as an employee would be between 54.436 and 73.964 units per hour. The width of this interval is 19.528 [i.e., 2(9.764)].

For purposes of illustration, we have used the same dexterity test score ($x = 15$) in determining both the 95% confidence interval for the mean and the 95% prediction interval for the productivity of an individual. Given that $x = 15$, notice that the 95% *confidence* interval (64.2 ± 4.281 units per hour) is much narrower than the 95% *prediction* interval (64.2 ± 9.764 units per hour). This reflects that we will have less uncertainty about the *mean productivity of all persons* who score $x = 15$ on the dexterity test than about the *productivity of one individual* who has scored $x = 15$ on the dexterity test.

Had the applicant scored $x = 14$ on the dexterity test, the prediction interval would be slightly narrower. This is because $\bar{x} = 14.0$, and there is less error in making interval estimates based on x values that are closer to the mean. For additional practice, you may wish to repeat the preceding calculations for $x = 14$. If so, you will find the interval to be 61.2 ± 9.610, for a total interval

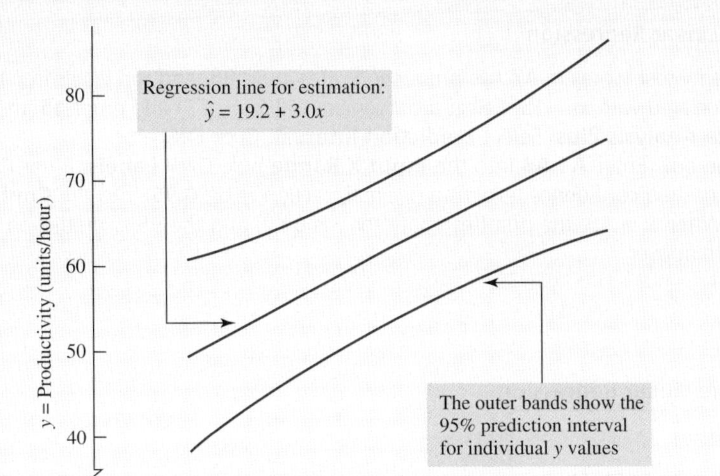

FIGURE 15.4

The 95% prediction interval for individual y values becomes slightly wider whenever the interval is based on x values that are farther away from the mean of x.

width of 2(9.610), or 19.220 units per hour. Figure 15.4 shows how the 95% prediction interval becomes wider whenever the specified value of x gets farther away from $\bar{x}$ in either direction.

Computer Solutions 15.2 shows the procedures and results when we use Excel and Minitab to obtain confidence and prediction intervals for productivity when the score on the dexterity test is $x = 15$. In each case, we also get a point estimate for productivity.

COMPUTER 15.2 SOLUTIONS

Interval Estimation in Simple Linear Regression

EXCEL

	A	B	C	D
1	**Prediction Interval**			
2				
3			Prodrate	
4	Predicted value		64.2	
5				
6	Prediction Interval			
7	Lower limit		54.44	
8	Upper limit		73.96	
9				
10	Interval Estimate of Expected Value			
11	Lower limit		59.92	
12	Upper limit		68.48	

(continued)

Excel Interval Estimation in Simple Linear Regression

1. Using Excel file **CX15DEX**, with dexterity test scores in A2:A6, production rates in B2:B6, and the labels in A1 and B1, we will obtain interval estimates associated with $x = 15$: Select a convenient cell (e.g., C3) and enter **15** into **C3.**
2. From the **Add-Ins** ribbon, click **Data Analysis Plus.** Select **Prediction Interval.** Click **OK.**
3. Enter **B1:B6** into the **Input Y Range** box. Enter **A1:A6** into the **Input X Range** box. Click **Labels.** Enter **C3** into the **Given X Range** box. Enter the desired confidence level as a decimal fraction (e.g., **0.95**) into the **Confidence Level (1 – Alpha)** box. Click **OK.** When $x = 15$, the point estimate for y is 64.2, and the 95% prediction and confidence intervals will be as shown above.

MINITAB

Minitab Interval Estimation in Simple Linear Regression

```
Predicted Values for New Observations
New
Obs     Fit   SE Fit       95% CI             95% PI
  1   64.20     1.35   (59.92, 68.48)   (54.44, 73.96)

Values of Predictors for New Observations
New
Obs   Dexscore
  1       15.0
```

Using Minitab file **CX15DEX**, follow the procedure in step 1 of Computer Solutions 15.1, but select **Options.** Enter **15** into the **Prediction intervals for new observations** box. Enter the desired confidence level as a percentage (e.g., **95**) into the **Confidence level** box. Click **OK.** Click **OK.** In addition to the printout in Computer Solutions 15.1, Minitab will generate the items shown above. When $x = 15$, the point estimate for y is 64.20, and the confidence and prediction intervals will be as shown here.

EXERCISES

15.15 What is the standard error of estimate, and what role does it play in simple linear regression and correlation analysis?

15.16 Is it possible for the standard error of estimate to be equal to zero? If so, under what circumstances?

15.17 Differentiate between a confidence interval and a prediction interval.

15.18 For a given value of x, which will be wider, a 95% confidence interval or a 95% prediction interval? Explain.

15.19 What happens to the width of a prediction interval for y as the x value on which the interval estimate is based gets farther away from the mean of x? Why?

15.20 For a set of 8 data points, the sum of the squared differences between observed and estimated values of y is 34.72. Given this information, what is the standard error of estimate?

15.21 For $n = 6$ data points, the following quantities have been calculated:

$$\Sigma x = 40 \quad \Sigma y = 76 \quad \Sigma xy = 400$$
$$\Sigma x^2 = 346 \quad \Sigma y^2 = 1160 \quad \Sigma(y - \hat{y})^2 = 52.334$$

a. Determine the least-squares regression line.
b. Determine the standard error of estimate.
c. Construct the 95% confidence interval for the mean of y when $x = 7.0$.
d. Construct the 95% confidence interval for the mean of y when $x = 9.0$.
e. Compare the width of the confidence interval obtained in part (c) with that obtained in part (d). Which is wider and why?

15.22 For the summary data provided in Exercise 15.21, construct a 95% prediction interval for an individual y value whenever
a. $x = 2$. b. $x = 3$. c. $x = 4$.

15.23 For the National Football League, ratings for the all-time leading passers were as shown below. Also shown for each quarterback is the percentage of passes that were interceptions, along with the percentage of passes that were touchdowns. Source: *Sports Illustrated 2009 Almanac*, p. 158.

Player	Rating	TD%	Inter%
Steve Young	96.8	5.6	2.6
Peyton Manning	94.7	5.7	2.8
Kurt Warner	93.2	5.1	3.4
Tom Brady	92.9	5.4	2.4
Joe Montana	92.3	5.2	2.6
Carson Palmer	90.1	5.1	3.1
Daunte Culpepper	89.9	4.9	3.2

a. Determine the least-squares regression line for estimating the passer rating based on the percentage of passes that were touchdowns.
b. A quarterback says 5.0% of his passes will be touchdowns next year. Assuming this to be true, estimate his rating for the coming season.
c. Determine the standard error of estimate.
d. For the quarterback in part (b), what is the 95% prediction interval for his passer rating next year?
e. For quarterbacks similar to the one in part (b), what is the 95% confidence interval for their mean rating next year?

15.24 Repeat parts (a)–(e) of Exercise 15.23, but using the interception percentage as the independent variable, and for a quarterback who claims his interception percentage will be just 3.0% next year.

15.25 For the data in Exercise 15.9, find the standard error of estimate, then construct the 95% prediction interval for the amount of stock owned by an individual employee who has been with the firm for 5 years.

15.26 For the data in Exercise 15.10, find the standard error of estimate, then construct the 95% prediction interval for boat trailer sales during a year in which the number of boats sold is 500,000.

(**DATA SET**) *Note:* Exercises 15.27–15.30 require a computer and statistical software.

15.27 Information describing 98 of the top automobile dealership groups in the United States is provided in computer database file **DEALERS**. For calendar year 1999, the data include the number of dealers, the average number of retail units sold per dealer, and the total revenue for the dealership group. Determine the regression equation estimating total revenue as a function of the number of dealers. For a dealership group consisting of 100 dealers, determine and interpret the 95% confidence and prediction intervals associated with total revenue. Source: *Automotive News*, May 1, 2000, pp. 54–60.

15.28 Repeat Exercise 15.27, but with average retail units sold per dealer as the independent variable and with the 95% interval estimates assuming a dealer group for which the average number of retail units sold per dealer is 1000.

15.29 Given the information in Exercise 15.13, determine and interpret the 90% confidence and prediction intervals associated with burned acreage in Florida when the January–May rainfall in the state is 20 inches.

15.30 Given the information in Exercise 15.12, determine and interpret the 90% confidence and prediction intervals associated with gross revenue when there are 200 equity partners.

CORRELATION ANALYSIS (15.4)

Regression analysis determines the nature of the linear relationship between two interval- or ratio-scale variables, while correlation analysis measures the strength of the linear relationship between them. In particular, correlation analysis provides us with two important measures of this strength: (1) the coefficient of correlation and (2) the coefficient of determination.

The Coefficient of Correlation

Briefly introduced in Chapter 3, the **coefficient of correlation** (r, with $-1 \leq r \leq +1$) is a number that indicates both the direction and the strength of the linear relationship between the dependent variable (y) and the independent variable (x):

- **Direction of the relationship** If r is positive, y and x are directly related—i.e., when x increases, y will tend to increase. If r is negative, y and x are inversely related—i.e., when x increases, y will tend to decrease.

- **Strength of the relationship** The larger the absolute value of r, the stronger the linear relationship between y and x. If $r = -1$ or $r = +1$, the regression line will actually include all of the data points and the line will be a perfect fit.

Calculating the coefficient of correlation for a set of data involves combining the same terms that appear in Table 15.1. The formula for r can be expressed as follows:

The coefficient of correlation:

$$r = \frac{n(\sum x_i y_i) - (\sum x_i)(\sum y_i)}{\sqrt{n(\sum x_i^2) - (\sum x_i)^2} \cdot \sqrt{n(\sum y_i^2) - (\sum y_i)^2}}$$

where r = coefficient of correlation, $-1 \leq r \leq +1$
n = number of data points

Referring to the dexterity test example, the coefficient of correlation between productivity (y) and dexterity test score (x) can be computed as

$$r = \frac{n(\sum x_i y_i) - (\sum x_i)(\sum y_i)}{\sqrt{n(\sum x_i^2) - (\sum x_i)^2} \cdot \sqrt{n(\sum y_i^2) - (\sum y_i)^2}}$$

$$= \frac{5(4362) - (70)(306)}{\sqrt{5(1006) - (70)^2} \cdot \sqrt{5(18{,}984) - (306)^2}} = \frac{390}{11.402 \cdot 35.833} = 0.9546$$

The coefficient of correlation ($r = 0.9546$) is positive, reflecting that productivity (y) is directly related to dexterity test score (x). In other words, persons scoring higher on the dexterity test tend to record higher levels of productivity. This is also reflected in the positive slope of the regression line, $\hat{y} = 19.2 + 3.0x$. Using Seeing Statistics Applet 19, at the end of the chapter, you can insert data points onto a blank grid and instantly see how the correlation coefficient and best-fit regression line change in response.

The Coefficient of Determination

Another measure of the strength of the relationship is the **coefficient of determination, r^2**. Its numerical value is the proportion of the variation in y that is explained by the regression line, $\hat{y} = b_0 + b_1x$. For the dexterity test example, $r = 0.9546$, $r^2 = 0.911$, and 91.1% of the variation in productivity is explained by the dexterity test scores.

The coefficient of determination can be described in terms of the *total* variation in y versus the *unexplained* variation in y. In turn, each type of variation can be described as a sum of squared deviations. In the denominator of the fraction in the following expression, each y value is compared to the mean of y in arriving at the total sum of squares (SST). In the numerator, each y value is compared to the value of y predicted by the regression line in obtaining the unexplained, or error, sum of squares (SSE):

The coefficient of determination, r^2:

$$r^2 = 1 - \frac{\textit{Error} \text{ variation, which is } \textit{not} \text{ explained by the regression line}}{\textit{Total} \text{ variation in the } y \text{ values}}$$

$$= 1 - \frac{\sum(y_i - \hat{y}_i)^2}{\sum(y_i - \bar{y})^2} = 1 - \frac{SSE}{SST}$$

In considering the variation of y, a natural question is, "Variation from what?" The logic involved in the calculations of *SST* and *SSE* warrants further explanation:

1. Without any knowledge of x, the best estimate of y is the *mean* of y. This is why the variation of y from $\bar{y}$ is used in calculating SST = total sum of squares, or total variation.
2. When knowledge about x is available, the best estimate of y is the *conditional mean* of y (i.e., $\hat{y}$ from the regression line). When an actual value of y differs from the value predicted by the regression line, this constitutes unexplained variation and it contributes to SSE = error sum of squares.

The error variation (*SSE*) is the sum of the squared residuals calculated earlier during our determination of the standard error of estimate. These calculations are repeated below, along with those for determining the total sum of squares (*SST*) for the y values. In the following calculations, recall that the mean of y is $\bar{y} = 61.2$ units per hour:

	Number of Units/Hour					
x_i = Test Score	Actual, y_i	Predicted, $\hat{y}_i$	$y_i - \hat{y}_i$	$(y_i - \hat{y}_i)^2$	$y_i - \bar{y}$	$(y_i - \bar{y})^2$
12	55	55.2	−0.2	0.04	−6.2	38.44
14	63	61.2	1.8	3.24	1.8	3.24
17	67	70.2	−3.2	10.24	5.8	33.64
16	70	67.2	2.8	7.84	8.8	77.44
11	51	52.2	−1.2	1.44	−10.2	104.04
				22.80		256.80

$\sum(y_i - \hat{y}_i)^2$, error, or residual, sum of squares (*SSE*)

$\sum(y_i - \bar{y})^2$, total sum of squares (*SST*)

and the coefficient of determination can be expressed as

$$r^2 = 1 - \frac{SSE}{SST} = 1 - \frac{22.80}{256.80} = 0.911$$

Total variation in y (*SST*) is the sum of the variation explained by the regression line (*SSR*) plus the variation not explained by the regression line (*SSE*), or

$$\begin{array}{ccccc} \text{Total variation} & & \text{Variation explained} & & \text{Variation } \textit{not} \text{ explained} \\ \text{in } y \text{ values} & = & \text{by regression line} & + & \text{by regression line} \\ (SST) & & (SSR) & & (SSE) \end{array}$$

We have calculated the total sum of squares as $SST = 256.80$ and the error sum of squares as $SSE = 22.80$. The variation explained by the regression line (*SSR*) is the difference between these two quantities, and $SSR = SST - SSE$, or $256.80 - 22.80$, or $SSR = 234.0$. All three of these quantities are typically included in the output whenever regression and correlation analyses are carried out with the assistance of a computer statistical package. You can use Seeing Statistics Applet 20, at the end of the chapter, to instantly see how explained and unexplained variation will change as you "try" different lines to fit the data in this chapter's primary example.

Some statistical software packages—Excel and Minitab included—provide an r^2 that has been "adjusted" for degrees of freedom. The "adjusted" r^2 approaches the size of the "unadjusted" value as the number of data points becomes greater. For our productivity example, the printouts in Computer Solutions 15.1 show the adjusted r^2 as 0.882. The formula for obtaining the adjusted r^2 is given in the footnote.[1]

Computer Solutions 15.3 shows the procedures and results when we use Excel and Minitab to obtain the coefficient of correlation between dexterity test score (x) and productivity (y). The procedures are comparable to their counterparts presented earlier in the text, in Computer Solutions 3.5. Minitab provides a p-value for the coefficient of correlation; if the true coefficient of correlation were 0, there would be a 0.012 probability of getting a sample r this far away from 0 just by chance. Testing whether the true coefficient of correlation could be 0 is discussed in Section 15.5.

COMPUTER 15.3 SOLUTIONS

Coefficient of Correlation

EXCEL

	A	B	C	D	E
1	Dexscore	Prodrate		*Dexscore*	*Prodrate*
2	12	55	Dexscore	1	
3	14	63	Prodrate	0.955	1
4	17	67			
5	16	70			
6	11	51			

1. Open the Excel data file **CX15DEX**. The labels and data are already entered, as shown in columns A and B in the printout above. From the **Data** ribbon, click **Data Analysis.**
2. In the **Analysis Tools** box, click **Correlation.** Click **OK.** Enter **A1:B6** into the **Input Range** box. In the **Grouped By** section, click next to **Columns.** Select **Labels in First Row.** Enter **C1** into the **Output Range** box. Click **OK.**

MINITAB

```
Correlations: Dexscore, Prodrate

Pearson correlation of Dexscore and Prodrate = 0.955
P-Value = 0.012
```

1. Open the Minitab data file **CX15DEX**. The dexterity test scores are in C1 and the production rates are in C2.
2. Click **Stat.** Select **Basic Statistics.** Click **Correlation.** Enter **C1** and **C2** into the **Variables** box. Select **Display P-values.** Click **OK.**

[1]Expressed as a proportion, the adjusted r^2 typically included in computer printouts of regression analyses is calculated as adjusted $r^2 = 1 - \dfrac{SSE/(n-2)}{SST/(n-1)}$.

EXERCISES

15.31 Differentiate between the coefficients of correlation and determination. What information does each one offer that the other does not?

15.32 The values of y and x are inversely related, and 64% of the variation in y is explained by the regression equation. What is the coefficient of correlation?

15.33 The coefficient of correlation between variables y and x is -0.90. Calculate and interpret the value of the coefficient of determination.

15.34 For the data in Exercise 15.23, and with y = touchdown percentage and x = interception percentage, determine and interpret the coefficients of correlation and determination.

15.35 In regression and correlation analysis, what are the two components of the total amount of variation in the y values? Define the coefficient of determination in terms of these components.

15.36 For a set of data, the total variation or sum of squares for y is $SST = 143.0$, and the error sum of squares is $SSE = 24.0$. What proportion of the variation in y is explained by the regression equation?

15.37 The Insurance Institute for Highway Safety has listed the following ratings based on collision and comprehensive claims for 12 makes of midsize four-door cars from the 2005–2007 model years. Higher numbers reflect higher claims in the collision and comprehensive categories of coverage. Source: iihs.org, August 8, 2009.

Collision	Comp	Collision	Comp
113	89	93	79
108	91	105	97
90	74	106	86
124	92	99	71
131	108	116	98
126	108	120	93

a. Determine the least-squares regression line for predicting the rate of collision claims based on comprehensive claim rating.
b. Calculate and interpret the values of r and r^2.
c. If a new model were to have a comprehensive claim rating of 90, what would be the predicted rating for collision claims?

15.38 The following table compares U.S. population (millions) with U.S. per-capita consumption of bottled water (gallons). Source: Bureau of the Census, *Statistical Abstract of the United States 2009*, pp. 10, 134.

Year	x = U.S. Population	y = Per-Capita Consumption of Bottled Water
2002	287.9 million	20.1 gallons
2003	290.4	21.6
2004	293.2	23.2
2005	295.9	25.5
2006	298.8	27.6

a. Determine the least-squares regression line and the value of r.
b. What proportion of the variability in per-capita bottled water consumption is explained by the regression equation?
c. During a year in which the U.S. population is 320 million, what would be the prediction for per-capita consumption of bottled water?

15.39 It has been reported that the average American male consumes 3774 calories per day and that 72.2% of American males are overweight. This information, along with data for seven other countries, is listed below. Source: "A Measure of How Much Countries Eat," *USA Today*, December 16, 2005, p. 4D.

Country	Calories per Day	% Overweight
Chad	2114	10.4
China	2951	27.5
Ecuador	2754	40.2
Egypt	3338	64.5
Great Britain	3412	62.5
India	2459	15.0
Mexico	3145	64.6
United States	3774	72.2

a. Determine the least-squares regression line for predicting the percentage of males who are overweight based on the number of calories consumed per day.
b. Calculate and interpret the values of r and r^2.
c. If a nation's males were to consume an average of 3000 calories per day, what would be the predicted percentage of the nation's males who are overweight?

(DATA SET) *Note:* Exercises 15.40–15.43 require a computer and statistical software.

15.40 Given the information in Exercise 15.13, determine and interpret the coefficients of correlation and determination for January–May burned acreage in Florida versus January–May rainfall in the state.

15.41 Given the information in Exercise 15.12, determine and interpret the coefficients of correlation and determination for gross revenue versus the number of equity partners.

15.42 For each of 10 popular prescription drugs, file **XR15042** lists the retail price (in U.S. dollars) for the drug in several different countries, including the United States, Canada, Great Britain, and Australia. Determine and interpret the coefficients of correlation and determination for U.S. prices versus Canadian prices. Source: "Prescription Drugs Cheaper in Other Nations," *USA Today,* November 10, 1999, p. 2A.

15.43 For each of the 20 top-grossing U.S. films made during the 1990s, file **XR15043** lists the domestic gross ticket sales versus foreign gross ticket sales (both in millions of U.S. dollars). Determine the linear regression equation relating the foreign gross as a function of the domestic gross. Identify and interpret the coefficients of correlation and determination. Source: "Overseas Grosses Outdoing North America," *USA Today,* December 13, 2000.

15.5 ESTIMATION AND TESTS REGARDING THE SAMPLE REGRESSION LINE

The coefficient of correlation (r) is based on sample data and estimates the population coefficient of correlation, ρ. A high absolute value for r may be due merely to chance, especially if the number of data points is small. Likewise, the slope (b_1) of the sample regression line is an estimate of the population slope (β_1). If $\beta_1 = 0$, the population regression line is horizontal and there is no linear relationship between the variables.

Testing H_0: $\rho = 0$ is equivalent to testing H_0: $\beta_1 = 0$. If one null hypothesis is true, there is no linear relationship between the variables and the other null hypothesis will also be true. The tests have the same p-value.

Testing the Coefficient of Correlation

In this test for the significance of the linear relationship, the null and alternative hypotheses pertain to the population coefficient of correlation, ρ:

***t*-test for the population coefficient of correlation, ρ:**

- **Null hypothesis**

H_0: $\rho = 0$ There is no linear relationship.

- **Alternative hypothesis**

H_1: $\rho \neq 0$ A linear relationship exists.

- **Test statistic**

$$t = \frac{r}{\sqrt{\frac{1 - r^2}{n - 2}}}$$

where r = sample coefficient of correlation
n = number of data points
t = t value for the level of significance of the test and $df = n - 2$

EXAMPLE

Testing the Coefficient of Correlation

For the employee dexterity test data, the coefficient of correlation was $r = 0.9546$. At the 0.05 level of significance, test the null hypothesis that the population coefficient of correlation (ρ) is really zero.

SOLUTION

For $n = 5$ persons in the dexterity test example, with $r = 0.9546$, the test statistic is

$$t = \frac{r}{\sqrt{\frac{1 - r^2}{n - 2}}} = \frac{0.9546}{\sqrt{\frac{1 - (0.9546)^2}{5 - 2}}} = 5.55$$

For a two-tail test at the 0.05 level and $df = 5 - 2$, or 3, the critical values are $t = -3.182$ and $t = +3.182$. The calculated test statistic ($t = 5.55$) falls outside these critical values, and the null hypothesis is rejected. At the 0.05 level, the sample coefficient of correlation differs significantly from zero, and we conclude that a linear relationship could exist between the dexterity score and the production rate. If the population coefficient of correlation were really zero, there is less than a 5% chance that a sample coefficient of correlation would be this large.

As shown in Computer Solutions 15.3, Minitab provides the coefficient of correlation along with the p-value for the test described here. The exact p-value is 0.012, considerably less than the 0.05 level of significance used in the test we've just carried out.

EXAMPLEEXAMPLEEXAMPLEEXAMPLEEXAMPLE

Testing and Estimation for the Slope

An equivalent method of testing the significance of the linear relationship is to examine whether the slope (β_1) of the population regression line could be zero.

t-test for the slope (β_1) of the population regression line:

- **Null hypothesis**

  H_0: $\beta_1 = \beta_{10}$ The population slope is β_{10}.

- **Alternative hypothesis**

  H_1: $\beta_1 \neq \beta_{10}$ The population slope is not β_{10}.

- **Test statistic**

  $$t = \frac{b_1 - \beta_{10}}{s_{b_1}}$$

  where b_1 = slope of the sample regression line
  β_{10} = a value that has been hypothesized for the slope of the population regression line

(*continued on page 574*)

STATISTICS IN ACTION 15.1

The Beta Coefficient and Investment Decisions

Beta, the second letter in the Greek alphabet, ascends to the number one position in the alphabet of financial decisions. In finance, million-dollar investments are made with the assistance of the Capital Asset Pricing Model (CAPM), a mathematical model in which beta plays a prominent role.

The CAPM is a lot easier to understand than its name might imply. This is because it is basically a simple linear regression model. For a given stock, and for $i = 1$ through n time periods, the model is

$$y_i = \beta_0 + \beta_1 x_i + \epsilon_i$$

where y_i = rate of return for the stock during the ith time period

x_i = rate of return for the overall stock market during the ith time period

For the stock being evaluated, the slope of the regression line (i.e., β_1, the beta coefficient) is a measure of how risky the stock is in terms of having volatile swings in return versus the market as a whole. A stock with $\beta_1 = 1$ is considered to be an average risk, while $\beta_1 < 1$ and $\beta_1 > 1$ reflect more conservative and more risky stocks, respectively.

Finance managers first set guidelines on how much risk they are willing to take in making their investments. The CAPM is then used in selecting individual investments that represent a mixture of risk levels, as represented by their beta coefficients, but for which the total amount of risk falls within the guidelines originally specified.

For a given stock, the slope of the sample regression equation (b_1) is used as an estimate of its population counterpart (β_1), and estimation and testing for β_1 can be done on the basis of the $i = 1$ through n time periods for which sample data have been collected. This is similar to the approach discussed within the chapter.

Source: For more information on the beta coefficient and the Capital Asset Pricing Model, interested readers may wish to refer to finance textbooks such as J. Fred Weston and Thomas E. Copeland, *Managerial Finance*, 9th ed. (Hinsdale, Ill.: The Dryden Press, 1992).

with

$$s_{b_1} = \frac{s_{y.x}}{\sqrt{\Sigma x_i^2 - n(\bar{x})^2}}$$

s_{b_1} = the estimated standard deviation of the slope

$s_{y.x}$ = the standard error of estimate

n = number of data points

t = t value for the level of significance of the test, and $df = n - 2$

Most typically, we are interested in testing whether β_1 could be equal to zero, and the numerator of the test statistic will simply be $(b_1 - 0)$. In this case, the hypothesized value will be $\beta_{10} = 0$. However, the method described above can also be used to test whether the slope of the population regression line might be any specific, nonzero value. (For example, in the Capital Asset Pricing Model discussed in Statistics in Action 15.1, an investor may wish to test whether the slope differs significantly from 1.0.)

An important component of the preceding test is the estimated standard deviation of the sample slope b_1. This is because s_{b_1} can also be used in the construction of a confidence interval for β_1.

Confidence interval for the slope (β_1) of the population regression line:

The interval is

$$b_1 \pm t s_{b_1}$$

where b_1 = slope of the sample regression line
t = t value for the confidence level desired, and $df = n - 2$
s_{b_1} = the estimated standard deviation of the slope, $s_{b_1} = \dfrac{s_{y.x}}{\sqrt{\Sigma x_i^2 - n(\bar{x})^2}}$

EXAMPLE

Testing and Estimation for the Slope

For the dexterity test data, the slope of the sample regression line was $b_1 = 3.0$.

1. Using the 0.05 level of significance, examine whether the slope of the population regression line could be zero.
2. Construct the 95% confidence interval for the slope of the population regression line.

SOLUTION

Testing H_0: $\beta_1 = 0$ versus H_1: $\beta_1 \neq 0$

For the $n = 5$ persons in the dexterity test example, we have already calculated $s_{y.x} = 2.757$, $\Sigma x_i^2 = 1006$, and $\bar{x} = 14.0$. The standard deviation of the slope will be

$$s_{b_1} = \frac{s_{y.x}}{\sqrt{\Sigma x_i^2 - n(\bar{x})^2}} = \frac{2.757}{\sqrt{1006 - 5(14.0)^2}} = 0.5407$$

and the observed value of the test statistic is

$$\text{Observed } t = \frac{b_1 - 0}{s_{b_1}} = \frac{3.0 - 0}{0.5407} = 5.55$$

For the 0.05 level of significance and $df = 5 - 2 = 3$, the critical values of t are -3.182 and $+3.182$, just as they were for the test of the coefficient of correlation. Notice that the calculated test statistic ($t = 5.55$) is also equal to the t statistic calculated during that test. As pointed out earlier, the tests are equivalent and yield the same conclusions and p-values. At the 0.05 level, the test statistic is outside the critical values, and we are able to reject the null hypothesis that the slope of the population regression line could be zero. In Computer Solutions 15.1, the exact p-value is 0.012—see the "Dexscore" row of either printout.

95% Confidence Interval for the Slope of the Population Regression Line

The 95% confidence interval is centered on the slope of the sample regression line, $b_1 = 3.0$, and s_{b_1} has already been calculated as 0.5407. For $n = 5$ data points, the number of degrees of freedom for the t distribution will be $df = n - 2$, or $df = 3$. Referring to the t distribution table, $t = 3.182$ will correspond to the 95% confidence interval:

$$b_1 \pm t s_{b_1} = 3.0 \pm 3.182(0.5407), \quad \text{or from } 1.279 \text{ to } 4.721$$

EXAMPLEEXAMPLEEXAMPLEEXAMPLEEXAMPLEEXAMPLEEXAMP

We have 95% confidence that the slope (β_1) of the population regression line is in the interval bounded by 1.279 and 4.721. This interval is also provided by Excel in Computer Solutions 15.1—see the "Dexscore" row of the printout. Since $\beta_1 = 0$ is not within the 95% confidence interval, we can conclude, at the 0.05 level, that a linear relationship might exist between productivity and the dexterity test score.

The Analysis of Variance Perspective

Finally, we can carry out an analysis of variance test based on the total amount of variation in y (SST) and the amount explained by the regression line (SSR). This test is equivalent to those for the coefficient of correlation and the slope conducted earlier, and its results are usually given when a computer statistical package is used in carrying out the analysis.

The test is based on the sum-of-squares values that were discussed in the context of the coefficient of determination, or

Total Variation in y values (SST)		Variation explained by regression line (SSR)		Variation *not* explained by regression line (SSE)
$\Sigma(y_i - \bar{y})^2$	=	$SSR = SST - SSE$	+	$\Sigma(y_i - \hat{y}_i)^2$

ANOVA test for the significance of a simple linear regression:

- **Null and alternative hypotheses**

  H_0: There is no linear relationship between x and y.
  H_1: A linear relationship exists between x and y.

- **Test Statistic**

  $$\text{Calculated } F = \frac{SSR/1}{SSE/(n-2)}$$

  where SSR = sum of squares explained by the regression line
  SSE = error, or residual, sum of squares
  n = number of data points

- **Decision Rule** Reject H_0 if calculated F exceeds the critical F for the level of significance of the test, with df (numerator) = 1 and df (denominator) = $n - 2$.

In the numerator of the calculated F-ratio, SSR is divided by 1, the number of degrees of freedom associated with the regression equation. In the denominator, SSE is divided by $(n - 2)$, the number of degrees of freedom for the error sum of squares. For the dexterity test example, our calculations in Section 15.4 showed that $SST = 256.80$, $SSR = 234.0$, and $SSE = 22.80$. We now substitute the SSR and SSE values into the expression for the calculated F-ratio:

$$\text{Calculated } F = \frac{SSR/1}{SSE/(n-2)} = \frac{234.0/1}{22.80/(5-2)} = 30.79$$

For df (numerator) = 1 and df (denominator) = $(n - 2) = 3$, the critical value of F for the 0.025 level is 17.44, while the critical F for the 0.01 level is 34.12. The calculated value ($F = 30.79$) exceeds the critical value for the 0.025 level but does not exceed that for the 0.01 level. Using the F-distribution tables, the most accurate statement we can make about the p-value for the test is that it is between 0.025 and 0.01. From Computer Solutions 15.1, the exact p-value is 0.012—see the ANOVA portion of either printout.

EXERCISES

15.44 Testing the null hypothesis that the slope of the true regression line equals zero is equivalent to testing whether the true coefficient of correlation could be zero. Why?

15.45 For $n = 15$ data points, $r^2 = 0.81$. At the 0.05 level of significance, can we conclude that the true coefficient of correlation could be zero?

15.46 For $n = 37$ data points, $r^2 = 0.29$. At the 0.02 level of significance, can we conclude that the true coefficient of correlation could be zero?

15.47 Based on sample data, the 90% confidence interval for the slope of the population regression line is found to be from −2.5 to 1.4. Based on this information, what is the most accurate statement that can be made about the p-value in testing H_0: $\beta_1 = 0$ versus H_1: $\beta_1 \neq 0$? Explain.

15.48 Based on sample data, the 95% confidence interval for the slope of the population regression line is found to be from 4.2 to 6.7. Based on this information, what is the most accurate statement that can be made about the p-value in testing H_0: $\beta_1 = 0$ versus H_1: $\beta_1 \neq 0$? Explain.

15.49 For the regression line developed in Exercise 15.37,

a. Use the 0.05 level in testing whether the population coefficient of correlation could be zero.
b. Use the 0.05 level in testing whether the population regression equation could have a slope of zero.
c. Construct the 95% confidence interval for the slope of the population regression equation.

15.50 For the regression line developed in Exercise 15.38,

a. Use the 0.05 level in testing whether the population coefficient of correlation could be zero.
b. Use the 0.05 level in testing whether the population regression equation could have a slope of zero.
c. Construct the 95% confidence interval for the slope of the population regression equation.

15.51 A computer analysis of 30 pairs of observations results in the least-squares regression equation $\hat{y} = 14.0 + 5.0x$, and the standard deviation of the slope is listed as $s_{b_1} = 2.25$.

a. At the 0.05 level of significance, can we conclude that no linear relationship exists within the population of x and y values?
b. Given the results in part (a), what conclusion would be reached if the same level of significance were used in testing whether the population coefficient of correlation could be zero?
c. Construct the 95% confidence interval for the population slope, β_1.

15.52 How is analysis of variance related to regression analysis?

15.53 In a regression analysis, the sum of the squared deviations between y and $\bar{y}$ is $SST = 120.0$. If the coefficient of correlation is $r = 0.7$, what are the values of SSE and SSR?

15.54 In a regression analysis, the sum of the squared deviations between y and $\bar{y}$ is $SST = 200.0$. If the sum of the squared deviations about the regression line is $SSE = 40.0$, what is the coefficient of determination?

15.55 Given the information below, calculate SST, SSR, and SSE, determine the coefficient of determination, then use ANOVA and the 0.05 level in testing whether r^2 is significantly different from zero.

x:	5	4	10	9	10	Least-squares equation:
y:	34	44	65	47	66	$\hat{y} = 20.66 + 4.02x$

15.56 Given the information below, calculate SST, SSR, and SSE, determine the coefficient of determination, then use ANOVA and the 0.05 level in testing whether r^2 is significantly different from zero.

x:	4	1	4	6	Least-squares equation:
y:	381	403	394	385	$\hat{y} = 404.94 - 3.78x$

(**DATA SET**) *Note:* Exercises 15.57–15.59 require a computer and statistical software.

15.57 The General Aviation Manufacturers Association has reported annual flying hours and fuel consumption for airplanes with a single, piston-driven engine as listed in file **XR15057**. Data are in millions of flying hours and millions of gallons of fuel, respectively. Determine the linear regression equation describing fuel consumption as a function of flying hours, then identify and interpret the slope, the coefficient of correlation, and the coefficient of determination. At the 0.05 level of significance, could the population slope and the population coefficient of correlation be zero? Determine the 95% confidence interval for the population slope. Source: General Aviation Manufacturers Association, *1999 Statistical Databook,* p. 25.

15.58 Given the information in Exercise 15.13, determine the linear regression equation describing January–May burned acreage in Florida as a function of January–May rainfall in the state, then identify and interpret the slope, the coefficient of correlation, and the coefficient of determination. At the 0.10 level of significance, could the population slope and the population coefficient of correlation be zero? Determine the 90% confidence interval for the population slope.

15.59 Computer database **GROWCO** describes the characteristics of 100 companies identified by *Fortune* as the fastest growing. Two of the variables listed are revenue and net income (millions of dollars) for the most recent four quarters. Determine the linear regression equation describing net income as a function of revenue, then identify and interpret the slope, the coefficient of correlation, and the coefficient of determination. At the 0.05 level of significance, could the population slope and the population coefficient of correlation be zero? Determine the 95% confidence interval for the population slope.
Source: "*Fortune's* 100 Fastest-Growing Companies," *Fortune,* September 4, 2000, pp. 142–158.

15.6 ADDITIONAL TOPICS IN REGRESSION AND CORRELATION ANALYSIS

Residual Analysis

If you wish, you can have your computer statistical package or spreadsheet list the residuals (observed minus estimated values of y) along with the regression printout. For example, Table 15.2 is Minitab's listing of the residuals for the dexterity data we've been examining throughout the chapter. It was generated using the Minitab procedure in Computer Solutions 15.4, which we will present shortly.

For each of the 5 data points in the dexterity example, Table 15.2 provides observed and estimated ("fitted") values of y, along with the difference between them (the residual). In Table 15.2, the residuals have also been "standardized," indicating how many standard deviation multiples each represents. The standard deviations for the estimated y values are listed in the "SE Fit" column. Note that the value differs from one row to the next—this is because estimates become more uncertain when they are based on x values that are farther away from the mean of x.

Besides simply listing the residuals, computer packages can readily provide an analysis of them. For example, some data points may be flagged as "unusual observations," or "outliers." These points are relatively distant from the regression line, and their distance may be a reflection of unique characteristics possessed by the sample members involved. For example, a person whose productivity is

TABLE 15.2

Minitab summary of the residuals and standardized residuals for the dexterity regression analysis using the procedure in Computer Solutions 15.4.

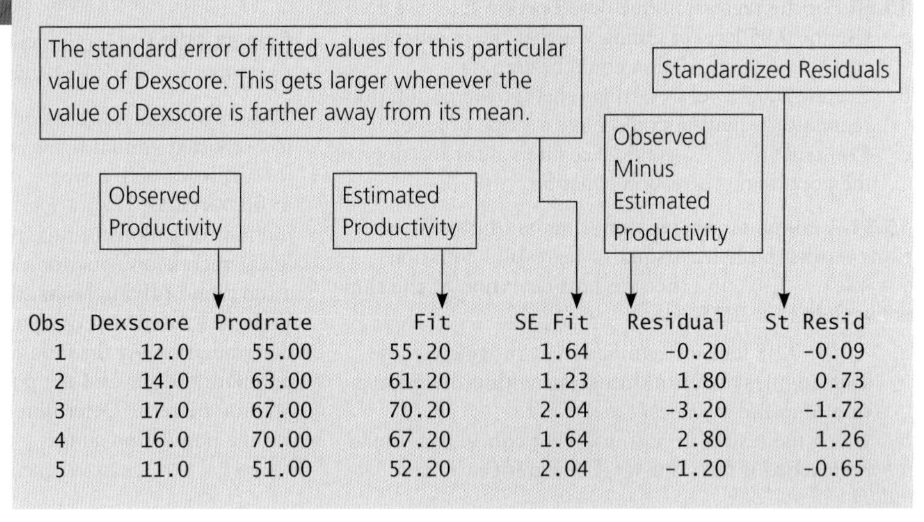

Obs	Dexscore	Prodrate	Fit	SE Fit	Residual	St Resid
1	12.0	55.00	55.20	1.64	-0.20	-0.09
2	14.0	63.00	61.20	1.23	1.80	0.73
3	17.0	67.00	70.20	2.04	-3.20	-1.72
4	16.0	70.00	67.20	1.64	2.80	1.26
5	11.0	51.00	52.20	2.04	-1.20	-0.65

extremely high may have done poorly on the dexterity test because he had a migraine headache on the test day. If the discrepancy between his observed and estimated productivity scores were great enough, he would be identified as an outlier.

Perhaps most important, residual analysis can also help us determine whether the assumptions of the regression model have been met. These assumptions (in terms of the errors or residuals) and the methods by which they can be examined are as follows:

Assumption 1: The population of ϵ values is normally distributed with a mean of zero.

a. Construct a histogram of the $y - \hat{y}$ values, or residuals. The histogram should be nearly symmetric, centered near or at zero, and have a shape that is approximately normal.

b. Use the computer to test for normality of the residuals. Whether the number of data points is small or large, we can use the Kolmogorov-Smirnov test for normality, introduced in Chapter 14. If the number of data points is sufficiently large ($n \geq 32$), Excel and the chi-square goodness-of-fit test for normality of Chapter 13 can be applied.

Excel and Minitab can also generate a normal probability plot for the residuals, another method introduced in Chapter 14. To the extent that the residuals are from a normal distribution, the points in the normal probability plot should fall approximately along a straight line.

c. Plot the residuals versus the x values. The points should be distributed about the horizontal line at zero, as shown in part (a) of Figure 15.5.

(*continued on page 582*)

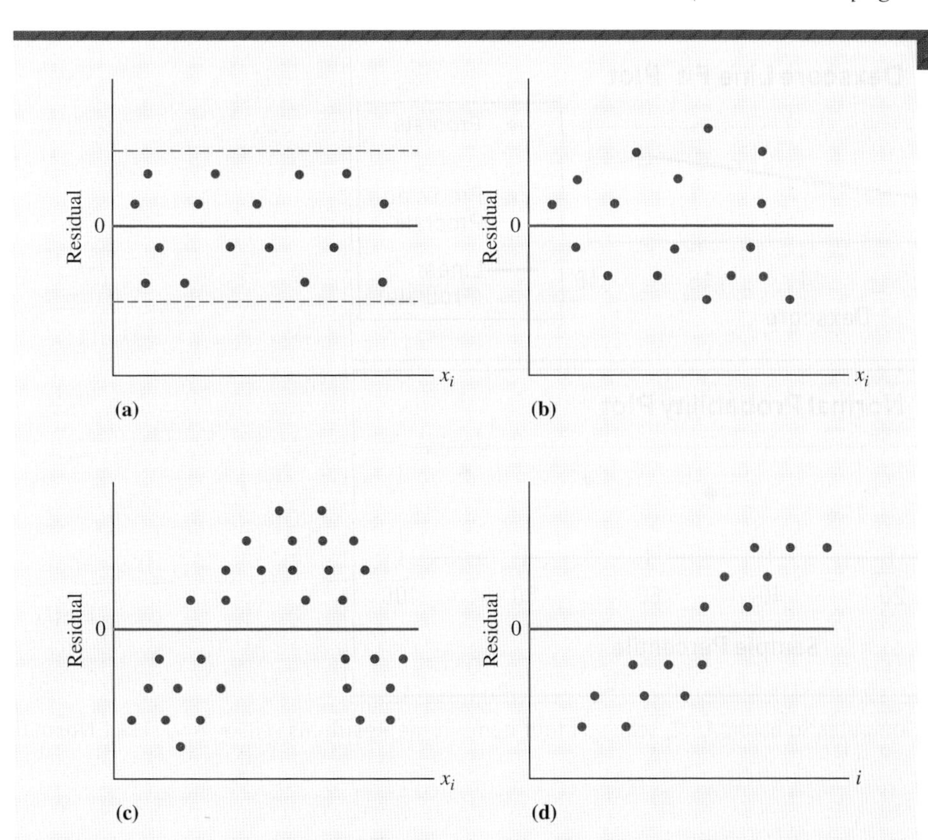

FIGURE 15.5
Selected residual plots; the interpretations are discussed in the text.

COMPUTER 15.4 SOLUTIONS

Residual Analysis

EXCEL

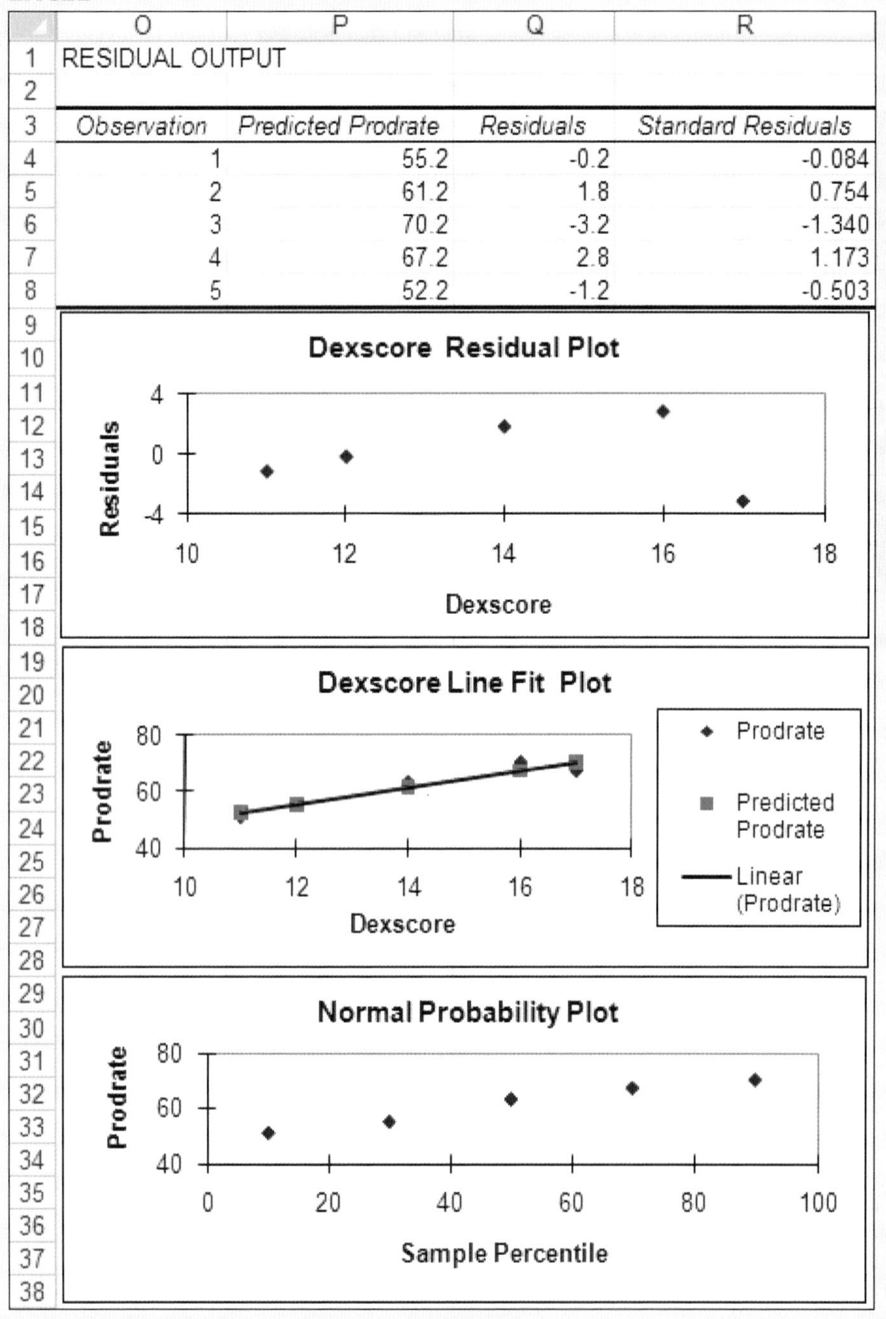

RESIDUAL OUTPUT

Observation	Predicted Prodrate	Residuals	Standard Residuals
1	55.2	-0.2	-0.084
2	61.2	1.8	0.754
3	70.2	-3.2	-1.340
4	67.2	2.8	1.173
5	52.2	-1.2	-0.503

In step 2 of the procedure in Computer Solutions 15.1, select all four items in the **Residuals** section and select **Normal Probability Plot.**

MINITAB

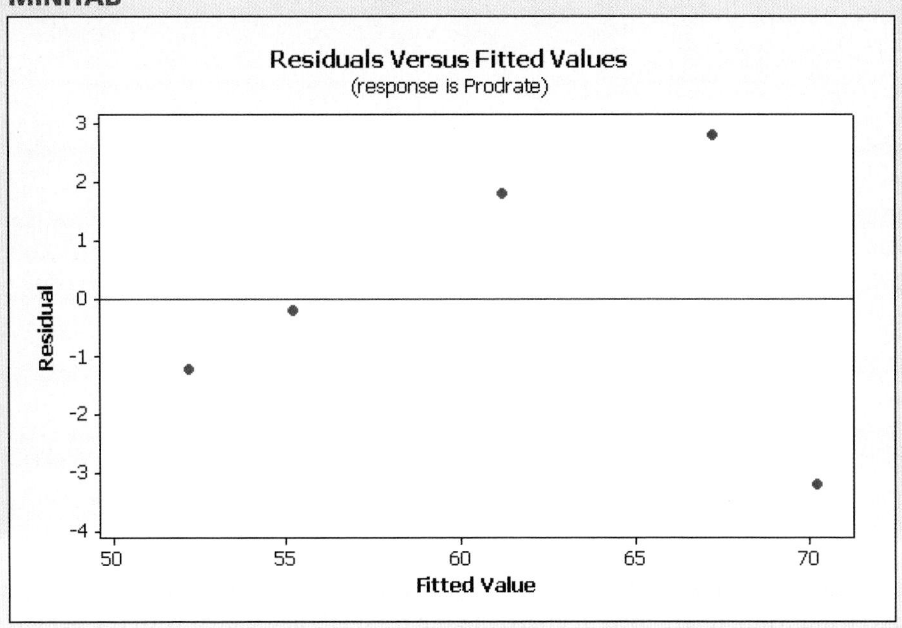

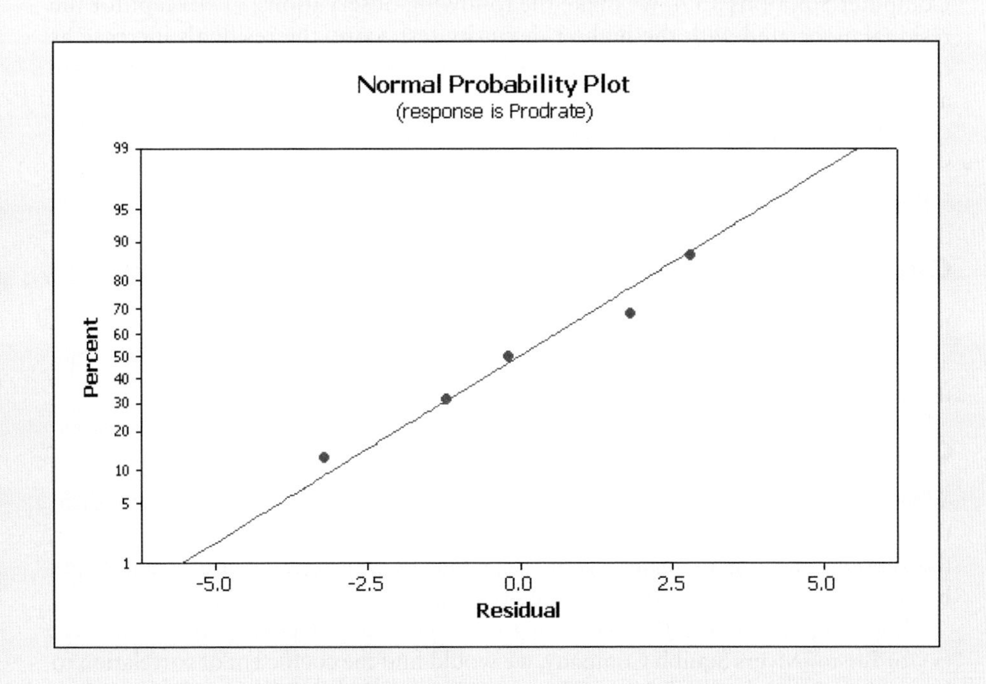

In step 2 of the procedure in Computer Solutions 15.1, click **Graphs.** In the **Residuals for plots** menu, select **Regular.** In the **Residual Plots** menu, select **Normal plot of residuals** and **Residuals versus fits.** A histogram of residuals and a plot of residuals versus order are also available, but not shown here. To obtain the printout of residuals shown in Table 15.2, in step 1 of Computer Solutions 15.1 click **Results,** then select **In addition, the full table of fits and residuals.**

Assumption 2: The standard deviation of the ϵ values is the same regardless of the given value of x.

Plot the residuals versus the x values. The amount of scatter should be approximately the same for all values of x, as shown in part (a) of Figure 15.5. In part (b) of Figure 15.5, the assumption of equal standard deviations has been violated, as the residuals tend to have a greater standard deviation for larger values of x. In part (c) of Figure 15.5, the scatter diagram of residuals is in a band that is not horizontal and centered on zero for all values of x, reflecting that the relationship between y and x may not be linear.

Assumption 3: The residuals are independent of each other.

Plot $(y_i - \hat{y}_i)$ versus the value of i. If successive observations have errors, or residuals, that are correlated, a phenomenon known as autocorrelation exists. For example, if the scatter diagram in this plot tends to have a negative slope or a positive slope, as shown in part (d) of Figure 15.5, this would suggest the presence of autocorrelation. The topic of autocorrelation is especially relevant to regression models fitted to a time series, and is discussed in Chapter 18.

Computer Solutions 15.4 describes the procedures and shows the results when Excel and Minitab are used in analyzing the residuals associated with the dexterity regression analysis. Referring to the first and third plots in the Excel portion of Computer Solutions 15.4, we make the following observations: (1) Except for the residual associated with the highest dexterity test score, the residuals increase as the dexterity test score increases; (2) The points in the normal probability plot are approximately in a straight line, which supports the possibility that the residuals could have come from a normal distribution. The number of data points is rather small and, despite the tendency noted in item (1), we are unable to conclude that the assumptions of the simple linear regression model have not been met.

Cautionary Notes

In applying regression and correlation analysis, it is not enough just to plug in y and x values, then carry out the calculations or assign them to the computer. Two major pitfalls can undermine our good intentions.

Causation Confusion

When analysis results in a high value for the coefficient of determination, the indication is that changes in x go a long way toward explaining changes in y. However, this does not mean that x causes y. In fact, y may be causing x, or both y and x may be caused by one or more other variables that have not been included in the analysis.

If we were to analyze 1990–1999 net sales (y) versus advertising and promotion (x) for Bristol-Myers Squibb Company, we would find the coefficient of correlation to be $r = 0.982$ and the coefficient of determination to be $r^2 = 0.965$ (see Figure 15.6). Statistically, changes in advertising and promotion "explain" 96.5% of the changes in net sales levels. This supports the widespread notion that more advertising tends to result in greater sales. However, higher sales allow for a larger advertising budget, and many companies actually budget their advertising as a fraction of their actual or anticipated sales. In such a case, y and x are actually causing each other!

Sometimes variables that are highly correlated have almost nothing at all to do with each other. For example, an analysis of BMW sales versus the average

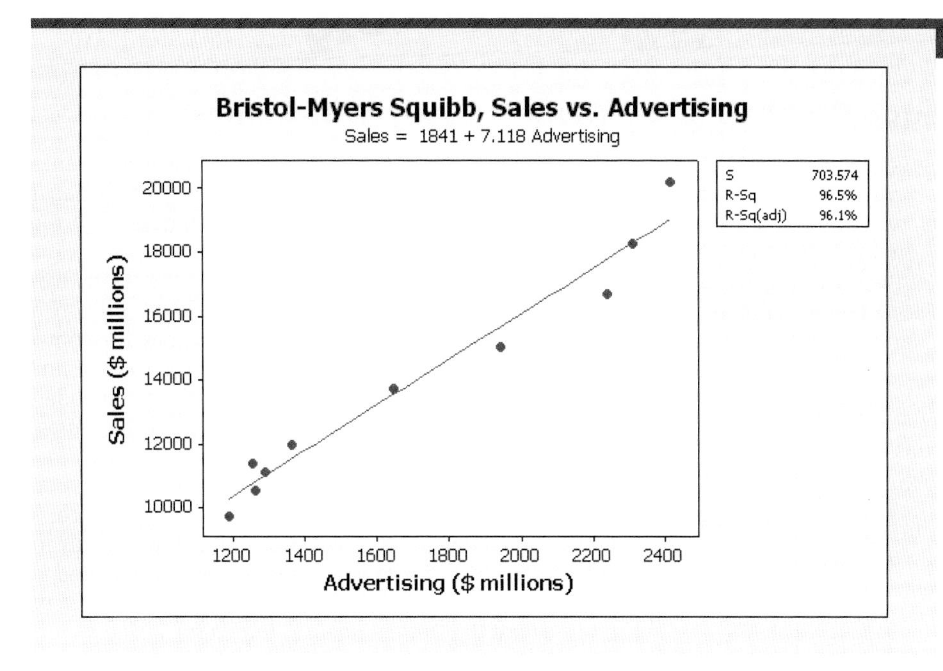

FIGURE 15.6

Minitab scatter diagram and fitted line for annual sales (y) versus annual expenditures on advertising and promotion (x) for Bristol-Myers Squibb Company for the period 1990–1999. While the coefficient of correlation is very high ($r = +0.982$), the question remains as to whether (1) higher levels of advertising cause greater sales, or (2) higher levels of sales lead to a larger advertising budget. *Source of underlying data:* Bristol-Myers Squibb Company, *1999 Annual Report,* pp. 50, 51.

salary of secondary school teachers results in a very high correlation:[2]

For the years 2002–2006:

y = Sales of BMW automobiles in the United States
x = Average salary, secondary school teachers

The regression equation is $\hat{y} = -375{,}997 + 14.126x$

$$r = +0.959 \quad \text{and} \quad r^2 = 0.920$$

Without further consideration, it might seem as though all these BMWs were being purchased by secondary school teachers, since the teachers' salary levels "explain" 92.0% of the variation in BMW sales. However, it is more likely that both y and x have simply increased largely *as a function of time.* If two variables are linearly related to time, they will also be linearly related to each other. Whenever two variables are highly correlated, but there is little evidence to suggest that either causes the other (e.g., annual flounder catch in the North Atlantic and annual bachelor's degrees granted in physics), this is known as a *spurious*, or nonsense, correlation.

Extrapolating Beyond the Range of the Data

When regression and correlation are based on a given set of data, it can be dangerous to make estimates for y based on x values outside the range of these data. Figure 15.7 shows the scatter diagram and regression line for EPA city miles per gallon (y) versus curb weight (x) for a sampling of automobiles from the 2009 model year. As expected, heavier cars tend to get fewer miles per gallon. However, if we go beyond the range of x values from which the regression line was derived, we find that *a car weighing 6591 pounds would get zero miles per gallon on the EPA city test,* a physical impossibility. Statistics in Action 15.2 discusses both causation and extrapolation beyond the data for salaries versus team performance in professional baseball.

[2]Source: autointell.com, August 8, 2009; and Bureau of the Census, *Statistical Abstract of the United States 2009,* p. 158.

STATISTICS IN ACTION 15.2

Major League Baseball: Does Money Really Win Games?

In the chapter's opening vignette, we discussed the possible relationship between how well a team is paid and how well they play. Using Excel, we've performed a simple regression analysis of the performance of Major League Baseball teams versus the size of their team payrolls during the 2006 season. Using Excel File **CX15BASE** and the methods in Computer Solutions 15.1 and 15.4, we obtained the regression equation

$$\text{Wins} = 67.980 + 0.167 \text{ Payroll}$$
(payroll in millions of dollars)
$$r = +0.536 \text{ with } r^2 = 0.287$$

Not unexpectedly, teams with higher payrolls tended to win more games. The slope (+0.167) indicates that an extra one million dollars in payroll tends to increase the number of wins by 0.167. Thus, on average, each additional win costs 1/0.167, or $5.99 million. Table 15.3 shows the payroll and number of wins for each team, the number of wins predicted by the regression equation, and the residual (actual wins minus predicted wins). Especially interesting in Table 15.3 is a comparison of two teams that were not among the highest-paid in the league: the Minnesota Twins and the Chicago Cubs. The Twins got a lot of wins for the money, 17.4 *more* than the equation predicted, while the Cubs were 17.79 wins *below* the performance predicted for a team with their payroll. (Which manager would *you* prefer to be if you're invited to meet with the team president the morning after the season ends?)

The upper portion of Figure 15.8 shows the scatter diagram and fitted regression line, and the lower portion is the plot of residuals (vertical axis) versus payrolls (horizontal axis). At approximately $63 million and $94 million payroll levels, respectively, the overachieving Twins and the underachieving Cubs are exceptionally distant from the regression line, but in opposite directions.

With their league-high $194.6 million payroll, one would expect the New York Yankees to have had a successful season. They did, in a way. Nobody had more regular-season wins than the Yankees, but they were eliminated in the post-season playoffs and the St. Louis Cardinals went on to defeat the Detroit Tigers in the World Series.

A final observation: If there were a team whose members received *zero* salary, the equation would nevertheless predict them to win nearly 68 games! This is misleading, however, because such an estimate would be based on a payroll that lies unrealistically far below those in the underlying data.

Data sources: usatoday.com and mlb.com, June 21, 2009.

FIGURE 15.7

Making estimates of y based on x values beyond the data is a dangerous proposition. This Excel scatter diagram represents EPA city gas mileage (y) versus vehicle weight (x) for a sample of 2009 auto models. According to the regression line, a car weighing 6591 pounds would get *zero* miles per gallon in the test. *Source of underlying data: Consumer Reports New Car Ratings & Reviews 2009.*

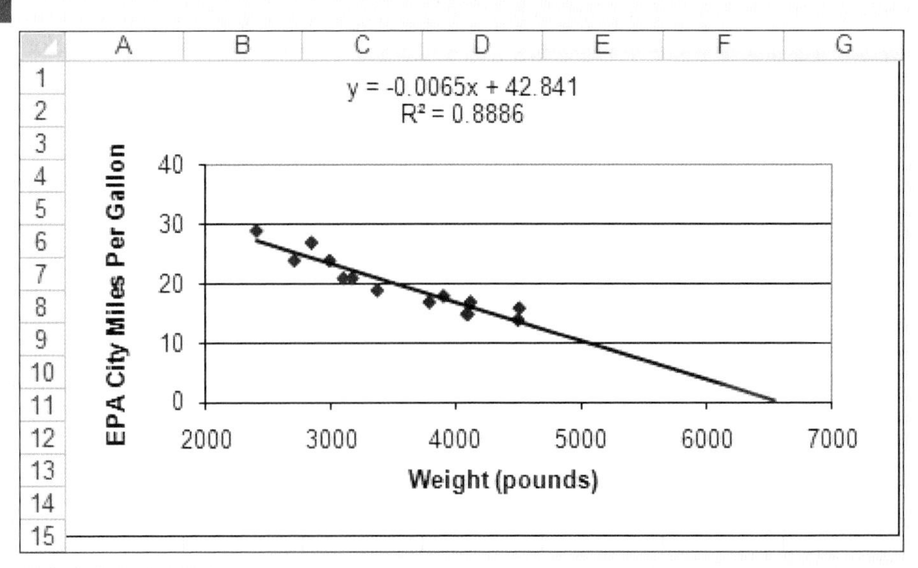

	A	B	C	D	E
1	Team	Payroll	Wins	Predicted Wins	Residual
2	Arizona Diamondbacks	59.684	76	77.97	-1.97
3	Atlanta Braves	90.157	79	83.08	-4.08
4	Baltimore Orioles	72.586	70	80.13	-10.13
5	Boston Red Sox	120.100	86	88.09	-2.09
6	Chicago Cubs	94.424	66	83.79	-17.79
7	Chicago White Sox	102.751	90	85.19	4.81
8	Cincinnati Reds	60.910	80	78.18	1.82
9	Cleveland Indians	56.032	78	77.36	0.64
10	Colorado Rockies	41.233	76	74.88	1.12
11	Detroit Tigers	82.613	95	81.81	13.19
12	Florida Marlins	14.999	78	70.49	7.51
13	Houston Astros	92.552	82	83.48	-1.48
14	Kansas City Royals	47.294	62	75.90	-13.90
15	Los Angeles Angels	103.472	89	85.31	3.69
16	Los Angeles Dodgers	98.447	88	84.46	3.54
17	Milwaukee Brewers	57.568	75	77.62	-2.62
18	Minnesota Twins	63.396	96	78.60	17.40
19	New York Mets	101.085	97	84.91	12.09
20	New York Yankees	194.633	97	100.57	-3.57
21	Oakland Athletics	62.243	93	78.40	14.60
22	Philadelphia Phillies	88.273	85	82.76	2.24
23	Pittsburgh Pirates	46.718	67	75.80	-8.80
24	St. Louis Cardinals	88.891	83	82.86	0.14
25	San Diego Padres	69.896	88	79.68	8.32
26	San Francisco Giants	90.056	76	83.06	-7.06
27	Seattle Mariners	87.960	78	82.71	-4.71
28	Texas Rangers	68.229	80	79.40	0.60
29	Tampa Bay Devil Rays	35.418	61	73.91	-12.91
30	Toronto Blue Jays	71.915	87	80.02	6.98
31	Washington Nationals	63.143	71	78.55	-7.55

TABLE 15.3

For the 2006 Major League Baseball season discussed in Statistics in Action 15.2, team payrolls, wins, wins predicted by the regression equation, and residuals.

EXERCISES

15.60 In regression analysis, what is a "residual"?

15.61 What is residual analysis, and what information can it provide?

15.62 What is spurious correlation? Provide a real or hypothetical example where two variables might exhibit such a relationship.

15.63 A firm finds the coefficient of correlation between y = annual sales and x = annual expenditure on research and development to be $r = +0.90$. Comment on the likely direction of causation, if any, between these variables.

15.64 "The Ford Edsel was sold only in the 1957–1959 model years, but the company made a big mistake in dropping it. If they had just fitted a regression line to the sales for these 3 years and extended it to 2006, they could have seen that it would have been one of the ten sales leaders of that year." Comment on the appropriateness of this statement.

(DATA SET) *Note:* Exercises 15.65–15.68 require a computer and statistical software.

15.65 For the y and x values listed in file **XR15065**, obtain the simple linear regression equation, then analyze

FIGURE 15.8

Excel scatter diagram, fitted regression equation, and residuals plot for the regression analysis of the payroll and wins data in Table 15.3.

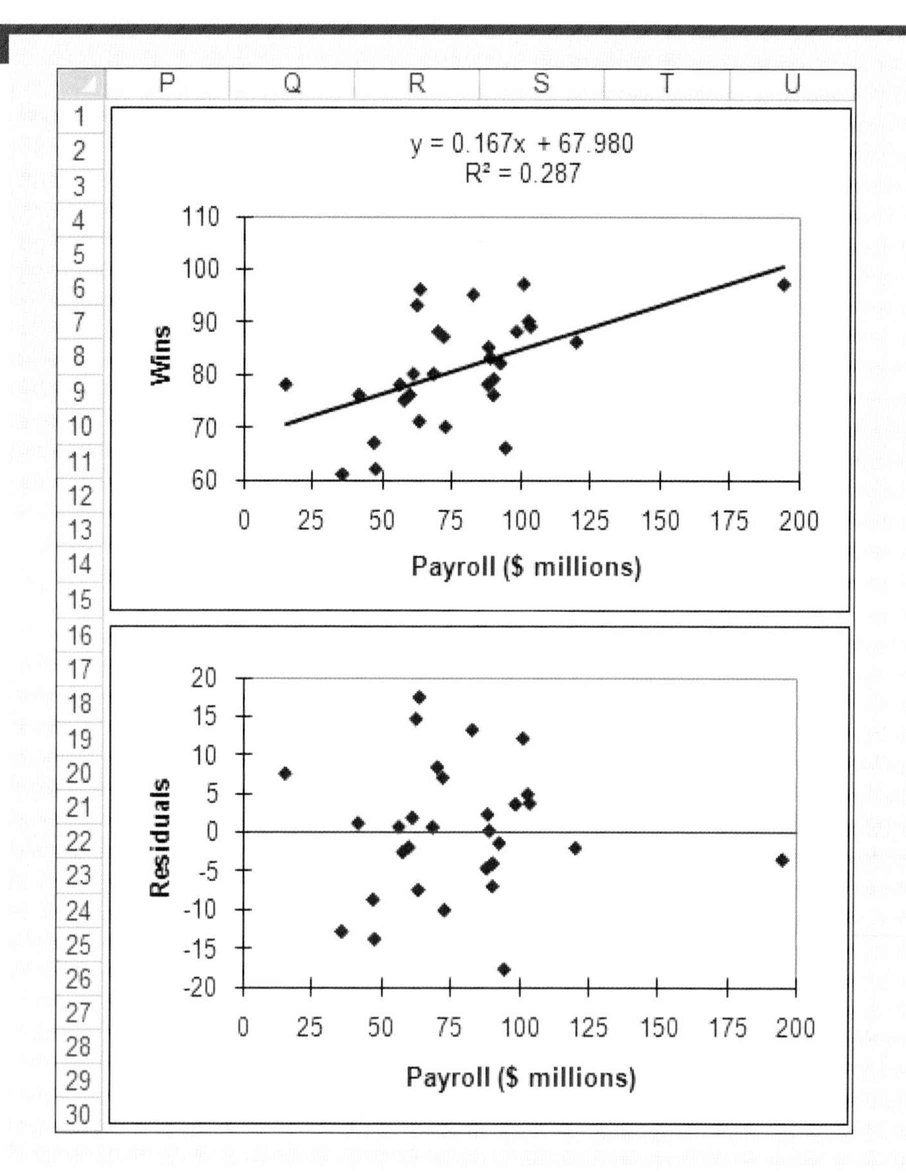

the residuals by (a) constructing a histogram, (b) using a normal probability plot, (c) plotting the residuals versus the x values, and (d) plotting the residuals versus the order in which they were observed. Do any of the assumptions of the simple linear regression model appear to have been violated?

15.66 For the y and x values listed in file **XR15066**, obtain the simple linear regression equation, then analyze the residuals by (a) constructing a histogram, (b) using a normal probability plot, (c) plotting the residuals versus the x values, and (d) plotting the residuals versus the order in which they were observed. Do any of the assumptions of the simple linear regression model appear to have been violated?

15.67 For the regression equation obtained in Exercise 15.57, analyze the residuals by (a) constructing a histogram, (b) using a normal probability plot, (c) plotting the residuals versus the x values, and (d) plotting the residuals versus the order in which they were observed. Do any of the assumptions of the simple linear regression model appear to have been violated? Comment on whether any of the years were associated with especially high or low fuel consumption for the number of flying hours during the year.

15.68 For the regression equation obtained in Exercise 15.59, analyze the residuals by (a) constructing a histogram, (b) using a normal probability plot, (c) plotting the residuals versus the x values, and (d) plotting the residuals versus the order in which they were observed. Do any of the assumptions of the simple linear regression model appear to have been violated? Comment on whether any of these high-growth companies had an especially high or low income for a firm with their level of revenue.

SUMMARY 15.7

• Linear regression model and its assumptions

In simple linear regression and correlation, regression analysis provides a "best-fit" linear equation for the data, and correlation analysis measures the strength of the linear relationship. In the linear model, $y_i = \beta_0 + \beta_1 x_i + \epsilon_i$, y and x are the dependent and independent variables, respectively. The simple linear regression model assumes the following: (1) for any given value of x, the y values are normally distributed with a mean that is on the regression line, or $\mu_{y.x} = \beta_0 + \beta_1 x_i$; (2) regardless of the value of x, the standard deviation of the y values about the regression line is the same; and (3) the y values are statistically independent of each other. In multiple linear regression and correlation (Chapter 16), there is one dependent variable, but two or more independent variables.

Variables are directly related when they increase or decrease together and inversely related when an increase in one is accompanied by a decrease in the other. The relationship may also be linear or curvilinear, depending on whether a straight line or a curve is best for describing their relationship. The emphasis of the chapter is on linear relationships.

• Determining the least-squares linear regression equation

The scatter diagram visually summarizes the data points and consists of vertical (y) and horizontal (x) axes. A ruler and the "eyeball" method can be used in fitting an approximate straight line to the data. A superior approach is to calculate the regression line, $\hat{y} = b_0 + b_1 x$, that satisfies the least-squares criterion. This criterion requires that the y-intercept (b_0) and slope (b_1) be such that the sum of the squared vertical distances between data points and the line is minimized.

Formulas are provided for the calculation of b_0 and b_1 for the least-squares regression line, $\hat{y} = b_0 + b_1 x$. Because the regression line is based on sample data, the y-intercept (b_0) and slope (b_1) are really estimates of their respective (and unknown) population values, β_0 and β_1. For a given value of x, the point estimate of y will be $\hat{y}$, calculated from the regression line. Interval estimates for y can also be developed, and the first step is to calculate the standard error of estimate, $s_{y.x}$. The standard error of estimate is the extent to which y values are scattered or dispersed above and below the regression line. For a given x value, the difference between the observed and estimated values of y is the error, or residual.

• Point and interval estimates

Given a specific value of x, two kinds of interval estimates can be made regarding y: (1) a confidence interval for the (unknown) true mean of y, and (2) a prediction interval for an individual observation of y. Each approach relies on a correction formula that takes into account the standard error of estimate, the number of data points in the sample, and the distance between $\bar{x}$ and the specific value of x on which the interval is based.

• Coefficients of correlation and determination

Two important measures of the strength of the linear relationship between y and x are the coefficient of correlation (r) and the coefficient of determination (r^2). The coefficient of correlation will be no less than $r = -1$ and no greater than $r = +1$, and the larger the absolute value of r, the stronger the linear relationship between the variables. Whenever $r = 0$, there is no linear relationship at all, and the slope of the regression line will be $b_1 = 0$. The coefficient of determination (r^2) is the proportion of the variation in y that is explained by the regression line, $\hat{y} = b_0 + b_1 x$, and r^2 will always be ≤ 1.

- **Testing the linear relationship**

The significance of the linear relationship can be tested by examining whether the coefficient of correlation differs significantly from zero, or by testing whether the slope differs significantly from zero. The tests are equivalent to each other as well as to an analysis of variance test that examines the amount of variation in y that is explained by the equation.

- **Computer-assisted regression analysis**

Computer statistical packages provide detailed regression and correlation analysis, and their output typically includes the descriptors, tests, and interval estimates discussed in the chapter. Both Excel and Minitab Computer Solutions procedures are provided that correspond to the principal example used throughout the chapter.

With statistical software, it is easy to analyze the residuals and examine the assumptions of the model. Computer statistical packages also identify data points that are relatively distant from the regression line. Such "unusual" observations may reflect the possession of unusual characteristics by the sample members that these points represent. Residual analysis is especially useful in examining whether various assumptions of the regression model may have been violated.

- **Potential problems with simple linear regression**

In applying regression and correlation analysis, it is important to avoid confusing a strong linear relationship between y and x with the conclusion that y is actually caused by x. In reality, y could be causing x, or both y and x might be heavily influenced by another variable that has not been included in the analysis. Another precaution is to avoid extrapolating beyond the range of the data. It can be extremely misleading to make point or interval estimates of y based on an x value that lies beyond the range of data points from which the regression line has been derived.

EQUATIONS

The Simple Linear Regression Model

$$y_i = \beta_0 + \beta_1 x_i + \epsilon_i$$

where y_i = a value of the dependent variable, y
x_i = a value of the independent variable, x
β_0 = the y-intercept of the regression line
β_1 = the slope of the regression line
ϵ_i = random error, or residual

1. For a given value of x, the expected value of y is given by the linear equation, $\mu_{y.x} = \beta_0 + \beta_1 x_i$. The term $\mu_{y.x}$ can be stated as "the mean of y, given a specific value of x."
2. The difference between the actual value of y and the expected value of y is the error, or residual, $\epsilon_i = y_i - (\beta_0 + \beta_1 x_i)$.

The Sample Regression Line

$$\hat{y} = b_0 + b_1 x$$

where $\hat{y}$ = the estimated value of the dependent variable (y) for a given value of x
b_0 = the y-intercept; this is the value of y where the line intersects the y-axis whenever $x = 0$
b_1 = the slope of the regression line
x = a value for the independent variable

Least-Squares Criterion for the Best-Fit Regression Line

The equation must be such that $\Sigma(y_i - \hat{y}_i)^2$ is minimized

where y_i = the observed value of y for the given value of x
$\hat{y}_i$ = the predicted value of y for the x value, as determined from the regression equation

Least-Squares Regression Line, $\hat{y} = b_0 + b_1x$

- Slope:

$$b_1 = \frac{\left(\Sigma x_i y_i\right) - n\bar{x}\bar{y}}{\left(\Sigma x_i^2\right) - n\bar{x}^2}$$

where n = number of data points

- y-intercept:

$$b_0 = \bar{y} - b_1\bar{x}$$

Standard Error of Estimate

$$s_{y.x} = \sqrt{\frac{\Sigma(y_i - \hat{y}_i)^2}{n - 2}}$$

where y_i = each observed value of y in the data
$\hat{y}_i$ = the value of y that would have been estimated from the regression line
n = the number of data points

Confidence Interval for the Mean of *y*, Given a Specific Value of *x*

$$\hat{y} \pm t(s_{y.x})\sqrt{\frac{1}{n} + \frac{(x\text{ value} - \bar{x})^2}{\left(\Sigma x_i^2\right) - \left(\Sigma x_i\right)^2/n}}$$

where $\hat{y}$ = the estimated value of y for the given value of x
n = the number of data points
$s_{y.x}$ = the standard error of estimate
t = t value for confidence level desired and $df = n - 2$
x value = the given value of x

Prediction Interval for an Individual *y*, Given a Specific Value of *x*

$$\hat{y} \pm t(s_{y.x})\sqrt{1 + \frac{1}{n} + \frac{(x\text{ value} - \bar{x})^2}{\left(\Sigma x_i^2\right) - \left(\Sigma x_i\right)^2/n}}$$

where $\hat{y}$ = the estimated value of y for the given value of x
n = number of data points
$s_{y.x}$ = the standard error of estimate
t = t value for the confidence level desired and $df = n - 2$
x value = the given value of x

The Coefficient of Correlation

$$r = \frac{n\left(\Sigma x_i y_i\right) - \left(\Sigma x_i\right)\left(\Sigma y_i\right)}{\sqrt{n\left(\Sigma x_i^2\right) - \left(\Sigma x_i\right)^2} \cdot \sqrt{n\left(\Sigma y_i^2\right) - \left(\Sigma y_i\right)^2}}$$

where r = coefficient of correlation, $-1 \le r \le +1$
n = number of data points

The Coefficient of Determination, r^2

$$r^2 = 1 - \frac{\textit{Error}\text{ variation, which is }\textit{not}\text{ explained by the regression line}}{\textit{Total}\text{ variation in the }y\text{ values}}$$

$$= 1 - \frac{\Sigma(y_i - \hat{y}_i)^2}{\Sigma(y_i - \bar{y})^2} = 1 - \frac{SSE}{SST}$$

t-Test for the Population Coefficient of Correlation, ρ

- Null hypothesis

H_0: $\rho = 0$ There is no linear relationship.

- Alternative hypothesis

H_1: $\rho \neq 0$ A linear relationship exists.

- Test statistic

$$t = \frac{r}{\sqrt{\dfrac{1 - r^2}{n - 2}}}$$

where r = sample coefficient of correlation
n = number of data points
t = t value for the level of significance of the test and $df = n - 2$

(Testing H_0: $\rho = 0$ is equivalent to testing H_0: $\beta_1 = 0$, described below.)

t-Test for the Slope (β_1) of the Population Regression Line

- Null hypothesis

H_0: $\beta_1 = \beta_{10}$ The population slope is β_{10}.

- Alternative hypothesis

H_1: $\beta_1 \neq \beta_{10}$ The population slope is not β_{10}.

- Test statistic

$$t = \frac{b_1 - \beta_{10}}{s_{b_1}}$$

where b_1 = slope of the sample regression line
β_{10} = a value that has been hypothesized for the slope of the population regression line

with

$$s_{b_1} = \frac{s_{y.x}}{\sqrt{\sum x_i^2 - n(\bar{x})^2}}$$

where s_{b_1} = the estimated standard deviation of the slope
$s_{y.x}$ = the standard error of estimate
n = the number of data points
t = t value for the level of significance of the test, and $df = n - 2$

Confidence Interval for the Slope (β_1) of the Population Regression Line

The interval is

$$b_1 \pm t s_{b_1}$$

where b_1 = slope of the sample regression line
t = t value for the confidence level desired, and $df = n - 2$
s_{b_1} = the estimated standard deviation of the slope, $s_{b_1} = \dfrac{s_{y.x}}{\sqrt{\sum x_i^2 - n(\bar{x})^2}}$

Components of the Total Variation in the Dependent Variable, *y*

Total variation in *y* values (SST)	=	Variation explained by regression line (SSR)	+	Variation *not* explained by regression line (SSE)
$\sum(y_i - \bar{y})^2$		$SSR = SST - SSE$		$\sum(y_i - \hat{y}_i)^2$

ANOVA Test for the Significance of a Simple Linear Regression

- **Null and alternative hypotheses**

H_0: There is no linear relationship between x and y.
H_1: A linear relationship exists between x and y.

- **Test statistic**

$$\text{Calculated } F = \frac{SSR/1}{SSE/(n-2)}$$

where SSR = sum of squares explained by the regression line
SSE = error, or residual, sum of squares
n = number of data points

- **Decision rule** Reject H_0 if calculated F exceeds the critical F for the level of significance of the test, with df (numerator) = 1 and df (denominator) = $n - 2$.

CHAPTER EXERCISES

15.69 For households in a community, many different variables can be observed or measured. If y = monthly mortgage payment, x = annual income would probably be directly related to y. For y = monthly mortgage payment, provide an example of an x variable likely to be (a) directly related to y, (b) inversely related to y, and (c) unrelated to y.

15.70 For high school seniors entering college, and for y = grade point average for the freshman year, x = Scholastic Aptitude Test score would tend to be directly related to y. Provide an example of another personal variable that might be directly related to y along with an example of a personal variable that you think would tend to be inversely related to y.

15.71 McDonald's Corporation has reported the following values for total revenues and net income during the 1998 to 2005 period. All data are in billions of dollars:
Source: McDonald's Corporation, *2005 Annual Report*.

	1998	1999	2000	2001
Net Income	1.55	1.95	1.98	1.64
Total Revenues	12.42	13.26	14.24	14.87

	2002	2003	2004	2005
Net Income	0.89	1.47	2.28	2.60
Total Revenues	15.41	17.14	19.07	20.46

Determine the least-squares regression equation line for estimating net income and interpret its slope. For a year in which total revenues are \$18.0 billion, estimate the net income for that year.

15.72 Prior to being hired, the five salespersons for a computer store were given a standard sales aptitude test. For each individual, the score achieved on the aptitude test and the number of computer systems sold during the first 3 months of their employment are shown here.

	John	Cora	Sam	Cindy	Henry
x = Score on aptitude test	80	70	45	90	20
y = Units sold in 3 months	25	15	10	40	5

Determine the least-squares regression line and interpret its slope. Estimate, for a new employee who scores 60 on the sales aptitude test, the number of units the new employee will sell in her first 3 months with the company.

15.73 As of mid-June 2009, five U.S. Government Bond mutual funds were reported as having generated the 1-year and 3-year (annualized) rates of return shown below. Source: *Money*, August 2009, p. 113.

U.S. Government Bond Fund	x = 3-Year Annualized Rate of Return	y = 1-Year Rate of Return
American Funds, AMUSX	6.3%	7.1%
Fidelity, FGOVX	7.4	8.4
Wells Fargo, STVSX	6.5	7.3
Morgan Stanley, USGBX	2.4	1.5
MFS Govt. Securities, MFGSX	7.2	9.0

a. Determine the least-squares regression line and interpret its slope.
b. For a U.S. Government Bond mutual fund with a 3-year annualized rate of return of 5.0%, estimate the 1-year rate of return.
c. For a U.S. Government Bond mutual fund with a 3-year annualized rate of return of 7.0%, estimate the 1-year rate of return.

15.74 Peoples Energy, Inc. reports the following data for operating revenue and net income for 2001–2005. Source: Peoples Energy, *2005 Annual Report*, pp. 83–84.

Year	x = Operating Revenue (Millions)	y = Net Income (Millions)
2001	2270.2	96.9
2002	1482.5	89.1
2003	2138.4	103.9
2004	2260.2	81.6
2005	2600.0	78.1

Determine the least-squares regression line and interpret its slope. Estimate the net income if the operating revenue figure is $2500 million.

15.75 For 2004 through 2008, Coca-Cola annual operating revenues and net incomes were as shown below. Source: *Coca-Cola 2008 Annual Review*, p. 7.

Year	x = Operating Revenue	y = Net Income
2004	$21.74 billion	$4.85 billion
2005	23.10	4.87
2006	24.09	5.08
2007	28.86	5.98
2008	31.94	5.81

Determine the least-squares regression line and interpret its slope. Estimate net income if the operating revenue is $30 billion.

15.76 The data in the table describe retail furniture store sales ($ billions) versus housing starts over a 5-year time period. Source: *Statistical Abstract of the United States 2009*, pp. 594, 644.

	2001	2002	2003	2004	2005
Retail Furniture Store Sales ($ billions)	50.6	51.3	52.1	56.5	59.8
Housing Starts (millions)	1.27	1.36	1.50	1.61	1.72

a. Determine the least-squares regression equation for predicting retail furniture store sales on the basis of housing starts.
b. Determine and interpret the coefficients of correlation and determination.
c. If there were 1.80 million housing starts in a year, what would be the prediction for retail furniture store sales?

15.77 In the early 1900s, population, prices, wages, and individual life insurance policy face amounts were all incredibly low compared to their counterparts of today. The following data show the number of life insurance policies (millions) and their total face amounts ($ millions) for five years selected from this long-ago era that was about to experience the Great Depression. Source: American Council of Life Insurers, *Life Insurers Fact Book 2008*, p. 71.

	1900	1905	1910	1915	1920
Individual Life Insurance Policies (millions)	14	22	29	41	64
Total Individual Life Insurance Face Amounts ($ millions)	7573	11,863	14,908	20,929	38,966

a. Determine the least-squares regression equation for predicting total value of individual life insurance on the basis of the number of individual life insurance policies in force. Interpret the slope of the equation.
b. Determine and interpret the coefficients of correlation and determination.
c. If there were 30 million individual life insurance policies in force during a given year, what would be the prediction for the total face value of these policies?

15.78 For the regression analysis of the data in Exercise 15.77:
a. Use the 0.10 level in testing whether the population coefficient of correlation could be zero.
b. Use the 0.10 level in testing whether the population regression equation could have a slope of zero.
c. Construct the 90% confidence interval for the slope of the population regression equation. Discuss the interval in terms of the results of part (b).

15.79 The following data describe U.S. motor vehicle travel and fuel consumption from 2002 through 2006. The data represent billions of gallons consumed and billions of miles traveled, respectively. Source: *Statistical Abstract of the United States 2009*, p. 676.

	2002	2003	2004	2005	2006
Fuel Consumed	168.7	170.0	173.5	174.8	174.9
Miles Traveled	2856	2890	2965	2989	3014

a. Determine the least-squares equation line for predicting fuel consumption on the basis of driving miles.
b. Determine and interpret the coefficients of correlation and determination.
c. If motor vehicle travel during a year were 3100 billion miles, what level of fuel consumption would be predicted?

15.80 For the regression analysis of the data in Exercise 15.79:

a. Use the 0.05 level in testing whether the population coefficient of correlation could be zero.
b. Use the 0.05 level in testing whether the population regression equation could have a slope of zero.
c. Construct the 95% confidence interval for the slope of the population regression equation. Discuss the interval in terms of the results of part (b).

15.81 As part of its ongoing testing and reporting program, the Insurance Institute for Highway Safety has crashed vehicles into a barrier intended to represent another vehicle's bumper, then recorded the cost of repairs necessary to fix whatever damage has occurred. In these tests, the simulated bumper was set at 16 inches from the ground in the corner tests, and at 18 inches from the ground in the full-width tests. In crash-testing a sample of mini and micro-cars at 6 mph, the repair costs were as shown in the table. Source: Insurance Institute for Highway Safety, *Status Report*, June 11, 2009, p. 2.

Vehicle Tested	Rear Full-Width	Rear Corner
Smart Fortwo	$631	$507
Chevrolet Aveo	1370	612
Mini Cooper	929	743
Toyota Yaris	3345	474
Honda Fit	3648	999
Hyundai Accent	2057	831
Kia Rio	3148	773

a. Based on these data, determine the least-squares regression equation for predicting rear full-width crash repair cost on the basis of rear corner crash repair cost.
b. Determine and interpret the coefficients of correlation and determination.
c. Based on these data, if a mini or micro-car like the ones tested were to incur $800 in damages in the rear corner test, what level of repairs would be predicted for the rear full-width test?

15.82 For the regression analysis of the data in Exercise 15.81:

a. Use the 0.10 level in testing whether the population coefficient of correlation could be zero.
b. Use the 0.10 level in testing whether the population regression equation could have a slope of zero.
c. Construct the 90% confidence interval for the slope of the population regression equation. Discuss the interval in terms of the results of part (b).

15.83 The following Minitab output describes the results of a production experiment in which $n = 10$ different steel bars were subjected to tempering times ranging from 6 to 10 seconds, then measured for tensile strength (thousands of pounds):

```
Regression Analysis

The regression equation is
STRENGTH = 60.0 + 10.5 TEMPTIME

Predictor        Coef       StDev        T        P
Constant        60.02       29.36     2.04    0.075
TEMPTIME       10.507       3.496     3.01    0.017

S = 14.03       R-Sq = 53.0%     R-Sq(adj) = 47.2%

Analysis of Variance

Source        DF        SS         MS        F        P
Regression     1    1777.4     1777.4     9.03    0.017
Error          8    1574.6      196.8
Total          9    3352.0
```

a. To the greatest number of decimal places in the printout, what is the least-squares regression line?
b. What proportion of the variation in strength is explained by the regression line?
c. At what level of significance does the slope of the line differ from zero? What type of test did Minitab use in reaching this conclusion?
d. At what level of significance does the coefficient of correlation differ from zero? Compare this with the level found in part (c) and explain either why they are different or why they are the same.
e. Construct the 95% confidence interval for the slope of the population regression line.

15.84 For the ANOVA portion of the printout shown in Exercise 15.83, explain the exact meaning of each number in the "SS" and "F" columns.

15.85 A tire company has carried out tests in which rolling resistance (pounds) and inflation pressure (pounds per square inch, or psi) have been measured for psi values ranging from 20 to 45. The regression analysis is summarized in the following Minitab printout:

```
Regression Analysis

The regression equation is
ROLRESIS = 9.45 - 0.0811 PSI

Predictor        Coef       StDev        T        P
Constant        9.450       1.228     7.69    0.000
PSI          -0.08113     0.03416    -2.38    0.029

S = 0.8808      R-Sq = 23.9%     R-Sq(adj) = 19.6%

Analysis of Variance

Source        DF        SS         MS        F        P
Regression     1    4.3766     4.3766     5.64    0.029
Error         18   13.9657     0.7759
Total         19   18.3422
```

a. To the greatest number of decimal places in the printout, what is the least-squares regression line?
b. What proportion of the variation in rolling resistance is explained by the regression line?
c. At what level of significance does the slope of the line differ from zero? What type of test did Minitab use in reaching this conclusion?
d. At what level of significance does the coefficient of correlation differ from zero? Compare this with the level

found in part (c) and explain either why they are different or why they are the same.

e. Construct the 95% confidence interval for the slope of the population regression line.

15.86 For the ANOVA portion of the printout shown in Exercise 15.85, explain the exact meaning of each number in the "SS" and "F" columns.

15.87 Given the data and regression analysis for Exercise 15.72, construct and interpret the 95% confidence and prediction intervals associated with a score of 60 on the sales aptitude test.

15.88 Given the data and regression analysis for Exercise 15.73, construct and interpret the 90% confidence and prediction intervals associated with a 3-year annualized rate of return of 7.0%.

15.89 Given the data and regression analysis for Exercise 15.76, construct and interpret the 95% confidence and prediction intervals associated with 1.4 million housing starts.

15.90 A physical-fitness researcher tests 25 professional football players, finding that the time required to run 40 yards (y, in seconds) is related to weight (x, in pounds) by the least-squares equation $\hat{y} = -2.0 + 0.038x$. According to this equation, a person weighing 200 pounds would run the 40-yard dash in an estimated time of 5.6 seconds. However, a child weighing 55 pounds would have an estimated time of just 0.09 seconds to run 40 yards. Briefly explain how this unlikely result could have been obtained.

(DATA SET) *Note:* Exercises 15.91–15.96 require a computer and statistical software.

15.91 A university administrator has collected data on total Scholastic Aptitude Test (SAT) score versus freshman grade point average (GPA) for a sample of students who recently finished their first year at the university. The data are listed in file **XR15091**. Given these data:

a. Determine the least-squares regression equation for estimating the freshman grade point average based on total SAT score, then estimate the freshman grade point average for an applicant who has scored a total of 1100 on his SAT examination.

b. What percentage of the variation in freshman GPA is explained by SAT total?

c. Given an SAT total of 1100, construct and interpret the 99% confidence and prediction intervals associated with freshman GPA.

15.92 For the regression equation obtained in Exercise 15.91, analyze the residuals by (a) constructing a histogram, (b) using a normal probability plot, (c) plotting the residuals versus the x values, and (d) plotting the residuals versus the order in which they were observed. Do any of the assumptions of the simple linear regression model appear to have been violated?

15.93 The National Association of Realtors has reported data on housing affordability for 1990–1999, and values for two of the variables are listed in file **XR15093**. Representing affordability is y = average monthly mortgage payment as a percentage of median household income. Representing the cost of a mortgage is x = the average mortgage rate, as a percentage. Source: *World Almanac and Book of Facts 2001*, p. 743.

a. Determine the least-squares regression equation for y as a function of x, then estimate y if the average mortgage rate is $x = 8.0\%$.

b. What percentage of the variation in affordability (y) is explained by the average mortgage rate (x)? At the 0.05 level, does the slope of the equation differ significantly from 0?

c. Given $x = 8.0\%$, construct and interpret the 95% confidence and prediction intervals associated with y.

15.94 For the regression equation obtained in Exercise 15.93, analyze the residuals by (a) constructing a histogram, (b) using a normal probability plot, (c) plotting the residuals versus the x values, and (d) plotting the residuals versus the order in which they were observed. Do any of the assumptions of the simple linear regression model appear to have been violated?

15.95 Computer database **GROWCO** describes the characteristics of 100 companies identified by *Fortune* as among the fastest growing. Two of the variables listed are (1) estimated stock price/earnings ratio next year and (2) annual revenue growth (as a percentage). Source: "*Fortune's* 100 Fastest-Growing Companies," *Fortune*, September 4, 2000, pp. 142–158.

a. Determine the least-squares regression equation for y = next year's price/earning ratio estimate as a function of x = annual revenue growth. What would be the predicted value of y for a company whose annual revenue growth has been 150%?

b. What percentage of the variation in the magazine's price/earnings estimates (y) is explained by the annual revenue growth percentage (x)? At the 0.05 level, does the slope of the equation differ significantly from 0?

c. Given $x = 150\%$, construct and interpret the 95% confidence and prediction intervals associated with y.

15.96 For the regression equation obtained in Exercise 15.95, analyze the residuals by (a) constructing a histogram, (b) using a normal probability plot, (c) plotting the residuals versus the x values, and (d) plotting the residuals versus the order in which they were observed. Do any of the assumptions of the simple linear regression model appear to have been violated?

INTEGRATED CASES

Thorndike Sports Equipment

Ted Thorndike's friend Mary Stuart teaches a statistics course at the local university, and she has asked Ted to stop in and talk to the class about statistics as it relates to sports. Among the topics that Ted thinks might be interesting is a discussion of how Olympic swimming performances for both men and women have improved over the years.

According to Ted, swimmers and other athletes have been getting bigger, faster, and stronger. In addition, training technologies and equipment have improved by leaps and bounds. Among the data that Ted has collected are the winning times for men and women in the 400-meter freestyle swimming event for Olympics held from 1924 through 2008. The times are shown here and are also listed in data file **THORN15**. Source: *Time Almanac 2009*, pp. 841, 842.

1. Using the women's winning time as the dependent variable and year as the independent variable, determine and interpret both the regression equation and the coefficient of correlation. For the year 2016, determine the point estimate and the 95% prediction interval for the winning time in the women's 400-meter freestyle.
2. Repeat question 1, using the men's winning time as the dependent variable.

400-Meter Freestyle Swimming, Winning Time for

Year	Women	Men
1924	6.037 minutes	5.070 minutes
1928	5.713	5.027
1932	5.475	4.807
1936	5.440	4.742
1948	5.297	4.683
1952	5.202	4.512
1956	4.910	4.455
1960	4.843	4.305
1964	4.722	4.203
1968	4.530	4.150
1972	4.324	4.005
1976	4.165	3.866
1980	4.146	3.855
1984	4.118	3.854
1988	4.064	3.783
1992	4.120	3.750
1996	4.121	3.800
2000	4.097	3.676
2004	4.089	3.718
2008	4.054	3.698

3. Using the women's winning time as the dependent variable and the men's winning time as the independent variable, determine and interpret both the regression equation and the coefficient of correlation. If the men's winning time were 3.600 minutes in a given Olympics, determine the point estimate and the 95% prediction interval for the winning time in the women's event.
4. For the regression equation obtained in question 3, use a test of your choice in examining the significance of the linear relationship between the variables.

Springdale Shopping Survey

This exercise involves the following variables from data file **SHOPPING**, the Springdale Shopping Survey:

- **Variables 7–9** Overall attitude toward the shopping area. The highest possible rating is 5. In each analysis, one of these will be the dependent variable.
- **Variables 18–25** The importance (highest rating = 7) placed on each of 8 attributes that a shopping area might possess. In each analysis, one of these will be the independent variable.

Information like that gained from this database exercise could provide management with useful insights into how respondents form their overall attitude toward a shopping area. For example, management can find out whether people who place more importance on "a lot of bargain sales" tend to have a higher perception or a lower perception of a shopping area.

1. With variable 7 (attitude toward Springdale Mall) as the dependent variable, perform two separate regression analyses—one for each of the independent variables listed below. In each case, determine and interpret the regression equation and the coefficients of correlation and determination, then use the 0.05 level of significance in reaching a conclusion on the significance of the linear relationship. If possible on your computer statistical package, retain the residuals for analysis in question 2.
   a. Variable 21 (good variety of sizes/styles).
   b. Variable 25 (a lot of bargain sales).
2. For each of the regression analyses in question 1, examine the residuals by using a histogram, a normal probability plot, and a plot of the residuals against the values of the independent variable. In each case, comment on whether an assumption of the linear regression model may have been violated.
3. Repeat questions 1 and 2, using variable 9 (attitude toward West Mall) as the dependent variable.

BUSINESS CASE

Pronto Pizza (B)

©Adam Crowley/Photodisc/Getty Images

In the Pronto Pizza case, visited previously in Chapter 10, the company was facing stiff competition in its pizza delivery service, and the owners were particularly concerned about being able to provide call-in customers with a guarantee as to how quickly their pizza would arrive. Manager Tony Scapelli had collected a month's worth of data that included such variables as preparation time, wait time, travel time, and delivery distance. In this chapter, we will use simple linear regression and correlation to help Tony examine the number of travel minutes required for a delivery versus the number of miles involved in making the delivery. Using the **PRONTO** data file and the variables TRAVEL_TIME and DISTANCE:

1. Determine the coefficient of correlation between TRAVEL_TIME and DISTANCE. Is the sign of the coefficient of correlation positive or negative? Is this the sign you would expect to be associated with this coefficient? What percentage of the variation in the driving time to deliver a pizza is explained by the number of miles for the trip?
2. Determine the best-fit linear regression equation for estimating TRAVEL_TIME on the basis of DISTANCE. Identify and interpret the slope of the equation in the context of this situation.
3. What would be the estimated time of travel for a pizza delivery that involved a 5-mile trip?
4. Determine and interpret the 95% confidence and prediction intervals associated with a pizza delivery that involved a 5-mile trip.
5. Were there any deliveries for which the time for travel was "flagged" by your computer statistical package as being unusually long or short compared to the time that would have been estimated by the equation? What managerial implications could this have, especially considering the possibility of litigation if a delivery person were to be involved in an accident while trying to meet a delivery guarantee set by the company?

SEEING STATISTICS: APPLET 18

Regression: Point Estimate for *y*

When this applet loads, we see the best-fit linear regression equation for the dexterity test example featured throughout this chapter. Moving the slider at the bottom allows us to select a dexterity test score and immediately see the estimated productivity for someone with that score.

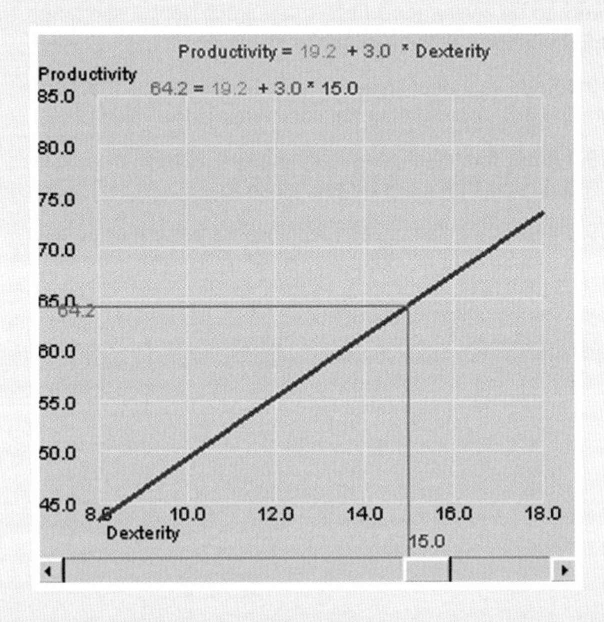

Applet Exercises

18.1 Move the slider so that the dexterity test score is 10. What is the point estimate of the productivity for someone with this score on the dexterity test?

18.2 Move the slider so that the dexterity test score is 11. What is the point estimate of the productivity for someone with this score on the dexterity test?

18.3 In Applet Exercise 18.2, we increased the dexterity test score by 1. By how much did the point estimate for productivity increase? Compare this increase with the slope of the regression equation.

18.4 By how much should the point estimate for productivity increase if we were to increase the dexterity test score by 5 points? Verify your answer by comparing the point estimates for dexterity test scores of 10 and 15.

Try out the Applets: http://www.cengage.com/international

SEEING STATISTICS: APPLET 19

Point Insertion Diagram and Correlation

This applet loads with a blank grid. By clicking the mouse, we can insert points anywhere within the grid. For the points we have chosen, the applet automatically updates the coefficient of correlation and the best-fit regression line. To start over again with a blank grid, just click on the "Reset" button.

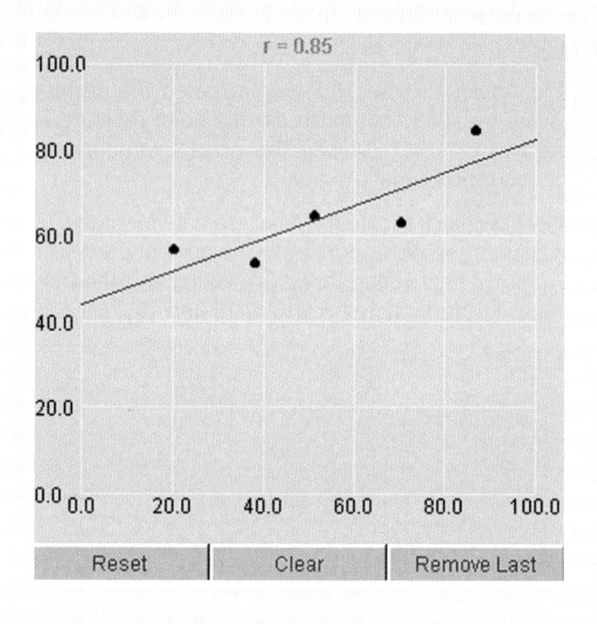

Applet Exercises

19.1 Insert any three points so that the best-fit line will slope upward. What is the sign of r?

19.2 Reset to a clear screen and insert a pattern so that r will be between $+0.7$ and $+0.9$. Does the regression line seem to be a very good fit to the points? Next, insert some additional points that will make $r < 0.7$.

19.3 Reset to a clear screen and insert two points that form an upward-sloping diagonal line. What is the value of r?

19.4 Reset to a clear screen and insert a pattern of at least 10 points so that r will be very close to -1.

19.5 Reset to a clear screen and insert a pattern of at least 10 points so that r will be very close to 0.

Try out the Applets: http://www.cengage.com/international

SEEING STATISTICS: APPLET 20

Regression Error Components

This applet loads with a scatter diagram containing the points in the dexterity test example used in this chapter. By clicking on the line, we can make it better or worse as an estimator of productivity. For each data point, the error is the vertical distance between that point and the line we have selected. In the display on the left, the sum of the squared errors is shown at the bottom and the value of r^2 is shown at the top. They are represented by the green and red rectangles, respectively. By clicking the "Find Best Model" button, we obtain the best-fit regression equation for the data.

Applet Exercises

20.1 Click on the regression line and move it so that it is as close as possible to horizontal. Identify the value of r^2 and interpret its meaning.

20.2 Click on the regression line and move it so that the slope of the equation is approximately 1.0. What effect has this had on r^2?

20.3 Click on the regression line and gradually move it so that the slope of the equation increases from approximately 1.0 to approximately 5.0. What happens to r^2 as you are shifting the line?

20.4 Click on the "Find the Best Model" button, then identify and interpret both the slope of the equation and the resulting r^2 value in the context of the dexterity test example in the text.

Try out the Applets: http://www.cengage.com/international

CHAPTER 16

Multiple Regression and Correlation

ACCIDENTS DON'T TAKE HOLIDAYS OFF

Statistics often represent positive events and pleasant experiences. For example, 44.0 million people visited Walt Disney World in 2007, and Americans bought 7.6 million new cars the same year.

On the other hand, it's often necessary to use statistics to keep track of things that are not so positive. A few examples from 2006: the 119 million visits to hospital emergency rooms (leading reason: stomach and abdominal pain), the 21,200 who died as the result of falls, and the 44,700 who perished in automobile accidents.

A disproportionate share of the auto accident fatalities tend to occur during major holiday periods, such as Thanksgiving, Christmas, and New Year's. Using multiple regression, an important technique covered in this chapter, we find a strong relationship exists between (1) the number of fatalities that occur during New Year's and (2) the numbers that occurred during the preceding Thanksgiving and Christmas periods.

As we'll examine more closely in Statistics in Action 16.1, a multiple regression analysis of holiday traffic fatalities for a 23–year period revealed that nearly 80% of the variation in New Year's fatalities was explained by a linear equation linking New Year's fatalities (y) with the numbers occurring during the preceding Thanksgiving and Christmas (x_1 and x_2, respectively). Although driving during all holiday periods tends to be a little more dangerous than everyday driving, holidays during some years would appear to be worse than holidays in others. If you're reading this either before or during a major holiday, be careful out there.

Sources: *TEA/ERA Attraction Attendance 2007*, p. vii, via parkworld-online.com; and *World Almanac and Book of Facts 2009*, pp. 122, 183, 216.

Some outcomes depend on many variables.

- Explain how the scatter diagram and least-squares concepts apply to multiple regression.
- Obtain and interpret the multiple regression equation; then make point and interval estimates regarding the dependent variable.
- Interpret the value of the coefficient of multiple determination and carry out a hypothesis test for its significance.
- Construct confidence intervals and carry out hypothesis tests involving the partial regression coefficients.
- Explain both the meaning and the applicability of a dummy variable.
- Use residual analysis in examining the appropriateness of the multiple regression model and the extent to which underlying assumptions are met.

LEARNING OBJECTIVES

After reading this chapter, you should be able to:

INTRODUCTION (16.1)

In Chapter 15, we examined the linear relationship between a dependent variable (y) and just one independent variable (x). More realistically, the value of y will be related to two or more independent variables (x_1, x_2, and so on). The inclusion of more than one independent variable is the reason for the *multiple* descriptor in the chapter title. **Multiple regression** and **correlation** analysis could prove useful with variables like these:

Dependent Variable (y)	Independent Variables (x's)
Annual unit sales	Advertising expenditure
	Size of sales force
	Price of the company's product
Selling price of a home	Number of rooms
	Square feet of living space
	Size of the surrounding lot

The purpose of this chapter is to determine, interpret, and examine the strength of linear relationships between a single y and two or more x's. As in Chapter 15, regression analysis is used in determining and interpreting the linear relationship, while correlation analysis measures its strength.

In previous chapters, the basics of a technique were presented through small-scale problems for which relevant calculations could easily be performed with the pocket calculator. The calculations in multiple regression and correlation analysis are especially intensive. Computer statistical packages are readily available for reducing the number-crunching drudgery, and the computer will be our partner through much of the chapter.

We'll begin by discussing the multiple regression model and examining the sample regression equation for an example that will continue throughout the chapter. The complete Excel and Minitab printouts will be presented and interpreted; then selected segments of these printouts will be featured as they become relevant to the topics under discussion. Later in the chapter, we'll revisit the full printouts for an overview of their format and interpretation. Once again, for multiple regression and correlation, the term "computer assisted" is a distinct understatement. It's like saying a jet pilot is "aircraft assisted."

As in the preceding chapter, point and interval estimates for y are discussed, as are tests of significance involving the regression and correlation analysis itself. The basic principles are similar to their counterparts from Chapter 15. However, specific topics and methods of analysis and interpretation will differ slightly in order to accommodate the greater number of independent variables.

EXERCISES

16.1 What is the difference between simple linear regression and multiple regression? Under what circumstances would a multiple regression analysis be preferable?

16.2 In multiple regression and correlation analysis, what is the purpose of the regression component? Of the correlation component?

16.3 For y = annual household expenditure for auto maintenance and repair, what independent variables could help explain the amount of money a household spends per year for this purpose?

16.4 A personnel director is using multiple regression analysis to examine employee absence records. He wants to find an estimation equation that will improve his understanding of variables that might explain the extent of such absences. For y = the number of days an employee called in sick last year, what x variables might be both useful to this analysis and likely to be available in each employee's personnel file?

16.2 THE MULTIPLE REGRESSION MODEL

Model and Assumptions

The multiple regression model is an extension of the simple linear regression model of Chapter 15. However, there are *two or more independent variables* instead of just one. As in Chapter 15, estimates of the population parameters in the model are made on the basis of sample data. Described in terms of its population parameters, the multiple regression model is as follows:

The multiple regression model:

$$y_i = \beta_0 + \beta_1 x_{1i} + \beta_2 x_{2i} + \cdots + \beta_k x_{ki} + \epsilon_i$$

where y_i = a value of the dependent variable, y

β_0 = a constant

$x_{1i}, x_{2i}, \ldots, x_{ki}$ = values of the independent variables, $x_1, x_2, \ldots, x_k$

$\beta_1, \beta_2, \ldots, \beta_k$ = partial regression coefficients for the independent variables, $x_1, x_2, \ldots, x_k$

ϵ_i = random error, or residual

For a given set of x values, the expected value of y is given by the regression equation, $E(y) = \beta_0 + \beta_1 x_{1i} + \beta_2 x_{2i} + \cdots + \beta_k x_{ki}$. For each x term in the equation, the corresponding β term is referred to as the **partial regression coefficient.** Each β_i ($i = 1, 2, \ldots, k$) is a slope relating changes in $E(y)$ to changes in one x variable whenever all of the other x's are held constant.

As in Chapter 15, the difference between an observed value of y and the value that was expected is the residual, ϵ_i. In terms of the residual component of the model, the following assumptions underlie multiple regression:

1. For any given set of values for the independent variables, the population of ϵ_i values will be normally distributed with a mean of zero and a standard deviation of σ.
2. The standard deviation of the ϵ_i values is the same regardless of the combination of values taken on by the independent variables.
3. The ϵ_i values are statistically independent of each other.

Determining the Sample Regression Equation

The sample regression equation is based on observed values for the dependent and independent variables. An extension of its counterpart from Chapter 15, it has the form $\hat{y} = b_0 + b_1x_1 + b_2x_2 + \cdots + b_kx_k$. The constants in the sample regression equation ($b_0, b_1, b_2, \ldots, b_k$) are estimates of their population counterparts ($\beta_0, \beta_1, \beta_2, \ldots, \beta_k$).

Determination of the "best-fit" multiple regression equation is according to the least-squares criterion, in which the sum of the squared deviations between observed and estimated values of y is minimized. The following information, and analyses related to it, will be used as our medium for discussing multiple regression and correlation in the sections to follow.

EXAMPLE

Multiple Regression Equation

The president of a large chain of fast-food restaurants has randomly selected 10 franchises and recorded for each franchise the following information on last year's net profit and sales activity. The data are also in file CX16REST.

Franchise Number	Net Profit, y	Counter Sales, x_1	Drive-Through Sales, x_2
1	$1.5 million	$8.4 million	$7.7 million
2	0.8	3.3	4.5
3	1.2	5.8	8.4
4	1.4	10.0	7.8
5	0.2	4.7	2.4
6	0.8	7.7	4.8
7	0.6	4.5	2.5
8	1.3	8.6	3.4
9	0.4	5.9	2.0
10	0.6	6.3	4.1

For these data, there will be one dependent variable (y = net profit) and two independent variables (x_1 = counter sales; x_2 = drive-through sales). The form of the sample regression equation will be $\hat{y} = b_0 + b_1x_1 + b_2x_2$.

TABLE 16.1

These portions of the Excel and Minitab printouts for the fast-food franchise analysis show the y-intercept and the partial regression coefficients for the least-squares multiple regression equation. The full printouts and the procedures for obtaining them are shown at the end of this section, in Computer Solutions 16.1.

	E	F
16		*Coefficients*
17	Intercept	-0.2159
18	$Counter	0.0855
19	$Driveup	0.1132

```
The regression equation is

$Profit = - 0.216 + 0.0855 $Counter + 0.113 $Driveup
```

SOLUTION

Table 16.1 shows Excel and Minitab printout segments describing the multiple regression equation for the restaurant data. The full printouts and the steps to obtain them are given in Computer Solutions 16.1, at the end of this section. In later sections of the chapter, we will examine other portions of the printouts. At this stage, our primary concern is the sample regression equation itself. To four decimal places, it is

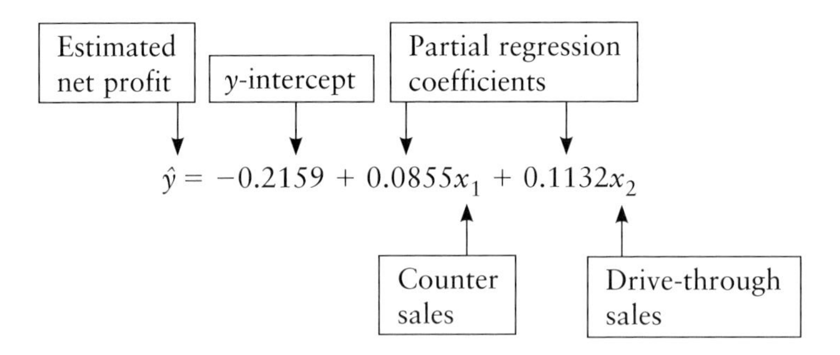

This least-squares regression equation is actually a two-dimensional plane, and for each possible value of x_1 and x_2, the estimated value of y lies on the surface of the plane. Imagine the data points in Figure 16.1 as helium-filled balloons in a room, with each balloon fastened to the floor at its particular (x_1, x_2) location. The length of the string holding each balloon represents the observed y value for that individual data point. The height where each string would intersect the regression plane is the estimated y value for that combination of x_1 and x_2 values. Some of the balloons (data points) will be above the plane, others below it. As in Chapter 15, the distances between observed and predicted positions are residuals, or errors of estimation. Because the least-squares criterion is also used in multiple regression, the sum of the squared residuals will be minimized.

When there are only two variables (y and x), the least-squares equation is a straight line. With (y, x_1, and x_2), it is a plane fitted to a three-dimensional scatter diagram like Figure 16.1. However, when we have more than three variables, there is no counterpart to the scatter diagram. With four or more variables, the regression equation becomes a mathematical entity called a **hyperplane.** Scatter diagrams in three dimensions are difficult to draw; with four or more dimensions, such visual summarization becomes impossible.

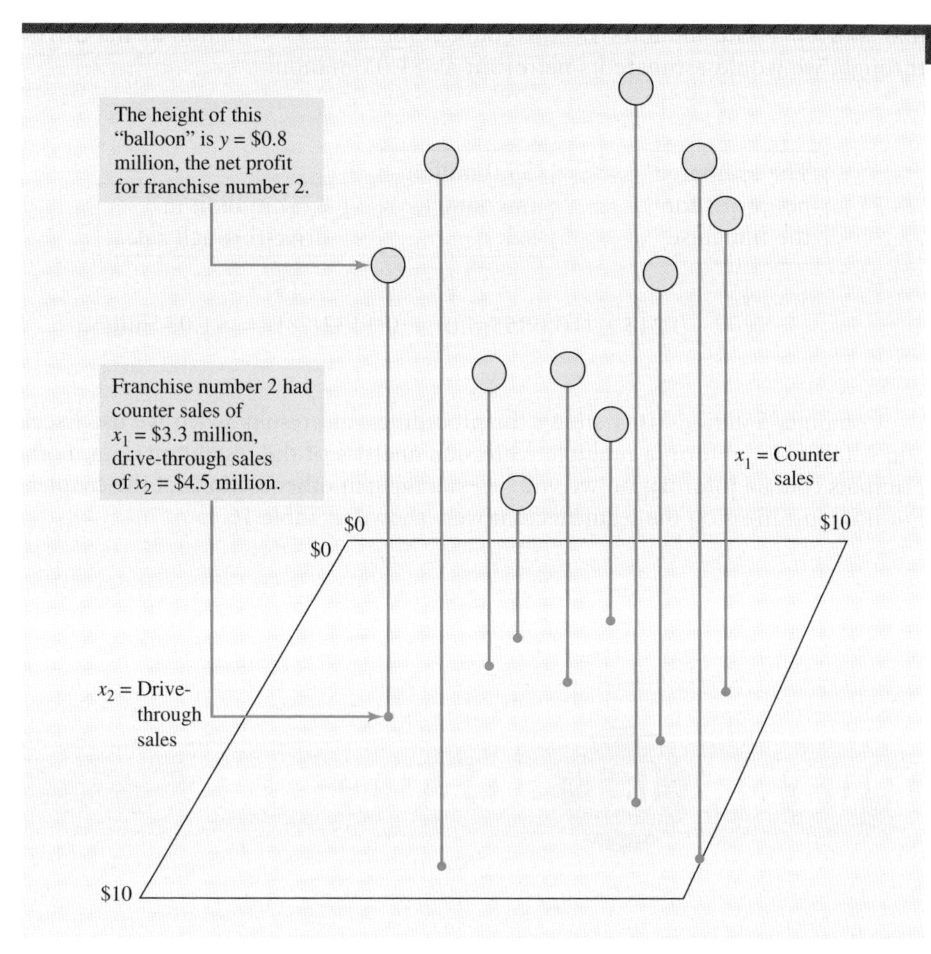

FIGURE 16.1

The scatter diagram for the fast-food data can be visualized as a room where each of 10 helium-filled balloons is held to the floor by a string. The length of each string is the observed value of *y* for that data point. The least-squares regression equation (not shown) passes through the data and takes the form of a two-dimensional surface, or plane.

Interpreting the Regression Equation

The *y*-intercept ($b_0 = -0.2159$) in the fast-food regression equation is the estimated value of *y* (net profit) whenever x_1 (counter sales) and x_2 (drive-through sales) are both zero. In other words, a franchise with no customers would be estimated as losing $215,900 for the year. (*Caution:* None of the 10 franchises in the study even approaches having zero sales in both departments, and we must be skeptical of estimates based on *x* values that lie beyond the range of the underlying data.)

The partial regression coefficient for x_1 ($b_1 = 0.0855$) indicates that, for a given level of drive-through sales, the estimated net profit will increase by $0.0855 for each additional $1.00 of counter sales.

The partial regression coefficient for x_2 ($b_2 = 0.1132$) shows that, for a given level of counter sales, the estimated net profit will increase by $0.1132 for each additional $1.00 of drive-through sales. Both partial regression coefficients are positive, reflecting that estimated net profit will increase regardless of whether an additional $1.00 of sales happens to occur over the counter or at the drive-through window.

Point Estimates Using the Regression Equation

Point estimates are made by substituting a set of *x* values into the regression equation and calculating the estimated value of *y*. For example, if a franchise had sold x_1 = \$5.0 million over the counter and x_2 = \$7.4 million at the drive-through, we would estimate its net profit as \$1.05 million:

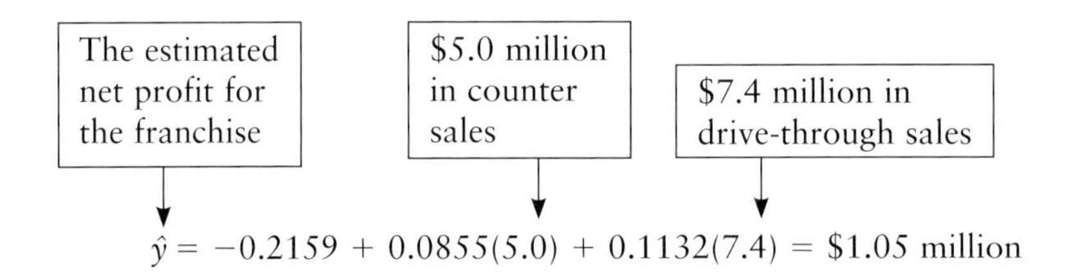

$$\hat{y} = -0.2159 + 0.0855(5.0) + 0.1132(7.4) = \$1.05 \text{ million}$$

Computer Solutions 16.1 shows the procedures and results when we use Excel and Minitab to perform a multiple regression analysis of the restaurant data. Both printouts contain information we will be examining in other sections of the chapter. For now, just focus on the segments that were shown in Table 16.1.

COMPUTER 16.1 SOLUTIONS

Multiple Regression

EXCEL

	E	F	G	H	I	J	K
1	SUMMARY OUTPUT						
2							
3	*Regression Statistics*						
4	Multiple R	0.8786					
5	R Square	0.7719					
6	Adjusted R Square	0.7067					
7	Standard Error	0.2419					
8	Observations	10					
9							
10	ANOVA						
11		*df*	*SS*	*MS*	*F*	*Significance F*	
12	Regression	2	1.3864	0.6932	11.8448	0.0057	
13	Residual	7	0.4096	0.0585			
14	Total	9	1.7960				
15							
16		*Coefficients*	*Standard Error*	*t Stat*	*P-value*	*Lower 95%*	*Upper 95%*
17	Intercept	-0.2159	0.2643	-0.8169	0.4409	-0.8409	0.4091
18	\$Counter	0.0855	0.0438	1.9516	0.0920	-0.0181	0.1890
19	\$Driveup	0.1132	0.0385	2.9368	0.0218	0.0220	0.2043

Excel Multiple Regression

1. The data for the primary example in this chapter are in Excel file **CX16REST**. The profits ($Profit) are in A2:A11, the counter sales ($Counter) are in B2:B11, the drive-through sales ($Driveup) are in C2:C11, and the labels are in the first rows. Note that the independent variables must be in adjacent columns. From the **Data** ribbon, click **Data Analysis.** Select **Regression.** Click **OK.**
2. Enter **A1:A11** into the **Input Y Range** box. Enter **B1:C11** into the **Input X Range** box. Select **Labels.** Enter **E1** into the **Output Range** box. Click **OK.** After slight adjustments to the column widths and the number of decimal places, the results are as shown in the printout.

MINITAB

```
Regression Analysis: $Profit versus $Counter, $Driveup

The regression equation is
$Profit = - 0.216 + 0.0855 $Counter + 0.113 $Driveup

Predictor      Coef   SE Coef      T      P
Constant    -0.2159    0.2643  -0.82  0.441
$Counter    0.08547   0.04380   1.95  0.092
$Driveup    0.11315   0.03853   2.94  0.022

S = 0.241912   R-Sq = 77.2%   R-Sq(adj) = 70.7%

Analysis of Variance
Source            DF       SS        MS      F      P
Regression         2  1.38635   0.69318  11.84  0.006
Residual Error     7  0.40965   0.05852
Total              9  1.79600

Source     DF   Seq SS
$Counter    1  0.88160
$Driveup    1  0.50475

Unusual Observations
Obs  $Counter  $Profit     Fit  SE Fit  Residual  St Resid
  8       8.6   1.3000  0.9039  0.1468    0.3961     2.06R

R denotes an observation with a large standardized residual.
```

Using Minitab file **CX16REST**, with $Profit in C1, $Counter in C2, and $Driveup in C3: Click **Stat.** Select **Regression.** Click **Regression.** Enter **C1** ($Profit) into the **Response** box. Enter **C2-C3** ($Counter and $Driveup) into the **Predictors** box. Click **OK.**

EXERCISES

16.5 Explain each of the terms in the multiple regression model.

16.6 What assumptions are required in using the multiple regression model?

16.7 In simple linear regression, the regression equation is a straight line. In multiple regression, what geometric form is taken by the regression equation when there are two independent variables? When there are three or more independent variables?

16.8 For the multiple regression equation $\hat{y} = 100 + 20x_1 - 3x_2 + 120x_3$:

a. Identify the y-intercept and partial regression coefficients.

b. If $x_1 = 12$, $x_2 = 5$, and $x_3 = 10$, what is the estimated value of y?
c. If x_3 were to increase by 4, what change would be necessary in x_2 in order for the estimated value of y to remain unchanged?

16.9 Last year's utility bill (in dollars) for a sample of homes is estimated by the multiple regression equation $\hat{y} = 300 + 7x_1 + 13x_2$, where x_1 is the number of occupants and x_2 is the number of rooms in the home.
a. Identify the y-intercept and partial regression coefficients.
b. If 3 people live in a 6-room home, what is the estimated bill?

16.10 A study describes operating costs for comparable manufacturing plants operating in 15 different U.S. metropolitan areas. From the costs obtained in the study, the following multiple regression equation was developed: $\hat{y} = 1.0 + 1.1x_1 + 2.8x_2$, with $\hat{y}$ = total operating cost, x_1 = labor cost, and x_2 = electric power cost. All costs are in millions of dollars and on an annual basis.
a. Interpret the y-intercept and the partial regression coefficients.
b. If labor costs $6 million and electric power costs $0.3 million, what is the estimated annual cost to operate the plant?

(**DATA SET**) *Note:* Exercises 16.11–16.14 require a computer and statistical software.

16.11 The owner of a large chain of health spas has selected eight of her smaller clubs for a test in which she varies the size of the newspaper ad and the amount of the initiation fee discount to see how this might affect the number of prospective members who visit each club during the following week. The results are shown in the table below.

Club	New Visitors, y	Ad Column-Inches, x_1	Discount Amount, x_2
1	23	4	$100
2	30	7	20
3	20	3	40
4	26	6	25
5	20	2	50
6	18	5	30
7	17	4	25
8	31	8	80

a. Determine the least-squares multiple regression equation.
b. Interpret the y-intercept and partial regression coefficients.
c. What is the estimated number of new visitors to a club if the size of the ad is 5 column-inches and a $75 discount is offered?

16.12 In testing 9 sedans, an automotive publication rated each on 13 different characteristics, including ride, handling, and driver comfort. Each vehicle also received an overall rating. Scores for each vehicle were as follows.

Car	y = Overall Rating	x_1 = Ride	x_2 = Handling	x_3 = Driver Comfort
1	85	8	7	7
2	88	8	8	8
3	85	6	8	7
4	85	8	7	9
5	97	9	9	9
6	86	8	8	9
7	90	9	6	9
8	84	7	8	7
9	94	8	9	8

a. Determine the least-squares multiple regression equation.
b. Interpret the y-intercept and partial regression coefficients.
c. What is the estimated overall rating for a vehicle that scores 6 on ride, 9 on handling, and 7 on driver comfort?

16.13 The makers of a pie crust mix have baked test pies and collected the following data on y = amount of crispness (a rating of 1 to 100), x_1 = minutes in the oven, and x_2 = oven temperature (degrees Fahrenheit).

y	x_1	x_2
68	6.0	460
76	8.9	430
49	8.8	360
99	7.8	460
90	7.3	390
32	5.3	360
96	8.8	420
77	9.0	350
94	8.0	450
82	8.2	400
97	6.4	450

a. Determine the least-squares multiple regression equation.
b. Interpret the y-intercept and partial regression coefficients.
c. What is the estimated crispness rating for a pie that is baked for 5 minutes at 200 degrees?

16.14 The following data have been reported for a sample of 10 major U.S. zoological parks:

City	y = Budget	x_1 = Attendance	x_2 = Acres	x_3 = Number of Species
1	$19.5 million	0.6 million	210	271
2	40.0	2.0	216	400
3	11.9	0.4	70	377
4	14.0	1.0	125	277
5	11.6	1.5	55	721
6	22.2	1.3	80	400
7	20.5	1.3	42	437
8	26.0	2.5	91	759
9	17.0	0.9	125	270
10	14.6	1.1	92	260

a. Determine the least-squares multiple regression equation.
b. Interpret the y-intercept and partial regression coefficients.
c. What is the estimated budget for a zoological park that draws an annual attendance of 2.0 million, occupies 150 acres, and has 600 species?

16.15 During its manufacture, a product is subjected to four different tests, each giving a score of 200 to 800. An efficiency expert claims the fourth test to be unnecessary since its results can be predicted based on the first three. The following Minitab printout is an analysis of scores received by a sample of 12 units subjected to all four tests.

```
Regression Analysis

The regression equation is
TEST04 = 12.0 + 0.274 TEST01 + 0.376 TEST02 + 0.326 TEST03

Predictor        Coef       StDev          T        P
Constant        11.98       80.50       0.15    0.885
TEST01         0.2745      0.1111       2.47    0.039
TEST02        0.37619     0.09858       3.82    0.005
TEST03        0.32648     0.08084       4.04    0.004

S = 52.72        R-Sq = 87.2%       R-Sq(adj) = 82.4%

Analysis of Variance

Source        DF         SS          MS          F         P
Regression     3     151417       50472      18.16     0.001
Error          8      22231        2779
Total         11     173648
```

a. Interpret the multiple regression equation.
b. An individual unit from the assembly area receives ratings of 350, 400, and 600 on the first three tests. Using the estimation equation in the printout, what is the unit's estimated score on test 4?

INTERVAL ESTIMATION IN MULTIPLE REGRESSION (16.3)

Given a specific combination of x values, the multiple regression equation provides our single best estimate for both the mean of y and an individual y value for that set of x values. However, as in simple linear regression, the estimate is subject to uncertainty. The first step in quantifying this uncertainty is to determine the multiple standard error of estimate.

The Multiple Standard Error of Estimate

The **multiple standard error of estimate (s_e)** is analogous to the standard error of estimate in Chapter 15, and it will be part of the output when the regression analysis is computer assisted. The numerical value of s_e reflects the amount of scatter, or dispersion, of data points about the plane (or hyperplane) represented by the multiple regression equation. For two independent variables, it describes the amount of scatter above and below the two-dimensional surface passing through data points like those shown in Figure 16.1. Regardless of the number of independent variables, the multiple standard error of estimate can be calculated as

The multiple standard error of estimate:

$$s_e = \sqrt{\frac{\Sigma(y_i - \hat{y}_i)^2}{n - k - 1}}$$

where y_i = each observed value of y in the data
$\hat{y}_i$ = the value of y that would have been estimated from the regression equation
n = the number of data points
k = the number of independent (x) variables

TABLE 16.2

For the sample of 10 fast-food franchises, the summary results from these calculations are important in determining both the multiple standard error of estimate and the coefficients of multiple correlation and determination. (Data are in millions of dollars.)

Counter Sales, x_1	Drive-Through Sales, x_2	Net Profit: Actual, y	Net Profit: Predicted, $\hat{y}$	Error, or Residual, $(y - \hat{y})$	Error, or Residual, Squared, $(y - \hat{y})^2$	Actual y versus the Mean of y: $(y - \bar{y})$	Actual y versus the Mean of y: $(y - \bar{y})^2$
$8.4	$7.7	$1.5	$1.3739	$0.1261	0.0159	$0.62	0.3844
3.3	4.5	0.8	0.5757	0.2243	0.0503	−0.08	0.0064
5.8	8.4	1.2	1.2309	−0.0309	0.0010	0.32	0.1024
10.0	7.8	1.4	1.5221	−0.1221	0.0149	0.52	0.2704
4.7	2.4	0.2	0.4576	−0.2576	0.0664	−0.68	0.4624
7.7	4.8	0.8	0.9858	−0.1858	0.0345	−0.08	0.0064
4.5	2.5	0.6	0.4519	0.1481	0.0219	−0.28	0.0784
8.6	3.4	1.3	0.9043	0.3957	0.1566	0.42	0.1764
5.9	2.0	0.4	0.5150	−0.1150	0.0132	−0.48	0.2304
6.3	4.1	0.6	0.7869	−0.1869	0.0349	−0.28	0.0784
		$\bar{y}$ = $0.88			0.4096		1.7960

$\Sigma(y - \hat{y})^2$; this is the residual (error) sum of squares, or *SSE*. It is the amount of variation in y that is *not* explained by the multiple regression equation.

$\Sigma(y - \bar{y})^2$; this is the total sum of squares, or *SST*. It is the *total* amount of variation in y, both explained and unexplained.

The numerator of the term beneath the square root symbol is the sum of the squared residuals, also known as the error sum of squares, or *SSE*. For the fast-food example and its multiple regression equation, the calculation of the error sum of squares (*SSE*) is demonstrated in Table 16.2. Each row of the table represents one of the franchises in the sample and compares the actual net profit (y) with the predicted value ($\hat{y}$). When the resulting residuals (errors) are squared and totaled, the error sum of squares is $SSE = 0.4096$.

For our example, there are $k = 2$ independent variables and $n = 10$ data points. Having calculated $\Sigma(y_i - \hat{y}_i)^2$, the error sum of squares, as $SSE = 0.4096$, the multiple standard error of estimate can be calculated as

$$s_e = \sqrt{\frac{\Sigma(y_i - \hat{y}_i)^2}{n - k - 1}} = \sqrt{\frac{0.4096}{10 - 2 - 1}} = \sqrt{0.0585} = \$0.2419 \text{ million}$$

Just as b_0, b_1, and b_2 are sample estimates of their counterparts in the "true" population regression equation, $E(y) = \beta_0 + \beta_1 x_1 + \beta_2 x_2$, s_e is our sample estimate of the actual multiple standard error of estimate, σ_e.

Confidence Interval for the Conditional Mean of *y*

In Chapter 15, the least-squares regression line passed through the point $(\bar{x}, \bar{y})$, and estimation intervals were shown to become wider when based on x values more distant from $\bar{x}$. The estimation formulas in Chapter 15 took into account the difference between x and $\bar{x}$, and increased the interval widths accordingly as x values got farther away from the mean of x.

As in simple regression, the least-squares multiple regression equation passes through the point containing the means of all the variables. (To verify this, substitute the means [$\bar{y}$ = \$0.88, $\bar{x}_1$ = \$6.52, and $\bar{x}_2$ = \$4.76] into the fast-food multiple regression equation.) Likewise, interval estimates become wider as the specified x values get farther away from their respective means.

Most computer statistical packages increase interval widths according to how far away the various x values are from their means. However, if your package does not have this capability, such calculations are nearly impossible to carry out by hand. Also, you may have only summary information (e.g., n, s_e, and the regression equation itself) from an analysis performed by someone else. In either case, all is not lost. You can still construct an *approximate* estimation interval using the methods in this section and the next.

Approximate confidence interval for the conditional mean of *y*:

$$\hat{y} \pm t\frac{s_e}{\sqrt{n}}$$

where $\hat{y}$ = the estimated value of y based on the set of x values provided

t = t value from the t distribution table for the desired confidence level and $df = n - k - 1$ (n = number of data points, k = number of x variables)

s_e = the multiple standard error of estimate

EXAMPLE

Confidence Interval

For franchises having counter sales of \$6.0 million and drive-through sales of \$4.0 million, what is the approximate 95% confidence interval for their average net profit?

SOLUTION

For x_1 = \$6.0 and x_2 = \$4.0, $\hat{y} = -0.2159 + 0.0855(6.0) + 0.1132(4.0)$, or \$0.7499 million. This is the midpoint of the approximate confidence interval for the mean of y for these x values. As in the previous example, the number of degrees of freedom will be $n - k - 1$, or $df = 7$. The multiple standard error of estimate is s_e = \$0.2419 million, and the t value for 95% confidence and $df = 7$ will be 2.365. The approximate confidence interval for the mean of y is

$$\hat{y} \pm t\frac{s_e}{\sqrt{n}} = 0.7499 \pm 2.365\frac{0.2419}{\sqrt{10}} = 0.7499 \pm 0.1809,$$

or from \$0.5690 to \$0.9308 million

We have 95% confidence that the average net profit for stores having \$6.0 million in counter sales and \$4.0 million in drive-through sales will be within the approximate confidence interval bounded by \$0.5690 and \$0.9308 million. The width of this interval is 0.9308 − 0.5690, or \$0.3618 million. Because the interval is based on x values that differ from their respective means, the exact 95% confidence interval will be wider than the \$0.3618 million calculated here.

EXAMPLEEXAMPLEEXAMPLEEXAMPLEEXAM

Prediction Interval for an Individual y Observation

As with the confidence interval, we can construct an approximate prediction interval for an individual y observation. Such intervals can be approximated by using the t distribution, the multiple standard error of estimate, and $\hat{y}$ as the point estimate for y.

Approximate prediction interval for an individual y observation:

$\hat{y} \pm ts_e$ where $\hat{y}$ = the estimated value of y based on the set of x values provided

t = t value from the t distribution table for the desired confidence level and $df = n - k - 1$

(n = number of data points, k = number of x variables)

s_e = the multiple standard error of estimate

EXAMPLE

Prediction Interval

The president of the fast-food chain believes that a new franchise being planned for construction in a suburban neighborhood would have experienced \$6.0 million in counter sales and \$4.0 million in drive-through sales had it been in operation during the preceding year. For these sales figures, what would be the approximate 95% prediction interval for the store's net profit?

SOLUTION

The midpoint of the approximate prediction interval will be the point estimate of net profit for the counter sales (x_1 = \$6.0 million) and drive-through sales (x_2 = \$4.0 million) specified. The point estimate is

$$\hat{y} = -0.2159 + 0.0855(6.0) + 0.1132(4.0) = \$0.7499 \text{ million}$$

For $n = 10$ data points and $k = 2$ independent variables, the number of degrees of freedom associated with t will be $df = n - k - 1$, or $df = 7$. Referring to the t distribution table, $t = \pm 2.365$ includes 95% of the area beneath the curve. The approximate 95% prediction interval is

$$\hat{y} \pm ts_e = 0.7499 \pm 2.365(0.2149) = 0.7499 \pm 0.5721,$$
$$\text{or from } \$0.1778 \text{ to } \$1.3220 \text{ million}$$

We have 95% confidence that this store's net profit would have been within the approximate prediction interval bounded by \$0.1778 and \$1.3220 million, an interval width of 1.3220 − 0.1778, or \$1.1442 million. Since $\bar{x}_1$ = \$6.52 million and $\bar{x}_2$ = \$4.76 million (as noted earlier), this approximate prediction interval is based on x values that differ from their respective means. As such, the exact prediction interval will be slightly wider than the \$1.1442 million calculated here.

Computer Solutions 16.2 shows the procedures and results when we use Excel and Minitab to generate confidence and prediction intervals like those obtained here. Because the values for x_1 and x_2 on which the intervals are based are not the sample means for the two independent variables, the exact intervals produced by Excel and Minitab are slightly wider than their more approximate counterparts that were determined using the methods described in this section.

COMPUTER 16.2 SOLUTIONS

Interval Estimation in Multiple Regression

EXCEL

	A	B	C	D
1	**Prediction Interval**			
2				
3			$Profit	
4	Predicted value		0.750	
5				
6	Prediction Interval			
7	Lower limit		0.146	
8	Upper limit		1.353	
9				
10	Interval Estimate of Expected Value			
11	Lower limit		0.558	
12	Upper limit		0.941	

Excel Interval Estimation in Multiple Regression

1. Using Excel file **CX16REST**, with data in the fields specified in Computer Solutions 16.1, select two convenient cells (e.g., A15:B15) and enter the specified values for the two independent variables—i.e., enter **6** into **A15** and **4** into **B15.**
2. From the **Add-Ins** ribbon, click **Data Analysis Plus.** Select **Prediction Interval.** Click **OK.**
3. Enter **A1:A11** into the **Input Y Range** box. Enter **B1:C11** into the **Input X Range** box. Select **Labels.** Enter **A15:B15** into the **Given X Range** box. Enter the desired confidence level as a decimal fraction (e.g., **0.95**) into the **Confidence Level (1 – Alpha)** box. Click **OK.**

MINITAB

Minitab Interval Estimation in Multiple Regression

```
Predicted Values for New Observations

New
Obs     Fit  SE Fit        95% CI              95% PI
  1  0.7496  0.0811  (0.5578, 0.9413)  (0.1462, 1.3529)

Values of Predictors for New Observations

New
Obs  $Counter  $Driveup
  1      6.00      4.00
```

(continued)

Using Minitab file **CX16REST**, follow the procedure in Computer Solutions 16.1, but select **Options.** Enter **6 4** (include the space between them) into the **Prediction intervals for new observations** box. Enter the desired confidence level as a percentage (e.g., **95**) into the **Confidence level** box. Click **OK.** Click **OK.** In addition to the printout in Computer Solutions 16.1, Minitab will generate the items shown here.

EXERCISES

16.16 Given the printout in Exercise 16.15, determine the approximate

a. 90% confidence interval for the mean rating on test 4 for units that have been rated at 300, 500, and 400 on the first three tests.

b. 90% prediction interval for the rating a unit will get on test 4 if its ratings are 300, 500, and 400 on the first three tests.

16.17 The regression equation $\hat{y} = 5.0 + 1.0x_1 + 2.5x_2$ has been fitted to 20 data points. The means of x_1 and x_2 are 25 and 40, respectively. The sum of the squared differences between observed and predicted values of y has been calculated as $SSE = 173.5$, and the sum of squared differences between y values and the mean of y is $SST = 520.8$. Determine the following:

a. The mean of the y values in the data.

b. The multiple standard error of estimate.

c. The approximate 95% confidence interval for the mean of y whenever $x_1 = 20$ and $x_2 = 30$.

d. The approximate 95% prediction interval for an individual y value whenever $x_1 = 20$ and $x_2 = 30$.

(**DATA SET**) *Note:* Exercises 16.18–16.22 require a computer and statistical software. If possible with your statistical software, determine the exact, rather than the approximate, intervals.

16.18 In testing eight value notebook computers, *PC World* reported various physical and performance measures of each. Data included overall rating, street price, PC WorldBench 4 performance score, and battery life, with the results shown here. Source: "Top 15 Notebook PCs," *PC World*, July 2003, p. 135.

Computer:	1	2	3	4
Overall Rating:	80	80	76	75
Street Price:	1024	949	1499	1478
Performance Score:	102	96	111	104
Battery Life (hours):	3.03	2.97	2.00	3.05

Computer:	5	6	7	8
Overall Rating:	75	74	72	72
Street Price:	1857	1865	1699	1298
Performance Score:	100	98	104	91
Battery Life (hours):	1.88	3.00	2.38	2.33

a. From these data, obtain the multiple regression equation estimating overall rating on the basis of the other three variables; then determine and interpret the approximate 95% confidence interval for the mean overall rating of notebook computers having a $1000 street price, a performance score of 100, and a 2.50-hour battery life.

b. Determine the 95% prediction interval for the overall rating of an individual notebook computer with a $1000 street price, a performance score of 100, and a 2.50-hour battery life.

16.19 In considering whether to implement a program where selected entering freshmen would be required to take a precalculus refresher course before moving on to calculus, a university's business college has given mathematical proficiency tests to a sample of entering freshmen. The college hopes to use this score and the Scholastic Aptitude Test (SAT) quantitative score in predicting how well a student would end up scoring on the standardized final exam in calculus later in his or her studies. Data are available for a sample of nine entering freshmen, and their scores are shown here.

Student:	1	2	3	4	5
Math Proficiency:	72	96	68	86	70
SAT, Quantitative:	462	545	585	580	592
Calculus Final:	71	92	72	82	74

Student:	6	7	8	9
Math Proficiency:	73	91	75	76
SAT, Quantitative:	516	638	615	596
Calculus Final:	71	100	87	81

a. From these data, obtain the multiple regression equation estimating calculus final exam score on the basis of the other two variables; then determine and interpret the approximate 90% confidence interval for the mean calculus final exam score for entering freshmen who score 70 on the math proficiency test and 500 on the quantitative portion of the SAT.
b. Determine the 90% prediction interval for the calculus final exam score for an entering freshman who scored 70 on the math proficiency test and 500 on the quantitative portion of the SAT.

16.20 For the multiple regression equation obtained in Exercise 16.11, determine the approximate
a. 95% confidence interval for the mean number of new visitors for clubs using a 5-column-inch ad and offering an $80 discount.
b. 95% prediction interval for the number of new visitors to a given club when the ad is 5 column-inches and the discount is $80.

16.21 For the multiple regression equation obtained in Exercise 16.12, determine the approximate
a. 95% confidence interval for the mean overall rating of cars that receive ratings of 8 on ride, 7 on handling, and 9 on driver comfort.
b. 95% prediction interval for the overall rating of a car that has received ratings of 8 on ride, 7 on handling, and 9 on driver comfort.

16.22 For the multiple regression equation obtained in Exercise 16.13, determine the approximate
a. 95% confidence interval for the mean crispness rating of pies that are baked 5.0 minutes at 300 degrees.
b. 95% prediction interval for the crispness rating of a pie that is baked 5.0 minutes at 300 degrees.

MULTIPLE CORRELATION ANALYSIS (16.4)

The purpose of the multiple correlation analysis is to measure the strength of the relationship between the dependent (y) and the set of independent (x) variables. This strength can be measured on an overall basis by the coefficient of multiple determination.

The Coefficient of Multiple Determination

The **coefficient of multiple determination** (R^2, with $0 \le R^2 \le 1$) is the proportion of the variation in y that is explained by the multiple regression equation. Its positive square root is the *coefficient of multiple correlation* (R), a measure that is less important than its counterpart from Chapter 15.

Computation of the coefficient of multiple determination is similar to that for the coefficient of determination for simple linear regression, and R^2 can be both defined and calculated in terms of the total variation in y versus the unexplained variation in y:

The coefficient of multiple determination:

$$R^2 = 1 - \frac{\text{Error variation, which is } \textit{not} \text{ explained by the regression equation}}{\textit{Total} \text{ variation in the } y \text{ values}}$$

$$= 1 - \frac{\Sigma(y_i - \hat{y}_i)^2}{\Sigma(y_i - \bar{y})^2} \quad \text{or} \quad 1 - \frac{SSE}{SST}$$

As in Chapter 15, the total variation in y is the sum of the explained variation plus the unexplained (error) variation, or

Total variation in y values (SST)	=	Variation explained by the regression equation (SSR)	+	Variation not explained by the regression equation (SSE)

For the fast-food multiple regression equation, both *SST* (the sum of the squared deviations of *y* from its mean) and *SSE* (the sum of the squared deviations between observed and predicted values of *y*) were calculated in Table 16.2. Substituting $SST = 1.7960$ and $SSE = 0.4096$ into the expression above, the coefficient of multiple determination is

$$R^2 = 1 - \frac{SSE}{SST} = 1 - \frac{0.4096}{1.7960} = 1 - 0.2281 = 0.7719$$

For the 10 franchises, 77.19% of the variation in net profit is explained by the multiple regression equation. This value of R^2 is listed in the computer output segments in Table 16.3. Note the inclusion of an "adjusted" R^2 that takes into account the number of data points and the number of independent variables. The formula used in obtaining the adjusted value is shown in the footnote.[1]

The printout segments in Table 16.3 list the value of the multiple standard error of estimate (s_e) calculated earlier. They also provide the regression, error (residual), and total sums of squares, along with the number of degrees of freedom associated with each. The remainder of the analysis of variance table will be discussed in the significance tests of Section 16.5.

TABLE 16.3

These portions of the Excel and Minitab printouts for the fast-food franchise analysis include the multiple standard error of estimate and the coefficient of multiple determination. The full printouts and the procedures for obtaining them are shown in Computer Solutions 16.1.

	E	F	G
3	*Regression Statistics*		
4	Multiple R	0.8786	
5	R Square	0.7719	
6	Adjusted R Square	0.7067	
7	Standard Error	0.2419	
8	Observations	10	
9			
10	ANOVA		
11		*df*	*SS*
12	Regression	2	1.3864
13	Residual	7	0.4096
14	Total	9	1.7960

```
S = 0.241912    R-Sq = 77.2%    R-Sq(adj) = 70.7%

Analysis of Variance

Source           DF       SS       MS      F      P
Regression        2  1.38635  0.69318  11.84  0.006
Residual Error    7  0.40965  0.05852
Total             9  1.79600
```

[1] Expressed as a proportion, the adjusted R^2 typically included in computer printouts of multiple regression analyses is calculated as:

$$\text{adjusted } R^2 = 1 - \frac{SSE/(n-k-1)}{SST/(n-1)}$$

EXERCISES

16.23 What is the coefficient of multiple determination, and what does it tell us about the relationship between y and the independent variables?

16.24 Among the results of a multiple regression analysis are the following sum-of-squares terms: *SST*, *SSR*, and *SSE*. What does each term represent, and how do the terms contribute to our understanding of the relationship between y and the set of independent variables?

16.25 Given the printout in Exercise 16.15, identify the coefficient of multiple determination and interpret its value in terms of that exercise.

(**DATA SET**) *Note:* Exercises 16.26 and 16.27 require a computer and statistical software.

16.26 For the multiple regression equation obtained in Exercise 16.12, what is the coefficient of multiple determination and exactly what does it mean?

16.27 For the multiple regression equation obtained in Exercise 16.11, what is the coefficient of multiple determination and exactly what does it mean?

SIGNIFICANCE TESTS IN MULTIPLE REGRESSION AND CORRELATION (16.5)

In the simple linear regression ($\hat{y} = b_0 + b_1x$) of Chapter 15, several tests were available regarding the relationship between the variables, but their results were the same. With just one independent variable, concluding that the population slope could be zero was equivalent to concluding that the population coefficient of correlation could be zero. However, in multiple regression there are two or more independent variables, and it is necessary to separately test (1) the overall significance of the multiple regression equation and (2) each of the partial regression coefficients in the equation.

Testing the Significance of the Regression Equation

The multiple regression equation, $\hat{y} = b_0 + b_1x_1 + b_2x_2 + \cdots + b_kx_k$, is based on sample data and is an estimate of the (unknown) population multiple regression equation, $E(y) = \beta_0 + \beta_1x_1 + \beta_2x_2 + \cdots + \beta_kx_k$. If there really is no relationship between y and any of the x variables, all of the partial regression coefficients ($\beta_1, \beta_2, \ldots, \beta_k$) in the population regression equation will be zero, and this is the basis on which we test the overall significance of our regression equation.

The test uses analysis of variance in examining the "explained" or regression sum of squares (*SSR*) versus the "unexplained" or error sum of squares (*SSE*). Some computer statistical packages carry out the entire test and report a p-value for its significance. Others may provide summary information (e.g., *SSR*, *SSE*, *SST*, and their degrees of freedom) that can be used in making the final calculations and interpreting the results. In any event, the procedure is as follows.

ANOVA test for the overall significance of the multiple regression equation:

- **Null hypothesis**

  H_0: $\beta_1 = \beta_2 = \cdots = \beta_k = 0$ The regression equation is not significant.

(*continued on page 618*)

- **Alternative hypothesis**

  H_1: One or more of the β_i ($i = 1, 2, ..., k$) $\neq 0$ — The regression equation is significant.

- **Test statistic**

$$F = \frac{SSR/k}{SSE/(n - k - 1)}$$

where SSR = regression sum of squares
SSE = error sum of squares
n = number of data points
k = number of independent (x) variables

- **Critical value of the F statistic**

  F for significance level specified and df (numerator) $= k$, and df (denominator) $= n - k - 1$

In the ANOVA test, the number of degrees of freedom associated with the numerator of the F statistic will be k, the number of independent variables used in estimating y. For the denominator, $df = n - k - 1$, or $n - (k + 1)$, since $(k + 1)$ quantities (b_0, along with $b_1, b_2, \ldots, b_k$) have been determined on the basis of sample data.

For the fast-food franchise data, we have calculated $SST = 1.7960$ and $SSE = 0.4096$. Since $SST = SSR + SSE$, SSR will be $1.7960 - 0.4096$, or $SSR = 1.3864$. There are $n = 10$ data points and $k = 2$ independent variables in our analysis, so the test statistic can be calculated as

$$F = \frac{SSR/k}{SSE/(n - k - 1)} = \frac{1.3864/2}{0.4096/(10 - 2 - 1)} = \frac{0.6932}{0.0585} = 11.85$$

Using the F distribution tables to identify the critical value for F, the number of degrees of freedom for the numerator will be $k = 2$; df for the denominator will be $(n - k - 1)$, or $(10 - 2 - 1) = 7$. Conducting the test at the 0.01 level of significance, the critical value is $F = 9.55$. The calculated test statistic ($F = 11.85$) exceeds the critical value, and H_0 is rejected. At the 0.01 level, the multiple regression equation is significant.

Table 16.4 shows the Excel and Minitab printout segments applicable to this test. The two printouts show the p-value for the overall significance of the equation as 0.0057 and 0.006, respectively.

Testing the Partial Regression Coefficients

In testing the significance of each partial regression coefficient (b_1, b_2, and so on), the null hypothesis is that the corresponding population value is really zero—i.e., that the observed value differs from zero merely by chance. The test is a two-tail t-test with the following format.

***t*-test for the significance of a partial regression coefficient, b_i:**

- **Null hypothesis**

  H_0: $\beta_i = 0$ — The population coefficient is 0.

- **Alternative hypothesis**

  H_1: $\beta_i \neq 0$ — The population coefficient is not 0.

- **Test statistic**

$$t = \frac{b_i - 0}{s_{b_i}}$$

where b_i = the observed value of the regression coefficient
s_{b_i} = the estimated standard deviation of b_i

TABLE 16.4

EXCEL

	E	F	G	H	I	J	K
10	ANOVA						
11		*df*	*SS*	*MS*	*F*	*Significance F*	
12	Regression	2	1.3864	0.6932	11.8448	0.0057	
13	Residual	7	0.4096	0.0585			
14	Total	9	1.7960				
15							
16		*Coefficients*	*Standard Error*	*t Stat*	*P-value*	*Lower 95%*	*Upper 95%*
17	Intercept	-0.2159	0.2643	-0.8169	0.4409	-0.8409	0.4091
18	$Counter	0.0855	0.0438	1.9516	0.0920	-0.0181	0.1890
19	$Driveup	0.1132	0.0385	2.9368	0.0218	0.0220	0.2043

MINITAB

```
Predictor      Coef   SE Coef      T      P
Constant    -0.2159    0.2643  -0.82  0.441
$Counter    0.08547   0.04380   1.95  0.092
$Driveup    0.11315   0.03853   2.94  0.022


S = 0.241912   R-Sq = 77.2%   R-Sq(adj) = 70.7%


Analysis of Variance

Source          DF       SS       MS      F      P
Regression       2  1.38635  0.69318  11.84  0.006
Residual Error   7  0.40965  0.05852
Total            9  1.79600
```

These portions of the Excel and Minitab printouts for the fast-food franchise data show the results of tests for (1) the overall significance of the regression equation and (2) the significance of each of the partial regression coefficients. The full printouts and the procedures for obtaining them are shown in Computer Solutions 16.1.

1. We can also test H_0: $\beta_i = \beta_{i0}$ versus H_1: $\beta_i \neq \beta_{i0}$ where β_{i0} is any value of interest. With the exception of β_{i0} replacing zero, the test is the same as described here.
2. Critical values of t are $\pm t$ for the level of significance desired and $df = n - k - 1$ (n = number of data points, k = number of independent variables).

The printout segments in Table 16.4 list the value and the standard deviation for each partial regression coefficient in the fast-food regression equation. In demonstrating these tests, we will use the number of decimal places shown in the Minitab portion of Table 16.4, and the 0.05 level of significance.

1. Test of the null hypothesis, H_0: $\beta_1 = 0$ versus H_1: $\beta_1 \neq 0$.

$$\text{Calculated test statistic, } t = \frac{b_1 - 0}{s_{b_1}} = \frac{0.08547 - 0}{0.04380} = 1.95$$

For $df = n - k - 1$, or $(10 - 2 - 1) = 7$, critical values for the 0.05 level of significance are $t = -2.365$ and $t = +2.365$. The calculated t (1.95) falls within these limits, and we are unable to reject the null hypothesis that $\beta_1 = 0$. At this level, $b_1 = 0.08547$ does not differ significantly from zero.

Had the test been done at the 0.10 level (with critical values of $t = -1.895$ and $t = +1.895$), the conclusion would be reversed, and the null hypothesis would have been rejected.

From our t distribution table, the most accurate statement we can make about the exact p-value for the test is that it is between 0.05 and 0.10. As shown in Table 16.4, the actual p-value is 0.092. If there were no linear relationship between profit and counter sales (i.e., $\beta_1 = 0$), there would be just a 0.092 probability of obtaining a sample b_1 this different from zero.

2. Test of the null hypothesis, H_0: $\beta_2 = 0$, versus H_1: $\beta_2 \neq 0$.

$$\text{Calculated test statistic, } t = \frac{b_2 - 0}{s_{b_2}} = \frac{0.11315 - 0}{0.03853} = 2.94$$

As with the preceding test, $df = n - k - 1$, or 7, and critical values for the 0.05 level of significance are $t = -2.365$ and $t = +2.365$. The calculated t exceeds these limits, and we are able to reject the null hypothesis that $\beta_2 = 0$. At this level, $b_2 = 0.11315$ differs significantly from zero. Had the test been done at the more demanding 0.02 level (having critical t values of $t = -2.998$ and $t = +2.998$), the conclusion would have been reversed, and the null hypothesis would not have been rejected.

From the t distribution table, the most accurate statement we can make about the exact p-value for this test is that it is between 0.05 and 0.02. As shown in Table 16.4, the actual p-value is 0.022. If there were no linear relationship between profit and drive-through sales (i.e., $\beta_2 = 0$), there would be just a 0.022 chance of obtaining a sample b_2 this different from zero.

We have just duplicated the procedure used by Excel and Minitab to test each of the partial regression coefficients. As expected, our results are consistent with those produced by the computer. The printouts in Table 16.4 go beyond our text discussion by including a t-test and p-value for the constant in the regression equation. The y-intercept, $b_0 = -0.2159$ does not differ very significantly from zero (p-value $= 0.441$). As noted in Chapter 15, testing the y-intercept is generally not of practical importance.

Interval Estimation for the Partial Regression Coefficients

In addition to playing a role in significance testing, the estimated standard deviation of b_i is used in constructing a confidence interval for its population counterpart β_i. The method, which is similar to the procedure used in Chapter 15, is described below.

Confidence interval for a partial regression coefficient, β_i:

The interval is

$$b_i \pm t(s_{b_i})$$

where b_i = **partial regression coefficient in the sample regression equation**

t = **t value for the confidence level desired, and $df = n - k - 1$**

s_{b_i} = **estimated standard deviation of b_i**

Both b_i and the estimated standard deviation of b_i will typically be included in the printout whenever the regression analysis is done with the assistance of a computer statistical package. Applying these to our fast-food example, we will construct the 95% confidence intervals for β_1 and β_2.

1. **95% confidence interval for β_1** Referring to the Minitab portion of Table 16.4, b_1 (the partial regression coefficient for x_1, or counter sales) is 0.08547, and the standard deviation of b_1 is 0.04380. There are $n = 10$ observations in the data and $k = 2$ independent variables. For $df = n - k - 1$, or 7 degrees of freedom, the value of t for a 95% confidence interval is $t = 2.365$. The confidence interval will be

$$b_1 \pm t(s_{b_1}) = 0.08547 \pm 2.365(0.04380)$$
$$= 0.08547 \pm 0.10359, \quad \text{or from } -0.01812 \text{ to } 0.18906$$

   We have 95% confidence that β_1 is somewhere from -0.01812 to 0.18906. Note that $\beta_1 = 0$ falls within the interval. This is consistent with our earlier finding that, at the 0.05 level, $b_1 = 0.08547$ is not significantly different from zero. As we've seen before, an x% confidence interval and a nondirectional significance test at the $(100 - x)/100$ level will give the same conclusion regarding a hypothesized value for a population parameter.

2. **95% confidence interval for β_2** Referring to the Minitab portion of Table 16.4, b_2 (associated with drive-through sales) is 0.11315, and the standard deviation of b_2 is 0.03853. With $n = 10$ observations and $k = 2$ independent variables, $df = n - k - 1$, or 7 degrees of freedom, and the value of t for the 95% confidence interval will be $t = 2.365$. The confidence interval is

$$b_2 \pm t(s_{b_2}) = 0.11315 \pm 2.365(0.03853)$$
$$= 0.11315 \pm 0.09112, \quad \text{or from } 0.02203 \text{ to } 0.20427$$

   We have 95% confidence that β_2 is somewhere from 0.02203 to 0.20427. Note that $\beta_2 = 0$ does not fall within the interval. This is consistent with our earlier finding that, at the 0.05 level, $b_2 = 0.11315$ differs significantly from zero.

If you're using Excel, confidence intervals for the partial regression coefficients are included in the output, as shown in Table 16.4. Unless you specify otherwise, 95% confidence intervals are provided. In the section exercises that follow, Excel users can readily obtain many of the requested confidence intervals without the need for further calculations. If you wish, you can use a pocket calculator and the techniques in this section to verify the confidence intervals your Excel spreadsheet package has so thoughtfully carried out for you.

EXERCISES

16.28 What is the relationship between (a) the results of the hypothesis test examining whether b_3 differs significantly from zero at the 0.02 level and (b) the 98% confidence interval for β_3?

16.29 For the multiple regression equation obtained in Exercise 16.11:

a. At the 0.05 level, is the overall regression equation significant?
b. Use the 0.05 level in concluding whether each partial regression coefficient differs significantly from zero.
c. Interpret the results of the preceding tests in the context of the variables described in that exercise.
d. Construct a 95% confidence interval for each partial regression coefficient in the population regression equation.

16.30 For the multiple regression equation obtained in Exercise 16.12:

a. At the 0.05 level, is the overall regression equation significant?
b. Use the 0.05 level in concluding whether each partial regression coefficient differs significantly from zero.
c. Interpret the results of the preceding tests in the context of the variables described in that exercise.
d. Construct a 95% confidence interval for each partial regression coefficient in the population regression equation.

16.31 Given the printout in Exercise 16.15, (a) determine the 90% confidence interval for each partial regression coefficient, and (b) interpret each significance test in the context of that exercise.

(**DATA SET**) *Note:* Exercises 16.32–16.35 require a computer and statistical software.

16.32 Referring to the least-squares regression equation and printout obtained in Exercise 16.18:

a. At the 0.10 level, is the overall regression equation significant?
b. Use the 0.10 level in concluding whether each partial regression coefficient differs significantly from zero.

16.33 Referring to the least-squares regression equation and printout obtained in Exercise 16.19:

a. At the 0.05 level, is the overall regression equation significant?
b. Use the 0.05 level in concluding whether each partial regression coefficient differs significantly from zero.

16.34 The computer database **GROWCO** describes the characteristics of 100 companies identified by *Fortune* as among the fastest growing. Three of the variables listed are (1) estimated stock price/earnings ratio next year, (2) annual revenue growth (as a percentage), and (3) annual earnings per share growth (as a percentage). Source: "*Fortune's* 100 Fastest-Growing Companies," *Fortune*, September 4, 2000, pp. 142–158.

a. Determine the least-squares regression equation for y = next year's price/earnings ratio estimate as a function of x_1 = annual revenue growth and x_2 = annual earnings per share growth. Interpret each of the partial regression coefficients.
b. At the 0.05 level, is the overall regression equation significant?
c. Use the 0.05 level in concluding whether each partial regression coefficient differs significantly from zero.
d. Construct a 95% confidence interval for each partial regression coefficient in the population regression equation.

16.35 Information describing 98 of the top automobile dealership groups in the United States is provided in computer database file **DEALERS**. For each group, the data include (1) total group revenue during 1999, (2) the number of retail vehicles sold that year, and (3) the number of dealerships in the group. Source: *Automotive News*, May 1, 2000, pp. 54–60.

a. Determine the least-squares regression equation for y = total group revenue as a function of x_1 = number of retail vehicles sold and x_2 = number of dealers in the group. Interpret each of the partial regression coefficients.
b. At the 0.02 level, is the overall regression equation significant?
c. Use the 0.02 level in concluding whether each partial regression coefficient differs significantly from zero.
d. Construct a 98% confidence interval for each partial regression coefficient in the population regression equation.

16.6 OVERVIEW OF THE COMPUTER ANALYSIS AND INTERPRETATION

A Summary of the Results

In the preceding sections, it was useful to segment the Excel and Minitab printouts and discuss each portion as it became relevant. We'll now reverse the process and put the pieces back together.

Table 16.5 is the Excel multiple regression printout for the fast-food example. In addition to the standard Excel printout shown in Computer Solutions 16.1, it includes the confidence and prediction interval estimates that were generated in Computer Solutions 16.2.

The Minitab printout in Table 16.6 includes the segments in Computer Solutions 16.1 and 16.2, plus two standard Minitab items we have not discussed. The first one is the "Unusual Observations" comment at the bottom, where franchise number 8 is identified as a data point that is especially distant from the regression equation. We'll expand on this shortly.

TABLE 16.5

Excel printout for the multiple regression analysis of the fast-food franchise data. In addition to the standard Excel printout from Computer Solutions 16.1, it includes the confidence and prediction intervals generated in Computer Solutions 16.2.

	E	F	G	H	I	J	K
1	SUMMARY OUTPUT			**Prediction Interval**			
2						$Profit	
3	*Regression Statistics*			**Predicted value**		0.750	
4	Multiple R	0.8786		**Prediction Interval**			
5	R Square	0.7719		Lower limit		0.146	
6	Adjusted R Square	0.7067		Upper limit		1.353	
7	Standard Error	0.2419		**Interval Estimate of Expected Value**			
8	Observations	10		Lower limit		0.558	
9				Upper limit		0.941	
10	ANOVA						
11		*df*	*SS*	*MS*	*F*	*Significance F*	
12	Regression	2	1.3864	0.6932	11.8448	0.0057	
13	Residual	7	0.4096	0.0585			
14	Total	9	1.7960				
15							
16		*Coefficients*	*Standard Error*	*t Stat*	*P-value*	*Lower 95%*	*Upper 95%*
17	Intercept	-0.2159	0.2643	-0.8169	0.4409	-0.8409	0.4091
18	$Counter	0.0855	0.0438	1.9516	0.0920	-0.0181	0.1890
19	$Driveup	0.1132	0.0385	2.9368	0.0218	0.0220	0.2043

TABLE 16.6

Minitab printout for the multiple regression analysis of the fast-food franchise data. It includes the standard Minitab printout in Computer Solutions 16.1 plus the confidence and prediction intervals in Computer Solutions 16.2.

```
Regression Analysis: $Profit versus $Counter, $Driveup

The regression equation is
$Profit = - 0.216 + 0.0855 $Counter + 0.113 $Driveup

Predictor      Coef   SE Coef      T      P
Constant    -0.2159    0.2643  -0.82  0.441
$Counter    0.08547   0.04380   1.95  0.092
$Driveup    0.11315   0.03853   2.94  0.022

S = 0.241912   R-Sq = 77.2%   R-Sq(adj) = 70.7%

Analysis of Variance
Source          DF       SS        MS      F      P
Regression       2  1.38635   0.69318  11.84  0.006
Residual Error   7  0.40965   0.05852
Total            9  1.79600

Source    DF   Seq SS
$Counter   1  0.88160
$Driveup   1  0.50475

Unusual Observations
Obs  $Counter  $Profit     Fit  SE Fit  Residual  St Resid
  8       8.6   1.3000  0.9039  0.1468    0.3961     2.06R
R denotes an observation with a large standardized residual.

Predicted Values for New Observations
New
Obs     Fit  SE Fit        95% CI             95% PI
  1  0.7496  0.0811  (0.5578, 0.9413)  (0.1462, 1.3529)

Values of Predictors for New Observations
New
Obs  $Counter  $Driveup
  1      6.00      4.00
```

The other portion of Table 16.6 that has not been previously discussed is the information in the "SEQ SS" column near the bottom. In this section, Minitab shows how much the error sum of squares (*SSE*) is reduced by each x variable when they are sequentially introduced into the regression equation. The first row in the "SEQ SS" column indicates that, compared to an "equation" in which only a constant is used as a predictor for y, the inclusion of x_1 reduces *SSE* by 0.88160. Once x_1 has been included, bringing x_2 into the regression equation reduces *SSE* by another 0.50475. Note that $0.88160 + 0.50475 = 1.38635$, the regression (or explained) sum of squares.

In describing our overall conclusions for the fast-food franchise regression analysis, we will use the number of decimal places shown in Table 16.5. This is to avoid confusion, since the two printouts do not always present findings to the same number of decimal places.

- **Dependent and independent variables**

  y = net profit $\quad$ x_1 = counter sales $\quad$ x_2 = drive-through sales

- **Multiple regression equation**

$$\hat{y} = -0.2159 + 0.0855x_1 + 0.1132x_2$$

- **Partial regression coefficients**

$$b_1 = 0.0855 \quad \text{and} \quad b_2 = 0.1132$$

  - For a given level of drive-through sales, another $1.00 of counter sales increases the estimated net profit by $0.0855.
  - For a given level of counter sales, another $1.00 of drive-through sales increases the estimated net profit by $0.1132.

- **Coefficient of multiple determination**

$$R^2 = 0.7719$$

  Changes in x_1 and x_2 explain 77.19% of the variation in y.

- **Significance test for the overall regression equation** The regression equation is quite significant. For the analysis of variance test of H_0: $\beta_1 = \beta_2 = 0$, the p-value is just 0.0057.

- **Significance of the individual partial regression coefficients**
  - b_1 (0.0855) differs from 0 at the p-value = 0.0920 level of significance
  - b_2 (0.1132) differs from 0 at the p-value = 0.0218 level of significance

- **Confidence intervals for the partial regression coefficients** These are provided by Excel but not by Minitab. We have 95% confidence that β_1 is between -0.0181 and 0.1890, and that β_2 is between 0.0220 and 0.2043.

- **Confidence and prediction intervals for the dependent variable** These intervals are associated with x_1 = $6 million in counter sales and x_2 = $4 million in drive-through sales. When x_1 = $6 million in counter sales and x_2 = $4 million in drive-through sales, the point estimate for y = profit is $0.750 million. The corresponding interval estimates are:
  - Confidence interval: For all franchises having x_1 = $6 million in counter sales and x_2 = $4 million in drive-through sales, we are 95% confident that *the mean profit for all such franchises* is between $0.558 and $0.941 million.
  - Prediction interval: For any individual franchise having x_1 = $6 million in counter sales and x_2 = $4 million in drive-through sales, we are 95% confident that *the profit for this single franchise* is between $0.146 and $1.353 million.

STATISTICS IN ACTION 16.1

Predicting Holiday Traffic Fatalities

Among other data, the National Highway Traffic Safety Administration (NHTSA) reports the number of U.S. traffic fatalities during holiday periods. To see if we might be able to predict New Year's fatalities based on the number during each of the two major holidays preceding it, we used NHTSA figures and multiple regression to analyze Thanksgiving, Christmas, and New Year's fatalities during the period 1983–2005. A portion of the Minitab printout is shown.

```
Regression Analysis: New_Yrs versus Thanksgiving, Christmas

The regression equation is
New_Yrs = - 47 + 0.291 Thanksgiving + 0.690 Christmas

Predictor          Coef   SE Coef       T      P
Constant          -46.6     140.2   -0.33  0.743
Thanksgiving     0.2912    0.2451    1.19  0.249
Christmas       0.69022   0.08054    8.57  0.000

S = 55.0154    R-Sq = 79.4%    R-Sq(adj) = 77.4%

Analysis of Variance
Source           DF      SS      MS      F      P
Regression        2  233898  116949  38.64  0.000
Residual Error   20   60534    3027
Total            22  294432
```

The overall regression equation is highly significant (the *p*-value, rounded, is 0.000), and x_2 = Christmas fatalities (*p*-value = 0.000) is easily the better of the two independent variables in predicting the number that will occur during the New Year's holiday. Over this time period, 79.4% of the variation in New Year's fatalities is explained by the regression equation.

Source: U.S. National Highway Traffic Safety Administration, *Traffic Safety Facts 2006*, p. 33.

Statistics in Action 16.1, mentioned in this chapter's opening vignette, describes a multiple regression analysis in which traffic fatalities during the New Year's holiday period are estimated based on the number of fatalities during the Thanksgiving and Christmas periods. The Minitab printout is also included, and it is similar in appearance to the one we've discussed for the fast-food franchise data.

Residual Analysis

Like most computer statistical software, Excel and Minitab give you an optional summary of the residuals, as shown in Table 16.7. As discussed in Chapter 15, this information can be used in finding data points that are especially distant from the regression equation as well as in identifying residual "patterns" that suggest our model may not be valid.

Refer to the Minitab portion of Table 16.7. For each of the 10 data points, observed and estimated ("fitted") values of y are printed, along with their difference (the residual). Each residual has been "standardized," or expressed in terms

TABLE 16.7

Using the procedures in Computer Solutions 16.3, we can obtain a listing of the residuals like the Excel and Minitab printouts shown here. Minitab and Excel differ slightly in their method of determining the values of the standardized residuals.

EXCEL

	E	F	G	H
23	RESIDUAL OUTPUT			
24				
25	*Observation*	*Predicted $Profit*	*Residuals*	*Standard Residuals*
26	1	1.373	0.127	0.594
27	2	0.575	0.225	1.053
28	3	1.230	-0.030	-0.142
29	4	1.521	-0.121	-0.569
30	5	0.457	-0.257	-1.206
31	6	0.985	-0.185	-0.869
32	7	0.452	0.148	0.696
33	8	0.904	0.396	1.857
34	9	0.515	-0.115	-0.538
35	10	0.787	-0.187	-0.874

MINITAB

Obs	$Counter	$Profit	Fit	SE Fit	Residual	St Resid
1	8.4	1.5000	1.3734	0.1279	0.1266	0.62
2	3.3	0.8000	0.5754	0.1564	0.2246	1.22
3	5.8	1.2000	1.2303	0.1756	-0.0303	-0.18
4	10.0	1.4000	1.5214	0.1592	-0.1214	-0.67
5	4.7	0.2000	0.4574	0.1157	-0.2574	-1.21
6	7.7	0.8000	0.9854	0.0919	-0.1854	-0.83
7	4.5	0.6000	0.4516	0.1172	0.1484	0.70
8	8.6	1.3000	0.9039	0.1468	0.3961	2.06R
9	5.9	0.4000	0.5147	0.1228	-0.1147	-0.55
10	6.3	0.6000	0.7865	0.0797	-0.1865	-0.82

R denotes an observation with a large standardized residual.

of the number of standard deviation multiples it represents. The standard deviations for the estimated y values, listed in the "SE Fit" column, differ for the various rows. This is because estimates become more uncertain when they are based on x_1 and x_2 values that are farther away from their respective means.

Franchise number 8 has been "flagged" as an outlier because its standardized residual is more than 2 standard deviation multiples away from zero. The net profit for this franchise was considerably higher than estimated, and the franchise may have some unique characteristic that explains its strong performance versus the regression equation. Perhaps the manager is an excellent cost cutter or employee motivator, or the franchise may have a very attractive location, such as near an office building or high school.

Among many other possibilities for residual analysis, we can (1) construct a histogram of the residuals as a rough check to see whether they are approximately normally distributed, (2) examine a normal probability plot to check for normality, (3) plot the residuals versus the predicted values of y, (4) plot the residuals versus

COMPUTER 16.3 SOLUTIONS

Residual Analysis in Multiple Regression

EXCEL

Excel Residual Analysis in Multiple Regression

Follow the procedure in Computer Solutions 16.1, but in step 2 select all four items in the **Residuals** section and select **Normal Probability Plots.** The tabular results will be as shown in the Excel portion of Table 16.7. Three of the resulting Excel residual analysis plots are shown in Figure 16.2.

MINITAB

Minitab Residual Analysis in Multiple Regression

With two exceptions, the procedure is as described in Computer Solutions 16.1. (1) To get the printout of residuals shown in the Minitab portion of Table 16.7, click **Results,** then select **In addition, the full table of fits and residuals.** (2) To get the plots shown in Figure 16.3 and discussed in the text: Click **Graphs.** In the **Residuals for plots** menu, select **Regular.** In the **Residual Plots** menu, select all of the options. Enter **C2 C3** (include the space between them) into the **Residuals versus the variables** box.

the order in which the observations are arranged, (5) plot the residuals versus x_1 to see whether they exhibit some kind of cycle or pattern with respect to this variable, and (6) repeat plot 5 for each of the other independent variables.

In testing for normality, we can also apply the chi-square test that was introduced in Chapter 13 (n must be ≥ 32) or the Kolmogorov-Smirnov test that was introduced in Chapter 14. In possibility number 4 of the preceding paragraph, we are concerned about the possible presence of *autocorrelation,* in which the value of a residual is related to the values of those that precede or follow it. The topic of autocorrelation is especially relevant to regression models fitted to a time series, and is discussed further in Chapter 18.

Computer Solutions 16.3 shows the Excel and Minitab procedures for getting residuals listings like those in Table 16.7, as well as the procedures involved in generating Excel and Minitab residual analysis plots like those in Figures 16.2 and 16.3 (pages 628–630). The Minitab residual analysis plots in Figure 16.3 include the following features:

- *Histogram of residuals.* There are only 10 franchises. Given this histogram, it looks like the 10 residuals could have come from a normal distribution with a mean of 0.
- *Normal probability plot.* The normal probability plot is not terribly nonlinear, again supporting the assumption that the residuals could have come from a normal distribution with a mean of 0.
- *Residuals versus predicted y values.* The residuals seem to be somewhat randomly scattered, so it does not look like there is any relationship between the residuals and the predicted profits for the 10 franchises.
- *Residuals versus order of the observations.* There seems to be a bit of a U shape here. It would appear that the order in which the franchises are listed

FIGURE 16.2

Several of the plots that Excel provides for analysis of the residuals, generated using the procedures in Computer Solutions 16.3.

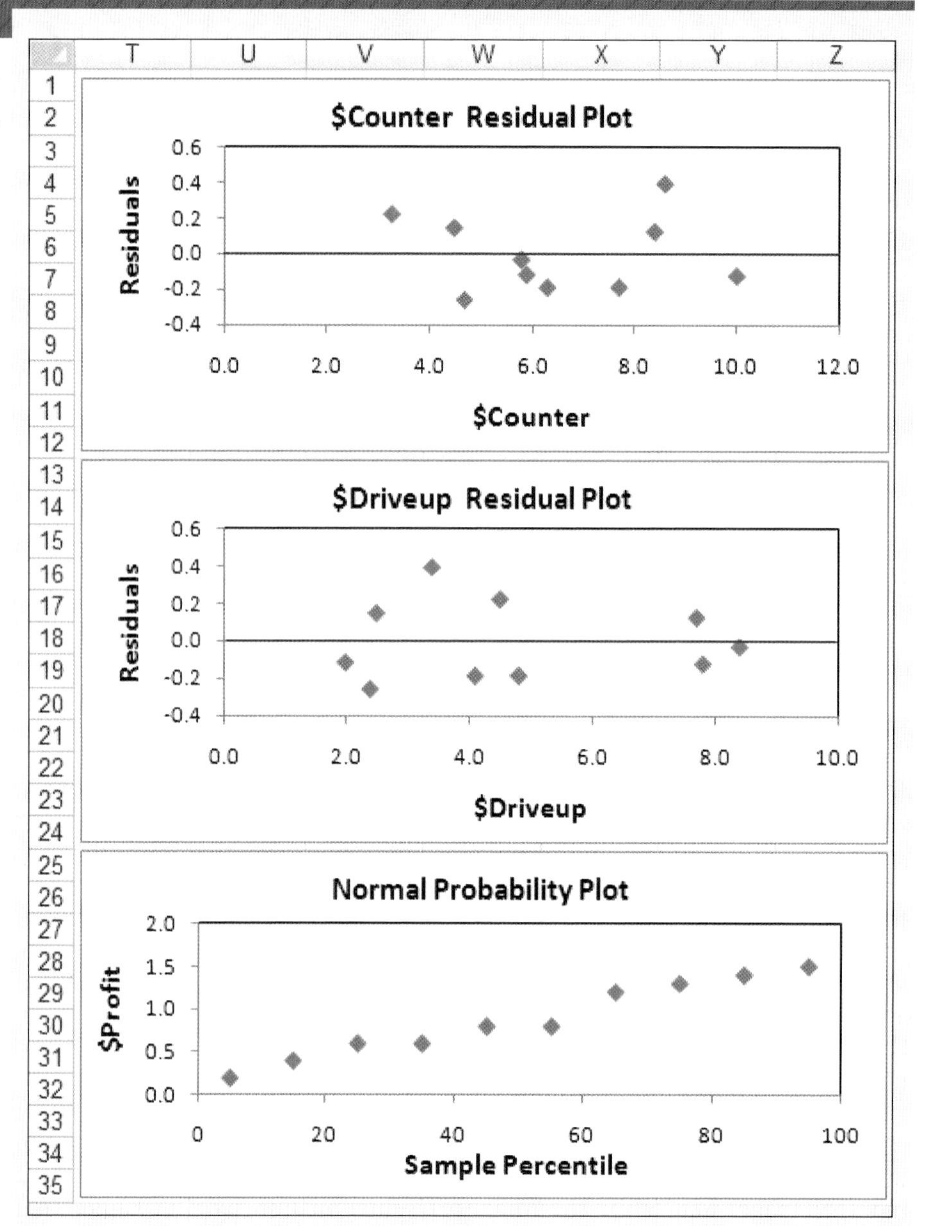

in the original data may have something to do with whether they have made (a) a higher profit than the equation predicts or (b) a lower profit than the equation predicts. For example, franchise #5 has "underperformed," earning a much lower profit ($0.2 million) than the $0.4574 million that the equation predicts, a residual of −$0.2574 million.

- *Residuals versus* x_1. This plot looks somewhat like a U-shaped curve, too. The residuals might not be independent of the level of counter sales.
- *Residuals versus* x_2. Although the residuals seem to be a bit more scattered for franchises with lower levels of drive-through sales, there are not very many data points, and this slight pattern could have happened by chance.

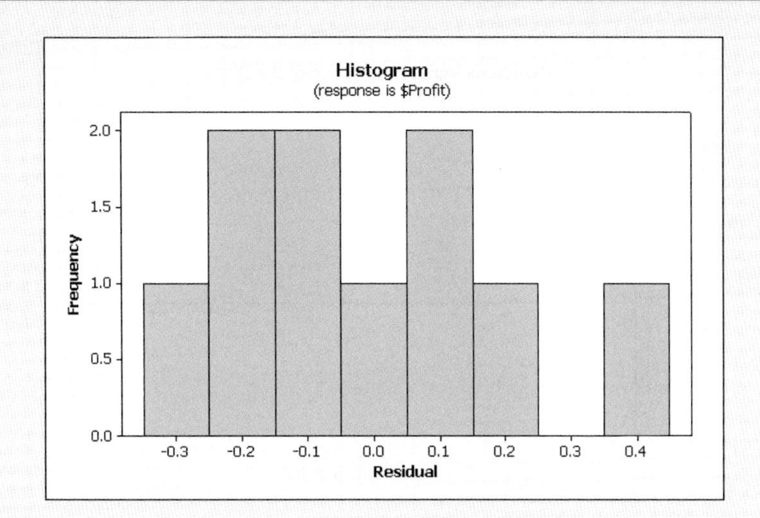

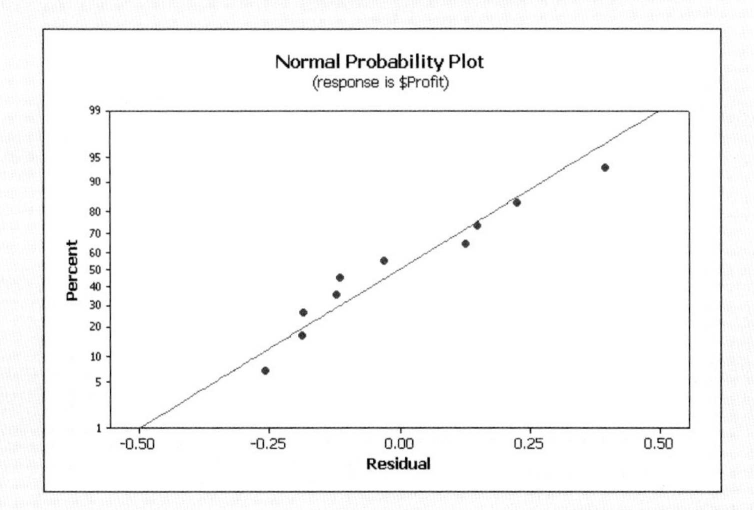

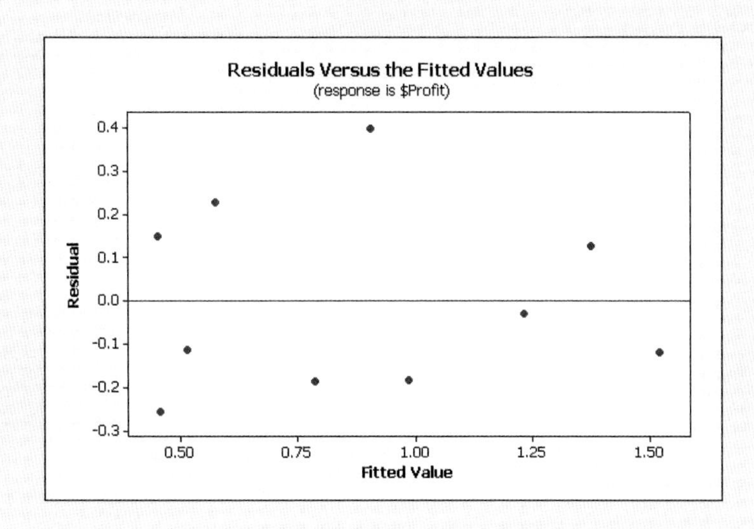

FIGURE 16.3

Minitab plots for analysis of the residuals, discussed in the text and generated using the procedures in Computer Solutions 16.3.

FIGURE 16.3
(continued)

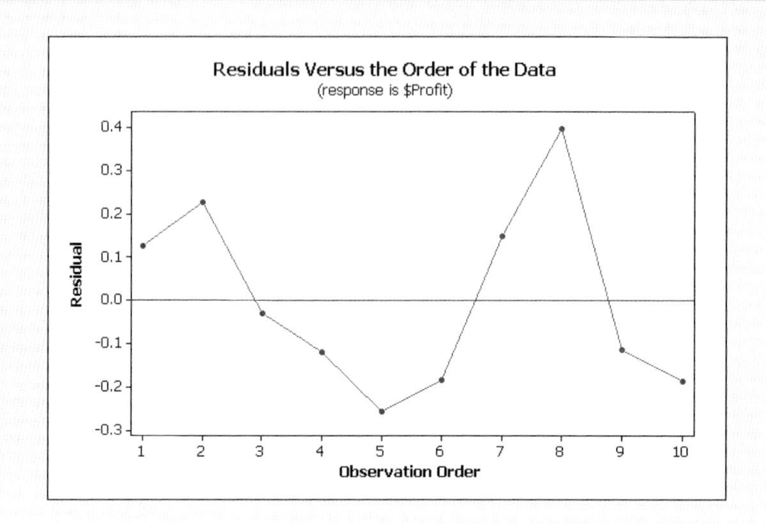

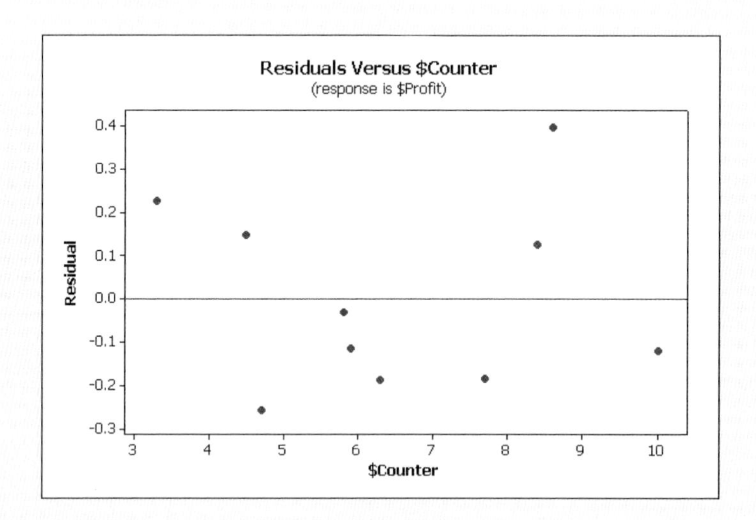

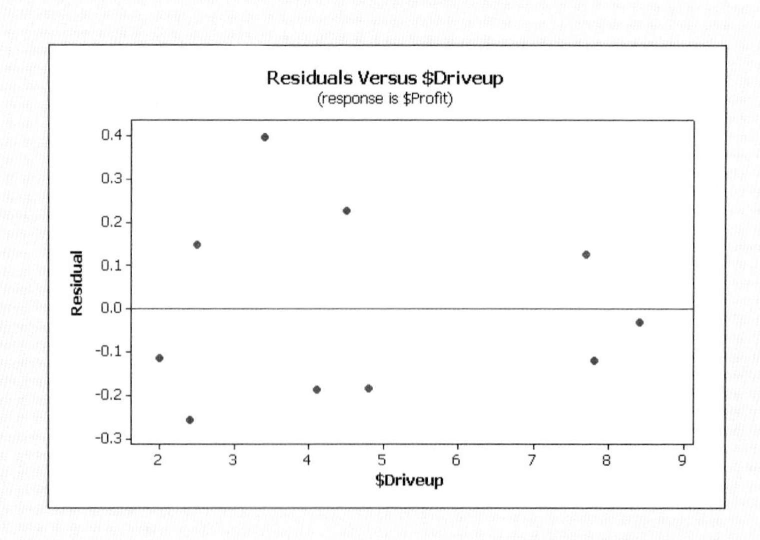

EXERCISES

16.36 What role does the normal probability plot play in examining whether the multiple regression model is appropriate for a given set of data?

16.37 Discuss residual analysis in terms of the assumptions required for using the multiple regression model.

16.38 Given the printout in Exercise 16.15, briefly summarize the overall results in the context of that exercise:

a. Identify and interpret each of the partial regression coefficients.
b. What proportion of the variation in y is explained by the equation?
c. At what p-value is the overall regression equation significant?
d. For each of the partial regression coefficients, at what p-value does the coefficient differ from zero? Which independent variables appear to be most and least useful in explaining changes in y?

(**DATA SET**) *Note:* Exercises 16.39–16.45 require a computer and statistical software.

16.39 For the regression equation obtained in Exercise 16.11, analyze the residuals by (a) constructing a histogram, (b) utilizing the normal probability plot, and (c) plotting the residuals versus each of the independent variables. Do any of the assumptions of the multiple regression model appear to have been violated?

16.40 For the regression equation obtained in Exercise 16.12, analyze the residuals by (a) constructing a histogram, (b) utilizing the normal probability plot, and (c) plotting the residuals versus each of the independent variables. Do any of the assumptions of the multiple regression model appear to have been violated?

16.41 For the regression equation obtained in Exercise 16.13, analyze the residuals by (a) constructing a histogram, (b) utilizing the normal probability plot, and (c) plotting the residuals versus each of the independent variables. Do any of the assumptions of the multiple regression model appear to have been violated?

16.42 For the regression equation obtained in Exercise 16.18, analyze the residuals by (a) constructing a histogram, (b) utilizing the normal probability plot, and (c) plotting the residuals versus each of the independent variables. Do any of the assumptions of the multiple regression model appear to have been violated?

16.43 For the regression equation obtained in Exercise 16.19, analyze the residuals by (a) constructing a histogram, (b) utilizing the normal probability plot, and (c) plotting the residuals versus each of the independent variables. Do any of the assumptions of the multiple regression model appear to have been violated?

16.44 An efficiency expert wants to determine whether the time an assembly worker requires to perform a standard task might be related to (1) the number of years on the job and (2) the score on a pre-employment aptitude test. The following data have been collected.

y = Time, seconds:	52	48	56	52	48	45	42
x_1 = Years on job:	7	14	8	6	10	6	10
x_2 = Test score:	73	78	71	75	78	78	80

y = Time, seconds:	51	51	53	47	45	49
x_1 = Years on job:	11	10	6	8	12	12
x_2 = Test score:	72	79	77	79	84	72

a. Determine the multiple regression equation and interpret the partial regression coefficients.
b. Determine the 95% confidence interval for the population partial regression coefficients, β_1 and β_2.
c. Interpret the coefficient of determination and the result of each significance test done by the computer.
d. Analyze the residuals. Does the analysis support the applicability of the multiple regression model to these data?

16.45 Of all U.S. credit unions, 67% have a "courtesy overdraft" policy that automatically covers overdrafts, with the median fee for each overdraft being approximately $25. For 14 federal credit unions with a courtesy overdraft policy, the accompanying table shows their fee income, total assets, and average fee income per member for 2008. Source: Kathy Chu, "Credit Unions Hit Customers with Fees, Too," *USA Today*, August 4, 2009, p. 2B.

Credit Union	Fee Income, y	Total Assets, x_1	Average Fee Income per Member, x_2
1	$0.5 million	$19.0 million	$140.7
2	2.8	128.2	140.6
3	1.1	53.0	165.1
4	0.6	42.1	107.4
5	0.3	29.0	93.9
6	1.8	147.7	69.4
7	2.2	211.2	97.9
8	0.6	58.4	90.8
9	4.0	443.3	86.3
10	3.4	395.4	85.6
11	9.0	1249.8	54.6
12	6.8	986.0	93.3
13	2.5	417.5	76.6
14	0.2	178.4	18.1

a. Determine the multiple regression equation and interpret the partial regression coefficients.
b. Determine the 95% confidence interval for the population partial regression coefficients, β_1 and β_2.
c. Interpret the coefficient of determination and the result of each significance test done by the computer.
d. Analyze the residuals. Does the analysis support the applicability of the multiple regression model to these data?

16.7 ADDITIONAL TOPICS IN MULTIPLE REGRESSION AND CORRELATION

Although we can't cover all facets of multiple regression and correlation analysis, two additional topics are deserving of our attention. These include (1) using a dummy variable to incorporate qualitative data into the analysis, and (2) a discussion of the potential problem known as multicollinearity.

Including Qualitative Data: The Dummy Variable

There may be situations where some of our data do not meet the requirements of either the interval or the ratio scales of measurement. In such cases, we can still include the data in our analysis by using one or more **dummy variables.** A dummy variable will have a value of either one or zero, depending on whether a given characteristic is present or absent.

For example, if some of the 10 franchise managers were college graduates and others were not, we could include an x_3 variable reflecting this information. Depending on the presence or absence of a college degree, x_3 could be recorded as either one or zero:

$x_3 = 1$ if the franchise is managed by a college graduate

and

$x_3 = 0$ if the manager is not a college graduate

The dummy variable has a partial regression coefficient, in this case b_3. As an example of its interpretation, if $b_3 = 0.300$, the estimated profit will be \$0.300 million higher if a franchise is managed by a college graduate.

Dummy variables can represent information such as marital status, gender, city versus urban location, trained versus untrained employees, or even whether someone wears contact lenses. As with other independent variables, computer analysis tests the significance of the partial regression coefficient for a dummy variable. For example, if b_3 were to differ significantly from zero, we would conclude that whether a manager is a college graduate is an important predictor for the net profit of a franchise. If b_3 were positive *and* significant, this would indicate that franchises managed by college graduates are significantly more profitable than those managed by nongraduates.

To further demonstrate how dummy variables can be employed, we will use actual data for a sample of 20 personal printers.[2] Some of the printers can print in color, while others are monochrome, and the models vary in how fast they can generate text-only materials. The data are in file **CX16PRIN** and represent the following variables:

PRICE = The list price for the printer
SPEED = Printing speed for text, pages per minute
COLOR = A dummy variable reflecting color versus monochrome printer (1 = color, 0 = monochrome)

[2]Source: "Top 10 Monochrome Laser Printers," *PC World,* August 2009, p. 70; and "Top 10 Color Laser Printers," *PC World,* March 2009, p. 48.

Using PRICE as the dependent variable, the estimation equation is:

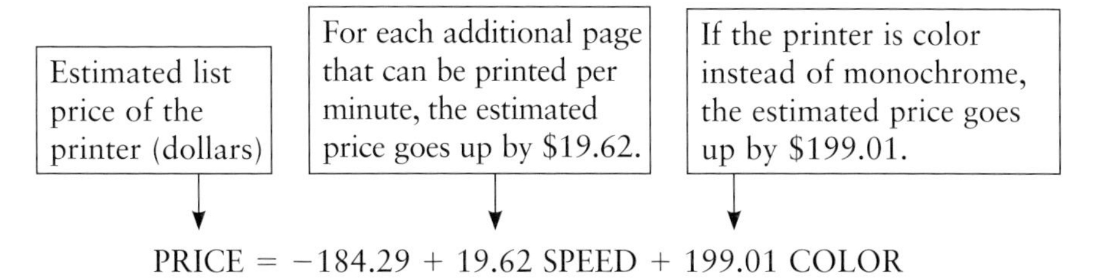

$$\text{PRICE} = -184.29 + 19.62\ \text{SPEED} + 199.01\ \text{COLOR}$$

Point and interval price estimates, confidence intervals, significance testing, and the coefficient of multiple determination are treated the same way as they were in earlier portions of the chapter. For example, a machine that prints 20 text pages per minute (SPEED = 20) and is capable of color printing (COLOR = 1) would have an estimated price of $-184.29 + 19.62(20) + 199.01(1)$, or $407.12.

Multicollinearity

Multicollinearity is a situation in which two or more of the independent variables are highly correlated with each other. When this happens, the partial regression coefficients become both statistically unreliable and difficult to interpret. This is because they are not really "saying" entirely different things about y. If our fast-food example had included x_3 = number of vehicles using the drive-through, this would be very highly correlated with x_2 = drive-through sales. As a result, these two variables would be the mathematical equivalent of the old-time comedians Laurel and Hardy carrying a piano. They would be very unsure of how much of the "load" each should carry, and the piano would tend to be quite unstable.

Multicollinearity is *not* a problem when we simply substitute x values into the regression equation in order to obtain an estimated value for y. However, it *is* a problem when we're trying to interpret the partial regression coefficients for the x variables that are highly correlated.

When multicollinearity is present, strange things can happen. Along with a high correlation between two or more independent variables, the following are clues to its presence:

1. An independent variable known to be an important predictor ends up having a partial regression coefficient that is not significant.
2. A partial regression coefficient that should be positive turns out to be negative, or vice versa.
3. When an independent variable is added or removed, the partial regression coefficients for the other independent variables change drastically.

One way to deal with multicollinearity is to avoid having a set of independent variables in which two or more variables are highly correlated with each other. In the fast-food example used throughout this chapter, the inclusion of a third independent variable representing the number of vehicles using the drive-through would likely cause multicollinearity problems, because x_2 = drive-through sales would probably be highly correlated with x_3 = drive-through traffic count—each variable is essentially describing the "bigness" of the drive-through aspect of the restaurant.

In Chapter 17, we will discuss **stepwise regression,** a regression model in which the independent variables enter the regression analysis one at a time. The first x variable to enter is the one that explains the greatest amount of variation in y. The second x variable to enter will be the one that explains the greatest amount of the *remaining* variation in y, and so on. At each "step," the computer decides which, if any, of the remaining x variables should be brought in. Every step results in a new regression equation and an updated printout of this equation and its analysis. The general idea of stepwise regression is the balancing act of trying to (1) explain the most possible variation in y, while (2) using the fewest possible x variables.

EXERCISES

16.46 What is a dummy variable, and how is it useful to multiple regression? Give an example of three dummy variables that could be used in describing your home town.

16.47 A multiple regression equation has been developed for y = daily attendance at a community swimming pool, x_1 = temperature (degrees Fahrenheit), and x_2 = weekend versus weekday ($x_2 = 1$ for Saturday and Sunday, and 0 for the other days of the week). For the regression equation shown below, interpret each partial regression coefficient.

$$\hat{y} = 100 + 8x_1 + 150x_2$$

16.48 For the regression equation in Exercise 16.47, the estimated number of persons swimming on a zero-degree weekday would be 100 persons. Since this level of pool attendance is unlikely on such a day, does this mean that an error was made in constructing the regression equation? Explain.

16.49 What is multicollinearity, and how can it adversely affect multiple regression analysis? How can we tell whether multicollinearity is present?

(**DATA SET**) *Note:* Exercises 16.50 and 16.51 require a computer and statistical software.

16.50 For 12 recent clients, a weight-loss clinic has collected the following data. (Session is coded as 1 for day, 0 for evening. Gender is coded as 1 for male, 0 for female.) Using a computer statistical package, carry out a multiple regression analysis, interpret the partial regression coefficients, and discuss the results of each significance test.

Client	Pounds Lost	Months as a Client	Session	Gender
1	31	5	1	1
2	49	8	1	1
3	12	3	1	0
4	26	9	0	0
5	34	8	0	1
6	11	2	0	0
7	4	1	0	1
8	27	8	0	1
9	12	6	1	1
10	28	9	1	0
11	41	6	0	0
12	16	6	0	0

16.51 A safety researcher has measured the speed of automobiles passing his vantage point near an interstate highway. He has also recorded the number of occupants and whether the driver was wearing a seat belt. The data are shown in the following table with seat belt users coded as 1, nonusers as 0. Using a computer statistical package, carry out a multiple regression analysis on these data, interpret the partial regression coefficients, and discuss the results of each significance test.

y = speed (mph):	61	63	55	59	59	52	61
x_1 = occupants:	2	1	2	1	3	1	2
x_2 = belt usage:	1	1	1	0	0	1	0

y = speed (mph):	51	57	55	54	49	61	73
x_1 = occupants:	2	2	2	3	1	1	1
x_2 = belt usage:	1	1	1	0	1	1	0

16.8 SUMMARY

• Multiple regression model

Unlike its counterpart in Chapter 15, multiple regression and correlation analysis considers two or more independent (x) variables. The model is of the form $y_i = \beta_0 + \beta_1 x_{1i} + \beta_2 x_{2i} + \cdots + \beta_k x_{ki} + \epsilon_i$ and, for a given set of x values, the expected value of y is given by the regression equation, $E(y) = \beta_0 + \beta_1 x_{1i} + \beta_2 x_{2i} + \cdots + \beta_k x_{ki}$. For each x term in the estimation equation, the corresponding β is referred to as the *partial regression coefficient*. Each β_i ($i = 1, 2, \ldots, k$) is a slope relating changes in $E(y)$ to changes in one x variable when all other x's are held constant. The multiple standard error of estimate expresses the amount of dispersion of data points about the regression equation.

- **Estimation and computer assistance**

Calculations involving multiple regression and correlation analysis are tedious and best left to a computer package such as Excel or Minitab. The chapter includes formulas for determining the approximate confidence interval for the conditional mean of y as well as the approximate prediction interval for an individual y value, given a set of x values. The exact intervals can be obtained using Excel or Minitab.

- **Coefficient of multiple determination**

Multiple correlation measures the strength of the relationship between the dependent variable and the set of independent variables. The coefficient of multiple determination (R^2) is the proportion of the variation in y that is explained by the regression equation.

- **Significance testing and residual analysis**

The overall significance of the regression equation can be tested through the use of analysis of variance, while a t-test is used in testing the significance of each partial regression coefficient. The results of these tests are typically included when data have been analyzed using a computer statistical package. Most computer packages can also provide a summary of the residuals, which may be analyzed further in testing the regression model and identifying data points especially distant from the regression equation.

- **Dummy variables and multicollinearity**

A dummy variable has a value of either one or zero, depending on whether a given characteristic is present, and it allows qualitative data to be included in the analysis. If two or more independent variables are highly correlated with each other, multicollinearity is present, and the partial regression coefficients may be unreliable. Multicollinearity is not a problem when the only purpose of the equation is to predict the value of y.

- **Stepwise regression**

In stepwise regression, independent variables enter the equation one at a time, with the first one entered being the x variable that explains the greatest amount of the variation in y. At each step, the variable introduced is the one that explains the greatest amount of the remaining variation in y. Stepwise regression will be discussed more completely in Chapter 17.

EQUATIONS

The Multiple Regression Model

$$y_i = \beta_0 + \beta_1 x_{1i} + \beta_2 x_{2i} + \cdots + \beta_k x_{ki} + \epsilon_i$$

where y_i = a value of the dependent variable, y

β_0 = a constant

$x_{1i}, x_{2i}, \ldots, x_{ki}$ = values of the independent variables, $x_1, x_2, \ldots, x_k$

$\beta_1, \beta_2, \ldots, \beta_k$ = partial regression coefficients for the independent variables, $x_1, x_2, \ldots, x_k$

ϵ_i = random error, or residual

The Sample Multiple Regression Equation

$$\hat{y} = b_0 + b_1x_1 + b_2x_2 + \cdots + b_kx_k$$

where $\hat{y}$ = the estimated value of the dependent variable, y for a given set of values for $x_1, x_2, \ldots, x_k$

k = the number of independent variables

b_0 = the y-intercept; this is the estimated value of y when all of the independent variables are equal to 0

b_i = the partial regression coefficient for x_i

The Multiple Standard Error of Estimate

$$s_e = \sqrt{\frac{\sum(y_i - \hat{y}_i)^2}{n - k - 1}}$$

where y_i = each observed value of y in the data

$\hat{y}_i$ = the value of y that would have been estimated from the regression equation

n = the number of data points

k = the number of independent (x) variables

Approximate Confidence Interval for the Conditional Mean of *y*

$$\hat{y} \pm t\frac{s_e}{\sqrt{n}}$$

where $\hat{y}$ = the estimated value of y based on the set of x values provided

t = t value from the t distribution table for the desired confidence level and $df = n - k - 1$ (n = number of data points, k = number of x variables)

s_e = the multiple standard error of estimate

Approximate Prediction Interval for an Individual *y* Observation

$$\hat{y} \pm ts_e$$

where $\hat{y}$ = the estimated value of y based on the set of x values provided

t = t value from the t distribution table for the desired confidence level and $df = n - k - 1$ (n = number of data points, k = number of x variables)

s_e = the multiple standard error of estimate

Coefficient of Multiple Determination

$$R^2 = 1 - \frac{\textit{Error}\text{ variation, which is }\textit{not}\text{ explained by the regression equation}}{\textit{Total}\text{ variation in the } y \text{ values}}$$

$$= 1 - \frac{\sum(y_i - \hat{y}_i)^2}{\sum(y_i - \bar{y})^2} \quad \text{or} \quad 1 - \frac{SSE}{SST}$$

As in Chapter 15, the total variation in y is the sum of the explained variation plus the unexplained (error) variation, or

Total variation in y values (SST)	=	Variation explained by the regression equation (SSR)	+	Variation not explained by the regression equation (SSE)

ANOVA Test for the Overall Significance of the Multiple Regression Equation

- **Null hypothesis**

H_0: $\beta_1 = \beta_2 = \cdots = \beta_k = 0$ The regression equation is not significant.

- **Alternative hypothesis**

  H_1: One or more of the β_i $(i = 1, 2, \ldots, k) \neq 0$ The regression equation is significant.

- **Test statistic**

$$F = \frac{SSR/k}{SSE/(n - k - 1)}$$

where SSR = regression sum of squares
SSE = error sum of squares
n = number of data points
k = number of independent (x) variables

- **Critical value of the F statistic**

  F for significance level specified and df (numerator) $= k$, and df (denominator) $= n - k - 1$

t-Test for the Significance of a Partial Regression Coefficient, b_i

- **Null hypothesis**

  H_0: $\beta_i = 0$ The population coefficient is 0.

- **Alternative hypothesis**

  H_1: $\beta_i \neq 0$ The population coefficient is not 0.

- **Test statistic**

$$t = \frac{b_i - 0}{s_{b_i}}$$

where b_i = the observed value of the regression coefficient
s_{b_i} = the estimated standard deviation of b_i

1. We can also test H_0: $\beta_i = \beta_{i0}$ versus H_1: $\beta_i \neq \beta_{i0}$, where β_{i0} is any value of interest. With the exception of β_{i0} replacing zero, the test is the same as described here.
2. Critical values of t are $\pm t$ for the level of significance desired and $df = n - k - 1$ (n = number of data points, k = number of independent variables).

Confidence Interval for a Partial Regression Coefficient, β_i

The interval is $b_i \pm t(s_{b_i})$

where b_i = partial regression coefficient in the sample regression equation
t = t value for the confidence level desired, and $df = n - k - 1$
s_{b_i} = estimated standard deviation of b_i

Excel provides these confidence intervals as part of the regression analysis printout.

CHAPTER EXERCISES

(**DATA SET**) *Note:* Exercises 16.52–16.58 require a computer and statistical software.

16.52 Interested in the possible relationship between the size of his tip versus the size of the check and the number of diners in the party, a food server has recorded the following for a sample of 8 checks:

Observation Number	y = Tip	x_1 = Check	x_2 = Diners
1	$7.5	$40	2
2	0.5	15	1
3	2.0	30	3
4	3.5	25	4
5	9.5	50	4
6	2.5	20	5
7	3.5	35	5
8	1.0	10	2

a. Determine the multiple regression equation and interpret the partial regression coefficients.
b. What is the estimated tip amount for 3 diners who have a $40 check?
c. Determine the 95% prediction interval for the tip left by a dining party like the one in part (b).
d. Determine the 95% confidence interval for the mean tip left by all dining parties like the one in part (b).
e. Determine the 95% confidence interval for the partial regression coefficients, β_1 and β_2.
f. Interpret the significance tests in the computer printout.
g. Analyze the residuals. Does the analysis support the applicability of the multiple regression model to these data?

16.53 Annual per-capita consumption of all fresh fruits versus that of apples and grapes from 1998 through 2003 was as shown in the table. Source: Bureau of the Census, *Statistical Abstract of the United States 2006*, p. 137.

Year	y = All Fresh Fruits	x_1 = Apples	x_2 = Grapes
1998	128.9 lb/person	19.0 lb/person	7.1 lb/person
1999	129.8	18.7	8.1
2000	128.0	17.6	7.4
2001	125.7	15.8	7.7
2002	126.9	16.2	8.7
2003	126.7	16.7	7.5

a. Determine the multiple regression equation and interpret the partial regression coefficients.
b. What is the estimated per-capita consumption of all fresh fruits during a year when 17 pounds of apples and 6 pounds of grapes are consumed per person?
c. Determine the 95% prediction interval for per-capita consumption of fresh fruits during a year like the one described in part (b).
d. Determine the 95% confidence interval for mean per-capita consumption of fresh fruits during years like the one described in part (b).
e. Determine the 95% confidence interval for the population partial regression coefficients, β_1 and β_2.
f. Interpret the significance tests in the computer printout.
g. Analyze the residuals. Does the analysis support the applicability of the multiple regression model to these data?

16.54 A university placement director is interested in the effect that grade point average (GPA) and the number of university activities listed on the résumé might have on the starting salaries of this year's graduating class. He has collected these data for a sample of 10 graduates:

Graduate	y = Starting Salary (Thousands)	x_1 = Grade Point Average	x_2 = Number of Activities
1	$40	3.2	2
2	46	3.6	5
3	38	2.8	3
4	39	2.4	4
5	37	2.5	2
6	38	2.1	3
7	42	2.7	3
8	37	2.6	2
9	44	3.0	4
10	41	2.9	3

a. Determine the multiple regression equation and interpret the partial regression coefficients.
b. Dave has a 3.6 grade point average and 3 university activities listed on his résumé. What would be his estimated starting salary?
c. Determine the 95% prediction interval for the starting salary of the student described in part (b).
d. Determine the 95% confidence interval for the mean starting salary for all students like the one described in part (b).
e. Determine the 95% confidence interval for the population partial regression coefficients, β_1 and β_2.
f. Interpret the significance tests in the computer printout.
g. Analyze the residuals. Does the analysis support the applicability of the multiple regression model to these data?

16.55 An admissions counselor, examining the usefulness of x_1 = total SAT score and x_2 = high school class rank in predicting y = freshman grade point average (GPA) at her university, has collected the sample data shown here and listed in file **XR16055**. Rank in high school class is expressed as a cumulative percentile, i.e., 100.0% reflects the top of the class.

Freshman GPA	SAT, Total	High School Rank	Freshman GPA	SAT, Total	High School Rank
2.66	1153	61.5	2.45	1136	45.5
2.10	1086	84.5	2.50	966	60.2
3.33	1141	92.0	2.29	1023	74.0
3.85	1237	94.4	2.24	976	86.4
2.51	1205	89.5	1.81	1066	73.0
3.22	1205	97.0	2.99	1076	55.2
2.92	1163	95.9	3.14	1152	72.1
1.95	1121	64.1	1.86	955	51.0

a. Determine the multiple regression equation and interpret the partial regression coefficients.
b. What is the estimated freshman GPA for a student who scored 1100 on the SAT exam and had a cumulative class rank of 80%?
c. Determine the 95% prediction interval for the GPA of the student described in part (b).
d. Determine the 95% confidence interval for the mean GPA of all students similar to the one described in part (b).
e. Determine the 95% confidence interval for the population partial regression coefficients, β_1 and β_2.
f. Interpret the significance tests in the computer printout.
g. Analyze the residuals. Does the analysis support the applicability of the multiple regression model to these data?

16.56 Data file **XR16056** lists the following information for a sample of local homes sold recently: selling price (dollars), lot size (acres), living area (square feet), and whether the home is air conditioned (1 = A/C, 0 = no A/C).
a. Determine the multiple regression equation and interpret the partial regression coefficients.
b. What is the estimated selling price for a house sitting on a 0.9-acre lot, with 1800 square feet of living area and central air conditioning?
c. Determine the 95% prediction interval for the selling price of the house described in part (b).
d. Determine the 95% confidence interval for the mean selling price of all houses like the one described in part (b).
e. Determine the 95% confidence interval for the population partial regression coefficients, β_1, β_2, and β_3.
f. Interpret the significance tests in the computer printout.
g. Analyze the residuals. Does the analysis support the applicability of the multiple regression model to these data?

16.57 In Exercise 16.56, what would be the estimated selling price of a house occupying a 0.01-acre lot, with 100 square feet of living area and no central air conditioning? Considering the nature of this "house," does this selling price seem reasonable? If not, might the computer have made an error in obtaining the regression equation? Explain.

16.58 Data file **XR16058** lists the following data for a sample of automatic teller machine (ATM) customers: time (seconds) to complete their transaction, estimated age, and gender (1 = male, 0 = female).
a. Determine the multiple regression equation and interpret the partial regression coefficients.
b. What is the estimated time required by a female customer who is 45 years of age?
c. Determine the 95% prediction interval for the time required by the customer described in part (b).
d. Determine the 95% confidence interval for the mean time required by customers similar to the one described in part (b).
e. Determine the 95% confidence interval for the population partial regression coefficients, β_1 and β_2.
f. Interpret the significance tests in the computer printout.
g. Analyze the residuals. Does the analysis support the applicability of the multiple regression model to these data?

INTEGRATED CASES

Thorndike Sports Equipment

For several years, Thorndike Sports Equipment has been a minority stockholder in the Snow Kingdom Ski Resort and Conference Center. The Thorndikes visit Snow Kingdom several times each winter to meet with management and find out how business is going. In addition, the visits give them a chance for informal discussions with customers and potential customers of Thorndike ski clothing and equipment. Luke claims that many good product ideas have been inspired by a warm drink in the Snow Kingdom lodge.

On their current visit, Ted and Luke are asked by Snow Kingdom management to lend a hand in analyzing some data that might help in predicting how many customers to expect on any given day. Overall business has been rather steady over the past several years, but the daily customer count seems to have ups and downs. The people at Snow Kingdom are curious as to what factors might be causing the seemingly random levels of patronage from one day to the next.

In response to Ted's request, management supplies data for a random sample of 30 days over the past two seasons. The information includes the number of skiers, the high temperature (degrees Fahrenheit), the number of inches of snow on the ground at noon, and whether the day fell on a weekend (1 = weekend, 0 = weekday). Data are shown here and are also in file **THORN16**.

Using multiple regression and correlation analysis, do you think this information appears to be helpful in explaining the level of daily patronage at Snow Kingdom? Help Ted in interpreting the associated computer printout for an upcoming presentation to the Snow Kingdom management.

Day	Skiers	Weekend	Snow (Inches)	Temperature (Degrees)	Day	Skiers	Weekend	Snow (Inches)	Temperature (Degrees)
1	402	0	22	24	16	648	0	14	8
2	337	0	17	25	17	540	0	17	11
3	471	0	17	28	18	614	1	36	6
4	610	0	39	21	19	796	1	25	29
5	620	0	11	18	20	477	0	13	5
6	545	1	24	17	21	532	0	35	30
7	523	0	25	29	22	732	0	36	15
8	563	0	34	17	23	618	0	18	11
9	873	1	18	11	24	728	1	19	28
10	358	0	28	6	25	620	0	29	8
11	568	0	14	9	26	551	0	18	6
12	453	0	12	18	27	816	0	31	13
13	485	0	27	27	28	765	0	24	19
14	767	1	37	16	29	650	1	22	27
15	735	1	12	6	30	732	0	11	24

Springdale Shopping Survey

This case involves the variables shown here and described in data file **SHOPPING**. Information like that gained from this case could provide management with useful insights into how two or more variables (including dummy variables) can help describe consumers' overall attitude toward a shopping area.

- **Variables 7–9** Overall attitude toward the shopping area. The highest possible rating is 5. In each analysis, one of these will be the dependent variable.
- **Variables 18–25** The importance (highest rating = 7) placed on each of 8 attributes that a shopping area might possess. In each analysis, these will be among the independent variables.
- **Variables 26 and 28** The respondent's gender and marital status. These independent variables will be dummy variables. Recode them so that (1 = male, 0 = female) and (1 = married, 0 = single or other).

The necessary commands will vary, depending on your computer statistical package. Be careful not to save the recoded database, because the original values will be overwritten if you do.

1. With variable 7 (attitude toward Springdale Mall) as the dependent variable, perform a multiple regression analysis using variables 21 (good variety of sizes/styles), 22 (sales staff helpful/friendly), 26 (gender), and 28 (marital status) as the four independent variables. If possible, have the residuals and the predicted y values retained for later analysis.
   a. Interpret the partial regression coefficient for variable 26 (gender). At the 0.05 level, is it significantly different from zero? If so, what does this say about the respective attitudes of males versus females toward Springdale Mall? Interpret the other partial regression coefficients and the results of the significance test for each.
   b. At the 0.05 level, is the overall regression equation significant? At exactly what p-value is the equation significant?
   c. What percentage of the variation in y is explained by the regression equation? Explain this percentage in terms of the analysis of variance table that accompanies the printout.

d. If possible with your computer package, generate a plot of the residuals (vertical axis) versus each of the independent variables (horizontal axis). Evaluate each plot in terms of whether patterns exist that could weaken the validity of the regression model.
e. If possible with your computer statistical package, use the normal probability plot to examine the residuals.

2. Repeat question 1, but with variable 8 (attitude toward Downtown) as the dependent variable.
3. Repeat question 1, but with variable 9 (attitude toward West Mall) as the dependent variable.
4. Compare the regression equations obtained in questions 1, 2, and 3. For which one of the shopping areas does this set of independent variables seem to do the best job of predicting shopper attitude? Explain your reasoning.

BUSINESS CASE

Easton Realty Company (A)

Digital Vision/Getty Images

Sam Easton started out as a real estate agent in Atlanta ten years ago. After working two years for a national real estate firm, he transferred to Dallas, Texas, and worked for another realty agency. His friends and relatives convinced him that with his experience and knowledge of the real estate business, he should open his own agency. He eventually acquired his broker's license and before long started his own company, Easton Realty Company, in Fort Worth, Texas.

Two salespeople at the previous company agreed to follow him to the new company. Easton currently has eight real estate agents working for him. Before the real estate slump, the combined residential sales for Easton Realty amounted to approximately $15 million annually.

Recently, the Dallas–Fort Worth (DFW) metroplex and the state of Texas have suffered economic problems from several sources. Much of the wealth in Texas was generated by the oil industry, but the oil industry has fallen on hard times in recent years. Many savings and loan (S&L) institutions loaned large amounts of money to the oil industry and to commercial and residential construction. As the oil industry fell off and the economy weakened, many S&Ls found themselves in difficulty as a result of poor real estate investments and the soft real estate market that was getting worse with each passing month. With the lessening of the Cold War, the federal government closed several military bases across the country, including two in the DFW area. Large government contractors, such as General Dynamics, had to trim down their operations and lay off many workers. This added more pressure to the real estate market by putting more houses on an already saturated market. Real estate agencies found it increasingly difficult to sell houses.

Two days ago, Sam Easton received a special delivery letter from the president of the local Board of Realtors. The board had received complaints from two people who had listed and sold their homes through Easton Realty in the past month. The president of the Board of Realtors was informing Sam of these complaints and giving him the opportunity to respond. Both complaints were triggered by a recent article on home sales that appeared in one of the local newspapers. The article contained the table shown below.

Typical Home Sale, DFW Metroplex

Average Sales Price	$104,250
Average Size	1860 sq. ft.

Note: Includes all homes sold in the Dallas–Fort Worth metroplex over the past 12 months.

The two sellers charged that Easton Realty Company had underpriced their homes in order to accelerate the sales. The first house is located outside of the DFW area, is four years old, has 2190 square feet, and sold for $88,500. The second house is located in Fort Worth, is nine years old, has 1848 square feet, and sold for $79,500. Both houses in question are three-bedroom houses. Both sellers believe that they would have received more money for their houses if Easton Realty had priced them at their true market value.

Sam knew from experience that people selling their homes invariably overestimate the value. Most sellers believe they could have gotten more money from the sale of their homes. But Sam also knew that his agents would not intentionally underprice houses. However, in these bad economic times, many real estate companies, including Easton Realty, had large inventories of houses for sale and needed to make sales. One quick way to unload these houses is to underprice them. On a residential sale, an agent working under a real estate broker typically makes about 3% of the sales price if he originally listed the property. Dropping the sales price of a $100,000 home down to $90,000 would speed up the sale and the agent's commission would fall only from $3,000 to $2,700. Some real estate agents might consider sacrificing $300 in order to get their commission sooner, but it is unethical because the agent is supposed to be representing the seller and acting in the seller's best interests. Sam had to convince the two sellers and the Board of Realtors that there was no substance to the complaints. The question was, How was he going to do it?

First, he needed to obtain recent residential sales data. Unfortunately, the local multiple listing service (MLS) did not contain actual sales prices of homes. However, Pat McCloskey, a local real estate appraiser, did maintain a database that had the sales information Sam needed. Phoning Pat, Sam found that she indeed had the data he required, but she would have to merge her personal database with data downloaded from the MLS in order to give Sam the necessary information. Fortunately, this was a relatively simple task, and Pat could get the data to Sam the next day.

Sam had asked Pat to give him all the data she had on home sales that had taken place in the DFW area over the previous three months. Although Pat's database did not contain all home sales in the DFW metroplex over that period of time, she felt the data she had were representative of the entire population. The data for each home sold included the sale month, the sale price, the size of the home (in square feet of heated floor space), the number of bedrooms, the age of the house, whether the house was located in Dallas versus Fort Worth or elsewhere within the metroplex, and the real estate company that made the sale.

Assignment

The real estate data compiled by Pat McCloskey for Sam Easton are contained in the file **EASTON**. The Data Description section provides a partial listing of this data file along with definitions of the variables. Using this data set and other information given in the case, help Sam Easton respond to the underpricing claims of his former clients. The Case Questions will assist you in your analysis of the data. Use important details from your analysis of the data to support your recommendation.

Data Description

The **EASTON** file contains data on home sales over the past three months in the DFW metroplex. A partial listing of the data is shown at the top of the next page.

The variables are defined as follows:

MONTH: Month in which the sale took place:
- 4, if April
- 5, if May
- 6, if June

PRICE: Sale price of the house in dollars

SQFEET: Square feet of heated floor space

BEDROOMS: Number of bedrooms in the house

AGE: Age of the house in years

DALLAS: Area within the DFW metroplex where the house is located:
- 1, if in Dallas
- 0, if in Fort Worth or elsewhere in the metroplex

EASTON:
- 1, if Easton Realty Company sold the house
- 0, otherwise

Case Questions

1. Considering the claim that the two houses in question did not sell for their fair market value:
   a. Compare the selling prices of the two houses to the average selling price of all houses sold in the most recent three-month period. Does the difference appear to be substantial?
   b. Provide at least two reasons why the comparison in part (a) is not fair—i.e., describe at least two pricing factors that are not being considered in drawing the comparison in part (a).
   c. In making their argument, the complaining sellers are relying heavily on the average selling price ($104,250) stated in the article for all homes sold in the area during the previous *twelve* months during a weakening housing market. Compare the $104,250 average with that for all houses sold in the area during the most recent three months, then comment on how this comparison might affect the validity of their argument.
2. Use a multiple regression model to estimate PRICE as a function of SQFEET, BEDROOMS, AGE, DALLAS, and EASTON.
   a. Interpret the partial regression coefficients in the equation. Specifically, comment on the sign and numerical value of the partial regression coefficient for EASTON in terms of the claim that Easton Realty Company has been underpricing its residential properties relative to other real estate companies.
   b. For each of the two houses that are the subject of complaints, construct and interpret the 95% prediction interval for a comparable house that is being sold by a realtor other than Easton.
   c. On the basis of your analyses in question 1 and in parts (a) and (b) of this question, prepare a brief, convincing response to the claims of underpricing.

Month	Price	SqFeet	Bedrooms	Age	Dallas	Easton
4	$73,800	1284	2	6	1	0
4	69,200	1919	3	8	0	0
4	98,500	2316	3	7	0	0
4	82,200	1821	4	4	0	0
4	100,300	1703	3	4	1	0
⋮	⋮	⋮	⋮	⋮	⋮	⋮

BUSINESS CASE

Circuit Systems, Inc. (C)

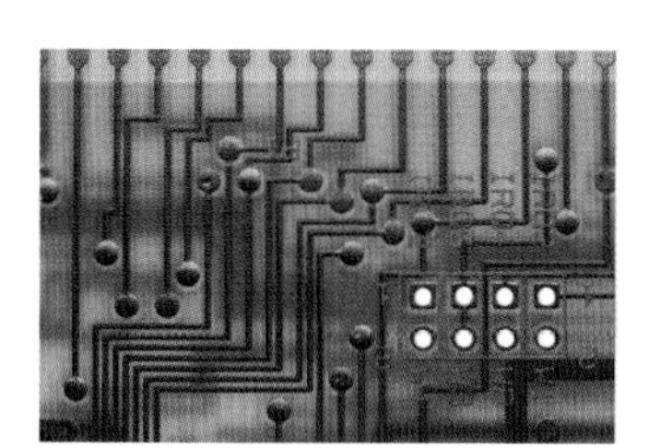

Photodisc/Getty Images

In Chapters 11 and 14, we visited Circuit Systems, Inc., a company that was concerned about the effectiveness of its new program for reducing the cost of absenteeism among hourly workers. Besides allowing employees to be paid for unused sick leave, the program also pays $200 toward membership in a local health club for those employees who wish to join. Thomas Nelson, who introduced the new program, has collected data describing the 233 employees who have been with the company for both the year preceding the new program and the year following its implementation. The information Mr. Nelson has collected is in the **CIRCUIT** data file, and the variables and description of his program are as described in Circuit Systems, Inc. (A), in Chapter 11.

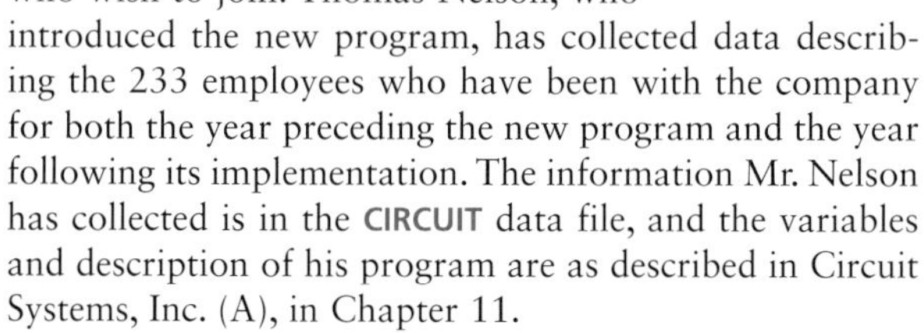

1. Use a multiple regression model to estimate the number of days of sick leave this year as a function of two variables: days of sick leave taken last year and whether the employee is a participant in the exercise program. Interpret the partial regression coefficients in the equation. Specifically, comment on the sign and numerical value of the partial regression coefficient for the dummy variable representing participation in the exercise program. On this basis, does the exercise program seem worthy of continuation?
2. What percentage of the variation in days of sick leave this year is explained by the regression equation you generated for question 1? To the extent that there is unexplained variation in the number of sick days taken this year, what other variables—including possible variables not in the database—might influence the number of days of sick leave taken this year?

CHAPTER 17

Model Building

THE NFL PASSER RATING

To many fans, one of the more mysterious facets of the National Football League (NFL) is its numerical rating of quarterbacks. It's something the sports announcers and "Monday morning quarterbacks" like to talk about, but they spend little or no time talking about how this rating is determined. Media listings of passer ratings are usually accompanied by other performance statistics, such as passing yards, completion percentage, touchdowns, interceptions, and so on.

In this chapter, we learn about stepwise regression, a method by which the candidate predictor variables are mathematically evaluated before being invited into the equation. As in the NFL, not all of them make the team. At each round of the evaluations, just one is invited in, and the invitations stop when we run out of variables that are able to make a meaningful contribution to estimating the dependent variable.

Later in the chapter, you can use Exercise 17.73 and stepwise regression in examining the NFL passer rating. The idea is to find out which variables are actually used, and how they might be combined. Although the passer rating is an interesting and useful measure, the official NFL site is quick to add this caveat: "Judging a quarterback based solely on statistical success overlooks critical intangibles such as leadership and decision making. While an elite quarterback routinely demonstrates his ability to put up big numbers, the great ones deliver those performances while leading their teams to victories."

Source: Bucky Brooks, "Game's Best Quarterbacks Produce Stats as Well as Victories," NFL.com, August 18, 2009.

Models represent the real thing, but are a lot easier to work with.

- Build polynomial regression models to describe curvilinear relationships.
- Apply qualitative variables representing two or more categories.
- Use logarithmic transformations in constructing exponential and multiplicative models.
- Identify and compensate for multicollinearity in the data.
- Apply stepwise regression in selecting which variables to use in a model.
- Determine which of several competing models might be most suitable for the data and the situation

LEARNING OBJECTIVES

After reading this chapter, you should be able to:

(17.1) INTRODUCTION

In Chapter 15, we applied the simple linear regression model in estimating the value of a dependent variable on the basis of a single independent variable. In Chapter 16, we expanded this model to include two or more independent variables. Both models assume a linear relationship, and the multiple regression model assumes that the independent variables are relatively independent from each other. In this chapter, we will examine more sophisticated models for the estimation of y. These will include curvilinear models as well as models in which qualitative variables are considered. We will also consider models in which the independent variables are related to each other. We will introduce a regression method that allows us to strategically choose a subset of the independent variables that has nearly the predictive power of the original, larger set.

The final section focuses on an important concept that is considered throughout the chapter: the importance of selecting a model that is appropriate to both the data and our analysis objectives. As an introductory example, note the scatter diagrams in parts (a) and (b) of Figure 17.1. As x increases, y initially increases, then it decreases, and then it increases again. As part (a) indicates, a linear model is not a very good fit to the data. In part (b), we use one of the models that will be presented in Section 17.2, and it does a much better job of estimating y for a given value of x.

(17.2) POLYNOMIAL MODELS WITH ONE QUANTITATIVE PREDICTOR VARIABLE

By extending the simple linear regression model of Chapter 15 to include terms that include the independent (or *predictor*) variable x to a power greater than 1, we can construct models to fit data that are decidedly nonlinear. Such models can fit even data like those shown in Figure 17.1, where y responds to increases in x by initially increasing, then decreasing, then increasing again. The equation shown in part (b) of Figure 17.1 includes both an x^2 term and an x^3 term, and is a member of the family of polynomial models discussed in this section.

The General Polynomial Model

The polynomial model includes p terms that contain the independent variable x, with the size of the exponent increasing in consecutive terms from 1 to a maximum of p. There is only one independent, or predictor variable (x), but each x

FIGURE 17.1

It is important to select an appropriate model for a given set of data. In part (a), the simple regression model of Chapter 15 is applied to the data. The model in part (b), to be introduced in Section 17.2, is a much better fit.

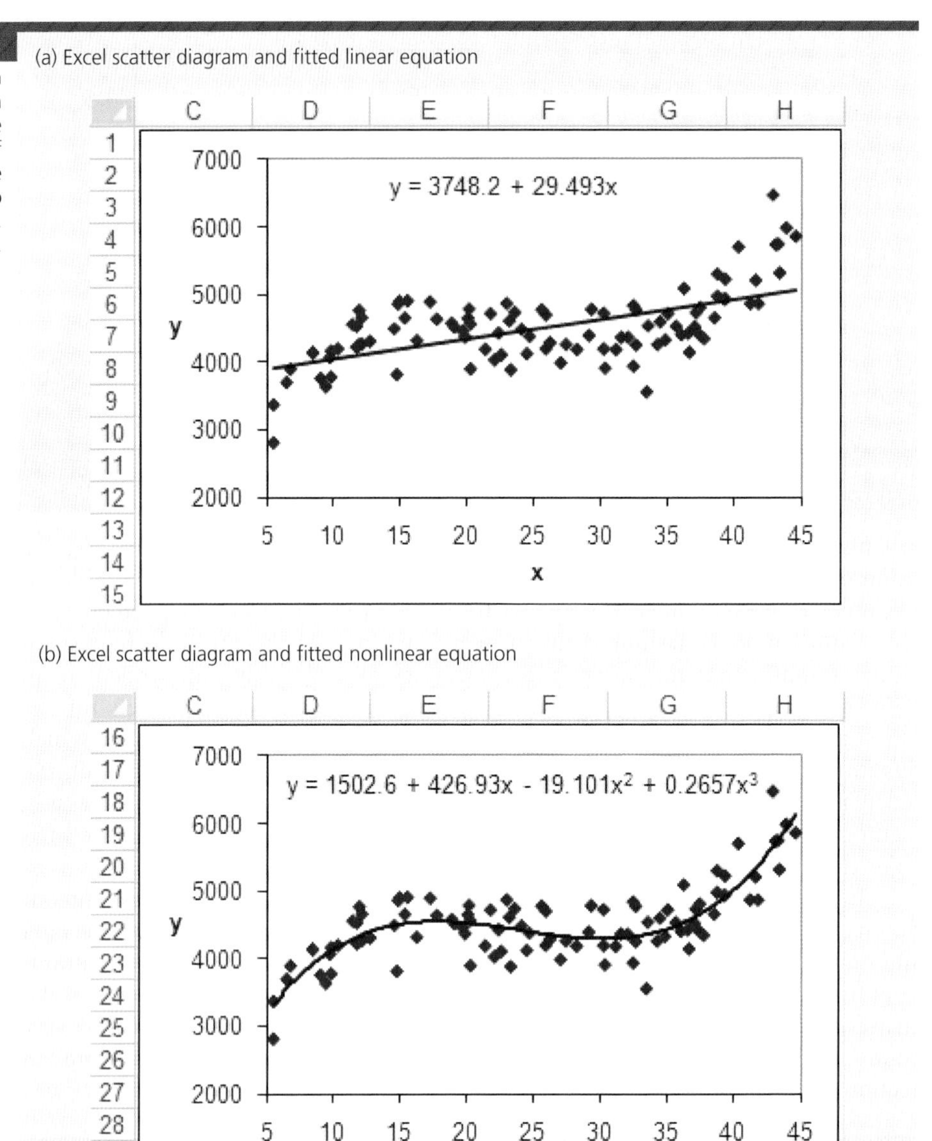

term in the equation is actually being treated as if it were a separate independent variable. The general form of the model is shown here:

The general polynomial model:

$$y_i = \beta_0 + \beta_1 x_i + \beta_2 x_i^2 + \beta_3 x_i^3 + \cdots + \beta_p x_i^p + \epsilon_i$$

where y_i = a value of the dependent variable, y

β_0 = a constant

x_i = a value of the independent (predictor) variable, x

$\beta_1, \beta_2, \ldots, \beta_p$ = partial regression coefficients

ϵ_i = random error, or residual

In the general polynomial model, $y_i = \beta_0 + \beta_1 x_i + \beta_2 x_i^2 + \beta_3 x_i^3 + \cdots + \beta_p x_i^p + \epsilon_i$, the *order* of the polynomial is the integer p. As in simple and multiple regression, the constant (β_0) and the partial regression coefficients ($\beta_1, \beta_2, \ldots, \beta_p$) must be estimated on the basis of sample data. Our discussion in the remainder of this section pertains to polynomial models in which $p = 1, 2$, or 3, because practical situations rarely call for a model that is higher than the third order. The first-, second-, and third-order polynomial models will be discussed individually.

First-Order Polynomial Model

The first-order polynomial model is the same as the linear regression model presented in Chapter 15, and is appropriate when there is a linear relationship between y and x. With $E(y)$ as the expected value of y for a given value of the predictor variable x, the format is

First-order polynomial model:

$$E(y) = \beta_0 + \beta_1 x$$

As shown in part (a) of Figure 17.1, this model is not suitable when the relationship between y and x is nonlinear. When the relationship does happen to be linear, the equation can have either a positive slope ($\beta_1 > 0$) or a negative slope ($\beta_1 < 0$), as shown in Figure 17.2. If you were driving along the plotted line in a car, you would always be driving straight.

Second-Order Polynomial Model

The second-order polynomial model has a parabolic shape in that the curve can resemble a "U" that is either upright or upside-down. The format for this model is

Second-order polynomial model:

$$E(y) = \beta_0 + \beta_1 x + \beta_2 x^2$$

In this model, the $\beta_1 x$ term provides a linear component while the $\beta_2 x^2$ term provides a nonlinear component that allows the line to curve. The shape of the curve will be determined by the relative sizes of β_1 and β_2 and on their signs, and typical curves for this model are shown in Figure 17.3. To the extent that

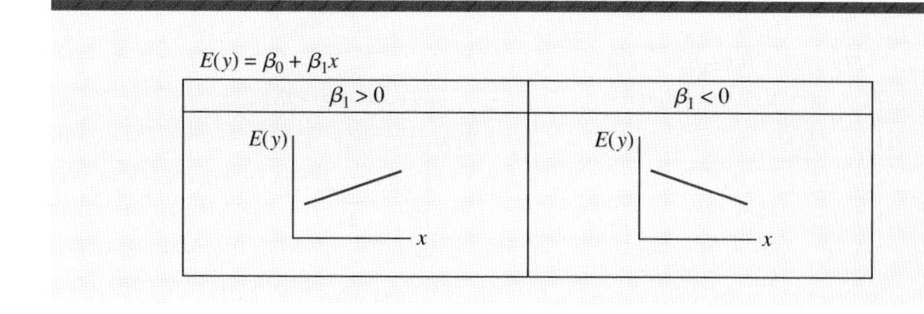

FIGURE 17.2

In the first-order polynomial model, the expected value of y is $E(y) = \beta_0 + \beta_1 x$. This is the same as the simple regression model in Chapter 15, and the slope can be either positive ($\beta_1 > 0$) or negative ($\beta_1 < 0$).

FIGURE 17.3

For the second-order polynomial model, the expected value of y will be $E(y) = \beta_0 + \beta_1 x + \beta_2 x^2$, and the shape of the curve will depend on both the signs and the relative sizes of the coefficients β_1 and β_2. With $x > 0$, combinations of positive and negative values of these coefficients can result in curves like these.

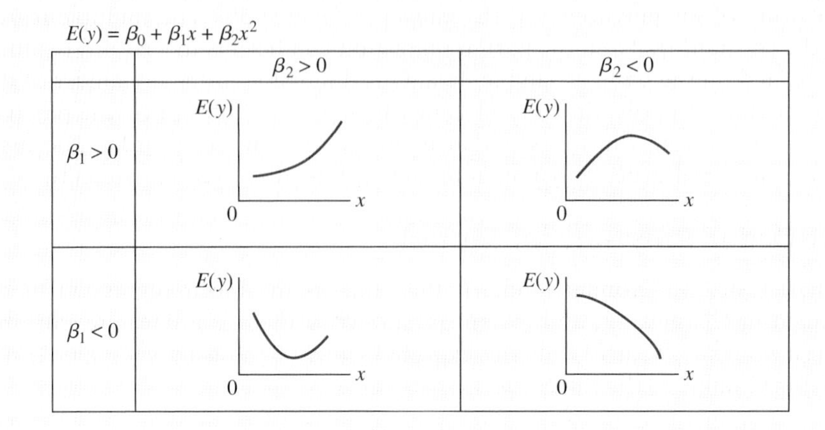

any given second-order polynomial exhibits a curve, it will always curve in the same direction.

Third-Order Polynomial Model

As with the second-order model, the third-order polynomial model allows the plotted line to curve. However, it makes it possible for the line *to reverse the direction in which it is curving*. The format for this model is

Third-order polynomial model:

$$E(y) = \beta_0 + \beta_1 x + \beta_2 x^2 + \beta_3 x^3$$

In this model, the $\beta_1 x$ term provides a linear component, the $\beta_2 x^2$ term provides a nonlinear component allowing the line to curve, and the $\beta_3 x^3$ term provides a nonlinear component enabling the line to reverse the direction of curvature. The sign of β_1 determines the initial slope of the line for small positive values of x. If β_2 and β_3 have opposite signs, the line will be able to reverse the direction it is curving. If β_2 and β_3 have the same sign, the effect of the $\beta_3 x^3$ term will be to make the line curve more strongly in the direction it is already curving. Figure 17.4 shows example plots of the third-order model when β_2 and β_3 have opposite signs.

FIGURE 17.4

In the third-order polynomial model, the expected value of y is $E(y) = \beta_0 + \beta_1 x + \beta_2 x^2 + \beta_3 x^3$. When β_2 and β_3 have opposite signs, the fitted line can reverse its direction of curvature. For positive values of x and β_1, these are examples of curves when $\beta_2 > 0$ and $\beta_3 < 0$, and vice versa.

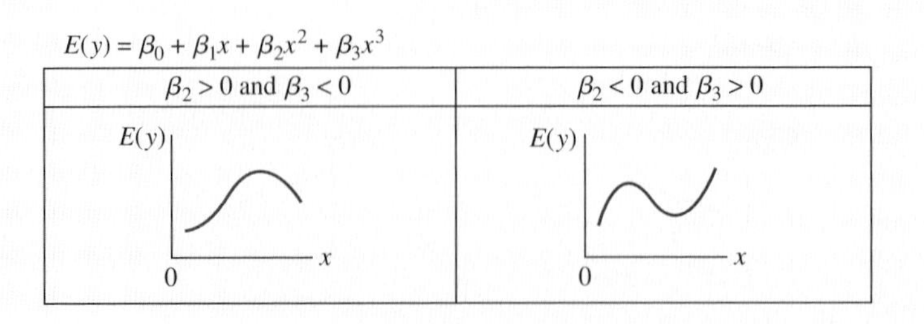

EXAMPLE

Fitting a Polynomial Model

U.S. News & World Report compiles information each year regarding leading U.S. graduate schools in a variety of fields. Data file CX17MBA includes the magazine's overall score for each of 25 leading MBA programs along with the average starting salary and bonus for the program's most recent graduating class.[1] In our analysis, we will use overall score as a predictor variable to estimate starting salary and bonus.

SOLUTION

Computer Solutions 17.1 describes the procedures and shows the results when we use Excel and Minitab in fitting a polynomial model to a set of data. With either software package, it is very easy to select whether the model will be first, second, or third order. Note that Minitab refers to these as linear, quadratic, or cubic, respectively. In Computer Solutions 17.1, we fit a second-order polynomial model to the data. With x = overall score and y = average starting salary and bonus (thousands of dollars), the estimation equation and coefficient of determination, R^2, are

$$\hat{y} = -39.678 + 2.887x - 0.0109x^2, \quad \text{with } R^2 = 0.885$$

EXAMPLEEXAMPLEEXAMPLEEXAM

COMPUTER 17.1 SOLUTIONS

Fitting a Polynomial Regression Equation, One Predictor Variable

EXCEL

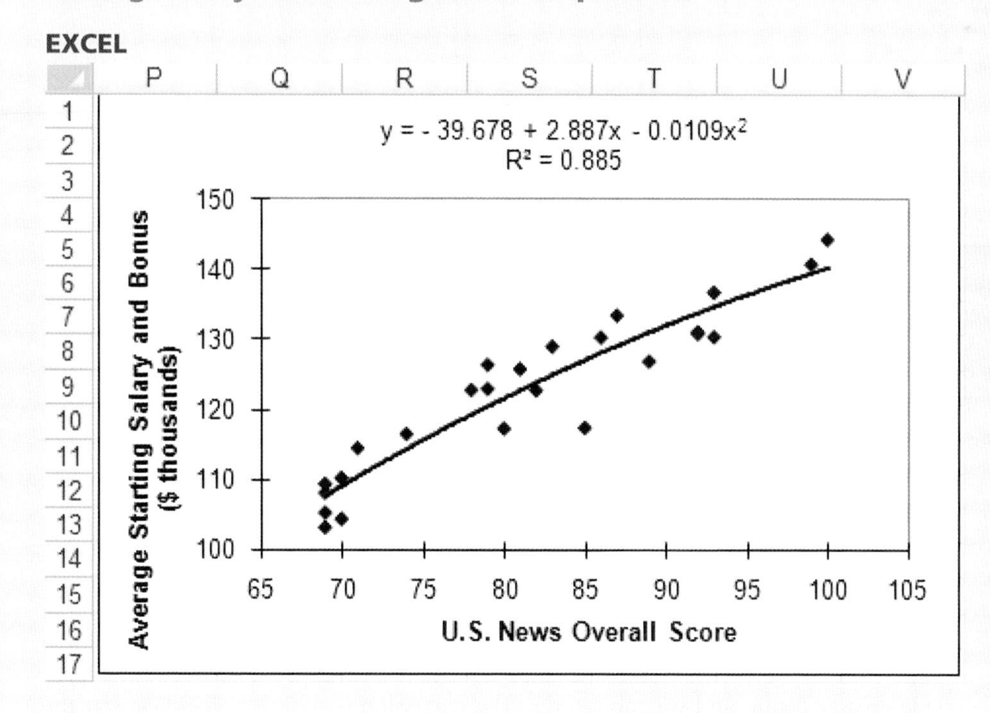

(continued)

[1]Source: *U.S. News & World Report: America's Best Graduate Schools 2009*, p. 14.

Fitting a Polynomial Regression Equation with Excel

1. Using Excel file **CX17MBA**, with MBA average starting salary and bonus data (in thousands of dollars) in A2:A26, *U.S. News & World Report* overall scores in F2:F26 and labels in A1 and F1: From the **Data** ribbon, click **Data Analysis.** Select **Regression.** Click **OK.**
2. Enter **A1:A26** into the **Input Y Range** box. Enter **F1:F26** into the **Input X Range** box. Select **Labels.** Enter **H1** into the **Output Range** Box. Select **Line Fit Plots.** Click **OK.**
3. Right-click within the legend descriptions to the right of the chart and click **Delete.** Right-click on any one of the linearly-arranged squares describing predicted values for the data points and click **Delete.**
4. Right-click on any one of the data points in the scatter diagram and click **Add Trendline.** In the **Trendline Options** menu, select **Polynomial** and enter **2** into the **Order** box. Click to select the **Display Equation on chart** and **Display R-squared value on chart** options. Click **Close.** Make formatting adjustments to the chart.

Note: To obtain the Excel printout in Table 17.1: In step 2, above, enter **F1:G26** into the **Input X Range** box and do not select **Line Fit Plots.** Omit steps 3 and 4. (The squares of the overall scores must be in column G.)

MINITAB

Fitting a Polynomial Regression Equation with Minitab

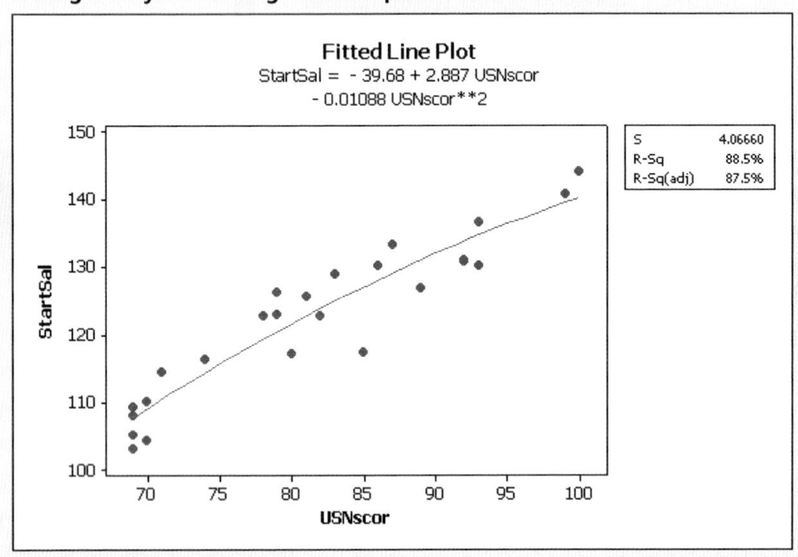

Using Minitab file **CX17MBA**, with MBA average starting salary and bonus data (in thousands of dollars) in C1 and *U.S. News & World Report* overall scores in C6: Click **Stat.** Select **Regression.** Click **Fitted Line Plot.** Enter **C1** into the **Response (Y)** box. Enter **C6** into the **Predictor (X)** box. We will be using the second-order polynomial model in this example. Within the **Type of regression model** box, select **Quadratic.** Click **OK.** The Minitab plot, fitted equation, and R^2 items appear as shown here.

Note: The Minitab printout in Table 17.1 is automatically generated when we carry out the Fitted Line Plot procedure described above.

EXAMPLE

Interpreting $R^2 = 0.885$, we can say that 88.5% of the variation in average starting salary and bonus is explained by the predictor variable, overall score, through the fitted equation. Table 17.1 shows the printouts obtained from performing a multiple regression analysis with x and x^2 treated as separate variables. These printouts provide greater detail, and we can see that the regression equation is significant at the 0.000 level (rounded). The procedures for obtaining the printouts in Table 17.1 are described in the Excel and Minitab portions of Computer Solutions 17.1.

TABLE 17.1

Regression printouts corresponding to the plots in Computer Solutions 17.1. In each case, we have created a new variable: the square of the *U.S. News & World Report* overall score. The *p*-value for the regression equation is 0.000 (rounded), and Excel reports it in scientific notation (4.64×10^{-11}).

EXCEL

	H	I	J	K	L	M	N
1	SUMMARY OUTPUT						
2							
3	*Regression Statistics*						
4	Multiple R	0.941					
5	R Square	0.885					
6	Adjusted R Square	0.875					
7	Standard Error	4.067					
8	Observations	25					
9							
10	ANOVA						
11		*df*	*SS*	*MS*	*F*	*Significance F*	
12	Regression	2	2800.90	1400.45	84.68	4.64E-11	
13	Residual	22	363.82	16.54			
14	Total	24	3164.71				
15							
16		*Coefficients*	*Standard Error*	*t Stat*	*P-value*	*Lower 95%*	*Upper 95%*
17	Intercept	-39.6777	62.6340	-0.6335	0.5329	-169.5726	90.2172
18	USNscor	2.8870	1.5309	1.8858	0.0726	-0.2879	6.0619
19	USNscor_sq	-0.0109	0.0093	-1.1756	0.2523	-0.0301	0.0083

MINITAB

```
Polynomial Regression Analysis: StartSal versus USNscor

The regression equation is
StartSal = - 39.68 + 2.887 USNscor - 0.01088 USNscor**2

S = 4.06660   R-Sq = 88.5%   R-Sq(adj) = 87.5%

Analysis of Variance

Source      DF       SS       MS      F      P
Regression   2  2800.90  1400.45  84.68  0.000
Error       22   363.82    16.54
Total       24  3164.71
```

If we use the procedures in Computer Solutions 17.1 to fit first-order and third-order polynomial models, the resulting equations and their R^2 values are as shown below.

First-order polynomial:

$$\hat{y} = 33.505 + 1.09x, \quad \text{with } R^2 = 0.878$$

Third-order polynomial:

$$\hat{y} = -1552.294 + 58.031x - 0.675x^2 + 0.0026x^3, \quad \text{with } R^2 = 0.910$$

The third-order polynomial, with $R^2 = 0.910$, explains a greater percentage of the variation in average starting salary and bonus than either of the other models. This is not surprising, because it includes two more terms than the first-order model and one more term than the second-order model. However, its R^2 advantage over the second-order model is relatively slight ($R^2 = 0.910$ versus $R^2 = 0.885$, respectively), and the greater simplicity of the second-order model may make it the most preferable of the polynomial models. In Section 17.8, we will further discuss the process of selecting the appropriate model for a set of data.

EXERCISES

17.1 The models in this section include the first-, second-, and third-order polynomials. What would be the form of a fifth-order polynomial? Would such a polynomial model tend to be practical for use with business data?

17.2 As x increases, y increases, but at a decreasing rate. If a second-order polynomial model were fitted to the scatterplot of the data, what would be the signs of the partial regression coefficients in the model?

17.3 As x increases, y decreases, and at an increasing rate. If a second-order polynomial model were fitted to the scatterplot of the data, what would be the signs of the partial regression coefficients in the model?

17.4 As the temperature increases, an experimental lubricant initially decreases in viscosity (resistance to flow), then increases, then decreases again. Of the polynomial models discussed in this section, which one would tend to be most applicable for such data? What would be the signs of the partial regression coefficient(s) in the model?

17.5 A communications firm employs a large number of technical sales representatives who make personal visits to potential customers. For data collected on y = average amount of sales per call and x = number of years of experience, the accompanying scatter diagram has been constructed. If a third-order polynomial model were fitted to these data, what would be the signs of the partial regression coefficients in the model?

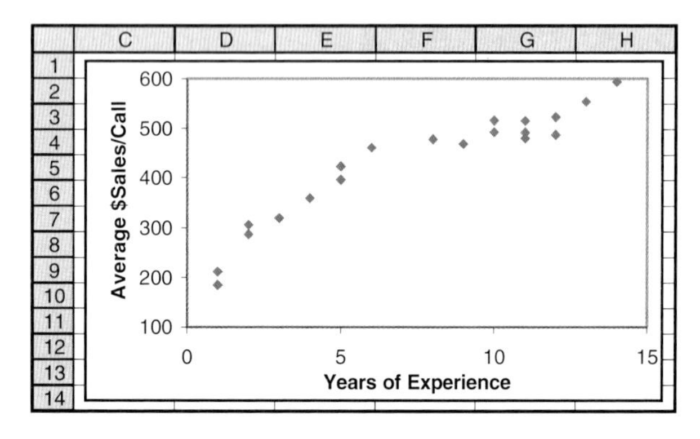

17.6 As the number of cable television subscribers has increased, the number of subscribers to satellite television services has increased as well. For y = number of satellite subscribers and x = number of cable subscribers, the scatter diagram shown here has been constructed. Of the polynomial models discussed in this section, which one would tend to be most applicable for such data, and what would be the signs of the partial regression coefficient(s) in the model? Source: *USA Today,* May 3, 2000, p. 26A.

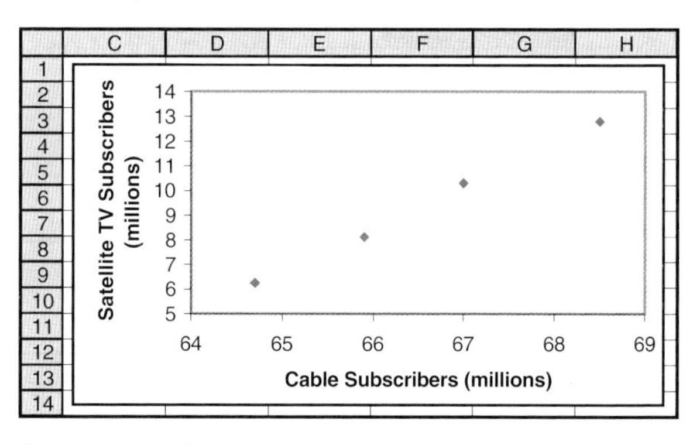

(**DATA SET**) *Note:* Exercises 17.7–17.10 require a computer and statistical software.

17.7 A travel research firm has reported average occupancy and room rates for hotels in 26 U.S. hotel markets. The data are also in file **XR17007**. For y = average room rate (dollars) and x = average occupancy rate (percent), fit a second-order polynomial model to the data and determine the percentage of the variation in average room rates that is explained by the model. Given the nature of these variables, comment on causation possibilities that might be present.

17.8 During the years 2005–2009, the world record for the 100-meter dash was repeatedly broken as never before. The years and corresponding record times are in data file **XR17008** and shown in the table. Using $x = 1$ through 5 to represent years 2005–2009, fit a second-order polynomial model to the data and estimate the world record time for the 100-meter dash in 2014.
Source: Karen Rosen, "9.58: Bolt Strikes Again at Worlds," *USA Today,* August 17, 2009, p. 1C.

Year	x = Year code	World Record (seconds)
2005	1	9.77
2006	2	9.77
2007	3	9.74
2008	4	9.69
2009	5	9.58

17.9 An early edition of *Car and Driver* magazine reported the engine horsepower, top speed (mph), curb weight (pounds), and 0–60 mph acceleration time (seconds) for a selection of automobiles that ranged from large American "muscle cars" to small economy imports. The data are in file **XR17009**. Using y = 0–60 time and x = horsepower, fit a second-order polynomial model and predict the 0–60 time for a car with 300 horsepower. At the 0.05 level of significance, is the regression equation significant? Source: *1964 Car and Driver Yearbook,* pp. 14–63.

17.10 Repeat Exercise 17.9, but for a third-order model and using y = top speed (mph) as the dependent variable. During the 1960s, manufacturers of the so-called "muscle cars" were accused of deliberately understating the massive amount of horsepower these engines produced. Examining the curve for the third-order model, would it appear that these accusations might have been justified?

17.11 For the 20 top-grossing U.S. films of the 1990s, data have been reported for y = foreign gross ticket sales and x = domestic gross ticket sales, both in millions of dollars. The printout here shows the results for a second-order polynomial model fitted to the data. What amount of foreign gross ticket sales would be predicted for a film that generated \$400 million in domestic gross ticket sales? At the 0.05 level of significance, is this model significant? Source: "Overseas Grosses Outdoing North America," *USA Today,* December 13, 2000.

	D	E	F	G	H	I	J
1	SUMMARY OUTPUT						
2							
3	*Regression Statistics*						
4	Multiple R	0.874					
5	R Square	0.764					
6	Adjusted R Square	0.736					
7	Standard Error	112.603					
8	Observations	20					
9							
10	ANOVA						
11		*df*	*SS*	*MS*	*F*	*Significance F*	
12	Regression	2	697099.23	348549.61	27.49	4.71E-06	
13	Residual	17	215551.88	12679.52			
14	Total	19	912651.11				
15							
16		*Coefficients*	*Standard Error*	*t Stat*	*P-value*	*Lower 95%*	*Upper 95%*
17	Intercept	860.8444	244.9641	3.5142	0.0027	344.0146	1377.6743
18	DomGross	-4.1517	1.4392	-2.8848	0.0103	-7.1882	-1.1153
19	DomGross_sq	0.0077	0.0019	4.0038	0.0009	0.0036	0.0117

POLYNOMIAL MODELS WITH TWO QUANTITATIVE PREDICTOR VARIABLES

17.3

In this section, we consider polynomial models in which two quantitative predictor variables are used in estimating the value of the dependent variable. Our discussion will include three such models.

First-Order Model with Two Variables

The first-order polynomial model with two quantitative predictor variables is the same as the two-variable multiple regression model of Chapter 16. Accordingly, the format is simply

First-order model with two predictor variables:

$$E(y) = \beta_0 + \beta_1 x_1 + \beta_2 x_2$$

In this model, the expected value of y is based on x_1 and x_2, as in the multiple regression models of Chapter 16. The model is applicable when there is *no interaction* between x_1 and x_2. Interaction between x_1 and x_2 is absent whenever the following conditions are met:

1. For a given value of x_1, a 1-unit increase in x_2 will cause $E(y)$ to increase by β_2.
2. For a given value of x_2, a 1-unit increase in x_1 will cause $E(y)$ to increase by β_1.

In other words, interaction between the two predictor variables is absent when the effect on $E(y)$ of a 1-unit increase in x_2 does not depend on the value of x_1, and vice versa. Figure 17.5 shows a situation in which interaction is absent. For the estimation equation $\hat{y} = 10 + 5x_1 + 3x_2$, we have plotted three separate linear relationships between $\hat{y}$ and x_1: one for each of the assumed values of x_2. It is very important to note that the lines are *parallel*—regardless of the value of x_2, the slope is the same.

FIGURE 17.5

In this first-order polynomial model with two quantitative predictor variables, there is no interaction between x_1 and x_2. Regardless of the value of x_2, the slope of the line relating $\hat{y}$ to x_1 remains the same.

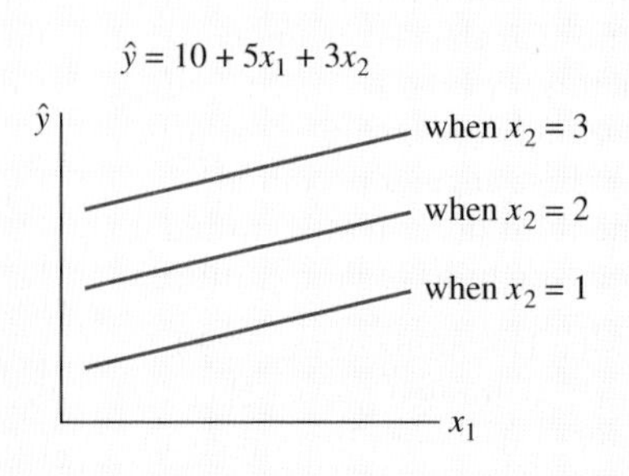

FIGURE 17.6

In this first-order polynomial model with an interaction term included, there is interaction between x_1 and x_2. The slope of the relationship between $\hat{y}$ and x_1 depends on the value of x_2.

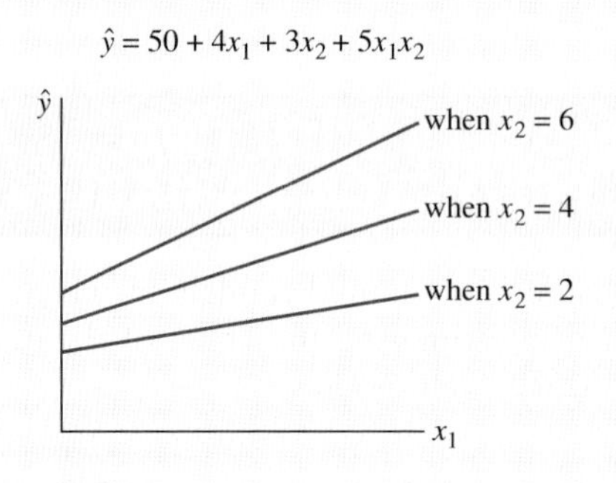

First-Order Model with an Interaction Term

When we believe there may be interaction between the two predictor variables, we can enhance the first-order model just discussed by adding an interaction term. The format for this model is

First-order model with two predictor variables and interaction:

$$E(y) = \beta_0 + \beta_1 x_1 + \beta_2 x_2 + \beta_3 x_1 x_2$$

In this model, the interaction term is $\beta_3 x_1 x_2$. The term includes the partial regression coefficient β_3 and the cross product $x_1 x_2$. Because the sum of the exponents in the cross product is 2, this model can technically be considered a second-order model. However, to avoid confusion with what is typically referred to as the "full" second-order model discussed next, we will consider this to be a first-order model that has had an interaction term added to it.

In this model, interaction is present because the effect on $E(y)$ of a 1-unit increase in x_2 will depend on the value of x_1, and vice versa. Figure 17.6 shows a situation in which interaction is present. For the estimation equation $\hat{y} = 50 + 4x_1 + 3x_2 + 5x_1x_2$, we have plotted three separate linear relationships between $\hat{y}$ and x_1: one for each of the assumed values of x_2. Note that the lines are not parallel and that the slope of the relationship between $\hat{y}$ and x_1 depends on the value of x_2.

Second-Order Model with Interaction

When, in addition to the possibility of interaction between the predictor variables, we believe there could be a nonlinear relationship between y and each of the predictor variables, we can use the second-order model with interaction. The format for this model is

Second-order model with interaction:

$$E(y) = \beta_0 + \beta_1 x_1 + \beta_2 x_2 + \beta_3 x_1^2 + \beta_4 x_2^2 + \beta_5 x_1 x_2$$

In addition to the interaction term $\beta_5 x_1 x_2$, this model includes the quadratic terms $\beta_3 x_1^2$ and $\beta_4 x_2^2$. As mentioned previously, this model is typically referred to as the "full" second-order model—not only do the exponents in the interaction term sum to 2, but each predictor variable is in a term where it is squared.

EXAMPLE

Polynomial Model with Two Predictor Variables

Residents of warm climates are familiar with the heat index that estimates how hot it feels outside. This index is based on two variables: air temperature and relative humidity. File **CX17HEAT** contains data showing the heat index for various combinations of air temperature and relative humidity.[2] The file also includes, as separate variables, the temperature-squared, relative humidity-squared, and (temperature × relative humidity) values. In Computer Solutions 17.2, we use Excel and Minitab to obtain a second-order polynomial model with two predictor variables and an interaction term. The Minitab equation is

$$\begin{aligned}\text{FeelsLike} = {} & 127.74 - 1.76\text{AirTemp} - 1.95\text{RelHumid} + 0.0134\text{AirTemp\_sq} \\ & + 0.0011\text{RelHumid\_sq} + 0.0259(\text{AirTemp} \times \text{RelHumid})\end{aligned}$$

SOLUTION

Interpreting the R^2 portion of the printouts, we can say that 98.0% of the variation in heat index is explained by this model. However, if we apply either of the two simpler models discussed in this section, we notice that one of them ends up obtaining nearly the same R^2 value for these data. The procedures for applying these models are similar to the steps described in Computer Solutions 17.2, and the resulting Minitab equations and R^2 values are

First-order polynomial:

$$\text{FeelsLike} = -55.13 + 1.52\text{AirTemp} + 0.341\text{RelHumid}, \quad \text{with } R^2 = 0.855$$

First-order polynomial, with interaction term added:

$$\begin{aligned}\text{FeelsLike} = {} & 2.54 + 0.867\text{AirTemp} - 1.467\text{RelHumid} + \\ & 0.0212(\text{AirTemp} \times \text{RelHumid}), \quad \text{with } R^2 = 0.968\end{aligned}$$

[2]*Time Almanac 2009*, p. 176.

EXAMPLEEXAMPLEEXAMPLEEXAMPLEEXAMPLEEXAMPLEEXA

Compared to the first-order model with no interaction, including the interactive term improves R^2 quite substantially, from 0.855 to 0.968. However, if we also include the square terms, R^2 goes up just slightly, to 0.980. Because the simpler interactive model does practically as well as the more complex one in its ability to explain variation in the heat index values, it would be the preferred model to fit to these data.

COMPUTER 17.2 SOLUTIONS

Fitting a Polynomial Regression Equation, Two Predictor Variables

EXCEL

	G	H	I	J	K	L	M	N
1	SUMMARY OUTPUT							
2								
3	*Regression Statistics*							
4	Multiple R	0.990						
5	R Square	0.980						
6	Adjusted R Square	0.979						
7	Standard Error	3.317						
8	Observations	93						
9								
10	ANOVA							
11		*df*	*SS*	*MS*	*F*	*Significance F*		
12	Regression	5	46350.120	9270.024	842.617	0.00000		
13	Residual	87	957.127	11.001				
14	Total	92	47307.247					
15								
16		*Coefficients*	*Standard Error*	*t Stat*	*P-value*	*Lower 95%*	*Upper 95%*	*Lower 95.0%*
17	Intercept	127.74311	17.8207	7.1683	0.00000	92.32260	163.16362	92.3226038
18	AirTemp	-1.75589	0.3701	-4.7449	0.00001	-2.49142	-1.02037	-2.4914155
19	RelHumid	-1.95291	0.1301	-15.0099	0.00000	-2.21151	-1.69430	-2.2115116
20	AirTemp_sq	0.01341	0.0019	7.0907	0.00000	0.00965	0.01717	0.009654
21	RelHumid_sq	0.00108	0.0005	2.2802	0.02504	0.00014	0.00202	0.00013838
22	AirTempxRelHumid	0.02586	0.0012	21.1306	0.00000	0.02342	0.02829	0.02342466

Fitting a Polynomial Regression Equation, Two Predictors, with Excel

1. The heat index data are in Excel file **CX17HEAT**, with heat index values in A2:A94, temperatures (degrees Fahrenheit) in B2:B94, relative humidity values in C2:C94, and their labels in A1:C1. The squares of the temperature and relative humidity values are in columns D and E, respectively, and the cross-products (temperature × relative humidity) are in column F.
2. We will generate the regression equation for the second-order model with interaction: From the **Data** ribbon, click **Data Analysis.** Select **Regression.** Click **OK.**
3. Enter **A1:A94** into the **Input Y Range** box. Enter **B1:F94** into the **Input X Range** box. Select **Labels.** Enter **G1** into the **Output Range** box. Click **OK.** After slight adjustments to the column widths and the number of decimal places, the results are as shown in the printout.

MINITAB

Fitting a Polynomial Regression Equation, Two Predictors, with Minitab

```
Regression Analysis: FeelsLike versus AirTemp, RelHumid, ...

The regression equation is
FeelsLike = 128 - 1.76 AirTemp - 1.95 RelHumid + 0.0134 AirTemp_sq
            + 0.00108 RelHumid_sq + 0.0259 AirTempxRelHumid

Predictor                 Coef     SE Coef       T      P
Constant                127.74       17.82    7.17  0.000
AirTemp                -1.7559      0.3701   -4.74  0.000
RelHumid               -1.9529      0.1301  -15.01  0.000
AirTemp_sq            0.013414    0.001892    7.09  0.000
RelHumid_sq          0.0010784   0.0004730    2.28  0.025
AirTempxRelHumid      0.025857    0.001224   21.13  0.000

S = 3.31685   R-Sq = 98.0%   R-Sq(adj) = 97.9%

Analysis of Variance

Source           DF        SS      MS       F      P
Regression        5   46350.1  9270.0  842.62  0.000
Residual Error   87     957.1    11.0
Total            92   47307.2
```

The heat index data are in Minitab file **CX17HEAT**, with heat index values in C1, temperatures (degrees Fahrenheit) in C2, relative humidity values in C3, temperature-squared values in C4, relative humidity-squared values in C5, and the cross-products (temperature × relative humidity) in C6: Click **Stat.** Select **Regression.** Click **Regression.** Enter **C1** into the **Response** box. Enter **C2-C6** into the **Predictors** box. Click **OK.**

EXERCISES

17.12 A researcher suspects that the dependent variable y is linearly related to both x_1 and x_2, but believes there is little or no interaction between the two predictor variables. Of the models in this section, which one would likely be most suitable for this situation?

17.13 In Exercise 17.12, suppose the researcher wishes to use one of the models in this section, but is uncertain as to whether y might be nonlinearly related to x_1 and/or x_2. Also, the researcher doesn't have any idea as to whether interaction might exist between x_1 and x_2. Of the models in this section, which one would result in the highest R^2 value while at the same time reducing the possibility that the researcher might "miss" an important aspect of the relationship between the dependent and predictor variables?

17.14 The regression model $\hat{y} = 200 + 5x_1 - 2x_2$ has been fitted to a set of data. If $x_2 = 10$, what will be the effect on $\hat{y}$ if x_1 increases by 1 unit? Will your answer change if $x_2 = 20$? Given the preceding, comment on whether interaction appears to exist between x_1 and x_2.

(DATA SET) *Note:* Exercises 17.15–17.20 require a computer and statistical software.

17.15 The personnel director of the communications firm in Exercise 17.5 has just provided another variable describing the company's technical sales representatives: their scores on a sales aptitude test they were recently given. Using data file **XR17015**, with y = average dollar sales per call, x_1 = years of experience, and x_2 = score on the sales aptitude test, fit a second-order model with interaction to the data. Interpret the R^2 value for the model and indicate whether, at the 0.05 level, the model is significant.

17.16 *Auto Rental News* has reported U.S. operations data for 15 top rental-car companies. Using data file **XR17016**, with y = rental revenue (millions of dollars), x_1 = thousands of cars in service, and x_2 = number of rental locations: Source: *Wall Street Journal Almanac 1999*, p. 649.

a. Fit a first-order model of the form $E(y) = \beta_0 + \beta_1x_1 + \beta_2x_2$. Interpret the R^2 value for the model and indicate whether, at the 0.01 level, the model is significant.
b. Repeat part (a), but with an interaction term included. Interpreting the R^2 value, to what extent has inclusion of the interaction term improved the model's explanatory power?

17.17 The Air Transport Association has reported the following kinds of information describing 23 individual U.S. airlines: total operating revenue, number of aircraft, number of employees, and number of aircraft departures. Using data file **XR17017**, with y = total operating revenue (millions of dollars), x_1 = number of employees, and x_2 = number of aircraft departures: Source: *Wall Street Journal Almanac 1999*, p. 644.

a. Fit a first-order model of the form $E(y) = \beta_0 + \beta_1x_1 + \beta_2x_2$. Interpret the R^2 value for the model and indicate whether, at the 0.01 level, the model is significant.
b. Repeat part (a), but with an interaction term included. Interpreting the R^2 value, to what extent has inclusion of the interaction term improved the model's explanatory power?

17.18 Repeat Exercise 17.17, using the data in file **XR17017**, but with y = total operating revenue (millions of dollars), x_1 = number of aircraft, and x_2 = number of employees.

17.19 For the *Car and Driver* information described in Exercise 17.9, it would seem reasonable that the number of seconds required to accelerate to 60 mph would depend on both a car's weight and its horsepower. Further, it would be logical to assume that horsepower and weight might interact. Using data file **XR17009**, with y = 0–60 mph acceleration time (seconds), x_1 = horsepower, and x_2 = curb weight (pounds), fit a model of the form $E(y) = \beta_0 + \beta_1x_1 + \beta_2x_2 + \beta_3x_1x_2$. Interpret the R^2 value for the model and indicate whether, at the 0.02 level, the model is significant.

17.20 Repeat Exercise 17.19, but using a second-order model with interaction. Interpreting the R^2 value, to what extent has inclusion of the squared terms improved the model's explanatory power?

17.4 QUALITATIVE VARIABLES

In Chapter 16, we briefly discussed the idea of including a binary qualitative variable in the regression analysis. Such a variable has two possible values, typically coded as 1 and 0. In examining home prices, whether a home has central air conditioning can be an informative variable. In studying employee salaries and wages, using gender as a qualitative variable can help in determining whether discrimination might exist within a company's workforce. In this section, we begin by reviewing the inclusion of a qualitative variable representing two possible categories, then go on to consider qualitative variables that happen to have more than two possible values. In regression analysis, qualitative variables are sometimes referred to as *dummy variables*.

Qualitative Variable Representing Two Possible Categories

One of the most important variables in determining a vehicle's highway fuel economy is weight. However, body style could also be important, since some types of vehicles are not very aerodynamic and would tend to get fewer miles per gallon at highway speeds. In a multiple regression of highway fuel economy on

(1) weight and (2) whether the vehicle is a sports-utility vehicle, the format of the fitted equation is

$\text{HwyMpg} = b_0 + b_1\text{Weight} + b_2\text{SUV?}$, where HwyMpg = highway miles per gallon
Weight = curb weight (pounds)
SUV? = sports-utility vehicle dummy variable (1 if vehicle is an SUV, 0 if not)

EXAMPLE

Qualitative Variable with Two Categories

For a sample of vehicles described in *Consumer Reports' New Car Preview 2001,* data include highway fuel economy test results, vehicle weights, and vehicle types.[3] Data file **CX17FUEL** describes 53 sports-utility vehicles, large or family sedans, and small cars in terms of their highway fuel economy and their curb weight. With the variable SUV? coded as 1 for sports-utility vehicles and 0 for other vehicles, we obtain the following fitted regression equation and R^2:

$$\text{HwyMpg} = 52.3 - 0.00534\text{Weight} - 6.01\text{SUV?}, \quad \text{with } R^2 = 0.784$$

SOLUTION

Interpreting this equation, we can say that the estimated highway fuel economy will decrease by 0.00534 mpg for each additional pound of curb weight. The coefficient for the SUV? variable indicates that, for a given curb weight, the estimated highway fuel economy will be 6.01 mpg lower if the vehicle is an SUV. The regression equation explains 78.4% of the variation in highway fuel economy for the vehicles in the sample.

EXAMPLEEXAMPLEEXAMPLEEXAMPLE

Qualitative Variables Representing More Than Two Possible Categories

With slight modification, the strategy we've just employed can also be used when a qualitative variable happens to represent more than two possible categories. For the new-car data described previously, we have $n = 3$ types of vehicles in data file **CX17FUEL**, but we will need only $(n - 1) = 2$ qualitative variables to represent

[3]Consumer Reports, *New Car Preview 2001.*

them. Using SUV? and FamSedan? as the two qualitative variables, our coding scheme is as follows:

SUV? =	FamSedan? =	Represents:
1	0	sports-utility vehicle
0	1	large/family sedan
0	0	small car

A portion of the data from file CX17FUEL is shown here; it represents the first five vehicles in the listing.

HwyMpg	Weight	SUV?	FamSedan?	
26	3225	1	0	← sports-utility vehicle
33	3455	0	1	← large/family sedan
30	4070	0	1	
38	3325	0	1	
39	2780	0	0	← small car

EXAMPLE

Qualitative Variable with More Than Two Categories

For the sample of vehicles in data file CX17FUEL, and using the coding scheme described above, we carry out a multiple regression with HwyMpg as the dependent variable, and Weight, SUV?, and FamSedan? as the predictor variables. The procedures and results for Excel and Minitab are shown in Computer Solutions 17.3. The equation can be expressed as

$$\text{HwyMpg} = 50.54 - 0.00428\text{Weight} - 8.62\text{SUV?} - 2.58\text{FamSedan?},$$
$$\text{with } R^2 = 0.799$$

SOLUTION

Interpreting this equation, we can say that the estimated highway fuel economy will decrease by 0.00428 mpg for each additional pound of curb weight. The estimated highway fuel economy will be 8.62 mpg lower if the vehicle is an SUV and 2.58 mpg lower if it is a large or family sedan. If the vehicle is a small car (i.e., SUV? = 0 and FamSedan? = 0), the estimated highway fuel economy will be 50.54 − 0.00428Weight − 8.62(0) − 2.58(0), or simply 50.54 − 0.00428Weight. The equation explains 79.9% of the variation in highway fuel economy.

Figure 17.7 describes the linear estimation equations when we consider the three types of vehicles individually. For the equation estimating highway fuel economy only for small cars, SUV? = 0, FamSedan? = 0, and the constant in the equation is 50.54. The constants in the corresponding sports-utility vehicle and large/family sedan equations will be smaller. As shown in Figure 17.7, the

EXAMPLEEXAMPLEEXAMPLEEXAMPLEEXAMPLEEXAMP

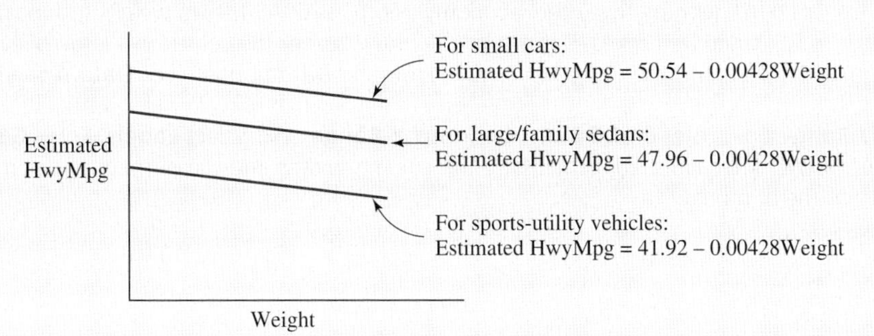

FIGURE 17.7

Given the multiple regression equation developed in Computer Solutions 17.3, these are the highway fuel-economy estimation equations for the individual types of vehicles in the analysis.

constant in the estimation equation for sports-utility vehicles is 50.54 − 8.62, or 41.92, and the constant in the equation for large/family sedans is 50.54 − 2.58, or 47.96.

EXAMPLE

COMPUTER 17.3 SOLUTIONS

Multiple Regression with Qualitative Predictor Variables

EXCEL

	G	H	I	J	K	L	M
1	SUMMARY OUTPUT						
2							
3	*Regression Statistics*						
4	Multiple R	0.894					
5	R Square	0.799					
6	Adjusted R Square	0.787					
7	Standard Error	3.089					
8	Observations	53					
9							
10	ANOVA						
11		*df*	*SS*	*MS*	*F*	*Significance F*	
12	Regression	3	1856.24	618.75	64.86	0.0000	
13	Residual	49	467.46	9.54			
14	Total	52	2323.70				
15							
16		*Coefficients*	*Standard Error*	*t Stat*	*P-value*	*Lower 95%*	*Upper 95%*
17	Intercept	50.5445	2.506	20.168	0.0000	45.508	55.581
18	Weight	-0.0043	0.001	-4.738	0.0000	-0.006	-0.002
19	SUV?	-8.6198	1.764	-4.885	0.0000	-12.166	-5.074
20	FamSedan?	-2.5828	1.348	-1.915	0.0613	-5.292	0.127

(continued)

Excel Multiple Regression with Qualitative Predictor Variables

1. The new-car data are in Excel file **CX17FUEL** with highway fuel economy (mpg) in A2:A54, curb weight (pounds) in B2:B54, whether the vehicle is an SUV (coded as 1 if yes, 0 if no) in C2:C54, whether the vehicle is a large or family sedan (coded as 1 if yes, 0 if no) in D2:D54, and their labels in A1:D1. From the **Data** ribbon, click **Data Analysis.** Select **Regression.** Click **OK.**
2. Enter **A1:A54** into the **Input Y Range** box. Enter **B1:D54** into the **Input X Range** box. Select **Labels.** Enter **G1** into the **Output Range** box. Click **OK.** After slight adjustments to the column widths and the number of decimal places, the results are as shown in the printout.

MINITAB

Minitab Multiple Regression with Qualitative Predictor Variables

```
Regression Analysis: HwyMpg versus Weight, SUV?, FamSedan?

The regression equation is
HwyMpg = 50.5 - 0.00428 Weight - 8.62 SUV? - 2.58 FamSedan?

Predictor          Coef     SE Coef      T      P
Constant         50.545       2.506  20.17  0.000
Weight       -0.0042766   0.0009026  -4.74  0.000
SUV?             -8.620       1.764  -4.89  0.000
FamSedan?        -2.583       1.348  -1.92  0.061

S = 3.08868   R-Sq = 79.9%    R-Sq(adj) = 78.7%

Analysis of Variance

Source          DF       SS      MS      F      P
Regression       3  1856.24  618.75  64.86  0.000
Residual Error  49   467.46    9.54
Total           52  2323.70
```

Using Minitab file **CX17FUEL**, with highway fuel economy (mpg) in C1, curb weight (pounds) in C2, whether the vehicle is an SUV (coded as 1 if yes, 0 if no) in C3, and whether the vehicle is a large or family sedan (coded as 1 if yes, 0 if no) in C4: Click **Stat.** Select **Regression.** Click **Regression.** Enter **C1** into the **Response** box. Enter **C2–C4** into the **Predictors** box. Click **OK.**

EXERCISES

17.21 For a sample of 100 recent credit-card customers of a company's on-line store, the following data are available: the dollar amount of merchandise ordered during the visit, how much time (minutes) the customer spent at the site during the purchase visit, and whether the customer used a Visa credit card. Describe the format of a linear regression equation for estimating the dollar amount of the order on the basis of the other variables described above. If any qualitative variables are used, indicate the strategy for coding them.

17.22 Repeat Exercise 17.21, but assume that more information is available regarding the specific type of credit card that was used for the purchase, that is, Visa, MasterCard, or other.

17.23 Data to be used in a linear regression analysis involving clients of an investment counselor include information regarding the risk level each client has specified: low, medium, or high. Determine the number of qualitative independent variables that will be required to represent this information and describe the coding scheme for each.

17.24 Repeat Exercise 17.23, but for clients whose risk levels have been specified as low, medium, high, or very high.

17.25 The manager of an ice cream store has developed a regression equation for estimating the number of customers on the basis of two independent variables: x_1 = the high temperature (degrees Fahrenheit) for the day, and x_2 = whether the day is during the weekend (coded as 1) or is a weekday (coded as 0). For the estimation equation $\hat{y} = 150 + 5x_1 + 50x_2$, interpret the partial regression coefficients, then estimate the number of customers on an 80-degree Saturday.

17.26 The dean of a law school has developed a regression equation for estimating the starting salary (thousands of dollars) of a new graduate on the basis of two independent variables: x_1 = the score achieved on the Law School Admission Test (LSAT) at the time of application, and x_2 = whether the new graduate's position is in the private sector (coded as 1) or the public sector (coded as 0). For the estimation equation $\hat{y} = 50 + 0.1x_1 + 30x_2$, interpret the partial regression coefficients, then estimate the starting salary for a new graduate with an LSAT score of 160 who is entering the private sector.

(DATA SET) *Note:* Exercises 17.27–17.30 require a computer and statistical software.

17.27 In a *PC World* comparison of computer hard drives, data describing the drives included the following variables: price (dollars), disk capacity (gigabytes), and rotational speed (RPM). The data are in file **XR17027**. With y = price, x_1 = disk capacity, and x_2 = rotational speed (coded as 1 = 7200 RPM and 0 = 5400 RPM), determine the linear regression equation for estimating price as a function of the two predictor variables and interpret the partial regression coefficients. For a given storage capacity, how much extra would one expect to pay for a hard drive with the higher rotational speed? Source: Stan Miastkowski, "Livin' Large," *PC World,* March 2001, p. 109.

17.28 A researcher has observed 50 automatic teller machine (ATM) users outside the gift shop of a museum, and the data are in file **XR17028**. With y = amount withdrawn (dollars), x_1 = time required for the transaction (seconds), and x_2 = gender (1 = female, 0 = male), determine the linear regression equation for estimating the withdrawal amount as a function of the two predictor variables and interpret the partial regression coefficients. According to the equation, how much difference does gender tend to make in the amount withdrawn?

17.29 An efficiency expert has studied 12 employees who perform similar assembly tasks, recording productivity (units per hour), number of years of experience, and which one of three popular assembly methods the individual has chosen to use in performing the task. Given the data shown here, determine the linear regression equation for estimating productivity based on the other variables. For any qualitative variables that are used, be sure to specify the coding strategy employed for each. Comment on the explanatory power of the regression equation and interpret the partial regression coefficients.

Worker	Productivity units/hour	Years of Experience	Assembly Method
1	75	7	A
2	88	10	C
3	91	4	B
4	93	5	B
5	95	11	C
6	77	3	A
7	97	12	B
8	85	10	C
9	102	12	C
10	93	13	A
11	112	12	B
12	86	14	A

17.30 For a small sample of suburban homes recently sold, the selling price ($), living space (square feet), and suburban area (Sunny Haven, Easy Acres, or Rolling Hills) have been recorded. Given the data shown here, determine the linear regression equation for estimating selling price based on the other variables. For any qualitative variables that are used, be sure to specify the coding strategy employed for each. Comment on the explanatory power of the regression equation and interpret the partial regression coefficients.

Home	Price	Space	Area
1	167,900	2860	Easy
2	184,800	2050	Sunny
3	161,000	2110	Rolling
4	160,800	2950	Easy
5	148,200	1570	Sunny
6	128,600	1860	Rolling
7	145,700	2620	Rolling
8	178,500	2010	Sunny
9	146,100	2710	Easy
10	146,300	2200	Rolling

DATA TRANSFORMATIONS (17.5)

When the relationship between the dependent variable and one or more predictor variables is not linear, one of the polynomial models discussed in Sections 17.2 and 17.3 will often provide satisfactory results. However, we sometimes encounter variables for which the underlying relationship is better described by other models that can be converted to linear form. This is possible through *data transformation*, which is what we did in Section 17.2 when we complemented x by forming the x^2 and x^3 terms in the third-order model. Many other types of transformations are possible, such as converting x to its square root, its inverse, or its logarithm. We will examine two nonlinear models that rely on logarithmic transformations to achieve linearity.

For the **exponential model** with estimation equation $\hat{y} = b_0 b_1^x$, conversion to a linear form involves taking the logarithms of both sides of the equation. Either natural logarithms (base $e = 2.71828\ldots$) or common logarithms (base 10) can be used, as long as we are consistent. We will be using common logarithms. The exponential estimation equation is converted to linear form as follows:

Conversion of the exponential equation, $\hat{y} = b_0 b_1^x$, to linear form:

$$\log \hat{y} = \log b_0 + x \log b_1, \text{ or } \log \hat{y} = \log b_0 + (\log b_1)x$$

As a brief review of logarithms, recall that a number is converted to a common logarithm by expressing it in terms of the power to which 10 must be raised. Also, multiplying two numbers is equivalent to adding their logarithms, and raising a quantity to a power is the same as multiplying its logarithm by that power. For example,

- **Logarithms and multiplication**

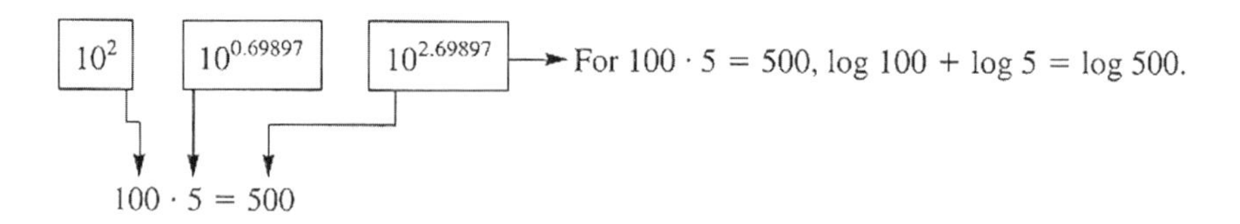

- **Logarithms and exponentiation**

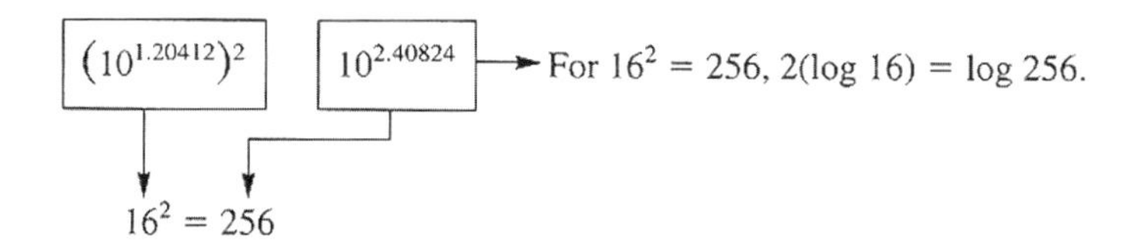

The second model we will consider is the **multiplicative model.** If there are two predictor variables, the estimated value of the dependent variable will be $\hat{y} = b_0 x_1^{b_1} x_2^{b_2}$. The multiplicative equation is converted to linear form as shown next.

Conversion of the multiplicative equation, $\hat{y} = b_0 x_1^{b_1} x_2^{b_2}$, to linear form:

$$\log \hat{y} = \log b_0 + b_1 \log x_1 + b_2 \log x_2$$

As an example of the logarithmic transformation of data, we will use data file CX17ACC, which lists the following variables for a sample consisting of sports cars and sedans described in an early issue of the *Car and Driver Yearbook:*[4]

$$y = \text{0–60 mph acceleration time (seconds)}$$
$$x_1 = \text{engine horsepower}$$
$$x_2 = \text{curb weight (pounds)}$$

Considering the possibility that the time to reach 60 mph might be directly proportional to horsepower and inversely proportional to curb weight (in essence, a multiplicative model with estimation equation $\hat{y} = b_0 x_1^{b_1} x_2^{b_2}$), we can use logarithms in converting this model to linear form. This can be done with either Excel or Minitab, and the procedures and results are shown in Computer Solutions 17.4. The equation and the result when it is converted back to multiplicative form are shown here:

Logarithmic form: $\text{Log\_0–60} = -0.0528 - 0.8607 \text{ log\_hp} + 0.8691 \text{ log\_wt}$

Multiplicative form: $\text{0–60} = 0.8855 \times \text{hp}^{-0.8607} \times \text{curbwt}^{0.8691}$

In this conversion, the antilog of -0.0528 is $10^{-0.0528} = 0.8855$. The estimated 0–60 mph time for a 100-horsepower car weighing 2000 pounds is calculated as $0.8855 \times 100^{-0.8607} \times 2000^{0.8691} = 12.44$ seconds.

In applying the models in this section, all of the original data values must be greater than zero. Also, because the dependent variable is log y instead of y, a comparison of R^2 values with those from regression models having y as the dependent variable may not be appropriate because of the differing units of measurement.[5]

COMPUTER 17.4 SOLUTIONS

Transformation of the Multiplicative Model

EXCEL

	G	H	I	J	K	L	M
1	SUMMARY OUTPUT						
2							
3	*Regression Statistics*						
4	Multiple R	0.923					
5	R Square	0.852					
6	Adjusted R Square	0.836					
7	Standard Error	0.066					
8	Observations	21					
9							
10	ANOVA						
11		*df*	*SS*	*MS*	*F*	*Significance F*	
12	Regression	2	0.455	0.228	52.014	0.000	
13	Residual	18	0.079	0.004			
14	Total	20	0.534				
15							
16		*Coefficients*	*Standard Error*	*t Stat*	*P-value*	*Lower 95%*	*Upper 95%*
17	Intercept	-0.0528	0.5531	-0.0954	0.925	-1.2148	1.1092
18	Log_hp	-0.8607	0.1062	-8.1022	0.000	-1.0839	-0.6375
19	Log_wt	0.8691	0.2170	4.0042	0.001	0.4131	1.3251

(continued)

[4]1964 *Car and Driver Yearbook*, pp. 14–63.
[5]T. Dielman, *Applied Regression Analysis*, Pacific Grove, CA: Duxbury Press, 1996, p. 240.

Excel Transformation of the Multiplicative Model

1. The data are in Excel file **CX17ACC**, with 0–60 mph times (seconds) in A2:A22, horsepower ratings in B2:B22, curb weights (pounds) in C2:C22, and their labels in A1:C1.
2. Click to select cell D2. From the **Formulas** ribbon, click the f_x **(Insert Function)** symbol. In the **Category** box, select **Math & Trig.** Within the **Function** box, select **LOG10.** Click **OK.** Enter **A2** into the **Number** box. Click **OK.** The logarithm of the first 0–60-mph time will now be in cell D2.
3. Click the lower right corner of cell D2 (cursor turns to "+" sign), then drag and fill to F2. Similarly, click the lower right corner of F2, then drag and fill to F22. Enter the column labels (**Log_0-60, Log_hp,** and **Log_wt,** respectively) into D1:F1.
4. From the **Data** ribbon, click **Data Analysis.** Select **Regression.** Click **OK.**
5. Enter **D1:D22** into the **Input Y Range** box. Enter **E1:F22** into the **Input X Range** box. Select **Labels.** Enter **G1** into the **Output Range** box. Click **OK.** After slight adjustments to the column widths and the number of decimal places, the results are as shown above.

MINITAB

Minitab Transformation of the Multiplicative Model

```
Regression Analysis: Log_60 versus Log_hp, Log_wt

The regression equation is
Log_60 = - 0.053 - 0.861 Log_hp + 0.869 Log_wt

Predictor        Coef    SE Coef        T       P
Constant      -0.0528     0.5531    -0.10   0.925
Log_hp        -0.8607     0.1062    -8.10   0.000
Log_wt         0.8691     0.2170     4.00   0.001

S = 0.0661493     R-Sq = 85.2%     R-Sq(adj) = 83.6%

Analysis of Variance

Source            DF        SS        MS       F       P
Regression         2   0.45520   0.22760   52.01   0.000
Residual Error    18   0.07876   0.00438
Total             20   0.53396
```

1. Using Minitab file **CX17ACC**, 0–60 times in C1, horsepower ratings in C2, and curb weights in C3: Click **Calc.** Click **Calculator.** In the **Functions** box, select **Log base 10.** Within the parentheses in the **Expression** box, enter **C1.** Enter **C4** into the **Store result in variable** box. Click **OK.**
2. Repeat this procedure two times, placing the logarithms for C2 into column C5 and placing the logs for C3 into C6.
3. Enter the **Log_0-60, Log_hp,** and **Log_wt** labels at the top of columns **C4–C6,** respectively. Click **Stat.** Select **Regression.** Click **Regression.** Enter **C4** into the **Response** box. Enter **C5–C6** into the **Predictors** box. Click **OK.**

The transformations in this section are but two of the many that can be done. For a more complete discussion of the possibilities and their applications, the interested reader may refer to one of the more advanced treatments cited in the footnote.[6]

[6]See Dielman or J. Neter, M. Kutner, C. Nachtsheim, and W. Wasserman, *Applied Linear Statistical Models*, Homewood, IL: Irwin, 1996.

EXERCISES

17.31 Following data transformation, regression analysis results in the estimation equation $\log \hat{y} = -0.179 + 0.140x$. Transform this equation to the equivalent exponential model with estimation equation $\hat{y} = b_0 b_1^x$. (The base of the logarithms is 10.)

17.32 Following data transformation, regression analysis results in the estimation equation $\log \hat{y} = 3.15 + 0.473 \log x_1 + 0.354 \log x_2$. Transform this equation to the equivalent multiplicative model with estimation equation $\hat{y} = b_0 x_1^{b_1} x_2^{b_2}$. (The base of the logarithms is 10.)

(**DATA SET**) *Note:* Exercises 17.33–17.36 require a computer and statistical software. Use logarithms with base 10.

17.33 For the situation and variables in Exercise 17.7, fit an exponential model and use it to estimate the average room rate for a market in which the average occupancy rate is 70%. The underlying data are in file **XR17007**.

17.34 For the situation and variables in Exercise 17.8, fit an exponential model and use it to estimate the world record time for the 100-meter dash in the year 2014. The underlying data are in file **XR17008**.

17.35 For the situation and variables in Exercise 17.17, fit a multiplicative model and use it to estimate the total operating revenue for an airline with 2000 employees and 10,000 departures. The underlying data are in file **XR17017**.

17.36 For the situation and variables in Exercise 17.16, fit a multiplicative model and use it to estimate the rental revenue for a company with 80,000 cars in service and 2500 rental locations. The underlying data are in file **XR17016**.

MULTICOLLINEARITY (17.6)

In Chapter 16, we briefly discussed *multicollinearity,* a condition in which two or more of the independent variables are highly correlated with each other. When this happens, partial regression coefficients can become both statistically unreliable and difficult to interpret. This is because they are not really "saying" different things about y. We likened multicollinearity to a situation where old-time comedians Laurel and Hardy are carrying a piano, and each comedian is unsure as to how much of the load he should carry. As a result, the piano will tend to be very unstable. Multicollinearity is *not* a problem if we are simply substituting x values into the equation and attempting to estimate y. However, it *is* a problem when we're trying to interpret the partial regression coefficients for the x variables that are highly correlated.

Besides the high correlation between two or more independent variables, there can be other clues to the presence of multicollinearity: (1) an independent variable known to be an important predictor ends up having a partial regression coefficient that is not significant; (2) a partial regression coefficient that should be positive turns out to be negative, or vice versa; or (3) when an independent variable is added or removed, the partial regression coefficients for the other independent variables change drastically.

A practical way to identify multicollinearity is to generate a **correlation matrix.** This is a matrix showing the correlation of each variable with each of the other variables. Computer Solutions 17.5 shows the correlation matrix for the following y and x variables for a sample consisting of 34 large automobile dealerships.[7] The data are in file **CX17DEAL**.

y = TOTSALES, total sales generated during the year (thousands of dollars)
x_1 = PARTSERV, parts and service sales for the year (thousands of dollars)
x_2 = RETUSED, number of used cars sold at retail

[7]Crain Communications, Inc., *Automotive News 1988 Market Data Book,* pp. 95–108.

x_3 = RETNEW, number of new cars sold at retail
x_4 = FLEET, number of cars sold to fleet customers
x_5 = YRSINBUS, number of years the dealership has been in business

COMPUTER 17.5 SOLUTIONS

The Correlation Matrix

EXCEL

	G	H	I	J	K	L	M
1		*TOTSALES*	*PARTSERV*	*RETUSED*	*RETNEW*	*FLEET*	*YRSINBUS*
2	TOTSALES	1.000					
3	PARTSERV	0.794	1.000				
4	RETUSED	0.780	0.386	1.000			
5	RETNEW	0.759	0.333	0.835	1.000		
6	FLEET	0.358	0.322	0.238	0.201	1.000	
7	YRSINBUS	0.190	0.275	-0.085	-0.130	0.336	1.000

Excel Correlation Matrix

1. The dealership data are in Excel file CX17DEAL, with total sales in A2:A35, parts and service sales in B2:B35, used retail units in C2:C35, new retail units in D2:D35, fleet units in E2:E35, years in business in F2:F35, and their labels in A1:F1. From the **Data** ribbon, click **Data Analysis.** Select **Correlation.** Click **OK.**
2. Enter **A1:F35** into the **Input Range** box. In the **Grouped By** menu, select **Columns.** Select **Labels in First Row.** Enter **G1** into the **Output Range** box. Click **OK.** After adjusting the column widths and the number of decimal places, the results are as shown here.

MINITAB

Minitab Correlation Matrix

```
Correlations: TOTSALES, PARTSERV, RETUSED, RETNEW, FLEET, YRSINBUS

           TOTSALES  PARTSERV   RETUSED    RETNEW     FLEET
PARTSERV      0.794
RETUSED       0.780     0.386
RETNEW        0.759     0.333     0.835
FLEET         0.358     0.322     0.238     0.201
YRSINBUS      0.190     0.275    -0.085    -0.130     0.336

Cell Contents: Pearson correlation
```

Using Minitab file CX17DEAL, with total sales in C1, parts and service sales in C2, used retail units in C3, new retail units in C4, fleet units in C5, and years in business in C6: Click **Stat.** Select **Basic Statistics.** Click **Correlation.** Enter **C1–C6** into the **Variables** box. Click **OK.**

With a total of six variables, there are $(6 \cdot 5)/2 = 15$ possible pairs of variables. As shown in the first column of the correlation matrix, several of the independent variables are highly correlated with the dependent variable, TOTSALES. However, there is a very high correlation (0.835) between the number of used (RETUSED) and new (RETNEW) cars sold at retail. This high correlation between two of the independent variables is an example of multicollinearity.

We can reduce the impact of multicollinearity by breaking up "partners" like the RETUSED and RETNEW variables and not including both of them in the equation at the same time. For example, we could decide to include only the "partner" with the higher correlation with y. An alternative approach involves the stepwise regression method described in the next section.

EXERCISES

17.37 What is a correlation matrix, and what role does it play in multiple regression?

17.38 One of the indicators that multicollinearity is present is when a variable that is known to be an important predictor ends up having a partial regression coefficient that is not significant. What are some of the other signals for the presence of multicollinearity?

17.39 In linear regression models, for what types of uses or applications would multicollinearity tend to be a problem? Are there any applications or uses for which multicollinearity would *not* be a problem?

17.40 A linear regression analysis was performed using x_1 as a predictor variable. The p-value associated with the regression coefficient for x_1 was 0.02, and the sign of the coefficient was negative. Another analysis was performed, this time using both x_1 and x_2 as predictor variables. In the second analysis, the partial regression coefficient for x_1 had a positive sign and the p-value associated with it was 0.39. Given these results, would multicollinearity appear to be present?

17.41 Four predictor variables are being considered for use in a linear regression model. Given the accompanying correlation matrix, does it appear that multicollinearity could be a problem?

	E	F	G	H	I
1		x1	x2	x3	x4
2	x1	1.000			
3	x2	0.483	1.000		
4	x3	0.952	0.516	1.000	
5	x4	-0.164	-0.202	-0.113	1.000

(DATA SET) *Note:* A computer and statistical software are required for Exercises 17.42–17.45.

17.42 For the situation and variables in Exercise 17.16, use data file **XR17016** and generate a correlation matrix for the independent variables. Examine the correlation matrix and comment on the possible presence of multicollinearity.

17.43 For the situation and variables in Exercise 17.17, use data file **XR17017** and generate a correlation matrix for these independent variables: number of aircraft, number of employees, and number of aircraft departures. Examine the correlation matrix and comment on the possible presence of multicollinearity.

17.44 Circulation (millions) for Sunday newspapers versus morning and evening daily papers for 1999–2007 are shown here. The data are also in file **XR17044**. With Sunday circulation as the dependent variable, generate the correlation matrix for the independent variables and comment on the possible presence of multicollinearity.
Source: Bureau of the Census, *Statistical Abstract of the United States 2006*, p. 738 and *Statistical Abstract of the United States 2009*, p. 697.

Year	Sunday	Daily, Morning	Daily, Evening
1999	59.9	46.0	10.0
2000	59.4	46.8	9.0
2001	59.1	46.8	8.8
2002	58.8	46.6	8.6
2003	58.5	46.9	8.3
2004	57.8	46.9	7.7
2005	55.3	46.1	7.2
2006	53.2	45.4	6.9
2007	51.2	44.5	6.2

17.45 For the years 1998–2008, production in the U.S. energy sector included the quantities shown here. The variables represent total energy production as well as energy production from nonrenewable fossil fuels. The units are in quadrillion Btu and the data are also in file **XR17045**. For these data, and with y = total energy production, generate the correlation matrix for the independent variables and comment on the possible presence of multicollinearity. Source: U.S. Energy Information Administration, *Annual Energy Review 2008*, pp. 5, 7.

Year	Total Energy Production	Crude Oil	Natural Gas	Coal
1998	66.39	13.24	22.03	24.05
1999	65.23	12.45	21.87	23.30
2000	65.23	12.36	22.27	22.74
2001	66.57	12.28	22.71	23.55
2002	65.03	12.16	22.00	22.73
2003	64.12	12.03	22.04	22.09
2004	64.13	11.50	21.56	22.85
2005	63.22	10.96	20.91	23.19
2006	64.18	10.80	21.38	23.79
2007	64.70	10.72	22.03	23.49
2008	66.41	10.52	23.57	23.86

17.7 STEPWISE REGRESSION

In **stepwise regression,** the variables enter the regression analysis one at a time. The first x variable to enter is the one that explains the greatest amount of variation in y. The second x variable to enter will be the one that explains the greatest amount of the *remaining* variation in y, and so on. At each "step," the computer decides which, if any, of the remaining x variables should be brought in. Every step results in a new regression equation and (if you choose) an updated printout of this equation and its analysis.

Stepwise regression is especially useful when there are a great many x variables. Under these conditions, a lot of independent variables may be highly correlated, and it can be difficult for us humans to sort out the confusion they represent. The general idea of stepwise regression is the balancing act of trying to (1) explain the most possible variation in y, while (2) using the fewest possible x variables.

In general, it's advisable to test a model that has already been hypothesized (e.g., a multiple linear regression between y and x_1, x_3, and x_6). If you don't have a specific model in mind, however, stepwise regression is very useful when you have a great many independent variables and wish to determine which linear model might best apply.

EXAMPLEEXAMPLEEXAMP

EXAMPLE

Stepwise Regression

As shown in Computer Solutions 17.5, two of the five independent variables (RETUSED and RETNEW) are highly correlated with each other. In viewing the application of stepwise regression to these data, we'll observe how the analysis is influenced by this situation in which multicollinearity is obviously present.

SOLUTION

Computer Solutions 17.6 shows the stepwise regression analysis of the underlying data for the six dealership variables. Although there are five independent variables, the procedure has gone through just four steps. The number of vehicles sold to fleet customers (FLEET) is denied entrance to the regression equation. Here is a brief discussion of the steps in the Computer Solutions 17.6 printout.

Step 1 The first variable invited into the equation is PARTSERV, sales revenue from parts and service. This is no surprise, since the correlation matrix shows this to be the independent variable most highly correlated with TOTSALES. At this step, the regression equation is

$$\text{TOTSALES} = -347.6 + 9.55\text{PARTSERV}$$

COMPUTER 17.6 SOLUTIONS

Stepwise Regression

EXCEL

	H	I	J	K	L	M	N
1	**Results of stepwise regression**						
2							
3	**Step 1 - Entering variable: PARTSERV**						
4							
5	Summary measures						
6		Multiple R	0.7940				
7		R-Square	0.6304				
8		Adj R-Square	0.6189				
9		StErr of Est	23143				
10							
11	ANOVA Table						
12		Source	df	SS	MS	F	p-value
13		Explained	1	29236853124	29236853124	54.59	0.000
14		Unexplained	32	17138622464	535581952		
15							
16	Regression coefficients						
17			Coefficient	Std Err	t-value	p-value	
18		Constant	-347.57	7685.61	-0.045	0.964	
19		PARTSERV	9.55	1.29	7.388	0.000	
20							

	H	I	J	K	L	M	N
21	**Step 2 - Entering variable: RETNEW**						
22							
23	Summary measures			Change	% Change		
24		Multiple R	0.9513	0.1573	%19.8		
25		R-Square	0.9050	0.2746	%43.6		
26		Adj R-Square	0.8989	0.2800	%45.2		
27		StErr of Est	11920	-11223	-%48.5		
28							
29	ANOVA Table						
30		Source	df	SS	MS	F	p-value
31		Explained	2	41971102084	20985551042	147.71	0.000
32		Unexplained	31	4404373504	142076565		
33							
34	Regression coefficients						
35			Coefficient	Std Err	t-value	p-value	
36		Constant	-6156.03	4005.73	-1.537	0.134	
37		PARTSERV	7.32	0.71	10.372	0.000	
38		RETNEW	10.81	1.14	9.467	0.000	
39							

(continued)

	H	I	J	K	L	M	N
40	Step 3 - Entering variable: RETUSED						
41							
42	Summary measures			Change	% Change		
43		Multiple R	0.9632	0.0119	%1.3		
44		R-Square	0.9279	0.0228	%2.5		
45		Adj R-Square	0.9206	0.0217	%2.4		
46		StErr of Est	10561	-1359	-%11.4		
47							
48	ANOVA Table						
49		Source	df	SS	MS	F	p-value
50		Explained	3	43029497732	14343165911	128.60	0.000
51		Unexplained	30	3345977856	111532595		
52							
53	Regression coefficients						
54			Coefficient	Std Err	t-value	p-value	
55		Constant	-5874.83	3550.30	-1.655	0.108	
56		PARTSERV	6.91	0.64	10.812	0.000	
57		RETNEW	6.46	1.74	3.724	0.001	
58		RETUSED	10.20	3.31	3.081	0.004	
59							

	H	I	J	K	L	M	N
60	Step 4 - Entering variable: YRSINBUS						
61							
62	Summary measures			Change	% Change		
63		Multiple R	0.9691	0.0058	%0.6		
64		R-Square	0.9391	0.0112	%1.2		
65		Adj R-Square	0.9307	0.0100	%1.1		
66		StErr of Est	9871	-690	-%6.5		
67							
68	ANOVA Table						
69		Source	df	SS	MS	F	p-value
70		Explained	4	43549942148	10887485537	111.74	0.000
71		Unexplained	29	2825533440	97432188		
72							
73	Regression coefficients						
74			Coefficient	Std Err	t-value	p-value	
75		Constant	-10235.89	3817.27	-2.681	0.012	
76		PARTSERV	6.41	0.64	10.100	0.000	
77		RETNEW	6.92	1.63	4.237	0.000	
78		RETUSED	10.42	3.10	3.362	0.002	
79		YRSINBUS	388.30	168.01	2.311	0.028	

Excel Stepwise Regression

1. Using Excel file **CX17DEAL**, with total sales in A2:A35, parts and service sales in B2:B35, used retail units in C2:C35, new retail units in D2:D35, fleet units in E2:E35, years in business in F2:F35, and their labels in A1:F1: Click on any cell within the data field. From the **Add-Ins** ribbon, click **Data Analysis Plus.** Select **Stepwise Regression.** Click **OK.**
2. Check to verify that the cursor is on a cell within the data field, then click **Continue.** If there are no missing values within the data, click **OK.** Click on **TOTSALES** as the response variable and click **OK.** Holding down the **Ctrl** key, click on each of the five explanatory variables to be used: **PARTSERV, RETUSED, RETNEW, FLEET,** and **YRSINBUS.** Click **OK.**
3. In the **Stepwise regression parameters** menu, click **p-values** and click **OK.** Accept the default **p-to-enter** and **p-to-leave** parameters (0.05 and 0.10, respectively) and click **OK.** Select any scatterplots that are desired, and click **OK.** In the **Location of results** menu, select **Same worksheet as the data.** Click **OK.** After adjusting the column widths and the number of decimal places, the results are as shown here.

MINITAB

Minitab Stepwise Regression

```
Stepwise Regression: TOTSALES versus PARTSERV, RETUSED, ...

  Alpha-to-Enter: 0.05  Alpha-to-Remove: 0.1

Response is TOTSALES on 5 predictors, with N = 34

Step              1        2        3        4
Constant     -347.6  -6156.0  -5874.8 -10235.9

PARTSERV       9.55     7.32     6.91     6.41
T-Value        7.39    10.37    10.81    10.10
P-Value       0.000    0.000    0.000    0.000

RETNEW                  10.8      6.5      6.9
T-Value                 9.47     3.72     4.24
P-Value                0.000    0.001    0.000

RETUSED                          10.2     10.4
T-Value                          3.08     3.36
P-Value                         0.004    0.002

YRSINBUS                                   388
T-Value                                   2.31
P-Value                                  0.028

S             23143    11920    10561     9871
R-Sq          63.04    90.50    92.79    93.91
R-Sq(adj)     61.89    89.89    92.06    93.07
Mallows Cp    140.1     15.7      7.2      4.0
```

1. Using Minitab file **CX17DEAL**, with total sales in C1, parts and service sales in C2, used retail units in C3, new retail units in C4, fleet units in C5, and years in business in C6: Click **Stat.** Select **Regression.** Click **Stepwise.** Enter **C1** into the **Response** box. Enter **C2–C6** into the **Predictors** box.
2. For consistency, we will use the same entry/removal parameters as the defaults in the Excel procedure: Click **Methods.** Select **Use alpha values.** Select **Stepwise (forward and backward).** Enter **0.05** into the **Alpha to enter** box. Enter **0.10** into the **Alpha to remove** box. Click **OK.** Click **OK.**

This regression equation explains 63.04% of the variation in TOTSALES, and the multiple standard error of estimate is listed as 23143, or $23.143 million.

Step 2 The next variable to enter is RETNEW, the number of new cars sold at retail. The regression equation now becomes

$$\text{TOTSALES} = -6156.0 + 7.32\text{PARTSERV} + 10.8\text{RETNEW}$$

The inclusion of RETNEW has shaken things up a bit by changing the constant from −347.6 to −6156.0. This is not too unusual, since improving the equation means changing the equation. The percentage of variation in TOTSALES that is explained rises from 63.04% to 90.50%, a hefty increase. The estimating ability of the equation has also improved: The multiple standard error of estimate is reduced to $11.920 million.

MPLEEXAMPLEEXAMPLEE

Step 3 Here comes the Laurel and Hardy piano act. Recall that RETNEW and RETUSED are highly correlated ($r = 0.835$). However, of the variables not yet included, RETUSED explains the greatest amount of remaining variability in TOTSALES, so it comes into the regression equation:

$$\text{TOTSALES} = -5874.8 + 6.91\text{PARTSERV} + 6.5\text{RETNEW} + 10.2\text{RETUSED}$$

The constant term and the partial regression coefficient for PARTSERV have changed slightly, but the partial regression coefficient for RETNEW has decreased dramatically. (In the Laurel and Hardy analogy, RETNEW was carrying most of the piano in step 2, but now RETUSED has taken up part of the load.) Notice how the t-ratio for RETNEW has gone down from 9.47 to 3.72, another reflection of the new sharing of the load. Although RETUSED has a high correlation with TOTSALES ($r = 0.780$), it has not added much to the predictive power of the regression equation. The percentage of variation explained has increased only slightly, from 90.50% to 92.79%. Likewise, the multiple standard error of estimate is reduced very little and is now $10.561 million.

If our sole purpose were the prediction of TOTSALES, there would be no problem in using the equations in this step and the next. On the other hand, if our purpose is to be able to interpret the partial regression coefficients, we should use only the equation that resulted from step 2. At step 3, multicollinearity has set in, and we have lost the ability to make meaningful interpretations of the partial regression coefficients.

Step 4 The next entrant is YRSINBUS, the number of years the dealership has been in business. The regression equation is now

$$\begin{aligned}\text{TOTSALES} = &-10{,}235.9 + 6.41\text{PARTSERV} + 6.9\text{RETNEW} \\ &+ 10.4\text{RETUSED} + 388.0\text{YRSINBUS}\end{aligned}$$

Now that RETNEW and RETUSED have the piano under control, their partial regression coefficients have not changed much from the previous step. However, the constant term is now $-10{,}235.9$. Bringing YRSINBUS into the equation increases the explained percentage of variation in TOTSALES to 93.91%, and the multiple standard error of estimate decreases to $9.871 million.

Once stepwise regression identifies a subset of the original independent variables and lists them in a multiple regression equation, it is useful to include only these independent variables in a conventional multiple regression analysis. The procedure described in this section is known as *forward selection stepwise regression,* since we begin with no predictor variables and add one predictor at each iteration. There is also a *backward elimination stepwise regression,* in which we begin with the entire set of predictor variables and eliminate one variable at each iteration.

EXERCISES

17.46 What is stepwise regression, and when is it desirable to make use of this multiple regression technique?

17.47 In general, on what basis are independent variables selected for entry into the equation during stepwise regression?

17.48 The two largest values in a correlation matrix are the 0.850 correlation between y and x_4, and the 0.790 correlation between y and x_9. During a stepwise regression analysis, x_4 is the first independent variable brought into the equation. Will x_9 necessarily be next? If not, why not?

17.49 The following Minitab printout is for a stepwise regression with 10 independent variables and 50 observations:

```
Stepwise Regression: y versus x1, x2, x3, x4,
x5, x6, x7, x8, x9, x10

 Alpha-to-Enter: 0.05  Alpha-to-Remove: 0.1
Response is    y    on 10 predictors, with N =  50
    Step          1          2          3
Constant      83.91     106.85     124.76

x5            -0.33      -0.35      -0.30
T-Value       -2.39      -2.65      -2.30
P-Value       0.021      0.011      0.026

x2                       -0.33      -0.38
T-Value                  -2.30      -2.70
P-Value                  0.026      0.010

x9                                  -0.27
T-Value                             -2.07
P-Value                             0.044

S              13.8       13.2       12.8
R-Sq          10.63      19.66      26.52
R-Sq(adj)      8.76      16.25      21.73
C-p             3.9        0.9       -0.9
```

a. List the order in which the independent variables have been introduced.
b. What is the multiple regression equation after two independent variables have been introduced?
c. As of step 3, three variables have been introduced. Which, if any, partial regression coefficients differ significantly from zero at the 0.05 level? At the 0.02 level?

(**DATA SET**) *Note:* Exercises 17.50–17.57 require a computer and statistical software. We suggest that Minitab users specify alpha-to-enter as 0.05 and alpha-to-remove as 0.10. (See Computer Solutions 17.6.)

17.50 For the data in Exercise 17.44, and using Sunday circulation as the dependent variable, perform conventional multiple regression and stepwise regression analyses on the data, then compare the results of the two approaches. The data are in file **XR17044**.

17.51 For the data in Exercise 17.45, and using total nonrenewable energy production as the dependent variable, perform conventional multiple regression and stepwise regression analyses on the data, then compare the results of the two approaches. The data are in file **XR17045**.

17.52 For the situation, data, and variables described in Exercise 17.16, perform conventional multiple regression and stepwise regression analyses on the data, then compare the results of the two approaches. The data are in file **XR17016**.

17.53 For the situation, data, and variables described in Exercise 17.15, perform conventional multiple regression and stepwise regression analyses on the data, then compare the results of the two approaches. The data are in file **XR17015**.

17.54 For the situation and data described in Exercise 17.17, perform conventional multiple regression and stepwise regression analyses on the data, then compare the results of the two approaches. Use total operating revenue as the dependent variable and number of aircraft, number of employees, and number of aircraft departures as the three independent variables. The data are in file **XR17017**.

17.55 An officer training program has 9 board members who provide ratings for the candidates. The high and low scores are dropped before the overall rating is computed. For a sample of 28 persons from the most recent class of candidates, the overall ratings and ratings by the individual board members are provided in data file **XR17055**.

a. Construct the correlation matrix for all 10 variables.
b. With overall rating as the dependent variable, perform a stepwise regression and interpret the results.
c. From the results in part (b), does it appear that the size of the board could be reduced without losing very much explanatory power? Explain.
d. Using only the data provided by the first two board members introduced in part (b), perform a multiple regression analysis. Compare the results with the stepwise regression as it appeared following the introduction of these board members.

17.56 Using the data file for Exercise 17.55:

a. Apply stepwise regression with the rating by board member 1 as the dependent variable and the ratings by the other 8 members as the set of independent variables from which to select. Interpret the results.

b. Using only the independent variables introduced during the stepwise regression, perform a multiple regression analysis.
c. Compare the results in parts (a) and (b).

17.57 Using the data file for Exercise 17.55:
a. Apply stepwise regression with the rating by board member 9 as the dependent variable and the ratings by the other 8 members as the set of independent variables from which to select, then interpret the results.
b. Including only the independent variables introduced during the stepwise regression, perform a multiple regression analysis.
c. Compare the results in parts (a) and (b).

17.8 SELECTING A MODEL

Given the variety of models discussed in this chapter, and in Chapters 15 and 16, it is natural to wonder which model is best for a given situation. However, the selection of which model to use is not quite so straightforward as the workings of the individual models themselves. Rather than feature a complex flow diagram and an elaborate discourse on each of its components, in this section we will consider a general sequence of steps and some of the more important considerations involved with each. For a more advanced discussion of the process of selecting the best model, the interested reader may wish to refer to a more advanced text, such as the excellent treatment given this topic by Kleinbaum, Kupper, Muller, and Nizam.[8]

1. **Specify the dependent variable and determine the goal for the analysis.** If the goal is simply to predict the value of the dependent variable based on a set of predictor variables, we need not worry about multicollinearity. Accordingly, we can use a large number of predictor variables without worrying about whether some of them might be highly correlated with each other. On the other hand, if we are concerned with interpreting partial regression coefficients in an attempt to better understand the relationship between the dependent variable and one or more of the predictor variables, it will be desirable to build a smaller model with predictor variables that are more independent from each other.
2. **Specify possible predictor variables.** Based on a combination of experience and knowledge about the dependent variable and the environment in which it exists, develop a list of predictor variables that could be related to the dependent variable. In doing so, keep in mind the data-collection step that comes next. It may not be practical or economically feasible to gather relevant data for all of the predictor variables that could be candidates for the model.
3. **Collect the data.** This may involve primary data, secondary data, or both. The kinds of methods and data sources described in Chapter 4 could be utilized. Consider how large the eventual set of predictor variables might be, and collect data accordingly. As a rule of thumb, the number of observations should be at least five times the number of predictor variables in the model.[9] This is due to degrees-of-freedom considerations. The larger the number of predictor variables, the easier it is for the model to fit any given set of data, even data that are completely random.
4. **Specify the candidate models.** The models under consideration could include any of those discussed in this chapter and the two that preceded, as well as models that are larger or more complex. In general, we should attempt to use the simplest possible model that satisfies the goal for which the model is being developed.

[8]Kleinbaum, D.G., L.L. Kupper, K.E. Muller, and A. Nizam, *Applied Regression Analysis and Multivariable Methods,* Pacific Grove, CA: Duxbury Press, 1998.
[9]Ibid., p. 389.

5. **For each candidate model, decide which of the available predictor variables will be included in the model.** Deciding which variable(s) to include can be done one variable at a time, as in the stepwise regression method presented in Section 17.7, and the approach can involve either forward selection or reverse elimination. However, with increasingly powerful computers and the greater sophistication of statistical software, more "brute force" methods are available. In an approach called *best subsets regression,* the computer considers all of the possible combinations or subsets of the available predictor variables, then reports which models are the best performers for each given number of predictors. Implicit in this process is the criterion by which one model is judged to perform better than another. In this chapter, our focus has been on the coefficient of determination (R^2) that indicates the proportion of variation in the dependent variable that is "explained" by the set of predictor variables, but many other criteria can be applied.
6. **Select a model and evaluate its suitability.** Once a model has been selected, we can evaluate whether it satisfies the assumptions underlying its use. This evaluation should include a residual analysis and related diagnostics.
7. **Test the reliability of the model by applying it to new or withheld data.** The model can be tested by applying it to new data as time goes by. Another possibility is to build the model on the basis of a portion of the available data, then test its performance on the "fresh" data that were initially withheld from the model.

SUMMARY (17.9)

Models, variables, and transformations

In this chapter, we go beyond the simple and multiple regression estimation models in Chapters 15 and 16 to include curvilinear models as well as models where a variable is qualitative and can represent two or more categories. For underlying relationships that are not linear, logarithmic transformation methods can be employed in constructing exponential and multiplicative models.

Multicollinearity

If two or more independent variables are highly correlated with each other, multicollinearity is present, and the partial regression coefficients may be unreliable. One way to identify multicollinearity is to generate a correlation matrix that shows the correlation of each variable with each of the other variables. Multicollinearity is not a problem when the only purpose of the equation is to predict the value of the dependent variable, but it can be a problem when we are trying to interpret the meaning of the partial regression coefficients.

Stepwise regression

Stepwise regression is introduced and the forward selection method is presented. In this approach, independent variables enter the equation one at a time, with the first one entering being the x variable that explains the greatest amount of the variation in y. At each step, the variable is introduced that explains the greatest amount of the remaining variation in y.

Model selection

Determining the best model for a given situation can be a complex process, and a sequence of general steps is presented for model selection.

EQUATIONS

Polynomial models with one quantitative predictor variable

General polynomial model: $y_i = \beta_0 + \beta_1 x_i + \beta_2 x_i^2 + \beta_3 x_i^3 + \cdots + \beta_p x_i^p + \epsilon_i$

First-order polynomial model: $E(y) = \beta_0 + \beta_1 x$

Second-order polynomial model: $E(y) = \beta_0 + \beta_1 x + \beta_2 x^2$

Third-order polynomial model: $E(y) = \beta_0 + \beta_1 x + \beta_2 x^2 + \beta_3 x^3$

Polynomial models with two quantitative predictor variables

First-order model with two predictor variables: $E(y) = \beta_0 + \beta_1 x_1 + \beta_2 x_2$

First-order model with two predictor variables and interaction:

$$E(y) = \beta_0 + \beta_1 x_1 + \beta_2 x_2 + \beta_3 x_1 x_2$$

Second-order model with interaction:

$$E(y) = \beta_0 + \beta_1 x_1 + \beta_2 x_2 + \beta_3 x_1^2 + \beta_4 x_2^2 + \beta_5 x_1 x_2$$

Conversion of the exponential equation, $\hat{y} = b_0 b_1^x$, to linear form

$$\log \hat{y} = \log b_0 + x \log b_1, \quad \text{or} \quad \log \hat{y} = \log b_0 + (\log b_1)x$$

Conversion of the multiplicative equation, $\hat{y} = b_0 x_1^{b_1} x_2^{b_2}$, to linear form

$$\log \hat{y} = \log b_0 + b_1 \log x_1 + b_2 \log x_2$$

CHAPTER EXERCISES

(**DATA SET**) *Note:* Exercises 17.58–17.73 require a computer and statistical software.

17.58 As reported by the Recording Industry Association of America, the percentage of recorded music dollar sales in CD format was as shown here for the years 2000 through 2007. The data are also in file **XR17058**. Using $x = 1$ through 8 to represent years 2000–2007, fit a second-order polynomial model to the data and estimate the percentage of music sales in CD format for the year 2015. Source: *The World Almanac and Book of Facts 2009*, p. 288.

Year	x = Year Code	CD Music Percentage
2000	1	89.3
2001	2	89.2
2002	3	90.5
2003	4	87.8
2004	5	90.3
2005	6	87.0
2006	7	85.6
2007	8	82.6

17.59 J.K. Rowling's popular *Harry Potter* books have tended to become thicker with each new edition. The publication year and page count for U.S. editions for the first seven books were as shown on the next page. The data are also in file **XR17059**. Using year codes so that $x = 1$ corresponds to 1998, the year the first book was published, fit a third-order polynomial model to the data and estimate the page count for a *Harry Potter* book published in 2015. Source: Barnes & Noble, bn.com, August 20, 2009.

17.60 For 1995 through the first half of 2009, PricewaterhouseCoopers and the National Venture Capital Association reported quarterly venture capital investments (trillions of dollars) and the number of investment deals as shown in data file **XR17060**. Generate a scatter diagram with the number of deals as the independent variable, then apply an appropriate polynomial model to fit the data. Does there appear to be any relationship between the variables? How much of the variation in quarterly venture capital investments is explained by the model? Source: PricewaterhouseCoopers, pwcmoneytree.com, August 20, 2009.

Data for Exercise 17.59

Title	Publication Year	x = Year Code	Page Count, U.S. Edition
The Sorcerer's Stone	1998	1	309
The Chamber of Secrets	1999	2	341
The Prisoner of Azkaban	1999	2	435
The Goblet of Fire	2000	3	734
The Order of the Phoenix	2003	6	870
The Half-Blood Prince	2005	8	652
The Deathly Hallows	2007	10	759

17.61 Each workstation in the final assembly area of a large manufacturing facility has a "backlog" area to hold units that are to be processed. A production engineer believes there could be a relationship between the number of units in the backlog area and the productivity of the worker. Specifically, the engineer thinks increasing the number of units in the backlog area might lead to workers being more in a hurry to "catch up," hence increasing their productivity. The data in file **XR17061** show the results of a test in which the engineer tried various backlog levels and measured productivity at 50 different workstations. Generate a scatter diagram relating y = productivity (units/hour) to x = number of units in the backlog area, then apply an appropriate polynomial model to fit the data. Does there appear to be any relationship between the variables? How much of the variation in productivity is explained by the model?

17.62 The data in the table show the number of years it took for selected technological innovations to spread to 25% of the U.S. population. The data are also in file **XR17062.** For y = years to spread and x = years past 1800 (e.g., the code for 1975 is 175), generate a scatter diagram and apply an appropriate polynomial model to fit the data. Does there appear to be any relationship between the variables? How much of the variation in dissemination time is explained by the model? Using your model, predict how long it will take for a major technological invention that is invented in 2015 to spread to 25% of the U.S. population. Source: *Wall Street Journal Almanac 1999,* p. 350.

Invention	Year Code	Years to Spread
Home electricity (1873)	73	46
Telephone (1875)	75	35
Automobile (1885)	85	55
Radio (1906)	106	22
Television (1925)	125	26
VCR (1952)	152	34
Microwave oven (1953)	153	30
Personal computer (1975)	175	15
Cellular phone (1983)	183	13

17.63 The data in file **XR17063** include the wind chill index (how cold it "feels") at different combinations of outside temperature (degrees Fahrenheit) and wind speed (miles per hour). For y = wind chill index, x_1 = actual outside temperature, and x_2 = wind speed, fit a first-order polynomial model to the data and interpret the partial regression coefficients and the R^2 value. Add an interaction term to the model. To what extent has the addition of the interaction term improved the explanatory power of the model? Source: *New York Times Almanac 2009,* p. 478.

17.64 For loans at insured commercial banks, the Federal Financial Institutions Examination Council has reported the percentage of various kinds of loans that were at least 30 days in delinquency during each of the years shown in the table. The data are also listed in file **XR17064.** For y = percent of all insured commercial bank loans overdue, x_1 = percent of consumer credit cards overdue, and x_2 = percent of residential loans overdue, fit a first-order polynomial model to the data and interpret the partial regression coefficients and the R^2 value. Add an interaction term to the model. To what extent has the addition of the interaction term improved the explanatory power of the model? Source: Bureau of the Census, *Statistical Abstract of the United States 2009,* p. 727.

Year	All Loans	Credit Cards	Residential Loans
2000	2.18%	4.50%	2.11%
2001	2.61	4.86	2.29
2002	2.69	4.87	2.11
2003	2.33	4.47	1.83
2004	1.80	4.11	1.55
2005	1.57	3.70	1.55
2006	1.57	4.01	1.73
2007	2.06	4.25	2.57

17.65 The production engineer for the manufacturing plant in Exercise 17.61 has just included another variable: the worker's gender, coded as 1 = female and 0 = male. File **XR17065** contains the expanded set of data. With y = productivity (units/hour), x_1 = number of units in the backlog area, and x_2 = gender, fit a first-order model with no interaction term. Interpret the partial regression coefficients and the R^2 value. At the 0.02 level, is the model significant?

17.66 The sales director for an industrial supplies firm has collected information describing the performance and personal characteristics of 80 members of her sales force. The data in file **XR17066** include the percent of sales quota the salesperson achieved last year, age, number of years with the firm, and personal appearance (very neat, about average, or a little sloppy). In the data file, a very neat person is coded as neat = 1, average = 0, a person of average neatness is coded as neat = 0, average = 1, and a sloppy person is coded as neat = 0, average = 0. With y = percent of sales quota, perform a linear regression using the other variables as predictors. Interpret the partial regression coefficients and the R^2 value for the model. At the 0.05 level, is the model significant?

17.67 For the loan delinquency data in Exercise 17.64 and listed in file **XR17064**, fit a multiplicative model and use it to estimate the delinquency rate for all insured commercial bank loans when the delinquency rates for credit cards and residential loans are 4.0% and 2.0%, respectively.

17.68 The Dow Jones Global Indexes describe over 3000 companies worldwide that represent more than 80% of the equity capital in the world's stock markets. Data file **XR17068** shows the quarterly indexes for the world as a whole, and indexes for the United States, the Americas, Europe, and the Asia/Pacific region over 23 consecutive quarters. Given that the dependent variable will be y = world index, construct a correlation matrix for the four possible predictor variables. Does multicollinearity seem to be present in the data? If so, would this present a problem if we were using the predictor variables to estimate the world index? If there is multicollinearity, would this present a problem if we were interpreting the partial regression coefficients in a linear model applied to the data? Source: *Wall Street Journal Almanac 1999*, p. 264.

Note: For Exercises 17.69–17.73, we suggest that Minitab users specify alpha-to-enter as 0.05 and alpha-to-remove as 0.10. (See Computer Solutions 17.6.)

17.69 Data file **XR17069** lists a variety of information for a selection of aircraft. The variables and their units are: operating cost (dollars per hour), number of seats, speed (mph), flight range (miles), and fuel consumption (gallons per hour). Considering operating cost per hour as the dependent variable, construct and interpret the correlation matrix for the possible predictor variables. Does it appear that multicollinearity could be present? For y = operating cost per hour, perform a stepwise regression analysis using the other variables as predictors and interpret the results. How much of the variation in operating costs is explained by the model? Source: *World Almanac and Book of Facts 2009*, p. 365.

17.70 Given the situation and variables described in Exercise 17.66 and data file **XR17066**, use y = percent of sales quota in performing a stepwise regression using the other variables as predictors. Interpret the results, including a commentary on which variables seem to be of greater importance in predicting sales performance. How much of the variation in sales performance is explained by the model?

17.71 The students in a training program have taken four tests along with a final exam, with the scores shown in the table. The data are also in file **XR17071**.

Student	Test 1	Test 2	Test 3	Test 4	Final Exam
1	59	65	53	65	62
2	69	77	57	76	78
3	79	74	81	86	82
4	87	88	83	90	96
5	92	89	97	92	94
6	65	69	60	66	68
7	71	70	68	81	78
8	93	87	85	95	94
9	83	92	89	90	91
10	71	68	69	74	77
11	64	61	74	74	73
12	76	82	79	78	86
13	65	75	67	77	79
14	90	88	75	94	89
15	69	68	74	76	72
16	85	81	99	79	96

a. Construct the correlation matrix for these variables. Does it appear that multicollinearity could be a problem if all of the independent variables are used in estimating y = final exam score? In a stepwise regression, which independent variable would be the first one introduced into the equation, and why?
b. With y = final exam score and the independent variables as shown above, apply stepwise regression analysis to these data.
c. Describe each step of the stepwise regression in terms of the variable introduced, the regression equation after its introduction, and the proportion of variation in y that has been explained.
d. Using only the independent variables introduced during the stepwise regression, carry out a conventional multiple regression analysis on the data. What, if any, additional information does your computer statistical package provide that was not given in the printout for the stepwise regression?

17.72 Repeat parts (b)–(d) of Exercise 17.71, with Test 1 as the dependent variable, and Test 2, Test 3, Test 4, and Final Exam as the independent variables.

17.73 As indicated in the chapter-opening vignette, the National Football League quarterback passer rating is somewhat of a mystery to most observers. Data file **XR17073** includes the passer rating for 32 NFL quarterbacks during the 2008 season, along with these ten performance measurements for each quarterback: passing attempts, completions, completion percentage, yards gained, average yards per attempt, number of touchdowns, percentage of passes that were touchdowns, longest pass completed, number of interceptions, and percentage of passes that were intercepted. With passer rating as the dependent variable, use stepwise regression to find out which predictor variables are especially important in determining the passer rating. Comment on the results, including the percentage of the variation in passer ratings that is explained by the model. Source: NFL.com, August 21, 2009.

INTEGRATED CASES

Thorndike Sports Equipment

Ted Thorndike has been asked to pay another visit to the statistics class taught by his friend, Mary Stuart. One of the university's track athletes has an excellent chance to make the Olympics in the 1500 meters, known as the "metric mile," and Ted has brought along his laptop computer, a statistical package, and data on the world record for the one-mile run.

In 1954, Great Britain's Roger Bannister became the first person to run a mile in less than four minutes (3:59.4). Since then, the record has been broken 18 times, most recently by Morocco's Hicham El Guerrouj, at 3:43.13. The following data describe the record runs and the years in which they occurred. The data are also in file **THORN17**. Source: *Sports Illustrated 2009 Almanac,* p. 526.

Note These times are not recorded to the same number of decimal places for the seconds, because officiating policies and timing technologies may have differed from one record run to the next. For convenience, you may wish to (1) code the times so that each is the number of seconds over 3 minutes and (2) code the years so that the year 1901 = 1.

Person	Time	Year
Roger Bannister	3:59.4	1954
John Landy	3:58	1954
Derek Ibbotson	3:57.2	1957
Herb Elliott	3:54.5	1958
Peter Snell	3:54.4	1962
Peter Snell	3:54.1	1964
Michael Jazy	3:53.6	1965
Jim Ryun	3:51.3	1966
Jim Ryun	3:51.1	1967
Filbert Bayi	3:51	1975
John Walker	3:49.4	1975
Sebastian Coe	3:49	1979
Steve Ovett	3:48.8	1980
Sebastian Coe	3:48.53	1981
Steve Ovett	3:48.4	1981
Sebastian Coe	3:47.33	1981
Steve Cram	3:46.32	1985
Noureddine Morceli	3:44.39	1993
Hicham El Guerrouj	3:43.13	1999

1. Fit a simple linear regression ($\hat{y} = b_0 + b_1x$) to the time series.
2. Fit a polynomial regression equation ($\hat{y} = b_0 + b_1x + b_2x^2$) to the time series.
3. Comparing the R^2 values for the equations from parts 1 and 2, which equation is the better fit to the data?
4. Using the equation identified in question 3, what is the predicted world record for the one-mile run in the year 2015?

Fast-Growing Companies

Computer database **GROWCO** describes the characteristics of 100 companies identified by *Fortune* as among the fastest growing. Source: "*Fortune's* 100 Fastest-Growing Companies," *Fortune*, September 4, 2000, pp. 142–158. The variables include:

1. Name
2. Earnings-per-share growth, 3-year annual rate (percent)
3. Net income, past four quarters (millions of dollars)
4. Revenue growth, 3-year annual rate (percent)
5. Revenue, past four quarters (millions of dollars)
6. Total return, 3-year annual rate (percent)
7. Stock price change in past year (percent)
8. Stock price (dollars)

9. Estimated price/earnings ratio a year from now
10. Category (1 = technology, 2 = industrial, 3 = retail, 4 = health care, 5 = financial services, 6 = telecommunications, 7 = other)

Variables 11–16 are dummy variables representing the categories of variable 10. They are coded according to the scheme described in Section 17.4. For a technology company, technology = 1 and the other dummy variables will be 0. For a financial services company, financial = 1 and the others will be 0. If a company is "other," all six values will be 0. Thus, we need only six variables to describe the seven possible categories.

1. Generate a correlations matrix that includes variables 2–9 and 11–16. (Variable 10 would be meaningless here because its values are 1–7 for the 7 company categories.) Do the correlations "make sense" in terms of which correlations are positive and which are negative?
2. Using y = stock price (variable 8) as the dependent variable, carry out a conventional multiple regression analysis using variables 2–7, 9, and 11–16. Excel users, see note 1, which follows. Examine the partial regression coefficients to see whether their signs match the signs of the corresponding correlation coefficients associated with stock price. What percentage of the variation in stock prices is explained by the set of predictor variables?
3. Repeat step 2, but this time perform a stepwise regression analysis. Minitab and Excel users should refer to the following notes. Identify the variables that were introduced and interpret the printout. What percentage of the variation in stock prices is explained by the reduced set of predictor variables?

Minitab users, note: In the solutions materials, we use alpha-to-enter = 0.05 and alpha-to-remove = 0.10 for stepwise regression analyses. This is to provide consistency with Excel and the Data Analysis Plus add-in used with the text. See Computer Solutions 17.6.

Excel users, note 1: Excel requires that the predictor variables be arranged in adjacent columns when we carry out a multiple regression analysis. The same is true for all variables for which a correlation matrix is being generated. Depending on the predictor variables being used in a given analysis, you may need to rearrange some of the columns in the database. If you save the revised database, it is advisable to change the name of the file when doing so.

Excel users, note 2: The estimated price/earnings ratio a year from now was not available for four of the companies, so these are "missing data." If you are using the Data Analysis Plus add-in for the stepwise regression, you may wish to either omit this variable or omit these four companies from the analysis.

BUSINESS CASE

Westmore MBA Program

Photodisc/Getty Images

The MBA program at Westmore University has undergone several dramatic changes over the past five years. During this time, the goal of the business school was to recruit as many students as possible into the MBA program in order to build up their student base and credit hour production. A massive campaign was launched five years ago in order to attract more applicants to the program. Special brochures containing information about the program were printed and mailed to prospective students as well as to other colleges and universities that were likely to have undergraduate students who might be interested in coming to Westmore. Mailings were also sent to students who indicated an interest in Westmore on their GMAT exam. (The GMAT exam is a national standardized test used by most business schools in making admissions decisions for applicants to their graduate programs.) Representatives from the Westmore School of Business began attending regional "MBA fairs," where MBA program representatives meet with prospective MBA students and share information.

In the beginning, the number of students applying to the Westmore MBA program was small, but eventually the advertising campaign began to work and the number of qualified applicants each year increased to the target value of 150 initially set by the dean of the business school and the director of the MBA program. The yield—the percentage of admitted applicants who actually enroll and attend Westmore—is typically around 70%. Admitted students who do not enroll either attend other MBA programs or accept job offers. The accompanying table shows the admissions and enrollment figures for the five years of the plan to build the MBA student base at Westmore.

Year	1	2	3	4	5
Admissions	86	108	134	141	154
Enrollment	54	77	91	96	106

Wayne McDonald, the director of the program, is currently putting the second phase of the plan into action. He knows that in order for the MBA program at Westmore to attain national recognition, administrators must become more selective in the admissions process. The number of applicants is now large enough to do this without falling below a critical mass of 60 enrolled students each year.

The major issue facing Wayne and the MBA program is how to go about selecting the best students for the program. Wayne recently met with the MBA Admissions Committee, which consists of himself and two faculty members, Dr. Susan Thompson, who is a finance professor, and Dr. Hector Gonzalez, who is a marketing professor.

Wayne: "Thanks for coming to the meeting today. As you both know, our recruiting effort over the past five years has been extremely successful. We were able to exceed our original enrollment goal last year. While many of our students have been outstanding and have given our program visibility in the business community, we have had a number of weak performers. Some professors have had to water down their courses to keep these people afloat in the program. If we are to have a nationally recognized, quality MBA program, we must become stricter in our admission policies. Fortunately, we are now at the point where we can be much more selective and still have our minimum critical mass of 60 enrolled students each year."

Susan: "Wayne is right. Our current admission standards require a minimum score of 400 on the GMAT and a minimum undergraduate grade point average of 2.0. Obviously, this is not much of a hurdle. Personally, I would like to see the minimum requirements set at 580 on the GMAT and 2.75 for the undergraduate grade point average."

Wayne: "Well, raising the minimums is one way of going about it, but there are many other factors that determine the degree of success a student has in our MBA program. We should consider including these factors in our decision making process."

Hector: "Too bad we don't know in advance which students are going to excel. Wayne, do you know what other schools are doing?"

Wayne: "From conferences that I've attended, I have found that many MBA programs put a lot of emphasis on the GMAT score and the undergraduate grade point average of the student. While some schools set a minimum entrance requirement on each of these criteria as we do currently, other schools combine these measures into a single overall score. For instance, there is a 'formula score' that many schools use which multiplies the undergraduate GPA by 200 and adds the result to the GMAT score. If the formula score is above a certain figure, say, 1000, then the student is considered to be admittable."

Susan: "But there are so many other factors to consider. Surely we don't want our admissions decisions to be based solely on a formula. There are many students who attend colleges with high grade inflation. Those applicants would have an unfair advantage over applicants from stronger schools with regard to undergraduate grade point average."

Hector: "Yes, I agree. There are also studies that have indicated the GMAT is not a strong predictor of success in graduate school for many reasons, including cultural bias."

Wayne: "I am not advocating that we go to a strictly mathematical basis for making our decisions. However, higher minimum standards than we currently have or some sort of formula involving GMAT and undergraduate grade point average might be a useful screening device for sifting through applicants."

Susan: "I'm not opposed to your suggestion. Such an approach could be used to identify those students with high potential for succeeding in our program. In a sense, many of these decisions could be automated."

Wayne: "That would certainly be a great timesaver. Our admissions committee would only have to meet to discuss those applicants with other strong characteristics or extenuating circumstances."

Susan: "Excellent idea! Now, if we go with raising the minimum requirements for GMAT and undergraduate GPA, how much should we raise them? Or if we go with a combined score approach, what formula should we use and how should we set its cutoff values?"

Wayne: "We could go with your earlier suggestion of a 580/2.75 minimum requirement or with the formula score I described earlier. I could talk with directors of other MBA programs to get some feel for how they set their cutoff criteria. After we gain some experience with future admissions, we could adjust the cutoff criteria."

Hector: "Why wait until then? We have five years worth of data already! We should be able to develop our own criteria based on our own past experience."

Susan: "We might even want to consider developing our own formula."

Wayne: "Great ideas! That's why I like working with the two of you on this committee. However, I would limit the data to the last two years because of several changes we made in the program a few years back. The data for the last two years are more reflective of our current program."

Hector: "In looking at these data, how are we going to measure the degree to which a student is successful in our program? Whether they graduate or not?"

Wayne: "Fortunately or unfortunately, depending on how you look at it, practically all of our students have eventually graduated. One of the things we are trying to accomplish is to make our program more rigorous and demanding, to raise the level of quality. If this is going to happen, we have to be more selective with the students we admit."

Susan: "Why not consider the grade point average of the student at the end of the program?"

Wayne: "The major problem there is that the students in our program do not take the same set of courses in their second year because they select different areas of concentration. Some go into marketing, some into finance, others into either accounting or management. There is a real lack of comparability in those final GPA figures. But what we might do is look at the first-year GPA in the MBA program. The courses taken by students in the first year are essentially the same because they are required core courses. It is not until their second year that they began taking elective courses in the different concentration areas. What first-year MBA grade point average would the two of you, as faculty members, define as indicating a successful first year?"

Hector: "Given the breadth of those first-year core courses, their level of difficulty, and our mild degree of grade inflation, I would say that any of our students in the past two years with at least a 3.2 average would be considered successful in the first year. Would you agree, Susan?"

Susan: "I believe most of the faculty would go along with that."

Wayne: "Don't set your goals too high! Remember, we need at least 60 students per year to even have a program. We probably need to look at the data to see what's possible."

Hector: "When can we get access to the past data? I really would like to get started."

Wayne: "I'll have one of the staff members write a database program to pull up the relevant information on all students who have completed the first year of the program in the past two years. I'll get the data to you as soon as I can. Let's plan to meet again in two weeks to see what we have discovered by then. I look forward to hearing your ideas."

Assignment

Wayne McDonald's assistant gathered the data for the past two years of experience with the Westmore MBA program and the information is stored in the **WESTMORE** data file. The Data Description section provides a partial listing of the data along with definitions of the variables. Using this data set and other information given in the case, help Wayne McDonald and the MBA Admissions Committee in their quest to develop better admissions guidelines for the Westmore MBA program. The case questions will assist you in your analysis of the data. Use important details from your analysis to support your recommendations.

Data Description

The data for the Westmore MBA admissions case are contained in the **WESTMORE** data file. The file contains data for the 202 students who completed their first year in the MBA program over the past two years. A partial listing of the data is shown below.

The variables are defined as follows:

ID_No: Student identification number.
GPA_Yr1: Grade point average for the first year of courses in the Westmore MBA program.
GMAT: Score on the GMAT test.
UG_GPA: Undergraduate grade point average.

Undergraduate Major is a qualitative measure broken down into two variables:

UG_Bus = 1 for undergraduate business major, otherwise = 0.
UG_Tech = 1 for undergrad science, engineering, or other technical field, otherwise = 0. (*Note:* A student with an undergraduate degree in a nonbusiness and nontechnical field is the "baseline" and will have both UG_Bus = 0 and UG_Tech = 0.)

Undergraduate School Rating is a qualitative measure broken down into four variables:

UG_1st5th = 1 if student is from the top 20% of undergraduate schools, otherwise = 0.
UG_2nd5th = 1 if student is from the second 20% of undergraduate schools, otherwise = 0.
UG_3rd5th = 1 if student is from the third 20% of undergraduate schools, otherwise = 0.
UG_4th5th = 1 if student is from the fourth 20% of undergraduate schools, otherwise = 0. (*Note:* A student with a degree from a school in the bottom fifth of undergraduate schools is the "baseline" and will have a 0 value for all four of the variables shown above.)

Age: Age of the student, in years.
US_Cit: 1, if student is a U.S. citizen, 0 if otherwise.

ID_ No	GPA_ Yr1	GMAT	UG_ GPA	UG_ Bus	UG_ Tech	UG_ 1st 5th	UG_ 2nd 5th	UG_ 3rd 5th	UG_ 4th 5th	Age	US_ Cit
4001	3.201	540	2.974	1	0	0	0	1	0	25	1
4002	2.964	540	2.529	1	0	0	1	0	0	23	1
4003	3.745	510	3.727	0	0	0	1	0	0	25	1
4004	3.290	570	2.905	1	0	0	0	1	0	23	1
4005	3.028	640	2.641	1	0	1	0	0	0	23	1
⋮	⋮	⋮	⋮	⋮	⋮	⋮	⋮	⋮	⋮	⋮	⋮

Case Questions

1. The committee wishes to use first-year MBA grade point average as its measure of the quality of a student. As a first step in assisting committee members, construct a correlation matrix and briefly discuss the information it contains that could be of help to the committee members. If they were to choose only one variable in predicting the first-year MBA GPA, which variable would you recommend? Do you think it would be a good idea to evaluate applicants on the basis of only this one variable?
2. Because the committee is very interested in using predicted first-year MBA grade point average in measuring the quality of applicants to its program, assist the members further by carrying out and interpreting the results of an appropriate stepwise regression of the data. (*Note to Minitab users:* For consistency with Excel results, you may wish to select the alpha-to-enter and alpha-to-remove values listed in the Minitab portion of Computer Solutions 17.6.) Discuss the results, including making and supporting any recommendations that you think the committee would find useful.
3. Based on your analysis in Case Question 2, and even considering the results from a complete regression that uses *all* of the possible predictive variables, does there seem to be very much variation in the first-year MBA grade point average that is *not* explained by the independent variables provided in the data? If so, provide the committee with some examples of other quantitative or nonquantitative applicant variables that you think could help them in evaluating the potential for academic success of persons applying for admission to the Westmore MBA program.

BUSINESS CASE

Easton Realty Company (B)

When we visited Easton Realty Company (A), in Chapter 16, Sam Easton was concerned about two former clients who complained that Easton Realty had underpriced their houses in order to get a quick sale and collect their realty fee. In defending his company against these charges, Mr. Easton obtained a set of residential sales data on recent home sales in the Dallas–Fort Worth metroplex area. The data included the sale month, the sale price, the size of the home in square feet, the number of bedrooms, the age of the house, whether the house was located in Dallas versus somewhere else in the Dallas–Fort Worth metroplex, and whether Easton was the real estate company that handled the sale. In this chapter, you will be asked to help Mr. Easton gain a better understanding of what determines the selling price of a house, and you will be using the same variables that were introduced earlier, in Easton Realty (A). The data are in the **EASTON** data file.

Case Questions

1. As a first step in helping Sam Easton better understand how the sales price of a residential property is determined, construct a correlation matrix for the variables PRICE, SQFEET, BEDROOMS, AGE, DALLAS, and EASTON. Examine the matrix and advise Mr. Easton as to which variable seems to be the most important in explaining the variation in residential housing prices. You can expect Mr. Easton to be a little skeptical, since he realizes that the sales price of a house depends on many other factors besides the one you have identified.

2. Provide further assistance to Sam Easton by carrying out and interpreting the results of an appropriate stepwise regression designed to indicate which of the predictive variables mentioned in Case Question 1 are most important in determining the sales price of a house. (*Note to Minitab users:* For consistency with Excel results, you may wish to select the alpha-to-enter and alpha-to-remove values listed in the Minitab portion of Computer Solutions 17.6.) Briefly discuss the results, including making and supporting any recommendations you believe could benefit Mr. Easton.

3. Based on your analysis in Case Question 2, and even considering the results from a complete regression that uses *all* of the possible predictive variables identified in Case Question 1, does there seem to be very much variation in sales price that is *not* explained by the independent variables provided in the data? If so, what other quantitative or nonquantitative variables associated with a house might cause its sales price to be especially high or low compared to the estimated value based on models such as the ones you applied here and in Case Question 2?

CHAPTER 18

Models for Time Series and Forecasting

TIME SERIES-BASED FORECASTING AND THE ZÜNDAPP

During the late 1950s, some very unusual, miniature-size economy cars were imported to the United States. One of these was the German Zündapp, a 9.5-foot, 14-horsepower four-seater with two doors: one at the front of the car, the other at the rear. The model designation was *Janus,* for the two-faced Roman god who guarded the gates with the help of peripheral vision that athletes and city pedestrians could only dream of. The Zündapp's driver and front passenger faced forward, while the two rear passengers faced (where else?) . . . to the rear.

Being a rear-seat passenger in the Zündapp was a lot like engaging in the time series-based forecasting approaches discussed in this chapter. The only thing you could see was where you'd been, and you often had little idea where you were going.

Imagine sitting in the rear of the Zündapp and having to give driving instructions to a blindfolded driver. Giving such directions would be no problem as long as no bridges were out, no busy intersections loomed ahead, and no freeway ramps were coming up. In other words, you could do a good job if the future were not much different from the past.

While forecasting isn't bothered by fallen bridges, intersections, and freeway ramps, unpredictable events such as technological advances and changes in either the economy or consumer demands can play havoc with the accuracy of a forecast. The moral of the story: When engaged in time series-based forecasting, remember the Zündapp.

The 1958 Zündapp Janus

Christine Berrie

LEARNING OBJECTIVES

After reading this chapter, you should be able to:

- Describe the trend, cyclical, seasonal, and irregular components of the classical time series model.
- Fit a linear or a polynomial trend equation to a time series.
- Smooth a time series with the centered moving average and exponential smoothing techniques.
- Determine seasonal indexes and use them to compensate for the seasonal effects in a time series.
- Use the trend extrapolation and the exponential smoothing forecast methods to estimate a future value.
- Use the mean absolute deviation (*MAD*) and mean squared error (*MSE*) criteria to compare how well fitted equations or curves fit a time series.
- Use the Durbin-Watson test to determine whether regression residuals are autocorrelated.
- Use index numbers to compare business or economic measurements from one period to the next.

(18.1) INTRODUCTION

When data are a sequence of observations over regular intervals of time, they are referred to as a **time series.** A time series may consist of weekly production output, monthly sales, annual rainfall, or any other variable for which the value is observed or reported at regular intervals.

Time series are examined for two reasons: (1) to understand the past and (2) to predict the future. By analyzing a time series, we can identify patterns and tendencies that help explain variation in past sales, shipments, rainfall, or any other variable of interest. Likewise, this understanding contributes to our ability to forecast future values of the variable.

(18.2) TIME SERIES

A time series can be visually portrayed as a graph in which the variable of interest (e.g., sales, costs, shipments) is the vertical (y) axis and time is the horizontal (x) axis. An example is shown in Figure 18.1, where U.S. hybrid vehicle sales are plotted against time for the years 2000–2005.

Components of a Time Series

Analysis of a time series involves identifying the components that have led to the fluctuations in the data. In what is known as the "classical" approach to time series, these components are as follows:

TREND (*T*) This is an overall upward or downward tendency. As Figure 18.1 shows, U.S. sales of hybrid vehicles tended to increase during the 2000–2005 period. To the extent that the trend component is present, a regression line fitted to the points on the time series plot will have either a positive or negative slope. The line fitted in Figure 18.1 has a positive slope.

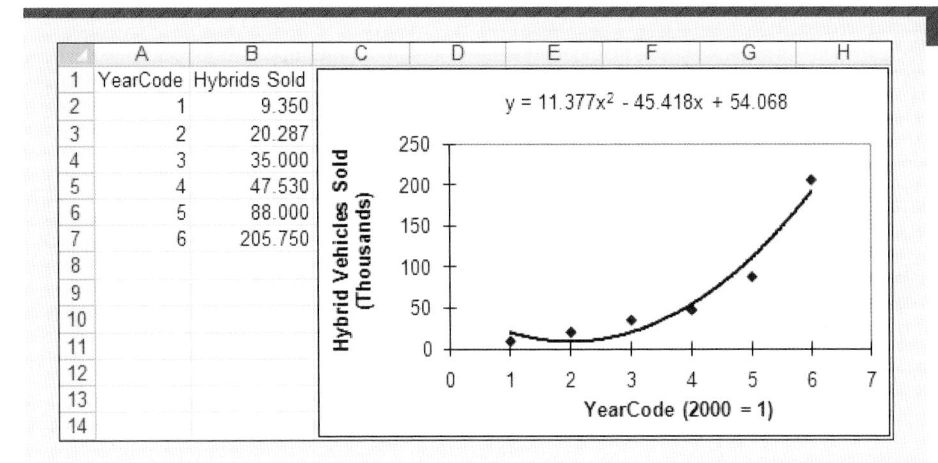

YearCode	Hybrids Sold
1	9.350
2	20.287
3	35.000
4	47.530
5	88.000
6	205.750

FIGURE 18.1

Excel plot with superimposed quadratic trendline for U.S. hybrid vehicle sales from 2000 through 2005. It was generated using the procedures in Computer Solutions 18.1, shown later in this section. The years are coded so that 2000 = 1.
Data source: HybridCars.com, August 25, 2006.

CYCLICAL (*C*) These are fluctuations that repeat over time, with the period being more than a year from one peak to the next. The sunspot and business cycles are examples of this type of fluctuation.

SEASONAL (*S*) These are also periodic fluctuations, but they repeat over a period of one year or less. For example, retail sales of bathing suits tend to be higher during the spring months and lower during the winter. Likewise, department stores typically do a great deal of their business during the Christmas period, vacation travel peaks during the summer months when children are out of school, and grocery stores may experience greater sales every other week, when employees at a local plant are paid.

IRREGULAR (*I*) This component represents random, or "noise" fluctuations that are the result of chance events, such as work stoppages, oil embargoes, equipment malfunction, or other happenings that either favorably or unfavorably influence the value of the variable of interest. Random variation can make it difficult to identify the effect of the other components.

In the classical time series model, the preceding components may be combined in various ways. In our discussion, we will use the *multiplicative* approach, described below and illustrated in Figure 18.2.

Classical time series model:

$y = T \cdot C \cdot S \cdot I$ where y = observed value of the time series variable
T = trend component
C = cyclical component
S = seasonal component
I = irregular component

The model assumes that any observed value of y is the result of influences exerted by the four components, and that these influences are multiplicative. The observed value of y is assumed to be the trend value, T, adjusted upward or downward by the combined influence of the cyclical, seasonal, and irregular components.

Several alternative models exist, including an additive model, with $y = T + C + S + I$. It can also be assumed that the value of y depends on the value of y

FIGURE 18.2

The components of the classical multiplicative time series model.

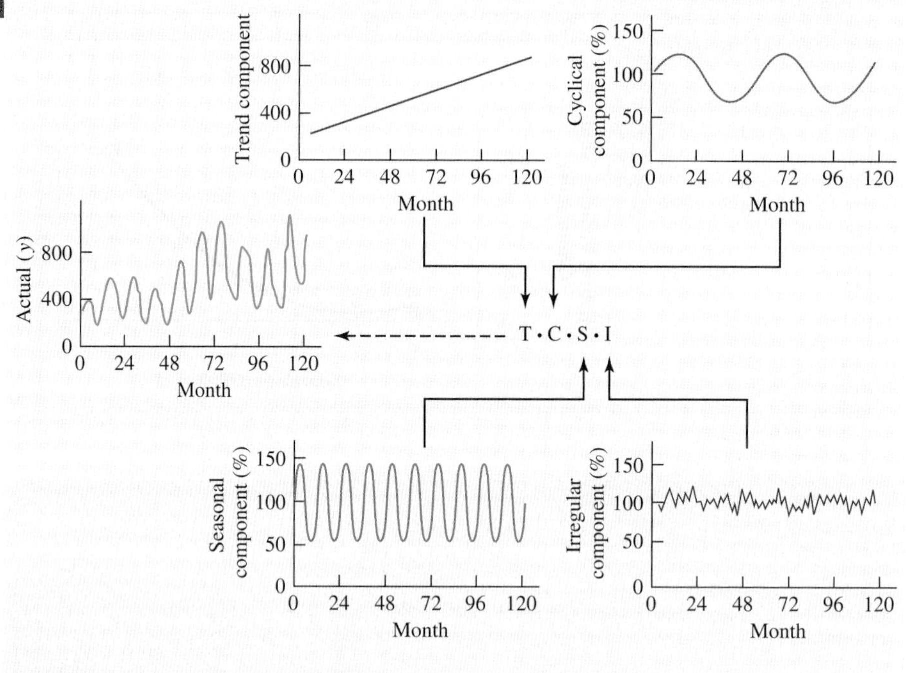

for the preceding period. In other words, y_t (the observed value for time period t) may be expressed as a function of T, C, S, I, and y_{t-1} (the value for the preceding time period). Also, the multiplicative and additive models may be combined, as in $y = (T + C) \cdot S \cdot I$.

In general, the most important component of most time series is the trend, which can be examined (1) by using regression techniques to fit a trend line to the data, or (2) by using the smoothing techniques described in the next section to moderate the peaks and valleys within the series.

In fitting a trend line with regression techniques, the line may be linear (e.g., $\hat{y} = b_0 + b_1x$) or nonlinear (e.g., $\hat{y} = b_0 + b_1x + b_2x^2$), where $\hat{y}$ is the trend line estimate for y and x = time. The nonlinear example is a quadratic equation of the type we discussed in Chapter 17.

Fitting a Linear Trend Equation

In this method, a least-squares linear equation is fitted to the set of observed data points. The approach is the same as in Chapter 15 and can be summarized as follows:

The linear trend equation:

$\hat{y} = b_0 + b_1x$ where $\hat{y}$ = the trend line estimate of y
x = time period

In determining this and other trend equations, we will code the time periods so that the first one has a value of 1, the second has a value of 2, and so on—this is the same approach taken by Excel and Minitab. Also, if you are not using the computer, this offers the advantage of having easier calculator computations because the x values have fewer digits than the four-digit years of the original data.

Using this coding technique and the procedures in Computer Solutions 18.1, we could have obtained the least-squares regression line ($\hat{y} = -52.114 + 34.219x$).

The slope of the equation is positive because vehicle sales tended to increase from 2000 to 2005. The slope, $b_1 = +34.219$, indicates that sales tended to increase by 34.219 thousand vehicles per year during the period. For any given year, we can substitute the appropriate value of x into the trend line equation to find the estimated value of y. For example, 2013 is coded as $x = 14$, and the trend estimate for that year is $\hat{y} = -52.114 + 34.219(14)$, or 426.95 thousand vehicles.

COMPUTER 18.1 SOLUTIONS

Fitting a Linear or Quadratic Trend Equation

EXCEL

Fitting a Linear or Quadratic Trend Equation with Excel (see Figure 18.1)

1. The data in Figure 18.1 are also in Excel file **CX18HYBRID**, with year codes in A2:A7, sales data (thousands of vehicles) in B2:B7, and labels in A1 and B1. From the **Data** ribbon, click **Data Analysis.** Select **Regression.** Click **OK.**
2. Enter **B1:B7** into the **Input Y Range** box. Enter **A1:A7** into the **Input X Range** box. Select **Labels.** Enter **J1** into the **Output Range** box. Select **Line Fit Plots.** Click **OK.**
3. Right-click within the legend descriptions to the right of the chart and click **Delete.** Right-click on any one of the linearly-arranged squares that describe predicted values for the data points and click **Delete.**
4. Right-click on any one of the data points in the scatter diagram and click **Add Trendline.** In the **Trendline Options** menu, select the **Polynomial** option and enter **2** into the **Order** box. Click to select **Display Equation on chart.** Click **Close.** Make formatting adjustments to the chart.

MINITAB

Fitting a Linear or Quadratic Trend Equation with Minitab

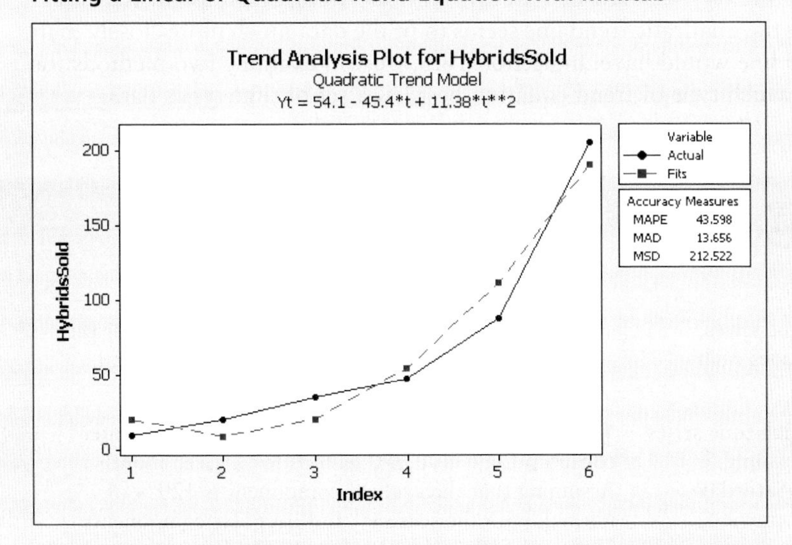

Using Minitab file **CX18HYBRID**, with the year codes in C1 and the vehicle sales data in C2: Click **Stat.** Select **Time Series.** Click **Trend Analysis.** Enter **C2** into the **Variable** box. Within the **Model Type** menu, select either **Linear** or **Quadratic.** We selected the latter. Click **OK.** The Minitab counterpart to Figure 18.1 appears as shown here.

Fitting a Quadratic Trend Equation

Another possibility for fitting an equation to a time series is to use a quadratic model, as shown in Figure 18.1. This involves an equation estimating y as a function of x, but the method treats x and x^2 as two independent variables instead of just one. The quadratic trend equation relies on the multiple regression techniques of Chapter 16 and is described as follows:

The quadratic trend equation:

$\hat{y} = b_0 + b_1x + b_2x^2$ where $\hat{y}$ = the trend line estimate of y
x = time period

(x and x^2 are treated as separate variables.)

The quadratic trend equation includes a constant component (b_0), a linear component (b_1x), and a nonlinear component (b_2x^2). As such, it is especially appropriate when an upward trend in the time series is followed by a downward trend, or vice versa. This type of equation can also be applied whenever the time series either increases or decreases at an increasing rate.

Computer Solutions 18.1 shows the procedures and results when Excel and Minitab are used in fitting a quadratic trend equation to the hybrid vehicle sales data for 2000 (coded as 1) through 2005 (coded as 6). Using the number of decimal places shown in Figure 18.1, the quadratic trend equation is

$$\hat{y} = 54.068 - 45.418x + 11.377x^2$$

As with the linear trend equation, the estimate for a given year is made using the x value corresponding to the code for that year. For 2013, coded as $x = 14$, the estimated number of hybrid vehicles sold in the United States during that year would be $\hat{y} = 54.068 - 45.418(14) + 11.377(14)^2 = 1648.11$ thousand vehicles.

Note that the quadratic trend line seems to fit the data much more closely than a linear trend line would have. In Section 18.6, we will examine two methods for determining which type of trend equation best fits a set of time series data.

EXERCISES

18.1 What are the four components of a time series, and to what kind of variation in the series does each refer?

18.2 For each of the following, indicate which time series component is represented. Explain your reasoning.

a. The proprietor of a movie theater finds Saturday is the most popular day of the week for movie attendance.
b. Political advertising in the United States peaks every 4 years.
c. Sales of a nonaspirin pain reliever decline after the product receives negative publicity in a medical journal article.

18.3 The estimated trend value of community water consumption is 400,000 gallons for a given month. Assuming that the cyclical component is 120% of normal, with the seasonal and irregular components 70% and 110% of normal, respectively, use the multiplicative time series model in determining the quantity of water consumed during the month.

18.4 The trend equation $\hat{y} = 1200 + 35x$ has been fitted to a time series for industry worker days lost due to job-related injuries. If $x = 1$ for 1995, estimate the number of worker days lost during 2012.

18.5 Fit a linear trend equation to the following data describing average hourly earnings of U.S. production workers. What is the trend estimate for 2013? Source: *World Almanac and Book of Facts 2009*, p. 149.

Year:	2004	2005	2006	2007
Average Hourly Earnings:	$15.69	$16.13	$16.76	$17.42

18.6 Fit a linear trend equation to these data showing the value of shipments of food products in billions of dollars. What is the trend estimate for 2015? Source: Bureau of the Census, *Statistical Abstract of the United States 2009*, p. 624.

Year:	2004	2005	2006	2007
Shipments:	$512.3	$532.4	$537.8	$573.5

18.7 U.S. cellular phone subscribership has been reported as shown here for the 1996–2007 period. Source: *World Almanac and Book of Facts 2009*, p. 406.

Year	Subscribers (Millions)	Year	Subscribers (Millions)
1996	44.0	2002	140.8
1997	55.3	2003	158.7
1998	69.2	2004	182.1
1999	86.0	2005	207.9
2000	109.5	2006	233.0
2001	128.4	2007	255.4

a. Fit a linear trend equation to the time series. Using this equation, determine the trend estimate for 2013.
b. Fit a quadratic equation to the time series, then use the equation to determine the trend estimate for 2013.
c. Construct a graph of the time series along with the equations fitted in parts (a) and (b). Which equation appears to better fit the data?

18.8 From 2000 through 2007, the value of U.S. exports to Mexico was as shown below. Data are in billions of dollars. Source: *World Almanac and Book of Facts 2009*, p. 117.

Year:	2000	2001	2002	2003
Exports:	$111.3	$101.3	$97.5	$97.4
Year:	2004	2005	2006	2007
Exports:	$110.8	$120.4	$134.0	$136.1

a. Fit a linear trend equation to the time series. Using this equation, determine the trend estimate for 2015.
b. Fit a quadratic equation to the time series, then use the equation to determine the trend estimate for 2015.
c. Construct a graph of the time series along with the equations fitted in parts (a) and (b). Which equation appears to better fit the data?

SMOOTHING TECHNIQUES (18.3)

While the preceding techniques fit an actual equation to the time series, the methods covered here smooth out short-term fluctuations in the data. These techniques function much like the shock absorbers on an automobile, dampening the sudden upward and downward "jolts" that occur over the series.

The Moving Average

The *moving average* replaces the original time series with another series, each point of which is the center of and the average of N points from the original series. For this reason, this technique is also known as the **centered moving average.** The purpose of the moving average is to take away the short-term seasonal and irregular variation, leaving behind a combined trend and cyclical movement. The technique will be explained through analysis of a time series for a state's electricity purchases.

We will examine the state's electricity purchases during 1995–2009, and there will be a lot of fluctuations from one year to the next. To smooth these fluctuations, we'll use a three-year moving average. Using several of the earlier data points as an example:

Year	Electricity Purchases (thousand kwh)	Three-Year Moving Total	Centered Three-Year Moving Average
1995	815	—	—
1996	826	2386	795.3 (or 2386/3)
1997	745	2347	782.3
1998	776	2359	786.3
1999	838		

The first point in the centered moving average is (815 + 826 + 745)/3, or 795.3. It corresponds to 1996 and is the average of the three years for which 1996 is at the center. The basis for the average now "moves," dropping the 1995 value and picking up the 1998 value. This second point includes purchases for 1996 through 1998, is centered at 1997, and has a value of 2347/3, or 782.3. The moving average can be viewed as the average weight of the occupants of a lifeboat that can hold only three people at a time.

Computer Solutions 18.2 shows the raw data, the procedures, and the results when Excel and Minitab are used in generating a three-year centered moving average of the electricity purchase data. Note that the Excel procedure is applicable only when the centered moving average is based on an odd number of periods; Computer Solutions 18.3 describes how to use Excel when the basis is an even number of periods. The same Minitab procedure is used regardless of the number of periods on which the moving average is based.

COMPUTER 18.2 SOLUTIONS

Centered Moving Average for Smoothing a Time Series

This Excel procedure applies when the centered moving average is based on an odd number of time periods. Computer Solutions 18.3 is applicable for Excel when the centered moving average is based on an even number of time periods. The Minitab procedure is applicable regardless of the number of time periods on which the centered moving average is based.

EXCEL

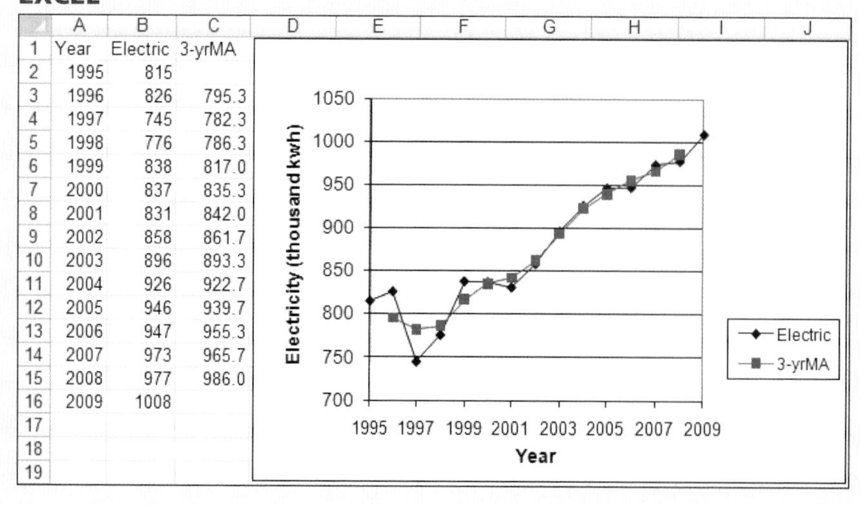

	A	B	C
1	Year	Electric	3-yrMA
2	1995	815	
3	1996	826	795.3
4	1997	745	782.3
5	1998	776	786.3
6	1999	838	817.0
7	2000	837	835.3
8	2001	831	842.0
9	2002	858	861.7
10	2003	896	893.3
11	2004	926	922.7
12	2005	946	939.7
13	2006	947	955.3
14	2007	973	965.7
15	2008	977	986.0
16	2009	1008	
17			
18			
19			

Excel Three-Period Centered Moving Average

1. The data in Table 18.1 are also in Excel file **CX18EPWR**, with years in A2:A16, electricity purchase data in B2:B16, and labels in A1 and B1. From the **Data** ribbon, click **Data Analysis.** Select **Moving Average.** Click **OK.**
2. Enter **B1:B16** into the **Input Range** box. Enter **3** into the **Interval** box. Select **Labels in First Row.** Enter **C1** into the **Output Range** box. (Do not select "Chart Output.") Click **OK.**
3. In column C, select the first two cells with "#NA" in them. Right-click and clear the contents of these cells. The first numerical value in column C should be C3, the centered moving average corresponding to 1996. Enter the label **3-yrMA** into **C1.**
4. Click on **B1** and drag to **C16** to select cells **B1:C16.** From the **Insert** ribbon, click **Line** from the **Charts** menu. From the **2-D Line** menu, click the fourth option (**Line with Markers**). Make formatting adjustments to the chart.

[*Excel tip:* Excel's default horizontal axis for this chart will show time periods 1, 2, and so on. *If you want to display the years, as shown here:* Right-click within the chart and click **Select Data.** In the **Select Data Source** menu, click the **Edit** button beneath **Horizontal (Category) Axis Labels.** When the **Axis Labels** box appears, click and drag to select cells **A2:A16.** As you click and drag, the years will appear within the box. Click **OK.** Click **OK.** *If you want to show every other year and have the years centered on the tick marks:* Right-click on the horizontal axis and click **Format Axis.** In the **Axis Options** menu, enter **2** into the **Interval between tick marks** box, enter **2** into the **Specify interval unit** box. In the **Position Axis** section, select **On Tick Marks.** Click **Close.**]

MINITAB

Minitab Three-Period Centered Moving Average

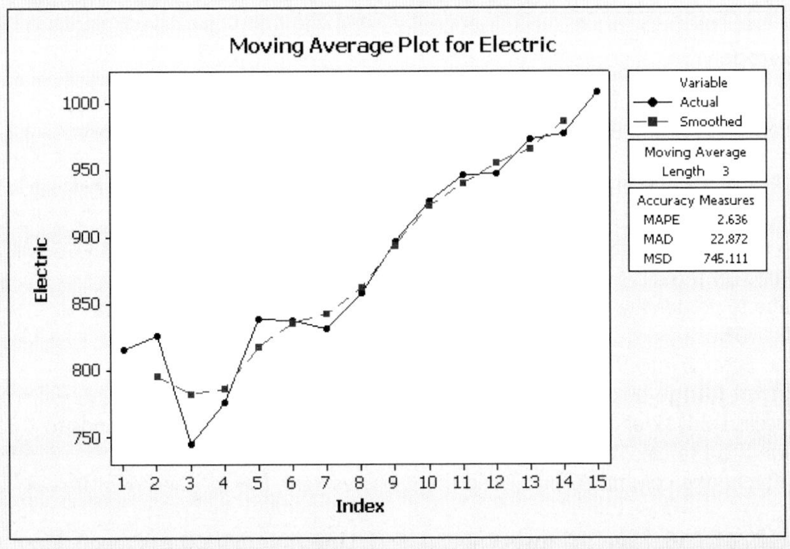

1. Using Minitab file **CX18EPWR**, with years in C1 and electricity purchase data in C2: Click **Stat.** Select **Time Series.** Click **Moving Average.**
2. Enter **C2** into the **Variable** box. Enter **3** into the **MA length** box. Select **Center the moving averages.**
3. Click **Graphs.** Select **Plot smoothed vs. actual.** Click **OK.** Click **OK.** The plot appears as shown here.

Constructing a Centered Moving Average with an Even Number of Periods

In the preceding example, the centered moving average had an *odd* number of periods as the base. With an *even* number of periods as the base, the method changes slightly so that each moving average point will correspond to an actual time period from the original series. The revised procedure is demonstrated in

COMPUTER 18.3 SOLUTIONS

Excel Centered Moving Average Based on Even Number of Periods

This Excel procedure is for centered moving averages based on an even number of periods.

EXCEL

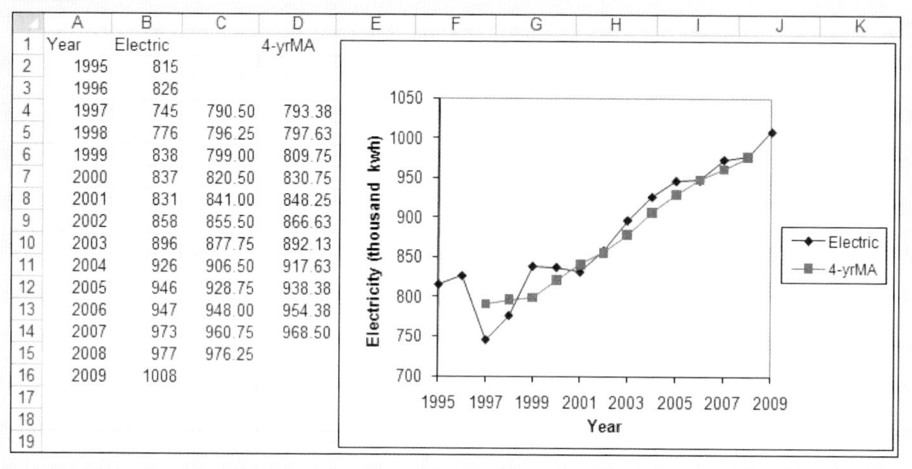

	A	B	C	D
1	Year	Electric		4-yrMA
2	1995	815		
3	1996	826		
4	1997	745	790.50	793.38
5	1998	776	796.25	797.63
6	1999	838	799.00	809.75
7	2000	837	820.50	830.75
8	2001	831	841.00	848.25
9	2002	858	855.50	866.63
10	2003	896	877.75	892.13
11	2004	926	906.50	917.63
12	2005	946	928.75	938.38
13	2006	947	948.00	954.38
14	2007	973	960.75	968.50
15	2008	977	976.25	
16	2009	1008		

Excel Four-Period Centered Moving Average

1. This procedure corresponds to the hand calculations in Table 18.1. Use Excel file **CX18EPWR**, with years in A2:A16, electricity purchase data in B2:B16, and labels in A1 and B1. From the **Data** ribbon, click **Data Analysis.** Select **Moving Average.** Click **OK.**
2. Enter **B1:B16** into the **Input Range** box. Enter **4** into the **Interval** box. Select **Labels in First Row.** Enter **C1** into the **Output Range** box. (Do not select "Chart Output.") Click **OK.**
3. In column C, select the cells with "#NA" in them. Right-click and clear the contents of these cells. Cells C4:C15 will contain smoothed values, but they are not centered.
4. Next we do a two-period moving average to center the smoothed values: From the **Data** ribbon, click **Data Analysis.** Select **Moving Average.** Click **OK.**
5. Enter **C4:C15** into the **Input Range** box. Enter **2** into the **Interval** box. (Do not select "Labels in First Row" or "Chart Output.") Enter **D1** into the **Output Range** box. Click **OK.**
6. Enter the label **4-yrMA** into **D1.** Select cells D2:D12 and move them downward so the first value corresponds to 1997. The results, shown here, correspond to the calculations demonstrated in Table 18.1.
7. For the plot shown here, temporarily interchange columns C and D, so the variables Year, Electric, and 4-yrMA will be in adjacent columns.
8. Click on **B1** and drag to **C16** to select cells **B1:C16.** From the **Insert** ribbon, click **Line** from the **Charts** menu. From the **2-D Line** menu, click the fourth option **(Line with Markers).** Make formatting adjustments to the chart. If you prefer years instead of the default period-number horizontal axis, follow the procedure in the note accompanying Computer Solutions 18.2.

Table 18.1. In proceeding from column (1), each of the four-year moving totals in column (2) falls between two of the actual time periods. In turn, each average in column (3) falls between two figures from column (2). The result of these consecutive "in-between" placements is a moving average in column (4) that corresponds exactly to a time period of the original series.

Computer Solutions 18.3 shows the yearly data, the procedure and the results when we use a four-period centered moving average to smooth the electricity

TABLE 18.1

When a centered moving average is based on an even number of periods, the procedure is as shown for the example years shown here. Otherwise, the moving averages will not correspond exactly to actual time periods from the original series.

Year	(1) Electricity Purchases	(2) Four-Year Moving Total	(3) Average of the Four-Year Moving Totals	(4) Four-Year Centered Moving Average
1995	815			
1996	826			
		3162		
1997	745		3173.5 ÷ 4	→ 793.38
		3185		
1998	776		3190.5	797.63
		3196		
1999	838		3239.0	809.75
		3282		
2000	837		3323.0	830.75
		3364		
2001	831		3393.0	848.25
⋮	⋮	3422	⋮	⋮
		⋮		

purchases data. As noted previously, Excel requires a different method when we use an even number of periods in the moving average, while the Minitab procedure does not change.

Selecting the Number of Periods for a Centered Moving Average

As the cumbersome operations in Table 18.1 suggest, if you're not using a computer, it's desirable to select an odd number of periods unless there is some compelling reason to do otherwise. Such a reason could arise in the case of quarterly ($N = 4$) or monthly ($N = 12$) data where seasonal effects are strong. For example, with a 12-month base, each point in the moving average will include one January, one February, one March, and so on. In the case of retailers who experience the annual Christmas shopping frenzy, this seasonal "blip" will be dampened so that the overall trend can be better observed. In general, a moving average with a base equal to the duration of a strong cyclical or seasonal pattern will be highly useful in dampening fluctuations caused by the pattern.

Selecting the number of periods (N) in the base of a centered moving average is a subjective matter. A larger base provides greater dampening, but is less responsive to directional changes. In addition, an extremely large base will make less efficient use of the data because a greater number of the original values at the beginning and end of the series will not be represented by moving average points. In the Excel portion of Computer Solutions 18.2, there are 15 original data values, but the three-year moving average has just 13 points. You may wish to verify that a nine-year centered moving average would include just 7 points.

Exponential Smoothing

The fluctuations in a time series can also be dampened by a popular and relatively simple technique known as **exponential smoothing.** In this approach, each "smoothed" value is a weighted average of current and past values observed in the series. It can be summarized as follows:

The exponential smoothing of a time series:

$E_t = \alpha \cdot y_t + (1 - \alpha) \cdot E_{t-1}$ where E_t = the exponentially smoothed value for time period t
E_{t-1} = the exponentially smoothed value for the previous time period
y_t = the actual y value in the time series for time period t
α = the smoothing constant, and $0 \leq \alpha \leq 1$

Each exponentially smoothed (E) value is a weighted average of (1) the actual y value for that time period and (2) the exponentially smoothed value calculated for the previous time period. The larger the smoothing constant (α), the more importance is given to the actual y value for the time period.

To demonstrate how time series fluctuations can be dampened by exponential smoothing, we'll apply the technique to the electricity purchases that were the topic of the preceding section. The procedure is as follows:

1. Select the value of the smoothing constant, α. For this example, we'll arbitrarily select $\alpha = 0.3$.
2. Identify an initial value for the smoothed curve. Since each smoothed value will be (0.3 · the actual value) + ([1 − 0.3] · [the previous smoothed value]), we're stuck at the beginning unless we have a smoothed value with which to start the process. It's common practice to "prime the pump" by assuming that $E_1 = y_1$, so this will be our approach. As shown below, the smoothed value for time period 1 (1995) is 815 thousand kilowatt-hours. This initial value will become less important as its effect is "diluted" in subsequent calculations.
3. The calculations through 1998 are shown below.

Year	Electricity Purchases	For $\alpha = 0.3$, Exponentially Smoothed Purchases
1995	815	815
1996	826	818.3 = 0.3(826) + 0.7(815.0)
1997	745	796.3 = 0.3(745) + 0.7(818.3)
1998	776	790.2 = 0.3(776) + 0.7(796.3)

For 1995, E is set as $E = y = 815$ so the smoothed curve will have a starting point.

The fitted value for 1998 is 0.3(the actual purchases for 1998) + 0.7(the smoothed purchases for 1997).

The results of the remaining calculations are shown in the Excel portion of Computer Solutions 18.4, along with a set of exponentially smoothed values based on a larger smoothing constant, $\alpha = 0.6$. Notice that the curve with the smaller smoothing constant ($\alpha = 0.3$) more strongly dampens the fluctuations in the original series. However, it does so at the cost of lagging further behind when the series increases, decreases, or changes direction. In the Excel plots of Computer Solutions 18.4, both exponentially smoothed curves lag behind actual sales during the growth years from 2001 to 2009, much more so than either of the moving averages shown in Computer Solutions 18.2.

COMPUTER 18.4 SOLUTIONS

Exponentially Smoothing a Time Series

EXCEL

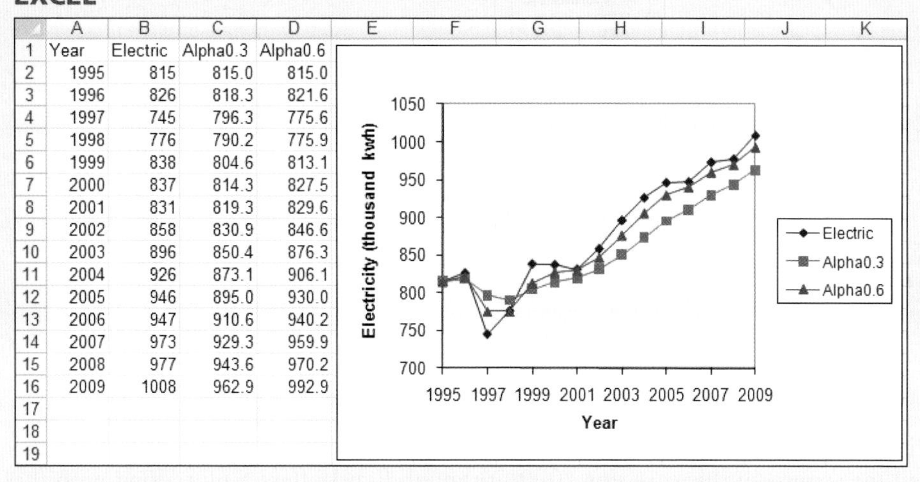

	A	B	C	D
1	Year	Electric	Alpha0.3	Alpha0.6
2	1995	815	815.0	815.0
3	1996	826	818.3	821.6
4	1997	745	796.3	775.6
5	1998	776	790.2	775.9
6	1999	838	804.6	813.1
7	2000	837	814.3	827.5
8	2001	831	819.3	829.6
9	2002	858	830.9	846.6
10	2003	896	850.4	876.3
11	2004	926	873.1	906.1
12	2005	946	895.0	930.0
13	2006	947	910.6	940.2
14	2007	973	929.3	959.9
15	2008	977	943.6	970.2
16	2009	1008	962.9	992.9
17				
18				
19				

Excel Exponential Smoothing with Smoothing Constant, $\alpha = 0.3$

1. Using Excel file CX18EPWR, with years in A2:A16, electricity purchase data in B2:B16, and labels in A1 and B1: From the **Data** ribbon, click **Data Analysis.** Select **Exponential Smoothing.** Click **OK.**
2. Enter **B1:B16** into the **Input Range** box. Enter **0.7** into the **Damping factor** box. (*Note:* Excel's "damping factor" corresponds to $1 - \alpha$.) Select **Labels.** Enter **C1** into the **Output Range** box. (Do not select "Chart Output.") Click **OK.**
3. C1 will have "#NA" in it. Right-click and clear the contents of this cell. The first numerical value in column C should be C2. Enter the **Alpha0.3** label into **C1.** Select **C15** and, clicking at the lower right corner to produce the AutoFill "+" sign, **AutoFill** down to **C16.**
4. (Optional) We can also generate exponentially smoothed values corresponding to $\alpha = 0.6$: In step 2, enter **0.4** (i.e., $1 - \alpha$) into the **Damping factor** box, and enter **D1** into the **Output Range** box. In step 3, select and clear the "#NA" from D1, then enter the **Alpha0.6** label into **D1.** Select **D15** and, clicking at the lower right corner to produce the AutoFill "+" sign, **AutoFill** down to **D16.**
5. Click on **B1** and drag to **D16** to select cells **B1:D16.** From the **Insert** ribbon, click **Line** from the **Charts** menu. From the **2-D Line** menu, click the fourth option **(Line with Markers).** Make formatting adjustments to the chart. If you prefer years instead of the default period-number horizontal axis, follow the procedure in the note accompanying Computer Solutions 18.2.

MINITAB

Minitab Exponential Smoothing with Smoothing Constant, $\alpha = 0.3$

1. Using Minitab file CX18EPWR, with years in C1 and electricity purchase data in C2: Click **Stat.** Select **Time Series.** Click **Single Exp Smoothing.**
2. Enter **C2** into the **Variable** box. Select **Use:** and enter the smoothing constant **0.3** into the box. Click **Graphs.** Select **Plot smoothed versus actual.** Click **OK.** Select **Options** and enter **1** into the **Set initial smoothed value: Use average of first K observations** box. Click **OK.** Click **OK.** The plot appears as shown here.

(*continued*)

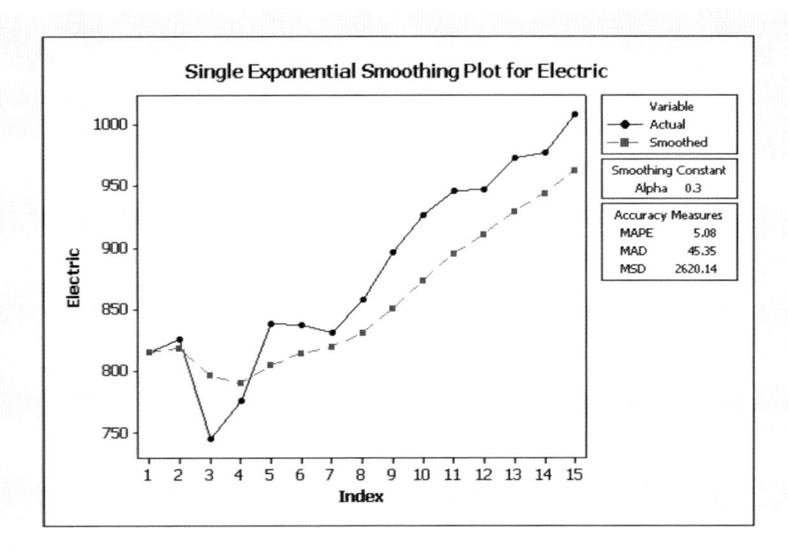

In the Excel portion of Computer Solutions 18.4, we must specify a "damping factor," calculated as $(1 - \alpha)$, then AutoFill to get the final smoothed value in the series—in this case, for 2009. Although Excel offers a plot along with the smoothed series, it assumes you are forecasting instead of just smoothing, and its plot automatically shifts the smoothed series one time period forward, leaving out the final smoothed value. The application of exponential smoothing to forecasting is covered in Section 18.5.

Minitab users can proceed relatively swiftly through the procedure described in Computer Solutions 18.4. However, to match the calculations in this section, be sure to specify that the first period be used in determining the initial smoothed value.

Selecting the value for the smoothing constant (α) is subjective and depends on the extent to which we wish to reduce the fluctuations in the data. At the extremes, a value of $\alpha = 0$ yields a horizontal line with no useful information, while $\alpha = 1$ results in an exponentially smoothed curve identical to the original time series. Naturally, neither extreme would serve the purpose of satisfactorily smoothing the series.

EXERCISES

18.9 The following data show residential and commercial natural gas consumption (quadrillion Btu) from 1992 through 2007. Source: Bureau of the Census, *Statistical Abstract of the United States 2009*, p. 567.

Year	Consumption	Year	Consumption
1992	29.6	2000	37.7
1993	30.6	2001	37.3
1994	30.8	2002	38.2
1995	33.3	2003	38.6
1996	33.0	2004	38.8
1997	33.0	2005	39.6
1998	32.8	2006	38.6
1999	36.0	2007	40.2

a. Construct a graph of the time series, then superimpose a three-year centered moving average over the original series.

b. Superimpose a five-year centered moving average over the original series. Why is the moving average "smoother" when $N = 5$?

18.10 For the data in Exercise 18.9, determine the centered moving average for $N = 2$; for $N = 4$.

18.11 According to the U.S. Energy Information Administration, a heating degree-day is a relative measurement of outdoor temperature that indicates how far the average daily temperature is below 65° Fahrenheit. The higher the number of heating degree-days, the greater the national demand for heating energy from sources like natural gas and electricity. The following monthly data show the number of heating degree-days from 2005 through 2008. Source: U.S. Energy Information Administration, *Annual Energy Review 2008*, p. 17.

	J	F	M	A	M	J
2005	859	676	648	305	186	25
2006	687	731	600	264	137	23
2007	841	853	502	372	111	24
2008	892	741	617	319	183	26

	J	A	S	O	N	D
2005	3	6	39	236	466	866
2006	2	9	82	304	467	690
2007	5	7	44	175	521	800
2008	5	13	52	281	534	831

a. Construct a graph of the time series. Does the overall trend appear to be upward or downward?
b. Determine the three-month and five-month centered moving average curves and superimpose each on the original time series.

18.12 For the data of Exercise 18.11:
a. Obtain the centered moving average for $N = 4$; for $N = 12$.
b. When $N = 12$ months, the centered moving average is relatively smooth compared to when $N = 4$. Aside from the fact that it includes a greater number of months at a time, is there any other reason why the moving average with $N = 12$ should be so smooth?

18.13 The following data describe the monthly volume for an urban courier service. Data are in thousands of pieces handled.

	J	F	M	A	M	J
2007	4.5	4.7	4.7	5.0	5.0	4.4
2008	5.4	4.9	5.3	5.6	5.4	5.4
2009	5.8	5.8	6.6	6.5	6.2	6.7

	J	A	S	O	N	D
2007	4.7	4.9	4.6	5.4	4.8	5.2
2008	5.5	5.4	5.7	6.3	5.3	6.3
2009	6.7	6.7	6.9	7.4	6.6	7.4

a. Construct a graph of the time series. Does the overall trend appear to be upward or downward?
b. Determine the three-month and five-month centered moving average curves and superimpose each on the original time series.

18.14 For the data of Exercise 18.13:
a. Obtain the centered moving average for $N = 4$; for $N = 12$.
b. When $N = 12$ months, the centered moving average is relatively smooth compared to $N = 4$. Aside from the fact that it includes a greater number of months at a time, is there any other reason why the moving average with $N = 12$ should be so smooth?

18.15 The following data describe U.S. manufacturers' shipments of general aviation jet aircraft from 1989 through 2008. Source: *2008 General Aviation Statistical Databook & Industry Outlook*, p. 20.

Year	Shipments	Year	Shipments
1989	157	1999	517
1990	168	2000	588
1991	186	2001	600
1992	171	2002	524
1993	198	2003	384
1994	222	2004	403
1995	246	2005	522
1996	233	2006	604
1997	342	2007	815
1998	413	2008	955

a. Construct a graph of the time series. Does the overall trend appear to be upward or downward?
b. Determine the three-year and five-year centered moving average curves and superimpose each on the original time series.

18.16 For the data of Exercise 18.11, fit an exponentially smoothed curve with smoothing constant $\alpha = 0.4$; with smoothing constant $\alpha = 0.7$. Which constant provides more "smoothing" and why? Which curve is quicker to "catch up" whenever the original series changes direction and why?

18.17 For the data of Exercise 18.13, fit an exponentially smoothed curve with smoothing constant $\alpha = 0.4$; with smoothing constant $\alpha = 0.7$. Which constant provides more "smoothing" and why? Which curve is quicker to "catch up" whenever the original series changes direction and why?

18.18 For the data of Exercise 18.15, determine the centered moving average for $N = 2$; for $N = 4$.

18.19 For the data of Exercise 18.15:
a. Fit an exponentially smoothed curve with smoothing constant $\alpha = 0.6$.
b. For these data, describe the appearance of the exponentially smoothed curve when the smoothing constant is $\alpha = 0.0$; when $\alpha = 1.0$. Why is it not useful to assign α either of these extreme values?

18.4 SEASONAL INDEXES

Depending on the nature of the time series variable, strong seasonal patterns may be present for certain months or quarters. The **seasonal index** is a factor that inflates or deflates a trend value to compensate for the effect of a pattern that repeats over a period of one year or less. Besides identifying the relative influence of various months or quarters, seasonal indexes are useful in the forecasting techniques of the next section.

To demonstrate the calculation of seasonal indexes, we will use a time series that has some very strong seasonal patterns. Shown in Figure 18.3, this is the quarterly sales of natural gas to customers in a northern city from 2005 through 2009. As the time series suggests, home heating and other natural gas demands are especially heavy during the first quarter of each year and tend to be relatively light during the third quarter.

The Ratio to Moving Average Method

The effect of seasonal fluctuations is quantified by using a technique called the **ratio to moving average** method. This method is explained in the context of the underlying data for Figure 18.3, with the following steps:

1. **Construct a centered moving average of the time series.** This is shown in columns (1)–(4) of Table 18.2; note that the procedure is similar to the moving average calculated earlier, in Table 18.1. Since data are quarterly, the moving average is based on four time periods. Had the data consisted of months, the moving average would have been based on 12 periods.
2. **Divide the original time series values by the corresponding centered moving average.** This yields a series of percentages of the moving average. The result of this operation is column (5) of Table 18.2. For example, the actual sales in

FIGURE 18.3

Time series plot of utility companies' sales of natural gas to customers in a northern city from 2005 through 2009.

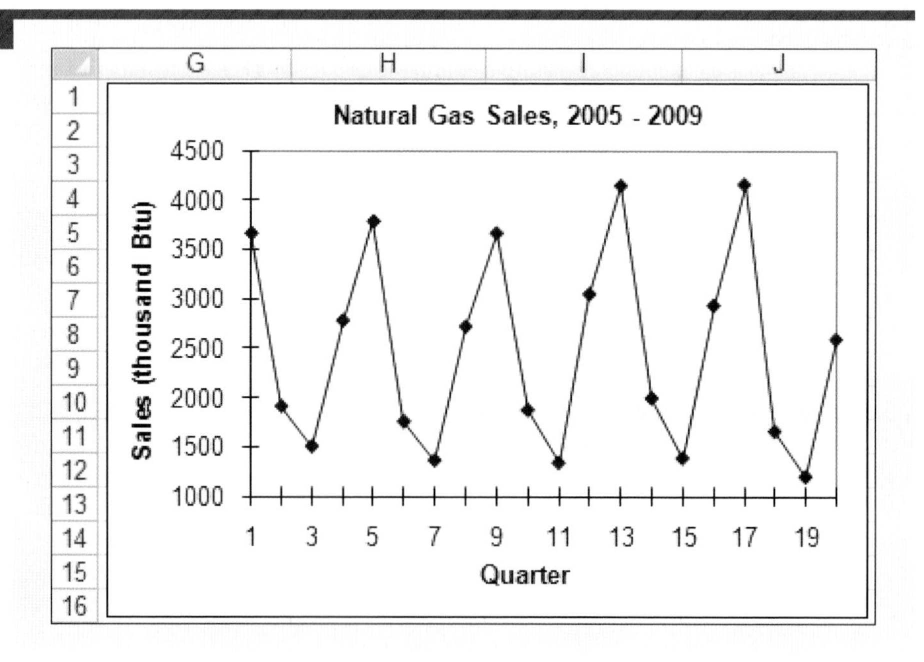

TABLE 18.2

The first stage of calculations in determining the seasonal indexes for the quarterly data of Figure 18.3. The remaining calculations and seasonal indexes are shown in Table 18.3.

Year	Quarter	(1) Sales	(2) Four-Quarter Moving Total	(3) Average of Moving Totals	(4) Centered Four-Quarter Moving Average	(5) Sales as a Percentage of Moving Average
2005	I	3661				
	II	1907				
			9,843			
	III	1500		9,903.0	2475.750	60.588%
			9,963			
	IV	2775		9,884.0	2471.000	112.303
			9,805			
2006	I	3781		9,734.5	2433.625	155.365
			9,664			
	II	1749		9,633.0	2408.250	72.625
			9,602			
	III	1359		9,541.5	2385.375	56.972
			9,481			
	IV	2713		9,541.0	2385.250	113.741
			9,601			
2007	I	3660		9,589.0	2397.250	152.675
			9,577			
	II	1869		9,741.5	2435.375	76.744
			9,906			
	III	1335		10,146.5	2536.625	52.629
			10,387			
	IV	3042		10,448.0	2612.000	116.462
			10,509			
2008	I	4141		10,532.5	2633.125	157.266
			10,556			
	II	1991		10,498.0	2624.500	75.862
			10,440			
	III	1382		10,446.5	2611.625	52.917
			10,453			
	IV	2926		10,284.0	2571.000	113.808
			10,115			
2009	I	4154		10,020.5	2505.125	165.820
			9,926			
	II	1653		9,755.0	2438.750	67.781
			9,584			
	III	1193				
	IV	2584				

the first quarter of 2006 were 3781 thousand Btu, which is 155.365% of the centered moving average (2433.625) for that quarter.

3. **For each quarter (or month), identify the mean percentage from those obtained in step 2.** This is done to reduce the effect of the irregular component of the time series, and the results are shown in the first box of Table 18.3. Other possibilities include using either the median or a trimmed mean that is calculated after the highest and lowest values are dropped.
4. **If the average value of the means is not 100, multiply each one by the same adjustment factor so that the average of 100 is obtained.** In other words, for quarterly data, the means should total 400. If this is not the case, inflate or

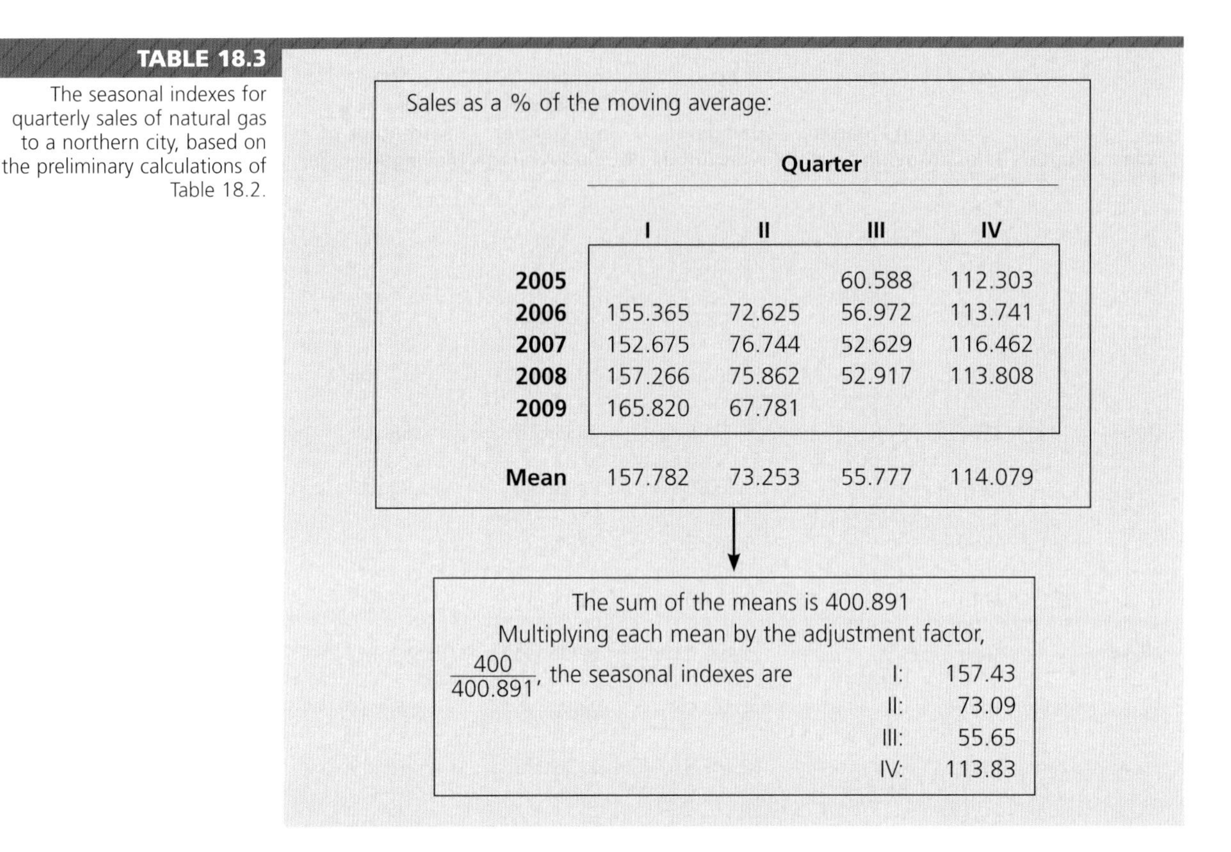

TABLE 18.3

The seasonal indexes for quarterly sales of natural gas to a northern city, based on the preliminary calculations of Table 18.2.

Sales as a % of the moving average:

	Quarter			
	I	**II**	**III**	**IV**
2005			60.588	112.303
2006	155.365	72.625	56.972	113.741
2007	152.675	76.744	52.629	116.462
2008	157.266	75.862	52.917	113.808
2009	165.820	67.781		
Mean	157.782	73.253	55.777	114.079

The sum of the means is 400.891
Multiplying each mean by the adjustment factor, $\frac{400}{400.891}$, the seasonal indexes are

I:	157.43
II:	73.09
III:	55.65
IV:	113.83

deflate the means accordingly. In Table 18.3, this is accomplished by multiplying each of the means by 400/400.891.

5. **Following the adjustment (if necessary) in step 4, the resulting values are the seasonal indexes.** For our example, these are shown in Table 18.3. The larger the seasonal index, the more influence that quarter has in terms of natural gas sales. As expected, the first quarter, and its heating requirements, has resulted in the highest seasonal index, 157.43. This means that 157.43% of the average quarterly gas purchases occur during the first quarter. Likewise, the third quarter accounts for only 55.65% of the average quarterly purchases.

Deseasonalizing the Time Series

Having determined the seasonal indexes, we can go back and *deseasonalize* the original time series. In this process, we remove the seasonal influences and generate a time series that is said to be *seasonally adjusted.*

From the standpoint of the classical time series model, the original and the *deseasonalized time series* can be summarized as follows:

1. **Original time series** Trend (T), cyclical (C), seasonal (S), irregular (I) components, or $T \cdot C \cdot S \cdot I$.
2. **Deseasonalized time series** Only trend, cyclical, and irregular components, or $T \cdot C \cdot I$.

For any time period, the deseasonalized value can also be expressed as $(T \cdot C \cdot S \cdot I)/S$, and dividing by S removes the seasonal influence from the series. In the denominator of the following formula, this seasonal influence (S) is represented as a proportion, or the seasonal index divided by 100:

Deseasonalized time series:

For each time period, the deseasonalized value is

$$D = \frac{\text{Value in the original time series}}{(\text{Seasonal index for that period})/100}$$

The deseasonalized time series for the natural gas data is determined by the preceding formula and shown in Table 18.4. Throughout the series, each first quarter's sales are divided by 1.5743, each second quarter's sales by 0.7309, and so on. The deseasonalized sales in column (3) are the actual sales in column (1) divided by the (seasonal index)/100 in column (2). For example, the deseasonalized sales for the first quarter of 2009 are

$$D = \frac{\text{Actual sales}}{(\text{Seasonal index})/100} = \frac{4154}{157.43/100} = 2638.6$$

The Excel portion of Computer Solutions 18.5 shows this and the other deseasonalized quarterly sales superimposed on the original time series of Figure 18.3. The end result of deseasonalizing the time series is a line that gives us a clearer view of the trend and cyclical components of the series.

Excel does not have a module dedicated to seasonal indexes, but we have provided Excel worksheet templates for determining seasonal indexes and deseasonalized values for quarterly and monthly series, and the Data Analysis Plus add-in can also be applied. Minitab does have a dedicated module but, as with Data Analysis Plus, the methodology differs from the technique described in this section and leads to seasonal index values that are slightly different from those obtained here.

TABLE 18.4

The deseasonalized time series for the natural gas sales of Table 18.2.

Year	Quarter	(1) Actual Sales	(2) (Seasonal Index)/100	(3) Deseasonalized Sales
2005	I	3661	1.5743	2325.5
	II	1907	0.7309	2609.1
	III	1500	0.5565	2695.4
	IV	2775	1.1383	2437.8
2006	I	3781	1.5743	2401.7
	II	1749	0.7309	2392.9
	III	1359	0.5565	2442.0
	IV	2713	1.1383	2383.4
2007	I	3660	1.5743	2324.8
	II	1869	0.7309	2557.1
	III	1335	0.5565	2398.9
	IV	3042	1.1383	2672.4
2008	I	4141	1.5743	2630.4
	II	1991	0.7309	2724.0
	III	1382	0.5565	2483.4
	IV	2926	1.1383	2570.5
2009	I	4154	1.5743	2638.6
	II	1653	0.7309	2261.6
	III	1193	0.5565	2143.8
	IV	2584	1.1383	2270.1

COMPUTER 18.5 SOLUTIONS

Determining Seasonal Indexes

EXCEL, USING WORKSHEET TEMPLATE

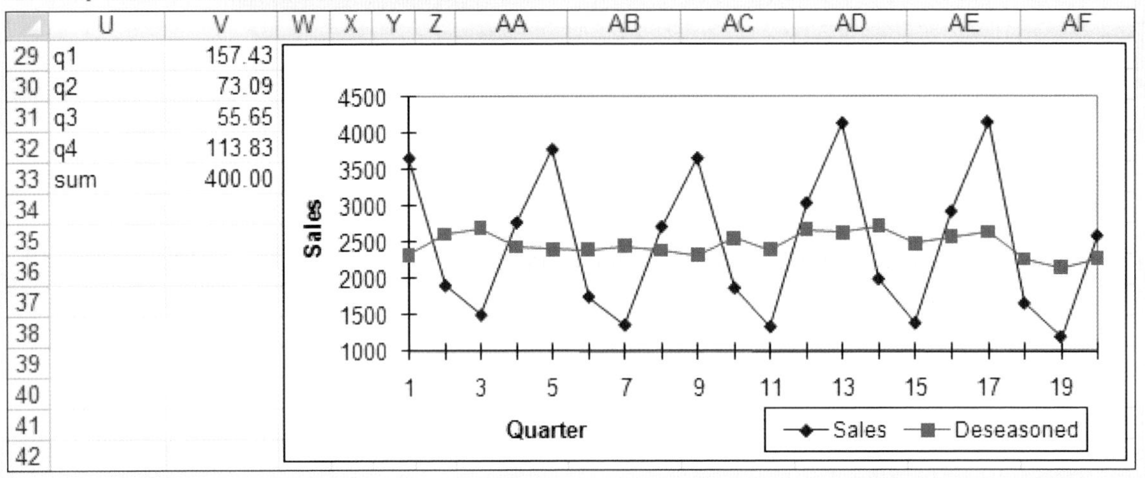

Seasonal Indexes for Quarters

1. Open Excel file **CX18GAS**, with quarters in B2:B21, gas sales data in C2:C21, and labels in B1 and C1. Simultaneously open Excel worksheet template **TMQUARTS**.
2. Working in **TMQUARTS**: Click on the row button for row 14, drag downward to select rows 14 through 37, then right-click and delete rows 14 through 37. There are now 20 rows in which to paste the 20 quarters of gas sales data.
3. Working in **CX18GAS**, click on C2 and drag to C21 to select cells C2:C21. Right-click and select **Copy.**
4. Working in **TMQUARTS**, click to select C6. Right-click and select **Paste.** The sales data are now in cells C6:C25. Click on cell D8 and, using the AutoFill "+" at the lower right corner, AutoFill downward over all existing values in column D. In column B, ensure that the quarters are numbered in sequence, from 1 to 20.
5. Within **TMQUARTS**, the seasonal indexes are shown in column V. Scroll downward in column R to view plots for the original time series and the deseasonalized time series. With slight formatting modifications, the deseasonalized plot appears as shown here. It has been moved next to the seasonal indexes for purposes of this Computer Solutions description.

Notes

1. In general, the specific rows for the locations of the indexes and plots will vary, depending on how many years' worth of quarterly data are being examined.
2. For monthly data, use Excel worksheet template **TMMONTHS**. The procedure will be similar to the one described here.

EXCEL, USING DATA ANALYSIS PLUS

Seasonal Indexes for Quarters

1. Open Excel file **CX18GAS**. The quarterly data are in C2:C21. Enter the label **Quarter** into cell **D1.** Enter the sequence of digits **1** through **4** into **D2:D5.** Continue repeating sequences of 1 through 4 into column D until you've put the sequence of digits 1 through 4 into cells D18:D21. Note that the repeating sequence of quarter identifications should be in the column immediately to the *right* of the column containing the data. If there is no blank column to the right, you will need to insert one.
2. From the **Add-Ins** ribbon, click **Data Analysis Plus.** Select **Seasonal Indexes.** Click **OK.**

3. Enter **C1:D21** into the **Input Range** box and select **Labels.** Click **OK.** The seasonal indexes are shown in the printout.

	A	B
1	Seasonal Indexes	
2		
3	Season	Index
4	1	1.5633
5	2	0.7411
6	3	0.5492
7	4	1.1463

Notes
1. The Data Analysis Plus methodology differs slightly from the one described in the text and used in the Excel worksheet template method described previously, and it will result in slightly different seasonal index values.
2. For monthly data, use the same procedure shown here, but repeat the sequence of digits 1 through 12 in the column immediately to the right of the column containing the data.

MINITAB

Minitab Seasonal Indexes, Quarters

1. Using Minitab file **CX18GAS**, with quarters in C2 and gas sales data in C3: Click **Stat.** Select **Time Series.** Click **Decomposition.**
2. Enter **C3** into the **Variable** box. Enter **4** into the **Seasonal length** box. Under **Model Type,** select **Multiplicative.** Under **Model Components,** select **Trend plus seasonal.** Click **Options** and enter **1** into the **First obs is in seasonal period:** box. Click **OK.** Click **OK.** The summary printout shown here is among the standard printout items, and a variety of tables and plots can be obtained as well.

```
Time Series Decomposition for Sales

Multiplicative Model

Data      Sales
Length    20
NMissing  0

Fitted Trend Equation
Yt = 2513.8 - 4.50850*t

Seasonal Indices
Period     Index
     1  1.56598
     2  0.74378
     3  0.55044
     4  1.13980
```

Notes
1. The Minitab methodology differs from the one described in the text and used in the Excel method above, and will result in slightly different seasonal index values.
2. For monthly data, use the same procedure, but in step 2 enter **12** into the **Seasonal length** box.

EXERCISES

18.20 In determining seasonal indexes for the months of the year, a statistician finds the total of the "unadjusted" seasonal indexes is 1180 instead of 12 · 100, or 1200. What correction must be made to arrive at the adjusted seasonal indexes?

18.21 The seasonal indexes for a convention center's bookings are 83, 120, 112, and 85 for quarters 1–4, respectively. What percentage of the center's annual bookings tend to occur during the second quarter?

18.22 The state of Texas has reported the following quarterly gross sales for the in-state accommodation and food services industry for the years 2004–2008.
Source: Texas Comptroller of Public Accounts site, ecpa.cpa.tx.us, August 22, 2009.

Gross Sales ($ Billions)

	I	II	III	IV
2004	$7.014	7.330	7.260	7.521
2005	7.441	7.904	7.716	8.228
2006	8.166	8.390	8.321	8.751
2007	8.642	9.322	9.136	9.599
2008	9.412	9.879	9.578	9.995

a. Construct the four-quarter centered moving average for these data and determine the percentages of the moving average for the quarters.
b. Determine the seasonal indexes for the quarters.
c. Use the seasonal indexes to deseasonalize the original time series.

18.23 A major amusement park has the following number of visitors each quarter from 2005 through 2009:

Number of Visitors (Thousands)

	I	II	III	IV
2005	155	231	270	105
2006	182	255	315	294
2007	160	250	280	297
2008	210	310	365	335
2009	225	325	384	386

a. Construct the four-quarter centered moving average for these data and determine the percentages of the moving average for the quarters.
b. Determine the seasonal indexes for the quarters.
c. Use the seasonal indexes to deseasonalize the original time series.

18.24 Given the degree-day data of Exercise 18.11, (a) determine the monthly seasonal indexes for January through December, then (b) use these indexes to deseasonalize the original time series.

18.25 Given the data of Exercise 18.13, (a) determine the monthly seasonal indexes for January through December, then (b) use these indexes to deseasonalize the original time series.

18.26 A mail-order firm has found the following seasonal indexes for the number of calls received each month. Below the seasonal indexes is the actual number of calls (thousands) per month during 2009.
a. Using the seasonal indexes, deseasonalize the data for 2009.
b. The president of the firm is quite pleased that the number of calls went up each month during the May through December portion of 2009. Based on the results obtained in part (a), is the firm's performance really on the increase?

Month:	J	F	M	A	M	J	J	A	S	O	N	D
Seasonal index:	110	105	90	60	70	75	85	95	100	105	140	165
Calls in 2009:	7.2	5.9	5.5	3.2	4.0	4.1	5.3	5.4	5.7	6.1	7.9	8.2

18.5 FORECASTING

In forecasting, we use data from the past in predicting the future value of the variable of interest, such as appliance sales, natural gas consumption, or the thickness of the ozone layer. The time series and seasonal index methods discussed earlier are often employed for this purpose. Using past data to predict future data is very much like facing backward in a car, then trying to figure out where you are going by looking at where you've been. There actually was such a car in the 1950s, and the analogy between it and time series-based forecasting is discussed in the opening vignette to this chapter.

Forecasts Using the Trend Equation

Once we've fitted a trend equation to the data, a forecast can be made by simply substituting a *future* value for the time period variable. The procedure is the same whether the fitted trend equation is linear or nonlinear.

For example, in Figure 18.1, we fitted a quadratic trend equation to hybrid vehicle sales in the United States from 2000 through 2005. Using this equation to make a forecast for 2013:

The fitted linear trend equation is

$$\hat{y} = 54.068 - 45.418x + 11.377x^2$$

where $\hat{y}$ = the trend line estimate of hybrid vehicle sales (thousands)
x = year code, with 2000 coded as 1.

↓

The forecast of hybrid vehicle sales for 2013 would be

$\hat{y} = 54.068 - 45.418(14) + 11.377(14)^2$, or 1648.11 thousand vehicles

Forecasting with Exponential Smoothing

Besides smoothing a time series, exponential smoothing can also be used in making a forecast for one time period into the future. The approach is similar to the smoothing method of Section 18.3, except that F (for forecast value) replaces E (exponentially smoothed value) for each time period, and the time period (t) "shifts" one unit into the future:

Forecasting with exponential smoothing:

$$F_{t+1} = \alpha \cdot y_t + (1 - \alpha) \cdot F_t$$

where F_{t+1} = forecast for time period $t + 1$
y_t = actual y value for time period t
F_t = the value that had been forecast for time period t
α = the smoothing constant, and $0 \leq \alpha \leq 1$

As in our previous application of exponential smoothing, there is a smoothing constant, $0 \leq \alpha \leq 1$. If our forecast involves annual sales, the forecast for next year is a weighted average of the *actual* sales this year and the sales level that had been *forecast* for this year. If we choose $\alpha = 1$, we are essentially saying "next year will be the same as this year."

To illustrate the application of exponential smoothing to forecasting, we will use the electric power purchases data. These data included purchases from 1995 through 2009, and exponential smoothing will give us a forecast for 2010. For our example, we will use $\alpha = 0.6$ as the smoothing constant.

As shown below, exponential smoothing requires that we assume an initial forecast value in order to get things started. For this reason, the forecast for 1995 is assumed to have been 815, the same as the actual purchases for that year. Proceeding toward 2010, the influence of this choice will decrease, since its effect is diluted as more recent forecast values are introduced. For each year, the forecast is a weighted average of the actual and the forecasted level of purchases for the preceding year, and the weighting is according to the smoothing constant, $\alpha = 0.6$:

Year	y = Actual purchases	F = Purchases forecast for the year, with smoothing constant $\alpha = 0.6$
1995	815	815.0 ←
1996	826	815.0 = 0.6(815) + 0.4(815.0)
1997	745	821.6 = 0.6(826) + 0.4(815.0)
1998	776	775.6 = 0.6(745) + 0.4(821.6)
⋮	⋮	⋮
2008	977	959.9 = 0.6(973) + 0.4(940.2)
2009	1008	970.2 = 0.6(977) + 0.4(959.9)
2010	—	992.9 = 0.6(1008) + 0.4(970.2)

To get the process started, the forecast for 1995 is assumed to have been 815.

The forecast for 2010 is a weighted average of the actual and the forecasted values for 2009.

The Excel portion of Computer Solutions 18.6 lists these and the other forecasted values for electricity purchases. In addition to the forecast and plot shown in Computer Solutions 18.6, Minitab can also generate a complete

COMPUTER 18.6 SOLUTIONS

Forecasting with Exponential Smoothing

EXCEL

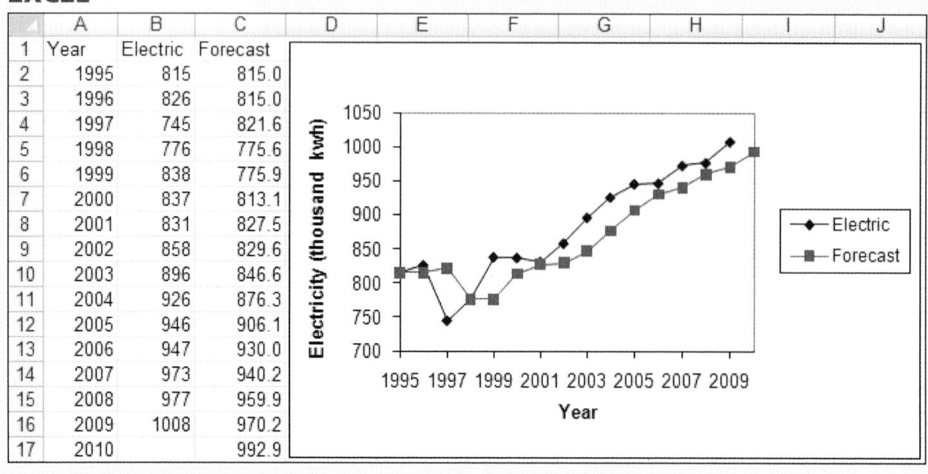

	A	B	C
1	Year	Electric	Forecast
2	1995	815	815.0
3	1996	826	815.0
4	1997	745	821.6
5	1998	776	775.6
6	1999	838	775.9
7	2000	837	813.1
8	2001	831	827.5
9	2002	858	829.6
10	2003	896	846.6
11	2004	926	876.3
12	2005	946	906.1
13	2006	947	930.0
14	2007	973	940.2
15	2008	977	959.9
16	2009	1008	970.2
17	2010		992.9

Excel Forecasting with Exponential Smoothing, $\alpha = 0.6$

1. Using Excel file **CX18EPWR**, with years in A2:A16, electricity purchase data in B2:B16, and labels in A1 and B1: From the **Data** ribbon, click **Data Analysis.** Select **Exponential Smoothing.** Click **OK.**

2. Enter **B1:B16** into the **Input Range** box. Enter **0.4** into the **Damping factor** box. (*Note:* Excel's "damping factor" corresponds to $1 - \alpha$.) Select **Labels.** Enter **C2** into the **Output Range** box. (Do not select "Chart Output.") Click **OK.**
3. Select **C2** and enter **815** (this replaces the "#NA" and provides a forecast for the first time period). Select **C16** and, clicking at the lower right corner to produce the AutoFill "+" sign, **AutoFill** down to **C17.** Enter **2010** into **A17.** Enter the **Forecast** label into **C1.**
4. Click on **B1** and drag to **C17** to select cells **B1:C17.** From the **Insert** ribbon, click **Line** from the **Charts** menu. From the **2-D Line** menu, click the fourth option (**Line with Markers**). Make formatting adjustments to the chart. If you prefer years instead of the default period-number horizontal axis, follow the procedure in the note accompanying Computer Solutions 18.2.

MINITAB

Minitab Exponential Smoothing with Smoothing Constant, $\alpha = 0.6$

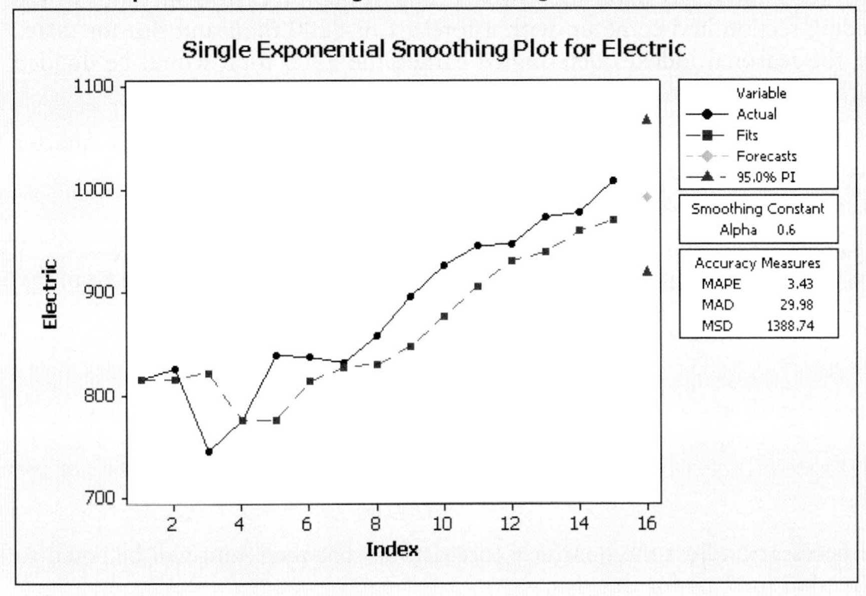

1. Using Minitab file **CX18EPWR**, with years in C1 and electricity purchase data in C2: Click **Stat.** Select **Time Series.** Click **Single Exp Smoothing.**
2. Enter **C2** into the **Variable** box. Select **Use:** and enter the smoothing constant **0.6** into the box. Select **Generate forecasts** and enter **1** into the **Number of forecasts** box. (Just leave the "Starting from origin" box blank and Minitab will make a forecast for the one period immediately following the last data value.)
3. Click **Graphs.** Select **Plot predicted vs. actual.** Click **OK.** Click **Options** and enter **1** into the **Set initial smoothed value: Use average of first K observations** box. Click **OK.** Click **OK.** The plot appears as shown here, with the forecast displayed as a single point beyond the data series.

table that includes time series values, smoothed values, forecast values, and residuals.

When exponential smoothing is used for forecasting, the forecast will tend to be too low for time series that are increasing (positive trend) and too high for time series that are decreasing (negative trend). This is because the forecast for a given period is always the weighted average of two values associated with the immediately preceding period. To correct this problem, the upward or downward trend of the time series can be introduced as an additional variable. Further details of this method, called the Holt-Winters forecasting model, can be found in more advanced texts on forecasting.

Seasonal Indexes in Forecasting

Once we have arrived at a forecast for a given year, we can use seasonal indexes to come up with a forecast for each month or quarter of the year. In doing so, we divide up the total forecast according to the relative sizes of the seasonal indexes. The procedure is as follows:

Dividing an annual forecast into seasonal components:

$$\text{Forecast for a season} = \frac{\text{Forecast for the year}}{\text{Number of seasons}} \cdot [(\text{Seasonal index})/100]$$

For the natural gas data, suppose that one of the forecasting methods of the preceding section had come up with a forecast of 9200 thousand Btu for 2015. Using the seasonal indexes determined earlier, the 9200 total would be divided up into the four quarters as follows:

1. If the forecast for the entire year = 9200 thousand Btu, the average quarter will be 9200/4, or 2300.
2. "Inflate" or "deflate" quarters according to their seasonal indexes:

Quarter:	I	II	III	IV
(Seasonal index)/100:	1.5743	0.7309	0.5565	1.1383

Quarter	
I	2300(1.5743) = 3620.89 thousand Btu
II	2300(0.7309) = 1681.07
III	2300(0.5565) = 1279.95
IV	2300(1.1383) = 2618.09
	9200.00

3. If necessary, adjust the quarterly forecasts so that their sum will be equal to the total forecast for the year.

The sum of these quarterly forecasts (9200.00) is equal to the 9200.00 forecasted for the entire year, so no further adjustment is necessary. (As an example, if the quarterly forecasts had totaled 9198.00, we would have had to multiply each of the quarterly estimates in step 2 by 9200/9198.00 so that the total for the year would be 9200.00.)

EXERCISES

18.27 The linear trend equation $\hat{y} = 120 + 4.8x$ has been developed, where $\hat{y}$ = estimated sales (thousands of dollars) and $x = 1$ for the year 1999. What level of sales would be forecast for 2013; for 2015?

18.28 An appliance repair shop owner has fitted the quadratic trend equation $\hat{y} = 90 + 0.9x + 3x^2$ to a time series of annual repair orders, with y = the number of repair orders and $x = 1$ for 2005. Forecast the number of repair orders for 2013; for 2015.

18.29 Based on annual truck rentals over the years, a rental firm has developed the quadratic trend equation $\hat{y} = 450 + 20x + 7.2x^2$, with $x = 1$ for 2005. Forecast y for 2013; for 2015.

18.30 When exponential smoothing is used in fitting a curve to a time series, the approach is slightly different from its application to forecasting. Compare the appropriate formulas and point out how they differ.

18.31 The U.S. trade deficit with Canada (billions of dollars) from 2000 through 2007 is reported as shown in the table. Using exponential smoothing and the smoothing constant $\alpha = 0.7$, what deficit would have been forecast for 2008? Source: *World Almanac and Book of Facts 2009*, p. 117.

Year:	2000	2001	2002	2003	2004	2005	2006	2007
Deficit:	$51.9	52.8	48.2	51.7	66.5	78.5	71.8	68.2

18.32 For the data of Exercise 18.7, use the weighting constant $\alpha = 0.5$ and exponential smoothing to determine the forecast for 2008.

18.33 For the data of Exercise 18.9, use the weighting constant $\alpha = 0.6$ and exponential smoothing to determine the forecast for 2008.

18.34 If a time series is such that sales are consistently increasing from one year to the next, will exponential smoothing tend to produce forecasts that are (a) over or (b) under the sales that are actually realized for the forecast period? Why?

18.35 A firm predicts it will use 560,000 gallons of water next year. If the seasonal index of water consumption is 135 for the second quarter of the year, what is the forecast for water consumption during that quarter?

18.36 The quarterly seasonal indexes for a firm's electricity consumption are 115, 92, 81, and 112 for quarters I–IV. It has been forecast that electricity consumption will be 850,000 kilowatt-hours during 2014. Forecast electricity consumption for each quarter of 2014.

EVALUATING ALTERNATIVE MODELS: *MAD* AND *MSE* 18.6

When two or more models are fitted to the same time series, it's useful to have one or more criteria on which to compare them. In this section, we'll discuss two such criteria.

According to the **mean absolute deviation (*MAD*)** criterion, the best-fit model is the one having the lowest mean value for $|y_t - \hat{y}_t|$. The direction of an error, or deviation, is not considered, only its absolute value. The *MAD* criterion is as follows:

The mean absolute deviation (*MAD*) criterion:

From a given set of models or estimation equations fit to the same time series data, the model or equation that best fits the time series is the one with the lowest value of

$$MAD = \frac{\sum |y_t - \hat{y}_t|}{n}$$

where y_t = an observed value of y
$\hat{y}_t$ = the value of y that is predicted using the estimation equation or model
n = the number of time periods

Another criterion is **mean squared error (*MSE*),** in which the best-fit model or equation is the one having the lowest mean value for $(y_t - \hat{y}_t)^2$. This is analogous to the least-squares criterion used in determining the regression equations of Chapters 15–17. The MSE approach is described below:

The mean squared error (*MSE*) criterion:

From a given set of models or estimation equations fit to the same time series data, the model or equation that best fits the time series is the one with the lowest value of

$$MSE = \frac{\sum (y_t - \hat{y}_t)^2}{n}$$

where y_t = an observed value of y
$\hat{y}_t$ = the value of y that is predicted using the estimation equation or model
n = the number of time periods

Compared to the *MAD* criterion, the *MSE* approach places a greater penalty on estimates for which there is a large error. This is because *MSE* is the mean of the *squared* errors instead of simply the mean of the absolute values. Because of this, the *MSE* criterion is to be preferred whenever the cost of an error in estimation or forecasting increases in more than a direct proportion to the amount of the error.

Table 18.5 shows the application of the *MAD* and *MSE* criteria to the linear and quadratic estimation equations that were developed for the hybrid vehicle sales data in Section 18.2 of this chapter.

TABLE 18.5

MAD and *MSE* measures of fit to the hybrid vehicle sales data of Section 18.2.

A. Fit for the Linear Estimation Equation, $\hat{y} = -52.114 + 34.219x$

Year	Year Code, x_t	Actual Sales, y_t	Estimated Sales, $\hat{y}_t$	Error, $(y_t - \hat{y}_t)$	Absolute Error, $\lvert y_t - \hat{y}_t \rvert$	Squared Error, $(y_t - \hat{y}_t)^2$
2000	1	9.350	−17.895	27.245	27.245	742.290
2001	2	20.287	16.324	3.963	3.963	15.705
2002	3	35.000	50.543	−15.543	15.543	241.585
2003	4	47.530	84.762	−37.232	37.232	1386.222
2004	5	88.000	118.981	−30.981	30.981	959.822
2005	6	205.750	153.200	52.550	52.550	2761.503
					167.514	6107.127

$$MAD = \frac{\sum |y_t - \hat{y}_t|}{n} = \frac{167.514}{6} = 27.919 \qquad MSE = \frac{\sum (y_t - \hat{y}_t)^2}{n} = \frac{6107.127}{6} = 1017.854$$

B. Fit for the Quadratic Estimation Equation, $\hat{y} = 54.068 - 45.418x + 11.377x^2$

Year	Year Code, x_t	Actual Sales, y_t	Estimated Sales, $\hat{y}_t$	Error, $(y_t - \hat{y}_t)$	Absolute Error, $\lvert y_t - \hat{y}_t \rvert$	Squared Error, $(y_t - \hat{y}_t)^2$
2000	1	9.350	20.027	−10.677	10.677	113.998
2001	2	20.287	8.740	11.547	11.547	133.333
2002	3	35.000	20.207	14.793	14.793	218.833
2003	4	47.530	54.428	−6.898	6.898	47.582
2004	5	88.000	111.403	−23.403	23.403	547.700
2005	6	205.750	191.132	14.618	14.618	213.686
					81.936	1275.133

$$MAD = \frac{\sum |y_t - \hat{y}_t|}{n} = \frac{81.936}{6} = 13.656 \qquad MSE = \frac{\sum (y_t - \hat{y}_t)^2}{n} = \frac{1275.133}{6} = 212.522$$

Regardless of the criterion, the quadratic equation is a better fit to these data. This is not surprising, since the quadratic equation has the advantage of including both a linear and a nonlinear term.

EXERCISES

18.37 How do the *MAD* and *MSE* criteria differ in their approach to evaluating the fit of an estimation equation to a time series?

18.38 The following data are the wellhead prices for domestically produced natural gas, in dollars per thousand cubic feet, from 2001 through 2008. Given these data and the trend equations shown here, use the *MAD* criterion to determine which equation is the better fit. Repeat the evaluation, using the *MSE* criterion. Source: U.S. Energy Information Administration, *Annual Energy Review 2008*, p.199.

Year	$x =$ Year Code	$y =$ Price
2001	1	$4.00
2002	2	2.95
2003	3	4.88
2004	4	5.46
2005	5	7.33
2006	6	6.39
2007	7	6.37
2008	8	8.07

Linear:
$\hat{y} = 2.896 + 0.6189x$

Quadratic:
$\hat{y} = 2.543 + 0.8305x - 0.02351x^2$

18.39 The following data are the number of Harley-Davidson motorcycles sold worldwide for 2003 through 2006 (thousands). Given these data and the trend equations, use the *MAD* criterion to determine which equation is the better fit. Repeat the evaluation, using the *MSE* criterion. Source: Harley-Davidson, Inc., *2008 Annual Report*, p. 20.

Year	$x =$ Year Code	$y =$ Sales
2003	1	291.1
2004	2	317.3
2005	3	329.0
2006	4	349.2

Linear:
$\hat{y} = 275.15 + 18.60x$

Quadratic:
$\hat{y} = 267.65 + 26.10x - 1.50x^2$

18.40 Using the *MAD* criterion, determine which one of the equations developed in Exercise 18.7 is the better fit to the data in that exercise. Repeat the evaluation, using the *MSE* criterion.

18.41 Using the *MAD* criterion, determine which one of the equations developed in Exercise 18.8 is the better fit to the data in that exercise. Repeat the evaluation, using the *MSE* criterion.

AUTOCORRELATION, THE DURBIN-WATSON TEST, AND AUTOREGRESSIVE FORECASTING (18.7)

Autocorrelation and the Durbin-Watson Test

As we've discussed in earlier chapters, one of the assumptions required by the regression model is that the residuals (errors) be independent of each other. When this assumption is not met, the condition known as *autocorrelation* (or *serial correlation*) is present. Regression models based on time series data are especially susceptible to this condition because: (1) key independent (x) variables that have not been included in the model are likely to be related to time, and (2) the omission of one or more of these independent variables will tend to yield residuals (errors) that are also related to time.

Autocorrelation will tend to cause the calculated standard error of estimate (s_e) to underestimate the true value (σ_e). The partial regression coefficients (b_1, b_2, and so on) will also have standard errors that are understated, and the t values in their respective tests of significance will tend to be exaggerated. As a result, confidence and prediction intervals will be too narrow, and the respective independent variables will be exaggerated in their significance.

A common method of testing for autocorrelation is the *Durbin-Watson* test. Our discussion of this test will be in terms of its ability to detect the presence of *first-order* autocorrelation. This is the correlation between adjacent residuals or errors—that is, between e_t and e_{t-1}. (By extension, *second-order* autocorrelation would be the correlation between e_t and e_{t-2}. When we refer to autocorrelation in the remainder of this discussion, we will be referring to first-order autocorrelation.) The Durbin-Watson test treats the population of residuals as shown below.

Durbin-Watson test, description of residual (error) for time period t:

$$\varepsilon_t = \rho\varepsilon_{t-1} + z_t$$

where ε_t = residual associated with time period t
ε_{t-1} = residual associated with time period $t - 1$
ρ = the population correlation coefficient between values of ε_t and ε_{t-1}, with $-1 \le \rho \le +1$
z_t = a normally distributed random variable with $\mu = 0$

If ρ is zero, there is no autocorrelation and $\varepsilon_t = z_t$—that is, ε_t is nothing more than random variation with a mean of zero. Likewise, a ρ that is greater than 0 reflects positive autocorrelation and a ρ that is less than zero reflects negative autocorrelation. Examples of positive and negative autocorrelation are shown in Figure 18.4. The Durbin-Watson test is carried out as shown below.

Durbin-Watson test for autocorrelation:

- **Null and alternative hypotheses**

The test is

for positive autocorrelation	for negative autocorrelation	a two-tail test
H_0: $\rho \le 0$ H_1: $\rho > 0$	H_0: $\rho \ge 0$ H_1: $\rho < 0$	H_0: $\rho = 0$ H_1: $\rho \ne 0$

- **Test statistic**

$$d = \frac{\sum_{t=2}^{t=n}(e_t - e_{t-1})^2}{\sum_{t=1}^{t=n} e_t^2}$$

where e_t = observed residual (error) associated with time period t
e_{t-1} = observed residual (error) associated with time period $t - 1$
n = number of observations, with $n \ge 15$

- **Critical values and decision zones** Look up the appropriate d_L and d_U values in the Durbin-Watson tables in Appendix A. The actual critical values and decision zones are based on (1) the type of test (positive, negative, or two-tail), (2) α = the level of significance for the test, and (3) k = the number of independent variables in the regression model. The decision zones and appropriate conclusions are shown in Figure 18.5.

FIGURE 18.4

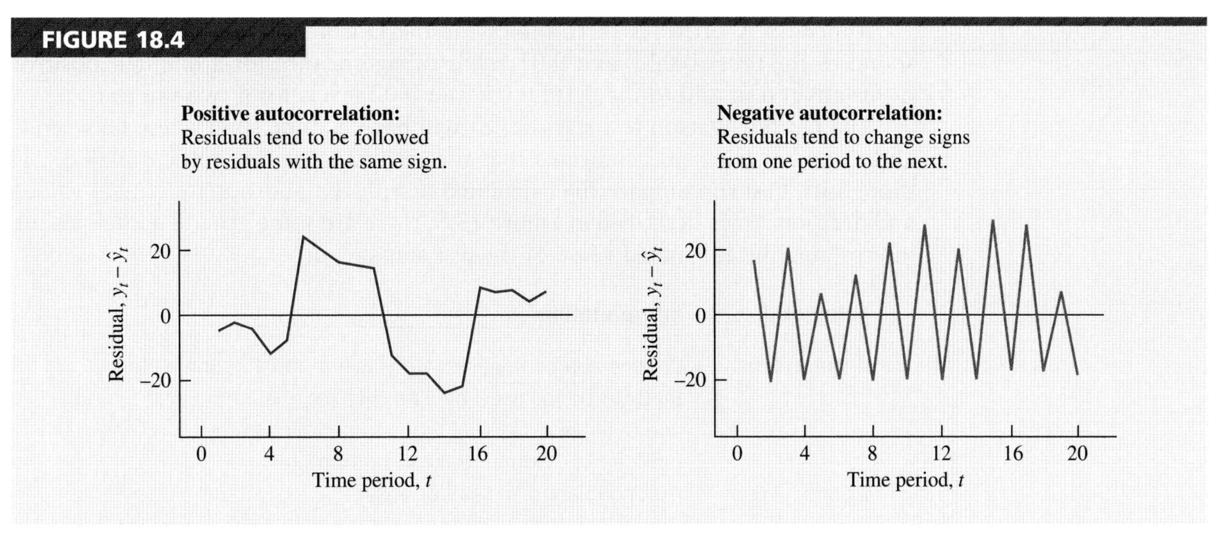

Examples of positive and negative autocorrelation of residuals

FIGURE 18.5

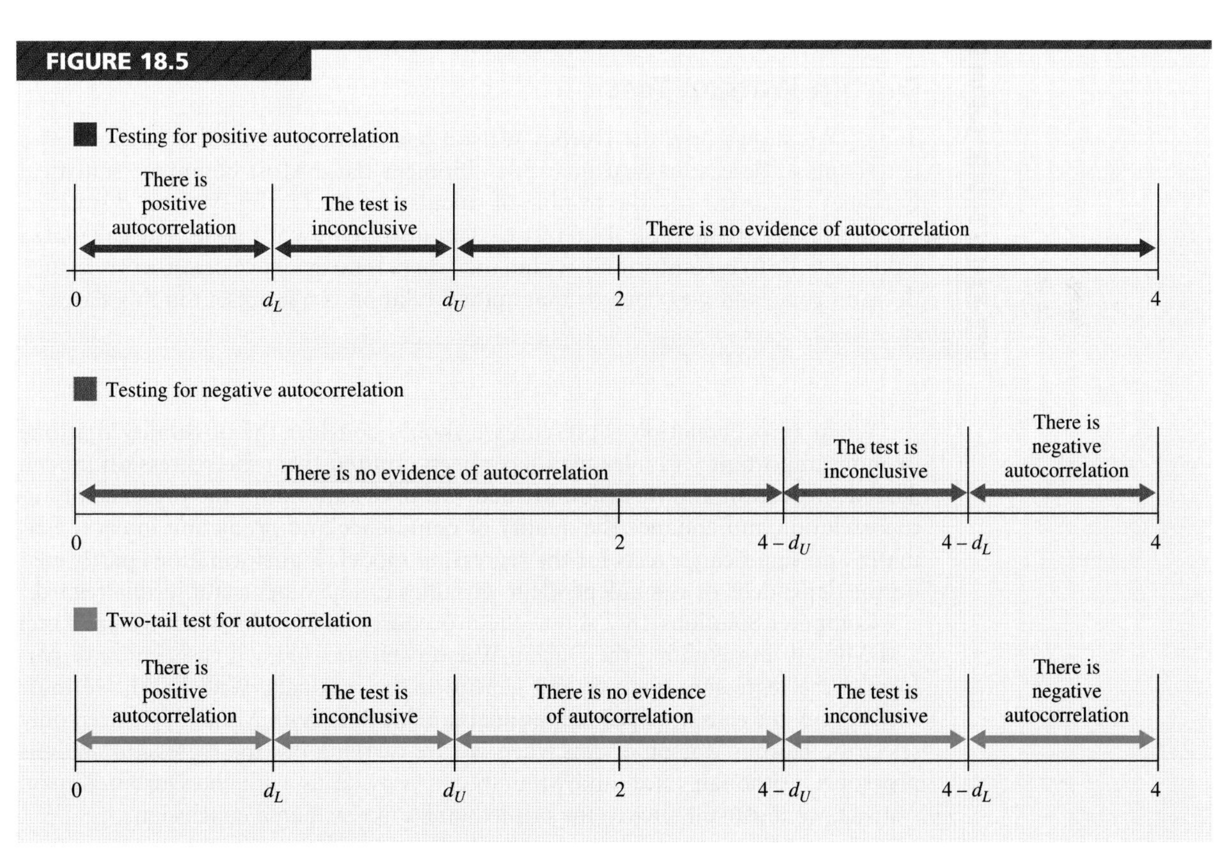

When proceeding with a Durbin-Watson test for autocorrelation, the calculated d statistic is compared to one of three sets of critical values and decision zones shown.

Appendix A provides three tables for the Durbin-Watson test. They correspond to the 0.05, 0.025, and 0.01 levels of significance for a one-tail (positive or negative) test, and to the 0.10, 0.05, and 0.02 levels for a two-tail test.

The Durbin-Watson test statistic (d) will have a value somewhere between 0 and 4, and a value of 2 indicates that the residuals are not autocorrelated. Although it's best to compare the calculated d with the appropriate critical values and decision zones described in Figure 18.5, there are some general guidelines for interpreting the calculated value of the test statistic, d:

Value of d	Interpretation
Close to 0.0	Positive autocorrelation: A positive residual in one period tends to be followed by a positive residual in the next period, and a negative residual tends to be followed by a negative residual. See Figure 18.4 for an example.
Approx. 2.0	Autocorrelation is weak or absent.
Close to 4.0	Negative autocorrelation: Residuals tend to change signs from one period to the next, as shown in Figure 18.4.

EXAMPLEEXAMPLE

EXAMPLE

Durbin-Watson Test

Table 18.6 shows how the Durbin-Watson test could be applied to a time series consisting of deseasonalized quarterly shipment data. Most computer statistical packages can spare you the trouble of calculating the Durbin-Watson d statistic, but the details are in Table 18.6 just to demonstrate how it's done. The calculated value ($d = 1.90$) is comfortably within the "there is no evidence of autocorrelation" decision zone, and we conclude that autocorrelation is not present in this series.

When autocorrelation is present, we should consider the possibility that one or more important variables may have been omitted from the regression model. The identification and inclusion of such variables can reduce the presence of autocorrelation and enhance the quality of confidence and prediction interval estimates and significance tests for the regression model. Transformations performed on the dependent or the independent variables can also be useful in this regard.

Computer Solutions 18.7 shows the procedures and results when we use Excel and Minitab in obtaining the Durbin-Watson statistic. In each case, we have performed the regression analysis and generated the residuals. With Excel, we must then specify the cells in which the residuals are located. With Minitab, we need only select the Durbin-Watson option when we are carrying out the regression analysis procedure. Although Excel and Minitab can provide us with the Durbin-Watson statistic, we must still refer to the printed tables in reaching a conclusion.

The Autoregressive Forecasting Model

When a time series exhibits autocorrelation, not only are the residuals correlated with each other, but the time series values themselves are correlated with each

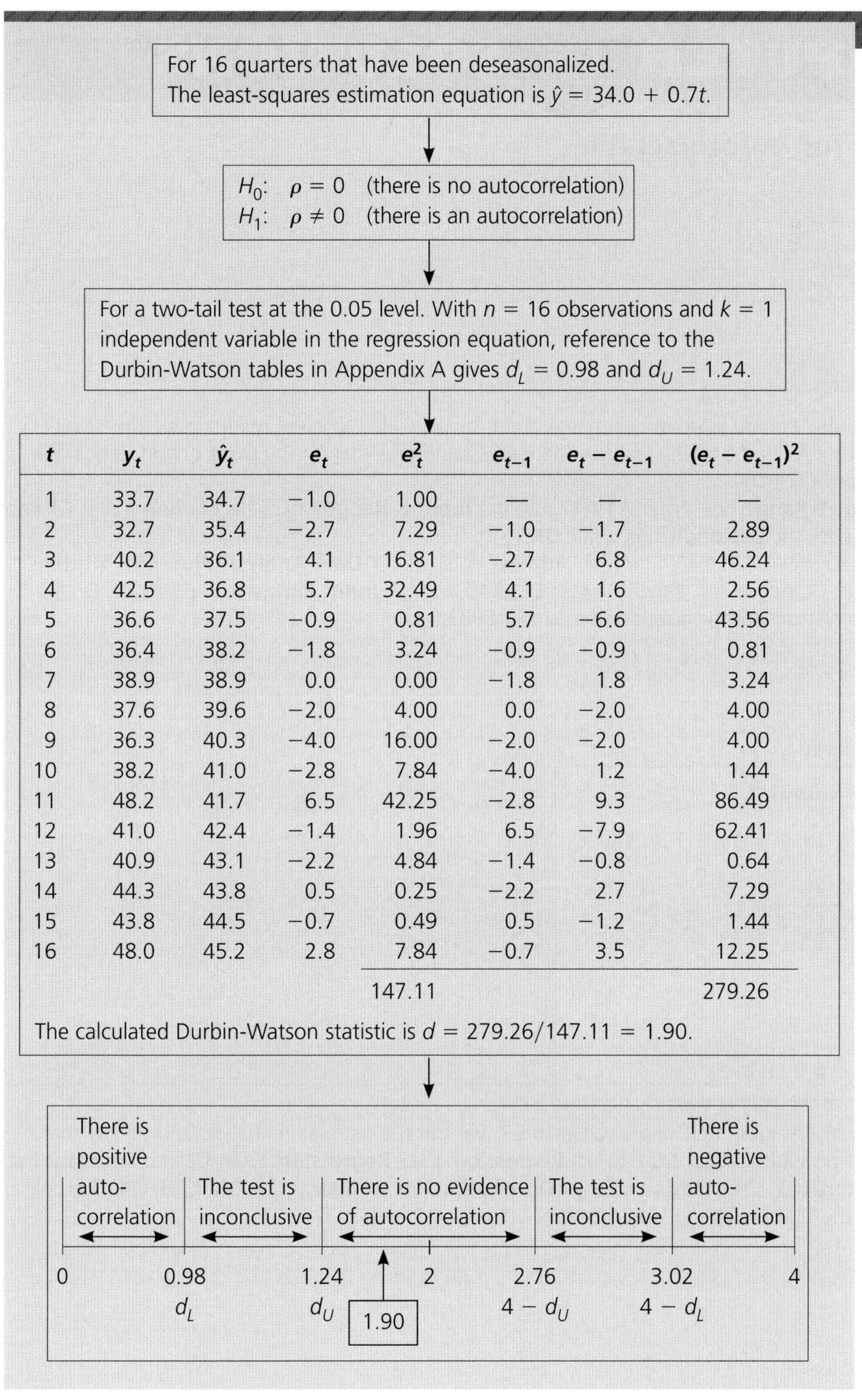

For 16 quarters that have been deseasonalized.
The least-squares estimation equation is $\hat{y} = 34.0 + 0.7t$.

H_0: $\rho = 0$ (there is no autocorrelation)
H_1: $\rho \neq 0$ (there is an autocorrelation)

For a two-tail test at the 0.05 level. With $n = 16$ observations and $k = 1$ independent variable in the regression equation, reference to the Durbin-Watson tables in Appendix A gives $d_L = 0.98$ and $d_U = 1.24$.

t	y_t	$\hat{y}_t$	e_t	e_t^2	e_{t-1}	$e_t - e_{t-1}$	$(e_t - e_{t-1})^2$
1	33.7	34.7	−1.0	1.00	—	—	—
2	32.7	35.4	−2.7	7.29	−1.0	−1.7	2.89
3	40.2	36.1	4.1	16.81	−2.7	6.8	46.24
4	42.5	36.8	5.7	32.49	4.1	1.6	2.56
5	36.6	37.5	−0.9	0.81	5.7	−6.6	43.56
6	36.4	38.2	−1.8	3.24	−0.9	−0.9	0.81
7	38.9	38.9	0.0	0.00	−1.8	1.8	3.24
8	37.6	39.6	−2.0	4.00	0.0	−2.0	4.00
9	36.3	40.3	−4.0	16.00	−2.0	−2.0	4.00
10	38.2	41.0	−2.8	7.84	−4.0	1.2	1.44
11	48.2	41.7	6.5	42.25	−2.8	9.3	86.49
12	41.0	42.4	−1.4	1.96	6.5	−7.9	62.41
13	40.9	43.1	−2.2	4.84	−1.4	−0.8	0.64
14	44.3	43.8	0.5	0.25	−2.2	2.7	7.29
15	43.8	44.5	−0.7	0.49	0.5	−1.2	1.44
16	48.0	45.2	2.8	7.84	−0.7	3.5	12.25
				147.11			279.26

The calculated Durbin-Watson statistic is $d = 279.26/147.11 = 1.90$.

TABLE 18.6

In this application of the Durbin-Watson test the calculated d (1.90) falls into the "there is no evidence of autocorrelation" decision zone of the two-tail test. The calculated d is available from most computer statistical packages, and Computer Solutions 18.7 shows how we can obtain it with Excel and Minitab.

other. Although this can undermine the independent-residuals assumption of the conventional regression approach to time series, it provides an excellent opportunity for what is known as the **autoregressive forecasting model.**

In autoregressive forecasting, the dependent variable (y) is treated the same way it is in other regression models within this chapter. However, each independent

COMPUTER 18.7 SOLUTIONS

Durbin-Watson Test for Autocorrelation

EXCEL

	A	B	C
1	Durbin-Watson Statistic		
2			
3	d = 1.8983		

Excel Durbin-Watson Statistic

1. The first two columns of Table 18.6 are listed in columns A and B of Excel file **CX18SHIP**, with t values in A2:A17, y values in B2:B17, and labels in A1 and B1. First, we run the regression analysis: From the **Data** ribbon, click **Data Analysis.** Select **Regression.** Click **OK.**
2. Enter **B1:B17** into the **Input Y Range** box. Enter **A1:A17** into the **Input X Range** box. Select **Labels.** Enter **C1** into the **Output Range** box. Check the **Residuals** box. Click **OK.**
3. The output lists the residuals, in cells E25:E40. From the **Add-Ins** ribbon, click **Data Analysis Plus.** Click **Durbin-Watson Statistic.** Click **OK.** Do not select "Labels." Enter **E25:E40** into the **Input Range** box. Click **OK.** On a separate sheet, the Durbin-Watson statistic is reported as d = 1.8983.

MINITAB

Minitab Durbin-Watson Statistic

```
Regression Analysis: y versus t

The regression equation is y = 34.0 + 0.700 t

Predictor     Coef  SE Coef      T      P
Constant    34.008    1.700  20.01  0.000
t           0.6999   0.1758   3.98  0.001

...
...
...

Durbin-Watson statistic = 1.89832
```

Using Minitab file **CX18SHIP**, with t values in C1 and y values in C2, we run the regression analysis and specify that the output include the Durbin-Watson statistic: Click **Stat.** Select **Regression.** Click **Regression.** Enter **C2** into the **Response** box. Enter **C1** into the **Predictors** box. Click **Options.** Select **Durbin-Watson statistic.** Click **OK.** Click **OK.** The partial printout shown here includes the Durbin-Watson statistic, reported as d = 1.89832.

variable represents *previous values of the dependent variable.* We will examine two levels of the autoregressive forecasting model:

1. **First-order autoregressive model** The forecast for a given time period is based on the value observed in the preceding time period:

$$\hat{y}_t = b_0 + b_1 y_{t-1} \qquad \text{with } y_{t-1} = \text{the value of } y \text{ in period } t - 1$$

2. **Second-order autoregressive model** The forecast for a given time period is based on the values observed in the two previous time periods:

$$\hat{y}_t = b_0 + b_1 y_{t-1} + b_2 y_{t-2} \qquad \text{with } y_{t-1} = \text{the value of } y \text{ in period } t-1$$
$$y_{t-2} = \text{the value of } y \text{ in period } t-2$$

Table 18.7 shows annual billings (millions of dollars) for new U.S. manufactured general aviation aircraft shipped from 1986 (coded as $t = 1$) through 2005

TABLE 18.7

Annual billings (millions of dollars) for new U.S. manufactured general aviation aircraft shipped from 1986 through 2005. In these autoregressive forecasting models, the estimated value for a given year is based either on the value observed in the preceding year (first-order model) or on the values in the two preceding years (second-order model). The underlying data are in file **CX18AIR**.

Year	Year Code (t)	Billings (y_t)	y_{t-1}	y_{t-2}
1986	1	\$1262		
1987	2	1364	\$1262	
1988	3	1918	1364	\$1262
1989	4	1804	1918	1364
1990	5	2008	1804	1918
1991	6	1968	2008	1804
1992	7	1840	1968	2008
1993	8	2144	1840	1968
1994	9	2357	2144	1840
1995	10	2842	2357	2144
1996	11	3048	2842	2357
1997	12	4580	3048	2842
1998	13	5761	4580	3048
1999	14	7843	5761	4580
2000	15	8558	7843	5761
2001	16	8641	8558	7843
2002	17	7719	8641	8558
2003	18	6434	7719	8641
2004	19	6816	6434	7719
2005	20	8667	6816	6434

First-order autoregressive model:
$\hat{y}_t = 443.9 + 0.987y_{t-1}$

Second-order autoregressive model:
$\hat{y}_t = 414.7 + 1.50y_{t-1} - 0.54y_{t-2}$

Data source: 2005 *General Aviation Statistical Databook & Industry Outlook*, p. 13.

(coded as $t = 20$). If we were to fit linear and quadratic trend equations to the time series, the results would be as follows:

Linear: The equation is $\hat{y}_t = -178.7 + 434.04t$ and the Durbin-Watson statistic = 0.47 (strong positive autocorrelation).

Quadratic: The equation is $\hat{y}_t = 557.3 + 233.3t + 9.558t^2$ and the Durbin-Watson statistic = 0.50 (strong positive autocorrelation).

In applying either of these estimation equations, we find very significant positive autocorrelation between consecutive values of the aircraft shipment billing data. A year in which billings exceed the model's estimate tends to be followed by *another* year in which billings are higher than estimated. Similarly, if the billings for a year are lower than the model has estimated, the billings for the next year also tend to be lower than estimated.

For each of the years in the time series, Table 18.7 shows the year code (t), the billing amount (y_t) for the year, the billing amount for the preceding year (y_{t-1}), and the billing amount for the period 2 years earlier (y_{t-2}). Notice how the billings in the fourth column are shifted downward one year from those in the third column, and how those in the fifth column are shifted downward two years from those in the third column. We simply apply either simple or multiple regression in obtaining the first-order and second-order autoregressive equations in the lower portion of the table.

The two equations at the bottom of Table 18.7 are the results when we apply the first-order and second-order autoregressive forecasting models to the aircraft billing data in the table. Using the mean absolute deviation criterion presented in Section 18.6, we find that MAD = \$603.1 million for the first-order model and \$575.0 million for the second-order model, indicating that the second-order model is a slightly better fit to the series. Given the available data and the autoregressive forecasting equations, the billings forecast for 2006 could be made as shown below:

Forecast for 2006, using the first-order autoregressive model:

$$\hat{y}_t = 443.9 + 0.987y_{t-1}$$

and

$$\hat{y}_{2006} = 443.9 + 0.987y_{2005} = 443.9 + 0.987(8667) = \$8998.2 \text{ million}$$

Forecast for 2006, using the second-order autoregressive model:

$$\hat{y}_t = 414.7 + 1.50y_{t-1} - 0.54y_{t-2}$$

and

$$\hat{y}_{2006} = 414.7 + 1.50y_{2005} - 0.54y_{2004} = 414.7 + 1.50(8667) - 0.54(6816)$$
$$= \$9734.6 \text{ million}$$

Because its *MAD* value is the lesser of the two, the forecast with the second-order autoregressive model would seem to be the preferred one. However, this does not mean we should automatically use a second-order autoregressive model instead of a first-order model. In fitting the second-order model, we must discard the initial *two* years in the series instead of just one. Referring to Table 18.7, note that years 1986 and 1987 would be considered as "missing data" in this model because no billing values were provided for years prior to 1986. Discarding the first two periods in a time series might not be acceptable if the series already happens to contain only a small number of periods.

Computer Solutions 18.8 describes the Excel and Minitab procedures for obtaining the first-order and second-order autoregressive equations shown at the bottom of Table 18.7. The Minitab procedure involves a special "Lag" module that makes it convenient to shift the data and create the y_{t-1} and y_{t-2} series that are lagged by one and two years, respectively.

COMPUTER 18.8 SOLUTIONS

Autoregressive Forecasting

EXCEL

Excel Autoregressive Forecasting [equations at the bottom of Table 18.7]

1. In Excel file **CX18AIR**, the billings time series in Table 18.7 is in C2:C21, with the label in C1. Select cells C2:C20 and copy them to cells D3:D21—this will be the series of (t – 1) values. Select cells C2:C19 and copy them to E4:E21—this will be the series of (t – 2) values.
2. The first-order autoregressive equation is obtained by performing a simple regression in which the Y input range is **C3:C21** and the X input range is **D3:D21.** When performing the regression, do not select "Labels." To obtain the residuals and predicted values, select **Residuals.**
3. The second-order autoregressive equation is obtained by performing a regression analysis in which the Y input range is **C4:C21** and the X input range is **D4:E21.** Do not select "Labels." To get the residuals and predicted values, select **Residuals.**

MINITAB

Minitab Autoregressive Forecasting [equations at the bottom of Table 18.7].

1. Using Minitab file **CX18AIR**, with y values in C3: Click **Stat.** Select **Time Series.** Click **Lag.** Enter **C3** into the **Series** box. Enter **C4** into the **Store lags in** box. Enter **1** into the **Lag** box. Click **OK.** The (t – 1) series will be stored in C4.
2. Repeat step 1, but enter **C5** into the **Store lags in** box and enter **2** into the **Lag** box. The (t – 2) series will be stored in C5.
3. To obtain the first-order autoregressive equation: Click **Stat.** Select **Regression.** Click **Regression.** Enter **C3** into the **Response** box. Enter **C4** into the **Predictors** box. Click **OK.**
4. To obtain the second-order autoregressive equation, carry out step 3, but enter **C4 C5** into the **Predictors** box.

EXERCISES

18.42 When autocorrelation of the residuals is present, what effect can this have on interval estimation and significance tests regarding the regression model involved?

18.43 What is the Durbin-Watson test for autocorrelation, and how can it be useful in evaluating the relevance of a given regression model that has been fitted to a set of time series data?

18.44 The least-squares regression equation $\hat{y} = 98.1 + 5.0t$ has been fitted to the shipment data given here. Calculate the Durbin-Watson d statistic and test for positive autocorrelation of the residuals at the 0.05 level of significance. Comment on the appropriateness of using this model for this set of data.

t	y_t	t	y_t	t	y_t
1	97.0	6	125.5	11	150.6
2	109.7	7	132.9	12	157.8
3	110.3	8	139.7	13	158.3
4	121.5	9	145.6	14	166.9
5	130.4	10	149.4	15	175.8

18.45 The least-squares regression equation $\hat{y} = 15.2 + 10.7t$ has been fitted to the shipment data in the table. Calculate the Durbin-Watson d statistic and test for positive autocorrelation of the residuals at the 0.05 level of significance. Based on your conclusion, comment on the appropriateness of using this model for this set of data.

t	y_t	t	y_t	t	y_t
1	65.6	7	68.3	13	141.3
2	56.7	8	79.1	14	149.6
3	79.5	9	104.2	15	177.1
4	53.4	10	104.2	16	193.6
5	59.7	11	115.4	17	238.9
6	71.1	12	86.6	18	259.2

18.46 A regression model has 3 independent variables and 20 observations, and the calculated Durbin-Watson d statistic is 0.91. What, if any, conclusion will be reached in testing for autocorrelation of the residuals with a:

a. two-tail test at the 0.02 level of significance?
b. one-tail test for positive autocorrelation at the 0.05 level?

18.47 A regression model has 2 independent variables and 16 observations, and the calculated Durbin-Watson d statistic is 2.78. What, if any, conclusion will be reached in testing for autocorrelation of the residuals with a:

a. two-tail test at the 0.05 level of significance?
b. test for negative autocorrelation at the 0.01 level?

18.48 Analysis of a time series consisting of weekly visits (in thousands) to a corporate website has led to the following autoregressive forecasting equation: $\hat{y}_t = 34.50 + 0.58y_{t-1} - 0.72y_{t-2}$. If there were 260 thousand visitors this week and 245 thousand last week, how many visitors would the forecast predict for next week?

18.49 For the shipment data listed in Exercise 18.45, construct a first-order autoregressive forecasting equation and make a forecast for period 19.

18.50 According to the Investment Company Institute, the number of U.S. mutual shareholder accounts (millions) were as shown in the table for years 1990 (coded as $t = 1$) through 2007. Construct a second-order autoregressive equation and determine what the forecast would have been for 2008. Source: Investment Company Institute, *2008 Investment Company Fact Book*, p. 110.

t	y_t	t	y_t	t	y_t
1	61.9	7	149.9	13	251.1
2	68.3	8	170.3	14	260.7
3	79.9	9	194.0	15	269.5
4	94.0	10	226.2	16	275.5
5	114.4	11	244.7	17	288.6
6	131.2	12	248.7	18	299.0

18.51 Over the past 20 years, inventory carrying costs for a large tire manufacturing facility have been as shown. Data are in thousands of dollars. Construct a first-order autoregressive forecasting equation for these data, calculate the mean absolute deviation (MAD) for the fit of the forecast values to the actual values, then use the equation to forecast inventory carrying costs for time period 21.

t	y_t	t	y_t
1	4710.0	11	2828.5
2	4187.7	12	2734.8
3	4275.6	13	2743.9
4	4076.9	14	2789.6
5	3731.4	15	2939.4
6	3484.3	16	2770.6
7	3437.2	17	2815.0
8	3203.2	18	2808.4
9	3206.2	19	2938.1
10	2963.1	20	2918.6

18.52 Repeat Exercise 18.51, but for a second-order autoregressive forecasting equation. Compare the MAD values for forecasts generated by the two equations and indicate which one is the better fit to these data.

18.8 INDEX NUMBERS

In Section 18.4, we discussed seasonal indexes, measures intended to reflect the extent to which various parts of the year experience higher or lower levels of production, demand, or other kinds of economic activity. This section introduces another concept that includes the word "index," but this is where the similarity ends. Before introducing the notion of index numbers, let's consider a variety of settings in which they could be useful:

- A family of four has a household income of $80,000, up 15% from three years ago, but they still have a lot of trouble making ends meet. Is the family's purchasing power stronger or weaker than it was three years ago?

- Wishing to compare the "livability" of various metropolitan areas, a researcher collects data on crime, suicide, alcoholism, and unemployment rates in each.
- Critics point out that the $8000 per semester tuition at a state university has escalated the cost of a college education beyond the reach of the typical family. University administrators, pointing to inflation, respond that a college education is more accessible than ever before.

For applications like these, it is useful to convert information to the form of an **index number.** An index number is a percentage that expresses a measurement in a given period in terms of the corresponding measurement for a selected base period. The measurement in the base period is defined as 100, and the measurement in a given period will be greater than or less than 100, depending on whether the measurement was greater than or less than that of the base period. For example:

In 2000 a company's inventory
costs were $3.6 million. $\rightarrow I_{2000} = 100$ for the 2000 base period

In 2009 the company's inventory
costs were $5.9 million. $\rightarrow I_{2009} = 163.9 \quad \left(\text{i.e., } \frac{5.9}{3.6} \cdot 100\right)$

In other words, if inventory costs in 2000 are defined as 100, 2009 costs can be expressed as 163.9, or 63.9% higher. Like other index numbers, this one is unitless. Note that both the numerator and the denominator are described in millions of dollars, and these cancel out during the division.

The base period for an index number should be a relatively *typical* period during which extremely unusual phenomena have not occurred. For example, it wouldn't be desirable to construct an index of oil prices using a base year during which there was either a severe oil glut or a shortage.

Index numbers are used in measuring changes over time for such diverse entities as stock values, prices, wages, production, and output. Some of them even combine many different measurements into a single value. If you happen to live in one of the colder regions of the nation, you're familiar with "wind chill," the translation of outside temperature and wind velocity into a single number representing what would be an equally uncomfortable temperature if the wind were calm. This particular indicator, originally based on Antarctic experiments of the 1930s, is not derived by the methods through which index numbers are usually determined, but it's yet another example of providing information by condensing two or more numbers into just one.

Many popular statistical indexes are in use, and you may already be familiar with such indexes as the Dow Jones Industrial Average, the Standard & Poor's Index of 500 Stocks, and the Consumer Price Index. The latter can be used in determining such things as whether the household earning $80,000 has more purchasing power than three years ago.

The Consumer Price Index

The **Consumer Price Index (CPI)** describes the change in prices from one time period to another for a fixed "market basket" of goods and services. It is generated monthly by the U.S. Bureau of Labor Statistics and is probably the index with which both businesspersons and the general public are most familiar.

Derivation of the CPI

The CPI is based on periodic consumer expenditure surveys. At this writing, the base prices are for the 1982–1984 period. The change to the 1982–1984 base period was made on the basis of data from the 1982–1984 Consumer Expenditure Survey. From 1971 until 1987, the base period for the index had been 1967.

The "market basket" for the CPI includes items that reflect the price of bread, physicians' and dentists' fees, alcoholic and nonalcoholic beverages, clothing, transportation, shelter, and other goods and services purchased for day-to-day living. Every month, prices are obtained from 57,000 housing units and 19,000 establishments representing 85 areas across the United States.[1] The resulting data are used in generating price indexes for specific regions and cities as well as for various categories of products and services.

Applications

Economic Indicator. By reflecting changes in prices, the Consumer Price Index is the primary measurement of inflation in the United States and of the perceived success or failure of governmental policies to control inflation. Political candidates are eager to point out increases in the CPI when they wish to gain office and equally quick to call attention to any slowing in the growth of the CPI whenever they have been in office and are defending past policies.

Escalator. A common application of the Consumer Price Index is in the automatic adjustment of contracted wages or pension benefits through what are known as *escalator clauses* or *cost-of-living adjustments (COLAs)*. As prices go up, the wages or benefits are increased, either (1) in proportion to the change in the CPI, or (2) according to a formula that has been included in the collective bargaining agreement or set forth in the relevant legislation.

Following a 1987 work stoppage, 4600 workers at Rohr Industries' aircraft parts plants in California got a new three-year contract negotiated by the Machinists Union. Among other items, the contract called for hourly wage increases based on the Consumer Price Index. The U.S. Department of Labor summed up the contract details:

Three-year agreement negotiated 2/87 also provided: lump sum payments equal to 10 percent of pay 1st year and 6 percent in the 2nd and 3rd years; current COLA rolled into base rates and cost-of-living adjustment formula continued to provide 1 cent for each 0.3–point movement in the BLS-CPI; $24 (was $20) monthly pension benefit rate; 100 percent (was 95) employer payment for preferred provider option-health care; $1 million (was $250,000) lifetime major medical maximum; $15,000 (was $10,000) life insurance. . . .[2]

Deflator. Using the Consumer Price Index, current prices and wages can be deflated to *real prices* or *real wages*. Compared to their "current" counterparts, these let us more readily determine whether we have more earning power than in some previous period. If a person's real income has not kept pace with real prices,

[1] *The New York Times Almanac 2009*, p. 338.
[2] Bureau of Labor Statistics, U.S. Department of Labor, *Current Wage Developments*, April 1987, p. 10.

his or her economic condition and standard of living will have decreased. Current (nominal) income can be converted to real (constant) income as follows:

Converting current income to real income:

$$\text{Real income} = \frac{\text{Current income}}{\text{Consumer Price Index}} \cdot 100$$

In 2007 the median U.S. family income was \$61,355, compared to the 1970 figure of \$9867. With 1982–1984 = 100 as the base, the 2007 Consumer Price Index was 207.3, compared to 38.8 in 1970.[3] Had real income increased or decreased from 1970 to 2007? To get the answer, we need only compare the real income levels for the two years:

- 2007

  Median family income = \$61,355 and CPI (1982–1984 base) = 207.3

  $$\text{Real income (in 1982–1984 dollars)} = \frac{\$61{,}355}{207.3} \cdot 100 = \$29{,}597$$

- 1970

  Median family income = \$9867 and CPI (1982–1984 base) = 38.8

  $$\text{Real income (in 1982–1984 dollars)} = \frac{\$9867}{38.8} \cdot 100 = \$25{,}430$$

In terms of current dollars, the 2007 median family income was 6.22 times as great as in 1970. The 2007 CPI, however, was just 5.34 times as large. Because the current income growth exceeded the CPI growth, the median family had a higher real income in 2007 than in 1970. Comparing real income for the two years, we find the \$29,597 median family real income of 2007 was 16.4% greater than the \$25,430 of 1970. Statistics In Action 18.1 provides a glimpse of some prices from years ago and takes a "time-traveler" approach to the CPI.

Other Indexes

The federal government and other institutions employ a great many special-purpose indexes for specific applications. The following are typical examples:

- **The Producer Price Index (PPI)** Dating from 1980, this is the oldest continuous statistical series published by the U.S. Bureau of Labor Statistics. It reflects the prices of commodities produced or processed in the United States.
- **Index of Industrial Production** Provided by the Board of Governors of the Federal Reserve System since the 1920s, this index measures changes in the physical volume or quantity of output of manufacturing, mining, and electric and gas utilities.
- **Employment Cost Index (ECI)** Compiled by the U.S. Bureau of Labor Statistics, the ECI is a quarterly measure of the total cost of employing labor. It includes wages and salaries, plus employers' costs for employee benefits such as paid leave, severance pay, health and life insurance, pension and saving plans, Social Security, and unemployment insurance.
- **Standard & Poor's Composite Index of Stock Prices** From Standard & Poor's Corporation, the index reflects the composite price of 500 common stocks.
- **New York Stock Exchange (NYSE) Composite Index** Compiled by the New York Stock Exchange, this index measures changes in the aggregate value of all stocks listed on the NYSE.

[3] *The New York Times Almanac 2009*, pp. 336–337.

STATISTICS IN ACTION 18.1

The CPI Time Machine

How would it feel to be able to buy first-class postage for 3 cents, a quart of milk for 15 cents, a new car for $1649, and a new house for $12,638? Seems like a pretty good deal, at least until you consider that your salary is a mere $2359—per year.

It was 1946. Timex watches and Tide detergent were brand new, less than 1% of households had a TV, and the hit song of the year was "Zip-A-Dee-Doo-Dah." The milk, car, house, and salary figures above were the averages for that year. The minimum wage was 40 cents an hour, but would go up to 65 cents later in the year.

Technology hasn't yet made it possible for us to time-travel, but—using the Consumer Price Index (CPI) discussed in this section—we can translate those long-ago prices and salaries into their modern-day equivalents. For example, that $1649 car would be about $17,530 in terms of 2007 dollars, and the 15-cent quart of milk would have been the equivalent of $1.59 in 2007. The multiplication factor we're using is the ratio of the Consumer Price Indexes for 2007 and 1946: 207.3 and 19.5, respectively.

Source: "A Snapshot of What Life Was Like in '46," *USA Today*, February 12, 1996, p. 6D; *New York Times Almanac 2009*, p. 338.

Shifting the Base of an Index

Changing the base of an index can be desirable for two purposes:

1. **To better reflect the mix of public tastes or economic activities from which the index is compiled** In this application, the change to a more recent base year involves a reweighting of the components comprising the index.
2. **To convert the index to the same base period as another series so that the two may be more easily compared** For example, Table 18.8 lists 1990 through 2007 values for two indexes: (a) the Producer Price Index for commercial natural gas and (b) the Producer Price Index for heating equipment. The former has 1990 as its base; the latter, 1982.

The two indexes will be more comparable if they have a common base year, and the 1990 base year of the commercial natural gas price index has been arbitrarily selected for this purpose. As shown in column (3) of Table 18.8, the shift is made by simply dividing each index number by the index for the new base year, then multiplying the quotient by 100. Following the conversion, it is apparent that the natural gas prices increased 135.4% from 1990 to 2007, while the price of heating equipment went up by 48.4% during the same period.

TABLE 18.8

As reported, these indexes have base years of 1990 and 1982, respectively. By shifting the base year for heating equipment as shown here, the two series will be more comparable, with each having 1990 as its base year.

	(1) Producer Price Index, Commercial Natural Gas	(2) Producer Price Index, Heating Equipment	(3)
Year	(1990 = 100)	(1982 = 100)	(1990 = 100)
1990	100.0	131.6	100.0
2000	134.7	155.6	118.2
2005	232.5	179.9	136.7
2007	235.4	195.3	148.4

$$148.4 = \frac{195.3}{131.6} \cdot 100$$

Data source: Bureau of the Census, U.S. Department of Commerce, *Statistical Abstract of the United States 2009*, pp. 475, 588.

EXERCISES

18.53 What is an index number, and what is its value for the base period?

18.54 What is the Consumer Price Index and what does it reflect?

18.55 The Consumer Price Index and average annual wages in selected industries were as shown in the table. Source: Bureau of the Census, *Statistical Abstract of the United States 2009*, pp. 409, 465.

	Average Wages ($/Hour)			
	Manufacturing	Construction	Leisure and Hospitality	CPI (1982–1984 = 100)
1990	$10.78	$13.42	$6.02	130.7
2000	14.32	17.48	8.32	172.2
2007	17.26	20.95	10.41	207.3

From 1990 to 2007, the average wages went up for workers in all three industries. However, which wage-earning groups fared best and worst in terms of their percentage change in spending power during this period?

18.56 An executive's salary increased from $95,000 to $130,000 between 2000 and 2007, but the Consumer Price Index went from 172.2 to 207.3 during this time span. In terms of real income, what was the percentage increase or decrease in the executive's real earnings?

18.57 Use the Consumer Price Index to convert the 2000 salary of the executive in Exercise 18.56 to its 2007 equivalent.

18.58 The Producer Price Index values for plumbing fixtures and brass fittings from 2002 through 2007 are shown here. Convert the index numbers so that the base period will be 2002 instead of 1982. Source: Bureau of the Census, *Statistical Abstract of the United States 2009*, p. 588.

2002	181.9	2005	197.6
2003	183.4	2006	207.2
2004	188.3	2007	220.8

SUMMARY 18.9

● Time series model and its components

The analysis of time series data is useful in identifying past patterns and tendencies as well as helping to forecast future values. In the classical time series model, the time series is made up of trend (T), cyclical (C), seasonal (S), and irregular (I) components. Although other possibilities exist, a popular approach to combining these components is the multiplicative approach, or $y = T \cdot C \cdot S \cdot I$. The model assumes that any observed value of y is the result of influences exerted by the four components.

● Trend equations and smoothing techniques

Trend equations can be fitted to a time series, and two approaches are examined: (1) the linear trend equation, $\hat{y} = b_0 + b_1x$; and (2) the quadratic trend equation, $\hat{y} = b_0 + b_1x + b_2x^2$. The time series can also be smoothed by using a centered moving average or by exponential smoothing. When moving averages are used, the procedure will differ slightly, depending on whether the number of periods in the basis of the average is odd or even.

● Seasonal indexes

Strong seasonal patterns may be present for certain months or quarters, and seasonal indexes can be calculated in order to inflate or deflate a trend value to compensate for seasonal effects. Seasonal indexes can be determined through the ratio to moving average method, and a time series can be deseasonalized by using the seasonal indexes for the respective months or quarters of the year.

● Seasonal indexes and forecasting

In forecasting based on time series information, data from the past are used in predicting the future value of the variable of interest. Two approaches to time series forecasting are the extrapolation of a trend equation fitted to the series and the use of exponential smoothing. Once a forecast has been made for a given year, seasonal indexes can be used in determining the forecast for each month or quarter of the year.

- **Testing for fit and autocorrelation**

When two or more models are fitted to the same time series, it's useful to have one or more criteria on which to compare the closeness with which they fit the data. Two such criteria are the mean absolute deviation (*MAD*) and mean squared error (*MSE*). The Durbin-Watson test can be used to determine whether the residuals in a regression model are autocorrelated.

- **Autoregressive forecasting**

When values in a time series are highly correlated with each other, autoregressive forecasting models can be useful. In autoregressive forecasting, each independent variable represents the value of the dependent variable for a previous time period. First-order and second-order autoregressive forecasting models are discussed.

- **Index numbers**

An index number is a percentage that compares a measurement to its value for a selected base period. The Consumer Price Index describes the change in prices from one period to the next for a fixed "market basket" of goods and services. The CPI is used as a measurement of inflation, as a "cost of living" adjustment in labor contracts, and as a means of deflating current prices or wages to "real" prices or wages.

EQUATIONS

Classical Time Series Model

$$y = T \cdot C \cdot S \cdot I$$

where y = observed value of the time series variable
T = trend component
C = cyclical component
S = seasonal component
I = irregular component

Linear Trend Equation

$$\hat{y} = b_0 + b_1 x$$

where $\hat{y}$ = the trend line estimate of y
x = time period

Quadratic Trend Equation

$$\hat{y} = b_0 + b_1 x + b_2 x^2$$

where $\hat{y}$ = the trend line estimate of y
x = time period

(x and x^2 are treated as separate variables.)

Exponential Smoothing of a Time Series

$$E_t = \alpha \cdot y_t + (1 - \alpha) \cdot E_{t-1}$$

where E_t = the exponentially smoothed value for time period t
E_{t-1} = the exponentially smoothed value for the previous time period
y_t = the actual y value in the time series for time period t
α = the smoothing constant, and $0 \le \alpha \le 1$

Deseasonalized Time Series

For each time period, the deseasonalized value is

$$D = \frac{\text{Value in the original time series}}{(\text{Seasonal index for that period})/100}$$

Forecasting with Exponential Smoothing

$$F_{t+1} = \alpha \cdot y_t + (1 - \alpha) \cdot F_t$$

where F_{t+1} = forecast for time period $t + 1$
y_t = actual y value for time period t
F_t = the value that had been forecast for time period t
α = the smoothing constant, and $0 \le \alpha \le 1$

Dividing an Annual Forecast into Seasonal Components

$$\text{Forecast for a season} = \frac{\text{Forecast for the year}}{\text{Number of seasons}} \cdot [(\text{Seasonal index})/100]$$

Mean Absolute Deviation (*MAD*) Criterion for Evaluating Fit

$$MAD = \frac{\sum |y_t - \hat{y}_t|}{n}$$

where y_t = an observed value of y
$\hat{y}_t$ = the value of y that is predicted using the estimation equation or model
n = the number of time periods

Mean Squared Error (*MSE*) Criterion for Evaluating Fit

$$MSE = \frac{\sum (y_t - \hat{y}_t)^2}{n}$$

where y_t = an observed value of y
$\hat{y}_t$ = the value of y that is predicted using the estimation equation or model
n = the number of time periods

Durbin-Watson Test, Description of Residual (Error) for Time Period *t*

$$\varepsilon_t = \rho\varepsilon_{t-1} + z_t$$

where ε_t = residual associated with time period t
ε_{t-1} = residual associated with time period $t - 1$
ρ = the population correlation coefficient between values of ε_t and ε_{t-1}, with $-1 \le \rho \le +1$
z_t = a normally distributed random variable with $\mu = 0$

Durbin-Watson Test for Autocorrelation

- **Null and alternative hypotheses**

The test is

for positive autocorrelation	for negative autocorrelation	a two-tail test
H_0: $\rho \le 0$ H_1: $\rho > 0$	H_0: $\rho \ge 0$ H_1: $\rho < 0$	H_0: $\rho = 0$ H_1: $\rho \ne 0$

- **Test statistic**

$$d = \frac{\sum_{t=2}^{t=n} (e_t - e_{t-1})^2}{\sum_{t=1}^{t=n} e_t^2}$$

where e_t = observed residual (error) associated with time period t
e_{t-1} = observed residual (error) associated with time period $t - 1$
n = number of observations, with $n \ge 15$

- **Critical values and decision zones** Look up the appropriate d_L and d_U values in the Durbin-Watson tables in Appendix A. The actual critical values and decision zones are based on (1) the type of test (positive, negative, or two-tail), (2) α = the level of significance for the test, and (3) k = the number of independent variables in the regression model. The decision zones and appropriate conclusions are shown in Figure 18.5. Appendix A provides three tables for the Durbin-Watson test. They correspond to the 0.05, 0.025, and 0.01 levels of significance for a one-tail (positive or negative) test, and to the 0.10, 0.05, and 0.02 levels for a two-tail test.

Autoregressive Forecasting Model

- **First-order autoregressive model** The forecast for a given time period is based on the value observed in the preceding time period:

$$\hat{y}_t = b_0 + b_1 y_{t-1} \quad \text{with } y_{t-1} = \text{the value of } y \text{ in period } t-1$$

- **Second-order autoregressive model** The forecast for a given time period is based on the values observed in the two previous time periods:

$$\hat{y}_t = b_0 + b_1 y_{t-1} + b_2 y_{t-2} \quad \text{with } y_{t-1} = \text{the value of } y \text{ in period } t-1, \; y_{t-2} = \text{the value of } y \text{ in period } t-2$$

Converting Current Income to Real Income

$$\text{Real Income} = \frac{\text{Current income}}{\text{Consumer Price Index}} \cdot 100$$

CHAPTER EXERCISES

18.59 Fit a linear trend equation to the following data describing the number of active U.S. Army personnel from 2001 through 2007. What is the trend estimate for 2015? Source: Bureau of the Census, *Statistical Abstract of the United States 2009*, p. 328.

Year	Army Personnel (thousands)
2001	481
2002	487
2003	499
2004	500
2005	493
2006	505
2007	522

18.60 Given the time series in Exercise 18.59, and using exponential smoothing with $\alpha = 0.7$, what would have been the forecast for 2008?

18.61 SUBWAY has grown to include more than 31,000 restaurants worldwide since they opened their first one in 1965. The accompanying data show the number of SUBWAY restaurants worldwide from 1985 through 2002. Fit a quadratic equation to these data and estimate the number of SUBWAY restaurants worldwide in 2014. Source: SUBWAY.com, July 3, 2003.

Year	SUBWAY Restaurants	Year	SUBWAY Restaurants
1985	596	1994	9893
1986	991	1995	11,420
1987	1810	1996	12,516
1988	2888	1997	13,066
1989	4071	1998	13,600
1990	5144	1999	14,162
1991	6187	2000	14,662
1992	7327	2001	15,762
1993	8450	2002	17,482

18.62 Given the time series in Exercise 18.61, and using exponential smoothing with $\alpha = 0.8$, what would have been the forecast for 2003?

18.63 From 2001 through 2007, average monthly cable TV bills were as shown here. Source: Bureau of the Census, *Statistical Abstract of the United States 2009*, p. 703.

Year	Average Bill
2001	$32.87
2002	34.71
2003	36.59
2004	38.14
2005	39.63
2006	41.17
2007	42.72

a. Fit a linear trend equation to the time series. Using this equation, determine the trend estimate for 2015.
b. Fit a quadratic equation to the time series, then use the equation to determine the trend estimate for 2015.
c. Which of the preceding equations is the better fit to the time series if the *MAD* criterion is used? If the *MSE* criterion is used?

18.64 The annual number of international visitors to the United States from 2000 through 2007 is shown here. Source: *World Almanac and Book of Facts 2009*, p. 127.

Year	Visitors (millions)
2000	50.9
2001	44.9
2002	41.9
2003	41.2
2004	46.1
2005	49.4
2006	51.0
2007	56.0

a. Fit a linear trend equation to the time series. Using this equation, determine the trend estimate for 2015.
b. Fit a quadratic equation to the time series, then use the equation to determine the trend estimate for 2015.
c. Which of the preceding equations is the better fit to the time series if the *MAD* criterion is used? If the *MSE* criterion is used?

18.65 The following data show U.S. retail sales of canoes from 1991 through 2008, with data in thousands of boats. Source: National Marine Manufacturers Association, *2008 Recreational Boating Statistical Abstract*, p. 72.

Year	Sales (thousands)	Year	Sales (thousands)
1991	72.3	2000	111.8
1992	78.0	2001	105.8
1993	89.7	2002	100.0
1994	99.8	2003	86.7
1995	97.8	2004	93.9
1996	92.9	2005	77.2
1997	103.6	2006	99.9
1998	107.8	2007	99.6
1999	121.0	2008	73.7

a. Construct a graph of the time series. Does the overall trend appear to be upward or downward?
b. Construct a three-year centered moving average for this series.
c. Using the constant $\alpha = 0.3$, fit an exponentially smoothed curve to the original time series.
d. Repeat part (c) using $\alpha = 0.7$. How has the new value for the constant affected the smoothness of the fitted curve?

18.66 Given the time series in Exercise 18.65, and using exponential smoothing with $\alpha = 0.5$, what would have been the forecast for 2009?

18.67 The following data are quarterly sales for the J. C. Penney Company for 2000 through 2002. Determine the quarterly seasonal indexes, then use these indexes to deseasonalize the original data. Source: J.C. Penney, Inc., *2001 Annual Report*, p. 31, *2002 Annual Report*, p. 39.

	Gross Sales (Billions)			
	I	II	III	IV
2000	\$7.53	7.21	7.54	9.57
2001	7.52	7.21	7.73	9.54
2002	7.73	7.20	7.87	9.55

18.68 Construct three-period and four-period centered moving averages for the quarterly J. C. Penney sales in Exercise 18.67. Are there any other retailing companies for whom a four-period moving average might have such a great effect on smoothing quarterly sales data? If so, give a real or hypothetical example of such a company.

18.69 The following data describe quarterly residential natural gas sales from 2005 through 2008. Determine the quarterly seasonal indexes, then use these indexes to deseasonalize the original data. Source: U.S. Energy Information Administration, eia.doe.gov, August 22, 2009.

	Residential Gas Sales (Trillions of Cubic Feet)			
	I	II	III	IV
2005	2329	784	353	1360
2006	2042	700	349	1277
2007	2317	761	346	1294
2008	2351	762	345	1408

18.70 Construct four-period and five-period centered moving averages for the quarterly residential gas sales in Exercise 18.69. Are there any other energy industries for whom a four-period moving average might have such a great effect on smoothing quarterly sales data? If so, give a real or hypothetical example of such an industry.

18.71 The Walt Disney Company has reported the following quarterly revenues (in billions of dollars) for its Parks & Resorts operating segment for fiscal years 1998 through 2002. Determine the quarterly seasonal indexes, then use these indexes to deseasonalize the original data. Source: The Walt Disney Company, *Fact Book*, editions 1998–2002.

	Quarter			
	I	II	III	IV
1998	\$1.260	1.244	1.504	1.524
1999	1.434	1.408	1.711	1.553
2000	1.577	1.571	1.940	1.715
2001	1.729	1.650	1.946	1.684
2002	1.433	1.525	1.847	1.660

18.72 Construct four-period and five-period centered moving averages for the quarterly revenues in Exercise 18.71.

Are there any other entertainment or recreation companies for whom a four-period moving average might have such a great effect on smoothing quarterly revenue data? If so, give a real or hypothetical example of such a company.

18.73 The actual number of help-wanted ads in a local newspaper was 1682, 1963, 2451, and 3205 for quarters I through IV, respectively, of the preceding year. The corresponding deseasonalized values are 2320, 2362, 2205, and 2414, respectively. Based on this information, what seasonal index is associated with each quarter?

18.74 The trend equation $\hat{y} = 2050 + 12x$ has been developed for the number of vehicles using the ferry service at a national seashore each year. (The years have been coded so that 1997 corresponds to $x = 1$.) If the seasonal indexes for quarters I through IV are 55.0, 130.0, 145.0, and 70.0, respectively, estimate the usage volume for each quarter of 2014.

18.75 The trend equation $\hat{y} = 150 + 12x + 0.01x^2$ has been developed for annual unit sales of a filtering system component used by municipal water authorities. The years have been coded so that 2001 corresponds to $x = 1$. If the seasonal indexes for quarters I through IV are 85.0, 110.0, 125.0, and 80.0, respectively, estimate the unit sales for each quarter of 2015.

18.76 The U.S. Energy Information Administration reports the following prices for regular unleaded gasoline in Germany from 2003 through 2008. Source: U.S. Energy Information Administration, *Annual Energy Review 2008*, p. 321.

	2003	2004	2005	2006	2007	2008
Price per Gallon	$4.59	$5.24	$5.66	$6.03	$6.88	$7.75

For the price data given here:

a. Fit a linear trend equation to the time series. Using this equation, determine the trend estimate for 2014.

b. Fit a quadratic equation to the time series, then use the equation to determine the trend estimate for 2014.

c. Which of the preceding equations is the better fit to the time series if the *MAD* criterion is used? If the *MSE* criterion is used?

18.77 Ford Motor Company reports the average labor cost per hour (earnings plus benefits) for 1993–2002. Fit a quadratic trend equation to the data, fit a linear trend equation to the data, and then use the *MSE* criterion to determine which equation is the better fit. Source: Ford Motor Company, *2002 Annual Report*, p. 91.

Year	Labor Cost	Year	Labor Cost
1993	$39.06/hour	1998	45.72
1994	40.94	1999	47.37
1995	40.45	2000	48.44
1996	41.77	2001	47.73
1997	43.55	2002	52.65

18.78 A Caribbean cruise line booked the number of passengers (in thousands) described below for each quarter from 2005 through 2009.

	I	II	III	IV
2005	27	23	21	24
2006	32	26	18	29
2007	28	25	16	30
2008	37	31	21	34
2009	38	33	23	39

a. Combining the quarterly bookings for each year, fit a linear trend equation and forecast the number of bookings for 2013.

b. Determine the seasonal indexes and break the 2013 forecast in part (a) into its quarterly components.

c. Combining the quarterly bookings for each year, use exponential smoothing, with $\alpha = 0.3$, to forecast the number of bookings for 2010.

d. Use the seasonal indexes to break the 2010 forecast in part (c) into its quarterly components.

18.79 Use the seasonal indexes obtained in Exercise 18.78 to generate a deseasonalized series of quarterly bookings for those data.

18.80 Monthly loan applications for a local bank from 2006 through 2009 were as shown in the table.

	J	F	M	A	M	J
2006	30	28	32	33	32	32
2007	22	26	30	36	35	40
2008	19	22	30	31	31	36
2009	29	28	42	47	49	53

	J	A	S	O	N	D
2006	39	44	48	47	33	28
2007	41	36	32	34	23	19
2008	39	37	36	36	32	29
2009	51	55	50	46	36	33

a. Combining the monthly applications for each year, fit a linear trend equation and forecast the number of applications for 2012.

b. Determine the seasonal indexes and break the 2012 forecast in part (a) into its monthly components.

c. Combining monthly applications for each year, use exponential smoothing and $\alpha = 0.4$ to obtain a forecast for 2010.

d. Use the seasonal indexes to break the 2010 forecast in part (c) into its monthly components.

18.81 Use the seasonal indexes obtained in Exercise 18.80 to generate a deseasonalized series of monthly loan applications for the bank.

18.82 Apply the Durbin-Watson test and the 0.05 level of significance in examining the data series in Exercise 18.65 for positive autocorrelation. Based on your conclusion, comment on the appropriateness of the linear regression model for these data.

18.83 Apply the Durbin-Watson test and the 0.05 level of significance in examining the data series in Exercise 18.61 for positive autocorrelation. Based on your conclusion, comment on the appropriateness of the linear regression model for these data.

18.84 The Producer Price Index for finished consumer goods from 1980 through 2007 is in file **XR18084**, with 1982 = 100 and 1980 coded as $t = 1$. Construct a first-order autoregressive forecasting equation for these data, calculate the mean absolute deviation (MAD) for the fit of the forecast values to the actual values, then use the equation to forecast the PPI for 2008. Source: Bureau of the Census, *Statistical Abstract of the United States 2009,* p. 474.

18.85 The Consumer Price Index from 1961 through 2007 is in file **XR18085**, with 1982–1984 = 100 and 1961 coded as $t = 1$. Construct a first-order autoregressive forecasting equation for these data, calculate the mean absolute deviation (MAD) for the fit of the forecast values to the actual values, then use the equation to forecast the CPI for 2008. Source: *The New York Times Almanac 2009,* p. 338; and Bureau of the Census, *Statistical Abstract of the United States 1999,* p. 495.

18.86 Repeat Exercise 18.85, but for a second-order autoregressive forecasting equation. Compare the MAD values for forecasts generated by the two equations and indicate which one is the better fit to these data.

18.87 From 1996 to 2007, the median weekly earnings of workers represented by unions went from $610 to $857, and the Consumer Price Index (1982–1984 = 100) went from 156.9 to 207.3. What was the percentage increase or decrease in real wages for these workers? Source: *World Almanac and Book of Facts 2009,* pp. 85, 150.

18.88 Convert each of the following indexes so that the new base period will be 2004.

	A	B
2001	101.3	100.0
2002	100.0	102.9
2003	103.4	104.7
2004	105.6	103.2
2005	107.9	107.5
2006	110.3	126.4
2007	116.2	122.1
2008	125.3	129.0
	(2002 = 100)	(2001 = 100)

INTEGRATED CASE

Thorndike Sports Equipment (Video Unit Five)

Negotiations are taking place for the next labor agreement between Thorndike Sports Equipment and its manufacturing employees. The company and the union have agreed to consider linking hourly wages to the predicted Consumer Price Index (CPI) for each of the four years of the new contract. The latest proposal would specify an hourly wage that is 10% of the estimated CPI (e.g., if the estimated CPI for a given year were 230.5, the hourly wage for that contract year would be $23.05).

Table 18.9 shows the CPI for 1961 through 2007, along with the year and the year code. The data are also in file **THORN18**. The base period for the CPI figures is 1982–1984, for which the average is 100. Assuming that the new contract will be in force from 2011 through 2014, use a computer statistical package to help determine the hourly wage that would be paid in each year of the contract if the estimated CPI is an extrapolation based on the following:

1. A simple linear estimation equation ($\hat{y} = b_0 + b_1x$) fitted to the 1961–2007 CPI data.
2. A quadratic estimation equation ($\hat{y} = b_0 + b_1x + b_2x^2$) fitted to the 1961–2007 CPI data.

Several additional questions regarding the estimation equations and the contract negotiations:

3. Which equation would each negotiating party tend to prefer as the basis of the hourly wages to be paid during the contract years?
4. Using the mean absolute deviation (MAD) criterion, which equation is the better fit to the CPI time series data?
5. Assuming (a) the hourly wages are linked to the equation identified in the preceding question, (b) the estimated CPI for 2014 actually occurs, and (c) the average Thorndike worker made $12.40 per hour in 1987, will his or her real earnings increase or decrease from 1987 to 2014? Explain.

TABLE 18.9

Values of the Consumer Price Index (CPI) for 1961 through 2007. The base period is 1982–1984 = 100.

Year	Year Code	CPI	Year	Year Code	CPI
1961	1	29.9	1985	25	107.6
1962	2	30.2	1986	26	109.6
1963	3	30.6	1987	27	113.6
1964	4	31.0	1988	28	118.3
1965	5	31.5	1989	29	124.0
1966	6	32.4	1990	30	130.7
1967	7	33.4	1991	31	136.2
1968	8	34.8	1992	32	140.3
1969	9	36.7	1993	33	144.5
1970	10	38.8	1994	34	148.2
1971	11	40.5	1995	35	152.4
1972	12	41.8	1996	36	156.9
1973	13	44.4	1997	37	160.5
1974	14	49.3	1998	38	163.0
1975	15	53.8	1999	39	166.6
1976	16	56.9	2000	40	172.2
1977	17	60.6	2001	41	177.1
1978	18	65.2	2002	42	179.9
1979	19	72.6	2003	43	184.0
1980	20	82.4	2004	44	188.9
1981	21	90.9	2005	45	195.3
1982	22	96.5	2006	46	201.6
1983	23	99.6	2007	47	207.3
1984	24	103.9			

Sources: Bureau of the Census, U.S. Department of Commerce, *Statistical Abstract of the United States 1999,* p. 495; and *New York Times Almanac 2009*, p.338.

CHAPTER 19

Decision Theory

LAWSUITS—TAKE THE CASH OR TRY FOR THE CURTAIN?

After Jill became pregnant, her company decided it no longer needed her services. Upon being released, Jill accused the company of gender bias and filed a wrongful-termination suit against it. The company is now trying to settle out of court and has offered Jill $160,000.

Should Jill take the settlement offer, or might it be more attractive to go to court and take her chances with the trial? Jill's attorney believes she has a 60% chance of winning her case if it proceeds to the trial stage. He also informs her that the median jury award in employment-practices liability cases has been estimated as $250,000.

Essentially, Jill's alternatives can be viewed as (a) the sure $160,000 versus (b) the 60% chance at $250,000 (which includes a 40% chance at $0). Based on the expected payoff decision criterion discussed in this chapter, Jill should take the money and run. After reading that section, you can verify our advice to Jill for yourself. By the way, Jill should be willing to pay up to $54,000 for a perfect forecast as to what the jury would decide if she were to take the case to trial. That's known as the *expected value of perfect information,* and it's in this chapter, too.

Source: The median award value is from "Report: EPL Jury Awards Up 18%," *Insurance Journal,* insurancejournal.com, August 29, 2009.

"When you come to a fork in the road, take it."
—Yogi Berra

LEARNING OBJECTIVES

After reading this chapter, you should be able to:

- Express a decision situation in terms of decision alternatives, states of nature, and payoffs.
- Differentiate between non-Bayesian and Bayesian decision criteria.
- Determine the expected payoff for a decision alternative.
- Calculate and interpret the expected value of perfect information.
- Express and analyze the decision situation in terms of opportunity loss and expected opportunity loss.
- Apply incremental analysis to inventory-level decisions.

19.1 INTRODUCTION

While it's been said that nothing is certain except death and taxes, there is one more thing that's certain: uncertainty. A fish market owner isn't sure how many customers will decide to buy a flounder today, a pharmaceutical company isn't completely certain that a new drug will be safe for every one of 20 million users, and a computer firm doesn't have complete certainty that a new computer in the final design stages will still be competitive when it is finally manufactured. In each of these cases, a decision must be made. For the fish market owner, it's how many flounder to stock; for the pharmaceutical and computer firms, it's whether to proceed with the new product.

Both decision making and uncertainty are also discussed in other chapters. In hypothesis testing, we had to decide whether to reject the null hypothesis, and uncertainty was reflected in the Type I and Type II errors. In the next chapter, we will examine how control charts can help us decide whether a process is in control, and we will briefly discuss how acceptance sampling can assist in deciding whether an entire shipment of products or components should be accepted.

The biggest difference between these discussions and the topics of this chapter is that we will be concentrating here on decisions made *without* the benefit of knowledge gained from sampling. While the possibility of gathering further information will be considered, this will be presented in terms of, "What is the most we should consider paying for additional information to reduce our uncertainty in making the decision?"

We will begin by structuring the decision situation and expressing it as either a payoff table or a decision tree diagram. The discussion then turns to two general approaches by which the actual decision might be reached. The chapter concludes with a discussion of incremental analysis, a technique especially useful for inventory decisions.

19.2 STRUCTURING THE DECISION SITUATION

Before analyzing a decision situation, we must have a structure to which its various components can be fitted. These components, and the structure itself, are described in this section.

Alternatives

Decision alternatives are the choices available to the decision maker. For the computer firm, decision alternatives could be (1) plan to introduce the new computer, and (2) drop plans to introduce the new computer. Only one alternative can be selected from among those available.

STATISTICS IN ACTION 19.1

New Technologies Take Film out of the Picture

In June 2009, Kodak announced the end of the line for Kodachrome film, an American icon in family and professional photography for 74 years. As it neared the end of its product life, Kodachrome was less than 1% of the company's still-photography film sales.

The major culprit in the demise of this legend was the state of nature known as *technology*. Not only had digital photography become the dominant mode of capturing and storing images, other types of film required considerably less complexity in both their manufacture and their processing. At this writing, there is only one photographic laboratory in the entire United States that still processes Kodachrome film.

Kodak itself went through a troubled technological period during the early 2000s, as its very survival depended on a successful transition from the film age to the digital age. Today, digital photography sales account for more than 70% of Kodak's revenue, and the company has become a leading competitor in the digital photography movement that many believe had threatened its corporate existence.

Source: "Kodak Kills Kodachrome Film After 74 Years," Reuters.com, June 22, 2009.

States of Nature

After a decision alternative has been selected, the outcome is subject to chance and no longer under the control of the decision maker. For example, the new computer could be either competitive or noncompetitive in technology and features versus other brands on the market at the time of its introduction. Each of these possibilities is a **state of nature,** a future environment that has a chance of occurring. Statistics in Action 19.1 describes how the occurrence of a given state of nature spelled doom for a part of Americana that had been around for 74 years. The states of nature must be both mutually exclusive (only one will occur) and exhaustive (one of them must occur).

Having introduced the concept of states of nature, we can now describe three different levels of doubt regarding the decision situation itself:

RISK Both the possible states of nature and their exact probabilities of occurrence are known. This is typical of many gambling situations, such as playing roulette. This is an extreme rarity in business decisions, since the exact probabilities are generally not known.

UNCERTAINTY The possible states of nature are known, but their exact probabilities are unknown. Most business decision situations fall into this category, and the probabilities must be estimated based on past experience or executive judgment.

IGNORANCE Neither the possible states of nature nor their exact probabilities are known. This is another rarity, as only a fugitive from Mars would have absolutely *no* idea what states of nature might occur after a decision has been made.

The Payoff Table

The **payoff table** is a table in which the rows are decision alternatives, the columns are states of nature, and the entry at each intersection of a row and column is a numerical **payoff** such as a profit or loss. The table may also include estimated probabilities for the states of nature.

EXAMPLEEXAMPLEEXAMPLEEXAMPLEEXA

EXAMPLE

Payoff Table

A German automobile manufacturer currently produces its best-selling U.S. model in Germany, but the relative strength of the Euro versus the U.S. dollar has been making the car very expensive for the market in which it competes. To ensure a lower, more stable price, the company is considering the possibility of manufacturing cars for U.S. consumption at one of its South American plants.

SOLUTION

Table 19.1 is a payoff table that summarizes this decision situation. It includes the following components:

- **Alternatives** (1) Continue producing cars in Germany for the U.S. market; or (2) manufacture cars for U.S. consumption in the South American plant.
- **States of nature** Over the next two model years, the relative strength of the dollar versus the Euro will be (1) weak, (2) moderate, or (3) strong.
- **Probabilities** Corporate economists have estimated the probabilities for the relative strength of the dollar as 0.3 (weak), 0.5 (moderate), and 0.2 (strong).
- **Payoffs** The estimated profit for this model depends on which alternative is selected and which state of nature occurs. For example, if manufacturing is switched to the South American plant and the dollar is strong, the firm's profit is estimated as $16 million over the two-year period.

The same decision situation could also be described by a **decision tree,** a diagram in which alternatives, states of nature and their probabilities, and payoffs are displayed in graphical form. Figure 19.1 is the decision tree counterpart to the payoff table of Table 19.1. It provides the same information, but in a graphical form. The first two branches are the decision alternatives. Depending on which alternative is selected, either node 1 or node 2 will be encountered next, and the probabilities and the states of nature are the same for each, just as in the payoff table.

In the decision tree, each square node shows the making of a decision, and each circular node shows the occurrence of a state of nature. The payoff table will be our primary method for describing decision situations, but the decision tree approach will be revisited for some later applications.

TABLE 19.1

The payoff table includes the decision alternatives, the possible states of nature, and the payoff for each combination of the alternatives and states of nature. Estimated probabilities for the states of nature may also be included, as shown here. Profits are in millions of dollars.

	State of Nature		
	Compared to the Euro, the Strength of the U.S. Dollar Will Be		
Alternative	Weak (Prob. = 0.3)	Moderate (Prob. = 0.5)	Strong (Prob. = 0.2)
German Production	$10	$15	$25
South American Production	$20	$18	$16

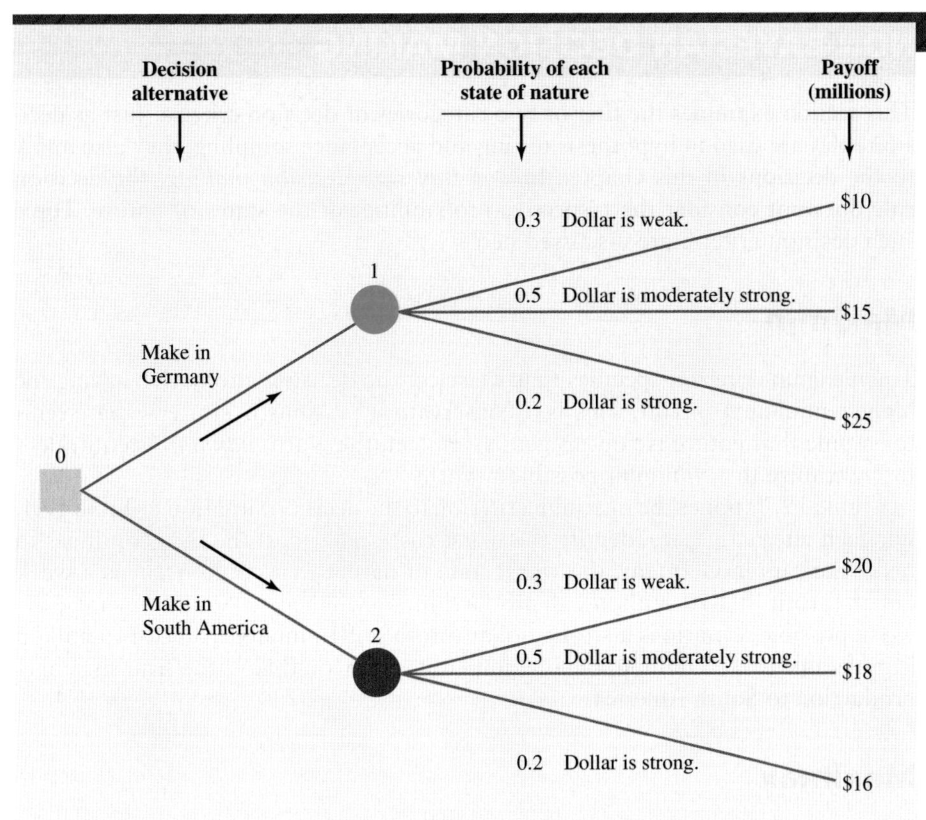

FIGURE 19.1

The decision tree counterpart to the payoff table of Table 19.1.

EXERCISES

19.1 What are the differences among risk, uncertainty, and ignorance? Into which category of decision situation do most business decisions fall?

19.2 What is meant when we say that the states of nature in the payoff table must be mutually exclusive and exhaustive?

19.3 The manager of a small local airport must extend a runway if business jets are to be accommodated. However, the airport is practically surrounded by a suburban housing development, and residents are already complaining to the city council about the volume and noise of existing air traffic at the facility. Identify the decision alternatives and the possible states of nature that might occur after the decision is made.

19.4 A management consultant, unable to locate receipts or records for several business trips during the past year, makes some rough estimates of these expenditures and considers claiming them as business expenses on his income tax form. If he makes the claims and is not audited, he will owe $12,000 in taxes. If he makes the claims and the IRS audits his return, he will have to pay an extra $8,000 in taxes and penalties. If he ignores the trips, he will owe $14,000 in taxes. Construct a payoff table and a decision tree diagram for this decision situation.

19.5 A trainee in a realty firm is being paid a straight salary, but has the opportunity to (a) continue with straight salary, (b) change to straight commission, or (c) change to a combination that includes a lower base salary and a commission on each sale. What states of nature might be useful in constructing a payoff table for this situation?

19.6 Having been injured while trying to use his rotary lawn mower as a snowblower, a consumer has sued the company and is offered the chance to settle out of court. If the case goes to trial, the jury could award him either nothing or a large amount in damages. Construct a payoff table for this consumer, including probabilities and payoff amounts that you feel might be realistic.

19.7 A Florida orange grower is considering the installation of a heating system to protect his crop in the event of frost during the coming winter. What would be the rows and columns of a payoff table applicable to this decision?

19.3 NON-BAYESIAN DECISION MAKING

This section examines the first of two categories of decision criteria. Just as decision rules are used in hypothesis testing and acceptance sampling, they also apply to the decisions in this chapter. In *non-Bayesian decision making*, the decision rule does not consider the respective probabilities of the states of nature. Three such decision criteria are discussed here.

Maximin

The **maximin** criterion specifies that we select the decision alternative having the highest minimum payoff. This is a conservative, pessimistic strategy that seems to assume that nature is "out to get us." In fact, the word *maximin* is equivalent to "*max*imize the *min*imum possible payoff."

Table 19.2 applies the maximin criterion to the decision situation in Table 19.1. For each alternative, we identify the worst-case scenario. If the firm continues to make the cars in Germany, the worst state of nature that could occur is a weak dollar (profit = \$10 million). If production is switched to South America, the worst possible scenario is a strong dollar (profit = \$16 million). Since \$16 million is preferable to \$10 million, the maximin criterion would have the firm switching production to South America.

Maximax

The **maximax** criterion specifies that we select the decision alternative having the highest maximum payoff. This is an optimistic strategy, making the assumption that nature will be generous to us. The word *maximax* is equivalent to "*max*imize the *max*imum possible payoff."

Table 19.3 applies the maximax criterion to the decision situation in Table 19.1. For each alternative, the best-case scenario is identified. If the firm continues German production for the United States, the best state of nature that could occur is a strong dollar (profit = \$25 million). If production is switched to South America, the best possible scenario is a weak dollar (profit = \$20 million). Because \$25 million is better than \$20 million, the maximax criterion would specify that German production continue.

TABLE 19.2

Using the maximin criterion, production of vehicles for the U.S. market would be switched to the South American plant. The maximin criterion pessimistically assumes that the worst possible state of nature will occur after the decision has been made.

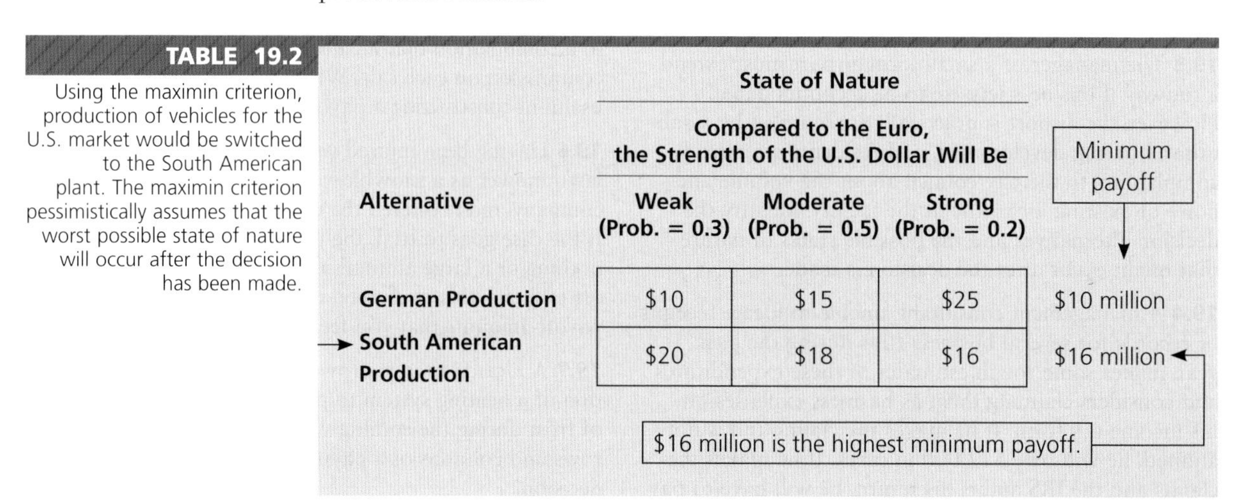

	State of Nature: Compared to the Euro, the Strength of the U.S. Dollar Will Be			
Alternative	Weak (Prob. = 0.3)	Moderate (Prob. = 0.5)	Strong (Prob. = 0.2)	Minimum payoff
German Production	\$10	\$15	\$25	\$10 million
→ South American Production	\$20	\$18	\$16	\$16 million

\$16 million is the highest minimum payoff.

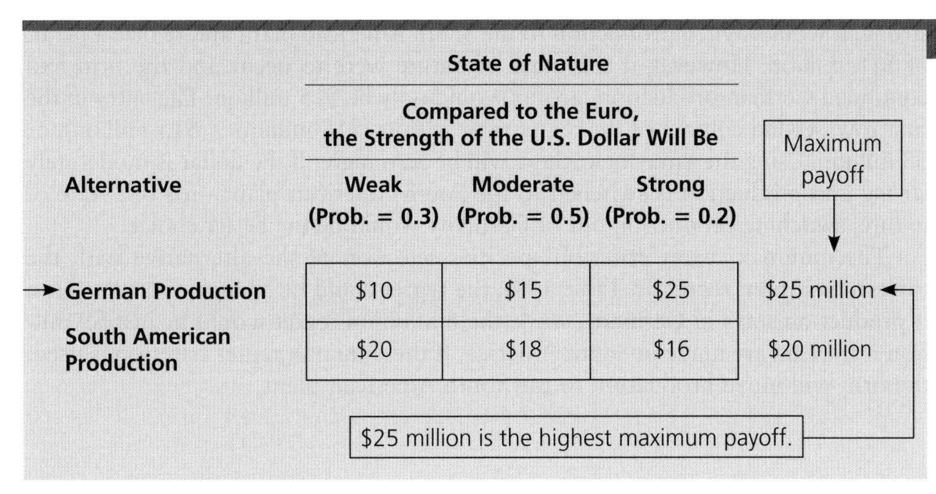

Alternative	State of Nature — Compared to the Euro, the Strength of the U.S. Dollar Will Be: Weak (Prob. = 0.3)	Moderate (Prob. = 0.5)	Strong (Prob. = 0.2)	Maximum payoff
→ German Production	$10	$15	$25	$25 million
South American Production	$20	$18	$16	$20 million

$25 million is the highest maximum payoff.

TABLE 19.3

Using the maximax criterion, production of vehicles for the U.S. market would remain in Germany. This criterion is optimistic, since it assumes that the best possible state of nature will occur after the decision has been made.

Minimax Regret

The **minimax regret** criterion is the equivalent of being paranoid about "Monday-morning quarterbacks" who, with the benefit of hindsight, come along later and tell you what you *should* have done. In using minimax regret, we must first construct a *regret table*. This is a table in which, for each state of nature, the entries are **regrets,** the differences between each payoff and the best possible payoff if that state of nature were to occur. The regret table is the same as the opportunity loss table, to be discussed in Section 19.5.

Table 19.4 is the regret table for the manufacturing decision described in Table 19.1. If the firm knew for sure that the dollar was going to be moderately

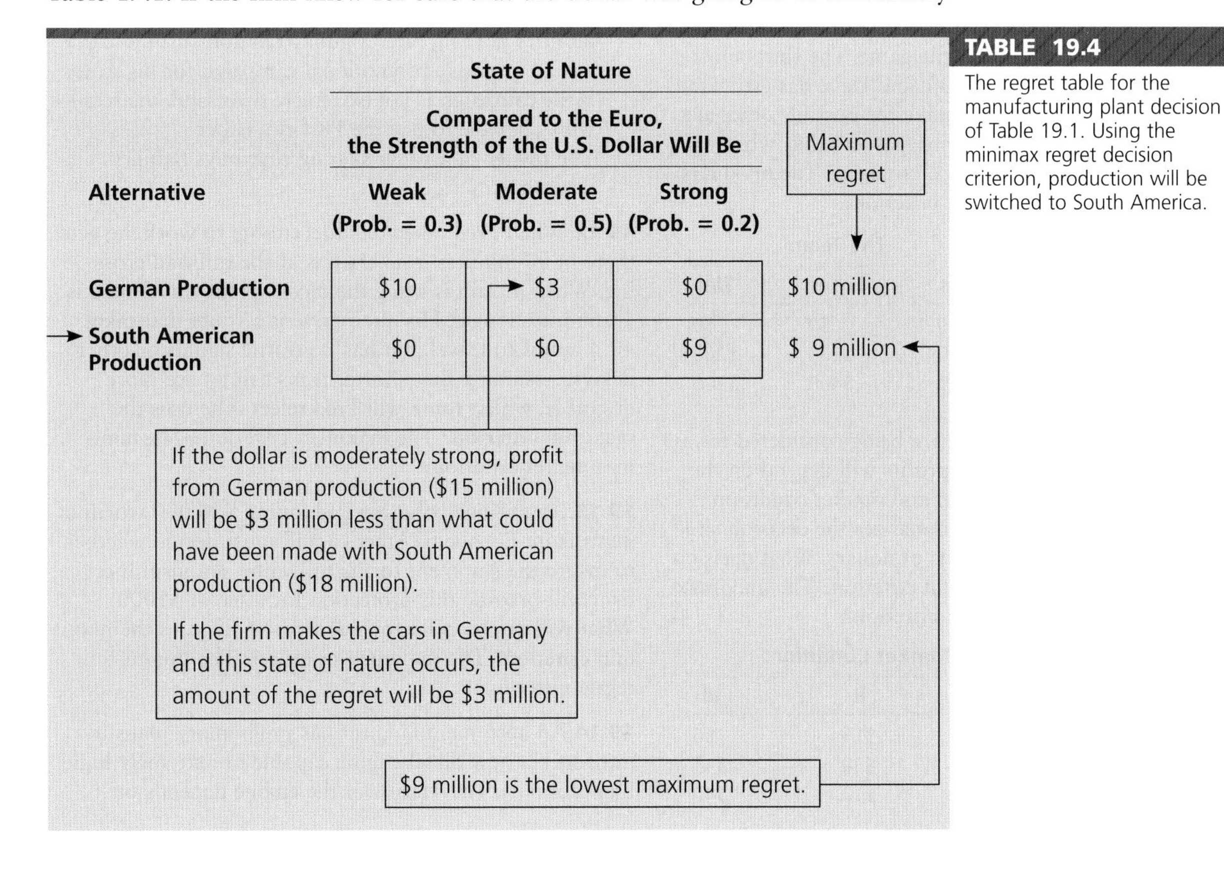

Alternative	State of Nature — Compared to the Euro, the Strength of the U.S. Dollar Will Be: Weak (Prob. = 0.3)	Moderate (Prob. = 0.5)	Strong (Prob. = 0.2)	Maximum regret
German Production	$10	$3	$0	$10 million
→ South American Production	$0	$0	$9	$ 9 million

If the dollar is moderately strong, profit from German production ($15 million) will be $3 million less than what could have been made with South American production ($18 million).

If the firm makes the cars in Germany and this state of nature occurs, the amount of the regret will be $3 million.

$9 million is the lowest maximum regret.

TABLE 19.4

The regret table for the manufacturing plant decision of Table 19.1. Using the minimax regret decision criterion, production will be switched to South America.

strong, it would switch production to the South American plant and realize a profit of $18 million. However, if this state of nature were to occur and the firm had continued German production, profits would only be $15 million. The entry in the first row, second column of the regret table is thus ($18 million − $15 million), or $3 million. Using the same logic, there will be zero regret if the dollar is moderately strong and production is switched to the South American plant—for this state of nature, switching production out of Germany would be the better choice.

The minimax regret criterion specifies selection of the alternative with the lowest maximum regret. In Table 19.4, the regret could be as high as $10 million if production stays in Germany, while the maximum regret would be just $9 million if the cars are made in South America. If the minimax regret criterion is used, the firm will move production to the South American plant.

EXERCISES

19.8 For Exercise 19.4, which decision alternative will be selected if the consultant uses the maximin criterion? The maximax criterion? The minimax regret criterion?

19.9 Dave's local college football team is playing for the national championship in next month's bowl game, and he's thinking about having "We're number one" T-shirts imprinted in honor of the team. It's too late to get the shirts before game day, but if he orders today, he can get them within a few days after the game. The shirts must be ordered in multiples of 1000, and Dave has estimated the amount of profit he will make for possible outcomes of the game. Given the following payoff table, what will Dave do if he uses the maximin criterion? The maximax criterion? The minimax regret criterion?

	The Team		
	Wins	Loses	Ties
Order no shirts	$0	$0	$0
Order 1000 shirts	$800	−$200	$200
Order 2000 shirts	$1600	−$400	$300

19.10 Three different designs are being considered for a new electronics product, and profits will depend on the combination of product design and market condition. The following payoff table summarizes the decision situation, with amounts in millions of dollars. What decision will be made using the maximin criterion? The maximax criterion? The minimax regret criterion?

	Market Condition		
	I	II	III
Design A	$25	$14	$6
Design B	$15	$30	$18
Design C	$5	$10	$40

19.11 For each of the following, indicate whether the behavior represents the maximin criterion or the maximax criterion. Explain your reasoning.

a. A trucker is behind schedule and knows he is approaching an area where police radar is often encountered, but he continues driving at 15 mph over the speed limit.

b. A struggling retailer, in danger of defaulting on next month's payroll, hires a public relations firm to carry out an intensive promotional campaign for his store. If the campaign is not extremely successful, the retailer won't be able to pay the PR firm, either.

c. The owner of a roller-skating rink buys liability insurance.

19.12 When Fred takes the direct route to work, he gets there in 30 minutes if no train is at the railroad crossing. When a train is using the crossing, his travel time is 15 minutes longer. His alternative is a route that takes 40 minutes to travel, but has no traffic signals or other delays. Assuming that shorter transit times are more desirable, which route will Fred select if he uses the maximin criterion? The maximax criterion? The minimax regret criterion?

19.13 An antique collector is shipping $50,000 worth of items from Europe to America and must decide whether to insure the shipment for its full value. An insurance firm will provide this protection for a fee of $1000. What will be the collector's decision if she uses the maximin criterion? The maximax criterion? The minimax regret criterion?

19.14 An auto manufacturer has preliminary plans to introduce a new diesel engine capable of extremely high fuel economy. The success of the engine depends on whether the U.S. Environmental Protection Agency goes ahead with stringent particulate emission limits that are scheduled for the coming model year. The payoff table, with amounts in millions of dollars for the corporation's profit, is shown here. What decision will be made using the maximin criterion? The maximax criterion? The minimax regret criterion?

	The Strict Emission Limits Are	
	Enacted	**Not Enacted**
Continue engine development	−\$10	\$30
Halt engine development	\$15	\$15

BAYESIAN DECISION MAKING (19.4)

Whereas the decision criteria just discussed ignore the probabilities for the respective states of nature, **Bayesian decision making** takes them into account. Specifically, the alternative chosen is the one with the best expected payoff.

Expected Payoff

The **expected payoff** of a decision alternative is the sum of the products of the payoffs and the state of nature probabilities. Each decision alternative is treated as a discrete probability distribution with an expected value that can be calculated using the method introduced in Chapter 6:

Expected payoff for a decision alternative:

Expected payoff = $\Sigma p_i \cdot m_i$ where p_i = the probability that state of nature i will occur

m_i = the payoff if this alternative is selected and state of nature i occurs

For the automobile production decision we've been following, the expected payoffs for the alternatives can be calculated as follows:

German production:

$$\text{Expected payoff} = \Sigma p_i \cdot m_i = 0.3(10) + 0.5(15) + 0.2(25) = \$15.5 \text{ million}$$

South American production:

$$\text{Expected payoff} = \Sigma p_i \cdot m_i = 0.3(20) + 0.5(18) + 0.2(16) = \$18.2 \text{ million}$$

As Table 19.5 shows, the alternative with the higher expected payoff is South American production, with an expected profit of \$18.2 million. Although \$18.2 million is not one of the entries in the payoff table, this would be the result if this decision situation were to be replayed a great many times—on the average, the profit would be \$18.2 million. Expected payoff is sometimes referred to as *expected monetary value* (*EMV*).

Figure 19.2 shows the decision tree diagram corresponding to Table 19.5. In using the decision tree, calculations proceed from right to left (i.e., from future to present), and the expected payoff for each alternative is obtained as in the calculations just presented. The "X" through the "German production" branch indicates that this alternative has an inferior expected payoff and would not be selected.

TABLE 19.5

Using the expected payoff criterion, production will be switched to South America and the expected profit will be $18.2 million.

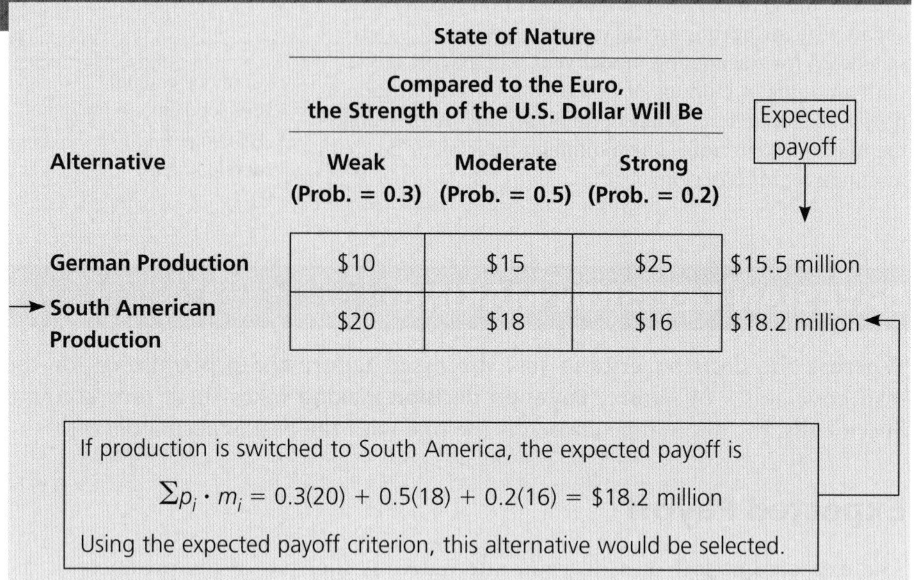

Alternative	State of Nature: Compared to the Euro, the Strength of the U.S. Dollar Will Be			Expected payoff
	Weak (Prob. = 0.3)	Moderate (Prob. = 0.5)	Strong (Prob. = 0.2)	
German Production	$10	$15	$25	$15.5 million
→ South American Production	$20	$18	$16	$18.2 million

If production is switched to South America, the expected payoff is

$$\Sigma p_i \cdot m_i = 0.3(20) + 0.5(18) + 0.2(16) = \$18.2 \text{ million}$$

Using the expected payoff criterion, this alternative would be selected.

FIGURE 19.2

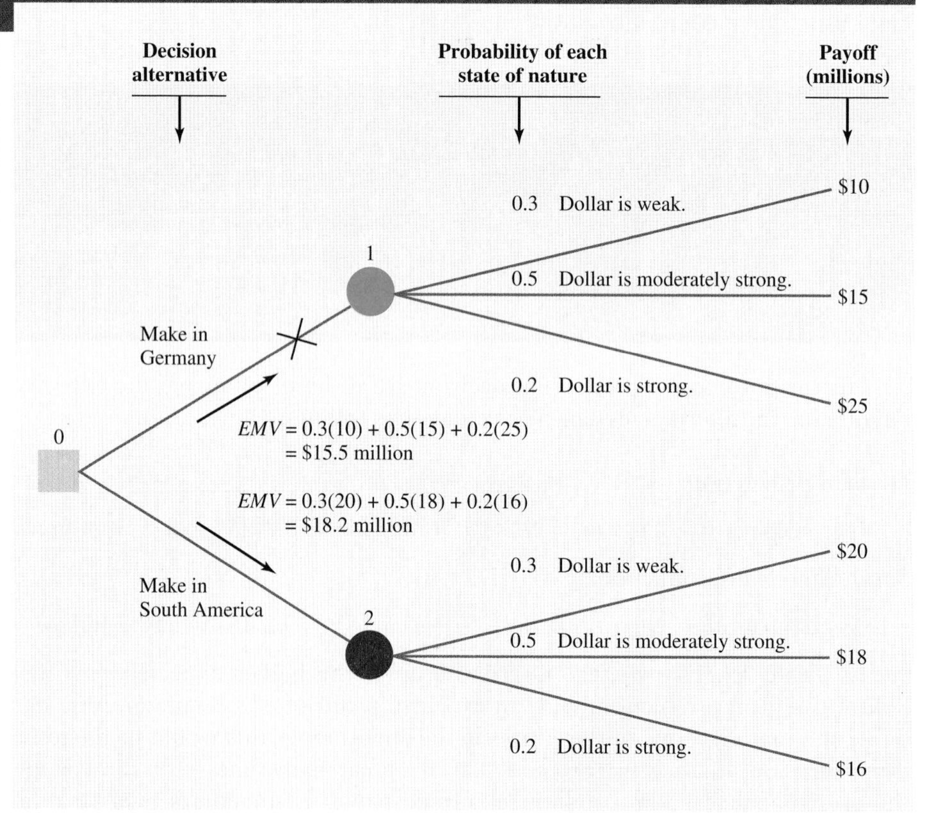

The decision tree summary for the table and expected payoff calculations of Table 19.5. The "X" through the German production alternative reflects that its expected payoff makes it inferior to the other alternative.

The Expected Value of Perfect Information

In decision making under uncertainty, it's often possible to invest in research that will give us a better idea as to which one of the states of nature will actually occur. For example, before deciding whether to continue making cars for the U.S.

market in Germany, the automobile company might like to consult with outside economic experts and forecasters regarding the predicted strength of the U.S. dollar versus the Euro.

When we are considering whether to collect additional information before making the decision, an important question arises: How much should we pay for such information? To help answer this question, we can determine the **expected value of perfect information (*EVPI*).** This is the difference between (1) the expected payoff if we are guaranteed perfect knowledge about the future and (2) the expected payoff under uncertainty:

The expected value of perfect information, *EVPI*:

$$EVPI = \text{Expected payoff with perfect information} - \text{Expected payoff with present information}$$

For the auto company, the expected payoff with present information has already been calculated as \$18.2 million. As Table 19.5 showed, this is the expected payoff for the alternative with the greatest expected payoff—that is, switching production of the car to South America.

If perfect information were available on the future relative strength of the U.S. dollar versus the Euro, the company could make its decision under *certainty*. It would know which state of nature was going to occur and would make the best choice for this state of nature. For example, if the firm had 100% confidence that the "weak dollar" state were going to occur, it would switch production to South America and have a profit of \$20 million.

In calculating the expected value *of* perfect information, we must first determine the expected payoff *with* perfect information. The perfect information has not yet been received, so we operate under the assumption that it might predict any one of the three states of nature. The probabilities of these states of nature have been estimated as 0.3, 0.5, and 0.2:

- **Probability = 0.3 that perfect information will say "weak dollar."** In this case, the firm switches production to South America and makes a profit of \$20 million.
- **Probability = 0.5 that perfect information will say "moderately strong dollar."** In this case, the firm switches production to South America and makes a profit of \$18 million.
- **Probability = 0.2 that perfect information will say "strong dollar."** In this case, the firm continues German production and makes a profit of \$25 million.

Then the expected payoff with perfect information is

$$0.3(20) + 0.5(18) + 0.2(25) = \$20.0 \text{ million}$$

The expected payoff with present information has already been calculated as \$18.2 million. This is the expected payoff for the alternative with the highest expected payoff under uncertainty (i.e., switch production to South America). Now we can calculate the expected value of perfect information:

$$EVPI = \text{Expected payoff with perfect information} - \text{Expected payoff with present information}$$
$$= \$20.0 \text{ million} - \$18.2 \text{ million} = \$1.8 \text{ million}$$

FIGURE 19.3

For the decision situation of Table 19.5, the expected value of perfect information is $1.8 million. This is the upper limit the firm should consider spending for any information to reduce uncertainty.

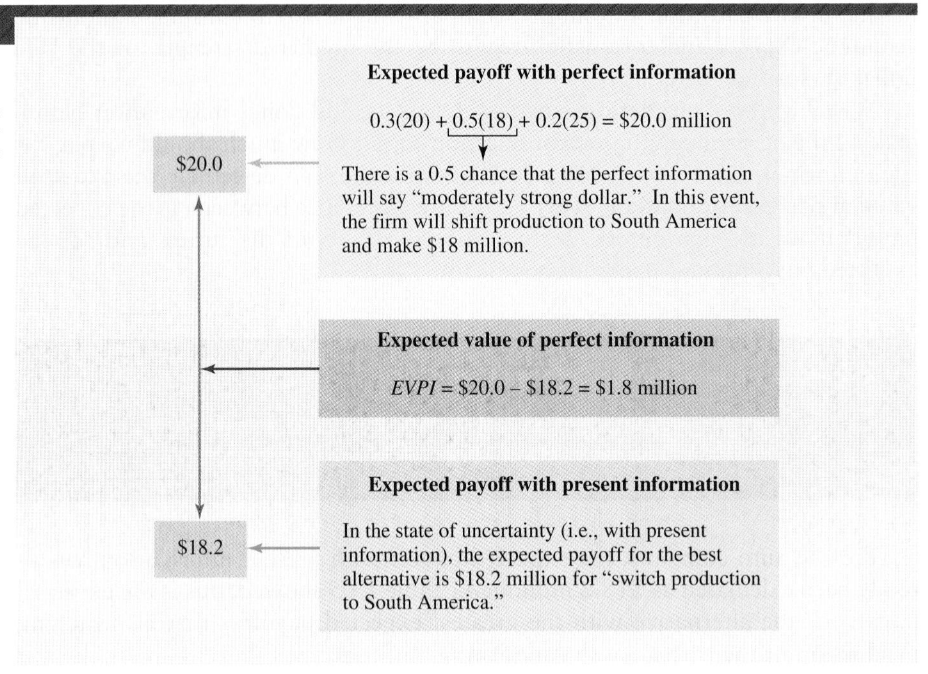

As calculated here and shown in Figure 19.3, the expected value of perfect information is $1.8 million. This is the upper limit the firm should consider spending for the purpose of reducing uncertainty.

EXERCISES

19.15 What is the most important difference between Bayesian and non-Bayesian decision criteria?

19.16 A financial institution is evaluating the loan application of an entrepreneur who has come up with a revolutionary product for which he needs developmental funding. The entrepreneur has practically no assets for collateral, but offers the institution a portion of the profits if the product is successful. The loan officer has constructed the following payoff table, with institution profits in thousands of dollars and the probabilities estimated for three states of nature.

	The Product's Level of Success Is		
	Low (0.1)	Moderate (0.3)	High (0.6)
Loan the full amount requested	–$25	$20	$50
Loan half the amount requested	–$10	$10	$20
Do not grant the loan	$0	$0	$0

Using the expected payoff criterion, which alternative will be selected and what is its expected monetary value? Determine the expected value of perfect information and interpret its meaning.

19.17 In Exercise 19.4, if the probability that his return will be audited is 0.1, which alternative will the consultant choose if he maximizes expected monetary value? What is the most he should be willing to pay for more information regarding his likelihood of being audited?

19.18 In Exercise 19.9, assume that the team has a 0.6 probability of winning, a 0.3 probability of losing, and a 0.1 probability for a tie game. Which alternative should Dave select if he wishes to maximize his expected payoff, and what is the expected value of perfect information?

19.19 In Exercise 19.10, the probabilities for the three market conditions are estimated as 0.1, 0.6, and 0.3, respectively. Which design should be selected in order to maximize the firm's expected profit? What is the most the firm should be willing to pay for a research study designed to reduce its uncertainty about market conditions?

19.20 For the following payoff table, determine the expected value of perfect information. Does the result seem unusual? If so, why did this occur?

	State of Nature	
	I (0.6)	II (0.4)
Alternative A	$80	$30
Alternative B	$12	$20

19.21 In Exercise 19.12, there is a 0.1 chance that a train will be using the railroad crossing. Given this information, which route should Fred select, and what is the expected number of minutes his trip will require? What is the expected number of minutes Fred would save if he could predict with certainty whether a train will be at the crossing?

19.22 In Exercise 19.13, if x is the probability that the shipment is lost, for what value of x will the alternatives have the same *EMV*? Assuming this value for x, what is the expected value of perfect information?

19.23 In business research, it's extremely rare that any study will provide a perfect assessment of the future. Why then should we bother determining the expected value of perfect information when such information is practically never available?

THE OPPORTUNITY LOSS APPROACH (19.5)

Another way to use probabilities in identifying the best alternative and finding the expected value of perfect information is to structure the decision in terms of opportunity loss. **Opportunity loss** is another term for regret, discussed earlier in the chapter, and expected opportunity loss is calculated in a manner similar to that of expected payoff:

Expected opportunity loss (*EOL*) for a decision alternative:

$$EOL = \sum p_i \cdot l_i$$

where p_i = the probability that state of nature i will occur

l_i = the opportunity loss if this alternative is selected and state of nature i occurs

The results of the opportunity loss approach will be the same as those obtained in the expected payoff method of the preceding section. In particular,

1. the alternative having the maximum expected payoff will also have the minimum expected opportunity loss (*EOL*), and
2. the expected opportunity loss for this alternative is the expected value of perfect information.

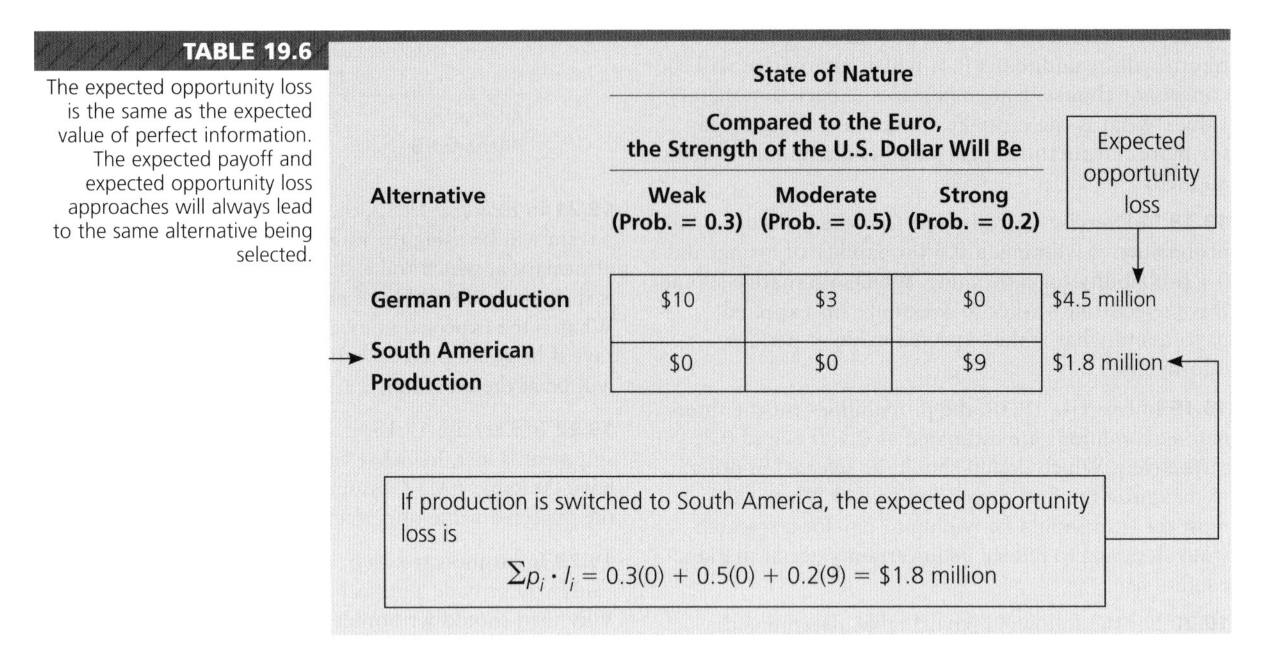

Alternative	State of Nature: Compared to the Euro, the Strength of the U.S. Dollar Will Be			Expected opportunity loss
	Weak (Prob. = 0.3)	Moderate (Prob. = 0.5)	Strong (Prob. = 0.2)	
German Production	\$10	\$3	\$0	\$4.5 million
→ South American Production	\$0	\$0	\$9	\$1.8 million

If production is switched to South America, the expected opportunity loss is

$$\Sigma p_i \cdot l_i = 0.3(0) + 0.5(0) + 0.2(9) = \$1.8 \text{ million}$$

TABLE 19.6

The expected opportunity loss is the same as the expected value of perfect information. The expected payoff and expected opportunity loss approaches will always lead to the same alternative being selected.

Table 19.6 is the opportunity loss approach to the auto maker's decision situation. Note that the table entries are the same as for the regret table of Table 19.4—this is because opportunity loss and regret are the same thing.

In Table 19.6, "South American production" has the minimum expected opportunity loss, \$1.8 million. This indicates that (1) "South American production" is the better alternative and (2) the absence of perfect information can be expected to cost the company \$1.8 million. Again, for any decision situation the expected payoff and expected opportunity loss approaches will always lead to the same alternative being selected. Just as opportunity loss and regret are the same thing, expected opportunity loss and the expected value of perfect information are equivalent.

EXERCISES

19.24 In what way are the minimax regret criterion and the expected opportunity loss (*EOL*) criterion similar? In what ways do they differ?

19.25 An investor intends to purchase stock in one of three companies. Company A is very sensitive to the overall national economy, company C is relatively insensitive to the economic environment, and company B is about average. The investor has constructed the following payoff table for his profit, in thousands of dollars.

	The State of the Economy Will Be		
	Weak (0.2)	Moderate (0.4)	Strong (0.4)
Invest in company A	−\$40	\$30	\$60
Invest in company B	−\$20	\$20	\$45
Invest in company C	\$24	\$28	\$30

Using the opportunity loss approach, which alternative will be selected, and what is its expected opportunity loss? What is the most the investor should be willing to pay for perfect information about the future state of the economy?

19.26 In Exercise 19.10, with the probabilities for the three market conditions estimated as 0.1, 0.6, and 0.3, respectively, what is the expected opportunity loss associated with each of the three decision alternatives? In what way is the lowest of these expected opportunity losses related to the expected value of perfect information?

19.27 In Exercise 19.12, with a probability of 0.1 that a train will be using the crossing, compute the expected opportunity loss for each of Fred's alternatives. Explain the meaning of the *EOL* for each alternative.

INCREMENTAL ANALYSIS AND INVENTORY DECISIONS (19.6)

In some cases, the number of decision alternatives and states of nature may be extremely large. This can occur, for example, if a retailer can stock anywhere from 0 to 1000 units and demand might also be anywhere from 0 to 1000. A payoff table could end up having 1000 rows and 1000 columns, and calculations to find the best decision alternative would be extremely tedious. In such applications, we can use **incremental analysis,** where units are sequentially considered, with each unit in the sequence being stocked only if the expected payoff for stocking it exceeds the expected payoff for not stocking it. Incremental analysis involves the following key terms:

MARGINAL PROFIT The profit if one more unit is stocked and sold.

MARGINAL LOSS The cost if one more unit is stocked and *not* sold.

EXAMPLE

Incremental Analysis

A supermarket receives weekly milk shipments from its dairy products supplier. The weekly demand for milk can be approximated by a normal distribution having a mean of 2000 gallons and a standard deviation of 300 gallons. The supermarket pays $2.10 per gallon and makes a profit of $0.50 on each gallon sold. At the end of the week, the supplier repurchases all unsold milk for $0.60 per gallon.

SOLUTION

A payoff table for this situation could have over 1000 rows and columns, with each one representing a single gallon of milk. Using incremental analysis, we can determine the alternative with the maximum expected payoff (or minimum expected opportunity loss) much more easily.

EXAMPLEEXAMPLEEXAMPLEEXAMP

TABLE 19.7

In incremental analysis, item i is stocked if the expected payoff from stocking it exceeds the expected payoff from not stocking it.

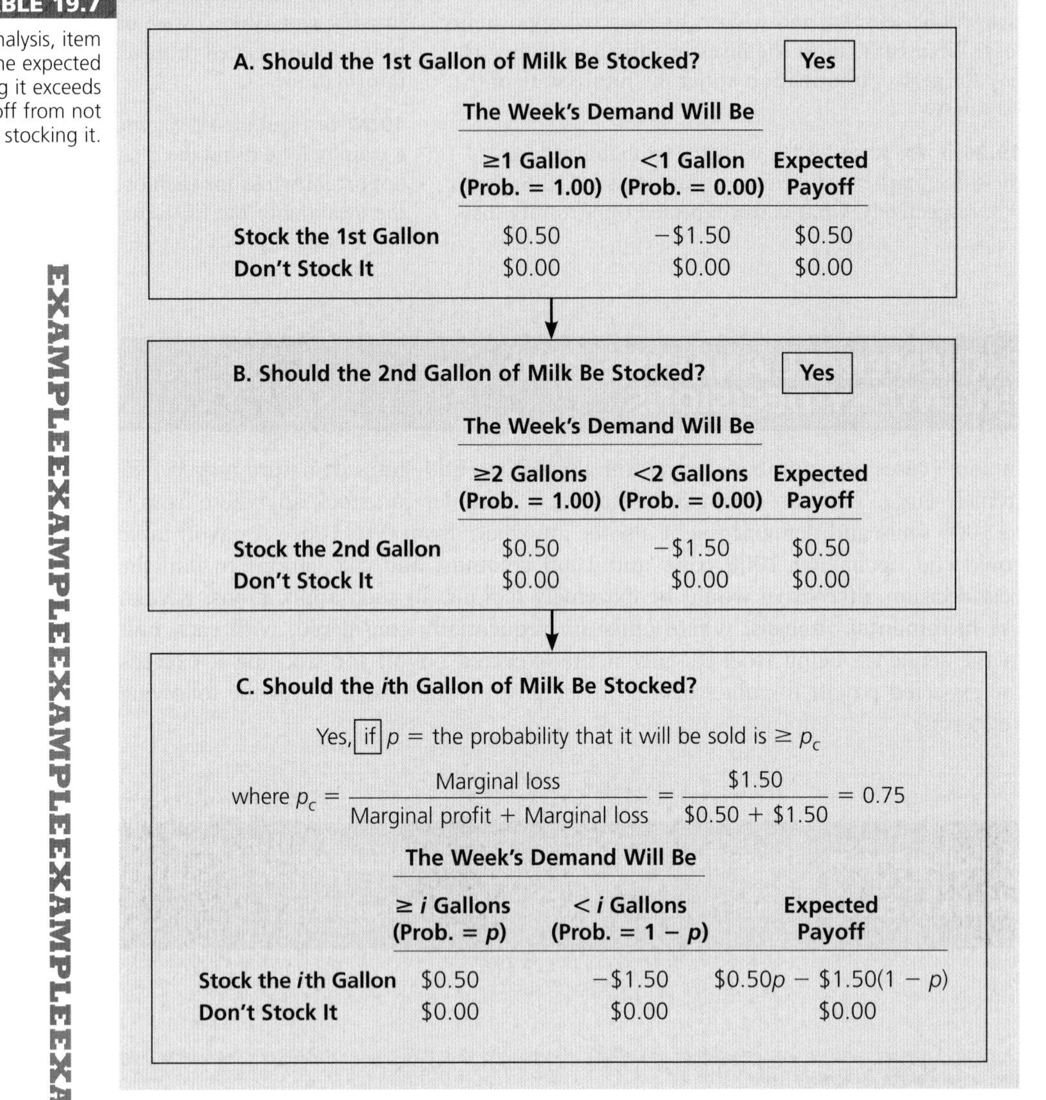

A. Should the 1st Gallon of Milk Be Stocked? **Yes**

	The Week's Demand Will Be		
	≥1 Gallon (Prob. = 1.00)	<1 Gallon (Prob. = 0.00)	Expected Payoff
Stock the 1st Gallon	\$0.50	−\$1.50	\$0.50
Don't Stock It	\$0.00	\$0.00	\$0.00

B. Should the 2nd Gallon of Milk Be Stocked? **Yes**

	The Week's Demand Will Be		
	≥2 Gallons (Prob. = 1.00)	<2 Gallons (Prob. = 0.00)	Expected Payoff
Stock the 2nd Gallon	\$0.50	−\$1.50	\$0.50
Don't Stock It	\$0.00	\$0.00	\$0.00

C. Should the *i*th Gallon of Milk Be Stocked?

Yes, if p = the probability that it will be sold is $\geq p_c$

$$\text{where } p_c = \frac{\text{Marginal loss}}{\text{Marginal profit} + \text{Marginal loss}} = \frac{\$1.50}{\$0.50 + \$1.50} = 0.75$$

	The Week's Demand Will Be		
	≥ i Gallons (Prob. = p)	< i Gallons (Prob. = $1 - p$)	Expected Payoff
Stock the *i*th Gallon	\$0.50	−\$1.50	$\$0.50p - \$1.50(1 - p)$
Don't Stock It	\$0.00	\$0.00	\$0.00

To show how it works, let's first consider whether the *first* gallon of milk should be stocked. If it's stocked and sold, the marginal profit is \$0.50. If it's stocked and *not* sold, the marginal loss is \$1.50. For a normal distribution with $\mu = 2000$ and $\sigma = 300$, it's virtually certain that at least one gallon of milk will be sold. Table 19.7 is a series of payoff tables for this decision setting, and part A shows that the expected payoff from stocking the first gallon easily exceeds the expected payoff if it is not stocked.

In part B of Table 19.7, we move on to consider the second gallon. As with the first gallon, the probability that demand will be at least two gallons is virtually certain. If the second gallon is stocked, the expected payoff will be \$0.50 higher than if it is not stocked.

Since it would be inefficient to continue evaluating one gallon at a time, let's introduce a time-saver. The key is the answer to the question, "What probability of selling one more gallon results in the expected payoff for stocking this gallon being at least as high as the expected payoff for not stocking it?" Given the marginal profit if the *i*th gallon is sold and the marginal loss if it is not, this probability can be determined as follows:

Critical probability for incremental analysis:

Stock the *i*th unit if Prob(demand $\geq i$) is at least p_c, with

$$p_c = \frac{ml}{mp + ml}$$

where p_c = the critical probability
mp = the marginal profit if the unit is stocked and sold
ml = the marginal loss if the unit is stocked and not sold

With marginal profit mp = \$0.50 and marginal loss ml = \$1.50, the critical probability for this decision situation will be

$$p_c = \frac{ml}{mp + ml} = \frac{\$1.50}{\$0.50 + \$1.50} = 0.75$$

Thus, as long as there is at least a 75% chance that the *i*th gallon will be sold, that gallon should be stocked. This can also be found by setting the expected payoff in the first row of Table 19.7, part C, to be ≥ 0, or

$$\$0.50p - \$1.50(1 - p) \geq \$0$$
$$2.00p - 1.50 \geq 0$$

and p must be ≥ 0.75, the value of p_c

To find which gallon of milk corresponds to a 0.75 probability of being sold, we use the standard normal distribution as shown in Figure 19.4. For a right-tail area of 0.75, the cumulative area to the left will be 0.25. This corresponds to $z = -0.67$ standard deviation units, and

$$z = \frac{x - \mu}{\sigma} \quad \text{or} \quad -0.67 = \frac{x \text{ gallons} - 2000 \text{ gallons}}{300 \text{ gallons}}$$

and

$$x = 1799 \text{ gallons}$$

The probability is 0.75 that the demand will be high enough to include the 1799th gallon, and this is the quantity the supermarket should stock. Stocking the 1800th gallon would not be advisable, since the probability of a demand at least this high would be slightly less than 0.75 and the expected payoff for stocking the 1800th gallon would be less than the expected payoff for not stocking it.

EXAMPLEEXAMPLEEXAMPLEEXAMPLEEXAMPLE

FIGURE 19.4

The probability is 0.75 that demand will include the 1799th gallon of milk. If the 1800th gallon is stocked, there is a less than 75% chance it will be sold. Using incremental analysis, expected profit is maximized if 1799 gallons are stocked.

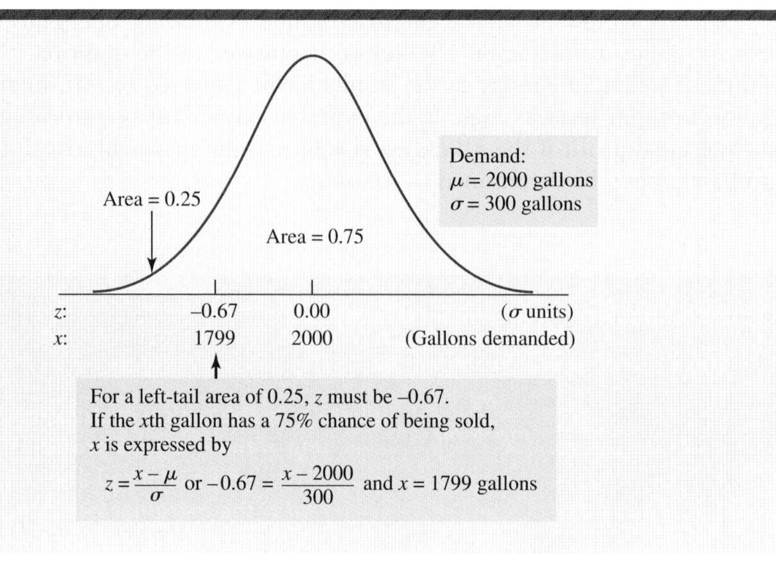

EXERCISES

19.28 A newsstand manager must decide how many daily papers to stock. Past experience has shown that demand can be approximated by a normal distribution with a mean of 400 papers and a standard deviation of 80 papers. The manager pays 30 cents for each paper, sells it for 50 cents, and any papers not sold are picked up by the distributor, with the manager receiving 5 cents for each of them. If the manager wants to maximize his expected profit, how many newspapers should be stocked?

19.29 The manager of a fish market pays 80 cents each for cod and sells them for $1.50 each. Fish left over at the end of the day are discarded. The daily demand can be approximated by a normal distribution having a mean of 300 and a standard deviation of 50. To maximize his expected profit, how many cod should the manager buy before the market opens each day?

19.30 For Exercise 19.29, how many cod should be purchased if leftover fish can be sold to a fertilizer company for 30 cents each?

19.31 An athletic director must order football programs well in advance of game day. She pays $1.00 for each program, and the selling price to fans is $5.00. Programs not sold at the game can be sold to the college bookstore for $0.30 each. Demand at the game is normally distributed, with $\mu = 3000$ and $\sigma = 500$. To maximize expected payoff, how many programs should the athletic director order?

19.32 Repeat Exercise 19.31, under the assumption that the bookstore manager has refused to purchase unsold programs for future games.

19.7 SUMMARY

- **Structuring the decision situation**

The decision situation can be structured according to the decision alternatives and the possible states of nature. The states of nature are both exhaustive (one of them must occur) and mutually exclusive (only one can occur). The decision situation may involve risk, uncertainty, or ignorance, depending on how much is known about the states of nature and their exact probabilities. Most business decisions are made under uncertainty, where the states of nature are known, but their exact probabilities are not.

The payoff table rows represent the decision alternatives, the columns show the states of nature, and the entries are payoffs for the various intersections of alternatives and states of nature. The decision situation can also be described by a decision tree, in which alternatives and states of nature are displayed in graphical form. Both the payoff table and the decision tree can include estimated probabilities for the states of nature.

- **Bayesian and non-Bayesian decision criteria**

In non-Bayesian decision making, there is no attempt to consider the probabilities of the respective states of nature. Three such decision rules are maximin, maximax, and minimax regret. Bayesian decision making takes these probabilities into account, and the alternative with the highest expected payoff is selected. If it is possible to obtain information that will reduce uncertainty, the expected value of perfect information (*EVPI*) can be calculated. This is the amount by which the expected payoff could be improved if perfect information were available, and it is the most that should be paid for any information designed to reduce uncertainty.

- **Opportunity loss approach**

In the opportunity loss table, each entry is the difference between a payoff and the best possible payoff for that state of nature. The results of the opportunity loss approach are the same as for the expected payoff method, and the alternative with the minimum expected opportunity loss will have the maximum expected payoff. Also, the minimum expected opportunity loss is equivalent to the expected value of perfect information.

- **Incremental stocking decisions**

For some situations, such as inventory stocking decisions, the number of alternatives and states of nature may be extremely large. In such cases, incremental analysis can be applied. In this method, units are sequentially considered, with each unit in the sequence being stocked only if the expected payoff for stocking it exceeds the expected payoff for not stocking it.

EQUATIONS

Expected Payoff for a Decision Alternative

$$\text{Expected payoff} = \Sigma p_i \cdot m_i$$

where p_i = the probability that state of nature i will occur

m_i = the payoff if this alternative is selected and state of nature i occurs

Expected Value of Perfect Information, *EVPI*

$$EVPI = \text{Expected payoff with perfect information} - \text{Expected payoff with present information}$$

Expected Opportunity Loss (*EOL*) for a Decision Alternative

$$EOL = \Sigma p_i \cdot l_i$$

where p_i = the probability that state of nature i will occur

l_i = the opportunity loss if this alternative is selected and state of nature i occurs

Critical Probability for Incremental Analysis

Stock the *i*th unit if Prob(demand $\geq i$) is at least p_c, with

$$p_c = \frac{ml}{mp + ml}$$

where p_c = the critical probability
mp = the marginal profit if the unit is stocked and sold
ml = the marginal loss if the unit is stocked and not sold

CHAPTER EXERCISES

19.33 For a given perishable product, a retailer pays $5 for each unit, then sells them for $10 each. At the end of the day, units not sold at the store are disposed of, and the retailer receives just $1 for each. Given the following probability distribution describing daily demand, how many units should be stocked?

x	Prob. (Demand = *x*)	*x*	Prob. (Demand = *x*)
0	0.01	7	0.18
1	0.02	8	0.14
2	0.04	9	0.09
3	0.07	10	0.05
4	0.09	11	0.02
5	0.12	12	0.01
6	0.15	13	0.01

19.34 Dennis Dunkwright is a former NBA player and free spirit whose face and unusual antics are well known to young Americans across the land. A marketer of athletic shoes is considering the possibility of retaining Dennis to appear in a series of high-profile television ads over the coming fiscal year. If Dennis is associated with their company and stays out of legal and substance abuse difficulties during the year, the company estimates its profits will be $4.4 million for the year. However, if Dennis runs afoul of the law or does anything else that might make the company look bad, the estimated profit would be just $0.5 million. If the company remains with its current spokesperson, who is lesser known but more stable, it estimates its annual profit for the year to be $2.2 million. If the company believes there to be a 0.20 probability that Dennis will get into trouble during the year, what strategy should be selected if the company wishes to maximize its expected profit? What would be the expected value of a perfect forecast for the future behavior of their potential new spokesperson?

19.35 In Exercise 19.34, what strategy would the company select if it were using the maximin criterion? The maximax criterion? The minimax regret criterion?

19.36 An investor is trying to decide between two promising technology stocks to round out his portfolio. The remainder of his portfolio consists of relatively conservative companies and will not enter into his decision as to which technology company will be selected. DiskWorth is a well-established company engaged in the design and manufacture of data storage disks, while ComTranDat is a start-up firm in the midst of developing what could be a wireless communication and data transmission system superior to those currently used or under development by the leaders in its industry. Depending on whether the technology sector weakens, remains the same, or strengthens over the next 10 years, the investor estimates that his profits with DiskWorth would be $20,000, $30,000, and $40,000, for the respective scenarios. The corresponding profits for ComTranDat are estimated to be −$20,000, $10,000, and $90,000, respectively. If the technology sector has a 20% probability of weakening over the next 10 years, a 50% probability of remaining the same, and a 30% probability of strengthening, which stock purchase should the investor make if he wishes to maximize his expected profits over the 10-year period? What is the most he should be willing to pay for a perfect forecast for the future strength of the technology sector?

19.37 In Exercise 19.36, which company would the investor select if it were using the maximin criterion? The maximax criterion? The minimax regret criterion?

19.38 A ski resort operator must decide before the winter season whether he will lease a snow-making machine. If he has no machine, he will make $20,000 if the winter is mild, $30,000 if it is typical, and $50,000 if the winter is severe. If he decides to lease the machine, his profits for these conditions will be $30,000, $35,000, and $40,000, respectively. The probability of a mild winter is 0.3, with a 0.5 chance of a typical winter and a 0.2 chance of a severe winter. If the operator wants to maximize his expected profit, should he lease the machine? What is the most he should be willing to pay for a perfect forecast?

19.39 In Exercise 19.38, the operator says that he has no idea regarding probabilities for the various weather conditions for the coming winter. What are some other criteria by which he might reach a decision, and what decision would be reached using each of them?

19.40 A driver in the Indianapolis 500 auto race has the lead but is nearly out of fuel with eight laps remaining in the race. The second-place car is just 10 seconds behind and has plenty of fuel. The driver in the lead car elects to try for the finish line without making a fuel stop. Does this behavior seem to be a maximin choice or a maximax choice? Explain.

INTEGRATED CASE

Thorndike Sports Equipment (Video Unit Seven)

For several years, Thorndike Sports Equipment has been one of the suppliers of refreshments to the Titans of Tennis national tournament. Luke Thorndike doesn't really like to call the company's product a "refreshment," since it's been scientifically designed to replenish the fluids and vital body chemicals lost through perspiration and exercise. According to the elder Thorndike, Thornado is not "just another sugar-fluffy soft drink. It helps athletes perform at their best."

Naturally, each competing athlete is given a complimentary carton of Thornado. However, the Thorndikes must decide how many cans of the product to send to the tournament for fan consumption. The Thorndikes make a profit of $0.25 on each can of Thornado purchased at the tournament. Each can that is not purchased is distributed to local sports clubs in the tournament's host city. Luke Thorndike believes in helping those in need, but is quick to point out that it costs him a total of $0.80 to produce and distribute each of these unsold cans of Thornado.

Luke believes in charity, but he is also rather frugal in making it available. He's asked Ted to figure out how many cans of Thornado to send to the Titans of Tennis tournament in order to make as much profit as possible. Ted Thorndike, having analyzed past sales patterns for Thornado at this tournament, has found annual tournament sales of the product to be closely approximated by a normal distribution with $\mu = 4500$ cans and $\sigma = 300$ cans. In order to satisfy his grandfather's dictate that profit be maximized, how many cans of Thornado should Ted plan on shipping to the Titans of Tennis?

CHAPTER 20

Total Quality Management

"THE TOYOTA WAY"

In early 2010, Toyota was in the midst of the largest safety recall in its history. The company had shut down production of vehicles that could be affected, and those already in dealer lots were sporting "Not for Sale" signs instead of festive balloons and ribbons. The recall involved enormous safety, monetary, and legal concerns, both for Toyota and its customers.

Why, then, does this chapter-opening vignette exist and why is it important? First, as both the Toyota experience and this chapter emphasize, it's much better to do something right the first time than to suffer the costs of tracking down a problem and fixing it afterwards. Second, although Toyota's safety recall was massive, the company nevertheless has a legendary history, reputation, and philosophy focused on quality. It is with this philosophy that our vignette is primarily concerned, and our medium lies in the title and contents of a book titled *The Toyota Way*, by Dr. Jeffrey K. Liker. In the foreword to the book, Gary Convis, Managing Officer and President of Toyota Motor Manufacturing, Kentucky, describes this philosophy:

The Toyota Way can be briefly summarized through the two pillars that support it: "Continuous Improvement" and "Respect for People." Continuous improvement, often called kaizen, *defines Toyota's basic approach to doing business. Challenge everything. More important than the actual improvements that individuals contribute, the true value of continuous improvement is in creating an atmosphere of continuous* learning *and an environment that not only accepts, but actually embraces* change.

As you know from your personal life, there are few things that we or our organization could not do better if we continually made even small, incremental improvements in the process through which we do them—this is the "kaizen" concept mentioned above, and it will be among the topics discussed within the chapter. If your course schedule is too tight for this chapter to be included in your syllabus, I encourage you to at least skim it, and I sincerely believe that doing so and embracing its spirit will help you greatly, both personally and professionally.

[Note: Regardless of what you drive or are considering driving, learn more about vehicle safety recalls and investigations from the National Highway Traffic Safety Administration (nhtsa.gov) and other online sources, such as Edmunds.com and Cars.com.]

Sources: cnn.com, February 3, 2010; Quoted material from Liker, Jeffrey K., *The Toyota Way* (New York: McGraw-Hill), 2004, p. xi.

Quality requires attention to detail.

- Discuss the concept of total quality management.
- Understand the nature and importance of the process orientation that is central to total quality management.
- Differentiate between defect prevention and defect detection strategies for the management of quality.
- Discuss the Kaizen concept of ongoing improvement.
- Explain the difference between random and assignable process variation.
- Describe, in general, how random and assignable process variation can each be identified and reduced.
- Explain what is meant by statistical process control and describe how it contributes to the quality of manufactured products.
- Construct and interpret statistical process control charts for variables and attributes.
- Determine and interpret the value of the process capability index.

LEARNING OBJECTIVES

After reading this chapter, you should be able to:

INTRODUCTION (20.1)

The quest for quality is probably more widespread and more intense than at any other time in our history. This pursuit is shared by producers and consumers alike, and is fueled by many factors, including global competition, consumer expectations, technological advances, and evolving techniques and philosophies for quality measurement and enhancement. More than ever before, the successful enterprise must focus on the quality of its product offerings.

What Is Quality?

If we look up "quality" in the dictionary, we will find it defined by conceptual descriptors such as "excellence" or "first-class." However, for the firm offering products to an increasingly demanding marketplace, the most important definition of quality is the one held by the customers the firm is trying to reach.

Products will differ greatly in terms of the quality dimensions that are most important to their consumers. People who buy roofing nails will tend to be most interested in the performance and durability dimensions, while buyers of fashion attire may place more value on the aesthetic and perceived-quality dimensions. Regardless of the product, it is increasingly important that all activities, from original design, to production, to after-sales service, have quality to the consumer as a central focus.

Why Is Quality Important?

Perhaps the most obvious reason quality is important to producers is the fact that quality is important to consumers; a producer who has been abandoned by his or her customers will not stay in business very long. Quality is also important because quality—and the lack of it—both cost money. For this reason, we can also look at the importance of quality in terms of two kinds of costs:[1]

1. Costs of preventing poor quality. These are also known as *costs of conformance,* and they are associated with making the product meet requirements or expectations.

   Some of the activities related to this objective include worker training and involvement, coordination with suppliers, customer surveys, and revisions to product design and the production process.
2. Costs that occur as the result of poor quality. These are also known as *costs of nonconformance,* and they are associated with the failure to meet requirements or expectations.

   Defective products may need to be scrapped, reworked, repaired, and retested, any of which can be both costly and disruptive to the production process. If defective products end up being purchased and used by consumers, this can raise warranty costs, lead to reduced goodwill and lost business, require image-damaging product recalls, and even result in expensive product liability lawsuits.

A large and growing school of thought argues that even great amounts of money spent on *preventing* poor quality are a bargain when compared to the enormous visible and invisible costs that can occur when poor quality is allowed to occur. Today's trend is heavily toward the "stitch in time" (spending on conformance) to "save the nine" (spending as a result of nonconformance).

Process Variation and Product Quality

The quality of products is closely related to variation in the process that has produced them. As in dart throwing and bowling, variation is inherent in manufacturing and other processes. The process (and the resulting product) will tend to be at least a little different from one unit of output to the next. There are two sources of process variation:

1. *Random variation* occurs as the result of chance. This type of variation is inherent in practically any process, and random variation in the process leads to random variation in the product. For example, each of a truckload of window air conditioners may have been designed to cool at the rate of 5000 Btu/hour. However, some may be capable of 5100 Btu/hour while others might reach only 4900. The pitch of the fan blades, the friction within the electric motor, the fit of the insulation, and the amount of system refrigerant injected are just a few of the process parameters that are not perfectly identical from one assembled unit to the next. Whether we're producing air conditioners, steel rods, or ball bearings, the output, tensile strength, or diameter will naturally tend to exhibit at least minor variations during the production run. This type of variation is not generally considered to be under the control of the individual worker. Industry often refers to random variation as *common-cause* variation.

   Random variation, while inherent, can often be reduced by using more expensive machines or materials. As long as the amount of random variation is tolerable, with the end product falling within acceptable specifications, neither its presence nor its sources are causes of alarm.

[1]Norman Gaither, *Production and Operations Management* 6/e. Fort Worth: The Dryden Press, 1994, ch. 17.

STATISTICS IN ACTION 20.1

And Here We Are Walking on the Moon. . . Whoops!

We've all accidentally taped or recorded over a video event we wanted to keep. Maybe it was a high school graduation, a wedding reception, or a vacation at the beach. But the original video recording of the first humans ever to set foot on the moon, a situation in which actual rocket scientists were involved? Yes, it happened. The original National Aeronautics and Space Administration (NASA) videotapes of U.S. astronauts Neil Armstrong and Buzz Aldrin walking on the surface of the moon on July 20, 1969, were, historical significance be hanged, somehow magnetically erased and reused to save money. Fortunately, all was not lost, as available tapes of the original TV broadcast have been found, restored, digitized, and enhanced, and they are said to look even sharper than the original version that was so foolishly trashed. The moral of the story is that, when you're erasing tapes of immense historical importance, you should perhaps consider the moral of the story.

Source: Maggie Fox, "Moon Landing Tapes Got Erased, NASA Admits," reuters.com, July 16, 2009.

2. *Assignable variation* occurs due to identifiable causes that have changed the process. For example, a worker might commit an error, a tool may break, a machine may be "drifting" out of adjustment, operators may become fatigued toward the end of the work day, contact surfaces may be getting worn, or an extrusion machine may accumulate deposits during its period of operation. Some even feel it's unwise to purchase an automobile that was made on a Friday or a Monday, since substitute workers may be filling in for some of the more highly skilled individuals who decided to enjoy an extended weekend. This type of variation is generally considered to be under the control of the individual worker. Industry often refers to assignable variation as *special-cause* variation.

   Compared to random variation, assignable variation tends to be directional. For example, as a cutting tool wears down, the thickness of the bars, rods, or disks being cut may increase over time. While the exact causes of random variation can be very expensive to track down and reduce, assignable variation is both more easily identified and more efficiently corrected.

One of the great challenges in industry is to be able to tell when process variation is random and when it is assignable; then, if it is assignable, to do something about it. Assignable variation is the primary concern of much of the chapter. Statistics in Action 20.1 describes a variation that was both assignable and unbelievable, especially considering the historical magnitude of the error involved.

EXERCISES

20.1 Given that a production process may experience both random and assignable variation, for which type of variation is the source generally easier to identify and eliminate? Why?

20.2 Differentiate between conformance and nonconformance in product quality. What are some of the costs associated with each?

20.3 In a manufacturing process, what is random variation? If cereal boxes are being filled by a machine, describe how random variation might affect the content weight of the individual boxes.

20.4 Repeat Exercise 20.3, but substitute "assignable variation" for "random variation."

20.2 A HISTORICAL PERSPECTIVE AND DEFECT DETECTION

Human concern with quality is at least as old as the wheel, since the discovery of this useful object was surely followed by the realization that it would work better if it were more perfectly round. As wheels and many other products came to be fabricated, control over their quality was typically the responsibility of the individual craftsman, and it was not unusual for each individual item to be the subject of its maker's scrutiny.

As our economy became more mass-production oriented, such individual inspection became less practical, and the 1920s marked the beginning of what is now known as *statistical quality control,* in which quality is monitored on the basis of *samples* of items taken from the production line. Two techniques introduced during this era are especially noteworthy:

1. *The Statistical Process Control (SPC) Chart* Developed by Walter Shewhart, this is a graphical procedure for determining whether a production process is operating properly. In general, as long as assignable variation is absent and random variation remains within acceptable limits, the process is deemed to be stable, or in statistical control. A production process that is unstable will tend to generate products that are of unpredictable and/or unsatisfactory quality.

   By monitoring the stability of the process as products are being made, the SPC method represents a *defect-prevention* strategy for quality management. Because it is employed *during* the manufacture of a stream of components or products, this method enables midstream corrections to be made if the process should begin to "jump" or "drift" due to an assignable cause. Today's quality orientation is heavily toward defect prevention, and the SPC method enjoys more popularity than ever. Statistical process control is discussed further in Sections 20.6–20.9.
2. *Acceptance Sampling* Developed largely through the efforts of Harold Dodge and Harry Romig, this is a set of procedures whereby, based on a random sampling of the items within, an entire production lot or shipment is (1) accepted, (2) rejected, or (3) subjected to further sampling before the final decision to accept or reject.

   Acceptance sampling involves inspection of a sample of products *after* an entire production run or shipment lot has been produced. As such, it represents a *defect-detection* approach to quality management. The methods of acceptance sampling remain useful and continue to be applied.

Although both statistical process control and acceptance sampling make inferences on the basis of samples, there is a very important difference in their purposes: (1) SPC uses the sampled products to make inferences about *the process* from which they came, while (2) acceptance sampling uses sampled products to make inferences about a larger population of *the products.*

Although statistical process control methods have much to contribute to defect *prevention,* they had the misfortune of appearing during an era that was dominated by defect *detection,* with its requisite array of quality inspectors in white coats whose responsibility was to ferret out defective products and prevent them from leaving the factory. For some companies, that era may have not yet ended, as they continue to place nearly sole reliance on weeding out defective products instead of trying to improve the design and production process that allowed the products to be defective in the first place.

While well intentioned, the traditional philosophy of defect detection has generally involved a corporate atmosphere in which quality is a separate entity within the firm. Not coincidentally, suppliers and workers have tended to be viewed as adversaries instead of partners. Those who aspire to global competitiveness and world-class quality now realize that quality can no longer be considered as an isolated function—rather, it must be integrated throughout the organizational structure and be made central to corporate thinking in all matters.

20.5 Differentiate between statistical process control and acceptance sampling. What is the purpose of each?

20.6 Statistical process control and acceptance sampling both rely on a sampling of products to make inferences. In what important way do the types of inferences differ?

20.7 Contrast the concepts of defect prevention and defect detection as they might apply to each of the following "defects" in society:
a. automobile accidents
b. juvenile delinquency
c. alcoholism

THE EMERGENCE OF TOTAL QUALITY MANAGEMENT (20.3)

The 1950s saw the beginnings of a movement toward *total quality management* (*TQM*), a philosophy that (1) integrates quality into all facets of the organization and (2) focuses on improving the quality of the product by systematically improving the process through which the product is designed, produced, delivered, and supported. The conviction of its believers and the success of its adopters combined to intensify the momentum of the TQM movement.

TQM Pioneers

W. Edwards Deming is the best known and probably most influential pioneer of TQM. During the years immediately following World War II, industry in the United States was prospering and Dr. Deming's ideas did not attract an audience in this country. On the other hand, Japan—with "made in Japan" perceived by many as "cheap," "shoddy," and "junk"—was anxious for any help it could get. In 1950 the Japanese invited Dr. Deming to visit their country and discuss the quality issue. He visited, they listened, and the quality of Japanese products was on its way to becoming the envy of the world. It was not until the 1980s that American companies began giving serious attention to Deming's ideas, and some observers would argue that they did so just in the nick of time.

Dr. Deming popularized a continual-improvement model developed earlier by Walter Shewhart. Dr. Deming's version involved a very slight change in terminology and has become more well known, but the core concept has remained the same: a four-stage repetitive cycle stressing the importance of continuous and systematic quality-improvement efforts at all levels of the organization. The following stage descriptors refer to the Deming adaptation, sometimes referred to as the *Deming cycle* or the PDCA (Plan-Do-Check-Act) cycle.[2] However, in our references to this cycle, we will give the originator his due and refer to it as the *Shewhart-Deming* cycle.

Plan Develop a plan for a possible improvement to the product or the process through which the product is made.

[2]Henry R. Neaves, *The Deming Dimension.* Knoxville: SPC Press, 1990, p. 142.

Do Apply the potential improvement in a small-scale test.

Check Evaluate the change to see whether it was successful.

Act Based on the results: (1) adopt the plan, (2) abandon the plan, or (3) use the information gained to revise the plan and go through the cycle again.

Among the other key pioneers of the TQM movement are Joseph M. Juran, Kaoru Ishikawa, Phillip B. Crosby, and Armand V. Feigenbaum.[3] Like Deming, Juran was instrumental in helping the postwar Japanese improve their product quality. Juran also authored the *Quality Control Handbook*. Ishikawa, author of *Guide to Quality Control,* was an innovator who developed the *cause-and-effect diagram* and the *quality circle,* concepts that are discussed later in the chapter. Crosby is the author of *Quality Is Free,* in which he argues that poor quality is so expensive (see Section 20.1 and the costs of nonconformance) to the firm that even great amounts of money could profitably be spent on moving toward the goal of *zero defects* and doing it right the first time. Feigenbaum's book, *Total Quality Control,* stresses a concept called *quality at the source,* where responsibility for quality is with the individual worker. In this concept, production workers can stop production whenever they spot quality problems.

The Process Orientation

Where traditional quality control tended to be *result oriented,* with primary emphasis on the end product and defect detection, total quality management is *process oriented,* stressing the prevention of defects through continual attention to, and improvements in, the processes through which the product is designed, produced, delivered, and supported. Figure 20.1 illustrates the emphasis that TQM places on understanding and reducing the sources of process variation that were discussed earlier. (*Note:* In TQM, the "satisfied customer" of Figure 20.1 is not necessarily the final retail customer—it could be the next workstation or procedure in the manufacturing sequence.)

Six Sigma and Other Quality Systems

TQM is being complemented and supplemented by other approaches to quality improvement. Perhaps the most popular and best known of these is the **Six Sigma** approach that was developed at Motorola in the 1980s. The name is inspired by the normal distribution and the incredible area beneath the curve between the -6σ and the $+6\sigma$ locations. The remaining area is so miniscule that it reflects only 3.4 defects per million products or opportunities—the result is a process or environment that is practically defect-free.

According to Greg Brue, author of *Six Sigma for Managers,* Six Sigma has a strong profitability and project/problem orientation, and involves five phases: (1) define the project or goal and the internal and external customers involved; (2) measure the current performance of the process; (3) analyze the root causes of the defects generated by the process; (4) improve the process so as to eliminate the defects; and (5) control the future performance of the process. There is also a strong orientation to using available data and the statistical tools for their analysis.[4]

[3]Norman Gaither, *Production and Operations Management* 6/e. Fort Worth: The Dryden Press, 1994, ch. 17.

[4]Greg Brue, *Six Sigma for Managers,* McGraw-Hill, 1988, pp. 6–8.

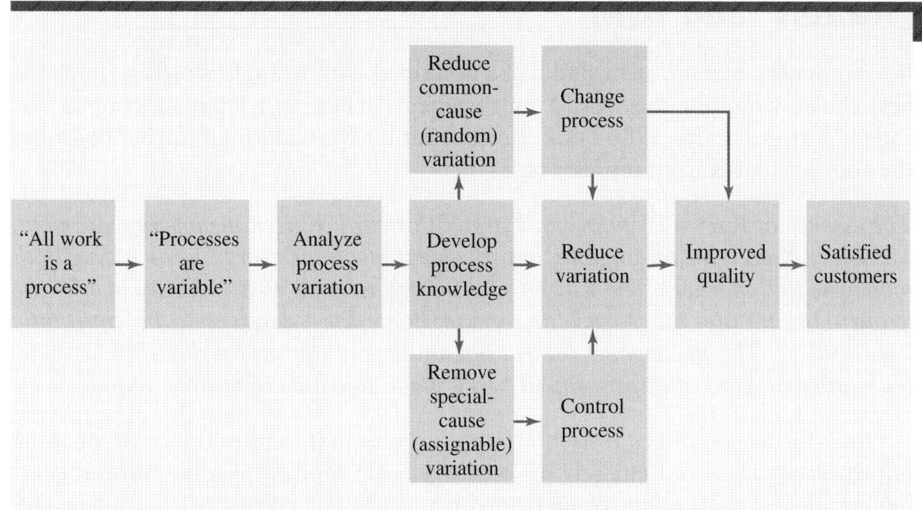

FIGURE 20.1

Total quality management stresses the importance of understanding and reducing variation within the process by which a product or service is generated.

Source: R. Snee, "Statistical Thinking and Its Contribution to Total Quality," *American Statistician,* Vol. 44, pp. 116–121.

A detailed comparison of current systems, whether the Toyota Production System (TPS), TQM, or Six Sigma, would lie beyond the scope of this text. Each system has its proponents and detractors, but they all have the same admirable goal of improving the processes involved in the quality of the goods and services we purchase, and they tend to utilize many of the same statistical tools. See the references in the footnote, or Google "Six Sigma versus TQM" for more information.[5] In the meantime, our emphasis will continue to be on the traditional TQM methods and on the spirit that tends to characterize the chapter-opening vignette, which features a corporation whose quality has been among the most respected in all of industry.

20.8 What is the Shewhart-Deming PDCA cycle, and how is it applicable to total quality management?

20.9 What is the basic difference between total quality management and the traditional approach to quality management?

20.10 What is meant by the statement that total quality management has a "process orientation"?

20.11 What are the major phases involved in the Six Sigma quality system?

PRACTICING TOTAL QUALITY MANAGEMENT (20.4)

As both concept and practice, total quality management is rapidly evolving. As such, it is increasingly difficult to differentiate between (1) strategies and practices that *comprise* TQM and (2) strategies and practices that *are conducive to* TQM. Accordingly, the topics that follow may contain elements of each.

[5]In addition to the work cited in reference 4, see Jeffrey K. Liker, *The Toyota Way,* pp. 135, 252–253, and 295–297; and Mekongcapital.com, "Introduction to Six Sigma," October 15, 2004.

"Kaizen" and TQM

If total quality management had to be described with a single word, perhaps the best choice would be the word "improvement." There is a Japanese counterpart called "Kaizen." In his 1986 book, *Kaizen,* author Masaaki Imai further describes the concept that Kaizen represents:

The essence of Kaizen is simple and straightforward: Kaizen means improvement. Moreover, Kaizen means ongoing improvement involving everyone, including both managers and workers. The Kaizen philosophy assumes that our way of life—be it our working life, our social life, or our home life—deserves to be constantly improved.... The message of the Kaizen strategy is that not a day should go by without some kind of improvement being made somewhere in the company.[6]

Imai further describes Kaizen as a philosophy that focuses on small, ongoing improvements rather than relying on large-scale (and, typically, more abrupt) innovations, such as might come from advances in technology or equipment. According to Imai, Western management has historically relied much more heavily on the latter than on the former, but that is being reversed as more companies move toward total quality management.

Deming's 14 Points

At the core of Dr. Deming's teachings is a set of guidelines that are known as "Deming's 14 points." There were fewer in the 1950s, when he began spreading his message, and the number gradually evolved to the present 14. Table 20.1 gives a condensed version of the points—a complete volume would be necessary to do true justice to all of them, and the interested reader may wish to refer to Henry R. Neaves' *The Deming Dimension* (1990) and William W. Scherkenbach's *The Deming Route to Quality and Productivity* (1988). Each book includes a foreword by Deming and a complete chapter on each one of his 14 points.[7]

Deming was adamant that the interdependence of the 14 points is such that all of them must be adopted as a package if success is to be achieved. The 14 points have an underlying theme of cooperation, teamwork, and the belief that workers truly want to perform their jobs in a quality fashion.

The Quality Audit

It is not unusual today for a company to carry out a quality audit of prospective suppliers in addition to evaluating its own quality efforts. According to Richard J. Schonberger, author of *World Class Manufacturing,* the most important element in an audit of suppliers is *process capability*—that is, the prospective supplier must provide evidence that it can meet the required product specifications. Evidence to this effect sometimes takes the form of studies involving statistical process control charts, and the process must be shown (1) to be in statistical control, with assignable process variation absent, and only random variation remaining, and (2) to have random variation that does not exceed the limits allowed by the product

[6]Masaaki Imai, *Kaizen.* New York: McGraw-Hill, 1986, pp. 3, 6.
[7]Neaves, *The Deming Dimension,* and William W. Scherkenbach, *The Deming Route to Quality and Productivity.* Washington: CEEPress Books, 1988.

TABLE 20.1
Deming's 14 points are at the core of his teachings on quality improvement.

1. "Create constancy of purpose for continual improvement of products and service to society . . . ."
2. "Adopt the new philosophy . . . . We can no longer live with commonly accepted levels of delays, mistakes, defective materials, and defective workmanship."
3. "Eliminate the need for mass inspection as the way of life to achieve quality by building quality into the product in the first place."
4. "End the practice of awarding business solely on the basis of price tag. Instead, require meaningful measures of quality along with price."
5. "Improve constantly and forever every process for planning, production, and service."
6. "Institute modern methods of training on the job for all, including management, to make better use of every employee."
7. "Adopt and institute leadership aimed at helping people to do a better job."
8. "Encourage effective two-way communication and other means to drive out fear throughout the organization so that everybody may work effectively and more productively for the company."
9. "Break down barriers between departments and staff areas."
10. "Eliminate the use of slogans, posters and exhortations for the work force, demanding Zero Defects and new levels of productivity without providing methods."
11. "Eliminate work standards that prescribe quotas for the work force and numerical goals for people in management . . . . Substitute aids and helpful leadership in order to achieve continual improvement of quality and productivity."
12. "Remove the barriers that rob hourly workers, and people in management, of their right to pride of workmanship."
13. "Institute a vigorous program of education, and encourage self-improvement for everyone. What an organization needs is not just good people; it needs people that are improving with education. Advances in competitive position will have their roots in knowledge."
14. "Clearly define top management's permanent commitment to ever-improving quality and productivity, and their obligation to implement all of these principles."

Source: W. Edwards Deming, in Henry R. Neaves, *The Deming Dimension,* Knoxville: SPC Press, 1990, pp. 287–413.

specifications. Statistical process control and process capability are discussed in Sections 20.6–20.9.

In addition to evaluating process capability, quality audits may involve the evaluation of other areas, such as employee quality training, the extent and nature of quality improvement efforts, the extent to which the firm applies statistical tools for TQM (discussed in Sections 20.5–20.9), and has an award structure to recognize and reward employees who make special contributions to quality improvement. On a more general level, the goal of a quality audit might simply be described as the assessment of the firm's strategies and successes regarding defect prevention.

Competitive Benchmarking

One popular approach to quality improvement is to engage in *benchmarking*. This involves studying, and attempting to emulate, the strategies and practices of organizations already known to generate world-class products and services. Depending on the specific type of excellence in which a firm happens to be interested, there may be one or several companies recognized as quality leaders.

Here are just a few examples: Institutions of higher education may wish to compare themselves against leading institutions within their category as listed in the annual *America's Best Colleges* and *America's Best Graduate Schools* publications from *U.S. News & World Report*. Companies wishing to improve their performance in employee enthusiasm and training would do well to compare themselves to the Walt Disney Company, and those engaged in Internet and mail-order activities might wish to study the excellence long exhibited by L.L. Bean.

Just-in-Time Manufacturing

Just-in-time (JIT) manufacturing systems operate with a minimum of inventory on hand, with materials and components literally arriving "just in time" to enter the production process. Some have the misconception that JIT is simply an inventory philosophy—actually, it is a *production* philosophy that happens to have an impact on inventory. Unlike systems where there are large inventory stocks, when a problem occurs in the JIT system, production is stopped until the problem is fixed. This approach fits in well with the philosophy held by Deming and others that the consumer must always be considered as the very next person or workstation in the manufacturing sequence.

Although JIT systems save money by reducing in-process inventory costs, there are quality benefits as well. Workers and suppliers become very conscious of product quality—with zero or minimal parts on hand to buffer irregularities in the production pace, the worker, machine, or supplier generating poor quality has nowhere to hide. Because production comes to a halt until quality has been restored, any problem that arises will be the object of serious efforts not only to momentarily solve it, but to permanently eliminate it. JIT also encourages small production lots, and mistakes that do occur are more quickly identified and corrected than in conventional systems that have large production runs before shipment or transfer to either the final customer or the next workstation. Through its effectiveness in preventing defects, the JIT system helps reduce the costs of nonconformance discussed in Section 20.1.

Worker Empowerment

Also called *quality at the source,* worker empowerment (1) makes the individual worker *responsible* for the quality of output from his or her own workstation, and (2) provides the worker with the *authority* to make changes deemed necessary for quality improvement. When workers have both responsibility and authority for quality, they can apply the Shewhart-Deming PDCA cycle at their own level, and statistical tools like those discussed in Sections 20.5–20.9 may be used or adapted for the worker's use at the workstation. The worker also has the authority to halt production when he spots a quality problem either in his own output or in components coming from a workstation earlier in the production sequence. Quality circles (see Statistics in Action 20.2) often play an integral role in the application of the worker empowerment philosophy.

STATISTICS IN ACTION 20.2

Quality Circles and the "Suggestion Box"

The typical "suggestion box" has included ideas that improved quality or productivity, ideas that were impractical to implement, and ideas that would have been physiologically impossible for the boss to perform. Although the box has not completely disappeared, its more useful functions have been taken over, and improved, by the quality circle.

The quality circle helps bring workers into the task of improving and maintaining product quality. An especially valued technique in Japan, the quality circle consists of about a dozen workers from the same department. Typically volunteers, they meet periodically to discuss issues and ideas pertaining to the quality of both their workplace and its output.

Ideas generated by the quality circle contribute to the reduction of both random and assignable variation in a process. Although they possess neither MBA nor engineering degrees, workers are closer to the production process than anyone else. When an intelligent person spends this much time in proximity to a machine or a process, familiarity with its capabilities and quirks is bound to develop, and it isn't surprising that workers can often recognize or solve problems that have evaded the efforts of those who are more highly educated in the conventional sense.

The quality circle is more than just a talk session held twice a month. The group may initiate specific studies and have access to company records pertinent to the problem at hand. This gives the worker both enhanced importance and an enhanced *sense* of importance. The quality circle helps raise workers from the simple status of "hired help" to a position where they are valued members of a quality team that has the enthusiastic support of management at the highest levels.

Awards and the Recognition of Quality

Besides enjoying the more direct benefits from offering world-class products and services, companies practicing TQM are increasingly being motivated by the challenge of being *recognized* for their quality achievements. One of the first such awards was the Deming Prize, established by the Japanese government in 1951 and awarded annually to companies and individuals that have distinguished themselves in the quality-management arena. The Malcolm Baldrige National Quality Award is awarded annually to U.S. companies that have demonstrated excellence in quality achievement and management, with award categories including manufacturing, service, small business, education, healthcare, and nonprofit.

EXERCISES

20.12 What is "Kaizen," and how is it related to total quality management?

20.13 In the Kaizen philosophy, what is the difference between "ongoing improvement" and "innovation"?

20.14 Give some specific examples of how the Kaizen concept might be applied (a) by a clerk at a convenience store, and (b) by the manager of an apartment that rents to university students.

20.15 What is a quality audit, and for what purposes might one be performed?

20.16 What is process capability, and how can its presence be ascertained?

20.17 What is competitive benchmarking, and how can it be used?

20.18 What is just-in-time manufacturing, and how does it contribute to total quality management?

20.19 "JIT manufacturing is just a gimmick for reducing inventory costs; it has nothing to do with quality." Discuss the preceding statement.

20.20 In what ways can large production runs and high inventory levels adversely affect product quality?

20.21 What two elements are necessary for worker empowerment to be a successful tool in total quality management?

20.5 SOME STATISTICAL TOOLS FOR TOTAL QUALITY MANAGEMENT

Many statistical tools can be applied to total quality management, and several are briefly described in this section. For the interested reader, more complete discussion of these and other tools can be found in relevant works, such as Kaoru Ishikawa's *Guide to Quality Control* and William W. Scherkenbach's *The Deming Route to Quality and Productivity*. Sections 20.6–20.9 are devoted to an especially popular and important statistical tool for total quality management: the statistical process control chart.

The Process Flow Chart

The *process flow chart* visually depicts the operations (or subprocesses) involved in a process, showing the sequential relationships between them. The process flow chart contributes to our understanding of the process as well as our efforts to improve it. By showing the operations in the context in which they are active in the overall process, the flow diagram facilitates discussion and data collection for each individual operation—in essence, each operation can be considered as a process itself, and the object of efforts employing the Shewhart-Deming PDCA cycle for continuing improvement. Figure 20.1, presented earlier in the context of the process orientation, can be viewed as a process flow chart for the study and reduction of sources of variation in a process. Figure 20.2 shows a process flow chart for visiting a video rental store and renting a movie. Note that activities are shown in rectangular boxes, while decisions are represented by diamond-shaped boxes.

The Cause-and-Effect Diagram

The *cause-and-effect diagram*, developed by Kaoru Ishikawa, assists problem solving by visually depicting effects and their possible causes. Because of its appearance, it is sometimes called a *fishbone diagram*. An example of a cause-and-effect diagram for an automotive tire manufacturer is shown in Figure 20.3, and

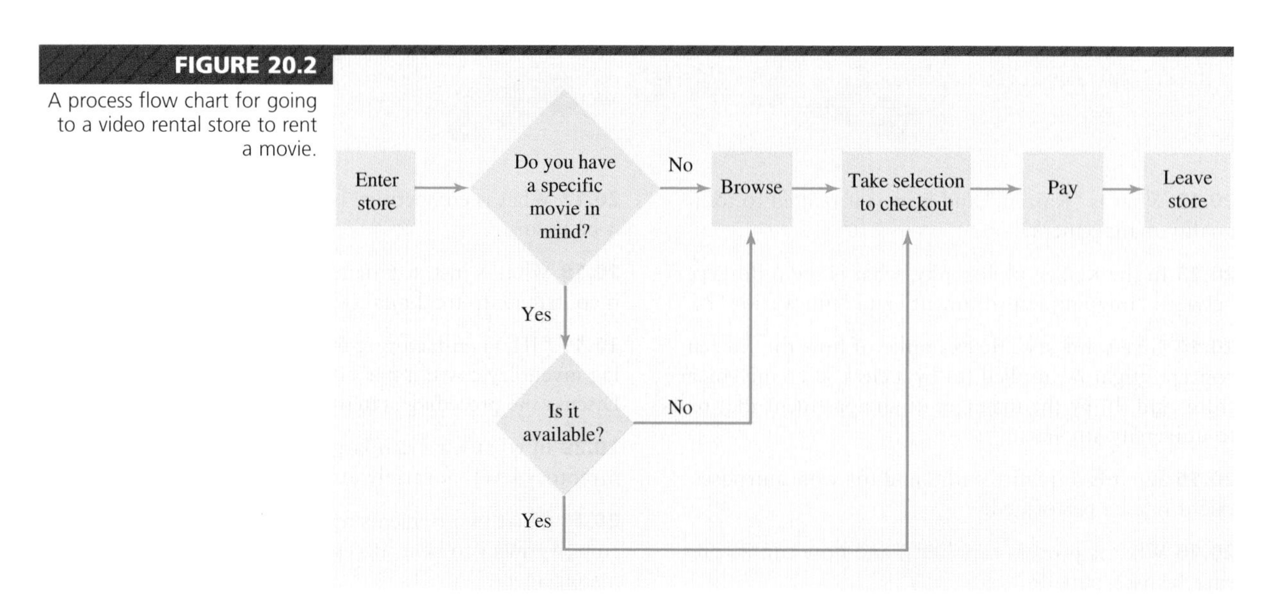

FIGURE 20.2
A process flow chart for going to a video rental store to rent a movie.

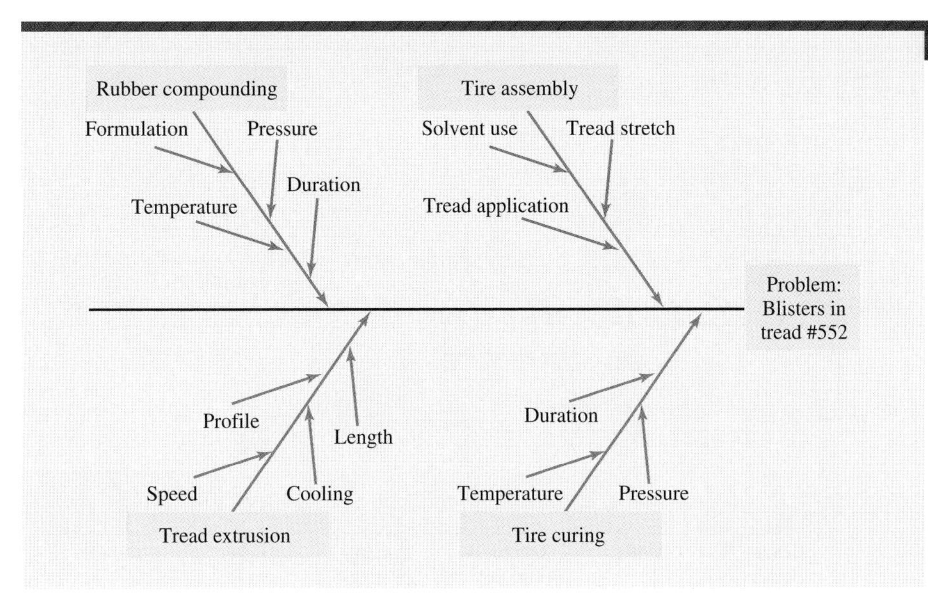

FIGURE 20.3

Because of its shape, the cause-and-effect diagram is sometimes called a *fishbone* diagram.

Source: Norman Gaither, *Production and Operations Management,* 6/e, Fort Worth: The Dryden Press, 1994, ch. 17.

it includes possible causative factors leading to blistering in the tread portion of the tire.

As Figure 20.3 shows, the cause-and-effect diagram depicts various causes at the left that feed directionally toward the ultimate effect at the far right. The causes will typically be associated with people, methods, materials, machines, or work environment, and the diagram can include categories such as the four subprocesses in Figure 20.3.

Either the diagram or the problem-solving sessions that it facilitates can become more complex when we consider that causes themselves have causes. By continually asking "*why?*" in a procedure similar to peeling back the layers of an onion, it is often possible to identify a root cause that lies at the core of the visible problem. The following series of "why?" questions is attributed to Taiichi Ohno, former vice president of Toyota Motor.[8]

Question 1: Why did the machine stop?
Answer 1: Because the fuse blew due to an overload.

Question 2: Why was there an overload?
Answer 2: Because the bearing lubrication was inadequate.

Question 3: Why was the lubrication inadequate?
Answer 3: Because the lubrication pump wasn't working properly.

Question 4: Why wasn't the lubrication pump working properly?
Answer 4: Because the pump axle was worn out.

Question 5: Why was the pump axle worn out?
Answer 5: Because sludge got in.

The permanent solution to this machine-stoppage problem ended up being a strainer to prevent dirt from getting into the lubrication pump. Had the question "why?" not been asked five times, the solution might have been a stopgap measure such as a replacement fuse.

[8]Masaaki Imai, *Kaizen.* New York: McGraw-Hill, 1986, p. 50.

FIGURE 20.4

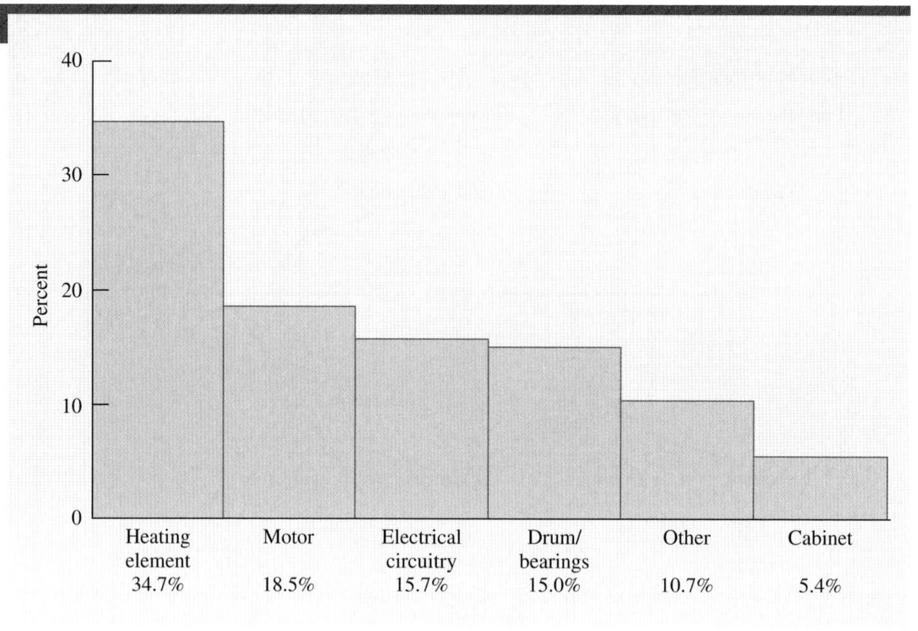

This Pareto diagram for warranty repairs for electric clothes dryers indicates that the heating element has the greatest single potential for quality improvement.

The Pareto Diagram

The *Pareto diagram,* also known as the *Juran diagram,* is a bar chart describing the relative frequencies with which various kinds of defects, or nonconformities, have occurred. The diagram contributes to the efficiency and effectiveness of quality management efforts by helping management and workers concentrate their limited resources on areas offering the greatest potential for improvement.

The usefulness of the Pareto diagram is underscored by a guideline that some refer to as the "80/20 rule"—that 80% of the problems come from only 20% of the problem areas. Naturally, the exact breakdown will not always be 80/20, but the general idea is to use the Pareto diagram for guidance in attacking those few areas where most of the problems have occurred. Figure 20.4 shows a Pareto diagram of warranty repairs for electric clothes dryers. The greatest potential for gains in quality appear to be associated with improving the heating element and the electric motor.

The Check Sheet

The *check sheet* can be likened to the "foot soldier" in the military. Often no more than a sheet of paper on which to tally frequency counts or record measurements, it is simply a mechanism for entering and retaining data that may be the subject of further analysis by another statistical tool. The continuing improvements sought by the Shewhart-Deming PDCA cycle and the Kaizen concept rely heavily on quantitative guidance, and companies that practice TQM listen most closely whenever advice or proposals are accompanied by supporting data.

Taguchi Methods

Traditional methods of evaluating quality are of a binary, yes/no nature—either a product is within specifications or it is not. For example, if a specification requires that a product be within 0.016 inches of 14.000 inches, an individual unit will be

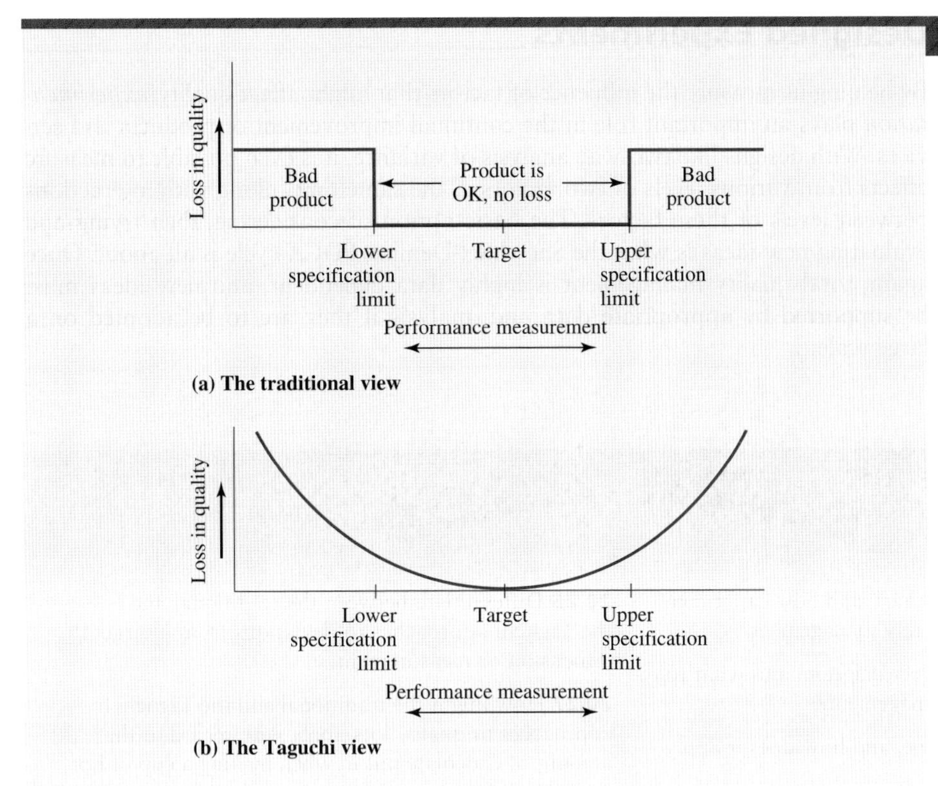

FIGURE 20.5

While traditional specification limits (a) view a product as either "good" or "bad," the Taguchi method (b) goes beyond this to recognize loss in quality as a dimension or measurement deviates farther from the target.

judged acceptable as long as it is no shorter than 13.984 inches and no longer than 14.016 inches. In this situation, 14.000 inches is the *target* and ±0.016 inches is the *tolerance*.

Engineer Genichi Taguchi suggested that the target measurement is an ideal or optimum, and that greater deviations from the target represent greater decreases in quality—in other words, one individual unit can be higher in quality than another, even though both are within the traditional specifications. Parts (a) and (b) of Figure 20.5 illustrate how the *Taguchi method* differs from the traditional approach to quality evaluation. The curve in part (b) is referred to as the *Taguchi loss function,* and it shows how quality is *lost* as the measurement gets farther away from the target, or optimum value.

As an example of how the Taguchi method applies, consider the temperature of a swimming pool. If the most comfortable water temperature is 80° Fahrenheit and a tolerance of ±5° is deemed acceptable to swimmers, a pool manager might tell personnel to make sure the temperature stays within the 75° to 85° range. On the other hand, swimmers do not suddenly become uncomfortable as the temperature drops from 75.1° to 74.9°—75.1° is a bit uncomfortable as well. In this situation, the Taguchi method would recognize that people experience greater discomfort (greater quality loss) as the temperature gets farther away from the optimum.

The Taguchi method can be applied to individual components as well as to manufactured products consisting of a great many components. A product is the totality of its parts, and some products can be of higher quality than others—even though all of their component parts might be within their respective traditional specifications. For example, chances are you will be much happier with your car if each of its approximately 15,000 parts is close to its ideal specification instead of just barely falling within the limits of acceptability.

Designed Experiments

By helping us measure the influence of factors that might affect quality, *experimentation* plays an important role in the continual improvement of products and services. With designs like two-way analysis of variance, it is even possible to measure effects from various levels of two factors at the same time, along with interactions between levels of these factors. The basic notion of considering, then trying and evaluating new ideas is what the Shewhart-Deming PDCA cycle is all about. Once again, total quality management is highly data dependent, and new ideas must be supported by appropriate data and analysis if they are to be adopted on a large scale.

EXERCISES

20.22 What is a process flow chart, and how can it contribute to the goals of total quality management?

20.23 What is a cause-and-effect diagram, and what role can it play in total quality management?

20.24 What is the "80/20" rule, and how does it apply to total quality management?

20.25 What is a "check sheet," and what is its role in total quality management?

20.26 Differentiate between the traditional and the Taguchi approaches to deviations from a target dimension or measurement.

20.27 How might the traditional and the Taguchi approaches to quality loss apply with regard to the amount of chocolate put in when making a cup of hot chocolate?

20.28 How can designed experiments facilitate the application of the Shewhart-Deming PDCA cycle?

20.6 STATISTICAL PROCESS CONTROL: THE CONCEPTS

On the highway, you and your car have probably been shaken by "rumble strips," harsh grooves across the road surface that warn you when a construction area or a dangerous intersection is about to appear. Imagine a long, straight road with two continuous and narrow sets of rumble strips—one at each edge of the lane in which you're traveling. If you were blindfolded, you could still stay in your lane. The strategy would be simple: adjust to the left when you touch the right-hand rumble strip, adjust to the right when you encounter the strip on the left. Such a highway would be analogous to the control chart.

Control charts were developed to keep track of ongoing processes and to signal when such a process had reached the point of going "out of control," much like the parallel set of rumble strips on our hypothetical highway. Instead of rumble strips, the control chart has upper and lower **control limits** for the sample statistics provided by each of successive samples taken over time. When nonrandom, or directional, variation causes these measurements to become unusually high or low, the upper or the lower control limit will be violated. In this event, the control chart will constitute a warning signal that the process may need corrective attention. The control chart cannot *make* the correction. It serves only as a warning that something needs to be done.

Types of Control Charts

Control charts apply to either variables or attributes, depending on the type of process and how it is being monitored. The distinction is as follows:

CONTROL CHART FOR VARIABLES **This type of control chart is appropriate whenever we are interested in monitoring measurements, such as weights, diameters, thicknesses, or temperatures. The next section of the chapter deals with control charts for variables, and includes those for sample means and for sample ranges.**

CONTROL CHART FOR ATTRIBUTES **This type of control chart applies when we are *counting* something associated with the process output. Section 20.8 discusses control charts for attributes, including those for the proportion of defectives (the *p*-chart) and the number of defects per unit of output (the *c*-chart).**

The Control Chart Format

Figure 20.6 shows the typical format of a control chart. As successive samples are drawn from the process, the results are plotted and the chart indicates whether the process is in control. While random variation may occur from one sample to the next, as long as sample results stay within the upper and lower control limits, there is deemed to be no cause for corrective action. However, if a sample result does not fall within the upper and lower control limits, the process is said to be *out of control*. In other words, we conclude that the sample statistic is too extreme to have been merely the result of random, or sampling, error.

If all this seems familiar, it's because the control chart is closely related to the concept of hypothesis testing. The upper and lower control limits are analogous to the upper and lower critical values that we determined when conducting a two-tail test involving a sample mean or proportion. From the standpoint of hypothesis testing, the null hypothesis is "The process is in control." As in hypothesis testing, there can be errors. For example, a process may be deemed to be in control when in fact it is not (Type II error), or it may be incorrectly judged to be out of control (Type I error).

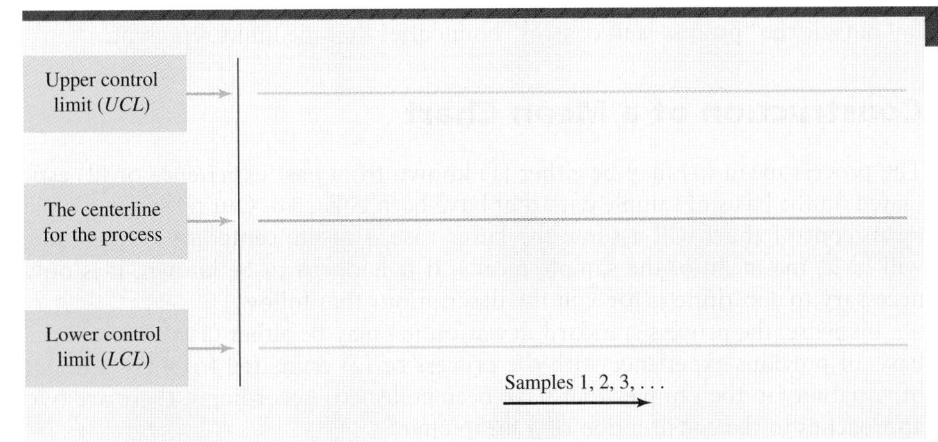

FIGURE 20.6

The general format of a control chart. If, in the succession of samples, the sample statistic (e.g., the sample mean) does not fall within the upper and lower control limits, the process is considered to be out of control.

EXERCISES

20.29 What is a control chart, and what purpose does it serve?

20.30 Differentiate between a variable and an attribute.

20.31 What function is performed by a control chart's upper and lower limits?

20.32 In terms of random variation and assignable variation, what is meant by the statement, "The process is in control"?

20.33 With respect to the null hypothesis, "H_0: The process is in control," explain the distinction between a Type I error and a Type II error.

(20.7) CONTROL CHARTS FOR VARIABLES

The *mean* and *range* charts monitor the central tendency and variability, respectively, of a process. These control charts are interrelated, and the range in successive samples can provide useful information for constructing the mean chart for the same process. The mean chart is more closely linked to our previous work with hypothesis testing, and it will be examined first.

Mean Charts

The **mean chart,** or "$\bar{x}$ chart," relies on the normal distribution and the central limit theorem. Introduced in Chapter 8, the central limit theorem holds that the means of simple random samples from any population will tend to be approximately normally distributed, with the distribution of sample means more closely approaching the normal distribution as the sample size is increased. Also, when the population is normally distributed, the sample means will tend to be normally distributed even for relatively small sample sizes. This latter point, which was covered in Chapter 8, is especially important in statistical process control, where it is common practice to assume that x (i.e., the width, diameter, weight, or thickness) is normally distributed.

In the mean chart, upper and lower control limits are usually selected as "3-sigma," so called because they represent the area under the normal curve that corresponds to ± 3 standard deviation units from the mean. This area is 0.9974, so there is only a $(1.0000 - 0.9974) = 0.0026$ chance the mean of a given sample will fall outside the "process is in control" boundaries that the limits represent.

Construction of a Mean Chart

The process mean (μ) may be either (1) known from past experience or (2) estimated on the basis of sample data that have been collected. Our treatment of the mean control chart will assume the latter case, and the **centerline** of the chart will be $\bar{\bar{x}}$, the mean of the sample means. If μ happens to be known, it is only necessary to substitute μ for $\bar{\bar{x}}$ in the descriptions that follow.

Likewise, the process standard deviation (σ) may be either (1) known on the basis of previous experience with the process or (2) estimated from the amount of variability in the samples that have been collected. Accordingly, there are two approaches to the construction of a mean chart:

The mean chart when the process standard deviation (σ) is known:

- Centerline = $\bar{\bar{x}}$, the mean of the sample means.
- Upper and lower control limits (*UCL* and *LCL*):

$$\bar{\bar{x}} \pm 3\left(\frac{\sigma}{\sqrt{n}}\right)$$

where σ = the process standard deviation
n = the sample size
$\sigma/\sqrt{n}$ = the standard error of the sample means

EXAMPLE

Mean Chart, Sigma Known

A machine produces aluminum cylinders, the diameters of which are approximately normally distributed. Twenty samples, each with $n = 5$, have been taken, and the mean of the sample means is a diameter of $\bar{\bar{x}} = 2.1350$ inches. Based on past experience, the process standard deviation is known to be $\sigma = 0.002$ inches.

SOLUTION

- If further samples are taken, what are the upper and lower 3-sigma control limits for determining whether the process is in control?

For $\sigma = 0.002$ inches and $n = 5$, the upper and lower control limits will be

$$\bar{\bar{x}} \pm 3\left(\frac{\sigma}{\sqrt{n}}\right) = 2.1350 \pm 3\left(\frac{0.002}{\sqrt{5}}\right) = 2.1350 \pm 3(0.0009)$$

or 2.1350 ± 0.0027, or 2.1323 (*LCL*) and 2.1377 (*UCL*) inches.

Provided that a sample mean is no less than 2.1323 inches and no greater than 2.1377 inches, the process is considered to be in control.

- If the process is in control, what is the probability that the next sample taken will falsely signal that the process is out of control?

The 3-sigma limits include 99.74% of the area under the normal curve. As long as the process remains in control, the probability is $1.0000 - 0.9974$, or 0.0026, that random variation alone will result in a given sample mean incorrectly indicating that the process has gone out of control.

EXAMPLEEXAMPLEEXAMPLEEXAMPLEEXAMPLE

If the process standard deviation is not known, the approach is similar, except that the upper and lower control limits are determined by using the average range ($\bar{R}$) of the samples as a measure of process variability. For simplicity, the procedure also involves the application of a special factor (A_2) obtained from the 3-sigma control chart factors table in Appendix A of the text. The table is reproduced here as Table 20.2. Note that the value of A_2 depends on n, the size of each

TABLE 20.2

The 3-sigma control chart factors table from Appendix A. A_2 depends on the size of the individual samples and is used in constructing the mean chart. Factors D_3 and D_4, also dependent on the value of n, are used in making the range chart.

Number of Observations in Each Sample, n	Factor for Determining Control Limits, Control Chart for the Mean, A_2	Factors for Determining Control Limits, Control Chart for the Range D_3	D_4
2	1.880	0	3.267
3	1.023	0	2.575
4	0.729	0	2.282
5	0.577	0	2.115
6	0.483	0	2.004
7	0.419	0.076	1.924
8	0.373	0.136	1.864
9	0.337	0.184	1.816
10	0.308	0.223	1.777
11	0.285	0.256	1.744
12	0.266	0.284	1.716
13	0.249	0.308	1.692
14	0.235	0.329	1.671
15	0.223	0.348	1.652

Source: From E. S. Pearson, "The Percentage Limits for the Distribution of Range in Samples from a Normal Population," *Biometrika* 24 (1932): 416.

sample, and that A_2 becomes smaller as n increases. Table 20.2 also includes two additional factors. These are used in constructing the range chart, covered in the next section.

The mean chart when the process standard deviation (σ) is not known:

- **Centerline = $\bar{\bar{x}}$, the mean of the sample means.**
- **Upper and lower control limits (*UCL* and *LCL*):**

$$\bar{\bar{x}} \pm A_2\bar{R} \quad \text{where } \bar{R} = \text{the average range for the samples}$$
$$A_2 = \text{value from the 3-sigma control chart factors table}$$

EXAMPLE

Mean Chart, Sigma Unknown

During the first few hours of a production run, 20 samples have been drawn from the output of a machine that produces steel rivets for industrial construction. The samples were taken at ten-minute intervals, and each sample consisted of $n = 4$ rivets, the lengths of which were measured. The results are shown in Table 20.3 and are also listed in file CX20RIVS.

EXAMPLEEXAMPLEEXA

Sample #	Item 1	Item 2	Item 3	Item 4	$\bar{x}$	Range
1	1.405	1.419	1.377	1.400	1.400	0.042
2	1.407	1.397	1.377	1.393	1.394	0.030
3	1.385	1.392	1.399	1.392	1.392	0.014
4	1.386	1.419	1.387	1.417	1.402	0.033
5	1.382	1.391	1.390	1.397	1.390	0.015
6	1.404	1.406	1.404	1.402	1.404	0.004
7	1.409	1.386	1.399	1.403	1.399	0.023
8	1.399	1.382	1.389	1.410	1.395	0.028
9	1.408	1.411	1.394	1.388	1.400	0.023
10	1.399	1.421	1.400	1.407	1.407	0.022
11	1.394	1.397	1.396	1.409	1.399	0.015
12	1.409	1.389	1.398	1.399	1.399	0.020
13	1.405	1.387	1.399	1.393	1.396	0.018
14	1.390	1.410	1.388	1.384	1.393	0.026
15	1.393	1.403	1.387	1.415	1.400	0.028
16	1.413	1.390	1.395	1.411	1.402	0.023
17	1.410	1.415	1.392	1.397	1.404	0.023
18	1.407	1.386	1.396	1.393	1.396	0.021
19	1.411	1.406	1.392	1.387	1.399	0.024
20	1.404	1.396	1.391	1.390	1.395	0.014
					$\bar{\bar{x}} = 1.398$	$\bar{R} = 0.022$

TABLE 20.3

Results from 20 samples, each with $n = 4$. The mean of the sample means is 1.398 inches and the mean of the sample ranges is 0.022 inches. This information, also listed in file **CX20RIVS**, is the basis for the Excel and Minitab mean charts generated in Computer Solutions 20.1.

SOLUTION

In determining the upper and lower control limits, several preliminary calculations are needed. As Table 20.3 shows, the mean of the 20 sample means is $\bar{\bar{x}} = 1.398$ inches. Since the process standard deviation is not known, we will use the average range and the appropriate factor from Table 20.2 to construct the mean control chart. The column at the far right in Table 20.3 shows the range for each sample. For example, the range in sample number 3 was the difference between the largest and smallest rivet, or $(1.399 - 1.385) = 0.014$ inches. At the bottom of the table, the average range for these samples has been calculated as $\bar{R} = 0.022$ inches.

Given that the size of each of the samples was $n = 4$ rivets, we refer to Table 20.2 and find the appropriate factor to be $A_2 = 0.729$. Having already calculated $\bar{\bar{x}} = 1.398$ inches and $\bar{R} = 0.022$ inches, the upper and lower control limits will be

$$\bar{\bar{x}} \pm A_2\bar{R} = 1.398 \pm (0.729)(0.022) = 1.398 \pm 0.016$$

and the control limits are 1.382 inches (LCL) and 1.414 inches (UCL).

EXAMPLEEXAMPLEEXAMPLEEXAMPLEEXAMPLEEXAMPLE

Computer Solutions 20.1 describes the procedures and shows the results when Excel and Minitab are used to generate the mean chart for the data in Table 20.3. Each plot shows the upper and lower control limits, the centerline ($\bar{\bar{x}} = 1.398$ inches), and the sequence of 20 sample means. All of the sample means are within the control limits, and the process is judged to be under control. If this were not the case, it would be necessary to identify the source of the assignable, or nonrandom, variation and take corrective action.

COMPUTER 20.1 SOLUTIONS

Mean Chart

In these procedures, the process standard deviation is assumed to be unknown and is estimated on the basis of the sample ranges.

EXCEL

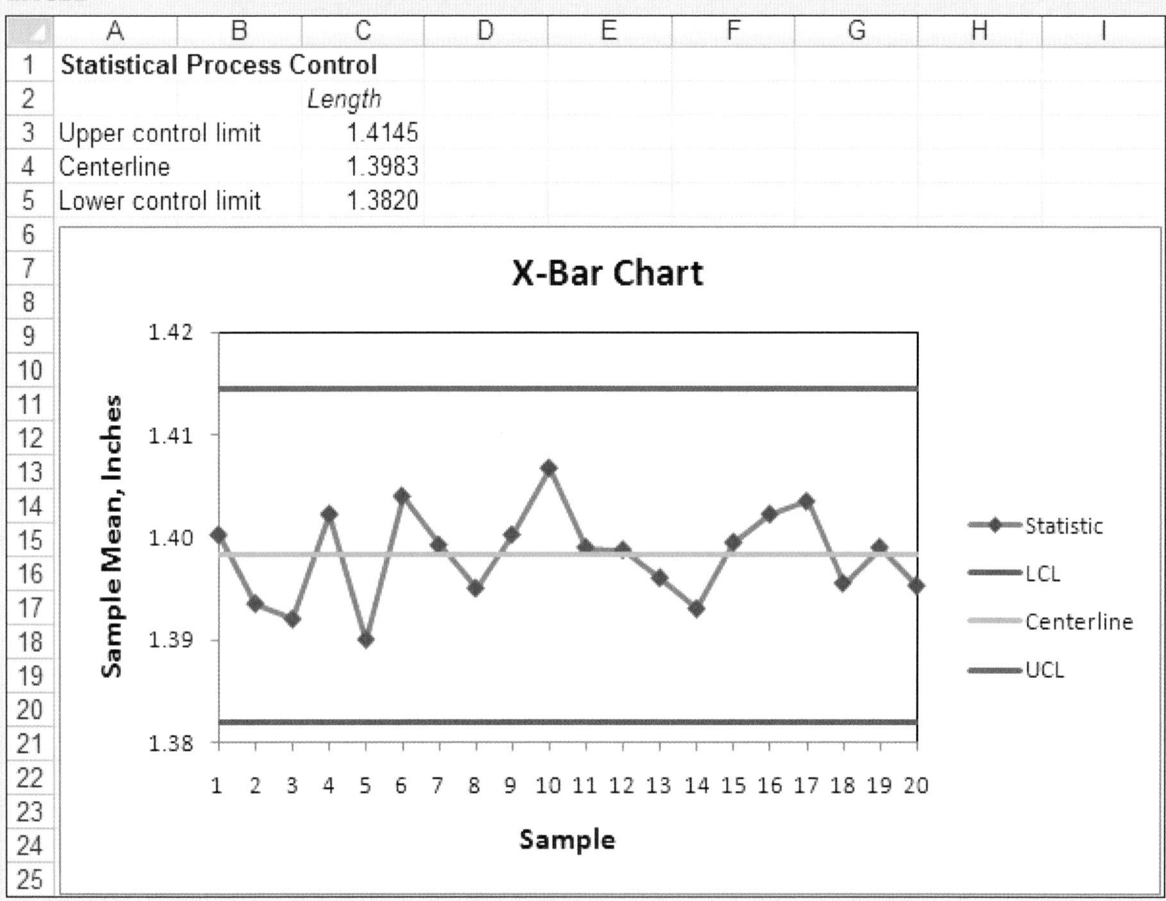

Excel Mean Chart, Sigma Unknown

1. In Excel file **CX20RIVS**, the label is in A1 and the sequence of rivet lengths is in A2:A81. From the **Add-Ins** ribbon, click **Data Analysis Plus.** Select **Statistical Process Control.** Click **OK.**
2. Enter **A1:A81** into the **Input Range** box. Click **Labels.** Enter **4** into the **Sample size** box. Select **XBAR (Using R).** Click **OK.** This provides the mean chart for the 20 samples in Table 20.3. (*Note:* As an alternative, you can use Excel worksheet template **TMSPCMR**. The instructions are within the template.)

MINITAB

Minitab Mean Chart, Sigma Unknown

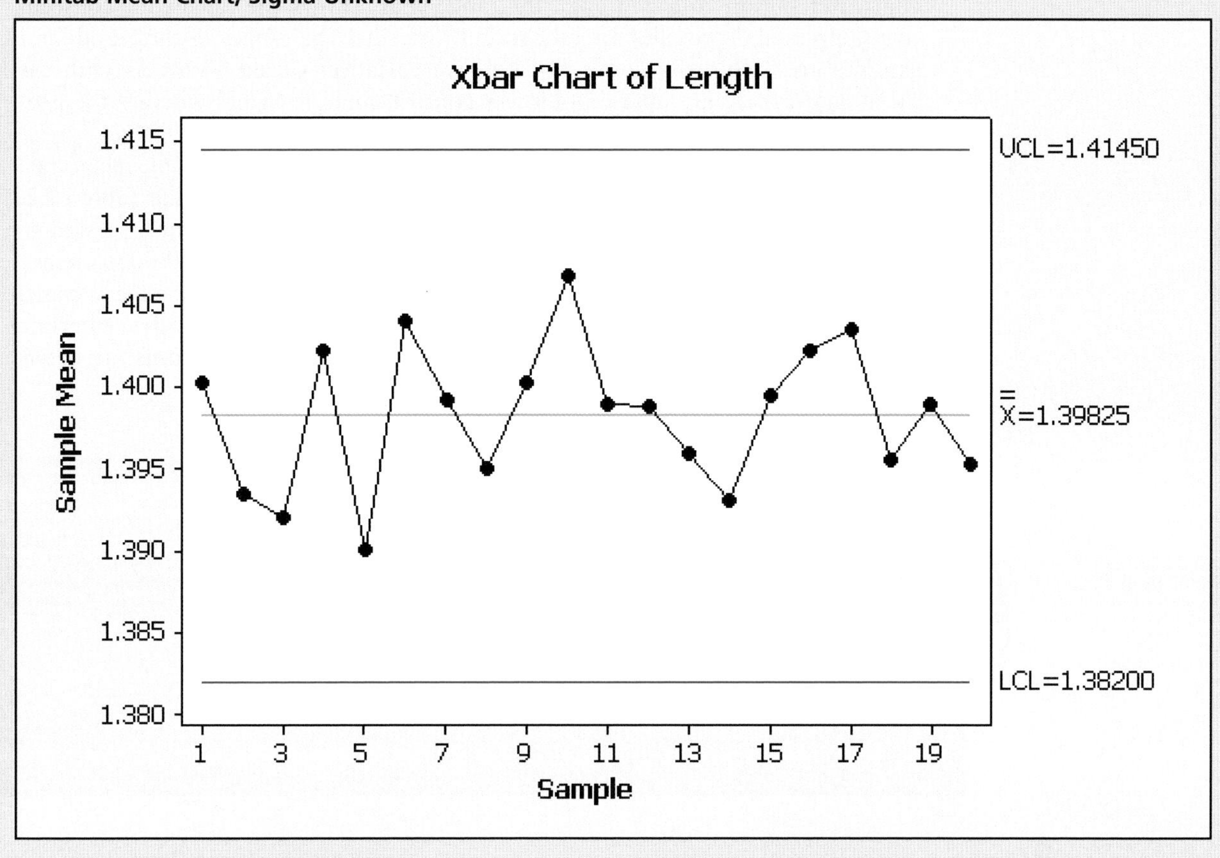

1. Using Minitab file CX20RIVS, with the 80 rivet lengths in C1: Click **Stat.** Select **Control Charts.** Select **Variables Charts for Subgroups.** Click **Xbar.**
2. Select **All observations for a chart are in one column** and enter **C1** into the box. Enter **4** (the size of each sample) into the **Subgroup sizes** box.
3. Click **Xbar Options.** Click **Estimate** and select **Rbar** in the **Method for Estimating Standard Deviation** menu. Click **OK.** Click **OK.** The mean chart appears as shown here.

Note: If the process standard deviation, σ, is known, the mean chart with known σ can be easily generated. In step 2, above, just click **Xbar Options,** click **Parameters,** and enter the known process standard deviation into the **Historical standard deviation** box.

An important note: The procedures in Computer Solutions 20.1 make no assumptions about the process standard deviation, and estimate it based on the sample ranges. In this chapter, our emphasis is on the so-called unknown-sigma mean chart available from both Excel and Minitab. However, as pointed out in the Minitab portion of Computer Solutions 20.1, Minitab is also capable of generating a mean chart when we do happen to know the process standard deviation.

Range Charts

The purpose of the **range chart** (or "R chart") is similar to that of the mean chart. Relying on collected data, the range chart shows the ranges for consecutive samples and provides an indication of whether the ranges of these and later samples are within the limits that random variation would allow. As with the mean chart, there are upper and lower control limits to signal whether the process is in control.

Although there is a counterpart to the standard error of the sample mean (i.e., σ_R instead of $\sigma_{\bar{x}}$), it is again simpler to refer to a statistical table like Table 20.2. Using the appropriate factors from this table, the result will again be a "3-sigma" chart. If the process is in control, any given sample will have just a 0.0026 chance of violating the chart's limits due to random variation alone. Like the mean chart, the range chart is based on successive samples of the same size, but the centerline will be the average range ($\bar{R}$), and the upper and lower control limits are determined as follows:

The range chart:

- **Centerline = $\bar{R}$, the average of the sample ranges.**
- **Upper and lower control limits:**

$UCL = D_4\bar{R}$ where $\bar{R}$ = the average range for the samples

$LCL = D_3\bar{R}$ D_3 and D_4 = values from 3-sigma control chart factors table

As an example of the construction of a range chart, we'll begin with the information from the 20 samples shown in Table 20.3. The average value for the range has already been calculated as $\bar{R} = 0.022$ inches, and this will be the centerline for the control chart. Referring to the "$n = 4$" row of Table 20.2, we find the factors for determining the upper and lower control limits to be $D_4 = 2.282$ and $D_3 = 0$, respectively. The control chart limits are

$$UCL = D_4\bar{R}, \text{ or } 2.282(0.022) = 0.050 \text{ inches}$$

$$LCL = D_3\bar{R}, \text{ or } 0.000(0.022) = 0.000 \text{ inches}$$

Computer Solutions 20.2 describes the procedures and shows the results when Excel and Minitab are used to generate a range chart for the data in Table 20.3. Each plot shows the upper and lower control limits, the centerline ($\bar{R} = 0.022$ inches), and the sequence of 20 sample ranges. All of the sample ranges are within the control limits, and the process appears to be in control.

In the range charts in Computer Solutions 20.2, note that the lower control limit is zero. As shown in Table 20.2, the value of D_3 is zero whenever $n \leq 6$, so the lower control limit will always be zero for these sample sizes. As noted in the Minitab portion of Computer Solutions 20.2, Minitab can also generate a range chart if the process standard deviation happens to be known.

For a given series of samples, the mean and range charts should be used together in determining whether the process is in control. Either chart (or both) may signal an "out of control" condition. In part (a) of Figure 20.7, the sample

COMPUTER 20.2 SOLUTIONS

Range Chart

In these procedures, the process standard deviation is assumed to be unknown and is estimated on the basis of the sample ranges.

EXCEL

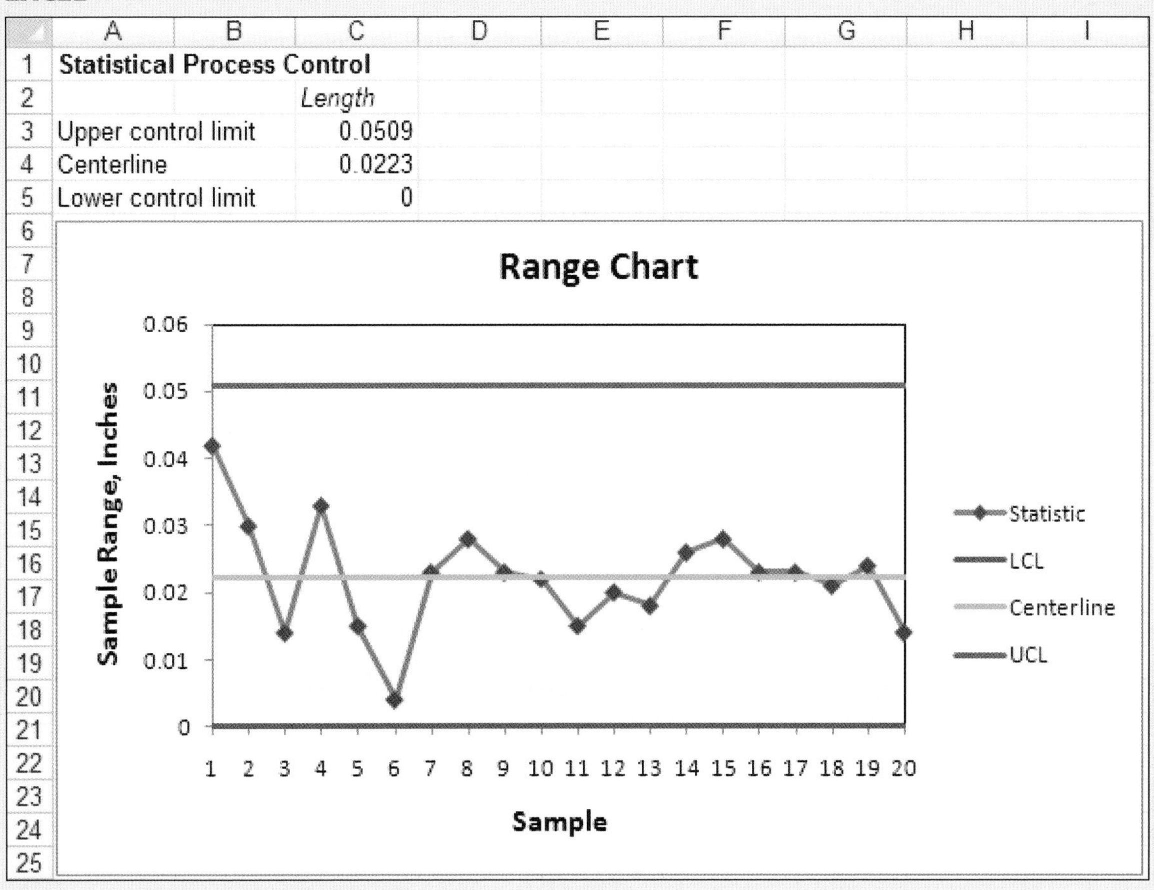

Excel Range Chart, Sigma Unknown

1. In Excel file CX20RIVS, the label is in A1 and the sequence of rivet lengths is in A2:A81. From the **Add-Ins** ribbon, click **Data Analysis Plus.** Select **Statistical Process Control.** Click **OK.**
2. Enter **A1:A81** into the **Input Range** box. Click **Labels.** Enter **4** into the **Sample size** box. Select **R.** Click **OK.** This provides the range chart for the 20 samples in Table 20.3. (*Note:* As an alternative, you can use Excel worksheet template TMSPCMR. The instructions are within the template.)

MINITAB

Minitab Range Chart, Sigma Unknown

1. Using Minitab file CX20RIVS, with the 80 rivet lengths in C1: Click **Stat.** Select **Control Charts.** Select **Variables Charts for Subgroups.** Click **R.**

(*continued*)

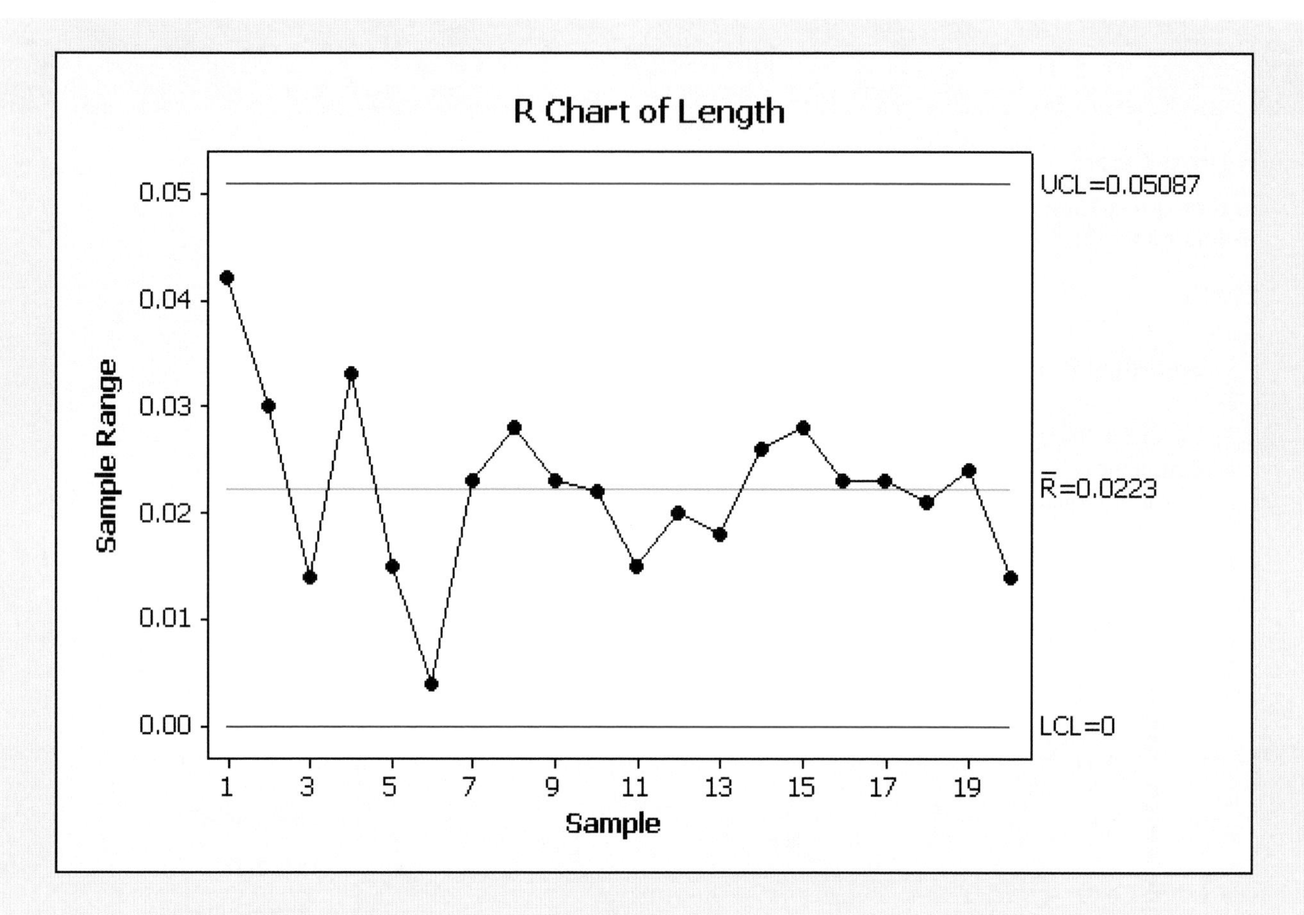

2. Select **All observations for a chart are in one column** and enter **C1** into the box. Enter **4** (the size of each sample) into the **Subgroup sizes** box.
3. Click **R Options.** Click **Estimate** and select **Rbar** in the **Method for Estimating Standard Deviation** menu. Click **OK.** Click **OK.** The range chart appears as shown here.

Note: If the process standard deviation, σ, is known, the range chart with known σ can be easily generated. In step 2, above, just click **R Options,** click **Parameters,** and enter the known process standard deviation into the **Historical standard deviation** box.

ranges are within their control limits, but the means have drifted upward and the process is no longer in control. The trend of the mean chart in part (a) could have occurred, for example, if a cutting tool were gradually wearing down during the production process. Part (b) of Figure 20.7 shows a series of samples in which the reverse has occurred. While the sample means remain within the control limits, the range chart has detected a shift in the variability of the process. Either mechanical wear or the need for maintenance may be increasing the dimensional variation in the output.

A process that is out of control can also be identified by a lone sample statistic that falls outside the control limits. For example, in the range chart in part (b) of Figure 20.7, the range for the seventh sample was unusually high. Even if the

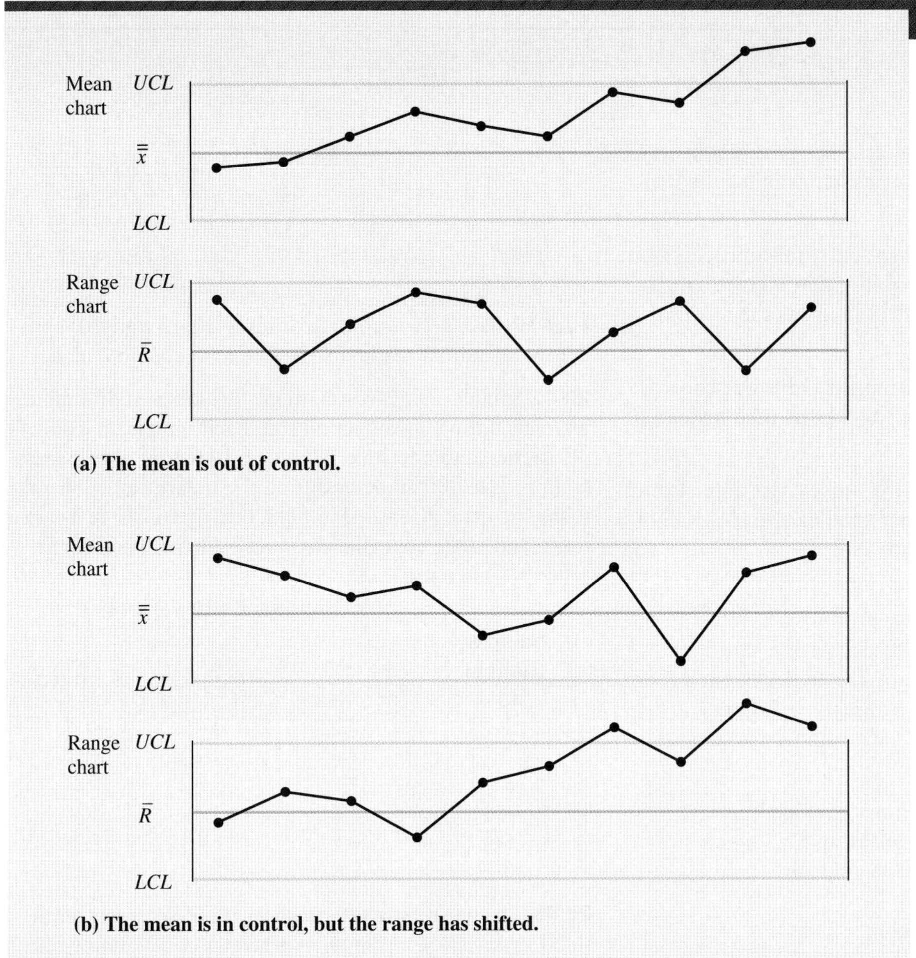

FIGURE 20.7

The mean chart and the range chart should be used together, since both are important in determining whether a process is in control. In part (a), the variability is within limits, but the mean is out of control. In part (b), the mean is in control, but the range control chart has detected a shift in the variability of the process.

ranges had settled down toward $\bar{R}$ after that sample was taken, the existence of the exceptionally large range for that sample would be a warning signal that the process might not have been in control at that time. A single sample may exhibit unusual characteristics due to an influence that is relatively temporary, such as an electrical surge.

In Figure 20.7, note that $\bar{\bar{x}}$ and $\bar{R}$ are not necessarily the same as the average mean or average range for the samples shown. In practice, the centerlines and the upper and lower control limits may be based on an earlier series of twenty or more samples during which the process was deemed to be in control. For example, the sample values shown in either segment of Figure 20.7 might represent samples 21–30 for the process examined by the control charts in Computer Solutions 20.1 and 20.2.

By using Seeing Statistics Applet 21, at the end of the chapter, you can easily select consecutive samples and generate two different mean control charts. The first chart represents a "well-behaved" process; the second one represents a process that tends to go out of control in some fashion. Notice how even the well-behaved process can occasionally give an indication that it might be out of control, but keep in mind that this is just due to random variation.

EXERCISES

20.34 In the mean chart, what assumption is made about the distribution of individual measurements for the process that is being monitored?

20.35 In a 3-sigma mean control chart for a process that is in control, what is the probability that a given sample mean will fall outside the control limits as the result of random variation alone?

20.36 In general, how will the upper and lower control limits change if a 2-sigma control chart is used instead of a 3-sigma control chart?

20.37 How does the purpose of a range chart differ from that of a mean chart? Why is it a good idea to use both of them in monitoring the same process?

20.38 When the robot welder is functioning properly, the natural gas cylinders being produced have an average burst strength of 3400 psi, with a standard deviation of 100 psi. Determine the centerline and the upper and lower control limits for a 3-sigma mean control chart in which each sample consists of 4 cylinders.

20.39 Use the control chart limits developed in Exercise 20.38 and the following sample results to evaluate whether the process is in control:

	Sample					
	1	2	3	4	5	6
Mean burst strength	3390	3447	3410	3466	3440	3384

20.40 Measured at the lid center, the thickness of the zinc coating applied to container lids averages 3.000 mil, with a process standard deviation of 0.300 mil. Determine the centerline and the upper and lower control limits for a 3-sigma mean control chart in which each sample consists of 4 lids.

20.41 Use the control chart limits developed in Exercise 20.40 and the following sample results to evaluate whether the process is in control:

	Sample					
	1	2	3	4	5	6
Mean thickness	2.95	2.97	3.03	3.11	2.35	3.19

(**DATA SET**) *Note:* For Exercises 20.42 and 20.43, a computer and statistical software are recommended.

20.42 Samples of 12-volt batteries are taken at 30-minute intervals during a production run. Each sample consists of 3 batteries, and a technician records how long each battery will produce 400 amperes during a standard test. Given the following data, also listed in file **XR20042**, construct 3-sigma mean and range control charts and evaluate whether the process is in control.

	Minutes, Battery Number		
Sample Number	**1**	**2**	**3**
1	13.3	9.4	12.1
2	12.2	13.4	8.5
3	11.2	8.2	9.2
4	7.8	9.7	10.0
5	10.1	11.4	13.8
6	9.9	11.7	8.5

20.43 The tensile strength of sheet metal screws is monitored by selecting a sample of 4 screws at 30-minute intervals, then seeing how many pounds of force are required to break each screw. Given the following data, also listed in file **XR20043**, construct 3-sigma mean and range control charts and evaluate whether the process is in control.

	Pounds, Screw Number			
Sample Number	**1**	**2**	**3**	**4**
1	760	843	810	706
2	771	800	747	742
3	710	809	876	754
4	882	875	873	763
5	718	770	728	939
6	988	927	902	956
7	843	739	792	787

20.8 CONTROL CHARTS FOR ATTRIBUTES

When we wish to *count* something regarding the output of a process, a control chart for attributes can be used. If we're counting the number of acceptable versus defective items in each sample, the *p*-chart is used. This would be the case if output consisted of metal containers that either leaked or did not leak. An attribute such as leakage is either present or it is not. Other processes involve output where the number of defects in a unit of production can be counted—for example, the number of surface flaws in a 30-foot length of fabric or the number of air bubbles in a sheet of glass. For such processes, the *c*-chart may be used in determining whether the process is in control.

p-Charts

The ***p*-chart** monitors a process by showing the proportion of defectives in a series of samples. It is also referred to as a "fraction-defective," "percent-defective," or "$\bar{p}$ chart." Like the mean and range charts, the *p*-chart has a centerline, and upper and lower control limits. The normal approximation to the binomial distribution is used in constructing a *p*-chart, and sample sizes must be larger than with the mean or range charts. In other respects, the concept is similar to the charts of the preceding section.

The *p*-chart:

- **Centerline: $\bar{p}$, the average of the sample proportions, or**

$$\frac{\Sigma p}{\text{Number of samples}}$$

- **Upper and lower control limits (*UCL* and *LCL*):**

$$\bar{p} \pm z\sqrt{\frac{\bar{p}(1-\bar{p})}{n}}$$

where $\bar{p}$ = the average of the sample proportions
n = the size of each sample
z = standard error units, normal distribution

In the expression for the control limits above, the quantity represented by the square root symbol is our estimate of σ_p, the standard error of the sample proportion for samples of this size. A 3-sigma ($z = 3$) control chart is common practice, and 99.74% of the sample proportions should fall within the limits if the process is in control. Both $n\bar{p}$ and $n(1 - \bar{p})$ should be ≥ 5 for the normal approximation to the binomial distribution to apply. If a process generates either a very high or a very low proportion of defectives, the samples may need to be quite large to satisfy this requirement.

EXAMPLE

p-Chart

As a test of filament integrity, lightbulbs coming off the manufacturing line are sampled and subjected to a voltage 30% higher than that for which they are rated. Each sample consists of 100 bulbs, and a bulb is categorized as "defective" if it burns out during a 2-second exposure to this higher voltage. Sixteen samples from yesterday's production run provided the data shown in Table 20.4 and listed in file CX20FILS.

EXAMPLEEXAMPLEEXA

TABLE 20.4

Results from 16 samples, each with $n = 100$ items. The mean of the proportion defective values is $\bar{p} = 0.135$. These data, also listed in file **CX20FILS**, are the basis for the Excel and Minitab *p*-charts in Computer Solutions 20.3.

Sample #	Defectives	Nondefectives	Proportion Defective, *p*
1	8	92	0.080
2	14	86	0.140
3	16	84	0.160
4	14	86	0.140
5	15	85	0.150
6	20	80	0.200
7	19	81	0.190
8	12	88	0.120
9	17	83	0.170
10	6	94	0.060
11	10	90	0.100
12	15	85	0.150
13	9	91	0.090
14	17	83	0.170
15	13	87	0.130
16	11	89	0.110
			$\bar{p} = 0.135$

SOLUTION

For each of the 16 samples described in Table 20.4, the numbers of acceptable and defective bulbs have been counted. The samples are of equal size, and the average of the 16 sample proportions is the total of the proportion-defective column divided by the number of samples, or $\bar{p} = 0.135$. This is the centerline for the control chart.

Constructing a 3-sigma control chart for these data, z is 3, the average proportion has been calculated as $\bar{p} = 0.135$, and each sample consists of $n = 100$ bulbs. The upper and lower control limits for the sample proportion are

$$\bar{p} \pm z\sqrt{\frac{\bar{p}(1-\bar{p})}{n}} = 0.135 \pm 3\sqrt{\frac{0.135(1-0.135)}{100}} = 0.135 \pm 0.103$$

and the control limits are

$$UCL = 0.135 + 0.103 = 0.238$$
$$LCL = 0.135 - 0.103 = 0.032$$

In the event that calculations result in a lower control limit that is negative, we use $LCL = 0$, because it is not possible to have either a negative number or a negative proportion of defectives. Like the mean and range charts, *p*-charts for later samples may employ the centerline and control limits that were determined for a *p*-chart constructed during a period when the process was judged as being in control.

Computer Solutions 20.3 describes the procedures and shows the results when Excel and Minitab are used to generate a p-chart for the data in Table 20.4. Each plot shows the upper and lower control limits, the centerline ($\bar{p}$ = 0.135), and the sequence of 16 sample proportions. All of the sample proportions are within the control limits, and the process appears to be in control.

COMPUTER 20.3 SOLUTIONS

p-Chart

EXCEL

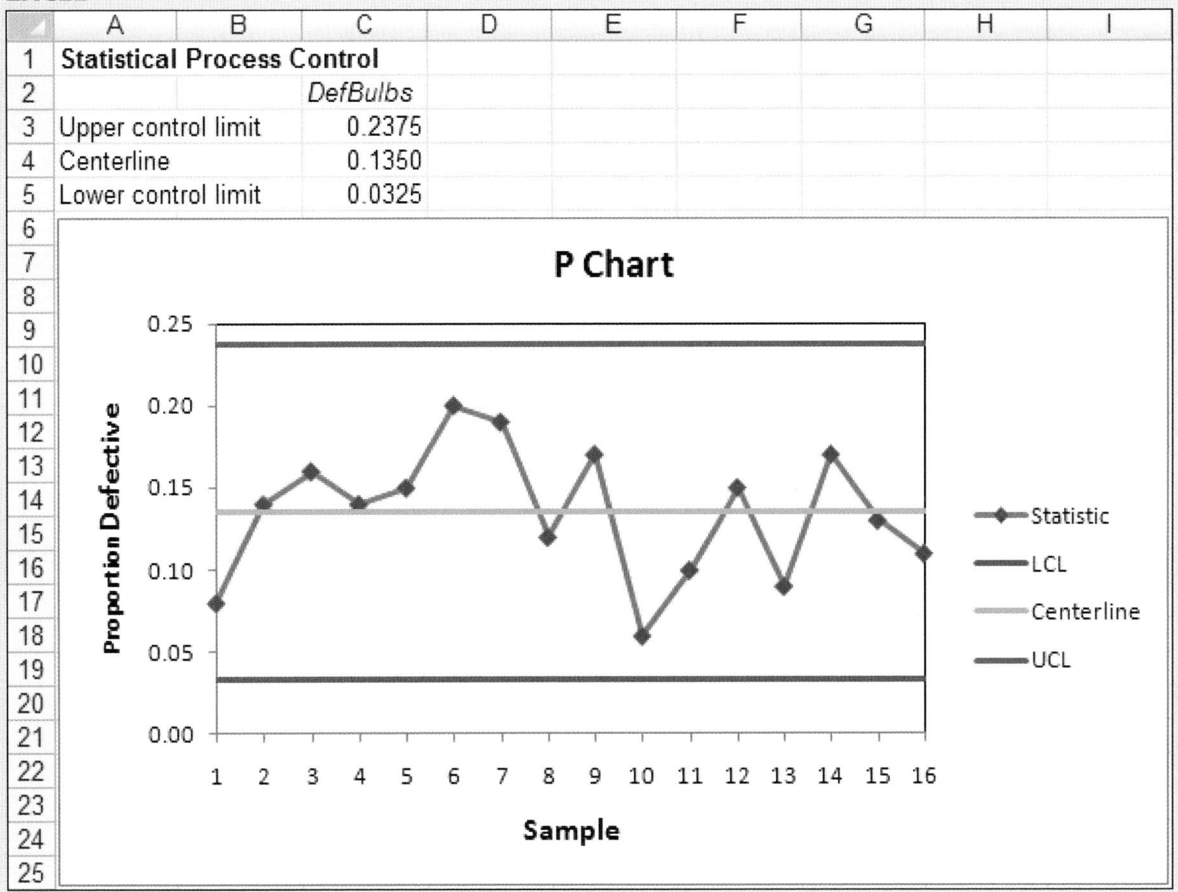

Excel *p*-Chart

1. In Excel file CX20FILS, the label is in A1 and the sequence containing the number of defectives in each sample is in A2:A17. From the **Add-Ins** ribbon, click **Data Analysis Plus.** Select **Statistical Process Control.** Click **OK.**
2. Enter **A1:A17** into the **Input Range** box. Click **Labels.** Enter **100** into the **Sample size** box. Select **P.** Click **OK.** This provides the p-chart for the 16 samples in Table 20.4. (*Note:* As an alternative, you can use Excel worksheet template TMSPCPCH. The instructions are within the template.)

(*continued*)

MINITAB

Minitab *p*-Chart

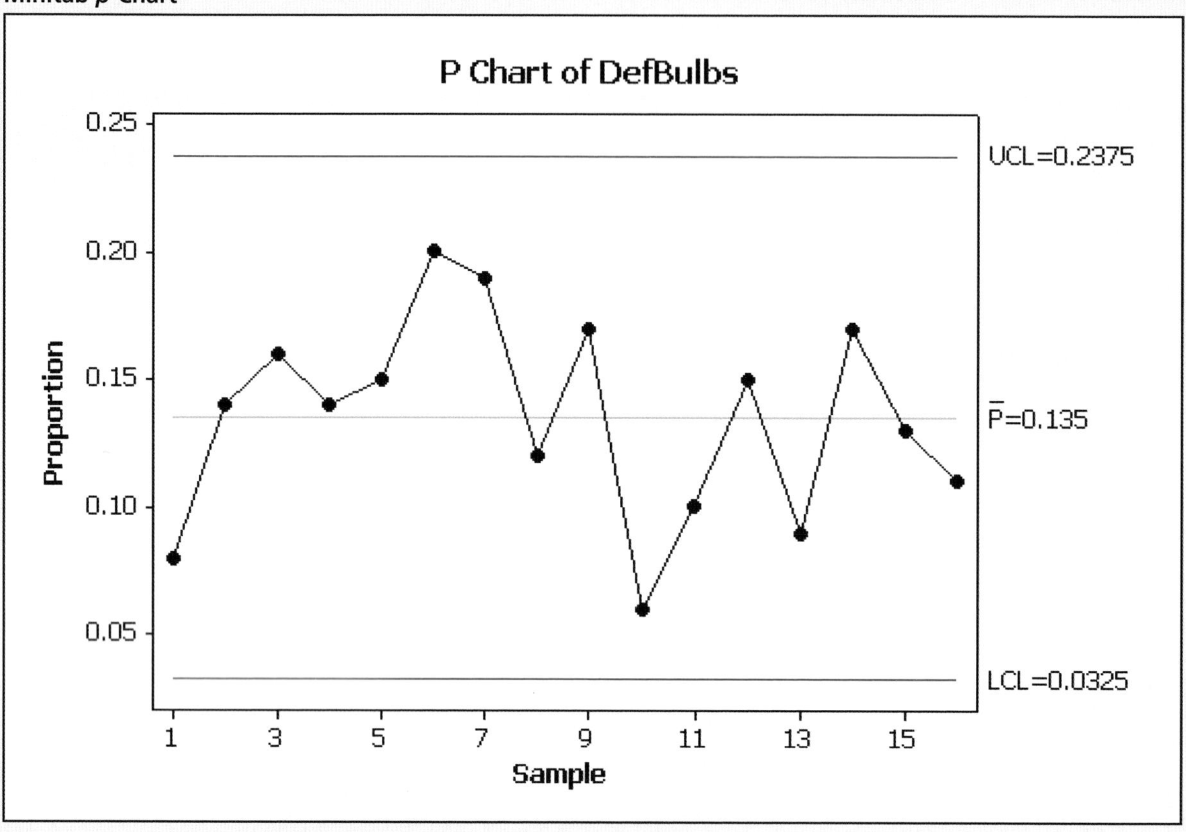

1. Using Minitab file **CX20FILS**, with the sequence containing the number of defectives in each sample in C1: Click **Stat.** Select **Control Charts.** Select **Attributes Charts.** Click **P.**
2. Enter **C1** into the **Variables** box. Enter **100** (the size of each sample) into the **Subgroup sizes** box. Click **OK.** The *p*-chart appears as shown here.

(*Note:* If we either know the process proportion defective or wish to specify an estimate that is based on past data, just select **P Chart Options,** click **Parameters,** and enter the known or specified process proportion defective into the **Proportion** box.)

c-Charts

If the objective of statistical process control is to monitor the number of defects from each sample, the **c-chart** is used. Also referred to as the "$\bar{c}$ chart," it can apply to such observations as the number of flaws in 10 feet of tubing, the number of scratches on a panel, or the number of "fit-and-finish" defects on an automobile body. Each sample consists of a unit of output, the statistic is c = the number of defects or flaws, and the distribution of the number of such defects tends to be Poisson. The c-chart is constructed using a normal approximation to the Poisson distribution.

The c-chart:

- Centerline: $\bar{c}$, the average number of defects per sample, or

$$\frac{\sum c}{\text{Number of samples}}$$

- Upper and lower control limits (*UCL* and *LCL*):

$\bar{c} \pm z\sqrt{\bar{c}}$ where $\bar{c}$ = the average number of defects per sample
z = standard error units, normal distribution

EXAMPLE

c-Chart

During the production of continuous sheet steel, periodic samples consisting of 10-foot lengths are examined to determine the number of surface scratches, blemishes, or other flaws. Twenty samples have been inspected at periodic intervals during a production run, with results as shown in Table 20.5 and listed in file CX20DEFS.

SOLUTION

For each of the 20 samples in Table 20.5, the number of surface defects has been counted. The average number of defects per sample is the total number of defects divided by the number of samples, or $\bar{c} = 106/20 = 5.30$.

Constructing a 3-sigma control chart for these data, the upper and lower control limits for c are

$$\bar{c} \pm z\sqrt{\bar{c}} = 5.30 \pm 3\sqrt{5.30} = 5.30 \pm 6.91$$

EXAMPLEEXAMPLEEXAMPLE

Sample #	Number of Defects, c	Sample #	Number of Defects, c
1	3	11	10
2	4	12	5
3	6	13	7
4	6	14	8
5	4	15	3
6	7	16	3
7	9	17	1
8	8	18	4
9	3	19	6
10	6	20	3
			$\bar{c} = 5.30$

TABLE 20.5

Each of the 20 samples consists of a 10-foot length of sheet steel. The measurement for each is c = the number of scratches, blemishes, and other defects over the 10-foot length. The mean number of defects is $\bar{c} = 5.30$. These data, also listed in file CX20DEFS, are the basis for the Excel and Minitab c-charts in Computer Solutions 20.4.

and the control limits are

$$UCL = 5.30 + 6.91 = 12.21$$
$$LCL = 5.30 - 6.91 = -1.61, \quad \text{which reverts to } 0$$

Since it's impossible to have a negative number of defects, the lower control limit (calculated as -1.61) is replaced by zero. The same reasoning was introduced during our discussion of the p-chart and its control limits. As with other control charts, later samples may be evaluated on the basis of c-chart centerline and control limit values determined during a period when the process was deemed to be in control.

Computer Solutions 20.4 describes the procedures and shows the results when Excel and Minitab are used to generate a c-chart for the data in Table 20.5. Each plot shows the upper and lower control limits, the centerline ($\bar{c} = 5.30$), and the number of defects in each of the 20 samples. Each of the c values falls within the upper and lower control limits, and the process is judged to be in control.

COMPUTER 20.4 SOLUTIONS

c-Chart

EXCEL

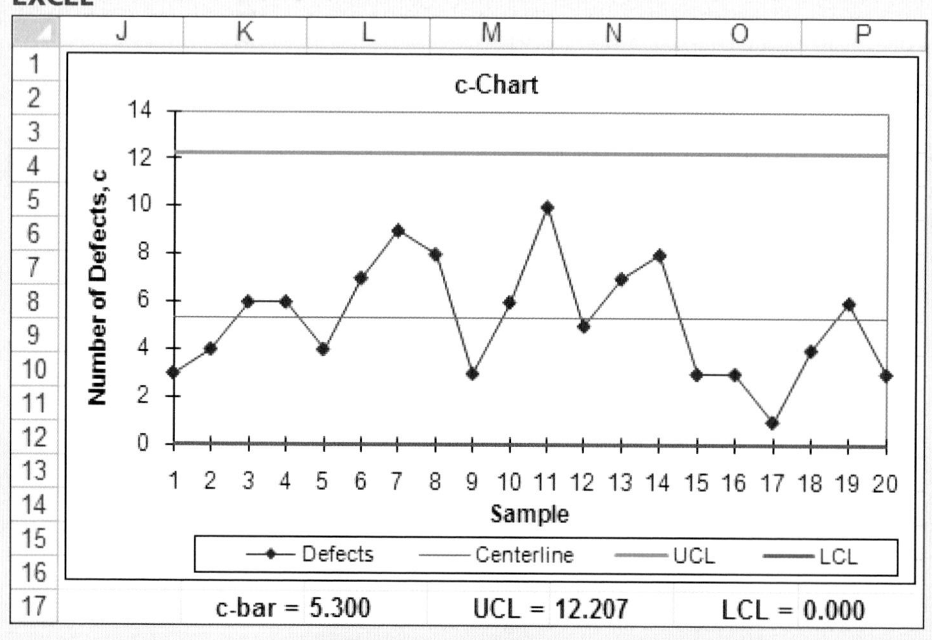

Excel *c*-Chart

1. Open Excel worksheet template **TMSPCCCH**.
2. With the template file remaining open, open file **CX20DEFS**, which contains the data shown in Table 20.5. Follow the short instruction set in the template. The centerline, control limits, and c-chart plot for the 20 samples will appear as shown here.

MINITAB

Minitab *c*-Chart

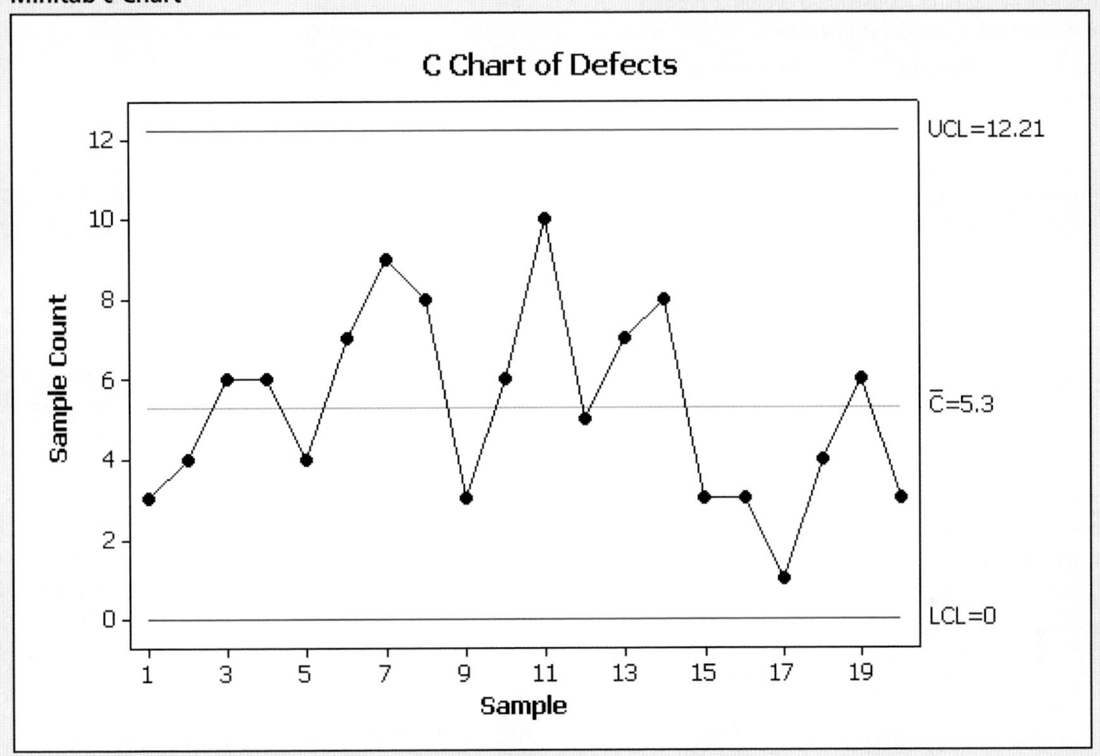

1. Using Minitab file **CX20DEFS**, with the sequence containing the number of defects in each sample in C1: Click **Stat.** Select **Control Charts.** Select **Attributes Charts.** Click **C.**
2. Enter **C1** into the **Variables** box. Click **OK.** The c-chart appears as shown here.

(*Note:* If we either know the process mean or wish to specify an estimated mean that is based on past data, just select **C Chart Options,** click **Parameters,** and enter the known or specified process mean into the **Mean** box.)

EXERCISES

20.44 Differentiate between the *p*-chart and the *c*-chart. Under what circumstances would each be used?

20.45 When a given process is operating properly, the percentage of defectives is 4%. What will be the 3-sigma upper and lower control limits for a *p*-chart in which each sample consists of 150 items?

20.46 When a given process is operating properly, the percentage of defectives is 15%. For a 3-sigma *p*-chart in which the upper and lower control limits are 4 percentage points apart, how large will each sample have to be?

20.47 In the data collected for construction of a *c*-chart, the average number of defects per unit of output is $\bar{c} = 7.5$. What are the 3-sigma upper and lower control limits for the control chart?

(**DATA SET**) *Note:* For Exercises 20.48–20.53, a computer and statistical software are recommended.

20.48 Twelve samples have been taken from a production run. Given the following data, construct a 3-sigma *p*-chart and evaluate whether the process is in control.

Sample Number	Sample Size	Number of Defectives
1	100	6
2	100	12
3	100	13
4	100	5
5	100	6
6	100	6
7	100	7
8	100	16
9	100	9
10	100	5
11	100	14
12	100	5

20.49 Twenty samples have been taken from a production run. Given the following data, construct a 3-sigma *p*-chart and evaluate whether the process is in control.

Sample Number	Sample Size	Number of Defectives
1	200	38
2	200	26
3	200	37
4	200	42
5	200	48
6	200	45
7	200	38
8	200	42
9	200	46
10	200	45
11	200	32
12	200	37
13	200	38
14	200	37
15	200	36
16	200	39
17	200	50
18	200	39
19	200	38
20	200	36

20.50 Using the control limits determined in Exercise 20.49, evaluate whether the process was in control when the following data were collected for samples 21–32.

Sample Number	Sample Size	Number of Defectives
21	200	55
22	200	45
23	200	45
24	200	56
25	200	54
26	200	53
27	200	53
28	200	56
29	200	53
30	200	57
31	200	42
32	200	53

20.51 After the stamping operation, door panels are periodically selected for a detailed examination in which the number of surface scratches, dents, and other flaws are counted. During today's production run, 20 panels were inspected. Given the data in the table, construct a 3-sigma *c*-chart, then reach a conclusion regarding whether the process is in control.

Panel Number	c = Number of Surface Defects	Panel Number	c = Number of Surface Defects
1	1	11	8
2	2	12	5
3	4	13	5
4	6	14	4
5	1	15	2
6	5	16	6
7	1	17	2
8	3	18	3
9	11	19	4
10	12	20	4

20.52 Automobiles are randomly selected for a close examination for fit-and-finish defects, including such items as dirt specks in the paint and panels that are not properly aligned. Based on data collected for the following vehicles that were so inspected, construct a 3-sigma *c*-chart, then reach a conclusion regarding whether the process is in control.

Vehicle Number	c = Number of Defects	Vehicle Number	c = Number of Defects
1	4	11	6
2	7	12	10
3	4	13	5
4	9	14	10
5	9	15	7
6	8	16	6
7	7	17	8
8	6	18	5
9	7	19	7
10	5	20	11

20.53 Using the control limits determined in Exercise 20.52, evaluate whether the process was in control when the following data were collected for vehicles 21–30.

Vehicle Number	c = Number of Defects	Vehicle Number	c = Number of Defects
21	6	26	4
22	4	27	8
23	4	28	13
24	3	29	7
25	10	30	8

ADDITIONAL STATISTICAL PROCESS CONTROL AND QUALITY MANAGEMENT TOPICS (20.9)

More Statistical Process Control Chart Possibilities

Besides the mean, range, *p*- and *c*-charts we've discussed, many other kinds of statistical process control charts can be constructed, including those for standard deviations, individual observations, moving averages, and moving ranges. In the latter two cases, the vertical axis represents either the average or the range for a "sample" of fixed size that changes in composition each time a new observation is made. This is analogous to the concept of the moving average discussed in Chapter 18. Minitab offers a variety of control charts like those just described, but Excel (even with Data Analysis Plus) is more limited in the variety and flexibility of the control charts it can generate.

Diagnostic Tests for the State of Control of a Process

Both Minitab and Excel (with Data Analysis Plus) provide a complete battery of tests on the state of control of a process that has been charted. This diagnosis takes the form of looking for and reporting any of eight conditions and patterns. Excel automatically performs all available diagnostic tests. With Minitab, we must select the "Test" option and specify "Select all tests." Shown below are the available diagnostic tests provided by Excel and Minitab. For mean and range charts, all eight tests are available; for *p*- and *c*-charts, only the first four.

1. A point is more than 3 sigmas away from the centerline.
2. Nine points in a row are on the same side of the centerline.
3. Six points in a row are all headed either upward or downward.
4. Fourteen points in a row alternate upward and downward (sawtooth pattern).
5. Two out of three consecutive points are on the same side of the centerline and more than 2 sigmas away from it.
6. Four out of five consecutive points are on the same side of the centerline and more than 1 sigma away from it.
7. Fifteen points in a row are within 1 sigma of the centerline.
8. Eight points in a row are more than 1 sigma from the centerline.

Process Capability Index

Earlier, in our discussion of Taguchi methods, we briefly introduced the concept of upper and lower specification limits. Such specifications may be necessary so an individual item will fit or function properly with other components in a finished product. For example, it may be required that the lengths of aluminum rods in an industrial application be no greater than 1.575 inches and no less than

1.525 inches. In this case, the upper specification limit (*USL*) would be 1.575 inches and the lower specification limit (*LSL*) would be 1.525. Thus, any rod resulting from the production process would be deemed acceptable if it fell within these limits. Think of these limits as the goalposts on an American football field; a kicked ball must go between the vertical posts in order to be successful and earn points.

We can express the *potential* ability of a process to generate products falling within the required specification limits with a measure known as the **Process Potential Index,** or **C_p.** It can be calculated as shown below.

Process Potential Index, C_p:

$$C_p = \frac{USL - LSL}{6\sigma},$$

where C_p = Process Potential Index
USL = upper specification limit
LSL = lower specification limit
σ = process standard deviation (If σ is unknown, use s.)

Assumptions: The process is in control, the process mean is midway between the upper and lower specification limits, and the output is at least approximately normally distributed.

This index is essentially the *allowable* spread in the output divided by the *actual* spread in the output. Higher numerical values for C_p are preferable to lower ones, a value of at least 1.0 is necessary for a process to be judged as potentially capable, and the ideal C_p will be as high as possible.[9]

Note that this measure assumes that the process mean is exactly midway between the upper and lower specification limits. In actuality, the process mean might be closer to the upper limit, closer to the lower limit, or even lie *outside* the range designated by the limits. Because of these possibilities, we will focus on the more realistic measure discussed below.

The **Process Capability Index (C_{pk})** takes into consideration that the process mean might be closer to one of the specification limits than to the other. In addition, we generally do not know the actual process mean and standard deviation, so we must typically estimate them based on the mean and standard deviation of a sample of output, such as the entire group of measurements we may have used in constructing a statistical process control chart for variables. In this case, we can calculate C_{pk} as shown below.

Process Capability Index, C_{pk}:

$$C_{pk} = \text{the lower of } \left(\frac{\bar{\bar{x}} - LSL}{3s}\right) \text{ or } \left(\frac{USL - \bar{\bar{x}}}{3s}\right)$$

where C_{pk} = Process Capability Index
$\bar{\bar{x}}$ = overall mean for all units sampled
s = standard deviation for all units sampled
USL = upper specification limit
LSL = lower specification limit

Assumptions: The process is in control and the output is at least approximately normally distributed.

[9] "Process Capability, Minding Your C_{pk}'s," qualitydigest.com, September 1, 2009.

EXAMPLE

Determination of Process Capability Index

In Computer Solutions 20.1, we were provided with the lengths (inches) of 80 rivets from a manufacturing process. These lengths are in file **CX20RIVS**, and we can use Minitab or Excel to calculate the sample mean and sample standard deviation for all 80 rivets. In either case, the results are as shown below.

$$\text{Overall mean for all units sampled: } \bar{\bar{x}} = 1.3983 \text{ inches}$$
$$\text{Sample standard deviation for all units sampled: } s = 0.0103 \text{ inches}$$

We have already concluded that the process is in control, and we have reasonably assumed that the distribution of process output measurements is at least approximately normal. The preceding items describe our knowledge of this production process. (Note that we are using the *actual* sample standard deviation for the sample of 80, not the estimated value that was used in constructing the statistical process control chart in Computer Solutions 20.1.)

Suppose we are the producer involved, and that the contract with our rivet customer requires that acceptable rivets have a length no greater than 1.4900 inches (USL) and no less than 1.3300 inches (LSL). Given this contractual requirement and the sample information from our production process, what is the Process Capability Index (C_{pk}) of our rivet production line?

SOLUTION

In determining our Process Capability Index (C_{pk}), we must first compute how well we're doing with regard to the lower (LSL) and the upper (USL) specification limits, respectively, then identify which value is the lower of the two.

Pertaining to the lower specification limit (LSL):

$$\left(\frac{\bar{\bar{x}} - LSL}{3s}\right) = \left(\frac{1.3983 - 1.3300}{3(0.0103)}\right) = 2.21$$

Pertaining to the upper specification limit (USL):

$$\left(\frac{USL - \bar{\bar{x}}}{3s}\right) = \left(\frac{1.4900 - 1.3983}{3(0.0103)}\right) = 2.97$$

The lower of these two values is 2.21, so our Process Capability Index is $C_{pk} = 2.21$. Both we and our customer should be impressed with a C_{pk} this high. In order to officially qualify as a Six-Sigma process, the C_{pk} performance must be at least 2.0.

EXAMPLEEXAMPLEEXAMPLEEXAMPLEEXAMPLEEXAMPLEEXAM

The Process Capability Index should be $C_{pk} \geq 1.0$ to demonstrate that a process is capable, and higher values are even better. However, diminishing returns may set in—for example, a C_{pk} greater than 2.5 or 3.0 might suggest that we are paying for levels of precision we don't really need. For processes involving both upper and lower specification limits, as we've discussed here, one expert has recommended minimum C_{pk} values of 1.33 for existing processes, 1.50 for new processes or for existing processes involving a safety or critical parameter,

1.67 for new processes involving a safety or critical parameter, and 2.00 for a Six-Sigma quality process.[10]

Because the C_{pk} evaluates how effectively a process meets the intended specifications for its output, it helps managers design and refine the operations generating the products on which their company's profits and reputations rely. Of necessity, our coverage of this key descriptor has been relatively brief, and the interested reader should refer to more advanced works for additional insights into the Process Capability Index and its managerial applications.

EXERCISES

20.54 Differentiate between the Process Potential Index (C_p) and the Process Capability Index (C_{pk}). In what way(s) are they similar? In what way(s) do they differ?

20.55 Can the Process Capability Index (C_{pk}) ever be greater than the Process Potential Index (C_p)? If so, how? If not, why not?

20.56 A process is in control and the results from a series of samples show the overall mean and standard deviation of all the units sampled to be $\bar{\bar{x}} = 23.50$ grams and $s = 0.21$ grams. The upper and lower specification limits for the output are $USL = 24.40$ grams and $LSL = 22.80$ grams. Based on this information, determine and interpret the value of the Process Capability Index for this process.

20.57 A process is in control and the results from a series of samples show the overall mean and standard deviation of all the units sampled to be $\bar{\bar{x}} = 2.320$ inches and $s = 0.003$ inches. The upper and lower specification limits for the output are $USL = 2.328$ inches and $LSL = 2.272$ inches. Based on this information, determine and interpret the value of the Process Capability Index for this process.

(**DATA SET**) *Note:* Exercises 20.58–20.64 require a computer and statistical software.

20.58 Repeat Exercise 20.42 and apply all of the diagnostic tests available with your statistical software. Has applying these diagnostics to the mean and range charts raised any concerns regarding the state of control of the process? If so, in what way(s)?

20.59 Repeat Exercise 20.43 and apply all of the diagnostic tests available with your statistical software. Has applying these diagnostics to the mean and range charts raised any concerns regarding the state of control of the process? If so, in what way(s)?

20.60 Repeat Exercise 20.52 and apply all of the diagnostic tests available with your statistical software. Has applying these diagnostics to the *c*-chart raised any concerns regarding the state of control of the process? If so, in what way(s)?

20.61 The data in file **XR20061** represent 30 samples, each with $n = 3$ items measured in inches. For the associated mean and range charts, have your statistical software apply all of the diagnostic tests of which it is capable. Has applying these diagnostics to the mean and range charts raised any concerns regarding the state of control of the process? If so, in what way(s)?

20.62 The data in file **XR20062** represent the number of anonymous tips phoned in during each of 50 consecutive days to a local law enforcement hotline. Use your statistical software in generating an appropriate control chart for these data and diagnosing unusual conditions and patterns. Have the diagnostics cast any doubt on whether the process is in control? If so, in what way(s)?

20.63 A process has generated the output (in centimeters) listed in file **XR20063**.

a. For a series of samples, each with $n = 4$, construct and interpret the $\bar{x}$ and range charts for this process. Does it appear to be in control?
b. Given your response to part (a), is it appropriate to compute the Process Capability Index (C_{pk}) for this process? If so, determine the overall mean and standard deviation for all the units measured, then compute and interpret the Process Capability Index if the specification limits are $USL = 5.700$ centimeters and $LSL = 5.100$ centimeters.

20.64 A process has generated the output (in ounces) listed in file **XR20064**.

a. For a series of samples, each with $n = 5$, construct and interpret the $\bar{x}$ and range charts for this process. Does it appear to be in control?

[10]Douglas Montgomery, *Introduction to Statistical Quality Control*, New York: John Wiley & Sons, 2004, p. 776.

b. Given your response to part (a), is it appropriate to compute the Process Capability Index (C_{pk}) for this process? If so, determine the overall mean and standard deviation for all the units measured, then compute and interpret the Process Capability Index if the specification limits are $USL = 45.400$ ounces and $LSL = 43.800$ ounces.

SUMMARY (20.10)

• TQM and the process orientation

Organizations seeking to improve the quality of their goods and services are increasingly adopting the principles and practices of total quality management (TQM). Companies are finding a close parallel between the adage, "a stitch in time saves nine" and the TQM concept of dealing with defects through a process orientation that concentrates on preventing them from occurring in the first place.

• Process variation and defect prevention

Much of the emphasis in TQM is on understanding and reducing variation within the process through which the product is produced, and samples of output are used in reaching inferences about the stability of the process. The two sources of process variation are random (common-cause) variation and assignable (special-cause) variation.

The traditional philosophy of defect detection generally involved a corporate atmosphere where quality was a separate entity within the firm. The TQM philosophy of defect prevention has quality integrated throughout the organizational structure and central to corporate thinking in all matters. The emergence of TQM saw pioneers like W. Edwards Deming spreading the teachings of TQM and its process orientation.

• Kaizen and TQM tools

The practices of TQM are closely related to the Kaizen philosophy of continuing improvement and include guidelines such as Deming's 14 points, quality audits, competitive benchmarking, just-in-time manufacturing, and worker empowerment. The chapter discusses several of the statistical tools used in applying TQM, including the process flow chart, the cause-and-effect diagram, the Pareto diagram, the check sheet, and the Taguchi view of quality loss.

• Statistical process control charts

A popular and important tool for applying TQM is statistical process control, where sampled products are used in making inferences about the *process* from which they came. The control charts used in monitoring the process can provide a warning signal that the process may have become unstable due to the presence of assignable variation. In this event, the process should receive attention and corrective action. A process that is unstable will tend to generate products that are of unpredictable and/or unsatisfactory quality.

• Control charts for variables versus attributes

Control charts may be constructed for either variables (measurements) or attributes (counts). The mean chart and the range chart apply to variables and monitor central tendency and variability, respectively. Applicable to attributes, the *p*-chart shows the proportion of defects in successive samples, while the *c*-chart reflects a count of the number of defects found in units of the output. In each of these charts, the upper and lower control limits are the boundaries for the amount of variation that can be attributed to chance alone. These and other control charts can be generated with the use of a computer statistical package.

Process Capability Index

The process capability index measures the capability of a process to generate output falling within specified upper and lower specification limits. It is an essential component of the Six-Sigma quality system.

EQUATIONS

Mean Chart When the Process Standard Deviation Is Known

- Centerline = $\bar{\bar{x}}$, the mean of the sample means.
- Upper and lower control limits (UCL and LCL):

$$\bar{\bar{x}} \pm 3\left(\frac{\sigma}{\sqrt{n}}\right)$$

where σ = the process standard deviation
n = the sample size
$\sigma/\sqrt{n}$ = the standard error of the sample means

Mean Chart When the Process Standard Deviation Is Not Known

- Centerline = $\bar{\bar{x}}$, the mean of the sample means.
- Upper and lower control limits (UCL and LCL):

$$\bar{\bar{x}} \pm A_2\bar{R}$$

where $\bar{R}$ = the average range for the samples
A_2 = value from the 3-sigma control chart factors table

Range Chart

- Centerline = $\bar{R}$, the average of the sample ranges.
- Upper and lower control limits:

$$UCL = D_4\bar{R}$$
$$LCL = D_3\bar{R}$$

where $\bar{R}$ = the average range for the samples
D_3 and D_4 = values from 3-sigma control chart factors table

p-Chart

- Centerline: $\bar{p}$, the average of the sample proportions, or

$$\frac{\Sigma p}{\text{Number of samples}}$$

- Upper and lower control limits (UCL and LCL):

$$\bar{p} \pm z\sqrt{\frac{\bar{p}(1-\bar{p})}{n}}$$

where $\bar{p}$ = the average of the sample proportions
n = the size of each sample
z = standard error units, normal distribution

c-Chart

- Centerline: $\bar{c}$, the average number of defects per sample, or

$$\frac{\Sigma c}{\text{Number of samples}}$$

- Upper and lower control limits (UCL and LCL):

$$\bar{c} \pm z\sqrt{\bar{c}}$$

where $\bar{c}$ = the average number of defects per sample
z = standard error units, normal distribution

Process Potential Index, C_p

$$C_p = \frac{USL - LSL}{6\sigma}$$

where C_p = Process Potential Index
USL = upper specification limit
LSL = lower specification limit
σ = process standard deviation (If σ is unknown, use s.)

Process Capability Index, C_{pk}

$$C_{pk} = \text{the lower of } \left(\frac{\bar{\bar{x}} - LSL}{3s}\right) \text{ or } \left(\frac{USL - \bar{\bar{x}}}{3s}\right)$$

where C_{pk} = Process Capability Index
$\bar{\bar{x}}$ = overall mean for all units sampled
s = standard deviation for all units sampled
USL = upper specification limit
LSL = lower specification limit

CHAPTER EXERCISES

20.65 For a person who eats hamburgers with ketchup, how might the traditional and the Taguchi approaches to quality loss apply with regard to the amount of ketchup that is put on the hamburger?

20.66 For each of the following, identify which kind of variation is at work—random or assignable—and explain your reasoning.

a. During the noon hour, there is an electrical power disruption and nuts machine-tightened during the five minutes that emergency power is on have 15% less torque than normal.
b. Although they are all labeled "9–D," the shoes on a retailer's shelf are not exactly the same size. Any given shoe is just as likely to be slightly larger than 9–D as it is to be slightly smaller.
c. The number of cars leaving downtown Cleveland during a Friday rush hour is greater than during other days of the week.
d. In a production run of curtain rods, the lengths are normally distributed, with a mean of 24.135 inches and a standard deviation of 0.142 inches.

20.67 A university is starting a new college of engineering, has extensive funding, and wants to use competitive benchmarking. What U.S. universities might be useful as objects of the competitive benchmarking effort?

20.68 As a machine produces steel rods, accumulated deposits and mechanical wear cause the rod diameters to increase gradually over time. What type of control chart should be used in determining when the process has gone out of control? Why?

20.69 In producing bulk rolls of paper for sale to newspapers, a firm wishes to control the quality of the paper by keeping track of how many discolored spots or other flaws are present in continuous strips of a given length. What type of control chart should be used and why?

20.70 A process is in control and the results from a series of samples show the overall mean and standard deviation of all the units sampled to be $\bar{\bar{x}} = 21.4500$ inches and $s = 0.0969$ inches. The upper and lower specification limits for the output are $USL = 21.8600$ inches and $LSL = 21.0600$ inches. Based on this information, determine and interpret the value of the Process Capability Index for this process.

20.71 A process is in control and the results from a series of samples show the overall mean and standard deviation of all the units sampled to be $\bar{\bar{x}} = 17.35$ kilograms and $s = 0.95$ kilograms. The upper and lower specification limits for the output are $USL = 20.50$ kilograms and $LSL = 14.30$ kilograms. Based on this information, determine and interpret the value of the Process Capability Index for this process.

(DATA SET) *Note:* For Exercises 20.72–20.80, a computer and statistical software are recommended.

20.72 In a production process for canvas, a sample consisting of 25 square feet contains 4 discolored spots that must be touched up before shipping. Data for this and subsequent samples, each consisting of 25 square feet of canvas, are shown here. Use an appropriate 3-sigma control chart in determining whether the process is in control.

Sample Number	Number of Spots	Sample Number	Number of Spots
1	4	16	13
2	6	17	10
3	3	18	9
4	9	19	8
5	9	20	12
6	8	21	10
7	3	22	12
8	5	23	9
9	8	24	12
10	12	25	10
11	15	26	12
12	8	27	12
13	10	28	10
14	8	29	11
15	10	30	16

20.73 The number of defective products in each of 40 samples taken from a production run are shown here. In each sample, 200 units were examined and classified as either good or defective, depending on whether dimensional tolerances had been met. Use an appropriate 3-sigma control chart in commenting on the state of control of the process.

Sample Number	Number of Defectives	Sample Number	Number of Defectives
1	12	21	10
2	9	22	10
3	15	23	7
4	11	24	8
5	15	25	12
6	8	26	9
7	14	27	12
8	13	28	16
9	10	29	12
10	12	30	11
11	7	31	11
12	12	32	9
13	12	33	7
14	14	34	12
15	8	35	8
16	9	36	8
17	7	37	12
18	8	38	9
19	10	39	6
20	12	40	11

20.74 Following their manufacture, plastic shims are sampled at periodic intervals and their thickness (in millimeters) measured. Each sample consists of $n = 3$ shims, and the following data have been collected for the past 20 samples. Given these measurements, use one or more appropriate 3-sigma control charts in evaluating whether the process is in control.

Sample Number	Measurement (mm)		
	1	2	3
1	47.7	51.8	42.7
2	55.4	47.9	53.9
3	51.6	47.9	47.6
4	45.3	45.5	57.4
5	48.9	49.4	47.0
6	46.5	46.3	47.5
7	55.0	44.3	51.1
8	49.5	50.3	43.9
9	53.7	47.6	49.2
10	48.4	49.2	41.8
11	46.7	51.1	43.9
12	48.8	49.8	48.7
13	48.1	48.8	50.9
14	55.6	52.5	42.8
15	54.3	54.6	44.4
16	48.4	48.3	47.6
17	46.2	46.3	50.3
18	52.4	51.8	49.1
19	48.1	51.7	53.5
20	52.7	46.5	49.2

20.75 An adhesives manufacturer periodically selects a sample of its product and glues 6 toothpicks together, in pairs. A measurement is then made of the number of pounds required to separate the toothpicks. The following data represent 15 samples, each consisting of 3 pairs of toothpicks, and the number of pounds required to separate each pair. Given these measurements, use one or more appropriate 3-sigma control charts in evaluating whether the process is in control.

Sample Number	Pounds to Separate Pair		
	1	2	3
1	28	24	27
2	29	27	27
3	27	26	28
4	28	31	30
5	31	25	25
6	22	27	29
7	30	27	26
8	29	25	28
9	28	29	34
10	32	28	32
11	26	25	28
12	22	27	29
13	26	28	34
14	27	30	24
15	26	24	28

20.76 During the production of 15-ampere circuit breakers, samples are taken at periodic intervals and tested for the exact amperage at which they break a circuit. Each sample consists of $n = 4$ circuit breakers. For the following data, representing the results from the past

20 samples, construct one or more appropriate control charts and comment on the state of control of the process.

Sample Number	Measurement Number 1	2	3	4
1	14.8	15.1	15.2	14.7
2	15.1	14.8	15.1	15.2
3	15.0	15.0	15.0	15.2
4	15.1	15.2	14.7	15.1
5	15.2	15.2	15.2	14.7
6	15.1	15.0	15.2	15.2
7	15.6	14.7	14.8	15.2
8	15.2	14.9	14.7	15.0
9	14.9	14.9	15.4	14.9
10	15.1	15.0	14.6	15.2
11	15.1	14.7	14.7	14.8
12	15.3	15.1	15.0	14.8
13	15.3	14.6	14.5	14.9
14	15.2	15.1	15.1	15.0
15	15.2	14.1	14.9	15.4
16	15.3	15.9	15.4	15.6
17	16.2	15.9	15.5	15.1
18	15.9	15.0	14.9	14.9
19	14.7	16.6	15.8	14.7
20	15.1	15.2	16.9	15.6

20.77 Telemarketers for a mail-order firm are assigned to make a total of 200 calls each day, then record the number of potential customers who are interested in receiving a catalog and first-order discount coupon in the mail. For each of the past 50 days, the number of catalog/coupon packages sent is as shown here. Construct one or more appropriate control charts and comment on the state of control of the process.

Days	Successes							
1–15	21	16	23	16	22	19	23	25
	15	22	17	21	21	23	15	
16–30	18	20	14	24	18	18	22	27
	15	13	29	18	21	21	25	
31–45	20	24	15	20	21	11	19	17
	22	16	17	28	23	22	16	
46–50	22	18	19	15	21			

20.78 A cabinet manufacturer randomly selects cabinets from the production line and examines them for defects. For each of the 30 cabinets sampled during today's production, the number of defects is shown below. Given these data, construct one or more appropriate control charts and comment on the state of control of the process.

Cabinet Number	Number of Defects	Cabinet Number	Number of Defects
1	4	16	3
2	3	17	5
3	2	18	4
4	3	19	3
5	3	20	3
6	2	21	1
7	2	22	3
8	3	23	9
9	3	24	2
10	4	25	2
11	6	26	5
12	1	27	4
13	4	28	4
14	2	29	4
15	1	30	3

20.79 Given the conclusion reached in the solution for Exercise 20.75, with the data in file **XR20075**, is it appropriate to compute the Process Capability Index (C_{pk}) for this process? If so, determine the overall mean and standard deviation for all the units measured, then compute and interpret the Process Capability Index if the specification limits are $USL = 33.700$ pounds and $LSL = 21.700$ pounds.

20.80 Given the conclusion reached in the solution for Exercise 20.74, with the data in file **XR20074**, is it appropriate to compute the Process Capability Index (C_{pk}) for this process? If so, determine the overall mean and standard deviation for all the units measured, then compute and interpret the Process Capability Index if the specification limits are $USL = 61.000$ millimeters and $LSL = 37.000$ millimeters.

INTEGRATED CASES

Thorndike Sports Equipment

Like any other large firm, Thorndike Sports Equipment receives daily telephone calls in which customers either praise or complain about a Thorndike product. Luke Thorndike is especially sensitive about the number of complaint calls received daily. Twelve weeks ago, he asked the customer service department to keep a daily log of the number of telephoned complaints received each day. Just six weeks ago, Thorndike ads began carrying a toll-free number through which customers could inquire or comment about Thorndike products.

The customer service department has just handed Luke a summary of the daily complaints over the past 60 business days. Luke has just handed it to Ted, who finds the daily totals shown in the table. The data are also in file **THORN20**.

1. Assuming the daily number of calls to be Poisson distributed, construct a c-chart for the first 30 days. Using 3-sigma limits, were there any days in which the volume of complaints fell outside the limits?
2. If any unusually high or low days were identified in part 1, construct a new pair of 3-sigma limits based on the days that remain from the initial 30-day period.
3. Using the upper and lower limits identified in part 2, extend the c-chart to include days 31–60. Based on the 3-sigma limits in this chart, does it appear that the toll-free number has caused a shift in the daily number of complaints received?

Day	Number of Complaints	Day	Number of Complaints	Day	Number of Complaints	Day	Number of Complaints
1	2	16	3	31	4	46	2
2	4	17	1	32	6	47	2
3	1	18	1	33	4	48	6
4	7	19	0	34	3	49	7
5	1	20	2	35	5	50	2
6	3	21	1	36	5	51	5
7	0	22	5	37	3	52	9
8	2	23	7	38	5	53	3
9	3	24	6	39	2	54	4
10	2	25	1	40	4	55	2
11	2	26	0	41	5	56	9
12	3	27	3	42	8	57	3
13	0	28	3	43	3	58	3
14	3	29	3	44	5	59	10
15	9	30	2	45	2	60	7

Willard Bolt Company

At regular intervals during the last three work shifts, technicians at the Willard Bolt Company have taken a total of 40 samples, each with $n = 5$ bolts. The measurement taken is the bolt diameter, in inches, and the data are listed in file **CDB20**.

1. Using only the first 20 samples, construct 3-sigma control charts for the mean and range. For each chart, plot all 20 sample means or ranges. Using the mean and range charts together, does the process appear to have been in control when each of these samples was taken?
   a. If the process was in control for samples 1–20, use the upper and lower control limits from these samples in a second set of control charts that includes all 40 samples. On this basis, was the process in control during the collection of samples 21–40? Use the mean chart and the range chart together in evaluating process control.
   b. If the process was not in control for each of samples 1–20, assume that an assignable cause has been identified for those that were outside the limits. Use the remaining samples as the basis for UCL and LCL values with which to evaluate samples 21–40.

Note: The preceding possibilities would be likely to occur in practice, since data that have already been gathered through sample 20 might be employed in monitoring the process through samples 21–40. In part 2, all of the samples will be considered at once.

2. Considering all 40 samples, construct 3-sigma control charts for the mean and range. For each chart, plot all 40 sample means or ranges. Using the mean and range charts together, does the process appear to have been in control when each of these samples was taken?

   In general, compared to the charts constructed in part 1, do the sample means and ranges for samples 21–40 fit more easily or less easily into the control limits? Would you expect that they *should* fit more easily or less easily into the control limits? Why?
3. If your computer statistical package provides additional tests for determining whether a process is in control, carry these out for the entire series of 40 sample means and sample ranges, then interpret the results. Did the computer identify any "out of control" clues that were not caught in part 2?

SEEING STATISTICS: APPLET 21

Mean Control Chart

This applet has two different mean charts:

1. The first mean chart represents a "well-behaved" process that tends to be under control, with samples of $n = 5$ taken from a production process known to have $\mu = 2.135$ inches and $\sigma = 0.002$ inches. Each time we press the "New Sample" button, a new sample is selected and its mean is displayed on the chart. We can select up to 100 samples in a row before it's necessary to reload the applet and try again.
2. The second mean chart is similar to the first, but is not so well behaved: The process has a tendency to go out of control in various ways somewhere between the 15th sample and the 35th sample.

Examples of the two applets are shown here.

1. The "well-behaved" process that tends to be in control:

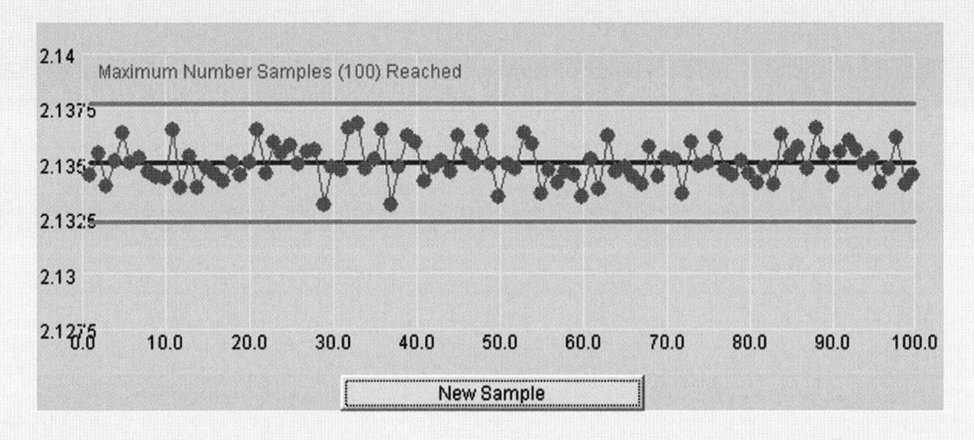

2. The process that has a tendency to go out of control:

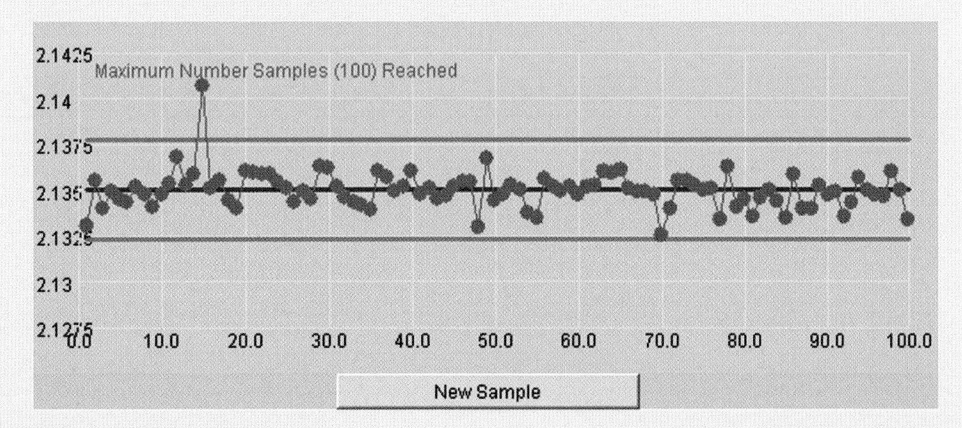

(*continued*)

Applet Exercises

21.1 Using the first applet, press the "New Sample" button 10 times. You have now generated 10 samples from the process and the means are shown on the chart. Does the process appear to be in control?

21.2 Continuing with the first applet, press the "New Sample" button 90 more times to reach the limit of 100 consecutive samples. Does it appear that the process was in control during the 100 samples?

21.3 Using the second applet, press the "New Sample" button 35 times. You have now generated 35 samples from the process and the means are shown on the chart. Does the process appear to be in control? If not, describe the condition(s) in the chart that led to your "not in control" conclusion.

21.4 Continuing with the second applet, press the "New Sample" button 65 more times to reach the limit of 100 consecutive samples. Describe the condition of the process during these last 65 samples. If the process was out of control earlier, did it seem to "settle down" and regain control during these later samples, or did it continue to have control problems?

21.5 Reload the two mean charts and repeat Applet Exercises 1 through 4. In what way(s) were the results similar and in what way(s) were they different the second time around?

Try out the Applets: http://www.cengage.com/international

APPENDIX A

STATISTICAL TABLES

TABLE A.1

Binomial Distribution, Individual Probabilities for x = number of successes in n trials, prob($x = k$)

$n = 2$

π:	0.1	0.2	0.3	0.4	0.5	0.6	0.7	0.8	0.9
k: 0	0.8100	0.6400	0.4900	0.3600	0.2500	0.1600	0.0900	0.0400	0.0100
1	0.1800	0.3200	0.4200	0.4800	0.5000	0.4800	0.4200	0.3200	0.1800
2	0.0100	0.0400	0.0900	0.1600	0.2500	0.3600	0.4900	0.6400	0.8100

$n = 3$

π:	0.1	0.2	0.3	0.4	0.5	0.6	0.7	0.8	0.9
k: 0	0.7290	0.5120	0.3430	0.2160	0.1250	0.0640	0.0270	0.0080	0.0010
1	0.2430	0.3840	0.4410	0.4320	0.3750	0.2880	0.1890	0.0960	0.0270
2	0.0270	0.0960	0.1890	0.2880	0.3750	0.4320	0.4410	0.3840	0.2430
3	0.0010	0.0080	0.0270	0.0640	0.1250	0.2160	0.3430	0.5120	0.7290

$n = 4$

π:	0.1	0.2	0.3	0.4	0.5	0.6	0.7	0.8	0.9
k: 0	0.6561	0.4096	0.2401	0.1296	0.0625	0.0256	0.0081	0.0016	0.0001
1	0.2916	0.4096	0.4116	0.3456	0.2500	0.1536	0.0756	0.0256	0.0036
2	0.0486	0.1536	0.2646	0.3456	0.3750	0.3456	0.2646	0.1536	0.0486
3	0.0036	0.0256	0.0756	0.1536	0.2500	0.3456	0.4116	0.4096	0.2916
4	0.0001	0.0016	0.0081	0.0256	0.0625	0.1296	0.2401	0.4096	0.6561

$n = 5$

π:	0.1	0.2	0.3	0.4	0.5	0.6	0.7	0.8	0.9
k: 0	0.5905	0.3277	0.1681	0.0778	0.0313	0.0102	0.0024	0.0003	0.0000
1	0.3281	0.4096	0.3601	0.2592	0.1562	0.0768	0.0284	0.0064	0.0005
2	0.0729	0.2048	0.3087	0.3456	0.3125	0.2304	0.1323	0.0512	0.0081
3	0.0081	0.0512	0.1323	0.2304	0.3125	0.3456	0.3087	0.2048	0.0729
4	0.0005	0.0064	0.0284	0.0768	0.1562	0.2592	0.3601	0.4096	0.3281
5	0.0000	0.0003	0.0024	0.0102	0.0313	0.0778	0.1681	0.3277	0.5905

Source: Probabilities generated by Minitab, then compiled as shown here.

TABLE A.1
(continued)

$n = 6$

π:	0.1	0.2	0.3	0.4	0.5	0.6	0.7	0.8	0.9
k: 0	0.5314	0.2621	0.1176	0.0467	0.0156	0.0041	0.0007	0.0001	0.0000
1	0.3543	0.3932	0.3025	0.1866	0.0937	0.0369	0.0102	0.0015	0.0001
2	0.0984	0.2458	0.3241	0.3110	0.2344	0.1382	0.0595	0.0154	0.0012
3	0.0146	0.0819	0.1852	0.2765	0.3125	0.2765	0.1852	0.0819	0.0146
4	0.0012	0.0154	0.0595	0.1382	0.2344	0.3110	0.3241	0.2458	0.0984
5	0.0001	0.0015	0.0102	0.0369	0.0937	0.1866	0.3025	0.3932	0.3543
6	0.0000	0.0001	0.0007	0.0041	0.0156	0.0467	0.1176	0.2621	0.5314

$n = 7$

π:	0.1	0.2	0.3	0.4	0.5	0.6	0.7	0.8	0.9
k: 0	0.4783	0.2097	0.0824	0.0280	0.0078	0.0016	0.0002	0.0000	
1	0.3720	0.3670	0.2471	0.1306	0.0547	0.0172	0.0036	0.0004	0.0000
2	0.1240	0.2753	0.3177	0.2613	0.1641	0.0774	0.0250	0.0043	0.0002
3	0.0230	0.1147	0.2269	0.2903	0.2734	0.1935	0.0972	0.0287	0.0026
4	0.0026	0.0287	0.0972	0.1935	0.2734	0.2903	0.2269	0.1147	0.0230
5	0.0002	0.0043	0.0250	0.0774	0.1641	0.2613	0.3177	0.2753	0.1240
6	0.0000	0.0004	0.0036	0.0172	0.0547	0.1306	0.2471	0.3670	0.3720
7		0.0000	0.0002	0.0016	0.0078	0.0280	0.0824	0.2097	0.4783

$n = 8$

π:	0.1	0.2	0.3	0.4	0.5	0.6	0.7	0.8	0.9
k: 0	0.4305	0.1678	0.0576	0.0168	0.0039	0.0007	0.0001	0.0000	
1	0.3826	0.3355	0.1977	0.0896	0.0313	0.0079	0.0012	0.0001	
2	0.1488	0.2936	0.2965	0.2090	0.1094	0.0413	0.0100	0.0011	0.0000
3	0.0331	0.1468	0.2541	0.2787	0.2187	0.1239	0.0467	0.0092	0.0004
4	0.0046	0.0459	0.1361	0.2322	0.2734	0.2322	0.1361	0.0459	0.0046
5	0.0004	0.0092	0.0467	0.1239	0.2187	0.2787	0.2541	0.1468	0.0331
6	0.0000	0.0011	0.0100	0.0413	0.1094	0.2090	0.2965	0.2936	0.1488
7		0.0001	0.0012	0.0079	0.0313	0.0896	0.1977	0.3355	0.3826
8		0.0000	0.0001	0.0007	0.0039	0.0168	0.0576	0.1678	0.4305

$n = 9$

π:	0.1	0.2	0.3	0.4	0.5	0.6	0.7	0.8	0.9
k: 0	0.3874	0.1342	0.0404	0.0101	0.0020	0.0003	0.0000		
1	0.3874	0.3020	0.1556	0.0605	0.0176	0.0035	0.0004	0.0000	
2	0.1722	0.3020	0.2668	0.1612	0.0703	0.0212	0.0039	0.0003	0.0000
3	0.0446	0.1762	0.2668	0.2508	0.1641	0.0743	0.0210	0.0028	0.0001
4	0.0074	0.0661	0.1715	0.2508	0.2461	0.1672	0.0735	0.0165	0.0008
5	0.0008	0.0165	0.0735	0.1672	0.2461	0.2508	0.1715	0.0661	0.0074
6	0.0001	0.0028	0.0210	0.0743	0.1641	0.2508	0.2668	0.1762	0.0446
7	0.0000	0.0003	0.0039	0.0212	0.0703	0.1612	0.2668	0.3020	0.1722
8		0.0000	0.0004	0.0035	0.0176	0.0605	0.1556	0.3020	0.3874
9			0.0000	0.0003	0.0020	0.0101	0.0404	0.1342	0.3874

TABLE A.1

(continued)

n = 10

π:	0.1	0.2	0.3	0.4	0.5	0.6	0.7	0.8	0.9
k: 0	0.3487	0.1074	0.0282	0.0060	0.0010	0.0001	0.0000		
1	0.3874	0.2684	0.1211	0.0403	0.0098	0.0016	0.0001	0.0000	
2	0.1937	0.3020	0.2335	0.1209	0.0439	0.0106	0.0014	0.0001	
3	0.0574	0.2013	0.2668	0.2150	0.1172	0.0425	0.0090	0.0008	0.0000
4	0.0112	0.0881	0.2001	0.2508	0.2051	0.1115	0.0368	0.0055	0.0001
5	0.0015	0.0264	0.1029	0.2007	0.2461	0.2007	0.1029	0.0264	0.0015
6	0.0001	0.0055	0.0368	0.1115	0.2051	0.2508	0.2001	0.0881	0.0112
7	0.0000	0.0008	0.0090	0.0425	0.1172	0.2150	0.2668	0.2013	0.0574
8		0.0001	0.0014	0.0106	0.0439	0.1209	0.2335	0.3020	0.1937
9		0.0000	0.0001	0.0016	0.0098	0.0403	0.1211	0.2684	0.3874
10			0.0000	0.0001	0.0010	0.0060	0.0282	0.1074	0.3487

n = 11

π:	0.1	0.2	0.3	0.4	0.5	0.6	0.7	0.8	0.9
k: 0	0.3138	0.0859	0.0198	0.0036	0.0005	0.0000			
1	0.3835	0.2362	0.0932	0.0266	0.0054	0.0007	0.0000		
2	0.2131	0.2953	0.1998	0.0887	0.0269	0.0052	0.0005	0.0000	
3	0.0710	0.2215	0.2568	0.1774	0.0806	0.0234	0.0037	0.0002	
4	0.0158	0.1107	0.2201	0.2365	0.1611	0.0701	0.0173	0.0017	0.0000
5	0.0025	0.0388	0.1321	0.2207	0.2256	0.1471	0.0566	0.0097	0.0003
6	0.0003	0.0097	0.0566	0.1471	0.2256	0.2207	0.1321	0.0388	0.0025
7	0.0000	0.0017	0.0173	0.0701	0.1611	0.2365	0.2201	0.1107	0.0158
8		0.0002	0.0037	0.0234	0.0806	0.1774	0.2568	0.2215	0.0710
9		0.0000	0.0005	0.0052	0.0269	0.0887	0.1998	0.2953	0.2131
10			0.0000	0.0007	0.0054	0.0266	0.0932	0.2362	0.3835
11				0.0000	0.0005	0.0036	0.0198	0.0859	0.3138

n = 12

π:	0.1	0.2	0.3	0.4	0.5	0.6	0.7	0.8	0.9
k: 0	0.2824	0.0687	0.0138	0.0022	0.0002	0.0000			
1	0.3766	0.2062	0.0712	0.0174	0.0029	0.0003	0.0000		
2	0.2301	0.2835	0.1678	0.0639	0.0161	0.0025	0.0002	0.0000	
3	0.0852	0.2362	0.2397	0.1419	0.0537	0.0125	0.0015	0.0001	
4	0.0213	0.1329	0.2311	0.2128	0.1208	0.0420	0.0078	0.0005	
5	0.0038	0.0532	0.1585	0.2270	0.1934	0.1009	0.0291	0.0033	0.0000
6	0.0005	0.0155	0.0792	0.1766	0.2256	0.1766	0.0792	0.0155	0.0005
7	0.0000	0.0033	0.0291	0.1009	0.1934	0.2270	0.1585	0.0532	0.0038
8		0.0005	0.0078	0.0420	0.1208	0.2128	0.2311	0.1329	0.0213
9		0.0001	0.0015	0.0125	0.0537	0.1419	0.2397	0.2362	0.0852
10		0.0000	0.0002	0.0025	0.0161	0.0639	0.1678	0.2835	0.2301
11			0.0000	0.0003	0.0029	0.0174	0.0712	0.2062	0.3766
12				0.0000	0.0002	0.0022	0.0138	0.0687	0.2824

TABLE A.1
(continued)

n = 13

π:	0.1	0.2	0.3	0.4	0.5	0.6	0.7	0.8	0.9
k: 0	0.2542	0.0550	0.0097	0.0013	0.0001	0.0000			
1	0.3672	0.1787	0.0540	0.0113	0.0016	0.0001	0.0000		
2	0.2448	0.2680	0.1388	0.0453	0.0095	0.0012	0.0001		
3	0.0997	0.2457	0.2181	0.1107	0.0349	0.0065	0.0006	0.0000	
4	0.0277	0.1535	0.2337	0.1845	0.0873	0.0243	0.0034	0.0001	
5	0.0055	0.0691	0.1803	0.2214	0.1571	0.0656	0.0142	0.0011	0.0000
6	0.0008	0.0230	0.1030	0.1968	0.2095	0.1312	0.0442	0.0058	0.0001
7	0.0001	0.0058	0.0442	0.1312	0.2095	0.1968	0.1030	0.0230	0.0008
8	0.0000	0.0011	0.0142	0.0656	0.1571	0.2214	0.1803	0.0691	0.0055
9		0.0001	0.0034	0.0243	0.0873	0.1845	0.2337	0.1535	0.0277
10		0.0000	0.0006	0.0065	0.0349	0.1107	0.2181	0.2457	0.0997
11			0.0001	0.0012	0.0095	0.0453	0.1388	0.2680	0.2448
12			0.0000	0.0001	0.0016	0.0113	0.0540	0.1787	0.3672
13				0.0000	0.0001	0.0013	0.0097	0.0550	0.2542

n = 14

π:	0.1	0.2	0.3	0.4	0.5	0.6	0.7	0.8	0.9
k: 0	0.2288	0.0440	0.0068	0.0008	0.0001	0.0000			
1	0.3559	0.1539	0.0407	0.0073	0.0009	0.0001			
2	0.2570	0.2501	0.1134	0.0317	0.0056	0.0005	0.0000		
3	0.1142	0.2501	0.1943	0.0845	0.0222	0.0033	0.0002		
4	0.0349	0.1720	0.2290	0.1549	0.0611	0.0136	0.0014	0.0000	
5	0.0078	0.0860	0.1963	0.2066	0.1222	0.0408	0.0066	0.0003	
6	0.0013	0.0322	0.1262	0.2066	0.1833	0.0918	0.0232	0.0020	0.0000
7	0.0002	0.0092	0.0618	0.1574	0.2095	0.1574	0.0618	0.0092	0.0002
8	0.0000	0.0020	0.0232	0.0918	0.1833	0.2066	0.1262	0.0322	0.0013
9		0.0003	0.0066	0.0408	0.1222	0.2066	0.1963	0.0860	0.0078
10		0.0000	0.0014	0.0136	0.0611	0.1549	0.2290	0.1720	0.0349
11			0.0002	0.0033	0.0222	0.0845	0.1943	0.2501	0.1142
12			0.0000	0.0005	0.0056	0.0317	0.1134	0.2501	0.2570
13				0.0001	0.0009	0.0073	0.0407	0.1539	0.3559
14				0.0000	0.0001	0.0008	0.0068	0.0440	0.2288

n = 15

π:	0.1	0.2	0.3	0.4	0.5	0.6	0.7	0.8	0.9
k: 0	0.2059	0.0352	0.0047	0.0005	0.0000				
1	0.3432	0.1319	0.0305	0.0047	0.0005	0.0000			
2	0.2669	0.2309	0.0916	0.0219	0.0032	0.0003	0.0000		
3	0.1285	0.2501	0.1700	0.0634	0.0139	0.0016	0.0001		
4	0.0428	0.1876	0.2186	0.1268	0.0417	0.0074	0.0006	0.0000	
5	0.0105	0.1032	0.2061	0.1859	0.0916	0.0245	0.0030	0.0001	
6	0.0019	0.0430	0.1472	0.2066	0.1527	0.0612	0.0116	0.0007	
7	0.0003	0.0138	0.0811	0.1771	0.1964	0.1181	0.0348	0.0035	0.0000
8	0.0000	0.0035	0.0348	0.1181	0.1964	0.1771	0.0811	0.0138	0.0003
9		0.0007	0.0116	0.0612	0.1527	0.2066	0.1472	0.0430	0.0019
10		0.0001	0.0030	0.0245	0.0916	0.1859	0.2061	0.1032	0.0105
11		0.0000	0.0006	0.0074	0.0417	0.1268	0.2186	0.1876	0.0428
12			0.0001	0.0016	0.0139	0.0634	0.1700	0.2501	0.1285
13			0.0000	0.0003	0.0032	0.0219	0.0916	0.2309	0.2669
14				0.0000	0.0005	0.0047	0.0305	0.1319	0.3432
15					0.0000	0.0005	0.0047	0.0352	0.2059

TABLE A.1

(continued)

$n = 16$

π:	0.1	0.2	0.3	0.4	0.5	0.6	0.7	0.8	0.9
k: 0	0.1853	0.0281	0.0033	0.0003	0.0000				
1	0.3294	0.1126	0.0228	0.0030	0.0002	0.0000			
2	0.2745	0.2111	0.0732	0.0150	0.0018	0.0001			
3	0.1423	0.2463	0.1465	0.0468	0.0085	0.0008	0.0000		
4	0.0514	0.2001	0.2040	0.1014	0.0278	0.0040	0.0002		
5	0.0137	0.1201	0.2099	0.1623	0.0667	0.0142	0.0013	0.0000	
6	0.0028	0.0550	0.1649	0.1983	0.1222	0.0392	0.0056	0.0002	
7	0.0004	0.0197	0.1010	0.1889	0.1746	0.0840	0.0185	0.0012	0.0000
8	0.0001	0.0055	0.0487	0.1417	0.1964	0.1417	0.0487	0.0055	0.0001
9	0.0000	0.0012	0.0185	0.0840	0.1746	0.1889	0.1010	0.0197	0.0004
10		0.0002	0.0056	0.0392	0.1222	0.1983	0.1649	0.0550	0.0028
11		0.0000	0.0013	0.0142	0.0667	0.1623	0.2099	0.1201	0.0137
12			0.0002	0.0040	0.0278	0.1014	0.2040	0.2001	0.0514
13			0.0000	0.0008	0.0085	0.0468	0.1465	0.2463	0.1423
14				0.0001	0.0018	0.0150	0.0732	0.2111	0.2745
15				0.0000	0.0002	0.0030	0.0228	0.1126	0.3294
16					0.0000	0.0003	0.0033	0.0281	0.1853

$n = 17$

π:	0.1	0.2	0.3	0.4	0.5	0.6	0.7	0.8	0.9
k: 0	0.1668	0.0225	0.0023	0.0002	0.0000				
1	0.3150	0.0957	0.0169	0.0019	0.0001	0.0000			
2	0.2800	0.1914	0.0581	0.0102	0.0010	0.0001			
3	0.1556	0.2393	0.1245	0.0341	0.0052	0.0004	0.0000		
4	0.0605	0.2093	0.1868	0.0796	0.0182	0.0021	0.0001		
5	0.0175	0.1361	0.2081	0.1379	0.0472	0.0081	0.0006	0.0000	
6	0.0039	0.0680	0.1784	0.1839	0.0944	0.0242	0.0026	0.0001	
7	0.0007	0.0267	0.1201	0.1927	0.1484	0.0571	0.0095	0.0004	
8	0.0001	0.0084	0.0644	0.1606	0.1855	0.1070	0.0276	0.0021	0.0000
9	0.0000	0.0021	0.0276	0.1070	0.1855	0.1606	0.0644	0.0084	0.0001
10		0.0004	0.0095	0.0571	0.1484	0.1927	0.1201	0.0267	0.0007
11		0.0001	0.0026	0.0242	0.0944	0.1839	0.1784	0.0680	0.0039
12		0.0000	0.0006	0.0081	0.0472	0.1379	0.2081	0.1361	0.0175
13			0.0001	0.0021	0.0182	0.0796	0.1868	0.2093	0.0605
14			0.0000	0.0004	0.0052	0.0341	0.1245	0.2393	0.1556
15				0.0001	0.0010	0.0102	0.0581	0.1914	0.2800
16				0.0000	0.0001	0.0019	0.0169	0.0957	0.3150
17					0.0000	0.0002	0.0023	0.0225	0.1668

TABLE A.1
(continued)

$n = 18$

π:	0.1	0.2	0.3	0.4	0.5	0.6	0.7	0.8	0.9
k: 0	0.1501	0.0180	0.0016	0.0001	0.0000				
1	0.3002	0.0811	0.0126	0.0012	0.0001				
2	0.2835	0.1723	0.0458	0.0069	0.0006	0.0000			
3	0.1680	0.2297	0.1046	0.0246	0.0031	0.0002			
4	0.0700	0.2153	0.1681	0.0614	0.0117	0.0011	0.0000		
5	0.0218	0.1507	0.2017	0.1146	0.0327	0.0045	0.0002		
6	0.0052	0.0816	0.1873	0.1655	0.0708	0.0145	0.0012	0.0000	
7	0.0010	0.0350	0.1376	0.1892	0.1214	0.0374	0.0046	0.0001	
8	0.0002	0.0120	0.0811	0.1734	0.1669	0.0771	0.0149	0.0008	
9	0.0000	0.0033	0.0386	0.1284	0.1855	0.1284	0.0386	0.0033	0.0000
10		0.0008	0.0149	0.0771	0.1669	0.1734	0.0811	0.0120	0.0002
11		0.0001	0.0046	0.0374	0.1214	0.1892	0.1376	0.0350	0.0010
12		0.0000	0.0012	0.0145	0.0708	0.1655	0.1873	0.0816	0.0052
13			0.0002	0.0045	0.0327	0.1146	0.2017	0.1507	0.0218
14			0.0000	0.0011	0.0117	0.0614	0.1681	0.2153	0.0700
15				0.0002	0.0031	0.0246	0.1046	0.2297	0.1680
16				0.0000	0.0006	0.0069	0.0458	0.1723	0.2835
17					0.0001	0.0012	0.0126	0.0811	0.3002
18					0.0000	0.0001	0.0016	0.0180	0.1501

$n = 19$

π:	0.1	0.2	0.3	0.4	0.5	0.6	0.7	0.8	0.9
k: 0	0.1351	0.0144	0.0011	0.0001					
1	0.2852	0.0685	0.0093	0.0008	0.0000				
2	0.2852	0.1540	0.0358	0.0046	0.0003	0.0000			
3	0.1796	0.2182	0.0869	0.0175	0.0018	0.0001			
4	0.0798	0.2182	0.1491	0.0467	0.0074	0.0005	0.0000		
5	0.0266	0.1636	0.1916	0.0933	0.0222	0.0024	0.0001		
6	0.0069	0.0955	0.1916	0.1451	0.0518	0.0085	0.0005		
7	0.0014	0.0443	0.1525	0.1797	0.0961	0.0237	0.0022	0.0000	
8	0.0002	0.0166	0.0981	0.1797	0.1442	0.0532	0.0077	0.0003	
9	0.0000	0.0051	0.0514	0.1464	0.1762	0.0976	0.0220	0.0013	
10		0.0013	0.0220	0.0976	0.1762	0.1464	0.0514	0.0051	0.0000
11		0.0003	0.0077	0.0532	0.1442	0.1797	0.0981	0.0166	0.0002
12		0.0000	0.0022	0.0237	0.0961	0.1797	0.1525	0.0443	0.0014
13			0.0005	0.0085	0.0518	0.1451	0.1916	0.0955	0.0069
14			0.0001	0.0024	0.0222	0.0933	0.1916	0.1636	0.0266
15			0.0000	0.0005	0.0074	0.0467	0.1491	0.2182	0.0798
16				0.0001	0.0018	0.0175	0.0869	0.2182	0.1796
17				0.0000	0.0003	0.0046	0.0358	0.1540	0.2852
18					0.0000	0.0008	0.0093	0.0685	0.2852
19						0.0001	0.0011	0.0144	0.1351

TABLE A.1
(continued)

$n = 20$

π:	0.1	0.2	0.3	0.4	0.5	0.6	0.7	0.8	0.9
k: 0	0.1216	0.0115	0.0008	0.0000					
1	0.2702	0.0576	0.0068	0.0005	0.0000				
2	0.2852	0.1369	0.0278	0.0031	0.0002				
3	0.1901	0.2054	0.0716	0.0123	0.0011	0.0000			
4	0.0898	0.2182	0.1304	0.0350	0.0046	0.0003			
5	0.0319	0.1746	0.1789	0.0746	0.0148	0.0013	0.0000		
6	0.0089	0.1091	0.1916	0.1244	0.0370	0.0049	0.0002		
7	0.0020	0.0545	0.1643	0.1659	0.0739	0.0146	0.0010	0.0000	
8	0.0004	0.0222	0.1144	0.1797	0.1201	0.0355	0.0039	0.0001	
9	0.0001	0.0074	0.0654	0.1597	0.1602	0.0710	0.0120	0.0005	
10	0.0000	0.0020	0.0308	0.1171	0.1762	0.1171	0.0308	0.0020	0.0000
11		0.0005	0.0120	0.0710	0.1602	0.1597	0.0654	0.0074	0.0001
12		0.0001	0.0039	0.0355	0.1201	0.1797	0.1144	0.0222	0.0004
13		0.0000	0.0010	0.0146	0.0739	0.1659	0.1643	0.0545	0.0020
14			0.0002	0.0049	0.0370	0.1244	0.1916	0.1091	0.0089
15			0.0000	0.0013	0.0148	0.0746	0.1789	0.1746	0.0319
16				0.0003	0.0046	0.0350	0.1304	0.2182	0.0898
17				0.0000	0.0011	0.0123	0.0716	0.2054	0.1901
18					0.0002	0.0031	0.0278	0.1369	0.2852
19					0.0000	0.0005	0.0068	0.0576	0.2702
20						0.0000	0.0008	0.0115	0.1216

$n = 25$

π:	0.1	0.2	0.3	0.4	0.5	0.6	0.7	0.8	0.9
k: 0	0.0718	0.0038	0.0001						
1	0.1994	0.0236	0.0014	0.0000					
2	0.2659	0.0708	0.0074	0.0004	0.0000				
3	0.2265	0.1358	0.0243	0.0019	0.0001				
4	0.1384	0.1867	0.0572	0.0071	0.0004				
5	0.0646	0.1960	0.1030	0.0199	0.0016	0.0000			
6	0.0239	0.1633	0.1472	0.0442	0.0053	0.0002			
7	0.0072	0.1108	0.1712	0.0800	0.0143	0.0009	0.0000		
8	0.0018	0.0623	0.1651	0.1200	0.0322	0.0031	0.0001		
9	0.0004	0.0294	0.1336	0.1511	0.0609	0.0088	0.0004		
10	0.0001	0.0118	0.0916	0.1612	0.0974	0.0212	0.0013	0.0000	
11	0.0000	0.0040	0.0536	0.1465	0.1328	0.0434	0.0042	0.0001	
12		0.0012	0.0268	0.1140	0.1550	0.0760	0.0115	0.0003	
13		0.0003	0.0115	0.0760	0.1550	0.1140	0.0268	0.0012	
14		0.0001	0.0042	0.0434	0.1328	0.1465	0.0536	0.0040	0.0000
15		0.0000	0.0013	0.0212	0.0974	0.1612	0.0916	0.0118	0.0001
16			0.0004	0.0088	0.0609	0.1511	0.1336	0.0294	0.0004
17			0.0001	0.0031	0.0322	0.1200	0.1651	0.0623	0.0018
18			0.0000	0.0009	0.0143	0.0800	0.1712	0.1108	0.0072
19				0.0002	0.0053	0.0442	0.1472	0.1633	0.0239
20				0.0000	0.0016	0.0199	0.1030	0.1960	0.0646
21					0.0004	0.0071	0.0572	0.1867	0.1384
22					0.0001	0.0019	0.0243	0.1358	0.2265
23					0.0000	0.0004	0.0074	0.0708	0.2659
24						0.0000	0.0014	0.0236	0.1994
25							0.0001	0.0038	0.0718

TABLE A.2

Binomial Distribution, Cumulative Probabilities for x = number of successes in n trials, prob($x \leq k$)

$n = 2$

π:	0.1	0.2	0.3	0.4	0.5	0.6	0.7	0.8	0.9
k: 0	0.8100	0.6400	0.4900	0.3600	0.2500	0.1600	0.0900	0.0400	0.0100
1	0.9900	0.9600	0.9100	0.8400	0.7500	0.6400	0.5100	0.3600	0.1900
2	1.0000	1.0000	1.0000	1.0000	1.0000	1.0000	1.0000	1.0000	1.0000

$n = 3$

π:	0.1	0.2	0.3	0.4	0.5	0.6	0.7	0.8	0.9
k: 0	0.7290	0.5120	0.3430	0.2160	0.1250	0.0640	0.0270	0.0080	0.0010
1	0.9720	0.8960	0.7840	0.6480	0.5000	0.3520	0.2160	0.1040	0.0280
2	0.9990	0.9920	0.9730	0.9360	0.8750	0.7840	0.6570	0.4880	0.2710
3	1.0000	1.0000	1.0000	1.0000	1.0000	1.0000	1.0000	1.0000	1.0000

$n = 4$

π:	0.1	0.2	0.3	0.4	0.5	0.6	0.7	0.8	0.9
k: 0	0.6561	0.4096	0.2401	0.1296	0.0625	0.0256	0.0081	0.0016	0.0001
1	0.9477	0.8192	0.6517	0.4752	0.3125	0.1792	0.0837	0.0272	0.0037
2	0.9963	0.9728	0.9163	0.8208	0.6875	0.5248	0.3483	0.1808	0.0523
3	0.9999	0.9984	0.9919	0.9744	0.9375	0.8704	0.7599	0.5904	0.3439
4	1.0000	1.0000	1.0000	1.0000	1.0000	1.0000	1.0000	1.0000	1.0000

$n = 5$

π:	0.1	0.2	0.3	0.4	0.5	0.6	0.7	0.8	0.9
k: 0	0.5905	0.3277	0.1681	0.0778	0.0313	0.0102	0.0024	0.0003	0.0000
1	0.9185	0.7373	0.5282	0.3370	0.1875	0.0870	0.0308	0.0067	0.0005
2	0.9914	0.9421	0.8369	0.6826	0.5000	0.3174	0.1631	0.0579	0.0086
3	0.9995	0.9933	0.9692	0.9130	0.8125	0.6630	0.4718	0.2627	0.0815
4	1.0000	0.9997	0.9976	0.9898	0.9688	0.9222	0.8319	0.6723	0.4095
5		1.0000	1.0000	1.0000	1.0000	1.0000	1.0000	1.0000	1.0000

$n = 6$

π:	0.1	0.2	0.3	0.4	0.5	0.6	0.7	0.8	0.9
k: 0	0.5314	0.2621	0.1176	0.0467	0.0156	0.0041	0.0007	0.0001	0.0000
1	0.8857	0.6554	0.4202	0.2333	0.1094	0.0410	0.0109	0.0016	0.0001
2	0.9841	0.9011	0.7443	0.5443	0.3437	0.1792	0.0705	0.0170	0.0013
3	0.9987	0.9830	0.9295	0.8208	0.6563	0.4557	0.2557	0.0989	0.0159
4	0.9999	0.9984	0.9891	0.9590	0.8906	0.7667	0.5798	0.3446	0.1143
5	1.0000	0.9999	0.9993	0.9959	0.9844	0.9533	0.8824	0.7379	0.4686
6		1.0000	1.0000	1.0000	1.0000	1.0000	1.0000	1.0000	1.0000

Source: Probabilities generated by Minitab, then compiled as shown here.

TABLE A.2
(continued)

$n = 7$

π:	0.1	0.2	0.3	0.4	0.5	0.6	0.7	0.8	0.9
k: 0	0.4783	0.2097	0.0824	0.0280	0.0078	0.0016	0.0002	0.0000	
1	0.8503	0.5767	0.3294	0.1586	0.0625	0.0188	0.0038	0.0004	0.0000
2	0.9743	0.8520	0.6471	0.4199	0.2266	0.0963	0.0288	0.0047	0.0002
3	0.9973	0.9667	0.8740	0.7102	0.5000	0.2898	0.1260	0.0333	0.0027
4	0.9998	0.9953	0.9712	0.9037	0.7734	0.5801	0.3529	0.1480	0.0257
5	1.0000	0.9996	0.9962	0.9812	0.9375	0.8414	0.6706	0.4233	0.1497
6		1.0000	0.9998	0.9984	0.9922	0.9720	0.9176	0.7903	0.5217
7			1.0000	1.0000	1.0000	1.0000	1.0000	1.0000	1.0000

$n = 8$

π:	0.1	0.2	0.3	0.4	0.5	0.6	0.7	0.8	0.9
k: 0	0.4305	0.1678	0.0576	0.0168	0.0039	0.0007	0.0001	0.0000	
1	0.8131	0.5033	0.2553	0.1064	0.0352	0.0085	0.0013	0.0001	
2	0.9619	0.7969	0.5518	0.3154	0.1445	0.0498	0.0113	0.0012	0.0000
3	0.9950	0.9437	0.8059	0.5941	0.3633	0.1737	0.0580	0.0104	0.0004
4	0.9996	0.9896	0.9420	0.8263	0.6367	0.4059	0.1941	0.0563	0.0050
5	1.0000	0.9988	0.9887	0.9502	0.8555	0.6846	0.4482	0.2031	0.0381
6		0.9999	0.9987	0.9915	0.9648	0.8936	0.7447	0.4967	0.1869
7		1.0000	0.9999	0.9993	0.9961	0.9832	0.9424	0.8322	0.5695
8			1.0000	1.0000	1.0000	1.0000	1.0000	1.0000	1.0000

$n = 9$

π:	0.1	0.2	0.3	0.4	0.5	0.6	0.7	0.8	0.9
k: 0	0.3874	0.1342	0.0404	0.0101	0.0020	0.0003	0.0000		
1	0.7748	0.4362	0.1960	0.0705	0.0195	0.0038	0.0004	0.0000	
2	0.9470	0.7382	0.4628	0.2318	0.0898	0.0250	0.0043	0.0003	0.0000
3	0.9917	0.9144	0.7297	0.4826	0.2539	0.0994	0.0253	0.0031	0.0001
4	0.9991	0.9804	0.9012	0.7334	0.5000	0.2666	0.0988	0.0196	0.0009
5	0.9999	0.9969	0.9747	0.9006	0.7461	0.5174	0.2703	0.0856	0.0083
6	1.0000	0.9997	0.9957	0.9750	0.9102	0.7682	0.5372	0.2618	0.0530
7		1.0000	0.9996	0.9962	0.9805	0.9295	0.8040	0.5638	0.2252
8			1.0000	0.9997	0.9980	0.9899	0.9596	0.8658	0.6126
9				1.0000	1.0000	1.0000	1.0000	1.0000	1.0000

$n = 10$

π:	0.1	0.2	0.3	0.4	0.5	0.6	0.7	0.8	0.9
k: 0	0.3487	0.1074	0.0282	0.0060	0.0010	0.0001	0.0000		
1	0.7361	0.3758	0.1493	0.0464	0.0107	0.0017	0.0001	0.0000	
2	0.9298	0.6778	0.3828	0.1673	0.0547	0.0123	0.0016	0.0001	
3	0.9872	0.8791	0.6496	0.3823	0.1719	0.0548	0.0106	0.0009	0.0000
4	0.9984	0.9672	0.8497	0.6331	0.3770	0.1662	0.0473	0.0064	0.0001
5	0.9999	0.9936	0.9527	0.8338	0.6230	0.3669	0.1503	0.0328	0.0016
6	1.0000	0.9991	0.9894	0.9452	0.8281	0.6177	0.3504	0.1209	0.0128
7		0.9999	0.9984	0.9877	0.9453	0.8327	0.6172	0.3222	0.0702
8		1.0000	0.9999	0.9983	0.9893	0.9536	0.8507	0.6242	0.2639
9			1.0000	0.9999	0.9990	0.9940	0.9718	0.8926	0.6513
10				1.0000	1.0000	1.0000	1.0000	1.0000	1.0000

TABLE A.2
(continued)

n = 11

π:	0.1	0.2	0.3	0.4	0.5	0.6	0.7	0.8	0.9
k: 0	0.3138	0.0859	0.0198	0.0036	0.0005	0.0000			
1	0.6974	0.3221	0.1130	0.0302	0.0059	0.0007	0.0000		
2	0.9104	0.6174	0.3127	0.1189	0.0327	0.0059	0.0006	0.0000	
3	0.9815	0.8389	0.5696	0.2963	0.1133	0.0293	0.0043	0.0002	
4	0.9972	0.9496	0.7897	0.5328	0.2744	0.0994	0.0216	0.0020	0.0000
5	0.9997	0.9883	0.9218	0.7535	0.5000	0.2465	0.0782	0.0117	0.0003
6	1.0000	0.9980	0.9784	0.9006	0.7256	0.4672	0.2103	0.0504	0.0028
7		0.9998	0.9957	0.9707	0.8867	0.7037	0.4304	0.1611	0.0185
8		1.0000	0.9994	0.9941	0.9673	0.8811	0.6873	0.3826	0.0896
9			1.0000	0.9993	0.9941	0.9698	0.8870	0.6779	0.3026
10				1.0000	0.9995	0.9964	0.9802	0.9141	0.6862
11					1.0000	1.0000	1.0000	1.0000	1.0000

n = 12

π:	0.1	0.2	0.3	0.4	0.5	0.6	0.7	0.8	0.9
k: 0	0.2824	0.0687	0.0138	0.0022	0.0002	0.0000			
1	0.6590	0.2749	0.0850	0.0196	0.0032	0.0003	0.0000		
2	0.8891	0.5583	0.2528	0.0834	0.0193	0.0028	0.0002	0.0000	
3	0.9744	0.7946	0.4925	0.2253	0.0730	0.0153	0.0017	0.0001	
4	0.9957	0.9274	0.7237	0.4382	0.1938	0.0573	0.0095	0.0006	0.0000
5	0.9995	0.9806	0.8822	0.6652	0.3872	0.1582	0.0386	0.0039	0.0001
6	0.9999	0.9961	0.9614	0.8418	0.6128	0.3348	0.1178	0.0194	0.0005
7	1.0000	0.9994	0.9905	0.9427	0.8062	0.5618	0.2763	0.0726	0.0043
8		0.9999	0.9983	0.9847	0.9270	0.7747	0.5075	0.2054	0.0256
9		1.0000	0.9998	0.9972	0.9807	0.9166	0.7472	0.4417	0.1109
10			1.0000	0.9997	0.9968	0.9804	0.9150	0.7251	0.3410
11				1.0000	0.9998	0.9978	0.9862	0.9313	0.7176
12					1.0000	1.0000	1.0000	1.0000	1.0000

n = 13

π:	0.1	0.2	0.3	0.4	0.5	0.6	0.7	0.8	0.9
k: 0	0.2542	0.0550	0.0097	0.0013	0.0001	0.0000			
1	0.6213	0.2336	0.0637	0.0126	0.0017	0.0001	0.0000		
2	0.8661	0.5017	0.2025	0.0579	0.0112	0.0013	0.0001		
3	0.9658	0.7473	0.4206	0.1686	0.0461	0.0078	0.0007	0.0000	
4	0.9935	0.9009	0.6543	0.3530	0.1334	0.0321	0.0040	0.0002	
5	0.9991	0.9700	0.8346	0.5744	0.2905	0.0977	0.0182	0.0012	0.0000
6	0.9999	0.9930	0.9376	0.7712	0.5000	0.2288	0.0624	0.0070	0.0001
7	1.0000	0.9988	0.9818	0.9023	0.7095	0.4256	0.1654	0.0300	0.0009
8		0.9998	0.9960	0.9679	0.8666	0.6470	0.3457	0.0991	0.0065
9		1.0000	0.9993	0.9922	0.9539	0.8314	0.5794	0.2527	0.0342
10			0.9999	0.9987	0.9888	0.9421	0.7975	0.4983	0.1339
11			1.0000	0.9999	0.9983	0.9874	0.9363	0.7664	0.3787
12				1.0000	0.9999	0.9987	0.9903	0.9450	0.7458
13					1.0000	1.0000	1.0000	1.0000	1.0000

TABLE A.2
(continued)

n = 14

π:	0.1	0.2	0.3	0.4	0.5	0.6	0.7	0.8	0.9
k: 0	0.2288	0.0440	0.0068	0.0008	0.0001	0.0000			
1	0.5846	0.1979	0.0475	0.0081	0.0009	0.0001			
2	0.8416	0.4481	0.1608	0.0398	0.0065	0.0006	0.0000		
3	0.9559	0.6982	0.3552	0.1243	0.0287	0.0039	0.0002		
4	0.9908	0.8702	0.5842	0.2793	0.0898	0.0175	0.0017	0.0000	
5	0.9985	0.9561	0.7805	0.4859	0.2120	0.0583	0.0083	0.0004	
6	0.9998	0.9884	0.9067	0.6925	0.3953	0.1501	0.0315	0.0024	0.0000
7	1.0000	0.9976	0.9685	0.8499	0.6047	0.3075	0.0933	0.0116	0.0002
8		0.9996	0.9917	0.9417	0.7880	0.5141	0.2195	0.0439	0.0015
9		1.0000	0.9983	0.9825	0.9102	0.7207	0.4158	0.1298	0.0092
10			0.9998	0.9961	0.9713	0.8757	0.6448	0.3018	0.0441
11			1.0000	0.9994	0.9935	0.9602	0.8392	0.5519	0.1584
12				0.9999	0.9991	0.9919	0.9525	0.8021	0.4154
13				1.0000	0.9999	0.9992	0.9932	0.9560	0.7712
14					1.0000	1.0000	1.0000	1.0000	1.0000

n = 15

π:	0.1	0.2	0.3	0.4	0.5	0.6	0.7	0.8	0.9
k: 0	0.2059	0.0352	0.0047	0.0005	0.0000				
1	0.5490	0.1671	0.0353	0.0052	0.0005	0.0000			
2	0.8159	0.3980	0.1268	0.0271	0.0037	0.0003	0.0000		
3	0.9444	0.6482	0.2969	0.0905	0.0176	0.0019	0.0001		
4	0.9873	0.8358	0.5155	0.2173	0.0592	0.0093	0.0007	0.0000	
5	0.9978	0.9389	0.7216	0.4032	0.1509	0.0338	0.0037	0.0001	
6	0.9997	0.9819	0.8689	0.6098	0.3036	0.0950	0.0152	0.0008	
7	1.0000	0.9958	0.9500	0.7869	0.5000	0.2131	0.0500	0.0042	0.0000
8		0.9992	0.9848	0.9050	0.6964	0.3902	0.1311	0.0181	0.0003
9		0.9999	0.9963	0.9662	0.8491	0.5968	0.2784	0.0611	0.0022
10		1.0000	0.9993	0.9907	0.9408	0.7827	0.4845	0.1642	0.0127
11			0.9999	0.9981	0.9824	0.9095	0.7031	0.3518	0.0556
12			1.0000	0.9997	0.9963	0.9729	0.8732	0.6020	0.1841
13				1.0000	0.9995	0.9948	0.9647	0.8329	0.4510
14					1.0000	0.9995	0.9953	0.9648	0.7941
15						1.0000	1.0000	1.0000	1.0000

TABLE A.2
(continued)

$n = 16$

π:	0.1	0.2	0.3	0.4	0.5	0.6	0.7	0.8	0.9
k: 0	0.1853	0.0281	0.0033	0.0003	0.0000				
1	0.5147	0.1407	0.0261	0.0033	0.0003	0.0000			
2	0.7892	0.3518	0.0994	0.0183	0.0021	0.0001			
3	0.9316	0.5981	0.2459	0.0651	0.0106	0.0009	0.0000		
4	0.9830	0.7982	0.4499	0.1666	0.0384	0.0049	0.0003		
5	0.9967	0.9183	0.6598	0.3288	0.1051	0.0191	0.0016	0.0000	
6	0.9995	0.9733	0.8247	0.5272	0.2272	0.0583	0.0071	0.0002	
7	0.9999	0.9930	0.9256	0.7161	0.4018	0.1423	0.0257	0.0015	0.0000
8	1.0000	0.9985	0.9743	0.8577	0.5982	0.2839	0.0744	0.0070	0.0001
9		0.9998	0.9929	0.9417	0.7728	0.4728	0.1753	0.0267	0.0006
10		1.0000	0.9984	0.9809	0.8949	0.6712	0.3402	0.0817	0.0033
11			0.9997	0.9951	0.9616	0.8334	0.5501	0.2018	0.0170
12			1.0000	0.9991	0.9894	0.9349	0.7541	0.4019	0.0684
13				0.9999	0.9979	0.9817	0.9006	0.6482	0.2108
14				1.0000	0.9997	0.9967	0.9739	0.8593	0.4853
15					1.0000	0.9997	0.9967	0.9719	0.8147
16						1.0000	1.0000	1.0000	1.0000

$n = 17$

π:	0.1	0.2	0.3	0.4	0.5	0.6	0.7	0.8	0.9
k: 0	0.1668	0.0225	0.0023	0.0002	0.0000				
1	0.4818	0.1182	0.0193	0.0021	0.0001	0.0000			
2	0.7618	0.3096	0.0774	0.0123	0.0012	0.0001			
3	0.9174	0.5489	0.2019	0.0464	0.0064	0.0005	0.0000		
4	0.9779	0.7582	0.3887	0.1260	0.0245	0.0025	0.0001		
5	0.9953	0.8943	0.5968	0.2639	0.0717	0.0106	0.0007	0.0000	
6	0.9992	0.9623	0.7752	0.4478	0.1662	0.0348	0.0032	0.0001	
7	0.9999	0.9891	0.8954	0.6405	0.3145	0.0919	0.0127	0.0005	
8	1.0000	0.9974	0.9597	0.8011	0.5000	0.1989	0.0403	0.0026	0.0000
9		0.9995	0.9873	0.9081	0.6855	0.3595	0.1046	0.0109	0.0001
10		0.9999	0.9968	0.9652	0.8338	0.5522	0.2248	0.0377	0.0008
11		1.0000	0.9993	0.9894	0.9283	0.7361	0.4032	0.1057	0.0047
12			0.9999	0.9975	0.9755	0.8740	0.6113	0.2418	0.0221
13			1.0000	0.9995	0.9936	0.9536	0.7981	0.4511	0.0826
14				0.9999	0.9988	0.9877	0.9226	0.6904	0.2382
15				1.0000	0.9999	0.9979	0.9807	0.8818	0.5182
16					1.0000	0.9998	0.9977	0.9775	0.8332
17						1.0000	1.0000	1.0000	1.0000

TABLE A.2

(continued)

$n = 18$

π:	0.1	0.2	0.3	0.4	0.5	0.6	0.7	0.8	0.9
k: 0	0.1501	0.0180	0.0016	0.0001	0.0000				
1	0.4503	0.0991	0.0142	0.0013	0.0001				
2	0.7338	0.2713	0.0600	0.0082	0.0007	0.0000			
3	0.9018	0.5010	0.1646	0.0328	0.0038	0.0002			
4	0.9718	0.7164	0.3327	0.0942	0.0154	0.0013	0.0000		
5	0.9936	0.8671	0.5344	0.2088	0.0481	0.0058	0.0003		
6	0.9988	0.9487	0.7217	0.3743	0.1189	0.0203	0.0014	0.0000	
7	0.9998	0.9837	0.8593	0.5634	0.2403	0.0576	0.0061	0.0002	
8	1.0000	0.9957	0.9404	0.7368	0.4073	0.1347	0.0210	0.0009	
9		0.9991	0.9790	0.8653	0.5927	0.2632	0.0596	0.0043	0.0000
10		0.9998	0.9939	0.9424	0.7597	0.4366	0.1407	0.0163	0.0002
11		1.0000	0.9986	0.9797	0.8811	0.6257	0.2783	0.0513	0.0012
12			0.9997	0.9942	0.9519	0.7912	0.4656	0.1329	0.0064
13			1.0000	0.9987	0.9846	0.9058	0.6673	0.2836	0.0282
14				0.9998	0.9962	0.9672	0.8354	0.4990	0.0982
15				1.0000	0.9993	0.9918	0.9400	0.7287	0.2662
16					0.9999	0.9987	0.9858	0.9009	0.5497
17					1.0000	0.9999	0.9984	0.9820	0.8499
18						1.0000	1.0000	1.0000	1.0000

$n = 19$

π:	0.1	0.2	0.3	0.4	0.5	0.6	0.7	0.8	0.9
k: 0	0.1351	0.0144	0.0011	0.0001					
1	0.4203	0.0829	0.0104	0.0008	0.0000				
2	0.7054	0.2369	0.0462	0.0055	0.0004	0.0000			
3	0.8850	0.4551	0.1332	0.0230	0.0022	0.0001			
4	0.9648	0.6733	0.2822	0.0696	0.0096	0.0006	0.0000		
5	0.9914	0.8369	0.4739	0.1629	0.0318	0.0031	0.0001		
6	0.9983	0.9324	0.6655	0.3081	0.0835	0.0116	0.0006		
7	0.9997	0.9767	0.8180	0.4878	0.1796	0.0352	0.0028	0.0000	
8	1.0000	0.9933	0.9161	0.6675	0.3238	0.0885	0.0105	0.0003	
9		0.9984	0.9674	0.8139	0.5000	0.1861	0.0326	0.0016	
10		0.9997	0.9895	0.9115	0.6762	0.3325	0.0839	0.0067	0.0000
11		1.0000	0.9972	0.9648	0.8204	0.5122	0.1820	0.0233	0.0003
12			0.9994	0.9884	0.9165	0.6919	0.3345	0.0676	0.0017
13			0.9999	0.9969	0.9682	0.8371	0.5261	0.1631	0.0086
14			1.0000	0.9994	0.9904	0.9304	0.7178	0.3267	0.0352
15				0.9999	0.9978	0.9770	0.8668	0.5449	0.1150
16				1.0000	0.9996	0.9945	0.9538	0.7631	0.2946
17					1.0000	0.9992	0.9896	0.9171	0.5797
18						0.9999	0.9989	0.9856	0.8649
19						1.0000	1.0000	1.0000	1.0000

TABLE A.2
(continued)

n = 20

π:	0.1	0.2	0.3	0.4	0.5	0.6	0.7	0.8	0.9
k: 0	0.1216	0.0115	0.0008	0.0000					
1	0.3917	0.0692	0.0076	0.0005	0.0000				
2	0.6769	0.2061	0.0355	0.0036	0.0002				
3	0.8670	0.4114	0.1071	0.0160	0.0013	0.0000			
4	0.9568	0.6296	0.2375	0.0510	0.0059	0.0003			
5	0.9887	0.8042	0.4164	0.1256	0.0207	0.0016	0.0000		
6	0.9976	0.9133	0.6080	0.2500	0.0577	0.0065	0.0003		
7	0.9996	0.9679	0.7723	0.4159	0.1316	0.0210	0.0013	0.0000	
8	0.9999	0.9900	0.8867	0.5956	0.2517	0.0565	0.0051	0.0001	
9	1.0000	0.9974	0.9520	0.7553	0.4119	0.1275	0.0171	0.0006	
10		0.9994	0.9829	0.8725	0.5881	0.2447	0.0480	0.0026	0.0000
11		0.9999	0.9949	0.9435	0.7483	0.4044	0.1133	0.0100	0.0001
12		1.0000	0.9987	0.9790	0.8684	0.5841	0.2277	0.0321	0.0004
13			0.9997	0.9935	0.9423	0.7500	0.3920	0.0867	0.0024
14			1.0000	0.9984	0.9793	0.8744	0.5836	0.1958	0.0113
15				0.9997	0.9941	0.9490	0.7625	0.3704	0.0432
16				1.0000	0.9987	0.9840	0.8929	0.5886	0.1330
17					0.9998	0.9964	0.9645	0.7939	0.3231
18					1.0000	0.9995	0.9924	0.9308	0.6083
19						1.0000	0.9992	0.9885	0.8784
20							1.0000	1.0000	1.0000

n = 25

π:	0.1	0.2	0.3	0.4	0.5	0.6	0.7	0.8	0.9
k: 0	0.0718	0.0038	0.0001	0.0000					
1	0.2712	0.0274	0.0016	0.0001					
2	0.5371	0.0982	0.0090	0.0004	0.0000				
3	0.7636	0.2340	0.0332	0.0024	0.0001				
4	0.9020	0.4207	0.0905	0.0095	0.0005	0.0000			
5	0.9666	0.6167	0.1935	0.0294	0.0020	0.0001			
6	0.9905	0.7800	0.3407	0.0736	0.0073	0.0003			
7	0.9977	0.8909	0.5118	0.1536	0.0216	0.0012	0.0000		
8	0.9995	0.9532	0.6769	0.2735	0.0539	0.0043	0.0001		
9	0.9999	0.9827	0.8106	0.4246	0.1148	0.0132	0.0005		
10	1.0000	0.9944	0.9022	0.5858	0.2122	0.0344	0.0018	0.0000	
11		0.9985	0.9558	0.7323	0.3450	0.0778	0.0060	0.0001	
12		0.9996	0.9825	0.8462	0.5000	0.1538	0.0175	0.0004	
13		0.9999	0.9940	0.9222	0.6550	0.2677	0.0442	0.0015	
14		1.0000	0.9982	0.9656	0.7878	0.4142	0.0978	0.0056	0.0000
15			0.9995	0.9868	0.8852	0.5754	0.1894	0.0173	0.0001
16			0.9999	0.9957	0.9461	0.7265	0.3231	0.0468	0.0005
17			1.0000	0.9988	0.9784	0.8464	0.4882	0.1091	0.0023
18				0.9997	0.9927	0.9264	0.6593	0.2200	0.0095
19				0.9999	0.9980	0.9706	0.8065	0.3833	0.0334
20				1.0000	0.9995	0.9905	0.9095	0.5793	0.0980
21					0.9999	0.9976	0.9668	0.7660	0.2364
22					1.0000	0.9996	0.9910	0.9018	0.4629
23						0.9999	0.9984	0.9726	0.7288
24						1.0000	0.9999	0.9962	0.9282
25							1.0000	1.0000	1.0000

TABLE A.3

Poisson Distribution, Individual Probabilities for x = number of occurrences, prob($x = k$)

λ:	0.1	0.2	0.3	0.4	0.5	0.6	0.7	0.8	0.9	1.0
k: 0	0.9048	0.8187	0.7408	0.6703	0.6065	0.5488	0.4966	0.4493	0.4066	0.3679
1	0.0905	0.1637	0.2222	0.2681	0.3033	0.3293	0.3476	0.3595	0.3659	0.3679
2	0.0045	0.0164	0.0333	0.0536	0.0758	0.0988	0.1217	0.1438	0.1647	0.1839
3	0.0002	0.0011	0.0033	0.0072	0.0126	0.0198	0.0284	0.0383	0.0494	0.0613
4	0.0000	0.0001	0.0003	0.0007	0.0016	0.0030	0.0050	0.0077	0.0111	0.0153
5		0.0000	0.0000	0.0001	0.0002	0.0004	0.0007	0.0012	0.0020	0.0031
6				0.0000	0.0000	0.0000	0.0001	0.0002	0.0003	0.0005
7							0.0000	0.0000	0.0000	0.0001
8										0.0000

λ:	1.1	1.2	1.3	1.4	1.5	1.6	1.7	1.8	1.9	2.0
k: 0	0.3329	0.3012	0.2725	0.2466	0.2231	0.2019	0.1827	0.1653	0.1496	0.1353
1	0.3662	0.3614	0.3543	0.3452	0.3347	0.3230	0.3106	0.2975	0.2842	0.2707
2	0.2014	0.2169	0.2303	0.2417	0.2510	0.2584	0.2640	0.2678	0.2700	0.2707
3	0.0738	0.0867	0.0998	0.1128	0.1255	0.1378	0.1496	0.1607	0.1710	0.1804
4	0.0203	0.0260	0.0324	0.0395	0.0471	0.0551	0.0636	0.0723	0.0812	0.0902
5	0.0045	0.0062	0.0084	0.0111	0.0141	0.0176	0.0216	0.0260	0.0309	0.0361
6	0.0008	0.0012	0.0018	0.0026	0.0035	0.0047	0.0061	0.0078	0.0098	0.0120
7	0.0001	0.0002	0.0003	0.0005	0.0008	0.0011	0.0015	0.0020	0.0027	0.0034
8	0.0000	0.0000	0.0001	0.0001	0.0001	0.0002	0.0003	0.0005	0.0006	0.0009
9			0.0000	0.0000	0.0000	0.0000	0.0001	0.0001	0.0001	0.0002
10							0.0000	0.0000	0.0000	0.0000

λ:	2.1	2.2	2.3	2.4	2.5	2.6	2.7	2.8	2.9	3.0
k: 0	0.1225	0.1108	0.1003	0.0907	0.0821	0.0743	0.0672	0.0608	0.0550	0.0498
1	0.2572	0.2438	0.2306	0.2177	0.2052	0.1931	0.1815	0.1703	0.1596	0.1494
2	0.2700	0.2681	0.2652	0.2613	0.2565	0.2510	0.2450	0.2384	0.2314	0.2240
3	0.1890	0.1966	0.2033	0.2090	0.2138	0.2176	0.2205	0.2225	0.2237	0.2240
4	0.0992	0.1082	0.1169	0.1254	0.1336	0.1414	0.1488	0.1557	0.1622	0.1680
5	0.0417	0.0476	0.0538	0.0602	0.0668	0.0735	0.0804	0.0872	0.0940	0.1008
6	0.0146	0.0174	0.0206	0.0241	0.0278	0.0319	0.0362	0.0407	0.0455	0.0504
7	0.0044	0.0055	0.0068	0.0083	0.0099	0.0118	0.0139	0.0163	0.0188	0.0216
8	0.0011	0.0015	0.0019	0.0025	0.0031	0.0038	0.0047	0.0057	0.0068	0.0081
9	0.0003	0.0004	0.0005	0.0007	0.0009	0.0011	0.0014	0.0018	0.0022	0.0027
10	0.0001	0.0001	0.0001	0.0002	0.0002	0.0003	0.0004	0.0005	0.0006	0.0008
11	0.0000	0.0000	0.0000	0.0000	0.0000	0.0001	0.0001	0.0001	0.0002	0.0002
12						0.0000	0.0000	0.0000	0.0000	0.0001
13										0.0000

Source: Probabilities generated by Minitab, then compiled as shown here.

TABLE A.3

(continued)

λ:	3.1	3.2	3.3	3.4	3.5	3.6	3.7	3.8	3.9	4.0
k: 0	0.0450	0.0408	0.0369	0.0334	0.0302	0.0273	0.0247	0.0224	0.0202	0.0183
1	0.1397	0.1304	0.1217	0.1135	0.1057	0.0984	0.0915	0.0850	0.0789	0.0733
2	0.2165	0.2087	0.2008	0.1929	0.1850	0.1771	0.1692	0.1615	0.1539	0.1465
3	0.2237	0.2226	0.2209	0.2186	0.2158	0.2125	0.2087	0.2046	0.2001	0.1954
4	0.1733	0.1781	0.1823	0.1858	0.1888	0.1912	0.1931	0.1944	0.1951	0.1954
5	0.1075	0.1140	0.1203	0.1264	0.1322	0.1377	0.1429	0.1477	0.1522	0.1563
6	0.0555	0.0608	0.0662	0.0716	0.0771	0.0826	0.0881	0.0936	0.0989	0.1042
7	0.0246	0.0278	0.0312	0.0348	0.0385	0.0425	0.0466	0.0508	0.0551	0.0595
8	0.0095	0.0111	0.0129	0.0148	0.0169	0.0191	0.0215	0.0241	0.0269	0.0298
9	0.0033	0.0040	0.0047	0.0056	0.0066	0.0076	0.0089	0.0102	0.0116	0.0132
10	0.0010	0.0013	0.0016	0.0019	0.0023	0.0028	0.0033	0.0039	0.0045	0.0053
11	0.0003	0.0004	0.0005	0.0006	0.0007	0.0009	0.0011	0.0013	0.0016	0.0019
12	0.0001	0.0001	0.0001	0.0002	0.0002	0.0003	0.0003	0.0004	0.0005	0.0006
13	0.0000	0.0000	0.0000	0.0000	0.0001	0.0001	0.0001	0.0001	0.0002	0.0002
14					0.0000	0.0000	0.0000	0.0000	0.0000	0.0001
15										0.0000

λ:	4.1	4.2	4.3	4.4	4.5	4.6	4.7	4.8	4.9	5.0
k: 0	0.0166	0.0150	0.0136	0.0123	0.0111	0.0101	0.0091	0.0082	0.0074	0.0067
1	0.0679	0.0630	0.0583	0.0540	0.0500	0.0462	0.0427	0.0395	0.0365	0.0337
2	0.1393	0.1323	0.1254	0.1188	0.1125	0.1063	0.1005	0.0948	0.0894	0.0842
3	0.1904	0.1852	0.1798	0.1743	0.1687	0.1631	0.1574	0.1517	0.1460	0.1404
4	0.1951	0.1944	0.1933	0.1917	0.1898	0.1875	0.1849	0.1820	0.1789	0.1755
5	0.1600	0.1633	0.1662	0.1687	0.1708	0.1725	0.1738	0.1747	0.1753	0.1755
6	0.1093	0.1143	0.1191	0.1237	0.1281	0.1323	0.1362	0.1398	0.1432	0.1462
7	0.0640	0.0686	0.0732	0.0778	0.0824	0.0869	0.0914	0.0959	0.1002	0.1044
8	0.0328	0.0360	0.0393	0.0428	0.0463	0.0500	0.0537	0.0575	0.0614	0.0653
9	0.0150	0.0168	0.0188	0.0209	0.0232	0.0255	0.0281	0.0307	0.0334	0.0363
10	0.0061	0.0071	0.0081	0.0092	0.0104	0.0118	0.0132	0.0147	0.0164	0.0181
11	0.0023	0.0027	0.0032	0.0037	0.0043	0.0049	0.0056	0.0064	0.0073	0.0082
12	0.0008	0.0009	0.0011	0.0013	0.0016	0.0019	0.0022	0.0026	0.0030	0.0034
13	0.0002	0.0003	0.0004	0.0005	0.0006	0.0007	0.0008	0.0009	0.0011	0.0013
14	0.0001	0.0001	0.0001	0.0001	0.0002	0.0002	0.0003	0.0003	0.0004	0.0005
15	0.0000	0.0000	0.0000	0.0000	0.0001	0.0001	0.0001	0.0001	0.0001	0.0002
16					0.0000	0.0000	0.0000	0.0000	0.0000	0.0000

λ:	5.1	5.2	5.3	5.4	5.5	5.6	5.7	5.8	5.9	6.0
k: 0	0.0061	0.0055	0.0050	0.0045	0.0041	0.0037	0.0033	0.0030	0.0027	0.0025
1	0.0311	0.0287	0.0265	0.0244	0.0225	0.0207	0.0191	0.0176	0.0162	0.0149
2	0.0793	0.0746	0.0701	0.0659	0.0618	0.0580	0.0544	0.0509	0.0477	0.0446
3	0.1348	0.1293	0.1239	0.1185	0.1133	0.1082	0.1033	0.0985	0.0938	0.0892
4	0.1719	0.1681	0.1641	0.1600	0.1558	0.1515	0.1472	0.1428	0.1383	0.1339
5	0.1753	0.1748	0.1740	0.1728	0.1714	0.1697	0.1678	0.1656	0.1632	0.1606
6	0.1490	0.1515	0.1537	0.1555	0.1571	0.1584	0.1594	0.1601	0.1605	0.1606
7	0.1086	0.1125	0.1163	0.1200	0.1234	0.1267	0.1298	0.1326	0.1353	0.1377
8	0.0692	0.0731	0.0771	0.0810	0.0849	0.0887	0.0925	0.0962	0.0998	0.1033
9	0.0392	0.0423	0.0454	0.0486	0.0519	0.0552	0.0586	0.0620	0.0654	0.0688
10	0.0200	0.0220	0.0241	0.0262	0.0285	0.0309	0.0334	0.0359	0.0386	0.0413
11	0.0093	0.0104	0.0116	0.0129	0.0143	0.0157	0.0173	0.0190	0.0207	0.0225
12	0.0039	0.0045	0.0051	0.0058	0.0065	0.0073	0.0082	0.0092	0.0102	0.0113
13	0.0015	0.0018	0.0021	0.0024	0.0028	0.0032	0.0036	0.0041	0.0046	0.0052
14	0.0006	0.0007	0.0008	0.0009	0.0011	0.0013	0.0015	0.0017	0.0019	0.0022
15	0.0002	0.0002	0.0003	0.0003	0.0004	0.0005	0.0006	0.0007	0.0008	0.0009
16	0.0001	0.0001	0.0001	0.0001	0.0001	0.0002	0.0002	0.0002	0.0003	0.0003
17	0.0000	0.0000	0.0000	0.0000	0.0000	0.0001	0.0001	0.0001	0.0001	0.0001
18						0.0000	0.0000	0.0000	0.0000	0.0000

TABLE A.3
(continued)

λ:	6.1	6.2	6.3	6.4	6.5	6.6	6.7	6.8	6.9	7.0
k: 0	0.0022	0.0020	0.0018	0.0017	0.0015	0.0014	0.0012	0.0011	0.0010	0.0009
1	0.0137	0.0126	0.0116	0.0106	0.0098	0.0090	0.0082	0.0076	0.0070	0.0064
2	0.0417	0.0390	0.0364	0.0340	0.0318	0.0296	0.0276	0.0258	0.0240	0.0223
3	0.0848	0.0806	0.0765	0.0726	0.0688	0.0652	0.0617	0.0584	0.0552	0.0521
4	0.1294	0.1249	0.1205	0.1162	0.1118	0.1076	0.1034	0.0992	0.0952	0.0912
5	0.1579	0.1549	0.1519	0.1487	0.1454	0.1420	0.1385	0.1349	0.1314	0.1277
6	0.1605	0.1601	0.1595	0.1586	0.1575	0.1562	0.1546	0.1529	0.1511	0.1490
7	0.1399	0.1418	0.1435	0.1450	0.1462	0.1472	0.1480	0.1486	0.1489	0.1490
8	0.1066	0.1099	0.1130	0.1160	0.1188	0.1215	0.1240	0.1263	0.1284	0.1304
9	0.0723	0.0757	0.0791	0.0825	0.0858	0.0891	0.0923	0.0954	0.0985	0.1014
10	0.0441	0.0469	0.0498	0.0528	0.0558	0.0588	0.0618	0.0649	0.0679	0.0710
11	0.0244	0.0265	0.0285	0.0307	0.0330	0.0353	0.0377	0.0401	0.0426	0.0452
12	0.0124	0.0137	0.0150	0.0164	0.0179	0.0194	0.0210	0.0227	0.0245	0.0263
13	0.0058	0.0065	0.0073	0.0081	0.0089	0.0099	0.0108	0.0119	0.0130	0.0142
14	0.0025	0.0029	0.0033	0.0037	0.0041	0.0046	0.0052	0.0058	0.0064	0.0071
15	0.0010	0.0012	0.0014	0.0016	0.0018	0.0020	0.0023	0.0026	0.0029	0.0033
16	0.0004	0.0005	0.0005	0.0006	0.0007	0.0008	0.0010	0.0011	0.0013	0.0014
17	0.0001	0.0002	0.0002	0.0002	0.0003	0.0003	0.0004	0.0004	0.0005	0.0006
18	0.0000	0.0001	0.0001	0.0001	0.0001	0.0001	0.0001	0.0002	0.0002	0.0002
19		0.0000	0.0000	0.0000	0.0000	0.0000	0.0001	0.0001	0.0001	0.0001
20							0.0000	0.0000	0.0000	0.0000

λ:	7.1	7.2	7.3	7.4	7.5	7.6	7.7	7.8	7.9	8.0
k: 0	0.0008	0.0007	0.0007	0.0006	0.0006	0.0005	0.0005	0.0004	0.0004	0.0003
1	0.0059	0.0054	0.0049	0.0045	0.0041	0.0038	0.0035	0.0032	0.0029	0.0027
2	0.0208	0.0194	0.0180	0.0167	0.0156	0.0145	0.0134	0.0125	0.0116	0.0107
3	0.0492	0.0464	0.0438	0.0413	0.0389	0.0366	0.0345	0.0324	0.0305	0.0286
4	0.0874	0.0836	0.0799	0.0764	0.0729	0.0696	0.0663	0.0632	0.0602	0.0573
5	0.1241	0.1204	0.1167	0.1130	0.1094	0.1057	0.1021	0.0986	0.0951	0.0916
6	0.1468	0.1445	0.1420	0.1394	0.1367	0.1339	0.1311	0.1282	0.1252	0.1221
7	0.1489	0.1486	0.1481	0.1474	0.1465	0.1454	0.1442	0.1428	0.1413	0.1396
8	0.1321	0.1337	0.1351	0.1363	0.1373	0.1381	0.1388	0.1392	0.1395	0.1396
9	0.1042	0.1070	0.1096	0.1121	0.1144	0.1167	0.1187	0.1207	0.1224	0.1241
10	0.0740	0.0770	0.0800	0.0829	0.0858	0.0887	0.0914	0.0941	0.0967	0.0993
11	0.0478	0.0504	0.0531	0.0558	0.0585	0.0613	0.0640	0.0667	0.0695	0.0722
12	0.0283	0.0303	0.0323	0.0344	0.0366	0.0388	0.0411	0.0434	0.0457	0.0481
13	0.0154	0.0168	0.0181	0.0196	0.0211	0.0227	0.0243	0.0260	0.0278	0.0296
14	0.0078	0.0086	0.0095	0.0104	0.0113	0.0123	0.0134	0.0145	0.0157	0.0169
15	0.0037	0.0041	0.0046	0.0051	0.0057	0.0062	0.0069	0.0075	0.0083	0.0090
16	0.0016	0.0019	0.0021	0.0024	0.0026	0.0030	0.0033	0.0037	0.0041	0.0045
17	0.0007	0.0008	0.0009	0.0010	0.0012	0.0013	0.0015	0.0017	0.0019	0.0021
18	0.0003	0.0003	0.0004	0.0004	0.0005	0.0006	0.0006	0.0007	0.0008	0.0009
19	0.0001	0.0001	0.0001	0.0002	0.0002	0.0002	0.0003	0.0003	0.0003	0.0004
20	0.0000	0.0000	0.0001	0.0001	0.0001	0.0001	0.0001	0.0001	0.0001	0.0002
21			0.0000	0.0000	0.0000	0.0000	0.0000	0.0000	0.0001	0.0001
22									0.0000	0.0000

TABLE A.3
(continued)

λ:	8.1	8.2	8.3	8.4	8.5	8.6	8.7	8.8	8.9	9.0
k: 0	0.0003	0.0003	0.0002	0.0002	0.0002	0.0002	0.0002	0.0002	0.0001	0.0001
1	0.0025	0.0023	0.0021	0.0019	0.0017	0.0016	0.0014	0.0013	0.0012	0.0011
2	0.0100	0.0092	0.0086	0.0079	0.0074	0.0068	0.0063	0.0058	0.0054	0.0050
3	0.0269	0.0252	0.0237	0.0222	0.0208	0.0195	0.0183	0.0171	0.0160	0.0150
4	0.0544	0.0517	0.0491	0.0466	0.0443	0.0420	0.0398	0.0377	0.0357	0.0337
5	0.0882	0.0849	0.0816	0.0784	0.0752	0.0722	0.0692	0.0663	0.0635	0.0607
6	0.1191	0.1160	0.1128	0.1097	0.1066	0.1034	0.1003	0.0972	0.0941	0.0911
7	0.1378	0.1358	0.1338	0.1317	0.1294	0.1271	0.1247	0.1222	0.1197	0.1171
8	0.1395	0.1392	0.1388	0.1382	0.1375	0.1366	0.1356	0.1344	0.1332	0.1318
9	0.1256	0.1269	0.1280	0.1290	0.1299	0.1306	0.1311	0.1315	0.1317	0.1318
10	0.1017	0.1040	0.1063	0.1084	0.1104	0.1123	0.1140	0.1157	0.1172	0.1186
11	0.0749	0.0776	0.0802	0.0828	0.0853	0.0878	0.0902	0.0925	0.0948	0.0970
12	0.0505	0.0530	0.0555	0.0579	0.0604	0.0629	0.0654	0.0679	0.0703	0.0728
13	0.0315	0.0334	0.0354	0.0374	0.0395	0.0416	0.0438	0.0459	0.0481	0.0504
14	0.0182	0.0196	0.0210	0.0225	0.0240	0.0256	0.0272	0.0289	0.0306	0.0324
15	0.0098	0.0107	0.0116	0.0126	0.0136	0.0147	0.0158	0.0169	0.0182	0.0194
16	0.0050	0.0055	0.0060	0.0066	0.0072	0.0079	0.0086	0.0093	0.0101	0.0109
17	0.0024	0.0026	0.0029	0.0033	0.0036	0.0040	0.0044	0.0048	0.0053	0.0058
18	0.0011	0.0012	0.0014	0.0015	0.0017	0.0019	0.0021	0.0024	0.0026	0.0029
19	0.0005	0.0005	0.0006	0.0007	0.0008	0.0009	0.0010	0.0011	0.0012	0.0014
20	0.0002	0.0002	0.0002	0.0003	0.0003	0.0004	0.0004	0.0005	0.0005	0.0006
21	0.0001	0.0001	0.0001	0.0001	0.0001	0.0002	0.0002	0.0002	0.0002	0.0003
22	0.0000	0.0000	0.0000	0.0000	0.0001	0.0001	0.0001	0.0001	0.0001	0.0001
23					0.0000	0.0000	0.0000	0.0000	0.0000	0.0000

λ:	9.1	9.2	9.3	9.4	9.5	9.6	9.7	9.8	9.9	10.0
k: 0	0.0001	0.0001	0.0001	0.0001	0.0001	0.0001	0.0001	0.0001	0.0001	0.0000
1	0.0010	0.0009	0.0009	0.0008	0.0007	0.0007	0.0006	0.0005	0.0005	0.0005
2	0.0046	0.0043	0.0040	0.0037	0.0034	0.0031	0.0029	0.0027	0.0025	0.0023
3	0.0140	0.0131	0.0123	0.0115	0.0107	0.0100	0.0093	0.0087	0.0081	0.0076
4	0.0319	0.0302	0.0285	0.0269	0.0254	0.0240	0.0226	0.0213	0.0201	0.0189
5	0.0581	0.0555	0.0530	0.0506	0.0483	0.0460	0.0439	0.0418	0.0398	0.0378
6	0.0881	0.0851	0.0822	0.0793	0.0764	0.0736	0.0709	0.0682	0.0656	0.0631
7	0.1145	0.1118	0.1091	0.1064	0.1037	0.1010	0.0982	0.0955	0.0928	0.0901
8	0.1302	0.1286	0.1269	0.1251	0.1232	0.1212	0.1191	0.1170	0.1148	0.1126
9	0.1317	0.1315	0.1311	0.1306	0.1300	0.1293	0.1284	0.1274	0.1263	0.1251
10	0.1198	0.1210	0.1219	0.1228	0.1235	0.1241	0.1245	0.1249	0.1250	0.1251
11	0.0991	0.1012	0.1031	0.1049	0.1067	0.1083	0.1098	0.1112	0.1125	0.1137
12	0.0752	0.0776	0.0799	0.0822	0.0844	0.0866	0.0888	0.0908	0.0928	0.0948
13	0.0526	0.0549	0.0572	0.0594	0.0617	0.0640	0.0662	0.0685	0.0707	0.0729
14	0.0342	0.0361	0.0380	0.0399	0.0419	0.0439	0.0459	0.0479	0.0500	0.0521
15	0.0208	0.0221	0.0235	0.0250	0.0265	0.0281	0.0297	0.0313	0.0330	0.0347
16	0.0118	0.0127	0.0137	0.0147	0.0157	0.0168	0.0180	0.0192	0.0204	0.0217
17	0.0063	0.0069	0.0075	0.0081	0.0088	0.0095	0.0103	0.0111	0.0119	0.0128
18	0.0032	0.0035	0.0039	0.0042	0.0046	0.0051	0.0055	0.0060	0.0065	0.0071
19	0.0015	0.0017	0.0019	0.0021	0.0023	0.0026	0.0028	0.0031	0.0034	0.0037
20	0.0007	0.0008	0.0009	0.0010	0.0011	0.0012	0.0014	0.0015	0.0017	0.0019
21	0.0003	0.0003	0.0004	0.0004	0.0005	0.0006	0.0006	0.0007	0.0008	0.0009
22	0.0001	0.0001	0.0002	0.0002	0.0002	0.0002	0.0003	0.0003	0.0004	0.0004
23	0.0000	0.0001	0.0001	0.0001	0.0001	0.0001	0.0001	0.0001	0.0002	0.0002
24		0.0000	0.0000	0.0000	0.0000	0.0000	0.0000	0.0001	0.0001	0.0001
25								0.0000	0.0000	0.0000

TABLE A.4

Poisson Distribution, Cumulative Probabilities for x = number of occurrences, prob($x \leq k$)

λ:	0.1	0.2	0.3	0.4	0.5	0.6	0.7	0.8	0.9	1.0
k: 0	0.9048	0.8187	0.7408	0.6703	0.6065	0.5488	0.4966	0.4493	0.4066	0.3679
1	0.9953	0.9825	0.9631	0.9384	0.9098	0.8781	0.8442	0.8088	0.7725	0.7358
2	0.9998	0.9989	0.9964	0.9921	0.9856	0.9769	0.9659	0.9526	0.9371	0.9197
3	1.0000	0.9999	0.9997	0.9992	0.9982	0.9966	0.9942	0.9909	0.9865	0.9810
4		1.0000	1.0000	0.9999	0.9998	0.9996	0.9992	0.9986	0.9977	0.9963
5				1.0000	1.0000	1.0000	0.9999	0.9998	0.9997	0.9994
6							1.0000	1.0000	1.0000	0.9999
7										1.0000
λ:	**1.1**	**1.2**	**1.3**	**1.4**	**1.5**	**1.6**	**1.7**	**1.8**	**1.9**	**2.0**
k: 0	0.3329	0.3012	0.2725	0.2466	0.2231	0.2019	0.1827	0.1653	0.1496	0.1353
1	0.6990	0.6626	0.6268	0.5918	0.5578	0.5249	0.4932	0.4628	0.4337	0.4060
2	0.9004	0.8795	0.8571	0.8335	0.8088	0.7834	0.7572	0.7306	0.7037	0.6767
3	0.9743	0.9662	0.9569	0.9463	0.9344	0.9212	0.9068	0.8913	0.8747	0.8571
4	0.9946	0.9923	0.9893	0.9857	0.9814	0.9763	0.9704	0.9636	0.9559	0.9473
5	0.9990	0.9985	0.9978	0.9968	0.9955	0.9940	0.9920	0.9896	0.9868	0.9834
6	0.9999	0.9997	0.9996	0.9994	0.9991	0.9987	0.9981	0.9974	0.9966	0.9955
7	1.0000	1.0000	0.9999	0.9999	0.9998	0.9997	0.9996	0.9994	0.9992	0.9989
8			1.0000	1.0000	1.0000	1.0000	0.9999	0.9999	0.9998	0.9998
9							1.0000	1.0000	1.0000	1.0000
λ:	**2.1**	**2.2**	**2.3**	**2.4**	**2.5**	**2.6**	**2.7**	**2.8**	**2.9**	**3.0**
k: 0	0.1225	0.1108	0.1003	0.0907	0.0821	0.0743	0.0672	0.0608	0.0550	0.0498
1	0.3796	0.3546	0.3309	0.3084	0.2873	0.2674	0.2487	0.2311	0.2146	0.1991
2	0.6496	0.6227	0.5960	0.5697	0.5438	0.5184	0.4936	0.4695	0.4460	0.4232
3	0.8386	0.8194	0.7993	0.7787	0.7576	0.7360	0.7141	0.6919	0.6696	0.6472
4	0.9379	0.9275	0.9162	0.9041	0.8912	0.8774	0.8629	0.8477	0.8318	0.8153
5	0.9796	0.9751	0.9700	0.9643	0.9580	0.9510	0.9433	0.9349	0.9258	0.9161
6	0.9941	0.9925	0.9906	0.9884	0.9858	0.9828	0.9794	0.9756	0.9713	0.9665
7	0.9985	0.9980	0.9974	0.9967	0.9958	0.9947	0.9934	0.9919	0.9901	0.9881
8	0.9997	0.9995	0.9994	0.9991	0.9989	0.9985	0.9981	0.9976	0.9969	0.9962
9	0.9999	0.9999	0.9999	0.9998	0.9997	0.9996	0.9995	0.9993	0.9991	0.9989
10	1.0000	1.0000	1.0000	1.0000	0.9999	0.9999	0.9999	0.9998	0.9998	0.9997
11					1.0000	1.0000	1.0000	1.0000	0.9999	0.9999
12									1.0000	1.0000
λ:	**3.1**	**3.2**	**3.3**	**3.4**	**3.5**	**3.6**	**3.7**	**3.8**	**3.9**	**4.0**
k: 0	0.0450	0.0408	0.0369	0.0334	0.0302	0.0273	0.0247	0.0224	0.0202	0.0183
1	0.1847	0.1712	0.1586	0.1468	0.1359	0.1257	0.1162	0.1074	0.0992	0.0916
2	0.4012	0.3799	0.3594	0.3397	0.3208	0.3027	0.2854	0.2689	0.2531	0.2381
3	0.6248	0.6025	0.5803	0.5584	0.5366	0.5152	0.4942	0.4735	0.4532	0.4335
4	0.7982	0.7806	0.7626	0.7442	0.7254	0.7064	0.6872	0.6678	0.6484	0.6288
5	0.9057	0.8946	0.8829	0.8705	0.8576	0.8441	0.8301	0.8156	0.8006	0.7851
6	0.9612	0.9554	0.9490	0.9421	0.9347	0.9267	0.9182	0.9091	0.8995	0.8893
7	0.9858	0.9832	0.9802	0.9769	0.9733	0.9692	0.9648	0.9599	0.9546	0.9489
8	0.9953	0.9943	0.9931	0.9917	0.9901	0.9883	0.9863	0.9840	0.9815	0.9786
9	0.9986	0.9982	0.9978	0.9973	0.9967	0.9960	0.9952	0.9942	0.9931	0.9919
10	0.9996	0.9995	0.9994	0.9992	0.9990	0.9987	0.9984	0.9981	0.9977	0.9972
11	0.9999	0.9999	0.9998	0.9998	0.9997	0.9996	0.9995	0.9994	0.9993	0.9991
12	1.0000	1.0000	1.0000	0.9999	0.9999	0.9999	0.9999	0.9998	0.9998	0.9997
13				1.0000	1.0000	1.0000	1.0000	1.0000	0.9999	0.9999
14									1.0000	1.0000

Source: Probabilities generated by Minitab, then compiled as shown here.

TABLE A.4

(continued)

λ:	4.1	4.2	4.3	4.4	4.5	4.6	4.7	4.8	4.9	5.0
k: 0	0.0166	0.0150	0.0136	0.0123	0.0111	0.0101	0.0091	0.0082	0.0074	0.0067
1	0.0845	0.0780	0.0719	0.0663	0.0611	0.0563	0.0518	0.0477	0.0439	0.0404
2	0.2238	0.2102	0.1974	0.1851	0.1736	0.1626	0.1523	0.1425	0.1333	0.1247
3	0.4142	0.3954	0.3772	0.3594	0.3423	0.3257	0.3097	0.2942	0.2793	0.2650
4	0.6093	0.5898	0.5704	0.5512	0.5321	0.5132	0.4946	0.4763	0.4582	0.4405
5	0.7693	0.7531	0.7367	0.7199	0.7029	0.6858	0.6684	0.6510	0.6335	0.6160
6	0.8786	0.8675	0.8558	0.8436	0.8311	0.8180	0.8046	0.7908	0.7767	0.7622
7	0.9427	0.9361	0.9290	0.9214	0.9134	0.9049	0.8960	0.8867	0.8769	0.8666
8	0.9755	0.9721	0.9683	0.9642	0.9597	0.9549	0.9497	0.9442	0.9382	0.9319
9	0.9905	0.9889	0.9871	0.9851	0.9829	0.9805	0.9778	0.9749	0.9717	0.9682
10	0.9966	0.9959	0.9952	0.9943	0.9933	0.9922	0.9910	0.9896	0.9880	0.9863
11	0.9989	0.9986	0.9983	0.9980	0.9976	0.9971	0.9966	0.9960	0.9953	0.9945
12	0.9997	0.9996	0.9995	0.9993	0.9992	0.9990	0.9988	0.9986	0.9983	0.9980
13	0.9999	0.9999	0.9998	0.9998	0.9997	0.9997	0.9996	0.9995	0.9994	0.9993
14	1.0000	1.0000	1.0000	0.9999	0.9999	0.9999	0.9999	0.9999	0.9998	0.9998
15				1.0000	1.0000	1.0000	1.0000	1.0000	0.9999	0.9999
16									1.0000	1.0000

λ:	5.1	5.2	5.3	5.4	5.5	5.6	5.7	5.8	5.9	6.0
k: 0	0.0061	0.0055	0.0050	0.0045	0.0041	0.0037	0.0033	0.0030	0.0027	0.0025
1	0.0372	0.0342	0.0314	0.0289	0.0266	0.0244	0.0224	0.0206	0.0189	0.0174
2	0.1165	0.1088	0.1016	0.0948	0.0884	0.0824	0.0768	0.0715	0.0666	0.0620
3	0.2513	0.2381	0.2254	0.2133	0.2017	0.1906	0.1800	0.1700	0.1604	0.1512
4	0.4231	0.4061	0.3895	0.3733	0.3575	0.3422	0.3272	0.3127	0.2987	0.2851
5	0.5984	0.5809	0.5635	0.5461	0.5289	0.5119	0.4950	0.4783	0.4619	0.4457
6	0.7474	0.7324	0.7171	0.7017	0.6860	0.6703	0.6544	0.6384	0.6224	0.6063
7	0.8560	0.8449	0.8335	0.8217	0.8095	0.7970	0.7841	0.7710	0.7576	0.7440
8	0.9252	0.9181	0.9106	0.9027	0.8944	0.8857	0.8766	0.8672	0.8574	0.8472
9	0.9644	0.9603	0.9559	0.9512	0.9462	0.9409	0.9352	0.9292	0.9228	0.9161
10	0.9844	0.9823	0.9800	0.9775	0.9747	0.9718	0.9686	0.9651	0.9614	0.9574
11	0.9937	0.9927	0.9916	0.9904	0.9890	0.9875	0.9859	0.9841	0.9821	0.9799
12	0.9976	0.9972	0.9967	0.9962	0.9955	0.9949	0.9941	0.9932	0.9922	0.9912
13	0.9992	0.9990	0.9988	0.9986	0.9983	0.9980	0.9977	0.9973	0.9969	0.9964
14	0.9997	0.9997	0.9996	0.9995	0.9994	0.9993	0.9991	0.9990	0.9988	0.9986
15	0.9999	0.9999	0.9999	0.9998	0.9998	0.9998	0.9997	0.9996	0.9996	0.9995
16	1.0000	1.0000	1.0000	0.9999	0.9999	0.9999	0.9999	0.9999	0.9999	0.9998
17				1.0000	1.0000	1.0000	1.0000	1.0000	1.0000	0.9999
18										1.0000

TABLE A.4
(continued)

λ:	6.1	6.2	6.3	6.4	6.5	6.6	6.7	6.8	6.9	7.0
k: 0	0.0022	0.0020	0.0018	0.0017	0.0015	0.0014	0.0012	0.0011	0.0010	0.0009
1	0.0159	0.0146	0.0134	0.0123	0.0113	0.0103	0.0095	0.0087	0.0080	0.0073
2	0.0577	0.0536	0.0498	0.0463	0.0430	0.0400	0.0371	0.0344	0.0320	0.0296
3	0.1425	0.1342	0.1264	0.1189	0.1118	0.1052	0.0988	0.0928	0.0871	0.0818
4	0.2719	0.2592	0.2469	0.2351	0.2237	0.2127	0.2022	0.1920	0.1823	0.1730
5	0.4298	0.4141	0.3988	0.3837	0.3690	0.3547	0.3406	0.3270	0.3137	0.3007
6	0.5902	0.5742	0.5582	0.5423	0.5265	0.5108	0.4953	0.4799	0.4647	0.4497
7	0.7301	0.7160	0.7017	0.6873	0.6728	0.6581	0.6433	0.6285	0.6136	0.5987
8	0.8367	0.8259	0.8148	0.8033	0.7916	0.7796	0.7673	0.7548	0.7420	0.7291
9	0.9090	0.9016	0.8939	0.8858	0.8774	0.8686	0.8596	0.8502	0.8405	0.8305
10	0.9531	0.9486	0.9437	0.9386	0.9332	0.9274	0.9214	0.9151	0.9084	0.9015
11	0.9776	0.9750	0.9723	0.9693	0.9661	0.9627	0.9591	0.9552	0.9510	0.9467
12	0.9900	0.9887	0.9873	0.9857	0.9840	0.9821	0.9801	0.9779	0.9755	0.9730
13	0.9958	0.9952	0.9945	0.9937	0.9929	0.9920	0.9909	0.9898	0.9885	0.9872
14	0.9984	0.9981	0.9978	0.9974	0.9970	0.9966	0.9961	0.9956	0.9950	0.9943
15	0.9994	0.9993	0.9992	0.9990	0.9988	0.9986	0.9984	0.9982	0.9979	0.9976
16	0.9998	0.9997	0.9997	0.9996	0.9996	0.9995	0.9994	0.9993	0.9992	0.9990
17	0.9999	0.9999	0.9999	0.9999	0.9998	0.9998	0.9998	0.9997	0.9997	0.9996
18	1.0000	1.0000	1.0000	1.0000	0.9999	0.9999	0.9999	0.9999	0.9999	0.9999
19					1.0000	1.0000	1.0000	1.0000	1.0000	1.0000

λ:	7.1	7.2	7.3	7.4	7.5	7.6	7.7	7.8	7.9	8.0
k: 0	0.0008	0.0007	0.0007	0.0006	0.0006	0.0005	0.0005	0.0004	0.0004	0.0003
1	0.0067	0.0061	0.0056	0.0051	0.0047	0.0043	0.0039	0.0036	0.0033	0.0030
2	0.0275	0.0255	0.0236	0.0219	0.0203	0.0188	0.0174	0.0161	0.0149	0.0138
3	0.0767	0.0719	0.0674	0.0632	0.0591	0.0554	0.0518	0.0485	0.0453	0.0424
4	0.1641	0.1555	0.1473	0.1395	0.1321	0.1249	0.1181	0.1117	0.1055	0.0996
5	0.2881	0.2759	0.2640	0.2526	0.2414	0.2307	0.2203	0.2103	0.2006	0.1912
6	0.4349	0.4204	0.4060	0.3920	0.3782	0.3646	0.3514	0.3384	0.3257	0.3134
7	0.5838	0.5689	0.5541	0.5393	0.5246	0.5100	0.4956	0.4812	0.4670	0.4530
8	0.7160	0.7027	0.6892	0.6757	0.6620	0.6482	0.6343	0.6204	0.6065	0.5925
9	0.8202	0.8096	0.7988	0.7877	0.7764	0.7649	0.7531	0.7411	0.7290	0.7166
10	0.8942	0.8867	0.8788	0.8707	0.8622	0.8535	0.8445	0.8352	0.8257	0.8159
11	0.9420	0.9371	0.9319	0.9265	0.9208	0.9148	0.9085	0.9020	0.8952	0.8881
12	0.9703	0.9673	0.9642	0.9609	0.9573	0.9536	0.9496	0.9454	0.9409	0.9362
13	0.9857	0.9841	0.9824	0.9805	0.9784	0.9762	0.9739	0.9714	0.9687	0.9658
14	0.9935	0.9927	0.9918	0.9908	0.9897	0.9886	0.9873	0.9859	0.9844	0.9827
15	0.9972	0.9969	0.9964	0.9959	0.9954	0.9948	0.9941	0.9934	0.9926	0.9918
16	0.9989	0.9987	0.9985	0.9983	0.9980	0.9978	0.9974	0.9971	0.9967	0.9963
17	0.9996	0.9995	0.9994	0.9993	0.9992	0.9991	0.9989	0.9988	0.9986	0.9984
18	0.9998	0.9998	0.9998	0.9997	0.9997	0.9996	0.9996	0.9995	0.9994	0.9993
19	0.9999	0.9999	0.9999	0.9999	0.9999	0.9999	0.9998	0.9998	0.9998	0.9997
20	1.0000	1.0000	1.0000	1.0000	1.0000	1.0000	0.9999	0.9999	0.9999	0.9999
21							1.0000	1.0000	1.0000	1.0000

TABLE A.4
(continued)

λ:	8.1	8.2	8.3	8.4	8.5	8.6	8.7	8.8	8.9	9.0
k: 0	0.0003	0.0003	0.0002	0.0002	0.0002	0.0002	0.0002	0.0002	0.0001	0.0001
1	0.0028	0.0025	0.0023	0.0021	0.0019	0.0018	0.0016	0.0015	0.0014	0.0012
2	0.0127	0.0118	0.0109	0.0100	0.0093	0.0086	0.0079	0.0073	0.0068	0.0062
3	0.0396	0.0370	0.0346	0.0323	0.0301	0.0281	0.0262	0.0244	0.0228	0.0212
4	0.0940	0.0887	0.0837	0.0789	0.0744	0.0701	0.0660	0.0621	0.0584	0.0550
5	0.1822	0.1736	0.1653	0.1573	0.1496	0.1422	0.1352	0.1284	0.1219	0.1157
6	0.3013	0.2896	0.2781	0.2670	0.2562	0.2457	0.2355	0.2256	0.2160	0.2068
7	0.4391	0.4254	0.4119	0.3987	0.3856	0.3728	0.3602	0.3478	0.3357	0.3239
8	0.5786	0.5647	0.5507	0.5369	0.5231	0.5094	0.4958	0.4823	0.4689	0.4557
9	0.7041	0.6915	0.6788	0.6659	0.6530	0.6400	0.6269	0.6137	0.6006	0.5874
10	0.8058	0.7955	0.7850	0.7743	0.7634	0.7522	0.7409	0.7294	0.7178	0.7060
11	0.8807	0.8731	0.8652	0.8571	0.8487	0.8400	0.8311	0.8220	0.8126	0.8030
12	0.9313	0.9261	0.9207	0.9150	0.9091	0.9029	0.8965	0.8898	0.8829	0.8758
13	0.9628	0.9595	0.9561	0.9524	0.9486	0.9445	0.9403	0.9358	0.9311	0.9261
14	0.9810	0.9791	0.9771	0.9749	0.9726	0.9701	0.9675	0.9647	0.9617	0.9585
15	0.9908	0.9898	0.9887	0.9875	0.9862	0.9848	0.9832	0.9816	0.9798	0.9780
16	0.9958	0.9953	0.9947	0.9941	0.9934	0.9926	0.9918	0.9909	0.9899	0.9889
17	0.9982	0.9979	0.9977	0.9973	0.9970	0.9966	0.9962	0.9957	0.9952	0.9947
18	0.9992	0.9991	0.9990	0.9989	0.9987	0.9985	0.9983	0.9981	0.9978	0.9976
19	0.9997	0.9997	0.9996	0.9995	0.9995	0.9994	0.9993	0.9992	0.9991	0.9989
20	0.9999	0.9999	0.9998	0.9998	0.9998	0.9998	0.9997	0.9997	0.9996	0.9996
21	1.0000	1.0000	0.9999	0.9999	0.9999	0.9999	0.9999	0.9999	0.9998	0.9998
22			1.0000	1.0000	1.0000	1.0000	1.0000	1.0000	0.9999	0.9999
23									1.0000	1.0000

λ:	9.1	9.2	9.3	9.4	9.5	9.6	9.7	9.8	9.9	10.0
k: 0	0.0001	0.0001	0.0001	0.0001	0.0001	0.0001	0.0001	0.0001	0.0001	0.0000
1	0.0011	0.0010	0.0009	0.0009	0.0008	0.0007	0.0007	0.0006	0.0005	0.0005
2	0.0058	0.0053	0.0049	0.0045	0.0042	0.0038	0.0035	0.0033	0.0030	0.0028
3	0.0198	0.0184	0.0172	0.0160	0.0149	0.0138	0.0129	0.0120	0.0111	0.0103
4	0.0517	0.0486	0.0456	0.0429	0.0403	0.0378	0.0355	0.0333	0.0312	0.0293
5	0.1098	0.1041	0.0986	0.0935	0.0885	0.0838	0.0793	0.0750	0.0710	0.0671
6	0.1978	0.1892	0.1808	0.1727	0.1649	0.1574	0.1502	0.1433	0.1366	0.1301
7	0.3123	0.3010	0.2900	0.2792	0.2687	0.2584	0.2485	0.2388	0.2294	0.2202
8	0.4426	0.4296	0.4168	0.4042	0.3918	0.3796	0.3676	0.3558	0.3442	0.3328
9	0.5742	0.5611	0.5479	0.5349	0.5218	0.5089	0.4960	0.4832	0.4705	0.4579
10	0.6941	0.6820	0.6699	0.6576	0.6453	0.6329	0.6205	0.6080	0.5955	0.5830
11	0.7932	0.7832	0.7730	0.7626	0.7520	0.7412	0.7303	0.7193	0.7081	0.6968
12	0.8684	0.8607	0.8529	0.8448	0.8364	0.8279	0.8191	0.8101	0.8009	0.7916
13	0.9210	0.9156	0.9100	0.9042	0.8981	0.8919	0.8853	0.8786	0.8716	0.8645
14	0.9552	0.9517	0.9480	0.9441	0.9400	0.9357	0.9312	0.9265	0.9216	0.9165
15	0.9760	0.9738	0.9715	0.9691	0.9665	0.9638	0.9609	0.9579	0.9546	0.9513
16	0.9878	0.9865	0.9852	0.9838	0.9823	0.9806	0.9789	0.9770	0.9751	0.9730
17	0.9941	0.9934	0.9927	0.9919	0.9911	0.9902	0.9892	0.9881	0.9870	0.9857
18	0.9973	0.9969	0.9966	0.9962	0.9957	0.9952	0.9947	0.9941	0.9935	0.9928
19	0.9988	0.9986	0.9985	0.9983	0.9980	0.9978	0.9975	0.9972	0.9969	0.9965
20	0.9995	0.9994	0.9993	0.9992	0.9991	0.9990	0.9989	0.9987	0.9986	0.9984
21	0.9998	0.9998	0.9997	0.9997	0.9996	0.9996	0.9995	0.9995	0.9994	0.9993
22	0.9999	0.9999	0.9999	0.9999	0.9999	0.9998	0.9998	0.9998	0.9997	0.9997
23	1.0000	1.0000	1.0000	1.0000	0.9999	0.9999	0.9999	0.9999	0.9999	0.9999
24					1.0000	1.0000	1.0000	1.0000	1.0000	1.0000

TABLE A.5

Table of Random Digits

Row	1–10		11–20		21–30		31–40		41–50	
1	37220	84861	59998	77311	87754	04899	75699	22423	01200	79606
2	31618	06840	45167	13747	74382	39250	37038	42810	81183	27226
3	53778	71813	50306	47415	21682	74984	69038	80269	18955	47908
4	82652	58162	90352	10751	49968	06611	23573	29523	78542	49685
5	40372	53109	76995	24681	42711	06899	14208	63897	60834	46618
6	24952	30711	22535	51397	37277	04343	84920	11970	28098	41367
7	94953	96367	87143	71467	31194	39897	56605	21810	72064	11296
8	98972	12203	90759	56157	40644	16283	80816	07379	87171	89016
9	61759	32900	05299	56687	92355	11684	23278	82492	00366	81502
10	14338	44401	22630	06918	86157	02229	68995	88455	35210	84775
11	03990	64484	50620	04134	04132	35323	39948	53446	13150	01522
12	04003	21357	21560	26820	97543	15607	75220	27896	29722	45389
13	75942	36578	85584	55496	88476	01255	75094	26800	78580	86252
14	29945	17580	17770	56883	90165	54800	47262	14471	66496	66618
15	05675	07093	47456	58942	13685	58422	50987	46420	60965	07193
16	38770	60956	63383	54619	28131	43657	25686	30869	44495	68087
17	93578	84503	84911	86482	45267	23662	59118	35926	11854	12602
18	54526	35249	68501	02206	77793	34752	67528	51706	94351	65383
19	66019	99264	83410	16479	55742	75660	24750	16698	15630	33348
20	30995	81658	22563	34561	29004	64725	80137	97704	63540	29097
21	54873	00592	16042	46776	31730	87258	15861	03115	95167	17410
22	69350	23570	89681	00936	22521	88911	10908	04029	21081	42224
23	43775	21148	77549	86081	02143	29803	17758	81909	09509	64255
24	94238	08655	67935	26234	17011	44426	18663	93938	12767	38048
25	90329	65940	30771	99446	08749	86843	72979	03701	72777	17442
26	49606	51896	13569	58172	37655	22835	26546	08220	22221	62889
27	85876	62869	92898	06662	89382	20703	70312	34032	40788	07235
28	99021	17433	31031	92399	28930	46651	81288	32707	50799	59045
29	43672	46199	83184	90585	32998	82557	82994	70297	79910	86498
30	77915	14650	48304	10282	16356	50667	23469	12116	42141	20876
31	62815	20732	36276	19416	09552	52494	85690	61579	99879	80023
32	35431	28351	10559	52307	49924	97564	68685	49850	62185	36563
33	26212	06143	84840	76247	08208	68191	73511	51271	28635	33390
34	03772	70603	28430	72720	70275	87396	54825	60139	48572	90762
35	22831	69776	07506	32013	51906	49757	41867	15688	53729	44526
36	91527	15810	88468	47664	28231	75104	24370	63076	53040	01302
37	98889	36471	23606	86428	36786	79805	17891	92903	52891	03626
38	80596	42027	72309	71805	39628	59368	15410	41685	05732	87997
39	06160	85702	50617	52953	85886	99013	52135	35314	13276	64765
40	07145	80861	37672	03747	76227	52880	65131	92937	32659	44959
41	25756	94616	75611	70478	82600	89252	99473	75367	36786	08165
42	31698	84269	82241	05714	68314	46971	27440	12794	51833	88879
43	65292	54687	65808	38799	44635	83610	83257	79134	29222	11974
44	55947	27034	40437	52112	94440	36443	89401	58176	74072	57605
45	21916	48584	21519	80149	38375	27493	94284	94171	65668	45875
46	38532	95321	76039	97791	57165	33428	37024	73345	32198	48764
47	33462	61546	54421	41760	68011	47564	77265	91710	89205	04352
48	13526	46548	09115	99063	45898	59162	70038	88042	85648	47177
49	34828	14870	70575	24312	79631	75697	72404	31103	00016	42082
50	89895	82390	89421	56029	61854	05548	21777	51553	17064	33859

Source: Random digits generated by Minitab.

TABLE A.5
(continued)

Row	1–10		11–20		21–30		31–40		41–50	
51	96018	27032	35948	81017	87632	68779	98468	82882	58835	61583
52	08657	49477	20792	55367	27957	21669	25807	01687	06853	87214
53	05183	80239	20082	32806	84896	36574	99899	02874	58987	07423
54	30558	18243	96761	22184	05203	62794	49547	62406	82189	38127
55	85574	37307	36048	72513	49070	82728	86719	96008	85102	66717
56	30554	02449	74975	54343	32643	06124	18521	63560	51187	23974
57	08408	15911	13377	74630	53089	11912	93262	10961	33576	60466
58	62944	23357	49036	41265	88109	02667	75931	89523	53776	14953
59	93160	32676	91167	63054	46459	76406	02739	44198	16693	85504
60	99581	93196	04941	28933	30480	81685	14525	34978	68211	18591
61	71317	85120	37427	57745	78399	49835	67713	10736	70576	81235
62	88100	72776	19450	63109	23254	45376	89233	40445	09325	73602
63	70782	68998	78042	28590	94522	49559	77357	16808	45458	35337
64	35340	66670	34011	47582	58480	32166	22558	69670	72272	26486
65	96757	46833	55325	73157	57438	18733	95369	55382	97468	19294
66	86797	63659	61773	64818	35218	19597	21190	18369	84201	85388
67	69259	17468	69838	56975	02929	21112	16636	61211	15809	48807
68	83837	17354	59419	24965	90989	83896	39712	78988	95621	81415
69	43638	54811	63054	08267	02125	17892	50132	01293	90524	30664
70	67436	14892	65654	88989	63574	39532	26906	72282	03567	54279
71	08336	51472	04056	15208	42517	97140	87426	55647	76220	70525
72	72678	31240	47192	49329	64702	25685	72651	25217	65463	65810
73	13063	46035	35899	22682	86499	28475	89917	11856	11202	45920
74	70022	46043	37033	04781	53014	68215	46567	83996	51586	00305
75	72803	61399	52929	75693	38989	14261	39257	38112	24227	15175
76	11566	26680	54965	42491	37382	11800	49989	96604	71890	62676
77	63085	07137	63898	54353	44912	00010	04212	51192	86710	38272
78	43029	84646	21389	71859	85939	95973	79164	23010	21267	39741
79	25172	64085	98417	45034	55105	15375	50465	75316	92495	93023
80	22565	88294	35872	45035	69125	46818	73424	10302	89830	66504
81	08342	34507	14599	76497	42542	52684	56830	80974	79301	25622
82	32972	02526	37798	66645	79124	60382	84256	55650	85417	83653
83	67256	76680	09063	90514	16663	29278	75623	96264	96627	21865
84	32925	24366	96567	21296	53412	46323	35922	56796	82847	43730
85	61270	37743	04870	55927	32722	51403	88391	35180	18657	14786
86	49847	35977	16510	95436	68933	06504	13832	58070	25616	00710
87	71292	25605	48086	25880	62403	90032	37255	49115	91855	85141
88	00829	65292	74516	37134	04001	09070	74628	70384	14075	85153
89	59329	18382	27521	73483	37372	94615	89458	56168	99200	58918
90	33251	02545	18701	39495	48746	23063	44545	99331	37707	11114
91	12488	88006	10476	66065	31729	75517	76418	49074	46531	71802
92	87379	02837	55836	46426	04337	20093	73444	93438	98875	92489
93	87793	28895	36704	80744	60137	62342	61414	31190	37768	17153
94	20540	32139	16496	72337	38768	82413	02398	89143	54376	16104
95	34086	44938	99218	17223	92393	39053	93918	56496	28382	77369
96	10403	83656	78289	65316	56677	99757	51479	03639	05957	62608
97	37124	75750	63234	00172	17453	87288	99592	98549	30527	03473
98	51218	08866	01031	39402	97757	19459	19458	52282	34979	82757
99	95060	16896	75433	40816	50161	87246	69040	64414	14555	27653
100	68291	35003	94288	95139	60876	61419	00615	49678	68123	55444

TABLE A.6

Critical Values of F

$\alpha = 0.05$; $F(\alpha, \nu_1, \nu_2)$

$v_1 = df$, numerator

$v_2 = df$, denominator	1	2	3	4	5	6	7	8	9	10	12	15	20	24	30	40	60	120	∞
1	161.4	199.5	215.7	224.6	230.2	234.0	236.8	238.9	240.5	241.9	243.9	245.9	248.0	249.1	250.1	251.1	252.2	253.3	254.3
2	18.51	19.00	19.16	19.25	19.30	19.33	19.35	19.37	19.38	19.40	19.41	19.43	19.45	19.45	19.46	19.47	19.48	19.49	19.50
3	10.13	9.55	9.28	9.12	9.01	8.94	8.89	8.85	8.81	8.79	8.74	8.70	8.66	8.64	8.62	8.59	8.57	8.55	8.53
4	7.71	6.94	6.59	6.39	6.26	6.16	6.09	6.04	6.00	5.96	5.91	5.86	5.80	5.77	5.75	5.72	5.69	5.66	5.63
5	6.61	5.79	5.41	5.19	5.05	4.95	4.88	4.82	4.77	4.74	4.68	4.62	4.56	4.53	4.50	4.46	4.43	4.40	4.36
6	5.99	5.14	4.76	4.53	4.39	4.28	4.21	4.15	4.10	4.06	4.00	3.94	3.87	3.84	3.81	3.77	3.74	3.70	3.67
7	5.59	4.74	4.35	4.12	3.97	3.87	3.79	3.73	3.68	3.64	3.57	3.51	3.44	3.41	3.38	3.34	3.30	3.27	3.23
8	5.32	4.46	4.07	3.84	3.69	3.58	3.50	3.44	3.39	3.35	3.28	3.22	3.15	3.12	3.08	3.04	3.01	2.97	2.93
9	5.12	4.26	3.86	3.63	3.48	3.37	3.29	3.23	3.18	3.14	3.07	3.01	2.94	2.90	2.86	2.83	2.79	2.75	2.71
10	4.96	4.10	3.71	3.48	3.33	3.22	3.14	3.07	3.02	2.98	2.91	2.85	2.77	2.74	2.70	2.66	2.62	2.58	2.54
11	4.84	3.98	3.59	3.36	3.20	3.09	3.01	2.95	2.90	2.85	2.79	2.72	2.65	2.61	2.57	2.53	2.49	2.45	2.40
12	4.75	3.89	3.49	3.26	3.11	3.00	2.91	2.85	2.80	2.75	2.69	2.62	2.54	2.51	2.47	2.43	2.38	2.34	2.30
13	4.67	3.81	3.41	3.18	3.03	2.92	2.83	2.77	2.71	2.67	2.60	2.53	2.46	2.42	2.38	2.34	2.30	2.25	2.21
14	4.60	3.74	3.34	3.11	2.96	2.85	2.76	2.70	2.65	2.60	2.53	2.46	2.39	2.35	2.31	2.27	2.22	2.18	2.13
15	4.54	3.68	3.29	3.06	2.90	2.79	2.71	2.64	2.59	2.54	2.48	2.40	2.33	2.29	2.25	2.20	2.16	2.11	2.07
16	4.49	3.63	3.24	3.01	2.85	2.74	2.66	2.59	2.54	2.49	2.42	2.35	2.28	2.24	2.19	2.15	2.11	2.06	2.01
17	4.45	3.59	3.20	2.96	2.81	2.70	2.61	2.55	2.49	2.45	2.38	2.31	2.23	2.19	2.15	2.10	2.06	2.01	1.96
18	4.41	3.55	3.16	2.93	2.77	2.66	2.58	2.51	2.46	2.41	2.34	2.27	2.19	2.15	2.11	2.06	2.02	1.97	1.92
19	4.38	3.52	3.13	2.90	2.74	2.63	2.54	2.48	2.42	2.38	2.31	2.23	2.16	2.11	2.07	2.03	1.98	1.93	1.88
20	4.35	3.49	3.10	2.87	2.71	2.60	2.51	2.45	2.39	2.35	2.28	2.20	2.12	2.08	2.04	1.99	1.95	1.90	1.84
21	4.32	3.47	3.07	2.84	2.68	2.57	2.49	2.42	2.37	2.32	2.25	2.18	2.10	2.05	2.01	1.96	1.92	1.87	1.81
22	4.30	3.44	3.05	2.82	2.66	2.55	2.46	2.40	2.34	2.30	2.23	2.15	2.07	2.03	1.98	1.94	1.89	1.84	1.78
23	4.28	3.42	3.03	2.80	2.64	2.53	2.44	2.37	2.32	2.27	2.20	2.13	2.05	2.01	1.96	1.91	1.86	1.81	1.76
24	4.26	3.40	3.01	2.78	2.62	2.51	2.42	2.36	2.30	2.25	2.18	2.11	2.03	1.98	1.94	1.89	1.84	1.79	1.73
25	4.24	3.39	2.99	2.76	2.60	2.49	2.40	2.34	2.28	2.24	2.16	2.09	2.01	1.96	1.92	1.87	1.82	1.77	1.71
26	4.23	3.37	2.98	2.74	2.59	2.47	2.39	2.32	2.27	2.22	2.15	2.07	1.99	1.95	1.90	1.85	1.80	1.75	1.69
27	4.21	3.35	2.96	2.73	2.57	2.46	2.37	2.31	2.25	2.20	2.13	2.06	1.97	1.93	1.88	1.84	1.79	1.73	1.67
28	4.20	3.34	2.95	2.71	2.56	2.45	2.36	2.29	2.24	2.19	2.12	2.04	1.96	1.91	1.87	1.82	1.77	1.71	1.65
29	4.18	3.33	2.93	2.70	2.55	2.43	2.35	2.28	2.22	2.18	2.10	2.03	1.94	1.90	1.85	1.81	1.75	1.70	1.64
30	4.17	3.32	2.92	2.69	2.53	2.42	2.33	2.27	2.21	2.16	2.09	2.01	1.93	1.89	1.84	1.79	1.74	1.68	1.62
40	4.08	3.23	2.84	2.61	2.45	2.34	2.25	2.18	2.12	2.08	2.00	1.92	1.84	1.79	1.74	1.69	1.64	1.58	1.51
60	4.00	3.15	2.76	2.53	2.37	2.25	2.17	2.10	2.04	1.99	1.92	1.84	1.75	1.70	1.65	1.59	1.53	1.47	1.39
120	3.92	3.07	2.68	2.45	2.29	2.17	2.09	2.02	1.96	1.91	1.83	1.75	1.66	1.61	1.55	1.50	1.43	1.35	1.25
∞	3.84	3.00	2.60	2.37	2.21	2.10	2.01	1.94	1.88	1.83	1.75	1.67	1.57	1.52	1.46	1.39	1.32	1.22	1.00

TABLE A.6

(continued)

$\alpha = 0.025$

$F(\alpha, \nu_1, \nu_2)$

$v_1 = df$, numerator

$v_2 = df$, denominator	1	2	3	4	5	6	7	8	9	10	12	15	20	24	30	40	60	120	∞
1	647.8	799.5	864.2	899.6	921.8	937.1	948.2	956.7	963.3	968.6	976.7	984.9	993.1	997.2	1001	1006	1010	1014	1018
2	38.51	39.00	39.17	39.25	39.30	39.33	39.36	39.37	39.39	39.40	39.41	39.43	39.45	39.46	39.46	39.47	39.48	39.49	39.50
3	17.44	16.04	15.44	15.10	14.88	14.73	14.62	14.54	14.47	14.42	14.34	14.25	14.17	14.12	14.08	14.04	13.99	13.95	13.90
4	12.22	10.65	9.98	9.60	9.36	9.20	9.07	8.98	8.90	8.84	8.75	8.66	8.56	8.51	8.46	8.41	8.36	8.31	8.26
5	10.01	8.43	7.76	7.39	7.15	6.98	6.85	6.76	6.68	6.62	6.52	6.43	6.33	6.28	6.23	6.18	6.12	6.07	6.02
6	8.81	7.26	6.60	6.23	5.99	5.82	5.70	5.60	5.52	5.46	5.37	5.27	5.17	5.12	5.07	5.01	4.96	4.90	4.85
7	8.07	6.54	5.89	5.52	5.29	5.12	4.99	4.90	4.82	4.76	4.67	4.57	4.47	4.42	4.36	4.31	4.25	4.20	4.14
8	7.57	6.06	5.42	5.05	4.82	4.65	4.53	4.43	4.36	4.30	4.20	4.10	4.00	3.95	3.89	3.84	3.78	3.73	3.67
9	7.21	5.71	5.08	4.72	4.48	4.32	4.20	4.10	4.03	3.96	3.87	3.77	3.67	3.61	3.56	3.51	3.45	3.39	3.33
10	6.94	5.46	4.83	4.47	4.24	4.07	3.95	3.85	3.78	3.72	3.62	3.52	3.42	3.37	3.31	3.26	3.20	3.14	3.08
11	6.72	5.26	4.63	4.28	4.04	3.88	3.76	3.66	3.59	3.53	3.43	3.33	3.23	3.17	3.12	3.06	3.00	2.94	2.88
12	6.55	5.10	4.47	4.12	3.89	3.73	3.61	3.51	3.44	3.37	3.28	3.18	3.07	3.02	2.96	2.91	2.85	2.79	2.72
13	6.41	4.97	4.35	4.00	3.77	3.60	3.48	3.39	3.31	3.25	3.15	3.05	2.95	2.89	2.84	2.78	2.72	2.66	2.60
14	6.30	4.86	4.24	3.89	3.66	3.50	3.38	3.29	3.21	3.15	3.05	2.95	2.84	2.79	2.73	2.67	2.61	2.55	2.49
15	6.20	4.77	4.15	3.80	3.58	3.41	3.29	3.20	3.12	3.06	2.96	2.86	2.76	2.70	2.64	2.59	2.52	2.46	2.40
16	6.12	4.69	4.08	3.73	3.50	3.34	3.22	3.12	3.05	2.99	2.89	2.79	2.68	2.63	2.57	2.51	2.45	2.38	2.32
17	6.04	4.62	4.01	3.66	3.44	3.28	3.16	3.06	2.98	2.92	2.82	2.72	2.62	2.56	2.50	2.44	2.38	2.32	2.25
18	5.98	4.56	3.95	3.61	3.38	3.22	3.10	3.01	2.93	2.87	2.77	2.67	2.56	2.50	2.44	2.38	2.32	2.26	2.19
19	5.92	4.51	3.90	3.56	3.33	3.17	3.05	2.96	2.88	2.82	2.72	2.62	2.51	2.45	2.39	2.33	2.27	2.20	2.13
20	5.87	4.46	3.86	3.51	3.29	3.13	3.01	2.91	2.84	2.77	2.68	2.57	2.46	2.41	2.35	2.29	2.22	2.16	2.09
21	5.83	4.42	3.82	3.48	3.25	3.09	2.97	2.87	2.80	2.73	2.64	2.53	2.42	2.37	2.31	2.25	2.18	2.11	2.04
22	5.79	4.38	3.78	3.44	3.22	3.05	2.93	2.84	2.76	2.70	2.60	2.50	2.39	2.33	2.27	2.21	2.14	2.08	2.00
23	5.75	4.35	3.75	3.41	3.18	3.02	2.90	2.81	2.73	2.67	2.57	2.47	2.36	2.30	2.24	2.18	2.11	2.04	1.97
24	5.72	4.32	3.72	3.38	3.15	2.99	2.87	2.78	2.70	2.64	2.54	2.44	2.33	2.27	2.21	2.15	2.08	2.01	1.94
25	5.69	4.29	3.69	3.35	3.13	2.97	2.85	2.75	2.68	2.61	2.51	2.41	2.30	2.24	2.18	2.12	2.05	1.98	1.91
26	5.66	4.27	3.67	3.33	3.10	2.94	2.82	2.73	2.65	2.59	2.49	2.39	2.28	2.22	2.16	2.09	2.03	1.95	1.88
27	5.63	4.24	3.65	3.31	3.08	2.92	2.80	2.71	2.63	2.57	2.47	2.36	2.25	2.19	2.13	2.07	2.00	1.93	1.85
28	5.61	4.22	3.63	3.29	3.06	2.90	2.78	2.69	2.61	2.55	2.45	2.34	2.23	2.17	2.11	2.05	1.98	1.91	1.83
29	5.59	4.20	3.61	3.27	3.04	2.88	2.76	2.67	2.59	2.53	2.43	2.32	2.21	2.15	2.09	2.03	1.96	1.89	1.81
30	5.57	4.18	3.59	3.25	3.03	2.87	2.75	2.65	2.57	2.51	2.41	2.31	2.20	2.14	2.07	2.01	1.94	1.87	1.79
40	5.42	4.05	3.46	3.13	2.90	2.74	2.62	2.53	2.45	2.39	2.29	2.18	2.07	2.01	1.94	1.88	1.80	1.72	1.64
60	5.29	3.93	3.34	3.01	2.79	2.63	2.51	2.41	2.33	2.27	2.17	2.06	1.94	1.88	1.82	1.74	1.67	1.58	1.48
120	5.15	3.80	3.23	2.89	2.67	2.52	2.39	2.30	2.22	2.16	2.05	1.94	1.82	1.76	1.69	1.61	1.53	1.43	1.31
∞	5.02	3.69	3.12	2.79	2.57	2.41	2.29	2.19	2.11	2.05	1.94	1.83	1.71	1.64	1.57	1.48	1.39	1.27	1.00

TABLE A.6

(continued)

$\alpha = 0.01$

0 $F(\alpha, \nu_1, \nu_2)$ F

$v_1 = df$, numerator

$v_2 = df$, denominator	1	2	3	4	5	6	7	8	9	10	12	15	20	24	30	40	60	120	∞
1	4052	4999.5	5403	5625	5764	5859	5928	5982	6022	6056	6106	6157	6209	6235	6261	6287	6313	6339	6366
2	98.50	99.00	99.17	99.25	99.30	99.33	99.36	99.37	99.39	99.40	99.42	99.43	99.45	99.46	99.47	99.47	99.48	99.49	99.50
3	34.12	30.82	29.46	28.71	28.24	27.91	27.67	27.49	27.35	27.23	27.05	26.87	26.69	26.60	26.50	26.41	26.32	26.22	26.13
4	21.20	18.00	16.69	15.98	15.52	15.21	14.98	14.80	14.66	14.55	14.37	14.20	14.02	13.93	13.84	13.75	13.65	13.56	13.46
5	16.26	13.27	12.06	11.39	10.97	10.67	10.46	10.29	10.16	10.05	9.89	9.72	9.55	9.47	9.38	9.29	9.20	9.11	9.02
6	13.75	10.92	9.78	9.15	8.75	8.47	8.26	8.10	7.98	7.87	7.72	7.56	7.40	7.31	7.23	7.14	7.06	6.97	6.88
7	12.25	9.55	8.45	7.85	7.46	7.19	6.99	6.84	6.72	6.62	6.47	6.31	6.16	6.07	5.99	5.91	5.82	5.74	5.65
8	11.26	8.65	7.59	7.01	6.63	6.37	6.18	6.03	5.91	5.81	5.67	5.52	5.36	5.28	5.20	5.12	5.03	4.95	4.86
9	10.56	8.02	6.99	6.42	6.06	5.80	5.61	5.47	5.35	5.26	5.11	4.96	4.81	4.73	4.65	4.57	4.48	4.40	4.31
10	10.04	7.56	6.55	5.99	5.64	5.39	5.20	5.06	4.94	4.85	4.71	4.56	4.41	4.33	4.25	4.17	4.08	4.00	3.91
11	9.65	7.21	6.22	5.67	5.32	5.07	4.89	4.74	4.63	4.54	4.40	4.25	4.10	4.02	3.94	3.86	3.78	3.69	3.60
12	9.33	6.93	5.95	5.41	5.06	4.82	4.64	4.50	4.39	4.30	4.16	4.01	3.86	3.78	3.70	3.62	3.54	3.45	3.36
13	9.07	6.70	5.74	5.21	4.86	4.62	4.44	4.30	4.19	4.10	3.96	3.82	3.66	3.59	3.51	3.43	3.34	3.25	3.17
14	8.86	6.51	5.56	5.04	4.69	4.46	4.28	4.14	4.03	3.94	3.80	3.66	3.51	3.43	3.35	3.27	3.18	3.09	3.00
15	8.68	6.36	5.42	4.89	4.56	4.32	4.14	4.00	3.89	3.80	3.67	3.52	3.37	3.29	3.21	3.13	3.05	2.96	2.87
16	8.53	6.23	5.29	4.77	4.44	4.20	4.03	3.89	3.78	3.69	3.55	3.41	3.26	3.18	3.10	3.02	2.93	2.84	2.75
17	8.40	6.11	5.18	4.67	4.34	4.10	3.93	3.79	3.68	3.59	3.46	3.31	3.16	3.08	3.00	2.92	2.83	2.75	2.65
18	8.29	6.01	5.09	4.58	4.25	4.01	3.84	3.71	3.60	3.51	3.37	3.23	3.08	3.00	2.92	2.84	2.75	2.66	2.57
19	8.18	5.93	5.01	4.50	4.17	3.94	3.77	3.63	3.52	3.43	3.30	3.15	3.00	2.92	2.84	2.76	2.67	2.58	2.49
20	8.10	5.85	4.94	4.43	4.10	3.87	3.70	3.56	3.46	3.37	3.23	3.09	2.94	2.86	2.78	2.69	2.61	2.52	2.42
21	8.02	5.78	4.87	4.37	4.04	3.81	3.64	3.51	3.40	3.31	3.17	3.03	2.88	2.80	2.72	2.64	2.55	2.46	2.36
22	7.95	5.72	4.82	4.31	3.99	3.76	3.59	3.45	3.35	3.26	3.12	2.98	2.83	2.75	2.67	2.58	2.50	2.40	2.31
23	7.88	5.66	4.76	4.26	3.94	3.71	3.54	3.41	3.30	3.21	3.07	2.93	2.78	2.70	2.62	2.54	2.45	2.35	2.26
24	7.82	5.61	4.72	4.22	3.90	3.67	3.50	3.36	3.26	3.17	3.03	2.89	2.74	2.66	2.58	2.49	2.40	2.31	2.21
25	7.77	5.57	4.68	4.18	3.85	3.63	3.46	3.32	3.22	3.13	2.99	2.85	2.70	2.62	2.54	2.45	2.36	2.27	2.17
26	7.72	5.53	4.64	4.14	3.82	3.59	3.42	3.29	3.18	3.09	2.96	2.81	2.66	2.58	2.50	2.42	2.33	2.23	2.13
27	7.68	5.49	4.60	4.11	3.78	3.56	3.39	3.26	3.15	3.06	2.93	2.78	2.63	2.55	2.47	2.38	2.29	2.20	2.10
28	7.64	5.45	4.57	4.07	3.75	3.53	3.36	3.23	3.12	3.03	2.90	2.75	2.60	2.52	2.44	2.35	2.26	2.17	2.06
29	7.60	5.42	4.54	4.04	3.73	3.50	3.33	3.20	3.09	3.00	2.87	2.73	2.57	2.49	2.41	2.33	2.23	2.14	2.03
30	7.56	5.39	4.51	4.02	3.70	3.47	3.30	3.17	3.07	2.98	2.84	2.70	2.55	2.47	2.39	2.30	2.21	2.11	2.01
40	7.31	5.18	4.31	3.83	3.51	3.29	3.12	2.99	2.89	2.80	2.66	2.52	2.37	2.29	2.20	2.11	2.02	1.92	1.80
60	7.08	4.98	4.13	3.65	3.34	3.12	2.95	2.82	2.72	2.63	2.50	2.35	2.20	2.12	2.03	1.94	1.84	1.73	1.60
120	6.85	4.79	3.95	3.48	3.17	2.96	2.79	2.66	2.56	2.47	2.34	2.19	2.03	1.95	1.86	1.76	1.66	1.53	1.38
∞	6.63	4.61	3.78	3.32	3.02	2.80	2.64	2.51	2.41	2.32	2.18	2.04	1.88	1.79	1.70	1.59	1.47	1.32	1.00

Source: Standard Mathematical Tables, 26th ed., William H. Beyer (ed.), CRC Press, Inc., Boca Raton, FL, 1983.

TABLE A.7

The Chi-Square Distribution

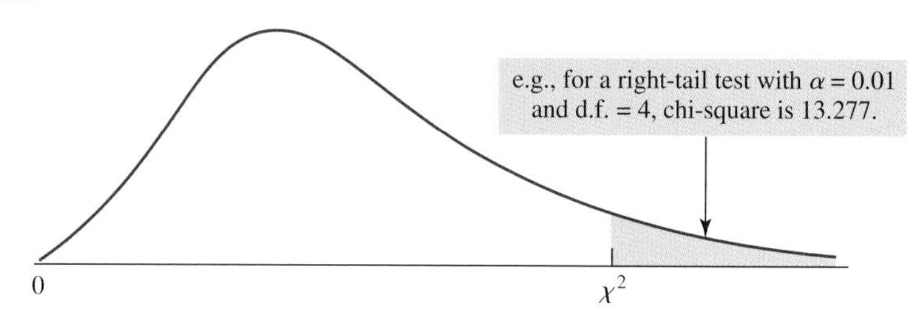

For α = Right-Tail Area of

d.f.	0.99	0.975	0.95	0.90	0.10	0.05	0.025	0.01
1	0.00016	0.00098	0.00039	0.0158	2.706	3.841	5.024	6.635
2	0.0201	0.0506	0.103	0.211	4.605	5.991	7.378	9.210
3	0.115	0.216	0.352	0.584	6.251	7.815	9.348	11.345
4	0.297	0.484	0.711	1.064	7.779	9.488	11.143	13.277
5	0.554	0.831	1.145	1.610	9.236	11.070	12.833	15.086
6	0.872	1.237	1.635	2.204	10.645	12.592	14.449	16.812
7	1.239	1.690	2.167	2.833	12.017	14.067	16.013	18.475
8	1.646	2.180	2.733	3.490	13.362	15.507	17.535	20.090
9	2.088	2.700	3.325	4.168	14.684	16.919	19.023	21.666
10	2.558	3.247	3.940	4.865	15.987	18.307	20.483	23.209
11	3.053	3.816	4.575	5.578	17.275	19.675	21.920	24.725
12	3.571	4.404	5.226	6.304	18.549	21.026	23.337	26.217
13	4.107	5.009	5.892	7.042	19.812	22.362	24.736	27.688
14	4.660	5.629	6.571	7.790	21.064	23.685	26.119	29.141
15	5.229	6.262	7.261	8.547	22.307	24.996	27.488	30.578
16	5.812	6.908	7.962	9.312	23.542	26.296	28.845	32.000
17	6.408	7.564	8.672	10.085	24.769	27.587	30.191	33.409
18	7.015	8.231	9.390	10.865	25.989	28.869	31.526	34.805
19	7.633	8.907	10.117	11.651	27.204	30.144	32.852	36.191
20	8.260	9.591	10.851	12.443	28.412	31.410	34.170	37.566
21	8.897	10.283	11.591	13.240	29.615	32.671	35.479	38.932
22	9.542	10.982	12.338	14.042	30.813	33.924	36.781	40.290
23	10.916	11.689	13.091	14.848	32.007	35.172	38.076	41.638
24	10.856	12.401	13.848	15.659	33.196	36.415	39.364	42.980
25	11.524	13.120	14.611	16.473	34.382	37.652	40.647	44.314
26	12.198	13.844	15.379	17.292	35.563	38.885	41.923	45.642
27	12.879	14.573	16.151	18.114	36.741	40.113	43.195	46.963
28	13.565	15.308	16.928	18.939	37.916	41.337	44.461	48.278
29	14.256	16.047	17.708	19.768	39.087	42.557	45.722	49.588
30	14.953	16.791	18.493	20.599	40.256	43.773	46.979	50.892
40	22.164	24.433	26.509	29.051	51.805	55.759	59.342	63.691
50	29.707	32.357	34.764	37.689	63.167	67.505	71.420	76.154
60	37.485	40.482	43.188	46.459	74.397	79.082	83.298	88.381
70	45.442	48.758	51.739	55.329	85.527	90.531	95.023	100.42
80	53.540	57.153	60.391	64.278	96.578	101.88	106.63	112.33
90	61.754	65.647	69.126	73.291	107.57	113.15	118.14	124.12
100	70.065	74.222	77.930	82.358	118.50	124.34	129.56	135.81

Source: Chi-square values generated by Minitab, then rounded as shown.

TABLE A.8

Wilcoxon Signed Rank Test, Lower and Upper Critical Values

Two-Tail Test:	$\alpha = 0.20$	$\alpha = 0.10$	$\alpha = 0.05$	$\alpha = 0.02$	$\alpha = 0.01$
One-Tail Test:	$\alpha = 0.10$	$\alpha = 0.05$	$\alpha = 0.025$	$\alpha = 0.01$	$\alpha = 0.005$
$n = 4$	1, 9	0, 10	0, 10	0, 10	0, 10
5	3, 12	1, 14	0, 15	0, 15	0, 15
6	4, 17	3, 18	1, 20	0, 21	0, 21
7	6, 22	4, 24	3, 25	1, 27	0, 28
8	9, 27	6, 30	4, 32	2, 34	1, 35
9	11, 34	9, 36	6, 39	4, 41	2, 43
10	15, 40	11, 44	9, 46	6, 49	4, 51
11	18, 48	14, 52	11, 55	8, 58	6, 60
12	22, 56	18, 60	14, 64	10, 68	8, 70
13	27, 64	22, 69	18, 73	13, 78	10, 81
14	32, 73	26, 79	22, 83	16, 89	13, 92
15	37, 83	31, 89	26, 94	20, 100	16, 104
16	43, 93	36, 100	30, 106	24, 112	20, 116
17	49, 104	42, 111	35, 118	28, 125	24, 129
18	56, 115	48, 123	41, 130	33, 138	28, 143
19	63, 127	54, 136	47, 143	38, 152	33, 157
20	70, 140	61, 149	53, 157	44, 166	38, 172

Source: Adapted from Roger C. Pfaffenberger and James H. Patterson, *Statistical Methods for Business and Economics* (Homewood, Ill.: Richard D. Irwin, Inc., 1987), p. 110, and R. L. McCornack, "Extended Tables of the Wilcoxon Matched Pairs Signed Rank Statistics," *Journal of the American Statistical Association* 60 (1965), 864–871.

TABLE A.9

Wilcoxon Rank Sum Test, Lower and Upper Critical Values

$\alpha = 0.025$ (one-tail) or $\alpha = 0.05$ (two-tail)

n_1:	3	4	5	6	7	8	9	10
n_2: 3	5, 16	6, 18	6, 21	7, 23	7, 26	8, 28	8, 31	9, 33
4	6, 18	11, 25	12, 28	12, 32	13, 35	14, 38	15, 41	16, 44
5	6, 21	12, 28	18, 37	19, 41	20, 45	21, 49	22, 53	24, 56
6	7, 23	12, 32	19, 41	26, 52	28, 56	29, 61	31, 65	32, 70
7	7, 26	13, 35	20, 45	28, 56	37, 68	39, 73	41, 78	43, 83
8	8, 28	14, 38	21, 49	29, 61	39, 73	49, 87	51, 93	54, 98
9	8, 31	15, 41	22, 53	31, 65	41, 78	51, 93	63, 108	66, 114
10	9, 33	16, 44	24, 56	32, 70	43, 83	54, 98	66, 114	79, 131

(Note: n_1 is the smaller of the two samples—i.e., $n_1 \leq n_2$.)

$\alpha = 0.05$ (one-tail) or $\alpha = 0.10$ (two-tail)

n_1:	3	4	5	6	7	8	9	10
n_2: 3	6, 15	7, 17	7, 20	8, 22	9, 24	9, 27	10, 29	11, 31
4	7, 17	12, 24	13, 27	14, 30	15, 33	16, 36	17, 39	18, 42
5	7, 20	13, 27	19, 36	20, 40	22, 43	24, 46	25, 50	26, 54
6	8, 22	14, 30	20, 40	28, 50	30, 54	32, 58	33, 63	35, 67
7	9, 24	15, 33	22, 43	30, 54	39, 66	41, 71	43, 76	46, 80
8	9, 27	16, 36	24, 46	32, 58	41, 71	52, 84	54, 90	57, 95
9	10, 29	17, 39	25, 50	33, 63	43, 76	54, 90	66, 105	69, 111
10	11, 31	18, 42	26, 54	35, 67	46, 80	57, 95	69, 111	83, 127

(Note: n_1 is the smaller of the two samples—i.e., $n_1 \leq n_2$.)

Source: F. Wilcoxon and R. A. Wilcox, *Some Approximate Statistical Procedures* (New York: American Cyanamid Company, 1964), pp. 20–23.

TABLE A.10

Critical Values of D for the Kolmogorov–Smirnov Test of Normality

Sample Size n	Significance Level α 0.20	0.15	0.10	0.05	0.01
4	.300	.319	.352	.381	.417
5	.285	.299	.315	.337	.405
6	.265	.277	.294	.319	.364
7	.247	.258	.276	.300	.348
8	.233	.244	.261	.285	.331
9	.223	.233	.249	.271	.311
10	.215	.224	.239	.258	.294
11	.206	.217	.230	.249	.284
12	.199	.212	.223	.242	.275
13	.190	.202	.214	.234	.268
14	.183	.194	.207	.227	.261
15	.177	.187	.201	.220	.257
16	.173	.182	.195	.213	.250
17	.169	.177	.189	.206	.245
18	.166	.173	.184	.200	.239
19	.163	.169	.179	.195	.235
20	.160	.166	.174	.190	.231
25	.142	.147	.158	.173	.200
30	.131	.136	.144	.161	.187
Over 30	$\frac{.736}{\sqrt{n}}$	$\frac{.768}{\sqrt{n}}$	$\frac{.805}{\sqrt{n}}$	$\frac{.886}{\sqrt{n}}$	$\frac{1.031}{\sqrt{n}}$

Source: From H. W. Lilliefors, "On the Kolmogorov–Smirnov Test for Normality with Mean and Variance Unknown," *Journal of the American Statistical Association*, 62 (1967), pp. 399–402. As adapted by Conover, *Practical Nonparametric Statistics* (New York: John Wiley, 1971), p. 398.

TABLE A.11

Critical Values of Spearman's Rank Correlation Coefficient, r_s, One-Tail Test. (For a two-tail test, the listed values correspond to the 2α level of significance.)

n	$\alpha = 0.05$	$\alpha = 0.025$	$\alpha = 0.01$	$\alpha = 0.005$
5	0.900	—	—	—
6	0.829	0.886	0.943	—
7	0.714	0.786	0.893	—
8	0.643	0.738	0.833	0.881
9	0.600	0.683	0.783	0.833
10	0.564	0.648	0.745	0.794
11	0.523	0.623	0.736	0.818
12	0.497	0.591	0.703	0.780
13	0.475	0.566	0.673	0.745
14	0.457	0.545	0.646	0.716
15	0.441	0.525	0.623	0.689
16	0.425	0.507	0.601	0.666
17	0.412	0.490	0.582	0.645
18	0.399	0.476	0.564	0.625
19	0.388	0.462	0.549	0.608
20	0.377	0.450	0.534	0.591
21	0.368	0.438	0.521	0.576
22	0.359	0.428	0.508	0.562
23	0.351	0.418	0.496	0.549
24	0.343	0.409	0.485	0.537
25	0.336	0.400	0.475	0.526
26	0.329	0.392	0.465	0.515
27	0.323	0.385	0.456	0.505
28	0.317	0.377	0.448	0.496
29	0.311	0.370	0.440	0.487
30	0.305	0.364	0.432	0.478

Source: E. G. Olds, "Distribution of Sums of Squares of Rank Differences for Small Samples," *Annals of Mathematical Statistics* 9 (1938).

TABLE A.12

Values of d_L and d_U for the Durbin–Watson Test for $\alpha = 0.05$

n = number of observations
k = number of independent variables

	$k = 1$		$k = 2$		$k = 3$		$k = 4$		$k = 5$	
n	d_L	d_U	d_L	d_U	d_L	d_U	d_L	d_U	d_L	d_U
15	1.08	1.36	0.95	1.54	0.82	1.75	0.69	1.97	0.56	2.21
16	1.10	1.37	0.98	1.54	0.86	1.73	0.74	1.93	0.62	2.15
17	1.13	1.38	1.02	1.54	0.90	1.71	0.78	1.90	0.67	2.10
18	1.16	1.39	1.05	1.53	0.93	1.69	0.82	1.87	0.71	2.06
19	1.18	1.40	1.08	1.53	0.97	1.68	0.86	1.85	0.75	2.02
20	1.20	1.41	1.10	1.54	1.00	1.68	0.90	1.83	0.79	1.99
21	1.22	1.42	1.13	1.54	1.03	1.67	0.93	1.81	0.83	1.96
22	1.24	1.43	1.15	1.54	1.05	1.66	0.96	1.80	0.86	1.94
23	1.26	1.44	1.17	1.54	1.08	1.66	0.99	1.79	0.90	1.92
24	1.27	1.45	1.19	1.55	1.10	1.66	1.01	1.78	0.93	1.90
25	1.29	1.45	1.21	1.55	1.12	1.66	1.04	1.77	0.95	1.89
26	1.30	1.46	1.22	1.55	1.14	1.65	1.06	1.76	0.98	1.88
27	1.32	1.47	1.24	1.56	1.16	1.65	1.08	1.76	1.01	1.86
28	1.33	1.48	1.26	1.56	1.18	1.65	1.10	1.75	1.03	1.85
29	1.34	1.48	1.27	1.56	1.20	1.65	1.12	1.74	1.05	1.84
30	1.35	1.49	1.28	1.57	1.21	1.65	1.14	1.74	1.07	1.83
31	1.36	1.50	1.30	1.57	1.23	1.65	1.16	1.74	1.09	1.83
32	1.37	1.50	1.31	1.57	1.24	1.65	1.18	1.73	1.11	1.82
33	1.38	1.51	1.32	1.58	1.26	1.65	1.19	1.73	1.13	1.81
34	1.39	1.51	1.33	1.58	1.27	1.65	1.21	1.73	1.15	1.81
35	1.40	1.52	1.34	1.58	1.28	1.65	1.22	1.73	1.16	1.80
36	1.41	1.52	1.35	1.59	1.29	1.65	1.24	1.73	1.18	1.80
37	1.42	1.53	1.36	1.59	1.31	1.66	1.25	1.72	1.19	1.80
38	1.43	1.54	1.37	1.59	1.32	1.66	1.26	1.72	1.21	1.79
39	1.43	1.54	1.38	1.60	1.33	1.66	1.27	1.72	1.22	1.79
40	1.44	1.54	1.39	1.60	1.34	1.66	1.29	1.72	1.23	1.79
45	1.48	1.57	1.43	1.62	1.38	1.67	1.34	1.72	1.29	1.78
50	1.50	1.59	1.46	1.63	1.42	1.67	1.38	1.72	1.34	1.77
55	1.53	1.60	1.49	1.64	1.45	1.68	1.41	1.72	1.38	1.77
60	1.55	1.62	1.51	1.65	1.48	1.69	1.44	1.73	1.41	1.77
65	1.57	1.63	1.54	1.66	1.50	1.70	1.47	1.73	1.44	1.77
70	1.58	1.64	1.55	1.67	1.52	1.70	1.49	1.74	1.46	1.77
75	1.60	1.65	1.57	1.68	1.54	1.71	1.51	1.74	1.49	1.77
80	1.61	1.66	1.59	1.69	1.56	1.72	1.53	1.74	1.51	1.77
85	1.62	1.67	1.60	1.70	1.57	1.72	1.55	1.75	1.52	1.77
90	1.63	1.68	1.61	1.70	1.59	1.73	1.57	1.75	1.54	1.78
95	1.64	1.69	1.62	1.71	1.60	1.73	1.58	1.75	1.56	1.78
100	1.65	1.69	1.63	1.72	1.61	1.74	1.59	1.76	1.57	1.78

Source: From J. Durbin and G.S. Watson, "Testing for Serial Correlation in Least Squares Regression," *Biometrika*, 38 June, 1951.

TABLE A.12

(continued)

Values of d_L and d_U for the Durbin–Watson Test for $\alpha = 0.025$

n = number of observations
k = number of independent variables

	$k = 1$		$k = 2$		$k = 3$		$k = 4$		$k = 5$	
n	d_L	d_U	d_L	d_U	d_L	d_U	d_L	d_U	d_L	d_U
15	0.95	1.23	0.83	1.40	0.71	1.61	0.59	1.84	0.48	2.09
16	0.98	1.24	0.86	1.40	0.75	1.59	0.64	1.80	0.53	2.03
17	1.01	1.25	0.90	1.40	0.79	1.58	0.68	1.77	0.57	1.98
18	1.03	1.26	0.93	1.40	0.82	1.56	0.72	1.74	0.62	1.93
19	1.06	1.28	0.96	1.41	0.86	1.55	0.76	1.72	0.66	1.90
20	1.08	1.28	0.99	1.41	0.89	1.55	0.79	1.70	0.70	1.87
21	1.10	1.30	1.01	1.41	0.92	1.54	0.83	1.69	0.73	1.84
22	1.12	1.31	1.04	1.42	0.95	1.54	0.86	1.68	0.77	1.82
23	1.14	1.32	1.06	1.42	0.97	1.54	0.89	1.67	0.80	1.80
24	1.16	1.33	1.08	1.43	1.00	1.54	0.91	1.66	0.83	1.79
25	1.18	1.34	1.10	1.43	1.02	1.54	0.94	1.65	0.86	1.77
26	1.19	1.35	1.12	1.44	1.04	1.54	0.96	1.65	0.88	1.76
27	1.21	1.36	1.13	1.44	1.06	1.54	0.99	1.64	0.91	1.75
28	1.22	1.37	1.15	1.45	1.08	1.54	1.01	1.64	0.93	1.74
29	1.24	1.38	1.17	1.45	1.10	1.54	1.03	1.63	0.96	1.73
30	1.25	1.38	1.18	1.46	1.12	1.54	1.05	1.63	0.98	1.73
31	1.26	1.39	1.20	1.47	1.13	1.55	1.07	1.63	1.00	1.72
32	1.27	1.40	1.21	1.47	1.15	1.55	1.08	1.63	1.02	1.71
33	1.28	1.41	1.22	1.48	1.16	1.55	1.10	1.63	1.04	1.71
34	1.29	1.41	1.24	1.48	1.17	1.55	1.12	1.63	1.06	1.70
35	1.30	1.42	1.25	1.48	1.19	1.55	1.13	1.63	1.07	1.70
36	1.31	1.43	1.26	1.49	1.20	1.56	1.15	1.63	1.09	1.70
37	1.32	1.43	1.27	1.49	1.21	1.56	1.16	1.62	1.10	1.70
38	1.33	1.44	1.28	1.50	1.23	1.56	1.17	1.62	1.12	1.70
39	1.34	1.44	1.29	1.50	1.24	1.56	1.19	1.63	1.13	1.69
40	1.35	1.45	1.30	1.51	1.25	1.57	1.20	1.63	1.15	1.69
45	1.39	1.48	1.34	1.53	1.30	1.58	1.25	1.63	1.21	1.69
50	1.42	1.50	1.38	1.54	1.34	1.59	1.30	1.64	1.26	1.69
55	1.45	1.52	1.41	1.56	1.37	1.60	1.33	1.64	1.30	1.69
60	1.47	1.54	1.44	1.57	1.40	1.61	1.37	1.65	1.33	1.69
65	1.49	1.55	1.46	1.59	1.43	1.62	1.40	1.66	1.36	1.69
70	1.51	1.57	1.48	1.60	1.45	1.63	1.42	1.66	1.39	1.70
75	1.53	1.58	1.50	1.61	1.47	1.64	1.45	1.67	1.42	1.70
80	1.54	1.59	1.52	1.62	1.49	1.65	1.47	1.67	1.44	1.70
85	1.56	1.60	1.53	1.63	1.51	1.65	1.49	1.68	1.46	1.71
90	1.57	1.61	1.55	1.64	1.53	1.66	1.50	1.69	1.48	1.71
95	1.58	1.62	1.56	1.65	1.54	1.67	1.52	1.69	1.50	1.71
100	1.59	1.63	1.57	1.65	1.55	1.67	1.53	1.70	1.51	1.72

TABLE A.12

(continued)

Values of d_L and d_U for the Durbin–Watson Test for $\alpha = 0.01$

n = number of observations
k = number of independent variables

	$k = 1$		$k = 2$		$k = 3$		$k = 4$		$k = 5$	
n	d_L	d_U	d_L	d_U	d_L	d_U	d_L	d_U	d_L	d_U
15	0.81	1.07	0.70	1.25	0.59	1.46	0.49	1.70	0.39	1.96
16	0.84	1.09	0.74	1.25	0.63	1.44	0.53	1.66	0.44	1.90
17	0.87	1.10	0.77	1.25	0.67	1.43	0.57	1.63	0.48	1.85
18	0.90	1.12	0.80	1.26	0.71	1.42	0.61	1.60	0.52	1.80
19	0.93	1.13	0.83	1.26	0.74	1.41	0.65	1.58	0.56	1.77
20	0.95	1.15	0.86	1.27	0.77	1.41	0.68	1.57	0.60	1.74
21	0.97	1.16	0.89	1.27	0.80	1.41	0.72	1.55	0.63	1.71
22	1.00	1.17	0.91	1.28	0.83	1.40	0.75	1.54	0.66	1.69
23	1.02	1.19	0.94	1.29	0.86	1.40	0.77	1.53	0.70	1.67
24	1.04	1.20	0.96	1.30	0.88	1.41	0.80	1.53	0.72	1.66
25	1.05	1.21	0.98	1.30	0.90	1.41	0.83	1.52	0.75	1.65
26	1.07	1.22	1.00	1.31	0.93	1.41	0.85	1.52	0.78	1.64
27	1.09	1.23	1.02	1.32	0.95	1.41	0.88	1.51	0.81	1.63
28	1.10	1.24	1.04	1.32	0.97	1.41	0.90	1.51	0.83	1.62
29	1.12	1.25	1.05	1.33	0.99	1.42	0.92	1.51	0.85	1.61
30	1.13	1.26	1.07	1.34	1.01	1.42	0.94	1.51	0.88	1.61
31	1.15	1.27	1.08	1.34	1.02	1.42	0.96	1.51	0.90	1.60
32	1.16	1.28	1.10	1.35	1.04	1.43	0.98	1.51	0.92	1.60
33	1.17	1.29	1.11	1.36	1.05	1.43	1.00	1.51	0.94	1.59
34	1.18	1.30	1.13	1.36	1.07	1.43	1.01	1.51	0.95	1.59
35	1.19	1.31	1.14	1.37	1.08	1.44	1.03	1.51	0.97	1.59
36	1.21	1.32	1.15	1.38	1.10	1.44	1.04	1.51	0.99	1.59
37	1.22	1.32	1.16	1.38	1.11	1.45	1.06	1.51	1.00	1.59
38	1.23	1.33	1.18	1.39	1.12	1.45	1.07	1.52	1.02	1.58
39	1.24	1.34	1.19	1.39	1.14	1.45	1.09	1.52	1.03	1.58
40	1.25	1.34	1.20	1.40	1.15	1.46	1.10	1.52	1.05	1.58
45	1.29	1.38	1.24	1.42	1.20	1.48	1.16	1.53	1.11	1.58
50	1.32	1.40	1.28	1.45	1.24	1.49	1.20	1.54	1.16	1.59
55	1.36	1.43	1.32	1.47	1.28	1.51	1.25	1.55	1.21	1.59
60	1.38	1.45	1.35	1.48	1.32	1.52	1.28	1.56	1.25	1.60
65	1.41	1.47	1.38	1.50	1.35	1.53	1.31	1.57	1.28	1.61
70	1.43	1.49	1.40	1.52	1.37	1.55	1.34	1.58	1.31	1.61
75	1.45	1.50	1.42	1.53	1.39	1.56	1.37	1.59	1.34	1.62
80	1.47	1.52	1.44	1.54	1.42	1.57	1.39	1.60	1.36	1.62
85	1.48	1.53	1.46	1.55	1.43	1.58	1.41	1.60	1.39	1.63
90	1.50	1.54	1.47	1.56	1.45	1.59	1.43	1.61	1.41	1.64
95	1.51	1.55	1.49	1.57	1.47	1.60	1.45	1.62	1.42	1.64
100	1.52	1.56	1.50	1.58	1.48	1.60	1.46	1.63	1.44	1.65

TABLE A.13

Factors for Determining 3-Sigma Control Limits, Mean and Range Control Charts

Number of Observations in Each Sample	Factor for Determining Control Limits, Control Chart for the Mean	Factors for Determining Control Limits, Control Chart for the Range	
n	A_2	D_3	D_4
2	1.880	0	3.267
3	1.023	0	2.575
4	0.729	0	2.282
5	0.577	0	2.115
6	0.483	0	2.004
7	0.419	0.076	1.924
8	0.373	0.136	1.864
9	0.337	0.184	1.816
10	0.308	0.223	1.777
11	0.285	0.256	1.744
12	0.266	0.284	1.716
13	0.249	0.308	1.692
14	0.235	0.329	1.671
15	0.223	0.348	1.652

Source: From E. S. Pearson, "The Percentage Limits for the Distribution of Range in Samples from a Normal Population," *Biometrika* 24 (1932): 416.

APPENDIX B

SELECTED ANSWERS

Answers to Selected Exercises

Chapter 2

2.3 a. 45.59 million **b.** lower limit is 35, upper limit is under 45 **c.** 10 years **d.** 40 years **2.5 a.** 183.53 thousand **b.** lower limit is 45, upper limit is under 55 **c.** 10 years **d.** 50 years **2.45 a.** $y = 323.53 + 0.8112x$ **b.** yes, direct **2.57 a.** 1342 cities **b.** 2815 cities **c.** 253 cities; 8.77% **d.** 175,000 **2.66 a.** yes **b.** lawjudge = 0.788 + 0.840*acad **2.67 c.** highest: mechanical 2 and electrical 1; lowest: mechanical 1 and electrical 3 **2.71 a.** yes **b.** Canada = 1.089 + 1.754*U.S.

Chapter 3

3.1 $\bar{x}$ = \$20.449, median = \$20.495 **3.3** $\bar{x}$ = 57.05 visitors, median = 57.50, mode = 63 **3.5** $\bar{x}$ = 6.93, median = 7.095 **3.7** $\bar{x}$ = 38.1, median = 36.5 **3.9** 83.2 **3.11 a.** mean **b.** median **3.13** $\bar{x}$ = 398.86, median = 396.75 **3.15** females: $\bar{x}$ = 40.62, median = 39.00; males: $\bar{x}$ = 41.08, median = 41.50 **3.17** range = 39, MAD = 10.35 visitors, s = 12.40, s^2 = 153.84 **3.19 a.** μ = 29.11 million, median = 22.4 million, range = 44.5 million, midrange = 39.45 million **b.** MAD = 12.99 million **c.** σ = 15.44 million, σ^2 = 238.396 **3.21 a.** $\bar{x}$ = 27.2 mpg, median = 28 mpg, range = 30 mpg, midrange = 25 mpg **b.** MAD = 5.6 mpg **c.** s = 8.052, s^2 = 64.84 **3.23** $Q_1 = 7$, $Q_2 = 18$, $Q_3 = 30$, interquartile range = 23, quartile deviation = 11.5 **3.25 a.** $\bar{x}$ = 23.35, median = 22.86, range = 26.13, midrange = 26.465 **b.** MAD = 4.316 **c.** s = 5.49, s^2 = 30.14 **3.27 a.** $\bar{x}$ = 90.771, median = 91.4, range = 40.4, midrange = 92.0 **b.** MAD = 6.70 **c.** s = 8.36, s^2 = 69.90 **3.29 a.** 84% **b.** 88.89% **c.** 96% **3.31** 90%, yes **3.33 a.** 68% **b.** 2.5% **c.** 84% **d.** 13.5% **3.35** Barnsboro **3.39 a.** approximately 7.00 and 1.504 **3.41** approximately 21.75 and 15.65 **3.43** $r = -0.8$ **3.45** lawjudge = 0.7880 + 0.8404*acad, r^2 = 0.9344, r = 0.967 **3.47** generic = −3.5238 + 0.377*brand, r^2 = 0.7447, r = 0.863 **3.48** \$4.69 **3.50** $\bar{x}$ = 117.75, median = 117.5, no **3.52 a.** $\bar{x}$ = 2.099, median = 2.115, range = 0.46, midrange = 2.08 **b.** MAD = 0.137 **c.** s = 0.156, s^2 = 0.02446 **3.56 a.** $\bar{x}$ becomes 3.1 lbs., s remains 0.5 lbs. **b.** 4.1 lbs. **3.59** greater variation for data in exercise 3.60 **3.60** no, positively skewed **3.63** 120, 116, 124, 20, symmetrical

Chapter 4

4.3 a. secondary **b.** secondary **4.11** response error **4.13** telephone **4.39** systematic **4.43 a.** sample **b.** sample **c.** sample **d.** census **4.68 a.** 54 **b.** 146

Chapter 5

5.1 subjective **5.7** decrease for men, increase for women **5.9** 0.36 **5.13 a.** 0 **b.** 196 **c.** 147 **d.** 372 **5.19** 0.945, 0.035 **5.21** 0.35, 0.65 **5.23** 0.797 **5.25 b.** 0.13 **c.** 0.30 **d.** 0.89 **5.27** 0.91 **5.29 a.** 0.947 **b.** 0.710 **c.** 0.351 **d.** 0.992 **5.31** no, no **5.33 a.** 0.16 **b.** 0.31 **c.** 0.12 **d.** 0.41 **5.35** 0.851 **5.37 a.** 0.025 **b.** 0.709 **c.** 0.266 **5.39** 0.175 **5.43** 0.154 **5.45 a.** 0.51 **b.** 0.85 **c.** 0.77 **5.47 a.** 0.5 **b.** 0.455 **5.49 a.** 0.1 **b.** 0.233 **5.51** 256 **5.53** 36 **5.57** 20 **5.59** $7.9019*10^{27}$ **5.62 a.** 0.000000005 **b.** 0.0625 **c.** 0.062500005 **d.** 0 **5.63 a.** 0.005 **b.** 0.855 **c.** 0.14 **d.** 0.995 **5.64 a.** 0.7 **b.** 0.9 **c.** 0.667 **5.67 a.** 3 **b.** 3 **c.** 1 **5.68 a.** 0.85 **b.** 0.278 **c.** 0.046 **5.71 a.** 0.001 **b.** classical **c.** yes **5.73** 120 **5.74** 22 **5.77 a.** 0.9 **b.** 0.429 **5.78 a.** 0.973 **b.** $4.741*10^{-12}$ **c.** 10 **5.81** 0.296, 0.963 **5.84** 10,000 **5.85** 1320

Chapter 6

6.3 a. discrete **b.** continuous **c.** discrete **d.** continuous **6.5** μ = 7.2, σ = 2.18, σ^2 = 4.76 **6.7** μ = 1.9, σ = 1.14, σ^2 = 1.29 **6.9** \$10 **6.11** $E(x) = 3.1$ **6.13** \$0.50 **6.15** \$37.9 million, \$41.9 million, \$40.3 million, minor **6.19 a.** 3.6 **b.** 1.59 **c.** 0.2397 **d.** 0.9133 **e.** 0.5075 **6.21 a.** 0.0102 **b.** 0.2304 **c.** 0.3456 **d.** 0.0778 **6.23 a.** 0.9703 **b.** 0.0294 **c.** 0.0003 **d.** 0.0000 **6.25 a.** 0.9375 **b.** 0.6875 **6.27** 0.1250, 0.1250, 0.3750 **6.29** 0.3439 **6.31 a.** 0.0014 **b.** 0.0139 **c.** yes **6.35 a.** 0.1000 **b.** 0.6000 **c.** 0.3000 **6.37** 0.9000 **6.39 a.** 0.1353 **b.** 0.2707 **c.** 0.8571 **d.** 0.5940 **6.41 a.** 4.9 **b.** 0.1460 **c.** 0.1753 **d.** 0.9382 **e.** 0.8547 **6.43 a.** 1.8 **b.** 0.2975 **c.** 0.0723 **d.** 0.9974 **e.** 0.5268 **6.45** 0.0047, should consider slight decrease **6.47 a.** 10.0 **b.** 0.0901 **c.** 0.1251 **d.** 0.7916 **e.** 0.7311 **6.49** 0.2275, \$45,000 **6.51** not merely a coincidence **6.58** 87.95% **6.59** no **6.61** 0.9098 **6.62** 0.3684 **6.65** 0.1904 **6.68** 0.3125 **6.69** 0.0839 **6.70** 0.6065 **6.73** 0.9872 **6.74** 0.1837 **6.75** 0.7748 **6.78** 0.0012, not believable

Chapter 7

7.9 a. 0.5 **b.** approx 0.683 **c.** approx 0.8415 **d.** 0 **e.** approx 0.4775 **f.** approx 0.9775 **7.11 a.** 0.5 **b.** approx 0.955 **c.** approx 0.683 **d.** approx 0.9985 **7.13 a.** approx 0.1585 **b.** approx 0.683 **c.** approx 0.0225 **d.** approx 0.819 **7.15 a.** approx 0.0015 **b.** approx 0.1585 **7.17 a.** −0.67, 0.67 **b.** −1.28, 1.28 **c.** −0.74, 0.74 **7.19 a.** −2.00 **b.** −0.80 **c.** 0.00 **d.** 3.40 **e.** 4.60 **7.21 a.** 0.3643 **b.** 0.1357 **c.** 0.9115 **7.23 a.** 0.8730 **b.** 0.2272 **c.** 0.1091 **7.25 a.** 0.52 **b.** −1.28 **c.** 0.10 **d.** 0.52 **7.27 a.** 0.0874 **b.** 0.8790

c. 0.8413 **7.29 a.** 0.4207 **b.** 0.3050 **c.** 0.3446 **7.31** $409,350 **7.33 a.** 0.4207 **b.** $9545 **7.35** 3.22 minutes **7.37** 70,500 **7.41 a.** 10.0, 2.739 **b.** 0.1098, 0.2823, 0.3900, 0.1003 **7.43 a.** 12.0, 1.549 **b.** 0.2501 **c.** 0.2510 **d.** 0.9463 **7.45** 0.4350 **7.47** 0.4191 **7.53 a.** 0.5488 **b.** 0.4493 **c.** 0.3679 **d.** 0.3012 **7.55** 0.2865 **7.57** 0.619, 0.383, 8779.7 hours **7.61** 11.51% **7.64** 0.3911 **7.65** 0.8023 **7.66** no **7.69** not credible **7.70** 0.0668, 0.0228 **7.73** 0.6604 **7.74** 0.1587 **7.77 a.** 0.4628 **b.** 0.7772 **c.** 12,900 **7.80** 0.0918, 0.0446 **7.81** 20.615 **7.82** 0.4307 **7.86** 0.4584, 0.2101 **7.87** 0.4724

Chapter 8

8.3 0.18, 0.15, 1000 **8.7 a.** 25 **b.** 10 **c.** 5 **d.** 3.162 **8.9 a.** 0.8413 **b.** 0.6853 **c.** 0.9938 **8.11** 0.8849, 0.8823 **8.13** concern is justified **8.15** 0.1056 **8.17** 0.1949 **8.19** 0.9976 **8.21** $\pi = 0.46, \sigma_p \leq 0.0895$ **8.23 a.** 0.9616 **b.** 0.5222 **c.** 0.9616 **8.25 a.** 0.40 **b.** 0.35 **c.** 0.035 **d.** 0.9236 **8.27** 0.9938 **8.29** 0.0000; district's claim is much more credible **8.35** 0.1170 **8.37** 0.8023 **8.42 a.** 0.0183 **b.** not credible **8.45 a.** 2.40 **b.** 0.0082 **c.** no **8.49 b.** 0.0446 **8.50 a.** 0.10 **b.** 0.0062 **c.** no **8.53** 0.0000 **8.56** 0.0228 **8.57** 0.0643

Chapter 9

9.7 a. 2.625 **b.** 4.554 **9.11 a.** 0.45 **b.** (0.419, 0.481) **c.** 95%, 0.95 **9.15** 90%: (236.997, 243.003); 95%: (236.422, 243.578) **9.17** 90%: (82.92, 87.08); 95%: (82.52, 87.48) **9.19** (149.006, 150.994) **9.23** (99.897, 100.027); yes **9.27** 1.313 **9.29 a.** 1.292 **b.** −1.988 **c.** 2.371 **9.31** 95%: (46.338, 54.362); 99%: (44.865, 55.835) **9.33 a.** (22.16, 27.84) **b.** yes **9.35** (28.556, 29.707) **9.37** (1520.96, 1549.04) **9.39** (17.43, 21.97); no **9.41** (14.646, 15.354); yes **9.43** (0.429, 0.491) **9.45** (0.153, 0.247); not credible **9.47** (0.449, 0.511) **9.49** (0.579, 0.625) **9.51 a.** (0.527, 0.613) **9.53** (0.649, 0.711) **9.55** (0.566, 0.634); no; may not succeed **9.57** (0.311, 0.489) **9.61** 92 **9.63** 1068 **9.65** 863 **9.67** 601 **9.71** 95%: (0.522, 0.578); 99%: (0.513, 0.587) **9.73** 458 **9.75** 92 **9.77** 462 **9.79** 1226 **9.81** 246 **9.82** 90%: (44.91, 49.96); 95%: (44.39, 50.47) **9.85** (135.211, 138.789) **9.86** 411 **9.89** 0.01 **9.90** $1200 **9.93** 3736 **9.96** 722 **9.97** 1110 **9.98** 4161 **9.101** 1068 **9.102** 95%: (0.347, 0.433); 99%: (0.334, 0.446) **9.106** (0.018, 0.062) **9.108** 90%: (0.375, 0.425); 95%: (0.370, 0.430) **9.109** ($95.56, $98.45) **9.111** (64.719, 68.301); funds are not endangered

Chapter 10

10.3 a. no **b.** yes **c.** yes **d.** no **e.** yes **f.** no **10.5** type I **10.7** type II **10.13** type I **10.17 a.** numerically high **b.** numerically low **10.23** no; do not reject H_0 **10.25 a.** 0.0618 **b.** 0.1515 **c.** 0.0672 **10.27** reject H_0; p-value = 0.021 **10.29** do not reject H_0; p-value = 0.035 **10.31** no; do not reject H_0; p-value = 0.052 **10.33** p-value = 0.316; do not reject H_0 **10.35 a.** do not reject H_0 **b.** reject H_0 **c.** do not reject H_0 **d.** reject H_0 **10.37** (1.995, 2.055); do not reject H_0; same **10.41** do not reject H_0 **10.43** no; do not reject H_0 **10.45** no; reject H_0 **10.47** yes; reject H_0 **10.49** do not reject H_0 **10.51** yes; reject H_0 **10.53** (86.29, 92.71); no; yes **10.55** (35.657, 37.943); no; yes **10.57** 0.03; reject H_0 **10.61** reject H_0 **10.63** reject H_0 **10.65** no; reject H_0; 0.003 **10.67** yes; reject H_0; p-value = 0.005 **10.69** reject H_0; p-value = 0.014 **10.71** yes; p-value = 0.220 **10.73** no; do not reject H_0; p-value = 0.079 **10.75** (0.735, 0.805); yes; yes **10.77** (0.402, 0.518); yes; yes **10.81** alpha unchanged, beta decreases **10.83** 0.9871 **10.87 a.** 2.33 **b.** 0.036 **10.91** reject H_0, has increased **10.92 a.** 0.05 level: reject H_0 **b.** 95% CI: (236,279; 255,321) **10.96** yes **10.99 b.** e.g., 0.005 **c.** e.g., 0.02 **d.** 0.024 **10.100** reject H_0; statement not credible **10.103 a.** 0.9505 **b.** 0.8212 **c.** 0.5753 **d.** 0.2946 **e.** 0.1020 **10.104** reject H_0 **10.107** do not reject H_0; claim is credible **10.109** yes; reject H_0; p-value = 0.015 **10.111** no; do not reject H_0; p-value = 0.059 **10.113** do not reject H_0; p-value = 0.282

Chapter 11

11.3 do not reject H_0, > 0.20 **11.5** yes; reject H_0 **11.7** do not reject H_0; between 0.05 and 0.10; (−8.872, 0.472) **11.9** reject H_0; between 0.10 and 0.05; 90% CI: (−1.318, −0.082) **11.11** claim could be valid; reject H_0; between 0.01 and 0.025 **11.13** yes; p-value = 0.031 **11.15** (−24.23, 1.03); yes; yes **11.17** (−3.33, 31.20); yes; yes **11.19** do not reject H_0 **11.21** yes, do not reject H_0; > 0.20; 95% CI: (−23.43, 6.43) **11.23** do not reject H_0; 0.1400; (−117.18, 17.18); yes; yes **11.25** do not reject H_0; (−2.03, −1.37); yes; yes **11.27** reject H_0; 0.000 **11.29** reject H_0; 0.006 **11.33** reject H_0 **11.35** yes, do not reject H_0; 0.2538; 95% CI: (−23.10, 6.10) **11.37** do not reject H_0; 0.1967; (−7.77, 37.77) **11.39** reject H_0; 0.0122 **11.41** reject H_0; 0.001 **11.43** do not reject H_0; 0.12 **11.45** dependent **11.47** reject H_0 **11.49** reject H_0; between 0.005 and 0.01 **11.51** do not reject H_0; 0.170; (−4.663, 1.774) **11.53** reject H_0 **11.55** no; do not reject H_0 **11.57** do not reject H_0 **11.59** yes; reject H_0; 0.0607; (−0.150, −0.010) **11.61** reject H_0; 0.0583 **11.63** do not reject H_0; 0.057; (−0.031, 0.211) **11.65** do not reject H_0; 0.1866; (−0.055, 0.005) **11.67** do not reject H_0; no, no **11.69** yes; do not reject H_0; no **11.71** yes; reject H_0 **11.73** reject H_0 **11.75** suspicion confirmed; reject H_0; between 0.025 and 0.05 **11.77** do not reject H_0; (−505.32, 65.32) **11.79** reject H_0; < 0.005 **11.81** yes; reject H_0; 0.0141 **11.83** reject H_0; table: < 0.01; computer: 0.0075; (0.13, 6.67) **11.85** do not reject H_0; (−1.13, 0.13) **11.87** yes; reject H_0 **11.89** no; do not reject H_0 **11.91** yes; reject H_0; < 0.005 **11.93** yes; reject H_0; 0.001 **11.95** yes; do not reject H_0; no **11.97** do not reject H_0; 0.140; (−1.95, 12.96) **11.99** not supported; do not reject H_0; 0.135 **11.101** do not reject H_0; 0.414; (−0.1204, 0.0404)

Chapter 12

12.7 designed **12.21** do not reject H_0; > 0.05 **12.23** do not reject H_0; > 0.05 **12.25** do not reject H_0 **12.27 b.** reject H_0 **12.29 b.** do not reject H_0 **c.** μ_1: (48.914, 61.086); μ_2: (42.714, 54.886); μ_3: (39.514, 51.686) **12.35** μ_1: (16.271, 19.009); μ_2: (13.901, 16.899); μ_3: (15.690, 18.060) **12.47** reject H_0; between 0.025 and 0.05 **12.49** do not reject H_0; do not reject H_0 **12.51** do not reject H_0; do not reject H_0 **12.53** reject H_0 **12.55** reject H_0 **12.59** yes; reject H_0 **12.69** Factor A, do not reject H_0; Factor B, reject H_0; Interaction, reject H_0

12.71 Factor A, do not reject H_0; Factor B, reject H_0; Interaction, do not reject H_0 **12.73** Factor A, reject H_0; Factor B, reject H_0; Interaction, reject H_0 **12.75** Assembly, do not reject H_0; Music, reject H_0; Interaction, reject H_0; Method 1, (35.846, 41.654); Method 2, (34.096, 39.904); Classical, (29.096, 34.904); Rock, (40.846, 46.654) **12.77** Bag, do not reject H_0; Dress, reject H_0; Interaction, reject H_0; Carry, (24.798, 30.536); Don't Carry, (27.298, 33.036); Sloppy, (41.736, 48.764); Casual, (17.736, 24.764); Dressy, (16.736, 23.764) **12.79** Keyboard, reject H_0; Wordpack, do not reject H_0; Interaction, reject H_0 **12.83** randomized block **12.85** independent: faceplate design; dependent: time to complete task; designed **12.87** no, randomized block procedure should be used **12.89** reject H_0 **12.91** do not reject H_0 **12.93** reject H_0 **12.95** do not reject H_0 **12.97** reject H_0 **12.99** Style, do not reject H_0; Darkness, reject H_0; Interaction, do not reject H_0; Style 1, (25.693, 30.307); Style 2, (23.693, 28.307); Light, (27.425, 33.075); Medium, (22.175, 27.825); Dark, (22.925, 28.575) **12.101** Position, reject H_0; Display, do not reject H_0; Interaction, reject H_0; Position 1, (42.684, 47.982); Position 2, (47.129, 52.427); Display 1, (42.089, 48.577); Display 2, (43.923, 50.411); Display 3, (46.923, 53.411)

Chapter 13

13.7 a. 3.490 **b.** 13.362 **c.** 2.733 **d.** 15.507 **e.** 17.535 **f.** 2.180 **13.9 a.** A = 8.547, B = 22.307 **b.** A = 7.261, B = 24.996 **c.** A = 6.262, B = 27.488 **d.** A = 5.229, B = 30.578 **13.13** 4 **13.15 a.** 6 **b.** 12.592 **c.** do not reject H_0 **13.17** yes; reject H_0 **13.19** do not reject H_0 **13.21** do not reject H_0 **13.23** do not reject H_0 **13.25** reject H_0 **13.29 a.** 6 **b.** 2 **c.** 12 **d.** 8 **e.** 12 **f.** 4 **13.31 a.** 12.833 **b.** 15.507 **c.** 16.812 **d.** 7.779 **13.33** no, reject H_0; between 0.025 and 0.01 **13.35** no; reject H_0; between 0.05 and 0.025 **13.37** no; reject H_0; less than 0.01 **13.39** yes; do not reject H_0 **13.41** no; reject H_0 **13.43** do not reject H_0 **13.45** do not reject H_0 **13.47** reject H_0 **13.49** yes; do not reject H_0 **13.53** (15.096, 43.011) **13.55** (2.620, 9.664) **13.57** (0.1346, 0.2791) **13.59** do not reject H_0 **13.61** do not reject H_0; between 0.05 and 0.025 **13.63** (0.0329, 0.0775) **13.65** do not reject H_0 **13.67** reject H_0 **13.69** yes; do not reject H_0 **13.71** do not reject H_0; between 0.10 and 0.05 **13.73** no; do not reject H_0; between 0.10 and 0.05 **13.75** (0.02001, 0.05694) **13.77** reject H_0 **13.79** do not reject H_0 **13.81** reject H_0 **13.83** reject H_0 **13.85** do not reject H_0 **13.87** probably not; reject H_0

Chapter 14

14.7 reject H_0; < 0.005 **14.9** yes; reject H_0; 0.017 **14.11** do not reject H_0; 0.137 **14.13** reject H_0 **14.15** yes, reject H_0 **14.17** do not reject H_0; between 0.05 and 0.10 **14.19** no; do not reject H_0; 0.137 **14.23** do not reject H_0 **14.25** yes; do not reject H_0; 0.2443 **14.29** reject H_0 **14.31** no; reject H_0 **14.33** no; reject H_0; 0.058 **14.37** yes; do not reject H_0; between 0.10 and 0.90 **14.39** no; do not reject H_0; 0.393 **14.43** do not reject H_0 **14.45** reject H_0 **14.49** do not reject H_0 **14.51** do not reject H_0 **14.53** do not reject H_0 **14.55** 0.83, yes **14.57** 0.343, no **14.59** reject H_0; 0.0287 **14.61** no, 0.619 **14.63** do not reject H_0 **14.65** yes; reject H_0; 0.0056 **14.67** no; do not reject H_0; between 0.05 and 0.10 **14.69** yes; do not reject H_0; between 0.025 and 0.05 **14.71** yes; do not reject H_0; between 0.025 and 0.05 **14.75** claim not credible, reject H_0; between 0.05 and 0.025 **14.77** 0.868, yes **14.79** no; do not reject H_0; 0.343 **14.81** do not reject H_0; 0.210 **14.83** they can tell the difference; reject H_0; 0.016 **14.85** do not reject H_0; 0.486

Chapter 15

15.5 second **15.7** second **15.9 a.** Shares = 44.3 +38.756*Years **b.** 431.9 **15.11** totgross = 110.02 + 1.0485*2wks; $214.87 million **15.13** Acres = 349,550 − 7851*Rain; 208,223 acres **15.21 a.** $\hat{y} = 21.701 - 1.354x$ **b.** 3.617 **c.** (8.107, 16.339) **d.** (4.647, 14.383) **e.** interval d is wider **15.23 a.** Rating = 54.97 + 7.166*TD% **b.** 90.796 **c.** 1.386 **d.** (86.728, 94.863) **e.** (88.833, 92.758) **15.25** 91.48; (−33.9, 510.1) **15.27** Revenue = $-1.66*10^8$ + 64,076,803*Dealers; CI: ($5.88*10^9$, $6.60*10^9$); PI: ($4.92*10^9$, $7.57*10^9$) **15.29** CI: (89,209; 295,832); PI: (−155,578; 540,619) **15.33** 0.81 **15.37 a.** Coll = 24.48 + 0.955*Comp **b.** 0.860, 0.740 **c.** 110.44 **15.39 a.** %Over = −82.19 + 0.04236*Cals **b.** 0.931, 0.866 **c.** 44.89% **15.41** $r = 0.800$; $r^2 = 0.640$ **15.43** Forgross = −69.382 + 1.5266*Domgross; $r = 0.7356$; $r^2 = 0.5411$ **15.45** no **15.47** > 0.10 **15.49 a.** reject H_0 **b.** reject H_0 **c.** (0.556, 1.354) **15.51 a.** reject H_0 **b.** reject H_0 **c.** (0.392, 9.608) **15.53** 61.2, 58.8 **15.55** 774.80, 536.01, 238.79; 0.692; do not reject H_0 **15.57** Gallons = 5,921,560.92 + 10.44806*Hours; $r = 0.993$; $r^2 = 0.986$; no; (8.746, 12.150) **15.59** NetIncome = 59.8006 + 0.0382*Revenue; $r = 0.7492$; $r^2 = 0.5612$; no; (0.031, 0.045) **15.71** NetIncome = 0.211 + 0.0999*TotRev; $2.009 billion **15.73 a.** OneYr% = −2.01 +1.45*ThreeYr% **b.** 5.264% **c.** 8.172% **15.75** NetIncome = 2.301 + 0.116*OpRev; $5.789 billion **15.77 a.** Amount = −2366.0 + 623.93*Policies **b.** $r = 0.99$; $r^2 = 0.984$ **c.** $16352 million **15.79 a.** Fuel = 48.065 + 0.042244*Miles **b.** 0.992, 0.984 **c.** 179.02 billion gallons **15.81 a.** RearFull = 318 + 2.612*RearCorner **b.** 0.397, 0.157 **c.** $2408 **15.83 a.** Strength = 60.02 + 10.507*Temptime **b.** 53.0% **c.** 0.017 **d.** 0.017 **e.** (2.445, 18.569) **15.85 a.** Rolresis = 9.450 − 0.08113*psi **b.** 23.9% **c.** 0.029 **d.** 0.029 **e.** (−0.15290, −0.00936) **15.87** CI: (8.26, 28.87); PI: (−6.66, 43.78) **15.89** CI: (49.710, 54.647); PI: (46.776, 57.581) **15.91 a.** GPA = −0.6964 + 0.0033282*SAT; 2.965 **b.** 69.5% **c.** CI: (2.527, 3.402); PI: (1.651, 4.278) **15.93 a.** Pay% = 7.020 + 1.516*Rate%; 19.148% **b.** 98.0%; yes **c.** CI: (18.9874, 19.3084); PI: (18.6159, 19.6799) **15.95 a.** Estp/e = 52.56 − 0.0959*Revgrow%; 38.17 **b.** 0.2%; no **c.** CI: (−0.74, 77.07); PI: (−94.67, 171.01)

Chapter 16

16.9 a. 300, 7, 13 **b.** 399 **16.11 a.** $\hat{y} = 10.687 + 2.157x_1 + 0.0416x_2$ **c.** 24.59 **16.13 a.** $\hat{y} = -127.19 + 7.611x_1 + 0.3567x_2$ **c.** −17.79 **16.15 b.** 454.42 **16.17 a.** 130.0 **b.** 3.195 **c.** (98.493, 101.507) **d.** (93.259, 106.741)

16.19 a. CalcFin = −26.6 + 0.776*MathPro + 0.0820*SATQ; 90% CI: (64.01, 73.46) **b.** 90% PI: (59.59, 77.88) **16.21 a.** (81.588, 90.978) **b.** (77.565, 95.002) **16.27** 0.716 **16.29 a.** yes **b.** β_1, reject H_0; β_2, do not reject H_0 **d.** β_1, (0.54, 3.77); β_2, (−0.07, 0.15) **16.31** β_1, (0.0679, 0.4811); β_2, (0.1928, 0.5596); β_3, (0.1761, 0.4768) **16.33 a.** yes **b.** each is significant **16.35 a.** $\hat{y} = -40{,}855{,}482 + 44{,}281.6x_1 + 152{,}760.2x_2$ **b.** yes **c.** β_1, reject H_0; β_2, do not reject H_0 **d.** β_1, (41,229.4, 47,333.8); β_2, (−4,446,472, 4,751,992) **16.45 a.** $\hat{y} = -0.8271 + 0.007163x_1 + 0.01224x_2$ **b.** β_1, (0.00615, 0.00818); β_2, (0.00198, 0.02249); **c.** 0.959 **16.51** Speed = 67.6 − 3.21*Occupants − 6.63*Seatbelt **16.53 a.** $\hat{y} = 99.865 + 1.236x_1 + 0.822x_2$ **b.** 125.816 lbs. **c.** (124.194, 127.438) **d.** (124.439, 127.193) **e.** β_1, (0.92, 1.55); β_2, (0.09, 1.56) **16.55 a.** gpa = −1.984 + 0.00372*sat + 0.00658*rank **b.** 2.634 **c.** (1.594, 3.674) **d.** (2.365, 2.904) **e.** β_1, (0.000345, 0.007093); β_2, (−0.010745, 0.023915) **16.57** $137,289

Chapter 17

17.3 negative, negative **17.5** positive, negative, positive **17.7** $Avgrate = 336.094 − 8.239*%Occup + 0.07709*%Occup^2; 49.1% **17.9** 0to60 = 26.8119 − 0.153866*hp + 0.0003083*hp^2; 8.396 seconds; yes **17.11** Forgross = 860.8 − 4.152*Domgross + 0.007689* Domgross^2; $430.4 million; yes **17.13** second-order with interaction **17.15** $percall = 61.2 + 25.63*yrs + 6.41*score −1.82*yrs^2 − 0.058*score^2 + 0.29*yrs*score; $R^2 = 0.949$; yes **17.17 a.** oprev = −231.2 + 0.129*employs + 0.00565*departs; yes **b.** oprev = 399.2 + 0.0745*employs + 0.00087*departs + 0.00000014*employs*departs; R^2 increases from 0.958 to 0.986 **17.19** 0to60 = 25.4 − 0.161*hp − 0.00030*curbwt + 0.000028*hp*curbwt; $R^2 = 0.734$; yes **17.23** two **17.25** 600 customers **17.27** price = −30.77 + 4.975*gb + 54.20*highrpm; $54.20 **17.29** productivity = 75.4 + 1.59*yrsexp − 7.36*metha + 9.73*methb; $R^2 = 0.741$ **17.31** $\hat{y} = 0.66(1.38)^x$ **17.33** $\hat{y} = 42.668371(1.0169419)^x$; $R^2 = 0.496$; $138.30 **17.35** log revenue = −0.1285 + 1.0040 log employs − 0.1121 log departs; revenue = 0.7439*employs$^{1.0040}$*departs$^{-0.1121}$; $546.2 million **17.41** yes **17.43** may be present **17.45** will not be a problem **17.49 a.** x_5, x_2, x_9 **b.** $\hat{y} = 106.85 - 0.35x_5 - 0.33x_2$ **c.** 0.05 level: x_5, x_2, x_9; 0.02 level: x_2 **17.59** Pages = $-141.5 + 422.5x - 61.01x^2 + 2.749x^3$; 3727 **17.61** $\hat{y} = 10.705 + 0.974x - 0.015x^2$; yes; 83.6% **17.63** Wchill = −11.296 + 1.320 Temp − 0.456 Wind; Wchill = −11.296 + 1.185 Temp − 0.456 Wind + 0.00542 TempxWind; from 0.995 to 0.997 **17.65** productivity = 19.09 + 0.211*backlog + 0.577*female; $R^2 = 0.676$; yes **17.67** Log_AllLoans = −1.0130 + 2.0567Log_CredCard + 0.0635Log_Resid; AllLoans = 0.09705*CredCard$^{2.0567}$*Resid$^{0.0635}$; 1.76%. **17.69** yes; OpCost/Hr = 525.8 +3.21*Gal/Hr − 0.68*Range; 90.68% **17.71 a.** yes, test 1 **b.** final = 14.79 + 0.885*test1; R^2 = 0.8568 **17.73** Rating = 2.225 + 4.1793*YdsperAtt − 4.1763*Int_pct +3.3231*TD_pct + 0.8305*Comp_pct; 100.00% (rounded)

Chapter 18

18.3 369,600 gallons **18.5** with $x = 1$ for 2004, Earnings = $15.045 + 0.582x$; $20.865. **18.7 a.** Subs = $13.371 + 19.357x$; 361.8 million **b.** Subs = $29.993 + 12.233x + 0.54798x^2$; 427.7 million **c.** quadratic **18.17** the 0.4 curve; the 0.7 curve **18.21** 30% **18.23 b.** I , 74.72; II, 103.98; III, 123.76; IV, 97.54 **18.25 a.** J, 100.50; F, 94.33; M, 103.16; A, 103.60; M, 98.06; J, 100.55; J, 98.17; A, 98.38; S, 96.86; O, 108.84; N, 93.20; D, 104.35 **18.27** $192.0 thousand, $201.6 thousand **18.29** 1213.2; 1541.2 **18.31** $69.43 billion **18.33** 39.655 quadrillion Btu **18.35** 189,000 gallons **18.39** quadratic, quadratic **18.41** quadratic, quadratic **18.45** 0.58, positive autocorrelation **18.47 a.** inconclusive **b.** inconclusive **18.49** $y_t = -1.51 + 1.119y_{t-1}$; 288.5 **18.51** $y_t = 525.6 + 0.812y_{t-1}$; 2895.5; MAD = 100.5 **18.53** 100 **18.55** best: leisure and hospitality; worst: construction **18.57** $114,364. **18.59** Army = $476.29 + 5.4643x$; 558.25 thousand **18.61** Restaurants = $-1408.0 + 1206.50x - 9.92905x^2$; 25,851 restaurants **18.63 a.** AvgBill = $31.4743 + 1.62536x$; $55.855 **b.** AvgBill = $31.0071 + 1.93679x - 0.038929x^2$; $51.300 **c.** quadratic, quadratic **18.67** I, 95.14; II, 89.69; III, 95.64; IV, 119.53 **18.69** I, 192.88; II, 63.87; III, 30.14; IV, 113.11 **18.71** I, 94.68; II, 93.54; III, 111.05; IV, 100.73 **18.73** I, 72.50; II, 83.11; III, 111.16; IV, 132.77 **18.75** I, 71; II, 91; III, 104; IV, 66 **18.77** Cost = $38.3268 + 0.794341x + 0.0538258x^2$; Cost = $37.1427 + 1.38642x$; quadratic **18.83** DW statistic = 0.48; positive autocorrelation **18.85** $CPI_t = 2.0216 + 1.01812 \cdot CPI_{t-1}$; MAD = 1.388; 213.08 **18.87** 6.36% increase

Chapter 19

19.9 0, 2000, 2000 shirts **19.11 a.** maximax **b.** maximax **c.** maximin **19.13** purchase, do not purchase, purchase **19.17** make claims, $600 **19.19** design B, $7.6 million **19.21** direct, 31.5 minutes, 0.5 minutes **19.25** A or C, $12,800, $12,800 **19.27** direct, 0.5 minutes; longer, 9.0 minutes **19.29** 296 cod **19.31** 3520 programs **19.33** 7 units **19.35** current; Dennis; Dennis **19.37** DiskWorth; ComTranDat; ComTranDat

Chapter 20

20.35 0.0026 **20.39** in control **20.41** out of control **20.43** out of control **20.45** LCL = 0, UCL = 0.088 **20.47** LCL = 0, UCL = 15.72 **20.49** centerline = 0.197; LCL = 0.113; UCL = 0.282; in control **20.51** centerline = 4.450; LCL = 0; UCL = 10.779; out of control **20.53** in control **20.55** no **20.57** 0.89; not capable **20.59** yes; mean chart failed test #1 at sample 6 **20.61** yes; mean chart failed tests #1 and #5 at sample 5 **20.63 a.** in control **b.** yes, 0.65; not capable **20.69** *c*-chart **20.71** 1.07; capable **20.73** centerline – 0.05225; LCL – 0.00504; UCL – 0.09946; in control **20.75** in control **20.77** centerline = 0.0988; LCL = 0.0355; UCL = 0.1621; in control **20.79** yes, 0.74; not capable

INDEX/GLOSSARY

A

Acceptance Sampling The use of a sampling plan to determine whether a shipment should be accepted or rejected, 762

Accuracy The difference between the observed sample statistic and the actual value of the population parameter being estimated. This may also be referred to as estimation error or sampling error, 275

Addition Rules Formulas for determining the probability of the union of two or more events, 143–145

Alpha (α) The probability of making a Type I error by rejecting a true null hypothesis, 315–316

Alternative Hypothesis (H_1) An assertion that holds if the null hypothesis (H_0) is false. If the null hypothesis is not rejected, H_1 must be rejected, and vice versa, 313
  formulating, 313–315, 317

Analysis of Variance (ANOVA) A procedure for comparing the means of two or more independent samples at the same time. Also used in testing the overall strength of the relationship between the dependent variable and the independent variable(s) in regression analysis, 411–454
  basic concepts of, 412–415
  between-group variation in, 414–415
  experimentation and, 412–414
  multiple regression analysis and, 617–618
  one-way, 416–426
  randomized block, 429–439
  regression analysis and, 576–577
  two-way, 441–454
  within-group variation in, 414–415

ANOVA, *See* Analysis of variance

A priori probability, *See* Classical probability

Arithmetic Mean Also called the average or mean, this measure of central tendency is the sum of data values divided by the number of observations, 7, 59–60
  from grouped data, 83–85

Assignable Variation In a process, variability that occurs due to nonrandom causes. Also known as *special-cause variation,* 761

Attributes, *See* Control charts for attributes

Autocorrelation Correlation between successive error terms, or residuals, in a regression and correlation analysis, 582, 715–718

Autoregressive Forecasting A forecasting model in which each independent variable represents previous values of the dependent variable, 718–723

Average, 59–60

B

Balanced Experiment A design in which an equal number of persons or test units receives each treatment, 414

Bar Chart For qualitative variables, a graphical method in which the number of values in a category is proportional to the length of a rectangle. Unlike the histogram, there is a gap between adjacent rectangles, 28

Bayesian Decision Making The use of a decision criterion in which the probabilities of the states of nature are considered, with the alternative having the greatest expected payoff being selected, 745–748
  expected payoff and, 745–746
  expected value of perfect information and, 746–748

Bayes' Theorem A theorem dealing with sequential events, using information obtained about a second event in order to revise the probability that a first event has occurred, 150–155

Bernoulli Process A process in which there are two or more consecutive trials, each of which has just two possible outcomes (success or failure), and in which the probability of a success remains the same from one trial to the next; i.e., the trials are statistically independent from each other, 175, 223

Beta (β) The probability of making a Type II error by failing to reject a false null hypothesis, 315–316

Between-group variation, in analysis of variance, 414–415

Bimodal Distribution A distribution in which two values occur with the greatest frequency, 62

Binomial Distribution A discrete probability distribution involving consecutive trials of a Bernoulli process, 175–181
  Bernoulli process and, 175
  description and applications of, 175–178
  mean of, 175, 224
  normal approximation to, 223–226, 338
  Poisson approximation to, 191–193
  standard deviation of, 224
  variance of, 175

Block In the randomized block ANOVA design, a set of persons or test units having similar characteristics in terms of the blocking variable, 429

Box-and-Whisker Plot A graphical method that simultaneously displays the median, quartiles, and highest and lowest values within the data, 77–78

Business Statistics The collection, summarization, analysis, and reporting of numerical findings relevant to a business decision or situation, 2–14
- attitudes toward, 2
- decision making and, 11
- history of, 3–4
- as tool, 11–12
- *See also* Statistics

C

Calculated Value A value of the test statistic that is calculated from sample data. It is compared to the critical value(s) in deciding whether or not to reject the null hypothesis, 317–319

Causal Research An investigation for the purpose of determining the effect of one variable on another; typically carried out by means of experimentation, 103

Cause-and-Effect Diagram A diagram that assists the solution of quality problems by visually depicting various causes at the left that feed directionally toward the ultimate effect at the far right. Because of its appearance, it is also known as a *fishbone diagram,* 770–771

Cause-and-effect problems, in experiments, 109–110

***c*-Chart** A control chart displaying the number of defects per unit of output, 790–793

Census The actual measurement or observation of all possible elements from the population, this can be viewed as a "sample" that includes the entire population, 7, 117
- sample vs., 118–119

Centered Moving Average A smoothing technique in which the original time series is replaced with another series, each point of which is the center of and the average of N points from the original series, 693–697

Centerline A control chart line representing the central tendency of a process when it is in control, 775–776

Central Limit Theorem For large, simple random samples from a population that is not normally distributed, the sampling distribution of the mean will be approximately normal, with $\mu_{\bar{x}} = \mu$ and $\sigma_{\bar{x}} = \sigma/\sqrt{n}$. The sampling distribution will more closely approach the normal distribution as the sample size is increased, 208, 252–253

Central tendency, measures of. *See* Measures of central tendency

Charts, *See* Control charts; Visual representation of data

Chebyshev's Theorem For either a sample or a population, the percentage of observations that fall within k (for $k > 1$) standard deviations of the mean will be at least $[1 - (1/k^2)](100)\%$, 78–79

Check Sheet In total quality management, a sheet or other mechanism for recording frequency counts or product measurements, 772

Chi-Square Distribution A family of continuous probability distributions, with a different curve for each possible value of the number of degrees of freedom, *df*. It is the sampling distribution of $(n - 1)s^2/\sigma^2$ whenever samples of size n are repeatedly drawn from the same, normal population, 468–471
- confidence interval for population variance and, 489–491
- hypothesis tests for population variance and, 491–496

Chi-square test for equality of proportions, 486–488

Chi-square test for goodness of fit, 471–476
- procedure for, 472–473
- purpose of, 471

Chi-square test for normality, 476–477

Chi-square test of variable independence, 479–484
- procedure for, 479–480
- purpose of, 479

Class Each category of a frequency distribution, 18
- exhaustive, 18
- mutually exclusive, 18
- open-end, 18

Classical Probability Probability defined in terms of the proportion of times that an event can be theoretically expected to occur, 136

Classical time series model, 689

Class Interval The width of each class in a frequency distribution; this is the difference between the lower class limit of a class and the lower class limit of the next higher class, 18

Class Limits The boundaries for each class in a frequency distribution; these determine which data values are assigned to the class, 18

Class Mark For each class in a frequency distribution, the midpoint between the upper and lower class limits, 18

Cluster analysis, 114

Cluster Sample A probability sampling technique in which the population is divided into groups, then members of a random selection of these groups are collected by either a sample or a census, 124

Coefficient of Correlation Expressed as r (with $-1 \leq r \leq +1$), a number describing the strength and direction of the linear relationship between the dependent variable (y) and the independent variable (x), 86–89, 567–568
- testing, 572–573

Coefficient of Determination, r^2 In simple regression, the proportion of the variation in y that is explained by the regression line $\hat{y} = b_0 + b_1x$, 88, 568–570

Coefficient of Multiple Determination Expressed as R^2, the proportion of the variation in y that is explained by the multiple regression equation. Its positive square root is the coefficient of multiple correlation, R, 615

Coefficient of Variation (*CV*) The relative amount of dispersion in a set of data; it expresses the standard deviation as a percentage of the mean, 82

COLAs (cost-of-living adjustments), 726

Combinations The number of unique groups for n objects, r at a time. The order of arrangement is not considered, 158–159

Common-Cause Variation In a process, variation that occurs by chance. Also known as *random variation,* 760
Competitive Benchmarking The practice of studying, and attempting to emulate, the strategies and practices of organizations already known to generate world-class products and services, 767–768
Compiled List A list or sample consisting of persons who share a common characteristic, such as age, occupation, or state of residence, 107
Complement Either an event will occur (A) or it will not occur (A'). A' is the *complement* of A, and $P(A) + P(A') = 1$, 135
Conditional Probability The probability that an event will occur, given that another event has already happened, 146
Confidence Coefficient For a confidence interval, the proportion of such intervals that would include the population parameter if the process leading to the interval were repeated a great many times, 275
Confidence Interval An interval estimate for which there is a specified degree of certainty that the actual value of the population parameter will fall within the interval, 271, 274
  with finite population, 298–299
  hypothesis testing and, 329
  for mean, with population standard deviation known, 276–280
  for mean, with population standard deviation unknown, 281–286
  multiple regression and, 610–611
  for population proportion, 288–291
  for population variance, 489–491
  regression analysis and, 562–563, 610–611
  for slope of regression line, 573–575
Confidence Level Like the confidence coefficient, this expresses the degree of certainty that an interval will include the actual value of the population parameter, but it is stated as a percentage. For example, a 0.95 confidence coefficient is equivalent to a 95% confidence level, 275
Consumer Price Index (CPI) An index produced by the U.S. Bureau of Labor Statistics, it describes the change in prices from one time period to another for a fixed "market basket" of goods and services, 725–728
Contingency Table A tabular method showing either frequencies or relative frequencies for two variables at the same time, 43–45, 140, 479
Continuity, correction for, 224. *See also* Correction for continuity
Continuous Probability Distribution A probability distribution describing probabilities associated with random variables that can take on any value along a given range or continuum, 206–232
  exponential distribution, 228–232
  normal approximation to binomial distribution, 223–226
  normal distribution, 208–226
  simulating observations from, 233–235
  standard normal distribution, 212–221
Continuous quantitative variables, 9
Continuous Random Variable A random variable that can take a value at any point along an interval. There are no gaps between the possible values, 168, 206
Control Chart A chart displaying a sample statistic for each of a series of samples from a process. It has upper and lower control limits and is used in determining whether the process is being influenced by assignable variation, 762, 774–793
  format of, 775
  types of, 775
Control Chart for Attributes A control chart in which the measure of product quality is a count, 775, 786–793
  c-charts, 790–793
  p-charts, 787–790
Control Chart for Variables A control chart in which the measure of product quality is a measurement, 776–785
  mean, 776–781
  range, 782–785
Control group, 109
Control Limit In a control chart, a boundary beyond which the process is judged to be under the influence of nonrandom, or assignable, variation, 774
Convenience Sample A nonprobability sampling technique where members are chosen primarily because they are both readily available and willing to participate, 125
Correction for Continuity In the normal approximation to the binomial distribution, the "expanding" of each possible value of the discrete variable, x, by 0.5 in each direction. This correction becomes less important when n is large and we are determining the probability that x will lie within or beyond a given range of values, 224
Correlation Analysis The measurement of the strength of the linear relationship between two variables, 552, 567–570
  causation confusion and, 582–583
  coefficient of correlation and, 86–89, 567–568
  coefficient of determination and, 88, 568–570
  extrapolating beyond range of data in, 583–584
  Kolmogorov-Smirnov test and, 579
  normal probability plot and, 579
  residual analysis and, 578–582
  types of relationships between variables and, 37–38, 86–87, 552
  See also Multiple correlation analysis
Correlation Matrix A matrix listing the correlation of each variable with each of the other variables, 667–668
Cost-of-living adjustments (COLAs), 726

Costs of Conformance In total quality management, the costs associated with making the product meet requirements or expectations—i.e., the costs of *preventing* poor quality, 760
Costs of Nonconformance In total quality management, the costs associated with the failure of a product to meet requirements or expectations—i.e., the costs that occur as *the result of* poor quality, 760
Counting, 156
Critical Value A value of the test statistic that serves as a boundary between the nonrejection region and a rejection region for the null hypothesis. A hypothesis test will have either one (one-tail test) or two (two-tail test) critical values, 319
Cross-tabulation Also known as the crosstab or contingency table, shows how many people or items are in combinations of categories, 43–45, 140, 479
Cumulative Frequency Distribution A frequency distribution showing the number of observations within or below each class, or above or within each class, 19
Cumulative relative frequency distribution, 19
Curvilinear Relationship A relationship between variables that is best described by a curved line, 38
Cyclical Variation in a time series due to fluctuations that repeat over time, with the period being more than a year from one peak to the next, 689

D

Data Mining Analyzing a large collection of data to discern patterns and relationships, 114–115
Data Warehouse A large collection of data from inside and outside a company to be analyzed for patterns and relationships, 114–115
Deciles Quantiles dividing data into 10 parts of equal size, with each comprising 10% of the observations, 70
Decision Alternatives The choices available to the decision maker, with one choice to be selected, 738
  expected opportunity loss for, 749–750
  expected payoff for, 745–746
  payoff table and, 739–740
Decision making, 737–754
  Bayesian, 745–748
  business statistics and, 11
  in hypothesis-testing procedure, 319
  incremental analysis and, 751–754
  inventory, 751–754
  non-Bayesian, 742–744
  opportunity loss approach to, 749–750
  payoff table in, 739–740
  regret table and, 743–744
  structuring decision situation in, 738–744
Decision Rule A statement specifying calculated values of the test statistic for which the null hypothesis should or should not be rejected (e.g., "Reject H_0 if $z > 1.96$, otherwise do not reject."), 319
Decision Tree A graphical method of describing the decision situation, including alternatives, states of nature, and probabilities associated with the states of nature, 740–741
Degrees of Freedom The number of values that remain free to vary once some information about them is already known, 281
Deming Cycle *See* Shewhart-Deming cycle
Deming's 14 Points, 766–767
Dependent Events Events for which the occurrence of one affects the probability of occurrence for the other, 146
Dependent Samples Samples for which the selection process for one is related to the selection process for the other—e.g., before–after measurements for the same people or other test units, 366
  analysis of variance and, 439
  hypothesis testing and, 385–389
  sign test for comparing, 530–535
  Wilcoxon signed rank test for comparing, 513–516
Dependent Variable In an experiment, the variable for which a value is measured or observed. In regression analysis, the y variable in the regression equation, 37, 109, 413, 552
Descriptive Research A study for the purpose of describing some phenomenon, 103
Descriptive Statistics A branch of statistics in which the emphasis is on the summarization and description of data that have been collected, 5
  from grouped data, 83–85
Deseasonalized time series A time series from which seasonal influences have been removed. Such a series is also described as seasonally adjusted, 704–705
Designed Experiment An experiment in which treatments are randomly assigned to the participants or test units, 413
Determination, coefficient of, 88, 568–570
Dichotomous Question In a questionnaire, a question having just two alternatives from which the respondent can choose, 105
Directional testing, 313. *See also* One-tail tests
Direct Relationship A relationship in which the variables y and x increase and decrease together, 37
Discrete Probability Distribution The probability distribution for a discrete random variable, 169–173
  binomial distribution, 175–181
  hypergeometric distribution, 183–186
  Poisson distribution, 187–193
  uniform, 173
Discrete quantitative variables, 8
Discrete Random Variable A random variable that can take on only certain values along an interval, with these values having gaps between them, 168
Discriminant analysis, 114
Dispersion, 58, 67–82
  Chebyshev's theorem and, 78–79
  coefficient of variation and, 82
  relative, 82
  standardized data and, 79–80

Distribution shape, and measures of central tendency, 63–64
Dotplot A visual display representing each data value with a dot, and allowing us to readily see the shape of the distribution as well as the high and low values, 27
Dummy Variable A variable that takes on a value of either 1 or 0, depending on whether a given characteristic is present or absent. It allows qualitative data to be included in a regression analysis, 632–633, 658–662
Durbin-Watson Test A test for autocorrelation of the residuals in a regression analysis. It is especially applicable when the data represent a time series, 715–718

E

Empirical Rule For distributions that are bell-shaped and symmetrical, about 68% of the observations will fall within 1 standard deviation of the mean, with about 95% falling within 2 standard deviations, and practically all of the observations falling within 3 standard deviations of the mean, 79
EMV (expected monetary value), 745
EOL (expected opportunity loss), 749–750
Equality of proportions. *See* Chi-square test for equality of proportions
Errors
  estimation, 272
  in hypothesis testing, 315–316, 346–347
  multiple standard error of estimate, 609
  nonresponse, 108
  nonsampling, 119
  nonsystematic, 108
  power of test and, 346–347
  response, 108
  sampling, 108, 119, 293, 419
  standard error of estimate, 561–562
  standard error of the mean, 248
  systematic, 108
  Type I, 315–316
  Type II, 315–316
Estimation error, 272
Estimator, unbiased, 272
Event One or more of the possible outcomes of an experiment; a subset of the sample space, 135
  dependent events, 146
  exhaustive events, 140
  independent events, 146
  intersection of events, 140
  mutually exclusive events, 140
  union of events, 140
EVPI (expected value of perfect information), 746–748. *See also* Expected value of perfect information
Excel applications, *See* front endsheet of text
Exhaustive In a frequency distribution, the condition in which the set of classes includes all possible data values. In an experiment, the condition in which a set of outcomes includes all of the possible outcomes, 18, 140
Exhaustive events, 140
Expected Frequency In chi-square analysis, E_{ij} is the frequency of occurrence that would be expected for a given cell in the table if the null hypothesis were really true, 472
Expected monetary value (*EMV*), 745. *See also* Expected payoff
Expected opportunity loss (*EOL*), 749–750
Expected Payoff For a decision alternative, the sum of the products of the payoffs and the state of nature probabilities, 745–746
Expected value, 171
Expected Value of Perfect Information (*EVPI*) The difference between the expected payoff with perfect information and the expected payoff of the best alternative with current information, 746–748
Experiment An activity or measurement that results in an outcome. Also, a causal study for determining the influence of one or more independent variables on a dependent variable, 135, 412–414
  balanced, 414
  designed, 413
  factorial, 442
Experimental group, 109
Experimentation, and analysis of variance, 412–414
Exploratory Research An investigation for the purpose of gaining familiarity with the problem situation, identifying key variables, and forming hypotheses that can be tested in subsequent research, 102–103
Exponential Distribution A continuous distribution that, for a Poisson process, describes probabilities associated with the continuous random variable, x = the amount of time, space, or distance between occurrences of the events of interest, 228–232
  description and applications of, 228
  Poisson distribution and, 228
Exponential model, 664
Exponential Smoothing A method that dampens the fluctuations in a time series; each "smoothed" value is a weighted average of current and past values observed in the series, 697–700
  forecasting with, 709–711
External Secondary Data Secondary data that have been generated by someone outside the researcher's firm or organization, 104, 112–116
External Validity In experimental research, the extent to which the results can be generalized to other individuals or settings, 110
Extraneous variables, 109

F

Factor An independent variable in an experiment, 413
Factor analysis, 114
Factorial For a positive integer, n, the factorial ($n!$) is the product of $n \times (n-1) \times (n-2) \times \cdots \times 1$. The value of 0! is defined as 1, 157
Factorial Experiment A design in which there are two or more factors and the treatments represent all possible combinations of their levels. The two-way ANOVA is the most basic form of this design, 442
Factorial rule of counting, 157

Factor Level A specific value for a factor. When an experiment involves just one factor, each factor level can be referred to as a treatment, 413

Finite Population Correction When sampling is without replacement and from a finite population, the use of a correction term is used to reduce the standard error according to the relative sizes of the sample and the population, 298–301

Finite population correction factor, 258

Fishbone diagram, 770–771. *See also* Cause-and-effect diagram

Focus group interview, 103

Forecasting, 708–723
- autoregressive, 718–723
- exponential smoothing and, 709–711
- seasonal indexes in, 712
- trend equation and, 709

Frequency The number of data values falling within each class of a frequency distribution, 18
- expected, 472
- observed, 472

Frequency Distribution A display table dividing observed values into a set of classes and showing the number of observations falling into each class, 16–19
- constructing, 16–19
- cumulative, 19
- guidelines for, 18
- relative, 19

Frequency Polygon A graphical display with line segments connecting points formed by intersections of class marks with class frequencies. May also be based on relative frequencies or percentages, 20

Friedman Test For the randomized block design, a nonparametric test that compares more than two dependent samples at the same time. It is the nonparametric counterpart to the randomized block analysis of variance, 525–528

F statistic
- for comparing variances of two independent samples, 397–400
- in one-way analysis of variance, 419
- in randomized block analysis of variance, 433
- in two-way analysis of variance, 446, 449–450

G

Goodness-of-Fit Test A test that examines whether a sample could have been drawn from a population having a specified probability distribution—e.g., the normal distribution, 471–477
- chi-square test for, 471–477
- Kolmogorov-Smirnov test for, 537–540, 579

Grand Mean The mean of all observations or measurements, 418

Graphs, *See* Visual representation of data

Grouped Data Data summarized in a frequency distribution, 83–85

H

Histogram For quantitative variables, a graphical method that describes a frequency distribution by expressing either the frequency or relative frequency within each class as the length of a rectangle. In a histogram, adjacent rectangles share a common side, 20–22

Homoscedasticity The assumption of equal standard deviations of *y* values about the population regression line, regardless of the value of *x*, 553

Hypergeometric Distribution A discrete probability distribution in which there are consecutive trials, but in which the probability of success is not the same from one trial to the next; i.e., the trials are not statistically independent from each other. 183–186

Hyperplane A mathematical counterpart to the two-dimensional surface fitted to data points when there are just two *x* variables. The multiple regression equation is either a plane (two *x* variables) or a hyperplane (more than two *x* variables), 604

Hypothesis, *See* Alternative hypothesis; Null hypothesis

Hypothesis tests, 312–344, 363–400
- basic procedures of, 317–319
- comparing means of two dependent samples, 385–389
- comparing means of two independent samples, 366–384
- comparing two sample proportions, 391–395
- comparing variances of two independent samples, 397–400
- confidence intervals and, 329
- errors in, 315–316, 346–347
- for population variance, 491–496
- nature of, 313–315
- of a mean with population standard deviation known, 320–323
- of a mean, with population standard deviation unknown, 330–336
- of a proportion, 338–444
- power of a test, 315–316, 346–350
- *p*-value and, 324–327
- *t*-tests, 330–336, 366–378
- *z*-tests, 320–323, 380–384, 391–395

I

Ignorance A decision situation in which neither the possible states of nature nor their exact probabilities are known, 739

Incremental Analysis An inventory decision method where items are sequentially considered, with each unit in the sequence being stocked only if the expected payoff for stocking it exceeds the expected payoff for not stocking it, 751–754

Independent Events Events for which the occurrence of one has no effect on the probability of occurrence for the other, 146

Independent Samples Samples for which the selection process for one is not related to the selection process for the other, 365
- comparing variances of, 397–400
- Kruskal-Wallis test for comparing more than two, 521–523
- *t*-test for comparing means of, 366–378
- Wilcoxon rank sum test for comparing, 517–520
- *z*-test for comparing means of, 380–384

z-test for comparing proportions of, 391–395

Independent Variable In regression analysis, an x variable in the regression equation. In an experiment, a variable that is observed or controlled for the purpose of determining its effects on the value of the dependent variable, 37, 109, 413, 552

Indexes, 724–728
- Consumer Price, 725–728
- Employment Cost, 727
- Industrial Production, 727
- New York Stock Exchange Composite, 727
- Process Capability, 796–798
- Producer Price, 727
- shifting base of, 728
- Standard & Poor's Composite, 727

Index Number A percentage expressing a measurement or combination of measurements in a given period compared to a selected base period, 724–728

Inferential Statistics A branch of statistics that goes beyond mere description and, based on sample data, seeks to generalize from the sample to the population from which the sample was drawn, 5–6, 102, 245, 271

Interactive Effect In a two-factor experiment, an effect that results from the combination of a level for one factor with a level for the other. In the randomized block design, interaction is present when the effect of a treatment depends on the block to which it has been administered, 431, 442–443

Internal Secondary Data Secondary data that have been generated from within the researcher's firm or organization, 104, 112

Internal Validity In experimental research, the extent to which the independent variable was actually responsible for changes measured in the dependent variable, 109–110

Internet data sources, 115–116

Interquartile Range The distance between the first and third quartiles, 71

Intersection Two or more events occur at the same time; e.g., "*A* and *B*", 140

Interval Estimate A range of values within which the actual value of the population parameter may fall, 271, 273–291
- with finite population, 298–299
- in multiple regression, 609–614
- in simple linear regression, 561–566, 573–575

Interval Limits The lower and upper values of the interval estimate, 274

Interval Scale of Measurement A scale that has a constant unit of measurement, but in which the zero point is arbitrary, 10

Interview
- focus group, 103
- personal, 105
- telephone, 105

Inverse Relationship A relationship in which y and x increase or decrease in opposite directions, 37

Irregular Variation in a time series due to random fluctuations or unusual events, 689

J

Joint Probability The probability that two or more events will all occur, 146

Judgment Sample A nonprobability sampling technique where the researcher believes the members to be representative of the population, 126

Juran diagram, 772

Just-in-Time Manufacturing A manufacturing system in which materials and components literally arrive "just in time" to enter the production process. It is a production philosophy that has favorable effects on both inventory costs and product quality, 768

K

Kaizen A philosophy that focuses on small, ongoing improvements rather than relying on large-scale (and typically, more abrupt) innovations, such as might come from advances in technology or equipment, 766

Kolmogorov-Smirnov Test A nonparametric test to examine whether a set of data may have come from a given distribution. It is applied in the text to test whether data or residuals could have come from a normal population, 537–540, 579

Kruskal-Wallis Test A nonparametric test that compares the medians of more than two independent samples at the same time, 521–523

L

Law of Large Numbers Over a large number of trials, the relative frequency with which an event occurs will approach the probability of its occurrence for a single trial, 136

Leaf In a stem-and-leaf display of raw data, the rightmost digit on which classification of an observation is based, 25

Least-Squares Criterion For fitting a regression line, a criterion which requires that the sum of the squared deviations between actual y values and y values predicted by the equation be minimized. The resulting equation is the least squares regression line, 554–558

Least-squares regression line, 554–558. *See also* Regression line

Left-Tail Test A one-tail test in which the rejection region for the null hypothesis is in the left tail of the sampling distribution, 313

Linear regression, *See* Regression analysis

Linear Relationship A relationship between variables that is best described by a straight line, 37–38

Linear trend equations, 690–691

Line Graph A display in which lines connect points representing values of one quantitative variable (y, or vertical axis) versus another (x, or horizontal axis). Two or more y variables may be plotted for the same x, 29

M

MAD (mean absolute deviation), 71–72, 713–714. *See also* Mean absolute deviation

Mailing lists, 107

Mail Survey A survey technique in which the medium of communication is a mail questionnaire, 105
  errors in, 108
  questionnaire design for, 105–106
  sampling process for, 107
Main Effect An effect caused by one of the factors in an experiment, 443
Marginal Loss In incremental analysis, the cost if one more unit is stocked and *not* sold, 751
Marginal Probability The probability that a given event will occur, with no other events taken into consideration, 146
Marginal Profit In incremental analysis, the profit if one more unit is stocked and sold, 751
Matched pairs, 385
Maximax A non-Bayesian criterion specifying selection of the decision alternative having the highest maximum payoff, 742
Maximin A non-Bayesian criterion specifying selection of the decision alternative having the highest minimum payoff, 742
Mean, 7, 59–60
  arithmetic, 7, 59–60
  from grouped data, 83–85
  grand, 418
  median and mode vs., 62–63
  sampling distribution of, 244, 244–253
  standard error of, 248
  trimmed, 66
  weighted, 61
Mean Absolute Deviation (*MAD*) As a measure of dispersion, this is the average of the absolute values of the differences between observations and the mean. In forecasting, a criterion for measuring the fit of an estimation equation to an actual time series. The best-fit model is the one having the lowest mean value for $|y_i - \hat{y}_i|$, 71–72, 713–714
  time series and, 713–714
Mean Chart A control chart displaying a series of sample means from the output of a process, 776–781
Mean of a Probability Distribution Expressed as μ or $E(x)$, this is the expected value of the random variable, x. It is a weighted average in which the possible values of x are weighted according to their probabilities of occurrence, 171–172
Mean Square A sum of squares value that has been divided by the number of degrees of freedom associated with the source of variation being examined, 419
Mean Squared Error (*MSE*) In forecasting, a criterion for measuring the fit of an estimation equation to an actual time series. The best-fit model is the one having the lowest mean value for $(y_i - \hat{y}_i)^2$, 713–714
Measurement, scales of, 9–10
Measure of Central Tendency A numerical measure describing typical values in the data, 58–66
  arithmetic mean, 7, 59–60
  distribution shape and, 63–64
  median, 7, 61–62
  mode, 7, 62
  weighted mean, 61
Measure of Dispersion A numerical measure describing the amount of scatter, or spread, in the data values, 58, 67–82
  mean absolute deviation (*MAD*), 71–72, 713–714
  quantiles, 70–71
  range, 7, 68
  standard deviation, 7, 72–75
  variance, 72–74
Median A value that has just as many observations that are higher as it does observations that are lower. It is the second quartile, 7, 61–62
Midrange The average of the lowest and highest data values, 70
Minimax Regret A non-Bayesian criterion specifying selection of the decision alternative having the lowest maximum regret, 743–744
Minitab applications, *See* front endsheet of text
Mode The value that occurs with the greatest frequency in a set of data, 7, 62
Moving average, centered, 693–697
MSE (mean squared error), 713–714
Multicollinearity A condition where two or more of the independent variables are highly correlated with each other, making the partial regression coefficients both statistically unreliable and difficult to interpret, 633, 667
Multidimensional scaling, 114
Multiple Choice Question In a questionnaire, a question having several alternatives from which the respondent may choose, 105
Multiple correlation analysis, 615–616
  coefficient of multiple determination and, 615
  dummy variable and, 632–633, 658–662
  multicollinearity and, 633, 667
  significance tests in, 617–620
  stepwise regression and, 670–674
Multiple determination, coefficient of, 615
Multiple Regression Regression analysis in which there is one dependent variable and two or more independent variables, 600–634
  assumptions for, 602–603
  confidence intervals and, 610–611
  dummy variable and, 632–633, 658–662
  interval estimation in, 610–614
  model for, 602–603
  multicollinearity and, 633, 667
  multiple standard error of estimate and, 609
  prediction intervals and, 612–614
  residual analysis and, 625–630
  significance tests in, 617–620
  stepwise regression and, 633–634, 670–674
Multiple Standard Error of Estimate The amount of dispersion of data points about the plane (or hyperplane) represented by the multiple regression equation; this is also the standard deviation of the residuals, 609
Multiplication, principle of, 156
Multiplication Rules Formulas for determining the probability of the intersection of two or more events, 146–149
Multiplicative model, 664–665
Mutually Exclusive Classes The classes in a frequency distribution are mutually exclusive when each

data value can fall into only one category, 18
Mutually Exclusive Events The condition in which two events cannot occur at the same time. An event (A) and its complement (A') are always mutually exclusive, 140

N

Negatively Skewed Distribution A nonsymmetrical distribution that tails off to the left; it has a mean that is less than the median, 64
Nominal Scale of Measurement The most primitive scale of measurement; it employs numbers only to identify membership in a group or category, 9
Non-Bayesian decision making, 742–744
Nondirectional testing, 313
  confidence intervals and, 329
  See also Two-tail tests
Nonparametric Test A hypothesis test that makes no assumptions about the specific shape of the population from which a sample is drawn, 505–543
  advantages and disadvantages of, 506–507
  corresponding parametric tests for, 507
  Friedman test, 525–528
  Kolmogorov-Smirnov test, 537–540, 579
  Kruskal-Wallis test, 521–523
  runs test, 535–537
  sign test, 530–535
  Wilcoxon rank sum test, 517–520
  Wilcoxon signed rank test, 508–516
Nonprobability Sampling Any sampling technique where not everyone in the population has a chance of being included in the sample. The selection process will involve at least some degree of personal subjectivity, 120, 124–126
Nonresponse Error In survey research, error that occurs when persons who respond to the survey are different from those who don't, 108
Nonsampling Error The error, or difference between the population parameter and its corresponding sample statistic, that occurs because of a directional tendency, or bias, in the measuring process, 119
Nonsystematic Error Also referred to as "random," an error that occurs when a measured value differs from the actual value of the population parameter, but is just as likely to be too high as too low, 108
Normal Distribution A family of distributions, each member of which is bell-shaped and symmetrical, and has the mean, median, and mode all located at the midpoint. The curve approaches the horizontal axis at both ends, but never intersects with it. The specific member of the family depends on just two descriptors: the mean (μ) and the standard deviation (σ), 208–221
  approximation to binomial distribution, 223–226, 338
  areas beneath normal curve, 209–221
  description and applications of, 208–209
  standard, 212–221
Normal probability plot, 541, 579
Null Hypothesis (H_0) A statement about the value of a population parameter, H_0 is put up for testing in the face of numerical evidence. It is not necessarily identical to the claim or assertion that led to the test, 312–313
  errors with, 315–316
  formulating, 312–313, 317

O

Observation Method of primary data collection in which behavior is observed and measured, 110–111
Observed Frequency In chi-square analysis, O_{ij} is the observed, or actual, frequency of occurrence for a given cell in the table, 472
Odds A method in which a ratio is used as a way of expressing the likelihood that an event will happen versus the likelihood that it will not, 138–139
Ogive A graphical display providing cumulative values for frequencies, relative frequencies, or percentages. May be either "less than" or "greater than", 21
One-Tail Test Test in which the null hypothesis can be rejected by an extreme result in one direction only. The test arises from a directional claim or assertion about the population parameter, 313
  directional assertions and, 313
One-Way Analysis of Variance An ANOVA procedure that examines two or more independent samples to determine whether their population means could be equal. Also referred to as the one-factor, completely randomized design, 416–426
  assumptions for, 416–417
  between-sample variation in, 417
  model for, 416–417
  pooled-variances t-test and, 426
  procedure for, 417–420
  purpose of, 416
  within-sample variation in, 417
Open-End Class In a frequency distribution, a class having either no lower limit or no upper limit, 18
Open-Ended Question In a questionnaire, a question that allows the respondent to answer in his own words and to express whatever thoughts he feels are relevant to the subject matter of the question, 105
Operating Characteristic (OC) Curve The complement to the power curve, the OC curve plots the probability (β) that the hypothesis test will make a Type II error (fail to reject a false null hypothesis) for a range of values of the population parameter for which the null hypothesis is false, 350
Opportunity Loss Equivalent to regret, 749–750
Ordinal Scale of Measurement A scale that allows the expression of "greater than" and "less than" relationships, but which has no unit of measurement, 9–10
Outlier A data value that is very different from most of the other data values, 25

P

Paired Observations Data collected from dependent samples, such as before-and-after measurements recorded for the same individuals or test units, 385
Paired samples
sign test for comparing, 530–535
Wilcoxon signed rank test for comparing, 513–516
See also Dependent samples
Parameter A characteristic of the population, such as the population mean (μ), standard deviation (σ), or proportion (π), 7, 117
Parametric tests, corresponding to nonparametric tests, 507. *See also* names of specific parametric tests
Pareto Diagram Also known as the *Juran diagram,* a bar chart describing the relative frequencies with which various kinds of defects, or nonconformities, have occurred, 772
Partial Regression Coefficient In the multiple regression equation, the coefficient for each independent variable. It is the *slope* of the equation when all other x variables are held constant, 602
interval estimation for, 620–621
testing, 618–620
Payoff The outcome at the intersection of a decision alternative and a state of nature, 739–740
expected, 745–746
Payoff Table A method of structuring the decision situation. The rows of the table are decision alternatives, the columns are states of nature, and the entry at each intersection of a row and column is a numerical consequence such as a profit or loss. The table may also include estimated probabilities for the states of nature, 739–740
***p*-Chart** A control chart that monitors a process by showing the proportion of defectives in each of a series of samples, 787–790
Percentiles Quantiles dividing data into 100 parts of equal size, with each comprising 1% of the observations, 70
Perfect information, expected value of, 746–748
Periodicity A potential problem with the systematic sample; this can occur when the order in which the population appears happens to include a cyclical variation in which the length of the cycle is the same as the value of k being used in selecting the sample, 123
Permutations The number of unique arrangements for n objects, r at a time, 157–158
Personal Interview A survey technique in which an interviewer personally administers the questionnaire to the respondent, 105
Pictogram A visual display using symbols to represent frequencies or other values, 32
Pie Chart A circular display divided into sections based on either the number of observations within, or the relative values of, the segments, 31
Point Estimate A single number (e.g., $\bar{x}$ or p) that estimates the value of a population parameter (e.g., μ or π), 271–272
using multiple regression equation, 605
using regression line, 558
Poisson Distribution A discrete probability distribution applied to events for which the probability of occurrence over a given span of time, space, or distance is extremely small. It can be used to approximate the binomial distribution when the number of trials is relatively large and the probability of a success in any given trial is very small, 187–193
approximation to binomial distribution, 191–193
description and applications of, 187–193
exponential distribution and, 228
Polynomial models extend the linear regression model to include terms in which the independent variable(s) are raised to a power greater than 1. They enable the model to fit data from relationships that are curvilinear, 653–657
with one quantitative predictor variable, 645–651
with two quantitative predictor variables, 653–657
Pooled-variances t-test
and analysis of variance, 426
in hypothesis testing, 366–372
Population The set of all possible elements that could theoretically be observed or measured; sometimes referred to as the *universe,* 6, 117
Population Parameter The characteristic of the population that corresponds to the sample statistic. The true value of the population parameter is typically unknown, but is estimated by the sample statistic, 7, 117
Positively Skewed Distribution A nonsymmetrical distribution that tails off to the right; it has a mean that is greater than the median, 64
Posterior Probability A revised probability based on additional information, 151
Power Curve A graph in which the power $(1 - \beta)$ of the hypothesis test is plotted against a range of population parameter values for which the null hypothesis is false, 349–350
Power of a Test $(1 - \beta)$ The probability that the hypothesis test will correctly reject a null hypothesis that is false, 315–316, 346–350
Prediction Interval In regression analysis, the estimation interval for an individual y observation, 563–565
multiple regression and, 612–614
Predictive Research An investigation for the purpose of obtaining a forecast or prediction of some value that will occur in the future, 103–104
Primary Data Data that have been collected by the researcher for the problem or decision at hand, 104 105–111
from experimentation, 109–110
from observation, 110–111
from survey research, 105–108
Principle of Multiplication If, following a first event that can

happen in n_1 ways, a second event can then happen in n_2 ways, the total number of ways both can happen is n_1n_2. Alternatively, if each of k independent events can occur in n different ways, the total number of possibilities is n^k, 156
Prior Probability An initial probability based on current information, 151
Probability A number between 0 and 1 which expresses the chance that an event will occur, 135–155
addition rules for, 143–145
Bayes' theorem and, 150–155
classical, 136
conditional, 146
joint, 146
marginal, 146
multiplication rules for, 146–149
odds and, 138–139
posterior, 151
prior, 151
relative frequency, 136–137
subjective, 137–138
Probability Density Function Curve describing the shape of a continuous probability distribution, 206
Probability Distribution For a random variable, a description or listing of all possible values it can assume, along with their respective probabilities of occurrence, 168
mean of, 171–172
variance of, 171–172
See also Continuous probability distributions; Discrete probability distributions
Probability Sampling Any sampling technique where each member or element in the population has some (nonzero) known or calculable chance of being included in the sample. This type of sample is necessary if it is desired to statistically generalize from the sample to the population. 107, 120–124
Process Capability Index The ability of a process to generate products falling within required specifications, 795–798
Process Potential Index The potential ability of a process to generate products falling within required specification limits, 796
Process Flow Chart A chart visually depicting the operations (or subprocesses) involved in a process, showing the sequential relationships between them, 770
Process orientation, 764
Proportion, 7
hypothesis testing of
one sample proportion, 338–344
two sample proportions, 391–395
sampling distribution of, 254–256
See also Chi-square test for equality of proportions; Population proportion estimation
Proportionate stratified sample, 123–124
Purposive Sample A nonprobability sampling technique where members are chosen specifically because they're not typical of the population, 126
***p*-value** The level of significance where the calculated value of the test statistic is exactly the same as a critical value. It is the lowest level of significance at which the null hypothesis can be rejected, 324–327

Q

Quadratic trend equations, 692
Qualitative Variable A variable that indicates whether a person or object possesses a given attribute, 8
Quality Audit The evaluation of the quality capabilities of a company or a supplier, 766–767
Quantiles Values that divide the data into groups containing an equal number of observations, 70–71
Quantitative Variable A variable that expresses how much of an attribute is possessed by a person or object; may be either discrete or continuous, 8
Quartile Deviation Half of the interquartile range, 71
Quartiles Quantiles dividing data into 4 parts of equal size, with each comprising 25% of the observations, 70–71
Questionnaire In survey research, the data collection instrument, 105–106
design of, 105–106
errors in, 108
Quota Sample A nonprobability sampling technique that is similar to the stratified sample, but where members of the various strata are not chosen through the use of a probability sampling method, 125

R

Random digit dialing, 107
Randomized Block Design An ANOVA procedure in which persons or test units are first arranged into similar groups (or *blocks*) before the treatments are randomly assigned. This reduces error variation by ensuring that treatment groups will be comparable in terms of the blocking variable, 429–439
assumptions for, 430–431
dependent-samples *t*-test and, 439
Friedman test for, 525–528
model for, 430–431
procedure for, 431–436
purpose of, 429–430
Randomness, runs test for, 535–537
Random Variable A variable that can take on different values depending on the outcome of an experiment, 168–169, 206
Random Variation In a process, variation that occurs by chance. Also known as *common-cause variation,* 760
Range The difference between the highest and lowest values in a set of data, 7, 68
interquartile, 71
Range Chart A control chart displaying a series of sample ranges from the output of a process, 782–785
Ratio Scale of Measurement A scale that has a constant unit of measurement and an absolute zero value, 10
Ratio to Moving Average A method for determining seasonal indexes, 702–704
Raw Data Data that have not been manipulated or treated in any way beyond their original collection, 16

Regression Analysis Fitting an equation to a set of data in order to describe the relationship between the variables, 553–566
analysis of variance perspective and, 576, 617–618
assumptions for, 553–554
causation confusion and, 582–583
confidence intervals and, 562–563
extrapolating beyond range of data in, 583–584
interval estimation and, 561–566
Kolmogorov-Smirnov test and, 579
model, 553–554
normal probability plot and, 579
prediction intervals and, 563–565
residual analysis and, 578–582
types of relationships between variables and, 37–38, 86–87, 552
See also Multiple regression
Regression coefficient, partial. *See* Partial regression coefficient
Regression Line In simple regression this is $\hat{y} = b_0 + b_1x$, the equation for estimating the value of y based on a known or assumed value of x, 554–558
Regret For a given state of nature, the difference between each payoff and the best possible payoff if that state of nature were to occur. Regret is the same as opportunity loss, 743–744
Relationships between variables, 37–38, 86–88, 552
Relative Frequency Distribution A display table showing the proportion or percentage of total observations falling into each class, 19
Relative Frequency Probability Probability defined in terms of the proportion of times an event occurs over a large number of trials, 136–137
Representative Sample A sample in which members tend to have the same characteristics as the population from which they were selected, 7
Research, 105–116
causal, 103
descriptive, 103
experimentation, 109–110
exploratory, 102–103
observational, 110–111
predictive, 103–104
primary data in, 105–111
process of, 104
secondary data in, 104, 112–116
survey, 105–108
Residual The difference between an observed y value and the estimated value ($\hat{y}$) predicted by the regression equation, 553
Residual analysis
multiple regression and, 625–630
regression analysis and, 578–582
Response Error In survey research, error that occurs whenever respondents report information that is not truthful, 108
Response List A list or sample consisting of persons who share a common activity or behavior, such as subscribing to a magazine, making a mail-order purchase, contributing to a charity, or applying for a credit card, 108
Right-Tail Test A one-tail test in which the rejection region for the null hypothesis is in the right tail of the sampling distribution, 314
Risk A decision situation in which both the possible states of nature and their exact probabilities of occurrence are known, 739
Runs Test A nonparametric test that evaluates the randomness of a series of observations by analyzing the number of runs it contains. A run is the consecutive appearance of one or more observations that are similar in some way, 535–537

S

Sample A smaller number, or subset, of the people or objects that exist within the population, 6, 117
census vs., 118–119
cluster, 124
convenience, 125
dependent, 366. *See also* Dependent samples
independent, 365. *See also* Independent samples
judgment, 126
paired, 385. *See also* Dependent samples
purposive, 126
quota, 125
simple random, 120–121
stratified, 123–124
systematic, 122–123
variance for, 72–74
Sample size
estimating population mean and, 293–294
estimating population proportion and, 295–296
with finite population, 300–301
selecting test statistic for hypothesis tests and, 317–318
Type I and Type II errors and, 350–351
Sample Space All possible outcomes of an experiment, 135
Sample Statistic A characteristic of the sample that is measured; it is often a mean, median, mode, proportion, or a measure of dispersion such as the standard deviation, 7
Sampling, 117–126
nonprobability, 120, 124–126
probability, 107, 120–124
process in research, 107
Sampling Distribution The probability distribution of a sample statistic, such as the sample mean ($\bar{x}$) or proportion (p), 244–259
with finite population, 257–259
of the mean, 244–253
of the proportion, 254–256
Sampling Error The error, or difference between the population parameter and its corresponding sample statistic, that occurs because a sample has been taken instead of a census. This is a nondirectional, or random, error. This type of error is reduced by increasing the size of a probability sample, 108, 119, 293, 419
Sampling with Replacement Sampling in which the item selected from the population is observed or measured, then put back into the population, 176
Sampling without Replacement Sampling in which the item selected from the population is not returned to the population after being observed or measured, 149, 176
Scales of measurement, 9–10

Scatter Diagram A visual summarization of the data in which each point is a pair of observed values for the dependent and independent variables, 37–41
Seasonal Variation in a time series due to periodic fluctuations that repeat over a period of one year or less, 689
Seasonal Index The relative strength of each season (week, month, quarter) in contributing to the values of annual observations within the time series, 702–707
deseasonalized time series, 704–705
in forecasting, 712
ratio to moving average method, 702–704
Secondary Data Data that have been collected by someone other than the researcher for some purpose other than the problem or decision at hand, 104, 112–116
evaluating, 116
external, 104, 112–116
internal, 104, 112
Seeing Statistics applets, *See* front endsheet of text
Shewhart-Deming Cycle Also known as the *Deming Wheel* and the *PDCA* (plan-do-check-act) *Cycle,* this model stresses the importance of continuous and systematic quality-improvement efforts at all levels of the organization, 763–764
Significance Level (α) The probability that the test will commit a Type I error by rejecting a true null hypothesis, 315–316
p-values and, 324–327
selecting, 317
Sign Test Used for the same purposes as the Wilcoxon signed rank test, this nonparametric test can be applied to ordinal data as well as interval or ratio. The sign test expresses each difference only in terms of being a plus ("+") or a minus, 530–535
Simple linear regression model, 553–566. *See also* Regression analysis
Simple Random Sample A probability sample in which every person or element in the population has an equal chance of being included in the sample. The more theoretical definition is that, for a given sample size, each possible sample constituency has the same chance of being selected as any other sample constituency, 120–121
Simulations
from continuous probability distributions, 233–235
from discrete probability distributions, 194–198
Sketch A relevant symbol the size of which reflects a frequency or other numerical descriptor, 33
Skewness The condition in which a distribution tails off to either the left or the right, exhibiting a lack of symmetry, 64
Smoothing techniques
exponential smoothing, 697–700
moving average, 693–697
Spearman Coefficient of Rank Correlation When paired observations are at least of the ordinal scale, this nonparametric method measures the strength and direction of the relationship between them, 540–543
Special-Cause Variation In a process, variation that occurs due to identifiable causes that have changed the process. Also known as assignable variation, 761
Standard Deviation A measure of dispersion, calculated as the positive square root of the variance, 7, 72–75
from grouped data, 85
Standard Error of Estimate A measure of the amount of scatter, or dispersion, of observed data points about the regression equation, 561–562
multiple, 609
Standard Error of the Mean The standard deviation of the sampling distribution of the mean, 248
Standardized data The expression of each data value in terms of its distance from the mean in standard deviation units. The resulting distribution will have a mean of 0 and a standard deviation of 1, 79–80
Standard Normal Distribution Also called a standardized normal distribution, a normal distribution that results from the original x values being expressed in terms of their number of standard deviations away from the mean. The distribution has a mean of 0 and a standard deviation of 1. Areas beneath the curve may be determined by computer statistical package or from the standard normal distribution table, 212–221
description and applications of, 212
z-score for, 212
Standard normal distribution table, *See* rear endsheet of text
State of Nature A future environment that has a chance of occurring; it is not under the control of the decision maker, 739
Statistic A characteristic of the sample, such as the sample mean ($\bar{x}$), standard deviation (s), or proportion (p), 7, 117
Statistical process control, 762, 774–793. *See also* Control chart
control charts for attributes, 775, 786–793
control charts for variables, 775, 776–785
Statistics
descriptive, 5, 102
inferential, 5–6, 102, 245, 271
sample, 6
See also Business statistics
Stem In a stem-and-leaf display of raw data, the leftmost digits on which classification of an observation is based, 25
Stem-and-Leaf Display A variant of the frequency distribution that uses a subset of the original digits in the raw data as class descriptors (stems) and class members (leaves), 24–26
Stepwise Regression A method in which independent variables enter the regression analysis one at a time. The first x variable to enter is the one explaining the most variation in y. At each step, the variable entered explains the greatest amount of the remaining variation in y, 633–634, 670–674

Stratified Sample A probability sample in which the population is divided into layers, or strata, then a simple random sample of members from each stratum is selected. Strata members have the same percentage representation in the sample as they do in the population, 123–124
Subjective Probability Probability based on personal judgment or belief, 137–138
Survey Research An investigation in which individuals are contacted by mail, telephone, or personal interview and their responses measured through the use of a questionnaire or data collection instrument, 105–108
errors in, 108
questionnaire design in, 105–106
sampling process in, 107–108
types of, 105
Symmetrical A distribution in which the left half and the right half are mirror images of each other, 64
Systematic Error Also called directional error, or bias, a tendency to either overestimate or underestimate the actual value of a population parameter, 108
Systematic Sample Probability sample in which a random starting point between 1 and k is selected, then every kth element in the population is sampled, 122–123

T

Tabulation
simple, 43–44
cross-tabulation, 43–44
Taguchi Methods In total quality management, a method that surpasses traditional specification limits to recognize the loss in quality that occurs as a dimension or measurement deviates farther from the target, 772–773
***t* Distribution** A family of continuous, unimodal, bell-shaped distributions, the t distribution is the probability distribution for the random variable $t = (\bar{x} - \mu)/(s/\sqrt{n})$. It has a mean of zero and a shape that is determined by the number of degrees of freedom (df). The t distribution tends to be flatter and more spread out than the normal distribution, especially for small sample sizes, but it approaches the shape of the normal curve as the sample size (and df) increases, 281–284
confidence intervals and, 284–286
See also t-tests
Telephone Interview A survey research technique in which the telephone serves as the medium of communication between interviewer and respondent, 105
Test for Equality of Proportions A chi-square application for examining, based on independent samples from two or more populations, whether the population proportions could be the same. This is a special case of the chi-square test for variable independence, 486–488
Test for Variable Independence A chi-square application for examining whether two nominal-scale (category) variables might be related, 479–484
Test Statistic Either a sample statistic or based on a sample statistic, a quantity used in deciding whether a null hypothesis should be rejected. Typical test statistics in the text are z and t, 317–318
Time Series A sequence of observations over regular intervals of time, 688–692
components of, 688–689
deseasonalized, 704–705
mean absolute deviation (MAD) criterion and, 713–714
mean squared error (MSE) criterion and, 713–714
ratio to moving average method and, 702–704
seasonal indexes and, 702–707
smoothing techniques and
exponential smoothing, 697–700
moving average, 693–697
trend equations and
linear, 690–691
quadratic, 692
Total Quality Management A philosophy that (1) integrates quality into all facets of the organization and (2) focuses on improving the quality of the product by systematically improving the process through which the product is designed, produced, delivered, and supported, 759–798
emergence, 763–765
pioneers, 763–764
practicing, 765–769
process capability and, 795–798
process orientation, 764
Six Sigma, 764–765
statistical tools, 770–774
Treatment When an experiment involves two factors, the combination of a level for one factor with a level for the second factor. When there is just one factor, each of its levels can be referred to as a treatment, 109, 413
Tree Diagram A diagram that visually depicts sequential events and their marginal, conditional, and joint probabilities, 146, 740–741
Trend Variation in a time series due to the overall upward or downward tendency of the series, 688–689
Trend equations
forecasts using, 709
linear, 690–691
quadratic, 692
Trimmed Mean A mean calculated after dropping an equal number or percentage of data values from both the high end and the low end of the data values, 66
t-distribution table, *See* pages preceding rear endsheet of the text
***t*-Test** A hypothesis test using t as a test statistic and relying on the t distribution for the determination of calculated and critical values.
for coefficient of correlation, 572–573
comparing means of two dependent samples, 385–389
comparing means of two independent samples, 366–378
of mean, population standard deviation unknown, 330–336
for partial regression coefficients, 618–620
pooled-variances, 366–372
for slope of regression line, 573–575
unequal-variances, 374–378

Two-Tail Test A test in which the null hypothesis can be rejected by an extreme result in either direction. The test arises from a nondirectional claim or assertion about the population parameter, 313
Two-Way Analysis of Variance An ANOVA procedure in which there are two factors, each operating on two or more levels. This design is able to examine both main effects and interactive effects, 441–454
assumptions of, 442–443
hypotheses tested, 443
model for, 442–443
procedure for, 443–446
purpose of, 441–442
Two-way stratified sample, 123–124
Type I Error The rejection of a null hypothesis that is true, 315–316
power of test and, 346–347
sample size and, 350–351
Type II Error The failure to reject a null hypothesis that is false, 315–316
power of test and, 346–347
sample size and, 350–351

U

Unbiased Estimator An estimator for which the expected value is the same as the actual value of the population parameter it is intended to estimate, 272
Uncertainty A decision situation in which the possible states of nature are known, but their exact probabilities are unknown, 739
Unequal-variances *t*-test, 374–378
Union At least one of a number of possible events occurs; e.g., "*A* or *B*", 140
Unpooled *t*-test. *See* Unequal-variances *t*-test

V

Validity
external, 110
internal, 109–110
Variable independence test. *See* Chi-square test of variable independence; Test of variable independence
Variables
continuous quantitative, 9
continuous random, 168, 206
control charts for, *See* Control charts, for variables
dependent, 37, 109, 413, 552
discrete quantitative, 8
discrete random, 168
dummy, 632–633, 658–662
extraneous, 109
independent, 37, 109, 413, 552
See also Chi-square test of variable independence
qualitative, 8
quantitative, 8
random, 168, 206
relationships between, 37–38, 86–89, 552
Variance A measure of dispersion based on the squared differences between observed values and the mean, 72–74
from grouped data, 85
of two independent samples, comparing, 397–400
Variation
assignable, 761
coefficient of, 82
random, 760
Venn Diagram For an experiment, a visual display of the sample space and the possible events within, 140–141
Visual representation of data, 16–47
abuse of, 34–36
bar chart, 28
dotplot, 27
frequency polygon, 20
histogram, 20–22
line graph, 29
pictogram, 32
pie chart, 31
scatter diagram, 37–41
sketch, 33

W

Weighted Mean Also called the weighted average, a measure of central tendency that weights each data value according to its relative importance, 61
Wilcoxon Rank Sum Test A nonparametric test comparing the medians of two independent samples, 517–520
Wilcoxon Signed Rank Test For one sample, a nonparametric test of the hypothesis that the population median could be a given value. For paired samples, it tests whether the population median of the difference between paired observations could be zero, 508–516
for comparing paired samples, 513–516
for one sample, 508–511
Within-group variation, in analysis of variance, 414–415
Worker Empowerment Also called *quality at the source,* a practice where the individual worker is (1) responsible for the quality of output from his or her own workstation, and (2) given authority to make changes he or she deems necessary for quality improvements, 768

Z

***z*-Score** Also referred to as the *z* value, the number of standard deviation units from the mean of a normal distribution to a given *x* value, 212
for sampling distribution of the mean, 249
for sampling distribution of the proportion, 254–256
for standard normal distribution, 212–221
***z*-Test** A hypothesis test using *z* as a test statistic and relying on the normal distribution for the determination of calculated and critical values, 320
approximation to Wilcoxon rank sum test, 519
approximation to Wilcoxon signed rank test, 510–511
comparing means of two independent samples, 380–384
comparing two sample proportions, 391–395
of mean, population standard deviation known, 320–323
of proportion, 338–344
Zündapp automobile, and time series-based forecasting, 687